July 2012

BRIDGE MAINTENANCE, SAFETY, MANAGEMENT, RESILIENCE AND SUSTAINABILITY

PROCEEDINGS OF THE SIXTH INTERNATIONAL CONFERENCE ON BRIDGE MAINTENANCE, SAFETY AND MANAGEMENT, STRESA, LAKE MAGGIORE, ITALY, 8–12 JULY 2012

Bridge Maintenance, Safety, Management, Resilience and Sustainability

Edited by

Fabio Biondini
Department of Structural Engineering, Politecnico di Milano, Milan, Italy

Dan M. Frangopol
ATLSS Engineering Research Center, Lehigh University, Bethlehem, PA, USA

CRC Press
Taylor & Francis Group
Boca Raton London New York Leiden

CRC Press is an imprint of the
Taylor & Francis Group, an **informa** business
A BALKEMA BOOK

Cover information:
View of the Europa Bridge, Parma, Italy
Owner: Authority for Urban Transformation, Parma, Italy
Contractor: Impresa Pizzarotti & C. and Codelfa S.p.A., Italy
Architectural Design: Vittorio Guasti Architettura S.r.l., Parma, Italy
Structural Design: Pier Giorgio Malerba, Paolo Galli and Marco Di Domizio, Milan, Italy
Construction Engineering: Paolo Sorba A.Ierre P&L Engineering, Parma, Italy
Plants Design: Pool Engineering, Mareno di Piave, Treviso, Italy
Environmental Planning: Ambiter S.r.l., Parma, Italy

CRC Press/Balkema is an imprint of the Taylor & Francis Group, an informa business

Typeset by MPS Limited, Chennai, India
Printed and bound in Great Britain by CPI Group (UK) Ltd, Croydon, CR0 4YY.

Published by: CRC Press/Balkema
P.O. Box 447, 2300 AK Leiden, The Netherlands
e-mail: Pub.NL@taylorandfrancis.com
www.crcpress.com – www.taylorandfrancis.com

ISBN: 978-0-415-62124-3 (Hbk)
ISBN: 978-0-203-10338-8 (eBook)

Bridge Maintenance, Safety, Management, Resilience and Sustainability – Biondini & Frangopol (Eds)
© 2012 Taylor & Francis Group, London, ISBN 978-0-415-62124-3

Table of Contents

Lifetime design, assessment, and maintenance of super long-span bridges

Organizers: H.-M. Koh, S. Shin, H.-K. Kim & N.-S. Kim

Strengthening of existing bridges with FRP composites

Organizer: C. Pellegrino

Steel bridge rehabilitation

Organizer: M. Sakano

Energy harvesting in bridges and transportation infrastructure networks
Organizer: K. Gkoumas

Vulnerability of bridges to fire and explosion
Organizer: L. Giuliani

Advances on structural robustness and redundancy of bridges
Organizers: F. Biondini & D.M. Frangopol

Risk-based and disaster resilience analysis of bridge systems and networked infrastructures under multiple hazards
Organizers: G.P. Cimellaro & L. Dueñas-Osorio

Advanced technologies in standard bridge structures – from research to implementation
Organizer: M. Saiid Saiidi

Design and seismic analysis of long span bridges – Case studies
Organizer: A.H. Malik

Recent advances in bridge health monitoring

Organizers: C.F. Cremona & A. Orcesi

Optical monitoring techniques for bridge maintenance and safety

Organizers: P.S. Sumitro & H. Matsuda

New developments on the bridge safety, maintenance and management in Mexico

Organizer: D. De León

Non deterministic schemes for structural safety and reliability of bridges

Organizer: S. Arangio

Structural control of bridges and footbridges: Extreme and every-day events

Organizers: L. Martinelli & M. Domaneschi

Integral bridges: Design and technological issues

Organizers: P.G. Malerba & V. Kristek

Management and preservation of long span historic bridges

Organizers: A.E. Aktan, F.L. Moon & D.S. Lowdermilk

Hybrid composite bridge system

Organizers: H. Furuta & S.-H. Kim

Understanding and enhancing bridge performance

Organizer: J.M. Hooks

Extending bridge life through industry academic partnerships

Organizer: E.J. OBrien

Bridge Maintenance, Safety, Management, Resilience and Sustainability – Biondini & Frangopol (Eds)
© 2012 Taylor & Francis Group, London, ISBN 978-0-415-62124-3

Conference organization

CONFERENCE CHAIRS

Fabio Biondini	*Politecnico di Milano, Milan, Italy*
Dan M. Frangopol	*Lehigh University, Bethlehem, PA, USA*

CONFERENCE HONORARY CHAIR

Pier Giorgio Malerba	*Politecnico di Milano, Milan, Italy*

INTERNATIONAL SCIENTIFIC COMMITTEE

Franco Bontempi (co-Chair)	*Sapienza University of Rome, Rome, Italy*
Airong Chen (co-Chair)	*Tongji University, Shanghai, China*
Hyun-Moo Koh (co-Chair)	*Seoul National University, Seoul, Korea*
Richard Sause (co-Chair)	*Lehigh University, Bethlehem, PA, USA*
Alessandro Palermo (Secretary)	*University of Canterbury, Christchurch, New Zealand*
Bryan T. Adey	*ETH Zürich, Zürich, Switzerland*
Anil K. Agrawal	*The City College of New York, New York, NY, USA*
Mitsuyoshi Akiyama	*Tohoku University, Sendai, Japan*
A. Emin Aktan	*Drexel University, Philadelphia, PA, USA*
Haluk Aktan	*Western Michigan University, Kalamazoo, MI, USA*
Sreenivas Alampalli	*New York State Department of Transportation, Albany, NY, USA*
Alfredo H.S. Ang	*University of California, Irvine, CA, USA*
Giuliano Augusti	*Sapienza University of Rome, Rome, Italy*
Konrad Bergmeister	*University of Natural Resources & Applied Life Sciences, Vienna, Austria*
Raimondo Betti	*Columbia University, New York, NY, USA*
Jan Bien	*Wroclaw University of Technology, Wroclaw, Poland*
Fabio Biondini	*Politecnico di Milano, Milan, Italy*
Túlio N. Bittencourt	*University of São Paulo, São Paulo, Brazil*
Geoff Boully	*VIC Roads, Camberwell, Victoria, Australia*
Fernando Branco	*Technical University of Lisbon, Lisbon, Portugal*
James Brownjohn	*The University of Sheffield, Sheffield, UK*
Eugen Bruehwiler	*EPFL, Lausanne, Switzerland*
Christian Bucher	*Vienna University of Technology, Vienna, Austria*
Peter G. Buckland	*Buckland & Taylor Ltd., North Vancouver, BC, Canada*
Harald Budelmann	*Technical University of Braunschweig, Braunschweig, Germany*
Joan R. Casas	*Technical University of Catalonia, Bercelona, Spain*
Necati Catbas	*University of Central Florida, Orlando, FL, USA*
Michael Chajes	*University of Delaware, Newark, DE, USA*
Chin-Chen Chang	*Hong Kong University of Science & Technology, Hong Kong, P.R.C.*
Kuo-Chun Chang	*National Taiwan University, Taipei, Taiwan, R.O.C.*
Steven Chase	*University of Virginia, Charlottesville, VA, USA*
Moe Cheung	*The Hong Kong University of Science & Technology, Hong Kong, P.R.C.*
Hyo-Nam Cho	*Hanyang University, Ansan, Korea*
Marios Chryssanthopoulos	*University of Surrey, Guilford, Surrey, UK*

NATIONAL ADVISORY COMMITTEE

NATIONAL ORGANIZING COMMITTEE

CONFERENCE SCIENTIFIC SECRETARIAT

Elena Camnasio & Andrea Titi
Department of Structural Engineering, Politecnico di Milano
Piazza Leonardo da Vinci, 32 – 20133 Milan, Italy

CONFERENCE ORGANIZING SECRETARIAT

Stella Pennini, Laura Manenti, Marta Padovan & Miriam Zanelli
Incentives e Congressi Via Crocifissa di Rosa, 15 – 25128 Brescia, Italy

CONFERENCE WEBSITE

http://www.iabmas2012.org

ORGANIZING ASSOCIATION

IABMAS
International Association for Bridge Maintenance and Safety

ORGANIZING INSTITUTION

POLIMI
Politecnico di Milano

SUPPORTING LOCAL AUTHORITIES (as of May 15th, 2012)

Regione Lombardia
Lombardy Regional Administration

Regione Piemonte
Piedmont Regional Administration

Consiglio Regionale del Piemonte
Regional Council of Piedmont

Provincia di Milano
Province of Milan

Provincia del Verbano Cusio Ossola
Province of Verbano Cusio Ossola

Comune di Verbania
Municipality of Verbania

Comune di Stresa
Municipality of Stresa

PATRONAGES AND SPONSORS (as of May 15th, 2012)

AASHTO
American Association of State Highway and Transportation Officials, USA

ACI
American Concrete Institute, USA

AICAP
Associazione Italiana Calcestruzzo Armato e Precompresso, Italy
Italian Association for Reinforced and Prestressed Concrete, Italy

AIOM
Associazione di Ingegneria Offshore e Marina, Italy
Association of Offshore and Marine Engineering, Italy

ARS.UNI.VCO
Associazione per lo sviluppo della cultura, degli studi universitari
e della ricerca nel Verbano Cusio Ossola, Italy

ASCE
American Society of Civil Engineers, USA

ASCP
The Portuguese Group of IABMAS, Portugal

ATLSS
Center for Advanced Technology for Large Structural Systems, USA

CERIC
Civil Engineering Research Information Center, Korea

CNI
Consiglio Nazionale degli Ingegneri, Italy
National Council of Engineers, Italy

COSEIK
Computational Structural Engineering Institute of Korea, Korea

CTE
Collegio dei Tecnici della Industrializzazione Edilizia, Italy
Italian Society of Building Engineers, Italy

FHWA
Federal Highway Administration, US Dept. of Transportation, USA

Fondazione Promozione Acciaio, Italy

IALCCE
International Association for Life-Cycle Civil Engineering

IASSAR
International Association for Structural Safety and Reliability

IBRACON
Brazilian Concrete Institute, Brazil

ISISE
Institute for Sustainability and Innovation in Structural Engineering, Portugal

JABMAS
Japanese Association for Bridge Maintenance And Safety, Japan

JRC
Joint Research Centre, European Commission, Italy

JSCE
Japan Society of Civil Engineers , Japan

JSMS
Japanese Society of Materials Science, Japan

KAIST
Korea Advanced Institute of Science and Technology, Korea

KBRC
Korea Bridge Design & Engineering Research Center, Korea

KCI
Korean Concrete Institute, Korea

KSSC
Korean Society of Steel Construction, Korea

KU
Kansai University, Japan

Kyoto University, Japan

RCEAS
P.C. Rossin College of Engineering and Applied Science,
Lehigh University, USA

ReLUIS
Rete dei Laboratori Universitari di Ingegneria Sismica, Italy
Laboratories University Network of Seismic Engineering, Italy

Sapienza University of Rome, School of Engineering, Italy

SISTEC
Smart Infra-Structure Technology Center, Korea

SNU
Seoul National University, Korea

Tongji University, China

TRB
Transportation Research Board of the National Academies, USA

T.Y. Lin International, USA

UPC
Technical University of Catalonia, Spain

UM
University of Minho, Portugal

TECHNICAL EXHIBITORS (as of May 15th, 2012)

Alpin Technik und Ingenieurservice GmbH, Leipzig, Germany
CSi Italia, Pordenone, Italy
DB System International, Cremona, Italy
DMT GmbH & Co. KG, Bochum, Germany
Intercomp, Medina, MN, USA
FiberSensing, Maia, Portugal
FORCE Technology, Brøndby, Denmark
IABMAS, International Association for Bridge Maintenance and Safety
IDS Ingegneria dei Sistemi S.p.A., Pisa, Italy
Intercomp, Medina, MN, USA
LUSAS, Kingston upon Thames, Surrey, UK
Mageba sa, Bülach, Switzerland
Maurer Söhne GmbH & Co. KG, Munchen, Germany
MISTRAS Group, Inc., Princeton Junction, NJ, USA
National Instruments, Austin, TX, USA
NEXCO Central Nippon Expressway Co.Ltd., Nagoya, Aichi, Japan
Pure Technologies Ltd., Calgary, Alberta, Canada
SINECO S.p.A., Milan, Italy
SMARTEC SA, Manno, Switzerland
Taylor & Francis Group, London, UK
Transpo Industries, Inc., New Rochelle, New York, NY, USA

Bridge Maintenance, Safety, Management, Resilience and Sustainability – Biondini & Frangopol (Eds)
© 2012 Taylor & Francis Group, London, ISBN 978-0-415-62124-3

Preface

The number of deteriorating bridges is increasing worldwide. Costs of maintenance, repair and rehabilitation of these bridges far exceed available budgets. Maintaining the safety and serviceability of existing bridges by making better use of available resources is a major concern for bridge management. Internationally, the bridge engineering profession continues to take positive steps towards developing more comprehensive bridge management systems. It was therefore considered appropriate to keep the tradition of the IABMAS conferences and bring together all of the very best work that has been done in the field of bridge maintenance, safety, management, resilience and sustainability at the Sixth International Conference on Bridge Maintenance, Safety and Management (IABMAS 2012), held in Stresa, Lake Maggiore, Italy, from July 8 to 12, 2012 (http://www.iabmas2012.org).

The First (IABMAS'02), Second (IABMAS'04), Third (IABMAS'06), Fourth (IABMAS'08), and Fifth (IABMAS 2010) International Conference on Bridge Maintenance, Safety and Management were held in Barcelona, Spain, July 14–17, 2002, Kyoto, Japan, October 18–22, 2004, Porto, Portugal, July 16–19, 2006, Seoul, Korea, July 13–17, 2008, and Philadelphia, PA, USA, July 11–15, 2010, respectively.

IABMAS 2012 has been organized on behalf of the International Association for Bridge Maintenance and Safety (IABMAS) under the auspices of Politecnico di Milano. IABMAS encompasses all aspects of bridge maintenance, safety and management. Specifically, it deals with: health monitoring and inspection of bridges; bridge repair and rehabilitation issues; bridge management systems; needs of bridge owners, financial planning, whole life costing and investment for the future; bridge safety and risk related issues, including economic and other implications. The objective of IABMAS is to promote international cooperation in the fields of bridge maintenance, safety, management, life-cycle performance and cost for the purpose of enhancing the welfare of society (http://www.iabmas.org).

The interest of the international bridge engineering community in the fields covered by IABMAS has been confirmed by the significant response to the IABMAS 2012 call for papers. In fact, over 800 abstracts from about 50 countries were received by the Conference Secretariat, and approximately 70% of them were selected for final publication as technical papers and presentation at the Conference within mini-symposia, special sessions, and general sessions. Compared to IABMAS 2010 the number of papers scheduled for presentation at IABMAS 2012 has increased from 511 to 555.

Contributions presented at IABMAS 2012 deal with the state of the art as well as emerging concepts and innovative applications related to all main aspects of bridge maintenance, safety, management, resilience and sustainability. Major topics covered include: advanced materials, ageing of bridges, assessment and evaluation, bridge codes, bridge diagnostics, bridge management systems, composites, damage identification, design for durability, deterioration modeling, earthquake and accidental loadings, emerging technologies, fatigue, field testing, financial planning, health monitoring, high performance materials, inspection, life-cycle performance and cost, load models, maintenance strategies, non-destructive testing, optimization strategies, prediction of future traffic demands, rehabilitation, reliability and risk management, repair, replacement, residual service life, resilience, robustness, safety and serviceability, service life prediction, strengthening, structural integrity, and sustainability, among others.

Bridge Maintenance, Safety, Management, Resilience and Sustainability contains the lectures and papers presented at IABMAS 2012. It consists of a book of extended abstracts and a DVD of full papers of 555 contributions, including the T.Y. Lin Lecture, nine Keynote Lectures, and 545 technical papers from 40 countries. This volume provides both and up-to-date overview of the field of bridge engineering and significant contributions to the process of making more rational decisions in bridge maintenance, safety, serviceability, resilience, sustainability, monitoring, risk-based management, and life-cycle performance using traditional and emerging technologies for the purpose of enhancing the welfare of society. The Editors hope that these Proceedings will serve as a valuable reference to all concerned with bridge structure and infrastructure systems, including students, researchers and engineers from all sections of bridge engineering.

Fabio Biondini and *Dan M. Frangopol*
Chairs, IABMAS 2012

Milan and Bethlehem, April 2012

Bridge Maintenance, Safety, Management, Resilience and Sustainability – Biondini & Frangopol (Eds)
© 2012 Taylor & Francis Group, London, ISBN 978-0-415-62124-3

Acknowledgments

The Editors are extremely grateful to all people who contributed to the organization of the IABMAS 2012 Conference and to the production of this volume. Particularly, the Editors would like to express their sincere thanks to all the authors for their contributions, to the members of the International Scientific Committee and the National Advisory Committee for their role in ensuring the highest scientific level of the Conference, and to the members of the National Organizing Committee for the time and efforts dedicated to make IABMAS 2012 a successful event.

Moreover, the Editors wish to thank all organizations, institutions, and authorities that offered their sponsorship. At the institutional level, a special acknowledgment has to be given to the Politecnico di Milano, for organizing and co-sponsoring this Conference along with the International Association for Bridge Maintenance and Safety (IABMAS), as well as to the Department of Structural Engineering for endorsing and supporting the Conference organization.

IABMAS 2012 has been conceived, planned, and developed in close consultation and cooperation with several individuals. Principally, the Editors wish to express their sincere gratitude to Pier Giorgio Malerba, Honorary Chair of the Conference, and Franco Bontempi, co-Chair of the International Scientific Committee, for their valuable and continuous support to the scientific organization of this Conference.

The Editors are also extremely thankful to Airong Chen, Hyun-Moo Koh and Richard Sause, co-Chairs of the International Scientific Committee; Marcello Ciampoli, Andrea Del Grosso and Claudio Modena, co-Chairs of the National Advisory Committee; Elsa Garavaglia and Luca Martinelli, co-Chairs of the National Organizing Committee; Alessandro Palermo, Secretary of the International Scientific Committee.

Finally, the Editors wish to express their warmest appreciation to Elena Camnasio and Andrea Titi, who led the Scientific Secretariat and provided a huge amount of effective teamwork, and Roberta Stucchi, who designed, developed and maintained the Conference website with praiseworthy dedication and technical skill. Special thanks are due to Stella Pennini, Laura Manenti, Marta Padovan, Miriam Zanelli, and all team of Incentives & Congressi, who professionally managed the Organizing Secretariat with outstanding expertise, commitment and enthusiasm which have been very important for the successful organization of this Conference.

T.Y. LIN LECTURE

Bridge Maintenance, Safety, Management, Resilience and Sustainability – Biondini & Frangopol (Eds)
 ISBN 978-0-415-62124-3

Precast segmental bridge construction in seismic zones

F. Seible
University of San Diego, California, La Jolla, California, USA

ABSTRACT: Following the inspiration and pioneering work of Professor T.Y. Lin, California is finally embarking on segmental precast concrete construction for major bridges in areas of high seismicity. For many years, uncertainty of the behavior of match-cast epoxy bonded joints between precast concrete segments in bridge superstructures during seismic excitation prohibited the use of precast concrete segmental bridge construction in California. While pre-stressed concrete is widely used in cast-in-place multi-cell box-girders and in cast-in-place segmental construction, the Skyway of the new San Francisco-Oakland Bay Bridge East Bay spans constitute the first modern concrete segmental bridge erected in precast segmental cantilever construction. A summary of the precast segmental bridge research for seismic zones conducted in the Powell Structural Research Laboratories at the University of California, San Diego will be presented together with the example of the new San Francisco – Oakland Bay Bridge East Bay Skyway design and construction.

DEDICATION

This paper is dedicated to and in memory of Professor T.Y. Lin (1912–2003) who pioneered pre-stressed concrete as well as pre-cast pre-stressed concrete in the United States. T.Y Lin was Professor at the University of California, Berkeley, for 30 years (1946–1976), where he inspired his students with his enthusiastic lecture style and simple and clear descriptions of the physics of pre-stressed concrete. He received the National Medal of Science in 1986.

Professor T.Y. Lin (1912–2003).

1 INTRODUCTION

Precast segmental construction has evolved over the years to cope with construction difficulties such as: deep valleys, irregular landscapes and congestion that prohibit conventional falsework erection, the desire for shorter construction times, and the ever-present need for higher quality and more efficient construction. Precast segmental construction, as we know it today began in Western Europe in the 1950s and was first implemented in the United States near Corpus Christi, Texas in the 1970s. Since then the use of segmental construction has steadily increased, however there is lack of acceptance in certain areas, specifically in seismic zones.

The precast construction process involves the segmental manufacturing of bridge components in precast yards or plants. The time in which precast segmental bridges are constructed can be substantially less than their cast-in-place counterparts as construction tasks can be completed simultaneously; superstructure segments can be manufactured off site while piers are constructed on site. Construction processes are more refined and efficient at established precast yards/plants than with the cast-in-place method. The quality of precast components is higher than cast-in-place components as precast yards/plants operate in a controlled environment with closely monitored control mix designs and curing conditions. Segments are normally match-cast, thus establishing a proper fit between segments.

Given all of the benefits of precast segmental construction and its advantages over cast-in-place construction there is still significant apprehension in utilizing this construction practice in seismic zones. The reason for this apprehension is the lack of information on the seismic response of segmental constructed precast bridges. The concern with precast segmental bridges is that under seismic motions the superstructure would undergo large deflections resulting in significant joint openings. There is added concern of the behavior of segmental superstructures in regions with both high moments and shears when joint opening is coupled with the possibility of vertical sliding between segments.

The AASHTO Guide Specifications for Design and Construction of Segmental Concrete Bridges permits the use of precast segmental construction in high seismic zones (Zones 3 and 4) provided that segment-to-segment joints are bonded by segmental bridge adhesive (epoxy). The same AASHTO specification also requires that external unbonded tendons should account for no more than 50 percent of the superstructure post-tensioning. Additional recommendations in seismic areas such as California suggest that mild reinforcement should cross the segment-to-segment joints of precast segmental bridge superstructures to avoid partial or complete superstructure collapse in case of tendon rupture. These recommendations are intended to be conservative but are based on little if any research investigating the seismic response of precast segmental bridges.

Research at the University of California, San Diego (UCSD) developed the basic understanding of how precast segmental bridges respond to seismic events and that precast construction is a feasible option for bridge construction in high seismic zones. A three-phase project was conducted by the American Segmental Bridge Institute (ASBI) and funded by the California Department of Transportation (Caltrans) for this study. Experimental and analytical studies focused on the seismic response of segment-to-segment joints in precast segmental superstructures with different ratios of internal to external post-tensioning under simulated seismic fully reversed cyclic loading simulating mid span type conditions of high moments and low shear. Next, the investigations also varied the ratio of internal to external post-tensioning but focused on segment-to-segment joints in regions with high negative moments and high shears. Finally, the research focused on the performance of segmental superstructures and columns under the combined effect of gravity loads and longitudinal and vertical seismic forces in a segmental cantilever constructed large scale bridge systems tests.

The goal of this project was to study the performance of segment-to-segment joints in bridge superstructures under simulated seismic fully reversed cyclic loading for varying ratios of internal to external post-tensioning. An additional objective was to study the seismic performance of segment-to-segment joints that have cast-in-place deck closures with mild reinforcement crossing the segment-to-segment joints, similar to the design proposed originally for the new East Span Skyway Structure of the San Francisco-Oakland Bay Bridge. The study of Phase I focused on superstructure joints close to mid-span where high moments and low shears are induced. An extension of this test phase comprised a proof-test of the modified SFOBB skyway design that replaced the originally proposed cast-in-place closure pour and mild reinforcement between segments with unbounded lightly stressed auxiliary tendons. The Phase II study focused on superstructure joints close to the bent cap where high negative moments and high shears are induced. Finally, Phase III studied the seismic performance of a precast pre-stressed segmental cantilever bridge column/superstructure system. The major objectives of this research were to investigate: (1) seismic behavior with respect to the opening and closing of joints under cyclic seismic loading, (2) crack development and propagation, (3) permanent deformations, and (4) failure modes. In this paper, only the tests directly relevant to the Skyway of the new San Francisco-Oakland Bay Bridge, namely Phase I, will be discussed in detail while general results from the other research findings will be mentioned.

2 THE NEW SAN FRANCISCO-OAKLAND BAY BRIDGE (SFOBB) EAST BAY SPAN SKYWAY

On October 17, 1989, an earthquake with a Richter magnitude of 7.1 struck the San Francisco-Oakland area, causing a 49 ft. (15 m) span of the current East Bay Bridge to collapse, resulting in one death and decommissioning the bridge for one month. This seismic event, known as the Loma Prieta earthquake, exposed weaknesses in the East Bay Bridge, which was designed in the 1930s, well before modern seismic design principles were introduced. In addition, the San Andreas and Hayward faults, both of which are located in the Bay Area, are capable of producing larger ground motions than those recorded during the Loma Prieta earthquake. With these factors in mind, Caltrans investigated the possibilities of retrofitting the existing bridge or constructing an entirely new East Bay Bridge. The design of an entirely new bridge was deemed more economically feasible than retrofitting the existing structure.

The new East Bay Bridge will consist of two main structures: a Self-Anchored Suspension (SAS) structure just east of Yerba Buena Island, and parallel Skyway structures that lead from the suspension span to the Oakland shore. The skyway structures consist of precast segmental concrete box girders of varying depth, see Fig. 1.

The precast segmental girders are post-tensioned with main tendons and auxiliary deck tendons. The auxiliary tendons will be stressed such that the effective pre-stress in the tendons after losses is 40 percent of their ultimate tensile strength. The addition of the auxiliary tendons was a major change from the initial design, which called for precast segmental girders

Figure 1. SFOBB skyway construction.

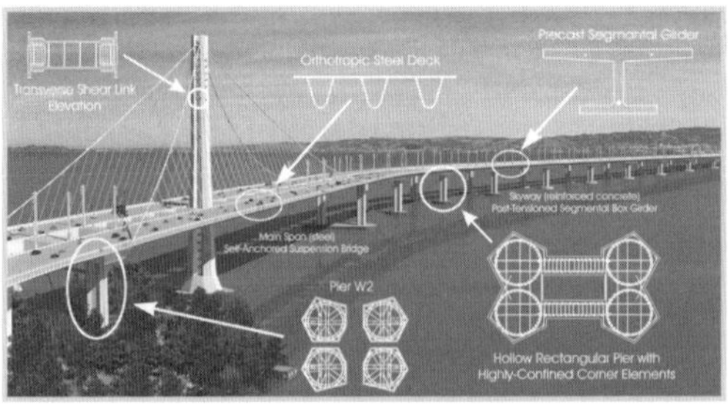

Figure 2. New east bay spans of the San Francisco Oakland Bay Bridge with location details for various testing programs.

with cast-in-place deck closures. Problems with cast-in-place deck closure joints include additional cost and construction time as well as the relative complexity of construction (vertical stirrups should be included in the closure joints to combat bucking of the longitudinal bars). The use of lightly stressed auxiliary tendons can avoid these added costs and construction difficulties while providing a comparable overall response.

Research programs that studied all structural components of the new East Bay Bridge that are expected to undergo inelastic deformation during an earthquake were conducted at UCSD. These programs included investigation of the seismic response of the shear links in the steel suspension tower, the skyway concrete piers (subjected to both uniaxial and biaxial bending), the orthotropic steel deck in the suspension span, and Pier W2, which contains the anchor for the suspension tower. Finally testing of the behavior of the precast segmental Skyway superstructure was conducted in the Charles Lee Powell Structural Systems Laboratory at UCSD. A rendering of the new East Bay Bridge, including details indicating the location on the bridge of the tests described above, is presented in Fig. 2. The research into the performance of pre-cast segmental bridge construction in seismic zones as applied to the Skyway is the subject of this paper.

3 RESEARCH OBJECTIVES FOR PRE-CAST SEGMENTAL CONSTRUCTION IN SEISMIC ZONES

A large-scale experimental research program on precast segmental bridge superstructures has been conducted in the Powell Structural Research Laboratories at UCSD. This test program consisted of three phases (Megally et al. 2002) (Densley et al. 2003) (Burnell et al. 2005). Phase I investigated segment-to-segment joints in regions of high moment and low shear (i.e the mid span region), and phase II tested the seismic performance of segment-to-segment joints in regions of high moments and high shears (i.e., joints close to the columns or bents) (Megally et al. 2002) (Densley et al. 2003). Finally, phase III examined the effects of seismic loading on a bridge system consisting of a cast-in-place (CIP) column with a precast segmental superstructure (Burnell et al. 2005). The research program in its entirety is summarized in (Megally et al. 2009). The research program presented herein specifically focuses on the phase I program with direct implications on the design and construction of the Skyway portion of the new SFOBB East Bay spans.

The original design of the SFOBB East Bay Span Skyway provided continuous deck reinforcement across the segment-to-segment joints in the form mild reinforcement and a 3 ft (1 m) closure pour of cast-in-place concrete that significantly compromised the erection schedule. These requirements were invoked following the 1994 Northridge earthquake, during which a post-tensioned bridge structure fully collapsed. Essentially, providing mild steel in addition to post-tensioning tendons adds redundancy to the precast segmental superstructure, preventing collapse in the event of full tendon rupture. Since cast-in-place deck joints between precast superstructure segments are counterproductive in segmental erection, alternative design options were investigated that would provide for redundancy in case of tendon rupture while allowing rapid cantilever erection. Lightly stressed auxiliary deck tendons provide such an alternative to cast-in-place closure joints with mild continuity reinforcement. Caltrans funded one additional test with a fully internal tendon configuration and lightly-stressed auxiliary tendons at the top flange location in order to draw direct comparisons between the behavior of this system and the cast-in-place deck closure joint configuration for the SFOBB Skyway.

4 TEST SPECIMEN DESIGN AND CONSTRUCTION, TEST SET-UP AND LOADING PROTOCOL

It is important to note that the test specimens were not a model of the actual SFOBB Skyway structure. Rather, the objective of this test was to obtain the general response of precast segmental bridge superstructures under seismic loads using auxiliary deck tendons; thus a generic bridge superstructure was tested. As a

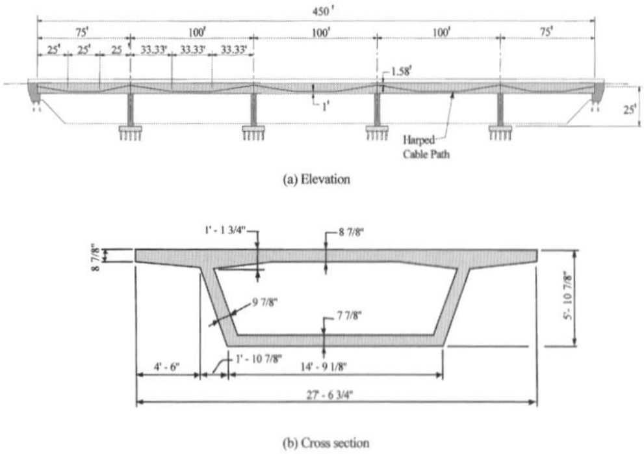

Figure 3. Prototype bridge structure.

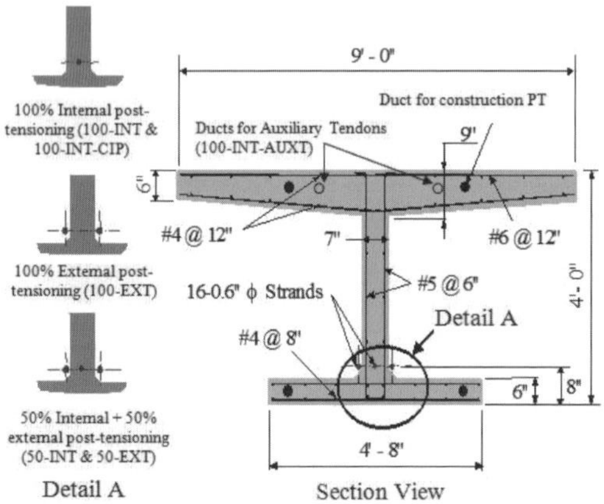

Figure 4. Test unit cross-section and reinforcement/ post-tensioning layout.

Table 1. Test matrix for phase I.

Unit	Description	Nomenclature
1	100% internal post-tensioning	100-int
2	100% internal post-tensioning with cast-in-place deck closure joints	100-int-cip
3	100% external post-tensioning	100-ext
4	50% internal and 50% external post-tensioning	50-int/50-ext
5	100% internal with auxiliary tendons	100-int-aux

consequence, test results are not specific to the SFOBB Skyway but can be used as guidelines for basic seismic performance of precast segmental bridges with lightly stressed auxiliary tendons in the deck slab. A prototype structure was designed initially for the first two phases of the project, and all the test units were designed at a 2/3-scale of a prototype structure (Megally et al. 2002), see Fig. 3. A single web with effective width flanges was isolated as test article and match-cast and assembled as shown in Fig. 4. The test matrix consisted of the 5 tests shown in Table 1.

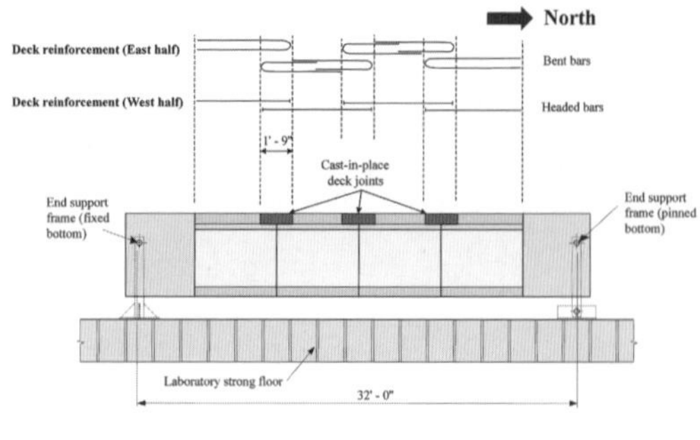

Figure 5. Test set-up with cast-in-place closure deck joints.

Figure 6. Construction and assembly of test units.

Test unit 1 (100-INT) featured 100% internal post-tensioning across segment-to-segment joints. Test unit 2 (100-INT-CIP) was modeled after the original SFOBB Skyway design with full internal post-tensioning but with cast-in-place deck closure joints with mild continuity reinforcement. The test setup and the closure joints/reinforcement are shown in Fig. 5. Test unit 3 (100-EXT) investigated the performance of 100% external post-tensioning. Test unit 4 (50-INT/50-EXT) featured combined 50% internal and 50% external post-tensioning as described as the limiting case by AASHTO, and finally test unit 5 (100-INT-AUX) investigated the proposed alternative for the SFOBB Skyway with lightly stressed auxiliary tendons. In lieu of the mild deck reinforcement across joints for test unit 2, test unit 5 contained two 3-strand tendons that ran through the deck region and across the segment-to-segment joints, see Fig. 4. Construction and assembly of the test units is documented in Fig. 6.

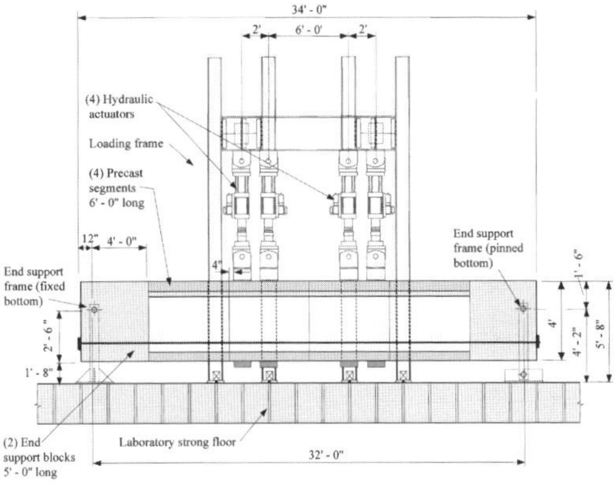

Figure 7. Elevation of test setup.

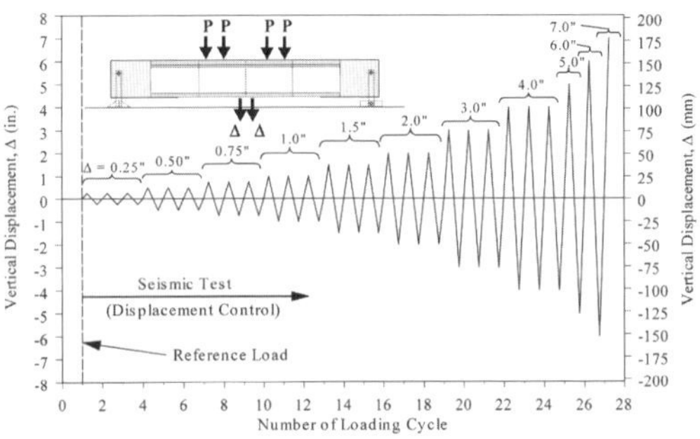

Figure 8. Loading protocol.

The schematic test setup for the match-cast epoxy bonded segment-to-segment joint tests is shown in Fig. 7, and the fully reversed cyclic loading protocol is depicted in Fig. 8.

5 PHASE I TEST RESULTS FOR INTERNAL VS. EXTERNAL TENDON ARRANGEMENT

The main objectives of this research phase were to study the behavior of joint opening and closing, crack patterns, and failure modes for different arrangement/combinations of internal (INT) and external (EXT) tendons. This research phase showed that superstructure segment-to-segment joints could reach significant openings without failure under seismic loading. In order to quantify the amount of joint opening, joint rotations were calculated for each specimen at each segment-to-segment joint under both upward and downward loading. Joint rotations were calculated by observing the displacements indicated by horizontal potentiometers that straddled each joint at the top and bottom surfaces of the test specimen; the difference between the potentiometers' readings was then divided by the height of the specimen. The results presented in the aforementioned report show that for each specimen, the center joint was the only joint that showed significant rotation. Table 2 shows only the maximum measured joint rotations at the center joint under both upward and downward loading.

Table 2. Maximum midspan joint rotations, phase i.

Test unit	Downward loading (Radians)
100-INT	0.035
100-EXT	0.071
50-INT/50-EXT	0.033

As can be seen from Table 2, for full external post-tensioning (100-EXT) for the same maximum vertical deflection the joint opening under downward loading at the center joint was twice the deformation occurring in the center joint of the other specimens. The other test specimens also showed opening in adjacent joints as can be expected from crack distribution with bonded reinforcement.

A second important aim of Phase I was to study the failure modes of each tendon configuration. The testing showed that the test units with full internally bonded tendons failed much more explosively than the fully externally post-tensioned unit. Although unit 100-EXT exhibited the greatest displacement capacity, unit 100-INT achieved higher strength levels. This comes from the from the tendon shift relative to the neutral axis for external un-bonded tendons. Unit 50-INT/50-EXT was found to be the least suitable for seismic regions due to its low displacement capacity initiated by rupture of the 50% internal bonded tendons. The test results show that while the total tendon/post-tensioning load capacity is the same, the 50/50 split cannot be assumed as additive but rather sequential as far as failure deformation levels are concerned. A quick comparison of the load-displacement behavior for units 100-INT, 100-EXT, and 50-INT/50-EXT is displayed in Fig. 9 where positive displacements correspond to downward loading. The total load in the figures is the total applied actuator load, and does not include beam self-weight. The reference load was defined as zero seismic loads, and was calculated such that the stress at the midspan joint was the same as the stress at the midspan joint of the prototype structure under dead loads and primary and secondary prestressing effects. Envelopes of total load-displacement behavior for downward loading are also provided in Fig. 10, so that clear comparisons of load carrying capacity and displacement capacity for each test unit can be made.

Permanent residual displacement is another concern in seismic regions. Although measuring residual displacements was not a primary objective in Phase I, they were noted during testing and presented in the report. Findings indicated that test unit 100-EXT exhibited the lowest residual displacement. Residual displacements were recorded during testing by unloading the specimen to zero applied seismic load. Shown below in Table 3 are the residual displacements recorded after loading to a 3 in. (76.2 mm) downward displacement and subsequent unloading.

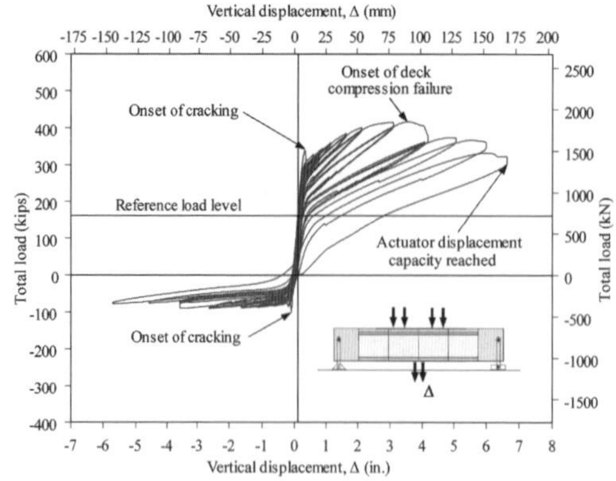

(a) Unit 100-Int

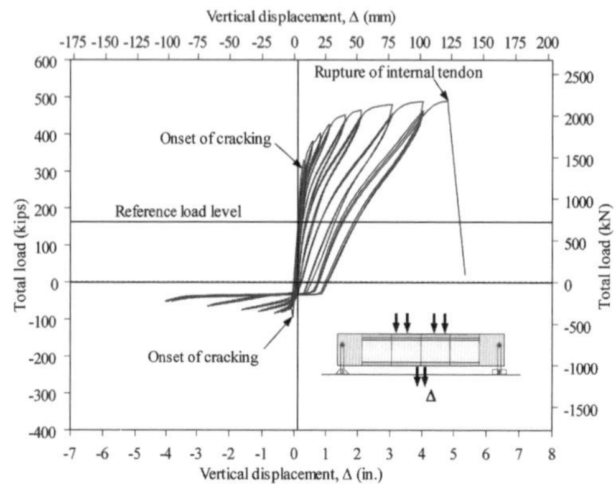

(b) Unit 100-EXT

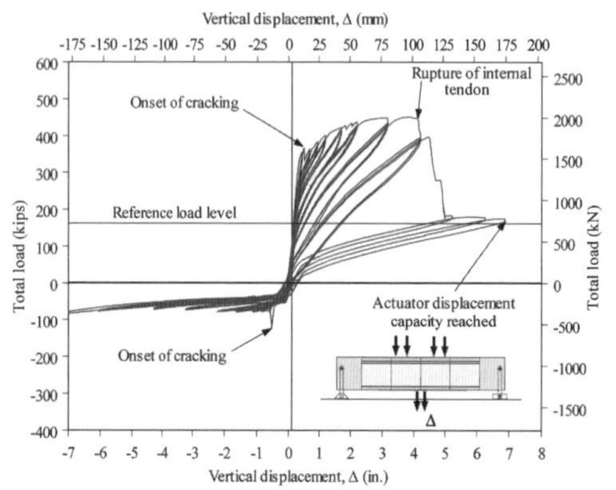

(c) Unit 50-INT/50-EXT

Figure 9. Load-displacement behavior for phase I test units with INT/EXT tendon arrangement.

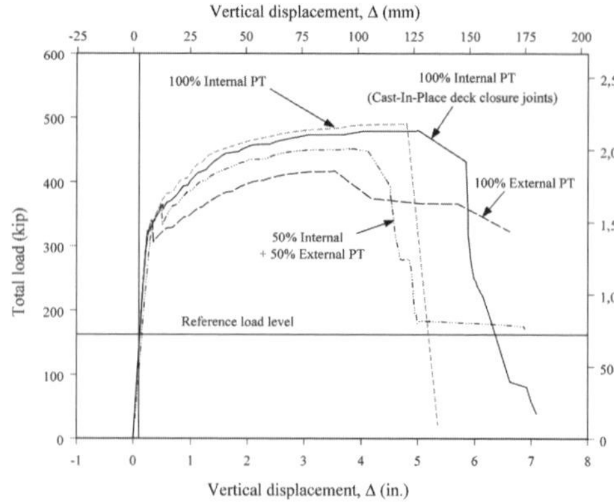

Figure 10. Envelopes of load versus displacement for downward loading, phase I.

Table 3. Maximum permanent residual displacements Δ_r during 3 in. (76.2 mm) cycle, Phase I.

	Δ_r	
Test unit	in	mm
100-INT	1.17	29.7
100-EXT	0.14	3.56
50-INT/50-EXT	0.82	20.8

6 TEST RESULTS FOR SKYWAY TYPE SPECIMENS WITH DECK CLOSURE POURS OR AUXILIARY TENDONS

The test unit with lightly stressed auxiliary deck continuity tendons (100-INT/AUX) can be directly compared to Units 100-INT and 100-INT-CIP of Phase I because all three units had the exact same dimensions and were all modeled based on the same prototype structure at midspan. The only variable that changed through these units was the technique employed to control joint opening behavior in the deck region. All units had the same 16-strand main tendon configuration, stressed to the same level. In the deck region, Unit 100-INT had no mild or prestressing steel crossing the joints, whereas Unit 100-INT-CIP had cast-in-place closure joints with mild steel reinforcement crossing each joint. The 100-INT-AUX Unit employed a system of two 40% stressed 3-strand tendons crossing each joint in the deck slab. In Unit 100-INT, joints were allowed to freely open during upward loading, while this opening was constrained in the other two units.

Figure 11 shows load-displacement envelopes for each test unit. Vertical deflections were measured at a location 6 in. (152 mm) from the midspan joint in all test units, so direct comparison of the load-displacement response is possible. Under downward loading, the response was comparable for all test units. Unit 100-INT-AUXT showed slightly more displacement capacity than Unit 100-INT-CIP. Unit 100-INT displayed high displacement capacity, but was not able to achieve the displacement capacities of the other two units without continuous deck reinforcement. Maximum load carrying capacity under downward loading was essentially identical for each of the three test units.

Although Units 100-INT-CIP and 100-INT-AUX showed similar results for both upward and downward

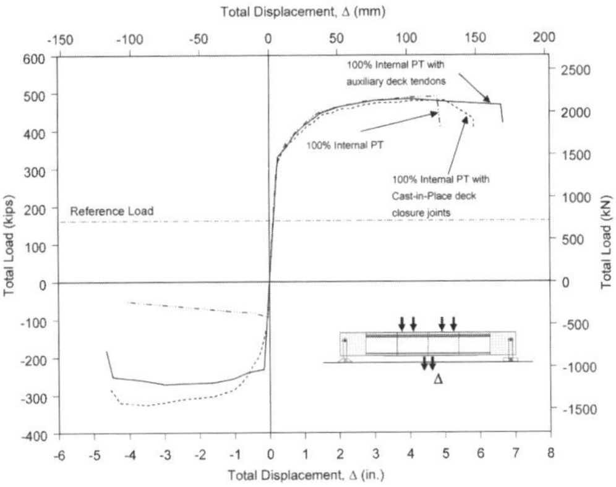

Figure 11. Load-Displacement Envelopes for Units 100-INT, 100-INT-CIP, and 100-INT-AUX.

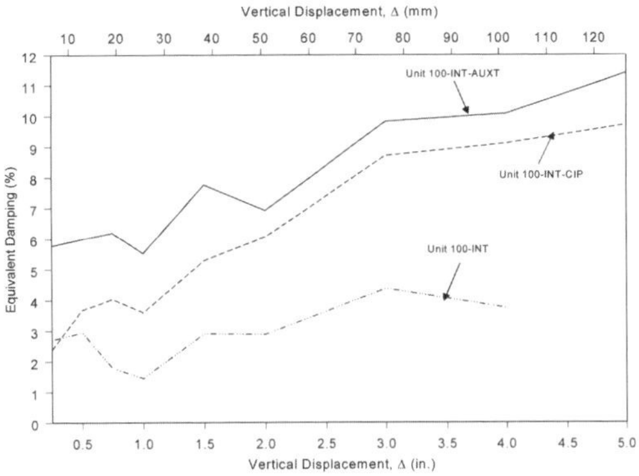

Figure 13. Energy dissipation versus displacement.

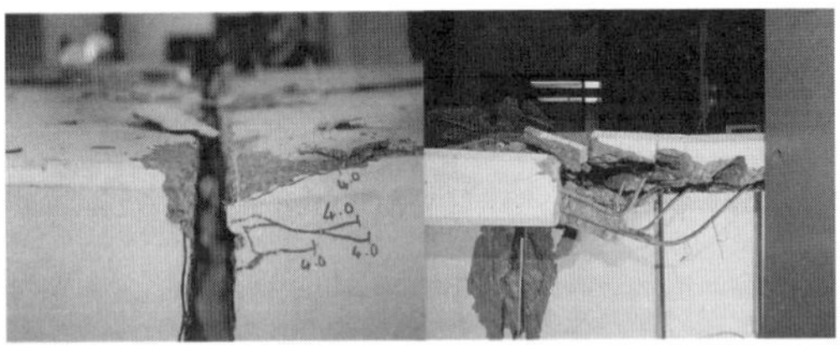

(a) Joint J3 After Failure, Unit 100-INT

(b) Joint J3 After Failure, Unit 100-INT-CIP

(c) Joint J3 at Failure in Unit 100-INT-AUX

(d) Concrete Crushing at Joint J3, Unit 100-INT-AUX

Figure 12. Joint failure modes.

loading, failure mechanisms were not similar, see Fig. 12. The main failure mechanism present in Unit 100-INT-CIP was compression failure of the deck under downward loading. At high downward displacement levels, the reinforcing bars that crossed each joint in the deck began to buckle and pushed against the concrete cover in the deck slab. Buckling of these bars prevented further testing from being conducted under upward loading. Unit 100-INT-AUX was tested to failure in both upward and downward loading directions as well, and tendon rupture occurred in each direction. No well-defined compression failure was observed in Unit 100-INT-AUX, with only moderate spalling occurring at the deck region and bottom surface of the specimen.

Relative energy dissipation characteristics for all test units were computed based on a direct comparison of equivalent hysteretic damping characteristics. Note that these values are not absolute measures of energy dissipation, but are intended only for relative comparisons between test units. Unit 100-INT exhibited virtually no energy dissipation during upward loading, as joints were allowed to freely open with no yielding of reinforcement or cracking of concrete. Units 100-INT-CIP and 100-INT-AUXT demonstrated well-defined energy dissipation under upward loading, especially at high levels of displacement. Although the energy dissipation characteristics of the superstructure may not be of concern because inelastic deformation will likely occur only in columns by means of column plastic hinging, it may be desirable for the superstructure to have the ability to obtain high levels of energy dissipation for reasons related to systems response and redundancy. Equivalent damping characteristics are depicted in Fig. 13.

7 SUMMARY OF RESULTS

Results obtained from these phase I tests and subsequently confirmed in the other testing phases for high shear regions and for combined systems response are as follows:

- At deformation level that can be expected in a segmental bridge under the Design Basis Earthquake (DBE) cracks will develop in the cover concrete adjacent to the epoxy bonded joint and not in the joint itself, independent of the post-tensioning method.
- After removal of the DBE load the cracks close completely and are no longer visible.
- No joint shear failures were observed in any test phase since the friction through the compression toe is always larger than the shear force.
- All internal (100-INT) or all external (100-EXT) tendon arrangements result in similar displacement capacities.
- All external (100-EXT) tendon arrangement require slightly more post-tensioning due to the tendon shift relative to the neutral axis of the cross-section.

- All external (100-EXT) tendon arrangements show significantly smaller residual deformations due to the non-linear elastic systems response.
- Mixed tendon arrangements (50-INT/50-EXT) show the lowest deformation capacity due to their sequential failure behavior with the internal bonded tendons rupturing prematurely.
- The failure mode of joints with bonded mild or grouted pre-stressed continuous deck reinforcement can be brittle due to bar buckling.
- The failure mode of joints without reinforcement through the joint is more gradual by concrete spalling.
- Ungrouted lightly stressed continuity tendons can provide the redundancy against superstructure collapse in the case of main tendon failure.
- No obvious reason was found why precast segmental construction could not be successfully used in regions of high seismicity.

This extensive research project showed that precast segmentally erected bridges can have excellent performance of the segment-to-segment joints in seismic zones. While cracks near joints will open during seismic motions, their deformation capacity is larger by an order of magnitude than the demands expected from the DBE. Cracks will completely close following the DBE.

The above findings and the validation via the phase III systems test showed that limited inelastic action in the joints of segmentally erected bridges can be an opportunity for new design approaches for seismic zones that can potentially reduce the amount of pre-stress required in the superstructure to force all inelastic action into the bridge columns.

ACKNOWLEDGMENTS

The following faculty, engineers, researchers, and graduate students contributed to the success of this research project: Sami Hanna Megally, Manu Garg, Robert K. Dowell, Jeffrey Densley, Kelly P. Burnell, and Jose I. Restrepo. Without their dedication and expertise this project would not have been possible. The research project was partially funded by ASBI (American Segmental Bridge Institute) and by Caltrans (California Department of Transportation.

REFERENCES

Burnell, K., Megally, S., Restrepo, J. & Seible, F., (2005) *Seismic Testing of Precast Segmental Bridges, Phase III: Bridge Systems Test*, Department of Structural Engineering, University of California, San Diego, CA, Structural Systems Research Project SSRP 2005/01.

Densley, D., Megally, S. & Seible, F., (2003) *Seismic Performance of Precast Segmental Bridge Superstructures with Lightly Stressed Continuity Tendons*, Department of Structural Engineering, University of California, San Diego, CA, Structural Systems Research Project SSRP 2003/10.

Megally, S., Garg, M., Seible, F. & Dowell, R., (2002) *Seismic Performance of Precast Segmental Bridge Superstructures*, Department of Structural Engineering, University of California, San Diego, Structural Systems Research Project SSRP-2001/24.

Megally, S., Veletzos, M., Burnell, K., Restrepo, J. & Seible, F., *Seismic Performance of Precast Concrete Segmental Bridges: Summary of Experimental Research on Segment-to-Segment Joints*, (2009) PCI Journal, Vol. 54, No. 2, pp 116–142.

KEYNOTE LECTURES

Bridge Maintenance, Safety, Management, Resilience and Sustainability – Biondini & Frangopol (Eds)
© 2012 Taylor & Francis Group, London, ISBN 978-0-415-62124-3

Renewal and rehabilitation of the Brazilian railway bridge infrastructure

T.N. Bittencourt, R.M. Teixeira & A.B. Colombo
Escola Politécnica – University of São Paulo, São Paulo, Brazil

ABSTRACT: This paper describes the challenges that Brazil faces in order to improve its transportation infrastructure to set the grounds for a continuous development and economic growth. The renewal and rehabilitation of the railway network is very important in this context. Plans for the expansion of the Brazilian railway system and its impact on the new and existing bridges are presented as well as research projects involving the assessment of some existing steel and reinforced concrete bridges through short term monitoring campaigns. The need for long term monitoring projects is also discussed.

1 INTRODUCTION

Recently Brazil has become the 6th most important economy in the World surpassing the United Kingdom in total numbers. Very soon the country is supposed to take the 5th position from France. Unfortunately, this impressive performance is terribly diminished when considered relatively to the total population and it can be easily realized that much needs to be done to improve the quality of life for the ordinary citizens. Education and infrastructure are major problems that need to be addressed urgently.

The enhancement and modernization of the transportation infrastructure is crucial to set the ground for a continuous development and economic growth in the coming years. In particular, the renewal and rehabilitation of the railway network is one the most important priorities. The modernization of the railways, and specifically of bridges and tunnels, is very important in this context.

In this paper an overview of this modernization process is presented together with the perspectives for the future and needs for the present. The impact of the expansion of the Brazilian railway system on the new and existing bridges. Railway bridges are facing an increase of axle loads to attend the internal and global demands for commodities, especially crops, sugar, oil, ethanol and minerals.

The existing bridge infrastructure and a description of some important and emblematic bridges is presented. Then the challenges in the renewal and rehabilitation processes are introduced. The main deterioration mechanisms are identified and the role of a bridge management system is established. The importance of monitoring and the application of some SHM (Structural Health Monitoring) techniques is shown through some examples of recent studies for steel and reinforced concrete bridges along the Vitoria-Minas Railway. Short term monitoring campaigns have been used to assess the performance of such bridges under controlled and service load conditions. Analyses are concentrated so far to fatigue problems, but other deterioration phenomena, like corrosion, carbonation, alkali-aggregate reaction and chloride attack, will have to be addressed in the near future.

To conclude some perspectives for the future are presented with special emphasis on the new plans for improving the regional and metropolitan railway network and for constructing high speed tracks connecting major cities in the country.

2 RECENT INVESTMENTS ON NEW BRIDGE INFRASTRUCTURE

In the last decade the railway industry in Brazil has received substantial amounts of new investments to improve the installed capacity of production of railway equipment and to introduce new technologies. These investments are reflected in the crescent numbers in the production of new wagons, locomotives and passenger cars over the last decades (Table 1).

Moreover, new technologies have been developed or introduced to improve the performance of the new and existing railway systems. Among these, it is important to notice the introduction of new wagons of higher capacity for iron ore transportation with 150 tons of weight and higher axle loads of 37.5 tons, new wagons for sugar and crops transportation with more efficient charge and discharge systems, double stack wagons for containers, 4400 HP diesel-electric locomotives, automated traffic control systems, comfortable passenger cars, more efficient noise attenuation systems and more sustainable concrete and plastic sleepers.

Currently, the major railways in Brazil are administrated by concession. The investment numbers in the existing railway network systems show an continuous increase since 1997 (Figure 1) with an important

Table 1. Production of new railway equipment in the last decades.

	1980-1990	1991-2000	2000-2010
New Wagons	3400	28200	40000
New Locomotives	69	175	2100
New Passenger Cars	640	1930	4000

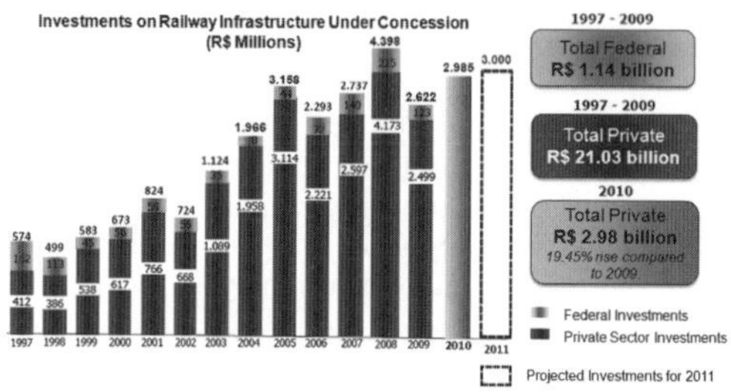

Figure 1. Private investment in the existing railway system.

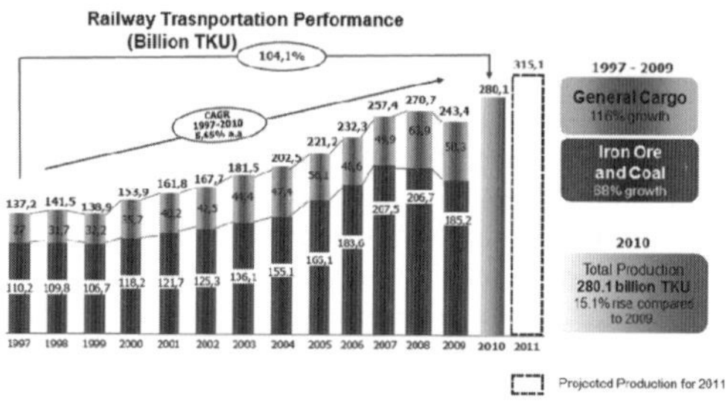

Figure 2. Transportation performance growth.

growth of 104.1% in the net transportation capacity per kilometer (TKU) and a compound annual growth rate of approximately 5.65% (Figure 2).

This scenario poses some new challenges to the railway infrastructure in Brazil. The existing railways will need proper maintenance and rehabilitation to face the demands of the economic situation while the new ones will need to be designed for an increasing environmental and loading demand. These challenges open attractive opportunities for the application and development of life cycle studies with the support of SHM techniques.

In the following sections the application of these studies for steel and reinforced concrete railway bridges will be emphasized.

3 EXISTING BRIDGE INFRASTRUCTURE

The Brazilian railway bridge infrastructure is composed of a few different bridge types. In order to give a general idea about this infrastructure, a few of the existing bridges will be shown as example.

Bridge design is greatly influenced by the physical obstacle that it is intended to span over. The Brazilian

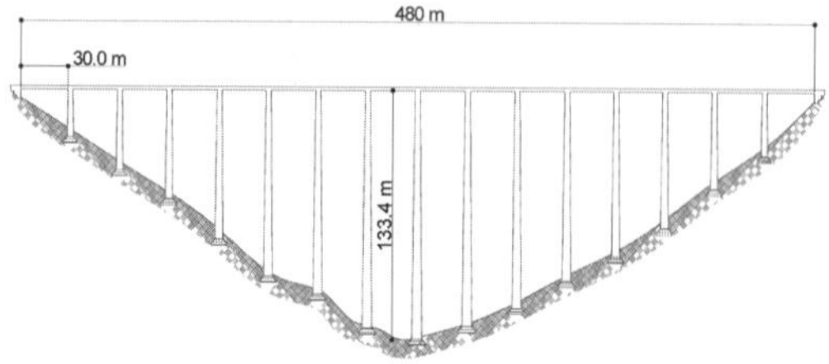

Figure 3. Elevation sketch of the Muçum Viaduct showing the principal dimensions of the structure and view of the structure (Vasconcelos 1993).

Figure 4. View of the Doce I River Bridge showing the three Warren truss spans with a total length of 103.10 m.

landscape is very diverse depending on the region. For example, northern Brazil has a vast area of relatively flat rainforest with large rivers, while southeastern Brazil has its largest mountain range with elevations just under 2,000 m. Brazil has also one of the largest river networks in the world, it is estimated to hold 13% of the world's fresh water reserve. This scenario shows how vital bridge engineering is in this country.

The Muçum Viaduct, in the southern state of Rio Grande do Sul, has the record for highest columns in Latin America, nearing 134 m. It is part of the railway mainly responsible for the transportation of grains from the region, being nicknamed "*The Wheat Railway*". This viaduct passes over a 480 m long valley having columns spaced in 30 m spans; it is composed of a reinforced concrete π-shaped deck supported by reinforced concrete columns.

Dynamic and wind load effects had to be considered in the design, especially because of the height of the columns. The studies included aerodynamic analysis of reduced scale models in wind tunnels, as reported by Blessman (1989).

In the crossing of the Vitoria Minas Railway with the Doce River, two bridges are found: the Doce I and Doce II bridges. Both bridges are composed of complete Warren trusses spans supported over reinforced concrete columns. The Doce I has three trusses over a total length of 103.10 m (Figure 4) and the Doce II has six spans constituting a bridge with total length of 247.60 m (Figure 5).

In northern Brazil the Carajas Railway is the main transportation mode for the iron ore from the town of Carajas to the harbor in São Luis. The OAE 40 bridge is located in this railway, it is a reinforced concrete bridge 593 m in length. The structure is composed of 22 spans with 25 m each (see Figure 6); two longitudinal beams are linked by transversal beams 300 mm in width in

Figure 5. View of the Doce II River Bridge showing the six spans with a total length of 247.60 m.

Figure 6. View of the OAE 40 Bridge (Teixeira et al. 2010).

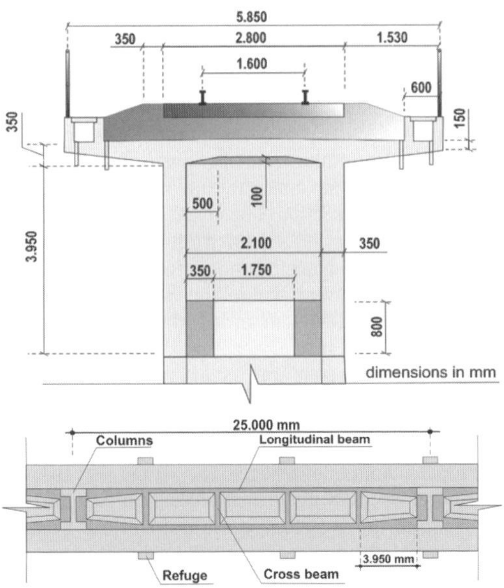

Figure 7. Typical cross section and span for the OAE 40 Bridge (Teixeira et al. 2010).

intermediary positions and 500 mm near the columns (see Figure 7).

Another typical example of a reinforced concrete railway bridge from the 1970's, is the Suaçui River Bridge (shown in Figure 8). This bridge is part of the Vitoria-Minas Railway in southeastern Brazil and its structure is currently under evaluation. The total length of this structure is 54.77 m divided in three isostatic

Figure 8. View of the Suaçui River Bridge.

spans of 18.26 m each, these spans are supported by reinforced concrete columns. A reinforced concrete slab over two longitudinal beams linked by transversal beams makes up this bridge's superstructure.

4 RENEWAL AND REHABILITATION OF EXISTING STRUCTURES

4.1 *Overview of needs*

The main degradation agents of existing railway bridges are linked to the growth of traffic and vehicle load. This is especially true for railways under heavy axle load (HAL). The traffic and load conditions imagined during the design and construction phases are usually different from the conditions found during the lifetime of the structure. These differences may be due to changes in design codes or changes in the use of the structure. In the particular case of HAL, the vehicle load proposed in design codes could prove to be unconservative. Recently in Brazil there has been a certain tendency to raise the axle load of rail vehicles in existing and future railways. This practice has direct impact in the bridge infrastructure, generating a need of an effective plan for the renewal and rehabilitation of this infrastructure.

4.1.1 *The need of a Bridge Management System*

Current inspection and maintenance programs also deserve some attention. New programs must be implemented within a Bridge Management System (BMS). The main Brazilian railway concessions are still dimensioning the funds necessary for the rehabilitation of the existing bridge infrastructure and do not have a solid and uniform management system, like the ones found in developed countries that use knowledge-based expert computer systems and artificial intelligence as management tool (Randomski 2001).

4.1.2 *The role of the field testing, evaluation techniques and SHM*

Currently, some field testing techniques have been extensively used in other countries. The application of these techniques with non-destructive inspections

(NDI) and Structural Health Monitoring (SHM) systems can improve the assessments and decision-making of the BMS.

Brazil has limited experience in field testing techniques and little research on the subject has been developed. Recently, some research projects have been underway in Brazilian universities with the intent of bring attention from scholar and companies linked to these fields. Different types of bridges have been under evaluation in two of the main railways responsible for the transportation of iron ore in the country, the Vitoria Minas Railway, with 153 bridges, and the Carajas Railway, with a total of 54 bridges. A similar program is being developed with the Central-Atlantic Railway concession which has 1,653 bridges.

The ongoing work consists basically of short-term monitoring and tests with controlled service loads with the monitoring of strains, displacements and accelerations in key points of the bridges. Non-destructive inspections are also used to determine the degradation of the structural elements using techniques such as the rebound hammer method and ultrasonic pulse velocity test in the qualitative evaluation of concrete uniformity.

Although there has been some effort towards implementing an SHM system in some bridges that are considered important for the iron ore railways, none has been implemented, that is of the author's knowledge. The need for such systems is well known but its implementation is difficult due to economic reasons. An efficient SHM system is directly linked to its capacity to identify a change in structural performance. This depends on a greater level of complexity, which in turn generates greater costs. Greater level systems depend on more efficient technologies with better performance for the sensor network and the acquisition system, which must transmit and process a large amount of data. The data processing must be able to give a diagnostic and prognostic with the remaining lifetime of the structure and present and adequate computer platform for the system management (Wenzel 2009).

4.2 *Recent assessments of Brazilian railway bridges*

As was mentioned earlier, some research projects in the assessment of railway bridges have been underway in Brazilian universities. A brief description of the problems found, the rehabilitation techniques and the short-term monitoring practices used will be shown in the subsequent items.

4.2.1 *Reinforced concrete bridges*

Two examples of the studies made that involve reinforced concrete bridges will be shown: the Suaçui River Bridge in the Vitoria Minas Railway and the OAE 40 in the Carajas Railway. Scaling, honeycombs, infiltration, exposed and corroded reinforcement and cracking with widths between 0.3 mm and 1.0 mm were found in both of these bridges before the repairs were made.

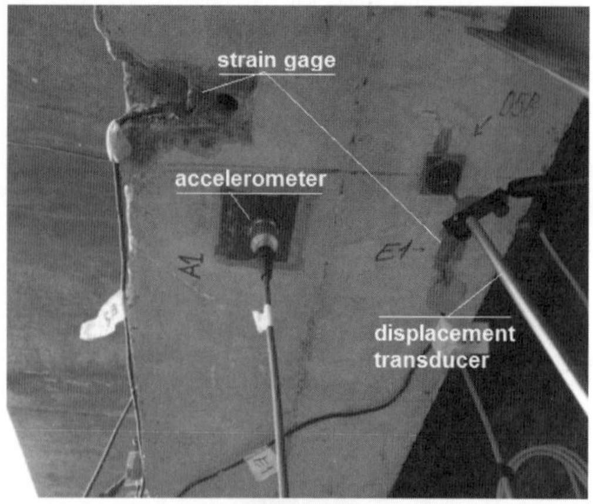

Figure 9. Instruments in longitudinal beam.

4.2.1.1 *Suaçui River Bridge*

A two-week short-term monitoring program was developed in the Suaçui River Bridge in 2011. In this program the curvature and stresses on the concrete and reinforcement of a cross-section were calculated. The displacements and strains were measured using inductive transducer and strain gauges, respectively. Figure 9 shows a typical monitoring setup of a longitudinal beam. Semi-destructive and non-destructive tests were also made at the site and in laboratory in order to characterize the concrete. With the collected data a fatigue analysis for the cyclic load was done, using as reference the load cycle of the monitored period.

Overall the results from the analysis showed low stress levels in the concrete and reinforcement for the studied cross section. The visual inspections, local tests and laboratory tests showed good quality concrete with uniform resistance among the tested regions. Crack openings were within the normative specifications in Brazil and in order to ensure a greater durability for the structure, a repair of the cracks and defects in the concrete was recommended.

4.2.1.2 *OAE 40 Bridge*

Teixeira (2009) presented a similar study of another reinforced concrete bridge, the OAE 40 Bridge. Structural integrity evaluations based on field testing and monitoring were made. Situated in the Carajas Railway, this bridge was normally submitted to axle loads of around 325 kN and was being tested with loads of 375 kN. This 25-year-old structure received repair in 2004 and 2006. The study presented a theoretical and experimental methodology to evaluate the effect of a 400 kN axle load on this bridge. An experimental program involving the concrete characterization with semi-destructive tests and non-destructive tests and, a monitoring program of the strains in cross sections of the superstructure and foundations was developed. The results from this experimental program were used in the calibration of the numerical models.

This study had conclusions similar to the ones found in the Suaçui River Bridge. The repairs made in 2006

Figure 10. Fatigue failure in the Correntes River Bridge.

left the structure in good conservation state. The reinforcement was treated with projected mortar, epoxy coating, polymer based mortar, cracks with openings larger than 0.3 mm were repaired, the drainage systems were repaired as well as the abutments, guardrails and the concrete cover were also repaired.

The non-destructive tests demonstrated an overall good quality concrete with distribution in the studied regions within acceptable limits. The structural design was checked for various load cases, including the 400 kN axle load, for all cases the structure showed to be adequate. These verifications were made using the Brazilian reinforced concrete code NBR 6118 (2003) recommendations.

4.2.2 *Steel Bridges*

Structural systems using Warren trusses, similar to the ones found in the Doce I (Figure 4) and Doce II (Figure 5) river bridges, are typical of the steel bridges found in the Vitoria Minas Railway.

Figure 11. Repair executed in the broken post in the Correntes River Bridge.

In 2006 a failure due to fatigue was detected in the Correntes River Bridge, which is a truss bridge with a 41 m span. A detailed view of the bridge, along with the point of failure and a detail of the failed element is shown in Figure 10.

The rehabilitation of the structural component was executed with urgency and a detail of the repair is shown in Figure 11. A new segment was welded to the remaining part of the failed element. The old connectors were also exchanged for high strength screws.

Following this incident in the Correntes River Bridge, the inspections were intensified and in 2007 some damaged elements that were classified as high risk were found. Other damages such as damages done by collisions, advanced corrosion and missing screws and rivets, classified as low risk, were also found. In all of these cases preventive maintenance was done.

Starting in 2009 a structural evaluation program of the truss bridges using short-term monitoring got underway. In these programs, the structural elements considered critical are instrumented and analyzed using traditional methods like the S-N curve as recommended by the AREMA (2009). The subsequent item shows one of these evaluation programs.

4.2.2.1 *Santa Maria River Bridge*

The Santa Maria River Bridge is a Warren truss bridge with spans with 30 m and 41 m supported by reinforced concrete columns. A longitudinal view of the truss structure is shown in Figure 12, in this figure the elements that were monitored are shown as M3 (central post), D2 (intermediate diagonal) and Bi3 (central

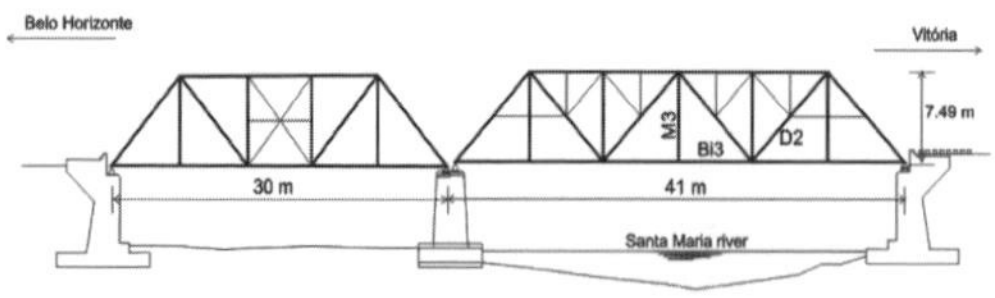

Figure 12. Overall profile of the Santa Maria Bridge.

Table 2. Fatigue analysis for monitored trainsets.

Component	Bi3		D2	M3
Detail category	B	D	B	D
Damage/block	–	1.46×10^{-6}	–	6.95×10^{-6}
Damage/year	–	0.0025	–	0.0122
Fatigue life (years)	–	392.17	–	82.16

bottom chord). The monitoring program that was realized in 2011 made it possible to gather a few samples of the different trainsets that pass through this bridge.

A fatigue analysis using the AREMA (2009) manual was done using the collected data. The results in terms of lifetime under fatigue, damage per trainset (block) and damage per year (considering approximately 24 trains per day) are shown in Table 2.

The safety factors for infinite lifetime under fatigue of these elements were found to be adequate so long as proper maintenance is done. Elements with riveted connections, that maintain the original design characteristics, present a finite lifetime under fatigue. Specifically the central post, M3, which dates back to 1944 and according to this analysis is in the final stage of its lifetime and therefore deserves special inspections in order to evaluate the need for its replacement.

5 FUTURE PERSPECTIVES AND CONCLUSION

Between 2005 and 2006 a new National Plan for Logistics and Transportation (PNLT – Plano Nacional de Logística e Transportes) was devised by the Brazilian Federal Government. This plan establishes new priorities for the Brazilian transportation system and selects 917 strategic projects for the modernization of roads, highways, railways, waterways, airports and ports all over the country, with a total necessary investment of R$ 428 billion (US$ 250 billion) until 2023. Among these, 96 projects are related to railways with an estimated investment of R$ 202 billion (US$ 120 billion).

All the projects in implementation or advanced phase of conception will permit an integration of the whole country and also an important link to Argentina, Chile and the Pacific Ocean through railways in 12 to 15 years, changing and improving the current transportation matrix of the country (Figure 13). It is clear that railways will have a significant share of investments and a more energetic efficient and sustainable network will be built. The expected benefits

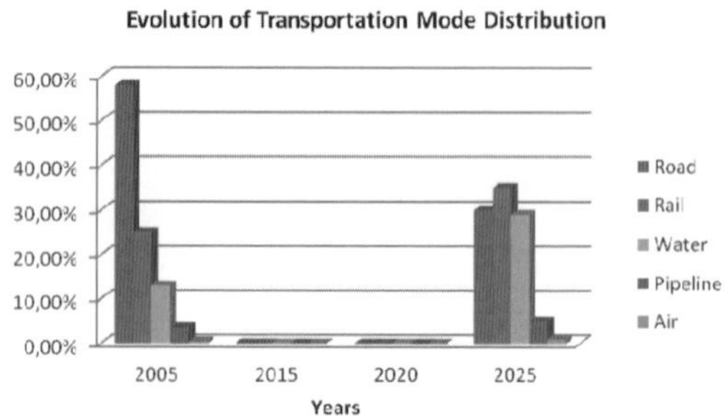

Figure 13. Current and future transportation modal percentage distribution.

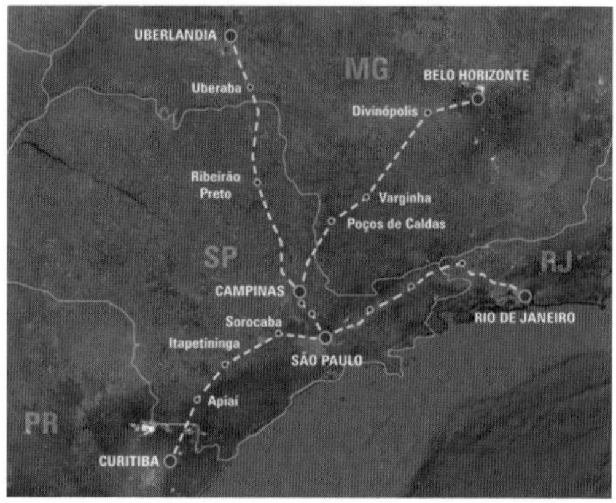

Figure 14. Future plans for the high speed rail tracks in Brazil.

of this change will lead to an increase of 850.9 to 1,510.4 billion TKU, to an increase of 38% in energetic efficiency, to 41% reduction of fuel consumption and 32% reduction of CO_2 emissions.

In this context, it is important to mention the projects related to high speed, regional and metropolitan trains. The high speed tracks are supposed to link the major Brazilian cities in the southeast portion of the country (Figure 14). In a first stage a railtrack of 511 will link Rio de Janeiro, São Paulo and Campinas. The construction of this line is supposed to start in 2012. Advanced sutdies are in place for the construction of more three high speed tracks connecting São Paulo to Curitiba (410 km), Campinas to Uberlândia (540 km) and Campinas to Belo Horizonte (530 km). And in the future two more high speed tracks are planned to link Curitiba to Porto Alegre (600 km) and Uberlândia to Brasíia (500 km). In total the Brazilian High Speed network will be 3,100 km long and it is supposed to be completed in the next 20 or 25 years. With respect to bridges, high speed rail tracks causes very particular structural behaviors with important consequences on their safety, service performance and durability.

Again all these plans and projects already in execution will bring a lot of new challenges and opportunities for the railway sector in Brazil in the coming years. And of course, they also bring to our attention the urgent necessity to develop and implement efficient and modern management and assessment systems not only for the railway bridges and tunnels, but also for the whole railway infrastructure (ballasted or unballasted), including all its components (rails, ties, connections, etc.).

ACKNOWLEDGEMENTS

Authors would like to acknowledge CNPq (Portuguese acronyms for Brazilian Ministry of Science and Technology Agency) and CAPES (Portuguese acronyms for Higher Education Improvement Agency) for providing an important part of the financial support needed to develop this project. The work described in this paper was supported by VALE Company. The opinions and conclusions presented in this paper are those of the authors and do not necessarily reflect the views of the sponsoring organizations.

REFERENCES

AREMA – American Railway Engineering and Maintenance-of-way Association. 2009. *Manual for railway engineering*. Washington: AREMA.

Blessmann, J. 1989. Aerodynamic instability in viaduct columns (in Spanish). *In Revista Sismodinámica* 1:1–17 (1989).

Frangopol, D. & Liu, M. 2007. Maintanance and management of civil infrastructure based on condition, safety, optimization and life-cycle cost. *Structure and Infrastructure engineering*, Taylor & Francis, Vol. 3, No. 1, p. 29–41.

Frangopol, D., Strauss, A. & Kim, S. 2008. Bridge reliability Assessment based on monitoring. Journal of Bridge Engineering, ASCE, Vol. 13, No. 3, p. 258–270.

Malite, M., Takeya, T., Gonçalves, R.M. & Sálses, J.J. 2000. Monitoring of the Paraná river bridge during launching. *In Structural Engineering International*, v. 10, n. 3, p. 193–196.

Miranda, M. 2010. New developments in Brazilian cable stayed bridges. *In 34th International Symposium on Bridge and Structural Engineering*, Venice.

National Confederation of the Transport. 2011. *Research CNT of Railroads 2011*. Brasília: CNT.

NBR 6118. 2003. *Design of structural concrete – Procedure*. Rio de Janeiro: Brazilian Association of Technical Codes.

NBR 7189. 1985. *Moving loads for structural design of railway bridges*. Rio de Janeiro: Brazilian Association of Technical Codes.

Protasio, F.C. 2008. Brazilian Bridge Management System: Comments on Axle-load Vehicle. *In XVII International Conference "In Situ Behaviour of Constructions"*, Bucharest.

Radomski, W. 2001. *Bridge Rehabilitation*. Warsaw: Imperial College Press.

Teixeira, R.M., Amador, S.D.R. & Oliveira, D.R.C. 2010. Static and dynamic analysis of a reinforced concrete rail bridge located in the Carajas Railroad. *In IBRACON Structures and Material Journal* v. 3 n. 3 p. 284–309.

Teixeira, R.M. 2009. *Study Theoretical-experimental of the Structural Behavior of a Rail Bridge in Reinforced Concrete*. Master thesis. Belem: Faculty of Civil Engineering of Federal University of Para.

Vasconcelos, A.C. 1993. *Brazilian bridges: notable viaducts and footbridges (in Portuguese)*. São Paulo: PINI.

Wenzel, H. 2009. *Health Monitoring of Bridges.*, Vienna: John Wiley & Sons.

Bridge Maintenance, Safety, Management, Resilience and Sustainability – Biondini & Frangopol (Eds)
© 2012 Taylor & Francis Group, London, ISBN 978-0-415-62124-3

Operational deformations in long span bridges

J.M.W. Brownjohn
Vibration Engineering Section, University of Sheffield, UK

K.-Y. Koo
School of Construction Engineering, Kyoung-il University, Korea

A. Scullion
Humber Bridge Board, UK

D. List
Tamar Bridge and Torpoint Ferry, UK

ABSTRACT: Being among the most flexible of civil engineering structures, long span bridges deform both dynamically and quasi-statically under a range of operational conditions including wind, traffic and thermal loads, in varying patterns, with different timescales and at different amplitudes. While external loads and internal forces can only rarely be measured, there are well developed and emerging technologies for measuring deformation. It is well known that some of these technologies can be used to validate or improve numerical simulations, which can then be used to estimate the external loads and internal forces. Changes in response patterns and relationships can also be used directly as a diagnostic tool, signaling unusual loading or structural changes, but excessive deformations themselves are a concern in terms of serviceability and allowable operational limits (e.g. of vibration, and bearing movement). This paper discusses the challenges of deformation measurement and applications in structural identification, performance diagnosis and load estimation, including observations of response to the most extreme loads.

1 INTRODUCTION

At a previous IABMAS conference we reviewed a number of technologies used in tracking deflection of bridge decks (Brownjohn & Meng, 2008). The imminent ASCE state of the art guidance on structural identification (ASCE, 2012) provides further information and references on these sensor technologies.

Our experience in structural health monitoring (SHM) of two major suspension bridges in the UK, Humber and Tamar, has demonstrated the important role of deformation measurements in checking performance. Unlike some SHM systems which emphasise dynamic response, the Humber and Tamar systems do not feature a high density of accelerometers. Instead they rely on previous modal surveys to permit modal superposition scaled by modal amplitudes recorded at strategically placed accelerometers. This technique misses only the usually modest changes in mode shapes that are occasionally observed under differing loading conditions.

Our greater concern, reflected in this paper, has been determining the slowly varying deformation patterns due to the range of operational loads such as wind, temperature and vehicles.

2 RELEVANCE OF BRIDGE DEFORMATION MEASUREMENTS, DYNAMIC AND STATIC

Deformation measurements play a critical role in bridge assessment in a number of ways. In the form of dynamic deformations summed from contributions of various vibration modes, they provide a direct view of the stiffness and mass properties of the structure. There are well documented procedures for vibration testing of long span bridges e.g. (Brownjohn et al., 2010), usually using accelerometers in an ambient vibration survey. Identifying modal frequencies and shapes and comparing with some form of analytical model allows for the verification that the bridge is behaving as reflected in the model, based on design assumptions.

High quality accelerometers are not limited to measurements of relatively high frequency vibrations. While they have traditionally not been used for quantifying the quasi-static deformations patterns due to slowly varying operational loads that do not engage inertia effects we have used them in some conditions to track movements with periods as long as 50 seconds.

For such purposes technologies for direct measurement of position, relative displacement (deformation) or rotation are used. While surveying technology is normally used for temporary snapshot measurements of position, our concern is with methods of continuously measuring and recording deformations that vary slowly with time, with frequency components ranging from several seconds to several months, usually in daily or seasonal cycles.

The earliest methods for measuring long span bridge deformations included the use of motion pictures e.g. to study the aero-elastic problems at Tacoma Narrows Bridge (University of Washington, 1954) and of hydrostatic leveling e.g. to check effects of structural changes to D. Luiz I bridge, Porto (Marecos, 1978). A sophisticated opto-mechanical system was the preferred solution for the deflection measurements of the Tagus River suspension bridge during a remarkably comprehensive exercise of instrumentation and proof load testing in 1966 (Marecos et al., 1969).

Before widespread use of GPS, opto-electronic technologies were developed. For example, lasers were used to track vertical and lateral movement in performance of the (new at the time) Foyle Bridge in Ireland (Sloan et al., 1992), and LED-based systems were used in Scandinavian bridges (Myrvoll et al., 1994). Two radically different technologies were employed in the Humber Bridge monitoring campaign of 1990/1991 (Stephen et al., 1993, Zasso et al., 1993). The motivation for the displacement measurements here was validation of numerical modeling procedures for predicting in-wind behavior of a super-long suspension bridge.

The Italian single axis 'optometer' (Zasso et al., 1993) used at Humber employed high contrast targets, CCD arrays and threshold detection, while the biaxial target tracking system developed at Bristol system was also used successfully for the Second Severn Crossing (Macdonald et al., 1997).

There have been experimental GPS applications in the UK, e.g. at Humber (Ashkenazi & Roberts, 1997) and Forth (Roberts et al., 2006), but until Humber (we believe), no permanent GPS installation. GPS is now the standard choice for deformation measurements, with wide application in the Far East, e.g. to the long span bridges operated by Hong Kong Highways Dept. (Wong, 2007).

All of these instruments have been used for direct measurement of translational deformations; complementary to these exist technologies already in use for several decades such as LVDTs and inclinometers for measuring small relative motions and inclinations. Also it is only recently that the combination of surveying technologies (theodolites and electronic distance measurement) found in robotic total stations has been considered for monitoring applications, in spite of their relatively slow and irregular sampling.

3 CURRENT VIABLE TECHNOLOGIES FOR MEASURING BRIDGE DEFORMATION

A number of deformation monitoring systems has been deployed, mainly in the Far East long span bridges in recent years. GPS is now regarded as a mature technology, alongside extensometers, inclinometers and lasers of various forms. Level sensors are now obsolete technology and like those on Tamar, those on the Lantau Fixed Crossing bridges are probably out of service.

The authors are not aware of robotic total stations (RTS) deployed for permanent instrumentation of long span bridges (or other types of bridge) although some short term applications have been reported (Erdoğan & Gülal, 2011, Psimoulis & Stiros, 2011). The system described in this paper seems to be unique.

This paper describes two currently operating long span bridge deformation monitoring systems where the technologies mentioned above have been or are being used. We start by describing the Tamar Bridge deformation measurement system as it has been operating for several years and has used less usual instrumentation. The system used presently at Humber is then described, with reference to methods used 20 years previously.

The main aim is to compare and contrast the ways in which these two bridges deform due to operational loads, to identify the advantages and limitations in the methods and to provide instructive examples of some extreme performance i.e. for the largest operational loads.

3.1 *Implementation on Tamar Bridge*

Tamar Bridge has provided a useful experience in tracking deformations. Built in 1961 to link Plymouth (east, in Devon) and Saltash (west, in Cornwall), the bridge was 'strengthened and widened' in 2000. The radical alterations included changing the deck, adding extra lanes cantilevered either side of the truss, adding a set of 16 stay cables to carry the extra load and restore the camber. Boundary conditions were also altered. To track the effects of the changes during and after the construction, Fugro Ltd installed a structural monitoring system' (SMS), a comprehensive array of sensors that included load cells in the stays, weather and structural temperature sensors and a hydrostatic leveling system. In late 2006, the Vibration Engineering Section (VES) installed a 16 channel dynamic data acquisition system (DAQ) to record deck and stay cable accelerations as well as relative movements (extensions) at the Saltash tower. Later, a robotic total station (RTS) and a small wireless sensor network were added. Components of these systems used for measuring deformations are described in the next sections.

3.1.1 *Total positioning system (TPS)*

From the information provided by the hydraulic leveling system and the Saltash tower extensometers extensometer during 2006 and 2007 it was clear that the vertical plane movement of the bridge, mostly due to thermal expansion, comprised an axial extension of the girder and a lowering of the deck. To study this two-dimensional motion (and to include any transverse motion) a TPS was installed, comprising a Leica

Figure 1. Tamar Bridge.

Figure 2. RTS unit on control room roof and reflectors on Saltash tower (top right) and Plymouth side tower (bottom right).

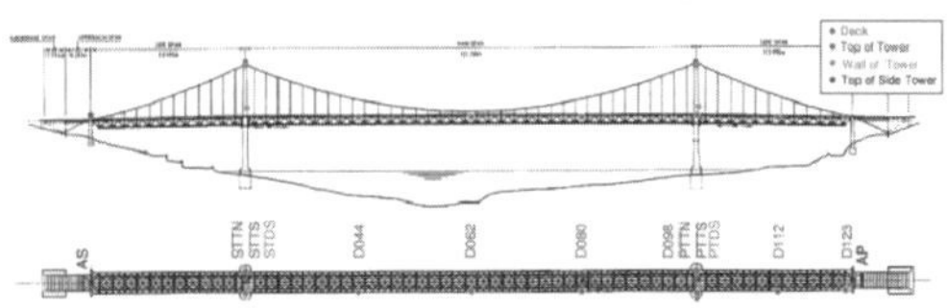

Figure 3. Reflector locations.

TCA1201M robotic total station (RTS) and a set of 15 reflectors (Figure 2). TPS was chosen in preference to GPS due to the expected small movements and the lower cost for multiple measurement positions, there being no need for 'dynamic' sample rates of 1 Hz or more offered by GPS.

The main problem with the TPS is the narrow range of angles available for fixed reference reflectors given the only secure location of the RTS on roof of the control room adjacent to the roadway. An experiment with the RTS unit located on the upper portal of the Plymouth tower demonstrated the destructive effect of the small tower longitudinal vibration on the precision of

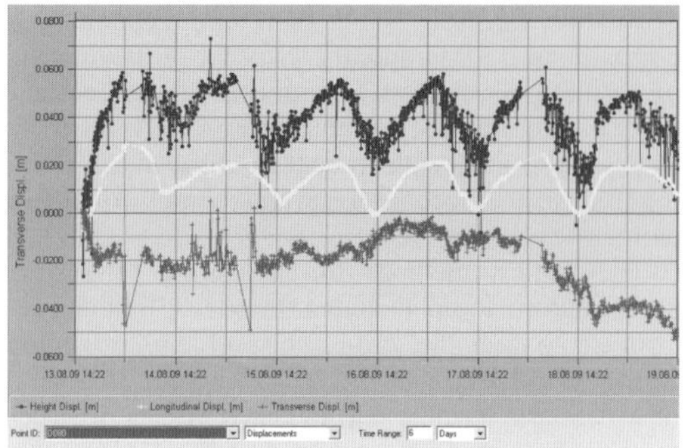

Figure 4. RTS raw data for reflector at 3/4 span position (Saltash end).

elevation measurements. Also the line of sight for distant reflectors around roadway level at the Saltash end sometimes results in failure to locate the data point. In poor visibility (e.g. drizzle) location capability is reduced, resulting in patchy data e.g. Figure 4 (actually these are good data). Finally the location on the roof surface proved to be unstable, so the RTS has been moved to the top of a structural column on the control room roof.

3.1.2 *Accelerometers and extensometers*

The 2006 DAQ system included a set of three accelerometers mounted close to midspan. Labelled VS, VN and H they sense vertical acceleration on the north and south sides of the truss and horizontal acceleration. They are primarily used for tracking response levels and modal parameters but being DC servo accelerometers (model Honeywell QA750) the H signal can also be used to track deck rotation. Since the RTS deck reflectors are located in a vertical plane on the south side (there is no unobstructed view of the north side from the RTS unit), accelerometers provide the only means of kinematically connecting the two sides of the bridge.

Three DAQ channels were originally used to record extensions at the only expansion joint in the structure, at the Saltash tower, using ASM WS12 potentiometers (Figure 5).

Signals from these units were available only for a short period, before the level sensors were repaired. The sensors were able to show that axial separation of Saltash tower and main span north and south sides (1st and 3rd channels) closely followed the separation between the spans i.e. that tower-sidespan motion was negligible. Earth loop problems and accidental damage to the wire threads (birds, workers) caused loss of all signals after March 2007.

From 2010, signals for the north and south extensometers have been available via a wireless network (de Battista, Westgate, Koo, & James M. W. Brownjohn, 2011). The reported study demonstrated that the clearest linear relationship of average longitudinal motion and temperature used main cable temperature as a reference.

Figure 5. Extensometer between Saltash and main spans.

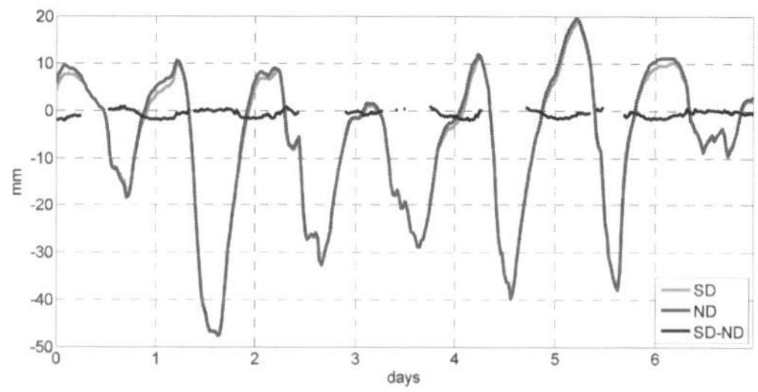

Figure 6. Extensometer data for July 2010.

Figure 6 shows signals for a week in July 2010. Daily patterns are clear; the range throughout the year exceeds 200 mm. North and south signals differ occasionally, and there is intermittent data loss due to a mechanical problem with one of the sensors.

3.1.3 *Data fusion*

RTS data are not sampled at regular intervals and data points are missed, likewise there are dropouts from extensometer data. Data streamed by the Fugro SMS are available at 10-second intervals, while accelerometer data are sampled at 8 Hz. All data streams are collated through the VES data management system then interpolated and averaged to produce 'summary' data (means and standard deviation values) at half-hourly intervals.

3.2 *Implementation on humber bridge*

The Humber Bridge, opened in 1981 has a main span of 1410 m and side-spans of 280 m and 530 m, all being 28.5 m wide 4.5 m deep steel box sections. Reinforced concrete towers are 151.5 m high and support twin 0.29 m^2 main cables with inclined hangers and sag of 115.5 m. It's now the fifth longest span in the world.

Humber has provided many opportunities and applications for deformation measurements, all with extensive support from Humber Bridge Board (HBB). The original ambient vibration study (Brownjohn et al., 1987) and the 1990/1991 monitoring (Brownjohn et al., 1994) served the purpose of validating simulation software to be used on other bridges. Specifically, the finite element analysis procedure validated in the 1985 test was used for studying seismic response of the two Bosporus crossings (Dumanoglu et al., 1992, Dumanoglu & Severn, 1987), while the 1990/1991 study validated procedures for simulating the in-wind performance of the proposed Messina Straits Bridge (Diana et al., 2003). For the bridge itself, the 1990/1 study identified relationships between loads and response presented in Table 1.

Table 1. Deflection/load relationships from 1990/1991.

Deflection component	Temperature T/°C	Wind speed U/m/sec
Vertical v/mm	$v \approx -60T$	$v \approx -0.81U^2$
Rotation θ/milli-rad		$\theta \approx 0.006\,U\vert U$
Lateral u/mm	$u \approx 15T$	$u \approx 1.18U\vert U$
Longitudinal w/mm	$w \approx 8T$	

Figure 7. Humber Bridge deformation monitoring system.

The present exercise extends the capabilities of the earlier system and provides useful management data for HBB, allowing them to make informed decisions on management and intervention.

3.2.1 *Present technology*

Figure 7 shows the present monitoring system.

Operational since 2010, the system comprises a GPS base station and two rovers mounted on the main cables at midspan, three accelerometers mounted at midspan in the same configuration as Tamar, four laser extensometers at the main span ends and an inclinometer at midspan. In addition temperatures are recorded at several points on the structure and environmental conditions (temperature and wind) are recorded at the site and from local weather stations.

As with Tamar, data fusion is major concern. Signals from the inclinometer and GPS antennae are recorded using the Leica GEOMOS software; at present RTK solutions are saved, as ASCII data. Accelerations and velocities from the RM Young anemometer are recorded using a National Instruments DAQ and temperatures by a Campbell Scientific logger. Data streams are also fed from HBB wind and temperature sensors and from local meteorological stations. As with Tamar, 30-minute summary values are saved to a common time-base, and can be viewed using a Google Apps viewer or via MATLAB database tools. Data are duplicated at the VES server by automated file transfer, and scheduled MATLAB routines process the data for summary files and also carry out operational modal analysis.

Not all sensors have been operational all of the time; acceleration data are recorded almost continuously, but there have been gaps from other deformation sensors, particularly the extensometers, which are a bespoke design. However, sufficient data for representative periods have been collected to characterize the bridge deformation mechanisms.

Compared to the 1990/1991 exercise, GPS is more reliable, using four extensometers has provided valuable information about the lateral movement of the deck and more comprehensive information are available about structural temperature. There has also been a greater focus on rotations about all axes

4 EXTREME RESPONSE MEASUREMENTS

The most revealing performance corresponds to extreme (operational) loading events e.g. strong winds, low or high temperatures, weak or strong sunlight and exceptional vehicle loads.

4.1 *Tamar*

For Tamar, (deck) temperature range recorded is between −5°C and 30°C, with largest daily range 13°C. Maximum wind speeds were generally lower than at Humber with maximum 51.5 mph (23 m/sec) on 10/3/2008. The only noticeable effect of wind on deformation has been the increased acceleration response in the lower vibration modes, when winds exceed 10m/sec.

The significant extreme response for this relatively short and bluff bridge is due to abnormal vehicle loading, specifically rush-hour traffic jams and passage of an abnormal load.

4.1.1 *Heavy vehicle crossing*

The single most significant loading event during the monitoring was the passage of a two-tractor low-loader carrying a power station component, with total vehicle weight of 350 tonnes. To track the vehicle the RTS was reconfigured to trace a single target at the fastest rate possible, while a second standalone RTS (TS30 unit) was deployed to track Saltash tower top deflections, also a single reflector. Other response parameters such as extension, cable tension and acceleration were simultaneously recorded.

Experimentation with the two RTS units determined that they could operate at sample rates varying in the range 2–3 Hz. The data collected during the vehicle passage were spliced using accurate time stamps and interpolated at 3 Hz to produce the plot of Figure 10; the whole event of the vehicle crossing the bridge lasted 3.5 minutes.

Peak midspan deflection of 300 mm corroborated our own and the consultant's predicted values.

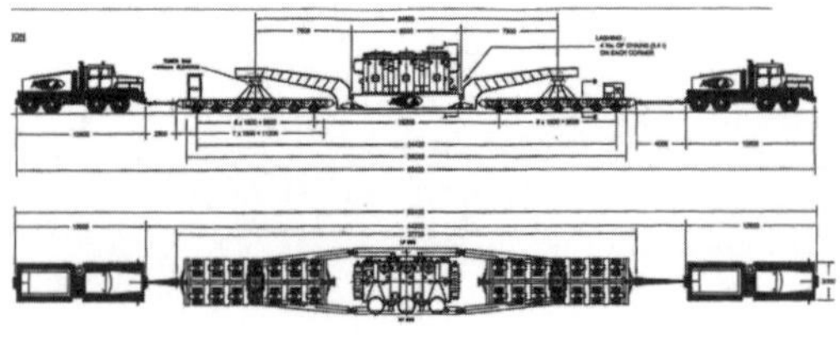

Figure 8. 350 tonne vehicle configuration, total length 65 m.

Figure 9. Westbound 350 ton vehicle.

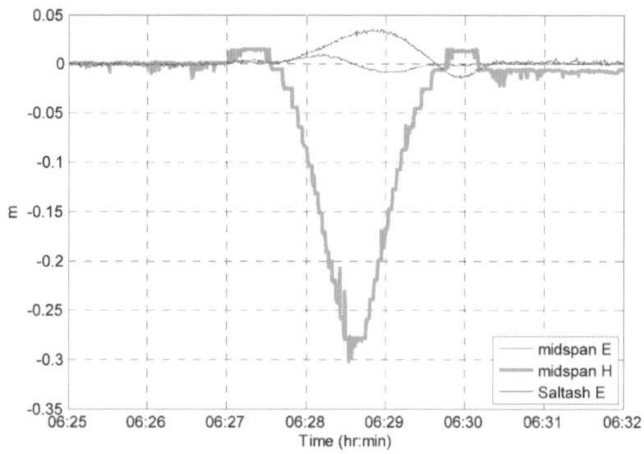

Figure 10. Vertical (H) and axial (E) deformation due to heavy vehicle.

4.2 *Humber Bridge*

Temperature ranges at Humber are greater than at Tamar. For air temperature the range is −10 to 30°C. Likewise, wind speeds are stronger, peaking at 27 m/sec (30 minute mean, 31st March 2011). The extreme response at Humber is wind-induced, specifically due to strong westerly wind. So far no digitized data are available on passage of a heavy vehicle, although there are plans to link weigh in motion (WIM) data to the monitoring system.

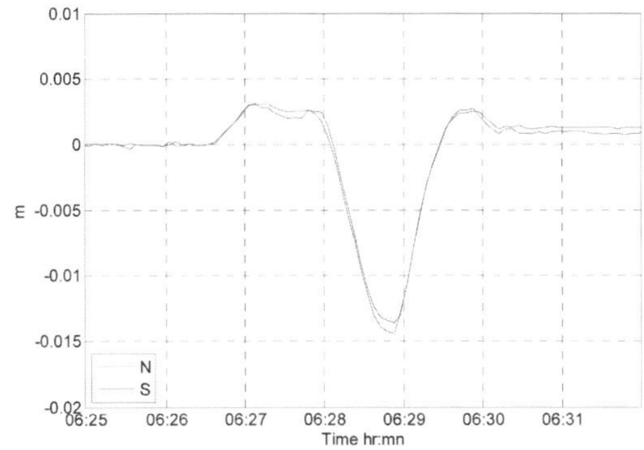

Figure 11. Extensometer readings at Saltash tower for north and sensors. Positive value is opening of the gap.

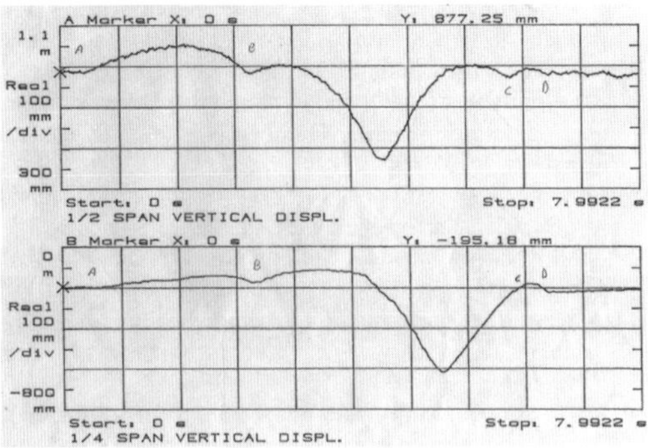

Figure 12. Humber Bridge midspan and quarter span deflection due to 172 tonne vehicle recorded by ISMES optometer.

4.2.1 *Heavy vehicle crossing*

With no specific data available for unusual vehicles the only information available (until WIM can provide it) is from a single measurement by the ISMES optometer on 17 May 1990, during passage of a 172 tonne vehicle. Figure 12 shows that this load generated a deflection of just over 400 mm, in fact smaller than deflections due to strong winds and normal temperature variations.

4.2.2 *Extreme traffic*

No abnormal vehicles are known to have crossed during the present monitoring but practically every conventional heavy vehicle can be identified from its signature in the response time series. Figure 13 is an example; winds are light, and there are two clear negative deflections in the GPS height signal. The middle plot shows difference in height from the two GPS receivers (west height – east height, on the main cables), height difference recovered from DC component of lateral acceleration (in g) scaled by chord and the same result obtained from the Nivel inclinometer. The vehicle at 609 minutes is northbound, that at 615 minutes is southbound. The lower plot indicates that a southerly movement of the deck followed by a northerly over-correction occurs for the northbound vehicle and vice-versa.

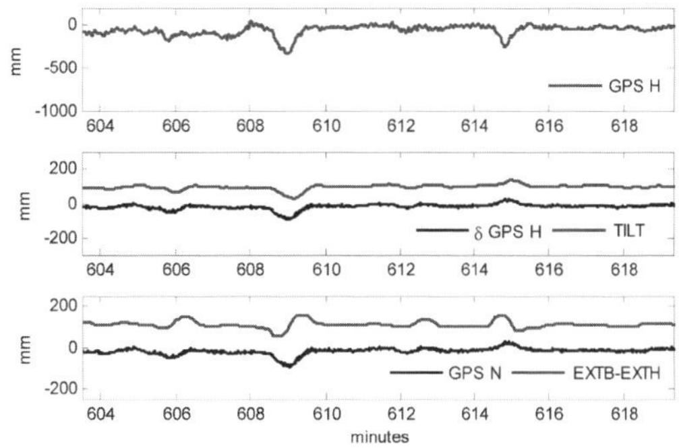

Figure 13. Heavy vehicle passage, 23/7/2011.

5 CAPTURING DEFORMATIONS BY ALTERNATIVE MEANS

Figure 14 shows the potential for estimating deformations using cheaper or more convenient sensors. For example the Nivel inclinometer provides practically the same data as the DC component of the lateral accelerometer. Rotation about the lateral axis (i.e. slope in the traffic direction) does not serve an obvious purpose. Similarly knowing (as a result of this exercise) that height differences between rovers either side of the bridge could be estimated using a single rover and the DC component of the lateral acceleration, the second rover appears to be redundant.

There is a strong correlation between lateral movement of the deck and rotation (about a vertical axis) at the deck bearings but the relationship depends on the deformed shape along the deck. In any case the GPS sensors have so far been more reliable.

Dynamic components of response can be recovered down to a frequency a little lower than the lateral mode frequency (0.05 Hz) so a relatively expensive GPS rover is an extravagant means for recovering dynamic response. It is possible to demonstrate that above a certain frequency (perhaps around 1 Hz) GPS cannot resolve dynamic signals, since typical bridge deck accelerations translate to mm level displacements for higher modes.

Apparently, vertical and lateral movement of a bridge deck at periods of 50 seconds or above requires either a TPS or GPS, and the choice depends on budget, required spatial resolution and good matching of deformation ranges to resolution capability of either system. GPS and TPS have very different types of random and systematic errors and despite having argued away the need for duplication, having some redundancy helps to validate signals and build confidence.

REFERENCES

ASCE. (2012). Structural Identification (St-Id) of Constructed Facilities. Approaches, Methods and Technologies for Effective Practice of St-Id (in press).

Ashkenazi, V., & Roberts, G. W. (1997). Experimental monitoring of the Humber Bridge using GPS. ICE Proceedings Civil Engineering, 120, 177–182.
de Battista, N., Westgate, R., Koo, K. Y., & Brownjohn, J. M. W. (2011). Wireless monitoring of the longitudinal displacement of the Tamar Suspension Bridge deck under changing environmental conditions. Proceedings of SPIE: San Diego., Vol. 7981.
Brownjohn, J. M. W., Dumanoglu, A. A., Severn, R. T., & Taylor, C. A. (1987). Ambient vibration measurements of the Humber Suspension Bridge and comparison with calculated characteristics. Proceedings Institution of Civil Engineers Part 2, 83, 561–600.
Brownjohn, J. M. W., & Meng, X. (2008). Methods for measuring structural deflection and applications to bridge deck performance monitoring. IABMAS08, Seoul.
Brownjohn, J. M. W., Magalhaes, F., Caetano, E., & Cunha, A. (2010). Ambient vibration re-testing and operational modal analysis of the Humber Bridge. Engineering Structures Vol. 32(8), 2003–2018.
Brownjohn, J. M. W., Bocciolone, M., Curami, A., Falco, M., & Zasso, A. (1994). Humber Bridge Full-Scale Measurements Campaigns 1990–1991. Journal of Wind Engineering and Industrial Aerodynamics, 52, 185–218.
Diana, G., Falco, M., Cheli, F., & Cigada, A. (2003). The aeroelastic study of the Messina Straits Bridge. Natural Hazards, 30, 79–106.
Dumanoglu, A. A., & Severn, R. T. (1987). Seismic response of modern suspension bridges to asynchronous vertical ground motion. Proceedings ICE Part 2, 83, 701–730.
Dumanoglu, A. A., Brownjohn, J. M. W., Severn, R.T., (1992). Seismic analysis of the Fatih Sultan Mehmet (Second Bosporus) Suspension Bridge. Earthquake Engineering and Structural Dynamics, 21, 881–906.
Erdoğan, H., & Gülal, E. (2011). Ambient Vibration Measurements of the Bosphorus Suspension Bridge by Total Station and GPS. Experimental Techniques, DOI: 10.1111/j.1747-1567.2011.00723.x
Macdonald, J. H. G., Dagless, E. L., Thomas, B. T., & Taylor, C. A. (1997). Dynamic measurements of the Second Severn Crossing. ICE Proceedings Bridge Transport, 123(4), 241–248.
Marecos, J. (1978). The measurement of vertical displacements through water levelling method. Materials and Structures, 11(5), 361–370.
Marecos, J., Castanheta, M., & Trigo, J. T. (1969). Field observation of Tagus River suspension bridge. ASCE Journal of Structural Engineering, 95(4), 555–583.
Myrvoll, F., DiBiagio, E., & Hansvold, C. (1994). Instrumentation for monitoring the Skarnsundet cable-stayed bridge. Third Symposium on Straits Crossings, Alesund, 207–215.
Psimoulis, P. A., & Stiros, S. C. (2012). Measuring deflections of a short-span railway bridge using a Robotic Total Station (RTS). Journal of Bridge Engineering (in press).
Roberts, G. W., Meng, X., Brown, C. J., & Andrew, A. (2006). Measuring the movements of the Forth Road Bridge by GPS; lorry trials. The First International Conference on Advances in Bridge Engineering. Bridges – Past, Present and Future, Uxbridge, UK.
Sloan, T. D., Kirkpatrick, J., Boyd, J. W., & Thomson, A. (1992). Monitoring the in service behaviour of the Foyle Bridge. The Structural Engineer, 70(7), 130–134.
Stephen, G. A., Brownjohn, J. M. W., & Taylor, C. A. (1993). Measurements of static and dynamic displacement from visual monitoring of the Humber Bridge. Engineering Structures, 15(3), 197–208.
Washington, University of, (1954). Aerodynamic stability of suspension bridges with special reference to the Tacoma Narrows bridge. Bulletin No. 116 University of Washington Engineering Experiment Station.
Wong, K.-Y. (2007). Design of a structural health monitoring system for long-span bridges. Structure and Infrastructure Engineering, 3(2), 169–185.
Zasso, A., Vergani, M., Bocciolone, M., & Evans, R. (1993). Use of a newly designed optometric instrument for long-term, long distance monitoring of structures, with an example of its application on the Humber Bridge. 2nd International Conference on Bridge Management, Guildford.

Bridge Maintenance, Safety, Management, Resilience and Sustainability – Biondini & Frangopol (Eds)
 ISBN 978-0-415-62124-3

Wind tunnel: A fundamental tool for long span bridges design

G. Diana
Department of Mechanical Engineering, Politecnico di Milano, Milan, Italy

ABSTRACT: An overview of the wind tunnel activities and methodologies to support the design of long span suspension bridges is proposed. The most important aspects of the wind-bridge interaction are investigated considering the aerodynamic phenomena affecting the different parts of the bridge (mainly deck and towers). The experimental activities and results are proposed in the framework of a synergic approach between numerical and experimental methodologies that represents the common practice in defining the full scale aeroelastic behavior of the bridge starting from scaled reproduction of the wind-bridge interaction. Static and dynamic wind loads, aeroelastic stability, vortex induced vibrations will be investigated.

1 INTRODUCTION

The definition 'long span bridge' is usually related to bridges with the main span of the order of one thousand meters or longer.

Actually the longest bridge is the Akashi in Japan, with its main span of 1990 m. Considering the bridges at the stage of the detailed final design, the longest is the Messina in Italy, with a 3300 m main span.

For these very long span bridges, as already said, the wind plays the most important role in the bridge design. In the following, we will make reference mainly to suspension bridges.

Many problems must be faced to guarantee the bridge performance to the wind action: the major are:

1) Static load, due to the average component of the wind blowing on the structure
2) Dynamic load, due to the incoming wind turbulence
3) Instability of the bridge
4) Vortex shedding on the tower, deck and cables.

The handling of these problems required the development of special analytical models. These models need tests in wind tunnel to identify the aerodynamic and aeroelastic behaviour of the different bridge components.

This paper is mainly aimed at the description of the types of tests generally performed in a wind tunnel in the design stage.

The present paper is organized in the following paragraphs:

1) Introduction
2) Description of the main problems related to the wind action
3) Analytical methods for the analysis of the bridge response to the wind
4) Wind Tunnel tests
   - Deck
   - Tower
   - Full bribge
5) Conclusions

2 DESCRIPTION OF THE MAIN PROBLEMS RELATED TO THE WIND ACTION

2.1 *Static load*

Static loads due to wind aerodynamic forces are very important for a long bridge. They are due to the average wind component and they condition all the bridge design process. These loads are applied to the towers, the deck and the main cables. The drag load of the deck plays the most important rule. The static wind load on the deck is transferred through the hangers to the main cables, since the deck lateral stiffness is negligible in comparison with that of the main cables. So the overall static load (deck + cables) is applied to the top of the towers, producing a very large flexural moment on the tower. This is one of the most important loads affecting the tower design.

2.2 *Dynamic load*

The incoming wind is turbulent. The wind speed is changing in space and time (Figure 1).

In any point of the structure the wind is changing in time around the average value V_m, with an horizontal turbulent component u and a vertical turbulent component w, as in Figure 2.

The turbulence effect produces a fluctuating load on all the bridge components.

The motion induced by this load causes fatigue problems on the structure.

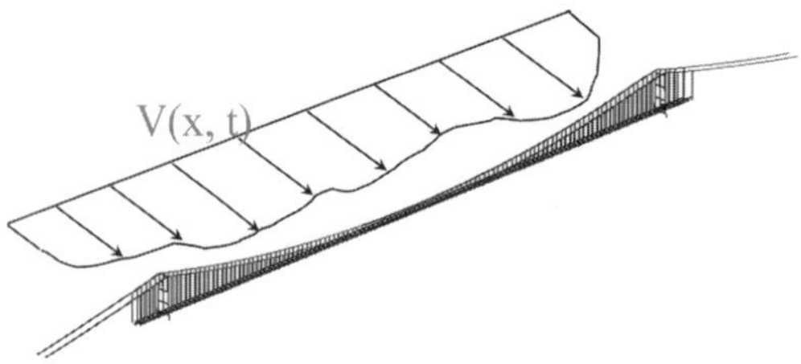

Figure 1. Wind turbulence: space variation of the wind speed along the span in the same instant of time.

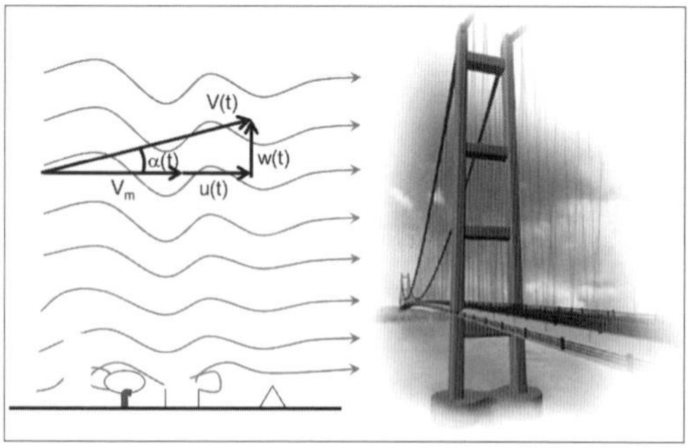

Figure 2. Average and turbulent component of the wind.

2.3 *Instability of the bridge*

Two forms of instability can be excited in a bridge:

- Single degree of freedom
- Two degree of freedom coupled flutter

These forms of instability are due to the motion-induced aerodynamic forces acting on the deck.

If these forces dissipate energy the system is stable, otherwise, if these forces introduce energy into the system the amplitude increase and the system becomes unstable. The single degree of freedom instability is due to the shape of the deck. Large frontal area of the deck can produce negative slopes of lift and moment aerodynamic coefficients that are responsible for this type of instability (see Figure 6, Tacoma). Deck sections with an airfoil-like geometry have positive slope of the lift and moment coefficients. But suffer of two degree of freedom flutter instability that is a coupling of one torsional mode of the deck with the corresponding vertical one.

2.4 *Vortex shedding*

Vortex shedding is produced in a large amount in bluff bodies, like the towers, but also in the deck due to the box girder corners and the wind barriers or any other device placed on the deck.

Vortex shedding produces a fluctuating load at the Strouhal frequency, related to the wind speed and the body dimensions, by the formula: $f_S = c_S V/D$, where c_S is the Strouhal number, V is the wind speed and D is a reference dimension of the body.

When the vortex shedding frequency equals one of the bridge natural frequencies, the vortex shedding induced motion is amplified.

3 ANALYTICAL METHODS FOR THE ANALYSIS OF THE BRIDGE RESPONSE TO THE WIND

It is common practice to develop finite element schematizations of the bridge (Figure 1) to identify its response to the different actions of interest: wind, earthquake and traffic.

For what concerns the wind action, generally beam elements are used for deck and tower, while taut string or tensioned beam elements are used for the cables. These models are generally used to compute the global behavior of the bridge to the different actions and they can be non linear or linear, as a function of the type of excitation.

Knowing the structure of the wind in the site surrounding the bridge, it is possible, by special tools (Chen & Kareem, 2001) (Ding, Zhu, & Xiang, 2006), to reproduce a space and time wind distribution, as shown in Figure 1.

As already said, the forces acting on the bridge are a function of the incoming turbulence.

On the other hand, the bridge motion produces aerodynamic forces which are motion dependent.

As a consequence, the aerodynamic forces acting on the bridge are functions both of the incoming turbulence and of the bridge motion.

In order to reproduce the over mentioned wind forces in the bridge FEM, the following equation have to be considered:

$$M_s\ddot{X} + R_s\dot{X} + K_sX = F_a\left(X, \dot{X}, s(t)\right) \quad (1)$$

where M_s, R_s, K_s are mass, damping and stiffness matrices of the bridge structure linearized in the neighbourhood of the static equilibrium position, X is a vector containing the bridge degrees of freedom.

F_a is the vector defined by the aerodynamic forces acting on the different sections of the bridge deck, towers, cables.

The identification of this non linear function through wind tunnel tests is a complex matter and it is for sure one of the most advanced topics of the research in this area (Diana, Resta, & Rocchi, 2008), (Diana, Rocchi, Argentini, & Muggiasca, 2010), (Wu & Kareem, 2011).

The most used approach consists in linearizing the aerodynamic forces in the neighbourhood of the static equilibrium position of the bridge X_0, identified by a given average wind V_m. This leads to:

$$M_s\ddot{X} + (R_s + R_a)\dot{X} + (K_s + K_a)X = A_m\left(s(t) - V_{m0}\right) (2)$$

where:

$$\left(\frac{\partial F_a}{\partial X}\right) = -K_a; \quad \left(\frac{\partial F_a}{\partial \dot{X}}\right) = -R_a; \quad \left(\frac{\partial F_a}{\partial s}\right) = A_m$$

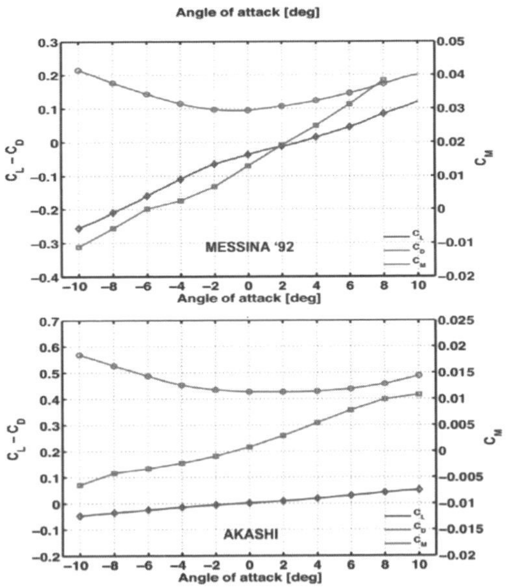

Figure 3. Aerodynamic coefficients of different deck geometry.

R_a and K_a are the equivalent damping and stiffness matrices due to the aerodynamic forces.

In other words, the linearization of the aerodynamic forces produces an equivalent elastic and damping system that is added to the structural one, changing the natural frequencies and the overall damping. An aeroelastic system is produced.

The K_a matrix, as well as the R_a matrix, are non symmetric and, as already mentioned, can give rise to flutter instability.

All the parameters needed to define the K_a , R_a and A_m matrices must be identified through wind tunnel tests in order to reproduce the bridge response to the turbulent wind.

4 WIND TUNNEL TESTS

This paragraph is aimed at the description of the types of wind tunnel tests performed in order to verify the bridge performances to the wind action.

4.1 *Deck*

The most important test for the deck, as well as for the other components, is that for the definition of the static aerodynamic coefficients as a function of the angle of attack.

This test is made in a wind tunnel on a sectional model of the deck (Figure 4).

The model can rotate around a deck longitudinal axis and the lift, drag and moment aerodynamic forces are measured as a function of the wind angle of attack, by changing the deck angle of rotation.

The output of this type of tests is shown in Figures 3,5,6,8 for different deck types (taken from (Brancaleoni, Diana et al., 2009)).

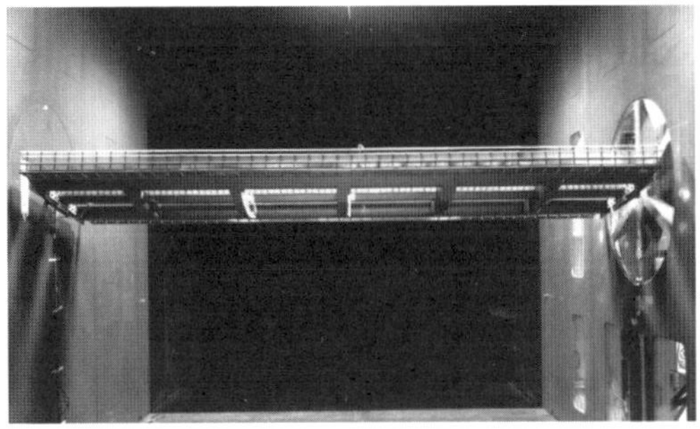

Figure 4. Deck sectional model in wind tunnel (external balance).

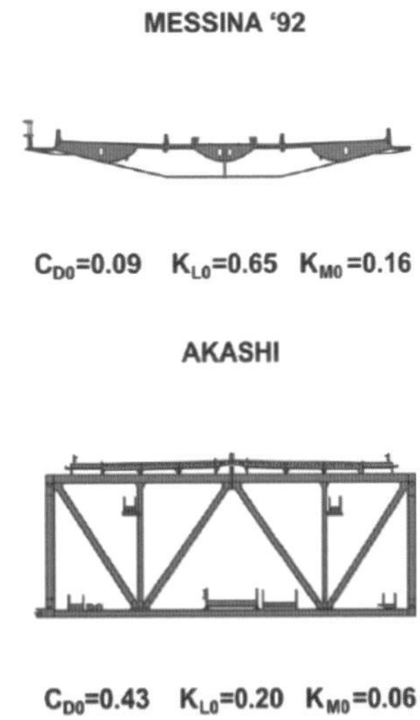

Figure 5. Deck geometry relevant to Figure 3.

4.2 *Forced motion method to identify the flutter derivatives*

A deck sectional model is forced to vibrate, as in Figure 7, in the three y, z and θ directions and the lift, drag and moment forces are measured for a given constant velocity V_m.

As already said, the forces are non linear functions of the motion.

$$
\begin{aligned}
Lift &= F_L = \frac{1}{2}\rho V^2 S C_L(X,\dot{X}) \\
Drag &= F_D = \frac{1}{2}\rho V^2 S C_D(X,\dot{X}) \qquad (3) \\
Moment &= F_M = \frac{1}{2}\rho V^2 S B C_M(X,\dot{X})
\end{aligned}
$$

being:

$$
X = \begin{Bmatrix} y \\ z \\ \vartheta \end{Bmatrix}
$$

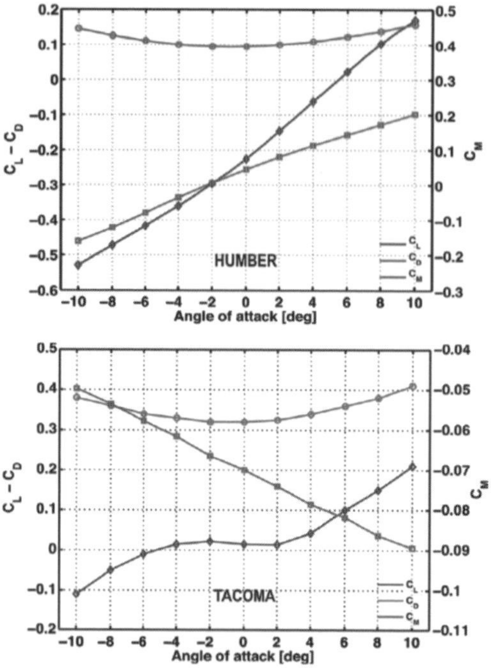

Figure 6. Aerodynamic coefficients of different deck geometry.

Figure 7. Sectional model on oil dynamic actuators driving its rigid vertical and torsional motion.

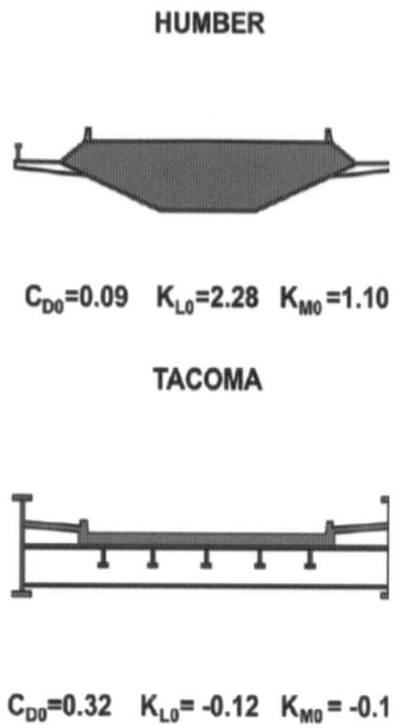

Figure 8. Deck geometry relevant to Figure 6.

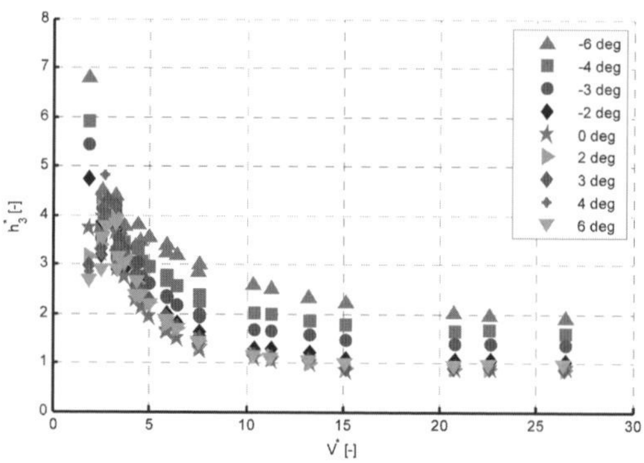

Figure 9. Flutter Derivatives (Messina bridge).

The identification of the C_L, C_D, C_M coefficients, non linear functions of $X, \dot{X}$, as already said, is not an easy task and represents a research topic.

Methods have been developed by some researchers and are reported in (Diana, Resta, & Rocchi, 2008) and (Diana, Rocchi, Argentini, & Muggiasca, 2010): for those based on rheological models and in (Wu & Kareem, 2011) for those based on neural network algorithms.

What is generally done is to linearize the forces around a static equilibrium position of the deck:

$$F = \begin{Bmatrix} F_D \\ F_L \\ F_M \end{Bmatrix} = \frac{1}{2}\rho V^2 S\left(\left[K_a\right]X + \left[R_a\right]\dot{X}\right) \qquad (4)$$

The 3×3 K_a and R_a matrices contains the flutter derivatives coefficients.

As an example, one of the K_a matrix coefficients, defined by h_3^*, is reported in Figure 9. The coefficients are function of the reduced velocity (reported in the abscissa) and of the angle of attack θ_0 around which the problem is linearized.

4.3 *Identification of the admittance matrix*

A method used in our wind tunnel and also by Chinese researcher (Niu, Chen, Liu, Han, & Hua, 2011) to identify the aerodynamic forces as a function of the incoming turbulence is to generate a well defined wind turbulence by an active device of the type shown in the Figure 10.

The aerodynamic forces are measured on a sectional deck model for a given average angle of attack θ_0 and

Figure 10. Active turbulence generator.

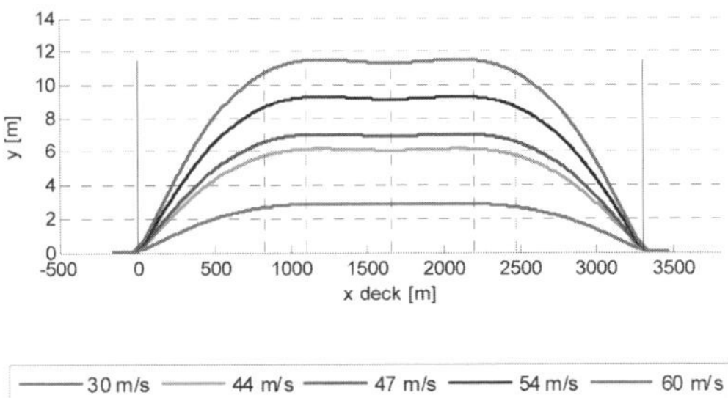

Figure 11. Static deformation under different mean wind speeds.

the coefficients of the admittance matrix are identified, in a similar way as the flutter derivatives.

Once the flutter derivatives and the admittance function are known, it is possible to compute the bridge response to the turbulent wind.

First of all the static equilibrium position of the bridge under the turbulent wind must be computed as already explained.

For each section of a bridge a time history of the turbulent wind must be generated. The spectrum of the single time history can be identified and the equation (2) can be solved, for instance, in the frequency domain, taking into account that the flutter derivatives and the admittance function are frequency dependent.

Some example of output result are reported in the following. Figure 11 reports the static deflection of the Messina bridge deck under the action of mean wind forces for different mean wind speeds.

Figure 12 reports RMS values of the vertical acceleration of the different deck section along the bridge axis of the Messina bridge deck under the action of turbulent wind.

5 VORTEX INDUCED VIBRATIONS

5.1 *Deck*

Vortex induced vibrations (VIVs) represent one of the most important aspects of the deck shape design. Even though the bridge deck of the last generation have an airfoil like geometry, flow separation occurs in correspondence of the deck sharp corners, in presence of adverse pressure gradients and of wind shields that are usually adopted to protect traffic. VIV is a serious problem that took also a part in the collapse of the Tacoma Narrow Bridge and has to be carefully analyzed. In fact, even if it can be separated from the

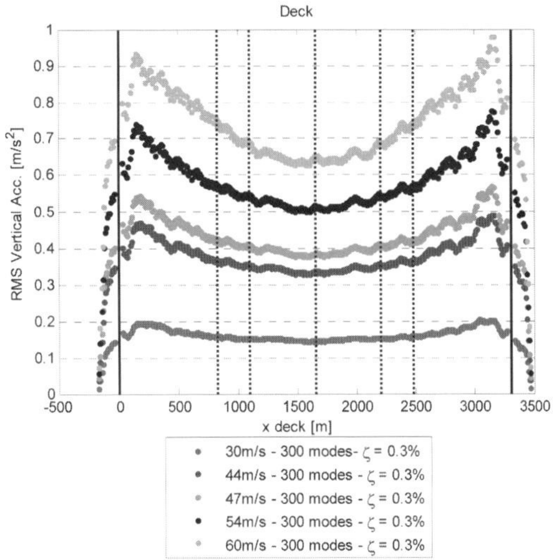

Figure 12. RMS values of the vertical acceleration of the different deck section along the bridge axis.

stability problem for the modern bridges, it can lead the deck to reach very large vibration amplitudes in case of low Scruton numbers

$$Sc = 2\pi \frac{mh_s}{\rho B^2} \tag{5}$$

where m is the linear mass, h_S is the structural damping, B is the deck chord and ρ is the air density, also at very low wind speeds having a high probability of occurrence.

Nevertheless the problem is widely known and studied, past and present experiences highlighted that it represents one of the major concerns in the bridge design.

The model response in the lock-in range is usually measured in order to define the oscillation amplitudes reached at regime condition throughout the lock-in range

As an example Figure 13 presents the two lock-in ranges for the flexural motion of the Messina Bridge multi box deck section for different Scrouton number.

In Figure 14, the vertical amplitudes of the deck divided by the deck chord B are reported as function of Sc. From this type of figure, it is possible to identify what is the damping of the structure able to control VIVs. Making reference to Figure 14, the 3 vertical dashed lines represent 3 different values of Sc that for the Messina bridge correspond to 3 different levels of non-dimensional ratios of structural damping equal to: $2 \cdot 10^{-3}$, $3 \cdot 10^{-3}$ and $5 \cdot 10^{-3}$. In order to avoid VIVs the structural damping of the Messina bridge should be greater than $3 \cdot 10^{-3}$.

5.2 *Tower*

The main problem that has to be addressed during the aerodynamic design of bridge towers is VIV. The usual wind activities on towers are

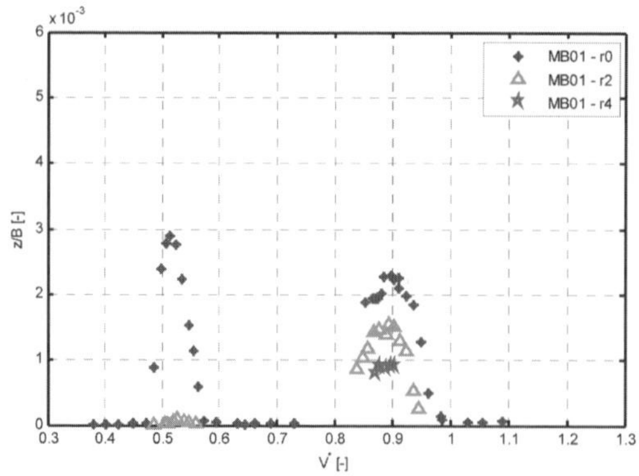

Figure 13. Steady state response: non dimensional oscillation amplitude, flexural motion as function of the reduced velocity varying the Scrouton number (r0 = 0.007; r2 = 0.022; r4 = 0.033).

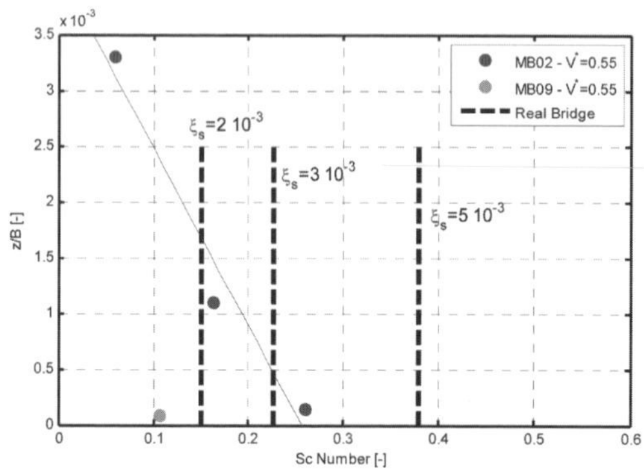

Figure 14. Steady state response: non dimensional oscillation amplitude, MB02 and MB09 configurations, flexural motion as function of Scruton Number.

1) The measurement of the static aerodynamic coefficients on tower sectional model;
2) The measurement of VIV response using elastically suspended tower sectional models
3) The measurement of VIV response using aeroelastic models.

Figure 15 shows a tower sectional model during the wind tunnel tests.

While the vortex shedding phenomenon is basically two-dimensional on the deck, on the contrary for the tower, it is affected by the three dimensional end effects at top and by the presence of the transversal beams along the height. Wind tunnel tests performed on aeroelastic models are therefore aimed to represent the full 3D tower geometry and the related aerodynamic effects (Figure 16).

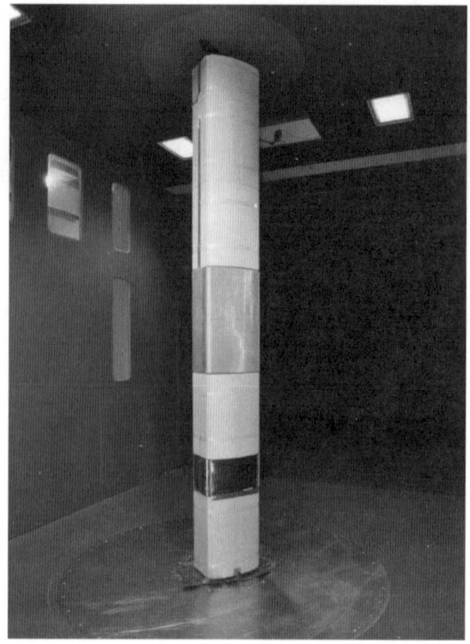

Figure 15. Cable stayed bridge tower sectional model in the wind tunnel test section: measure of static aerodynamic coefficients.

Figure 16. Free standing tower.

6 FULL AEROELASTIC MODELS

Full aeroelastic models are used to simulate the complex wind structure interaction of the completed or semi completed bridge. Once defined the aerodynamic behavior of each single component of the bridge by means of the over described wind tunnel tests the full bridge behavior may be predicted by numerical approaches relying on numerical models or by a scaled simulation of the full aeroelastic structure.

Figure 17. Full aeroelastic model on the turning table of Polimi Wind Tunnel. Length of the model 14 m.

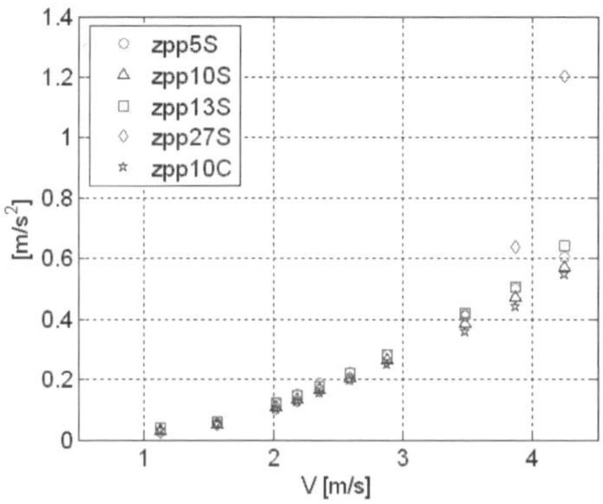

Figure 18. RMS value of the vertical acceleration of section positioned at different positions along the bridge axis varying the mean wind speed under turbulent conditions.

Full aeroelastic models are also used to check if undesired dangerous conditions may arise during the construction stage when the structure has not yet reached its final robustness.

Another key feature is the correct reproduction of the atmospheric boundary layer turbulent characteristics. The goal is reached through the adoption of spires and floor roughness elements that interacting with the wind tunnel flow produce a turbulent mixing of the incoming wind.

From the analysis of the bridge response measured under turbulent wind conditions, it is possible to study the buffeting response of the bridge for different mean wind.

Figure 18 plots the RMS value of the vertical acceleration on the different instrumented section of the deck versus the mean wind speed. The trend is parabolic, nevertheless vortex shedding phenomena occur, up to conditions where instability problem arise.

7 CONCLUSIONS

The common approach to design very long span bridges is to adopt a combined methodology relying on both wind tunnel tests on scaled models and numerical models of the wind-bridge interaction.

Wind tunnel tests on sectional models allows to measure:

- the aerodynamic static coefficients that are used for the definition of the static wind loads and for the check if one degree of freedom instability conditions may occur
- the aerodynamic coefficients (flutter derivatives) that are used for the definition of the buffeting response and for defining the flutter critical wind speed
- the response of deck and tower to vortex induced vibrations as a function of the Sc number.

Wind tunnel results with numerical models are able to drive in the early stage of the aerodynamic design of the deck and tower geometry the designer choice for the most important aeroelastic problems that are the aerodynamic stability of the bridge and the VIV of the deck, tower and cables.

The most promising solution is therefore analysed most in detail through aeroelastic full bridge models in wind tunnel.

The results of the full aeroelastic model are compared with the numerical simulations in order to control all the design procedure.

ACKNOWLEDGEMENTS

I want to express my first acknowledgements to the staff of the wind engineering group of the Mechanical Department of the Politecnico di Milano: Ferruccio Resta, Alberto Zasso, Fabio Fossati, Marco Belloli, Daniele Rocchi, Sara Muggiasca, Tommaso Argentini and to the Wind Tunnel team: Luca Ronchi, Stefano Giappino, Lorenzo Rosa.

Special acknowledgments are also due to Società Stretto di Messina, who promoted relevant investigations on bridge aero elasticity.

REFERENCES

Brancaleoni, F., Diana, G., Faccioli, E., Fiammenghi, G., Firth, P. T., Gimsing, N. J., et al. (2009). *The Messina Strait Bridge.* CRC Press.

Caracoglia, L., & Jones, N. P. (2003). Time domain vs. frequency domain characterization of aeroelastic forces for bridge deck sections. *Journal of Wind Engineering and Industrial Aerodynamics, 91*, 371–402.

Chen, X., & Kareem, A. (2001). Aeroelastic Analysis of Bridges under Multicorrelated Winds: Integrated State-Space Approach. *Journal of Engineering Mechanics*, 1124–1134.

Chen, X., & Kareem, A. (2001, December). Nonlinear response analysis of long-span bridges under turbulent winds. *Journal of Wind Engineering and Industrial Aerodynamics*, *89*(14–15), 1335–1350.

Chen, X., Kareem, A., & Matsumoto, M. (2000). Coupled flutter and buffeting response of bridges. *Journal of Engineering Mechanics*, 17–26.

Cigada, A., Falco, M., & Zasso, A. (2001, June). Development of new systems to measure the aerodynamic forces on section models in wind tunnel testing. *Journal of Wind Engineering and Industrial Aerodynamics, 89*(7–8), 725–746.

Diana, G., Bruni, S., & Rocchi, D. (2005). A numerical and experimental investigation on aerodynamic non linearities in bridge response to turbulent wind. *Proceedings of the EACWE4*. Prague.

Diana, G., Falco, M., Bruni, S., Cigada, A., Larose, G. L., Darnsgaad, A., et al. (1995, February). Comparisons between wind tunnel tests on a full aeroelastic model of the proposed bridge over Stretto di Messina and numerical results. *Journal of Wind Engineering and Industrial Aerodynamics, 54–55*, 101–103.

Diana, G., Resta, F., & Rocchi, D. (2008, October–November). A new numerical approach to reproduce bridge aerodynamic non-linearities in time domain. *Journal of Wind Engineering and Industrial Aerodynamics, 96*(10–11), 1871–1884.

Diana, G., Rocchi, D., Argentini, T., & Muggiasca, S. (2010, June–July). Aerodynamic instability of a bridge deck section model: Linear and nonlinear approach to force modeling. *Journal of Wind Engineering and Industrial Aerodynamics, 98*(6–7), 363–374.

Ding, Q., Zhu, L., & Xiang, H. (2006). Simulation of stationary Gaussian stochastic wind velocity field. *Wind and Structures*, 231–243.

Jain, A., Jones, N. P., & Scanlan, R. H. (1996). Coupled aeroelastic and aerodynamic response analysis of long-span bridges. *60*, 69–80.

Jain, A., Jones, N., & Scanlan, R. H. (1996). Coupled flutter and buffeting analysis of long-span bridges. *J. Struct. Engrg*, 716–725.

Katsuchi, H., Jones, N. P., & Scanlan, R. H. (1999). Multimode coupled flutter and buffeting analysis of the Akashi-Kaikyo Bridge. *J. Struct. Engrg*, 60–70.

Minh, N. N., Miyata, T., Yamada, H., & Sanada, Y. (2000, November). Numerical simulation of wind turbulence and buffeting analysis of long-span bridges. *Journal of Wind Engineering and Industrial Aerodynamics, 83*(1–3), 301–315.

Minh, N. N., Yamada, H., Miyata, T., & Katsuchi, H. (2000, December). Aeroelastic complex mode analysis for coupled gust response of the Akashi Kaikyo bridge model. *Journal of Wind Engineering and Industrial Aerodynamics, 88*(2–3), 307–324.

Niu, H. W., Chen, Z. Q., Liu, M. G., Han, Y., & Hua, X. G. (2011). Development of the 3-DOF Forced Vibration Device to Measure the Aerodynamic Forces on Section Models. *Proc. of ICWE*. Amsterdam.

Wu, T., & Kareem, A. (2011, April). Modeling hysteretic nonlinear behavior of bridge aerodynamics via cellular automata nested neural network. *Journal of Wind Engineering and Industrial Aerodynamics, 99*(4), 378–388.

Bridge Maintenance, Safety, Management, Resilience and Sustainability – Biondini & Frangopol (Eds)
© 2012 Taylor & Francis Group, London, ISBN 978-0-415-62124-3

Pathology, appraisal, repair and management of old prestressed beam and slab bridges

B.F. Godart
Université Paris-Est, IFSTTAR, Paris, France

ABSTRACT: After a presentation of the pathology of old prestressed beam and slab bridges (refered in France as VIPP), the paper reviews the causes and then presents the difficulties of the appraisal and the methodology developed to assess the residual load carrying capacity of this type of bridges. Then, the paper addresses the problems encountered with the management of such structures by emphasizing the interest of the risk analysis for helping owners to optimize the management of their patrimony, and finally reviews different solutions of repair used with success to maintain or to strengthen the capacity of these structures.

1 DESCRIPTION AND PATHOLOGY OF POST-TENSIONED BEAM AND SLAB BRIDGES

A great number of beam and slab bridges was built in France, after the end of the second world war (about 250 between 1945 and 1957, and 450 between 1957 and 1967). Most of these structures (including the two longest: the Saint-Waast bridge at Valenciennes, with a span of 64 m, and the Hippodrome bridge in Lille, with a span of 66 m) were built after 1947 in the absence of formal design rules.

1.1 *Description of beam and slab bridges*

The Beam and slab bridges (refered in France according to their acronym “VIPP”: Viaducs à travées Indépendantes à Poutres Précontraintes par post-tension) are viaducts with multiple single spans made of post-tensioned concrete beams. These bridges are made of simply supported spans, and each span is composed of precast post-tensioned concrete beams (generally precast on site) that are linked together by crossbeams and a slab on the top. According to the design the crossbeams and/or the slab may be prestressed transversally. The length of the spans are generally ranging between 30 and 50 metres (Fig. 1).

These bridges are slim and piecemeal structures, that often suffered from a poor grouting of their prestressing sheaths, from defects of sealing behind tendon anchorages and from a lack or failing waterproofing membrane. However, very often, the concrete of these decks was of good quality. We are now detailing the pathology of these VIPP according to their three possible causes of defects: design, construction and maintenance.

Figure 1. Example of a beam and slab bridge (VIPP bridge), the Merlebach bridge.

1.2 *Pathology of beam and slab bridges*

1.2.1 *Design defects*

The enthusiasm which prevailed when the first generation of prestressed structures was being built was reflected in a total confidence in full prestressing (i.e. lack of cracking) and in a belief that concrete under compression was watertight (HA, SETRA, TRL & LCPC 1999). This is why a number of details appeared such as:

- lack of waterproofing (deck waterproofing only became mandatory after 1966);
- lack of sealing behind tendon anchorages;
- lack of provision for drainage. (It is often beams located under drainage channels which are most affected by corrosion (Fig. 2);
- leaking expansion joints;
- piecemeal construction resulting in a large number of unprepared construction joints which could give rise to cracking due to restrained shrinkage;
- use of unprotected transverse tendons in grooves in the deck;

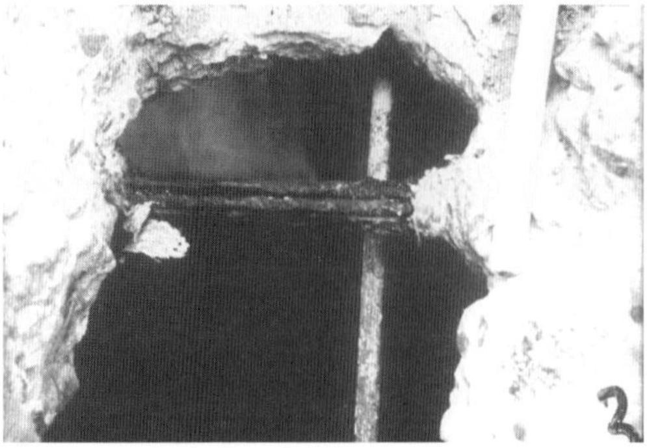

Figure 2. Condition of a transversal tendon located in the slab below the drainage channel.

Figure 3. Damages at the bottom flanges of VIPP beams due to poor water proofing.

- use of sheaths made from bitumen-coated Kraft paper wrapped around the tendons which made grouting impossible (metal sheaths were not used until the late 1950s);
- large numbers of tendons per span in older structures with deck anchorages (e.g. typically up to 15 out a total of 20 to 25) increasing the number of points of possible water ingress;
- use of lead lined steel ducts between 1950 and 1975 to reduce friction sometimes caused tendon corrosion as a result of bimetallic action between the ducts and the steel;
- use between 1950 and 1970 of prestressing steel which was susceptible to stress corrosion, which gave rise to the possibility of brittle fracture;
- insufficient concrete cover to the reinforcement, resulting in corrosion of the reinforcement and spalling of the concrete giving easier access to the prestressing tendons for aggressive agents.

1.2.2 *Construction defects*

The two main construction defects leading to corrosion are poor waterproofing (Fig. 3) and incomplete grouting of the ducts. It is common to find thin waterproofing layers which do not extend beneath the footways. It is also common to find partly empty or sand-filled ducts in structures built before 1960 when grouts used to contain sand. Grouting techniques have also been poorly applied, and evidence of blockages has frequently been found. In addition to these defects, poor sealing of beam end anchorages, deck anchorages and transverse anchorages are observed.

Although concrete has generally performed well in older structures, there may be areas where poor workmanship has given rise to honeycombing or shrinkage cracking. The most common locations are the flange soffits where concreting has been made difficult by the congestion of ducts, or where there has not been proper compaction, resulting in large areas where spalling of concrete may allow aggressive agents to penetrate.

1.2.3 *Defects linked to maintenance and operation*

During the 30 years which followed the end of the second world war, the absence of an inspection policy for structures and the failure to take the necessary action to make the structures watertight are the two main causes of maintenance problems.

Resurfacing the carriageway without investigating the condition of the waterproofing layer, failure to restore drainage to its required condition and clearing obstructed drainage pipes, failure to act to prevent water leaking through expansion joints and running over the ends of beams are some of the many maintenance defects which have encouraged the development of corrosion in the tendons.

In addition the increasing use of deicing salts on the carriageway surface exacerbated the aggressive nature of the water seeping into the structure.

1.3 *The corrosion of prestress*

The corrosion of prestress can be divided into three main types: electrolytic corrosion (rust, small pits, generalized corrosion, etc.), stress corrosion and hydrogen embrittlement corrosion.

Most corrosion defects are linked to the first type of corrosion (conventional corrosion) and are caused by water which seeps through cracks and which infiltrates into the structure through leaking seals or zones of porous concrete. Then water flows through the network of ducts which have been grouted to a greater or lesser extent ... Experience shows that problems due to corrosion are distributed randomly, in accordance with the random nature of the movement of water within a structure. Figure 4 taken under the soffit of a beam after removal of the concrete cover is a perfect illustration of this "ramdom" distribution; we can observe that healthy tendons are close to heavily corroded tendons in the same cross-section, and that healthy zones alternate with corroded zones on a same tendon. In general, relatively few chlorides have been found in steel corrosion products in France.

Brittle fractures attributed to stress corrosion cracking have been found in some cases. Investigations conducted by LCPC show that stress corrosion appears primarily in quenched hot rolled wires, and more often in such wires containing more than 0.1% of Cu. This type of prestressing steel was only used from 1950 to 1965. We are especially cautious with some

Figure 4. Illustration of the "random" distribution of corrosion.

type of wires named "sigma-oval" used with the KA prestressing process (Fig. 5).

The third type of corrosion has never been found in any of the surveys on French bridges. Likewise no fatigue fractures of tendons or wires have so far been discovered in existing prestressed concrete structures.

Although defective grouting of the ducts is a primary condition for the development of corrosion, this condition on its own is not sufficient to trigger the reaction. It has been found that ungrouted tendons in dry structures located in regions with a continental climate have not corroded at all.

Experience shows that neither the steel sheath, nor even a well compacted grout can form a sufficiently tight barrier against the percolation of an aggressive water through the more or less porous concrete of a beam. This was particularly well highlighted during the expertise of a viaduct made up of simply supported box girders of good quality, located in a port, whose bottom of the box girders was immersed in sea water only when very strong tides occured, and where the corrosion of tendons at mid-span was such that the bridge had to be rebuilt. This was also observed on another viaduct whose prestressed piers bathed permanently in sea water.

Figure 5. Brittle failure of wires of the KA prestressing kit.

Although the evidence of tendon corrosion arising from a deck waterproofing defect usually shows up as simple water leakage or the presence of seepage products such as efflorescence or traces of rust, there are a number of cases where tendons have been partly corroded without any external signs or visible defects.

Finally, the risk created by corrosion of prestressing tendons is greatest in twin-beam bridges without intermediate cross-beams. The risk is even higher in the areas subject to high shear forces where brittle fracture is likely because there is insufficient passive reinforcement to carry the tensile forces transferred to the steel as the concrete cracks.

2 THE METHODOLOGY OF DIAGNOSIS

Some of these VIPP present strong deficits of prestress related to the corrosion or the failure of tendons, not detected during traditional examinations, because no external sign makes it possible to reveal them. It is particularly the case of the prestressing steels sensitive to stress corrosion cracking or to hydrogen embrittlement, and more particularly the case of the KA post tensioning kit (sigma oval wires).

Moreover, in the first generation of VIPP, longitudinal reinforcement steels were very reduced (as well in area as in number), and vertical reinforcement steels close to the supports were insufficient. This weakness of reinforcement steels involve a lack of ductility of the structure during the failure of prestressing tendons, and in this case the ruin of this type of structure can then be fast and fragile.

Moreover, knowing that there is not any non destructive testing methods to evaluate the global state and the quantity of residual prestressing all along a span of VIPP, one can consider easily that the diagnosis methodology for the VIPP is difficult to conceive and to carry out. Nevertheless, we propose a methodology which is strongly inspired of the LCPC guide (LCPC 2001) and improved by the knowledge gathered since 2001. This methodology of diagnosis comprises the three following stages.

2.1 *Stage 1: Analysis of the bridge records*

The first stage of the diagnosis consists in identifying in the bridge file all the records that are useful for the diagnosis and in particular those which make it possible to carry out a preliminary analysis of the risk:

- construction year (a VIPP built before 1967 is *a priori* doubtful)
- design assumptions and calculation hypotheses
- post tensioning kit and type of unit
- grout composition
- grouting method used and incidents of injection
- waterproofing system

2.2 *Stage 2: Detailed inspection of the bridge*

This second stage aims to continue the preliminary analysis of the risk and especially to evaluate the risks of defects of integrity of the tendons. It consists in detecting the presence of disorders which can make fear a corrosion or a failure of tendons, such as for example:

- transverse cracks starting from the bottom part of the beam and generally going up in the web vertically (flexure) and sometimes with a slope close to the supports (shearing force)
- longitudinal cracks following the tendons in the webs, with traces of water circulation in the cracks
- longitudinal cracks following the tendons in the bottom flange with traces of water circulation in the cracks
- detachment of sealings behind anchorages of longitudinal or transversal prestressing

But as we already mentioned, the problems of these bridges are that it is not because none of these disorders is apparent that prestressing is healthy... Whatever the result of the detailed inspection, it is thus appropriate to go to the following stage which comprises 3 levels of investigations.

2.3 *Stage 3: Investigations*

The investigations mainly based on non destructive techniques, but also on partially destructive techniques, are graded according to 3 levels N1 to N3.

2.3.1 *Level of investigation N1*

This level allows to evaluate the quality of grouting of the prestressing ducts. This evaluation is done by gammagraphy or radiography examinations which make it possible to detect the presence or the lack of grout, the possible presence of wire or strands failures (Fig. 6) and the conformity of the reinforcement. But, these techniques can not detect the presence of corroded wires.

It should be noted that the Radar (Ground Penetrating Radar applied to investigations on structures) also makes it possible to evaluate the conformity of the layout of the ducts and reinforcement, but it does not allow in any way to evaluate the presence or the absence of grout in metal sheaths. It can make it possible to position the gamma or radiographic pictures. It may also be noted that the use of the impact-echo method allows to detect large grouting defects in metal sheaths, without being able, in the current state of the technique, to detect small voids.

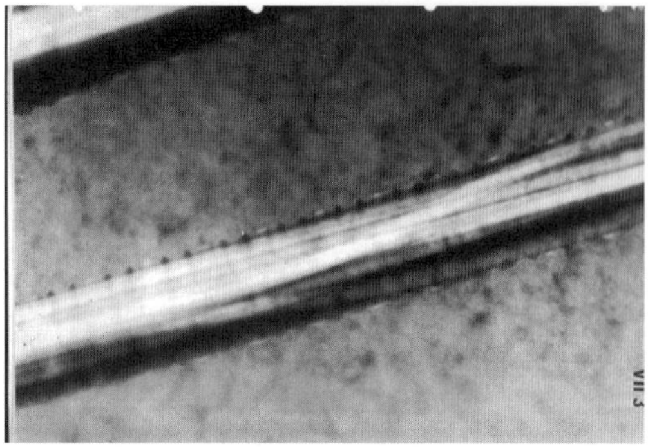

Figure 6. Gammagraphy showing a lack of grout and wire failures on a tendon in a corrugated duct.

2.3.2 *Level of investigation N2*

This level allows to evaluate qualitatively the prestress by opening access ports in order to inspect the state of the tendons. In particular, it allows to examine the type of corrosion : simple rust, pits of corrosion, craters, generalized corrosion, stress corrosion. It also makes it possible to take samples of grout, water, sheath, wires, strands, rust, etc ... for the purpose of laboratory analysis. The opening of windows also allows to practise the test known as "of the flat screwdriver" in order to detect a possible diminution of tension in the wires or strands.

If the results are satisfactory (absence of corrosion, lack of failures and water in the sheaths), the LCPC guide advises to proceed with a normal surveillance.

If the results show stress corrosion, or a corrosion generalized on at least 5 to 10 % of the prestress, or if the analysis of the materials shows risks of corrosion, then it recommends to pursue to level N3.

2.3.3 *Level of investigation N3*

This level aims at assessing the residual load carrying capacity of the bridge by practising various testing methods of the prestress and the structure, in order to allow a recalculation of the bridge. The various testing methods that can be used are the following ones:

- the measurement of tension of the prestressing steels by the crossbow method (LCPC 2009): this latter is particularly rich in terms of knowledge acquisition because it allows to assess the residual tension in wires or strands, and this whatever the degree of corrosion affecting the prestressing steel (Fig. 7). It also provides a very valuable average tension for the recalculation;
- the evaluation of the local longitudinal stresses by the stress release method: this method allows to

Figure 7. Measurement of the tension of a prestressing wire by the crossbow method.

estimate the stress gradient existing in the thickness of an element at a given point. It is possible to obtain the normal force existing in a cross section of a VIPP beam, with the condition of carrying out a minimum of 3 to 4 slots, taking into account the non-linearity of the stress diagram (because of the delayed strains of the concrete, the stresses are lowered in the slender zones of a cross section and increased in the thicker zones…).
- the curvaturemetry: this method allows to detect non-linearities of curvatures under bending moment indicating the beginning of a not yet visible cracking in the central zone of a beam.

The acoustic monitoring can also be used to survey a VIPP and to listen to the elementary wires or strands failures. If it does not provide directly useable results to assess a load carrying capacity, it however allows to confirm the existence of a prestress damage by corrosion and to follow it with time.

The final evaluation is done on the basis of recalculation integrating the results of the investigations and in particular the estimates of prestress losses. This recalculation can be done by traditional methods and if necessary on the basis of a reliability approach which requires to know the probabilistic distribution laws of certain calculation parameters.

According to the results of the recalculation and investigations of level N1, N2 and N3, one can proceed with various possible types of treatment (see chapter 5).

3 RISK ANALYSIS APPLIED TO THE MANAGEMENT OF VIPP

3.1 *Introduction*

Risk analysis is a powerful tool to achieve certain strategic choices when financial resources of the authority are not expandable. It is the case when the authority has to manage a great quantity of old VIPP that are difficult to appraise, and the authority has to do priorities to investigate among the bridges.

In France, the first application of risk analysis on bridges was conducted in 2007 by C. Cremona at LCPC on the VIPP of the conceded motorway network on behalf of the French Association of Motorway Companies (ASFA) (Dabert & Cremona 2009) (Guérard 2009). This association decided in 2005 to launch a bid to study the reliability of the knowledge of their 116 VIPP. LCPC (now IFSTTAR) then proposed to apply a standard approach founded on Operational Safety (or dependability).

3.2 *Operational safety analysis of the ASFA VIPP*

3.2.1 *Stages of the Operational Safety*

The Operational Safety analysis proposed by LCPC has been divided into 4 stages:

Stage 1: an Internal Functional Analysis (IFA) which covered the following aspects:

- the detailed inventory of the concerned structures, including their technical characteristics
- the recovery of data files on bridges and the inventory of the possible modes of exceptional operation modes of each bridge
- the identification and analysis of repairs already carried out (span attachment, change of bearings…)
- the analysis of the bridge environment (aggressiveness, network environment, winter viability…)
- the analysis of constituent materials and methods of construction,
- the analysis of ancient calculation methods
- the inventory and analysis of investigations conducted.

Stage 2: a Preliminary Risk Analysis (PRA) with the objective to:

- define families of structures based on relevant criteria (construction, failure, …);
- identify gaps and problems specific to each bridge and each family.

The preliminary risk analysis (PRA) is a deductive analysis based on information available in the bridge records that allows to have visibility on the shortcomings of bridges and provides a hierarchy of bridges in five classes. It allows to concentrate the analysis and detailed investigations on the most sensitive structures. It seeks both to identify the factors or events that could reduce the bridge performance, and to affect to bridges a grade of severity according to predetermined criteria.

The PRA index is obtained from severity notes given to some identified characteristics of the bridge and modified by weighting coefficients. Il allows to classify bridges into 5 classes of risk (Table 1).

Stage 3: a Risk Quantification (RQ) aimed at:

- checking by simple calculation, the ability of the current configuration of bridges to support the operational modes;
- defining the possible interim operating modes of the failed deck as a function of the probable visible defects.

Quantifying the mechanical criticality is therefore based on simple indicators to assess the serviceability

(SLS: service limit states) and the structural safety (ULS: ultimate limit states). In this goal, this criticality is represented by a criticality indicator I of a generic type:

I = *Resistance*/*Load*

This indicator is evaluated for a bridge used in its initial configuration, in its present situation, then with reduced lanes. The evolution of the indicators is estimated for deteriorated situations (losses of prestressing tendons that are assumed or effective according to the data inventory) or for limited traffic in weight.

Because of the complexity of a VIPP-type structure, a single indicator can not summarize the behaviour of the bridge. Therefore, one should look for specific remarkable elements that together contribute to the structural safety of the structure. It is the resistance of the slab receiving the traffic loads, the resistance of the longitudinal main beams, then the resistance of the supports. For beams that are in a single span, we consider that a good evaluation of their resistance is to control their flexural strength at mid-span and their shear strength on support.

The overall index RQ is defined for the resistance of beams with respect to bending and shear, and for the resistance of the deck slab by retaining the maximum of the different coefficients calculated. A bridge is then ranked among the 5 classes defined in Table 2.

Stage 4: a detailed analysis of criticality to:

- identify and propose further investigations that are necessary to characterize the condition of each bridge and each family
- define a method for evaluating the residual load carrying capacity and the remaining life according to different operating modes.

The qualitative index reflects the overall susceptibility to potential or actual risks of the bridge and reports primarily for the assessment of the risk of prestress loss. The quantitative index reflects the overall mechanical health condition assuming that the structure is sane. For the manager, these indexes should therefore be crossed to try to assess the mechanical condition given the likely condition of prestress. A matrix of decision making has been proposed and is illustrated in Table 3.

From this table, it is possible to define overall criticality classes to outline a method of classifying bridges by assigning an overall and unique rating to each bridge. The overall criticality rating crosses information on the apparent condition of the structure (mainly characterized by the PRA classes) with information on the theoretical structure condition characterized mainly by the RQ note (Table 4).

The criticality analysis can then determine what actions will be undertaken on each of the VIPP according to its ranking. VIPP with criticality D will be treated in priority and subjected to heavy investigations and thorough recalculations. VIPP with criticality A will just be subjected to a normal surveillance. VIPP with intermediate criticality B or C will be subjected to recalculation or targeted investigations as defined in the LCPC guide on assessment of VIPP.

4 THE REPAIR OF VIPP

Several solutions of repair or strengthening may be applied to the VIPP according to their condition and their residual load carrying capacity. The main solutions are summarized below.

If the corrosion of tendons is negligible, and that some spalling of concrete has happened due to the corrosion of the reinforcement, then a patch repair is sufficient provided that it has been checked that the carbonation of concrete and/or the ingress of chlorides

Table 1. Overall qualitative index of a bridge (PRA).

Classification of bridges	Note
No risk	0 to 20
Moderate Risk	20 to 40
Rather high Risk	40 to 60
High Risk	60 to 80
Very high Risk	80 to 100

Table 2. Overall quantitative index of the risk presented by a bridge.

RQ Index	Class
1	Good
2	Fair
3	Correct
4	Poor
5	Bad

Table 3. Percentages of VIPP for each pair (PRA, RQ).

PRA Index	RQ Index 1	2	3	4	5
0–20	0%	3%	4%	0%	0%
20–40	5%	8%	34%	7%	9%
40–60	3%	3%	14%	3%	5%
60–80	0%	0%	1%	0%	2%
>80	0%	0%	0%	0%	1%

Table 4. Global criticality of VIPP.

Overall criticality	
A	Healthy bridge
B	Little Deteriorated bridge
C	Deteriorated bridge
D	Very deteriorated bridge

have not penetrated beyond the rebars. Sprayed concrete can also be used in the case where a large amount of concrete spalling has occurred.

In the case where the penetration depth of aggressive agents is beyond the rebars, the problem becomes difficult because the removal of the contaminated concrete could be impossible in some regions of the beam like the toe in the middle of the beams or the shear resisting zone near the ends of the beam and the bearings; electrochemical treatment may be applied, but with extra care due to the possible damaging of the prestressing steels. The standard NF EN 12 696 (AFNOR 2000) gives some rules to avoid hydrogen embrittlement of the prestressing steels, and we must be very cautious with chloride extraction as compared with realkalinization.

If the corrosion of tendons is small, or if the reinforcing steels are insufficient because of initial underestimation or loss of cross-section due to corrosion, two repair solutions may be used:

- the first one is the gluing of steel plates: it is a proven technique that allows to add a cross-section of reinforcement which is strong enough. It should ensure the respect of the ultimate limit state (ULS), but the improvement of the Service Limit State (SLS) will be limited since it only affects the moments of live loads. This technique is subject to a good condition of the concrete surface (pull-off test). The implementation is quite delicate because sheets are heavy and they need to be laid on under the bottom flange of the beam with a certain pressure, and the traffic should be interrupted during the gluing phase and during the prolonged period of the resin hardening to prevent the plates from dropping...
- the second solution is the gluing of tissues or strips made of Carbon Fiber Reinforced Plastics (CFRP): this technique is increasingly used. Compared to the strengthening of a slab, the problem may be more difficult to solve on beams because the width of the flange is only 65 cm and it requires to glue several layers of tissues or thick bands. As for the steel plates glued, this technique requires a good strength of the concrete surface. The implementation is a bit easier than for the steel plates because of a lower weight and a setting by masking.

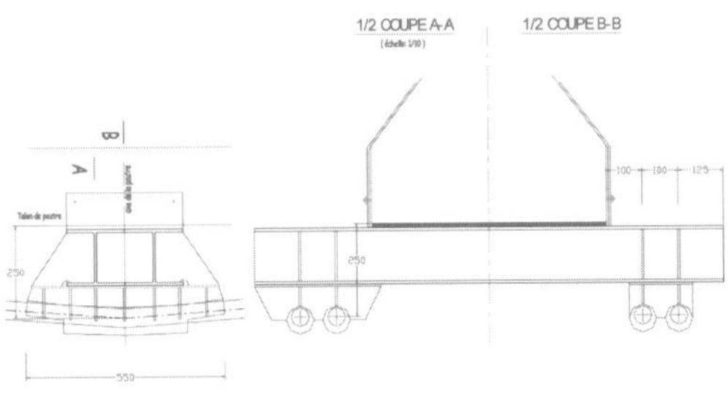

Figure 8. Cross-section of the deviator in the middle of each beam.

If the corrosion of tendons is important, then the only solution is to strengthen by additional prestressing. This latter can be straight or deviated. The deviation of the prestressing is generally more efficient because it allows a strengthening of the shear zones and has a better efficiency to add bending moment, but it necessitates to add a deviator in the middle of the beam (Fig. 8). As illustrated by the following example of the Merlebach bridge in the East of France, a current strengthening consists of adding two tendons symmetrically on each side of the beam. Each tendon comprising 2 strands of diameter 15,7 mm is able to replace one corroded existing tendon of cross-section 600 mm^2 (12 wires of diameter 8 mm and post-tensioned at 133 kg/m^2).

One of the main difficulties is to anchor the new tendons at their ends. The anchor blocks must be installed in a suitable zone and preferably in the upper corner of the beam. Because of the presence of a crossbeam at the end of the span, and because there is no place behind the crossbeam to install a prestressing jack, the anchor blocks have to be fixed before the crossbeam and at a certain distance of this latter in order to have enough space to install the prestressing jack. To improve the strengthening system and to diffuse the anchoring efforts in the web of the beam, it is advisable to alternate dead and active anchoring as shown on figure 9. Each tendon goes from a dead anchoring block located in a corner to an active anchoring block located at 3,84 metres from the crossbeam.

The blocks are preferably precast on site and are fixed to the beam with the help of very short prestressing bars that must be carefully post-tensioned to avoid a rapid loss of prestress.

The tendons are tensioned as planned at one end only and there is no conflict between them because the two tendons are deviated in different planes (at the deviator) see figure 8.

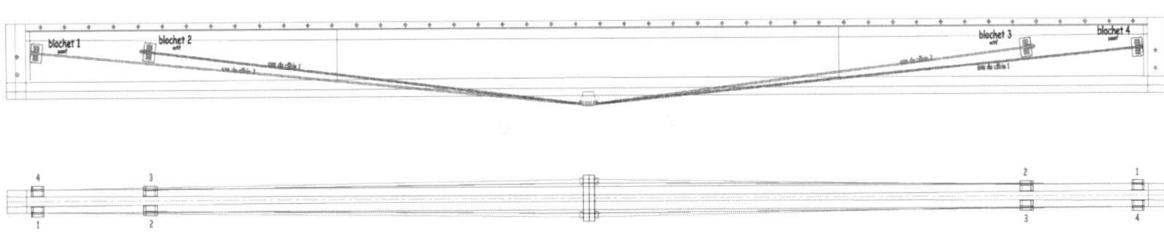

Figure 9. Installation of additional tendons with their dead and active anchoring block.

REFERENCES

SETRA, 1996. Guide de conception des ponts à poutres préfabriquées précontrainte par post-tension (VIPP), Ed. Service d'Etudes Techniques des Routes et Autoroutes.

Godart, B. & Chatelain, J. 1997. L'auscultation des Ouvrages d'Art. In *Maintenance et réparation des ponts*, sous la direction de Calgaro J.A. & Lacroix R., Presses de l'Ecole Nationale des Ponts et Chaussées, 666 p.

HA, SETRA, TRL, LCPC. 1999. Post-tensioned concrete bridges. Joint publication of Highways Agency, Service d'Etudes Techniques des Routes et Autoroutes, Transport Research Laboratory and Laboratoire Central des Ponts et Chaussées, 164 pages, ISBN: 0 7277 2760 5. (French and english). Thomas Telford.

AFNOR, 2000. Protection cathodique de l'acier dans le béton, Standard NF EN 12696.

LCPC, 2001. Viaducs à travées indépendantes à poutres précontraintes (VIPP), Guide méthodologique de surveillance et d'auscultation, Technical Guide, Ed. LCPC, 72 p.

Godart, B. 2003. L'évaluation et la maintenance des viaducs à travées indépendantes à poutres précontraintes (VIPP). In *Application des notions de fiabilité à la gestion des ouvrages existants*, sous la direction de C. Cremona, Presses de l'Ecole Nationale des Ponts et Chaussées.

Cremona, C. 2007. Suivi du comportement mécanique d'une poutre de VIPP sous chargement: Le cas du VIPP de Merlebach. – Etudes et Recherches des Laboratoires des Ponts et Chaussées, Série OA n° 56, Ed. LCPC, 154 p.

LCPC, 2009. Mesure de la tension des armatures de précontrainte à l'aide de l'Arbalète, Technical Guide, Ed. Laboratoire Central des Ponts et Chaussées, 24p.

Dabert, J.L. & Cremona, C. 2009. Analyse de la criticité du patrimoine VIPP autoroutier. In GC'2009, Association Française de Génie Civil, Cachan, 18–19 mars 2009, Ed. AFGC.

Guérard, H. 2009. Les VIPP du réseau ASF/Recalculs d'ouvrages : les VIPP d'ASF. In Journées GC'2009, Association Française de Génie Civil, Cachan, 18–19 mars 2009, Ed. AFGC.

SETRA, 2010. Analyse des risques appliquée aux viaducs à travées indépendantes en poutres précontraintes (VIPP), Guide technique, Ed. Sétra.

Bridge Maintenance, Safety, Management, Resilience and Sustainability – Biondini & Frangopol (Eds)
© 2012 Taylor & Francis Group, London, ISBN 978-0-415-62124-3

Assessment and retrofitting of existing bridges

C. Modena, C. Pellegrino, G. Tecchio, F. da Porto, R. Morbin & M. Grendene
Department of Civil, Environmental and Architectural Engineering, University of Padua, Italy

ABSTRACT: The existing stock of road and railway bridges in Italy and in most European countries frequently exhibits insufficient performances, both in terms of structural safety and functionality, compared to the current demands coming from modern structural codes, transportation systems and necessity of reducing the costs of maintenance. Quick and reliable methodologies are then needed to assess the specific vulnerability of any bridge typology. In the first part of the paper the application of such type of methodologies to both road and railway transportation infrastructure in Italy is described showing that, e.g., masonry arch bridges are usually quite robust structural systems, r.c. bridges generally present durability problems and are vulnerable to seismic actions, steel bridges, other than presenting durability problems, are particularly vulnerable to fatigue. Typical retrofitting interventions considered in these studies are also briefly presented. In the second part of the work significant case studies of retrofitting interventions are in more detail described, focusing on four existing r.c. bridges, which are examples of some of the most usual typologies adopted in post–World War II period, making their retrofitting description suitable for considerations of a general nature. The refurbishment interventions are presented outlining a methodological approach, which takes into account the typological characteristics of the structure, the state of maintenance, the functional requirements and the environmental aspects connected to the repair and strengthening system.

1 INTRODUCTION

A lot of in-service bridges usually show dimensional, structural and functional deficiencies caused by different factors: deterioration processes, increase of the traffic loads, updating of the site seismic classification. In many cases, the severe reduction of the structural safety level makes necessary a repair and/or retrofit intervention. In this paper, after discussing the condition appraisal and retrofitting of usual road and railway bridge typologies, four case histories of refurbishment interventions on existing rc bridges are presented in detail. In these examples, as usual, rehabilitation is coupled with refurbishment, in order to improve the safety and comfort to road users.

2 ASSESSMENT AND RETROFITTING OF USUAL ROAD AND RAILWAY BRIDGE TYPOLOGIES

A simple procedure for evaluation bridge conditions developed at the University of Padova (Pellegrino et al. 2011) is presented. The system of inspections is the visual survey as prescribed by the standards of most countries (BRIME 2001) and by the Italian Code.

As the greatest part of infrastructural authorities have applied (Ryall 2001), the Condition Value index has been chosen to express the functional condition of every element. For every element of the bridge the Condition Value index points out five possible levels of deterioration. These numerical values have been chosen according to the results of BRIME (2001), Ryall (2001) and UK local authority experience. Condition Value (CV) is then converted to a Condition Factor (CF) (Pellegrino et al. 2011). The bridge structure has been divided into its fundamental components: structural elements, fundamental for the structural capacity and the safety of the bridge against collapse; non-structural elements, that do not contribute to the structural capacity of the bridge, but are relevant for functionality and durability of the structure.

A different weight is assigned to every element of the bridge that must be evaluated. This weight (W) varies from 10 (maximum importance) to 5 (minimum importance) and contributes to the calculation of the global efficiency. A Location Factor (LF) corresponds to each weight (Pellegrino et al. 2011). LF and corresponding weights are defined according to Blakelock et al. (1998) and Ryall (2001).

The evaluation of the condition of the elements through the Condition Value (CV) is not enough to establish the priorities of maintenance/rehabilitation/strengthening interventions on the structure and does not allow providing a maintenance planning of the bridge stock investigated.

The definition of the Element Sufficiency Rating (ESR), as a "grade" of the single bridge element, considers both Project and Network Levels. Such index has been calculated starting from the CV index and taking into account that the elements of the bridge have not the same importance; for example, it is necessary to give a higher weight (W) to the maintenance of a principal structural component, i.e. a pier, than that of a secondary non-structural one, i.e. the parapets.

Finally the age of the bridge is also taken into account for the quantification of the Element Sufficiency Rating (ESR).

Therefore the calculation of the ESR has to be influenced by (Pellegrino et al. 2011): the condition of the element, through the Condition Factor (CF), linked to the CV; the importance of the element into the bridge, Location Factor (LF) linked to the weight of the element; the road type to which the bridge belongs, Road Type (RT); the traffic on the bridge, Traffic Index (TI), measured in ADTV (Average Daily Traffic Volume); the importance of the bridge into the network, Network Bridge Importance (NBI); the age of the bridge, Age Factor (AF).

According to the above considerations, the formula for the calculation of the Element Sufficiency Rating (ESR) can be expressed as follows:

$$ESR = CF \times LF \times (RF \times NBI \times AF) \tag{1}$$

where: CF = Condition Factor; LF = Location Factor; RF = Road Factor (RF = RT × TI); NBI = Network Bridge Importance; AF = Age Factor.

This index allows to define the degree of efficiency of the components of the bridge, establish a priority plan of intervention for the single structure (Project Level), and establish a priority plan of intervention for the whole network (Network Level).

Once defined Element Sufficiency Rating (*ESR*), the calculation of the efficiency of the whole structure starting from the efficiency of its components was developed. Considering the Network Level, the problem is to give a "grade" for every structure that allows the authority to have an overview on the general state of efficiency of the bridges of the stock. Such "grade", named Total Sufficiency Rating (*TSR*), is calculated with a weighted arithmetic average:

$$TSR_{real} = \left(\frac{\sum_{i=1}^{t} CF_i \cdot W_i}{\sum_{i=1}^{t} W_i} \right) \cdot PF \cdot 10 \tag{1}$$

where CF_i = Condition Factor of the i-th elements evaluated; W_i = weight of the i-th elements evaluated; PF = Penalty Factor = $(RF \times NBI \times AF)$; t = number of elements evaluated; TSR_{real} = index of total efficiency referred to the elements evaluated.

The final value of *TSR* is calculated starting from TSR_{real} also considering the elements not evaluated. Therefore it was introduced, the Confidence Factor (*CoF*), that must be superior to a threshold limit:

$$CoF = 100 \times \left(\frac{\sum_{i=1}^{t} W_i}{\sum_{i=1}^{n} W_i} \right) \tag{2}$$

where n is the total number of elements.

The criterion that has seemed to be the most appropriate for the calculation of the final value of the Total Sufficiency Rating (*TSR*) is to refer to a weighted arithmetic average between the real situation (TSR_{real}) and the worst situation that can happen (TSR_{min}). TSR_{min} has been evaluated assuming $CV = 5$ for all the elements that are not evaluated, except for the foundations for which the worst-case scenario is assumed when $CV = 3$ since the foundations are usually not visible elements.

The final expression of *TSR* is:

$$TSR = \left(\frac{TSR_{real} \times 100 + TSR_{min} \times CoF}{100 + CoF} \right) \tag{3}$$

2.1 *Rc bridges*

Proper assessment of rc bridges cannot disregard seismic issues. Fragility curves can be considered as one of the most performing tools to assess existing bridge seismic vulnerability (Padgett & DesRoches 2008, Monti & Nisticò 2002, Franchin et al. 2006, Shinozuka et al. 2000a). They are instruments describing the probability of a structure being damaged beyond a specific damage state for various levels of ground shaking.

Analytical fragility curves are developed through seismic analyses of the structure. Most of the analytical methods of the literature consist in three steps: simulation of ground motions, modelling of bridges taking into account the uncertainties and generation of fragility curves from the seismic response data of the bridges. In most studies the fragility curves have been evaluated considering only the pier of the bridge (for example Karim & Yamazaki 2001, Karim & Yamazaki 2003) as the most vulnerable element of the structure, but it is a simplified approach which, as shown in this work, can lead to overestimate or underestimate the vulnerability of the bridge.

The fragility curve becomes a lognormal cumulative distribution:

$$P_{f,PL}(a) = P\left[Da > d_{PL} \middle| a\right] \tag{4}$$

where the damage density probability function is defined by the lognormal distribution in the following equation:

$$f_D(d) = \frac{1}{\sqrt{2\pi}\varepsilon d} \exp\left[-\frac{1}{2} \left(\frac{\ln d - \lambda}{\varepsilon} \right)^2 \right] \tag{5}$$

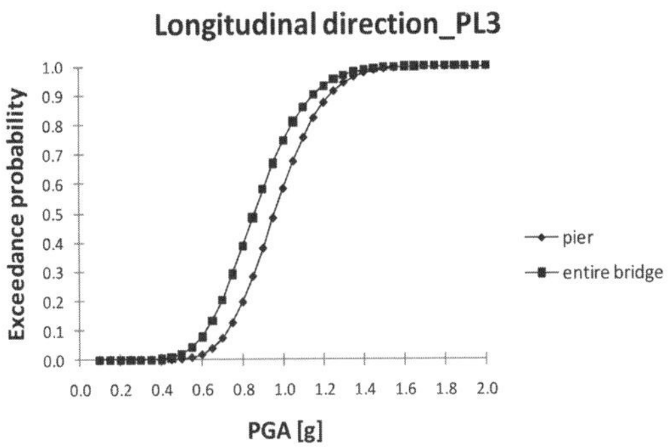

Figure 1. Fragility curves (longitudinal direction): comparison for PL3 between the analytical models of the single pier and the entire bridge.

being $\lambda = A + B \ln(IM)$ the average value related to a specific IM value and ε the dispersion of data.

Two modelling strategies with an increasing level of complexity have been developed. The pier modelled as a cantilever with fixed end (1) and the entire bridge (2), have been studied to generate and compare the fragility curves.

The seismic response was obtained by non-linear time history analysis (Morbin et al. 2010), carried out by OpenSees code (2009): piers are modelled with beam elements. Kent & Park (1971) constitutive law, has been considered for concrete elements and an elastic-hardening plastic law has been considered for reinforcing steel. The deck is modelled with beam elements having linear elastic behaviour. After having built fragility curves for each numerical model, a comparison between the fragility curves obtained with the two considered analytical models (single pier and entire bridge) is presented in Fig.1.

There are various strategies to reduce the seismic vulnerability of rc bridges. A typical intervention consists in concrete jacketing of piers eventually including new steel reinforcement with the aim of increasing strength and ductility of the piers.

FRP jacketing (CNR-DT 200 2004, fib bullettin n.14 2001, Pellegrino & Modena 2010, Tinazzi et al. 2003, Wang & Wu 2008) is commonly used to increase compressive strength and ductility of bridge piers.

In this study the model described in CNR-DT 200 (2004) is considered for the constitutive law of the FRP confined RC pier. According to this model and taking into account the above-mentioned bridge case study, fragility curves for the retrofitted bridge were calculated and compared with those of the existing bridge (Fig. 2).

2.2 *Steel bridges*

The effect of fatigue is particularly relevant on steel bridges since the influence of service load cycles on serviceability limit stress values is very high if compared to relatively low dead weight. Orthotropic steel decks of road bridges, directly subjected to traffic loads, are very sensitive to fatigue: in most cases,

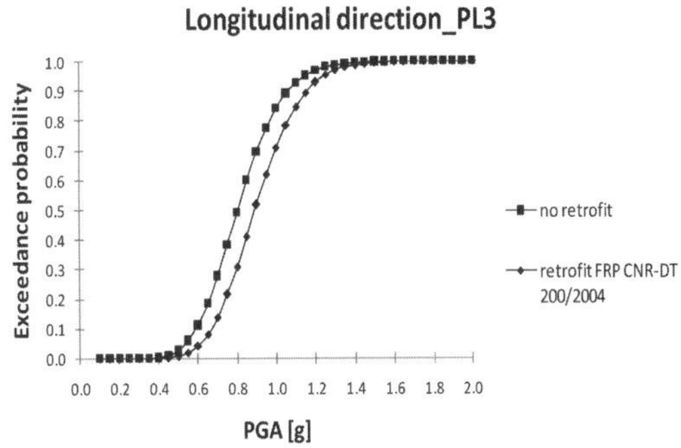

Figure 2. Fragility curves (longitudinal direction): comparison between entire bridge without retrofit and retrofitted bridge (CNR-DT 200 2004 is used for modelling the constitutive law of the bridge with confined piers).

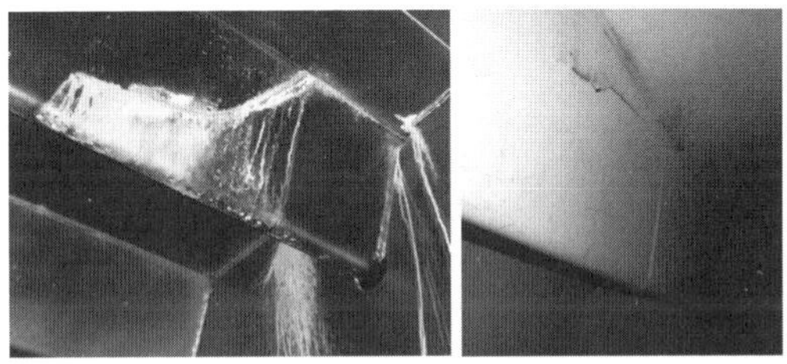

Figure 3. Fatigue cracks on the longitudinal ribs of the orthotropic plate of the A4-Peschiera Bridge.

fatigue defects appear as fatigue cracks, which affect the top plates, longitudinal ribs and cross-beams of the deck.

Generally speaking the poor fatigue performance exhibited by this kind of steel bridges built in the last 30 years is related to insufficient fatigue design (Caramelli, 1990) and lack of sensitivity to the problem (in some cases connected to a lack of code requirements).

In Fig. 3 typical examples of fatigue cracks on a orthotropic plate deck are illustrated: the structure is a 20 years old box girder bridge, which has three spans for a total length of 152 m, the central one being 70 m long. The cracks started on the longitudinal welding of the ribs, at the connection with the cross-beam, in correspondence to the slow traffic lanes, where the heavy truck loads are cyclically acting.

With regard to fatigue assessment of riveted historical metal bridges, many factors have found to play an important role, as also documented by several studies (Pipinato et al. 2009, 2011a,b).

A series of experiments on an old railway riveted metal bridge taken out of service and transported to a laboratory of the University of Padua (see Fig. 4). The focus was placed on main riveted connections. A material characterization of the aged constitutive materials was carried out.

Fatigue cracks are usually identified in the crack growth stage; in this case, if the cracks do not extend away from the weld into parent metal, a crack stop hole (10–15 mm in diameter) can be drilled at the crack tip

Figure 4. Bending tests on the structural elements taken for the railway bridge at the University of Padova.

to reduce stress concentration, and cracked welds can be effectively repaired by grinding out and rewelding the section (Transit New Zealand, 2001). Otherwise it might be necessary to reduce the stress in the area of the crack, introducing new load paths and redesigning the joint or connection, or to replace the entire component.

2.3 *Masonry bridges*

According to a recent survey approximately 40% of the 220000 railway bridges in Europe are masonry arch bridges (SB-ICA, 2007). Due to the natural ageing phenomena and poor maintenance, combined with the intrinsic weakness of some structural components increased traffic loads, many of these artefacts usually show one or multiple damages, which influence the bridge condition and reduce structural safety.

The main defects can roughly be divided into damages related to foundations and defects in the superstructure. Foundations can be generally affected by:

- local undermining;
- masonry dislocation due to loss of mortar.

As regards the superstructure, the most common defects detected are:

- deterioration of brick, loss of joint mortar or loss of bricks units, salt efflorescence from bricks (often due to insufficient waterproofing, freeze-thaw cycles, penetrating vegetation);
- deformation of the arch barrel with longitudinal or transverse cracking, joints of the arch opening;
- spandrel wall movements (sliding, bulging, detachment from the barrel);
- separation between brick rings in multi-ring arch barrel;
- fractures in piers and wing walls.

Figure 5. Railway masonry bridge on the Orte-Falconara route (km 226 + 221)

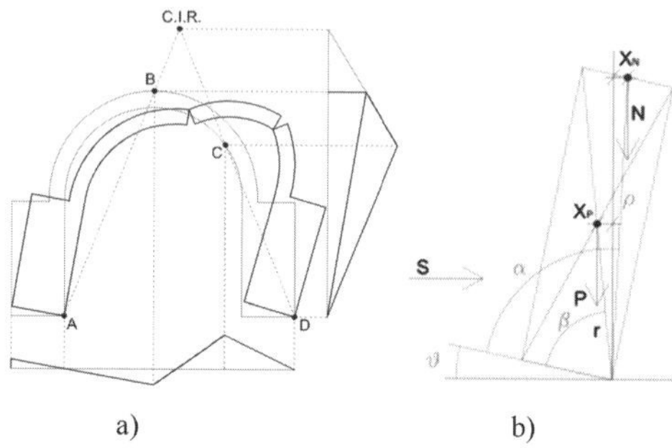

Figure 6. Collapse mechanisms: a) Arch and abutment global mechanism in longitudinal direction for arches with tall abutments b) spandrel wall out of plane overturning.

If the condition appraisal requires a static or seismic evaluation of the capacity of the arch bridge, the quickest approach is based on the kinematic analysis of collapse mechanism of the bridge, through the application of the principle of virtual work.

The equation of the principle of virtual work is the following:

$$\alpha\sum_i P_i \cdot \delta_{x,i} - \sum_i P_i \cdot \delta_{y,i} - \sum_j F_j \cdot \delta_j = 0 \qquad (6)$$

where α is the horizontal load multiplier that activates the mechanism, P_i represents the weight force of the arch voussoirs and of the filling voussoirs, F_j refers to the generic external force applied to the structure, δ stands for the virtual displacement of the load application point. The hypotheses of the method for a single span bridge are: the absence of sliding between voussoirs and infinite compressive strength of masonry (Heyman 1982, Clemente 1998), large displacements (for the evaluation of the ultimate displacement at collapse, being the problem affected by second order effects), and infinitely rigid abutments; this last hypothesis is removed for the particular typology of masonry arch with slender and tall abutments (Fig. 6).

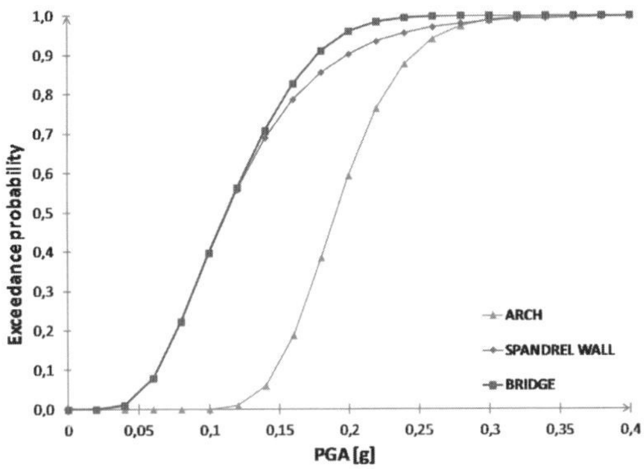

Figure 7. Fragility curves obtained for the bridge type SSMT (single span masonry arch bridge with tall abutments): fragility curve for longitudinal behaviour (arch), transverse behaviour (spandrel wall) and global curve for the overall system (bridge).

When considering the overall seismic vulnerability of a masonry bridge, not only the main components of the structure (for a single span bridge: arch barrel, abutment walls, foundations), but also secondary elements, like spandrel walls, should be verified. Generally the spandrels are the weakest elements in a seismic condition (on the contrary the arches exibit a higher resistance), because they are subjected to considerable out-of plane forces and also are often not properly connected with the lower barrel vault, so a low ground acceleration it's required for their out of plane overturning. Also for these elements the kinematic analysis can be used to calculate the structural capacity to seismic forces (see Fig. 7), and to determine the corresponding fragility curve, which is used to characterizes the seismic vulnerability.

Generally the repair intervention involves the rehabilitation of the structural integrity of the arches and the improvement of the element connections to upgrade seismic resistance; this means that the proposed intervention should fully rely on the intrinsic load-bearing capacity and design characteristics of the original structure, that has to be preserved, and even enhanced.

The repair and strengthening intervention can involve:

- enlargement and/or underpinning of the foundations with possible execution of micro-piles;
- repointing of the joints with hydraulic-lime based mortar;
- grout injection into the arch vault or vertical walls (piers or abutments made by stone with large cavities);
- patch repair: removal and substitution of the damaged bricks with new ones;
- removal and replacement of the backfill;
- thickening of the existing masonry arch at the extrados with new brick layers inserting suitable connectors;
- FRP application to the vault;
- application of sprayed reinforced concrete to the arch barrel intrados.

For the reduction of seismic vulnerability it is often cost-effective to insert transversal steel ties connecting the spandrel walls, to prevent them from overturning during the seismic event.

If this arrangement is executed, a substantial enhancement of the global seismic response of the bridge is generally obtained, being the other components generally less vulnerable.

3 RC BRIDGES' RETROFITTING: CASE HISTORIES

Four case histories of refurbishment interventions on rc existing bridges are presented in detail below (Fig. 8). The defects they evidenced after fifty years of service life were typical of these kind of structures, being often the consequences of a poor maintenance and the lack of durability rules in the original design.

- *Albaredo bridge.* The bridge has six spans with a total length of approx. 230 m entirely. Three spans, on the right bank of the Adige, have an effective span of approx. 23 m, and are composed of simply-supported beams, whereas the remaining three spans, approximately 53 m, are "Nielsen" type arches (Modena et al. 2004). The roadway was originally 6.00 m wide, flanked by two footpaths of slightly more than 1 meter wide.
- *Zevio bridge.* The bridge consists of seven spans of approx. 32 m and two terminal spans of approx. 14.3 m, for an overall length of 252.7 m.

The isostatic structure is formed by "Gerber" type beams (a statically determinate beam consisting of one or a greater number of beam members connected by hinged joints): the height of the transverse section varied from a minimum of 1.52 m at the abutments to a maximum of 2.67 m on the piers. The roadway was originally 6.00 m wide with two lateral 1.05 m wide footpaths.

- *Sega bridge.* This arch bridge has two spans, respectively 60 m and 25.45 m long; the deck, which originally was 8.34 m wide, is supported by transversal rc portals which rely on a couple of massive arches. The section of the main arches has variable height (from 1.2 to 1.9 mm in the longest span) and a width of 2.0 m.
- *San Francesco Bridge.* It is symmetric in plan, with three arched spans, the central one 41.45m long and the lateral 35.81 m. The upper structure is constituted by a 14 cm thick rc slab and a 25 cm high grillage of downstanding beams. These beams find their support on pillars distributed with a pitch of 2.0 m in longitudinal and trasverse direction, with a variable height from the crown (0.30 m) to the arch springing (aprox. $h_{max} = 4.66$ m, in the central span). The main structure is represented by the lower rc arch barrel, with a variable thickness of 40–70 cm.

a)

b)

c)

d)

e)

f)

g)

h)

Figure 8. View of the bridges before (on the left) and after the retrofitting (on the right): the Albaredo Bridge (a,b), the Zevio Bridge (c,d), the Sega Bridge (e,f), the San Francesco Bridge (g,h).

3.1 *Treatments of the deteriorated parts of the structure*

Deteriorated materials (see Figs 9,10) were systematically treated to stop the degradation process. The concrete cover was generally hydrodemolished in seriously damaged parts, while more light treatment by blast sanding was used for the well preserved concrete. The entire surface area was then dressed by pressurised

Figure 9. Hydro-demolition at the intrados of the arches of the S. Francesco Bridge.

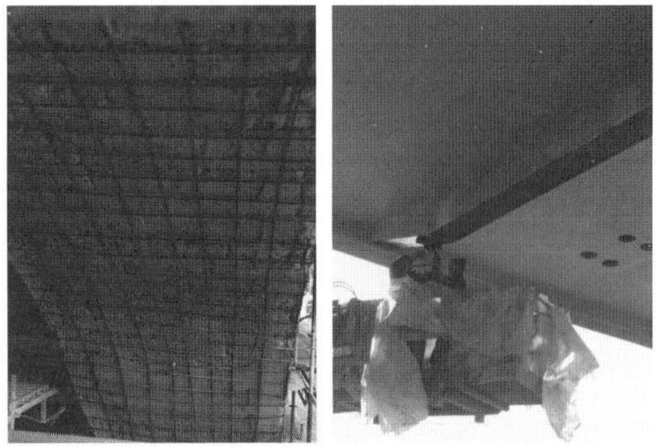

b) c)

Figure 10. b) insertion of additional bars and preparation of the surface of the arches of the Sega Bridge c) application of FRP strips to the intrados of cross beam of the Albaredo Bridge.

sanding, all the exposed rebars were sanded down to white metal, blown with pressurised air jets and treated with an anti-corrosive agent. New bars having the same diameter but with improved adhesion, were positioned next to the oxidised bars. Lastly, the new plastering was applied to the cover using thixotropic shrinkage-compensated cement.

3.2 *Elimination or substitution of joints*

Expansion joints are typically one of the most critical points of existing structures, in relation to the durability and viability conditions of the bridges. An intervention of fundamental importance from both the static and functional points of view consists in eliminating the deck joints, if it has no unfavorable static consequences for the existing structure. This intervention allows the consolidation of the entire deck in terms of the transmission of horizontal forces.

In the cases of Albaredo Bridge and Zevio Bridge, the joints between adjacent spans were eliminated by means of laying a continuous supplementary concrete slab along all the spans. Expansion joints were only placed at the ends of the bridges, to allow temperature expansion.

3.3 *Widening*

The managing authorities of the examined bridges had expressly asked for the bridges to be widened, especially in order to obtain new cycle paths and footpaths raised above the roadway and separated from it by safety barriers. The design choices were different in the four cases, taking into account both structural and architectural-functional considerations.

The more complex solution was adopted for Albaredo bridge (Fig. 8). The widening of the deck to 14.60 m, was obtained by cantilevered reticular structures in steel supporting the layers composed of corrugated metal sheet and a supplementary covering in light reinforced concrete. To achieve this it was necessary to install new steel tie-rods and cross-beams which allowed integration with the existing bridge and ease of installation of the metal structures used to build the new lanes reserved for pedestrians and cyclists.

Functional adaptation of the Zevio (Fig. 8) bridge was obtained, for both the main structures and the newly-built supported spans, using lateral cantilever concrete slabs supported by auxiliary metal structures in cor-ten steel. Deck width after the intervention is 14.00 m.

In the case of Sega Bridge (Fig. 8), taking into account the specific construction and geometric properties, the widening has been realized with post-tensioned r.c. cantilevers, obtaining a total width of the deck of 12.60 m instead of the initial width of 8.34 m.

In the S. Francesco Bridge the functional requirements regarded the necessity to obtain four traffic lanes instead of the existing three, to have wider lateral cyclepaths, and leave ample room for the passage of pipes and ducts, including high pressure gas and district heating for residential use; these requirement were particularly demanding, operating in a urban context, and led to a widening of 5.50 m of the road platform and to a total width of 19.90 m of the transverse section of the bridge.

4 CONCLUSIONS

Natural ageing, poor maintenance, severe environmental actions or changes in the use of a structure, as well as increased safety requirements, involve repair and/or strengthening for existing bridges. In the first part of the paper the assessment and typical retrofitting interventions of usual road and railway bridge typologies in Italy are described showing that, e.g., masonry arch bridges are usually quite robust structural systems, r.c. bridges generally present durability problems and are vulnerable to seismic actions, steel bridges, other than presenting durability problems, are particularly vulnerable to fatigue. In the second part of the work significant case studies of retrofitting interventions are in detail described, focusing on four existing r.c. bridges, which represent some of the most usual structural typologies adopted in the post–World War II period. The refurbishment interventions are presented outlining a methodological approach, which takes into account the typological characteristics of the structure, the state of maintenance, the functional requirements and the environmental aspects connected to the repair and strengthening system.

ACKNOWLEDGMENTS

The authors would like to thank the infrastructure design team of SM Ingegneria s.r.l. (www.smingegneria.it), for the continuous and decisive support.

REFERENCES

Blakelock R., Day W. and Chadwick R. 1998. Bridge condition index, The Management of Highway Structures, Great Britain Highways Agency.

Brime 2001. Bridge management in Europe, Final Report D14, IV Framework Programme, Brussels

Caramelli S., Croce P., Froli M., Sanpaolesi L. 1990. Orthotropic bridge decks: fatigue behavior, *Costruzioni Metalliche 6*: 378–411

Clemente P. 1998. Introduction to dynamics of stone arches. *Earthquake engineering and structural dynamics* 27: 513–522.

fédération internationale du béton. 2001. *Externally bonded FRP reinforcement for RC structures.* Bullettin No.14.

Franchin, P., Lupoi, A. & Pinto, P.E. 2006. On the role of road networks in reducing human losses after earthquakes. *Journal of Earthquake Engineering* 10 (2): 195–206.

Heyman J., 1982. *The masonry arch*, Chiester.

Italian Ministry of Infrastructures. 2008. *Norme Tecniche per le Costruzioni* DM 2008-1-14.

Karim, K.R. & Yamazaki, F. 2003. A simplified method of constructing fragility curves for highway bridges. *Earthquake Engineering & Structural Dynamics* 32 (10): 1603–1626.

Karim, K.R. & Yamazaki, F. 2001. Effect of earthquake ground motions on fragility curves of highway bridge piers based on numerical simulation. *Earthquake Engineering & Structural Dynamics* 30 (12): 1839–1856.

Kent, D.C. & Park, R. 1971. Flexural Members with Confined Concrete. *Journal of Structural Division* 97 (7): 1969–1990.

Modena C., Franchetti P., Grendene M. 2004. *Guidelines for design of bridges in seismic areas*, Mestre (VE): published by Veneto Strade spa.

Modena C., Laurenco P.B., Roca P. 2004. *Structural analysis of historical constructions* 2: 1091–1101, Balkema.

Monti, G. & Nisticò, N. 2002. Simple Probability-Based Assessment of Bridges under Scenario Earthquakes. *Journal of Bridge Engineering* 7 (2): 104–114.

Morbin, R., Pellegrino, C., Grendene & M., Modena, C. 2010. Strategies for seismic vulnerability evaluation of common RC bridges typologies. In *14th European Conference on Earthquake Engineering,* Ohrid, Republic of Macedonia: 30th August – 3rd September 2010.

National Research Council 2004. Guide for the Design and Construction of Externally Bonded FRP Systems for Strengthening Existing Structures. CNR-DT 200.

Padgett, J.E. & DesRoches, R. 2008. Methodology for the development of analytical fragility curves for retrofitted bridges. *Earthquake Engineering & Structural Dynamics* 37 (8): 1157–1174.

Pellegrino, C. & Modena, C. 2010. Analytical Model for FRP Confinement of Concrete Columns with and without Internal Steel Reinforcement. *Journal of Composites for Construction* 14 (6): 693–705.

Pellegrino C., Pipinato A., Modena C. 2011. A simplified management procedure for bridge network maintenance", *Structure and Infrastructure Engineering*, 7(5): 341–351.

Pipinato A., Pellegrino C., Bursi O.S., Modena C. 2009. High-cycle fatigue behavior of riveted connections for railway metal bridges. *Journal of Constructional Steel Research*, 65(12): 2167–2175.

Pipinato A., Pellegrino C., Modena C. 2011a. Fatigue assessment of highway steel bridges in presence of seismic loading. *Engineering Structures*, 33(1): 202–209.

Pipinato A., Molinari M., Pellegrino C., Bursi O.S., Modena C. 2011b. Fatigue tests on riveted steel elements taken from a railway bridge. *Structure and Infrastructure Engineering*, 7(12): 907–920.

Ryall, M.J. 2001. Bridge Management. Oxford: Butterworth, Heinemann.

SB-ICA 2007. Guideline for Inspection and Condition Assessment of Existing European Railway Bridges. *Prepared by Sustainable Bridges – a project within EU FP6.* Available from: www.sustainablebridges.net

Shinozuka, M., Feng, M.Q., Lee, J. & Naganuma, T. 2000a. Statistical analysis of fragility curves. *Journal of Engineering Mechanics-Asce* 126 (12): 1224–1231.

Tinazzi, D., Pellegrino, C., Cadelli, G., Barbato, M., Modena, C. & Gottardo, R. 2003. An experimental study of RC columns confined with FRP sheets. In *Structural Faults & Repair, 10th International Conference*. London, UK: 1–3 July 2003.

Transit New Zealand 2001. *Bridge and Inspection and Maintenance Manual*, Transit New Zealand.

Wang, L.M. & Wu, Y.F. 2008 Effect of corner radius on the performance of CFRP-confined square concrete columns: Test. *Engineering Structures* 30 (2): 493–505.

Bridge Maintenance, Safety, Management, Resilience and Sustainability – Biondini & Frangopol (Eds)

Innovative steel bridge girders with tubular flanges

R. Sause
ATLSS Engineering Research Center, Department of Civil and Environmental Engineering, Lehigh University, Bethlehem PA, USA

ABSTRACT: I-shaped steel girders with tubular flanges have been studied for application in highway bridges because of their large torsional stiffness compared to conventional I-shaped steel plate girders (I-girders). For straight girder bridges, the large torsional stiffness of a tubular flange girder (TFG) results in significantly greater lateral-torsional buckling strength compared to a corresponding I-girder. For horizontally curved girder bridges, the large torsional stiffness of a TFG results in much less normal stress, vertical displacement, and cross section rotation compared to a corresponding I-girder. The paper presents experimental and finite element analysis results for straight and horizontally curved TFG bridges. The results show the advantages of TFGs in comparison to conventional I-girders. A TFG demonstration bridge constructed in the USA is described.

1 INTRODUCTION

The potential advantages of I-shaped girders with tubular flanges (TFGs) for highway bridges, including large local buckling resistance and large torsional stiffness, were outlined by Wassef et al. (1997). Wimer & Sause (2004) studied TFGs with a rectangular concrete-filled steel tube as the compression flange and a flat steel plate as the tension flange, as shown in Figure 1(a). Kim & Sause (2005a, b) studied TFGs with a round concrete-filled steel tube as the compression flange and a flat steel plate as the tension flange, as shown in Figure 1(b). These TFGs, called concrete-filled tubular flange girders (CFTFG), were found to have several advantages relative to conventional I-shaped steel plate girders (I-girders) for straight bridges: (1) the concrete-filled tubular flange provides more strength, stiffness, and stability than a flat plate flange with same amount of steel, and (2) CFTFGs require fewer diaphragms (or cross frames) to maintain lateral-torsional stability compared to corresponding I-girders. Sause & Dong (2008) studied the application of steel TFGs in curved highway bridges, where the significant torsional stiffness of TFGs is the main advantage.

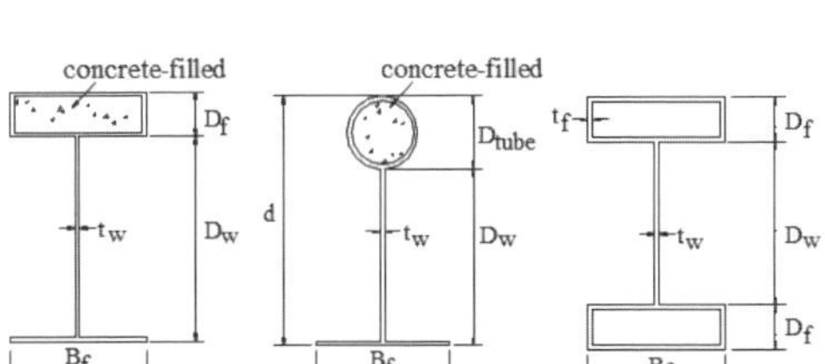

Figure 1. Girders with tubular flanges.

This paper reviews and synthesizes previous work on TFG bridges. This work includes tests, finite element analysis studies, the development of flexural strength formulas, and the design and construction of a demonstration bridge. Applications in straight and curved highway bridges are considered.

2 INITIAL DESIGN STUDY

CFTFGs with minimum steel weight were designed for a straight prototype bridge and compared with minimum weight conventional steel I-girders by Kim & Sause (2005b). The prototype bridge has a simply-supported single span of 40 m and four girders spaced at 3.8 m. Possible framing plans for the prototype bridge with the diaphragm arrangements are shown in Figure 2. Scheme 1, with diaphragms spaced at 8 m is typical for a conventional I-girder bridge. Schemes 8 and 9 have fewer interior diaphragms, resulting in reduced fabrication and erection costs. However, the designs of the conventional I-girders with a larger diaphragm spacing will be controlled by lateral-torsional buckling (LTB) under construction conditions (e.g., during construction of the concrete deck), and will require heavier I-girder cross sections. On the other hand, CFTFGs with their large torsional stiffness efficiently accommodate the larger diaphragm spacing.

The design study considered criteria for strength, stability, service, and fatigue. Figure 3 compares the total steel weight of girders for the prototype bridge for three cases: (1) conventional I-girders with the Scheme 1 diaphragm arrangement; (2) conventional I-girders

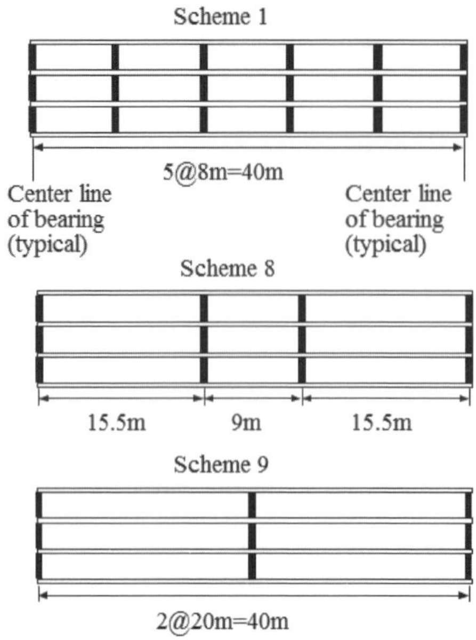

Figure 2. Framing plans for prototype bridge.

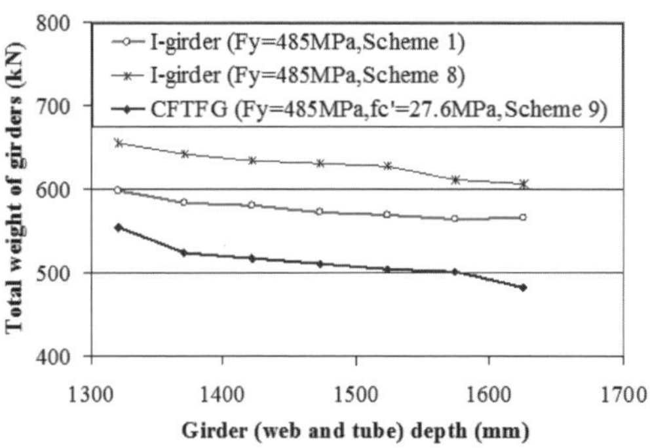

Figure 3. Total steel weight of girders for prototype bridge.

with the Scheme 8 diaphragm arrangement; and (3) CFTFGs with the Scheme 9 diaphragm arrangement. Figure 3 shows that the CFTFGs are significantly lighter than the I-girders, even when a large diaphragm spacing is used. In particular, the CFTFGs with Scheme 9 (with only one line of interior diaphragms) are more than 10% lighter than the I-girders with Scheme 1 (with four lines of interior diaphragms). Thus the CFTFGs have the advantages of decreased steel weight, and decreased diaphragm fabrication and erection effort.

3 TESTS OF TFGS FOR STRAIGHT BRIDGES

Tests of CFTFGs (Sause et al. 2008) for a straight bridge were conducted considering two different conditions: (1) critical construction conditions, which are the loads and support conditions that occur during construction of the concrete deck, where the flexural strength of the girders is controlled by LTB; and (2) under maximum loads in the final constructed condition during normal use of the bridge, where the flexural strength of the girders is controlled by the cross-section flexural capacity.

Figure 4. CFTFG test specimen without concrete deck.

The CFTFG test specimen was similar to the 40m span prototype bridge (Fig. 2), but only two girders were included, and the girder dimensions were scaled by 0.45 to reduce costs. The simply-supported test specimen (Fig. 4) included a girder with a rectangular tube and a flat web (denoted FWG) and a girder with a rectangular tube and a corrugated web (denoted CWG). The CFTFGs were made composite with a concrete deck in the final constructed condition using shear studs. The test specimen was tested in two different diaphragm configurations: (1) with two end diaphragms and one interior diaphragm (Scheme 9); and (2) with only the two end diaphragms (Scheme 10). The Stage 1 test used the Scheme 9 diaphragm arrangement. The Stage 2 and Stage 3 tests used the Scheme 10 diaphragm arrangement. The Stage 1 and Stage 2 tests simulated the deck construction stage. The Stage 3 test simulated maximum loads in the final constructed condition.

The CFTFG tests and comparisons with the analysis results demonstrated the composite action between the steel TFG cross section with the tube concrete infill. In addition, the Stage 1 and Stage 2 tests demonstrated the significant LTB capacity of TFGs. The tests showed that the CFTFGs can carry their design loads; but owing to the loading method employed in the tests, the CFTFGs could not be safely loaded to failure and the test results could not be used to validate flexural strength formulas for TFGs.

4 ANALYSIS OF TFG FLEXURAL STRENGTH

Design flexural strength formulas for TFGs considering LTB and cross section yielding were proposed by Kim & Sause (2005a, b). They used finite element (FE) analysis results for CFTGFs to develop and calibrate these design flexural strength formulas. Dong & Sause (2009) conducted additional FE studies to establish the flexural strength of TFGs with hollow tubes (HTFGs shown in Fig. 1(c)), and to show that these design flexural strength formulas can be used to determine the flexural strength of TFGs which do not have tube concrete infill. The FE models used by Dong & Sause (2009) considered material inelasticity, second-order effects, initial geometric imperfections, and residual stresses.

The FE studies by Dong & Sause (2009) included studies of the effect of cross section distortion on the lateral-torsional buckling strength of HTFGs. The results showed that three intermediate transverse web stiffeners (at the quarter points and mid-span) and two bearing stiffeners (one at each end of the span), with a thickness of 25 mm, are sufficient to eliminate any significant effect of cross-section distortion on the load capacity of HTFGs.

Dong & Sause (2009) also presented parametric studies on the effects of initial geometric imperfections and residual stresses on the flexural strength HTFGs. FE results showed that formulas from the AASHTO LRFD bridge design specifications (2004) overestimate the flexural strength of HTFGs, but the design flexural strength formulas proposed by Kim & Sause (2005a, b) provide an accurate but conservative estimate of the flexural strength of HTFGs

5 STRAIGHT TFG DEMONSTRATION BRIDGE

A two-span straight TFG demonstration bridge was designed and constructed to span Tionesta Creek in Pennsylvania, USA (Schoedel & Sause 2010). The TFGs have a concrete filled rectangular steel tube as the top flange and a flat steel plate as the bottom flange. The demonstration bridge has two 30.5 m spans and four CFTFGs spaced at 2.58 m (Fig. 5). The CFTFGs are made composite with the concrete deck using shear studs. The bridge was designed and constructed as simple spans for dead loads and continuous spans for superimposed dead loads and live loads.

Each span of the bridge has one line of interior diaphragms near mid-span. End diaphragms are located 0.91 m from the pier centerline in each span to permit construction of the initial 29.6 m simple spans. End diaphragms are located at each abutment.

The use of only a one line of interior diaphragms between the CFTFGs and the use of span-by-span construction enabled rapid erection of the steel girders. However, the span-by-span construction required a bolted continuity splice at the pier. The splice was completed after the dead load of the CFTFGs, diaphragms, and concrete deck were carried by the CFTFGs as simple spans. The splice made the girders continuous for superimposed dead loads and live loads (Schoedel & Sause 2010).

The successful completion of the demonstration bridge showed that CFTFGs are a viable steel bridge girder and demonstrated the improved stability of TFGs (compared to conventional I-girders) during steel erection and concrete deck construction, which eliminated many lines of diaphragms that would have been needed for a conventional I-girder bridge.

Figure 5. TFG demonstration bridge viewed from below.

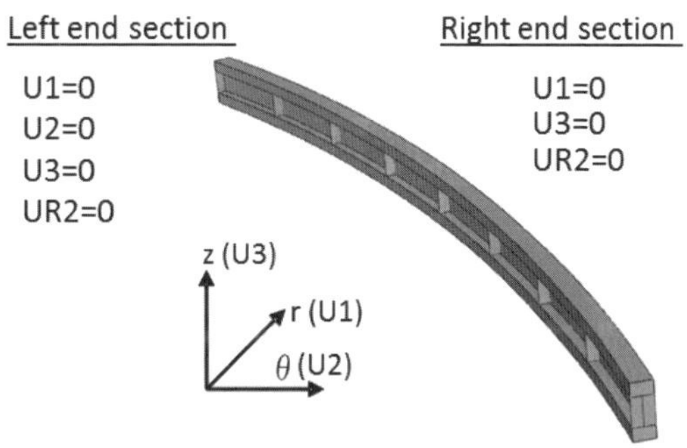

Figure 6. FE model of individual curved HTFG.

6 TFGS FOR CURVED BRIDGES

Sause & Dong (2008) studied the application of TFGs to curved highway bridges, where the horizontal curvature induces significant torsion to the bridge girder system. The significant torsional stiffness of TFGs, compared to conventional I-girders, is the main advantage of TFGs in this application.

Dong & Sause (2010a) presented FE studies of individual curved HTFGs (see Fig. 1(c) and Fig. 6). The FE models considered material inelasticity, second-order effects, initial geometric imperfections, and residual stresses. It was found that initial geometric imperfections and residual stresses had little effect on the load capacity of curved HTFGs. The FE models were used to study the effect of cross section distortion on the load capacity of curved HTFGs. The results showed that that cross section distortion could significantly reduce the load capacity of curved HTFGs. Further studies showed that transverse web stiffeners and tube diaphragms mitigate the effect of cross section distortion. Seven intermediate transverse web stiffeners within the span, a bearing stiffener at each end of the span, and tube diaphragms (within each tube) at each end of the span (all 25 mm thick) were sufficient to eliminate any significant effect of cross-section distortion.

The behavior of individual curved HTFGs was compared with the behavior of individual curved I-girders.

Table 1. Girder cross sections.

Girder	Flanges (mm)	Web (mm)	Depth (mm)	Area (mm^2)
TG1	508 × 101.6 × 12.7	1066.8 × 12.7	1282.7	44516
TG2	508 × 304.8 × 12.7	660.4 × 7.9	1282.7	46553
IG1	508 × 25.4	1257.3 × 14.9	1282.7	44516
IG2	508 × 27.9	1254.8 × 14.5	1282.7	46553

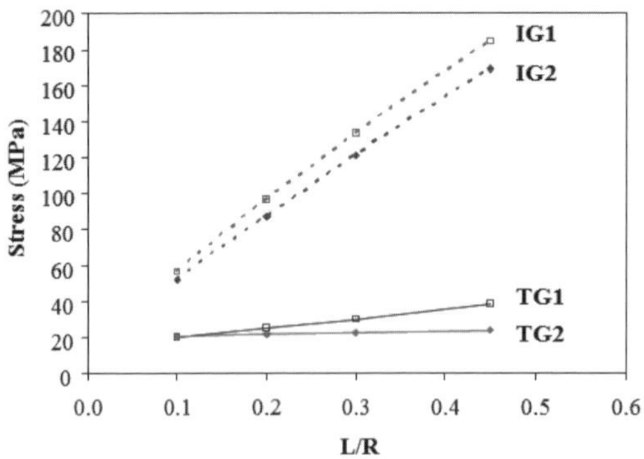

Figure 7. Maximum total longitudinal normal stress.

Curved I-girders (denoted IG1 and IG2) having the same girder weight (cross section area), cross section depth, and flange width as the curved HTFGs (denoted TG1 and TG2) were developed as shown in Table 1. The corresponding I-girders have slightly larger flexural rigidity but much smaller St. Venant torsional rigidity than the corresponding HTFGs (Dong & Sause 2010a). For the FE studies of these curved girders, the span L was constant and equal to 27 m, and the radius of curvature R was varied so that L/R varied from 0.1 to 0.45.

Figure 7 compares the maximum total longitudinal normal stress for the curved HTFGs and the corresponding curved I-girders under the girder self-weight. The results in Figure 7 indicate that because the warping normal stress is large for the curved I-girders, the maximum total normal stress for the curved I-girders is much larger than that for the curved HTFGs.

Figures 8 and 9 show the mid-span vertical displacement ($U3_M$) and the mid-span cross section rotation ($UR2_M$) for the curved HTFGs and the corresponding curved I-girders under the girder self-weight. The curved HTFGs develop much smaller displacement and cross section rotation. For example, for TG2 with $L/R = 0.45$, the mid-span vertical displacement $U3_M$ is 0.019 m which is 1/1500 of the span L, and the cross section rotation at mid-span $UR2_M$ is 0.005 rad. However, for the corresponding curved I-girder IG2, $U3_M$ is 0.394 m which is 1/70 of L, and $UR2_M$ is 0.296 rad. Therefore, under girder self-weight, the mid-span vertical displacement of the I-girder is about 20 times that of the HTFG and the mid-span cross section rotation of the I-girder is about 60 times that of the HTFG.

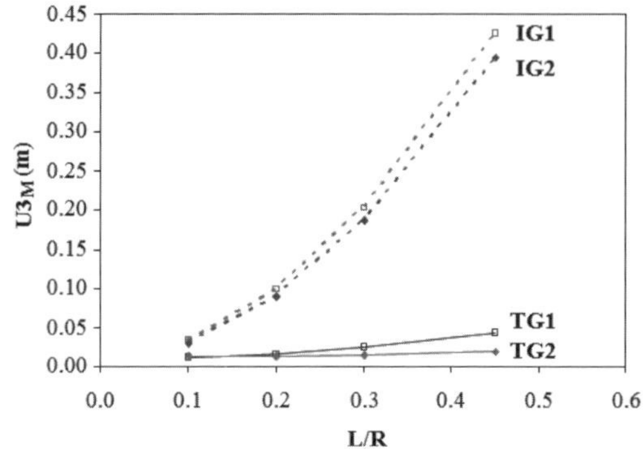

Figure 8. Mid-span vertical displacement.

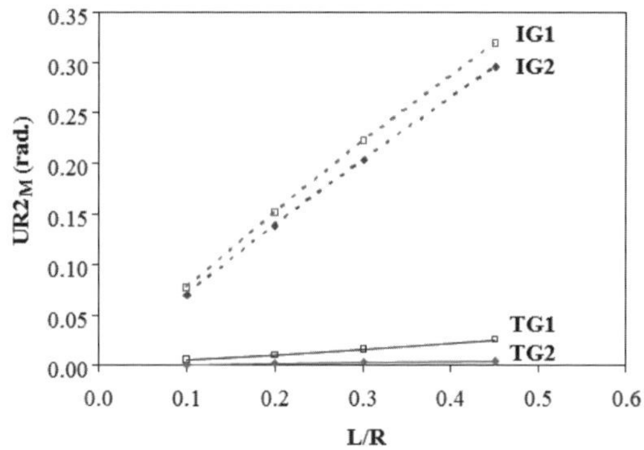

Figure 9. Mid-span cross section rotation.

The comparison of individual curved HTFGs with corresponding individual curved I-girders showed that curved I-girders are better at resisting primary bending, but they develop much larger warping normal and total normal stresses, and much larger vertical displacement and cross section rotation than corresponding curved HTFGs (i.e., individual curved HTFGs have much better structural behavior than individual curved I-girders). Under the girder self-weight, the normal stresses, vertical displacement, and cross section rotation of the curved I-girders were quite large, which suggests that temporary support within the span would be needed for these curved I-girders during erection, while the normal stresses, vertical displacement, and cross section rotation of the curved HTFGs were quite reasonable, which suggests that such temporary support within the span would not be needed during erection.

7 SYSTEMS OF TFGS FOR CURVED BRIDGES

Dong & Sause (2010b) studied framing systems for curved highway bridges composed of three HTFGs braced by diaphragms (cross frames). FE models for three-girder systems of curved HTFGs were presented. The FE models considered material inelasticity, second-order effects, initial geometric

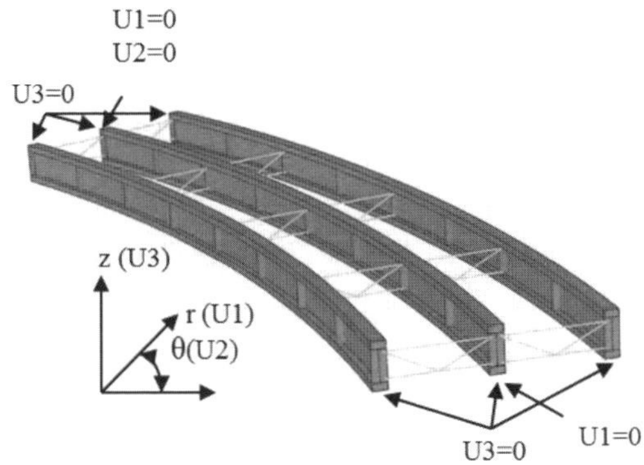

Figure 10. FE model of curved HTFG system.

imperfection, and residual stresses. These models were used in parametric studies, where the curved HTFG systems were compared with corresponding curved I-girder systems, and the effects of the curvature, cross section dimensions, number of cross frames, and a concrete deck were investigated. Two types of FE models were developed. The FE model denoted M1 (Fig. 10) included the three girders and cross frames, but did not include a composite concrete deck. This model was used to study the behavior of the three-girder systems during construction of the composite concrete bridge deck. The FE model denoted M2 included the three-girder system with a composite concrete deck. This model was used to study the behavior of the three-girder systems in the final constructed condition.

The dimensions of the HTFG cross sections are given in Table 2, where the cross sections are organized into various "cross section sets". Each cross section set includes the cross section dimensions for each girder in the three-girder system, where G1 is the girder on the inside of the curve, G2 is the middle girder, and G3 is the girder on the outside of the curve. Cross section sets TGA, and IGA are used for the curved HTFG systems and the curved I-girder systems, respectively. Each I-girder had the same girder weight (cross section area), cross section depth, and flange width as the corresponding HTFG. The span L of the middle girder G2 was equal to 27.4 m. The spacing between girders was 3.05 m. The radius of the curvature of each girder R was varied so the L/R ratio was the same for all three girders. The L/R ratio for the three-girder system was varied from 0.1 to 0.45. Three interior cross frames were used within the span, spaced evenly along the span between the cross frames at the bearings.

Dong & Sause (2010b) studied the effect of cross section distortion on the behavior of three-girder systems of curved HTFGs. The FE results showed that cross section distortion could reduce the load capacity of three-girder systems of curved HTFGs. Further results showed that using seven intermediate web transverse stiffeners, and bearing stiffeners and tube diaphragms (within the tube) at the end sections (all 25 mm thick) was sufficient to eliminate any significant effect of cross-section distortion Dong & Sause

Table 2. Girder cross section sets.

Set	Girder	Flange (cm)	Web (cm)	Area (cm^2)
TGA	G1	40.6 × 20.3 × 1.3	86.4 × 1.0	395.2
	G2	40.6 × 20.3 × 1.3	86.4 × 1.0	395.2
	G3	40.6 × 20.3 × 1.3	86.4 × 1.0	395.2
IGA	G1	40.6 × 2.5	125.7 × 1.5	395.2
	G2	40.6 × 2.5	125.7 × 1.5	395.2
	G3	40.6 × 2.5	125.7 × 1.5	395.2

(2010b) studied the effects of initial geometric imperfection and residual stresses on the load capacity of a curved HTFG system. The results showed that initial geometric imperfections and residual stresses had a small effect on the load capacity of curved HTFG systems.

Using FE analyses, the behavior of the curved HTFG systems were compared with the behavior of the curved I-girder system. Model M1 with cross sections set TGA and with cross section set IGA were analyzed under a reference load equal to the weight of the girders and the concrete deck.

The maximum total normal stress for each girder in M1 is plotted in Figure 11(a) for the curved HTFGs and the corresponding curved I-girders under the reference load. As the L/R ratio increases, more vertical load is carried by the outer girder G3, and less vertical load is carried by inner girder G1. Since the cross sections of these girders are constant as the L/R ratio increases, the normal stress increases in girder G3 and decreases in girder G1. For the curved HTFGs, the difference in the maximum normal stress between girders G1, G2, and G3 is much smaller than for the corresponding curved I-girders.

Figure 11(a) shows that I-girders G2 and G3 (I_G2 and I_G3) have greater total normal stress than HTFGs G2 and G3 (T_G2 and T_G3). Figure 11(b), however, shows that the vertical displacements of the HTFGs and I-girders are similar, since the HTFGs and I-girders have similar flexural rigidity, and the cross frames and girders work together to resist the torsion acting on the curved bridge system. Figure 11(c) shows that the maximum cross frame force in the curved I-girder system is significantly larger than in the curved HTFG system.

Dong and Sause (2010b) studied the effect of the number of interior cross frames (i.e., cross frames that are equally spaced between the end cross frames at the bearings). In this study, the number of interior cross frames was varied from zero to five. The results showed that once a single interior cross frame at mid-span was included in the curved HTFG system, the effect on the maximum total normal stress and the vertical displacement was small as more cross frames are added. Once the HTFG system had two interior cross frames, the effect of adding more cross frames on the maximum cross frame force was small. However, the effect of the number of interior cross frames

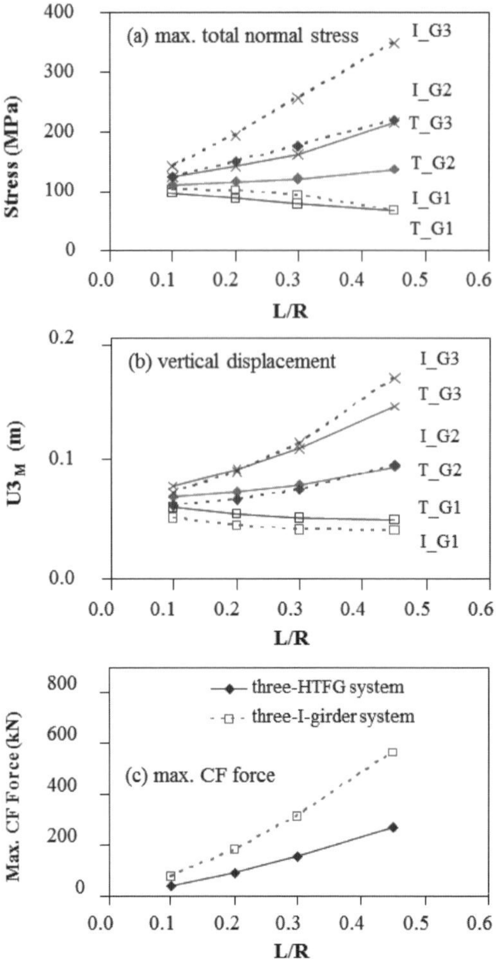

Figure 11. Results for M1 with section sets TGA and IGA.

on the curved I-girder system was significant, especially when three or fewer interior cross frames were used. In fact, the curved I-girder system without a single interior cross frame could not carry the weight of the girders and concrete deck, while, the HTFG system carried its own weight and the weight of the concrete deck without any interior cross frames.

Dong & Sause (2010b) also studied the behavior of curved girder systems with a composite concrete deck. FE analyses of model M2 of the three-girder system with a composite concrete deck under the reference load (i.e., the weight of the girders and the concrete deck) were conducted. The three-girder systems had three interior cross frames. The results showed that the composite concrete deck increased the stiffness of the system significantly. Compared to the results for model M1, the maximum normal stress, the vertical displacement, and the maximum cross frame force were reduced, and the load capacity was increased. The curved HTFG system with the composite concrete deck developed smaller maximum normal stress and maximum cross frame force than the corresponding

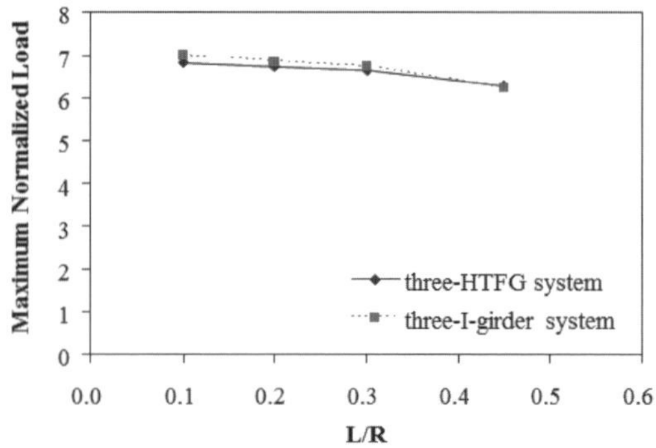

Figure 12. Load capacity for M2 with composite concrete deck.

system of curved I-girders. Figure 12 shows that with a composite concrete deck in place, the load capacities of the curved HTFG system and the curved I-girder system are similar.

In summary, comparing the behavior of curved HTFG systems with corresponding curved I-girder systems demonstrated the following advantages of the HTFG system: (1) under the same load, the HTFGs develop less total normal stress than the corresponding curved I-girders; (2) the forces in the cross frames of the HTFG systems are smaller than in the corresponding I-girder systems, and thus lighter cross frame members could be used for the HTFG systems; (3) fewer cross frames are needed for the HTFG systems; and (4) the HTFG systems can carry their own weight (plus the weight of a concrete deck) without any support within the span and without interior cross frames, and, therefore, temporary support for curved HTFG systems during construction may not be needed.

Overall, the results indicate that the curved HTFG systems are more structurally efficient than the corresponding curved I-girder systems. Note that the curved HTFG systems and corresponding curved I-girder systems compared in this paper have the same cross sectional area (and weight), and, therefore, the material costs for the two systems should be similar. In practice, curved I-girder flanges are often cut from rectangular plates producing wasted material. On the other hand, curved tubular flanges will be cold-bent from conventional hollow structural shapes, eliminating this wasted material. Fewer cross frames and lighter cross frame members are needed in HTFG systems. Other expected advantages of HTFG systems are easier handling, shipping, and erection of the individual curved girders, based on their significantly increased torsional stiffness, as well as less need for temporary support of HTFG systems before the concrete deck is composite with the girders, resulting in faster and easier bridge construction.

8 CONCLUSIONS

In conclusion, the paper presents and summarizes advantages of TFGs compared with conventional

I-girders. For straight bridges, the steel tube flange of a TFG provides more stability than a flat plate flange with same amount of steel, so that fewer diaphragms (or cross frames) and less temporary support are needed to maintain stability (during construction of the concrete deck) compared to corresponding I-girders.

For curved bridges, the significant torsional stiffness of individual curved TFGs, compared to corresponding I-girders, is a significant advantage. Individual curved TFGs develop significantly less warping normal stress, total normal stress, vertical displacement, and cross section rotation than corresponding individual I-girders. After bridge steel erection is completed, and the system of curved girders braced by diaphragms (or cross frames) is in place, the curved TFGs develop less normal stress and require fewer (and lighter) diaphragms and less temporary support than corresponding curved I-girders during construction, resulting in more efficient bridge construction.

ACKNOWLEDGEMENTS

This work was conducted at the Advanced Technology for Large Structural Systems (ATLSS) Engineering Research Center at Lehigh University, Bethlehem, PA, USA. Financial support provided by the Federal Highway Administration (FHWA), the Pennsylvania Department of Transportation (PennDOT), and the Pennsylvania Department of Community and Economic Development through the Pennsylvania Infrastructure Technology Alliance (PITA) is gratefully acknowledged. The author is grateful for the contributions of Bong-Gyun Kim, Mark Wimer, Zhou Fan, Jun Dong, Richard Schoedel, and Robert Bondi to the work reported herein. Michael Baker Corporation of Coraopolis, PA, USA engineered the superstructure of the TFG demonstration bridge. Opinions and conclusions given in the paper are the author's and do not necessarily reflect the views of those acknowledged herein.

REFERENCES

AASHTO LRFD bridge design specifications (1998, 2004). American association of state highway and transportation officials, Washington, DC, USA.

Dong, J., Sause, R. (2009). Flexural strength of tubular flange girders. Journal of constructional steel research 65(3): 622–630.

Dong, J., Sause, R. (2010a). Finite element analysis of curved tubular flange girders. Engineering structures 32(1): 319–327.

Dong, J., Sause, R. (2010b). Behavior of hollow tubular flange girder systems for curved bridges. Journal of structural engineering 136(2): 174–182.

Kim B.G., Sause R. (2005a). High performance steel girders with tubular flanges. ATLSS Report 05-15, ATLSS Engineering Research Center, Lehigh University, Bethlehem, PA, USA.

Kim B.G., Sause R. (2005b). High performance steel girders with tubular flanges. International journal of steel structures 5(3): 253–263.

Sause, R., Kim, B.G., Wimer, M.R. (2008). Experimental study of tubular flange girders. Journal of structural engineering 134(3): 384–392.

Sause, R., Dong, J. (2008). Finite-element analysis of curved hollow tubular flange girders. Proceedings, 25th annual international bridge conference, Pittsburgh, PA, USA.

Schoedel, R.M., Sause, R. (2010). Design of concrete filled tubular flange girders. Proceedings, 27th annual international bridge conference, Pittsburgh, PA, USA.

Wassef, W.G., Ritchie P.A., Kulicki J.M. (1997). Girders with corrugated webs and tubular flanges – an innovative bridge system. Proceedings, 14th annual international bridge conference, Pittsburgh, PA, USA.

Wimer, M., Sause, R. (2004). Rectangular tubular flange girders with corrugated and flat webs. ATLSS Report No. 04-18, ATLSS Engineering Research Center, Lehigh University, Bethlehem, PA, USA.

Bridge Maintenance, Safety, Management, Resilience and Sustainability – Biondini & Frangopol (Eds)
© 2012 Taylor & Francis Group, London, ISBN 978-0-415-62124-3

Remote monitoring: Concept and pilot study

M. Shinozuka, M. Torbol & K.G. Papakonstantinou
University of California, Irvine, CA, USA

ABSTRACT: This paper addresses the emerging technology here referred to as "remote structural monitoring and control in real time". The technology can improve resilience and sustainability of large-scale complex infrastructure systems under natural and manmade hazards. In this context, the paper attempts to initiate and promote the effort to expand the existing concept of structural monitoring into continuous, if not permanent, system performance monitoring and control, remotely and in real time. This is done in the spirit of advancing the existing SCADA system (Supervisory Control and Data Acquisition system), widely deployed by utility providers for steady operations of their systems, into the next generation of SCADA system. For this reason, the paper first deals with the next generation SCADA for water distribution networks as a test-bed and then it presents examples involving bridges and other systems.

1 INTRODUCTION

Deterioration of infrastructure systems throughout the world is a major concern. Based on ASCE's 2009 Report Card for America's Infrastructure, the following summarizes the current status of various water, transportation and structural systems: "Grades ranged from a high of C+ for solid waste to a low of D− for drinking water, inland waterways, levees, roads, and wastewater. U.S. surface transportation and aviation systems declined over the past four years, with aviation and transit dropping from a D+ to D, and roads dropping from a D to a nearly failing D−. Showing no significant improvement since the last report, the nation's bridges, public parks and recreation, and rail remained at a grade of C, while dams, hazardous waste, and schools remained at a grade of D, and drinking water and wastewater remained at a grade of D−. Levees, the newest category, debuted on the 2009 Report Card at a barely passing grade of D−. Just one category (energy) improved since 2005, raising its grade from D to D+." Note that scores of "D" refer to "poor" quality and scores of "C" to "mediocre" quality.

This paper presents the authors' most recent unpublished work on the subject matter and integrates earlier published works of theirs, which are referred to at the end. Duramote, part of next generation SCADA system, developed at UCI, is also presented.

2 PROBLEM STATEMENT

2.1 *Urgent needs for remote monitoring and damage assessment in real time*

Urban water delivery network system components, particularly under- as well as above-ground, such as pipes, are subject to damage due to earthquakes, corrosion, severely cold weather, heavy traffic loads on the ground surface, and other causes. In all these situations the damage could be disastrous; water leakage at high pressure might threaten the safety of nearby buildings due to scouring of their foundations, flooding could create a major traffic congestion if the pipe breaks under a busy street, and above all, after a severe earthquake, pipe damage may result in reduction in the water head, thus degrading post-earthquake fire fighting capability of the community and at the same time may force the human consumption of water below unacceptably low level. Yet, the current technology cannot detect the location and extent of damage easily, especially immediately after the damaging event. This will make it difficult to implement rapid and effective emergency response to minimize human miseries, inconveniences, and potential outbreak of diseases. These observations, particularly those related to severe earthquakes, strongly suggest that the technology supporting "remote structural monitoring and control in real time" is urgently needed.

2.2 *Next generation SCADA*

Utility networks are usually equipped with a SCADA system that encompasses and monitors the entire network, and centrally collects measured data of key operational parameters at a number of locations in the network for the purpose of efficient system operation and control. The key parameters for water networks include the flow rate, water head, temperature, and chemical quality data. A SCADA system as such can in principle be used for damage detection purposes. However, the usual sparseness of the data acquisition locations, which are concentrated only on key components such as pump stations, makes it difficult to detect pipe damage with the accuracy and rapidity needed for a timely repair. To overcome this difficulty, the next generation SCADA is developed for water distribution networks.

3 METHODOLOGY USED AND MONITORING SYSTEM

3.1 *Methodology*

The methodology for the development of next generation SCADA takes advantage of the network water flow that exhibits sharp transient behavior upon sudden local change in the hydraulic conditions, caused for example by a pipe break. This behavior makes it possible to develop a cost effective damage identification procedure for water distribution networks. The key signature in this procedure is the rate of change in the water head under sudden pipe breaks and it takes the form of absolute maximum WHG (water head gradient = time derivative of water head). WHG is an easily computable quantity in real time at each sensor node. Henceforth, this absolute maximum WHG is denoted by D for simplicity. In the methodology proposed here for fundamental damage localization, D value is obtained by sensors (in principle by pressure gauges) at all the joints in the network so that the two end joints of every link of the network are covered. The link represents a pipeline, possibly consisting of segmented pipes, connecting these two joints. The numerical simulation results in subsection 3.2 "*Numerical simulation for damage identification*" below show that pipe damage can be found in the link between the two end joints where the D values form local maxima. It is also shown in the same subsection that this can be easily achieved by automatically identifying such links on the contour map of D value plotted over the service area. The magnitude of the D value is expected to relate closely to the severity of the damage that caused the transient. Their exact relationship will be the subject of future research. The methodology described above, together with field experiments for verification, lays the foundation for establishment of an innovative cost effective monitoring of water delivery networks that can be most useful for effectively carrying out ensuing emergency response. The proposed procedure must integrate a dense sensor network with rapid and robust data transmission capability, with the software capable of recognition of damage, unique for each lifeline system.

The transient flow based damage localization technique described above can also be applied in principle to other lifeline systems with appropriate modifications. It is noted however that when a water network is considered, change in water pressure in the pipe can be translated to change in acceleration on the pipe. This makes it possible to use non-invasive acceleration sensors attached on pipe surface for detection of pressure change without using invasive pressure gauges. This is the critical innovation that made the proposed damage localization method practical. The existence of the clear correlation between the pressure and acceleration

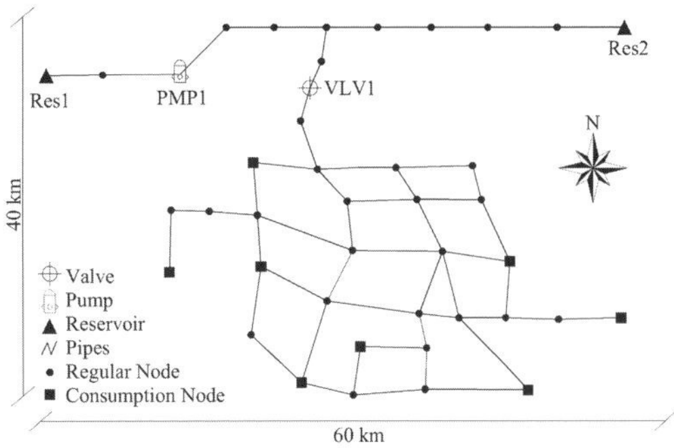

Figure 1. Water delivery system.

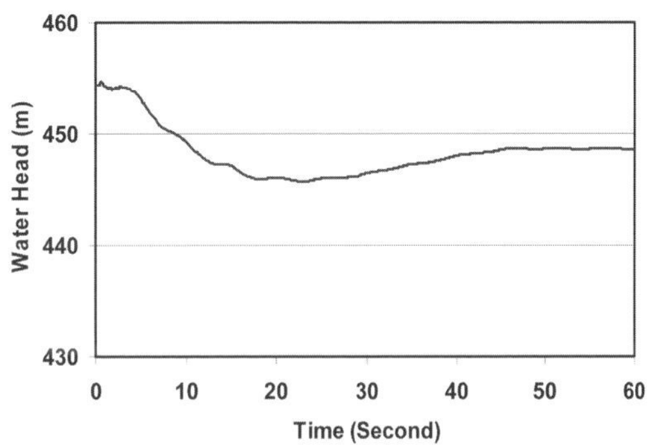

Figure 2. Water head time history at joint 9 under scenario 1.

is verified experimentally in subsection 3.3, and it is used in field tests for verifying the damage localization method.

3.2 *Numerical simulation for damage identification*

To validate the methodology described above, an industry standard computer code HAMMER for transient analysis of flow in water networks is used to simulate the transient behavior under sudden pipe breaks. A virtual hydraulic system which appears in HAMMER User's Guide (www.haestad.com) is used for this purpose. As shown in Figure 1, this water system consists of two reservoirs, one pump, one valve, thirty-eight nodes and 54 pipe links covering an area approximately 40 km × 60 km. The initial hydraulic boundary conditions are such that Reservoir 1 supplies water to the network and the water is consumed in steady state at consumption nodes. In this simulation analysis, two scenarios are considered. Scenario 1 in which one of the pipes breaks in Pipe 111, and scenario 2 in which two pipes rupture in the network, one in Pipe 111 and another in Pipe 24. In Figures 2 and 3 time history plots of water head at Joint 9 under scenarios 1 and 2, are shown respectively. It is noted that the water head shown in Figures 2 and 3 is inclusive of the elevation effect. The time history begins with a pressure value representing the initial steady state and converges downward to the value associated with

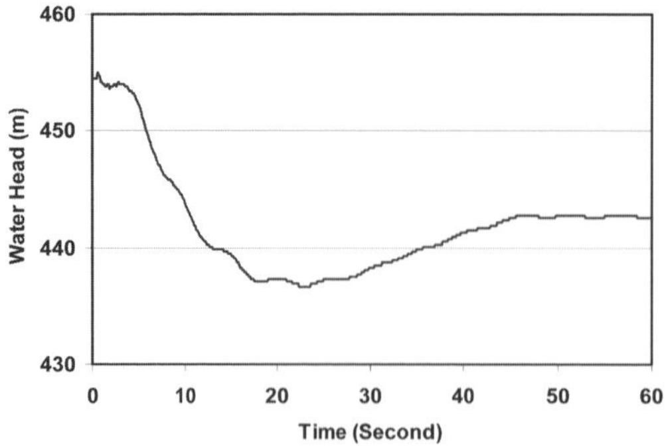

Figure 3. Water head time history at joint 9 under scenario 2.

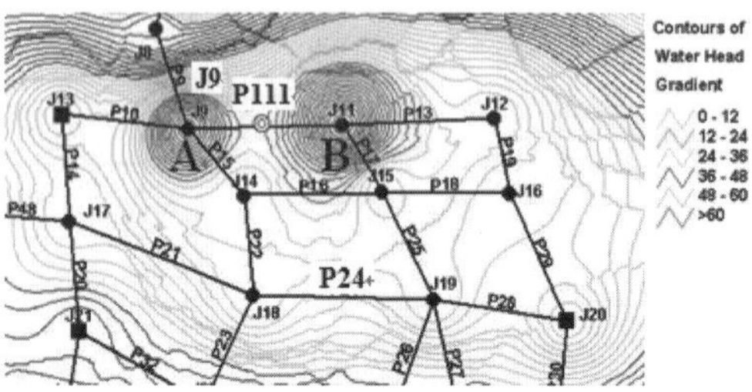

Figure 4. Contour map of D (absolute maximum water head gradient) due to a pipe break in P111.

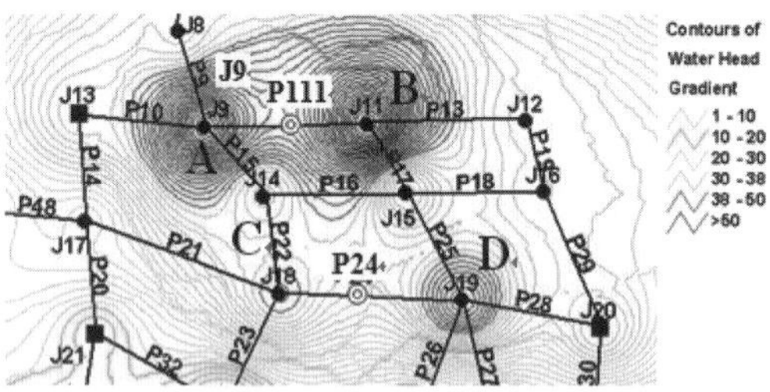

Figure 5. Contour map of D (absolute maximum water head gradient) due to two pipe breaks one in P111 another in P24.

the post-damage steady state as the system completes its adjustment to the change. The absolute maximum value of WHG computed over the first 30 seconds of record is used as D value at each joint, and its spatial distribution over the entire network is plotted for scenario 1 in Figure 4 and scenario 2 in Figure 5. In Figure 4, it is observed that D values at the two end nodes A (Joints 9) and B (Joint 11) of Pipe 111 are spatial local maxima identifying the location of pipe break (double circle) to be in this link. Similarly, Figure 5 identifies the locations of pipe breaks in Pipes 111 and 24 by recognizing two pairs of local maxima of D values at A (Joint 9) and B (Joint 11) for Pipe 111 and at C (Joint 18) and D (Joint19) for Pipe 24. The procedure for the damage identification thus boils down to the identification of the pipes in the map (as in Figures 4 and 5) bounded by a pair of joints with maximum D values. This selection can be achieved automatically in near real time.

Figure 6. Experimental setup.

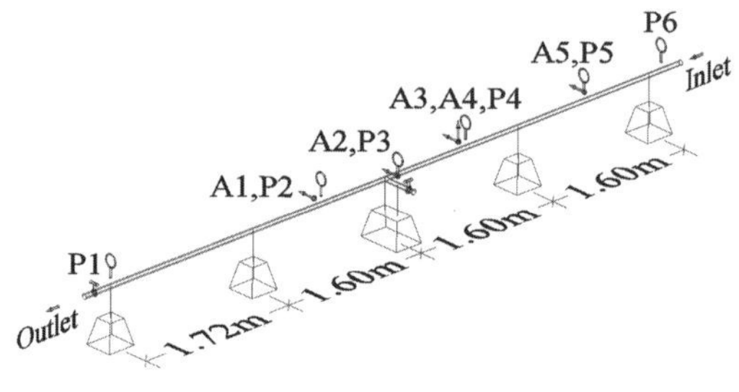

Figure 7. Sensor Layout.

It is noted that the inversion process involved in the localization procedure does not always produce unique solutions. For example, consider joints 1, 2, 3 and 4 represent corners of a rectangular network, with 1-2, 2-3, 3-4 and 4-1 being its links. Suppose pipes break in link 1-2 and link 3-4. This example will most probably produce maximum D's at all the joints, and hence damaged links cannot be determined uniquely by the simplistic logic used above. However, in practice, this is the situation where a cluster of link damages occurs in a service area strongly affected, for example by a severe earthquake. Even in such a case, information of many damaged links clustered (may not be uniquely identified) in a specific limited area is useful, from emergency response point of view, since it enables rapidly dispatching search and rescue teams and repair and restoration crews to the identified area.

3.3 *Experimental verification of the methodology*

Laboratory tests were conducted to check the correlation between the pressure drop in the pipe and the vibrations of the pipe wall. The pipe was 15 ft long. The experimental setup included six pressure sensors and five accelerometers installed along the pipe at different locations, as seen in Figure 6 and Figure 7.

The pipe was equipped with 2 valves and the inlet was connected to the water distribution system of the building. The pressure in the system was recorded between 80 to 90 psi. During the experiment, different pressure change conditions were observed for different

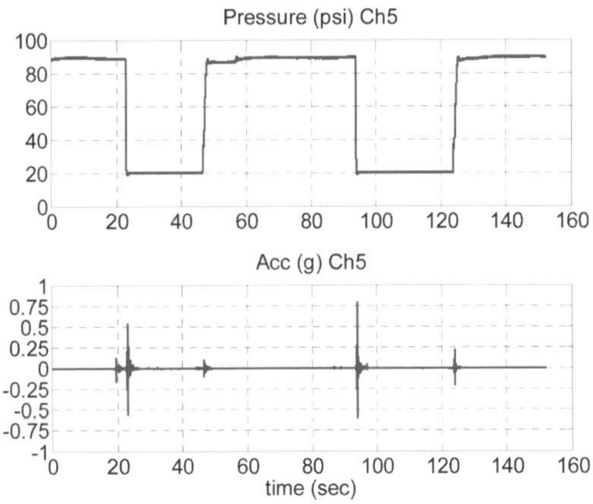

Figure 8. Pressure and acceleration time histories.

(a)

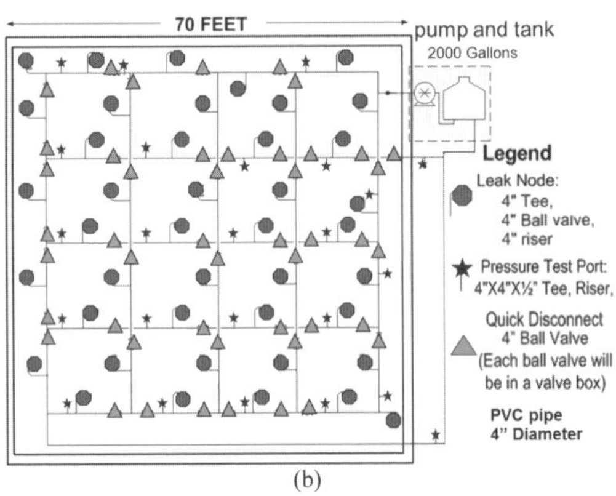

(b)

Figure 9. IRWD water distribution system: (a) model (b) layout.

induced boundary conditions, e.g. a valve opening or use of water inside the building. The pressure change was always coupled with an acceleration spike, indicating significant pipe vibrations. An example of the obtained results is seen in Figure 8.

Therefore, it is possible to find the location of a pressure drop in a water distribution system using a series of accelerometers throughout the network. Large scale model tests are also carried out, as shown in Figure 9,

-CAN bus connection: Daisy chain connection up to 120 nodes
-Low noise figure accelerometers
-Supports 4th interface for a commercial piezoelectric accelerometer
-Built in tilt sensor, humidity sensor, temperature sensor
-Supports pigtail-type MEMs accelerometers

Figure 10. Design and features of Gopher.

using the facilities constructed for this purpose in the campus of Irvine Ranch Water District. The experimental results obtained so far using this facility also confirmed the existence of close relationship between changes in water pressure and in pipe acceleration. In all these tests the accelerometer based Duramote system, together with standard industrial pressure gauges, were used to verify the correlation. The details of the Duramote system are described in section 4.

4 THE NEXT GENERATION SCADA SYSTEM

Duramote, part of the next generation SCADA system for the identification of leaks and pipe breaks has multiple components: the gopher, the roocas, a base station, and eventually routers.

The gopher is the main component of the system (Figure 10). Each gopher includes one or more MEMS accelerometers. One accelerometer is embedded on the board and two more can be added to cover all three acceleration directions. The gopher is enclosed in a water tight box and can be placed underground attached to the water pipe. As stated before the main advantage of using accelerometers over pressure gauges is that they are not invasive sensors. The gopher can be placed or replaced without drilling the pipe and without interrupting the service.

The roocas (Figure 11) is the second component of the system. Each roocas is connected with a wire to one or more gophers. The roocas is designed to be installed at ground level and its function is to power the gopher, to collect the acceleration data from it, and to transmit the data to a base station using the wireless network. Each roocas is capable of supporting multiple channels and multiple gophers. Currently the roocas is powered by a battery or by the external power. However, studies are underway in order to harvest energy locally.

The wireless network is supported by a certain number of inexpensive and commercially available Wi-Fi routers. Their functions are: to synchronize the roocas with the base station, to collect the signals from the

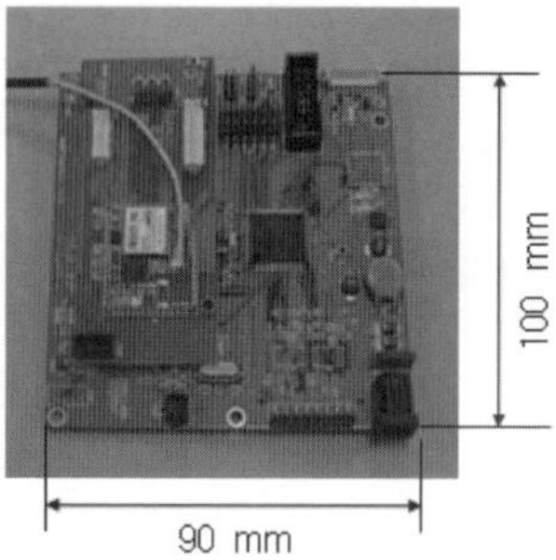

-Multi-Wireless Technologies:
 Wi-Fi/ Xstream/ XBee/ Eco/ Media converter (TX to FX)
-Different topologies: P to P, P to MP, 3G network, mesh topology
-Data logging: nonvolatile storage (FRAM, HCSD)
-Time synchronization modules: GPS, RTC, WWVB
-Supports real time monitoring system: Server/Client program

Figure 11. Design and features of Roocas.

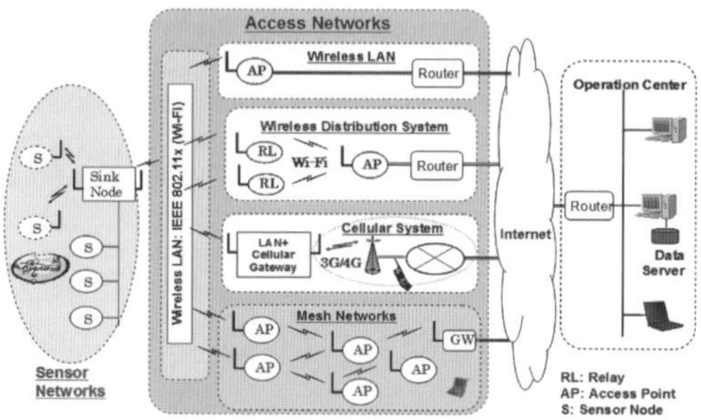

Figure 12. The next generation SCADA.

roocas, and to hop the signals between each other and the base station.

Figure 12 indicates the layout of the SCADA system from the gophers to the base station passing through roocas and routers.

5 THE PROTOTYPE DEVELOPMENT

This section describes recent efforts, performed at UCI campus, Vincent Thomas bridge, and Hwamyung bridge in South Korea, towards the development of the prototype Duramote system.

As stated, the whole system is mainly composed by 3 different components: gopher (Figure 10), which contains the accelerometers, roocas (Figure 11), which transmits the data, and the base station, which stores the data and post-processes the results. The connectivity between the roocas and the base station can be achieved using different technologies (Figure 12).

5.1 *UCI campus facilities*

The experiment was conducted to test the long-range wireless communication capability of the Duramote system. The experiment involved 3 campus facilities: the engineering tower, the parking structure, and the green bridge. The distance between the nodes ranges from 150 m to 520 m (Figure 13). The base station was placed on the roof of engineering tower while the sensors were placed on the roof of the parking structure and on the green bridge (Figure 14).

Figure 13. Layout of the experiment at UCI campus.

(a) (b) (c)

Figure 14. (a) Eng. tower – base station, (b) Parking structure – sensor, (c) Green bridge – sensor.

The devices required to build the long range Wi-Fi wireless network are:

- Wireless Access Points (AP): APs make it possible for various types of Wi-Fi clients to be associated with the existing network.
- Repeaters: Repeaters relay incoming data to the parent repeater or the access point.
- Bridges: Bridges provide point-to-point wireless communication as well as wired communication with other client devices.

Different network topologies were tested during the experiment since the Duramote system can also employ two different wireless communication technologies: Wi-Fi and modified Zigbee. Each node was equipped with either a repeater or a bridge. The experiment focused on the performance of the Wi-Fi technology including its bandwidth, which limits the number of nodes that constitute the network, the sampling frequency of the sensors, and the allowable number of hops between the sensors and the base station.

Acceleration records at each location can be observed in real time on the monitors at the base station

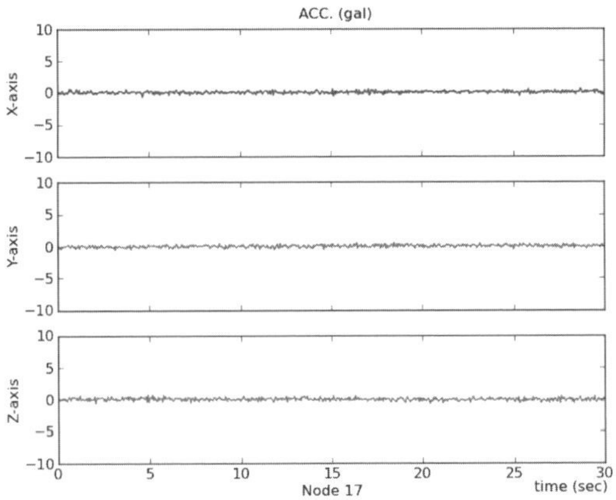

Figure 15. Real time acceleration record at green bridge.

in Figure 14a, and are also shown here in Figure 15, only for the green bridge structure.

5.2 *System versatility, Vincent Thomas bridge experiment*

In order to test the reliability and robustness of the wireless network in a noisy environment, such as the steel girders of a long bridge, and due to its relative closeness to UCI campus the Vincent Thomas bridge was used as a test bed. The Vincent Thomas bridge is a four lane suspension bridge located in San Pedro, California, spanning over the main channel of Los Angeles Harbor and part of Seaside Freeway and State Route 47. The bridge was designed by the bridge department of the California Department of Transportation in 1960 and opened to traffic in 1963. Currently, it is the 19th longest suspension bridge in the US and the third in California. The bridge's deck is supported by two 120 m tall towers resting on steel piles. The main span is 457 m long and the side spans are 154 m each.

The first layout of the experiment involved accelerometers placed on a steel water pipe that runs inside the steel box girder for the entire length of the bridge. The pipe runs parallel to the service walk way and it can be easily accessed throughout its length. 3 accelerometers were attached on the steel pipe. The 3 sensors were placed in a symmetrical configuration, with one in the middle and the other two approximately halfway between the towers and on the center line of the deck, Figure 16a.

The results of the test indicated various peaks in the frequency domain of the acceleration data, both in high and low range. High frequency peaks were associated with the vibrations of the pipe. The sensors were removed from the pipe and attached directly on the bridge, Figure 16b. After comparing the frequency domain of the acceleration data from the sensors on the pipe and the sensor on the deck, the low frequencies were nothing else than the natural frequencies of the bridge structure itself.

To identify the natural frequencies of the pipe and bridge the Frequency Domain Decomposition (FDD)

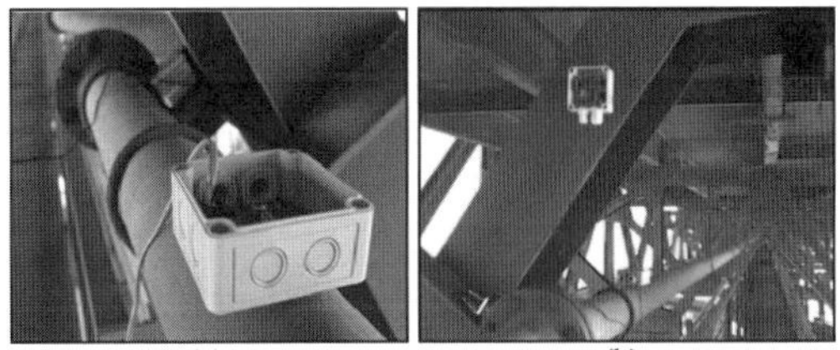

(a) (b)

Figure 16. Sensors placement at Vincent Thomas bridge.

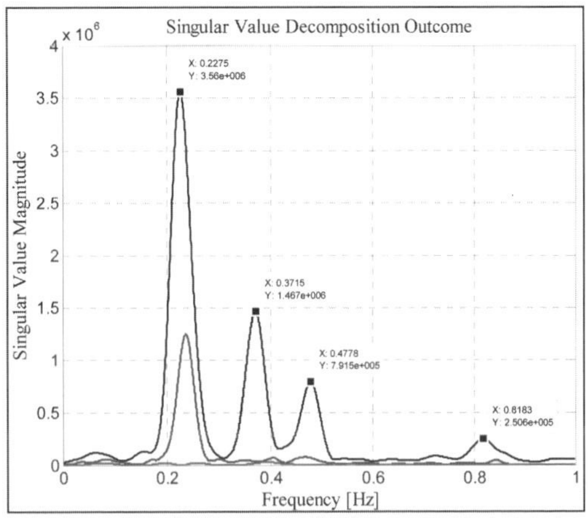

Figure 17. Singular Value Decomposition of the collected data

was used. FDD technique is a simple yet accurate output-only system identification technique based on decomposition of the response of the system into a number of SDOF systems. The FDD technique resolves some of the issues associated with the classical frequency domain approach, such as the extraction of natural frequencies close to each other, and it keeps the important features of simplicity and user friendliness. FDD can produce exact results in the case where loading is white noise, the structure is lightly damped and the mode shapes of the natural frequencies close to each other are geometrically orthogonal. Even if these assumptions are not well satisfied, the results are still significantly more accurate than the results of the classical approach. Some data analyzed using FDD showing natural frequencies of the bridge structure can be seen in Figure 17.

Based on the recorded data from the three sensors only four basic vertical modes were able to be identified. The difference between the bridge results and the pipe results is minor. From the obtained results it is concluded that if a pipe is placed on a structure special attention is needed while analyzing the frequency domain and the structure's natural frequencies should be filtered out. Overall, the results also proved that the Duramote system can be used successfully for structural health monitoring of large engineering structures, such as bridges. The capability of the system, specifically the high frequency range up to 450 Hz and over,

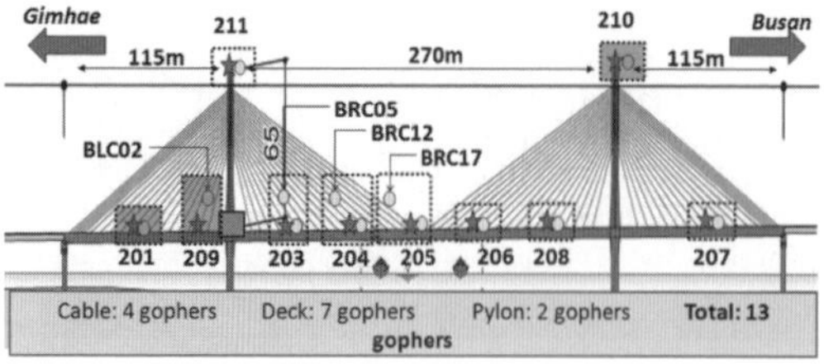

Figure 18. Hwamyung bridge geometry and sensor layout.

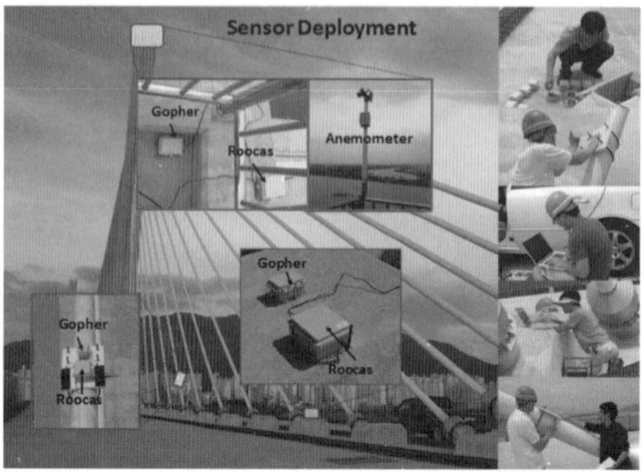

Figure 19. Hwamyung bridge sensor deployment.

was not fully utilized in this case but the system can clearly identify the behavior of the structure, the natural frequencies, the mode shapes, and other structural parameters.

5.3 *Hwamyung bridge test*

After the Vincent Thomas bridge experiment, the Duramote system was installed on the Hwamyung bridge, which is a cable stayed bridge between the city of Busan and the town of Gimhae, in South Korea. The scope of the experiment was to test the system's durability and performance.

The Hwamyung bridge, Figure 19, is a newly built cable stayed bridge with a main span of 270 m. The top of the towers are 65 m above the deck. The bridge was instrumented with 10 nodes (Figures 19 and 20) and each node has a different combination of sensors. 7 sensors were placed on the center line of the deck; each of them has 2 accelerometers to measure the vertical and transversal deck vibrations. 2 sensors were placed on top of the towers; each of them has 2 accelerometers to measure the longitudinal and transversal tower vibrations. 4 sensors were placed on the cables to measure their behavior.

The data were transmitted through Wi-Fi to the base station, which was placed inside the tower on the Gimhae side. The base station recorded data continuously and the sampling frequency was set at 450 Hz. While a much lower frequency would have been good enough to monitor the bridge, the 450 Hz frequency was used in order to produce heavy packet traffic on the Wi-Fi. The base station was connected to the internet through 3G connection and it could be accessed remotely.

The experiment lasted 3 months and during that time the system survived two major rain storms and continued to aggregate the acceleration data proving its advanced robustness. The natural frequencies and mode shapes were successfully identified and they are in good agreement with the analytical results, which were obtained from a Finite Element Model of the bridge, as seen in Table 1.

Table 1. Natural frequencies.

Experimental ω_n	Analytical ω_n	Direction
0.413 Hz	0.476 Hz	transversal
0.465 Hz	0.484 Hz	vertical
0.752 Hz	0.833 Hz	vertical
0.797 Hz	0.797 Hz	transversal
1.072 Hz	*	vertical
1.134 Hz	1.165 Hz	vertical
1.276 Hz	1.219 Hz	vertical
1.478 Hz	1.238 Hz	transversal
1.522 Hz	1.561 Hz	vertical
2.270 Hz	2.299 Hz	vertical
2.475 Hz	2.112 Hz	transversal+torsional

6 CONCLUSIONS

This study presents the progress made so far at UCI in developing next generation SCADA systems that remotely and in real time monitor and control the decaying and/or anomalous behaviors of infrastructure systems. The paper first described in some detail the concept and methodology of the next generation SCADA system using water distribution networks as test-bed. In this study, the next generation system consists of 3 components: (1) Duramote (MEMS-based sensor and transmission network, developed at UCI), (2) Access network providing optional ways of data transmission (acceleration in this case), and (3) SCADA operational center to where the data are sent and displayed in real time for observation, analysis and control. Performance of Duramote network and access network were verified analytically and in field experiments. The field experiments included, among others, a set of UCI campus facilities, the Vincent Thomas bridge, and the 500m long cable stayed Hwamyung bridge in South Korea. All systems performed very well, while the bridge systems demonstrated the robustness of the installed Duramote system as well.

ACKNOWLEDGMENTS

This study was carried out under National Institute of Standards and Technology (NIST) Grant. Their support is deeply appreciated. The partners on the NIST Project include Earth Mechanic Inc. (EMI), Irvine Ranch Water District (IRWD), Orange County Sanitation District (OCSD), Santa Ana Water Project Authority (SAWPA) and Pacific Advanced Civil Engineering (PACE). The authors thank all these partners for their support and contributions. The authors also acknowledge the support provided by NSF National Science Foundation for the early conceptual development of this study. The support of Hyundai for the experiment on the Hwamyung bridge is greatly appreciated. In addition, the authors would like to thank Dr. Sergio Carnalla for the sup-port during the experiments, as well as the research group from the Department of Electrical Engineering at UC Irvine including Professor Pai Chou, Dr. Sehwan Kim, Brett Chien and Eunbae Yoon for their contributions concerning the development of Duramote. The suggestion by Professor Fujiwara (Hakodate National College of Technology in Japan) to integrate Access network into the next generation SCADA system was very valuable.

REFERENCES

Kim, S., Yoon, E., Chien, T-C., Mustafa, H., Chou, P. H., Shinozuka, M., 2011. Smart wireless sensor system for lifeline health monitoring under a disaster event. *Proceedings of SPIE Conference on Smart Structures & Materials/NDE, San Diego, USA.*

Papakonstantinou, K. G., Shinozuka, M., and Beikae, M., 2011. Experimental and analytical study of water pipe's rupture for damage identification purposes. *Proceedings of SPIE Conference on Smart Structures & Materials/NDE, San Diego, USA.*

Shinozuka, M. and Dong, X., 2005. Evaluation of hydraulic transients and damage detection in water system under a disaster event. *Proceedings of the 3rd US-Japan Workshop on Water System Seismic Practices, Kobe, Japan.*

Shinozuka, M., Park, C., Chou, P. H. and Fukuda, Y., 2006. A sensor network for real-time damage location and assessment. *Proceedings of the 3rd International Workshop on Advanced Smart Materials and Smart Structures Technology, Lake Tahoe, USA.*

Shinozuka, M., Chou, P. H., Kim, S., Kim, H. R., Yoon, E., Mustafa, H., Karmakar, D. and Pul, S., 2010. Nondestructive monitoring of a pipe network using a MEMS-based wireless network. *Proceedings of SPIE Conference on Smart Structures & Materials/NDE, San Diego, USA.*

Shinozuka, M., Kim, S., Chou, P. H., Fei, L. and Kim, H. R., 2010. A sensor network for real-time damage location and assessment. *6th Taiwan-US-Japan Workshop on Water System Seismic Practices, Taipei.*

Shinozuka, M., Chou, P. H., Kim, S., Kim, H., Karmakar, D., and Lu, F., 2010. Non-invasive acceleration-based methodology for damage detection and assessment of water distribution system. *Smart Structures and Systems.*

Bridge Maintenance, Safety, Management, Resilience and Sustainability – Biondini & Frangopol (Eds)
© 2012 Taylor & Francis Group, London, ISBN 978-0-415-62124-3

The art of arches

M.-C. Tang
T.Y. Lin International, San Francisco, USA

ABSTRACT: People have been building beautiful arch bridges for more than 2000 years. From the spectacularly solid Roman arch to today's elegant and graceful styles, there are a large variety of arch shapes. The development of arches was influenced by the construction materials available at the time. Today, with more versatile materials such as steel and reinforced concrete, many beautiful signature arch bridges are being built.

1 ANCIENT ARCHES

The form of an arch is very natural. The inspiration behind arch construction may have come from nature. There are an abundance of examples of arch formations in our natural landscape.

Arch construction started before the Roman Empire. But the Romans were master builders. And while we do not know how good they were at building arch bridges with today's materials, the longest-lasting bridges that we can see today are all stone bridges. And, for obvious reasons, these can only be deck arch bridges, with the bridge deck placed above the arch.

Modern arches are much lighter and are mostly built using steel or reinforced concrete. These materials have tensile capacities. A modern arch rib using modern materials can handle compression and tension, as well as bending moment. They are much more slender. The advantages of slenderness and lightness are that they offer excellent opportunities for creating graceful arch forms. They also enable much larger spans. However, the disadvantage of slenderness and lightness is that the arch rib by itself, if not restrained, is usually not sufficiently stable under the required design loads. Because the arch rib is under high compression, it behaves like a beam-column. It can buckle in both in-plane and out-of-plane directions.

There are several ways to stabilize slender arch ribs. In the in-plane direction, the arch ribs are connected to the girder by hangers. Any vertical deformation of the arch rib will force the girder to follow. If the hangers do not elongate, the vertical deformation of the girder must be equal to that of the rib. Thus, the stiffness of both the rib and the girder contribute to the stability of the structure. The axial deformation, that is, the elongation or shortening of the hangers is typically small. Consequently, adding the stiffness of the girder and the stiffness of the arch rib to estimate the buckling load is a good approximation in most cases. As a result, in the case of a stiff girder, the arch ribs can be made more slender. The Fremont Bridge in Portland, Oregon, USA and the Guandu Bridge in Taiwan are good examples.

Another way to increase the in-plane stability of the arch rib is to use inclined hangers. This system is called "Network Arch Bridge." In actuality, the network arch bridge acts more like a truss than an arch bridge. The hangers are similar to the diagonals of a truss. Consequently, the arch rib can be made very slender in the in-plane direction.

In the out-of-plane direction, the hangers are usually in the same plane as the arch rib, so they cannot provide any assistance to strengthen the arch rib against lateral or out-of-plane buckling. The most common way to solve this problem is to have two arch ribs bracing against each other. The bracing can either be in a crisscross pattern, where the structure will act more like a truss, or be simply parallel struts, where the structure will act more like a Vierendeel truss.

Turning the arch and hanger combination into a spatial structure is another effective way of increasing the lateral stability of the arch rib. There are several alternative methods to achieve similar result: One simple way is to attach two planes of cables to each arch rib, thus making it into a three-dimensional or spatial structure.

Once the stability problem of an arch rib in both the in-plane and the out-of-plane directions is resolved, the arch can be a very versatile structure for medium- and long- span bridges. In the following section, we will explore a variety of configurations with real-life examples (Fig. 1).

2 CAIYUANBA BRIDGE

The Caiyuanba Bridge is located in Chongqing, China. It crosses the Yangtze River from the southern district of the City to the central district. It has a main span of 420 m, and is one of the largest arch bridge spans in the world featuring a highway and monorails. The upper deck is 36.50 m wide and the lower deck is 12.10 m

Figure 1. Top: Caiyuanba, Dagu Bridge, Huihai Road Bridge, Sanhao Bridge, Taijiang Bridge.

wide. This forms a trapezoidal-shaped girder cross section which appears more slender as it crosses the Yangtze River Valley. A rectangular box shape would have looked too bulky at this location.

Because of the aesthetic requirements, the arch ribs were made to look as slender as possible. A simple rectangular box section, 2.40 m wide by 4 m deep, was used for the arch rib. To assure the lateral stability of the arch ribs, a basket-handle arch system was chosen. This arch bridge has two ribs leaning towards each other and they are connected by transverse struts.

The basket-handle arch bridge is a good choice for long-span arch bridges. It can also be attractive for short-span bridges if the deck is narrow so that the inclination of the arch ribs is not too acute.

3 DAGU BRIDGE

The basket-handle arch configuration is effective, except that with the arches leaning inwards, travelers may feel boxed in while driving through the bridge. This is especially so if the bridge deck is wide and the span is small. To provide them an open view while driving on the bridge, the arch ribs were bent outwards, as in the case of the Dagu Bridge in Tianjin, China. Each arch rib of the Dagu Bridge has two planes of hangers, forming a spatial structural system. This arrangement provides the required lateral restraint for the arch ribs, which are very slender.

The Dagu Bridge is located in the center of Tianjin, a city of 11 million close to Beijing. It is one of the first bridges to be built under the Haihe Area Revitalization Plan. Because of its location, aesthetics was extremely important. The Owner envisioned a signature structure for the City. The bridge crosses the Haihe River, which traverses the entire city. The river is about 96 m wide at this point; so, the span of the bridge is 106 m to clear the entire river. It carries six lanes of city traffic and two pedestrian/bicycle paths, one on each side. The minimum width of the bridge deck is 32 m at both ends and varies to about 56 m at the mid-span including some openings in the deck. Due to navigational clearance requirements, the depth of the girder was limited to about 1.40 m at the centerline of the deck. The 1.40 m girder was not sufficient to span the 32 m width of the bridge in the transverse direction. Hence, the arch ribs were placed inside of the pedestrian/bicycle paths. They are 24 m apart transverse to the bridge axis. Both arch ribs are made of steel with trapezoidal cross sections. The girder is a steel box girder with an orthotropic steel deck paved with 50 mm of epoxy asphalt.

4 HUIHAI ROAD BRIDGE

By moving both arch ribs, like those of the Dagu Bridge, to the centerline of the deck, a new configuration has been derived for the Huihai Road Bridge in Lianyungang, China. The bridge, currently under construction, is located in the middle of a new development area of the City. Aesthetics for this bridge project is important. The bridge has a main span of 100 m. The deck is 39 m wide. Each of the two arch ribs is restrained by two planes of hangers.

Moving the two arch ribs together changed the behavior of the structural system. For the Dagu Bridge, the arches are about 24 m apart, and any eccentric load on the deck or any asymmetrical loads acting on the structure can be carried by the two arch ribs to the end supports. There is no need for the girder to be torsionally rigid. However, when the ends of the two arch ribs come together at the centerline of the main girder, as in the case of the Huihai Road Bridge, such a system can no longer carry any eccentric load on the deck, or any asymmetrical loads due to wind, earthquakes, etc. Consequently, the main girder must be torsionally stiff to carry all torsional moments induced by any asymmetric loading to the piers. To achieve this, a multi-cell box section is used for the girder to enable the transfer of all torsional moment created by eccentric loads to the supports at the ends of the bridge.

The bridge deck is widened in the middle section to offer pedestrians additional space to rest and view the beautiful surroundings. This bridge is still under construction.

5 SANHAO BRIDGE

The configuration for the Sanhao Bridge can be shown by turning the Huihai Road Bridge 90° transversely. The Sanhao Bridge is located in the City of Shenyang. It crosses the Hun River connecting two newly developed communities. The City requested a signature bridge at this location to complement the high-end residential developments on both sides of the river. The bridge has a twin main span of 100 m each and the deck is 34 m wide. The girder depth is limited to 2.40 m to satisfy the bridge's navigational requirements. However, the budget was limited and the Owner could only afford to build a concrete girder. In China, where costs are concerned, a steel girder is more than twice as expensive as a concrete girder; however, a steel girder is significantly lighter in weight. To employ a concrete girder, a new bridge concept or the "partially cable-supported girder bridge," was developed (Tang, 2007). This design concept achieves its economic efficiency by fully utilizing the carrying capacity of both the cable- stay system and the girder itself. The cables carry only a portion of the weight of the bridge so the towers can be made much lighter. This design concept also allows a more liberal selection of cable forces, which is a major advantage for such a tower shape. For the design of the Sanhao Bridge, the cables were assigned to carry only about 50% of the load.

The two arch leaves divide each cable into three sections. The force in the middle section of each cable was fine-tuned to eliminate any out-of-plane force in the arch ribs. The two arch ribs were assembled into their final shape by laying them flat on the completed bridge deck and then hoisting them up into their final position using strand jacks attached to a temporary steel frame installed between the two arch ribs.

6 TAIJIANG BRIDGE

Yet another new configuration was innovated by combining the two arch ribs, similar to those of the Sanhao Bridge, together into one single arch. This configuration was implemented for the Taijiang Bridge in the City of Sanming in Fujian Province, China. This bridge has two adjacent main spans of 110 m each. The maximum allowable girder depth is 2.60 m. Again, due to budget constraints, the Owner could only afford to build a concrete girder. The design was based on the concept of a partially cable- supported girder bridge. In this case, as in the Sanhao Bridge, the cables carry approximately 50% of the total load on the bridge. However, putting many cables along the top portion of the arch creates large bending moment in the arch rib. Therefore, a horizontal tie was used to balance the outward thrust, creating a tied arch from the upper part of the arch. This significantly reduces the bending moment in the tower. This horizontal tie consists of two parallel cables of PWS strands. A vertical-tie cable was used to tie these horizontal tie cables to the deck girder. The lower end of this vertical cable was anchored at the bottom of the main girder. The purpose of this vertical-tie cable is to control the force in the horizontal tie cables. The force in the horizontal-tie cables can be easily estimated after the exact geometry of the horizontal-tie cables is obtained (by surveying) and the force in the vertical-tie cable is calculated. In other words, the force in the horizontal-tie cables can be adjusted by adjusting the force in the vertical-tie cable.

7 CONCLUSION

A few examples of modern arches have been described to show that arch bridges can have a variety of forms. While the arch is the oldest form of long-span bridges, it also represents some of the newest developments in bridge design configurations.

REFERENCES

Tang, M.C., Rethinking Bridge Design – A New Configuration, *Civil Engineering*, July 2007, Volume 77, Number 7: 38–45.

Bridge Maintenance, Safety, Management, Resilience and Sustainability – Biondini & Frangopol (Eds)
© 2012 Taylor & Francis Group, London, ISBN 978-0-415-62124-3

System design and implementation of structural health monitoring and maintenance management system for marine viaduct bridges

K.Y. Wong
Bridges & Structures Division, Highways Department, The Government of the Hong Kong Special Administrative Region, Hong Kong

ABSTRACT: Structural health monitoring of cable-supported bridges in Hong Kong has been carried out since 1997 by on-structure instrumentation systems and associated data processing and analysis tools. Recently, such works have been extended to marine viaduct bridges with the combination of relevant application systems for maintenance management. This paper introduces the works of the system design and implementation of Structural Health Monitoring and Maintenance Management System (SHM&MMS) in seven aspects: (1) conventional approach of inspection and maintenance, (2) modern approach of inspection and maintenance, (3) degradation mechanisms in highway infrastructures, (4) identification of measurands for marine viaduct bridges, (5) system architecture and operation of SHM&MMS, (6) framework of SHM&MMS, and (7) implementation of SHM&MMS for marine viaduct bridges. Discussion of structural health monitoring systems in Hong Kong with typical examples of monitoring and evaluation results are also presented. Finally, conclusions are drawn regarding the future applications and developments of SHM&MMS.

1 CONVENTIONAL APPROACH OF INSPECTION AND MAINTENANCE

The service life of the highway infrastructures such as bridges, viaducts and tunnels will be decreased as time goes on due to degradation mechanisms induced by the combined actions of environmental attacks, operational loads and aging effects. The current trend of designing durable or sustainable structures for overcoming such degradation effects is to increase the service life and reduce the maintenance cost of the highway infrastructure by routine inspection and maintenance with early identification of potential problems/defects through the usage of long-term instrumentation systems and corresponding analytical tools.

Figure 1 illustrates the performance of a highway infrastructure as a function of time. In this figure, the vertical steps correspond to maintenance actions and the performance may be represented by the condition rating or the load carrying capacity. The two maintenance steps of preventive and corrective maintenance are considered as the conventional maintenance approach and are defined as follows:

- Corrective maintenance – When the component is identified or reported to be damaged, maintenance work will then be carried out.
- Preventive maintenance – Periodically checking the performance and material condition of the components to determine if their operating conditions and resulting degradation rate are within the expected limits. If they are not, a search for the reason of the rapid degradation shall be carried out so that the problems can be corrected, or at least mitigated, before component breaks down.

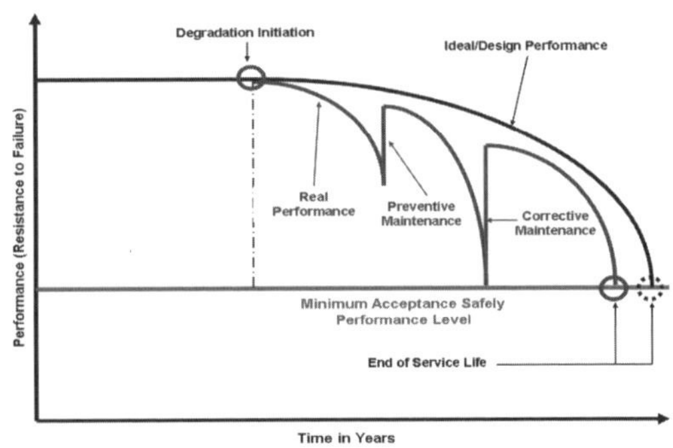

Figure 1. Performance of a highway infrastructure as a function of time.

Figure 2 illustrates the arrangement of different types of activities in conventional maintenance approach. In this figure, visual inspection (or routine inspection) is the key activity in inspection. The figure shows that when major structural problems are identified in routine inspection, in-depth inspection will be carried out. This type of inspection includes the strength and integrity ratings by vehicular load trials, laboratory material tests, on-structure vibrating tests, and modeling analysis. Such tests are all ad hoc and

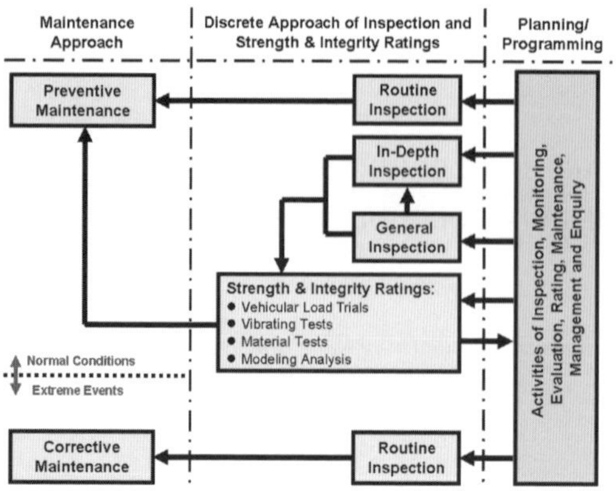

Figure 2. Conventional inspection and maintenance approach for highway infrastructure.

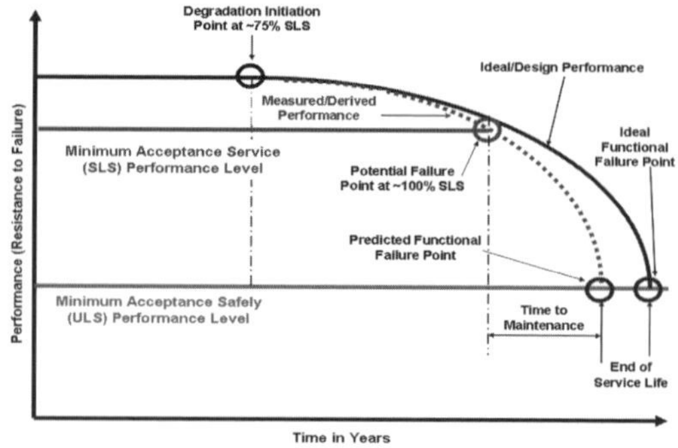

Figure 3. Condition-based Maintenance.

historical data regarding structural performance is not available, and the causes of problems/defects sometimes might not been properly identified/diagnosed. Without knowing the causes of problems/defects, the maintenance options so proposed or carried out in turn might not be effective.

2 MODERN APPROACH OF INSPECTION AND MAINTENANCE

Modern approach of inspection and maintenance of highway infrastructure shall normally involve the works in seven aspects, namely, (i) inspection – by visual inspection with systematic inventory system and structural condition rating system, (ii) monitoring – by on-structure instrumentation systems and high-definition video cameras monitoring system with appropriate data processing, analysis and reporting software tools (an alternative form of inspection), (iii) evaluation – by routine field calibrated finite element models and appropriate analytical methods, (iv) rating – by codified requirements of design and rehabilitation with MATLAB programming tools, (v) maintenance – by maintenance strategies, options, priorities and availability of resources, (vi) management of data and information – by data warehouse management system and on-line analytical processing tools and (vii) enquiry – by data buffer and network security tools.

With the introduction of the SHM&MMS, the inspection and maintenance of the marine viaduct bridges shall be improved and/or enhanced from conventional approach to modern approach. Such improvement/enhancement may be referred to as condition-based maintenance. The conditioned-based maintenance is based on the actual condition of the maintained component; that is the component that is consistently in good to superior condition does not need to be maintained as frequently as the component is deteriorating or has reached an age where deterioration is anticipated. In condition-based maintenance is composed of two parts, real-time monitoring and predictive maintenance. The former refers to the on-structure instrumentation systems and associated software systems for processing, analysis and report of measured data; whereas the latter refers to routine field-calibrated analytical tools such as finite element models for prognostic and/or diagnostic analysis of the potential problematic structural components that require maintenance/repair. The graphical illustration of condition-based maintenance with reference to structural health monitoring is given in Figure 3.

Figure 4 illustrates the arrangement of the different types of activities in modern inspection and maintenance approach. In this figure, it is shown that the SHM&MMS solely replaces the ad hoc works of strength and integrity ratings by the long-term and continuous monitoring system of SHM&MMS and at the same time enhance both the preventive maintenance under normal bridge operating conditions and corrective maintenance under abnormal operating conditions (such as ship impacting, historical typhoons, earthquakes, etc.) by respective predictive maintenance and condition-based maintenance with the output of: (i) scope and priority reports for inspection, and (ii) scope and priority reports for maintenance.

In Figure 4, the condition-based maintenance (prognostic) shall be advocated with measurements under normal operating condition by on-structure instrumentation systems, which aims at the early detection of degradation mechanisms by periodically updating of information on the performance of the structure under its designated functions of load-carrying capacity and durability resistance with the co-existence of degradation effects due to aging and environments; whereas the condition-based maintenance (diagnostic) shall immediately be activated for detection and diagnosis of structure after extreme events such as historical great typhoons, earthquakes, ship-impacts, etc. with the aims to provide, in near real-time, reliable information on: (i) the integrity of the structure for continuous operation, and (ii) the cause and scope of (repairable) damage for immediate maintenance action, if any. This final evolutionary step is the real key to optimizing the high valued and critical operational and maintenance process.

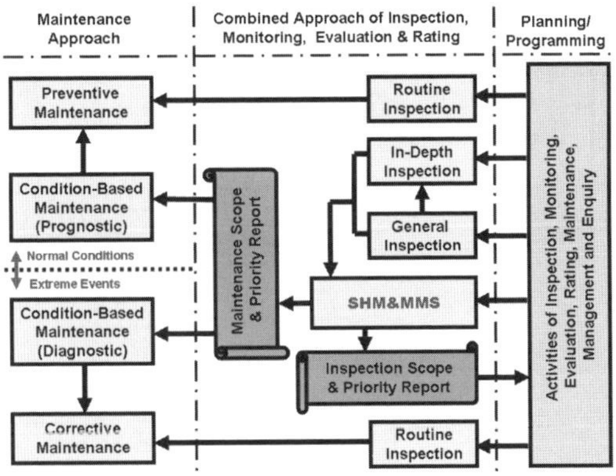

Figure 4. Modern inspection and maintenance approach for highway infrastructure.

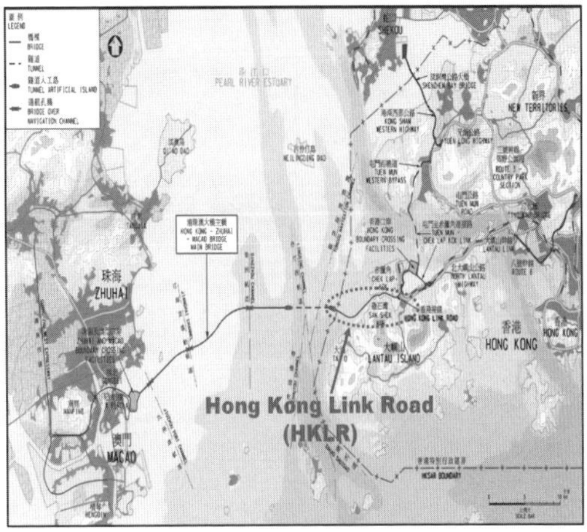

Figure 5. Location of HKLR HZBMP.

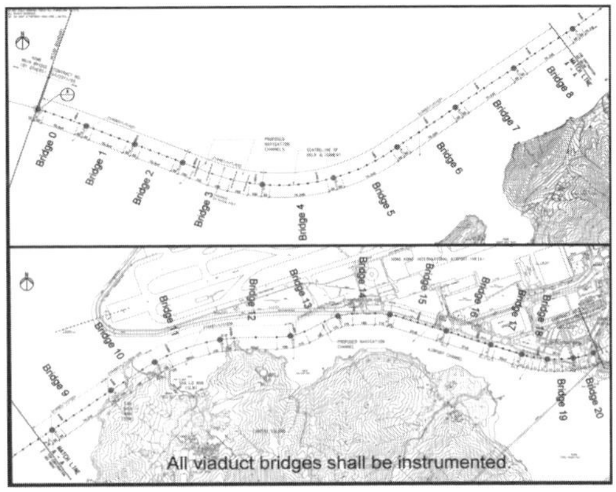

Figure 6. Layout of marine viaduct bridges.

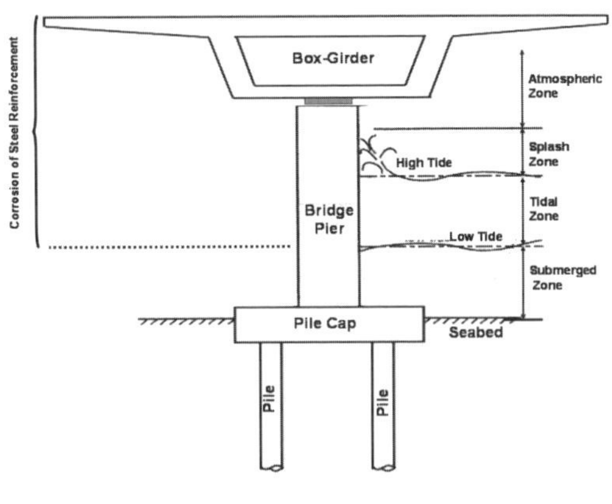

Figure 7. Exposure zone and possible locations of corrosion in marine viaduct bridges.

3 DEGRADATION MECHANISMS IN HIGHWAY INFRASTRUCTURES

A highway infrastructure project, called Construction of Hong Kong Link Road (HKLR) which is part of the Hongkong-Zhuhai-Macao Bridge Project (HZMBP), shall be constructed and instrumented. This project is composed of 21 numbers of marine viaduct bridges with a total span-length of ~9.3 km. The layout of the HKLR in HZMBP is shown in Figure 5, and the details of the 21 numbers of bridges are shown in Figure 6.

Deterioration in highway infrastructure is a commonly occurred phenomenon, particularly for those marine structures and such deterioration can lead to a number of deleterious effects such as loss of serviceability, loss of load carrying capacity, reduction in safety, increase in traffic restriction and loss of aesthetic value. Since the frequency and extent of maintenance and repair operations are dependent upon the rate of deterioration, the determination of the value of the rate of deterioration for marine viaduct bridges is thus essential to evaluate the effectiveness of the proposed maintenance/repairing option/options. As the rate of deterioration is a function of time, conventional inspection approach, due to its ad hoc nature and lacking of continuous time-history supporting data, cannot fulfill such requirement. Furthermore, the rate of deterioration, at which the structural components deteriorate, cannot be generalized because they are exposed to different macro and micro climates. Long-term real-time instrumentation systems and analytical tools are thus required for estimating the deterioration rate at key locations/components of each marine viaduct bridge under its in-service condition.

The possible environmental attacks and/or damages in marine concrete viaducts are illustrated in Figure 7. Such attacks will induce corrosion cracking cycles in reinforced concrete piers and decks exposed to seawater, as shown in Figure 8.

However, in prestressed bridge-decks, the environmental attacks when combined with intrinsic material defects of concrete and steel, construction defects and repeated highway traffic loads, will result in prestress loss in tendons, hence the formation of different types of deterioration phenomena, as shown in Figure 9.

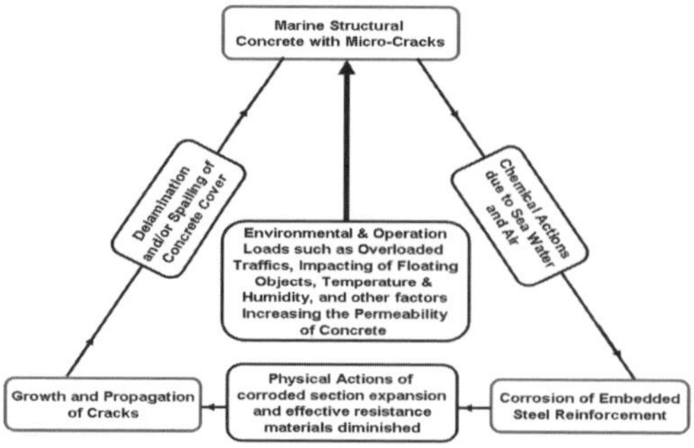

Figure 8. Diagram to show the cracking-concrete cycle in structural concrete piers and decks exposed to seawater.

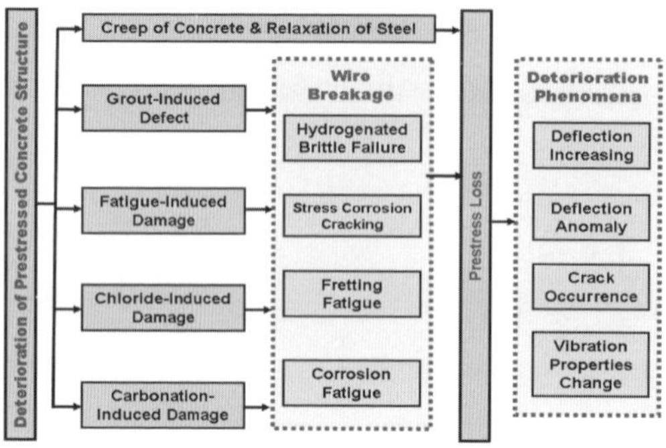

Figure 9. Possible defects in prestressed concrete structures.

4 IDENTIFICATION OF MEASURANDS FOR MARINE VAIDUCT BRIDGES

With reference to the degradation mechanisms in marine viaduct bridges, the measurands are: (i) potential risk of corrosion status of embedded steel reinforcement in structural concrete – for evaluation of corrosion on structural performance, (ii) creep of concrete and relaxation of steel reinforcement in prestressed concrete – for load-carrying capacity assessment, (iii) highway traffic loads and flows on marine viaduct bridges – for evaluation of highway traffic load-effects (i.e. jammed traffic load-effects and fatigue damage effects), (iv) static and dynamic features of marine viaduct bridges – for calibration of analytical models and for prediction of future loads and responses at locations/components without instrumentations; and (v) transient responses – for monitoring of transient loads such as ship impacting and earthquakes.

5 SYSTEM ARCHITECTURE AND OPERATION OF SHM&MMS

A system, named as "structural health monitoring and maintenance management system or SHM&MMS" is thus deployed to standardize and/or unify the seven aspects of maintenance works (as stated in Section 2

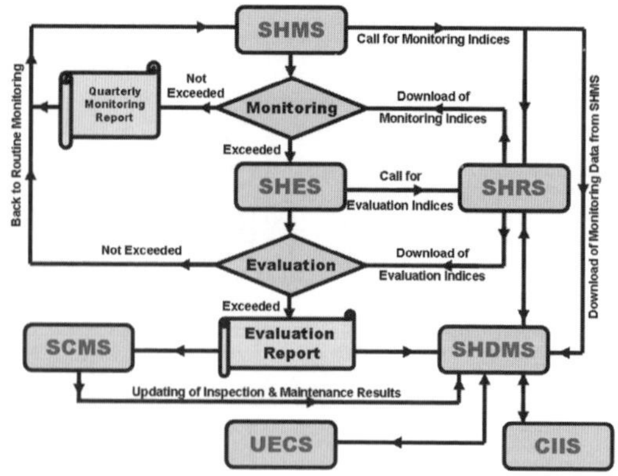

Figure 10. Architecture layout & flow diagram of SHM&MMS.

above) in continuous (real-time) and systematic manner so that the inspection and maintenance of the marine viaduct bridges can be executed efficiently and semi-automatically. This SHM&MMS, whose architecture layout and flow diagram are given in Figure 10, is composed of the following seven systems: (i) Condition Inspection and Inventory System (CIIS), (ii) Structural Health Monitoring System (SHMS), (iii) Structural Health Evaluation System (SHES), (iv) Structural Health Rating System (SHRS), (v) Structural Component Maintenance System (SCMS), (vi) User Enquiry and Communication System (UECS), and (vii) Structural Health Data Management System (SHDMS).

6 FRAMEWORK OF SHM&MMS

6.1 *Framework of CIIS*

The CIIS is deployed to carry out four functions: (i) to establish and maintain an inventory system for systematic storage and fast retrieval of inventory information and inspection records, (ii) to setup and operate an updatable structural condition rating system, (iii) to forward the structural condition rating results to SHRS to form synthetic rating by combining them with the results of criticality and vulnerability ratings, and to establish/update the priority orders for inspection, and (iv) to receive and display the synthetic rating results and the priority orders of components inspection from SHRS.

The CIIS is devised for facilitating the planning, scheduling, execution and recording of inspection activities. The architectural layout and flow diagram of CIIS is shown in Figure 11.

6.2 *Framework of SHMS*

The SHMS, which is the most important and costly system, has 3 major functions: (i) monitoring the structural performance of the marine viaduct bridges

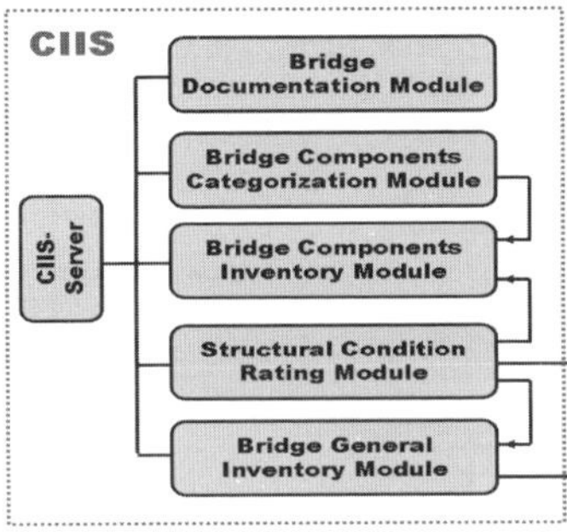

Figure 11. Architectural layout & flow diagram of CIIS.

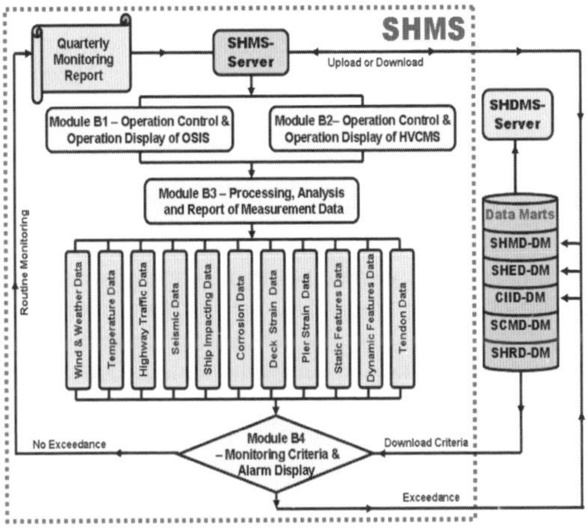

Figure 12. Architectural layout & flow diagram of SHMS.

and their surrounding environments against their designated performance limits and/or variation ranges at serviceability limit state, (ii) statistics analysis of the measured/derived data for pattern recognition and trend investigation; and (iii) features extraction of loads and responses.

Four categories of physical and chemical quantities are considered for monitoring, and they are: (i) environmental loads and attacks, which are composed of wind, temperature, seismic, corrosion, etc.; (ii) operation loads, which are composed of highway traffics, ship impacting, etc.; (iii) structure features, which are composed of static and dynamic features of the bridge; and (iv) structural responses, which are composed of displacements, strains, tendon forces, etc..

The SHMS is composed of 7 modular systems, namely, sensory systems (SS), local cabling network systems (LCNS), portable data acquisition system (PDAS), data acquisition system (DAS), global cabling network system (GCNS), data processing and control system (DPCS), and portable inspection and maintenance system (PIMS). The architectural layout and flow diagram of SHMS is given in Figure 12 and the schematic network layout of SHMS is given in Figures 13 for on-structure instrumentation systems (OSIS) and 14 for high definition video cameras

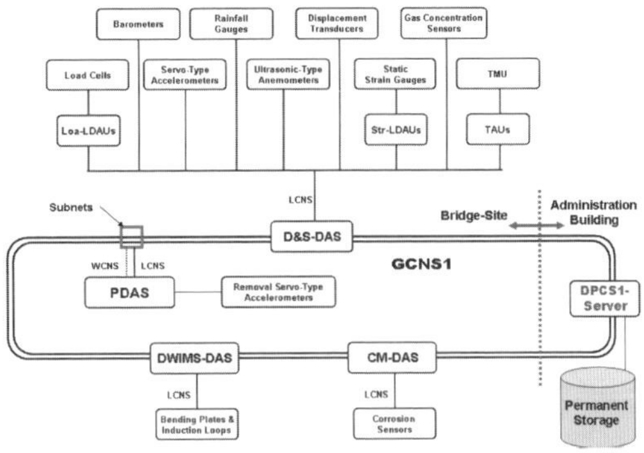

Figure 13. Layout of OSIS in SHMS.

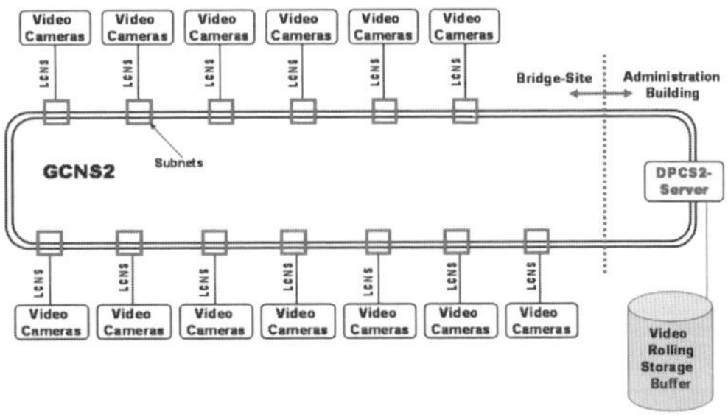

Figure 14. Layout of HVCMS in SHMS.

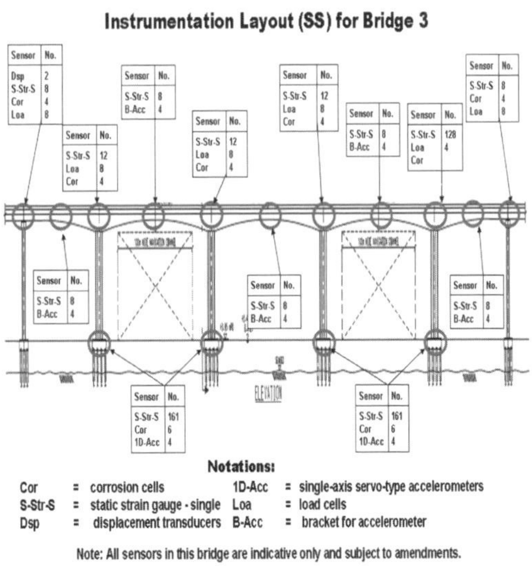

Figure 15. Instrumentation Layout of Sensory Systems in Bridge 3 (5-span bridge: 100 m + 150 m × 3 + 100 m = 650 m).

(HVCMS). The instrumentation layout for the longest bridge is shown in Figure 15.

6.3 *Framework of SHES*

The SHES is deployed as a computational system to carry out two major types of analysis, i.e., performance analysis (real-time or near real-time) and

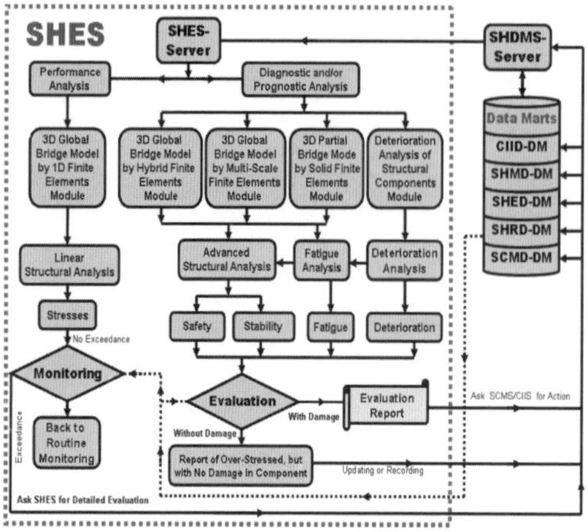

Figure 16. Architectural layout and flow diagram of SHES.

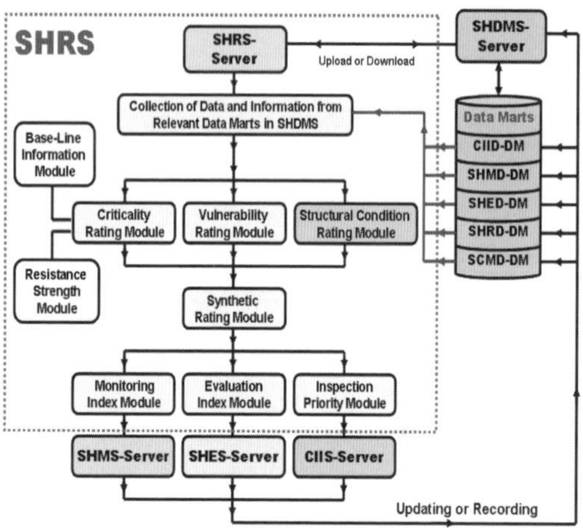

Figure 17. Architectural layout and flow diagram of SHRS.

diagnostic/prognostic analysis (off-time). The scope of evaluation, basing on the codified approach for viaduct bridges, is to analysis the structural performance of the bridge under the four limit states: (i) serviceability limit state, (ii) ultimate limit state, (iii) fatigue limit state, and (iv) durability limit state. The architectural layout and flow diagram of SHES is shown in Figure 16.

6.4 *Framework of SHRS*

The SHRS is deployed as a neuro-computing system to process and analyze the data and information received from CIIS, SHMS, SHES and SCMS for prioritization of component inspection. The designated functions of SHRS are: (i) collection and analysis of base-line information from surveys, vehicular load-trials and vibration measurements carried out before the opening of the completed viaduct bridges to public traffics for subsequent structural health ratings works, (ii) computation of the resistance strengths of structural

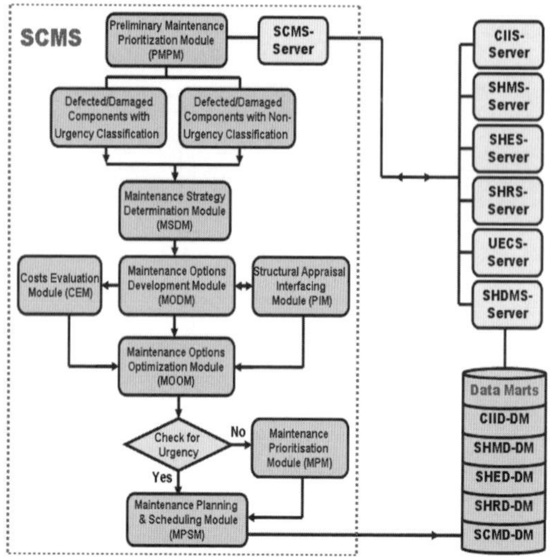

Figure 18. Architectural layout and flow diagram of SCMS.

components, (iii) criticality and vulnerability ratings of structural components, (iv) collection of structural condition ratings from CIIS, (v) synthetic rating and prioritization of structural component for inspection, and (vi) automatic and semi-automatic updating and forwarding the structural health monitoring and evaluation indices to respective SHMS and SHES. The architectural layout and flow diagram of the SHRS is illustrated in Figure 17.

6.5 *Framework of SCMS*

The SCMS is deployed as a maintenance decision making system to carry out the works: (i) confirmation of defects or damages in problematic component/components as reported by SHES, SHMS and CIIS; (ii) identification of defect/damage causes and qualification of defect/damage severity; (iii) preliminary maintenance prioritization based on maintenance urgency; (iv) determination of maintenance strategy; (v) development of maintenance options; (vi) structural appraisal of maintenance options; (vii) evaluation of direct and indirect maintenance costs; (viii) optimization of maintenance options; (ix) maintenance prioritization for problematic components that need non-urgent maintenance/repair actions; and (x) maintenance planning and scheduling for execution. The architectural layout and flow diagram of the SCMS is given in Figure 18 in which arrowless lines indicate the affiliation of items, while arrow lines only represent the directions of data transmission.

6.6 *Framework of UECS*

The UECS is deployed as a data buffer for user-access of data and information and also as a security wall to protect the SHM&MMS from virus attacks induced by users. The functions of the UECS are: (i) to provide internal users accessing data and information from the

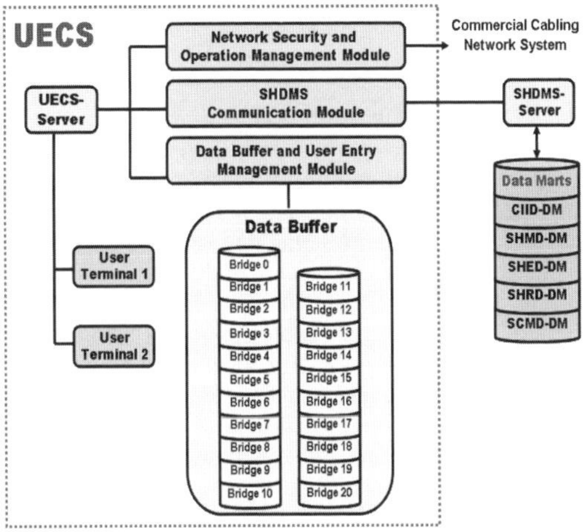

Figure 19. Architectural layout and flow diagram of UECS.

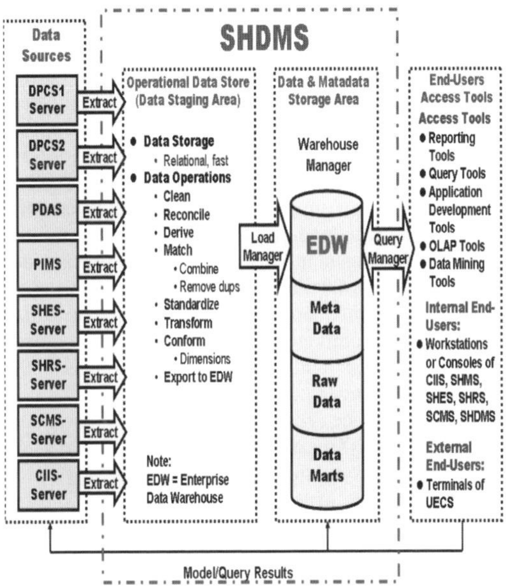

Figure 20. Architectural layout and flow diagram of SHDMS.

data buffer and communicate with SHDMS; (ii) to provide external communication with others maintenance authorities and universities via commercial network; and (iii) to protect the SHM&MMS from virus attacks induced by users. The architectural layout and flow diagram of the UECS is illustrated in Figure 19.

6.7 *Framework of SHDMS*

The SHDMS is deployed as a data and information management system to carry out five functions: (i) systematic storage and fast retrieval of data and information for automatic and semi-automatic processing, analysis and reporting of measured and analyzed data; (ii) data interfacing and analysis execution among SHMS, SHES, SHRS, CIIS, SCMS, UECS and SHDMS; (iii) formation of data marts for the storage and retrieval of highly summarized data and information such as bridge condition inspection reports, structural health monitoring reports, structural health evaluation reports, structural health rating reports, structural components maintenance reports, etc; (iv) multi-dimensional views of measured data and analyzed results, and (v) provision of data mining tools for extraction of additional information from measured data and analyzed results. The architectural layout and flow diagram of SHDMS are shown in Figure 20.

7 IMPLEMENTATION OF SHM&MMS

Upon completion of the system design of SHM&MMS, there remains five aspects to guarantee the success of implementing the SHM&MMS. These five aspects, from the operator's point of view, are: (i) performance specification of SHM&MMS, (ii) tender assessment, (iii) system installation supervision, (iv) deployment of staff for operation and maintenance, and (v) training for operation and maintenance.

8 DISCUSSION OF STRUCTURAL HEALTH MONITORING SYSTEMS IN HONG KONG

The limitations in data retrieval, processing, analysis and reporting that occur in the Wind and Structural Health Monitoring Systems (WASHMS) of Tsing Ma Bridge (TMB), Kap Shui Mun Bridge (KSMB), Ting Kau Bridge (TKB), and Hongkong-Shenzhen Western Corridor (SWC-HK) are discussed. Standardization and enhancements of such works in Stonecutters Bridge (SCB) are also introduced and graphical examples of improvements of wind and traffic data processing and analyzed results are illustrated in Figures 21, 22 and 23.

As the WASHMS in TMB, KSMB, TKB and SWC-HK include solely SHMS and SHES. In SCB, the WASHMS is upgraded to include two more system of SHRS and SHDMS. Such improvement is made on the WASHMS itself and does not take into account of facilitating inspection and maintenance works, and the communication between structural health monitoring team and inspection and maintenance team is by quarterly monitoring reports and monthly meetings.

In HKLR, the WASHMS is therefore further modified to have interfacing with inspection and maintenance directly and three more systems of CIIS, SCMS and UECS are introduced. Such an introduction provides the inspection and maintenance team with not only the first hand information regarding the current structural performance and/or condition of the bridges, but also the software tools and databases regarding inspection and maintenance. The former provision will minimize the frequency of inspection activities; whereas the latter provision will improve the effectiveness of preventive maintenance under normal

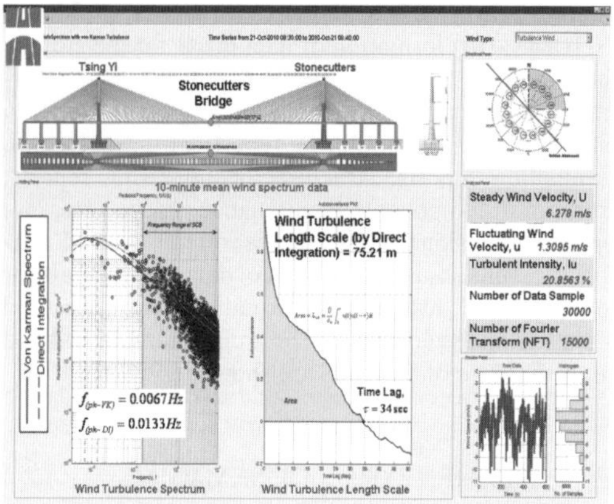

Figure 21. Wind parameters processing and analysis – Stonecutters Bridge.

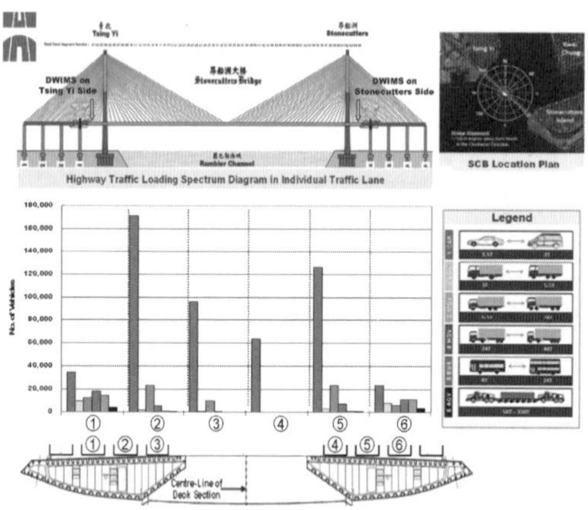

Figure 22. Highway traffic composition analysis at each traffic lane – Stonecutters Bridge (1-month data).

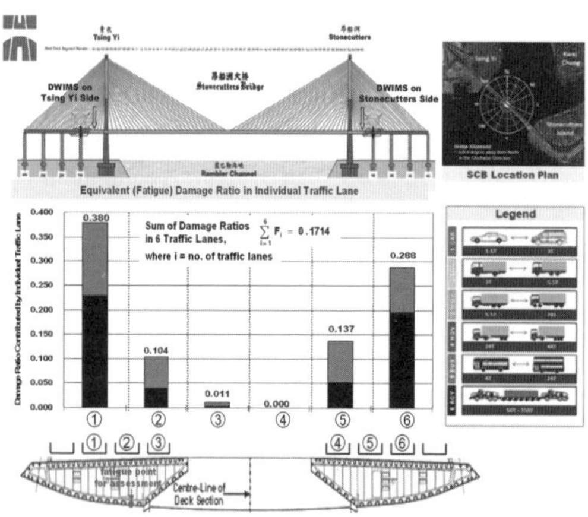

Figure 23. Equivalent standard fatigue vehicle loading spectrum at each traffic lane with reference to the deck-trough section under consideration – Stonecutters Bridge.

operation condition and corrective maintenance under extreme events.

9 CONCLUSIONS

Regarding the future applications and developments of SHM&MMS, the conclusions are drawn in three aspects, i.e. applications, maintenance and references.

In the application aspect, under the philosophy of condition-based maintenance, SHM&MMS could be applied to not only cable-supported bridges and marine viaduct bridges, but also to other types of highway infrastructures such as submerged tunnels, bored-tunnels, pavement structures, etc., when and where the cost of preventive maintenance is high and the effectiveness is uncertain.

In the maintenance aspect, for the purpose of improvement and/or enhancement of maintenance management with references to cost and effectiveness, SHM&MM should be devised to: (i) monitor structural/durability health conditions under the performance thresholds at serviceability limit condition (SLS); (ii) evaluate structural/durability safety when the SLS thresholds are exceeded; (iii) rate the inspection, monitoring and evaluation results basing on codified/designated criteria for inspection prioritization of structural components; (iv) identify and quantify problematic components for existence, causes and extend of defects and/or damages; (v) determine the required maintenance strategy and develop relevant maintenance options/methods; (vi) optimize the developed maintenance options/methods, (vii) prioritize the structural components (non-urgent type) required for maintenance; and (viii) plan and schedule the corresponding maintenance activities.

In the reference aspect, the SHM&MMS should be developed as: (i) an automatic system for provision of data and information in the operation, inspection and maintenance of highway infrastructures, and (ii) a database system for updating/calibrating relevant design codes, manuals and standards used in the design and construction of new highway infrastructures and in the maintenance and rehabilitation of old or existing highway infrastructures.

REFERENCES

Cadick, J. and Traugott, G. 2009, *Condition Based Maintenance – A White Paper Review of CBM Analysis Techniques*, CC Technical Bulletin – 017a, Cadick Corporation.

Connolly, T. M. and Begg, C.E. 2010, *Database Systems – A Practical Approach to Design, Implementation, and Management,* Pearson Education International, fifth edition, 2010.

Highways Department 1993a, *Wind and Structural Health Monitoring System*, Particular Specification for Electrical and Mechanical Services in Lantau Fixed Crossing, Highway Contract No. HY/1993/09, The Government of Hong Kong.

Highways Department 1993b, *Wind and Structural Health Monitoring System*, Particular Specification for the

Construction of Ting Kau Bridge and Approach Viaducts, Highway Contract No. HY/1993/38, The Government of Hong Kong

Highways Department 1997, Wind and Structural Health Monitoring System for Lantau Fixing Crossing and Ting Kau Bridge, Consultancy Agreement No. CE 75/94) – a consultancy agreement between Highways Department, The Government of Hong Kong and Flint & Neil Partnership, U.K.

Highways Department 2002a, *Wind and Structural Health Monitoring System*, Appendix W of the Particular Specification for the Construction of Hongkong-Shenzhen Western Corridor, Highway Contract No. HY/2002/21, The Government of the Hong Kong Special Administrative Region

Highways Department 2002b, *Wind and Structural Health Monitoring System*, Section 33 of the Particular Specification for the Construction of Stonecutters Bridge, Highway Contract No. HY/2002/26, The Government of the Hong Kong Special Administrative Region.

Highways Department 2011, *Performance Specification of Structural Health Monitoring and Maintenance Management System for Hong Kong Link Road*, Annex 4L1 of the Employer's Requirements for the Construction of Hongkong-Zhuhai-Macao Bridge, Highway Contract No. HY/2011/09 (HZMB- HKLR – Section between HKSAR Boundary and Scenic Hill.

Hoffer, J. A., Prescott, M. B. and McFadden, F. R. 2005, Modern Database Management, Pearson Education International, seventh edition, 2005.

Jarrell, Don 2000, *An information architecture for coping with aging infrastructure,* Proceedings of SPIE, Vol. 3995 (2000), 524–536pp.

Liu, T.T., Xu, Y.L., Zhang, W.S., Wong, K.Y., Zhou, H.J. and Han, K.W.Y. 2009, *Buffeting-induced stresses in a long suspension bridge: structural health monitoring oriented stress analysis*, An International Journal of Wind & Structures, Vol. 12, No. 6 (2009), 479–504.

Wai, C.S., Hui, M.C.H. and Wong, K.Y. 2007, *Structural Health Monitoring of Cable-Supported Bridges in Hong Kong*, Proceeding of Sino-US-Japan Forum on "Bridge Inspection & Safety Evaluation held by Chinese Highway Institutes, in Gui-Lian, China, 11–13 December 2007

Wong, K.Y. 2004, *Instrumentation and Health Monitoring of Cable-Supported Bridges*, Structural Control and Health Monitoring 2004, STC-033, Journal of Structural Control.

Wong, K.Y. 2006a, *Highway Traffic Loads Monitoring in Stonecutters Bridge,* Proceedings of International Conference on Bridge Engineering, held in Hong Kong, during 1–3 November 2006, organized by Hong Kong Institution of Engineers.

Wong, K.Y. 2006b, *Criticality and Vulnerability Analyses of Tsing Ma Bridge*, Proceedings of International Conference on Bridge Engineering, held in Hong Kong, during 1–3 November 2006, organized by Hong Kong Institution of Engineers.

Wong, K.Y. 2007, *Design of a structural health monitoring system for long-span bridges*, Structure and Infrastructure Engineering, Vol. 3, No. 2, June 2007, 169–185, Taylor & Francis Group.

Wong, K.Y., Chan, K.W.Y. and Chong, K.W. 2008, *Monitoring of Wind-Effects in Tsing Ma Bridge,* the Proceedings of 4th International Conference on Advances in Wind and Structures (AWAS'08), in Jeju, Korea, 29–31 May 2008.

Wong, K.Y. 2011, *Design of Wind and Structural Health Monitoring System for Stonecutters Bridge, an invited lecture,* Proceedings of the 5th Cross-Strait Conference on Structural and Geotechnical Engineering, 13–15 July 2011, Hong Kong, China, pp 243–289.

MINI-SYMPOSIA

Field tests for bridge assessment

Organizers: A. Miyamoto & I. Hakola

Bridge Maintenance, Safety, Management, Resilience and Sustainability – Biondini & Frangopol (Eds)
© 2012 Taylor & Francis Group, London, ISBN 978-0-415-62124-3

Bridge testing, monitoring and condition assessment in Finland

I. Hakola, P. Hradil & M. Halonen
VTT, Technical Research Centre of Finland, Espoo, Finland

ABSTRACT

The paper gives an overview of Finnish bridges, their conditions and methods to estimate the condition using loading tests and monitoring. The measured data from bridge loading tests and afterward from monitored data have been transferred to laboratory for detail analyzing. The analysis also includes Finite Element model about the bridge, which have been used to simulate the loading tests and also to calculate possible failure modes for the bridge. By analyzing methods heavy vehicles crossing the bridge are possible to find out and calculate the speeds and weights and also define the life age of welded connections. The bridge loading tests are introduced as a part of whole bridge management system and also part of whole infra management system.

According to The Finnish Traffic Agency there are about 14500 bridges in Finland. Most of the bridges have been built in the years of 1960 to 1990 and therefore quite many of them are at the age when the rehabilitation is approaching or is necessary to perform.

In Finland about 100 different kind bridges have been tested and analyzed during the last 30 years and e.g. in 2011 five bridges have been tested using heavy trucks, mobile cranes and heavy carriages. The bridges have been analyzed by FE (Finite Element) models and most of the tested bridges include monitoring devices and sensors.

The article describes mainly Kirjalansalmi suspension bridge and its field tests, monitoring and analyzing of condition. The bridge has been tested many times and the last two loadings were done before the global renovation to research the bridge deck composite behavior.

The monitoring system by Halonen (2006) has been put together with National Instruments hardware and software. Data from the strain gages and displacement transducers is measured with NI CompactDAQ, a modular data acquisition system for USB. It is equipped with two 4 channel strain gage modules one 16 channel analog voltage module. The software has been installed in a Windows XP based computer and it has been done with LabVIEW graphical programming environment.

Since the bridge is extensively used nowadays, the additional numerical simulations were carried out to study the structure behavior in case of accidental failure of one or several load-bearing elements.

Two methods for evaluating mass, speed and direction of passing vehicles (Bridge weighting-in-motion BWIM) were developed in order to provide approximate information about the traffic. Both methods are using the strains recorded at 100 Hz frequency for 30 seconds after the measurement was automatically triggered by elevated monitoring values.

At the first step, the speed and direction of vehicle is evaluated by identifying exact time, when maximum and minimum stress measured at the main girder is reached (e.g. the maximum stress is always obtained when the vehicle passes over the first quarter of the bridge span.

The second method that is not yet fully implemented in automatic signal evaluation is the optimization of estimated parameters from the previous method using measured stress on multiple places at the same time. A curve-fitting approach to achieve the smallest error in all monitoring points at the same time is used when the measured signal is compared to pre-calculated finite element models.

REFERENCES

European Committee for Standardization *Eurocode 3. Design of steel structures*. Brussels, Belgium: 2006.

Halonen, M. (2006). *Langaton siltojenmonitorointijärjestelmä,* Master's thesis at Helsinki University of Tehchnology.

Kirjalasalmen riippusillan koekuormitus, VTT report no. RTE4656/05, 22.12.2005.

Tirkkonen, T. FEM-laskennan soveltaminen riippusiltojen kantavuustarkasteluun, Lisensiattityö, TKK, 17.12.2001.

Bridge Maintenance, Safety, Management, Resilience and Sustainability – Biondini & Frangopol (Eds)

Field tests for remaining life and load carrying capacity assessment of concrete bridges

A. Miyamoto
Yamaguchi University, Ube, Japan

ABSTRACT

It is an important problem for maintenance and rehabilitation of existing bridges to develop a method of safety evaluation such as remaining life and load carrying capacity of bridges. This paper describes a method of safety evaluation of concrete bridges in service and the verification of the evaluated results by field tests. The field tests to evaluate structural safety were performed as static loading test by test trucks for application to the system identification(SI) method, and also as dynamic loading test under forced vibration caused by falling mass for application to the model analysis.

Both of the safety factors for flexural and shear failures and the change of dynamic behavior were evaluated form the field tests. These results were verified through the ultimate load test carried out in field on the reinforced concrete main girders isolated by cutting off from the bridge system. Finally, the remaining life of the bridge was predicted by application of fuzzy set theory which deal with the subjective information of bridge engineers. The fuzzy mapping which was determined based on questionnaire results performed on more than 20 experts was introduced for remaining life prediction of the existing bridges. A few concrete bridges on which on field data have been collected are analyzed to demonstrate the applicability of this method. Through the application to the cracked reinforced concrete bridge girders, reasonable results were obtained by tests.

The major part of bridge diagnosis which is the kernel of the systematization for bridge maintenance is to develop a method of safety evaluation on items such as remaining life and load carrying capacity. The main conclusions obtained in this study can be summarized as follows:

1 A few examples of the field testing and the measuring procedures, the evaluation methods of both the safety factors, including the safety index and the probability of failure, for flexural and shear failure and the verification concept of the evaluated results by field tests were performed on six reinforced concrete bridges.
2 The prediction of remaining life of concrete bridges in service, which is a major problem faced in the process of bridge maintenance, can be quantitatively evaluated based on the evaluated results for the judgment factors. The relations among the factors are obtained through the fuzzy relations in the questionnaire results performed on experts.

REFERENCES

Miyamoto, A. and Fujii, M., Safety Evaluation and Verification by Field Testing of Concrete Bridges. Proc. of ASCE Structures Congress'89, 1989.

Miyamoto, A. Morikawa H. and Yamamoto M., Sensitivity Analysis of Dynamic Behavior in Existing Concrete Bridges and Its Application to Damage Detection. Proc. of JSCE, No. 442, 1992.

Miyamoto, A. and Isoda, S., Sensitivity Analysis of Mechanical Behaviors for Bridge Damage Assessment, The 2011 World Congress on Advances in Structural Engineering and Mechanics (ASEM 2011 Plus), 2011.

Morikawa, H. and Miyamoto, A., Structural Safety Evaluation and Remaining Life Prediction of Existing Concrete Bridges. Proc. of the 17th Conference on Our World in Concrete & Structures, 1992.

Natke, H.G. and Yao, J.T.P. (ed.), Structural Safety Evaluation Based on System Identification Approaches. Proc. of the Workshop at Lambrecht/Pfalz, Friedr. Vieweg & Sohn, 1988.

Bridge Maintenance, Safety, Management, Resilience and Sustainability – Biondini & Frangopol (Eds)
© 2012 Taylor & Francis Group, London, ISBN 978-0-415-62124-3

Development of a damage detection system for expansion joints of highway bridges applying acoustic method

Y. Nishikawa & K. Taniguchi
Kokusai Shinon Keiso Company Limited, Osaka, Japan

L.H. Ichinose
Japan Industrial Testing Corporation, Osaka, Japan

S. Tsukamoto & T. Yamagami
Hanshin Expressway Engineering Company Limited, Osaka, Japan

ABSTRACT

Damages in expansion joints of expressways not only cause serious problems that can affect the driving performance, but can also be a source for unwanted noises and a trigger for accidents that can be of serious consequences. Thus, the early detection of damaged expansion joints in expressways is a matter of great importance.

It has been reported that experienced road inspectors, while driving during ordinary patrol routines, can detect abnormality in expansion joints by only hearing the sound emitted by the joint when the vehicle travels on it. However, the number of inspectors having such skills is decreasing and transferring their skills to younger generations is becoming more and more difficult. Therefore, the present study proposes a method that can emulate the skills of an experienced inspector through a computational detection system.

To develop the detection system, sound emitted by the expansion joints when a vehicle passes over them were measured through instruments installed inside and outside the vehicle. In the first stage, as a preliminary survey, sounds of joints already rated as damaged by road inspectors were measured. As a result, it was found that there are characteristic frequency bands, 200–500 Hz or 500–800 Hz, in the collected data. These measurements were carried out without road inspectors.

In the second stage, measurements we carried in a vehicle in movement, in cooperation with an experienced road inspector and confirmed the results of the preliminary surveys. In addition, it was found that partial overall levels outside the car also had higher levels in the band range of 1,000–5,000 Hz.

In the third stage, applying the characteristics found in the previous surveys, several methods, such as partial overall (POA) of sounds inside the car, product of the two POA levels which obtained from sound inside the car and sound outside car and 1/3 octave analysis, were tried. And the detection results of the experienced inspector were compared to that of the results for the different methods. The comparison results are shown in Table 1.

Table 1. Comparison between the detection results from the analyses and the inspector.

Method	Inspector	POA	POA product	1/3 octave
Detection rate	65% (28/43)	27% (11/41)	37% (15/41)	51% (21/41)
Rate to inspector	100%	42%	57%	78%

From the above results, it can be concluded that some basic knowledge to develop a detection system could be obtained. It is certain that each detection rate of methods is lower than inspector's detection rate. However, this problem can be solved by increasing number of measurement times. The purpose of the present study is to detect damaged joints without traffic regulation using moving vehicle. Therefore, detecting damaged joints at one measurement time (sweep) is not necessarily required. Thus, the method can be considered effective if a damaged joint can be detected during the repeated patrol routings.

REFERENCES

Hanshin Expressway Co. Ltd. 2005. *Guidelines for inspection of road structures*. Osaka: Hanshin Expressway Co. Ltd. (in Japanese).

Ichinose, L.H., Matsumura, et al. 2010. Proposal of an acoustic method to evaluate vertical misalignment between plates of steel expansion joints of finger type. *Proceeding of JSSC* 18 (69): 1–8 (in Japanese).

Randall, R.B. 1977. *Application of B & K Equipment to FREQUENCY ANALYSIS*. Naerum: Brüel & Kjaer.

Sopon, R., Abe, M., Fujino, Y., Miyashita, T. 2003. Quantitative health monitoring of bolted joints using piezoceramics acuator-sensor. *Proc. of SPIE Smart Structures and Materials* 5057: 126–137.

Tsukamoto, S., Yamagami, T., et al. 2011. Development of a system to detect abnormalities in expansion joints applying acoustic method. *Proceedings of the 66th Annual Meeting of JSCE*: 1097–1098 (in Japanese).

Bridge Maintenance, Safety, Management, Resilience and Sustainability – Biondini & Frangopol (Eds)
© 2012 Taylor & Francis Group, London, ISBN 978-0-415-62124-3

Smart system of bridge strain monitoring during construction and service

P. Olaszek
Road and Bridge Research Institute, Warsaw, Poland

ABSTRACT

The popularization of bridge structure monitoring is connected with a rapid development of electronics, optics and cordless communication and information processing technologies. On account of the costs the monitoring systems created nowadays are usually based on modular elements and cordless communication. On the basis of measurement and monitoring equipment market analysis it was concluded that well-known producers actually do not offer any ready and relatively cheap small measurement sets.

The only system of mini recorders intended for the analysis of bridge structure service that is known to the author was presented in a paper by Shenton (Shenton et al. 2010) but it was a single execution of the system which had not been implemented into serial production.

For this reason a system intended for bridge structure monitoring during construction and service was worked out. The system is designed to monitor bridge structures for which it is sufficient to measure strain/stress and acceleration in one or at the most several points. The developed conception assumes the creation of strain/stress, acceleration and temperature recorders which would work independently.

A single recorder enables the strain/stress measurement to be taken at two points (with the application of two full Wheatstone bridge strain gauges), acceleration measurement in one of three directions and measurement of temperature (Fig. 1). The recorders should be characterized first of all by high measurement reliability and a relatively low price. The first feature is intended to be worked out through the duplication of strain/stress measurement methods. Apart from an electric resistance wire strain gauge the recorder contains a photo-elastic element. The role it has to serve is to control and verify the measurement of constant component of strain. A special portable polariscope has been constructed for photo-elastic readings.

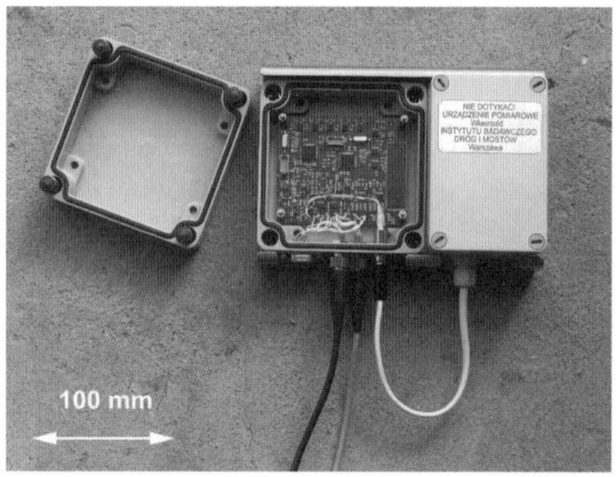

Figure 1. View of the recorder: on the top from the left – the main recorder lid, the recorder and the power supply module.

System service tests were carried out during construction and exploitation of bridge structures. The first instance concerns monitoring of structure creation, from system installation in steel structure factory through transport of particular elements, assembly of structure on a construction site and concreting of a deck to a disassembly of shuttering. The second example focuses on the monitoring of bridge construction via the process of launching steel girders with a composite reinforced concrete deck slab and during bridge load testing and service.

REFERENCE

Shenton III H.W., Connor K., Chajes M.J., Rakowski M., Brookes B. 2010. Analysis of in-service data collected during biennial inspections on typical bridges. *The Fifth International Conference on Bridge Maintenance, Safety, Management and Life-Cycle Optimization (IABMAS)*. Philadelphia 2010: 521–526.

Bridge Maintenance, Safety, Management, Resilience and Sustainability – Biondini & Frangopol (Eds)
© 2012 Taylor & Francis Group, London, ISBN 978-0-415-62124-3

Investigation of displacements of road bridges under test loads using radar interferometry – Case study

T. Owerko, Ł. Ortyl, R. Kocierz & P. Kuras
AGH University of Science and Technology, Kraków, Poland

M. Salamak
Silesian University of Technology, Gliwice, Poland

ABSTRACT

Reliable assessment of the bridge condition is a complex process which requires extensive knowledge of any tested structure, e.g. the response under loads (Lazinski & Salamak 2011). Techniques used to observe the response of structure under load are very diverse (Ko & Ni 2005). According to the type of supporting structure the following values may be measured: beam deflection, bridge support subsidence, pylon deflection, horizontal displacement. Dynamic measurements are performed by determination time series of vertical displacements and accelerations of the bridge girders.

Disadvantages of techniques used to above mentioned measurements are: necessity of direct access to the structure and only discrete information on the current position of the object. The paper presents the results of comparing the measurements carried out during test loads of a bridge, performed using conventional measurement methods and ground-based radar interferometer, which enables non-contact, direct and multi-point measurement of structure displacements. It enables static and dynamic tests of structures such as bridges (Gentile 2010).

Structure, for which the static and dynamic load tests were carried out, is the cable-stayed bridge over the Skawa River in Zembrzyce, Poland (Fig. 1). To perform measurements during the load tests, the following methods were applied: radar interferometry (IBIS-S system), satellite positioning (GPS/GNSS receivers) and levelling (precise digital level). Static loads were carried out according to seven schemes. Results of an exemplary load scheme are presented in Figure 2. During dynamic measurements IBIS-S system was working with 50 Hz sampling frequency. An exemplary dynamic response of a span during the truck jumping over the threshold is shown in Figure 3.

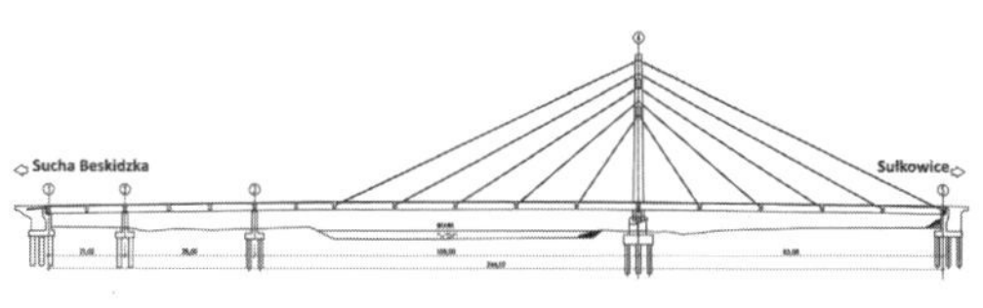

Figure 1. Longitudinal section of the structure.

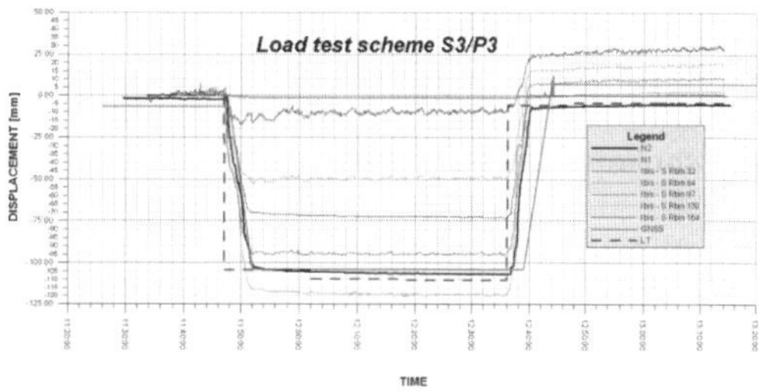

Figure 2. Static response of a span.

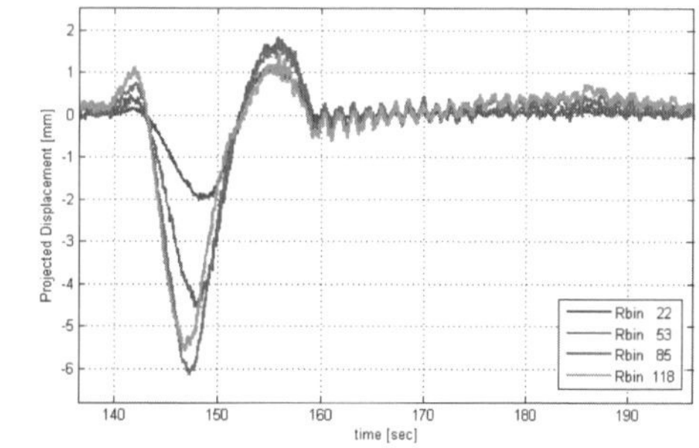

Figure 3. Dynamic response of a span.

The natural frequencies calculated on the basis of the theoretical modal analysis and observation performed by IBIS-S system and induction sensors have been summarized in the paper.

REFERENCES

Gentile, C. 2010. Application of Radar Technology to Deflection Measurement and Dynamic Testing of Bridges. In Guy Kouemou (ed.), Radar Technology: 141–162. Vukovar: InTech.

Ko, J.M. & Ni, Y.Q., 2005. Technology developments in structural health monitoring of large-scale bridges. *Engineering Structures* 27(12): 1715–1725.

Lazinski, P. & Salamak, M., 2011. Identification of computational models in load carrying structures of concrete bridges on the basis of making load tests. In *Proceedings of 7th Central European Congress on Concrete Engineering CCC2011, Baltonfured, 22–23 September 2011.*

Bridge Maintenance, Safety, Management, Resilience and Sustainability – Biondini & Frangopol (Eds)
© 2012 Taylor & Francis Group, London, ISBN 978-0-415-62124-3

Rehabilitation prioritization for Canadian Alaska Highway structures

P. Phillips
Delcan Corporation, Calgary, Alberta, Canada

R. El-Hacha
University of Calgary, Calgary, Alberta, Canada

ABSTRACT

The Alaska Highway serves as the only land route between mainland USA and the state of Alaska. This remote Northern highway passes through Canada's province of British Columbia and their Yukon territory, before entering into Alaska. The very first Alaska Highway bridge and culvert structures were built during WWII. Since then, many of these structures have been replaced or rehabilitated due to structural, functional, or condition type reasons. Some of the original structures, however, still remain. The aging infrastructure along this route is of great concern as it impacts the integrity of this sole land connection.

This paper outlines how bridge and culvert structure maintenance, repairs, rehabilitations, and/or strengthenings can be prioritized based on existing condition rating data. The sixty-three structures located along the British Columbian section of the Alaska Highway are used as examples. Results from four detailed visual condition evaluation sessions of these structures, completed over the last decade, are provided and compared and contrasted with time. Where deficiencies, structural or otherwise, were found in the field, strategies for rectifying such issues were recommended to the Owner of these structures. Then, how these recommendations can be prioritized, based on severities of the existing conditions, risk to public safety, cost, and possibility of total structure replacement verses repairs to deficient structures, is described.

Proof that an Owner well maintains their structure assets or not can be provided by plotting condition ratings of the structures as compared to the ages of the structures at the times of their inspections and observing if the overall conditions of the structures improve with time. Following this methodology it is shown that the sixty-three British Columbian Alaska Highway structures are well-maintained by their Owner.

REFERENCES

American Concrete Institute Committee 364. 2007. *Guide for Evaluation of Concrete Structures Before Rehabilitation*. Farmington Hills, Michigan, USA: ACI – 364.1R-07.

Canadian Standards Association. 2006. *Canadian Highway Bridge Design Code and Commentary*. Mississauga, Ontario, Canada: CAN/CSA S6-06.

Ministry of Transportation of Ontario. 2008. *Ontario Structure Inspection Manual*. 301 – St. Paul Street, St. Catharines, Ontario, Canada: MTO.

Phillips, P. 2011. *Comprehensive Bridge and Culvert Condition Evaluations and Rehabilitations, Alaska Highway, km 233 to km 956, British Columbia; M.Eng. Thesis*. Calgary, Alberta, Canada: University of Calgary.

Public Works and Government Services Canada. 2008. *Bridge Inspection Manual*. 11 Laurier Street, PDP III, Gatineau, Quebec, Canada: PWGSC.

Tracy, R. & Fling, R. 1989. Rehabilitation Strategies. *American Concrete Institute Compilation #10: Repair and Rehabilitation of Concrete Structures*: 3 to 7.

US Army Corps of Engineers. 1995. *Evaluation and Repair of Concrete Structures, Engineering Manual: 1110-2-2002*. Washington, DC, USA, 20314-1000: Department of the Army.

Vaysburd, A. & Emmons, P. 2007. Fitting the Pieces Together: A Complete System Philosophy for Concrete Repair. *Structural Engineer* (July): 30 to 33.

Bridge Maintenance, Safety, Management, Resilience and Sustainability – Biondini & Frangopol (Eds)
© 2012 Taylor & Francis Group, London, ISBN 978-0-415-62124-3

Bridge condition assessment for short and medium span bridges by vibration responses of city bus

A. Yabe
KOZO KEIKAKU Engineering Inc. Tokyo, Japan

A. Miyamoto
Yamaguchi University Yamaguchi, Japan

ABSTRACT

Bridges as one of the important infrastructures are suffering from many defects during their service lives. In our country, because there are a huge amount of existing short and medium span bridges in service, it is becoming one of major social concern how those bridges can be maintained in good condition during their whole lifetime. In this paper, a method of the condition assessment is proposed newly for existing medium & short span RC bridges based on monitoring data from public bus vibration. This paper describes the details of not only a prototype monitoring system using information technology and sensors which is capable of providing more accurate knowledge of bridge performance than traditional ways but also a few specific examples of bridge condition assessment by public bus vibration measured in operating on the bridge based on the data from the system. This paper also describes a sensitivity analysis for deteriorating bridges using the city bus acceleration response by using "substructure method" based on finite element model for verification of the above mentioned results of bridge performance. Main conclusions about foundational investigation for reasonable method of detecting sever damaged and sustainable bridge maintenance can be drawn as follows.

1 Because the vibration responses of a passenger vehicle such as city bus had a good liner relationship with the vibration responses of target bridges, it is possible to apply a practical monitoring system for bridge condition assessment.
2 The results of sensitivity analysis by using "substructure method" are shown that public bus vibration responses were useful for evaluation of target bridge performance.
3 The characteristic deflection value will be influenced by measurement conditions especially oncoming vehicles.

REFERENCES

Ayaho Miyamoto, Akito Yabe: Bridge Condition Assessment based on Vibration Responses of Passenger Vehicle, Journal of Physics(CS), Vol. 305(1), pp. 1–10, 2011.7.

Ishii, H., Fujino, Y., Mizuno, Y., and Kaito, K.: "Development of Train Intelligent Monitoring System (TIMS) using wheel vibration of in-service vehicle", Doboku Gakkai Ronbunshuu F, Vol. 64, No. 1, 44–61, February 2008.

Japan Ministry of Land, Infrastructure, Transport and Tourism, http://www.mlit.go.jp/road/sisaku/yobohozen/yobo2_3.pdf, 2011

Japan Society of Mechanical Engineering: Computer analysis in vibration engineering, Corona Publishing, 1987.

JSCE Subcommittee on Bridge Vibration Monitoring Committee of Structural Engineering, "Guidelines for Bridge Vibration Monitoring", 2000.9. (in Japanese)

Kawatani, M. and Shimada, R.: "Nonstationary random response and impact of girder bridges under moving vehicles", Proceedings of the Japan Society of Civil Engineers, No. 398/I-10, pp. 385–394, October 1988.

Kojima, T., Tsunashima, H., Matsumoto, A., and Ogata, M.: "Rail track defect detection by use of onboard measurement data", Transactions of the Japan Society of Mechanical Engineers (Series C), Vol. 72, No. 720, pp. 2447–2454, August 2006.

Lin C.W. and Yang Y.B. (2005) "Use of a passing vehicle to scan the fundamental bridge frequencies: an experimental verification. Engineering Structures, 27, 1865–1878.

Mizuno, Y., Fujino, Y., Kataoka, K., and Matsumoto, Y.: "Development of a mobile sensing unit and its prototype implementation", Tsinghua Science and Technology, Vol. 13, No. S1, pp. 223–227, October 2008.

Nakajima, A. and Tsumura, N.: "A study on simulation method for vehicle-bridge system", Journal of Structural Engineering, Vol. 38A, pp. 781–788, March 1992.

Sugiura, K., Oshima, Y., Yamaguchi, T., Kobayashi, Y., Okano, H., and Okashiro, S.: "Structural health assessment of road structures by use of multifunction inspection vehicle", Report on results of research and development of technology contributing to improvement of quality of road policy measures, No. 17–8, Committee on Advanced Road Technology, July 2008.

Yabe, A and Miyamoto, A. 2010. Bridge Maintenance, Safety, Management and Life-Cycle Optimization, CRC Press (A Balkema Book), IABMAS'2010, Pennsylvania, USA et al.

Bridge Maintenance, Safety, Management, Resilience and Sustainability – Biondini & Frangopol (Eds)
© 2012 Taylor & Francis Group, London, ISBN 978-0-415-62124-3

Rule-type knowledge discovery from field inspection data for highway bridges based on advanced data mining technique

H. Yagi
Ryo-sen engineer's Ltd. Hiroshima, Japan

A. Miyamoto
Yamaguchi University, Ube, Japan

N. Tsukamoto
Hanshin Expressway Engineering Company Limited, Osaka, Japan

ABSTRACT

In this study, an attempt is made to acquire rule-type knowledge from large volumes of data stored in a field inspection database on bridges in expressway networks by a data mining method for efficient bridge maintenance. As a data mining method, the decision table reduction method based on the concept of the rough set theory[1),2)] that enables the extraction of rule-type knowledge from a database of examples and is applicable to such problems as "discrimination" and "association". Reduction is made in a decision table that lists condition and decision attributes to extract rules. Then, data with the same condition attributes but different decision attributes is considered contradictory data and is removed when rules are extracted for decision table reduction. Actual field inspection data, however, sometimes contains numerous contradictory data, so most of the field inspection data, although collected in large quantities, becomes useless.

To cope with the above problem, contradictory data that is removed when extracting rules, may be minimized by adding a function for saving the large majority of contradictory data (an algorithm for rescuing contradictory data). In this study, a contradictory data rescuing algorithm was proposed that should be added to the rough set theory and its effectiveness was verified. Various studies were also made on the effects of the addition of the proposed algorithm on the extraction of rules.

The results obtained from this study can be summarized as follows:

(i) Rescuing deleted contradictory data was made possible by applying an algorithm for rescuing contradictory data to rough sets.
(ii) In the verification using data on actual bridges, contradictory data was rescued. Thus, the effectiveness of an algorithm for rescuing contradictory data was verified.
(iii) Lowering the percentage of decision attribute for rule extraction enables the rescue of contradictory data. Reducing the percentage to an extremely low level, however, means the extraction of rules based on ambiguous data and is likely to output false results.
(iv) When setting the percentage of decision attribute for rule extraction, the contradictory data contained in the data to be handled should be analyzed to verify the contents of contradictory data and the percentages of majority and minority contradictory data. Establishing an appropriate method for using the percentage for rule extraction is a future task.

REFERENCES

Pawlak, Z. (1982). "Rough sets", *International Journal of Computer and Information Sciences*, 11(5), 541–356.

Yokomori, T. and Kobayashi, S. (1994). "Rough Sets and Decision Making". *Mathematical Sciences*, No. 375, 76–83.

Bridge Maintenance, Safety, Management, Resilience and Sustainability – Biondini & Frangopol (Eds)

Application of electromagnetic testing to orthotropic steel deck

T. Yamada, A. Shiraishi & M. Okuno
Nihon Densokuki Co., Ltd., Fukuoka, Japan

H. Sugiyama & N. Kanjo
Hanshin Expressway Company Limited, Osaka, Japan

S. Tsukamoto & T. Yamagami
Hanshin Expressway Engineering Company Limited, Osaka, Japan

ABSTRACT

Road bridges with orthotropic steel decks are widely used for urban highways in coastal areas, because of lightweight and short construction time. In recent years, with increase in traffic, fatigue cracks at weld between deck plate and U-rib in the orthotropic steel deck have been reported. These damages may cause danger to the traffic on the bridges. Therefore, detection of the cracks in early stage is highly required. The early detection is also useful for saving on repair costs.

Based on these backgrounds, it is significant to construct an inspection method for fatigue cracks in the deck plate. Thus we have already proposed a 3-stage inspection method in cooperation with Hanshin Expressway group (Sugiyama et al. 2010). In this method, eddy current testing occupies an important place.

In this paper, we describe the eddy current testing technique in the second stage of the proposed method. This technique is based on the principle of electromagnetic induction and suitable for this application, because the deck plate is a conductor and simple in shape, and it is possible to make an inspection without physical contact between a sensor and the deck plate surface. In order to demonstrate the effectiveness of the scanning technique from pavement surface, not only experiments for artificial cracks but also field tests for actual through cracks are carried out.

Furthermore, a scanning system in Figure 1 is newly developed to improve inspection efficiency. In this system, zigzag movement of the sensors to scan a crack is realized mechanically. The system can measure a single lane along 80 m back and forth for a 4-hour shift. The 4 sensors measure 4 weld lines along a bridge axis, and each sensor covers one specific weld line. The single lane is about 3.5 m wide and contains 8 weld lines to be inspected. These 8 weld lines are located near loading points in the orthotropic steel deck. When the system is applied through the asphalt pavement of 100 mm thick, the minimum surface-breaking crack that can be detected is 100 mm long.

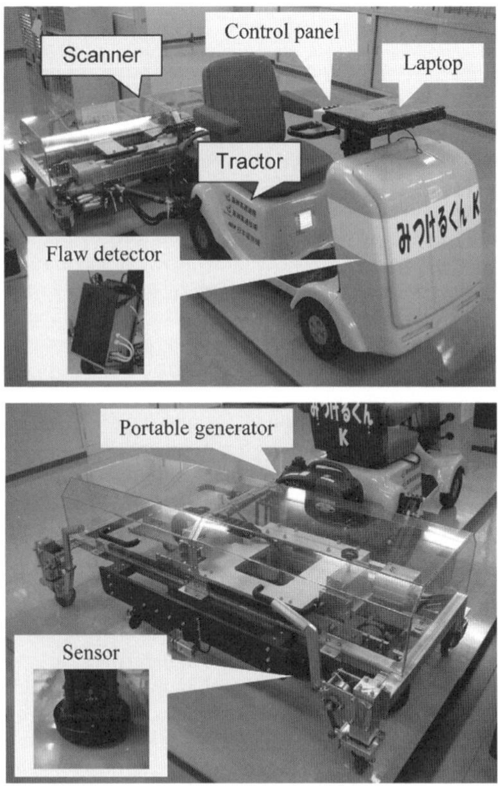

Figure 1. Newly-developed scanning system.

REFERENCES

Shull, P.J. (ed.) 2002. *Nondestructive evaluation: theory, techniques, and applications.* New York: CRC Press.

Sugiyama, H., Tabata, A., Nakajima, H., Yamagami, T., Tsukamoto, S., Uchima, M., Okuno, M. & Kohno, Y. 2010. Development of a compound inspection method to detect fatigue damages on orthotropic steel deck. *Proc. of the Fifth International Conference on Bridge Maintenance, Safety and Management (IABMAS2010).*

SmartEN ITN – Smart management for sustainable built environment including bridges and structural systems

Organizers: T. Onofriou & R. Helmerich

Bridge Maintenance, Safety, Management, Resilience and Sustainability – Biondini & Frangopol (Eds)
 ISBN 978-0-415-62124-3

Power-efficient wireless sensor reachback for SHM

D. Ampeliotis, N. Bogdanović & K. Berberidis
Department of Computer Engineering and Informatics, University of Patras & C.T.I RU-8, Rio – Patras, Greece

F. Casciati
Department of Structural Mechanics, University of Pavia, Italy

R. Al-Saleh
Architecture and Building Systems Engineering Department, AlZaytoonah University of Jordan, Jordan

ABSTRACT

Wireless Sensor Networks (WSN) offer tremendous promise for accurate and continuous Structural Health Monitoring (SHM) using a dense array of inexpensive sensors and possess many advantages over conventional wired systems, particularly for large civil infrastructure. The process of collecting the measurements acquired by a sensor network into a central sink node, constitutes one of the main challenges in this area of research and is often referred to as the sensor reachback problem.

In general, there are several difficulties in the sensor reachback problem in an SHM system. Firstly, the amount of data generated by the sensor nodes is immense, owing to the fact that structural monitoring applications need to transfer relatively large amounts of dynamic response measurement data with sampling frequencies as high as 1000 Hz. Moreover, the assumption that all sensors have direct, line-of-sight link to the sink does not hold in the case of these structures.

Regarding the former problem, the fact that sensor readings from nearby sensors in an SHM system may be correlated may be exploited so as to reduce the amount of information required to be transmitted to the sink node. For instance, the data collected by the sensors on each span of a bridge are correlated since they are measuring the vibration of the same part of the physical structure. Regarding the latter one, recent advances in the field of cooperative communications hold the promise to alleviate the problem.

In this work, we study a communication protocol that exploits the spatial and temporal correlations among the measurements of the sensor nodes by employing an adaptive filter at each node that predicts the current measurement using past measurements acquired from its neighbors. The sink node, keeps an exact replica of the filters that run on each sensor node. When a sensor node records a new measurement, it computes the prediction error. If the prediction error is small enough, the node sends the output of its filter to its neighbors, so that they can use this value as input for the prediction filters they operate. In the opposite case, when the prediction error is not that small, the node updates its filter coefficients and sends its actual measurement to its neighbors. In this case, the neighbors will transmit that measurement to the sink node in a cooperative fashion.

The new technique has been tested extensively via real acceleration measurements from the Canton Tower in China and it turns out that it may offer considerable savings in transmitted energy. Furthermore, the appropriate selection of cooperating sensor nodes is of great importance, as depicted in Figure 1.

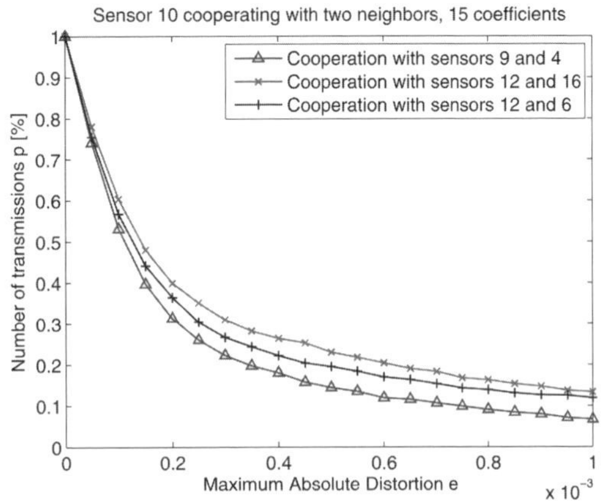

Figure 1. Sensor 10 cooperating with two neighbors.

REFERENCES

Hong, Y.-W., W.-J. Huang, F.-H. Chiu, & C.-C. Kuo (2007, May). Cooperative communications in resource-constrained wireless networks. Signal Processing Magazine, IEEE 24(3), 47–57.

Nagayama, T., P. Moinzadeh, K. Mechitov, M. Ushita, N. Makihata, M. Ieiri, G. Agha, J. B. F. Spencer, F. Y., & J. Seo (2010). Reliable multi-hop communication for structural health monitoring.

Sayed, A. H. (2008). Adaptive Filters. Wiley-IEEE Press.

Bridge Maintenance, Safety, Management, Resilience and Sustainability – Biondini & Frangopol (Eds)
© 2012 Taylor & Francis Group, London, ISBN 978-0-415-62124-3

Structural diagnostic via compressive sensing

F. Casciati & L. Faravelli
University of Pavia, Italy

R. Al-Saleh
Al-Zaytoonah University of Jordan, Jordan

K. Hinc
formerly at the University of Pavia, Italy

ABSTRACT

Wireless sensor networks (WSN) are built of spatially distributed nodes which are connected to one or more sensors. Each sensor node typically consists of several parts that are connected to an energy source like a battery, which limits the power resources of a wireless sensor. The major energy demanding parts in a sensor network node are the processor and the transmitter. Typically, those nodes that work continuously consume huge amounts of power. This suggested reducing the quantity of the obtained data to reduce the required power, of course without losing any important data. It means that the data should be sent through sensors with minimal additional operations performed by the processor. This is called the Compressive Sensing which is an innovative technique that fulfills the previously mentioned requirements. The aim of this research is to evaluate the reliability of compressive sensing when applied on signals obtained in real life. For this purpose; two types of signals are used with two different reconstruction algorithms. The obtained results are discussed and some conclusion remarks are presented.

REFERENCES

Al-Saleh, R. 2010.Verification of Wind Pressure and Wind Induced Response of a Supertall Structure using a Long Term Structural Health Monitoring System, Ph. D. thesis, Graduate School of Civil Engineering, University of Pavia, Pavia, Italy.

Candes, E. J. & Romberg, J. 2005. L1-MAGIC: Recovery of Sparse Signals via Convex Programming.

Candes, E. J.& Wakin, M. B. 2008. An Introduction To Compressive Sampling. Signal Processing Magazine, IEEE, vol. 25, no. 2, pp. 21–30.

Chrabiwala, Z. & Kim, Y. & Zahedi, S. & Friedman, J. 2009. Energy Efficient Sampling for Event Detection in Wireless Sensor Network, ISLPED'09; Proc. intern. symp. San Francisco, USA 19–21 August 2009.

Figueiredo, M.A.T. & Nowak, R.D. & Wright, S.J. 2007. Gradient Projection for Sparse Reconstruction: Application to Compressed Sensing and Other Inverse Problems. Selected Topics in Signal Processing, IEEE Journal of, vol. 1, no. 4, pp. 586–597.

Ni, Y. Q. & Xia, Y. & Liao, W. Y. & Ko, J. M. 2009. Technology innovation in developing the structural health monitoring system for Guangzhou New TV Tower. Structural Control and Health Monitoring, vol. 16, no. 1, pp. 73–98.

Romberg, J. 2008. Imagining via Compressive Sampling. Signal Processing Magazine, IEEE, vol. 25, no. 2, pp. 14–20.

Website 1: http://users.ece.gatech.edu/~justin/l1magic/

Website 2: http://www.lx.it.pt/~mtf/GPSR/

Yang, S. & Grela, M. 2010. Energy-Efficient Accelerometer Data Transfer for Human Body Movement Studies SUTC/UMC'2010; Proc. intern. symp., Newport Beach, USA, 7–9 June 2010. pp. 304–311.

Bridge Maintenance, Safety, Management, Resilience and Sustainability – Biondini & Frangopol (Eds)
© 2012 Taylor & Francis Group, London, ISBN 978-0-415-62124-3

Towards a SHM-based methodology for updating fatigue reliability of orthotropic steel decks

I.F. Alcover
COWI A/S, Kongens Lyngby, Denmark, and University of Surrey, Guildford, UK

J.E. Andersen
COWI A/S, Kongens Lyngby, Denmark

M.K. Chryssanthopoulos & M.I. Rafiq
University of Surrey, Guildford, UK

ABSTRACT

Orthotropic steel decks are, due to their light weight and high stiffness, a preferred structural element in long-span bridges. Despite their advantages, fatigue-related problems can develop over time at particular details, such as welded joints between various sub-elements (deck plate and longitudinal stiffeners or stiffeners and cross-beams), this being the main drawback in using this type of deck.

Fatigue assessment of orthotropic decks is a relatively complex phenomenon due to their intricate geometric features and the combined influence of traffic and temperature on the characteristics of the pertinent load effects.

Structural Health Monitoring (SHM) can play a role in reducing some of the inherent uncertainties while assessing fatigue performance using either the S-N or a fracture mechanics approach. However, it is still necessary to develop the theoretical framework for SHM-based methodologies in order to use effectively the outcomes provided by SHM systems.

The present paper reviews current approaches for assessing fatigue in orthotropic decks using SHM data in combination with an S-N approach. A model to assess fatigue damage considering pavement temperatures and traffic intensities is presented. The model can be seen as a probabilistic extension of the formulation developed by Guo et al. (2008) and is intended to serve as the basis towards an SHM-based methodology for updating fatigue reliability. The prior stochastic description of this *kernel* model can be based either on a reference monitored period, laboratory tests, analytical approaches or a combination of the above.

The underlying motivation in trying to develop such a methodology is to provide a framework and associated tools with which it will be possible to integrate the multiple sets of monitored data acquired during the lifetime of a structure, while formally accounting for the different sources of uncertainties. In essence, the aim is to avoid different sets of data which are being acquired by modern SHMS without much thought to their subsequent utilization and contribution to improved rational decision making.

Model parameter distributions can be updated according to the multiple (incomplete) sets of monitoring outcomes obtained during the service life of a structure through Bayesian inference. As more data is incorporated into the updating methodology, epistemic uncertainties shall be reduced. This will contribute towards assessing cost-effective monitoring durations and will be used to establish relevant time intervals between model updates, thus rationalizing the use of multiple SHM data. Systematic time-dependent effects could also be taken into account when considering long-term monitoring data, thus improving current assessment approaches, which tend to be based on a single monitoring reference period and, as a result, fail to capture time-dependent degradations or other trends (such as live load evolutions or climate effects).

Eventually, SHM-updated fatigue models should lead to more robust performance simulations, resulting in updated distributions of remaining fatigue life with reduced levels of uncertainties. This can be regarded as the primary goal in attempting to optimize bridge management strategies through the rational use of limited resources.

Although focus has been given to orthotropic steel decks, parallel approaches could be developed for other structural systems susceptible to fatigue damage.

Several research questions are currently being addressed in order to further develop the updating methodology and to evaluate, considering field data, its potential.

REFERENCE

Guo, T., Li, A. & Li, J. (2008). Fatigue Life Prediction of Welded Joints in Orthotropic Steel Decks Considering Temperature Effect and Increasing Traffic Flow. *Structural Health Monitoring*, 7(3), 189–202.

Bridge Maintenance, Safety, Management, Resilience and Sustainability – Biondini & Frangopol (Eds)
© 2012 Taylor & Francis Group, London, ISBN 978-0-415-62124-3

Low cost wireless sensor networks for continuous bridge monitoring

B. Han
CTiF, Aalborg University, Aalborg, Denmark & Athens Information Technology Athens, Greece

A. Kalis & P. Tragas
Athens Information Technology, Athens, Greece

R.H. Nielsen & R. Prasad
CTiF, Aalborg University, Aalborg, Denmark

Continuous monitoring wireless sensor networks (WSN) are considered as one of the most promising means to harvest information from large structures in order to assist in structural health monitoring and management. At the same time, continuous monitoring WSNs suffer from limited network lifetimes, since they need to propagate large amounts of data over regular time intervals towards a single destination in the network. Propagation of information is done through multiple hops, suffering from collisions, retransmissions and therefore high energy consumption. Moreover, since there is a bottleneck effect around the network sink, all routing layer algorithms will always deplete the power of the last tier before the fusion center. Finally, theory shows that in such networks scalability could become an issue since transport capacity per node is severely affected as the number of nodes within the network increases. Therefore, in order for WSNs to be considered as an efficient tool to monitor the health state of large structures, their energy consumption should be reduced to a bare minimum. In this work we consider a couple of novel techniques for increasing the lifetime of the sensor network, related to both node and network architecture. Namely, we consider new node designs that are of low cost, low complexity, and low energy consumption. Moreover, we present a new network architecture for such small nodes, that would enable them to reach a base station at large distances from the network, with minimal energy.

Smart management of large structures requires the acquisition of large amounts of data from multiple sensors placed on critical points of the structure, which are sampled continuously over very large periods of time. In order to access the sampled data, a network of sensors is formed, responsible for transferring information from the structure to a fusion centre. The use of wireless sensor networks (WSN) has already been proposed for this task, since the use of wireless links overcomes some significant problems (Bhardwaj & Chandrakasan, 2002) related to the cost of network placement and maintenance.

However, the use of WSNs does not come without problems. Traditional sensor networks, adopting standard wireless schemes to organize the network topology (e.g. Zigbee), are not optimal in terms of energy efficiency, Propagation of information is done through multiple hops, suffering from collisions, retransmissions and therefore high energy consumption. Moreover, since there is a bottleneck effect around the network sink, all routing layer algorithms will always deplete the power of the last tier before the fusion center. Finally, theory shows that in such networks scalability could become an issue since transport capacity per node is severely affected as the number of nodes within the network increases. Therefore, in order for WSNs to be considered as an efficient tool to monitor the health state of large structures, their energy consumption should be reduced to a bare minimum.

To overcome the aforementioned problems, low cost, energy efficient wireless sensor network techniques are proposed in this work, which can reduce the power consumption significantly. Namely, we propose the use of an energy efficient RF front end circuit on the sensor node, which is composed of an adaptive antenna scheme and an active reflector node architecture, in order to significantly increase network lifetime. In order to take full advantage of the improved sensor node design, a single hop method is also considered for increasing the energy efficiency of the whole sensor network.

REFERENCES

Bhardwaj, M & Chandrakasan, A.P. 2002 Bounding the lifetime of sensor networks via optimal role assignments", *IEEE Proceedings of INFOCOM 3*: 1587–1596.

Bridge Maintenance, Safety, Management, Resilience and Sustainability – Biondini & Frangopol (Eds)
© 2012 Taylor & Francis Group, London, ISBN 978-0-415-62124-3

Knowledge representation system about existing bridges

R. Helmerich
Federal Institute for Materials Research and Testing, Berlin, Germany

ABSTRACT

The paper presents an ontology-based system for retrieval and semantic organization of knowledge in the domain of existing bridges. Ontologies represent a novel information retrieval and processing system for simpler reuse, sharing and processing of information. Although basically developed in the domain of biomedicine, the free open source ontology editor Protégé (2012) was applied to design ontologies in the technical domain of existing bridges. Protégé follows worldwide standardized rules established by the www-consortium (W3C). The main goal of the presented ontology is to formalize knowledge about multidisciplinary aspects in bridge engineering, material sciences and testing in a set of ontologies. The addressed domain covers typical bridge structures, deterioration processes, typical defects, causes, materials and test (measurement) methods. The ontology is exemplarily populated with knowledge and data from the domain of historic iron and steel bridges retrieved from own tests and measurements. The considered full scale fatigue testing, material investigation and field measurements were carried out at the Federal Institute for Materials Research and Testing (Helmerich 2005). Typical causes, deterioration processes and defects are linked to typical structures, sensitive structural details and appropriate methods to detect these defects.

Additional advantage of semantic organization of knowledge is the availability of information not explicitly up-loaded to the system due to constraints and axiomatization of the concepts in the domain. All data is formalized to machine readable and processable information and at the same time understandable by humans. Open world assumption characterizes a system that is newer complete, open for update or amendments. An increasing number of case studies or new findings in the domain can enrich the ontology any time and allows the immediate access to the recent findings by other researchers, owners, bridge inspectors or users.

Fast knowledge extraction from the system allows inspectors owners and structural engineers to learn from available lessons about typical deterioration and defects to faster identify weak elements or details in historic steel bridges to optimize inspection and assessment strategies. Adding knowledge to the sub-ontologies on structures, materials, deterioration processes, causes, defects and methods to early detect these defects or deterioration processes in weak details, is symbolized in the Logo of BrOntEx (see Figure 1) BrOntEx is open for further extension to any bridge type and material in the future.

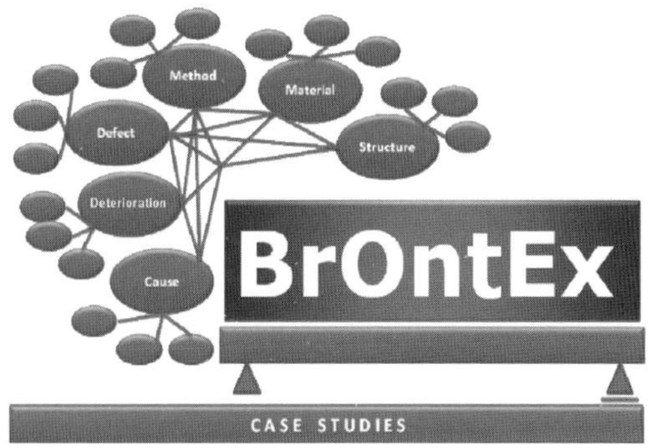

Figure 1. Symbol of the BrOntEx-ontology.

The support of BAM (Germany) to write down these results, the supervision by Professor Jan Bien from Wroclaw University of Technology (Poland) and the use of Protégé is kindly acknowledged.

REFERENCES

Helmerich, R. (2005). Old steel and steel structures- material tests, fatigue testing and field measurements of original bridge girsders between 1990 and 2003, *BAM Report 271*, 1–86, ISBN 3-86509-362-0; ISSN 0938-5533.

Helmerich, R. (2012). Historic Iron and Steel Bridges: Semantic Management of Knowledge, unpublished.

Protégé (2012). Information retrieved from the Protégé-website at http://protégé.stanford.edu, last accessed January 2012.

Bridge Maintenance, Safety, Management, Resilience and Sustainability – Biondini & Frangopol (Eds)
 ISBN 978-0-415-62124-3

NDT-based monitoring of accelerated steel corrosion in concrete

S.-X. Hong, W.L. Lai, R. Helmerich & B. Milmann
BAM, Federal Institute for Materials Research and Testing, Berlin, Germany

ABSTRACT

Reinforced concrete is used more than any other man-made material in the infrastructure. Corrosion of reinforcement is one of the main reasons for accelerated aging and sometimes even for failure of reinforced concrete structures.

Maintenance, restoration and replacement require an efficient inspection method to evaluate the current corrosion state and to early detect beginning corrosion processes. Our research attempts to develop a strategy to detect reinforcement corrosion in the concrete with different Nondestructive Testing (NDT) techniques, meeting one of the challenges proposed by SmartEN research project. The initial training network for early stage researchers SmartEN is funded by the European commission in its 7th Framework Programme.

In this study, the ground penetration radar (GPR), ultrasonic and half-cell potential (HCP) measurements are used separately to periodically monitor the corrosion process in a concrete slab with 150 cm × 149 cm × 30 cm in size. The slab was reinforced by two layers of rebars and partly immersed into a bath with a saline water solution. The accelerated corrosion process was induced by impressing constant direct current across a pair of the rebars on the first layer.

For GPR measurement, the general potential of high frequency GPR to detect corrosion was shown by Lai (Lai et al. 2010). This study follows and develops Lai's work from 1-D waveform GPR measurement into area scan on the slab. The GPR signals are processed in time domain and visualized with energy intensity mapping (Fig. 1) and C-scan. Also the chloride effect to GPR signal is discussed and suppressed by background removal.

During the accelerated corrosion, the possible change of acoustic properties was monitored using a dry coupling low frequency ultrasonic array. The

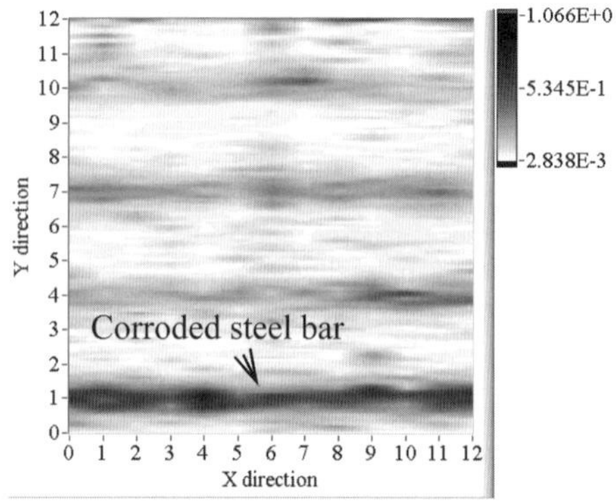

Figure 1. Visualized reinforcement corrosion in concrete by GPR signal energy intensity mapping.

whole surface was automatically scanned before corrosion started to get a basic picture of the inner structure. Images of the obtained 3-D data set visualize slices parallel to the concrete surface based on amplitude and phase analysis. The visualized results show the change of the received signal before and after corrosion. Ongoing work will analyze the different influences from concrete cracking due to corrosion and the steel corrosion.

Our work present an example of monitoring the corrosion process with NDT techniques and shows the way to the ultimate goal of finding objective tools to evaluate the corrosion state.

REFERENCE

Lai W.L., Kind T., Wiggenhauser H., 2010. Detection of accelerated reinforcement corrosion in concrete by ground penetrating radar. Ground Penetrating Radar (GPR), 13th International Conference, Leece, Italy.

Bridge Maintenance, Safety, Management, Resilience and Sustainability – Biondini & Frangopol (Eds)

Bi-objective layout optimization of a wireless sensor network for footbridge monitoring

K. Jalsan, K. Flouri & G. Feltrin
Laboratory for Structural Engineering, EMPA, Duebendorf, Switzerland

ABSTRACT

Wireless Sensor Networks (WSNs) have emerged as a potentially competitive structural monitoring technology. Even though literature is rich in cross-discipline applications for structural health monitoring (SHM) using WSNs, most approaches are validated from the WSN issues such as energy consumption, network connectivity, fault tolerance, etc. Monitoring application aspects are marginally considered. Nevertheless, the integration of the application aspects (structural engineering requirements on data quality) can provide further information for node placement in the network. For such cross-discipline aspects, on one hand, both civil engineering and computer science should focus on and contribute their own expertise. On the other hand, challenges in WSNs (prolonging network lifetime, reliability, etc.) often clash with the structural engineering requirements for maximizing the information quality of the acquired data. Balancing these two diverging objectives has an effect on sensor placement, monitoring system organization and operation.

Although intensive development continues on innovative sensor systems, there is still considerable uncertainty in deciding on the number of sensors required and their location in order to obtain adequate information on structural behavior. This paper studies the optimization of sensor placement for wireless monitoring systems combining WSN and monitoring application aspects. Sensor placement is analyzed within a framework that can best capture the structure properties. The issues related to WSNs such as topology control, data routing and energy consumption are considered.

In this paper, we discuss the trade off between network lifetime and sensor placement quality in the framework of the optimal sensor placement on a structure. We propose an objective function that provides more than one sensor placement solution, depending

Figure 1. The monitored footbridge in Mellingen, Switzerland.

on a given trade off between energy consumption and sensor placement quality. Simulation experiments on a bridge in Mellingen, Switzerland (Figure 1) exhibit that there is space for improvement in the overall systems performance by weighting the importance of energy efficiency and information quality depending on application demands.

REFERENCES

Feltrin, G., Meyer, J., Bischoff, R., & Motavalli, M. 2010. Long-term monitoring of cable stays with a wireless sensor network. *Structure and Infrastructure Engineering*, 6(5): 535–548.

Kammer, D. C. 1990. Sensor placement for on-orbit modal identification and correlation of large space structures. *American Control Conference*, 2984–2990.

Li, B., Wang, D., Wang, F., & Ni, Y. Q. 2010. High quality sensor placement for shm systems: Refocusing on application demands. *INFOCOM*, 1–9.

Lloyd, E. L., & Xue, G. 2007. Relay node placement in wireless sensor networks. *IEEE Transactions on Computers*, 56(1): 134–138.

Qin, W., Hempstead, M., & Woodward, Y. 2007. A realistic power consumption model for wireless sensor network devices. *Sensor, Mesh and Ad Hoc Communications and Networks (SECON)*, 286–295.

Bridge Maintenance, Safety, Management, Resilience and Sustainability – Biondini & Frangopol (Eds)
© 2012 Taylor & Francis Group, London, ISBN 978-0-415-62124-3

Wisespot, a novel approach for wireless localization of damages in bridges

A. Kounoudes & M. Milis
SignalGeneriX Ltd, Lemesos, Cyprus

T. Onoufriou & R.A. Votsis
Cyprus University of Technology, Lemesos, Cyprus

A. Kalis & P. Tragas
Athens Information Technology, Athens, Greece

A.G. Constantinides
Imperial College, London, UK

ABSTRACT

Despite the significant increase of applications for Wireless Sensor Networks (WSN) in various industries, their use in the infrastructure monitoring industry remains low, and this is mainly due to the limitations introduced by the current technology of wireless nodes. More specifically, wireless nodes currently used in WSN have significant limitations in energy consumption, communication range, cost, and localization efficiency. These limitations impede their wide integration in applications such as health monitoring of large infrastructures (bridges, tunnels, water supply systems, transport networks, etc.), environmental control and many other industries, requiring deployment of sensor nodes on a large scale. The WiseSPOT research project developed a new generation of compact low-power wireless sensor nodes that utilize advanced smart antenna technology for continuous monitoring, localization and tracking of events in the network environment (i.e, crack detection in large structures, land movements in landslide prediction etc.). Localization is a network function that requires the synergy of both hardware and software mechanisms, and is applied both for finding the relative and absolute locations of sensor nodes, as well as for localizing these events on the structure. An integral part of this research is the development of fast, energy efficient localization algorithms embedded on miniaturized sensor node hardware to be developed in the framework of the project This technology is intended to develop a powerful remote monitoring tool for structural civil engineers that can be used for continuous monitoring of the health of large structures such as bridges, dams, the transportation and utilities infrastructure.

WiseSPOT proposes innovative system architecture, illustrated in Figure 1, which utilizes four directional antennas able to be selected via antenna switching. This system architecture is motivated by the considerable advantages that the use of smart antennas has shown in wireless ad-hoc networks. These advantages, which include range increase, power consumption decrease, and interference cancellation at the network level, are quite attractive for WSN applications where range, node power consumption and interference largely dictate the overall lifetime of the network. Additionally, an advanced localisation algorithm which combines RSSI and AOA techniques was designed. The algorithm utilises the advantages provided by the four directional antennas while at the same time eliminates the need for additional hardware.

Both hardware and software components developed under the WiseSPOT project have been tested intensively in real world applications and experiments yielding to very promising results regarding the algorithms performance and damage localization accuracy.

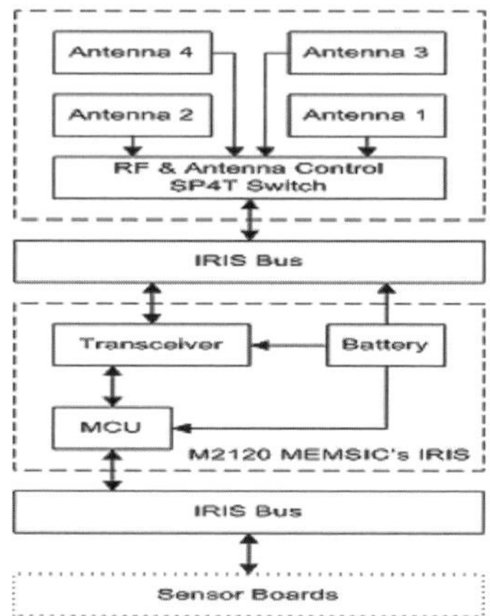

Figure 1. Wisespot System Architecture.

REFERENCE

Leang D., Kalis A., Smart Sensor DVBs: sensor network development boards with smart antennas, in: Proc. of IEEE International Conference on Communications, Circuits and Systems (ICCCAS 2004), June 2004, vol. 2, pp. 1476–1480.

Bridge Maintenance, Safety, Management, Resilience and Sustainability – Biondini & Frangopol (Eds)

Physical characterization of reinforcing bar corrosion in concrete

L. Llano, M.I. Rafiq & M.K. Chryssanthopoulos
University of Surrey, Guildford, Surrey, UK

ABSTRACT

Chloride induced corrosion, caused primarily by de-icing salts and/or salt spray from marine environments, is one of the most common deterioration processes in reinforced concrete. It is often characterized by localized loss of section, known as pitting, which can lead to a significant reduction of the structure's service life. In order to predict the impact of this phenomenon on the mechanical properties of the reinforcing bars in concrete a thorough analysis of its characteristics is needed.

A majority of the models found in literature uses the concept of equivalent uniform corrosion. Only a few models have accounted for the localized corrosion following a similar approach: modeling corrosion in terms of the cross-sectional area loss using maximum pit depth measurements (Rodriguez, et al. 1996; Val & Melchers, 1997). This may be attributed to the limitations of current non-automated and largely heuristic methods used in evaluating the corrosion characteristics on the surface of reinforcement. Automation of the evaluation method could allow the creation and development of comprehensive corrosion models which consider both systematic and random features of the deterioration process. The use of 3D computerized imaging has been considered as a powerful solution to the current status. Tentative first advances in the use of this 3D technique have been found in literature (Oyado & Sato, 2008). However, actual measurement and data analysis is still scarce.

The objective of the present paper is to present an automated procedure for the acquisition of localized corrosion data, together with the first steps involved in statistical data analysis.

Results are presented for longitudinal profiles of corrosion depth measurements, frequency distribution of corrosion depths and frequency distribution of maximum corrosion depths for rebars experiencing corrosion in the range of 2.5% to 4.5%.

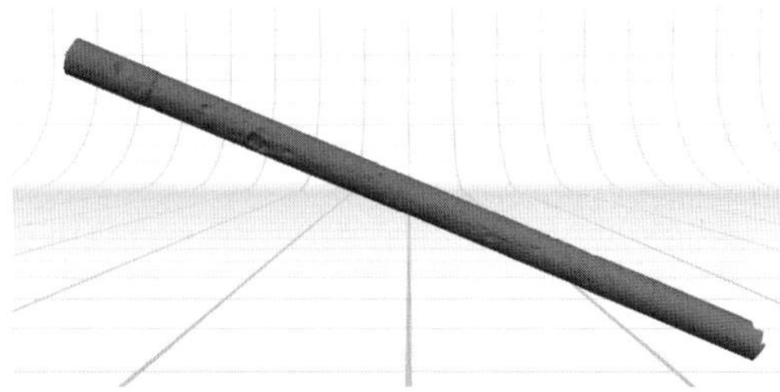

Figure 1. Example of scanned rebar.

It is concluded that the current models utilize only the maximum corrosion penetration depth to characterize the engineering mechanical properties of the corroded rebar are on the conservative side.

From the limited amount of data available so far, it can be suggested that the frequency distribution of corrosion depths for rebars experiencing corrosion in the range of 2.5% to 4.5% can be modeled with lognormal distributions.

Finally, the observed variability of maximum corrosion depths can be taken into account to improve previous models considering only the maximum pit depths.

REFERENCES

Oyado, M., & Sato, T. (2008). Bending performance of reinforced concrete member deteriorated by corrosion. In F. Biondini & D. M. Frangopol (Eds.), *Proceedings of the International Symposium on Life-Cycle Civil Engineering, IALCE* (pp. 415–420). Varenna, Lake Como, Italy: CRC Press.

Rodriguez, J., Ortega, L. M., Casal, J., & Diez, J. M. (1996). Corrosion of reinforcement and Service Life of Concrete Structures. *Durability of Building Materials* (pp. 117–121).

Val, D., & Melchers, R. (1997). Reliability of Deteriorating RC Slab Bridges. *Journal of Structural Engineering, 123*(12).

Bridge Maintenance, Safety, Management, Resilience and Sustainability – Biondini & Frangopol (Eds)
 ISBN 978-0-415-62124-3

Physical layer network coding for bridge wireless monitoring

A.M. Mejri, G.R. Rekaya & J.-C.B. Belfiore
Telecom-ParisTech, Paris, France

ABSTRACT

Structural Health Monitoring (SHM) of bridges has aroused the interest of many researchers on various disciplines.

Our purpose in this paper is to make an overview on SHM of bridges from two widely different research areas: Civil and Structural Engineering domain, and Electrical and Communication Engineering field.

For the Civil Engineering part, we aim to provide some insights into the objectives and the ways SHM of bridges is made by answering to five main questions: why, what, when, where and how to monitor bridges?

For the Electrical and Communication Engineering side, we provide some background on the used technologies in SHM of bridges. We distinguish the wired and wireless sensor networks used to inspect the performance of a bridge and provide proactive management of changes in such system. We focus on communication limitations in Wireless Sensor Networks, in particular the interference problem occurring when the sensors wirelessly broadcast their data measurements. This salient issue makes the whole network suffer from a loss in spectral efficiency and results in performance degradation and reduction of the system life time.

In order to ameliorate these drawbacks, we review basic tools existing in literature using the Physical Layer Network Coding approaches and we focus on the Compute-and-Forward protocol known to provide significant gains.

Our main contribution concerns the real implementation of this protocol and discussion of its practical challenges that need to be met to achieve its promised potential to interference mitigation.

REFERENCES

Dargie, W. and Poellabauer, C., 2010. Fundamentals of Wireless Sensor Networks: Theory and Practice. Wiley-Blackwell, Chichester.

Fragouli, C., Boudec, J.L. and Widmer, J., 2006. Network Coding: An instant Primer. ACM SIGCOMM Computer Communication Review, 63–68.

Harms, T., Segidh, S. and Bastianini, F., 2010. Structural Health Monitoring of Bridges Using Wireless Sensor Networks. IEEE Instrumentation and Measurement Magazine, 14–18.

Hu, P. and Ibnkahla, M., 2010. A Survey Of Physical Layer Network Coding in Wireless Networks. Biennial Symposium on Communications, 311–314.

Bridge Maintenance, Safety, Management, Resilience and Sustainability – Biondini & Frangopol (Eds)

SmartEN – Smart management for sustainable built environment including bridges, structures and infrastructure systems

T. Onoufriou & M.A. Kyriakides
Cyprus University of Technology, Cyprus

K. Berberidis
Computer Technology Institute & Press "DIOPHANTUS", Patras, Greece

M.K. Chryssanthopoulos
University of Surrey, UK

A. Kalis
Athens Information Technology, Greece

ABSTRACT

This paper presents the research and training focus of major European Marie Curie Initial Training Network research programme, SmartEN, which focuses on the development of smart proactive management methods for sustainable built and natural environment. The SmartEN research focuses on multi-disciplinary developments combining mainly the areas of proactive management, non-destructive evaluation methods from the civil engineering discipline, with wireless sensor networks, signal processing and communication technologies from the electrical engineering discipline to develop effective smart proactive management systems for the benefit of the built and natural environment.

This in an area of particular current interest worldwide and significant potential benefits, given the increasing concerns regarding the environmental impact of human actions, the use of the environment and climate changes. These are coupled with ageing infrastructure systems, continuously growing and changing demands on the built and natural environments and limited financial and depleting natural resources.

Recent emerging technologies in miniature wireless sensor platforms which utilize novel digital signal processing offer new opportunities for continuous monitoring of complex systems. These can be adopted to obtain large quantities of highly diverse sensor data that are continuously collected from multiple locations providing insight on the condition, demands and performance of the system. These developments open up a completely novel area for multidisciplinary research towards the 'smart' management of sustainable environment. Integration of such technologies within effective smart proactive management systems can offer significant benefits for the built and natural environment.

The SmartEN Marie Curie Programme aims to push innovation through effective development and integration of emerging technologies targeting key application areas of current interest to Europe and worldwide. Within SmartEN, four work packages have been established (Figure 1), namely: i) Wireless Sensor Networks, ii) Sensor Signal Processing, iii) Non Destructive Evaluation, and iv) Smart Proactive Management. Interaction and collaboration between the researchers from all four WPs is achieved through the four multi-disciplinary research projects (MDPs) developed that involve research themes from all WPs. The application areas of the MDPs are: 1) Structural systems with focus on tall buildings, 2) Heritage and infrastructure with focus on modern long span bridges and heritage bridges, 3) Transportation infrastructure systems with focus on infrastructure networks, and 4) Urban microclimate with focus on its effect on structure and infrastructure networks.

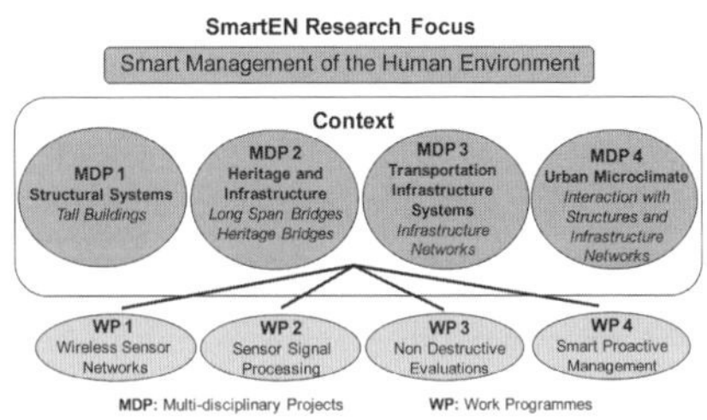

Figure 1. SmartEN research context.

The SmartEN research and training programmes provide a timely and most appropriate platform for interdisciplinary and intersectorial research and training in the aforementioned multi-disciplinary application areas. The SmartEN ITN network aims to equip the young researchers with world-class scientific knowledge and important complementary skills which will help them to become the future leaders in the area of research and development of IP, systems and processes for smart management of the sustainable human environment, an area of significance to Europe's competitiveness.

Bridge Maintenance, Safety, Management, Resilience and Sustainability – Biondini & Frangopol (Eds)
© 2012 Taylor & Francis Group, London, ISBN 978-0-415-62124-3

Performance indicators based on structural health monitoring for management of bridges

A. Orcesi
University Paris-Est, IFSTTAR, Paris, France

D.M. Frangopol
Lehigh University, Bethlehem, PA, USA

ABSTRACT

An important aim of structural health monitoring (SHM) is the development of reliable and robust indicators to detect, locate, quantify and predict damage. Timely detection of structural changes likely to become critical can avoid the occurrence of unserviceability and/or ultimate limit states including partial and even total collapse (Liu et al. 2009a,b,c, Frangopol 2011). In such a context, SHM should enable to determine whether a structure presents an abnormal behavior or not. A feature sensitive to damage is a quantity extracted from the measured system's response that is able to indicate the presence of structural changes. Identifying features that can accurately distinguish a damaged from an undamaged structure is the focus of most SHM techniques.

The objectives of this paper are to build performance indicators under uncertainties (in resistance and actions) for bridges by using a reliability-based approach, and ultimately to define efficient warning notifications. When identifying limit state(s), it is possible to calculate the probability of unsatisfactory performance of the structure (as a system) by using advanced methods (Orcesi et al. 2010). Combining this reliability approach with monitoring information is a great challenge and should help stakeholders determine their management strategies under uncertainties by considering the actual response of the structure to in situ loads (Orcesi & Frangopol 2011a,b).

The proposed framework provides probabilistic metrics to quantify the performance of a structure, based on long-term SHM information. The originality of the proposed approach consists in the use of a performance indicator in the structural reliability analysis, based on monitoring programs.

To reach the objective of warning notifications, a two-step procedure is proposed: (1) the uncertainties associated with material properties and the measured load effects are integrated in the structural model and a cumulative-time failure probability (CTFP) function, based on load occurrences, is built; and (2) a monitoring-based importance index is introduced and compares the responses for different components.

The monitoring-based performance index introduced in this paper enables to assess the contribution of each component to the overall performance under uncertainty, and control if the performance of components changes with time. A sudden change can be a warning meaning that some components have undergone degradation and that maintenance actions are required. The proposed approach is applied to an existing bridge to illustrate how the methodology can be implemented in a real case study.

REFERENCES

Frangopol, D.M. 2011. Life-cycle performance, management, and optimization of structural systems under uncertainty: Accomplishments and challenges, *Structure and Infrastructure Engineering*, Taylor & Francis, 7(6): 389–413.

Liu, C., DeWolfa, J.T. & Kima, J.-H. 2009a. Development of a baseline for structural health monitoring for a curved post-tensioned concrete box-girder bridge. *Engineering Structures*, Elsevier, 31(12): 3107–3115.

Liu, M., Frangopol, D.M. & Kim, S. 2009b. Bridge system performance assessment from structural health monitoring: A case study, *Journal of Structural Engineering*, ASCE 135 (6): 733–742.

Liu, M., Frangopol, D.M. & Kim, S. 2009c. Bridge safety evaluation based on monitored live load effects, *Journal of Bridge Engineering*, ASCE, 14(4): 257–269.

Orcesi, A.D., Frangopol, D.M. & Kim, S. 2010. Optimization of bridge maintenance strategies based on multiple limit states and monitoring, *Engineering Structures*, Elsevier, 32(3): 627–640.

Orcesi, A.D. & Frangopol, D.M. 2011a. Optimization of bridge maintenance strategies based on structural health monitoring information, *Structural Safety*, Elsevier, 33(1): 26–41.

Orcesi, A.D. & Frangopol, D.M. 2011b. Bridge performance monitoring based on traffic data, Submitted for publication.

Bridge Maintenance, Safety, Management, Resilience and Sustainability – Biondini & Frangopol (Eds)
© 2012 Taylor & Francis Group, London, ISBN 978-0-415-62124-3

Monitoring of bridge using a wireless sensor network based on network coding

J. Skulic & K.K. Leung
Electrical and Electronic Engineering Department, Imperial College, South Kensington Campus, London, UK

ABSTRACT

Wireless Sensor Network (WSN) can be deployed on bridges for Structural Health Monitoring (SHM). SHM applications tend to produce a very high amount of data for transfer which consumes much battery power and makes the lifetime of network very short. We propose to use network coding to improve network efficiency and thus prolong its lifetime. A frequently overlooked disadvantage of network coding in WSN is the link quality issue. After a message is lost or overhearing opportunity is missed due to insufficient received signal power or excessive interference, the decoding for network coding cannot be performed at the intermediate nodes, thus significantly degrading performance. For bridge monitoring, sensor nodes are likely to form a linear topology such as along both sides of the bridge. To ensure proper data transfer and decoding by use of network coding, we propose here to control transmission power as a means to adjust the number of nodes that can overhear a message transmission by a neighboring node. Network coding gain relies on such message overhearing. On the other hand, too much overhearing by high power transmission consumes too much limited battery energy. By simulation, we study the tradeoffs between overhearing and power consumption for the network-coding scheme, Figure 1. Specifically, we consider a bridge with fixed length and sensor nodes are deployed at a uniform distance along one or both sides of the bridge. Each radio link is characterized by an exponential path loss, shadowing and Rayleigh fading. Our numerical results reveal that appropriate choices of transmission power (thus the degree of communication connectivity) can achieve the optimal extent of overhearing for network coding gain, while minimizing the overall power consumption for the WSN.

REFERENCE

Skulic J. & Leung K. K. 2012. Application of Network Coding in Wireless Sensor Networks for Bridge Monitoring, submitted for publication.

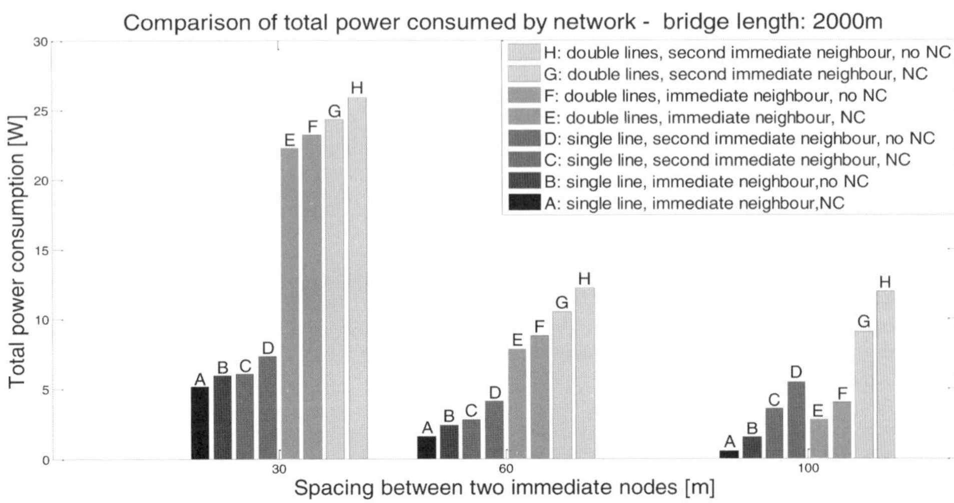

Figure 1. Total power consumption for WSN with and without network coding applied.

Bridge Maintenance, Safety, Management, Resilience and Sustainability – Biondini & Frangopol (Eds)
© 2012 Taylor & Francis Group, London, ISBN 978-0-415-62124-3

Optimization of wireless sensor locations for SHM based on application demands and networking limitations

R.N. Soman, T. Onoufriou, R.A. Votsis, C.Z. Chrysostomou & M.A. Kyriakides
Cyprus University of Technology, Limassol, Cyprus

ABSTRACT

Structural Health Monitoring (SHM) is the continuous evaluation of the strength and reliability of a structure. The aging infrastructure along with the more severe in-service conditions has made SHM a need rather than a luxury. The major advantage of SHM is that it can warn the owners early about the presence of damage in the structure so as to perform preventative maintenance and avoid any future disaster. Performing planned preventative maintenance results to lower maintenance cost along with increased safety.

Until recently, the monitoring was carried out using wired networks of sensors and data collectors. However, the cost of instrumentation, the labor cost and the cost of maintenance of the system proved to be greater than the savings in the maintenance of the structure. The weakest link of the system has been the co-axial wiring used to transfer data and power between the nodes and the data logger. This realization along with the improvement in the field of communications engineering paved the way for use of wireless sensors for SHM. The wireless sensor networks (WSN) allow similar levels of performance at a lower cost of installation, lower maintenance and give flexibility to the nodes of the network.

While WSN offers several advantages, there are still areas that can be improved in their use for SHM. For instance, the sensor nodes have a limited battery reserve, and thus limited network lifetime. As a result, there is a need to improve the longevity of the WSN. Efforts have been made to increase battery life through optimal sensor placement to reduce communication loads. However, restricting the node locations on the energy principle might affect the quality of information which can make the WSN ineffective from the SHM perspective.

This paper focuses on optimizing the number of sensors and their location to cater to the specific requirements of structural engineering while adhering to the energy limitations imposed due to the use of WSN. To fulfill this objective, a two step methodology is proposed. In the first step, the number of sensors is minimized using Auto-MAC matrix (Ewins, 2000) as the criterion. The minimum number of sensors which fulfills the criterion is taken as the lower limit for the number of sensors. In the second step a minimization problem is formulated and the Genetic Algorithm (GA) is used to optimize the sensor location. The GA employs a fitness function that combines the determinant of Fisher Information Matrix (FIM) – an indicator of the information quality and the maximum energy used by the sensor node.

The approach has been validated on a numerical model of a long span bridge (Grand- Mere Bridge, Quebec Canada) (Massicotte et al., 1994) and the results obtained are compared with other optimal sensor placement principles particularly the Fisher Information Matrix (FIM) alone, energy alone, evenly spaced placement and the Power-Aware Sensor Placement (p-SPEM) using Effective Independence (EFI) Method (Li et al., 2010) in order to ascertain its suitability and effectiveness. The results have shown that the developed methodology yields better sensor locations than the previously suggested methods given the connectivity limitations.

REFERENCES

Ewins, D.J., 2000, Modal Testing: Theory, practice and application, Research Studies Press

Li, B., Wand, D., Wang, F., Yi, Q-N., 2010, High Quality Sensor Placement for SHM systems: Refocussing on Application Demands, *IEEE INFOCOM 2010.*

Massicotte, B., Picard, A., Ouellet, C., Gaumond, Y., 1994, Strengthening of a long span post tensioned segmental box girder bridge, *PCI Journal* 29: 52–65

Research and applications in bridge health monitoring

Organizers: F.N. Catbas, J.R. Casas, H. Furuta & D.M. Frangopol

Bridge Maintenance, Safety, Management, Resilience and Sustainability – Biondini & Frangopol (Eds)
© 2012 Taylor & Francis Group, London, ISBN 978-0-415-62124-3

Experimental load rating of a steel girder bridge using structural health monitoring and modeling

E.S. Bell
University of New Hampshire, Durham, New Hampshire

P.J. Lefebvre
Michie Corporation, Henniker, New Hampshire

M. Sanayei, J.D. Sipple & M. Iplikcioglu
Tufts University, Medford, Massachusetts

B.R. Brenner
Fay, Spofford and Thorndike, INC, Burlington, Massachusetts

ABSTRACT

The state of highway infrastructure in the U.S. is in a critical state. While the design and construction of the next generation of US highway bridges is underway, the current fleet must remain operational through proper visual inspection and load rating. The current bridge evaluation paradigm is based on an inspection process which is by definition elemental and subjective, which could lead to overly conservative bridge load ratings. Recognizing the need to capitalize on advances in structural health monitoring and assessment technologies, the Federal Highway Administration (FHWA) began a 20-year study in 2007 entitled the Long-Term Bridge Performance (LTBP) program, which will assess bridge performance changes over time using instrumentation, structural modeling and advanced data management techniques (Ghasemi, 2009).

One of the goals of LTBP program is to improve inspection standards and condition assessment through non-destructive testing (NDT) and structural health monitoring (SHM) (Ghasemi, 2009). Concurrently, the International Society for Structural Health Monitoring of Intelligent Infrastructure (ISHMII) defines SHM as "a type of system that provides information on demand about any significant change or damage occurring in the structure" (ISHMII, 2010). SHM typically includes data collection through sensor based instrumentation, post-processing the collected responses with respect to predicted responses from both hand calculations and structural modeling and then monitoring of any changes to the structure. This process can be effectively integrated with a structural model of the instrumented bridge for objective condition assessment, design verification and bridge management.

In synergy with the goals of the LTBPP and ISHMII, the integration of instrumention and modeling into the initial design and construction of a bridge structure with with shift the design paradigm from a opening day focus to the long-term focus. This paradigm shift will create a quantitative performance measure and load rating for this new fleet of bridge age that will be availabe throughout the service life of the bridge.

This paper will present an integrated load rating protocol that includes 3D structural modeling, SHM and NDT beginning during the design phase and will create a quantitative, effective and efficient asset allocation program for long-term bridge decision-making for a steel girder bridge in Massachusetts, see Figure 1 (Santini-Bell and Sipple, 2009).

Figure 1. The Vernon Avenue Bridge (VAB).

REFERENCES

Ghasemi, H. (2009, January). FHWA Long-term Bridge Performance Program-A Flagship Initiative. Washington, D.C., USA.

ISHMII. (2010, March 31). International Society for Structural Health Monitoring of Intelligent Infrastructure. Retrieved March 31, 2010, from http://www.ishmii.org/Literature/SHMGlossaryDefinitions.html#S

Santini-Bell, E. M., & Sipple, J. D. (2009). Integrating Baseline Structural Modeling, Structural Health Monitoring and Intelligent Transportation Systems. In U. D. Security, Aging Infrastructure: Issue, Research and Technology. Washington, D.C.: Department of Homeland Security.

Bridge Maintenance, Safety, Management, Resilience and Sustainability – Biondini & Frangopol (Eds)
© 2012 Taylor & Francis Group, London, ISBN 978-0-415-62124-3

Bridge condition assessment using digital image correlation and structural modeling

E.S. Bell, J.T. Peddle & A. Goudreau
University of New Hampshire, Durham, New Hampshire

ABSTRACT

Bridge owners, especially municipalities, are becoming plagued with increasing maintenance costs and decreasing maintenance budgets. Limited available funds require that each maintenance dollar be efficiently allocated to the most critical bridge structure. Objectively determining the most critical bridge structure often requires non-destructive evaluation and testing, which can be costly in terms of equipment, traffic delay and money. The installation of traditional contact-based sensors can require special equipment for access to key bridge elements as well as wiring for power supply and data acquisition. Digital image correlation (DIC) can be used as an alternative to traditional bridge response measurement instruments such as strain gauges or linear variable differential transformers, commonly referred to as LVDTs. DIC is an optical measurement technique that has the ability to capture deformation, stress, and strain in both two and three dimensions through high-resolution digital photography. Because it is a non-contact, non-destructive means of measuring bridge responses, it is an attractive choice for rapid testing of in-service structures.

Researchers at the University of New Hampshire have conducted a series of laboratory and field experiments for verification of DIC application for civil structures in which the DIC system and LVDTs recorded deflection data simultaneously. Upon comparison, the two methods showed nearly identical results, with the DIC within .05 mm of LVDTs, See Figure 1. With a high confidence in DIC, the field-collected deflection data was used to verify design and analytical structural models of two tested bridges; a newly constructed three span continuous steel girder bridge (Sanayei, et al. 2012) and a short concrete slab culvert with fiber reinforced polymer reinforcement retrofit (Whittemore and Durfee 2011).

The collected DIC measurements were used for model verification and evaluation of the innovative retrofit program, respectively. Results from both field tests are presented in this paper. The ability to capture a bridge's behavior with DIC and calibrate a structural model with the collected data provides bridge designers and managers with an easy-to-collect objective measure of bridge performance.

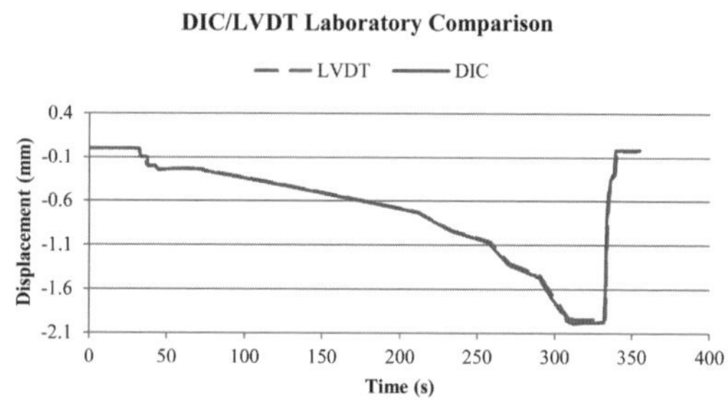

Figure 1. Laboratory verification test data.

REFERENCES

Sanayei, M., Phelps, J.E., Sipple, J.D. Bell, E.S., and Brenner, B.R. (2012). Instrumentation, Nondestructive Testing and FEM Updating for Bridge Evaluation Using Strains. *Journal of Bridge Engineering*. 17(1): 130–138.

Whittemore, M.D., and Durfee, R.H. (2011) Low-Cost Rehabilitation with FRP Strips. *International Bridge Conference*. Pittsburg.

Bridge Maintenance, Safety, Management, Resilience and Sustainability – Biondini & Frangopol (Eds)
 ISBN 978-0-415-62124-3

Implementation of Robust Regression Algorithm (RRA) to detect structural change using Fiber Bragg Grating (FBG) data

F.N. Catbas & M. Malekzadeh
University of Central Florida, Orlando, Florida, USA

I.-B. Kwon
Korea Research Institute of Standards and Science, Daejeon, South Korea

ABSTRACT

Fiber optic sensors (FOS) offer a number of advantages for the purpose of long term Structural Health Monitoring, such as distributed sensing capability, durability, stability and immunity to electrical noise. There are different FOS technologies with a wide range of performance metrics that define their suitability for different applications. One of the most commonly used fiber optic sensing technologies is point sensors with Fiber Bragg Gratings (FBG) sensors. It is also critical to couple such sensing capabilities with effective precise data analysis methods that can identify structural changes and detect possible damage. In this study, robust regression analysis (RRA) is employed to analyze strain data collected with FBG sensors that are installed on a laboratory 4-span bridge. In order to test the efficiency of this non-parametric data analysis approach, several tests are conducted with different damage scenarios in the laboratory environment. The efficiency of both FBG sensors and robust regression algorithm for detection and localizing damage is explored. Based on the findings presented in this paper, the RRA coupled with fiber bragg grating sensors can be deemed to deliver promising results to observe and detect both local and global damage implemented on the structure.

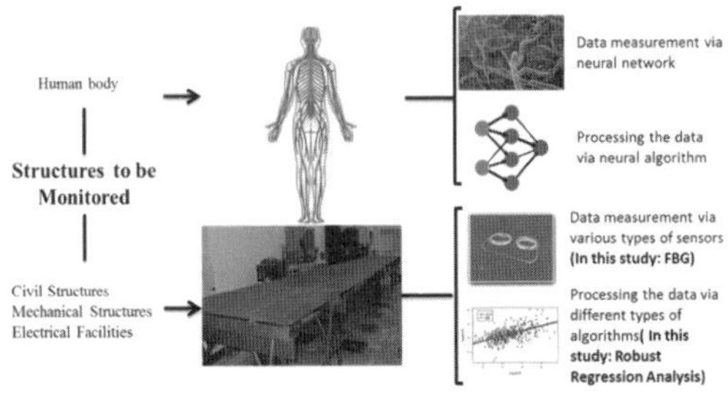

Figure 1. Comparing SHM with human body performance.

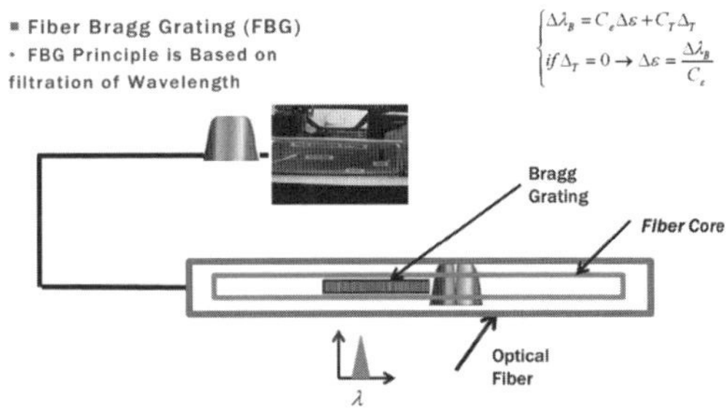

Figure 2. BG Fundamental Concept of FBG sensors.

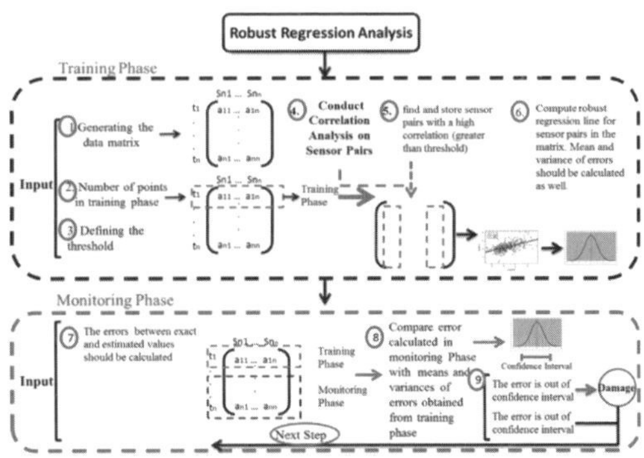

Figure 3. Damage detection procedure for Robust Regression Analysis.

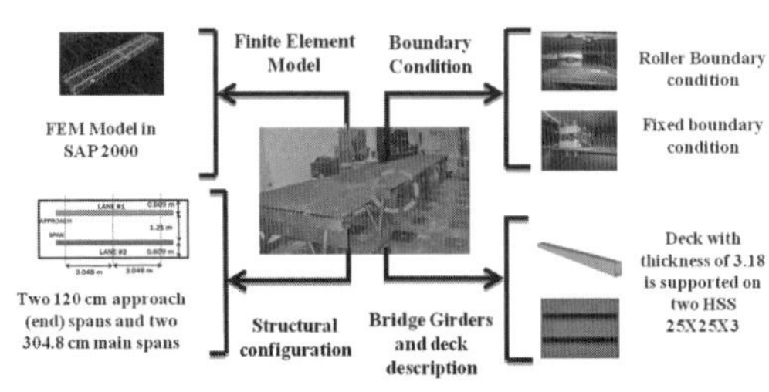

Figure 4. Structural description and instrumentation (UCF 4-span Bridge).

Bridge Maintenance, Safety, Management, Resilience and Sustainability – Biondini & Frangopol (Eds)
© 2012 Taylor & Francis Group, London, ISBN 978-0-415-62124-3

Indirect structural health monitoring in bridges: Scale experiments

F. Cerda, J. Garrett, J. Bielak, J. Barrera, Z. Zhuang, S. Chen, M. McCann & J. Kovačević
Carnegie Mellon University, Pittsburgh, Pennsylvania, USA

P. Rizzo
University of Pittsburgh, Pittsburgh, Pennsylvania, USA

ABSTRACT

In this paper, we use a scale model to experimentally validate an indirect approach to bridge structural health monitoring (SHM). In contrast to a traditional direct monitoring approach with sensors placed on a bridge, the indirect approach uses instrumented vehicles to collect data about the bridge. Indirect monitoring could offer a mobile, sustainable, and economical complementary solution to the traditional bridge SHM approach.

Acceleration signals were collected from a vehicle and bridge system in a laboratory-scale experiment. Figure 1(a) shows an overview of the complete system. It consists of an acceleration and deceleration ramp, a simply supported bridge, and a vehicle. One of the supports is detailed in Figure 1(b). A close-up of the vehicle is shown in Figure 1(c).

Acceleration signals were acquired with wireless sensors on the vehicle and at the midspan of the bridge. Damage was simulated by adding mass at the midspan of the bridge. Experiments covered four damage scenarios and five speeds.

These signals were classified using a simple short-time Fourier transform technique meant to detect shifts in the fundamental frequency of the bridge due to changes in its structure. Results in Figure 2 show near-perfect damage detection when applying this technique to signals collected from the bridge (direct monitoring), and promising damage detection using signals from the sensors on the vehicle (indirect monitoring), especially at the lowest speed.

These results further encourage the research into the indirect approach for bridge SHM. Future work will explore other detection and classification approaches, different vehicle characteristics, different and more subtle changes in the bridge structure, and smaller increments in the vehicle speeds.

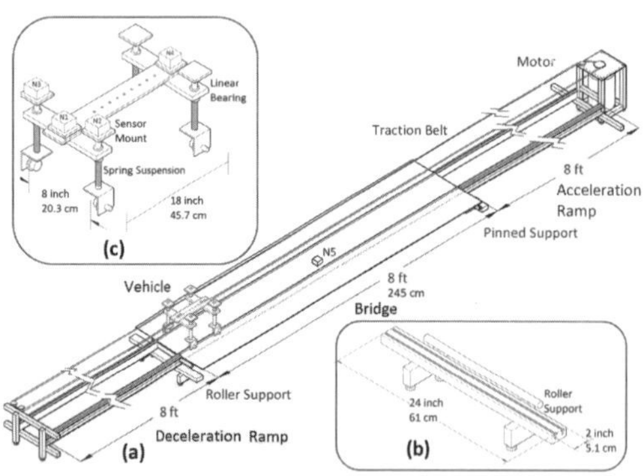

Figure 1. Overview (a), structural components (b) and vehicle (c) from experimental setup.

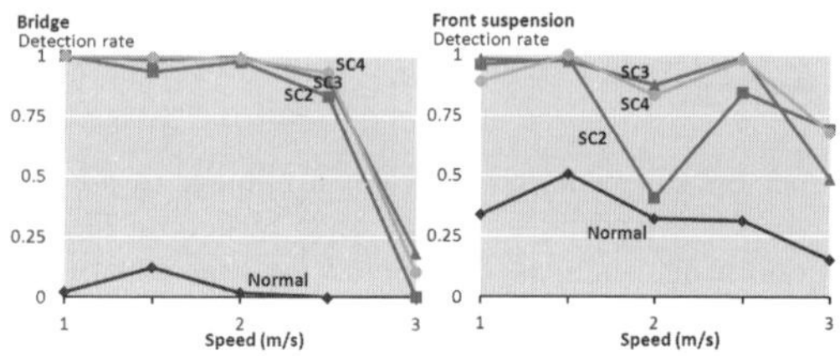

Figure 2. Detection rates from sensors on the bridge (left) and the vehicle (right). A perfect system has a detection rate of 1.0 for each damage scenario and 0.0 for the normal bridge.

REFERENCES

Cerda F., Garrett J., Bielak J., Bhagavatula R. & Kovačević J. 2010. Exploring Indirect Vehicle-Bridge interaction for SHM. *Proceedings of the Fifth International Conference on Bridge Maintenance, Safety and Management*, IABMAS2010, Philadelphia, USA, 696–702.

Kim, C.W. Kawatani, M. & Fujimoto T, (2010), Identifying bending stiffness change of a beam under a moving vehicle, *Proceedings of the Fifth International Conference on Bridge Maintenance, Safety and Management*, IABMAS2010, Philadelphia, USA, 761–767.

Lin, C.W., & Yang Y.B. 2005. Use of a passing vehicle to scan the fundamental bridge frequencies: An experimental verification. *Engineering Structures* 27, no. 13 (November): 1865–1878.

Yang, Y. B., Lin C. W., & Yau J. D. 2004. Extracting bridge frequencies from the dynamic response of a passing vehicle. *Journal of Sound and Vibration* 272, no. 3–5 (May 6):471–493.

Bridge Maintenance, Safety, Management, Resilience and Sustainability – Biondini & Frangopol (Eds)
© 2012 Taylor & Francis Group, London, ISBN 978-0-415-62124-3

Bridge pier scouring: A new approach for monitoring. A case in northern Italy

A. Cigada & S. Manzoni
Politecnico di Milano, Department of Mechanics, Milano, Italy

F. Ballio & G. Crotti
Politecnico di Milano, D.I.I.A.R, Milano, Italy

G. Rossi
Province of Mantova, Department of infrastructure design and traffic, Mantova, Italy

C. Someda
GHT Photonics srl, Padova, Italy

ABSTRACT

The major cause for bridge failure is local scour around piers and/or abutments (Melville & Coleman, 2000). During floods, real-time monitoring systems are the only way to assess bridge stability. Several methods have been developed allowing for in situ scour measurements but these tools present different advantages and disadvantages (NCHRP, 1997).

Using only a single sensor for scour evaluation is not just risky, but also reliability of data is a major system weakness. That is why, to improve the quality of risk assessment evaluation for a bridge crossing a river it has been thought that a sensor fusion approach was the solution to be adopted.

Thanks to Province of Mantova, a complex measurement set-up has been designed, laid down and validated on a real bridge, in Borgoforte (Mantova), on the river Po, Figure 1.

The important step forward in this monitoring system has been the design and production of a new fiber optic based sensor, which has been patented as BLESS (Cigada et al. 2008 and Manzoni et al. 2011). BLESS and echo-sounder measure in real time the local scour.

As discussing before, the potential of this monitoring system is the possibility to make a cross check between these two tools, Figure 2.

In the paper the monitoring system is fully showed and results of all devices are discussed. Finally, a discussion about the potential of the monitoring systems for assessing vulnerability of the bridge is presented.

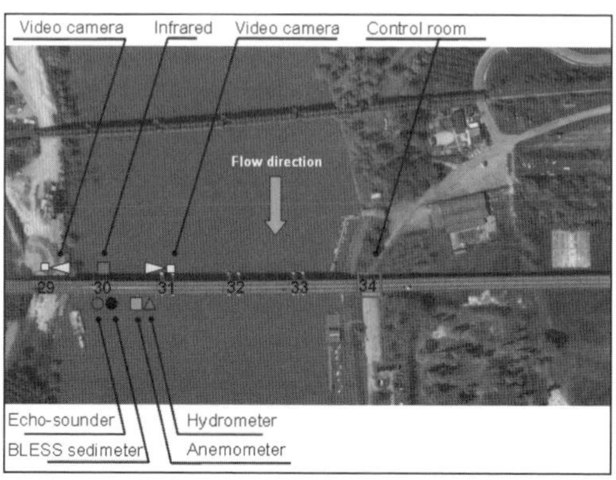

Figure 1. Layout of the monitoring system at Borgoforte bridge.

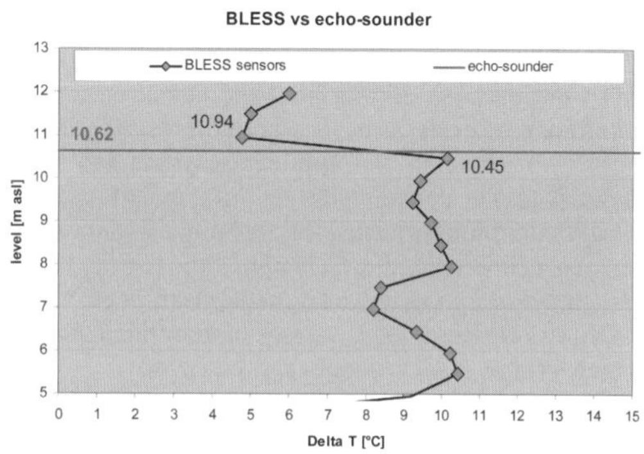

Figure 2. Level of the first 13 sensors (green). Echo-sounder measure (red line).

REFERENCES

Cigada A., Ballio F., Inzoli F., 2008. Hydraulic Monitoring Unit, application for international patent n. PCT/EP2008/059075.

Melville B.W. and Coleman S.E, 2000. "Bridge scour", Water Resources Publications, LLC, Highlands Ranch, Colo.

National Cooperative Highway Research Program (NCHRP), 1997. "Report 396: Instrumentation for Measuring Scour at Bridge Piers and Abutment".

Manzoni S. Crotti G., Ballio F., Cigada A., Inzoli F. and Colombo E., 2011. Bless: A fiber optic sedimeter, Flow Measurement and Instrumentation, doi number: 10.1016/j.flowmeasinst.2011.06.010.

Bridge Maintenance, Safety, Management, Resilience and Sustainability – Biondini & Frangopol (Eds)
© 2012 Taylor & Francis Group, London, ISBN 978-0-415-62124-3

On the static monitoring of bridges and bridge-like structures

A.E. Del Grosso
University of Genoa, Italy

ABSTRACT

The assessment of the actual safety conditions in bridges and similar structures, like harbor piers or other structural systems characterized by a one-dimensional development of the main structural component and subjected to bending and shear, is a very important issue in infrastructure management, as shown by so many papers presented in previous IABMAS and other relevant Conference Series.

To the purpose of assessing the actual structural conditions of bridges, in the recent 15 to 20 years a significant research and application effort has been devoted to the development of instrumental Structural Health Monitoring (SHM) techniques. Innovative sensing technologies like fiber optics and various types of contact and non-contact sensing have been introduced to monitor a large variety of quantities among which strains, displacements and vibration.

To attempt a rough categorization, two SHM approaches may be adopted, depending on the frequency and type of measurement campaigns:

- *periodic*, when the measurements are taken and interpreted only at relatively long time intervals (e.g. several months, one or more years);
- *continuous*, when the measurements are taken at intervals of typically a few hours and the interpretation takes place over time-series of data of appropriate length, theoretically in real-time.

As concerning the physical quantities that are measured by the monitoring system to characterize the structural response, a further distinction can be made between static and dynamic quantities (displacements or strains, accelerations).

When dynamic measurements are of concern, at every measurement campaign high frequency data streams from the different sensors are repeatedly collected in batches of several minutes while, in the static case, only single data values are collected from the sensors.

Assessment methods based on the interpretation of dynamic measurements have been used since a relatively long time and a very extensive literature is available on the subject.

Long-term static monitoring of structural systems has been developed quite recently. By static monitoring we intend the result of the installation of a permanent instrumentation system on the bridge structure. This instrumentation normally provides data streams comprising strain and displacement components at significant points on the structure, temperatures and other relevant environmental and electro-chemical parameters.

The paper is aimed at discussing some of the characteristic features of both methods and at presenting in a greater detail strength and weaknesses of the static monitoring approach.

For static monitoring, data processing and interpretation for damage detection can be performed according to the following methods:

- model based, input-output;
- non model based, output only.

According to the model based approach, strains and deflections recorded by the instrumentation systems are compared with the values obtained by computer simulations on finite element models.

A large number of different non model based algorithms is available to process monitoring data for damage detection, no one having definitely proven to be superior to the others.

Characteristic features that allow comparisons among the different algorithms are:

- length of observation for stabilizing the algorithm;
- length of observation needed after damage;
- minimum detectable damage;
- signal/noise ratio for a given damage level.

Very few comparative studies considering static monitoring have been conducted up to now considering real data.

It is believed that mixed dynamic/static approaches and multi-algorithm data processing tools should be preferred in real applications.

Bridge Maintenance, Safety, Management, Resilience and Sustainability – Biondini & Frangopol (Eds)
© 2012 Taylor & Francis Group, London, ISBN 978-0-415-62124-3

Reliability prediction based on family of models

H.B. Gokce & F.N. Catbas
University of Central Florida, Orlando, Florida, USA

D.M. Frangopol
Lehigh University, Bethlehem, Pennsylvania, USA

ABSTRACT

A family of models approach is proposed for predicting the performance of real life structures such as bridges under uncertainties. Uncertainties exists in the data collected by means of intermittent testing or monitoring, the model boundary conditions, material and geometric properties, non-stationary nature of behavior, and assumed system models, among others. Contrary to utilizing a single model, a family of models can better represent the non-deterministic behavior of civil structures. The methodology presented in the paper is summarized in Figure 1.

This study proposes a family of models calibrated with SHM data to predict more accurately the system reliability of a movable bridge. For the system reliability approach, the movable bridge is modeled with different system configurations and the analyses are conducted based on SHM data, which provides the extreme traffic stresses, temperature cycles and correlation of the random variables. In order to explore the effect of span lock failure on system reliability of the structure, damage is induced and the predicted system reliability indices for undamaged and damaged cases are presented in a comparative fashion (Figure 2).

The main observations and conclusions from this study are: (1) A single FEM that is calibrated using a set of monitoring data may reproduce that data set accurately within acceptable numerical limits. However, time dependent behavior of civil structures and other uncertainties may not necessarily be incorporated properly. Thus predictive analysis using this model might not be accurate; (2) The system reliability of the bridge is calculated for different system models and different correlations. A careful selection of the system model and correlation of variables are crucial for accurate performance predictions; (3) As a demonstration, a real life movable bridge performance is evaluated for undamaged and span-lock damage cases. It is found that the lower bound of reliability index is 4.45 for the undamaged case and 3.07 for the damaged case. This indicates the importance of the span lock component of movable bridges.

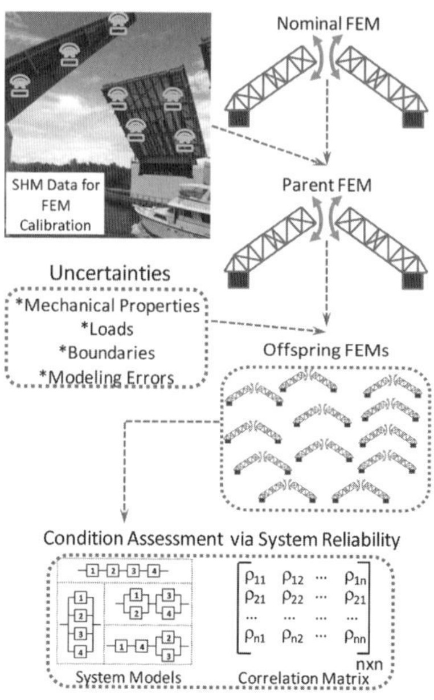

Figure 1. Family of model methodology.

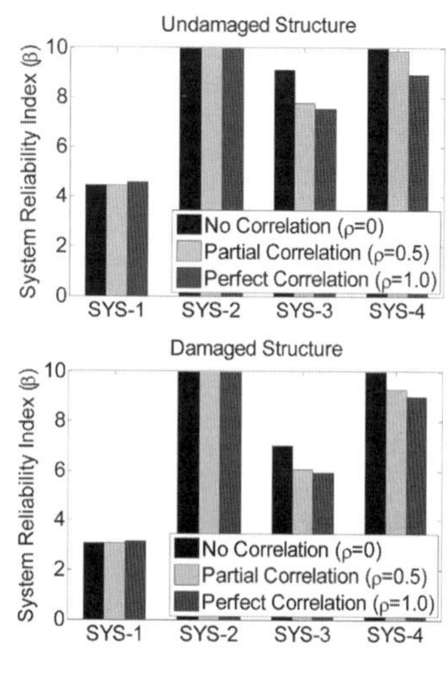

Figure 2. System reliability results for undamaged and damaged structure.

Bridge Maintenance, Safety, Management, Resilience and Sustainability – Biondini & Frangopol (Eds)
© 2012 Taylor & Francis Group, London, ISBN 978-0-415-62124-3

Automation of concrete bridge deck condition assessment and rehabilitation

N. Gucunski & J. Yi
Rutgers University, Center for Advanced Infrastructure and Transportation, Piscataway, NJ, USA

F.L. Moon
Drexel University, Civil, Architectural and Environmental Engineering, Philadelphia, PA, USA

ABSTRACT

Maintenance, rehabilitation, and reconstruction of concrete bridge decks substantially represent the highest expenditure in the overall cost of bridge management. However, the current practice of bridge deck condition assessment relies mostly on visual inspection and simple nondestructive evaluations using chain drag and hammer sounding. Advanced (NDE) techniques are being used, but their wide application has been held back due to the low speed and high cost of evaluation. In addition, significant expertise is needed in both the data collection and data analysis and interpretation. Two systems under development are presented. The first system is called ANDERS (Automated Nondestructive Evaluation and Rehabilitation System) and concentrates on early problem detection and rehabilitation. The second system is RD^2 (Robotic Deck Diagnostics) and concentrates on comprehensive condition assessment of concrete bridge decks.

ANDERS aims to provide a uniquely comprehensive tool that will transform the manner in which bridge decks are assessed and rehabilitated. This will be achieved through: 1) much higher evaluation detail and comprehensiveness of detection at an early stage of deterioration; 2) comprehensive condition and structural assessment at all stages of deterioration; and 3) integrated assessment and rehabilitation that will be minimally invasive, rapid, and cost-effective. ANDERS is composed of four systems. To perform assessments, ANDERS will be equipped with two complimentary, nondestructive approaches. The first, Multi-Modal Nondestructive Evaluation (MMNDE) System aims to identify and characterize localized deterioration with a high degree of resolution using acoustic and ground-penetrating radar (GPR) arrays. The second, Global Structural Assessment (GSA) System aims to capture global structural characteristics and identify any appreciable effects of deterioration on a bridge structure. Output from these two approaches will be merged through a novel Automated Structural Identification (Auto St-Id) approach that will construct, calibrate, and utilize simulation models to assess the overall structural vulnerability and capacity. These three systems comprise the assessment suite of ANDERS and will directly inform the Nondestructive Rehabilitation (NDR) System. The NDR System leverages robotics for the precision and rapid delivery of novel materials capable of halting the identified early-stage deterioration.

RD^2 concentrates on a more objective condition assessment of bridge decks using a complementary suite of NDE technologies from a robotic platform. The condition assessment has three main components: assessment of corrosive environment and corrosion processes, concrete degradation assessment, and assessment with respect to deck delamination. The NDE technologies used in the assessment include: electrical resistivity (ER), ultrasonic surface waves (USW), GPR, and impact echo (IE) method. Each of the four techniques can contribute to a more comprehensive condition assessment of a deck. For example, GPR can identify deteriorated bridge deck areas, while IE can accurately detect and characterize delaminations in the deck. USW, on the other hand, provides information about the material degradation through a measurement of concrete elastic moduli. Finally, ER will assess the potential for corrosive environment.

Bridge Maintenance, Safety, Management, Resilience and Sustainability – Biondini & Frangopol (Eds)
© 2012 Taylor & Francis Group, London, ISBN 978-0-415-62124-3

A bridge damage detection approach using vehicle-bridge interaction analysis and Neural Network technique

H. Hattori
Department of Civil and Environmental Engineering, Kyoto University, Japan

X. He
Faculty of Engineering, Hokkaido University, Japan

F.N. Catbas
Civil, Environmental and Construction Engineering Department, University of Central Florida, USA

H. Furuta
Faculty of Informatics, Kansai University, Japan

M. Kawatani
Graduate School of Engineering, Kobe University, Japan

ABSTRACT

In the maintenance scheme of bridge infrastructures nowadays in developed countries, effective health monitoring and diagnosis processes become more and more important because of the huge number of the structures. Currently, the overall health conditions of the bridges are mainly examined by visual inspections, which demand a large number of technicians and also a considerable cost. The bridge collapse accident in Minneapolis in 2007 give the bridge engineering a warning against bridge health monitoring, and there is an urgent demand to develop effective screening techniques for the global bridge health conditions. The dynamic characteristics of the bridge structure will change when subjected to deteriorations or damages, such as the changing of the structural mass, stiffness and damping effect. Therefore, the dynamic response of the bridge is considered effective to be used for damage detections. If the dynamic bridge responses induced by the running vehicles can be effectively used for the damage identification process, it will be an economical and convenient method.

In this study, a damage detection approach using traffic-induced bridge vibration data is proposed, employing vehicle-bridge coupled vibration analysis and Neural Network technique (NN). In recent years, the NN techniques have been introduced into the civil engineering filed for various kinds of problems, which are networks of highly interconnected neural computing elements that have the ability to respond to input stimuli and to learn to adapt to the environment. In the proposed approach, a NN for bridge damage identification is developed using the running vehicle-induced bridge vibration data as input and the structural damage condition as output. To develop such a system, the possible damage patterns of the bridge are assumed according to theoretical and empirical considerations at first. Then, the running vehicle-induced dynamic responses of the bridge under a certain damage pattern are calculated employing a developed vehicle-bridge interaction analysis procedure. The obtained structural responses are then used as training data to establish the NN system. The inputs of the neural network are the time history of acceleration response of the bridge, while the output values are the state of the structure, including the damage location and degreed of the members. In this study, the input data is created using numerical analysis, while in actual applications the field measurements of bridge vibration should be used.

In this paper, the basic process of the proposed approach and the evaluation of its feasibility are indicated using a two-dimensional simply supported girder bridge model and a 2-degree-of-freedom train model. The numerical results indicate that even a simple Multi-Layer Neural Network is capable to detect damages of the simply-supported bridge using traffic-induced vibration data. Moreover, the network can provide acceptable accuracy for damage identification even only by learning partial response data of the total damage cases, which is more applicable to actual engineering problems. Therefore, the future feasibility of the proposed damage identification approach is confirmed through the investigations of this paper.

Not to mention, toward practical application on actual structures, more detailed models and critical discussions are indispensable. Since the simulated bridge response is used as pseudo-measurement data and no noises are taken into account, further discussions are needed using more complicated models and field test data. Furthermore, demonstrations of the approach using laboratory experiment as well as field test data are also necessary in the further.

Bridge Maintenance, Safety, Management, Resilience and Sustainability – Biondini & Frangopol (Eds)
© 2012 Taylor & Francis Group, London, ISBN 978-0-415-62124-3

Structural health monitoring and damage detection using AdaBoost technique

H. Hattori
Graduate School of Engineering, Kyoto University, Japan

M. Gul
Department of Civil and Environmental Engineering, University of Alberta, Canada

F.N. Catbas
Department of Civil, Environmental and Construction Engineering, University of Central Florida, USA

H. Furuta
Faculty of Informatics, Kansai University, Japan

ABSTRACT

Recently, a vast amount of research has been conducted on health monitoring of existing structures such as buildings, bridges and other civil structures. Furthermore, in Japan, natural disasters like typhoons and earthquakes occur frequently increasing the importance of the damage assessment of the existing structures. In order to evaluate the damage state of structures, health monitoring technology is quite promising to provide useful information. However, there are still some research needs in modeling, analysis and experimental examination before routine applications of health monitoring systems. In this paper, an attempt is made to develop a damage detection approach system by the learning ability. This learning ability facilitates a monitoring paradigm without a need for preliminary investigation of the underlying structure and environment. In other words, it is not necessary to use the precise modeling and analysis methods before con-ducting the health monitoring. The proposed system learns the vibration response by using AdaBoost technique that uses fuzzy-neural networks as a weak learner. By using AdaBoost technique, the network can respond to various types of external forces and the prediction accuracy increases. The fuzzy reasoning predicts the next state of structural behavior such as displacement, velocity and acceleration from the current state of structural behavior and external force. Previously, a health monitoring system that can adapt to the structural systems and environments through the learning ability was developed with the recognition rate of over 80% using numerical simulations. However, experimental verification is needed before real life application of the proposed system. In this paper, results from laboratory experiments are presented to show the effectiveness of the methodology. It is observed that the proposed system can recognize the change of structural characteristics and condition states of a large scale steel grid type laboratory structure.

Table 1 shows the prediction error of the proposed system for ecah node. In the damage case, the deterioration is occurred at node 4 (node 3 and 5 are marked since node 4 is not used in the table).

From Table 1, it is confirmed that the prediction error for each node is bigger than the intact situation. Also, the prediction error of the node 3 and node 5 next to node 4 are bigger than other nodes. Thus, it is concluded that the systam can detect the deterioration and find the deterioration position.

In this paper, an attempt is made to develop a structural health monitoring system that can adapt to the structural systems and environments, by introducing the learning ability.

The proposed system learns the vibration response and predicts the next response by AdaBoost and Fuzzy neural network. The proposed system can recognize existence of deterioration. The model experiment is presented to demonstrate the applicability of the proposed method. It is proved that the proposed system can identify the existence and position of deterioration.

Table 1. The prediction error for damage case (*shows the damage location).

Node	Average	Maximum value
Node 2	0.000886	0.000752
Node 3*	0.007554	0.151900
Node 5*	0.011749	0.184994
Node 6	0.005612	0.093701
Node 9	0.000801	0.021463
Node 10	0.002816	0.050744
Node 12	0.007509	0.110725
Node 13	0.006618	0.109149

Bridge Maintenance, Safety, Management, Resilience and Sustainability – Biondini & Frangopol (Eds)
© 2012 Taylor & Francis Group, London, ISBN 978-0-415-62124-3

Development of a bridge damage detection approach using vehicle-bridge interaction analysis and soft computing methods

X. He
Hokkaido University, Sapporo, Japan

F.N. Catbas
University of Central Florida, USA

H. Hattori
Kyoto University, Japan

H. Furuta
Kansai University, Japan

M. Kawatani
Kobe University, Japan

T. Hayashikawa & T. Matsumoto
Hokkaido University, Sapporo, Japan

ABSTRACT

In developed countries nowadays, there are enormous numbers of bridge structures facing increasing deteriorations, including the Shinkansen lines in Japan that are constructed nearly 50 years agao. In the maintenance scheme of Shinkansen viaducts, the health monitoring and diagnosis of the structures becomes especially important because of the huge number of the structures. Currently, the overall health condition of the Shinkansen viaducts is mainly examined by visual inspections, which demand a large number of technicians and also a considerable cost. With the proceeding of a decreasing birthrate as well as an aging society in Japan, it is significant to develop more effective and economical health monitoring and diagnosis approaches to the civil infrastructures. It is already reported that the dynamic characteristics are possible to be used to identify the structural conditions. In fact, impact tests have been adopted to investigate the integrity of the bridge structure in Shinkansen system since 1991. However, the impact tests not only demand enormous man-power and cost, but also have the deficiency that it cannot be carried out during the operational hours. If the dynamic bridge response induced by the running trains can be effectively used for the health monitoring and diagnosis process, it will be an economical and convenient way because the traffic-induced vibrations of the Shinkansen viaducts have been continually recorded by not only the railway companies but also the local communities.

Some researches on structural identification and bridge health monitoring using traffic-induced vibration data have been initiated recently, which mostly need inverse analysis to identify the structural damage. The numerical errors caused by the inverse analysis can grow severely with the increasing structural members, thus bring serious difficulties in practical applications. In this study, a structural identification approach using train-induced vibration data of the Shinkansen viaducts is developed, employing only direct analyses by means of introducing soft computing methods. In this approach, different from identifying the structural damages using inverse analyses, the possible damage patterns of the bridge are assumed according to theoretical and empirical considerations at first. Then, the running train-induced dynamic responses of the bridge under a certain damage pattern are calculated employing a train-bridge interaction analysis procedure established by the authors. If the calculated responses are identical to the recorded ones, this damage pattern is the exact solution. However, owing to the large number of damage patterns, it will demand enormous computational work to identify the exact solution. Therefore in this approach, an effective optimization method is applied to identify the damage pattern including the damage locations and degrees. In this paper, the basic concepts and process of this approach are represented using a real bullet train model and simply-supported girder bridge model. Several damage scenarios, including various damage locations and degrees, are designed and the identification of the damages is carried out. The results indicate that the proposed approach can effectively detect the damages, therefore the feasibility of the approach can be confirmed.

This study laid a foundation toward developing a practical structural identification approach to actual bridge structures, although more detailed models and critical discussions are necessary.

Bridge Maintenance, Safety, Management, Resilience and Sustainability – Biondini & Frangopol (Eds)
© 2012 Taylor & Francis Group, London, ISBN 978-0-415-62124-3

Distributed sensing for damage localization

D. Inaudi
Roctest Group/SMARTEC SA, Manno, Switzerland

R. Belli & D. Posenato
SMARTEC SA, Manno, Switzerland

ABSTRACT

Structural health monitoring (SHM) is a process aimed at providing accurate and real-time information concerning structural condition and performance. Needs for structural health monitoring in the last decades ere rapidly increasing, and these needs stimulated developments of new sensing technologies. Distributed optical fiber sensing technology has opened new possibility in structural monitoring. Distributed deformation sensor (sensing cable) is sensitive at each point of its length to strain and temperature changes and cracks. Such a sensor is therefore able to record one-dimensional strain field and can be installed over the whole length of the structural members to be monitored (suspension cables, bridge girders, tunnel vaults, dam basis, etc.), and therefore provides for integrity monitoring, i.e. for direct detection, characterization (including recognition, localization, and quantification or rating), and report of local strain changes. These sensors are therefore not only able to measure strain (answering the "how much" question) but to localize damage areas (answering the "where" questions). This makes them ideal to monitor structure where the location of possible damage is a-priori unknown. As an example the sensor can detect and localize a fatigue crack appearing on a bridge girder or a leakage in a pipeline. Distributed sensing techniques and components based on Brillouin and Raman scattering are briefly introduced and their potential for integrity monitoring is discussed. Finally, several large-scale application examples are presented, including crack detection on steel girders, leakage detection on pipelines and crack localization in a tunnel.

The use of this distributed sensing technology is illustrated by several field application examples.

Figure 1. Fatigue crack detection and localization on Göta Bridge in Sweden.

Figure 2. Pipeline strain monitoring in a landslide area in Italy.

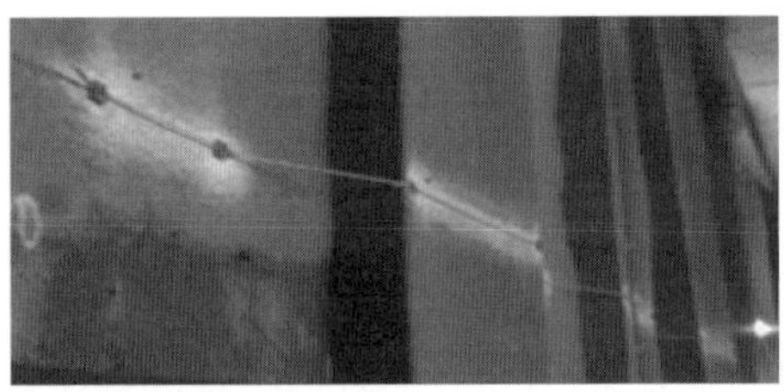

Figure 3. Crack detection in a tunnel lining, Spain.

Figure 4. Detection and localization of sinkholes, Kansas, USA.

Bridge Maintenance, Safety, Management, Resilience and Sustainability – Biondini & Frangopol (Eds)
© 2012 Taylor & Francis Group, London, ISBN 978-0-415-62124-3

A novel image-based approach for structural displacement measurement

Y.F. Ji & Q.W. Zhang
Department of Bridge Engineering, Tongji University, Shanghai, P.R. China

ABSTRACT

With the incessant advancement in optics, electronics and computer technologies during the last three decades, commercial digital video cameras have experienced a remarkable evolution, and can now be employed to measure complex motions of objects with sufficient accuracy, which render great assistance to structural displacement measurement in civil engineering. This paper proposes a novel image-based approach for dynamic measurement of structures. One digital camera is used to capture image sequences of planar targets mounted on vibrating structures. The mathematical relationship between image plane and real space is established based on computer vision theory. Then, the structural dynamic displacement at the target locations can be quantified using point reconstruction rules. Compared with other tradition displacement measurement methods using sensors, such as accelerometers, linear-variable-differential-transducers (LVDTs) and global position system (GPS), the proposed approach gives the main advantages of great flexibility, a non-contact working mode and ease of increasing measurement points. To validate, four tests of sinusoidal motion of a point, free vibration of a cantilever beam, wind tunnel test of a cross-section bridge model, and field test of bridge displacement measurement, are performed.

The proposed image-based approach for real time displacement measurement is composed of a commercial video camera and a planar pattern attached on object surface. The whole processing algorithms of the approach includes camera calibration, object tracking and point reconstruction, as illustrated in Figure 1.

Three experiments of vibration measurement have been performed to verify the reliability of the proposed image-based technique which is actually a post-experiment method.

Table 1 shows the means and the standard deviations of image-based measurement errors for the 1D sinusoidal motion.

A conclusion can be drawn that the proposed image-based approach can provide acceptable accuracy and hold great advantages of great flexibility, a non-contact working mode, ease of increasing measurement points and low cost.

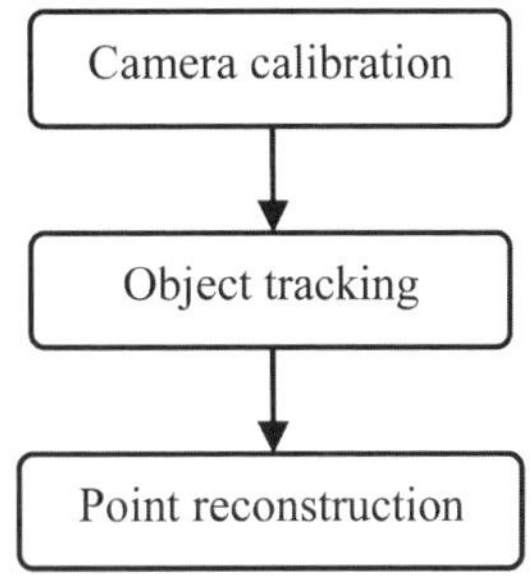

Figure 1. Procedures of the image-based approach.

Table 1. Means (μ) and standard deviations (σ) of measurement errors for one-dimensional 20-mm sinusoidal motion (in mm)

Focal length (mm)	Frequency (Hz) 0.5		1.0		2.0	
	Mean	Std	Mean	Std	Mean	Std
5.2	0.18	0.22	−0.20	0.26	0.22	0.29
15.6	−0.17	0.19	0.17	0.22	0.21	0.25
20.8	−0.12	0.18	0.16	0.22	−0.18	0.24

REFERENCES

Chang, C. C., and Ji, Y. F. (2007) "A flexible videogrammetric technique for three-dimensional structural vibration measurement," *J. of Eng. Mech.*, ASCE, 133(6), 656–664.

Hartley, R. and Zisserman, A. (2003). *Multiple View Geometry in Computer Vision*, 2nd Ed. Cambridge University Press, UK.

Chung, H., Liang, J., Kushiyama, S. and Shinozuka, M. (2004). "Digital image processing for non-linear system identification." *International Journal of Non-linear Mechanics*, 39(5), 691–707.

Bridge Maintenance, Safety, Management, Resilience and Sustainability – Biondini & Frangopol (Eds)
© 2012 Taylor & Francis Group, London, ISBN 978-0-415-62124-3

Testing and long term monitoring of a pre-cast pre-stressed concrete girder bridge

N.R. Johnson, S.M. Petroff, M.W. Halling & P.J. Barr
Utah State University, Logan, UT, USA

ABSTRACT

High quality data is required to develop management procedures, deterioration models, and life cycle cost models for bridge owners. Structural health monitoring (SHM) provides a means of collecting this data (Nagayama & Spencer 2007, Olund & DeWolf 2007). Researchers at Utah State University (USU) have conducted diagnostic testing and installed long term instrumentation on multiple bridge structures in order to obtain this data. Due to the substantial U.S. inventory of bridges, developing this database will be essential for increasing efficiency and advancing the understanding of bridge performance through uniform protocols and procedures.

As part of a larger research program USU has conducted testing and installed instrumentation on a bridge in Utah and in California. Live load and dynamic tests were conducted on both bridges. Long term instrumentation has been installed and is being analyzed and continuously monitored on both bridges.

This paper explains the SHM system that was installed on the bridge in Utah which is known as the Cannery Street Overpass. A quality SHM system consists of more than simply installing a number of long term instruments and collecting the data on a routine basis. Olund & DeWolf (2007) explained that initial benchmarks must be set. Field testing such as live load and dynamic tests not only provide these benchmarks but also allow for determination of optimum long term sensor location and calibration of a quality SHM after these sensors are installed. A quality SHM requires calibration from initial field testing such as live load and dynamic testing. The components and layout of the long term instrumentation are described. A summary of the live load and dynamic tests performed on the Cannery Street Overpass are also presented. This initial testing provided a baseline for subsequent testing done with long term instruments. The process of using the initial test data in conjunction with long term measurements on the same bridge is discussed. This comparison includes a correlation between the field and long term measurements as well as methods of using long term data to monitor a bridge's changing condition state.

Multiple sensors and data types are being used on the Cannery Street Overpass to monitor its structural condition. This paper limits the discussion to four categories of data. The categories discussed in this paper are strain, rotation, acceleration and temperature.

The strain readings from the permanent gauges were compared to the measured results from the live load test. Strain readings were also used to determine maximum values of strain the bridge experiences and compare them to traffic counts from a nearby Weigh-In-Motion (WIM) station. Vibration response from permanently installed velocity transducers were compared to the response measured during the dynamic test. Strain and rotation measurements were compared with corresponding temperature readings to analyze the bridge's thermal response.

Long term instrumentation is beneficial because the change of health over time can be measured continuously rather than just in isolated instances when field testing is performed (Balageas et al. 2006). This will allow for early detection of changes in the bridge condition as well as provide a better indication as to why the changes occurred.

Bridge Maintenance, Safety, Management, Resilience and Sustainability – Biondini & Frangopol (Eds)
© 2012 Taylor & Francis Group, London, ISBN 978-0-415-62124-3

Structural diagnosis of bridges using traffic-induced vibration measurements

C.W. Kim, R. Isemoto & K. Sugiura
Department of Civil & Earth Resources Engineering, Graduate School of Engineering, Kyoto University, Kyoto, Japan

M. Kawatani
Department of Civil Engineering, Graduate School of Engineering, Kobe University, Kobe, Japan

ABSTRACT

An important problem that must be solved in bridge health monitoring (BHM) using vibration measurements is how to excite the bridge economically, reliably and rapidly. For small and medium span bridges the traffic-induced vibration becomes dominant. A noteworthy point is that the traffic-induced vibration of bridges is a kind of non-stationary vibration (Kim et al. 2005). Despite of the non-stationary property of the traffic-induced vibration, the idea of using the traffic-induced vibration data of short span bridges in BHM is that the parameter identified repeatedly under moving vehicles can provide a pattern and give useful information to make a decision about the bridge health condition.

This study is intended to investigate the feasibility of bridge health monitoring using a linear system parameter of a time series model identified from traffic-induced vibration data of bridges through a moving vehicle experiment on scaled model bridges. In order to detect possible abnormality of bridges, this study adopts a parameter from autoregressive (AR) coefficients as damage indicator (DI). Consideration goes into diagnosis of the bridge from pattern change of identified system parameters due to damage. Mahalanobis-Taguchi system (MTS) (Taguchi and Jugulum, 2000) is applied to make a decision on bridge health condition.

It was observed that DI would be a useful indicator for diagnosis of short span bridges (see Fig. 1). MTS worked very well to recognized pattern changes in terms of BHM (see Fig. 2). In making a decision for bridges' health condition using the proposed approach, higher the vehicle speed was the better accuracy would be. Another interesting observation is that AIC could be utilized as a damage-sensitive feature.

In applying MTS in this study, however, the important assumption was that the observed DI values followed the normal distribution. If the assumption of the normal distribution is no more valid, accuracy of MTS might decrease. The next step for this study thus should be considering even data of non-normal distribution.

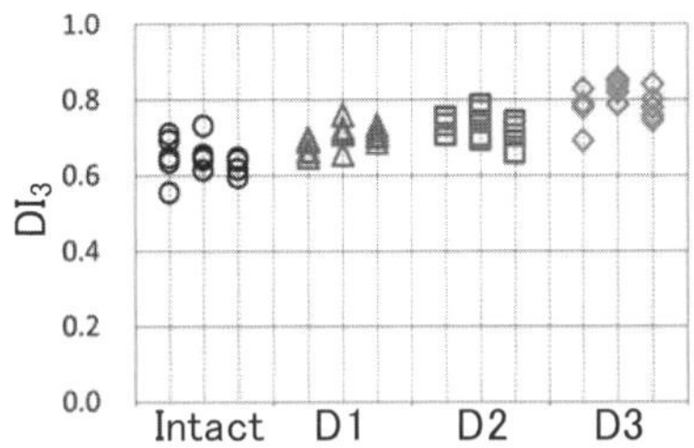

Figure 1. Variation of damage indivator (DI3) observed at the span center according to different damage patterns (severity of damage: D1 < D2 < D3).

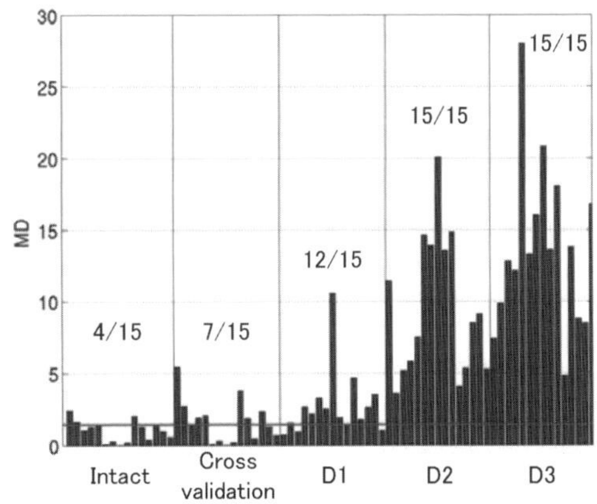

Figure 2. Mahalanobis distance with threshold.

REFERENCES

Kim, C.W., Kawatani, M., and Kim, K.B. (2005). "Three-dimensional dynamic analysis for bridge-vehicle interac tion with roadway roughness", *Computers and Structures*, 83(19–20), 1627–1645.

Taguchi, G., and Jugulum, R. (2000). "New trends in multivariate diagnosis", *Indian Journal of Statistics*, 62, Series B, pp.233–248.

Bridge Maintenance, Safety, Management, Resilience and Sustainability – Biondini & Frangopol (Eds)
 ISBN 978-0-415-62124-3

Modal parameters identification under multi-operational grades and its application to a cable-stayed bridge

Y. Liu
School of Transportation Science and Engineering, Harbin Institute of Technology, Harbin, China

H. Li
School of Civil Engineering, Harbin Institute of Technology, Harbin, China

ABSTRACT

With the installation of various sensors on brides, the SHM system can continuously acquire many kinds of information regarding structural acceleration response, the traffic load, environmental temperature and humidity, and wind load etc. As the SHM system operates for several years, a huge number of structural responses are collected, and these massive data provide a solid foundation for diagnosing the damage of bridges by using the data-driven method. For example, the long-term measured acceleration response of structures can be analyzed statistically, and then the continuous damage diagnosis of bridges may be realized by comparing the variation trend of the statistical modal parameters.

However, in practical situations, the bridge structures are inevitable to be suffered to varying environmental and operational conditions, which may mask the change of modal parameters induced by the damage of bridges. For example, researchers from Los Alamos National Laboratory found that the first three natural frequencies of Alamosa Canyon Bridge varied about 4.7%, 6.6%, 5.0% respectively (H Sohn, 1999) during 24 hours when the temperature of bridge deck changed about 22°C, whereas the natural frequencies of I-40 Bridge had little change as a significant artificial damage was occurred on this bridge (C.R. Farrar, 1997). Peeters and De Roeck (B. Peeters, 2001, J. Maeck, 2000) found that the first four natural frequencies of Z24 Bridge in Switzerland changed about 14%-18% during 10 months, and then a serial of damage test were conducted and the results shown that the vibration frequencies varied less than 10% as the final damage was generated. Therefore, the effects of operational load and time-variant environmental conditions have become the barrier to the development of damage diagnosis of bridges.

In this study, modal parameter identification under multi-operational grades was proposed to simultaneously eliminate the effect of environmental factors (traffic load, environmental temperature, environmental humidity and wind load etc,) on the modal parameters of bridges. A cluster method was applied to classify the monitoring modal parameters, obtained by using the long-term monitoring data of bridges, into different grades. In each grade, the relationship between modal parameters and environmental factors could be deemed as a simple linear relationship, and then the classified modal parameters could be applied to detect the damage of bridges under the effect of environmental factors. Finally, the long-term monitoring modal parameters of a practical bridge were applied to demonstrate the effectiveness of proposed method.

REFERENCES

H. Sohn, M. Dzwonczyk, E.G. Straser, A.S. Kiremidjian, Law K.H., T. Meng. An Experimental Study of Temperature Effect on Modal Parameters of the Alamosa Canyon Bridge. Earthquake Engineering and Structural Dynamics, 1999, 28(8), p: 879–897.

C.R. Farrar, S.W. Doebling, P.J. Cornwell, E.G. Straser. Variability of Modal Parameters Measured on the Alamosa Canyon Bridge. In: Proceedings of the 15th international modal analysis conference, 1997, p: 257–263.

B. Peeters, G. De Roeck. One-year Monitoring of the Z24-Bridge: Environmental Effects Versus Damage Events. Earthquake Engineering and Structural Dynamics, 2001, 30 p: 149–71.

J. Maeck, B. Peeters, G. De Roeck. Damage Identification on the Z24-bridge Using Vibration Monitoring Analysis. In: Proceedings of European COSTF3 conference on system identification and structural health monitoring, 2000, p: 233–242.

Bridge Maintenance, Safety, Management, Resilience and Sustainability – Biondini & Frangopol (Eds)
© 2012 Taylor & Francis Group, London, ISBN 978-0-415-62124-3

Theoretical testing of an empirical mode decomposition damage detection approach using a spatial vehicle-bridge interaction model

J. Meredith & A. González
University College Dublin, Dublin, Ireland

ABSTRACT

It is desirable that bridge damage detection techniques employ operating loads, such as vehicular loads, as excitation sources, which do not require controlled vibration conditions, a perfect knowledge of the excitation source or closure of the structure. This paper has proposed a health monitoring technique with the potential to be used with everyday operating loads, and to identify and locate damage with minimal use of resources.

Empirical mode decomposition (EMD) is used to detect and locate damage in a bridge using its acceleration response to the crossing of a vehicle. EMD is a technique that converts the measured signal into a number of basic functions that make up the original signal. These functions are obtained purely from the original signal in a sequential procedure, where lower order basic functions contain a range of high frequency components of the signal and higher order basic functions contain the low frequency components (Huang & Shen 2005). Damage is identified through a distinctive peak in the decomposed signal resulting from applying EMD. Recent studies have shown the potential of this tool to detect single and multiple damages when using the response of a one-dimensional beam model to the crossing of a constant load (Bradley et al. 2010, Meredith et al. 2011, Meredith & Gonzalez 2011).

In this paper, the technique is further developed using simulations from a quarter-car vehicle-bridge dynamic interaction finite element model. The vehicle model is composed of mass elements, which represent the tyre and body masses, and stiffness and damping elements, which represent tyre and suspension systems. The bridge deck is modelled using plate elements with typical properties found on site and the road profile is generated stochastically from power spectral density functions based on ISO standard guidelines. Different levels of damage are simulated as localised losses of stiffness at random locations within the bridge and a number of longitudinal and transverse locations are used as observation points. The ability of the EMD algorithm to detect damage is analysed for a variety of scenarios including two vehicle configurations (light and heavy), a range of speeds between 5 and 15 m/s, and smooth and rough road surfaces. The influence of the distance from the simulated acceleration points to the damage locations, on the accuracy of the predicted damage, is also discussed.

Results from testing show that the method is capable of detecting a losses in stiffness as low as 10%, however to detect this level the vehicle must cross the damage location. It was also found that an increase in road roughness made detection more difficult, requiring larger vehicle loads and an increase in vehicle speed has an effect on the accuracy of locating the damage. This approach, however, is shown to be a promising tool for damage detection of real structures with relatively low-costs and simple implementation.

REFERENCES

Bradley, M., González, A. & Hester, D. 2010. Analysis of the structural response of a moving load using Empirical Mode Decomposition. In *Proceedings of the 5th International Conference on Bridge Maintenance, Safety, Management and Life-Cycle* Optimization, *IABMAS2010*: 356–363. Philadelphia.

Huang, N.E. & Shen, S.S. 2005. *Hilbert-Huang transform and its applications*. London, World Scientific.

Meredith, J. & González, A. 2011. Damage identification in bridges using Empirical Mode Decomposition of the acceleration response to random traffic. In *Proceedings of Innovations on Bridges and Soil-Bridge Interaction*, IBSBI2011. Athens, Greece.

Meredith, J., González, A. & Hester, D. 2011. Empirical Mode Decomposition of the acceleration response of a prismatic beam subject to a moving load to identify multiple damage locations. In *Proceedings of the International Conference on Structural Engineering Dynamics, ICEDyn2011*. Tavira, Portugal.

Bridge Maintenance, Safety, Management, Resilience and Sustainability – Biondini & Frangopol (Eds)
© 2012 Taylor & Francis Group, London, ISBN 978-0-415-62124-3

Investigation of structural health of timber piles supporting aged bridge

T. Nishikawa, Y. Komatsu & S. Yumoto
Public Works Bureau, Osaka Municipal Government, ATC Building, Osaka, Japan

T. Yamaguchi
Osaka City University, Osaka, Japan

T. Mino & T. Matsumoto
CTI Engineering Co., Ltd., Osaka, Japan

ABSTRACT

In case of deciding whether a bridge should be replaced or enhanced its service life, it is important to evaluate the health state and seismic resistance performance of the foundation because of huge amount of expense for its retrofitting in Japan. [The Jyuso Bridge] is one of typical aged bridge in Osaka whose decks and piers are supported by aged timber piles. The health state and seismic safety of it are not clear. Because the asset value of [the Jyuso Bridge] is significantly high in Osaka City, it is urged to evaluate the intact state of timber piles to decide the maintenance policy of it on early stage (refer to Figure 1).

Accordingly, the experimental and analytical investigations conducted by authors in order to evaluate the structural health of an aged timber pile and its seismic resistance performance under the strong earthquake.

Structural health of an aged timber pile is evaluated based on the investigations and several experiments as follows:

- It is known that the timber where is under the water level would not decay. According from the research on fluctuation of ground water level in Osaka City and the boring data near the Jyuso Bridge, it has been found that the ground water level near the Jyuso Bridge is stabled around O.P. -0 m (O.P means Osaka peil). Since the head of the timber pile is settled in O.P. -2 m, it can be stated that possibility of timber's decaying is quite low.

· Jyuso Bridge
· 1932 Constructed
· Bridge Length is 681m (17 spans)

· Width of Road: 20m
· Traffics: 42,208 vehicles/day (heavy vehicles: 4,011vehicles /day)
· Foundation: Timber Pile, Caisson

Figure 1. Description of Jyuso Bridge.

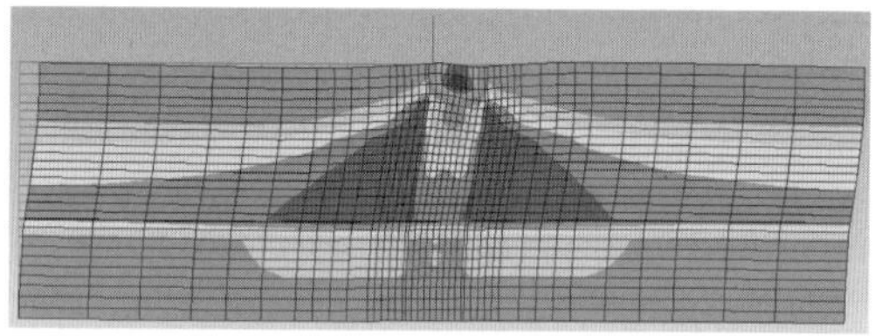

Figure 2. Shear Deformations of the Timber Pile.

- Several timber piles are taken from old bridges which have been replaced, the bending and compressive tests are conducted. Totally 48 and 10 specimens were tested for compressive and flexural strength, respectively. It is understood from the material tests that the strength of an aged timber material does not decrease significantly.

To evaluate the dynamic behavior of the multiple timber piles subjected to the strong earthquake, 2 dimensional finite element analyses have been carried out. The following conclusions are obtained as follows:

- Seismic performance of the multiple timber piles evaluated by the 2D-FEM shows they behave as an integrated structure rather than the single pile structure (refer to Figure 2).
- The peak rotations at the top of pile relative to the bottom were only 0.01 rad which satisfies the required capacity rotation specified in JSHB (=0.02 rad).

It is concluded that the timber piles which support the Jyuso Bridge are in healthy state and possess seismic required resistance in current design code.

REFERENCES

Watanabe, E., Furuta, H., Yokota, T., et al. 2010. Study For Bridge Renewal and Repair By Osaka Municipal Government, *International Association for Bridge Maintenance And Safety*, Philadelphia, USA.

Yokota, T., Komatsu, Y. and Nagahashi, S. 2008. Replacement plan of old bridges by Osaka Municipal Government. Bridge and Foundation Engineering, 42(8): 36–38 (in Japanese).

Bridge Maintenance, Safety, Management, Resilience and Sustainability – Biondini & Frangopol (Eds)
 ISBN 978-0-415-62124-3

Baseline-less structural health monitoring system based on recurrence quantification analysis

Y. Nomura, T. Kusaka & D. Morimoto
Ritsumeikan University, Shiga, Japan

H. Furuta
Kansai University, Osaka, Japan

ABSTRACT

The field of vibration-based structural health monitoring involves extracting a feature which robustly quantifies damage induced changes to the structure. Generally, the change of structural characteristics due to damage is analyzed by observing frequencies, mode shapes, damping, etc. However, the change of those modal features due to damage is minor, so that detecting the location of damage is difficult. Recently, an attractor-based monitoring system has been proposed. The study by Nichols et al. (2006) has demonstrated that the change of the attractor caused by damage is larger than that of the most sensitive frequency and mode of the vibrational response of the structure where the recurrence plot was applied successfully to the damage detection. In this study an attempt is made to develop a damage detection method without baseline by using recurrence quantification analysis.

This study investigates the damage situation in which the 10% deterioration of bending rigidity occurs at the element 17 of the structure shown in Figure 1. Figure 2(a) is the results in which chaotic excitation is given to the structure against the direction of x axis. % recurrence ratio obtained at the node 17 is the smallest among all. Figure 2(b) shows the % recurrence ratio when chaotic excitation is given to the structure against the direction of y axis. It is found from Figures 2(a) and 2(b) that the node 17 provides the smallest % recurrence ratio in both cases. From these results, it is predicted that the element 17 immediately below node 17 has something abnormal.

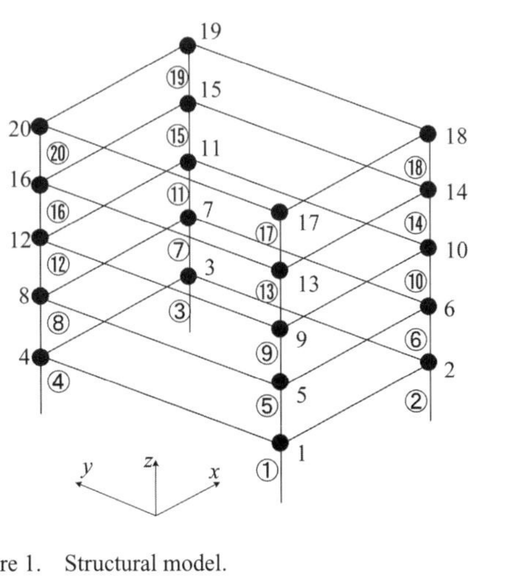

Figure 1. Structural model.

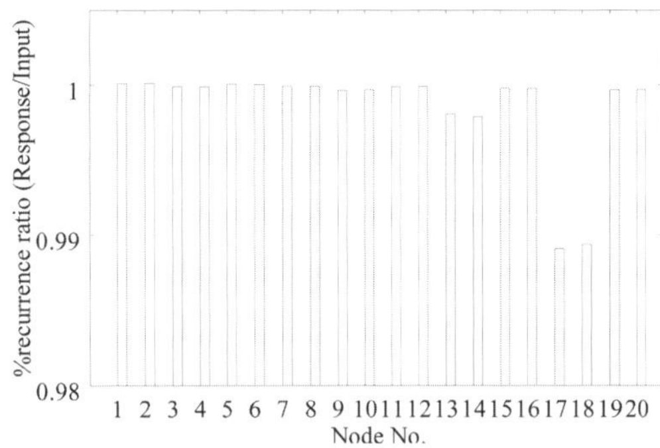

(a) Forcing to x axis direction

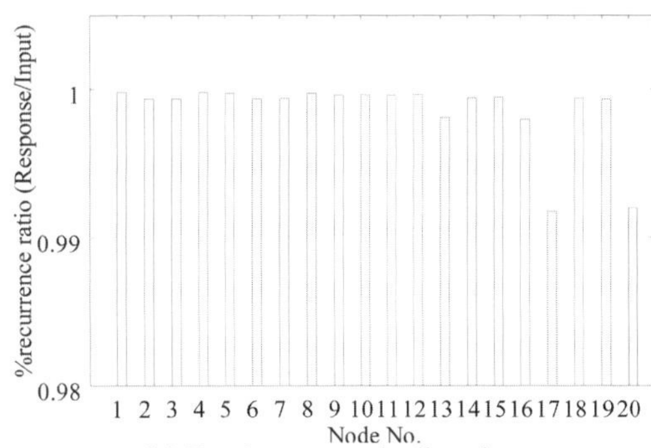

(b) Forcing to y axis direction

Figure 2. The results of % recurrence ratio.

REFERENCE

Nichols, J.M., Trickey, S.T., Seaver, M. 2006. Damage detection using multivariate recurrence quantification analysis, *Mechanical Systems and Signal Processing*, 20:421–437.

Bridge Maintenance, Safety, Management, Resilience and Sustainability – Biondini & Frangopol (Eds)
© 2012 Taylor & Francis Group, London, ISBN 978-0-415-62124-3

One year monitoring of bridge eigenfrequency and vehicle weight for SHM

Y. Oshima, S. Heng & H. Kawano
Kyoto University, Kyoto, Japan

ABSTRACT

Due to high sensitivity to environmental factors, the estimated eigenfrequency of a bridge is varied and blurred at every measurement. Especially for short-span bridges the effect of traffic becomes very serious. To assess the bridge condition based on vibration properties, the effect of such environmental factors has to be clarified.

Therefore, we conducted one year monitoring of bridge eigenfrequency and vehicle weight at the same time to find the relationship between them. The monitored bridge is a Geber-type composite girder with seven spans and the sensors including accelerometers, strain gauges and thermocouples are installed on suspended girders of the last span. In this monitoring, vehicle weight is detected by Bridge Weigh-in-Motion (B-WIM) based on the bending moment of main girders and reaction forces of supports.

In this monitoring, vehicle location is recognized by B-WIM system and then the time series during the vehicles being on the bridge can be extracted by its system. However, the vehicles recognized by B-WIM must include the vehicles together with the other vehicles on the monitored section. To eliminate such influence, the data of the vehicles associated with no other vehicle passing on the section at the same time are adopted for analysis. Forced vibration is defined as the vibration during the vehicle being on the section, while free vibration can be defined as the vibration for 5 seconds after the vehicle exits the section.

Figure 1 shows the relationship between vehicle weight and frequency in forced vibration. The data is also limited to the cases where the vehicles pass the bridge at a velocity of 55–56 km/h under the temperature 26–27 degree Celsius. From this figure, it was found that there is no significant trend in the region over 10 Hz, but some linear trend can be confirmed in the region from 2 Hz to 6 Hz. This trend is stronger in upstream than in downstream. Figure 2 shows the relationship in the data of forced vibration induced by upstream traffic at upstream locations. From this figure, when we look at the locations directly excited by traffic, linear trend can be found clearly. Thus it can be said that lower frequency detected in the forced vibration, which is close to vehicle eigenfrequency, may be affected by the vehicle weight.

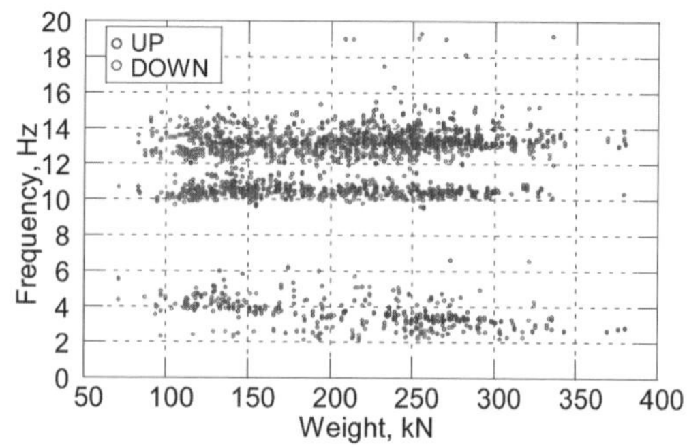

Figure 1. Vehicle weight vs. frequency (forced vibration).

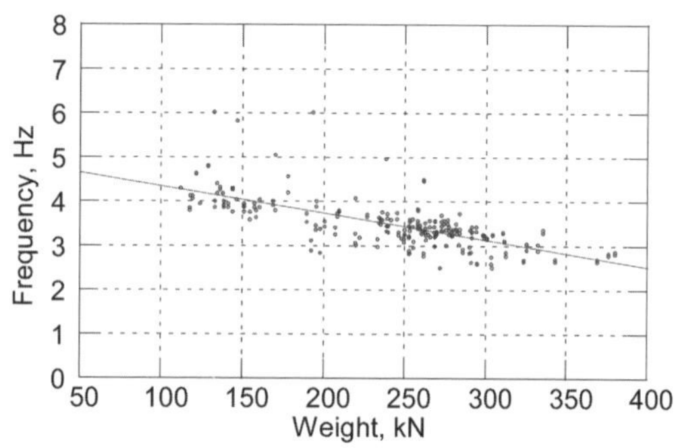

Figure 2. Vehicle weight vs. frequency at upstream locations (forced vibration induce by upstream traffic).

REFERENCES

Yoshinobu Oshima and Kunitomo Sugiura, 2008. Fatigue life estimation of a concrete slab using B-WIM and traffic census, Proc. of HVParis 2008, 359–368, 2008.5.

Bridge Maintenance, Safety, Management, Resilience and Sustainability – Biondini & Frangopol (Eds)
© 2012 Taylor & Francis Group, London, ISBN 978-0-415-62124-3

Monitoring applications providing long-term benefits to owners

T. Spuler, G. Moor & R. Berger
Mageba SA, Switzerland

ABSTRACT

Modern technology has much to offer those who build and maintain bridges and other structures, and the development from "traditional" methodology is especially marked in the case of performance assessment and investigation. For such purposes, structural health monitoring (SHM) now enables any aspect of a structure's condition to be measured, often more precisely, more efficiently and more reliably than was ever possible using traditional manual inspections and measurements.

The capabilities of the modern technology used are extensive. Engineering data can be measured and recorded with minimal effort, and made available in any desired format. Such systems can offer immediate transmission of information from the structure to be viewed online from anywhere in the world, at any time, and automatic notification by SMS or email to the structure's owner of the occurrence of predefined critical events. These systems can be installed in even the most remote locations, with power supply by battery or solar panel, and transmission of data using mobile telecommunications networks.

The purposes the technology can serve are very varied, and limited more by the imagination and knowledge of practitioners than by technical limitations. Purposes can include, for example, investigation of specific concerns which may arise during the life of a structure – for instance, to establish whether or not the surface cracking observed in the pre-stressed concrete deck of a bridge, as shown in Figure 1, is a sign of serious deterioration and reducing structural strength. Gaining a detailed understanding of the problem can enable the need for remedial works to be determined, and if necessary, the optimal solution to be defined. Another common purpose is safety monitoring, which can provide immediate notification of any deterioration in the condition or safety of a structure.

Examples of the use of monitoring systems to address such needs and others in practice are presented.

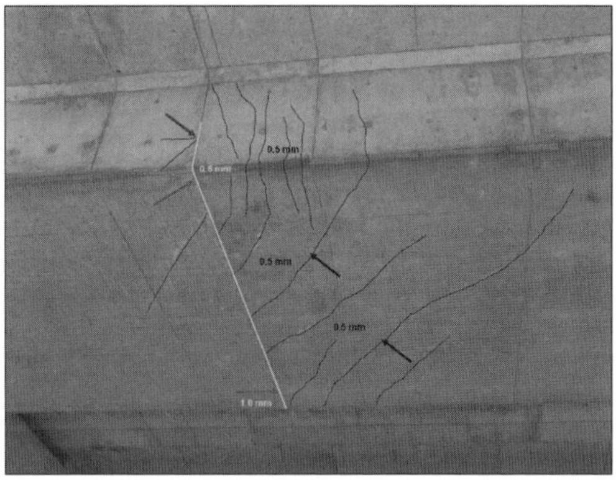

Figure 1. Surface cracking in a pre-stressed concrete bridge deck – the significance of which could be efficiently investigated using an automated monitoring system.

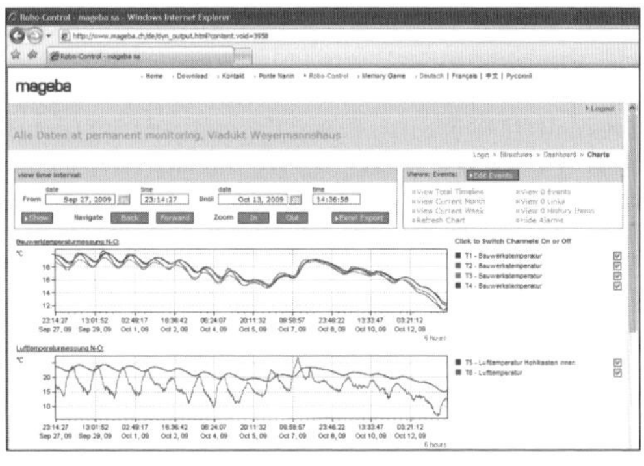

Figure 2. Graphical data presentation of detailed measurements.

REFERENCES

Hoffmann S. and Bergmeister K., Long-term monitoring the bearings of the Ponte 25 de Abril in Lisbon, IABSE Congress in Weimar, Germany, 2007.

Wenzel, H, and Pichler, D. 2005, Ambient Vibration Monitoring J. Wiley and Sons Ltd., Chichester – England, ISBN 0470024305.

Bridge Maintenance, Safety, Management, Resilience and Sustainability – Biondini & Frangopol (Eds)
© 2012 Taylor & Francis Group, London, ISBN 978-0-415-62124-3

First year data mining for vibration based condition monitoring of a cable stayed bridge

Z. Sun
State Key Laboratory on Disaster Reduction in Civil Engineering, Tongji University, Shanghai, China

T. Yan
Department of Bridge Engineering, Tongji University, Shanghai, China

ABSTRACT: To monitor bridge condition and alarm damage occurrence at the earliest possible stage using structural dynamical properties, one important task is to determine reliable damage detection thresholds. Concerning that bridge physical condition and dynamical properties are influenced by non-damage environmental factors (such as temperature, vehicle, and etc.), it is required to develop statistical models to calibrate those environmental factors before a reasonable threshold can be obtained. This paper presents such a work on setting up data-driven polynomial regression models using the first year data for online monitoring of Shanghai Yangtze River Main Navigation Channel Bridge (as shown in Figure 1). The concerned dependent variables of the model are the identified bridge natural frequencies of the first symmetrical and anti-symmetrical vertical bending modes. The independent variables are the measured air temperatures and RMS acceleration at the mid-point of bridge main span, which is adopted to evaluate the moving vehicle mass effect. Model degree is determined by checking the difference and similarity factors between different degrees of models. Model coefficients are estimated using least square method. Damage alarms are then designed according to the obtained nonlinear regression models. A 24-hours continuous monitoring verification study (as shown in Figure 2) tells that the presented method is efficient and applicable.

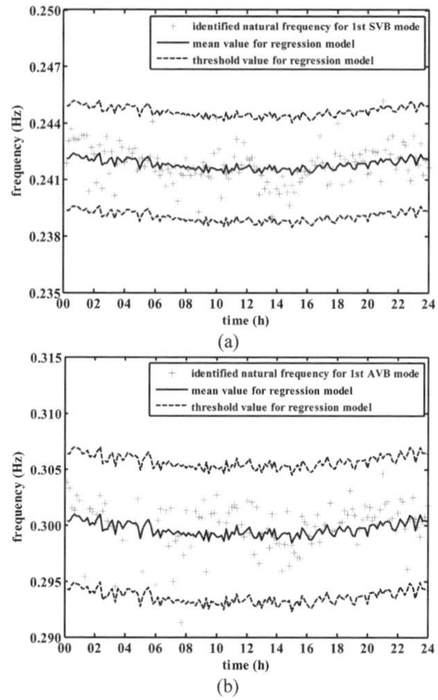

Figure 2. One day monitoring results concerning environmental variation by checking the natural frequency of (a) the 1st SVB mode and (b) the 1st AVB mode.

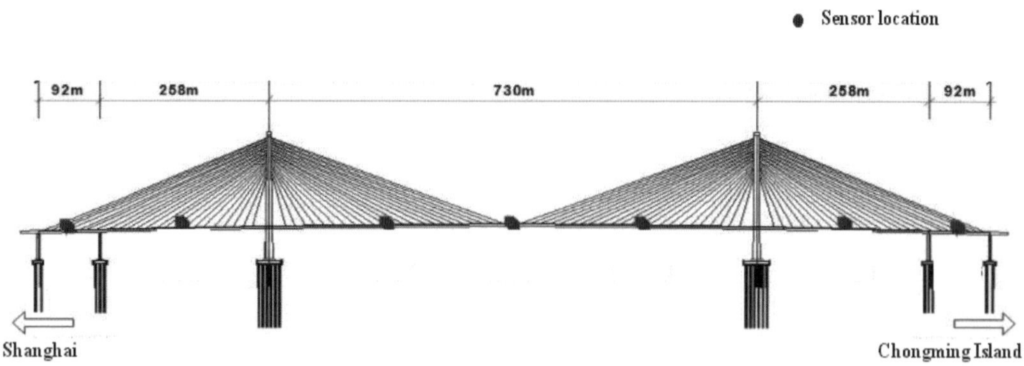

Figure 1. Layout and sensor location of Shanghai Yangtze River Main Navigation Channel Bridge.

Bridge Maintenance, Safety, Management, Resilience and Sustainability – Biondini & Frangopol (Eds)
© 2012 Taylor & Francis Group, London, ISBN 978-0-415-62124-3

Update on AAR bridge testing and monitoring

R.A.P. Sweeney
Modjeski and Masters, Montreal, Canada

D. Otter
Transportation Technology Center, Inc., Pueblo, CO, USA

ABSTRACT

The Transportation Technology Center Inc. (TTCI), a subsidiary of the Association of American Railroads (AAR) has been conducting tests relevant to North American Railroads for many years (Sweeney & Otter 2007). The AAR is supported by member railroads.

This facility includes a 4.4 km oval where heavy trains can be run continuously to test various track components, equipment and several bridges. Given the size of the loop, it is possible to generate the equivalent of a year's traffic on a normal main line in a month.

For the past 24 years, the test train has included several locomotives and a string of 143 tonne cars operating at 64 kph.

The steel bridge at the FAST Facility has a 19.8 m welded span and a 16.8 m Vintage Riveted Span.

Theoretical models predict stresses about 10 percent higher than measured but deflections about 20 percent lower than measured on both the welded and riveted spans.

Both spans have numerous cracks and different repair techniques are evaluated.

Other tests on a movable rail bridge joint, alternative ties (sleepers) and Ultrasonic Impact testing are reviewed.

Two concrete bridges are under load testing, a conventional prestressed concrete bridge and a State of the Art prestressed concrete bridge that includes a high performance concrete span. A number of tests are in progress.

When ballast depth was reduced from 12 inches to 8 inches the measured impact increased by 30%. Also fouled ballast increased impacts from 12% to 38%.

A full production span is under test. The lifting weight of this 12.8 m spans is roughly the same as a 9.14 m prestressed concrete span.

The new T-Rail reference bridge will be used to evaluate and/or develop an on-board bridge defect detection system to be placed in either a locomotive or a set of specially instrumented rail cars or wagons. The stiffness of the bridge can be changed relatively easily.

REFERENCE

Sweeney, R.A.P., Otter, D., "A Review of North American Railway Bridge Research, IABSE Symposium, Weimar, 2007 Report 93, IABSE, Zurich, Switzerland.

Bridge Maintenance, Safety, Management, Resilience and Sustainability – Biondini & Frangopol (Eds)
© 2012 Taylor & Francis Group, London, ISBN 978-0-415-62124-3

Ankara-Istanbul Railway High Speed Train Project, construction of viaduct V4 of 2400 meters

S. Uluöz, S. Düzbasan & M. Camcıoğlu
Ilgaz İnşaat Tic. Ltd. Şti. (Construction) Ankara, Turkey

E. Yakıt
Railone Ilgaz Demiryolu Sis.Ürt. İth. İhr. Ltd. Şti. (Concrete sleeper) Ankara, Turkey

ABSTRACT

Construction of Viaduct V4 which length is 2,400 meters, located on Ankara-Istanbul Railway High-Speed Train Project and one of longest viaducts in Turkey, has been accomplished on 2007.

175.000 m^3 of concrete in various classes and compatible with TS EN 206-1 standard criteria, 25.000 tons of iron and 1.000 tons of pre-stressing steel cord have been used on viaduct construction.

1,100 ea bored piles, 67 ea foundation, 67 ea piers and 792 ea pre-stressed precast beams have been constructed and executed within the scope of project. Sulfate resisting cement has been used for construction of bored piles and foundations, due presence of sulfate risk on main soil on the construction site. CEM I cement has been used for construction of piers, pre-stressed beams, capping beam, bearing, seismic block floor concrete layers, protective concrete layers, cable ducts and front elements, which do not have any risk to contact with main soil on the construction site. Concrete designs in C 25/30, C 30/37 and C 40/50 classes and envisaged to be used in the project. A fixed facility has been established for the manufacture of 792 ea I 195 type prestressed precast beam to be used within the scope of the project.

Samples taken continuously from the iron and concrete used within the scope of project were tested in

- Construction inspection authority's laboratory.
- Company's construction site's laboratory.
- Witness laboratories holding TS EN ISO/IEC 17025 Certificate, in terms of compatibility of fabrication quality with standards and contract's technical specifications.

Solution teams established during execution of the project, within the scope of the company's ISO EN 9001:2000 Quality Management System, to solve problems occurred and to prevent repetition of same incidents and carry out studies within this frame, have performed brain storm studiers.

Quality tests of bored piles have been checked by performing integrity test and bored pile loading test.

Quality tests of prestressed beams have been checked by loading test.

Since Top management of Ilgaz Construction:

- Had continuously monitored all phases of the project and had solved all bottlenecks on time.
- The mobilization had been completed before the scheduled date.
- Spares of equipments which have a direct affect over the production had been made available all the time.
- A high level of coordination had been established in between Project management and employees.
- Information transmission had been made on time.
- All personnel had worked in a devoted manner.

The project had been completed in a very short period of time as 7 months despite of adverse winter conditions. At 26.04.2007, test drives had been started over Ankara-Istanbul High Speed Train Project V4 Viaduct.

Figure 1,2,3. Pre-stressed precast beam production.

REFERENCES

Uluöz S., Düzbasan S., Yakıt E., (2007) Concrete Quality of Pre-stressed Precast Beams which have been produced at adverse weather conditions for Astana Ishim River Bridge. 7th National Concrete Congress Notifications Book, Istanbul Technical University, Istanbul, pg. 233–242

Uluöz S., Düzbasan S., Yakıt E., (2008) Sleeper Production and Quality. 2nd International Railway Symposium Notifications Book, Volume No 1, R.T. Ministry of Transportation, TCDD General Directorate, International Union Railways, Istanbul, pg. 567–577.

Bridge Maintenance, Safety, Management, Resilience and Sustainability – Biondini & Frangopol (Eds)
© 2012 Taylor & Francis Group, London, ISBN 978-0-415-62124-3

Application of OBR fiber optic technology in the structural health monitoring of the Can Fatjó Viaduct (Cerdanyola del Vallés – Spain)

V. Villalba
Structural Monitoring Manager SGS Tecnos
Department of Construction Engineering, Technical University of Catalonia (UPC), Barcelona, Spain

S. Villalba & J.R. Casas
Department of Construction Engineering, Technical University of Catalonia (UPC), Barcelona, Spain

ABSTRACT

The safest structures are those that are managed throughout their service life. Continuous monitoring plays a vital role in this management process. By way of analyzing the data, collected by optic fiber monitoring in real-time, an insight into actual service behavior of a structure is provided and, if necessary, enabling the development of a plan of action with respect to maintenance, repair work or the replacement of component structural elements.

Figure 1. Sample of a longitudinal fiber optic sensors.

Structural monitoring systems can be used for short or long periods of time, periodically or continuously. For some applications, periodic monitoring is satisfactory, but this implies that information is lost between the deferent readings. Only by way of continuous monitoring with OBR technology, during the life of a structure, can a full service history be recorded.

In 2010, a bridge load test was used to investigate the effectiveness of the OBR technology monitoring system. The load test was carried out on the central span of the new Can Fatjó viaduct on the BP-1413 highway, Cerdanyola del Vallés, Barcelona. The bridge monitoring set-up consisted of two 25 m long optic fibers and 50 m transverse optic fiber located along and across the soffit of the prefabricated bridge beams respectively.

The investigation confirms that the optic fiber can be connected to the bridge soffit with epoxy glue; this system can be easily applied and enabled the continuous monitoring of the bridge during the load test. Strain readings obtained from the optic fibers were compared to results from previous identical load tests applied to the bridge, confirming the high accuracy inherent in the optic fiber technology. Thanks to a strain accuracy of $+-1\ \mu\varepsilon$ and a spatial resolution of $10\ \mu\varepsilon$, the proposed OBR monitoring system has important advantages over other monitoring equipment that use optic fiber.

REFERENCES

Villalba, S., Casas, J.R., Fernández, A., Güemes, A., 2009. Feasibility of structural health monitoring of concrete structures by optical backscatter reflectometer. A: The 7th International Workshop on Structural Health Monitoring, Stanford University (USA), 1–8.

Villalba, V., Casas, J.R., 2005. Long term monitoring and updating of corrosion models in post-tensioned concrete structures". A: Advances in life-cycle analysis and design of civil infrastructure systems, Nebraska University, 349–357.

Bridge Maintenance, Safety, Management, Resilience and Sustainability – Biondini & Frangopol (Eds)
© 2012 Taylor & Francis Group, London, ISBN 978-0-415-62124-3

A comparison of different dynamic characterization methods for a truss bridge

T.R. Wank, E.V. Fernstrom & K.A. Grimmelsman
Department of Civil Engineering, University of Arkansas, Fayetteville, Arkansas, USA

ABSTRACT

Bridges are a vital part of the nation's infrastructure; however, the nation's transportation system continues to deteriorate. Faster and more accurate methods of global bridge characterization would be a useful tool in the process of allocating limited transportation repair and reconstruction funding.

The economic viability of characterization of bridges by dynamic testing is controlled by several factors, including the amount of time the field testing requires. As part of a larger study, this paper considers the reduction in testing time that can be realized by enhancing ambient input.

An existing truss bridge was characterized using typical ambient input from traffic and the environment. The same bridge was also characterized using a small APS shaker to supplement the ambient excitation, and another characterization was accomplished using a vibroseis truck on an adjacent span to supplement the excitation. The amount of modal data that could be captured in three different time periods was then compared.

Consideration was also given to whether any time reduction noted could be attributed to the added power or to the broadband nature of the signal. Obviously putting power into the natural frequencies is the best way to shorten data collection needs, but those frequencies are not known at the start of testing. Thus, broadband input is needed. It is assumed that natural ambient excitation is broadband; however, it may take significant time for adequate power to be imparted to each frequency.

The data collected in this study consisted of: four hours of purely ambient data which included occasional passenger vehicle crossings; ten minutes of data from the vibroseis truck producing burst random excitation; thirty-three minutes from the vibroseis truck producing sine sweep excitation; and fifty-four minutes from the APS shaker producing continuous random excitation.

The analysis showed that the enhancement of the ambient input via either the small APS shaker or the vibroseis truck allowed identification of natural frequencies and mode shapes with a shorter data collection period than with pure ambient excitation. This was the expected outcome.

The amount of power added by the auxiliary input devices was assumed to correspond to the average recorded accelerations. The underlying idea is to have a rough scale of how much more energy was being put into the system so that a relative statement could be made on how important the level of added power is. Based on this analysis, it appears that the intentional broadband nature of the additional input was more important than the level of added power in identifying more modes. Further work will be required to make this statement with greater certainty.

Overall, the research described herein indicates that it is beneficial from a time-savings perspective to add broadband excitation to a structure during dynamic characterization. The input can be relatively low in power and can be treated as unrecorded ambient input. Data processing for modal properties is then identical to a pure ambient testing program.

REFERENCES

Allemang, R.J. 2003. The Modal Assurance Criterion (MAC) – Twenty Years of Use and Abuse. *Sound and Vibration*, 37(8): 14–21.

American Association of State Highway and Transportation Officials (AASHTO). 2008. *Bridging the Gap: Restoring and Rebuilding the Nation's Bridges*, Washington DC.

Brincker, R. Zhang, L. Andersen, P. 2001. Modal identification of output-only systems using frequency domain decomposition. *Smart Materials and Structures*, (10): 441–445.

Phares, B.M., Washer, G.A., Rolander, D.D., Graybeal, B.A., and Moore, M. 2004. Routing Highway Bridge Inspection Condition Documentation Accuracy and Reliability. *J. Bridge Eng.*, 9(4): 403–413

Risk based bridge management

Organizer: L. Klatter

Bridge Maintenance, Safety, Management, Resilience and Sustainability – Biondini & Frangopol (Eds)
 ISBN 978-0-415-62124-3

Risk Based Inspection (RBI) at Rijkswaterstaat

J. Bakker & L. Klatter
Rijkswaterstaat, Ministry of Infrastructure and the Environment, Utrecht, The Netherlands

ABSTRACT

Inspection is an important tool for infrastructure asset management processes. Inspection gives the information on the condition of an object and its elements. The use of this information is various. The goal to be achieved determines to a large extend how useful results of inspections are. The change inspection results are useful is low when a manager has not defined these goals clearly. Inspection without a clear goal can be risky. The amount of information gathered may give the impression of being in control of all risks. In many cases the opposite is truth. The more information available, the more difficult it is to have the clear focus needed to be in control. The key is to have a focused process in obtaining information. The subject of this paper is risk based inspection (RBI). The aim is to recognize risks of undesired events in time by gaining the right information at the right time. RBI provides information fit for risk based planning of maintenance (van der Velde, J., Klatter, H.E., Bakker, J.D., 2012).

Definition of Risk Based Inspection: *Risk Based Inspection is a process of gathering and analyzing information with an aim to timely detect risks due to undesired events. An undesired event is an event with a negative effect on the required performance of objects.*

Risk Based Inspection (RBI) differs from most sorts of inspection because the structure itself is not the central theme, but the performance of the structure instead.

The definition of an undesired event depends on the goals of the infrastructure manager (required performance). A defect is not the only factor that determines the risk. Local conditions, for instance the loads and traffic intensity on a structure, are determining factors as well. The word "timely" in the definition is related to the level of acceptance. In general the chance of too late detection will increase with inspections being less frequent. Early recognition of risks can be achieved by prediction (assessment of future risks from inspection results) and/or by periodic review of the risks during inspections.

Risks are not always recognizable based on a visual or measurable defect. In RBI inspection is aimed at elimination of possible risks that could affect performance required, rather than at finding defects. Present defects are regarded as risk indicator. A risk consists of an undesired event with its consequences multiplied by the probability of occurrence (risk = probability × consequences). A damage detected with no risk of affecting performance required is not a defect according to the risk based inspection principles.

A risk based inspection cannot be executed without first carrying out a risk assessment and desk study. Here the process of elimination will be followed. Thus: a risk is present unless the contrary is proved by desk study or inspection. The risk of the half-tooth joint has to be recognized in the desk study (de Boer A, Booij, N. 2012). Much more is needed to quantify the risk than just a visual inspection. The depth of the inspection has to be determined based on what is needed to quantify or eliminate the risk. Elements with a low risk profile (little direct effect on the performance required) still have to be assessed in a risk based inspection. However the depth of the inspection and possibly the inspection frequency have to be tuned to the risk profile. The effects on performance have to be taken into account at all times.

REFERENCES

de Boer, A. & Booij, N: *Inspection method related to structural safety for RC structures*, to be published in the Proceedings of the 6th international conference on Bridge Maintenance, Safety and Managememnt, IABMAS 2012, Rotterdam, Balkema

Van der Velde, J., Katter, H.E., Bakker, J.D., *A holistic approach to asset management in the Netherlands*, to be published in February in Structure and Infrastructure Engineering, 2012. Taylor and Francis

Bridge Maintenance, Safety, Management, Resilience and Sustainability – Biondini & Frangopol (Eds)
© 2012 Taylor & Francis Group, London, ISBN 978-0-415-62124-3

Application of bridge management system to determine preservation and improvement budgets, meet condition targets, and manage risk for the City of Hamilton

A. Dalziel
Stantec Consulting Ltd., Kitchener, ON, Canada

R.M. Ellis
Stantec Consulting Ltd., Edmonton, AB, Canada

G. Moore & R. Andoga
City of Hamilton, Hamilton, ON, Canada

ABSTRACT

In 2005, the City of Hamilton produced a Life-Cycle State of the Infrastructure (SOTI) Report for Public Works Assets, as well as a State of the Infrastructure Report Card for those assets. This report and a 2009 updated report showed that many of the asset groups had deteriorated to undesirable levels and as a result City Council requested staff to report on the practical options for increasing the condition of these assets to a B+ rating over a ten-year period.

As with many similar agencies across Canada the funding available for the maintenance of the roads network is tied to property taxes, therefore, it is often difficult to secure the level of investment that is actually required to keep the roads and bridges operating at an optimal level of service. Having a well maintained transportation network is an essential component of any strategy that seeks to provide a sustainable community.

The objective of this paper is to review the results of the study that was completed on behalf of the City of Hamilton of the options available for increasing the condition of the roads network which includes roads, bridges and traffic from its current D− grade to a B+ grade. For bridge and major culvert structures this required the use of the advanced tools of a Bridge Management System (BMS 2010) which could determine prioritized needs and budgets, predict changes in future bridge condition index (BCI) for various funding levels, and perform risk analysis to develop a risk profile for the City's structures inventory. The analysis determined the effects on report card grade for various funding levels.

This paper shares the methodology and results of this analysis and will be of interest to agencies which manage bridges and any agency interested in preparing an infrastructure report card.

Bridge Maintenance, Safety, Management, Resilience and Sustainability – Biondini & Frangopol (Eds)
© 2012 Taylor & Francis Group, London, ISBN 978-0-415-62124-3

Inspection method related to structural safety of RC structures

A. de Boer & N. Booij
Ministry of Infrastructure and the Environment, Centre for Infrastructure, Utrecht, The Netherlands

ABSTRACT

Inspection techniques are not new in the world of the bridge management systems. Related to structural safety and risk oriented together is more and more wanted, while the budget for maintenance is growing. Till today it was quite common, that the inspection office contains two arts, the inspection part and an additional analysis part of the structure. The inspection part was foreseen and the calculation part needs an extra budget for the contractor. So a need was born to relate the inspection part to the calculation part based on risk analysis in order to minimize the extra costs after the inspection part. A weighted method related to the structural safety of the structure and the condition of the structure indicates a so-called risk index of the structure.

The first possible difference between the design parameters of the structure and the actual use or the wished use of the structure gives a first parameter of the weigthed method. A second difference is the hardening of behaviour of the concrete material in the structure. The concrete structures build before 1975 show a hardening behaviour, where sometimes the concrete has twice the strength of the old design circumstances. This means the structure can carry more load. The traffic load could be growing too, this item is an important negative parameter for the structure. Till now the scope is mentioned in a way that all parameters form the design period of the structure is known. However during all the years some information can be destroyed or lost, this indicates a higher risk for the safety of a structure because there are coming some uncertainties around the design parameters. All together a weigthed method is developed to get a ranking of the structures for extra calculations of the structure and the way of inspection of a structure. This method can be explained and discussed in a paper on a wide range of concrete structures build in the 20th century.

Not all the information of the structures is available. Which information is missing isn't known on forehand so a weighted process is born. Three main aspects can be drawn. The archive database with structural information like drawings, calculations, reinforcement schemes, foundations, birth-year, design recommendation information from the past are aspects coming from the design stage. The relation between this building information and the structural behaviour recommendation changes from the past till today show the uncertainties of structural reliability of the structure. This can be seen as the risque in the reliability index. Also the changes in the use of the structure are part of the uncertainties, but these changes can be controlled partly by measurements during the inspection process of the structure. The type of inspection method is strongly related to the unknown aspect, which are coming from the archive aspects and the structural risque process. If it is necessary an inspection is needed close to the structure area to look at shear cracks or chloride ingress areas. The inspection reports over the lifetime of the structure show the development of the related damage to a structure. The most actual condition of the structure is of course indicated on the last inspection tour. The summation of the scores of the three main aspects is signed as A.

The secondary parameters counts a lot of aspects and will be signed with scores from 1 till 6, depending the influence on the structural reliability. In the case of the archive process secondary aspects are a geometry drawing, a reinforcement drawing, a prestressed scheme, the type of design office, the material properties of the concrete at the construction stage, the type of cement, the quantity of the cement, the type of reinforcement bars, the traffic load recommendation, the load combination factors, etc. The summation of the scores of the secondary aspects is signed by the letter B.

The overall score of the structure is the multiplication of A × B. If the overall score is low, the risque is low and this means there are no further actions to be done.

Bridge Maintenance, Safety, Management, Resilience and Sustainability – Biondini & Frangopol (Eds)
© 2012 Taylor & Francis Group, London, ISBN 978-0-415-62124-3

SmartBMS – improving bridge inspection accuracy and efficiency using a bridge management system in a smartphone

R.M. Ellis & C. McElhinney
Stantec Consulting Ltd., Edmonton AB, Canada

K. Hong
Stantec Consulting Ltd., Kitchener ON, Canada

ABSTRACT

Bridge inspections are an important part of the bridge management cycle. In addition to being a fundamental part of the safety assurance of the road network, inspections collect important information for decision making on maintenance, repair, and rehabilitation needs. There is a continuous improvement process in bridge inspection. In the current study, we sought to address the issue of how to collect bridge inspection data more efficiently and accurately. In response to this need, a new bridge inspection tool has been developed. Dubbed SmartBMS, this inspection tool is a BMS on a hand-held touch screen smartphone using the Windows Mobility operating system. The BMS also synchronizes photographs and video from the inspection to the main BMS.

The SmartBMS tool was implemented for the PEI Department of Transportation and is designed to be compatible with Stantec's latest BMS, BMS 2010 (Evans & Ellis (2012), Ellis & Hong (2012), IABMAS Committee Report (2012)).

Bridge inspections are a key aspect of the bridge management cycle. In addition to being a fundamental part of the safety assurance of the road network, the inspection process collects important information that is entered into the bridge management system (BMS) to help manage bridge assets. This inspection data is used in decision making on rehabilitation, repair, and maintenance needs of bridges and is a critical part of the risk management process. Also, the project level and network level analyses in bridge management systems requires quantified data of the condition, defects, and appraisal/safety/vulnerability to hazards. It is therefore extremely important that all inspection information be collected and entered into BMS accurately and efficiently.

In this study, we sought to address the issue of how to collect bridge inspection data in a more efficient manner, and how to improve the accuracy of the data recording and reporting process. Our inspectors and many of our clients have been using laptop field computers for many years and durable tablet computers since they became available more recently. The most common complaint about these tools is their size and weight. They are also quite costly. As a result, we wished to develop a tool specifically using hand-held SmartPhone technology due to the small size, ability to fit in a pocket, and low cost of replacement. The smartphone also inherently brings the other advantages of the device such as mobile phone access, computer processing such as office tools, calculator etc.

In response to this need, a new bridge inspection and bridge management tool has been developed. The SmartBMS is a BMS on a hand-held touchscreen smartphone using the Windows Mobility operating system. The BMS user interface is nearly identical to the desktop BMS and uses a SQL database that is synchronized to the master database. All inspection data is recorded in the device through a touch screen interface and, when necessary, using a slide-out keyboard. The BMS also synchronizes photographs and video from the inspection to the main BMS, BMS 2010.

The focus of this paper is to outline the SmartBMS tool including the hardware and software used, as well as its applicability in modern bridge inspections.

REFERENCES

Ellis, R.M., Hong., K, Incorporating Risk and Criticality in Bridge Management Decision Making and Project Prioritization. *Proceedings IABMAS Conference*, Lake Como, Italy.

Evans, D., Ellis, R.M., Development, implementation and application of Bridge Management in Prince Edward Island. *Proceedings IABMAS Conference*, Lake Como, Italy.

IABMAS Bridge Management Committee, 2012. Report – Overview of Existing Bridge Management Systems. Available at IABMAS website.

Bridge Maintenance, Safety, Management, Resilience and Sustainability – Biondini & Frangopol (Eds)
© 2012 Taylor & Francis Group, London, ISBN 978-0-415-62124-3

Incorporating risk and criticality in bridge management decision making and project prioritization

R.M. Ellis
Stantec Consulting Ltd., Edmonton AB, Canada

K. Hong
Stantec Consulting Ltd., Kitchener ON, Canada

ABSTRACT

The current generation of full-function bridge management systems (BMS) assist the bridge manager/owner in managing inventory and inspection information, reporting this information and typically including some form of condition performance measure, as well as an overall sufficiency rating, and perhaps other management indices. These systems also include the functionality to recommend prioritized work programs and budgets for the network or subsets of the network. This consists of recommended treatments for the various bridge elements comprising each structure, the timing of recommended work, and the resulting forecasted condition after the work is complete. Most commonly, lifecycle cost analysis is used to determine the benefits of doing the work, and the ratio of benefit to cost (B/C) is used as the main ranking method (IABMAS 2010). The same approach can be used for preservation work, replacements, as well as functional improvements such as strengthening or widening.

A common shortcoming of these powerful systems is the lack of explicit consideration of risk in the prioritization and timing of recommended work. In fact condition indices are only a measure the relative condition of structures and cannot be used reliably to prioritise projects; structures can have the same condition index but very different criticality or urgency. Recent bridge collapses have reminded us that criticality and urgency of recommendations should consider element function and behaviour as well as condition severity and extent.

A new approach has been developed which incorporates risk into the traditional B/C analysis and project prioritising process. In addition to condition based inspection, explicit consideration is given to element function, behaviour, and criticality of inspection findings in the determination of likelihood of service interruption. Overall structure behaviour and importance are considered in the determination of consequences of service interruption.

A method to explicitly consider risk in modern bridge management systems is proposed. The method is relatively simple but considers the relevant factors that bridge engineers/managers consider important in ranking projects in addition to typical LCCA. The results can be expressed in a risk profile which is useful for communicating results, and as a possible target performance measure. The tool can also assist is determining structures of higher risk that require higher inspection frequencies than typical or structures that are of low risk and may be inspected less frequently (Dalziel et al. (2012), Evans & Ellis (2012)).

This paper describes this methodology and should be of interest to bridge inspectors, engineers and bridge managers.

REFERENCES

Dalziel et al. 2012, Application of a Bridge Management System to Determine Preservation and Improvement Budgets, Meet Condition Targets, and Manage Risk for the City of Hamilton. *Proceedings IABMAS Conference*, Lake Como, Italy.

Evans, D., Ellis, R.M. 2012, Development, implementation and application of Bridge Management in Prince Edward Island. Proceedings IABMAS Conference, Lake Como, Italy.

IABMAS Bridge Management Committee, 2010. Report – Overview of Existing Bridge Management Systems. Available at IABMAS website.

Bridge Maintenance, Safety, Management, Resilience and Sustainability – Biondini & Frangopol (Eds)
© 2012 Taylor & Francis Group, London, ISBN 978-0-415-62124-3

Development, implementation and application of bridge management in Prince Edward Island

D.J. Evans
Transportation and Infrastructure Renewal, Charlottetown, Prince Edward Island, Canada

R.M. Ellis
Stantec Consulting Ltd., Edmonton, Alberta, Canada

ABSTRACT

The Province of Prince Edward Island, Canada boasts the densest network of roadway and bridges per capita of any Province in the nation. Yet, as late as 2006, PEI Transportation and Infrastructure Renewal (TIR) had no formal process for bridge management. Inspections were carried out on an ad-hoc basis with no formal inspection method required and with local consultants having only basic bridge inspection experience. Further, the Department lacked a bridge management system to manage the province's structures.

In a very short period of only four years, TIR has been able to implement a bridge management strategy that has included adoption of an accepted standardized inspection methodology, implementation of a bridge management system, and development of an inspection training program for in-house staff and local engineering firms. It has also included adoption of new performance measures such as a bridge condition index (BCI), adoption of bridge asset valuation methods from the BMS, and application of new project level and network level analysis techniques for forecasting needs and developing prioritized work programs. For the first time, the Province has been able to forecast condition of structures based on specified budgets.

The Department wished to set a provincial standard for bridge inspections and provide a course to achieve those standards initially and in future as new inspectors require training certification. Since bridge inspections were not previously provided by the local consulting engineering community, it was necessary to provide comprehensive training to TIR and engineering consultants.

The consultant had developed bridge inspection training courses in other regions in Canada and internationally and was familiar with the successful national bridge inspection courses offered in the USA.

In concert with the inspection program, the Department decided that it would be beneficial to conduct regular refresher training courses for the inspectors including calibration bridges.

In 2008, the Department entered into an agreement with Stantec Consulting Ltd. to license and implement the OBMS for our inspection data collection. The system accepts the inspection data in the OSIM format and calculates a Bridge Condition Index (BCI). In early 2010, the Department licensed Stantec's BMS 2010 software, an advanced BMS that is compatible with the OSIM inspection method and offers many new enhancements over the OBMS which is now over 10 years old.

This paper describes from an owner's perspective the process of developing a complete strategy for bridge management in PEI and shares some of the challenges that needed to be overcome in attempting to achieve t his in a short period of time. This should be of interest to other provincial, state or municipal agencies facing some or all of these challenges.

REFERENCES

AASHTO. 1992. AASHTO Guidelines for Bridge Management Systems. Washington, D.C.,

BRIME, 2002. Bridge Management in Europe Final Report. European Commission Directorate General for Transport

Canadian Institute of Chartered Accountants. 1998. Public Sector Accounting and Auditing Handbook (PSAB), Canadian Institute of Chartered Accountants.

Federal Highway Administration, July 2007. Bridge Evaluation Quality Assurance Summary Report. FHWA International Technology Exchange Program, U.S. Department of Transportation, Washington, DC, USA.

Governmental Accounting Standards Board, 1999. Statement No. 34 of the Governmental Accounting Standard Board. Basic Financial Statements and Management's Discussion and Analysis for State and Local Governments. *Governmental Accounting Standard Series, No. 171-A*. Washington, D.C.

Ministry of Transportation of Ontario. 2003. Ontario Structure Inspection Manual, Queen's Printer, Toronto, ON, Canada.

Thompson, P.D., Ellis, R.M., Merlo, T. and Hong, K. 2003. Implementation of the Ontario Bridge Management System. 9th International Bridge Management Conference, TRB, National Research Council, Washington, D.C., USA.

Bridge Maintenance, Safety, Management, Resilience and Sustainability – Biondini & Frangopol (Eds)
 ISBN 978-0-415-62124-3

First results of the German BMS – Influence of data availability and quality

R. Holst
Federal Highway Research Institute (BASt), Bergisch Gladbach, Germany

ABSTRACT

Due to the increasing globalization of world, traffic grows steadily all around the World and especially in Europe. Germany represents, also following the enlargement of the European Union, a very strong transit country. In order to grow the economy in Europe, the security and ease of traffic be guaranteed at all times.

Besides the increase in traffic and here particularly of heavy vehicles, the total allowable weight of vehicles has increased very strong over time. Thereby the increasing overloading of the truck is not considered.

Furthermore, it is important to note that the majority of the trunk road network in Germany was built after the Second World War, especially in the 60s to the 80s of the last century, and this meant 30 to 50 years of the planned life time.

These factors have left significant impact over time on the roads and also on bridges. Due to limited budget funds available and the fact that in the past very often the construction of new roads was preferred before maintenance measures, in recent years (decades) has flowed too little money in the maintenance of bridges and engineering structures.

Foreseeing this development, the Federal Highway Research Institute and the Federal Ministry of Transport (formerly BMV) and highway authorities of the countries have conducted research projects for the 1997 first design of a Bridge Management System (BMS) for the area of federal trunk roads. After reviewing of management systems in the world, and considering the available information, e.g. from the bridge inspection in accordance with DIN 1076, has been decided to develop a system that builds on these data. This bottom-up approach enables, starting with the damages on the individual bridges, optimization on object- and network level.

The first results show that the algorithms lead both on the object- and on network-level to technically meaningful results. The consideration of construction and damage data offer the possibility to consider the individual circumstances of the respective bridge. However, the quality requirements for construction and damage data are very high and the quality of the results is directly related to data quality. Thus, the extents of the damages have a huge impact on the cost of repairs. In addition, these data must be continuously available in this very high quality. To reach this level and keep the long term is a demanding task for the future.

REFERENCES

Deutsches Institut für Normung, 1999. DIN 1076 "Highway Structures – Testing and Inspection", Berlin.

Haardt, P., 2002. Entwicklung eines Bauwerks-Management-Systems für das deutsche Fernstraßennetz Stufe 1 und 2, BASt Reports B43, Bergisch Gladbach.

Holst, R., 2005. Entwicklung eines Bauwerks-Management-Systems für das deutsche Fernstraßennetz Stufe 1 und 2, BASt Reports B50, Bergisch Gladbach.

Holst, R., 2010. Conventional/standard testing methods for concrete: an overview. In C. Maierhofer, H.-W. Reinhardt, G. Dobmann, *Non-destructive evaluation of reinforced concrete structures*: 117–136. Cambridge: Woodhead Publishing.

Bridge Maintenance, Safety, Management, Resilience and Sustainability – Biondini & Frangopol (Eds)
© 2012 Taylor & Francis Group, London, ISBN 978-0-415-62124-3

Updating bridge management system in Korea considering recent subjects in bridge management

J.H. Kim, H.-M. Park & J.S. Kong
School of Civil, Environmental and Architectural Engineering, Korea University, Seoul, Korea

K.-H. Park
Structural Engineering & Bridge Research Division, Korea Institute of Construction Technology, Goyang-Si, Korea

ABSTRACT

The bridge population in Korea has greatly increased since the 1990's. The major bridge authorities in Korea can be divided into three; the central government, the express way corporation and the local governments. The history of the BMS with the central government goes back to the middle of 1980, however, the application of the BMS has been limited very much because it did not include some important issues in bridge management such as structural safety, realistic life-cycle cost data and solution finding strategies. To overcome these restraints, the Korean government has launched a new project to revise the current BMS, including not only practical approaches of realistic cost data and condition status but also research finding bridge deterioration models in terms of safety and decision making processes which is appropriate for local conditions and the environment of Korea.

In this study, the current status of the bridge management system development in Korea is introduced including several key issues such as the analytical/computational framework for statistical regression and reliability based models for bridge member degradation and optimized decision making strategies considering preventive/demanded maintenance.

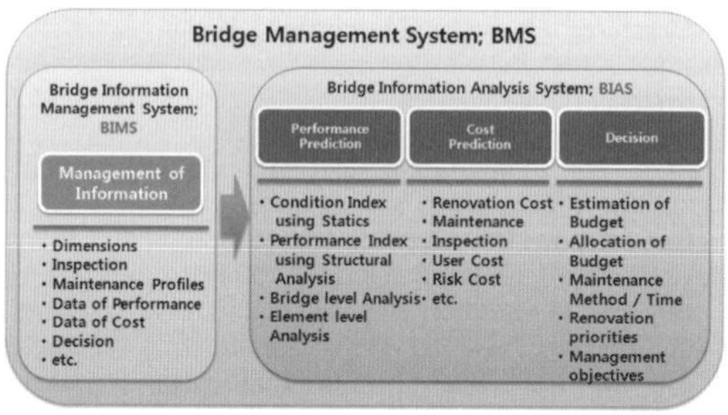

Figure 1. Main features of Bridge Management System.

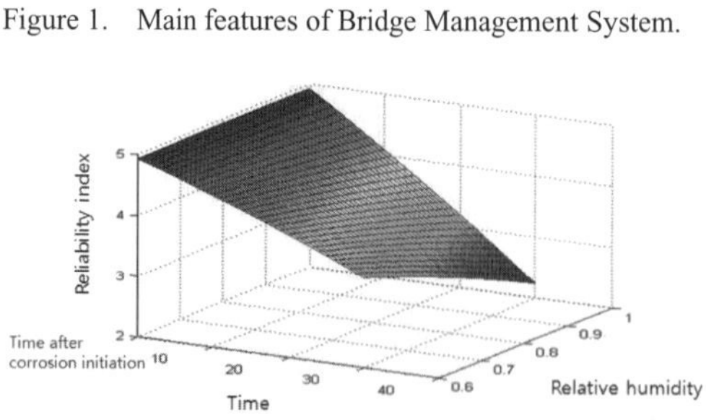

Figure 2. Performance Profile of Time-Relative Humidity-Reliability index.

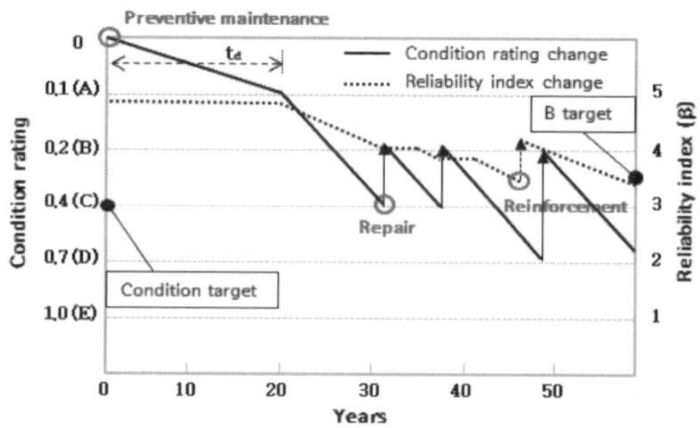

Figure 3. Preventive maintenance scenario of PSCI.

REFERENCES

Comité Euro-International du Béton (CEB). 1989. *Durable structures—CEB design guide*, London: Thomas Telford.

Kong, J. S., and Frangopol D.M. 2003. Life-Cycle Reliability-Based Maintenance Cost Optimization of Deteriorating Structures with Emphasis on Bridges. *Journal of Structural Engineering, ASCE* 129(6): 818–828.

Kong, J. S., and Frangopol D.M. 2004. Prediction of Reliability and Cost Profiles of Deteriorating Structures under Time- and Performance-Controlled Maintenance. *Journal of Structural Engineering, ASCE* 130(12): 1865–1874.

Mori, Y. 1992. Reliability-based condition assessment and life prediction of concrete structures. PhD thesis, *Johns Hopkins University*.

Mori, Y., and Ellingwood, B.R. 1994. Maintaining reliability of concrete structures II: optimum inspection/repair. *Journal of Structural Engineering, ASCE* 120(3): 846–862.

Bridge Maintenance, Safety, Management, Resilience and Sustainability – Biondini & Frangopol (Eds)
© 2012 Taylor & Francis Group, London, ISBN 978-0-415-62124-3

The use of MINLP to determine optimal preservation strategies for road links composed of pavement sections and bridges

N. Lethanh & B.T. Adey
Institute for Construction and Infrastructure Management, Swiss Federal Institute of Technology, Zurich, ETHZ, Switzerland

ABSTRACT

Effective management of road links composed of road sections and bridges requires the determination of preservation strategies that maximize stakeholder benefits. Solving such an optimization problem is challenging since the objective function and constraints are mostly non-linear, and the execution of interventions is discrete. This type of problem can, however, be solved using mixed-integer nonlinear programming (MINLP) methods (Bussieck & Pruessner 2003). In MINLP programs, the objective function, deterioration and constraints can all be modeled as non-linear continuous functions, and the control variables (or decision variables) can be modeled as integers. The use of MINLP programs are an improved representation of reality than earlier methods.

This paper proposes a MINLP model for determination of optimal preservation strategies for a road link composed of multiple infrastructure objects. The development of our model is in favor of the model developed by (Vipul & Ignacio 1998). In the cited paper, the author developed a MINLP model for solving the optimal time to clean the furnaces, which are used to produce Ethylene. The model presented in this paper is based on the model in the cited paper, but it is extended to allow the determination of the optimal time and type of intervention to be executed on a road link, where deterioration on individual object results in increasing user and public costs. A discount factor, which is not included in the cited paper, has also been added. The deterioration process of each object is modeled deterministically as the changing value of a performance indicator with respect to time.

The impacts on four groups of stakeholders (owner, users, directly affected public, and indirectly affected public) (Adey et al. 2012), all of which are dependent on the physical condition of the objects and intervention types are considered. The use of the MINLP methods is illustrated by determining the optimal preservation strategies for a road link consisting of 2 road sections and 1 bridge that are each subjected to gradual deterioration processes.

REFERENCES

Adey, B., Lethanh, N. & Phillippe, L., 2012. An Impact Hierarchy for the Evaluation of Intervention Strategies for Public Roads. In *4th European Pavement and Asset Management Conference*. Malmo, Sweden.

Bussieck, M.R. & Pruessner, A., 2003. Mixed-Integer Nonlinear Programming. *Memory*, 20007(1), pp. 1–7.

Vipul, J. & Ignacio, E.G., 1998. Cyclic Scheduling of Continuous Parallel-Process Units with Decaying Performance. *AIChE Journal*, 44(7), pp.1623-1636. Available at: http://onlinelibrary.wiley.com/doi/10.1002/aic.690440714/pdf.

Bridge Maintenance, Safety, Management, Resilience and Sustainability – Biondini & Frangopol (Eds)
 ISBN 978-0-415-62124-3

Bridge risk management: Back to basics

P.S. McCarten
Napier, New Zealand

ABSTRACT

Bridge Management Systems are designed to assist Road Controlling Authorities (RCA's) in developing preservation and improvement programmes for bridges in their networks. Bridge data (from both a visual inspection programme and special monitoring programmes) is used with analytical tools to develop these programmes.

Bush et al (2011) have recognized that while efficient application of current knowledge and learning is being achieved it may not be an effective use of resources and effort. It is also contended that sub-optimal outcomes result and this impacts on the value of the public infrastructure expenditure. RCA audits are showing that Bridge Managers need to move away from a maintenance and condition centric viewpoint to align with operational and strategic targets and should be moving toward an advanced asset management approach. Where advanced asset management is a process consisting of predictive modeling, risk management and optimized decision making to establish asset lifecycle treatment options and related long term cash flow predictions. With advanced asset management there will be clear links between delivered projects and strategic targets. Bush et al (2011) highlight the importance of collecting bridge data that allows the wider strategic vision to be understood and optimal decisions, to achieve a sustainable economic network, to be made. In order to understand the broader performance target requirements data related to environmental impact, safety related issues, asset history, project and RCA costs, service levels and performance targets is required.

The paper highlights the importance for using high quality data to determine bridge safety (identifying the conditions under which bridge failure could occur) and reliability (the likelihood that a bridge will fail to perform). The method used by Stewart et al (2001) to determine reliability is summarised.

The paper reinforces the importance for following a structured approach to risk assessment and outlines the activities presented in AS/NZS HB-89 (2011) being:

- Risk identification
- Control analysis
- Consequence
- Likelihood
- Level of risk
- Risk evaluation

The Cost Benefit Analysis technique is discussed with a bridge pier washout vulnerability presented.

The paper concludes that while managing maintenance and condition of bridges it is important for the RCA to ensure bridge stock management aligns with operational and strategic targets. Risk management is fundamental to achieving those targets. This paper reinforces the need for collecting high quality data in a cost effective manner and utilizing it to assess bridge safety, reliability and undertaking comprehensive risk assessment. Risk evaluation techniques, such as Cost Benefit Analysis, are available to transfer complex engineering issues into terms suitable for decision makers. The paper shows that bridges with aggregated probabilities of failure greater than 3% Annual Exceedance Probability are likely candidates for bridge replacement. The paper concludes that Bridge Asset Management Plans provide a suitable method for recording, tracking and reporting risk issues to stakeholders.

REFERENCES

AS/NZS HB-89, "Risk Management – Guidelines on Risk Assessment Techniques", 2011 (Working draft)

Bush, Omenzetter, Henning and McCarten "A risk and criticality based approach to bridge performance data collection". Structure and Infrastructure Engineering, UK, Taylor & Francis 2011

Stewart, Rosowsky and Val, "Reliability based bridge assessment using risk ranking decision analysis", Structural Safety, 23, pp. 397–405, 2001

Bridge Maintenance, Safety, Management, Resilience and Sustainability – Biondini & Frangopol (Eds)
© 2012 Taylor & Francis Group, London, ISBN 978-0-415-62124-3

Consistency of bridge deterioration rates across agencies

P.D. Thompson
Consultant, Seattle, WA, USA

ABSTRACT

Most agencies having analytical capabilities in their bridge management systems use inspections based on bridge elements and condition states, and use Markovian deterioration models. It is common to base these models on expert judgment elicitation, but increasingly agencies have sufficient data to develop deterioration models using past inspection and work accomplishment data.

In the United States, the American Association of State Highway and Transportation Officials maintains the Pontis bridge management system, used in 45 of the nation's 50 states, and offers a standardized inspection manual. This makes it possible to estimate and compare deterioration models among agencies, to investigate the level of consistency in model parameters and to understand how agency characteristics, policies, and climate can affect deterioration rates. In this paper, models developed for Florida and Virginia are compared, describing how the similarities and differences between these two agencies affect deterioration rates for pure Markovian and hybrid Markovian/Weibull models.

In order to estimate statistical models of deterioration and action effectiveness, it is necessary to separate the effect of deterioration from the effect of agency actions. These effects are not directly measured, but must be deduced from a limited amount of information in two snapshots of condition spaced 2 years apart, plus any available evidence of agency actions that may have been performed in between the two snapshots. Figure 1 shows the problem schematically. Methods of data processing and statistical analysis were developed to separate the effects of deterioration from those of preservation so these processes could be accurately measured.

In both Florida and Virginia, it was found that actual deterioration transition times, measured from inspection data, were at least twice as long as the original transition times estimated by expert judgment. This very large difference indicates that expert judgment is an unreliable source of deterioration models for bridge elements.

Another interesting finding is that traffic volume appears to affect deterioration rates for all elements,

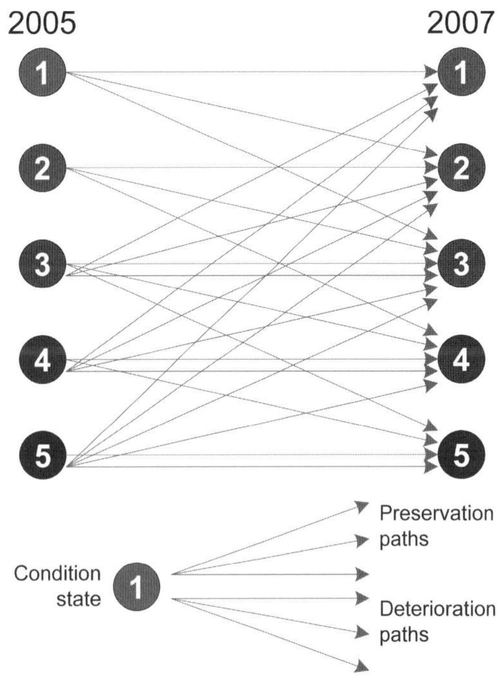

Figure 1. Changes in condition between inspections.

with faster transition times on bridges with higher traffic volume. In general, Virginia's bridges, with lower traffic volumes than Florida, experienced slower deterioration even though its climate is more aggressive.

In both Florida and Virginia, the researchers developed Weibull models for the onset of deterioration. These models were consistent between the two agencies, in that expansion joints and decks showed relatively rapid onset of deterioration, compared to superstructure and substructure elements. However, Virginia again showed a consistently longer time before the first signs of deterioration appeared.

Comparison of the two studies provided some valuable insights on data processing and analysis that will help practitioners improve future deterioration studies.

Bridge Maintenance, Safety, Management, Resilience and Sustainability – Biondini & Frangopol (Eds)
© 2012 Taylor & Francis Group, London, ISBN 978-0-415-62124-3

Risk based bridge planning in Minnesota

P.D. Thompson
Consultant, Seattle, WA, USA

H. Rogers
Michael Baker Jr., Inc., Harrisburg, PA, USA

D. Thomas
Minnesota Department of Transportation, Saint Paul, MN, USA

ABSTRACT

The Minnesota Department of Transportation (Mn/DOT) manages an inventory of more than 19,000 structures. In response to 2010 state legislation, the Bridge Replacement and Improvement Management (BRIM) system was developed as a tool to rank potential bridge projects by directly considering the risk of an interruption to service in Mn/DOT's long term planning process. As a functioning communicative tool, this database will help project stakeholders to understand how Mn/DOT prioritizes and programs bridge projects for future contracts.

The database and tool contain a risk assessment model to provide a consistent rating and ranking of Minnesota bridges using the principles of risk management. BRIM consists of a set of risk evaluation models that consider the major natural and man-made hazards affecting the bridge inventory: advanced deterioration of deck, superstructure, and substructure; scour; fracture criticality; fatigue; overweight and over-height trucks.

Risk assessment considers three main elements of risk for each structure:

- Likelihood of hazard. This element incorporates much of the external, probabilistic aspect of the hazard, such as the probability of a damaging flood or the number of overweight trucks in the traffic stream. This information is typically developed from network-level (as opposed to site-specific) research, such as hydrology and traffic studies. However, it describes the specific site where the bridge is located.
- Consequence to the structure. This element is the direct, deterministic measurement of characteristics of the structure that make up its resilience, its ability to resist damage and interruption to service; for example redundancy, condition, and operating rating.
- Impact on the public. This element incorporates characteristics of the network and environment that cause damage beyond what is experienced at the time and place that a hazard strikes. For example, a traffic accident on a bridge at rush hour can tie up traffic for miles around. A bridge that is posted due to advanced deterioration may force trucks and buses to use longer routes, incurring higher costs.

Likelihood and consequence are combined into a single indicator of bridge resilience. This is conventionally measured on a bounded scale of 0 to 100, making it easy to compare one bridge with another and to combine separate hazard scales. A separate table or formula computes a resilience score for each hazard (Figure 1). Then the scores are combined using a weighted average, to yield a Facility Level Resilience Score. Bridges can be compared and ranked using this score.

The potential impact of each bridge's resilience on the public is characterized by traffic volume and by the role of the bridge in the network, represented by an importance factor. The analysis is presented as a customizable spreadsheet to enable the engineer to quickly evaluate and compare bridges in terms of their relative risk, to assist in selecting and prioritizing bridge corrective and replacement actions.

		SUPERSTRUCTURE CONDITION		
		Smart flag reduction		
NBI	Condition	None	Case 1	Case 2
N	Not applicable	100	100	100
9	Excellent	100	90	80
8	Very good	95	85	80
7	Good	90	80	60
6	Satisfactory	75	60	40
5	Fair	55	40	25
4	Poor	35	25	10
3	Serious	15	10	0
2	Critical	5	5	0
1	Imminent fail	0	0	0
0	Failed	0	0	0

Figure 1. BPI (resilience) table for superstructure condition.

Bridge Maintenance, Safety, Management, Resilience and Sustainability – Biondini & Frangopol (Eds)
© 2012 Taylor & Francis Group, London, ISBN 978-0-415-62124-3

Railway bridge risk assessment in Finland

J. Wuorenjuuri
VR Track Ltd, Helsinki, Finland

ABSTRACT

There are several different kinds of risks on railway bridges. Only some of them are structurally related. The most significant reason is that the design age of these bridges is very old.

The design age means that, these bridges do not meet the standards of the current traffic needs. At the times of design and construction, speeds of trains were less than 100 km/h and the design load models of bridges were made of 18 or 20 ton axle loads. Now the current traffic needs 200 km/h speeds and axle loads up to 25 tons.

The bridges are too narrow and low for electrified lines, the geometry of tracks on bridges are often not optimal for the new trains. All these have a risk value for required service-level.

The risk of failure may cause either derailment of trains or greater impact loads on structures or their elements. This has to be managed with inspection programs. Often inspections have to be executed under train traffic, because without loading, possible defects cannot be detected. It is also important to collect maintenance data from these bridges and not rely on only inspection data.

Structural risks include the fact that the safety margins are already in use as the traffic loads have increased. Old steel bridges have several details that were not structurally calculated to carry loads. Fatigue was not taken into account during old designs as known today. There is very little documented knowledge of the old abutments of stone and wooden piles. Also the quality of rivets in steel structure joints is an unknown detail.

This risk has to be managed with specially trained inspectors whose experience and understanding is sufficient of old steel structures, their design and behavior under loads. Also monitoring can be used, but the problem for this often is, that monitoring can not be executed without understanding of the bridges. An experienced professional of these bridges needs to be involved in specifying what should be monitored, what are the limit values and how to interpret the results.

One of the greater risks of underpasses is the collisions of road vehicles. Especially steel girder bridges are very light and are easily moved out of place on impact. This risk is a major train traffic disturbance and for the vehicles under the bridge.

Safety related risks on railway bridges always include train safety, electrical safety and the safety of personnel work on the track. All maintenance work has to be done according to safety regulations. Inspection and maintenance aspects should be included into designs of bridges more often and maintenance contracts as well.

In addition the European Railway Agency and the demands for the interoperability (TSI) require that safety margins and all other safety issues are met on bridges. The approval processes of new and improved bridges require very high risk evaluation and documentation.

The risks of railway bridges differ of those of road bridges very much. In most cases of a collapse or other severe defect, there are no alternative routes. If something happens, the track is out of service for a long time. This is one of the reason safety margins and risk management need to be sufficient on railway bridges.

REFERENCE

Wuorenjuuri Janne 2010. Management Report of Railway Bridges in Finland 2010. Helsinki: The Finnish Transport Agency.

Reliability analysis of bridge structures

Organizers: F. Schoefs & F. Lanata

Bridge Maintenance, Safety, Management, Resilience and Sustainability – Biondini & Frangopol (Eds)
© 2012 Taylor & Francis Group, London, ISBN 978-0-415-62124-3

Updating the reliability of existing PC bridge girders by incorporating spatial variations

M. Akiyama
Waseda University, Tokyo, Japan

D.M. Frangopol
Lehigh University, Bethlehem, PA, USA

I. Yoshida
Tokyo City University, Tokyo, Japan

ABSTRACT

This paper presents a framework for estimating the time-dependent reliability of prestressed concrete (PC) bridge girders in a marine environment. The analysis considers the spatial variability associated with concrete cover c, surface chloride concentrations C_0, and coefficient of diffusion of chloride D_c over the entire area. Random variables for estimating the corrosion process of PC bridge girders are updated based on observational information such as chloride concentration distribution by coring test. Sequential Monte Carlo Simulation (SMCS) can update multiple random variables based on observational information (Akiyama et al 2010). Figure 1 shows the flowchart for estimating the time-dependent reliability of existing PC bridge girders by incorporating spatial distribution and updating using SMCS.

PC bridge girder is modeled as one-dimension spatial model. The bridge girder is divided into N elements, and a random variable is used to represent the field over each element. There are several methods associated with discretization of random fields. This study uses the midpoint method to model random field (e.g., Stewart & Suo 2009). Using the measured values of concrete cover, coefficient of diffusion of chloride, and surface chloride concentration provided by the field investigation of existing corroded concrete decks in a marine environment (Kato et al. 2006), correlation lengths are determined by empirical variogram describing the degree of spatial dependence of a spatial random field. Once the random field is defined, random variables are generated using MCS for each element.

In an illustrative example, the effect of the spatial interval of observational information given by coring test on the updated estimate of PC bridge girder reliability is investigated.

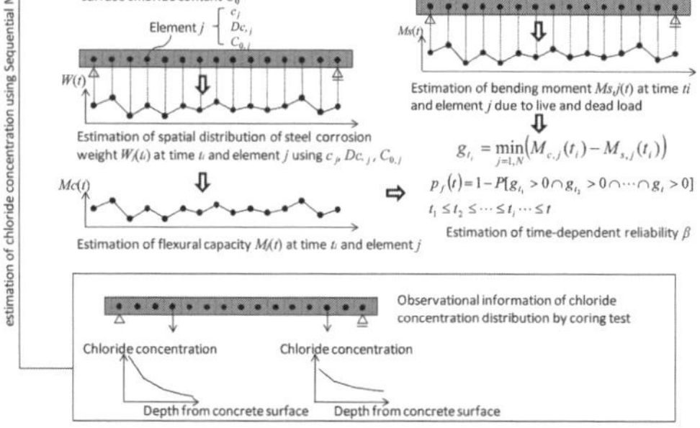

Figure 1. Flowchart for estimating the time-dependent reliability of existing PC bridge girders by incorporating spatial distribution and updating using Sequential MCS.

REFERENCES

Akiyama M., Frangopol D.M. and Yoshida I. 2010. Time-dependent reliability analysis of existing RC structures in a marine environment using hazard associated with airborne chlorides, *Engineering Structures* 32(11): 3768–3779.

Akiyama M., Frangopol D.M. and Matsuzaki H. 2011. Life-cycle reliability of RC bridge piers under seismic and air-borne chloride hazards. *Earthquake Engineering and Structural Dynamics* 40(15): 1671–1687.

Akiyama M., Frangopol D.M. and Suzuki M. 2012. Integration of the effects of airborne chlorides into reliability-based durability design of R/C structures in a marine environment. *Structure and Infrastructure Engineering*, 8(2): 259–270.

Kato E., Iwanami M., Yamaji T. & Yokota H. 2006. Structural performance and deterioration due to chloride attack of reinforced concrete deck of existing piers. *Technical note of the port and airport research institute* 114. (in Japanese)

Stewart M. G. 2009. Mechanical behavior of pitting corrosion of flexural and shear reinforcement and its effect on structural reliability of corroding RC beams. *Structural Safety* 31: 19–30.

Stewart M.G. & Suo Q. 2009. Extent of spatially variable corrosion damage as an indicator of strength and time-dependent reliability of RC beams. *Engineering Structures* 31(1): 198–207.

Bridge Maintenance, Safety, Management, Resilience and Sustainability – Biondini & Frangopol (Eds)
 ISBN 978-0-415-62124-3

Estimating the remaining service life of a historical railway bridge

P. Basso
University of Pavia, Pavia, Italy

S. Casciati
University of Catania, Siracusa, Italy

L. Faravelli
University of Pavia, Pavia, Italy

ABSTRACT

In 1889, an arch iron truss bridge (Figure 1) was realized by Gustave Eiffel along the "Oran-Tlemcen" railway line at 7 km from the east of Tlemcen city, in Algeria. In this paper, the age of the bridge and the variation of the loading conditions over the time are considered in order to estimate its remaining service life. All the data concerning the geometry of the structure and the materials were previously collected in (Boumechra and Hamdaoui, 2008), together with the historical data on the train traffic loading conditions. A summary of the most relevant data is reported.

The main structure of the bridge is made of puddled iron, whose fatigue behavior is characterized by the corresponding Wöhler curves. The influence of the mean stress level on the fatigue life is also considered in the analyses using both the models of Goodman (Goodman, 1899) and Morrow (as reported, for instance, in Darveaux et al., 1995). The Palmgren-Miner's linear rule (Miner, 1945) is adopted to model the process of damage accumulation into the material during the fatigue cycles. A damage index is accordingly defined.

The framework for the fatigue analysis is further described with reference to a critical element of the considered case-study. The stress time histories in a node of the element are simulated by a multi-step static analysis of a finite element model of the bridge under different load combinations, which are selected according to the train traffic data of the last century. Randomized artificial time histories are then created based on the spectral analysis of the former results. The rainflow-counting technique (Matsuishi & Endo, 1968) is used to evaluate the number of meaningful loading cycles to be considered in the fatigue analysis.

To assess the reliability of the bridge, a fully probabilistic approach is developed by formulating a limit state function based on the adopted damage accumulation criterion and by assuming the stress amplitude,

Figure 1. The El Ourit Bridge.

the damage threshold value, and the material fatigue strength as random variables. Finally, the achieved results are discussed in order to evaluate the residual service life of the bridge. A comparison with the reference values of the reliability index prescribed in the Eurocodes (Annexes B and C in UNI EN 1990:2006) is also provided.

REFERENCES

Boumechra N. and Hamdaoui K., 2008, Dynamic and Fatigue Analyses of an 18th Century Steel Arch Bridge, *Proceedings of the 2008 Seismic Engineering International Conference (MERCEA'08)*, July 08–11, Reggio Calabria, Italy, pp. 69–75.

Darveaux, R., Banerji, K., Mawer, A. and Dody, G., 1995. *Reliability of plastic ball grid array assemblies*, Chap. 13, Ball Grid Array Technology, ed. J. H. Lau, McGraw-Hill, pp. 379–442.

Goodman, J., 1899. *Mechanics Applied to Engineering*, Longman, Green & Company, London.

Miner, M.A., 1945. Coumulative Damage in Fatigue, *J. App. Mech*, ASME, 12: 159–164.

Matsuishi, M. & Endo, T., 1968. Fatigue of metals subjected to varying stress, *Japan Soc. Mech. Engineering*.

UNI EN 1990:2006 (Basis of structural design).

Bridge Maintenance, Safety, Management, Resilience and Sustainability – Biondini & Frangopol (Eds)
© 2012 Taylor & Francis Group, London, ISBN 978-0-415-62124-3

Segmentation and condition rating of concrete bridge decks using NDE for more objective inspection and rehabilitation planning

N. Gucunski & A. Maher
Rutgers University, Center for Advanced Infrastructure and Transportation, Piscataway, NJ, USA

H. Ghasemi & F.S. Ibrahim
Federal Highway Administration, Turner-Fairbank Highway Research Center, McLean, VA, USA

ABSTRACT

Bridge decks encompass by far the largest part of expenditures used in bridge maintenance, rehabilitation and repair. The dominant practice by state Departments of Transportation (DOTs) in evaluation of bridge decks is by visual inspection, and the use of simple nondestructive methods like chain drag and hammer sounding. Such an evaluation often provides a subjective condition assessment and is highly dependent on the experience of an inspector. The presented study concentrates on a more objective condition assessment of bridge decks using a complementary use of nondestructive evaluation (NDE) techniques.

The condition assessment has three main components: assessment of corrosive environment and corrosion processes, concrete degradation assessment, and assessment with respect to deck delamination. The NDE technologies used in the assessment include: half-cell potential (HCP), electrical resistivity (ER), ultrasonic surface waves (USW) ground penetrating radar (GPR), and impact echo (IE) method. The HCP is used to assess probability of active corrosion, while the ER method is used to characterize the concrete's corrosive environment. The USW method allows measurement of concrete modulus, and thus detection of possible concrete degradation. The GPR provides a qualitative assessment of the condition of the deck based on the measurement of signal attenuation on the top rebar level. Finally, the IE enables detection and characterization of delamination.

Results from NDE surveys are presented in terms of condition maps. The maps serve as the basis to obtain the deck's condition rating with respect to one of the three deterioration types. The condition rating is calculated using a weighted area approach. For example, the overall rating on a scale 0 to 100 (best) with respect to active corrosion is calculated from the percentages of area falling into three probability-of-active-corrosion zones identified by the ASTM C876. The area having the 90% probability of no active corrosion uses a weight factor of 100, the area having the 90% probability of active corrosion uses 0, and the area in the transition zone uses a weight factor of 50. Similar approaches are used to evaluate condition rating with respect to delamination and concrete degradation. Finally, the overall condition rating of a deck can be defined from a weighted average of condition ratings obtained from different NDE surveys and visual inspection. This is illustrated in Table 1 for a bridge in Virginia surveyed in 2009 and 2011. The results are also a good illustration of objective assessment of deterioration progression since the analysis is based on the use of quantitative NDE data.

Table 1. Condition rating with respect to active corrosion, delamination and concrete degradation, and the overall rating of Virginia LTBPP bridge.

	2009	2011
Active Corrosion	39.4	28.1
Delamination Assessment	70.0	57.2
Concrete Degradation	48.1	35.3
Combined Rating	52.5	40.2

Different condition rating schemes, guided by different objectives of their usage, are presented. Those include: 1) condition rating comparisons of bridges on the network level for condition monitoring and rehabilitation, and 2) segmentation and rating on the project level to identify areas of the deck that should have higher priority in rehabilitation. The NDE of bridge decks and condition rating is illustrated by a series of examples from bridge deck evaluation within the FHWA's Long Term Bridge Performance (LTBP) Program.

Bridge Maintenance, Safety, Management, Resilience and Sustainability – Biondini & Frangopol (Eds)
© 2012 Taylor & Francis Group, London, ISBN 978-0-415-62124-3

Effect of ASR on steel-concrete bond behavior in the lap-splice region of bridge columns

Q. Huang
The University of Akron, Akron, OH

P. Gardoni & A. Pagnotta
Texas A&M University, College Station, TX

D. Trejo
Oregon State University, Corvallis, OR

ABSTRACT

Columns in reinforced concrete (RC) bridges commonly contain a lap splice of the longitudinal reinforcement near the base of the column. It is important to ensure that the splice is of sufficient length to transfer internal forces so that the column can resist the applied lateral load. However, deterioration mechanisms, such as alkali silica reaction (ASR) may weaken the bond between the reinforcement and the concrete in the lap-splice region, which may subsequently lead to changes in the flexural capacity and demand of the column and a reduction in the bridge reliability.

This paper aims to develop and calibrate a finite element model (FEM) of RC columns subjected to ASR to study the effect of ASR on the bond behavior in the lap-splice region. Eight large-scale bridge column specimens were constructed. Figure 1 shows the specimen dimensions and rebar layout. Two specimens, C1 and C2, were the control specimens and did not experience any deterioration. The remaining six specimens, C3-C8, were subject to accelerated wet-dry cycles to accelerate the development of ASR. The columns were then tested destructively at different levels of ASR damage (two columns for each level) to determine the reduction in capacity due to the deterioration. The levels of deterioration were assessed qualitatively before the destructive testing based on the extent of concrete cracking as: no damage (C1 and C2), early stage (C3 and C4), middle/late stage (C5 and C6), and late stage (C7 and C8). Note that these stages were established based on the petrography analysis of concrete cores taken from specimens after structural testing and also from the surface and internal expansion measurements and cracking throughout the specimen prior to testing.

For each column specimen, a FEM is built in Abaqus, where a bond-slip model is explicitly used to model the interface of the concrete and reinforcing bar. The bond-slip model is then calibrated using the experimental force-displacement data obtained from four-point load test of the large-scale column specimens subject to different degrees of deterioration.

It is found that the post-cracking stiffness becomes greater when ASR deterioration becomes severe according to the experimental test results. To reflect this behavior, the initial slope of the bond model in the FEM needs to be increased. Therefore, with limited ASR deterioration, the bond condition actually can be improved due to the fact that the expansion resulting from ASR can cause more confinement effect on the bond. However, as the deterioration continues, the bond strength may start decreasing. When more specimens with higher level of ASR deterioration are tested, the bond behavior can be further studied.

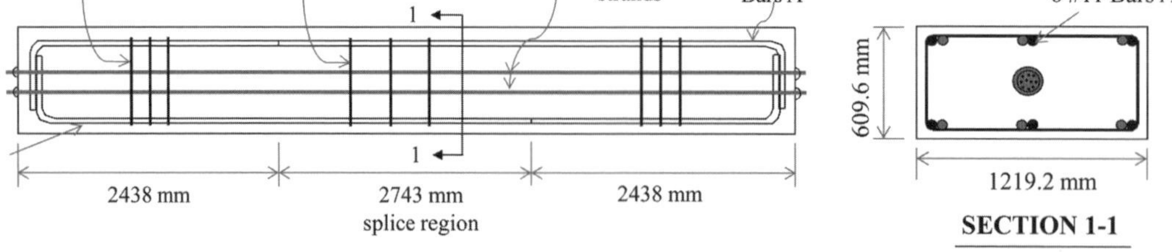

Figure 1. Dimensions and reinforcement layout of large-scale column specimens.

Bridge Maintenance, Safety, Management, Resilience and Sustainability – Biondini & Frangopol (Eds)
© 2012 Taylor & Francis Group, London, ISBN 978-0-415-62124-3

A probabilistic approach for the quantification of structural robustness

N. Kagho & A. Orcesi
University Paris-Est, IFSTTAR, Paris, France

C. Cremona
Technical Center for Bridge Engineering, SETRA, Sourdun, France

ABSTRACT

For civil engineering structures, robustness reflects the ability of the entire system, including the most critical elements, to withstand events without being damaged to an extent disproportionate to the original cause. Under uncertainties on loading effects and material/geometric properties, defining and quantifying robustness is a crucial issue.

The concept of robustness has been recognized as a major property of structural systems to prevent disproportionate failure due to accidental actions (Eurocode EN 1991-1-7, Knoll & Vogel 2008, Starossek & Haberland 2008, Biondini & Restelli 2008), and has become a significant aspect to consider in the design of structures. Several researches have been recently conducted (Faber et al. 2006, Mohammadkhani-Shali & Cremona 2007, Baker et al. 2008), in spite of which a global consensus for the consideration of the structural robustness during the phases of design and execution of structures still does not exist.

This paper presents a probabilistic framework to quantify the structural robustness in uncertain context, by considering uncertainties associated with structural resistance and loads. Under various uncertainties in modelling of structures and randomness in loading phenomena, the use of a probabilistic approach is justified. By definition, a measure of robustness should allow appreciating the gap between the extent of the damage and the cause, or, in other words, quantifying the impact of local failure on the overall system failure. Small and large impacts should then characterize robust and non robust structures, respectively. Since there might be a large number of failure paths for structures with a high degree of redundancy, stochastically dominant failure paths are identified by using structural system reliability methods (Thoft-Christensen & Murotsu 1986).

The proposed approach aims at quantifying the robustness of a structural system by means of reliability theory. The proposed robustness indices quantify the impact of a local damage on major structural damage. As disproportionate failure, progressive collapse, and structural robustness are tied, failure mechanism analysis of a structure can efficiently help assess the robustness of a structure by considering the stochastically most significant failure paths. The proposed methodology is applied to some examples and the results are discussed.

REFERENCES

Baker, J.W., Schubert, M. & Faber, M.H. 2008. On the assessment of robustness. *Elsevier, Structural Safety* 30(3): 253–267.

Biondini, F. & Restelli, S. 2008. Damage propagation and structural robustness. In Biondini & Frangopol (eds), *Life-Cycle Civil Engineering*. Taylor & Francis Group, London.

EN 1991-1-7. 2006. Actions on Structures: Part 1–7: Accidental Actions. *Eurocode 1: CEN 2006.*

Faber, M.H., Maes, M.A., Straub, D. & Baker, J. 2006. On the Quantification of robustness of Structures. *Offshore Mechanics and Arctic Engineering; Proc. Intern. Conf., Hamburg, 4–9 June 2006* . Germany.

Knoll, F. & Vogel, T. 2008. Design for Robustness. *Structural Engineering Documents: IABSE, AIPC, IVBH.* Zurich.

Madsen, H.O., Krenk, S. & Lind, N.C. 1986. Methods of structural safety. *Englewood Cliffs (NJ): Prentice-Hall, Inc.*

Mohammadkhani-Shali, S. & Cremona, C. 2007. Damage effects on system redundancy for existing reinforced concrete bridges using response surface methods. ESREL 2007, Safety and Reliability Conference, ESRA, Stavanger, Norway, June 2007.

Starossek, U. & Haberland, M. 2008. Approaches to measures of structural robustness. *Bridge Maintenance, Safety and Management; IABMAS'08, 4th Intern. Conf., Seoul, 13–17 July 2008.* Korea.

Thoft-Christensen, P. & Murotsu, Y. 1986. Application of structural Systems Reliability Theory. *Springer-Verlag*, Berlin.

Bridge Maintenance, Safety, Management, Resilience and Sustainability – Biondini & Frangopol (Eds)
© 2012 Taylor & Francis Group, London, ISBN 978-0-415-62124-3

Reliability analysis of highway bridge structures considering ultimate load effects

L.A. McCarthy & C.C. Caprani
Department of Civil & Structural Engineering, Dublin Institute of Technology, Ireland

ABSTRACT

In the reliability analysis of highway bridge structures, traffic loading is one of the most variable parameters and so bridge safety estimates are very sensitive to the model assumed for it. It is often assumed that the bridge responds elastically under extreme traffic loads. This may not be the case, since bridges are typically designed to be close to their load-carrying capacity under such loading events. Bridges whose strength is compromised through degradation are even more susceptible to extreme loading events. An elastic-perfectly plastic material model is often assumed, but the real moment-rotation relationship of common bridge sections is quite different. In reality therefore, the bridge may not be behaving elastically under extreme loading, but may be yielding in one or more locations. Whilst by the Lower-Bound Theorem, safety against collapse is ensured when using such models, their use in a reliability analysis may lead to incorrect reliability indices.

In this work, a nonlinear material response of a three span beam and slab bridge structure is assessed using a nonlinear finite element model. The bridge is subjected to a lifetime (100 years) of simulated traffic using Monte Carlo Simulation of Weigh in Motion (WIM) data. The annual maximum loading events are determined using elastic analysis: influence lines and lateral distribution factors are determined from a grillage analysis of the bridge. These 100 extreme loading events are then analyzed using the nonlinear finite element model.

The material nonlinearity of the structure is analyzed using the generalized Clough model (Li & Li, 2007) which traces the spread of plasticity through a structure using force recovery parameters and an incremental loading procedure. A load factor for failure for each of the 100 annual maximum loading events is established. Failure is defined as when the bending moment at any point in the structure exceeds an allowable bending moment.

Extreme value statistics are used to establish the characteristic elastic load effect. A simplistic linear

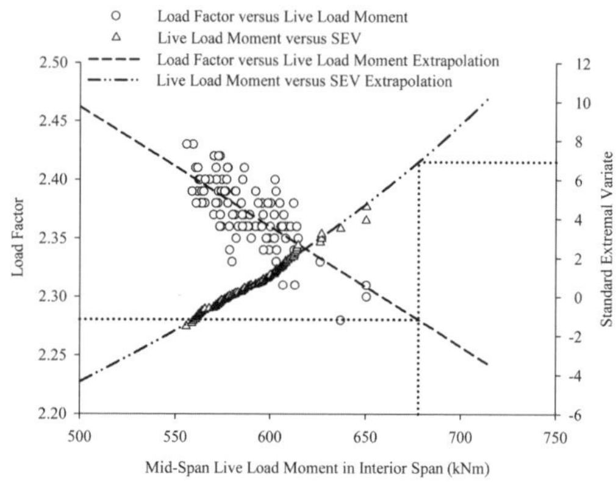

Figure 1. Extrapolation to 1000-year load factor and elastic live load effect.

correlation between elastic load effect and the load factor against failure is established and used to determine the load factor for a 1000-year return period. Figure 1 illustrates the extrapolation process for the interior span mid-span bending moment. The resulting load factor indicates the suitability of a nonlinear structural model in a reliability analysis.

For comparison, conventional reliability indices are also established using the First Order Reliability Method is also done. The load factor against collapse are compared to the design points corresponding to the maximum probability of failure. It is found that nonlinear modeling may be required for some load effects.

REFERENCES

Li, G.Q. & Li, J.J. 2007. *Advanced Analysis and Design of Steel Frames*. Sussex: John Wiley & Sons.

Rackwitz, R. & Fiessler, B. 1978. Structural Reliability under Combined Random Load Sequences, *Journal of Computer and Structures* 9:489–494.

Bridge Maintenance, Safety, Management, Resilience and Sustainability – Biondini & Frangopol (Eds)

Reliability-based analysis of the progressive collapse of bridges

F. Miao & M. Ghosn
Department of Civil Engineering, The City College of New York/CUNY

ABSTRACT

According to ASCE 10-05 (ASCE 2010), Progressive Collapse is defined as the spread of an initial local failure from element to element resulting in the collapse of an entire structure or a disproportionately large part of it. Although a redundant bridge may survive the damage to one member and continue to carry some load, the dynamic response of the system due to a sudden removal of the member may sometimes cause the collapse of the whole structure.

Existing criteria for reducing the risk of progressive collapse have been developed for buildings using traditional deterministic methods. However, the large uncertainties associated with estimating the strength as well as the dead and live loads applied on bridges require the use of structural reliability methods.

The safety evaluation of a bridge structure requires checking if the effects of the loads applied on the structure exceed the capacities of the individual members or the capacity of the whole system. The live load model used in this study is based on site-specific truck weight and traffic data collected using Weigh-In-Motion (WIM) systems. According to LRFD design code, the design load HL93 were developed using 1975 truck data from the Ontario Ministry of Transportation to project a 75-year live load occurrence. Because truck traffic volume and weight have increased and truck configurations have become more complex, the 1975 Ontario data do not represent U.S. traffic loadings. Recent observations made on truck weight data collected from Weigh-In-Motion (WIM) stations at representative New York State sites have shown that truck weights in New York can be significantly heavier than the generic truck weight data used during the calibration of the AASHTO specifications (Sivakumar et al, 2008 & Ghosn et al. 2010). Therefore, site-specific or state-specific live load models are necessary to be developed based on actual truck weight and traffic data collected at the site or within the state.

This paper uses a Markov-chain based advanced simulation technique to perform a probabilistic dynamic analysis of the mechanisms that may lead to the progressive collapse of bridge structures. The flexibility of the approach allows for the use of site-specific truck weight and traffic data collected using Weigh-In-Motion (WIM) systems. The application of the method is illustrated using one bridge example.

REFERENCES

ASCE, 2002. Minimum Design Loads for Buildings and Other Structures (ASCE 7), American Society of Civil Engineers, New York.

ASCE, 2005. Minimum design loads for buildings and other structures, Reston, Virginia.

Ellingwood, B. R., 2006. Mitigating Risk from Abnormal Loads and Progressive Collapse. Journal of Performance of Constructed Facilities, ASCE, November, 315–323.

Eurocode8, 1994. Design Provisions for structures, European Committee for Standardization.

FEMA, 1997. NEHRP guidelines for the seismic rehabilitation of buildings, FEMA-273, Washington, D.C.

Ghosn, M., Sivakumar, B. and Miao, F., 2010. Calibration of Load and Resistance Factor Rating Methodology in New York State. TRR, 2200, 81–89.

Miao, F. and Ghosn, M., 2011. Modified Subset Simulation Method for Reliability Analysis of Structural Systems. Structural Safety, 33(4–5), 251–260.

Sivakumar, B., Ghosn, M. and Moses, F., 2008. Protocols for Collecting and Using Traffic Data in Bridge Design. NCHRP Report 683, Transportation Research Board, Washington, DC.

Bridge Maintenance, Safety, Management, Resilience and Sustainability – Biondini & Frangopol (Eds)
© 2012 Taylor & Francis Group, London, ISBN 978-0-415-62124-3

Reliability assessment of concrete bridges

D. Novák & B. Teplý
Brno University of Technology, Brno, Czech Republic

R. Pukl
Cervenka Consulting, Prague, Czech Republic

A. Strauss
IKI BOKU University, Vienna, Austria

ABSTRACT

The paper describes a complex methodology for statistical, reliability and risk analyses of concrete bridges. It describes the virtual simulation used on the way from assessment of experimental results to reliability analysis. The whole approach is based on randomization of nonlinear fracture mechanics finite element analysis of concrete structures. Theoretical as well as practical application aspects are presented emphasizing the conceptual framework and key points of solution. Efficient techniques of both nonlinear numerical analysis of concrete structures and stochastic simulation methods of Monte Carlo type have been combined in order to offer an advanced tool for assessment of realistic behavior of concrete structures from statistical and reliability point of views. Degradation aspects, as carbonation of concrete, corrosion of reinforcement, are included too, enabling reliability prediction in time.

The authors combined efficient techniques of both nonlinear numerical analysis of engineering structures and stochastic methods to offer an advanced tool for the assessment of the realistic behavior of concrete structures from the reliability point of view. Within the framework of this complex system attention is also paid to the modeling of degradation phenomena, such as carbonation of concrete, corrosion of reinforcement, chloride attack, etc.

The procedure can be outlined as follows:

- experiment (laboratory, in situ)
- development of a deterministic computational model to capture the experiment
- inverse analysis to get parameters of the model
- deterministic computational model of a structure
- stochastic model of a structure
- statistical, sensitivity and reliability analyses

The combination of all parts (structural analysis, reliability assessment, inverse analysis and degradation modeling) is presented together in a package as the SARA software system.

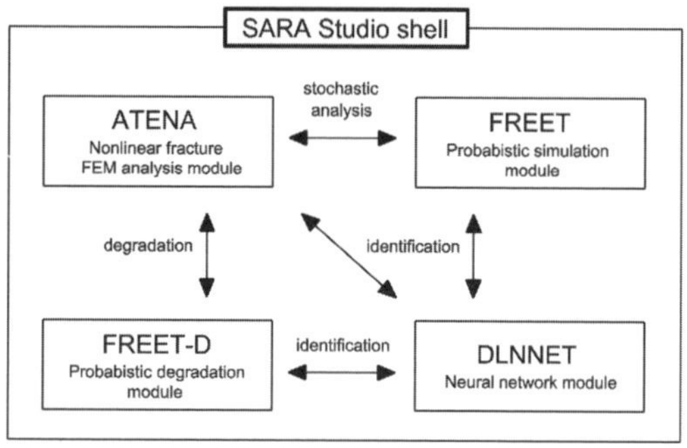

Figure 1. The program combination within SARA software.

A representation of the program combination within SARA software is presented in Fig. 1. It includes: SARA – a software shell which controls the communication between following individual programs: ATENA (Červenka et al., 2007) – FEM nonlinear analysis of concrete structures; FReET (Novák et al., 2011) – the probabilistic engine based on LHS simulation; DLNNET – artificial neural network software; FReET-D (Teplý et al., 2010) – degradation module.

Finally, the paper documents the most interesting application of this methodology and software tools for assessment of bridges from Austria and Italy.

REFERENCES

Červenka, V., Jendele, L. and Červenka, J. 2007. ATENA Program Documentation – Part 1: Theory. Cervenka Consulting, Prague, Czech Republic.

Novák, D., Vořechovský, M. and Rusina, R. 2011. FReET v.1.5 – program documentation. User's and Theory Guides. Brno/Červenka Consulting, Czech Republic, http://www.freet.cz.

Teplý, B., Chromá, M., Vořechovská, D. and Rovnaník, P. 2010. FReET-D. Program documentation, Part 1: Theory, Brno/Červenka Consulting, Czech Republic.

Bridge Maintenance, Safety, Management, Resilience and Sustainability – Biondini & Frangopol (Eds)
© 2012 Taylor & Francis Group, London, ISBN 978-0-415-62124-3

Probabilistic load-modelling and reliability-based load-rating for existing bridges

S.G. Reid
University of Sydney, School of Civil Engineering, NSW, Australia

N. Yaiaroon
Intercity Motorways Division, Department of Highways, Thailand

ABSTRACT

Whereas reliability-based design standards for new bridges are typically based on 'notional' reliabilities for anticipated (hypothetical) traffic loads and idealised structural conditions, the management of existing bridges requires assessments of 'realistic' reliabilities for realistic traffic loads and realistic assessments of the relevant structural conditions.

An important aspect of bridge management that can benefit from realistic reliability assessments is the process of safety assessment and load-rating for old bridges that were not designed for current traffic loads and that do not satisfy current design standards. To date, reliability-based methods have generally not been adopted for routine safety assessment and load-rating for old bridges, and conventionally trained structural engineers are generally not able to perform the relevant reliability calculations.

To facilitate reliability-based assessments, a calculation spreadsheet was developed, based on load and resistance parameters similar to those considered in conventional structural engineering calculations. The calculation spreadsheet gives the bridge reliability as a function of the load and resistance parameters that are entered by the structural engineer. The calculation spreadsheet can also account for the beneficial 'proof-loading' effect of historical traffic loading.

The calculation procedure and results are based on a load model developed to describe bounded 'overload' events. This model has been adapted to describe traffic overloading specifically for bridges with spans of 5–40 m (for which the load-effects are dominated by the effect of a single axle group) and this load model was calibrated in relation to some Australian traffic load data.

The reliability-based calculation procedures and the probabilistic traffic load model (including the relevant extreme value distribution) are reviewed in the paper and reassessed with regard to extensive analyses of Weigh-in-Motion (WIM) data from New South Wales (NSW) in Australia.

Based on analyses if the NSW WIM data, a consistent model of the distributions of peak load-effects is described, and a traffic load model based on the statistics of GVM overload factors is proposed for the purposes of reliability-based load-rating. Estimated distributions for the extreme peak load-effect from 5 million trucks show significant differences between the results based on the NSW WIM data for overloading of ST42.5 trucks (considering a load-rating of ST42.5) and overloading from general traffic accompanying ST42.5 trucks (and sharing common legal limits on axle loads). The choice of the appropriate model to describe the effects associated with a load-rating of ST42.5 depends on the influence that a load-rating of ST42.5 would have in relation to the overloading from other general traffic.

The distributions for overloading based on the NSW WIM data have also been compared with the original traffic load model based on different WIM data. The original traffic load model is shown to overestimate the load-effects for a load-rated bridge (because it is based on WIM data which focused on a relatively small number of special vehicles that are not representative of the traffic for bridges with reduced load-ratings). Accordingly, the load-rating procedure based on the original traffic load model gives conservative results.

However this does not affect the general conclusion that reliability indices are relatively low for structures with low deterministic load ratings and relatively high for structures with high deterministic load ratings. The enhancement of the reliability due to prior service loading can be significant, particularly for relatively weak bridges and bridges with unreliable strengths, and for a bridge with a low deterministic load rating, it is likely that prior service loading must be taken into account to satisfy the target reliability.

Bridge Maintenance, Safety, Management, Resilience and Sustainability – Biondini & Frangopol (Eds)
© 2012 Taylor & Francis Group, London, ISBN 978-0-415-62124-3

Reliability of bridge deck subject to random vehicular and seismic loads through subset simulation

D. Sen & B. Bhattacharya
Indian Institute of Technology Kharagpur, Kharagpur, India

C.S. Manohar
Indian Institute of Science, Bangalore, India

ABSTRACT

This paper presents a reliability analysis of a simply supported bridge deck subjected to random moving and seismic loads. Vehicle structure interactions effects are taken into account, which are affected by vehicle speed, vehicle and bridge masses, deck surface irregularities and material and geometric properties of the bridge all of which could be random in nature. The random field modeling of the bridge surface roughness and the stochastic process modeling of non-stationary seismic loads result in a reliability problem of very high dimensions and typically low probability of failure. Subset simulation incorporating Markov Chain Monte Carlo moves is adopted in this study for the reliability analysis. The bridge is modeled as a single span simply supported Euler-Bernoulli beam and the vehicle is modeled as a SDOF oscillator. Mid span deflection of the beam is computed using method of weighted residual and is used as the performance criteria.

Synthetic earthquake records using the Kanai-Tajimi power spectral density are generated. The records are assumed to be Gaussian zero mean processes. Surface roughness profile is a random process in space. It is generated using the power spectral density proposed by Dodds and Robson.

The random variables considered for the present study are mass of the oscillators, damping coefficient of the oscillators, stiffness of the oscillators, elastic modulus of the beam, second moment of area of the beam, the time at which the earthquake starts (with respect to entry time of first oscillator which is assumed to be $t = 0$), the time taken by each oscillator to travel over the beam (it is assumed that all oscillators travel with the same velocity), time at which the subsequent oscillators enter (with respect to the first vehicle), the intensity of the power spectrum (white noise) of the vibrations at the bedrock level (parameter of Kanai-Tajimi power spectral density) and the surface roughness coefficient (parameter of the road roughness power spectral density).

To begin with only one oscillator is assumed to be travelling over the beam subjected to seismic loading. Initially failure probability is estimated considering all the random parameters. Subsequently each parameter is considered to be deterministic (with mean value) keeping the others to be random. In this study it was seen that the surface profile may not be assumed as a random field. This reduced the dimension of the reliability problem by 20.

Next the failure probability is computed for the case when two oscillators pass over the beam. Again a similar sensitivity study described in the previous paragraph was done on the time between two vehicles. This showed that it cannot be assumed to be deterministic.

Failure probability is estimated for the cases when only one oscillator passes over the beam (no seismic loading) and two oscillators pass over the beam. Failure probability is also estimated for the case when only earthquake acts on the beam. Results for these cases show that even though the loading conditions taken into account for this study are mutually independent the result of the combined case does not reflect that. Interaction between the vehicle structure interactive forces and the earthquake forces leads to drastic increase in the failure probability.

REFERENCES

Au S.K., Beck J.L. (2001). "Estimation of small failure probabilities in high dimensions by subset simulation." *Probabilistic Engineering Mechanics* **16**(2001): 263–277.

Fan F.-G., Ahmadi G. (1990). "Nonstationary Kanai-Tajimi models for El Centro 1940 and Mexico City 1985 earthquakes." *Probabilistic Engineering Mechanics* **Vol. 5, No. 4**.

Fryba L. (2001). "A rough assessment of railway bridges for high speed trains." *Engineering Structures* **23**(2001): 548–556.

Palamas J., Coussy O., Bamberger Y. (1995). "Effects of surface irregularities upon the dynamic response of bridges under suspended moving loads." *Journal of Sound and Vibration* **99**(2): 235–245.

Bridge Maintenance, Safety, Management, Resilience and Sustainability – Biondini & Frangopol (Eds)
 ISBN 978-0-415-62124-3

Extreme value distribution model of vehicle loads incorporating de-correlated tail fitting and stationary gamma process

L. Shunlong, L. Hui, Z. Fujian & G. Yiming
Key Laboratory of Advanced Material & Intelligent Control Technology on Transportation Safety, Harbin Institute of Technology, Harbin, China
School of Transportation Science and Engineering, Harbin Institute of Technology, Harbin, China

Z. Guo
Administration Agency of Nanjing No. 3 Yangtze River Bridge, Nanjing, China

ABSTRACT

Vehicle traffic plays an important role in fatigue deterioration and overload leading to the collapse of bridges. The monitored data show that occurrences of vehicle loads are correlated. And the correlation of vehicle loads is excluded using a Peak Over Threshold (POT) method. Additionally, it is more reasonable to employ the tail region of a distribution when estimating extreme loads. To choose a proper threshold for the tail region, the mean exceedance function (MEF) of the GPD (Shi 2006) is employed and is defined as:

$$e(u) = E(y-u \mid y>u) = \frac{\sigma+\gamma u}{1-\gamma}$$

Considering that the stochastic process of the tail part may not follow the well-known Poisson or Erlang process, a generalized EV distribution function for any homogeneous renewal process is derived.

$$\mathrm{L}\left\{F_{Y_M}(y;t)\right\}(s) = \frac{F(y)\left(1-\mathrm{L}\left\{f(t)\right\}(s)\right)}{s\left(1-F(y)\mathrm{L}\left\{f(t)\right\}(s)\right)}$$

Generalized Pareto distribution is proposed to fit the tail distribution of vehicle loads, whereas the inter-arrival times of vehicle loads are fitted by gamma distribution.

The extreme value distribution function of vehicle weight can be obtained by inversion of the Laplace transform over $[0, T]$,

$$P(Y_m > y) = \mathrm{L}^{-1}\left\{\frac{\left((1+\theta s)^{\frac{1}{2}}-1\right)}{s\left((1+\theta s)^{\frac{1}{2}}-F(y)\right)}\right\}$$

$$= \frac{1-Erf\left(\sqrt{\frac{T}{\theta}}\right)+F(y)e^{\frac{\left(-1+F(y)^2\right)T}{\theta}}\left(1+Erf\left(F(y)\sqrt{\frac{T}{\theta}}\right)\right)}{1+F(y)}$$

where $Erf(\cdot)$ is the error function.

The validation of the proposed approach is conducted using the monitored vehicle loads from the Nanjing No.3 Yangtze River Bridge, and the results show the effectiveness and accuracy of the proposed approach.

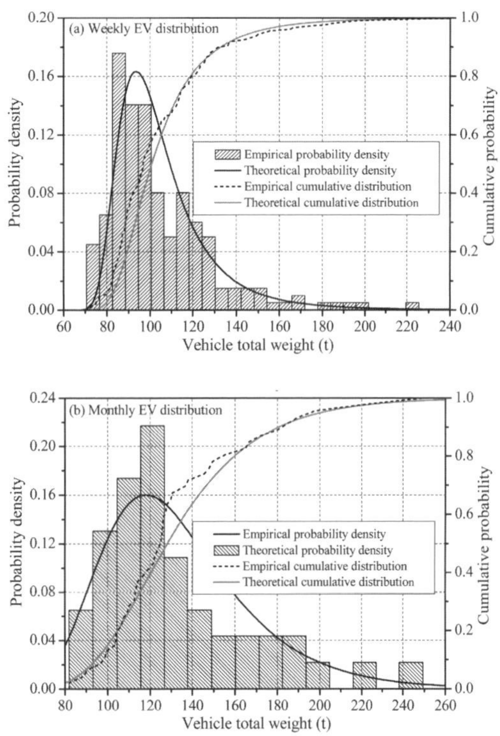

Figure 1. Weekly and monthly extreme value distribution estimation.

REFERENCES

Lan, C., Li, H. & Ou, J. 2011. Traffic load modelling based on structural health monitoring data. *Structure and Infrastructure Engineering* 7: 379–386.

O'Connor, A. & Eugene, J.O.B. 2005. Traffic load modelling and factors influencing the accuracy of predicted extremes. *Canadian Journal of Civil Engineering* 32(1): 270–278.

Shi, D. 2006. *Practical statistical methods for extreme values*. Tianjin: Tianjin science and technology press (PRC).

Bridge Maintenance, Safety, Management, Resilience and Sustainability – Biondini & Frangopol (Eds)
© 2012 Taylor & Francis Group, London, ISBN 978-0-415-62124-3

Probabilistic performance assessment of concrete structures subjected to corrosion process

A. Strauss, R. Wendner & K. Bergmeister
University of Natural Resources and Life Science, Vienna, Austria

B. Teplý & D. Novák
Brno University of Technology, Brno, Czech Republic

ABSTRACT

Influence factors for the service life of a structure are apart from its design, the quality of construction, environmental loads and maintenance during operation. Although in the design phase much consideration is given to durability and relevant loads, a high degree of uncertainty remains with respect to the future development of loads and mechanical resistance of materials. Ensuring proper safety levels of concrete bridges under their gradual deterioration over the entire service life is an expensive task leading to complex decisions regarding maintenance strategies and/or balancing the cost of possible repair or replacement (Petcherdchoo et al. 2008). Within financial constraints the optimal decision can only be made on the basis of effective probabilistic assessment methodologies, as e.g. developed by Stewart & Rosowsky (Stewart and Rosowsky 1998) or proposed in ISO 13822, ISO 13823 and fib.

This paper aims at the discussion and verification of practical approaches to the assessment and the prediction of an existing structures time variable performance with respect to the Serviceability Limit State (SLS) and the Ultimate Limit State (ULS) based on experimentally obtained chloride distribution data. In particular, different Non Linear Finite Element (NLFE) based mapping concepts between chloride ingress and the SLS and ULS performance are presented. Only probabilistic NLFE methods allow for the required incorporation of epistemic and aleatoric uncertainties in the aging material properties as well as relevant loads and accurate description of failure mechanism based on fracture mechanics and nonlinear material laws. The entire procedure was applied to a recently demolished highway bridge, which serves as a case study object in order to illustrate (the feasibility of) the proposed concept, see Fig. 1.

In particular, the interests were focused on the concrete stress performance at the top and bottom surface of the V-girders and the vertical deflection and chracking behavior during the gradually reduction of the pre-stressing cross sections. In addition, to the cross section reduction the variability of the yield strength

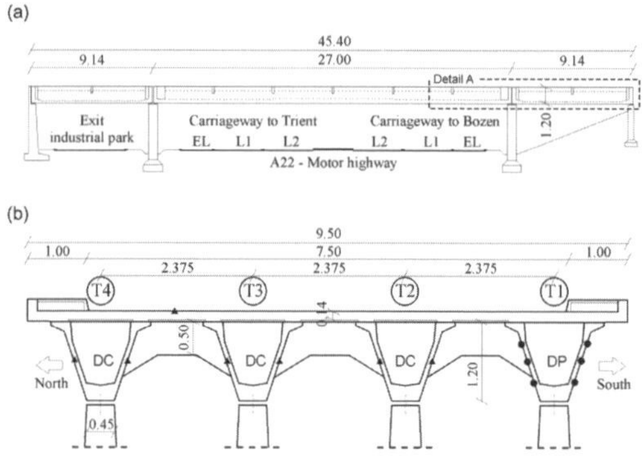

Figure 1. Neumarkt Bridge: (a) longitudinal cut, and (b) cross section.

had been taken into account according to Almusallam (2001). The deterministic and probabilistic studies allow an insight in the development of the reliability level and the globale and structural ductile or brittel performance of the system during the corrosion process.

REFERENCES

Almusallam, A. A. "Effect of degree of corrosion on the properties of reinforcing steel bars." *Construction and Building Materials* **15**: 361–368, 2001

Petcherdchoo, A., Neves, L. A. C. and Frangopol, D. M. "Optimizing lifetime condition and reliability of deteriorating structures with emphasis on bridges." *Journal of Structural Engineering* **134**(4): 554–552, 2008

Stewart, M. G. and Rosowsky, D. V. "Time dependent reliability of deteriorating reinforced concrete bridge decks." *Structural Safety* **20**(1): 129–139, 1998

Strauss, A., Hoffmann, S., Wendner, R. and Bergmeister, K. "Structural assessment and reliability analysis for existing engineering structures, applications for real structures." *Structure and Infrastructure Engineering* **5**(4): 277–286, 2009

Bridge Maintenance, Safety, Management, Resilience and Sustainability – Biondini & Frangopol (Eds)

Detailed comparison between ASR/LFR and LRFR for reinforced concrete highway bridges

H. Toutanji, D. Wang & R. Vuddandam
Department of Civil and Environmental Engineering, University of Alabama in Huntsville, Huntsville AL, USA

ABSTRACT

A detailed comparison study on three highway bridge rating procedures, Allowable Stress Rating (ASR), Load Factor Rating (LFR) and Load and Resistance Factor Rating (LRFR), is presented in this research. The latest rating method LRFR brings many new factors into account other than ASR and LFR. Not only the factors used in LRFR are based on probability theory, but also LRFR uses new live load truck models, considers the level of redundancy of the complete superstructure system, and separates dead loads into two types. The engineers involved in inspection and load rating work of existing bridges should recognize the difference between rating methods so that they can efficiently use new rating method with confidence.

The comparison of these procedures was conducted in early research based on Rating Factor (RF). Earlier studies show conflicting results. For example, one study indicated that LRFR produces lower rating factors than the LFR, while other studies showed that LRFR produces nearly equal or higher rating factors than LFR. The aim of this paper is to find the reasons for these different results found in the literature. In addition, the paper shows the difference in calculating the capacity of bridges, dead loads, live loads during rating factor calculation and determines the significant factors that are responsible for disparity among these rating methods. In this study, this is done by rating several bridges and the intermediate parameters that govern the rating factor are shown.

Data from previous research is collected and reorganized to support this study. Based on the data, it is easy to find that the rating factors from LRFR are all smaller than the rating factors from LFR. Figures 1–2 show the differences and variable trends of live load effects from two methods. It can be inferred from the figures that the differences between LFR and LRFR have opposite trends along with the increase of span length. The difference of live load moment gets smaller while the span length increases. On the contrary, for shear due to live load difference is bigger with increasing span length.

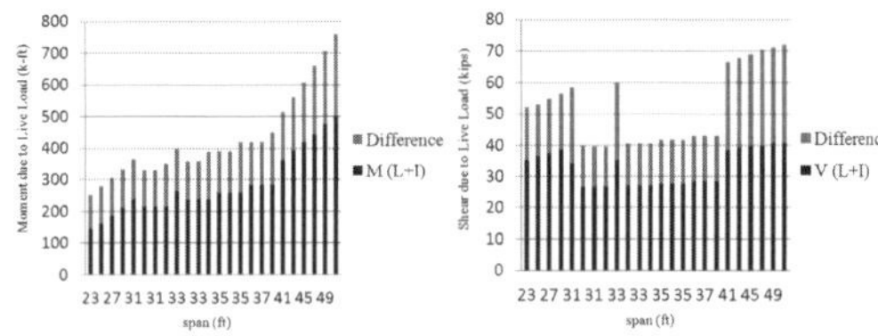

Figure 1. Moment (left) and Shear (right) due to Live Load from LFR and LRFR.

For further comparison study and to support the conclusion acquired above, one detailed bridge rating example is performed. The example considers moment and shear rating factors of the interior girders. For ASR, LFR the inventory and operating level ratings are performed using HS 20-44 live load model. For LRFR the design load level rating is performed using HL-93 live load model at strength I limit state. Further, results of three more examples having different span lengths are used to draw the conclusions.

REFERENCES

Dennis R. Mertz, 2005, *Load Rating by Load and Resistance Factor Evaluation Method*

Lichtenstein Consulting Engineers, Inc. and Paramus, NJ, 2001, *Manual for Condition Evaluation and Load Rating of Highway Bridges Using Load and Resistance Factor Philosophy*

Michael Murdock, 2009, *Comparative Load Rating Study under LRFR and LFR Methodologies for Alabama Highway Bridges*, MS Thesis, Auburn University, Alabama

Rogers, B.J., and Jáuregui, D.V., 2005, Load Rating of Prestressed Concrete Girder Bridges: Comparative Analysis of Load Factor Rating and Load and Resistance Factor Rating, *Transportation Research Record: Journal of the Transportation Research Board*, Transportation Research Board, Washington, D.C., Volume 1928/2005, pp. 53–63

Bridge Maintenance, Safety, Management, Resilience and Sustainability – Biondini & Frangopol (Eds)
© 2012 Taylor & Francis Group, London, ISBN 978-0-415-62124-3

Structural reliability analysis of deteriorating RC bridges considering spatial variability

T.V. Tran, E. Bastidas-Arteaga, F. Schoefs & S. Bonnet
LUNAM Université, Université de Nantes, Institute for Research in Civil and Mechanical Engineering, Nantes, France

A.J. O'Connor
Trinity College Dublin, Ireland

F. Lanata
Ecole Supérieure du Bois, Nantes, France

ABSTRACT

Most part of probabilistic studies of corroding RC structures focus on the reliability analysis on a single section of the structure. Recent works have demonstrated that deterioration processes and loading are highly space-variant. Therefore, the consideration of their spatial variability is essential for proper reliability assessment. Herein, we considered the spatial variability of two parameters: chloride surface concentration, C_s and diffusion coefficient D_o. These random fields will be represented by a Karhunen-Loève expansion (Schoefs et al, 2011). The characteristics of these random fields (mean, standard deviation and fluctuation parameter) are determined from real measurements by using the Maximum-Likelihood method.

At the beginning of the reliability analysis the structural component is discretized into m elements. Afterwards, we compute the failure probability for each element. Then, assuming that the probability of failure is constant for each element, the probability of failure of the structural component is estimated as the probability of failure of a series system:

$$P_{fS} = 1 - \Phi_m\left(\vec{\beta}, [\rho]\right) \quad (1)$$

Most part of studies assume that the probabilities of failure of the elements are independent – i.e. $\rho = 0$. However, as chloride surface concentration and diffusion coefficient are considered as random fields, there is a correlation between the probabilities of failure of the elements. In this paper, we use a Bi-normal model to consider this correlation (Yuan and Pandey, 2006).

The proposed methodology is illustrated with a numerical example that focuses on the reliability assessment of a RC bridge girder considering the spatial variability of the parameters C_s and D_0.

Figure 1 presents the influence of spatial variability on the probability of failure. These results show that the probability of failure obtained by considering spatial variability are higher than for the case without spatial variability. This means that the influence

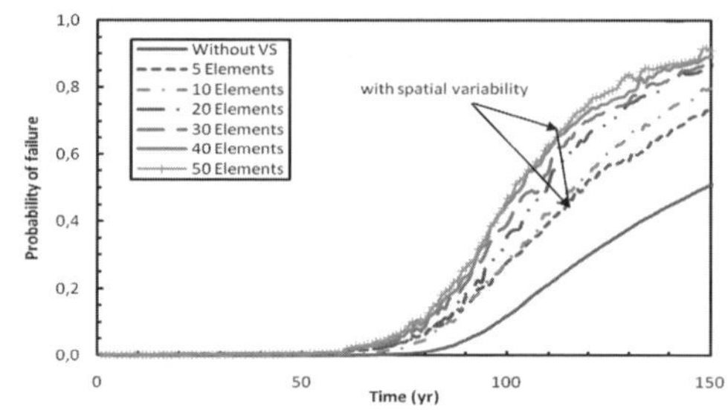

Figure 1. Influence of spatial variability on the probability of failure.

of spatial variability is very important in structural reliability assessment. For instance, it can be noted that for a given probability of failure ($P_f = 0.5$), the times to failure computed considering spatial variability are 60–80% smaller than the computed without considering spatial variability.

Figure 1 also shows that the reliability assessment is sensitive to the discretization – i.e., number of elements. We can observe that for a lower number of elements the results are close to the obtained for the case without considering spatial variability. On the contrary, when the number of elements increase the there is a convergence on the probability of failure assessment. Therefore, a comprehensive reliability assessment that considers spatial variability should define the number of elements for each specific case.

REFERENCES

Schoefs F, Tran TV, Bastidas-Arteaga E. Optimization of inspection and monitoring of structures in case of spatial fields of deterioration/properties. In: 11th International Conference on Applications of Statistics and Probability in Civil Engineering, ICAPS 11, Zurich, Switzerland; 2011. pp. 2249–2256.

Yuan X.-X. and Pandey M.D. 2006. Analysis of approximations for multinormale integration in system reliability computation. *Structural Safety*, 28, 361–377.

Bridge Maintenance, Safety, Management, Resilience and Sustainability – Biondini & Frangopol (Eds)
© 2012 Taylor & Francis Group, London, ISBN 978-0-415-62124-3

Probabilistic seismic response of a bridge-soil-foundation system under the combined effect of vertical and horizontal ground motions

Z. Wang, J.E. Padgett & L. Dueñas-Osorio
Department of Civil and Environmental Engineering, Rice University, Houston, TX, USA

ABSTRACT

Different levels of model sophistication have recently emerged to support risk assessment studies, but mostly at the expense of neglecting the influence of vertical ground motions (VGMs). Both observational field bridge damage and time history analyses have shown that the inclusion of VGMs has a great influence on the bridge seismic response. In this paper, the effect of adding VGMs on the seismic response is presented and the results of the seismic response are compared with the case of horizontal-only excitations. A detailed CBSF system with a three-dimensional bridge superstructure, two-dimensional soil domain and one-dimensional *p-y*, *t-z*, and *q-z* springs is investigated.

This study shows that the bending moment demand at the middle of the end span, the center of the bridge, and at the bents are all impacted significantly by the VGMs. Similar VGM effects are observed for the axial force of the columns, which impacts the flexural and shear capacity of the columns, and the normal force of the bearings that in turn affects their force-displacement relationship. However, VGMs do not have a great impact on the seismic demand of the pile cap displacement or the maximum axial force of the piles. On the issue of the bridge deck bending moments, the amplification of the moment demand in the negative direction may exceed their capacity. Although the shear capacity of the columns is influenced by the VGMs, this reduction is of less concern since the columns are still controlled by flexural behavior. For the bearings, it is found that VGMs have a great impact on the normal force of both the expansion and fixed bearings. Additionally, VGMs have a significant influence on the force-displacement relationship of the fixed bearing yet little influence on the force displacement relationship of the expansion bearing attributed to different vertical load distributions to each bearing type. VGMs may also result in deck uplift from the bearings leading to potential instability of bearings, particularly the high-type steel bearings considered in this study. However, isolating the liquefaction effects under the influence of VGMs remains ambiguous and needs to be investigated in the future.

In addition, a probabilistic seismic demand model is used to correlate the hazard intensity measure to structural engineering demand parameters. Moreover, fragility curves that consider capacity change with fluctuating axial force of the columns and the piles are derived. Results show that the presence of VGMs only has a minor effect on the failure probabilities of piles and expansion bearings, while it has a great influence on fixed bearings. Whether VGMs have an impact on column fragilities depends on the design axial load ratio.

Finally, first-order and second-order system reliability bounds are used to obtain the fragility curve of the CBSF system. Results show that the first-order and second-order reliability lower bound and upper bound of the CBSF system are very close. The reason is that the failure probability of the pile is much higher than other components and dominates the failure probability of the CBSF system. Therefore, first-order bounds are sufficient for this case study. In addition, VGMs do not have much impact on the overall first-order and second-order system reliability bounds of the selected CBSF system, since the failure probabilities of the piles are not influenced by VGMs. However, other bridge systems with other components dominating the failure could benefit from the inclusion of VGMs, given their significant impact on column and fixed bearing fragility.

This study reveals the important role that vertical ground motions can play due to the shift in probabilistic seismic demands on key structural components, the capacity and fragility representation of certain components of the selected CBSF systems. Therefore, the effects of vertical ground motions should be more explicitly considered in the design and analysis of critical components of the bridges, especially for bridges located in near fault regions.

Bridge Maintenance, Safety, Management, Resilience and Sustainability – Biondini & Frangopol (Eds)
© 2012 Taylor & Francis Group, London, ISBN 978-0-415-62124-3

Reliable damage detection and localization using direct strain sensing

Y. Yao & B. Glisic
Princeton University, Princeton, NJ, USA

ABSTRACT

Civil infrastructure in the U.S. is aging and has been identified as an area in critical need. It is important to determine and monitor its structural health in order to mitigate risks, prevent disasters, and plan maintenance activities in an optimized manner. The need for reliable, robust, and affordable Structural Health Monitoring (SHM) is thus rapidly increasing. In spite of its importance, however, SHM is rarely utilized on real structures. The main reason for this is the cost and limited reliability achievable by current monitoring technologies. The sensors currently available must be sparsely spaced and either provide severely insufficient spatial-resolution for early damage detection and location or rely on complex algorithms that degrade specificity against environmental and variable-load conditions.

The objectives of this research are two-fold: to investigate sensing system principles that provide affordable monitoring through a dense and expansive array of sensors enabled by two technologies, namely distributed fiber optic technology and technology called large-area electronics; and to experimentally study how the high-resolution sensing offered by such systems can overcome the robustness and reliability limitations affecting current SHM technologies. The main concepts are presented in this paper along with both reduced- and large-scale test results, which demonstrate that the proposed technologies and direct strain sensing approach are beneficial for reliable damage detection and localization of damage over large areas of a structure.

The first signs of damage to a structure often have local character and occur in the form of strain-field anomalies. The primary underlying challenge in current strain monitoring technologies is that their spatial resolution is limited, which, as a consequence, leads to unreliable damage detection at points distant from sensors. As an illustration, recent measurements performed on the Streicker Bridge on the Princeton University campus are given in Figure 1. A crack occurred at the early age of concrete at the locations P10h11U and P10h11D, and thus these sensors were directly activated. The crack created anomalies in the strain field that were reliably identified as a large change in the strain magnitude. The sensor P10h11L was installed less than one meter away, it did not cross the crack, and its measurements could not be reliably interpreted as a crack.

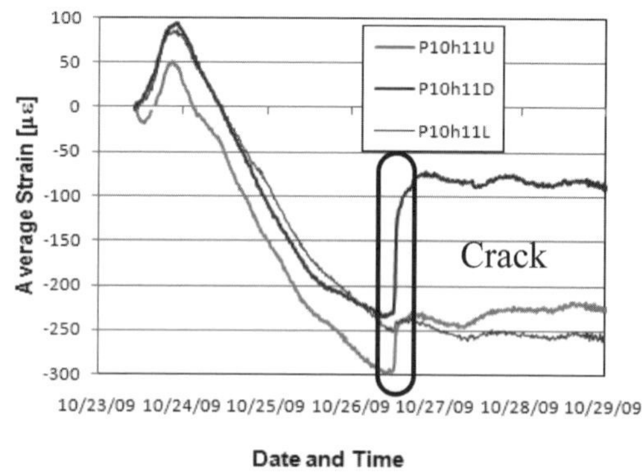

Figure 1. Direct crack detection vs. indirect crack detection in the Streicker Bridge.

Above example shows that damage detection and localization based on direct strain monitoring can be very reliable. However, uncertainty of location of the damage before it occurred requires a very dense sensor network, i.e., installation of large amount of sensors with a very high spatial resolution. Traditional sensing technologies based on short-gauge sensors would be too expensive and complicated to use in such application.

Two novel sensing technologies are identified as promising for damage detection and localization based on direct strain monitoring: distributed fiber optic sensors (DFOS) for one-dimensional (1D) strain monitoring and large area electronics (LAE) for 2D. These two technologies are at different levels of maturity. The concept of "direct" damage detection and localization, along with real applications of the DFOS and preliminary results of the LAE are presented and discussed in this paper.

Bridge Maintenance, Safety, Management, Resilience and Sustainability – Biondini & Frangopol (Eds)
© 2012 Taylor & Francis Group, London, ISBN 978-0-415-62124-3

Bayesian networks for post-earthquake assessment of bridges

Y.C. Yue
School of Human Settlement and Civil Engineering, Xi'an Jiaotong University, China

M. Pozzi, D. Zonta & R. Zandonini
DIMS, University of Trento, Italy

ABSTRACT

This paper presents a framework for Bayesian Network-based assessment of the seismic vulnerability of bridges. The proposed framework has three main parts: the demand model, the capacity model, and the fragility function which correlates the demand and the capacity as shown. The capacity model is constructed using Hazus (FEMA, 2003). The demand on the bridge is calculated using an attenuation function. The uncertainty terms are considered in both the capacity and the demand model. In seismic risk analysis, it is common to assume a lognormal distribution for component capacities (C) and component demands (D). If we assume that: $C \sim \ln N\ (\mu_C\ , \sigma_C^2)$; $D \sim \ln N\ (\mu_D, \sigma_D^2)$, then the probability of failure is:

$$F = \Pr[D - C > 0] = \Pr[\ln C - \ln D < 0]$$
$$= \Pr[g < 0] = \Phi(\frac{\mu_g}{\sigma_g}) = \Phi(\frac{\mu_D - \mu_C}{\sqrt{\sigma_D^2 + \sigma_C^2}}) \quad (1)$$

where Φ is the standard normal cumulative distribution function.

Fig. 1 is the conceptual BN representing the relationship between demand and capacity. g is the intermediate variable that is related to the probability of failure, $g = \ln(D/C)$. In order to facilitate the calculation, all the variables follow a normal distribution or lognormal distribution. In this simple BN, obviously we have $g \sim N\ (\mu_g, \sigma_g^2)$. The demand is calculated using an attenuation function depending on the local site conditions. The capacity is calculated based on some empirical or analytical models, depending on the characteristics of the bridges.

In order to apply the exact inference algorithm of Bayesian Networks with continuous variables in this framework, all the variables in the framework are assumed to follow Gaussian distribution and the continuous variables are not allowed to have discrete child

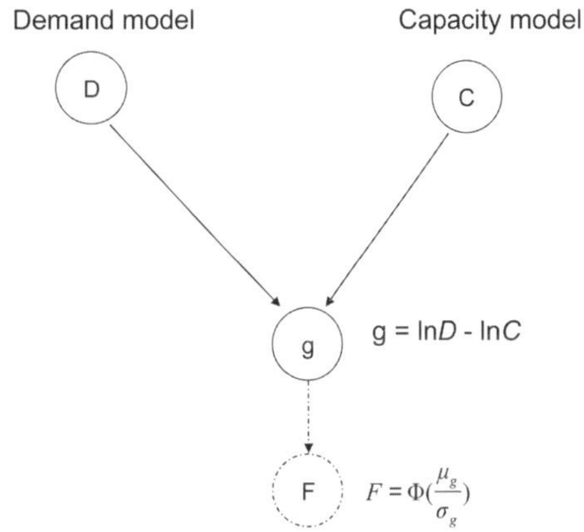

Figure 1. The conceptual BN that contains the three main parts in the framework.

variables (Lauritzen & Jensen 2001). Using this framework, the correlation between two bridges is analyzed, and the prior probabilities of the two bridges being in a damage state are calculated during the initialization procedures. When an earthquake occurs, if data such as magnitude and epicenter are known, or the damage state of one bridge is observed by on-site sensors, then the probability of another bridge being in a damage state can be predicted and updated. Two similar bridges in Trentino, Italy are used to illustrate the procedures.

REFERENCES

HAZUS-MH MR3 2003. Multi-hazard Loss Estimation Methodology (Earthquake Model). Technical Manual. Department of Homeland Security Emergency Preparedness and Response Directorate. FEMA. Washington, D.C.

Lauritzen, S. L., & Jensen, F. (2001). Stable local computation with conditional Gaussian distributions. *Statistics and Computing*, 11(2), 191–203.

Smart SHM and application to bridge condition assessment and maintenance

Organizers: Y. Zhang, H. Sohn, C. Wang & D. Zonta

Bridge Maintenance, Safety, Management, Resilience and Sustainability – Biondini & Frangopol (Eds)
 ISBN 978-0-415-62124-3

Remote guided wave imaging using wireless PZT excitation and laser vibrometer scanning for local bridge monitoring

Y.K. An, H.M. Song, H.J. Park, H. Sohn & C.-B. Yun
Department of Civil and environmental Engineering, KAIST, Daejeon, Korea

ABSTRACT

A number of structural health monitoring (SHM) techniques have been developed and applied to *in-situ* bridges for assessing their safety and integrity. For example, vibration-based SHM techniques are useful for characterizing global dynamic characteristics of bridges (Chen, Y. et al. 2009). However, these techniques are often insensitive to local incipient damage.

To overcome this limitation, local SHM techniques have been developed. In particular, the guided wave technique is one of the most promising techniques for local monitoring, because guided waves are sensitive to an incipient damage and can propagate a long distance with little attenuation (Rose 2002).

This study proposes a remote damage visualization technique where a surface mounted piezoelectric transducer (PZT) is wirelessly excited by a laser beam for guided wave generation and the corresponding response is scanned by a laser Doppler vibrometer (LDV) so that interaction of guided waves with a hidden defect can be identified from created guided wavefield images.

The proposed technique has the following advantages over conventional wired/wireless sensing techniques: (1) No wiring is necessary for both power and data transmission, (2) no need for battery replacement unlike conventional wireless sensors, (3) guided wavefield images with high temporal and spatial resolutions are obtained allowing instantaneous damage visualization, (4) no baseline image is necessary, making this technique less prone to false alarms due to ambient temperature and loading variations in field.

The effectiveness of the proposed technique is tested at a decommissioned Ramp-G Bridge site in South Korea with curved steel box girders as shown in Figure 1. Our monitoring efforts focus on visualization of a hidden notch inside the bridge box girder when a PZT placed outside the girder is wirelessly excited and the out of place velocity of the box girder is scanned

(a) (b)

Figure 1. Overview of Ramp-G bridge in Goyang, Gyunggi, Korea: (a) Curved 90 m-long Ramp-G bridge and monitored section and (b) Two steel box girders.

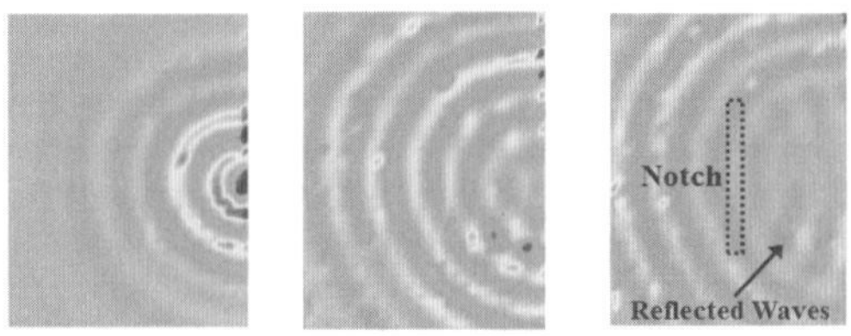

(a) (b) (c)

Figure 2. Snapshots of guided wave propagation images near the hidden notch: (a) 86 μs, (b) 107 μs and (c) 120 μs.

by LDV. The wave propagation images obtained from the box girder with a hidden notch are shown in Figure 2. Reflections from the hidden notch are shown in Figure 2(c). The associated implementation issues for field bridge testing are also discussed in this paper.

REFERENCES

Chen, Y. et al. 2009. Bridge Structural Condition Assessment based on Vibration and Traffic Monitoring. *Journal of Engineering Mechanics* 135(8): 747–758.

Rose, J. 2002. A baseline and vision of ultrasonic guided wave inspection potential. *ASME Journal of Pressure Vessel Technology* 124(3): 273–282.

Bridge Maintenance, Safety, Management, Resilience and Sustainability – Biondini & Frangopol (Eds)
 ISBN 978-0-415-62124-3

Damage detection for local components of long suspension bridges using influence lines

Z.W. Chen
School of Architecture and Civil Engineering, Xiamen University, Fujian Province, China

S. Zhu & Y.L. Xu
Department of Civil and Structural Engineering, The Hong Kong Polytechnic University, Hong Kong

Q. Li
Department of Bridge Engineering, Tongji University, Shanghai, China

ABSTRACT

Many long-span cable-supported bridges have been built throughout the world in recent years. They begin to deteriorate once built and continuously accumulate damage during their service life due to natural hazard and harsh environment such as typhoons, earthquakes, vehicles, temperature and corrosion. Meanwhile, structural health monitoring technology gains a rapid development recently. Comprehensive structural health monitoring systems (SHMS) have been designed and installed in a number of long-span bridges, and different types of sensors are used for monitoring the loading, responses and conditions of bridges. A well-known example is the Wind and Structural Health Monitoring System (WASHMS) installed in the Tsing Ma suspension bridge in 1997. It is a new trend to integrate SHMS and damage detection technology for real-time condition assessment of long-span bridges.

On the other hand, bridges failure often begins with minor local damage in bridge components. Therefore, an efficient and effective damage detection method sensitive to local damage is essential for bridge health monitoring. It has been recognized that conventional vibration-based damage detection techniques, commonly based on modal properties or their derivatives, are often insensitive to structural local damage and are significantly dependent on the change of operational environment, such as temperature. This paper explores a novel damage detection technique based on stress influence lines in representative bridge components, and its efficacy is validated through a case study of Tsing Ma Suspension Bridge. A mathematical method with regularization is first introduced to identify the stress influence line (SIL) based on the in-situ measurement of train information and train-induced stress responses in local bridge components. A good agreement between the identified and baseline SIL demonstrates the effectiveness of the proposed identification method. A damage index based on SIL is subsequently proposed and applied to a hypothetic

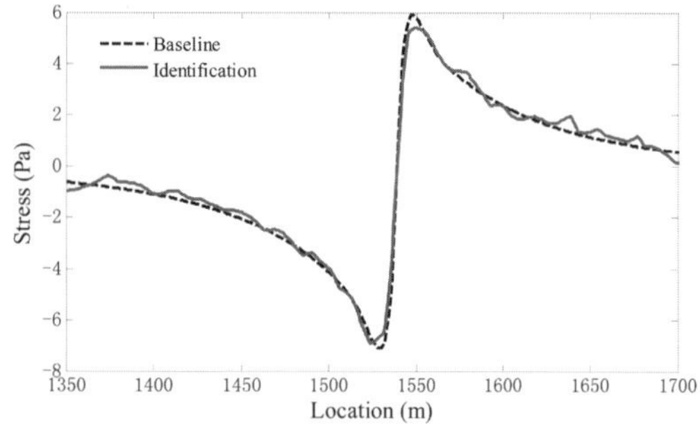

Figure 1. Comparison between the identified SIL and the baseline SIL.

damage scenario in which a typical local component is damaged. Train-induced dynamic stress response in the damage scenario is simulated numerically using coupled train-bridge dynamic method. The SIL in the damage scenario is identified from the simulated responses. The damage index is computed based on the SILs in the undamaged and damaged cases. The results indicate that the proposed method offers a promising technique for real-time damage detection of long-span cable-supported bridges equipped with comprehensive SHMS.

REFERENCES

Chen, Z.W., Xu, YL, Li, Q., & Wu, D.J. 2011a. Dynamic stress analysis of long suspension bridges under wind, railway and highway loading. *Journal of Bridge Engineering, ASCE* 16(3):383–391.

Chen, Z.W., Xu, Y.L., & Wong, K.Y. 2011b. Fatigue Analysis of Long-Span Suspension Bridges under Multiple Loading: Case Study. *Engineering Structures* 33: 3246–3256.

Wong, K.Y., Lau, C.K., & Flint, A.R. 2000. Planning and implementation of the structural health monitoring system for cable-supported bridges in Hong Kong. *Proceedings of Health Monitoring and Management of Civil Infrastructure Systems, SPIE*, 266–275.

Bridge Maintenance, Safety, Management, Resilience and Sustainability – Biondini & Frangopol (Eds)
© 2012 Taylor & Francis Group, London, ISBN 978-0-415-62124-3

Filtering environmental load effects to enhance novelty detection on cable-supported bridge performance

E.J. Cross & K. Worden
Dynamics research group, Department of Mechanical Engineering, University of Sheffield, UK

K.-Y. Koo & J.M.W. Brownjohn
Vibration Engineering Section, Department of Civil and Structural Engineering, University of Sheffield, UK

ABSTRACT

As dense and complex monitoring systems are becoming more common for cable-supported bridges, useful ways to utilise the large amounts of data collected must be imagined. Of great advantage to performance monitoring of these structures is to have a sound understanding of their normal condition, i.e. what constitutes a normal response to environmental and operational loadings. If this can be achieved anomalous performance data can obviously be easily identified. The idea of filtering the effects of environmental and operational loadings to enhance novelty detection is addressed here in the context of data collected from the Tamar Suspension Bridge in Southwest England. Data that will be studied in this paper from the Tamar monitoring campaign include modal properties extracted using a stochastic subspace (SSI) routine from accelerometer readings and deflection data from a state of the art total positioning system (TPS) recently installed.

From analysis carried out on data from the Tamar monitoring campaign it has been found that the lower natural frequencies of the deck are most influenced by the traffic loading, and hence display a daily variation as opposed to a seasonal one. The deflections of the bridge deck, as measured by the TPS system, are strongly correlated with the ambient temperature, and as such display daily and seasonal trends. Such trends as these complicate any novelty detection seeking to identify performance anomalies. If responses driven by, for example, a temperature differential, are not defined within the normal condition, then a novelty detection process will wrongly assign these responses as performance anomalies.

Two different approaches to deal with this problem are available; the first seeks to encompass responses due to environmental and operational variation into the definition of the normal condition. The second seeks to remove (or filter) the effects of environmental/operational conditions from response measurements before novelty detection is attempted. Both approaches are explored in this work in the context of natural frequency and deck deflection data from the Tamar monitoring campaign. The first approach of incorporating trends induced by environmental or operational conditions into the definition of a normal response is most applicable to the natural frequency data which demonstrates no strong seasonal trend. Where strong seasonal trends are evident, as is the case for deflection measurements, it may be necessary to remove these trends before novelty detection is attempted. In this work, Gaussian process regression is used to model natural frequency and deck deflection with respect to the external conditions that drive their variation. The use of Gaussian processes (GPs) is a sophisticated nonparametric Bayesian approach to regression and classification problems. Figure 1 shows the predictions of an implemented Gaussian process regression for the prediction of measured deck deflection. One advantage of GP regression visible in Figure 1 is that confidence intervals on predictions are available, which may be helpful for novelty detection, as discussed further in the paper.

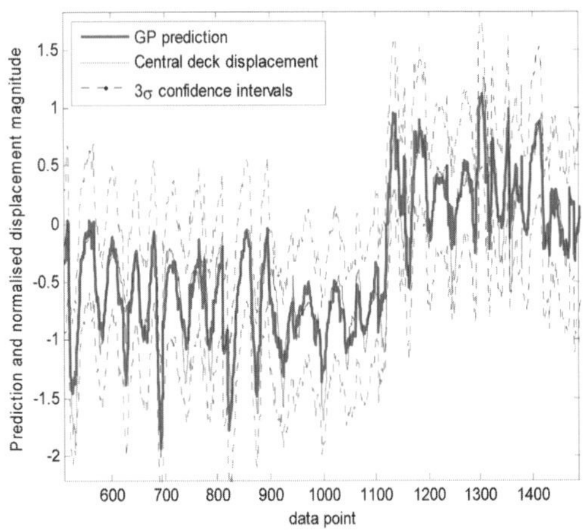

Figure 1. GP prediction of central easterly deck displacement.

Bridge Maintenance, Safety, Management, Resilience and Sustainability – Biondini & Frangopol (Eds)
 ISBN 978-0-415-62124-3

System identification using wirelessly acquired vehicle-bridge interaction data from a highway bridge excited by a moving vehicle

J. Kim
Korea Advanced Institute of Science and Technology, Daejeon, Korea

J.P. Lynch
University of Michigan, Ann Arbor, MI, USA

ABSTRACT

Highway bridges are vital infrastructure components that allow for the movement of people and goods over topological obstacles including bodies of water and mountain valleys. Most highway bridges are regularly loaded with both cars and trucks; however, trucks are known to be the more detrimental load in terms of the long-term health of the bridge. Recent studies on the condition of bridges in the United States has revealed a rapidly aging fleet of bridges whose structural conditions are worsening with time [1]. While repeated truck loading is known to lead to long-term deterioration, the precise causal relationship between truck loading and deterioration is not yet completely understood. As such, there has been a renewed interest in the study of vehicle-bridge interaction (VBI) within the structural health monitoring community.

Experimental investigation of how vehicles precisely load bridges is needed to further advance understanding of VBI. Due to the untethered nature of wireless communications, wireless sensors installed in a moving truck can be used to communicate data pertinent to the truck behavior to a static wireless sensor network installed in a bridge as a permanent monitoring system. In this study, a wireless monitoring system for VBI data acquisition is used to record the dynamic behavior of a moderate-span highway bridge intentionally loaded by a heavy truck. Using the time-history data obtained from the truck and the bridge, a novel system identification strategy is proposed to extract an analytical model of the vehicle-bridge system.

A two-stage system identification strategy is proposed to model the dynamic behavior of the bridge exposed to a moving truck. Dividing the system identification strategy into two stages allows for the time invariant properties of the bridge to be identified in one stage and the time variant loading properties to be identified in the other. In this study, subspace system identification methods [8] are adopted for the first stage of the system identification methodology.

In the first stage, the free-vibration response of the bridge is used to extract the linear time invariant properties of the system. A state-space model corresponding to the unloaded, time-invariant system is

$$\begin{aligned} \mathbf{x}(k+1) &= \mathbf{A}\,\mathbf{x}(k) + \mathbf{w}(k) \\ \mathbf{y}(k) &= \mathbf{C}\,\mathbf{x}(k) + \mathbf{v}(k) \end{aligned} \tag{1}$$

The identification of the optimal set of **A** and **C** is determined by stochastic subspace identification (SSI).

In the second stage, vehicle position and vibratory response time-histories are used to identify a linear, time-variant model that fully models the vehicle-bridge interaction occurring during loading. A key innovation of the second stage of the proposed system identification process is the use of a kernel approximation of the bridge loading based on the vehicle position. The estimates for **A** and **C** identified in Stage I ($\{\hat{\mathbf{A}}, \hat{\mathbf{C}}\}$) will be used as known model parameters in Stage II. Hence, the objective of Stage II is to find the optimal load matrix, $\mathbf{B}(k)$:

$$\begin{aligned} \mathbf{x}(k+1) &= \hat{\mathbf{A}}\,\mathbf{x}(k) + \mathbf{B}(k)u(k) + \mathbf{w}(k) \\ \mathbf{y}(k) &= \hat{\mathbf{C}}\,\mathbf{x}(k) + \mathbf{v}(k) \end{aligned} \tag{2}$$

The Yeondae Bridge is 180 m long and located in Icheon, Korea. The bridge is designed as a continuous steel box girder system. The bridge proves to be an idea testbed for experimental observation of VBI using a calibrated truck. Truck and bridge response data is shown to be sufficient to derive an accurate tine variant model of the bridge system.

Bridge Maintenance, Safety, Management, Resilience and Sustainability – Biondini & Frangopol (Eds)
© 2012 Taylor & Francis Group, London, ISBN 978-0-415-62124-3

Sensor driven prognosis scheme based on moment estimator

Z. Li & Y. Zhang
Department of Civil & Environmental Engineering, University of Maryland, College Park, MD, USA

ABSTRACT

In this paper, an extreme value theory (EVT) based structural health prognosis scheme is presented to estimate the remaining useful life (RUL) of monitored structure by taking advantage of abundant sensor data acquired from online monitoring system

In the proposed EVT-based prognosis scheme, hundreds of sensor data records are recorded within a short time period at each degradation stage. Moment Estimator (Dekkers et al. 1989) from EVT is introduced to fuse the RUL information derived from sensor data. EVT is a semi-parametric method. It utilizes a small percentage of total available data to predict the quantile value for the total data with unknown distribution. Only a small portion of the total measurements are needed to estimate the RUL information derived from full sensor data. Quantile value and end point estimation of the RUL can be realized. Figure 1 shows the flowcharts for the full sensor data based prognosis scheme using distribution fitting method and the proposed EVT-based prognosis scheme.

As a prerequisite, the monotonicity relation between measurements and estimated RUL is derived in this paper through proving the First-order stochastic dominance of the posterior distribution in Bayesian updating.

An example based on fatigue prognosis of the Yellow Mill Pond Bridge (Fisher et al. 1981 etc) is presented in this paper to demonstrate the proposed prognosis scheme. The fatigue crack development data from NDT on this bridge is used.

Utilizing abundant sensor data facilitates the RUL estimation. Results of this study show that EVT based sensor driven prognosis scheme is computationally efficient in performing prognosis without loss of much accuracy. as shown in Table 1. It is seen that the proposed EVT-based prognosis scheme is capable of obtaining an estimator for the end point value with a difference less than 5% and low quantile values with a difference less than 1% from the results obtained by using full sensor data set while the processed data amount is reduced to 20%.

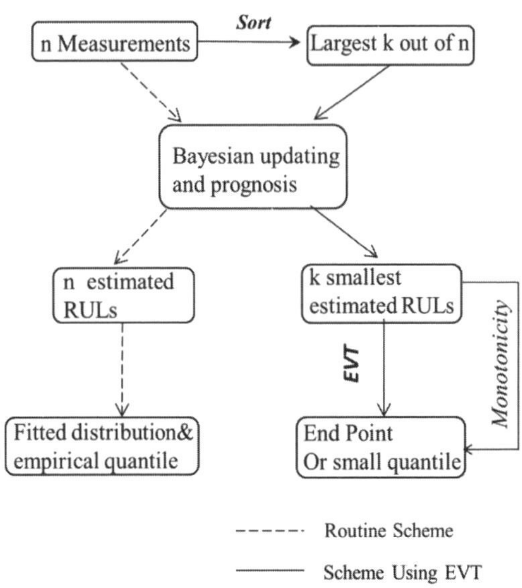

Figure 1. Prognosis Scheme based on EVT.

Table 1. Estimated RUL from moment estimator.

Quantile	0	0.002	0.01	0.05
RUL from Moment Estimator	2.990e7	3.790e7	4.020e7	4.340e7
RUL from distribution fit	3.135e7	3.780e7	4.010e7	4.330e7
Difference	4.7%	−0.26%	−0.24%	−0.23%

REFERENCES

Dekkers A.L.M., Einmahl J.H.J., De Haan L. 1989. A moment estimator for the index of an extreme-value distribution. *Ann Stat.* 17:1833–55.

Fisher J.W., Slockbower R.E., Hausammann H., Pense A.W. 1981. Long time observation of a fatigue damaged bridge. *J of Tech Counc ASCE*; 107(TC1): 55–71.

Bridge Maintenance, Safety, Management, Resilience and Sustainability – Biondini & Frangopol (Eds)
© 2012 Taylor & Francis Group, London, ISBN 978-0-415-62124-3

Develop on-line parameter estimation methods for bridges under changing environment

C.-H. Loh, M.-C. Chen, S.-H. Chao & C.-H. Li
Department of Civil Engineering, National Taiwan University, Taipei, Taiwan

ABSTRACT

For the vibration-based structural health monitoring, two important issues need to be addressed: one is the methodology for detect damage and feature extraction, and the other is to achieve successful novelty detection to distinguish the signal changes caused by abnormality from those caused by environmental and operational variation. In operational modal analysis the Stochastic Subspace Identification (SSI) technique is a well known multivariate identification technique by using the output-only measurements. The advantages of the method is robust against non-stationary excitation and thus are applicable to structural vibration study in the presence of ambient excitation, (d) SSI can directly apply to real-time monitoring the modal parameters.

In this paper application of RSSI-COV method is applied to identify the modal parameters of a bridge under operational condition. Time-varying environmental and operational conditions such as temperature and traffic loading induce changes in the identified modal parameters are examined. The bridge we considered is the 3-span steel arch bridge. The center span is 165. M and the other two spans are with length of 143. meter. Twenty-four hour ambient vibration data was collected from the 10 sensor nodes distributed along the bridge deck to measure the vertical vibration. One of the two identified time-varying system natural frequency (24 hours) is shown in Fig. 1, and the mode shapes for these two modes are shown in Fig. 2.

Since the modal parameters are subject to change due to the variation of temperature and the traffic loading conditions, novelty analysis needs to be employed to identify the correlation of modal frequencies and the environmental factors. Auto-associate Neural Networks (AANN) was applied to extract the nonlinear principal component (NPCA). As shown in Fig. 3, the principal component of the identified frequency (f = 2.575 Hz), temperature and mean response variance can be extracted.

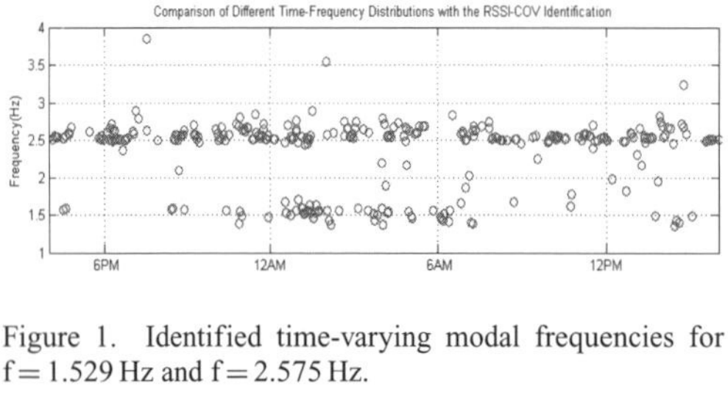

Figure 1. Identified time-varying modal frequencies for f = 1.529 Hz and f = 2.575 Hz.

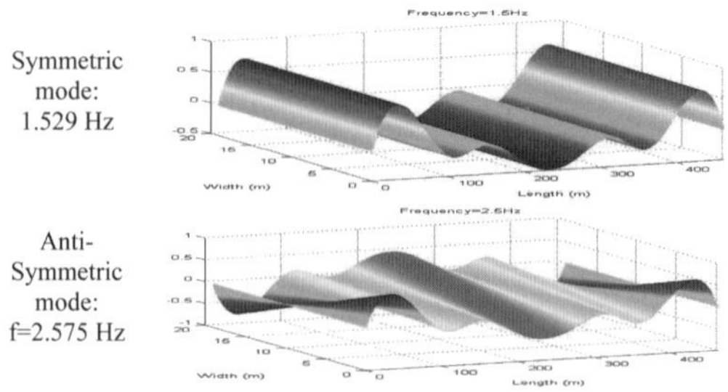

Figure 2. Identified symmetric and anti-symmetric modesf Kuan-Du bridge at f = 1.529 Hz and f = 2.575 Hz, respectively.

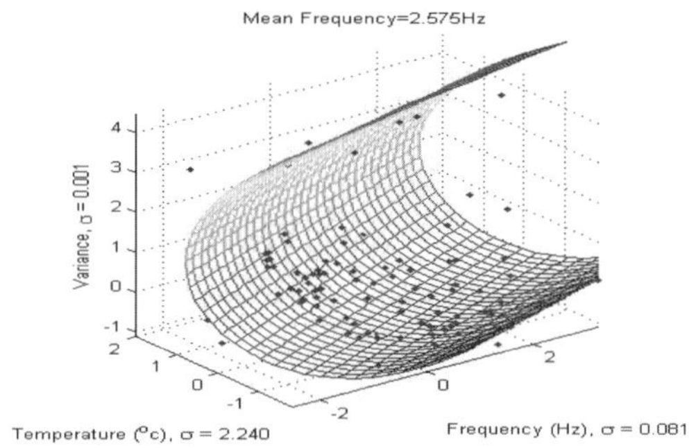

Figure 3. 3D plot of NPC among system natural frequency, mean response and temperature change.

REFERENCE

Loh, C.H. et al. Integrate on-line RSSA and RSSI-COV algorithm for operational modal analysis of bridge structure, 8th Int. Workshop on Structural Health Monitoring, Stanford Univ. 2011.

Bridge Maintenance, Safety, Management, Resilience and Sustainability – Biondini & Frangopol (Eds)

Long-term monitoring of composite girders using optical fiber sensor

Y. Oshima
Kyoto University, Kyoto, Japan

M. Kado
JGC corporation, Yokohama, Japan

ABSTRACT

In this paper, we report on the long term monitoring of a road bridge using optical fiber sensors. In this monitoring, the optical fiber sensor, OSMOS, is installed on the main girders of a steel-concrete composite girder bridge to monitor the vehicle weight as well as the neutral axis of the composite girder for more than three years.

The monitored bridge is located in a main road in the west area of Japan and fabricated more than 30 years ago. The bridge has been rehabilitated several times due to the heavy traffic which leads to fatigue cracks of main beam and concrete slab. The purpose of the monitoring is to assess the position of neutral axis of the composite girder and also to assess the traffic volume: the position of neutral axis may denote the state of composite between concrete slab and steel girder, and the neutral axis may be shifted if the concrete slab is deteriorated or the stiffness of the composite girder is changed. The traffic volume is a direct factor of fatigue and this information can be used for the prediction of fatigue deterioration.

As a result of the one-year monitoring, it is found that the traffic volume is periodically changed according to the day of the week and the almost identical amplitude is repeated in a weekday. Thus it can be said that one week monitoring is enough to represent the traffic characteristic of this road for one year. As for the neutral axis, it is found that the positions of the neutral axis of the outer girders at upstream lane (G4) and downstream lane (G1), which are designed as non-composite girders, are different from each other and the axis position of upstream line is closer to that of non-composite girder, while the axis position of downstream line is close to that of the composite. Normally the girder designed as non-composite girder becomes a composite girder in actual cases because of the binder of concrete and steel.

Therefore, the difference of the axis position may indicate that the girder of upstream are deteriorated more than that of downstream, which can be also confirmed by the fact that the traffic volume of heavy vehicles in the upstream is 1.5 times more than that in the downstream. However, any shift of the axis cannot

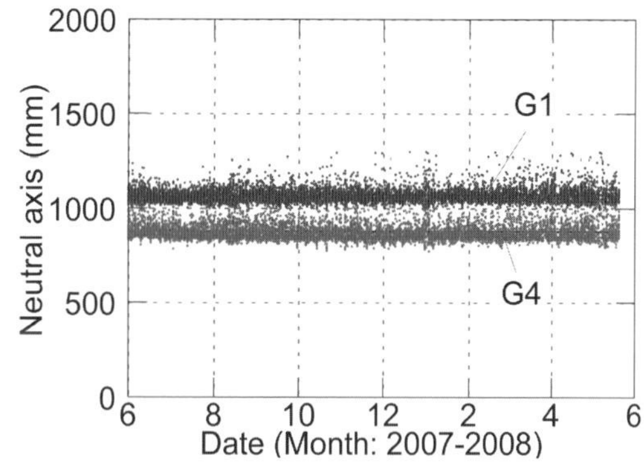

Figure 1. History of neutral axis for one year.

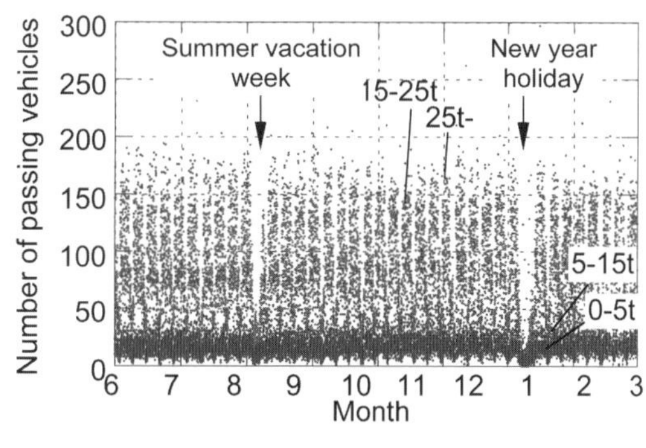

Figure 2. Yearly trend in upstream lane (G4).

be confirmed for three years. Finally, it can be concluded that once OSMOS is installed on the bridge the sporadic monitoring such as one week monitoring in a year is enough for the assessment of the composite girder.

REFERENCE

Braunstein J., Ruchala J., Hodac B., "SMART STRUCTURES: Fiber-optic deformation and displacement monitoring," International Association for Bridge Maintenance and Safety (IABMAS) conference, July 2002, Barcelona, Spain.

Bridge Maintenance, Safety, Management, Resilience and Sustainability – Biondini & Frangopol (Eds)
© 2012 Taylor & Francis Group, London, ISBN 978-0-415-62124-3

Assessing the value of alternative bridge health monitoring systems

M. Pozzi & A. Der Kiureghian
Department of Civil & Environmental Engineering, University of California, Berkeley, USA

ABSTRACT

Many sensing devices and damage detection methods are available for monitoring of civil structures, and selection of the "best" system is not a trivial problem. This selection can be guided by the Value of Information (VoI) principle of decision theory, which measures the quality of the information gained and the effect on reducing overall costs. VoI is essentially the difference between the expected costs of managing the structure without and with the monitoring system.

When the focus is on reducing the consequences of seismic events on infrastructure, it is only by considering the savings resulting from subsequent decisions (e.g., on inspection, repair, continued operation or closure of a structure) that the value of such a system can be determined.

In this paper, for an idealized probabilistic model, we compare the expected values of measuring the seismic excitation, the structural response or the damage experienced by the structure, depending on the precision of the observations. We also propose a numerical procedure for obtaining an approximate estimate of the VoI by a relatively small number of simulations. The procedure requires generating samples of all random variables involved in the decision problem using their conditional distributions, and makes use of non-parametric regression. The approach is applicable to a broad class of decision problems. We validate the approach on a simple case study and provide an assessment of the accuracy of the estimation: Figure 1 shows the evolution with the increasing number of simulations. The dashed line indicates the exact result, solid lines indicate the mean estimates, and the dash-dotted lines represent mean plus/minus one standard deviation estimates. The uncertainty in the estimation is progressively reduced, from a standard deviation of \$1.6 K for 100 simulations down to \$270 for 3,000 simulations.

Theoretically, the estimate we obtain is biased, however the prediction is consistent with the actual VoI (\$7,118) along the plot, indicating that the bias is not an issue, at least for this application.

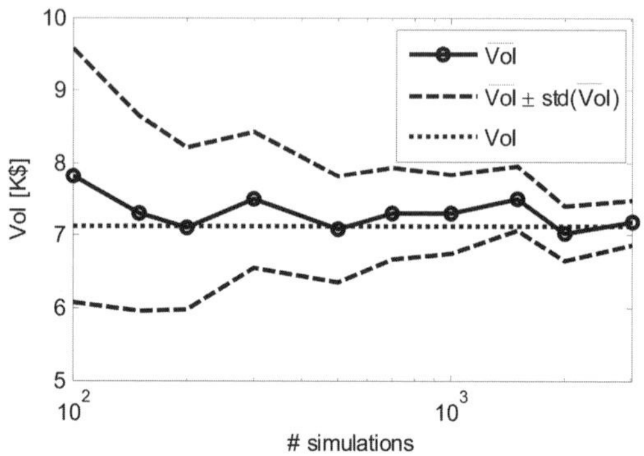

Figure 1. Estimated VoI *vs* number of simulations.

REFERENCES

Bensi, M.T., Der Kiureghian, A. & Straub, D. 2011. *A Bayesian Network Methodology for Infrastructure Seismic Risk Assessment and Decision Support*. PEER Report 2011/02.

Jensen, F.V. & Nielsen, T.D. 2007. *Bayesian Networks and Decision Graphs*. Springer.

Oakley, J.E., Brennan, A., Tappenden P. & Chilcott J. 2010. Simulation sample sizes for Monte Carlo partial EVPI calculations. Journal of Health Economics. 29(3), Pg. 468–477.

Pozzi, M. & Der Kiureghian, A. 2011a. Assessing the value of information for long-term structural health monitoring. In T. Kundu (ed.), *SPIE Smart Structures/NDE Conference, Health Monitoring of Structural and Biological Systems 2011*, Proceedings Vol. 7984.

Pozzi, M. & Der Kiureghian, A. 2011b. Computation of Life-time Value of Information for Monitoring Systems. In Fu-Kuo Chang (ed.), *Structural Health Monitoring 2011*, Proceedings of the 8th Int. Workshop on SHM, Stanford University.

USGS. (2008). *ShakeMaps. USGS Earthquake Hazard Program*. http://earthquake.usgs.gov/eqcenter/shakemap/

Wasserman L. 2005. *All of Nonparametric Statistics*. Springer.

Bridge Maintenance, Safety, Management, Resilience and Sustainability – Biondini & Frangopol (Eds)
 ISBN 978-0-415-62124-3

Streicker Bridge: A two-year monitoring overview

D.H. Sigurdardottir
Princeton University, USA

J.P.S. Afonso
Princeton University, USA/EPFL, Switzerland

D.L.K. Hubbell
Princeton University, USA/University of Toronto, Canada

B. Glisic
Princeton University, USA

ABSTRACT

New and advanced analysis methods allow modern engineers to design more daring structures than ever. Society demands that the development of new infrastructure strives for efficiency and sustainability resulting in even slender elements and less material usage. These modern designs require better understanding of structural behavior of individual elements and more importantly global structural systems. Structural Health Monitoring (SHM) is gaining increasing importance in facilitating such understanding and providing important information for assessment of structural safety.

The complete benefits of SHM have not been fully assessed and further research and real applications are needed to provide more experience. Streicker Bridge is a new pedestrian bridge built on the Princeton University campus (Fig. 1). Structural Health Monitoring (SHM) is applied with the aim of transforming the bridge into an on-site laboratory for various research and educational purposes. This allows for research in the area of SHM methods and data analysis as well as various damage detection techniques and system identification.

The Streicker Bridge project is also important in educating new generations of structural engineers. The project serves as a basis for an undergraduate and graduate course on SHM, allowing the students to work with real data and appreciate the potential of SHM. Furthermore, the project serves as a basis for several senior, master and Ph.D. theses.

Figure 1. Streicker Bridge. Right: Photograph. Left: Rendering, courtesy of Facilities of Princeton University.

The research focuses on addressing some general SHM challenges, and it also includes specific studies in the domain of monitoring approaches, methods, and instrumentation. Two fiber-optic sensing technologies are currently permanently deployed: discrete long-gage sensing technology based on Fiber Bragg-Gratings (FBG) and truly distributed sensing technology based on Brillouin Optical Time Domain Analysis (BOTDA). The sensors were embedded in the concrete during the construction (Fig. 2). The post-tensioning of each part of the bridge was performed about one week after the pouring.

This paper presents a two year overview including: monitoring systems, applied monitoring strategy, and comparison of numerical model with monitoring results: bridge behavior at early age and post-tensioning, under temperature load, and under various static and dynamic load tests. A study of the strain evolution over these two years has given indication of creep and shrinkage; these observations will be published at a later stage. The results showed that the selected monitoring strategy was suitable for monitoring of this complex bridge, and that the selected monitoring systems were able to capture the main features related to the real structural behavior of the bridge (e.g. early age cracking, post-tensioning etc.). SHM confirmed that the bridge is in a very good condition.

Figure 2. Photographs taken during sensor installation. Top left: Student installing sensor. Bottom left: Students installing extension cable. Right: Parallel discrete and distributed sensors installed on rebar.

Bridge Maintenance, Safety, Management, Resilience and Sustainability – Biondini & Frangopol (Eds)
© 2012 Taylor & Francis Group, London, ISBN 978-0-415-62124-3

Damage detection on a full-scale highway sign structure with a distributed wireless sensor network

Z. Sun, S. Krishnan & S.J. Dyke
School of Mechanical Engineering, Purdue University, USA

G. Hackmann & C. Lu
Department of Computer Science and Engineering, Washington University in St. Louis, USA

G. Yan
School of Engineering, University of Western Sydney, Australia

A. Irfanoglu
School of Civil Engineering, Purdue University, USA

ABSTRACT

Civil engineering structures such as bridges will suffer from environmental corrosion, wind loading and traffic impact in their lifetime. Monitoring the condition of these structures is important for prioritizing repair and replacement of our inventory and even for protecting structures from collapse or accelerated deterioration. Monitoring using distributed wireless sensor networks (WSNs) have become a novel and promising solution to the challenges of structural health monitoring (SHM) in civil engineering structures (Spencer & Yun, 2010), such as bridges, buildings and highways.

WSNs have numerous advantages over traditional sensor networks. Installation and maintenance costs are considerably reduced and energy consumption rate is relatively low (Zimmerman et al, 2008). Also, the embedded computational capabilities of wireless sensors can reduce the amount of data transmitted through the network significantly and, consequently, improve the damage detection efficiency (Hackmann et al, 2010).

While research projects using WSNs are ongoing worldwide, implementations of WSNs on full-scale structures are limited. In this paper, a WSN is implemented on a full-scale 17.3 m, 11-bay highway sign support structure (see Fig. 1) to investigate the ability to use vibration response data to detect damage induced in the structure. A multi-level damage detection strategy is employed: the Angle-between-String-and-Horizon (ASH) flexibility-based algorithm as the first level and the Axial Strain (AS) flexibility-based algorithm as the second level. For the proposed multi-level damage detection routine, level I damage detection will be conducted first to detect potential damaged regions. Subsequently, level II damage detection is conducted using the knowledge of potential damaged regions to locate the damaged elements. Several damage cases are created on the full-scale highway sign support structure to test the robustness of two methods involved in the multi-level damage detection strategy. The results are compared to the results from a numerical study. The strategy is shown to be successful in localizing damage in the structure in several cases.

Figure 1. Setup of the full-scale truss structure.

REFERENCES

Hackmann, G., Guo, W., Yan, G., Lu, C., & Dyke, S. (2010). Cyber-Physical Codesign of Distributed Structural Health Monitoring With Wireless Sensor Networks. *In Proceedings of the 1st ACM/IEEE International Conference on Cyber-Physical Systems (ICCPS '10).*, (pp. 119–128). New York.

Spencer, B., & Yun, C. (2010). *Wireless Sensor Advances and Applications for Civil Infrastructure Monitoring.* Newmark Structural Engineering Laboratory. University of Illinois at Urbana-Champaign.

Zimmerman, A. T., Shiraishi, M., Swartz, R. A., & Lynch, J. P. (2008). Automated Modal Parameter Estimation by Parallel Processing within Wireless Monitoring Systems. *Journal of Infrastructure Systems*, 14 (1), 102–113.

Bridge Maintenance, Safety, Management, Resilience and Sustainability – Biondini & Frangopol (Eds)
© 2012 Taylor & Francis Group, London, ISBN 978-0-415-62124-3

Decentralized damage diagnosis for beam-like truss structure considering modeling error

Z. Sun
State Key Laboratory on Disaster Reduction in Civil Engineering, Tongji University, Shanghai, China

B. Zhou
Department of Bridge Engineering, Tongji University, Shanghai, China

ABSTRACT

In this paper, a decentralized damage diagnosis method for planar beam-like truss is presented with a special emphasis on discussing the effect of modeling error on damage diagnosis algorithm. According to structural geometrical layout, the structure is divided into multiple local subgroups firstly. Based on vibration test within each subgroup, structural dynamic flexibility matrix of each measured subgroup is estimated and structural damage locating vector can be computed. The vector is taken as an input force vector imported into structural baseline FEM model to calculate the normalized cumulative stress of all components of the tested subgroup. Structural damage can then be detected and located by checking whether the obtained normalized cumulative stresses for the components in this subgroup are below a preset threshold value. The final decisions on structure damage occurrence and location are made by comparing the results obtained from different subgroups. Since the algorithm rely on an accurate baseline model during the local computing, parametrical and physical modeling error will induce some unpredictable effect on damage diagnosis and is carefully checked in this study. A numerical study on a cantilever beam-like planar truss is conducted to demo the procedure and inspire the discussion. The results tell that both the distribution and the level of the parametrical modeling error will have some influence on damage diagnosis conclusions (as shown in Figure 1). The physical modeling of the baseline structure as a frame model with rigid connection will not affect the result of damage diagnosis if the damage level is set to be 20% (as shown in Figure 2).

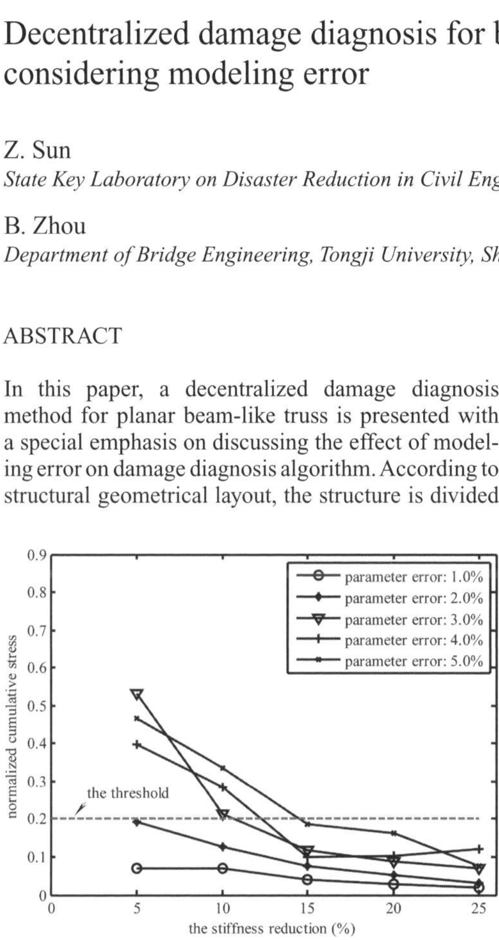

Figure 1. The normalized cumulative stress of element 8 for the damage scenario considering different levels of parametrical modeling error and 0% noise.

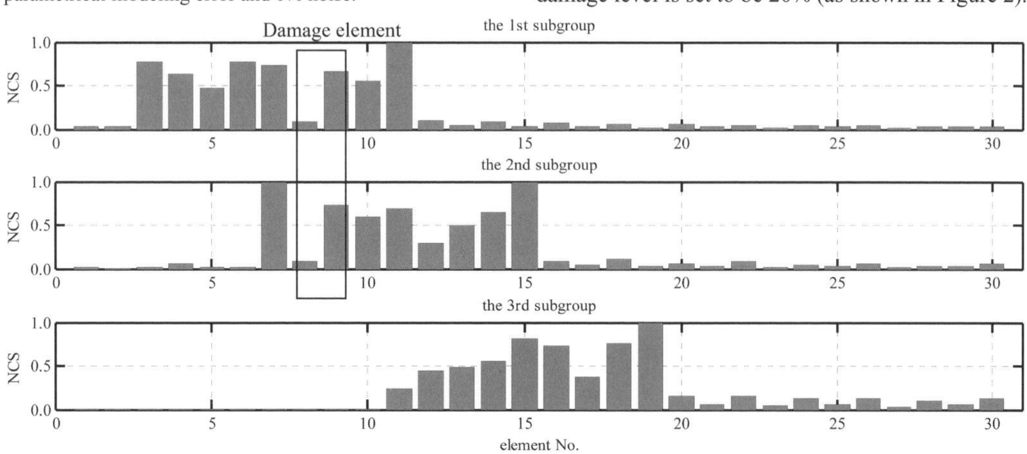

Figure 2. The normalized cumulative stress of all elements if element 8 is damaged considering physical modeling error.

Bridge Maintenance, Safety, Management, Resilience and Sustainability – Biondini & Frangopol (Eds)
 ISBN 978-0-415-62124-3

Fatigue safety assessment of existing railway steel bridges based on in-situ monitoring data

C. Wang, S. Yan & L. Hao
Key Laboratory for Bridge and Tunnel Engineering of Shaanxi Province, Highway College, Chang'an University, Xi'an, Shaanxi Province, China

ABSTRACT

The existing railway steel bridges are often required to carry an increasing volume of traffic and heavier freight trains in China, so bridge owners pay more attention to the actual fatigue life and service safety of such structures. In 2006 twenty two long fatigue cracks were found at the bottom flange connection plates between end cross-frames and main girders for each span of NO.1 Wei River Bridge, which is a railway bolted and welded girder bridge built in 1982. In order to prevent crack propagation, stop holes were drilled in front of the crack tips. Furthermore, the connection plates in the tenth span of NO.1 Wei River Bridge were strengthened by bonding steel plates in 2007, and the dynamic strains near the cracks were monitored before and after strengthening to evaluate the fatigue safety and rehabilitation effectiveness. In current paper, the fatigue damage initiation and propagation in connection plates were simulated using finite element models. Based on the fatigue and fracture safety evaluation results of NO.1 Wei River Bridge using three short term in-situ monitoring data, the following conclusions can be drawn: (1) The tension stresses induced by the out plane deformation at the connection plates were very high, which led to crack initiation and propagation, and very short fatigue life evaluated by 2006 field monitoring data. (2) Using stress data of 10DPU by in situ monitoring in December 2008 and April 2011 after retrofit, the remaining fatigue life of the new bonding connection plate was evaluated by linear elastic fracture mechanics model. The results indicate that the remaining fatigue life assessed using 2008 monitoring data is longer than 200 years, so the trial rehabilitation is very effectiveness and the other cracked connection plates can and should be retrofitted by adding bonding plates. The remaining fatigue life evaluated using 2011 monitoring data is about 150 years, which indicates that degeneration of strengthening joint has happened from 2008 to 2011, so it is necessary to inspect the damaged and strengthened connection plates periodically to ensure the bridge service safety.

ACKNOWLEDGEMENT

The writer gratefully acknowledges the financial support provided by National Natural Science Foundation of China (Grant No.50878025, 51078039), the Foundation for the Author of National Excellent Doctoral Dissertation of the P.R. China (Grant No. 2007B49), the Special Fund for Basic Scientific Research of Central Colleges of the P.R. China, Chang'an University (CHD2012ZD008), China West Transportation Development Research Projects (Grant No. 20113185191410) and the technical support provided by Xi'an Railway Bureau for the field tests.

REFERENCES

Barsom, J. M. & Rolfe, S. T. 1999. *Fracture and Fatigue Control of Structures: Applications and Fracture Mechanics*. West Conshohocken: ASTM.

Fisher, J. W. 1984. *Fatigue and Fracture in Steel Bridges*. New York: John Wiley & Sons.

Hobbacher, A. 1993. Stress intensity factors of welded joints. *Engineering Fracture Mechanics*, 46(2):173–182.

Newman, J. C. & Raju, I. S. 1981. An empirical stress-intensity factor equation for the surface crack. *Engineering Fracture Mechanics*, 15(1-2):185–192.

Wang, C.S., Chen, A.R., Chen, W.Z. & Xu, Y. 2006. Application of probabilistic fracture mechanics in evaluation of existing riveted bridges. *Journal of Bridge Structures*, 2(4): 223–232.

Wang, C.S., Chen, W.Z. & Chen, A.R. 2009. Fatigue Safety Assessment of Existing Steel Bridges in China. *Structural Engineering International*, 19(2), 174–179.

Bridge Maintenance, Safety, Management, Resilience and Sustainability – Biondini & Frangopol (Eds)
 ISBN 978-0-415-62124-3

Fatigue cracking monitoring and evaluation using smart sensors for steel bridge decks

C. Wang, L. Tian & B. Fu
Engineering Research Center for Large Highway Structure Safety of Ministry of Education, Highway College, Chang'an University, Xi'an, Shaanxi Province, China

Y. Zhang
Department of Civil & Environmental Engineering, University of Maryland, College Park, MD, USA

ABSTRACT

As the complex configurations and high stress concentration of orthotropic steel bridge decks, many details in orthotropic steel bridge decks are fatigue-sensitive under repeated vehicle loading. For predicting and extending the service life of steel bridge decks, it is necessary to monitor and evaluate the fatigue cracks in orthotropic steel decks. In this paper, piezo paint sensors and commercial acoustic emission sensors are used to monitor and evaluate the full-scale orthotropic steel bridge deck under cyclic loading. The piezo paint sensor has broad frequency bandwidth of interest to bridge monitoring, is applicable to curved surfaces and can monitor fatigue crack in close-range. During the fatigue test, fatigue crack monitoring was mainly conducted at the rib butt welding joints, and the fatigue crack growth information was acquired by smart acoustic emission sensors.

AE signals released by genuine hit, friction emission and PLB testing collected by commercial AE sensor are identified. The waveforms collected during friction emission and PLB testing provide the reference to distinguish the noise data from the emission caused by crack growth. By comparing with the friction emission and PLB testing signals, the genuine AE signals collected from the orthotropic steel bridge deck were compared and verified to be due to AE events associated with fatigue crack propagation. Acoustic emission monitoring results show that typical genuine AE signals can be collected by piezoelectric paint sensor as well as commercial AE sensor under the same cyclic loading. This indicated that piezoelectric paint sensor has good ability to detect AE events associated with fatigue crack propagation in the orthotropic steel bridge deck under cyclic loading.

ACKNOWLEDGEMENT

The writer gratefully acknowledges the financial support provided by National Natural Science Foundation of China (Grant No.50878025, 51078039), the Foundation for the Author of National Excellent Doctoral Dissertation of the P.R. China (Grant No. 2007B49) and the Special Fund for Basic Scientific Research of Central Colleges of the P.R. China, Chang'an University (CHD2012ZD008), China West Transportation Development Research Projects (Grant No. 20113185191410) and the technical support provided by Dr. Weiping Dong and Mr. Delian Kong of Physical Acoustics Company.

REFERENCES

Holford, K.M., Davies, A.W., Pullin R. & Carter, D.C. 2001. Damage Location in Steel Bridges by Acoustic Emission. *Journal of Intelligent Material Systems and Structures*, 12: 567–576.

Li, X. & Zhang, Y. 2008. Feasibility study of wide-band low-profile ultrasonic sensor with flexible piezoelectric paint. *Smart Structures and Systems*, 4(5): 1–10.

Nair, A. & Cai, C. S. 2010. Acoustic Emission Monitoring of Bridges: Review And Case Studies. *Engineering Structures*, 32(6): 1704–1714.

Physical Acoustics Corporation. 1996. *Inspecting Bridges with Acoustic Emission*. Technical Report TR-103-12, Physical Acoustics Corporation.

Wang, C.S. & Feng, Y.C. 2008. Review of the fatigue behaviors and finite element analysis of orthotropic steel bridge decks. *Proceeding of the first International Orthotropic Bridge Conference*. Sacramento: ASCE Sacramento Section: 290–304.

Xu, W. & Zhang, X.N. 2008. Analysis distress charactors and design of steel orthotropic bridge decks pavement in China. *Proceeding of the first International Orthotropic Bridge Conference*. Sacramento: ASCE Sacramento Section: 184–192.

Yu, J., Ziehl, P., Zárate, B. & Caicedo, J. 2011. Prediction of fatigue crack growth in steel bridge components using acoustic emission. *Journal of Construction Steel Research*, 67(8): 1254–1260.

Bridge Maintenance, Safety, Management, Resilience and Sustainability – Biondini & Frangopol (Eds)
© 2012 Taylor & Francis Group, London, ISBN 978-0-415-62124-3

Wireless crack sensing using an RFID-based folded patch antenna

X. Yi, Y. Wang & R.T. Leon
School of Civil and Environmental Engineering, Georgia Institute of Technology, Atlanta, GA, USA

J. Cooper & M.M. Tentzeris
School of Electrical and Computer Engineering, Georgia Institute of Technology, Atlanta, GA, USA

ABSTRACT

This paper describes the crack sensing performance of a wireless and passive smart-skin sensor, designed as a folded patch antenna (Yi *et al.* 2011). When stress concentration or cracks occur on the patch antenna, the antenna electrical length changes and its electromagnetic resonance frequency changes accordingly. The resonance frequency change can be wirelessly interrogated by a radiofrequency identification (RFID) reader (Finkenzeller 2003). An inexpensive off-the-shelf RFID chip is adopted in the sensor design for signal modulation and collision avoidance. The RF interrogation energy from the reader is captured by the patch antenna, and then used to activate the RFID chip that transmits modulated signal back to the reader. Therefore, the interrogation process is wireless and the antenna sensor is battery-free.

The sensing resolution and measurement limit of the RFID antenna sensor were investigated through extensive tensile tests. It was shown in previous studies that the prototype passive wireless strain sensor can detect small strain changes (<20 με), and perform well at large strains (>10,000 με). To validate crack sensing performance of the antenna sensor, a specially designed crack testing device is manufactured. Fig. 1 shows the experimental setup. By turning a fine-resolution displacement control screw, a rotation is imposed on the upper plate and a crack is opened between the upper and lower plates. A total of fourteen crack opening sizes are measured during the experiment, before the antenna sensor stopped functioning properly at a 32-mil crack opening. Fig. 2 shows the relationship between the resonance frequency of the antenna sensor and the applied crack size. The slope of −0.3013 MHz/mil represents the crack sensing sensitivity, which means a 1-mil crack size increase generates 0.3013 MHz decrease in antenna resonance frequency.

In summary, this research shows that the antenna sensor is capable of measuring milli-inch crack width and tracking its propagation. The crack sensitivity is reasonable and the equivalent strain sensitivity is consistent with the sensitivity reported in previous studies.

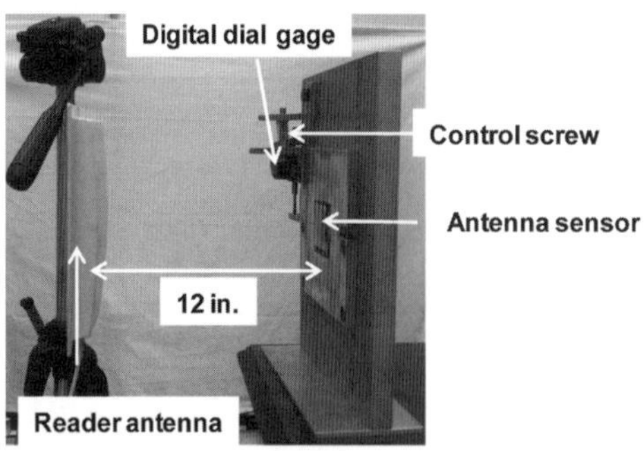

Figure 1. Experimental setup.

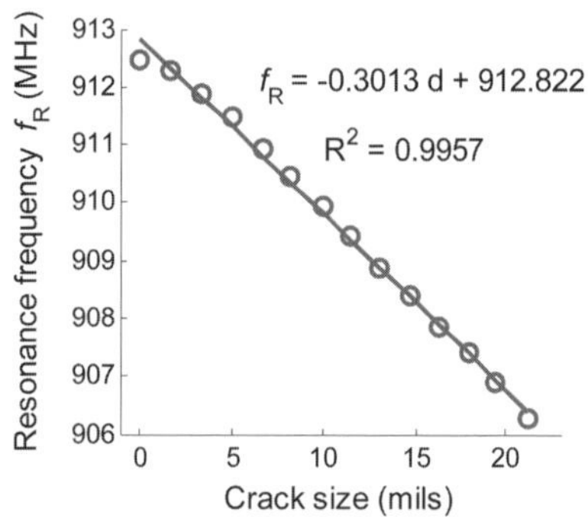

Figure 2. Relationship between antenna resonance frequency and crack size.

REFERENCES

Finkenzeller, K. 2003. *RFID Handbook*, John Wiley & Sons: New York.

Yi, X., Wu, T., Wang, Y., Leon, R. T., Tentzeris, M. M., and Lantz, G. 2011. Passive wireless smart-skin sensor using RFID-based folded patch antennas. *International Journal of Smart and Nano Materials* 2(1): 22–38.

Bridge Maintenance, Safety, Management, Resilience and Sustainability – Biondini & Frangopol (Eds)

Steel bridge fatigue crack monitoring with broadband thin-film acoustic emission sensor

C. Zhou & Y. Zhang
Department of Civil and Environmental Engineering, University of Maryland, College Park, MD, USA

ABSTRACT

This paper presents a fatigue crack monitoring strategy termed close range acoustic emission (AE) signal sensing enabled by the use of low-cost thin-film AE sensor. In particular, flexible piezo paint has been used to make broadband thin-film AE sensors. This close-range AE signal monitoring method offers a promising new approach to AE-induced stress-wave sensing in the ultrasonic frequency range of interest to engineering structure monitoring: low profile, conformable to curved surface, broad-band sensor that has less signal distortion in the specified frequency range, lower cost suitable for large amount use on bridges. In this work, the concept of close-range AE monitoring strategy, the thin-film sensor response and the analysis method are discussed.

Close-range AE monitoring strategy with thin-film AE sensor is explained in details. In close-range monitoring, the thin-film sensor works based on strain measurement but with a much higher sensitivity compared with normal strain gauges. The operational frequency in ultrasonic range makes itself could response to rapid stress release in nearby locations such as fatigue crack propagation. Its lower sensitivity compared with displacement sensor leads to environmental load-induced noise immunity. The most important is close-range monitoring is able to perform inverse analysis due to the much shorter traveling path of the induced stress wave. A test of thin-film AE sensor made of piezo paint material mounted on a test-bed bridge in South Korea has been conducted. The collected data pro-vides useful information for demonstrating the concept of the close-range AE signal monitoring.

The response of thin-film AE sensor to surface pulse on a half space has also been studied. Since there is no existing formula to calculate the strain response, formula to calculate the surface in plane strain due to a surface pulse on a half space is derived and the resulted surface strain considering aperture effect is also plotted. Surface pulse generated by breaking glass capillary on a large steel block is used as a step force source in experimental validation of strain-based AE sensor characteristics. Comparable performance with displacement based AE sensor in both amplitude and frequency response is obtained, which is shown in Figure 1. The derived formula for surface response and experiment setup might be adapted for strain-measurement based thin-film AE sensor calibration.

Waveform based signal analysis for thin-film AE sensor is also studied. Moment tensor theory and Green's function is utilized to form the general expression representing signals initiating from fatigue crack. With the possible broadband sensing provided by thin-film sensor, signal interpretation for identification of AE source parameters such as crack type and location can be carried out.

Future work for strain-based AE sensor will be focused on real fatigue crack source as well as obtaining some quantities from strict calibration procedure.

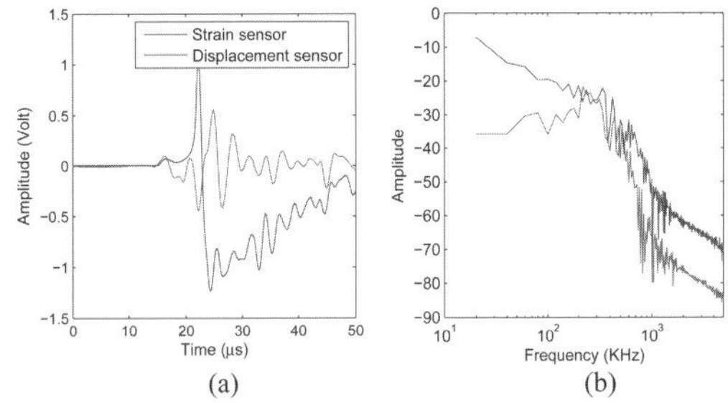

Figure 1. Sensor response to a breaking glass capillary on a large steel block: (a) response in time history; (b) frequency spectrum.

Bridge Maintenance, Safety, Management, Resilience and Sustainability – Biondini & Frangopol (Eds)
© 2012 Taylor & Francis Group, London, ISBN 978-0-415-62124-3

Piezoelectric-based crack detection techniques of concrete structures: Experimental study

J.S. Zhu, C.E. Gao & L.K. He
Tianjin University, Tianjin, P.R. China

ABSTRACT

Steel reinforced concrete (RC) is widely used in civil infrastructures. It can be seen everywhere in buildings, bridges, and highway systems. Health monitoring of civil infrastructures has achieved considerable importance in recent years, since the failure of these structures can cause immense loss of life and property. However, the large size and complex nature of the civil-structural systems render the conventional visual inspection very tedious, expensive, and sometimes unreliable, thus necessitating investigations for the development of automated structural health monitoring (SHM) techniques for this purpose are necessary. Most of the researches in the field of SHM for concrete structures have so far focused on the vibration method, using low frequency natural frequencies and mode shapes. The disadvantage of the method is that the first few global modes are less sensitive to localize damage in the structure. Therefore, these techniques tend to fail in detecting relatively small damage, and they lack precision. In addition, the vibration methods are usually operated in a passive manner, in which an exterior loading impact is needed.

In recent years, due to the advantages such as high sensitivity, active sensing, low cost, assorted shapes and simple implementation, piezoceramic materials have been successfully applied in SHM of concrete structures. Many researchers have attempted on piezoceramic-based health monitoring, and which are two major categories: (1) the impedance based technique and (2) the wave propagation-based health monitoring technique.

In their studies, there are two categories of PZT patches were applied for wave propagation sensing, the surface bonded PZT and the embedded PZT. Whereas, it is noted that no significant literature can be found involving experimentally comparative study on the different wave propagation modes in the concrete structures due to employing different PZT patches. Further studies are necessary on PZT-based wave propagation in concrete structures for the widely application.

This paper reports on the use and comparison of surface bonded PZT patches and embedded SAs for crack detection of RC structures experimentally. At first, the wave propagation mechanisms in concrete are analyzed. Then, an active sensing system with integrated actuators/sensors is constructed. In the experimental study, progressive cracked damage inflicted artificially on the plain concrete beam is assessed by using both lateral and thickness modes of the PZT patches. A relative voltage attenuation coefficient R_v in the voltage signals of the PZT patches is used as the damage indicator. The excitation central frequencies, stress wave propagation modes, the actuator-sensor distances and the crack quantities and depth are set as sensitive parameters, and their affects on R_v are discriminated.

REFERENCES

Arghadeep Laskar, Haichang Gu, YLMo, *et al.* 2009. Progressive collapse of a two-story reinforced concrete frame with embedded smart aggregates. *Smart Mater. and Struct.*, 18(7):75–84.

Ayres J.W., Lalande F., Chaudhry Z., *et al.* 1998. Qualitative impedance-based health monitoring of civil infrastructures. *Smart Mater. and Struct.*, 7(5):599–605.

F. Song, G.L. Huang, J.H. Kim and S. Haran. 2008. On the study of surface wave propagation in concrete structures using a piezoelectric actuator/sensor system. *Smart Mater. and Struct.*, 17(5):1–8.

Gangbing Song, Haichang Gu and Yi-Lung Mo. 2008. Smart aggregates: multi-functional sensors for concrete structures-a tutorial and a review. *Smart Mater. and Struct.*, 17(3):1–8.

Ou J.P., Li H. 2010. Structural health monitoring in mainland China: review and future trends. *Structural Health Monitoring*, 9(3):219–232.

R. Tawie, H.K. Lee. 2010. Monitoring the strength development in concrete by EMI sensing technique. *Constr. and Build. Mater.* 24(9):1746–1753.

R. Tawie, H.K. Lee. 2010. Piezoelectric-based nondestructive monitoring of hydration of reinforced concrete as an indicator of bond development at the steel-concrete interface. *Cement and Concrete Research*, 40(12): 1697–1703.

S. Park, S. Ahmad, C.-B. Yun, *et al.* 2006. Multiple crack detection of concrete structures using impedance-based structural health monitoring techniques. *Experimental Mechanics*, 46(5): 609–618.

Song G., Gu H., Mo Y.L., *et al.* 2007. Concrete structural health monitoring using embedded piezoceramic transducers. *Smart Mater. and Struct.*, 16:959–968.

Tseng K.K., Wang L. 2005. Impedance-based method for nondestructive damage identification. *ASCE J. Eng. Mech.*, 131(1):58–64.

Yuhang Hu, Yaowen Yang. 2007. Wave propagation modeling of the PZT sensing region for structural health monitoring. *Smart Mater. and Struct.*, 16(3):706–716.

Zhu Jinsong, He Likun. 2011. Study on piezoelectric actuator/sensor wave propagation based nondestructive active monitoring method of concrete structures. *Journal of Wuhan University of Technology-Mater. Sci. Ed.*, 26(3):541–547.

Bridge Maintenance, Safety, Management, Resilience and Sustainability – Biondini & Frangopol (Eds)

 ISBN 978-0-415-62124-3

Structural health monitoring-based finite element model of Stonecutters Bridge

S. Zhu, Y.L. Xu & Y. Zhen
Department of Civil and Structural Engineering, The Hong Kong Polytechnic University, Hong Kong

Z.W. Chen
School of Architecture and Civil Engineering, Xiamen University, Fujian Province, China

K.Y. Wong
Bridges & Structures Division, Highways Department, The Government of Hong Kong S.A.R., Hong Kong

S.L. Li
School of Transportation Science and Engineering, Harbin Institute of Technology, Harbin, China

ABSTRACT

The recently built Stonecutters Bridge in Hong Kong is a cable-stayed bridge with a main span of 1018 m, the second-longest cable-stayed span in the world. As the bridge is located in a highly urbanized area, its serviceability and safety in long service life becomes a public concern. Therefore, a comprehensive structural health monitoring system (SHMS) has been installed on Stonecutters Bridge by Highways Department of Hong Kong S.A.R., and the sensory system is composed of totally 1,571 sensors in 15 different types, namely, anemometers, barometers, rainfall gauges, temperature sensors, hygrometers, corrosion cells, accelerometers, dynamic strain gauges, static strain gauges, global positioning systems, tiltmeters, dynamics weigh-in-motion stations, video cameras, articulation sensors and tensio-magnetic gauges. The SHMS pro-vides very useful information for safety evaluation and regular maintenance of the bridge under in-service condition. After the instrumentation of SHMS in the bridge, Hong Kong's Highways Department has carried out a serious of in-situ measurement, including the measurement of as-built profiles, modal frequencies, cable forces, etc. In particular, the displacement and strain influence lines were obtained by sensors at multiple locations during the road testing.

It is noteworthy that the number of sensors is still very limited for such a large-scale structural system, and the locations of structural defects or degradation may not be directly monitored by sensors. In view of this fact, an accurate finite element model (FEM) that relates the bridge performance and condition to the measurement at limited locations becomes an essential tool for the effective health diagnosis and prognosis of the bridge. This paper presented the development of the SHM-oriented FEM for Stonecutters Bridge, including the modeling, updating and validation of the FEM. All the superstructures, substructures,

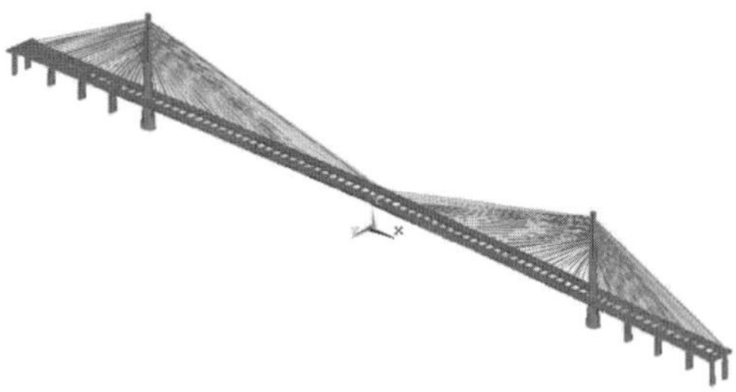

Figure 1. 3D space frame FEM of Stonecutters Bridge.

connections and boundary conditions of the bridge were properly modeled in 3D frame FEM. The model updating was carried out according to in-situ measured results (e.g. the deck profile, cable forces and modal frequencies). A good agreement between the computed results and in-situ measurement validates that the established FEM can well serve as a basis for bridge rating system and performance prediction. Furthermore, the static influence lines due to moving highway vehicles were analyzed using the 3D FEM, and the critical locations or members were identified through the deflection influence lines. It should be noted that the 3D global FEM presented in this paper is only the first stage in the development of health prognosis tool for Stonecutters Bridge. A more sophisticated model, 3D multi-scale FEM of Stonecutters Bridge, is being developed currently.

REFERENCE

Wong, KY. 2010. Structural health monitoring and safety evaluation of Stonecutters Bridge under the in-service condition. *Bridge Maintenance, Safety, Management and Life-cycle Optimization*, Frangopol, Sause & Kusko (eds), London: Taylor & Francis Group.

Bridge Maintenance, Safety, Management, Resilience and Sustainability – Biondini & Frangopol (Eds)
 ISBN 978-0-415-62124-3

Combination of sensing techniques to estimate tension and elongation in bridge cable-stays

D. Zonta, P. Esposito, M. Pozzi & R. Zandonini
University of Trento, Trento, Italy

M. Wang
Northeaster University, Boston, USA

Y. Zhao
Intelligent Instrument Systems, Chicago, USA

D. Inaudi & D. Posenato
Smartec SA, Manno, Switzerland

ABSTRACT

The motivation of this work is the installation of a monitoring system on a new cable-stayed bridge spanning the Adige River 10 km north of the town of Trento. This is a statically indeterminate structure, having a composite steel-concrete deck of length 260 m overall, supported by 12 stay cables, 6 per deck side. These are full locked steel cables of diameters 116 mm and 128 mm, designed for operational loads varying from 5000 to 8000 kN. The structural redundancy suggests that plastic load redistribution among the cables can be expected in the long term.

To monitor such load redistribution, the owner decided to install a monitoring system to measure cable stress; the precision specified was of the order of few MPa. However no cable release or any form of on-site calibration involving tension change was allowed. The solution found was a combination of built-on-site electromagnetic and fiber-optic elongation gauges, these appropriately distributed on both the cables and the anchorages.

We discuss how the set of gauges allows tension and elongation measurement with the appropriate precision, and compare the initial monitoring results with the tension estimates made using a non-destructive vibration test.

The preliminary results show that the strain acquired from FOS is consistent with the prediction and allows a precision equal or better than 5 $\mu\varepsilon$, corresponding to 5 kN in terms of elastic strain. However, because the structure is statically indeterminate and undergoes elastic redistribution on temperature changes, additional electromagnetic sensors are needed to directly detect the tension on the cables.

These sensors are built *in-situ*, and they require extensive calibration first in the laboratory and then on the bridge. After calibration, the standard deviation that can be achieved is of the order of 200 kN. This precision could be improved by comparing EM response with that of the fiber optics in a controlled load test, which has not been performed yet.

REFERENCES

Glisic, B. & Inaudi, D 2007. Fibre Optic Methods for Structural Health Monitoring. Chichester: John Willey & Sons.

Sumitro, S., Kurokawa, M. & Wang, M.L. 2002. Elastomagnetic sensor utilization on steel cable stress measurement. *Proc. of the first fib congress, Concrete Structures in the 21th Century, Osaka, 13–19 October 2002.*

a)

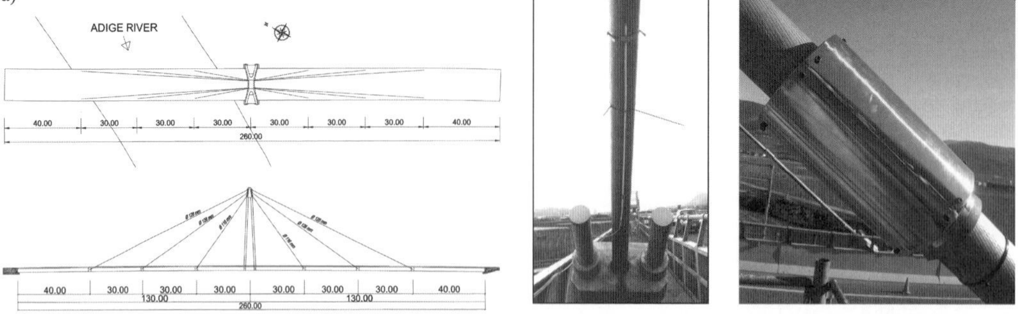

Figure 1. Plan view and elevation of Adige bridge; view of FOS strain sensor and EM load sensor on cable.

Monitoring and assessment of bridges using novel techniques

Organizers: A. Strauss & D.M. Frangopol

Bridge Maintenance, Safety, Management, Resilience and Sustainability – Biondini & Frangopol (Eds)
 ISBN 978-0-415-62124-3

Experimental study on bridge scour monitoring system

C.C. Chen & S.C. Wong
National Center for Research on Earthquake Engineering, Taipei, Taiwan

K.-C. Chang & C.Y. Lin
National Taiwan University, Taipei, Taiwan

ABSTRACT

The news of heavy rain and floods disasters with enormous cost loss is often reported in the world. Additionally, the raising scope of natural hazards associating with the global climate change threatens civil infrastructures on the increase. As for cross-water bridges, it is no doubt that scour effect and flood events could be accordingly expected to be a critical problem after the exposition of bridge foundations. Flood water intensifies the scour menace to bridges due to unsteady nature of the flow. Furthermore, floating or submerged debris induced by flood may deepen the scour depth of bridge foundations (Melville et al. 2000). In accordance with the research in the Unites States (Reddy 2006), earthquake, wind, corrosion, structural, accidental and construction related bridge failures combined do not exceed failures caused by scour. Therefore, the public pay more and more attention towards the bridge safety under natural disasters, and the nations and related organizations have put a lot of efforts and resources on such subjects which lead that bridge scour monitoring plays a more and more important role in the development of practical bridge safety evaluation technologies. Although the complete and actual field bridge failure due to scour effect during floods is still a difficult process to be measured and not be clarified enough yet.

In order to obtain the information of field bridges for engineering evaluation references, monitoring technologies would be generally applied for performing on-site measurements. Even though bridge failure cases are still observed during flood and typhoon events, which pushes researchers studying bridge engineering to develop the more reliable and more robust measuring system for bridge monitoring. Regarding to investigating the scour effect affecting bridge structural conditions, the most straight forward way is using field measured data for the evaluation of in-situ bridge safety. Thus, scour monitoring is the implementation of scour status detection for bridges. But budgetary constraints make it difficult to manage and maintain the existing bridges in reality. Moreover, facing the extreme difficulties in scouring, it is inefficient and dangerous to manually monitor bridge scour as it occurs.

By applying bridge monitoring technologies can help us to obtain the field information for engineering evaluation references. The purpose of this study not only focus on the verification of scour monitoring system developed for field application but also on the exploration of bridge failure condition as a result of local scour effect. Several experimental case studies have been planned and are conducting in the laboratory, the preliminary results shows that the failure mode of modeling bridge caisson specimen can be guided to insufficient soil bearing capacity as experimental expectation. The preliminary result in this paper indicates that studying the knowledge about bridge collapse due to scour by conducting experiment in the flume is a good way with feasibility. However, due to the highly variability for each experimental setup, further study cases should be conducted and able to provide more specific information regarding to the experimental results. Furthermore, the enhancement of reliability for bridge scour safety evaluation could be anticipated with experimental consequences.

Bridge Maintenance, Safety, Management, Resilience and Sustainability – Biondini & Frangopol (Eds)
© 2012 Taylor & Francis Group, London, ISBN 978-0-415-62124-3

Subsequent anchorage of transverse prestressing cables in bridge decks

Ch. Fust & P. Mark
Institute of Concrete Structures, Ruhr-University Bochum, Germany

M. Wolff
Grassl Consultant Engineers, Hamburg, Germany

ABSTRACT

Both, partial deconstruction as well as rehabilitation of prestressed bridge superstructures demand distinct technical knowledge often accompanied by economically reasoned restrictive time limits (Bergmeister 2009 & Strauß 2009 & Haveresch 2010). In case of road bridges, transverse and post-tensioned prestressing is widely used in solid deck slabs of bridges, e.g. with boxes or t-beams (Tue et al 2007). While nowadays prestressing cables usually include coupled wires, aged structures were often built using plain prestressing steel. In case of partial deconstructions tendon cuts involve an immediate release of prestressing forces. Hence, adequate ways are needed to preserve the prestressing forces for the remaining structure. Anchorage by rebonding via shear friction is often not applicable due to its demand on extended bond lengths. Moreover, grouting of ducts can be partially incomplete.

Due to usually short bond length available between cutting position and the section, where full action of prestressing in transverse direction is required again, subsequent mechanical anchorage devices are needed to sustain prestressing and maintain a well-defined safety level for a demanded duration (Tue et al 2007).

The contribution introduces an effective set-up for subsequent anchorage devices of transverse prestressing cables. The device consists of an concrete anchor assembled to a tendon cable that induces the released prestressing forces into a retained deck part. A mechanical steel connector embedded into high-strength concrete overtakes the released forces from the tendon by friction and transfers it via contact compression to the surrounding concrete. Thus, the steel device – consisting of two separated steel plates – is pressed to the tendon cable by well-defined prestressing forces of bolts. The high-strength concrete element itself activates compressive forces to finally induce the released prestressing into the existent bridge deck.

Main focus is laid on the general idea, its theoretical background as well as the design, detailing and experimental verification of the anchorage device.

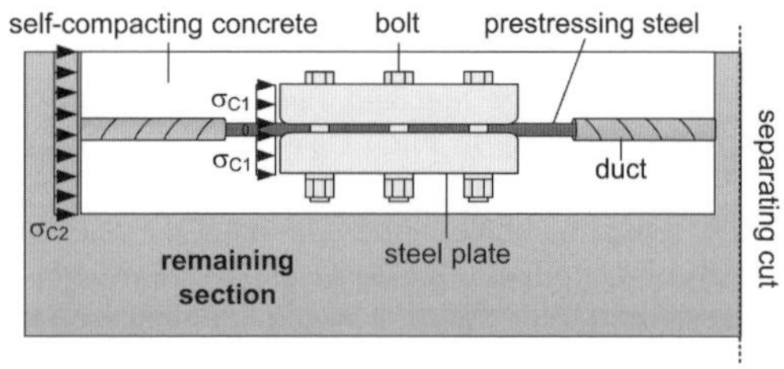

Figure 1. Schematic sketch of the developed anchorage device.

Furthermore, design and feasibility of the developed device are demonstrated at a reference bridge located in Hamburg/Germany.

The elaboration of the innovative anchorage device needed interactive input from Hamburg authorities, researchers and consultant engineers. Due to the uncomplicated and excellent collaboration of these three partners, a ready to use steel device could be developed in an efficient and economic way.

REFERENCES

Bergmeister, K., Novák, D., Pukl, R., Cervenka, V. 2009. Structural assessment and reliability analysis for existing engineering structures, theoretical background. In Structure and Infrastructure Engineering, 5 (4): 267–275.

Haveresch, K.H. 2010. Checking and Strengthening older Prestressed Concrete Bridges. In Beton- und Stahlbetonbau (106): 89–102.

Strauss, A., Hoffmann, R., Wendner, R., Bergmeister, K. 2009. Structural assessment and reliability analysis for existing engineering structures, applications for real structures. In Structure and Infrastructure Engineering. 5 (4): 277–286.

Tue, N., Küchler, M., Ma, J., Zink, M. & Nehrkorn, M. 2007. Anchorage of tendons for partly deconstruction of prestressed constructions – an innovation in UHFB. In Beton- und Stahlbetonbau 84 (11): 762–768.

Bridge Maintenance, Safety, Management, Resilience and Sustainability – Biondini & Frangopol (Eds)
 ISBN 978-0-415-62124-3

Bridge management system: Challenges of adopting a bridge management system appropriate to the needs of a local authority. Example from the United Kingdom

B. Kamya
Kamya Consulting Ltd, Aberdeen, UK

ABSTRACT

For many local bridge authorities, challenges continue to be experienced due to a backlog of maintenance and tight budgetary constraints. Many have tried to deal with these challenges, as covered in various research studies, by adopting some form of bridge management system to aid the decision making process. The bridge management systems BMS adopted range from a simple database inventory system to a more sophisticated database. The complex BMS often include a range of additional criteria, such as condition states, that interface with other road maintenance management tools thus aiding decision making in the process. Issues identified in similar studies in the past, are that the BMS in the UK were mostly developed overseas and often with application features incompatible to UK practice.

Some locally developed systems for example by the Transport Research Laboratory (TRL) mainly focused on whole life costing techniques with limited processing or decision making capabilities. Where systems had some functionality to incorporate condition states assessment or processing, the use of condition states did not always relate to the structures' assessed capacity. A European study in 2001 noted that European maintenance engineers, with exceptions of the Nordic countries such as Denmark, preferred reliance on human judgment than their American counterparts who were more prone to reliance on computerized tools. In 2010 at the IABMAS10 conference, a report was presented that followed a 2008 decision by IABMAS to compile a report on the bridge management systems of the world. This report was based on completed questionnaires sent to 15 countries. Unfortunately, the United Kingdom was not represented in the review of bridge management systems.

This paper in part attempts to deal with this gap of information representation of bridge management systems as applied in the United Kingdom.

The paper reviews the current bridge management systems through questionnaires submitted to a representative section of bridge management authorities in the United Kingdom. It then at-tempts to rationalize the Implementation of a bridge management system by a typical local authority with specific focus on the key drivers to the implementation of a bridge management system. It is hoped that this information would be useful to those with an interest of finding out the bridge management systems in application in the United Kingdom and also provide an opportunities of developing a standard platform that can be used as a guidance and reference to bridge management systems in the United Kingdom.

This paper, in conclusion, proposes further research to develop a system of regulating bridge management systems.

REFERENCES

Adey B.T., Klatter L. and Kong J.S. 2010 Overview of existing bridge management systems. The IABMAS Bridge Management Committee.

Flaig K.D. and Lark R.J. 2000 The development of UK bridge management systems Proc. Instn Civ Engnrs Transp 141, 99–106.

Woodword D.J. 2001 External review of existing bridge management systems and outline framework for a European System. Deliverable 13 BRIME PL97-2220. TRL.

Bridge Maintenance, Safety, Management, Resilience and Sustainability – Biondini & Frangopol (Eds)
© 2012 Taylor & Francis Group, London, ISBN 978-0-415-62124-3

Optimized monitoring concepts for arch bridges

A. Krawtschuk, A. Strauss & K. Bergmeister
Institute for Structural Engineering, BOKU University Vienna, Austria

R. Wendner
Department of Civil and Environmental Eng., NW University Evanston, IL, USA

ABSTRACT

Arch bridges made from nature stone nowadays are the oldest structures which are still in use on road and railway lines. With an average age of more than hundred years, these structures often are seen as historical important buildings. Most of them had been constructed during the great building period of roads and railways from the 1840ies to 1900. Lots of the considered nature stone bridges are constructed as circle or three center curve, some of them also in a parabolic form or catenaries or cycloide. The height of the apex cover varies in a large range. If masonry was appropriated, usually sand, chalkstone or clay bricks were used. For most bridges no observations of the material parameter are available, as a result the stone and the mortar strengths are unknown. Under the usage of the German railway company, there are more than 8000 arch bridges yet, although at local roads there is an additional unknown number of them. In Austria, the railway network, especially along the southern railway line has around 1000 arch bridges in usage. In whole Europe, the stock of masonry railway bridges is estimated with around 70.000. In the course of route expansion plans in the past especially arch bridges have been replaced by new steel or reinforced concrete structures. Considerations of preservation, the budgetary situation of the rail and road operators, as well as a sustainable, re-source efficient usage of resources and existing infrastructure are motivations to maintain and – if necessary – toughen up existing arch bridges. Therefore, the issues of sustainability, durability and serviceability become more important.

Practical considerations have shown that stone arches in combination with an appropriate structural state can have considerable reserves in their bearing capacity. Therefore they often reach the standards which are recommended nowadays. In this article the topic of monitoring based modeling for masonry structures is discussed as it has already been worked out for reinforced concrete structures by Strauss et al. (2009a) and Wendner et al. (2010). The monitoring based modeling consists of the input factors material, which can be seen as inhomogeneous at masonry structures as

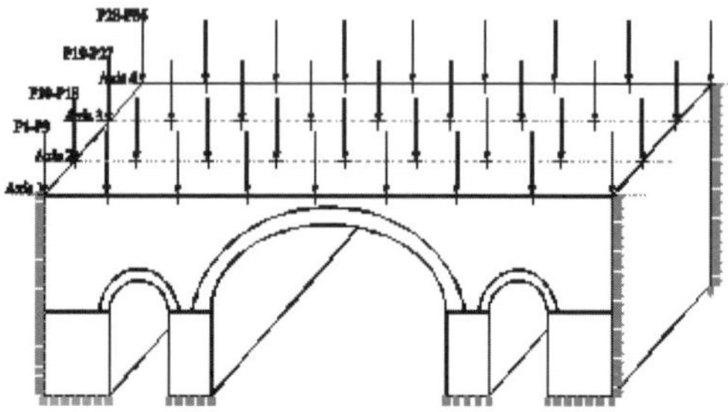

Figure 1. Definition of the axes and load positions at arch bridge object "Bernhardsthal km 75.702".

discussed by Zimmermann & Strauss (2011) and codes which lead to the relevant load models. Moreover the historical development of the load cases which have to be considered is discussed. The model setup includes also the static system and additional testing methods.

The nonlinear modeling of an existing arch bridge structure is done by the program SOFiSTiK and various loading scenarios should be considered. In the first modelling and simulation steps a unit load P was put on the system. The altogether 36 unit loads were applied on defined nodes at four predefined load axes 1 to 4, see Figure 1.

REFERENCES

Strauss, A., Bergmeister, K., Wendner, R. and Hoffmann, S. "System- und Schadensidentifikation von Betonstrukturen". Betonkalender 2009. K. Bergmeister, F. Fingerloos and J. D. Wörner. Berlin, Ernst und Sohn. 2: 55–125, 2009a

Wendner, R., Strauss, A., Bergmeister, K. and Frangopol, D. M. "Monitoring Based Evaluation of Design Criteria for Concrete Frame Bridges". IABSE Symposium 2010. Venice, Italy, IABSE, AIPC, IVBH. Report Vol. 97: 7, 2010

Zimmermann, T. and Strauss, A.: Masonry and Earthquakes: Material properties, Experimental testing and Design approaches. In: Moustafa, A. (Ed): Earthquake Research and Analysis/Book 4, ISBN 979-953-307-680-4, Intech, 2011.

Bridge Maintenance, Safety, Management, Resilience and Sustainability – Biondini & Frangopol (Eds)
 ISBN 978-0-415-62124-3

Advantages of radar interferometry for assessment of dynamic deformation of bridge

P. Kuras, T. Owerko, Ł. Ortyl, R. Kocierz & O. Sukta
AGH University of Science and Technology, Kraków, Poland

S. Pradelok
Silesian University of Technology, Gliwice, Poland

ABSTRACT

Among many sources of information used in the structure diagnosis, the results of load testing are particularly important. They are obtained by performing load tests or tests during operation (Cunha et al. 2007). Structures in service are exposed to different types of load. For this reason they may be damaged by fissures and cracks. In the presented research a railway lattice girder bridge was tested. Cracks appeared in the bottom chord on one of the lattice girders.

In order to explain the reasons for cracking calculation models were built. The results should be verified in empirical tests (Pradelok 2009). For selected points of structure the displacement values were recorded by means of radar interferometer and induction displacement sensors. The main aim of this paper is to test the properties of ground-based radar interferometry as a novel method of dynamic displacements measuring.

The IBIS-S version of radar is suitable for measuring displacement of structures of elongated shape, e.g. bridges. Advantages, which allow to use it in dynamic testing, are the following: 0.1 mm displacement accuracy, 200 Hz sampling frequency, 1000 m range and 0.5 m range resolution (Gentile 2010). The IBIS-S system operates basing on two radar techniques: microwave interferometry and stepped-frequency continuous wave modulation (Skolnik 2001).

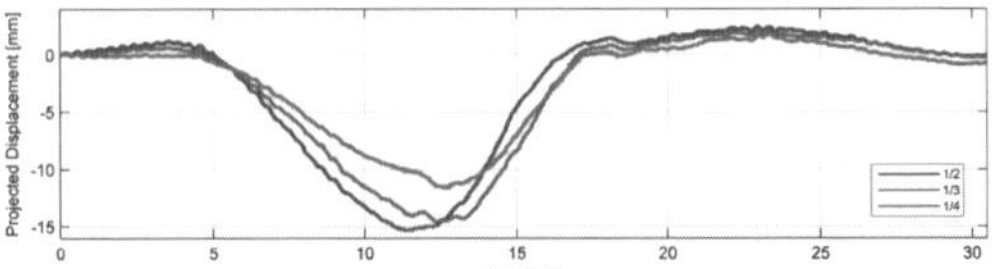

Figure 1. Displacement measurements results.

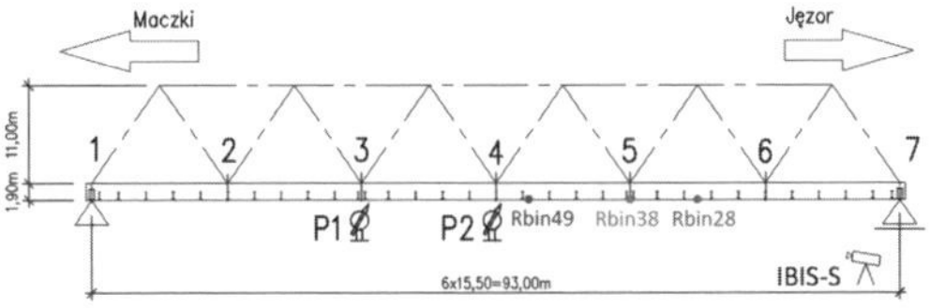

Figure 2. Layout of devices and measured points.

Table 1. Maximum deflection values *d* [mm].

Load no.	Radar		Sensors		Difference	
	1/3	1/2	1/3	1/2	1/3	1/2
1	14.62	15.37	14.05	15.37	−0.57	0.00
2	20.19	21.55	19.93	22.06	−0.26	0.51
3	20.58	22.48	22.74	25.50	2.16	3.02
4	14.62	16.32	16.25	17.63	1.63	1.31

Table 2. Dominant frequencies [Hz].

Radar		Sensors	
1/3	1/2	1/3	1/2
1.875	1.875	1.821	1.839

The subject of study was the 93 m long lattice girder. During tests the bridge was subjected to operational loads, caused by passing trains. An examplary displacement measurements results, recorded by the interferometric radar, are presented in Figure 1. Layout of devices is shown in Figure 2. Maximum deflection values, calculated for both methods, are summarized in Table 1. Dominant frequencies of vibration were determined on the basis of power spectral density and presented in Table 2.

REFERENCES

Cunha, Á., Caetano, E. & Magalhães, F. 2007. Output-only dynamic testing of bridges and special structures. *Structural Concrete* 8(2): 67–85.

Gentile, C. 2010. Application of Radar Technology to Deflection Measurement and Dynamic Testing of Bridges. In Guy Kouemou (ed.), Radar Technology: 141–162. Vukovar: InTech.

Pradelok, S. 2009. Tests of Certain Lattice Girder Railway Bridge Type Nodes. In *Proceedings of EVACES'09. International Conference on Experimental Vibration Analysis for Civil Engineering Structures, Wroclaw, 14–16 October 2009.*

Skolnik, M.I. 2001. *Introduction to radar systems. Third Edition*. McGraw-Hill.

Bridge Maintenance, Safety, Management, Resilience and Sustainability – Biondini & Frangopol (Eds)
© 2012 Taylor & Francis Group, London, ISBN 978-0-415-62124-3

Combined use of ground penetrating radar and laser scanner for bridge health assessment

M. Marelli
Codevintec Italiana srl, Milan, Italy

G. Lommori & S. Oppioni
Sineco spa, Milan-Tortona, Italy

ABSTRACT

Ground penetrating radar and laser scanner were used to quickly and effectively evaluate health state on a 40-years old highway bridge in Forte dei Marmi (Tuscany).

Static and dynamic lidar surveys of ultimate technology were undertaken and output a very accurate 3D model for the whole bridge, assuring the absence of major structural and geometrical faults (according to project specifications) and giving a reference for any possible future three-dimensional multi-temporal monitoring.

Yet, ground penetrating radar dense profiles were acquired both on the lane, underneath the bridge and on the top and base of the pillars by the use of a ground-coupled high frequency antenna.

For the extrados surveyed area (19.8×5.10 m) a detailed stratigraphy of the bridge was derived (confirmed by cores) locating the asphalt-concrete interface around 16 cm depth and identifying inner levels of rebars both transversal and longitudinal to traffic directions.

Their spacing and depth were exactly calculated and vary depending on lane portions, being the reinforcement deeper along the roadway (up to 28 cm) and much tighter as inter-distance on the hard shoulder than on other directions and regions (ca. 12 vs 25 cm).

Time slices and volume analysis were drawn out of the final GPR 3D processed models even for some restricted areas on the pillars and bridge intrados where reinforcements prove to be very shallow (2–3 cm deep) and generally dense, though not always regular, especially close to the top of the structure.

Moreover, GPR signal amplitude maps of the superficial layer and rebars were finally rendered, confirming the presence of light diffused cortical deteriorations, principally on the hard shoulder/driving lane limit and where some cracks are evident on the asphalt as also pointed out by thermal-camera data anomalies. This may be due to trucks traffic and slight slope of the roadway thus channeling stress and humidity over certain portions.

On completion, chemical and mechanical tests, both laboratory and in situ, testified the general good quality of the concrete and the absence of major defects while some small deteriorations have started on the pillars, coming with bulges, where no significant radar signal attenuation maps can be generated because of the restricted extension of the studied areas.

REFERENCES

Dobrin, M.B. & Savit, C.H., 1988, Introduction to Geophysical Prospecting. Singapore, McGraw-Hill.

Parrillo, R. et al., 2006, Bridge Deck Condition Assessment using Ground Penetrating Radar, ECNDT.

Rhazi, J. et al., 2007, A new application of the GPR technique to reinforced concrete bridge decks, 4th Middle Est NDT Conference and Exhibition, Kingdom of Bahrain.

Reynolds, J.M., 1997, An introduction to Applied and Environmental Geophysics.

Rhazi, J. et al., 2003, Non destructive health evaluation of concrete bridge decks by GPR and half cell potential techniques, NDTCE 2003–10.

Telford, W.M et al., 1990, Applied Geophysics. Cambridge, Cambridge University Press.

UNI 10174:1993, Instruction for the inspection of reinforced concrete structure exposed atmosphere by half potential mapping.

Bridge Maintenance, Safety, Management, Resilience and Sustainability – Biondini & Frangopol (Eds)
© 2012 Taylor & Francis Group, London, ISBN 978-0-415-62124-3

Modeling and structural health monitoring of a geriatric signature movable bridge

E. Minaie, F.L. Moon & A.E. Aktan
Intelligent Infrastructure Systems, a Division of Pennoni Associates Inc., Philadelphia, PA, USA

ABSTRACT

As the infrastructure of the United States age, many bridges are deteriorating into complex, geriatric structural systems. While sensing, imaging and information technology recorded significant advances, routine inspection and traditional load rating approaches by over-idealized analytical models are still serving as common practices available to the owners of geriatric bridges. Further, rehabilitation design and implementation is typically based on "as originally constructed" philosophy. Although in some cases these traditional empirical methods may yield reasonable outcomes due to the application of large safety margins, in the case of deteriorated geriatric systems repair and rehabilitation engineering require a mechanistic approach. Design should be based on fully appreciating the actual global and local behaviors of such systems in their deteriorated state. In addition, with today's economy, limited rehabilitation budgets should be allocated based on greater insight for public safety versus operational demands and based on informed and wise decisions. Structural identification (St-Id) may guide engineers and owners in decision making and budget allocation process by providing them mechanistic and reliable information regarding the probable influences of repairs and retrofits on the behavior and lifetime extension of a bridge. This paper discusses the application of structural identification in the evaluation of rehabilitation design for a deteriorated bridge. A finite element model was developed for simulating the "as-is" condition of the deteriorated bridge and was calibrated using the results from a controlled truck load test performed on the bridge. This finite element model was then used to evaluate and prioritize different rehabilitation plans, and to load rate the bridge for each different rehabilitation scenario. Through these evaluations, it was determined that even though the traditional repair plans may improve the performance of some regions, they may, in fact, adversely impact other members and connections of the system. Based on the analyses results, recommendations to modify the repair plans were submitted to the bridge owner, and a retrofit concept was developed which promised a more economical yet effective solution for providing the desired lifetime extension.

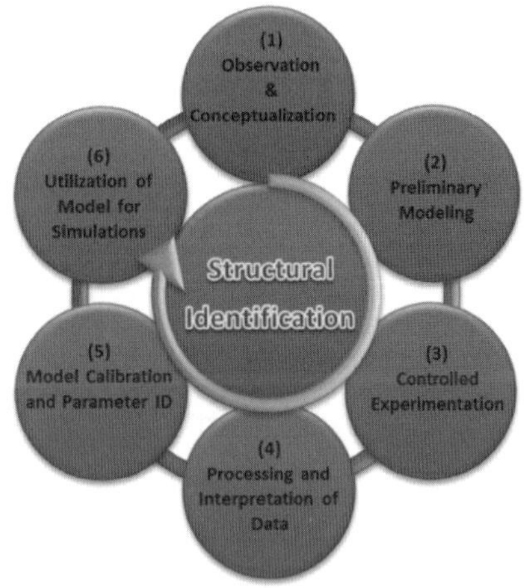

Figure 1. Six Stages of Structural Identification.

In addition, leveraging the FE model, a long-term health monitoring system was designed to capture the most critical responses of the bridge as it further ages in order to alert the owner of any potential hazards which may require immediate intervention. Figure 1 illustrates the six stages of structural identification (St-Id).

REFERENCES

BIRM 2006, FHWA-NHI-03-003 (2006), "Bridge Inspector's Reference Manual (BIRM)," Federal Highway Administration (FHWA)-National Highway Institute (NHI), October 2002 (Revised 2006).

Koglin, T. L. (2003), "Movable Bridge Engineering," John Wiley and Sons.

Lowdermilk D. S., L. Egan & J. Prader (2010), "Structural health monitoring of signature bridges: an engineer's Perspective," Proceedings of the Fifth international Conference on Bridge Maintenance, Safety and Management (IABMAS), Philadelphia, PA, USA, 11–15 July 2010.

Wallner, M. & M. Pircher (2007), "Kinematics of Movable Bridges," ASCE Journal of Bridge Engineering Vol 12, No. 2, pp. 147–153.

Bridge Maintenance, Safety, Management, Resilience and Sustainability – Biondini & Frangopol (Eds)
© 2012 Taylor & Francis Group, London, ISBN 978-0-415-62124-3

Advanced methods for estimating the natural frequency and the damping from monitoring data of structures

M. Österreicher, A. Strauss & K. Bergmeister
University of Natural Resources and Life Sciences, Vienna, Austria

ABSTRACT

The interest in developing approaches and possibilities of determining the condition and remaining lifetime of engineering structures, in particular bridges (both railway and road bridges) has considerably increased over the past decade. The development of approaches based on vibration measurements are of particular interest for the operators of such structures because they are easy to perform and therefore an economical solution. The present work introduces methods for automatically evaluating the natural frequencies and the damping of a structure from monitoring data. The advanced method to calculate natural frequencies is based on a FFT – summation.

The main difference is that before the sum is calculated, a trigger condition is introduced to check if a certain spectrum should be added to the sum. This gives the possibility to define special trigger functions so that f.g. just a spectrum from the ambient vibrations or exited vibrations is added. The results for the analyzed structure show that it is a suitable method to evaluate natural frequencies from large data sets automatically.

The second method introduced in the present work is an algorithm for estimating the damping from vibration data automatically. The evaluation of the damping value, the percentage of critical damping, is generally more complex to evaluate than the maximum deflection or the natural frequency of a structure. The results of the evaluation of the damping indicate, as expected, wide spread distributed results. Nevertheless the distribution of this data shows the damping clearly. To test the method of estimating the damping of the structure from the observation data it is necessary to decide for which natural frequency it should be calculated. To use this method for evaluating the damping a big amount of measurement data set is necessary to get reliable results.

The simulation with both methods indicates that one important thing to get reliable results is to evaluate some measurement data before manually.

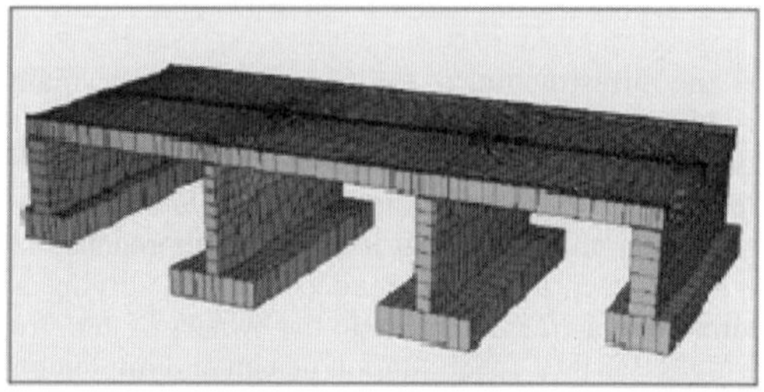

Figure 1. Structural system and fe-model of the railway bridge.

Especially to find suitable trigger conditions to evaluate the natural frequencies based on the TRWFT a good knowledge about the monitoring data is necessary.

Additionally this work shows the results of the introduced methods applied on a structure. The structural model is shown in Figure 1

Finally, both methods could help to get estimates for the damping and the natural frequencies respectively for the analyzed structure, but both methods need the user evaluate parts of the data manually to get the calculation parameters for the introduced methods.

REFERENCES

Geier, Österreicher, Pircher 2008. Langzeitmessungen an einer Eisenbahnbrücke. Beton und Stahlbetonbau, JG 103, Heft 6

Österreicher, Strauss, Geier 2009. Long Term Monitoring Activities for the Assessment of a Railway-Bridge Performance, 4th International Conference on Structural Health Monitoring on Intelligent Infrastructure (SHMII-4) 2009

Österreicher, Geier, Pircher 2008. RELAS – Reliability (Zuverlässigkeit)

Strauss 2010. Framework for code based regulations in monitoring of bridge structures, Proceedings IALCCE 2010

von Strukturen. FFG-Project 2008, Projektnummer 811878

Bridge Maintenance, Safety, Management, Resilience and Sustainability – Biondini & Frangopol (Eds)
 ISBN 978-0-415-62124-3

Stress measurement and material defect detection in steel strands by magneto elastic effect. Comparison with other non-destructive measurement techniques

A.M. Sarmento
LABEST, Faculty of Engineering, University of Porto, Portugal

A. Lage
Faculty of Engineering, University of Porto, Portugal

E. Caetano
ViBEST, Faculty of Engineering, University of Porto, Portugal

J. Figueiras
LABEST, Faculty of Engineering, University of Porto, Portugal

ABSTRACT

The efficiency of the use of magneto-elastic effect on the stress level of prestressed strands is evaluated by the comparison of results obtained by different measuring techniques: magnetic flux variation, vibration and cable elongation. The detection of corrosion and other surface defects by magneto-elastic effect is also investigated. Laboratory tests on seven-wire strands with ME sensors for dynamic and static force evaluation were conducted at the Laboratory of Civil Engineering Department of University of Porto. The results are compared with those measured by fiber optic sensors and electric strain gages installed on the strands and obtained indirectly from identification of natural frequencies associated with the strand vibration. The tensile load measured by a ring-load cell installed between the tension jack and the anchorage was used as reference. The instrumentation setup is shown in Figure 1.

Previous tests for load and temperature calibration of ME sensor with strand were conducted. The values obtained for magneto-elastic coefficient and temperature coefficient were of 50.417 kN/mWb and of 0.0045 mWb/°C, respectively. The force temperature sensitivity is up to 2.3 kN/°C.

Dynamic measurements were further conducted for the various stress levels based on a conventional accelerometer for identification of cable frequency

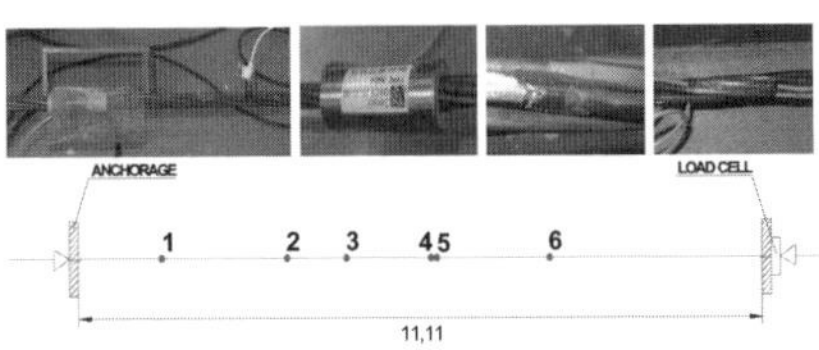

Figure 1. Experimental setup for instrumentation: 1: dynamic ME sensor; 2: accelerometer; 3: static ME sensor; 4,6: strain gages; 5: Bragg optical sensors.

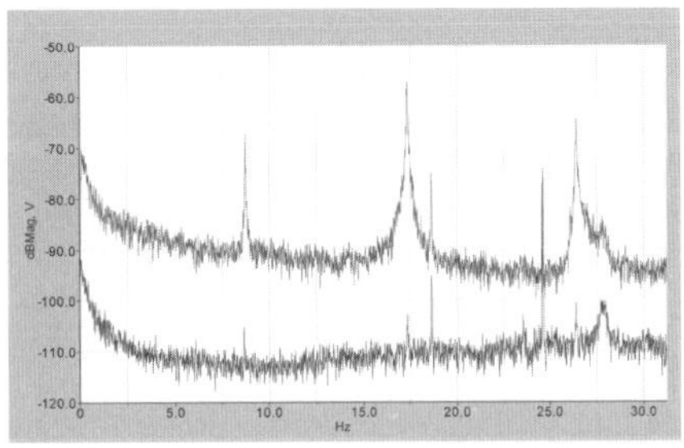

Figure 2. Average power spectra density estimates of time records registered with the accelerometer (top) and the ME sensor (bottom).

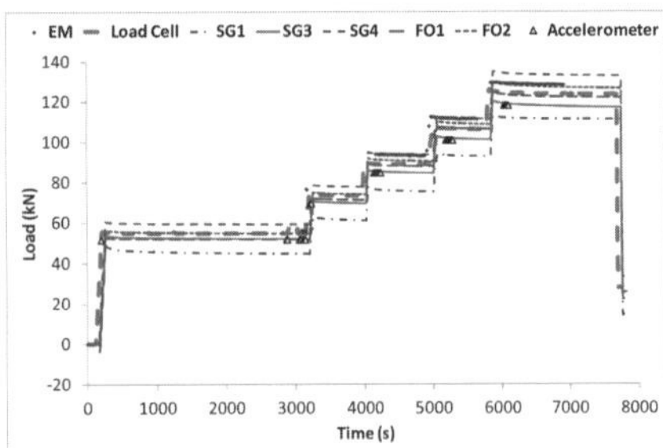

Figure 3. Comparison of strand force results.

and indirect assessment of force under ambient vibrations. The ME sensor was also tested in dynamic measurement mode. Corresponding frequency peaks were obtained for the two sensors, showed the ability of ME sensor to be used for the indirect assessment of force based on the vibration method

The strand load values obtained by static and dynamic measurements are compared in Figure 3.

The maximum error of 4.7% was observed for ME sensors showing their viability for force assessment in steel strands. ME sensors showed to be also suitable for detection of corrosion and surface defects.

Bridge Maintenance, Safety, Management, Resilience and Sustainability – Biondini & Frangopol (Eds)
© 2012 Taylor & Francis Group, London, ISBN 978-0-415-62124-3

Monitoring based assessment of a jointless bridge

A. Strauss, A. Krawtschuk & R. Wendner
Institute for Structural Engineering, BOKU University Vienna, Austria

D.M. Frangopol
Department of Civil and Environmental Engineering, ATLSS Engineering Research Center, Lehigh University, Bethlehem, PA, USA

K. Bergmeister
Institute for Structural Engineering, BOKU University Vienna, Austria

ABSTRACT

In traditional structural design, linear elastic behavior is typically assumed for the global structural response as well as for the soil-structure interaction. This assumption simplifies the design process, in particular with regard to the multitude of load conditions that have to be investigated. However, it is associated with a high degree of uncertainty. While errors resulting from simple superposition of load cases are of little significance, this is not true for some structures subjected to time variant loads resulting from creep, shrinkage, temperature processes and earth pressure. Within this paper, linearity assumptions that are typically made in the design of frame bridges and ultimately govern performance will be presented and discussed on the basis of available monitoring information for an existing bridge. In particular, the uncertain soil-structure interaction and the structural detail for the transition area between concrete deck and soil are investigated.

The so called S33.24 (the Marktwasser Bridge) is a bridge, part of an important highway connection and located in an environmentally sensitive area north of Vienna, Austria. This three-span continuous structure carries five lanes of highway traffic. The span lengths are 19.50 m, 28.05 m and 19.50 m.

The original linear Finite Element Model (FEM) see Figure 1 served for the adaptation of the slab-stiffness distribution and the boundary conditions by incorporating information from the installed monitoring system. The S33.24 Bridge was, already during its setting period, equipped with a system for integrative monitoring which focuses (a) on the structural behavior of the first lateral south/west field of the S33.24 Bridge, and (b) on the deformation–strain field performance of the pavement and the soil next to the approach slab and the abutment. In total, five different sensor systems with 54 sensors were installed (Strauss et al. 2010a; Strauss et al. 2010b; Wendner et al. 2010a).

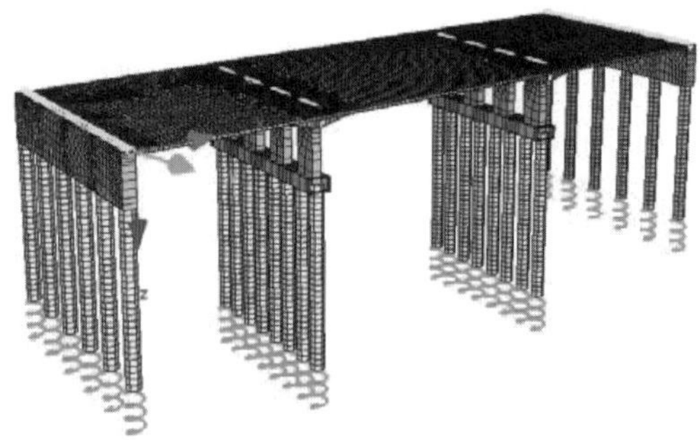

Figure 1. Side elevation of the Marktwasser Bridge.

REFERENCES

Strauss, A., Wendner, R., Bergmeister, K. and Frangopol, D. M. "Monitoring based analysis of an concrete frame bridge 'Marktwasser bridge'". *Third International fib Congress*. Washington, DC, USA: 11, 2010a

Strauss, A., Wendner, R., Bergmeister, K. and Reiterer, M. "Monitoringbasierte Analyse von Integralen Brücken am Beispiel der Marktwasserbrücke". *Schriftenreihe des Departments*. K. Bergmeister. Vienna, Department für Bautechnik und Naturgefahren, Universität für Bodenkultur. **20**, 2010b

Wendner, R., Strauss, A. and Bergmeister, K. "Monitoring based verification of the soil structure interaction of the Markwasser Bridge S33.24". *IABMAS 2010*. D. M. Frangopol, Sause and Kusko. Philadelphia, Pennsylvania, USA, Taylor & Francis, London: 540–547, 2010a

Bridge Maintenance, Safety, Management, Resilience and Sustainability – Biondini & Frangopol (Eds)
 ISBN 978-0-415-62124-3

Linearity assumptions in design: Soil-structure interaction

R. Wendner, A. Strauss & T. Zimmermann
Institute of Structural Engineering, University BOKU, Vienna, Austria

ABSTRACT

In structural engineering traditionally linear models are used to determine structural response and internal forces. The nonlinear characteristics of reinforced concrete are only considered on the cross-sectional level during its layout. Although, depending on the structural system, this assumption can be associated with significant error and thus lead to uncertain design solutions, it is widely accepted in practice and by codes, especially for statically determinate structures, e.g. small span bridges.

With regard to internal forces this assumption generally is valid for serviceability related load levels but becomes questionable for the ultimate limit state design. However, the ability to apply the principle of superposition significantly simplifies structural design, considering the multitude of load situations that are to be investigated. The soil structure interaction in practice is typically considered by means of linear elastic springs with appropriately selected stiffness values corresponding to the upper or lower percentile of the governing soil properties. Unfortunately, for more complex structures the choice for the overall most unfortunate assumption is not possible.

Due to liability reasons as well as for the sake of efficiency, design offices tend to investigate the limiting cases of the anticipated range of nonlinear behaviour by linear models; e.g. fully constrained frame for internal constraint forces versus unconstrained model for extreme deformations. This practice generally ensures sufficiently save design. However, the resulting (highly) conservative design is questionable from the point of economics.

Within this contribution linear design assumptions concerning the soil structure interaction of concrete frame bridges without expansion joints will be investigated. In particular, the linearity between temperature and structural expansion will be studied, as will be the linearity of the strain field in the soil body and pavement in the transition area between concrete bridge deck and soil dam.

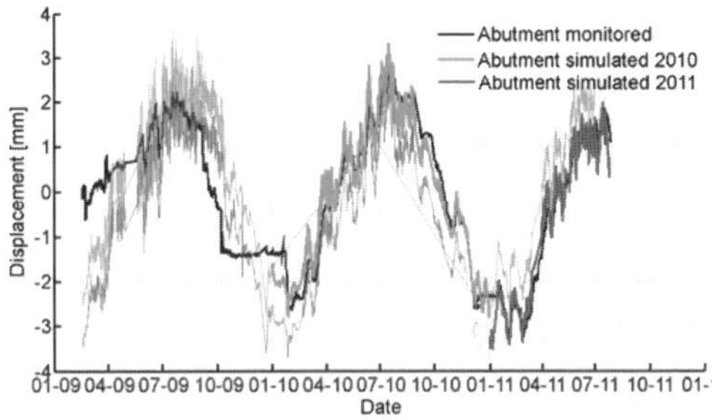

Figure 1. Monitored abutment movement versus prediction based on average concrete temperature for the 2010 and 2011 monitoring period.

The presented investigation of a recently constructed jointless bridge structure has shown that abutment movement seems to be highly correlated with the temperature in the bridge deck. Although this is in agreement with the design assumptions, the monitored structural response only amounts to roughly 59% of the value for free thermal expansion, which was assumed during design, thus leading to an overly conservative layout of the transition area between concrete deck and soil body.

The analysis of the monitoring data for the strain field in the soil body behind the abutment indicates a predominantly linear relationship with the abutment movement and in consequence with the temperature in the bridge deck. However, the encountered redistribution in strains (and trend in the linear relationship) during summer 2010 (time of maximum expansion) as well as winter 2010–2011 shows that further nonlinear analysis of this particular detail and respective optimization is necessary in order to ensure safe and efficient linear design practice.

Bridge Maintenance, Safety, Management, Resilience and Sustainability – Biondini & Frangopol (Eds)
© 2012 Taylor & Francis Group, London, ISBN 978-0-415-62124-3

Long-time health monitoring system of an in-service concrete cable-stayed bridge

Y. Zeng, H. Tan & X. Zhang
State Key Laboratory Breeding Base of Mountain Bridge and Tunnel Engineering, Chongqing Jiaotong University, Chongqing, China
Department of Bridge Engineering, Chongqing Jiaotong University, Chongqing, China

X. Jiang
CCCC First Highway Consultants Co.,LTD., Xi'an, China

X. Shen
Yunnan Chuxiong Highway&Bridge Engineering General CO. Chuxiong, China

ABSTRACT

Lijiatuo Yangtze River Bridge is located in Chongqing, China, built in 1997. Its main span is 169 m + 444 m + 169 m with 24-meter width of the deck. Lijiatuo Bridge is a pre-stressed concrete cable-stayed bridge with twin towers and double cable planes. Health monitoring system of an in-service bridge, which must consider the existing damage and disease, is different from a new bridge.

Health monitoring system (HMS) of Lijiatuo bridge is one integrated system of structural analysis, computer technology, communication technology, network technology, sensor technology and other high technology, which can be used in structural damage and condition assessment of maintenance to meet the needs of economic benefits.

Monitoring of main beam stress in the main girder structure of control parts and key parts of the main girder structure internal force monitoring, research on the internal force distribution, local structure and connection in response to various loads, structural damage identification, fatigue life assessment and structural state assessment provides the basis.

Pylons bear great axial pressures, and generate additional bending moment in the case of possible un-symmetric loads, which is unfavorable for pylons. Therefore it is necessary to monitoring the section stress of main towers.

Distribution monitoring of temperature field can provide the original basis of bridge design. Bridge working condition changes, such as deformation and stress change can be compared in different temperatures.

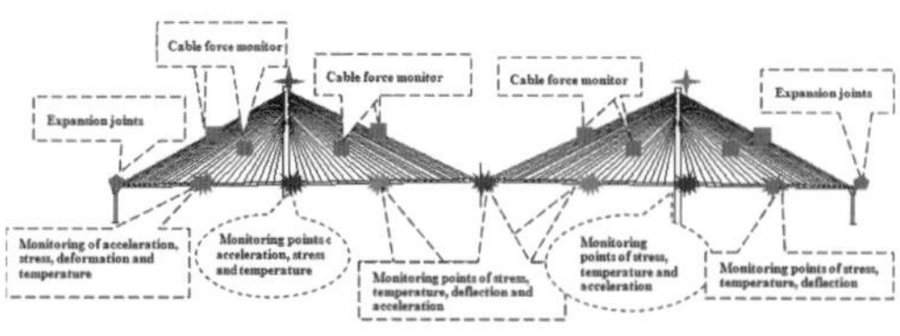

Figure 1. Overall arrangement of LHMS in Lijiatuo bridge.

Vibration can be measured by acceleration sensors. Monitoring of cable forces, not only provide a basis of the overall assessment of the safety and durability of the bridge to provide a basis, but also can detect the integrity and corrosion of cable anchoring system and the protection system.

Wind speed has a great impact on cable-stayed bridges. Wind is the main load of bridges, which can affect the normal operation of the bridge.

Data are collected by combining comprehensive methods of real-time monitoring, regular inspection and detection. Real-time monitoring is automatic data collection mode of sensors. Regular inspection is manual input-data mode.

Importance of operational conditions, namely safety, durability and serviceability of large span in-service bridges in China have aroused the attention of bridge management. Many of these bridges will or have installed long term structural health monitoring system.

Health monitoring system (HMS) of Lijiatuo bridge is one integrated system of structural analysis, computer technology, communication technology, network technology, sensor technology and other high technology, which can be used in structural damage and condition assessment of maintenance to meet the needs of economic benefits.

REFERENCES

Airong Chen, Yong Zeng. 2007. Life-circle based monitor, maintenance and management research of Jiangyin Yangtse River Bridge. Shanghai: Tongji University (in Chinese)

Xuesong ZHANG. 2010. Long term health monitor framework of Chongqing Lijiatuo Yangtze River Bridge. China: Chongqing Jiaotong University

Yong Zeng. 2007.Monitor, maintenance and management of large span suspension bridges in lifetiime. Shanghai: Tongji University (in Chinese)

Bridge Maintenance, Safety, Management, Resilience and Sustainability – Biondini & Frangopol (Eds)
© 2012 Taylor & Francis Group, London, ISBN 978-0-415-62124-3

Extreme value statistics for the life-cycle assessment of masonry arch bridges

T. Zimmermann, A. Krawtschuk & A. Strauss
Institute for Structural Engineering, BOKU University Vienna, Austria

R. Wendner
Department of Civil and Environmental Eng., NW University Evanston, IL, USA

ABSTRACT

Arch bridges made of natural stone masonry nowadays rank among the oldest, structures still in service in the road and railroad infrastructure. Some of them with ages of more than 100 years, represent a historically very valuable building fabric. In Austria's railroad network alone there are around 1,000 arch bridges still in operation. In whole Europe the estimated number of masonry railroad bridges amounts to 70,000. Nowadays increasingly more attention is paid to the maintenance and rehabilitation of existing arch bridges than on their reconstruction, as a result of financial reasons as well as monument conservation aspects. Since those bridges usually have been projected for very different loads, assessments concerning the current bearing capacity and the future utilization are required. A recalculation with conventional calculation methods is often insufficient as results can considerably deviate from the actual load capacity due to various influencing factors. The challenge therefore is to find and verify more accurate methods to determine the actual condition of such arch bridges. A point of interest is the combination of finite element modeling strategies of masonry with inspection and monitoring strategies of existing structures.

Inspections enable the detection of "damage", whereby damage is defined as changes in material as well as geometric properties of a structure. The applications of monitoring systems allow a continuous observation of a system over time. In case of long term observations, the output of this process is a periodically updated information regarding the ability of the structure to perform its tasks.

Within the proposed contribution a generally valid methodology for the performance and lifetime assessment of masonry arch bridges based on nonlinear finite element analysis and monitoring data will be presented. Further two different approaches for the assessment of extreme values and probability of failure will be discussed. The two approaches will be applied to a railroad arch bridge located in Austria.

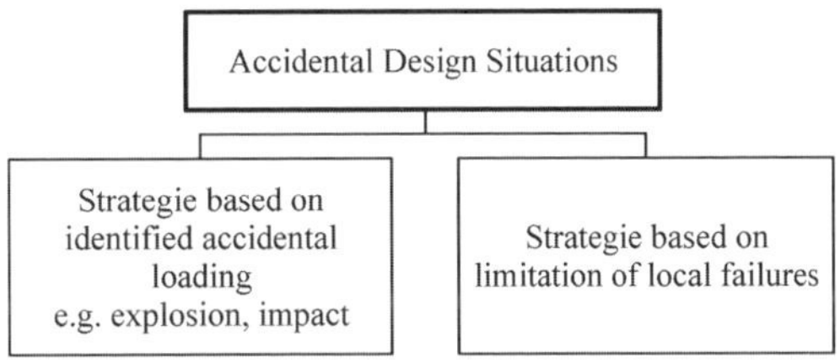

Figure 1. General strategies for structural assessment, according to Eurocode 1-1-7.

The knowledge of hazard-resisting structures has been accumulated over time by integrating the findings from catastrophic failures with the existing information. Although structures are designed to resist certain levels of hazards, it is not possible to completely ensure their safety due to the uncertainties in both hazards and structural behavior. Natural hazards can be classified as earthquake, flood, avalanches and landslides amongst others. Several studies are mentioned herein, related to structures under the effects of such events. For the structural assessment two general strategies can be used (a) based on the identification of accidental actions or (b) based on the limitation of local failures, see Figure 1

For both strategies a calculation of the bearing capacity for the accidental loading as well as an estimation of the level of safety is possible. But case (b) is only useful applicable for the dimensioning and design of new structures, whereby case (a) can be used for new as well as for existing ones.

The focus within this contribution is on horizontal impacts caused by mudflows, acting on existing masonry arch bridges. Hence the target strategy is the identification of the value of accidental loads and thereby the resulting level of safety of the investigated structure.

In particular for the accidental load case mudflow two assessment approaches can be distinguish for existing arch bridges: (a) forward and (b) backward approach.

Life-cycle design and assessment of bridges exposed to corrosion and other hazards

Organizers: F. Biondini, D.M. Frangopol, J. Padgett & A. Palermo

Bridge Maintenance, Safety, Management, Resilience and Sustainability – Biondini & Frangopol (Eds)
 ISBN 978-0-415-62124-3

Damage modeling and nonlinear analysis of concrete bridges under corrosion

F. Biondini & M. Vergani
Department of Structural Engineering, Politecnico di Milano, Milan, Italy

ABSTRACT

The structural performance of concrete structures should be considered time-dependent due to environmental damage. Possible damaging factors include the effects of the diffusive attack from aggressive agents, such as chlorides, which may involve corrosion of steel reinforcement. The direct and indirect costs associated to steel corrosion and related effects, in particular for concrete bridges and viaducts, are generally very high. It is therefore of major importance to promote a life-cycle design of durable structures and infrastructures (Biondini and Frangopol 2008).

Design for durability with respect to chemical-physical damage phenomena is usually based on simplified criteria associated with threshold values for concrete cover, water-cement ratio, amount and type of cement, among others, to limit the effects of damage induced by carbonation of concrete and corrosion of reinforcement. However, a life-cycle design cannot be based on indirect evaluations of the effects of damage, but needs proper methodologies to take into account the global effects of local damage phenomena on the overall system performance.

In this study a general approach to nonlinear analysis of concrete bridges under corrosion is presented. This approach is based on the formulation of a three-dimensional reinforced concrete beam finite element which accounts for both mechanical non-linearity, associated with the constitutive properties of the materials, and geometrical non-linearity, due to the second order effects (Malerba 1998, Biondini *et al.* 2004*a*, Biondini 2004). The formulation considers uniform and localized (pitting) corrosion and damage modeling includes the reduction of cross-sectional area of corroded bars, the reduction of ductility of reinforcing steel, the deterioration of concrete strength due to the development of longitudinal cracks induced by the corrosion products, and the spalling of concrete cover (Biondini and Vergani 2011). These effects are described in the structural model through damage indices and corrosion can selectively be applied to damaged structural elements with a different level of penetration in each reinforcing bar.

The proposed approach focuses on the evaluation of the effects of prescribed damage patterns and corrosion levels. However, the adopted formulation can be extended to include the time factor in a lifetime scale by modeling the diffusive process of the aggressive agents leading to corrosion initiation and damage propagation (Biondini *et al.* 2004*b*, 2006).

The reinforced concrete beam finite element is validated by comparison with the results of experimental tests carried out on beams with corroded reinforcement. Finally, the three-dimensional structural analysis of a reinforced concrete arch bridge under different damage scenarios is presented. The results show the effectiveness and application potentialities of the proposed formulation for life-cycle assessment and design of concrete bridges exposed to corrosion.

REFERENCES

Biondini, F. 2004. A Three-dimensional Finite Beam Element for Multiscale Damage Measure and Seismic Analysis of Concrete Structures. *13th World Conf. on Earthquake Engineering*, Vancouver, B.C., Canada, August 1–6, Paper No. 2963.

Biondini, F., Bontempi, F., Frangopol, D.M., Malerba, P.G. 2004*a*. Reliability of Material and Geometrically Nonlinear Reinforced and Prestressed Concrete Structures. *Computers & Structures* **82**(13–14), 1021–1031.

Biondini, F., Bontempi, F., Frangopol, D.M., Malerba, P.G., 2004*b*. Cellular Automata Approach to Durability Analysis of Concrete Structures in Aggressive Environments, *Journal of Structural Engineering*, ASCE, **130**(11), 1724–1737.

Biondini, F., Bontempi, F., Frangopol, D.M., Malerba, P.G., 2006. Probabilistic Service Life Assessment and Maintenance Planning of Concrete Structures, *Journal of Structural Engineering*, ASCE, **132**(5), 810–825.

Biondini, F., Frangopol, D.M. (Eds.), 2008. *Life-Cycle Civil Engineering*. CRC Press, Taylor & Francis Group.

Biondini, F. & Vergani, M., 2011. Modellazione del degrado di strutture in calcestruzzo armato soggette a corrosione. *Giornate AICAP 2011*, Padua, May 19–21 (in Italian).

Malerba, P.G., (Ed.), 1998. *Limit and Nonlinear Analysis of Reinforced Concrete Structures*. International Centre for Mechanical Sciences (CISM), Udine, Italy (in Italian).

Bridge Maintenance, Safety, Management, Resilience and Sustainability – Biondini & Frangopol (Eds)
© 2012 Taylor & Francis Group, London, ISBN 978-0-415-62124-3

Life-cycle analysis of bridges considering historic seismic damage and aging

J. Ghosh & J.E. Padgett
Rice University, Houston, TX

M. Sánchez-Silva
Universidad de Los Andes, Bogotá, Colombia

ABSTRACT

Throughout their lifetime civil engineering structures are subjected to progressive deterioration, as a result of processes such as aging, as well as to natural hazards such as earthquakes. Progressive degradation mechanisms are slow processes that accumulate damage in time while extreme events such as earthquakes lead to significant damage accumulation at specific points along the service life. Given such multiple threat scenarios, there is very limited work which analyzes the coupled effects of aging and damage accumulation along the service life of the structure.

Most of the work on repeated earthquake occurences consider very short durations between earthquake occurrences and hence does not reflect on the aspect of damage accumulation along the service life of the structure. Additionally, seismic damage accumulation can also be of critical concern during aftershock events before and after a large earthquake as identified by several researchers. With respect to aging, a significant volume of recent work exists on seismic fragility of highway bridges considering coupled effects of deterioration and single earthquake. Such models which typically neglect damage accumulation mechanisms have also been used to conduct life-cycle analysis of buildings and bridges in seismic regions by assuming that the damaged structures are re-constructed after every event.

Given the above mentioned drawbacks, the primary focus of this paper is twofold. Firstly, a framework is proposed to assess the probability of exceeding a certain level of total energy dissipation–considered as a proxy for structural damage–along the service life of the structure after considering multiple earthquake events. As an example case study, an integral frame concrete box girder bridge with single column is idealized as a lumped mass column model and studied under multiple shocks. It is found that even when no aging effects are considered, accounting for damage

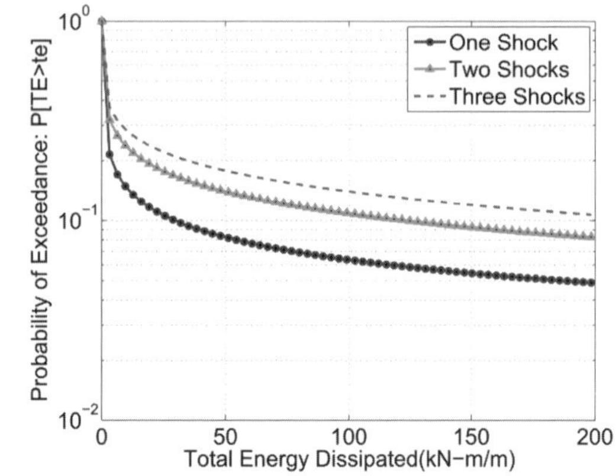

Figure 1. Increase in probability of failure with number of shocks.

accretion leads to higher chances of exceeding energy dissipation levels than when the accumulation effects are neglected. Introduction of aging and deterioration effects reveals that even for single shock scenarios the probabilities of exceeding different energy levels is significantly higher when compared to a pristine structure.

REFERENCES

Amadio, C., Fragiacomo, M., and Rajgelj, S. (2003). "The effects of repeated earthquake ground motions on the non-linear response of SDOF systems." *Earthquake Engineering & Structural Dynamics*, 32(2), 291–308.

Ghosh, J., and Padgett, J. (2010). "Aging Considerations in the Development of Time-Dependent Seismic Fragility Curves." *Journal of Structural Engineering*, 136(12), 1497–1511.

Jeong, G. D., and Iwan, W. D. (1988). "The effect of earthquake duration on the damage of structures." *Earthquake engineering & structural dynamics*, 16(8), 1201–1211.

Bridge Maintenance, Safety, Management, Resilience and Sustainability – Biondini & Frangopol (Eds)
© 2012 Taylor & Francis Group, London, ISBN 978-0-415-62124-3

The damage characteristic and repair of the concrete-bridges under severe environment

S. Kiso
Maintenance Management Team, Nagoya Branch, C-NEXCO, Nagoya City Aichi Prefecture, Japan

T. Tajiri
Maintenance Laboratory, Maintenance Department, RI-NEXCO, Machida City Tokyo metoropolitan, Japan

H. Miyauchi & A. Takeuchi
Environment/Engineering Department, C-NEXCO, Nagoya City Aichi Prefecture, Japan

ABSTRACT

This paper describes the deteriorating features of the expressway-bridges made of concrete and having been in service for more than 30 years under severe environment in cold region. It is estimated that this deterioration is caused by frost attack by freezing and thawing repeatedly in extremely cold climate and the chloride attack other than the longtime traffic load. Particularly the chloride attack is induced by "chloride ion" contained the anti-freezing agent which we scatter on the surface of road every winter for snow and ice control.

The crack and the corrosion of reinforcing bars in the reinforced concrete structure come into being as the initial deterioration and damage. The damage accelerates rapidly when this initial condition is observed at first. Furthermore, when the deterioration and damage of the concrete structure become advanced, the chloride ion concentration inside the concrete structures increases, and then, the reinforcing bars become corroded and expand. Finally, when the cross section of reinforcing bars decreases and the piece of cover concrete fall off, these phenomenon cause safety issues. Also, in this paper we conducted study with about the deterioration and damage of the substructure and superstructure of the bridges made of concrete, and the reinforced concrete deck slab on steel girder bridges.

Next, in this paper the repairing methods of the deterioration and damage on the high pier reaching a height of 50 m and the superstructure is indicated in detail. The major concrete repair and lining surface on the structure are applied after removing the damaged concrete parts.

The reinforced concrete deck slab on steel girder bridges is suffered under not only such an severe environmental damage, but also the impact of the wheel load directly. In such a condition, the deterioration and damage appears again several years later after it was repaired once. In this paper we analyzed this re-deterioration and re-damage of the reinforced concrete structure and proposes the improved repairing method. It is expected that the present study is useful in the future.

Figure 1. The situation of the broken prestressing strands in the deck slab.

Figure 2. Fatigue test of the repaired deck slab.

A further study of developing everlasting repairing or protecting method and materials should be conducted. Also considering life cycle at the design stage is important in order to manage road bridges.

Furthermore, the systematic, reasonable maintenance procedure and system of bridge and concrete structures should be established.

It is consider that the number of decrepit structures will grow even larger, so, preventive operation and maintenance technology will expected to be especially important in bridges and other road structures.

Bridge Maintenance, Safety, Management, Resilience and Sustainability – Biondini & Frangopol (Eds)
© 2012 Taylor & Francis Group, London, ISBN 978-0-415-62124-3

Effect of varying surface ageing on frost-salt scaling

H. Kuosa, M. Leivo & E. Holt
VTT Technical Research Centre of Finland, Espoo, Finland

R.M. Ferreira
University of Minho, C-TAC, Guimarães, Portugal

ABSTRACT

Surface scaling of bridge structures, caused by a moist and saline environment with frost attack, should be minimized to guarantee a long service life. For frost-salt scaling, near surface material pro-perties are critical. Thus it is also essential to know how ageing resulting from carbonation, drying and hydration will change durability performance of concrete. Laboratory studies on the effect of surface ageing on frost-salt scaling were included in the Finnish DuraInt project. The effect of five different ageing conditions on the frost-salt scaling performance of concrete is analysed, using natural or accelerated ageing methods. Concrete specimens were subject to one or two rounds of combined ageing – degradation mechanisms. The effect of binder type, w/b ratio, air pore structure is discussed. The results show that ageing with carbonation typically increased scaling. Binder material type and air entrainment quality were also decisive.

Three of the five different exposure cases (EC) are described briefly (more details in Leivo et al. 2011):

- *EC1*: Concrete specimens were subject to a standardized 50-cycle frost-salt test to measure the amount of scaling (kg/m^2);
- *EC3:* Concrete specimens were stored for approximately 1.6 years in the same conditions as in EC2 described in paper. A 10 mm concrete layer was sawed off to remove the carbonated layer. The concrete specimens are subject to the standardized frost-salt test;
- *EC4:* The concrete specimens from EC3 were re-tested after an additional ageing procedure, (3 months at 65% RH and 20°C + accelerated carbonation + 11 month storage wrapped in plastic at 65% RH and 20°C). This was followed by the standardized frost-salt test.

Figure 1 presents the results of scaling for EC1, EC3 and EC4. The concrete scaling for EC1 and EC3 is approximately the same. It can be considered that for these two exposure classes the effect of carbonation has not yet been felt, and therefore, the effect of frost-salt scaling is relatively identical, despite the age difference of the concretes.

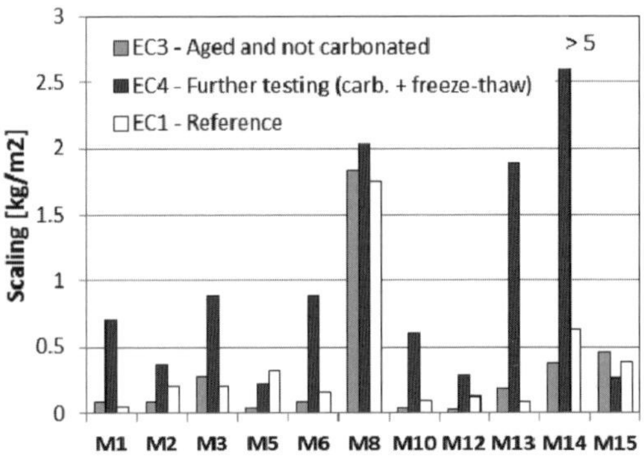

Figure 1. Comparison of the scaling results: first and second ageing cycle (EC3 & EC4) with the reference case (EC1).

In EC3 there was no natural long term carbonation. In EC4 there was first a 3 months natural carbonation at RH 65%, followed by accelerated carbonation at 1% CO_2 (at RH 60%, 56 days).

When the concrete is subject to drying and an intensive accelerated carbonation test prior to freeze-thaw scaling (EC4), a clear increase in the amount of scaling is visible. This seems to indicate that carbonation has a negative effect on the freeze-thaw scaling test.

REFERENCES

Leivo, M., *et al.* 2011. Effect of interacted deterioration parameters on service life of concrete structures in cold environments. Laboratory test results 2009–2010. VTT Research report VTT- Report VTT-R-04799-11.

Bridge Maintenance, Safety, Management, Resilience and Sustainability – Biondini & Frangopol (Eds)

Impact of corrosion on the seismic vulnerability of multi-span integral concrete bridges

M. Ní Choine & A.J. O'Connor
Trinity College Dublin, Ireland

J.E. Padgett
Rice University, Texas

ABSTRACT

Traditionally, it has been common practice to neglect the effects of deterioration when assessing the seismic vulnerability of bridges. As most bridges currently undergoing seismic retrofit are over 40 years old, levels of deterioration of bridge components are often significant. This paper evaluates the impact of the level of deterioration on the seismic vulnerability of multi-span concrete integral bridges. Chloride-induced corrosion of the steel reinforcement is considered, represented by a loss of steel cross section in the concrete columns. A finite element model of a 3 span reinforced concrete integral bridge was generated using the finite element platform OpenSees (Opensees). For this study only deterioration of the RC columns is considered. The deterioration mechanism considered is chloride-induced corrosion resulting in a loss of steel cross section and a reduction in the steel strength. Corrosion models are taken from work previously carried out by Ghosh and Padgett (2010) and Du et al. (2005a, b). A full probabilistic analysis is conducted to develop time-dependent fragility curves for the bridge in its pristine condition, after 25 years and after 50 years. These fragility curves give the probability of reaching or exceeding a defined damage limit state, for a given ground motion intensity measure taken as Peak Ground Acceleration (PGA). For this study fragility curves are only generated for the columns and the abutments. Figure 1 shows the fragility curves for the columns at the extensive damage state. The findings of this paper suggest that more research needs to be done in the area including the development of more accurate corrosion models and the inclusion of other deterioration mechanisms such as concrete spalling and cracking, a loss of bond strength and deterioration of the superstructure. A general conclusion from

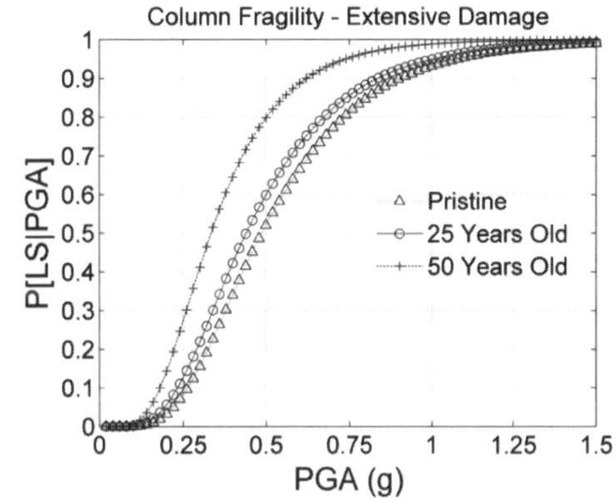

Figure 1. Column fragility curves: Extensive Damage state.

the findings in this paper and others (Ghosh and Padgett, 2010; Alipour et al. 2010) is that the inclusion of bridge age and exposure condition in seismic risk assessments is necessary to offer a more realistic estimate of the seismic vulnerability of the bridge. This in turn could lead to more accurate estimates of life cycle costs and more efficient rehabilitation strategies.

REFERENCES

Alipour, A., Shafei, B., and Shinozuka M. 2010. J. Bridge. Eng. 16(5), 597–611.

Ghosh, J., Padgett, J. E. 2010. Aging considerations in the development of time-dependent seismic fragility curves. Struct. Eng. 136(12), 1497–1511.

Opensees [Computer software]. OpenSees Development Team, Pacific Engineering Research Center, University of California, Berkeley, CA.

Bridge Maintenance, Safety, Management, Resilience and Sustainability – Biondini & Frangopol (Eds)
© 2012 Taylor & Francis Group, London, ISBN 978-0-415-62124-3

Probabilistic estimation of the initial time of corrosion of reinforced concrete components situated in a marine environment

R.A. Oliveira
Science and Technology Centre, Catholic University of Pernambuco, Brazil

T.N. Bittencourt
Escola Politécnica, University of São Paulo, Brazil

ABSTRACT

The work aims to make a probabilistic analysis of the initial time of corrosion of the reinforcement of reinforced concrete component due to the action of chloride ions. It is considered that the dissemination of chlorides on the concrete is in accordance with Fick's second law. It is assumed that the concentration of chlorides on the surface of the component, the critical concentration in the concrete located around the reinforcement, the concrete cover and the diffusion coefficient follow normal laws of probability. Application of the Monte Carlo method allows showing that the initial time of corrosion satisfies the Weibull distribution with three parameters. The reliability of the component in this moment is an indicator of its service life. The method is applied to the case of a reinforced concrete bridge located in marine area. The parameters were obtained from information of composition of concrete and design of structural elements analyzed.

Depending on the strength of concrete against chloride penetration and the thickness of the layer of concrete cover in a marine environment, it may take several years for the reinforced bar to be reached. Once the rebar is reached, due to its depassivation, cracks and traces of products resulting from oxidation appear in a few years. After this period, it may still take several years to compromise the safety of the structural element.

After the beginning of corrosion the anodic regions concentrate corrosion spots in the form of pitting. Subsequently, when the corrosion starts, there will be an evolution of the process that leads to a stage in which cracking occurs in the surrounding of the bars, after the pores of the concrete are filled with corrosion products. In the next stage, the physical manifestation is the appearance of visible cracks and stains caused by the products of corrosion. Thereafter the evolutionary process will lead to spalling with the detachment of pieces of concrete located in the surrounding of the rebar and, finally, the collapse of the concrete component.

Taking into account considerations of cost and safety, one can decide on the critical time to be reached to perform intervention through predictive or corrective maintenance. The time required to reach this stage will be considered service life.

After the construction of a concrete structure, chlorides penetrate with relative easiness through the concrete cover. During the first years, there will be an increase of the surface concentration of chloride and a decrease of the diffusion coefficient, due to both: the cement hydration, and chlorides that have penetrated. Later on, the surface layer of concrete may carbonate or be subject to cycles of wetting and drying, resulting in a change in the mechanism of penetration of chlorides. After the saturation of the superficial region of the concrete component, chloride ions penetrate the concrete by diffusion.

Bridge Maintenance, Safety, Management, Resilience and Sustainability – Biondini & Frangopol (Eds)
© 2012 Taylor & Francis Group, London, ISBN 978-0-415-62124-3

Life-cycle seismic evaluation of existing reinforced concrete bridges considering corrosion of steel reinforcement

Y.-C. Ou, A.B. Delima & L. Cute
Department of Construction Engineering, National Taiwan University of Science and Technology, Taipei, Taiwan

ABSTRACT

A life-cycle seismic evaluation method of existing reinforced concrete bridges considering the effects of corrosion of steel reinforcement is proposed in this paper. The evaluation method is based on the well-known ATC-40 nonlinear static pushover analysis procedure and is implemented in common commercial software familiar to engineers. The effects of steel corrosion are considered by modeling degradation of mechanical properties of steel reinforcement, softening of cover concrete in compression, and reduction of bond between concrete and steel reinforcement. Twelve existing reinforced concrete bridges in Taiwan are analyzed using the proposed method. By considering the effects of chloride and carbonation attacks, the life cycle seismic performance curves of these bridges are established and discussed.

REFERENCES

ATC-40. (1996). "Seismic Evaluation and Retrofit of Concrete Buildings", Volume 1, Applied Technology Council, Redwood City, California.

Bhargava, K., Ghosh, A. K., Yasuhiro, M., and Ramanujam, S. (2008). Suggested empirical model for corrosion-induced bond degradation in reinforced concrete. *J. Struct. Eng.*, 134(2): 221–230.

Coronelli, D., and Gambarova, P. (2004). Structural assessment of corroded reinforcement concrete beams: Modeling guidelines. *J. Struct. Eng.*, 138(8): 1214–1224.

Comite Euro-International du Beton (CEB). (1992). Model Code 1990. Case Postal 88, Laussane (CH), Switzerland, 437.

Computer and Structures, Inc., SAP2000. (2010). Integrated finite element analysis and design of structures, Version 14, Berkeley, California, USA.

Lee, H-S., and Cho, Y-S. (2009). Evaluation of mechanical properties of steel reinforcement embedded in concrete speciment as a function of the degree of reinforced corrosion. International Journal Fract., 157: 81–88.

Ou, Y-C., Chiewanichakorn, M., Aref, A. J., and Lee, G. (2007). Cyclic behavior of precast segmental concrete bridge column with high performance or conventional steel reinforcing bars as energy dissipation bars. Earthquake Engineering and Structural Dynamic.

Ou, Y-C., Tsai, M-S., Chang, K-C., and Lee, G. (2010). Large scale experimental study of precast segmental unbounded posttensioned concrete bridge columns for seismic region." *J. Struct. Eng.*, 133(11): 1636–1647.

Sung, Y-C., et al. (2005). A study on pushover analysis of reinforce concrete columns." Journal of Structure Engineering and Mechanics., 21(1): 35–52.

Bridge Maintenance, Safety, Management, Resilience and Sustainability – Biondini & Frangopol (Eds)

Life cycle assessment of existing steel bridges considering corrosion and fatigue coupled problems

A. Pipinato, C. Pellegrino & C. Modena
Department of Construction and Transportation, University of Padova, Italy

ABSTRACT

Corrosion and fatigue are important issues which affect the structural performance of steel bridges. This is particularly important for bridges and viaduct, for their key role in infrastructural nets. As the management of bridges net requires the prediction of the remaining life for all partially deteriorated structures, the development of assessment models for time-varying changes in structural performance due to the coupled problem of corrosion and fatigue is relevant and need to be investigated. In this paper, the accumulated load cycles causing cracking or even failure is analyzed together with different scenarios of corrosion propagation. The analysis is performed for the fatigue limit state function of shear in a detailed riveted connection. Prediction of the remaining fatigue life is illustrated on a typical example of existing bridge girders. Variations in the remaining life are evaluated as a function of time, covering the whole life-time of the bridge. In this context, issues as maintenance, assessment, rehabilitation and strengthening of existing bridges assume a significant importance (Pellegrino et al. 2009, Pipinato et al. 2009). The authors have developed some works concerning assessment and fatigue behavior of metal railway bridges by means of full scale experimental testing. In particular in Pipinato et al. (2009, 2011a) full scale tests on dismantled steel bridges have been developed whereas assessment of existing bridges and estimation of their remaining fatigue life are shown in Pipinato et al. (2010a, 2010b). Moreover a comprehensive method to assess the reliability of existing bridges taking fatigue into account has been recently published (Pipinato 2010). Other works have focused on the combined action of fatigue and seismic loading (Pipinato et al. 2011b), and on special structures as arch rail and road bridges (Pipinato et al. 2010). Finally, detailed studies on existing bridge nets have been presented in Pipinato et al. (2008) and in Pipinato et al. (2011). In this paper, the accumulated load cycles causing cracking or even failure is analyzed together with different scenarios of corrosion propagation. Variations in the remaining life are evaluated as a function of time, covering the whole life-time of the bridge.

REFERENCES

Pipinato A., Pellegrino C., Bursi O. S., Modena C. (2009). High-cycle fatigue behavior of riveted connections for railway metal bridges, Journal of Constructional Steel Research, Vol. 65, No. 12. (14 December 2009), pp. 2167–2175.

Pipinato A., Modena C. (2010). Structural Analysis and Fatigue Reliability Assessment of the Paderno Bridge, Practice Periodical on Structural Design and Construction, Vol. 15, No. 2. (2010), 109.

Pipinato A. (2010), Step Level Procedure for Remaining Fatigue Life Evaluation of One Railway Bridge, The Baltic Journal of Road and Bridge Engineering, Vol. 5, No. 1. (29 March 2010), pp. 28–37.

Pipinato A., Pellegrino C., Modena C. (2010a). Assessment procedure and rehabilitation criteria for the riveted railway Adige Bridge, Structure and Infrastructure Engineering (16 July 2010), pp. 1–18.

Pipinato A., Pellegrino C., Modena C. (2010b). Structural Analysis of Historical Metal Bridges in Italy, Advanced Materials Research, Vol. 133–134 (October 2010), pp. 525–530.

Pipinato Alessio, Marco Molinari, Carlo Pellegrino, Oreste S. Bursi, Claudio Modena (2011a). Fatigue tests on riveted steel elements taken from a railway bridge, Structure and Infrastructure Engineering, Vol. 7, No. 12. (24 May 2011), pp. 907–920.

Pipinato A., Pellegrino C., Modena C. (2011b). Fatigue assessment of highway steel bridges in presence of seismic loading, Engineering Structures, Vol. 33, No. 1.

Pipinato A. (2011). Assessment of existing bridges: Safety and security issues, Ingegneria Ferroviaria, Vol. 66, No. 4. (2011), pp. 355–371, by Alessio Pipinato, edited by Collegio Ingegneri Ferroviari Italian.

Stiglat, K. (1997). Brücken am Weg, Ernst & Sohn, Berlin (Germany) , ISBN 3433012997, 1997; pp. 86.

Sustainable Bridges – Assessment for future traffic demands and longer lives (2006). SB-4.2, SB-LRA. Background document D4.2: Guideline for Load and Resistance Assessment of Existing European Railway Bridges. Advices on the use of advanced methods e Annex 3.7: Fatigue as-sessment of structural details and components in steel using probabilistic methods. pp. 371–380.

Bridge Maintenance, Safety, Management, Resilience and Sustainability – Biondini & Frangopol (Eds)
© 2012 Taylor & Francis Group, London, ISBN 978-0-415-62124-3

Structural modeling of corroded reinforced concrete bridge columns

A.S. Rao, M.D. Lepech & A.S. Kiremidjian
Stanford University, Stanford, California, USA

ABSTRACT

Corrosion of the embedded reinforcement in concrete represents one of the primary causes of damage observed in concrete bridges across the world. The U.S. Federal Highway Administration (FHWA) estimates the annual direct cost of corrosion of bridges in the United States to be around $8.3 billion (Koch et al. 2002). In recent years, advanced levels of deterioration have been observed in reinforced concrete highway bridges exposed to seawater or spray in coastal locations and due to the application of deicing salts in northern locations (Hartt et al. 2004).

Consequently, several researchers have investigated the influence of reinforcement corrosion on the time-dependent reliability of reinforced concrete structures subjected to primarily dead and live gravity loads (Vu and Stewart 2000; Stewart and Rosowsky 1998; Frangopol et al. 1997). These studies have found that the corrosion of reinforcing steel can lead to substantial reductions in reliability over time in reinforced concrete members and structural systems.

While previous researchers have shown the significant impact that time-dependent corrosion can have on the probability of failure of structures under seismic loads, the simple fashion in which structural deterioration due to corrosion has been modeled reduces the usefulness of these results for either structural reliability prediction or life cycle cost or impact estimation.

This study improves upon previous research by explicitly incorporating the effects of corrosion in the structural model, which is implemented using Open System for Earthquake Engineering Simulation (OpenSees) (Mazzoni et al. 2009), a finite element method based object oriented software framework for simulating the seismic response of structural systems. A simplified approach for modeling reinforced concrete bridge columns that have undergone corrosion deterioration is presented and the approach is validated by comparison with experimental tests on columns subjected to accelerated corrosion at the State Key Laboratory of Coastal and Offshore Engineering at Dalian University of Technology in China (Li and Gong 2008). To the best of the authors' knowledge, the validation of the structural modeling of corrosion in bridge columns with large-scale accelerated corrosion and cyclic loading of experimental specimens is the first of its kind.

The simplified non-linear analytical model which accounts for the reduction in the steel cross-section due to corrosion showed excellent agreement with the experimental tests. Results from this modeling show that the degree of corrosion and the axial load ratio can have a significant effect on the behavior of corroded concrete elements. The presented model can thus be used in conjunction with time-dependent corrosion models to assess the seismic risk of a bridge at various points during its lifetime.

REFERENCES

Frangopol, D. M., Lin, K.-Y., and Estes, A. C. (1997). "Reliability of reinforced concrete girders under corrosion attack." *Journal of Structural Engineering*, 123(3), 286–297.

Hartt, W. H., Powers, R. G., Leroux, V., and Lysogorski, D. K. (2004). *Critical Literature Review of High-Performance Corrosion Reinforcements in Concrete Bridge Applications, FHWA-HRT-04-093*. McLean, VA.

Koch, G. H., Brongers, P. H., Thompson, N. G., Virmani, Y. P., and Payer, J. H. (2002). *Corrosion costs and preventive strategies in the United States, FHWA-RD-01-156*. Washington, DC.

Li, J., and Gong, J. (2008). "Influences of Rebar Corrosion on Seismic Behavior of Circular RC Columns." *China Journal of Highway and Transport*, 21(4), 55–60.

Mazzoni, S., McKenna, F., Scott, M. H., and Fenves, G. L. (2009). "Open System for Earthquake Engineering Simulation User Command-Language Manual." Pacific Earthquake Engineering Research Center, Berkeley, California.

Stewart, M. G., and Rosowsky, D. V. (1998). "Time-dependent reliability of deteriorating reinforced concrete bridge decks." *Structural Safety*, 20(1), 91–109.

Vu, K. A. T., and Stewart, M. G. (2000). "Structural reliability of concrete bridges including improved chloride-induced corrosion models." *Structural Safety*, 22(4), 313–333.

Bridge Maintenance, Safety, Management, Resilience and Sustainability – Biondini & Frangopol (Eds)

Maintenance optimization of suspender ropes of suspension bridges

K. Sakai, S. Kusuhara, A. Moriyama & K. Ogihara
Honshu-Shikoku Bridge Expressway Company Limited, Kobe, Japan

ABSTRACT

Two types of suspender ropes, Center Fit Rope Core (CFRC) ropes and Parallel Wire Strand (PWS) ropes are used in the Honshu-Shikoku suspension bridges. CFRC ropes are painted and are expected to be replaced periodically, and PWS ropes are fully protected by polyethylene covers and are expected to be maintenance-free for long time. Honshu-Shikoku Bridge Expressway Co., Ltd. (HSBE) maintains ten suspension bridges. Since CFRC suspender ropes are installed in seven of ten bridges, huge cost is necessary for the replacement of CFRC suspender ropes. Therefore HSBE is tackling optimization of maintenance management of CFRC suspender ropes in order to reduce the Life Cycle Cost (LCC) of the bridges.

Honshu Shikoku Bridge Expressway Co., Ltd. (HSBE) has conducted the visual inspection of the ropes once per year. The inspection result showed not only paint coating deterioration and slight corrosion at the general part of the ropes of several bridges but also heavy corrosion at the anchoring part of the ropes of one bridge. Since the visual inspection cannot show the internal corrosion situation, HSBE developed a nondestructive inspection system using the electromagnetic method (main flux method) in order to know the internal corrosion situation appropriately. Moreover, the relation between the data measured by the main flux method and the tensile strength reduction rate was found.

Considering the relation between the sectional area reduction rate and the load which is actually acting on the rope, we developed a maintenance management flow of suspender ropes (Figure 1) based on the control criteria of the sectional area reduction rate.

According to the management flow, the repair method or the repainting method is selected for each corroded part of the ropes. Generally, the dipping painting method with flexible fluorine resin paint is selected as a repainting method for the general part. When the corrosion is in early stage, the internal filling up method is applied for the anchoring part. When the corrosion is heavy, the partial replacement method or the replacement of a suspender rope is applied for the anchoring part.

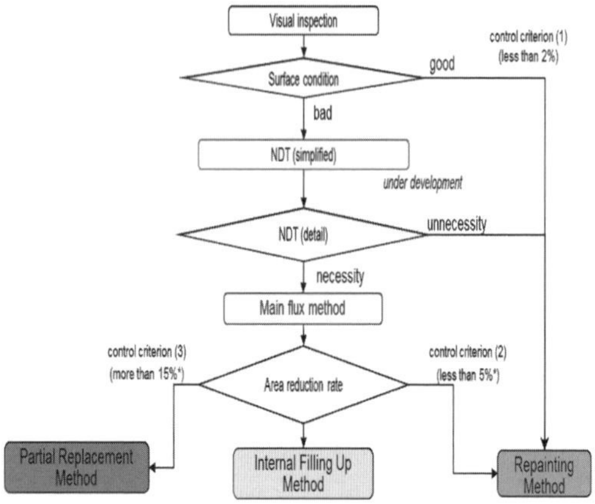

Figure 1. Maintenance management flow of suspender rope.

REFERENCES

T. Takazawa et al, 2002. Corrosion Protection System of Main Cable and Non-Destructive Inspection of Suspender Ropes in a Suspension Bridge, IABMAS.

A. Moriyama et al, 2003. Nondestructive Inspection of the Suspender Ropes in a Suspension Bridge, 11th REAAA Conference.

A. Moriyama et al, 2004. Establishment of the Non-destructive Inspection Technique for Suspension Ropes of Suspension Bridge, 4th International Cable Supported Bridge Operators' Conference.

M. Nakamura et al, 2004. Soundness investigation of suspender ropes of the suspension bridge which has passed 20 years, IABMAS.

Y. Hasegawa et al, 2009. Preventive Maintenance of Suspender Ropes on the Honshu-Shikoku Suspension Bridges, 13th REAAA Conference.

Bridge Maintenance, Safety, Management, Resilience and Sustainability – Biondini & Frangopol (Eds)

Multi-objective cost analysis for bridges considering disasters, bridge form and driving comfort

Y.Q. Wang
Research Institute of Highway Ministry of Transport, Beijing, China & Department of Bridge Engineering, Tongji University, Shanghai, China

A. Chen & X. Ruan
Department of Bridge Engineering, Tongji University, Shanghai, China

Y. Li
China Communication Construction Company Highway Consultants CO., Ltd. Beijing, China

ABSTRACT

This article mainly studies on how to quantify disaster, bridge form and driving comfort related objectives, which are also very difficult to quantify.

(1) For the four outburst disasters, namely wind, flood, earthquake, and collision, firstly determine their typical parameters, and then make calculations as shown in Fig. 1. Extreme Value type I Distribution is selected to simulate yearly maximum average wind speed in 10 minutes at the height of 10 m. Pearson III Distribution is adopted to describe the annual peak discharge of flood. Earthquake magnitude is selected as the typical parameter to quantify earthquake induced damage, while Gutenberg Liszt's law is adopted to simulate the minor earthquakes and characteristic earthquake method for the major ones. After determination of the types of vehicles or vessels, the collision times of each type can be estimated based on traffic volume and incidence rate of collision.

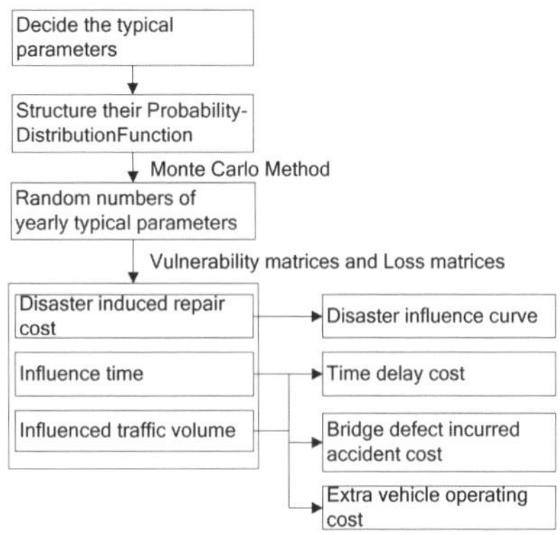

Figure 1. The method of outburst disaster simulation and related cost calculation.

(2) Damages from fatigue, salt and overload are long-term and gradually increasing, thus extra reductions of bridge resistance index are used to model their affect on bridges. It is suggested to make an assumption of the initial resistance deteriorating curve, namely the function reflecting the variation of bridge resistance index in calculation period, and the converting function between cost and resistance index. Then simulate the reduced resistance index vectors resulted respectively from fatigue, ice-removal salt and overload, take the vectors as the disaster influential curves and add them on the initial resistance deteriorating curve. Convert maintenance costs to the increase of bridge resistance index according to the converting function to form the maintenance influential curve, and also add them on the resistance deteriorating curve. Arrange overhaul and rehabilitation by the maintenance strategy and the resistance deteriorating curve considering disasters and the planned maintenance. Finally we can obtain the overhaul and rehabilitation cost by the converting function. Besides, calculation of the time delay cost, bridge defect incurred accident cost and extra vehicle operating cost can be carried out on the base of the influence time and influenced traffic volume of extra overhaul and rehabilitation.

(3) For unquantifiable objectives as driving comfort, bridge form, etc., it is suggested to adopt qualitative descriptions and then quantify them. Expert investigation can be taken to get the quantified values; then pretreatment and statistical analyses of those data are suggested to carry out in order to obtain the final attribute values. The construction cost or Willingness to Pay compared to the referenced alternative can also be taken as the substitutive attribute.

REFERENCES

Ehlen, M.A. BRIDGELCC 2.0 users manual [R]. NISTGCR 03-853, 2003.

Hawk, H. Bridge life-cycle cost analysis [R]. NCHRP Report 483, 2003.

Joint Committee on Structural Safety (JCSS). Guideline for Reliability Based Assessment of Existing Structures [S], Rilem: The Joint Committee on Structural Safety, 2000.

WANG Yuqian. Cost Analysis Based Decision Method and Process of Bridges under Given Structural Life [D]. Shanghai: Tongji University, 2011.

Bridge Maintenance, Safety, Management, Resilience and Sustainability – Biondini & Frangopol (Eds)
© 2012 Taylor & Francis Group, London, ISBN 978-0-415-62124-3

Effect of corrosion of reinforcement on the coupled shear and bending behaviour of reinforced concrete beam

W.J. Zhu & R. François
Université de Toulouse; UPS, INSA; LMDC (Laboratoire Matériaux et Durabilité des Constructions), Toulouse, France

D. Coronelli
Politecnico di Milano, Dipartimento di Ingegneria Strutturale, Milano, Italy

ABSTRACT

The influence of corrosion on the shear behaviors of highly corroded short-span beams was studied in this paper.

An experimental investigation into the shear behaviors of the corroded reinforced concrete beams in chloride environment for a period of 26 years under service loading was presented.

Cracking maps of the corroded beams were drawn. The corrosion maps and the loss of the diameter of the tensile bars were also described after having broken in the concrete. The shear behaviors of the beams are discussed by the experimental results and the cracking maps.

Two seriously corroded short-span beams were tested until broken by three-point load system. Another two beams of the same age but without corrosion were also tested as the control ones. The relationship of load and deflection was recorded so as to better understand the influence of corrosion on shear behaviors of the short-span beams.

The failure processes of the shear tests on the corroded beams and the control beams were totally different (Figure 1). For the two corroded beams, the first inclined crack appeared in the web of the beam. Some other cracks developed gradually, all the cracks propagated in both the width and the length. One main

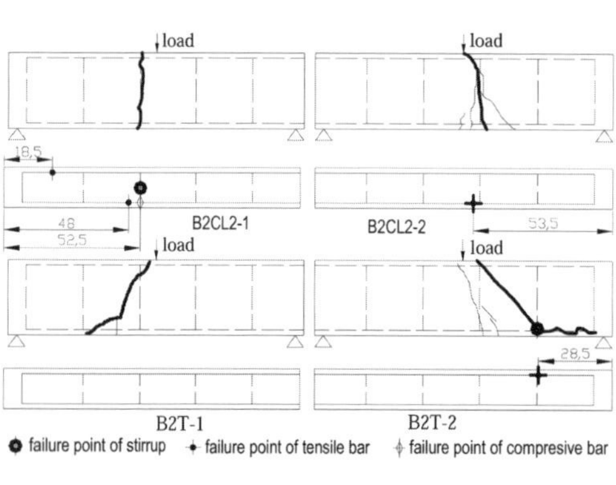

Figure 1. The failure points and the main cracks of the beams.

inclined crack was formed, stretching to the loading point and the bottom of the beam. With the crack wider and wider, the concrete crushed suddenly, which typically belonged to the shear failure.

But for the corroded beams, when the beams were loaded, the main cracks also appeared at the location of the nearest stirrups, but compared with the main cracks of the control beams, the cracks almost at the location of the stirrups, and then the corroded beams were broken with the failure of the concrete and the failure of the tensile bars. The beams were broken in bending failure mode.

The mid-span deflections of four short-span beams were recorded together with load, the load-deflection curves were plotted (Figure 2). Compared with the control beams that reached failure status in a more brittle mode, there is an obvious difference of the ultimate deflection between the two control beams (B2T-1 and B2T-2) which is due to the fact that the failure path intercepted a stirrup for B2T-2, then the ductile failure of stirrup increase the global ductility of short beams. For corroded beams, there are a degradation of concrete cross-section (spalling, cracking), a loss of cross-section for longitudinal bars (tensile and compressed) and for stirrups and usually the corrosion leads to a more brittle behavior of corroded RC elements. But as shown in Figure 2, the important degradation of short beams due to corrosion lead to an increase of ductility and a quite small reduction of ultimate capacity. The change in failure mode from shear to bending for B2CI2-2 or mixture between shear and bending for B2CI2-1 explained this very surprising result.

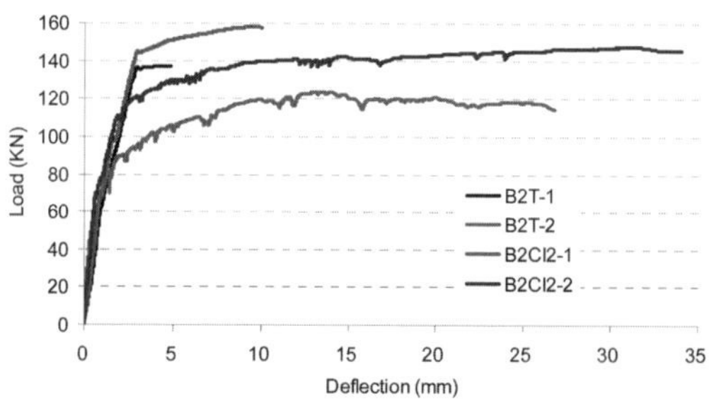

Figure 2. Load-deflection curves of the shear tests.

Brick and stone masonry bridge safety and durability

Organizers: A. Benedetti & L. Jurina

Bridge Maintenance, Safety, Management, Resilience and Sustainability – Biondini & Frangopol (Eds)
 ISBN 978-0-415-62124-3

Patch loading of longitudinally stiffened webs

A. Benedetti & F. Dall'Aglio
Department DICAM, University of Bologna, Italy

ABSTRACT

The paper deals with the case of patch loading of steel I-shaped girders with two longitudinal stiffeners.

The patch loading or partial edge loading of steel girder webs is a loading condition that occurs when a concentrated force acting perpendicular to a girder's flange. This usually leads to a local buckling of the web near the loaded flange.

The patch loading is a loading conditions that occurs during bridges launching when a girder section can be subjected to repeated support reactions provided by the slides (or rollers).

Studies conducted by Lagerqvist and those conducted by Graciano, with a few changes, have led to the design criteria contained in Eurocode 3 (EN 1993-1-5) that are valid for girder without or with only one longitudinal stiffener. The configuration with two longitudinal stiffeners is often an excellent solution for girders higher than 3 meters but has not yet been discussed in EN 1993-1-5 with regard to patch loading resistance.

It is proposed a model of resistance to the ultimate limit state for steel I-shaped girders with two longitudinal open stiffeners based on the works cited previously by Lagerqvist and Graciano, and on the more recent works by Clarin and Gozzi, which treats the problem of yield and buckling, with an approach harmonized with the methods used in Eurocode for the other problems of buckling. The model contains three significant

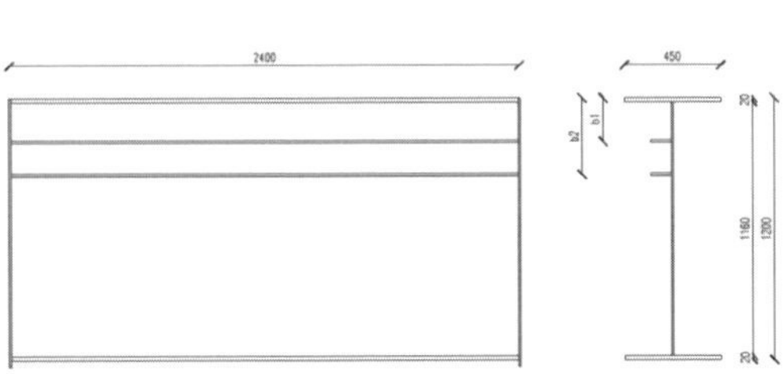

Figure 1. Girder geometry.

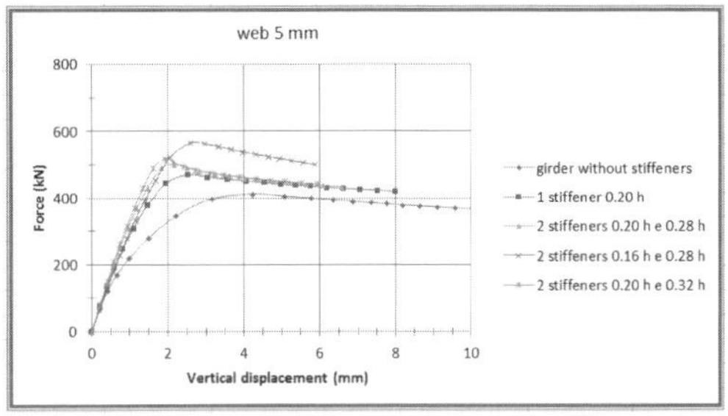

Figure 2. Influence of stiffeners position on the resistance.

parts: the yield resistance, the elastic critical load used to determine the slenderness parameter and a reduction factor that relates the resistance to the slenderness, but uses the same equations regardless of the prevalent cause of collapse (yield or buckling).

The presented work aims to propose a prediction of patch loading resistance of double stiffened girders, which is able to evaluate the influence of longitudinal stiffeners vertical position too.

REFERENCES

Clarin, M. (2007). Plate buckling resistance – Patch loading of longitudinally stiffened webs and local buckling, Doctoral thesis 2007:31, Lulea University of Technology, ISRN:LTU-DT—07/31—SE.

Lagerqvist, O. (1994). Patch loading – Resistance of steel girders subjected to concentrated forces, Doctoral thesis 1994:159D, Luleå University of Technology, ISRN: HLU-TH-T-159-D—SE.

Graciano, C. (2002). "Patch Loading – Resistance of Longitudinally Stiffened Steel Girder Webs", Doctoral Thesis 2002:18, Division of Steel Structures, Luleå University of Technology, Luleå. ISSN: 1402–1544.

Gozzi, J. (2007). Patch Loading Resistance of Plated Girders – Ultimate and serviceability limit state, Doctoral thesis 2007:30, Lulea University of Technology, ISRN: LTU-DT—07/30—SE.

Bridge Maintenance, Safety, Management, Resilience and Sustainability – Biondini & Frangopol (Eds)
© 2012 Taylor & Francis Group, London, ISBN 978-0-415-62124-3

Masonry bridges: Static and dynamic response through reduced scale models

A. Brencich
DICAT – Department of Civil, Environmental and Architectural Engineering, University of Genoa, Italy

ABSTRACT

Masonry bridges are often considered as an arch to which non-structural elements (spandrels, fill, etc.) are added. This is in contrast with several experimental and theoretical evidences showing that this kind of bridges behaves as a complex structure in which all the elements of the bridge take part in the load bearing structure. To this aim, some hidden elements may play a crucial role, such as internal spandrels; neglecting these elements may lead to completely unreliable structural models.

To this extent, dynamic testing is a powerful technique to get a detailed insight in the mechanical response of this kind of bridges comparing the natural frequencies of the real bridge to the ones expected on the basis of structural models. As an example, a great contribution to the bridge stiffness, and thus to the natural frequencies and mode shapes, is provided by the arch-spandrel interaction and by internal spandrels, if present.

Besides, it is already well known that the actual load bearing structure of masonry bridges is not only the arch but a complex system consisting of arch + spandrels + fill, figure 1. It is not yet well understood the specific role of each of these elements. To this aim, static monotonic and cyclic tests may allow detailed considerations into the role of arch-fill interaction and arch-spandrel interaction.

In this paper, a series of laboratory tests are used to identify and quantify the contributions to the different elements of the bridge to its static and dynamic response. 1:3 prototypes have been used retaining the similarity to the prototype reducing either the geometric quantities and the material strength. The arches have been tested both statically and dynamically at different levels of load-induced damage and in different settings: bare arch; arch + fill, arch + spandrel, any at different levels of damage.

A quite large database is thus obtained providing some data for the static and dynamic identification

Figure 1. Shallow ($r/s = 0.2$) ARCH + FILL. a) global mechanism; b) + c): details (final stage).

of masonry bridges. It is showed that the structural elements, and their effective contribution to the load bearing structure, may be identified on the basis of dynamic identification techniques. Damages, mainly due to material degradation or overload, on the contrary, may be identified on the basis of dynamic testing only if their extent is large enough to be detected also visually. Therefore, at present, the condition state of a bridge cannot entirely rely on dynamic identification procedures, while dynamic testing appears to be a crucial test to assess the reliability of structural models.

REFERENCES

Brencich, A. and Sabia, D. 2008. Experimental identification of a multi-span masonry bridge: the Tanaro Bridge. *Constr. & Build. Mat.s*, **22**, 2087–2099

Riotto G. 2010. *Experimental analysis of static and dynamic behavior of masonry bridges* (in Italian). Genoa: University of Genoa, Department of Civil, Env. & Arch. Eng.ng.

Bridge Maintenance, Safety, Management, Resilience and Sustainability – Biondini & Frangopol (Eds)
© 2012 Taylor & Francis Group, London, ISBN 978-0-415-62124-3

Strengthening effectiveness of ancient masonry bridges

N. Gattesco & R. Franceschinis
Department of Civil Engineering and Archtecture, University of Trieste, Trieste, Italy

V. Kristek, A. Kravtsov & J. Řimal
Faculty of Civil Engineering, Czech Technical University, Prague, Czech Republic

ABSTRACT

Masonry arch bridges represent a major part of the bridge population in Europe, including more than 200,000 masonry bridges. They comprise railway bridges built in the 19th century, and ancient roman, medieval and Renaissance bridges which have survived along the centuries and nowadays represent relevant landmarks around Europe.

These bridges are normally made through masonry barrel vaults supported by strong piers. The spandrels are filled with a backing and a filling that plays a significant role in the stability of the structure. Backing is normally made with a conglomerate of rubble stones and lime mortar; the filling is frequently made with incoherent material as coarse gravel. Sometimes the backing and filling are made with the same material. Lateral spandrel walls are used to contain the filling.

At present, most of masonry bridges need to be strengthened and repaired because of ageing of the component materials, the increased service loads and the poor performance under extreme loading conditions. In fact, masonry arches were not designed to carry the current traffic loads as well as spandrel walls were not designed to withstand high out-of-plane actions due to filling thrust or seismic excitation. Moreover significant differential temperature among parts of the structure may cause considerable tensile stresses (Rimal 2003, Zeman et al. 2008).

Therefore, transversal cracks in the vaults as well as the separation between the spandrel wall and the arch are frequently present due mainly to the filling thrust on the spandrel walls and/or to the significant temperature gradient from external and inner parts of the structure in hot and cold seasons. Moreover, the lateral filling thrust and the earthquake excitation cause significant out-of-plane actions in the spandrel walls, which frequently lead to the occurrence of extensive vertical cracks. Vertical cracks may also be caused by longitudinal tensile stresses in long bridges due to temperature seasonal variations.

In the study two new techniques to strengthen bridge spandrel walls were proposed (Gattesco & Dudine 2010, Borri et al. 2010). They are based on the application on the masonry surface of either a GFRP reinforced lime and cement mortar coating or a reinforced repointing with a net of stainless strands.

Some diagonal compressive tests on specimens made with rubble stone and solid brick masonry evidenced a significant increase in shear resistance due to the application of GFRP reinforced coating on both surfaces of the wall. In particular the shear resistance of strengthened samples was tripled in stone masonry and 60% greater in solid brick masonry with respect to unstrengthened samples.

Numerical simulations of the experimental tests allowed determining the values of the parameters to be used in the material model. The numerical simulation of four point bending tests allowed evidencing a significant effectiveness of the strengthening techniques also for out-of-plane actions. In particular the loading increase was from 3.5 to 5.5 times, for stone masonry, and from 2 to 4 times, for solid brick masonry.

REFERENCES

Borri, A., Corradi, M., Speranzini, E., Giannantoni, A. 2010. Reinforcement of Historic Masonry with High Strength Steel Cords, *Masonry International*: 23/3, 79–90.

Gattesco, N., Dudine, A. 2010. Effectiveness of a masonry strengthening technique made with a GFRP-mesh-reinforced mortar coating, *Proceedings of the Eighth International Masonry Conference*: 1881–1890. Dresden.

Rimal, J. 2003. Charles Bridge in Prague – Measurements of temperature fields, *International Journal for Restoration of Buildings and Monuments*: 9(6):585–602.

Zeman, J., Novák, J., Šejnoha, M., Šejnoha, J. 2008. Pragmatic multi-scale and multi-physics analysis of Charles Bridge in Prague, *Engineering Structures*: 30, 3365–3376.

Bridge Maintenance, Safety, Management, Resilience and Sustainability – Biondini & Frangopol (Eds)
© 2012 Taylor & Francis Group, London, ISBN 978-0-415-62124-3

Strengthening of arch masonry bridges with "RAM" – Reinforced Arch Method

L. Jurina
Politecnico di Milano, Italy

ABSTRACT

In order to increase the safety of masonry arch bridges, a new technique, called "RAM" – Reinforced Arch Method, has been recently developed by the author. The different blocks of an arch transmit compression forces each other, and as long as the force stays within the central "core" of the section, all the stresses across the section will be compressive. If the resulting load moves out of the central zone, the masonry bricks detach themselves, giving rise to hinge-points, as the mortar joints are not able to resist tensile stresses. While a three-pin arch is still a statically determinate structural form, the fourth hinge converts the arch into a mechanism and collapse occurs. It is possible to prevent the formation of almost one family of hinges (the extrados or the intrados ones), by introducing post-tensioned cables, able to stand the

Figure 1. 1996-Tests with different consolidation techniques.

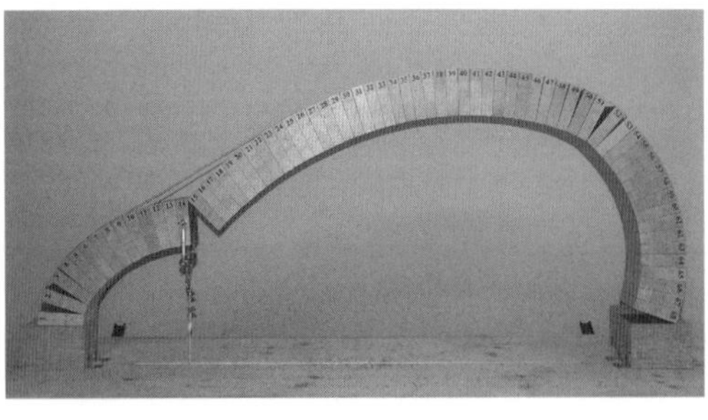

Figure 2. 2008-Tests with vertical loads.

Figure 3. 2011-Tests with horizontal (seismic) loads.

tensile stresses. Loading the arch in radial direction, the cables increase the average compression between the bloks and improve their resistance to pressure-flexure. The same tension applied to the cables, in fact, is transferred to the masonry, that will work mainly in compression. More than 500 small and medium scale tests on arches of different shape and 12 real cases have been performed, with results of absolute interest. The tests show that it is possible to strongly increase the collapse load of the arch, even preserving reversibility and respecting the original static behavior of the structure. No additional weight is needed (that is a relevant factor in seismic areas) and no shape modification of the arch is required. Moreover the confining actions can be re-calibrated where and how it is necessary. The "RAM Method" represents an innovative, easy, quick and cheap instrument for a respectful consolidation of masonry arch bridges.

Three different experimental campaign have been conducted in 1996, 2008 and 2011 (Figure 1,2,3).

REFERENCES

Bonfigliuoli, S., 2011, Consolidamento strutturale e antisismico di archi e volte in muratura: una sperimentazione sulla tecnica dell'arco armato, Tesi di Laurea

Giglio, S.M., 2008, Consolidamento di archi e volte in muratura mediante la tecnica dell'arco armato. Approccio sperimentale, Tesi di Laurea

Jurina, L., 1999, Una tecnica di consolidamento attivo per archi e volte in muratura, in Int. Symp. on Seismic Performance of Built Heritage in Small Historic Centers, Assisi, Italia

Bridge Maintenance, Safety, Management, Resilience and Sustainability – Biondini & Frangopol (Eds)
© 2012 Taylor & Francis Group, London, ISBN 978-0-415-62124-3

Remedial works and repairs of Prague's historical Charles Bridge

V. Křížek, V. Kristek & J. Římal
Czech Association of Civil Engineers & Czech Technical University, Prague, Czech Republic

ABSTRACT

The Gothic Charles Bridge built in the 14th century with its total length of 516 m, 31 statues and 16 arches is undoubtedly one of the biggest Prague's attractions. In the 600 years of its existence it survived substancial flood-breakdowns, rainwater leak and trafic damages, or disintegration of sandstone blocks. Around 1980 it was clear that an immediate general reconstruction is vital. However, the explorations and studies leading to the "right concept" were the subject of endless engineers' and conservationists' disputes and took 20 years.

Critical evaluation of the existing unsatisfactory course of technical preparation of the Charles Bridge repairs made the investor take extraordinarily vigorous steps:

- Nullify the already approved project documentation of the "radical" concept of repairs
- Replace the authority of direct investor, design and engineering organization
- Based on the dispute findings order to newly elaborate a comprehensive concept of optimal repair of the Charles Bridge carried away in a friendly and economical way and performed in separate stages

The time schedule of repairs was divided into three major stages:

Stage I: Ensuring the static stability of the bridge during the flood emergency by protecting the shallow-based pillars against the undermining of bedrock foundation joints (see Figure 1).

Stage II: Protecting of the upper bridge structure against the destructive effects of rainwater leakage, climate and emissions. Establishing durable waterproof system all over the bridge.

Stage III: Repair or replacement of sandstone blocks of the facial wall of the bridge threatened by the loss of load carrying capacity and durability through staling.

Figure 1. Deepening of pillar fundaments.

In addition to describing repair of the Charles Bridge, the paper presents the general lessons learned: multi-decade response of a masonry structure, assessment of damages, failures and deterioration processes related to materials, environment, climate, structural arrangement and detailing, assessment of the masonry bridge with respect to heritage, historic and structural aspects, discussion on appropriate strategies for refurbishment, restoration, conservation and preservation corresponding to location and significance of the historical bridge and ways to increase durability and long service life of historical bridges ensuring that they will be protected against harmful effects for many years to come.

REFERENCES

Petrák, J., Tvrzník, V. 2011: The Charles Bridge Repair – the end of the second phase of repairs – Roads and Railways, 3

Batal, Z., Kara, D., Krízek, V. 2011: The Prague Charles Bridge – the successful repair of the upper construction – Road Horizon 10

Tvrzník, V. 2009: The Prague Charles Bridge Repair – preparation and implementation – Civil Engineering, 1

Bridge Maintenance, Safety, Management, Resilience and Sustainability – Biondini & Frangopol (Eds)
© 2012 Taylor & Francis Group, London, ISBN 978-0-415-62124-3

Structural assessment of the railway masonry arch bridge crossing the Reno river in Bologna

C. Mazzotti, S. de Miranda, G. Castellazzi & F. Carrea
DICAM, Department of Structural Engineering, University of Bologna, Bologna, Italy

ABSTRACT

A large number of historical railway masonry arch bridges built in the nineteenth and twentieth centuries are still in service; nevertheless, the train loads and the traffic amount increased tremendously in this last century. For this reason and due to environmental aggressive conditions, a good number of these bridges suffered important mechanical deteriorations. It is thus becoming more and more important the definition of an effective process of evaluation of the actual structural safety of these fundamental infrastructures, both with respect to serviceability and ultimate limit state conditions (see, for example, Brencich & Sabia, 2008 and the references therein).

In the present paper, the procedure applied for structural assessment of the railway masonry arch bridge crossing the Reno river in Bologna (Italy), see Figure 1, is presented, together with the obtained results. The addition of a new railway track on the existing bridge, in fact, required a complete evaluation of the actual structural conditions to be performed in order to find out possible criticality.

The bridge, built in 1852, has 15 archspans and an overall length of 360 m. Arches have a 20 m free-span and piers are 2 m thick and 10 m high. The bridge was originally designed for two railway tracks, with a barrel width of 9 m. Later, the bridge was enlarged building a second, 6 m width, barrel in order to increase the number of tracks. Visual inspection of the intrados reveals that the barrels seem to be separated (percolation from the rail deck are visible along all the arches), so slips between the two parts are possible. The evaluation of the degree of collaboration of the two parts is one of the key aspect of the investigation.

A number of experimental tests have been carried out in order to find out both the material properties of the masonry constituting the arches and the piers (Mazzotti & Sassoni 2011) and the structural behavior of the bridge. In particular, accurate static and dynamic load tests have been performed on a very limited number of arches whereas simplified dynamic tests have been repeated on all the spans of the bridge in order to verify the homogeneity of their structural behavior.

Figure 1. Longitudinal view of the multi-span masonry arch bridge.

The experimental findings have been used to calibrate a refined finite element model of the whole bridge, able to describe the static and dynamic behavior of the structure under service conditions (train traffic). The very good agreement obtained between experimental and numerical results confirms the accuracy of the assumptions made in setting-up the finite element model, which could also be used in the evaluation of the structural effects of an additional, new railway track and for the definition of a proper strengthening intervention, if required.

REFERENCES

Brencich, A. & Sabia, D., 2008. Experimental identification of a multi-span masonry bridge: the Tanaro bridge. *Construction and building Materials*, **22**, 2087–2099.

Sassoni E., Mazzotti C., 2011. A new semi-destructive method for evaluating the compressive strength of brick masonries, Proceedings of the "5th International Congress on Science and Technology for the Safeguard of Cultural Heritage in the Mediterranean Basin", Istanbul, Turkey 22–25 November 2011.

Bridge Maintenance, Safety, Management, Resilience and Sustainability – Biondini & Frangopol (Eds)
© 2012 Taylor & Francis Group, London, ISBN 978-0-415-62124-3

Investigation and upgrading of a historical multispan arch masonry bridge

An. Paeglitis & Ai. Paeglitis
Riga Technical University, Riga, Latvia

ABSTRACT

The aim of this research was the investigation of the properties of clay brick masonry arch bridge materials with a purpose to find the best methods for reconstruction and renovation of the structure. The restoration, reconstruction and upgrading of historical heritage bridges require the careful investigation of materials and causes of damages. In many cases, the use of incorrect composition of joint mortar and clayed brick could lead until an unfavorable result. It is important to ensure the natural water migration in historic masonry could not be interrupted after restoration or reconstruction measures. As an example had been analyzed the results of investigations and upgrading of a historical clayed brick masonry bridge over Venta River in Kuldiga town in Latvia (figure 1). During the long lifetime and pro-active maintenance polity, a bridge had a lot of damages that could influence the further service life. For geometrical data collection was used laser scanning method that ensures enough accurate data for reconstruction design and information for architectural investigation. Three-dimensional scanning of heritage structure by 3D laser scanners allow the further transformation of information into the surface mesh model. Paper presents results of investigation and design of restoration and reconstruction works.

Figure 2 show the bridge after reconstruction.

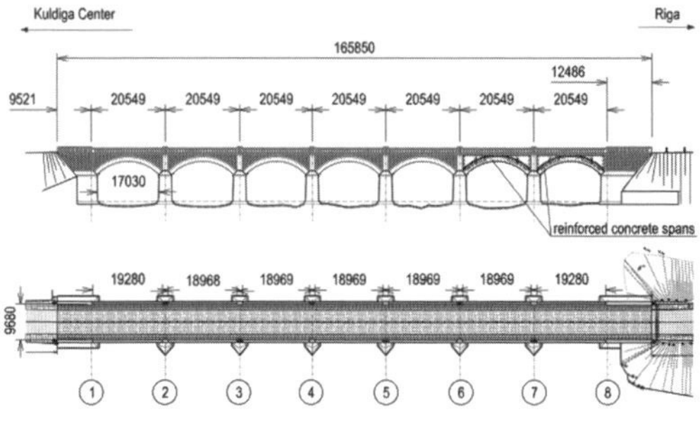

Figure 1. Elevation and plan of the bridge.

Figure 2. View on bridge after reconstruction.

REFERENCES

Brencich A., Sabia D. 2008. Experimental identification of a multi-span masonry bridge: The Tanato Bridge. *Construction and building materials*, 22: 2087–2099.

Brencich A., de Francesco U. 2004. Assessment of multi-span masonry arch bridges: Examples and applications. *Journal of Bridge Engineering* (11/12): 591–598.

Domede N., Pons G., Sellier A., Fritih Y. 2009. Mechanical behavior of ancient masonry. *Materials and structures* 42: 123–133.

Mammino A., Tonon F., Tonon L., 2006. Renovation of the 17-th century Ponte Lungo bridge in Chioggia, Italy, *Journal of Bridge Engineering* (1/2): 13–20.

LVS EN 196-2 Methods of testing cement: chemical analyses of cement.

LVS EN 196-21 Methods of testing cement: determination of chloride, carbon dioxide and alkali content of cement.

LVS EN 623-2 Advanced technical ceramics – Monolithic ceramics – General and textural properties – Part 2: Determination of density and porosity.

Perret S., Khayat K. H., Gagnon E., Rhazi J, 2002. Repair of 130-Year old masonry bridge using high performance cement grout. *Journal of Bridge Engineering* (1/2): 31–38.

Ritter W. 1877. Die Windaubrücke bei Goldingen", *RIGASCHE*: 49–53

Teutonico J.M.A. 1988. Laboratory Manual for Architectural Conservators, *ICCROM*, ISBN 92-9077-083-X.

Bridge Maintenance, Safety, Management, Resilience and Sustainability – Biondini & Frangopol (Eds)
© 2012 Taylor & Francis Group, London, ISBN 978-0-415-62124-3

Static and seismic retrofit of masonry arch bridges: Case studies

G. Tecchio, F. da Porto, P. Zampieri & C. Modena
Department of Structural and Transportation Engineering, University of Padova, Italy

C. Bettio
S.M. Ingegneria srl, Via longhin, Padova, Italy

ABSTRACT

Thousands of road and railway masonry arch bridges are still in operation in the Italian transportation network: most of them need being improved in their carrying capacity and to be upgraded to the standards of the current seismic code. In this paper three case-studies of the static and seismic retrofit of historical masonry arch bridges are presented, outlining some methodological approaches to the renewal intervention according to the different typological characteristics of the bridges and their state of maintenance. The main phases of work, are described and appropriate analytical models have been carried out to prove the effectiveness of the strengthening techniques.

The upgrading of the "Sandro Gallo" brick masonry arch bridge, in Venice, consisted in the substantial conservation of the bridge structure and the enhancement of the actual load bearing capacity: the central part of the arch span has been thickened by the insertion of one more layer of bricks; locally, the connection between the new and the old masonry has been improved by means of brick units and metallic dowels, used as shear connectors. Moreover, the intrados of the masonry arch have been restored in a "traditional" way with excavation of the deteriorated part of the mortar joints and repointing with proper hydraulic-lime based mortar. At the level of the abutments, a new foundation structure on micro-piles has been constructed to bear the increase of the horizontal and vertical loads transmitted by lateral friction from the existing foundations.

The bridge over the Rio Moline appeared in a very poor condition, and a provisional wooden shoring with steel ties had already been put in place: the masonry was significantly affected by loss of mortar joint, the surfaces at the intrados of the vaults were also partially damaged by the presence of vegetation which deepened its roots between the joints. Many refurbishment techniques, derived from the historical heritage restoration field, have been used for the repair intervention, with injections of grout

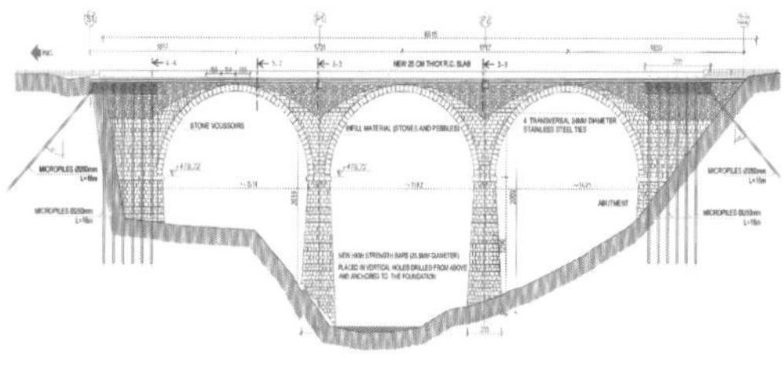

Figure 1. Design longitudinal section of the Gresal Bridge.

based on hydraulic lime, repointing of the stone joints with proper hydraulic lime mortar, and local masonry reconstruction by manual methods. New internal spandrel brick walls connected to the extrados of the vaults have been built to share some of the load and enhance the seismic resistance.

In the case of the Gresal Bridge the project interventions have been decided on the basis of a preliminary seismic analysis and in the light of the results obtained from structural investigations. The bridge was rated highly vulnerable to seismic action, mostly due to the slenderness of its high piers: the seismic vulnerability has been reduced by creating a new structural arrangement through a new rc slab anchored to the piers with vertical ties and restrained at the abutments, collaborating with the existing structure in carrying horizontal loads. The r.c. slab is anchored to the abutments, where new reinforced concrete plinths on micro-piles have been positioned outside the existing masonry foundations.

REFERENCES

Clemente, P., Occhiuzzi, A., Raithel, A., 1995. Limit Behaviour of Stones Arch Bridges. *Journal of Structural Engineering*, pp. 1045–1050.

Da Porto, F., Franchetti, P., Grendene, M., Valluzzi, M., Modena, C. 2007, Structural capacity of masonry arch bridges to horizontal loads, *Proc. 5th Int. Conference on Arch Bridges (ARCH'07)*, p. 451–458.

Lifetime design, assessment, and maintenance of super long-span bridges

Organizers: H.-M. Koh, S. Shin, H.-K. Kim & N.-S. Kim

Bridge Maintenance, Safety, Management, Resilience and Sustainability – Biondini & Frangopol (Eds)
 ISBN 978-0-415-62124-3

Modeling of truck traffic for long span bridges

E.-S. Hwang, K.-T. Lee & D.-Y. Kim
Kyung Hee University, Yongin, Korea

ABSTRACT

The purpose of this study is to develop rational design live load model for long span bridges with span length up to 2,000 m.

New live load model reflects real truck traffic data in Korea using Weigh-In-Motion system that was installed on the road with the heaviest truck traffic volume. By using collected WIM data, various truck traffic scenarios are assumed based on congestion condition. To analyze the load effects, typical long span bridge such as a suspension bridge and a cable stayed bridge are modeled by the structural analysis program. Based on traffic scenarios, equivalent uniformly distributed load (EUDL) are calculated and loaded to the selected bridge. The results of load effects are compared with other load model in internationally renowned design code such as AASHTO LRFD (2007), Honshu-Shikoku bridge design code (1989) and current Korea Bridge Design Code (2010) as shown in Figure 1.

Based on the results of analyses, new live load model is proposed as a combination of design truck (Figure 2) and design lane load (Equation (1)) which the magnitude is decreased as influence line length become longer. From the comparison with other design load model in typical cable stayed and suspension bridge, the load effect by proposed model is smaller than AASHTO LRFD or current bridge design code in Korea and greater than Honshu-Shikoku bridge design code.

$$L \le 60m : w = 12.7\ kN/m$$
$$L > 60m : w = 12.7\left(\frac{60}{L}\right)^{n}\ kN/m \qquad (1)$$

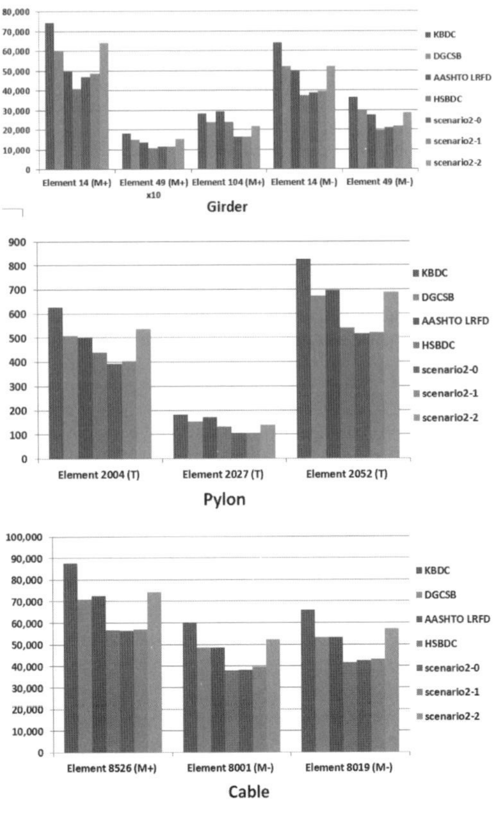

Figure 1. Effects of truck traffic.

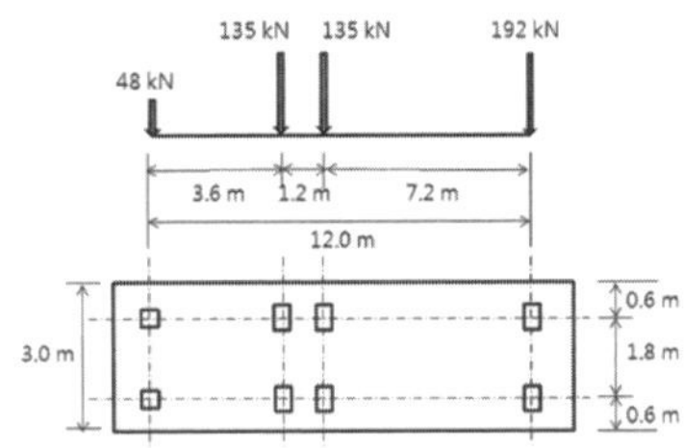

Figure 2. Proposed design truck model.

Bridge Maintenance, Safety, Management, Resilience and Sustainability – Biondini & Frangopol (Eds)
© 2012 Taylor & Francis Group, London, ISBN 978-0-415-62124-3

Analytical prediction of lateral-torsional buckling of long-span suspension bridge

H. Katsuchi & H. Yamada
Yokohama National University, Yokohama, Japan

K. Hasegawa
JFE Engineering Corporation, Yokohama,Japan

ABSTRACT

As the main span length of a suspension bridge becomes much longer, it was pointed out (Nakamura, et al., 1998) that not only aerodynamic instability like flutter but also aerostatic instability like torsional lateral buckling would be concerns of wind-resistant design. Dr. Hirai (1942) of the University of Tokyo of those days firstly proposed a calculation formula of the critical wind speed of torsional lateral buckling of a suspension bridge. It is believed that this study was initiated by the collapse of Tacoma Narrows Bridge. This formula included some simplifications such as only consideration of force equilibrium of a girder and exclusion of geometric nonlinearity. Recently, owing to the availability of powerful advanced FEM analysis software, sophisticated structural analysis including nonlinearities and wind-structure interactions can be possible. Authors have tried to analyze torsional lateral buckling of suspension bridges (Yamada, et al., 2009). However difficulty of distinguish torsional lateral buckling from divergence was encountered. Because they are both large wind-induced deflections suddenly occurring at a high wind speed. In this study, a Mode Tracing Flutter Analysis method (Nguyen, et al., 1998), which was originally developed for tracing modal parameters of pre-selected particular vibration modes in flutter analysis, was used to identify and distinguish torsional lateral buckling from divergence. The paper presents the analytical methodology and demonstration results for torsional lateral buckling of a long-span suspension bridge.

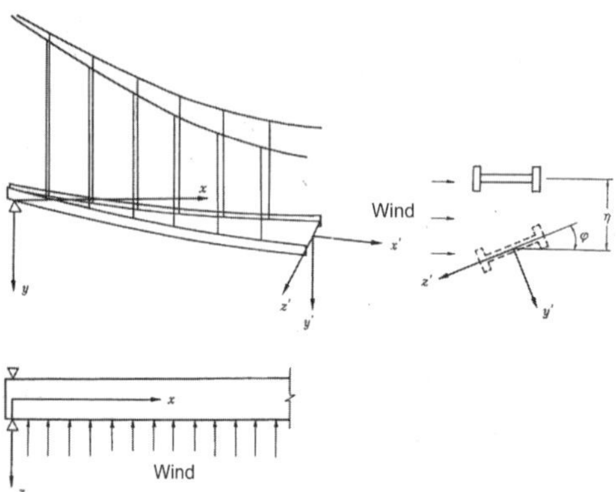

Figure 1. General plan and FEM model of Akashi Kaikyo Bridge.

REFERENCES

Hirai, A. 1942. On stability of suspension bridges against torsional vibration, *Journal of Japan Society of Civil Engineers*, JSCE, vol. 28, No. 9, pp. 769–786.

Nakamura, T., Miyata, T. and Yamada, H. 1998. Study on wind-induced static instability and its analytical method of long-span suspension bridge, *Proc. of 53rd annual meeting of JSCE*, I-B62, pp. 124–125.

Nguyen, N. D., Miyata, T., Yamada, H. and Ngyuen, N. M., 1998. Flutter responses in long span bridges with wind induced displacement by the mode tracing method, *Journal of Wind Engineering and Industrial Aerodynamics*, Vol. 77&78, pp. 367–379.

Yamada, H., Katsuchi, H. and Sasaki, E. and Ishihara, D. 2009. Wind-induced lateral-torsional buckling analysis of long-span suspension bridge, *Proc. of 10th International Conference on Structural Safety and Reliability*, Osaka, Japan, pp. 818–823.

Bridge Maintenance, Safety, Management, Resilience and Sustainability – Biondini & Frangopol (Eds)
 ISBN 978-0-415-62124-3

Application of vision-based monitoring system to stay cables

S.W. Kim & N.S. Kim
Pusan National University, Busan, South Korea

Y.-M. Kim
DAEWOO Institute of Construction Technology, Suwon, South Korea

ABSTRACT

Stay cables as a critical member of cable-stayed bridges are an important element for supporting the entire structure. They would be vibrated due to various reasons and such vibration has been often investigated in many existing bridges. Thus, current studies on verification of its mechanism and measures for vibration control have been actively progressed. It is also required to verify the influence of cable vibration by health monitoring because cable vibration leads cable fatigue and visual anxiety of drivers. Accelerometers, as one of the conventional sensors being used to measure stay cable vibration of existing bridges, may not easy to install at stay cables and require the considerable cabling work to facilitate a direct connection between each sensor and instrument. For this reason, a technique using digital image processing, which is one of non-contact sensing systems, may be needed to detect cable vibration affecting bridge structures. In this study, a method is suggested to measure vibration of multiple cables using image processing techniques and a vision-based monitoring system, as a sensing system to measure cable vibration remotely considering convenience and economic aspects for use, has been developed.

Figure 1 shows the structure of the developed system. In general, the image storage server was located at the position of the dynamic and static data loggers; and as the wire between the image storage server and the camera was 100–500 m long, optical cables were used for the data transfer. A housing device was installed in the system for waterproof, heatproof, and dustproof, so that the system can be operated even in bad weather. In addition, a pan-tilt drive that can be remotely controlled was installed to measure multiple cables with one camera, and a feature for storing 60 locations in the program was added for long-term cable monitoring. The camera resolution for remote cables was attained with a 20× optical zoom lens. The control signals for the pan-tilt drive and the zoom lens were sent through an RS485 cable. Figure 2 shows the camera, housing, and TX control box installed at the cross-beam 2 of Busan-Geoge fixed link, a three-span steel-concrete composite cable-stayed bridge in South Korea.

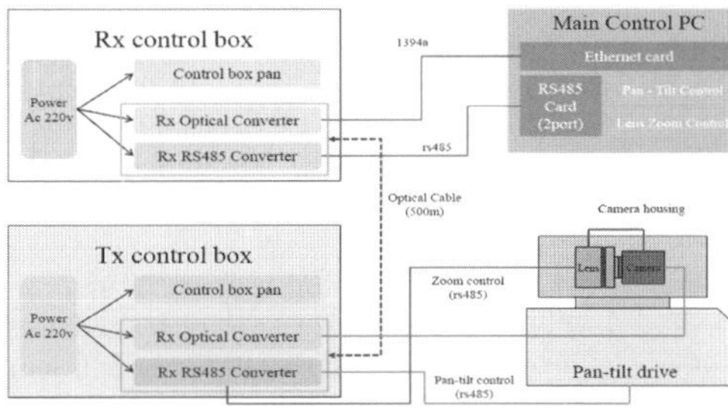

Figure 1. Structure of vision-based monitoring system.

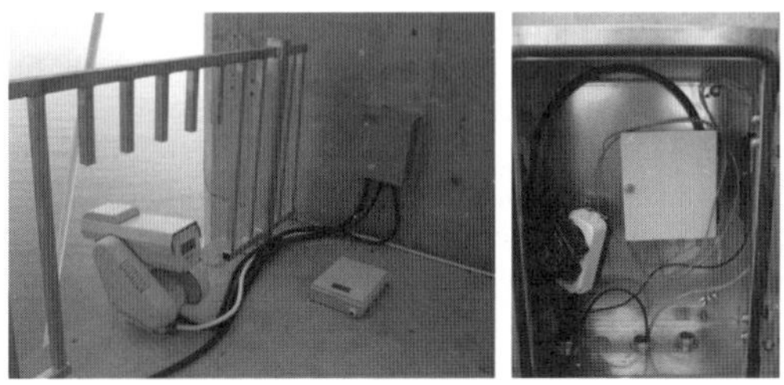

Figure 2. Camera, housing, and TX control box.

In this study, a system for detecting the vibration that affects a structure by monitoring stay cables with a digital image device, which will overcome the limitations (high price, fixed measurement, and durability) of and replace the conventional sensors, was developed.

REFERENCES

Ji, Y.F. and Chang, C.C. 2008. Nontarget Image-based Technique for Small Cable Vibration Measurement, *Journal of Bridge Engineering*, 13(1):34–42.

Kim, S.W. and Kim, N.S. 2011. Multi-point Displacement Response Measurement of Civil Infrastructures Using Digital Image Processing, *The Twelfth East Asia-Pacific Conference on Structural Engineering and Construction*, Hong Kong, China, January 26–28.

Bridge Maintenance, Safety, Management, Resilience and Sustainability – Biondini & Frangopol (Eds)
© 2012 Taylor & Francis Group, London, ISBN 978-0-415-62124-3

Wireless impedance sensor node for structural health monitoring of cable-anchorage subsystem

J.-T. Kim, K.-D. Nguyen & P.-Y. Lee
Department of Ocean Engineering, Pukyong National University, Busan, Korea

ABSTRACT

In this study, wireless impedance sensor node that includes smart PZT-interface is proposed to monitor prestress-loss in tendon anchorage. In order to achieve the goal, the following approaches are implemented.

Firstly, a smart PZT-interface is newly designed to sensitively monitor electro-mechanical impedance changes in tendon-anchorage subsystem. To analyze the effects of prestress force, an analytical model of tendon-anchorage is described regarding to relationship between prestress force and structural parameters of anchorage contact region as shown in Fig. 1. Based on the analytical model, impedance-based monitoring of prestress-loss is conducted using the impedance-sensitive PZT-interface.

Secondly, wireless impedance sensor node working on Imote2 platform, which is interacted with the PZT-interface, is outlined as shown in Fig. 2.

Finally, an experiment on lab-scale tendon-anchorage of a prestressed concrete girder is performed to evaluate the performance of the smart PZT-interface along with the wireless impedance sensor node on prestress-loss detection. The monitoring results are shown in Fig. 3.

From the experimental verification, implementation of the smart PZT-interface was successful in indicating various prestress-losses in the tendon-anchorage connection. Electro-mechanical impedance of the PZT-interface is sensitive to prestress-loss even the examined frequency range is rather low, smaller than 100 kHz. For wireless application, the impedance sensor node designed in this study showed the good performance for impedance-based monitoring of prestress-loss.

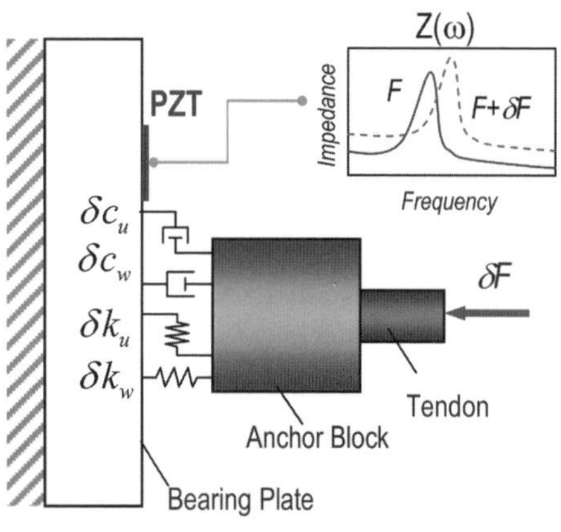

Figure 1. Schematic of prestress-force monitoring by PZT sensor.

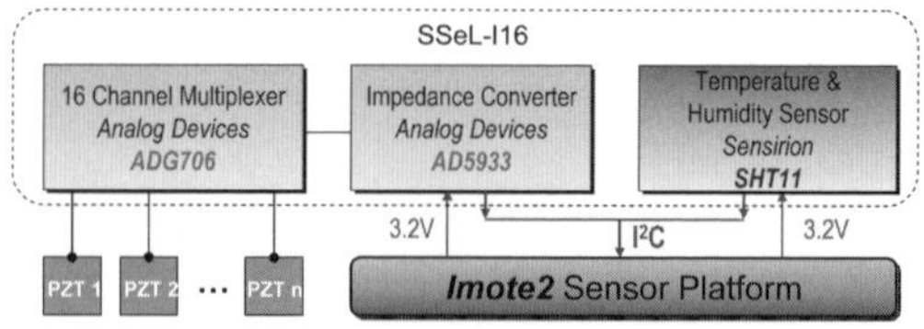

Figure 2. Design of impedance sensor node.

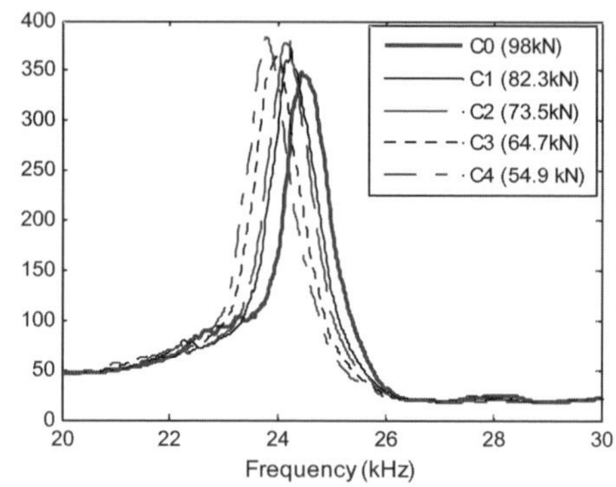

Figure 3. Impedances of PZT-interface under various prestress forces.

REFERENCES

Mascarenas, D.L., Todd, M.D., Park, G. & Farrar, C.R. 2007. Development of an impedance-based wireless sensor node for structural health monitoring. *Smart Materials and Structures*, **16**(6), 2137–2145.

Park, J.H., Kim, J.T., Hong, D.S., Mascarenas, D. & Lynch, J.P. 2010. Autonomous smart sensor nodes for global and local damage detection of prestressed concrete bridges based on accelerations and impedance measurements. *Smart Structures and Systems*, **6**, 711–730.

Bridge Maintenance, Safety, Management, Resilience and Sustainability – Biondini & Frangopol (Eds)
© 2012 Taylor & Francis Group, London, ISBN 978-0-415-62124-3

Field loading test for evaluation of load bearing capacity of a cable-stayed bridge

Y.-M. Kim & S.-H. Shin
Daewoo Institute of Construction Technology, Suwon-si, Kyunggi-do, Republic of Korea

J. Lee
Dankook University, Yongin-si, Kyunggi-do, Republic of Korea

J.-H. Jang
TM E&C, Seongnam-si, Kyunggi-do, Republic of Korea

This paper describes the procedure and results of a set of initial field loading tests performed on a 2-pylon cable-stayed bridge, Geoga Bridge which is a part of the Busan–Geoje fixed link project. The objective bridge is located between two islands, Gaaduk Island and Geoje Island, in South coast of Korea. During the field loading test, a series of static and dynamic responses were measured from pre-installed instruments, and the measured responses were compared with the results of numerical analyses. Major responses considered were static strains and natural frequencies.

Measured strain responses were compared with corresponding calculated strains, and the results are listed in Table 1. Here, measured strains were obtained from strain gauges installed on lower surface of the steel girder at the corresponding section.

As shown in Table 1, all response ratios of experimental and analytical strain results are below unity, and this means that the constructed bridge possesses higher stiffness than design value.

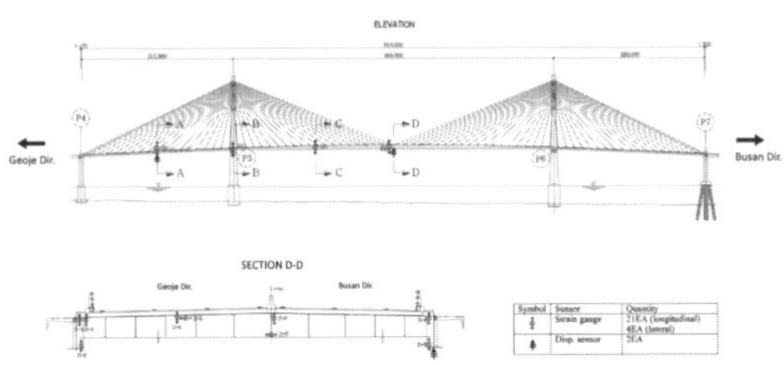

Figure 1. Location of static sensors on Geoga Bridge (unit: m).

Table 1. Comparison of static strain (με).

Sec.	Load Case	Experiment (1)	Numerical analysis (2)	Ratio (2)/(1)
A-A	LC5	−110	−88.5	0.80
C-C	LC10	−94	−86.5	**0.92**
D-D	LC15	−200	−140.5	0.70

Table 2. Comparison of natural frequencies (Hz).

Mode	Experiment	Numerical analysis
1st Vertical bending	0.232	0.221
1st Torsional	0.623	0.578

Measured and calculated natural frequencies of 1st bending and torsional mode are compared in Table 2. Also, measured natural frequencies are higher than the calculated values, therefore higher stiffness of the constructed bridge is confirmed again.

Bridge Maintenance, Safety, Management, Resilience and Sustainability – Biondini & Frangopol (Eds)
© 2012 Taylor & Francis Group, London, ISBN 978-0-415-62124-3

A vision-based damage detection of cable exterior in cable-stayed bridges

J.-J. Lee, K.-D. Kim & H.-N. Ho
Department of Civil & Environmental Engineering, Sejong University, Seoul, Republic of Korea

ABSTRACT

This study presents an effective vision-based system for surface cable bridge damage detection, shown in Figure 1. Generally, the cable should be inspected for surface damage as well as inside defect. Starting from August 2010, a new research project supported by Korea Ministry of Land, Transportation Maritime Affairs (MLTM) was initiated focusing on the damage detection of cable system. In this study, only the vision-based surface damage detection system based on image processing will be focused on.

Principally, the damage detection algorithm combines some image enhancement techniques with principal component analysis (PCA) to detect damages on cable surface. In more detail, the input image from a camera is wirelessly transferred to server PC located at the cable support, then it is processed with image enhancement method together with noise removal technique to improve overall image quality, next it is projected into PCA sub-space. Finally, the Mahalanobis square distances of the projected image to all sample patterns are calculated, then the smallest distance is found to be the match for the input image. Many laboratory tests showed that this system could effectively be applied to inspect real bridge cables.

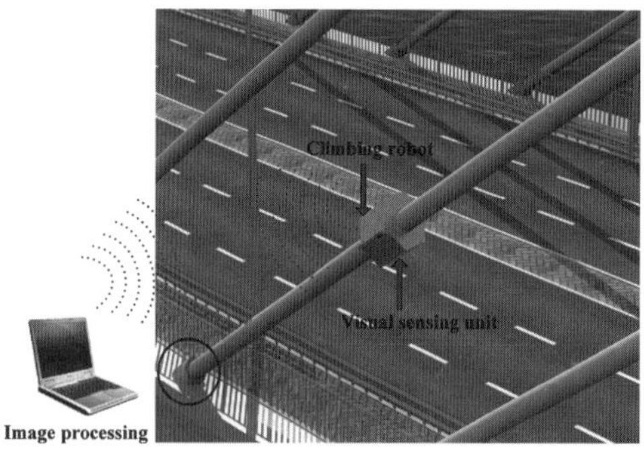

Figure 1. An overview of cable inspection system.

REFERENCES

Artac, M., Jogan, M., and Leonardis,. A., 2002. Incremental PCA for On-line Visual Learning and Recognition, *16th International Conference on Pattern Recognition (ICPR'02)* – Vol. 3.

Boyle, R.D., 1998. Scaling additional contributions to principal component analysis, *Pattern Recognition*, 31: 2047–2053.

Huang, D., Yi, Z. and Pu, X., 2009. A new incremental PCA algorithm with application to Visual Learning and Recognition, *Neural Processing Letters, Springer*, 30:171–185.

Tisarelli, M., and Grosso, E., 2000. Active vision-based face authentication, *Image and Vision Computing*, 18:299–314.

Turk, M., and Pentland, A., 1991. Face recognition using Eigenfaces. *Proc. IEEE Conference on Computer Vision & Pattern Recognition*, pp.586–591.

Bridge Maintenance, Safety, Management, Resilience and Sustainability – Biondini & Frangopol (Eds)
© 2012 Taylor & Francis Group, London, ISBN 978-0-415-62124-3

Long-term monitoring for dynamic properties on a suspension bridge under wind-induced vibration

D.U. Park, N.S. Kim & J.H. Cheung
Pusan National University, Busan, South Korea

H.-K. Kim
Seoul National University, Seoul, South Korea

ABSTRACT

Long span bridge market has being increased since local connecting enterprises in North Europe, early 1980s. As time passed, techniques of maintenance and health monitoring for long span bridges have been developed because of those infrastructure owner's requirement. The state evaluation with natural frequencies and damping properties of long span bridges are ones of most important factors for stability and serviceability evaluation.

In this study, changing of natural frequencies and damping properties according to environmental factors have been evaluated. The objective for this study was the suspension bridge part of Sorok Bridge located in Goheung, South Korea. Measured signals were ambient vibration signals of vertical acceleration, wind speed, and wind direction on center of middle span which were collected at every 10 P.M. for a year. Daily minimum temperatures which were reported at online site of Korea Meteorological Administration were used for temperature data. Ambient vibration signals had been mostly collected under wind-induced vibration condition without traffic loading. It was possible to observing relationship between dynamic properties and environmental factors due to a lack of traffic.

FFT (Fast Fourier Transform) and PP (peak picking) methods were used for analyzing natural frequencies. It could be possible to find a periodical changing trend of first natural frequency according to temperature although the variation of first natural frequency was small as 2.5 percents for a year. Natural frequencies for a year are shown in figure 1. It was possible to evaluate change of natural frequencies according to seasons and ambient temperatue.

Meanwhile, extended Kalman filter as one of nonlinear system identification method was used for estimating damping properties. According to previous studies, damping ratio could be changed by wind speed and amplitude of acceleration. In this study, the relationships among estimated damping ratio, wind speed and amplitude of acceleration have been investigated by regression analysis. At results, estimated damping ratio seems to be more related with amplitude of acceleration than average of normal component of wind speed. Figure 2 shows the relationship between estimated damping ratio and amplitude of acceleration.

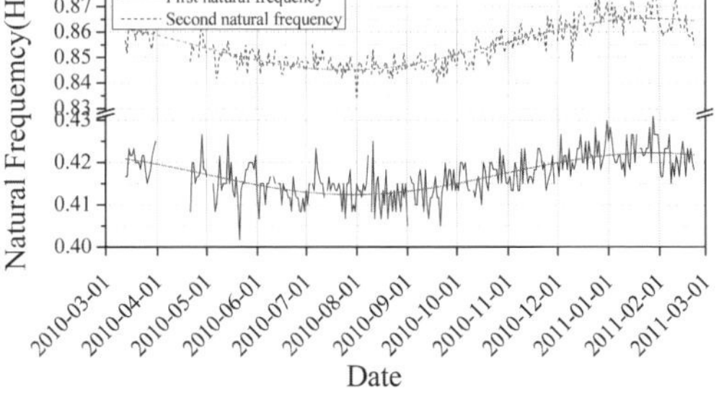

Figure 1. First and second natural frequency of Sorok suspension Bridge.

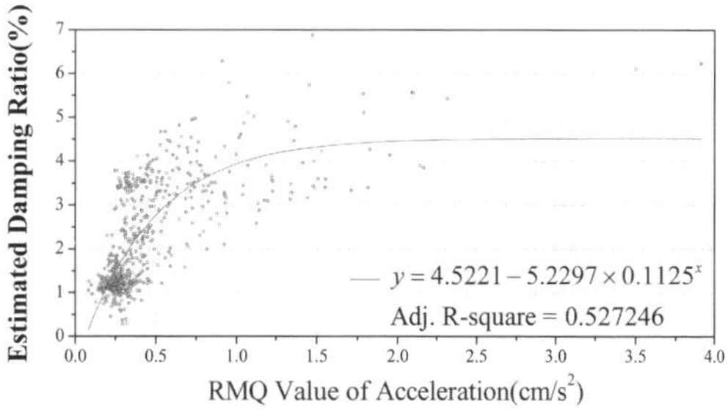

Figure 2. Relationship between estimated damping ratio and RMQ value of acceleration.

REFERENCES

Park, D.U., Kim, N.S., and Kim, S.I., 2009, *Damping Estimation of Railway Bridges Using Extended Kalman Filter*, Transactions of the Korean Society for Noise and Vibration Engineering, 19(3):294–300.

Simiu, E. and Scanlan, R.H., 1986, *Wind Effects on Structures*, Canada; John Wiley & Sons, Inc.

Yi, J.H., Koo, K.Y., Kim, N.S. and Yun, C.B., *Estimation of cable characteristics using extended Kalman filter*, The Magazine of the Korean Society of Civil Engineers, 24(6A):1267~1274.

Bridge Maintenance, Safety, Management, Resilience and Sustainability – Biondini & Frangopol (Eds)
© 2012 Taylor & Francis Group, London, ISBN 978-0-415-62124-3

Real-time steel cable NDE for corrosion defects using E/M sensors installed in a cable climbing robot

S. Park
Department of Civil and Environmental Engineering, Sungkyunkwan University, Suwon, Korea

J.-W. Kim & M.-J. Nam
Department of U-City Design and Engineering, Sungkyunkwan University, Suwon, Korea

J.-J. Lee & H.-N. Ho
Department of Civil and Environmental Engineering, Sejong University, Seoul, Korea

ABSTRACT

The steel cables are frequently used for various infrastructures. Especially, the steel cables in long span bridges such as cable stayed bridges and suspension bridges are critical members which suspend dead loads due to both the main girders and floor slabs. Damage at cable members can occur in the form of cross-sectional loss caused by corrosion and/or fracture, which can lead to structural failure due to concentrated stress of the cable. Therefore, nondestructive evaluation (NDE) of steel cables is needed so any cross-sectional damage can be detected. This study proposes a steel cable health monitoring technique using an elasto-magnetic (E/M) sensor installed in a cable climbing robot. The E/M sensor is applied to detect the cross-sectional loss in this study while it was originally developed for measuring the tensile force in the previous works.

To verify the feasibility of the proposed damage detection technique, a series of experimental studies were performed. Firstly, a steel bars with 4-different diameters were fabricated and their output voltage values were obtained by the E/M sensor. The optimal input voltage and working point were chosen to ensure the linearity and resolution of the results through the repeated experiments. And then, the E/M sensor measured the output voltage values at the damage points of the steel bar specimen under the applied external and

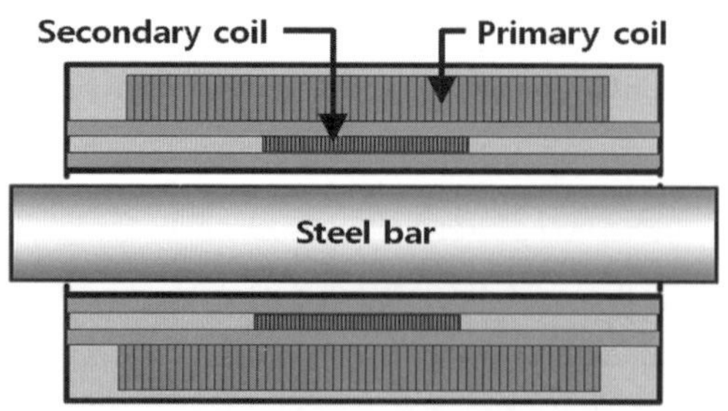

Figure 1. The concept of an elasto-magnetic sensor.

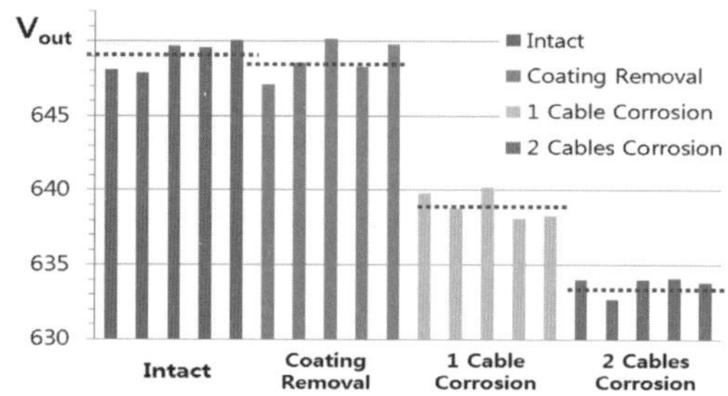

Figure 2. The variation in the output voltage values according to the corrosion damages.

inner damage conditions based on the selected optimal experimental conditions. A final experiment was carried out with a steel cable which has corrosion damages to examine the capabilities of corrosion damage detection using E/M sensor, likewise experiments with steel bar specimen. This series of experimental studies verified the feasibility of the proposed technique, and confirmed the following facts. 1. The measured output voltage decreased according to the decrease in the diameter of the steel bar specimen. 2. The linearity and resolution of the measurement were varied as to the working points were changed. 3. The output voltage values were decreased according to the increase of the damage severity. 4. The output voltage value was also changed due to the occurrence of inner damage. 5. Even small cross-sectional area loss due to the corrosion could be detected by the E/M sensor.

Conclusively, these results demonstrated that the steel cable monitoring technique using the E/M sensor can be effectively used to detect cross-sectional area loss damage.

REFERENCES

Joule, J. P. 1842. On a new class of magnetic forces, *Annals of Electricity Magnetism and Chemistry*, 8: 219–224

Wang, M. L., Wang, G. and Zhao, Y. 2005. *Sensing issues in civil structural health monitoring*, Springer, Dordrecht

Bridge Maintenance, Safety, Management, Resilience and Sustainability – Biondini & Frangopol (Eds)
© 2012 Taylor & Francis Group, London, ISBN 978-0-415-62124-3

Buffeting responses of a cable-stayed bridge during the typhoon Kompasu

J. Park, H.-K. Kim, H.S. Lee & H.-M. Koh
Seoul National University, Seoul, Korea

S. Cho
University of Illinois at Urbana-Champaign, Urbana, IL, USA

ABSTRACT

An international test-bed study on the application of smart wireless sensor network is on-going for the 2nd Jindo bridge, a cable-stayed bridge with the main span length of 344 m. Total 113 sensor nodes were deployed on the bridge including couple of ultrasonic anemometers.

Just after the installation of the wireless sensor network, the typhoon Kompasu struck the Korean peninsula in August, 2010, and the wind speed and synchronized acceleration data were measured. These sorts of data are very uncommon and provide a challenging chance for engineers to examine the effectiveness of the state-of-the-arts buffeting analysis approaches used in the aerodynamic design of the cable-stayed bridge.

The measured wind data can be used to estimate a simplified wind spectrum for analysis. Power spectra of wind velocity fluctuation which are used for analysis are represented in Fig. 1. Best-fit curves are more suitable in the 0.3–3 Hz frequency range.

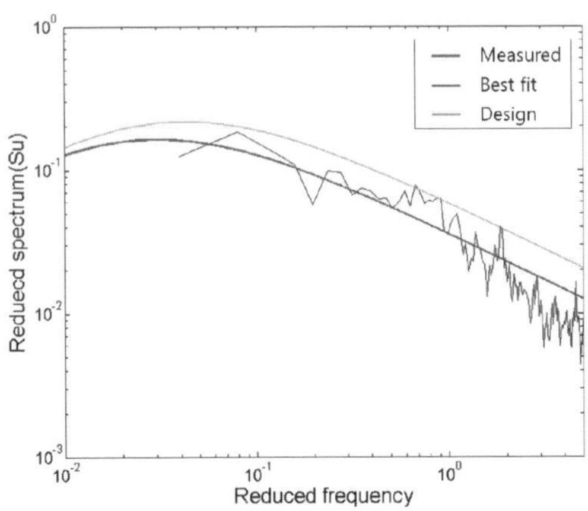

Figure 1. Spectra of longitudinal wind fluctuation.

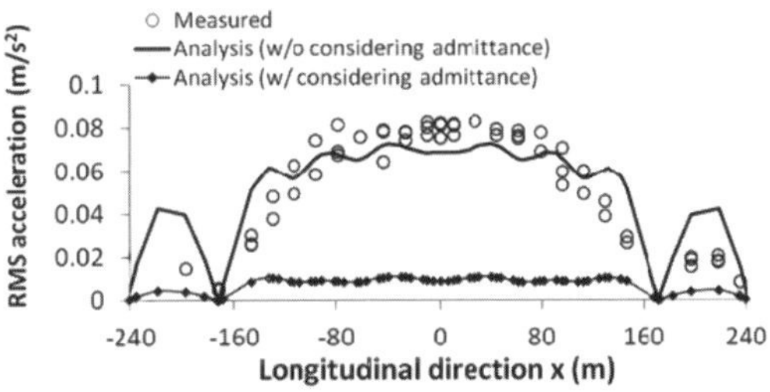

Figure 2. RMS lateral responses along the bridge direction.

The analysis results are represented in Fig. 2. The calculated buffeting responses are correctly fitted with measured responses without the consideration of admittance effect and are underes-timated under the admittance effect. However, it doesn't mean that the aerodynamic admittance function should not be considered because still several uncertainties are underlined for the comparison between buffeting analysis and field.

The monitoring system of the bridge is still working and accumulating data with various wind conditions, and further refinements in comparison of analysis and field measurement are expected.

REFERENCES

John H.G. Macdonald, 2003, Evaluation of buffeting predictions of a cable-stayed bridge from full-scale measurements, *Journal of Wind Engineering and Industrial Aerodynamics*, 91, pp 1465–1483.

John H.G. Macdonald & Wendy E. Daniell, 2005, Variation fo modal parameters of a cable-stayed bridge identified from ambient vibration measurements and FE modelling, *Engineering Structures*, 27, pp 1916–1930.

Korean Society of Civil Engineers (KSCE), 2006, Design guidelines for steel cable-supported bridges.

Z.Q. Chen et al., 2009, Investigation on influence factors buffeting response of bridges and its aeroelastic model verification for Xiaoguan Bridge, Engineering Structures, 31, pp 417–431.

Bridge Maintenance, Safety, Management, Resilience and Sustainability – Biondini & Frangopol (Eds)
© 2012 Taylor & Francis Group, London, ISBN 978-0-415-62124-3

Electromechanical impedance-based health diagnosis for tendon and anchorage zone

H. Shim, J. Min & C.-B. Yun
Korea Advanced Institute of Science and Technology, Daejeon, Korea

ABSTRACT

The tendon and anchorage block are responsible to deliver high stress as key members of a prestressed structure. Thus, the deterioration of tendon in carrying tensile force or damages on the anchorage block may lead to failure of the structure so that they should be monitored continuously. In this paper, a simple impedance-based method was proposed to estimate the tensile force and monitor the condition of the anchorage zone. Experimental studies were carried out on two sizes of models of an anchorage system. The relationship between the impedance signature and the tensile force was first constructed for the tensile force estimation. Simultaneously, the damage diagnosis was performed using the relationship among impedance signatures of multiple sensors attached to the structure. When the loss of the tensile force occurred, the root mean square deviation (RMSD) index increased. Hence an empirical curve can be obtained for the estimation of tension loss based on the experimental data. On the other hand, when the structure got damaged, the RMSD index abruptly increased over this curve so that the status of the damage was able to be identified. The experimental results indicated a big potential of the proposed method for monitoring and diagnosis of the tendon and the anchorage zone.

REFERENCES

Bartoli, I., Phillips, R., Scalea, F. L., Salamone S., Coccia, S., and Sikorsky, C.S. (2008), "Load Monitoring in Multiwire Strands by Interwire Ultrasonic Measurements", Proc. of SPIE, Vol. 6932.

Bati, S.B., and Tonietti, U. (2001), "Experimental methods for estimating in situ tensile force in tie-rods", Journal of Engineering Mechanics, ASCE, Vol. 127, No. 12, pp. 1275–1283.

Giurgiutiu, V. (2008), Structural health monitoring with piezoelectric wafer active sensors, Amsterdam: Elsevier/Academic Press.

Grisso, B.L. and Inman, D.J. (2005), "Developing an autonomous on-orbit impedance-based SHM system for thermal protection systems", Proceedings of 5th Int'l Workshop on Structural Health Monitoring, Stanford, CA, September.

Kim, J.T., Park, J.H., and Hong, D.S. (2011), "Sensitivity Analysis of Impedance Sensor Node for Structural Health Monitoring in Cable-Anchor System", Advanced Science Letters, Vol. 4, pp. 1342–1347.

Kim, M.K., Lim, H.J., Sohn, H., and Park, C.Y. (2010), "Impedance-based bolt loosening detection under varying temperature and loading", Asian Pacific Workshop on Structural Health Monitoring, Tokyo, Japan.

Kim, B.H., Jang, J.B., Lee, H.P., and Lee, D.H., (2010), "Effect of prestress force on longitudinal vibration of bonded tendons embedded in a nuclear containment", Nuclear Engineering and Design, 240(6), pp. 1281–1289.

Liang C, Sun FP, and Rogers CA. (1996), "Electromechanical impedance modeling of active material systems", Smart Materials and Structures, Vol. 5, pp. 171–86.

Min, J. (2012). "Structural Health Monitoring for Civil infrastructure Using Wireless Impedance Sensor Nodes and Smart Assessment Techniques", Doctoral Dissertation.

Min, J., Park, S., and Yun, C-B. (2010), "Impedance-based structural health monitoring using neural networks for autonomous frequency range selection", Smart Materials and Structures, Vol. 19.

Nguyen, K.D., and Kim, J.T. (2011), "Numerical Simulation of Electro-Mechanical Impedance Response in Cable-Anchor Connection Interface:, Journal of the Korean Society for Nondestructive Testing, Vol. 31, pp. 11–23.

Overly, T.G., Park, G., Farrar, C.R., and Allemang, R.J. (2007), "Compact hardware development for structural health monitoring and sensor diagnostics using admittance measurements", Proceedings of IMAC-XXV: A Conference & Exposition on Structural Dynamics, Orlando, FL, February.

Park G., Sohn H., Farrar C.R., and Inman D.J. (2003), "Overview of piezoelectric impedance-based health monitoring and path forward", Shock Vibration Digest, Vol. 35, pp. 451–63.

Park, S., Yun, C.-B., Roh, Y., and Lee, J.-J. (2005), "Health monitoring of steel structures using impedance of thickness modes at PZT patches", *Smart Struct. and Syst.*, **1**, 339–353.

Park, S., Shin, H.-H., and Yun, C.-B. (2009), "Wireless impedance sensor nodes for functions of structural damage identification and sensor self-diagnosis", *Smart Mater. Struct.*, **18**, 1–11.

Taylor, S.G., Farinholt, K.M., Park, G., and Farrar, C.R. (2009), "Wireless impedance device for electromechanical impedance sensing and low-frequency vibration data acquisition", Proc. of SPIE, San Diego, CA, March.

Bridge Maintenance, Safety, Management, Resilience and Sustainability – Biondini & Frangopol (Eds)

ISBN 978-0-415-62124-3

Vibration-based BHMS for long-span bridges considering environmental actions

S. Shin, H. Kim & Y. Kim
Inha University, Incheon, Korea

J.C. Park
Korea Expressway, Co., Kyunggi-do, Korea

ABSTRACT

A BHMS (Bridge Health Monitoring System) algorithm is proposed. It contains three sequential systematic steps: (a) application of numerical schemes to recover or correct missing data, (b) application of some statistical methods to eliminate environmental effects or outliers from the data, and (c) computation of health index to evaluate the health condition of the bridge.

For the recovery of missing data, the current study used the concept of the NN (Neural Network) approach. Since previous studies have demonstrated that a bridge structure usually has a good correlation between natural frequencies and temperature variation, it is clear that a conspicuous pattern between them may be easily recognized by a NN approach.

To eliminate environmental effects from the measured data, the current study applies PCA (Principal Component Analysis) method. The main idea of the PCA method is to reduce a matrix of natural frequencies $\mathbf{Y}_{(n \cdot N)}$ with n modes and N samples to a matrix $\mathbf{X}_{(m \cdot N)}$ with only m principal components by multiplying a loading matrix $\mathbf{T}_{(m \cdot n)}$ as expressed in Eq. (1) (Yan et al. 2005).

$$\mathbf{X} = \mathbf{T}\ \mathbf{Y} \tag{1}$$

where $\mathbf{X}$ = a linearly projected matrix onto a domain $\Re^m$ from a domain $\Re^n$ of $\mathbf{Y}$ where $n > m$.

To evaluate the health condition of a bridge, Mahalanobis norm of Eq. (1) is selected as a health index (Bellino et al. 2010).

$$MD_k = \sqrt{\widetilde{\mathbf{Y}}_k^{\mathrm{T}}\ \mathbf{R}^{-1}\ \widetilde{\mathbf{Y}}_k \mathrm{Y}} \tag{2}$$

where $\mathbf{R} = (1/N)\,\mathbf{Y}\mathbf{Y}^{\mathrm{T}}$ = the covariance matrix of the matrix $\mathbf{Y}$ and k = the time step.

The developed BHMS algorithm has been applied to field data measured from a cable-stayed bridge in Korea. The bridge is located on a highway crossing over a sea channel and the total length of the cable-stayed bridge part is 990 m. Various types of sensors were installed to measure the bridge behaviors and environmental actions.

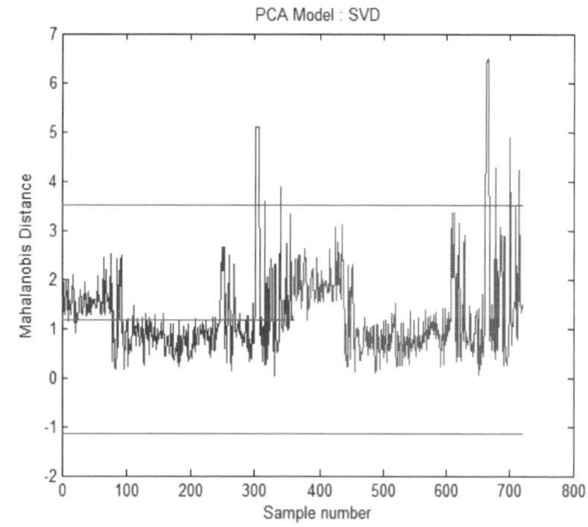

Figure 1. Health index by Mahalanobis distance.

The health indices of the Mahalanobis distance defined by Eq. (1) were computed for two years with all the revised data and are drawn in Figure 1. The first year data was used for training the NN model and the second year data was predicted using the defined model. Figure 1 shows that some peaks of the computed health index around the same season in a year were out of the upper limit line. However, overall, the computed health indices were within the limit boundaries. Since the patterns of the health index for the first year and the second year are quite similar, we can conclude that the health condition of the bridge could be retained in the second year without serious change or damage to it.

REFERENCES

Bellino, A., Fasana, A., Garibaldi, L. & Marchesiello, S. 2010. PCA-based detection of damage in time-varying systems. *Mechanical Systems and Signal Processing*, 24: 2250–2260.

Yan, A.-M., Kerschen, G., Boe, P. De & Golinval, J.-C. 2005. Structural damage diagnosis under varying environmental conditions – part I: A nonlinear analysis. *Mechanical Systems and Signal Processing*, 19(4): 847–864.

Bridge Maintenance, Safety, Management, Resilience and Sustainability – Biondini & Frangopol (Eds)
© 2012 Taylor & Francis Group, London, ISBN 978-0-415-62124-3

Fatigue analysis of steel box stiffening girders for large-span suspension bridge in its lifetime

Y. Zeng & H. Tan
Key Laboratory of Bridge-Structure Engineering Ministry of Communications, Chongqing Jiaotong University, Chongqing, China
College of Civil and Architectural Engineering, Chongqing Jiaotong University, Chongqing, China

X. Li
The Second Harbour Engineering Company, China Communications Construction Company Ltd., Wuhan, China

Z. Xiang
Department of Bridge Engineering, Chongqing Jiaotong University, Chongqing, China

F. Yu
Engineering Company Ltd. of CCCC First Harbor Engineering Company Ltd., Qingdao, China

ABSTRACT

Orthotropic steel decks are especially important to box girder. It is vulnerable and sensitive to damage fatigue induced by heavy trunks, particularly in the vicinity of welds where cracks and voids appear. Ensuring the safety of main girder under different load conditions is of great importance to engineers and managers. Occurrence of fatigue cracks and corrosion in suspension bridge is always an important concern for bridge managers, and efforts should been devoted in order to ensure safety of the bridges during the service life. In this paper, traffic investigation, corrosion maintenance strategies for the girders and main cable, and fatigue assessment for steel box girder will be presented respectively.

Main girder of Qingcaobei bridge is a streamlined shape with 6 lanes and the width of each lane is 7.75 meters. There is a 1.5-meters repair lane on each side of main girder. The deck plates and stiffeners are welded structures.

If $D(n)$ is Miner's damage accumulation index, the limit state function of the fatigue reliability is defined as follows: $g(X) = D(n) - D_C$ (2) where D_C is critical damage accumulation index; n is the number of stress cycles. If $g(X) \geq 0$, it implies that the girder is safe; failure due to fatigue is assumed to occur when $g(X) < 0$.

The actual fatigue reliability of a steel box girder, calculated by above-mentioned method, will be compared with target reliability to yield information that can help in scheduling of inspections and maintenance. If the fatigue reliability of a girder is higher than the target reliability during the service life, only routine inspections are needed. If the fatigue reliability of the detail falls below the target reliability level, carefully scheduled and detailed inspections may be required.

The fatigue category of steel box girder of Qingcaobei bridge is C category of AASHTO, conservatively. Since stress range of steel box girder is so small, the fatigue life of steel box girder is enough safe. Steel box girder of large-span suspension bridge is flexible, lightweight and sensitive to fatigue damage caused by passing trucks. Attention should be paid to fatigue damage of girder. Fatigue reliability assessment of girder is presented in this paper, which can be used to the scheduling of inspections and maintenance.

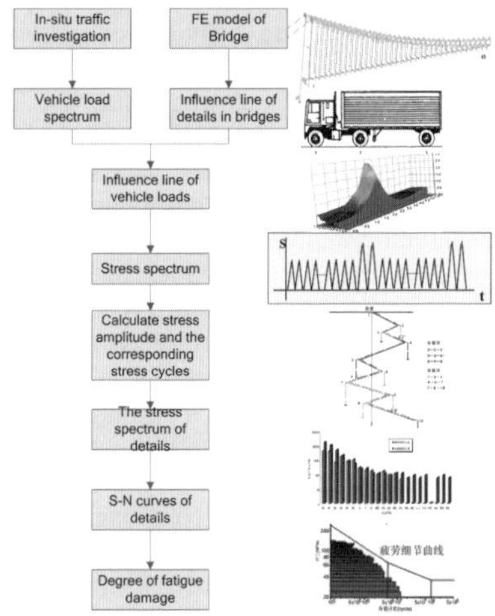

Figure 1. Fatigue assessment of steel bridges.

REFERENCES

Airong Chen, Yong Zeng. 2007. Life-circle based monitor, maintenance and management research of Jiangyin Yangtse River Bridge. Shanghai: Tongji University (in Chinese)

Yong Zeng. 2007. Monitor, maintenance and management of large span suspension bridges in lifetiime. Shanghai: Tongji University (in Chinese)

Strengthening of existing bridges with FRP composites

Organizer: C. Pellegrino

Bridge Maintenance, Safety, Management, Resilience and Sustainability – Biondini & Frangopol (Eds)
© 2012 Taylor & Francis Group, London, ISBN 978-0-415-62124-3

Bond of FRP strengthening systems for concrete structures: A round Robin test

M. Guadagnini & A. Serbescu
The University of Sheffield, Sheffield, UK

A. Palmieri & S. Matthys
Ghent University, Ghent, Belgium

A. Bilotta & E. Nigro
University of Naples, Naples, Italy

C. Mazzotti
University of Bologna, Bologna, Italy

G. Sas & B. Taljsten
University of Lulea, Lulea, Sweden

F. Ceroni
University of Sannio, Sannio, Italy

C. Czaderski & S. Olia
Laboratory EMPA, Dübendorf, Switzerland

Z. Szabo & G. Balazs
University of Budapest, Budapest, Hungary

J. Barros & I. Costa
University of Minho, Guimarães, Portugal

V. Tamuzs
University of Latvia, Riga, Latvia

ABSTRACT

Over the last decade, the interest in using FRP materials for the rehabilitation of structurally deficient concrete bridges has increased dramatically. FRPs can be used to increase the flexural, shear and torsional capacity of RC elements (e.g. concrete box girders) and a variety of systems are commercially available, including externally bonded and near surface mounted solutions. Unless specific anchorage systems are used, the bond of the FRP system to the concrete substrate can limit the effectiveness of the adopted strengthening solution and lead to poor material utilization.

Current design models to predict the load causing debonding of FRP strengthening systems are based either on purely empirical approaches (e.g. Chen & Teng 2001) or rely on observations and results obtained from small scale testing (e.g. Dai et al. 2005). A standard test methodology, however, has yet to be generally accepted. With these issues in mind, an extensive international exercise was initiated within the framework of the European funded Marie Curie Research Training Network, EN-CORE, and completed with the support of Task Group 9.3 of the International Federation for Structural Concrete (*fib*).

Ten laboratories and eight manufacturers and suppliers participated in this testing programme aimed at examining the reliability of the test methods that are most commonly used for the characterization of FRP strengthening systems. Only the tests on the bond performance of selected strengthening systems are presented in this paper. Six different FRP externally bonded strengthening systems and eight NSM bars/strips were tested. A minimum of 2 tests per strengthening system were carried out at each laboratory for a total of 180 tests (89 on EBR and 91 on NSM reinforcement). The tests were designed to subject the concrete specimens to a tension-tension state of stress, and a double shear test (DS) or a single shear test (SS) setup was adopted according to given specifications.

From the analysis of the experimental results it was found that, although the COV of the data obtained at an individual laboratory was typically around 10%, this could increase to more than 35% when all of the results across the participating laboratories are taken into account. This can be attributed to a series of factors, including variability of concrete, surface preparation, load misalignment. In addition, the preparation of specimens for DS tests has proven to be problematic, and concerns have been raised at several of the participating laboratories regarding the ability of effectively ensure alignment of the various components and avoid the development of undesired bending effects. If similar conditions are ensured, the use of a DS or an SS configuration does not seem to affect considerably overall performance and behavior. The number of specimens exhibiting concrete splitting related failures, however, was comparatively higher for SS tests than for DS tests. This could be attributed to the possible effects of load eccentricity, but also to the lower concrete strength used for the specimens tested in this configuration.

When examining the performance of externally bonded laminates, the quality of the concrete substrate, primarily in terms of surface finish, was found to affect the bond behaviour of the tested systems to a large extent.

REFERENCES

Chen, J.F., & Teng, J.G.. 2001. Anchorage strength models for FRP and steel plates bonded to concrete. *J. Struct. Eng.*, 127(7): 784–791.

Dai, J.G., Ueda, T. & Sato, Y. 2005. Development of the nonlinear bond stress-slip model of fiber reinforced plastics sheet-concrete interfaces with a simple method. *Journal of Composites for Construction*, 9(1): 52–62.

Bridge Maintenance, Safety, Management, Resilience and Sustainability – Biondini & Frangopol (Eds)
© 2012 Taylor & Francis Group, London, ISBN 978-0-415-62124-3

Behaviour of FRP confined concrete cylinders under different temperature exposure

M.N.S Hadi
University of Wollongong, Wollongong, New South Wales, Australia

B.A. Louk Fanggi
The State Polytechnic of Kupang, Kupang, Nusa tenggara timur, Indonesia

ABSTRACT

Fibre Reinforced Polymer (FRP) is known as an excellent material for retrofitting, repairing and strengthening structural members. It is characterised by the following features: light weight material, corrosion resistant; available in various forms for field application, such as bar, sheet, strips and plate, and also available in long length thus eliminating the need for joints and splices; and can be cured within 24 hours when applied in the field.

Moreover, application of FRP to confine concrete members is widely used all over the world, due its high strength to weight ratio and ease of installation. When FRP is applied, it will lead to improving the strength of concrete members and it will increase their ductility dramatically, and reduce their maintenance compared to other methods such as attachment of steel to concrete and concrete jacket method.

However the behaviour of FRP confined concrete members has not been explored extensively especially in a situation of different temperature environment where solar gain could lead to a high concrete surface temperature. The maximum reported environment temperature applied for FRP confined concrete member is +45°C. It is also noticed that the global temperature tend to increase as a result of climate changes.

Therefore if the behaviour of FRP confined concrete members under different temperatures is known, the application of FRP confined concrete members can be used more confidentially in a wid range of temperature situation. This study investigates the effect of different temperatures on the behaviour of FRP confined concrete cylinders under axial compression loading.

Based on findings of this study, the following conclusions are drawn.

1. The ultimate strength of cylinders wrapped with one and two layers of CFRP did not significantly change after being exposed to 20°C to 70°C and 70°C. Meanwhile, their ductility increased. The maximum increase was observed on the cylinder after being exposed to 70°C.
2. The ultimate strength of cylinders wrapped with one layer of GFRP decreased after being exposed to 20°C to 70°C and 70°C. The maximum decrease was observed on the cylinder after being exposed to 70°C. Meanwhile, the ductility did not significantly change after being exposed to 20°C to 70°C but increased after being exposed to 70°C.
3. The ultimate strength of cylinders wrapped with two layers of CFRP decreased but it did not significantly change. On the other hand their ductility increased especially on the cylinder after being exposed to 70°C.

REFERENCES

El-Hacha, R., Green, M. F. and Wight, G. R, 2010. “Effect of severe environmental exposure on CFRP wrapped concrete columns”, *Journal of composites for construction*, 14(1): 83–88.

Green, M. F., Bisby, L. A., and Kodur, V. K. R. 2006. FRP confined concrete columns: behavior under extreme condition, *Cement and concrete composites*, 28: 928–937.

Karbhari, V. M. and Eckel, D. A,1994. “Effect of cold regions climate on composite jacketed concrete columns”, *Journal of cold region engineering*, 8(3): 73–86.

Karbhari, V. M, 2002. “Response of fiber reinforced polymer confined concrete exposed to freeze and thaw regimes”, *Journal of composites for construction*, 6(1): 35–40.

Bridge Maintenance, Safety, Management, Resilience and Sustainability – Biondini & Frangopol (Eds)
© 2012 Taylor & Francis Group, London, ISBN 978-0-415-62124-3

Bond behavior and failure mechanisms of EBR made of UHM carbon fibers

F. Jesse
Chair of Concrete Structures, BTU Cottbus, Cottbus, Germany

W. Stremmel
Hessisches Landesamt für Straßen- und Verkehrswesen, Wiesbaden, Germany

M. Curbach
Institute of Concrete Structures, TU Dresden, Dresden, Germany

ABSTRACT

An attractive method for strengthening reinforced concrete (RC) and pre-stressed concrete (PC) members are externally bonded reinforcements (EBR) made of carbon fiber reinforced polymers (CFRP). CFRP are supplied in form of lamellas or sheets.

From perspective of mechanical properties ultra high modulus (UHM) carbon fiber sheets are very interesting materials for structural strengthening of RC structures with EBR. Their very high elastic modulus ($E_f = 640{,}000\ \text{N/mm}^2$) promises immediate action at very low deformation. Especially shear strengthening should benefit from UHM carbon fibers, because small crack widths foster aggregate interlock as a main contributor to shear capacity.

To recheck design values for this special type of EBR two test series has been carried out. RC strain specimens with UHM carbon fiber sheet strengthening has been tested under uni-axial tension. Series A was intended to study interaction between internal reinforcement made from un-deformed low strength steel and EBR made from UHM carbon fiber sheets. Surprisingly, these tests show no increase of ultimate capacity due to double-sided strengthening. Because of a very inhomogeneous strain distribution in the sheets it was not possible to measure sheet load share with sufficient accuracy.

Series B was intended to study required anchoring length and had only one sheet applied. It was found that ultimate sheets capacity is equivalent to only 1/3 of nominal strength. From strain measurements with strain gauges and digital photogrammetrie we concluded, that waviness and inhomogeneous delamination leads to a very heterogeneous strain distribution (figure 1). High stressed areas fail first and trigger a collapse like failure type. Thus the combination of waviness, heterogeneous delamination and UHM carbon fibers leads to low ultimate capacity.

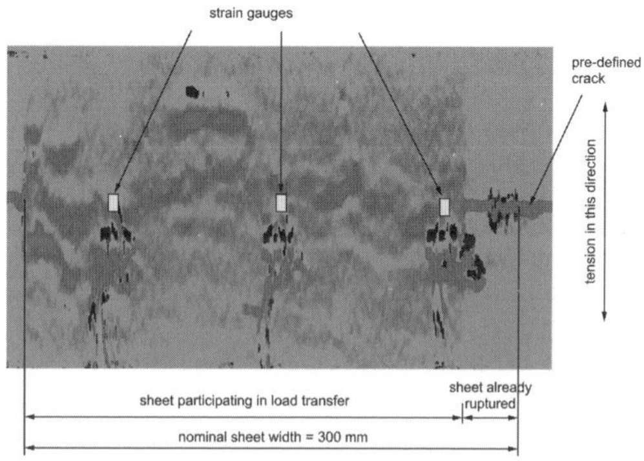

Figure 1. Qualitative strain distribution from digital photogrammetry measurements (spots directly below strain gauges are artefacts from cables).

Waviness of UHM carbon fiber sheets has at least two sources. There is waviness induced by in-situ lamination that supposedly cannot be eliminated. And there is waviness that comes from residual stresses when pre-impregnated sheets are rolled off during application. The latter source of waviness seems to be the dominating one for waviness and can probably be avoided by modifying the production process. Waviness and residual stresses seem directly related to the pre-impregnation.

UHM carbon fibers sheets with low ultimate capacity can even fail before internal steel reinforcement begins to yield. As a consequence, individual contributions of internal reinforcement and EBR are not correctly considered by simply adding their maximum load capacity but by taking their interaction into account. If EBR fails to early, no strengthening effect will show.

The presented results show that materials used for strengthening should be carefully selected and that this requires detailed knowledge on their interaction with internal reinforcement in structural members. Further on, material properties derived from small laboratory test on raw materials—e.g. strength tests of carbon fibers—are not transferable to structural applications in case of in-situ laminated UHM carbon fiber sheets.

Bridge Maintenance, Safety, Management, Resilience and Sustainability – Biondini & Frangopol (Eds)
© 2012 Taylor & Francis Group, London, ISBN 978-0-415-62124-3

Strengthening of multi-storey parking by bridge engineering means

G. Lagoda
Warsaw University of Technology, Warsaw, Poland

M. Lagoda
Road & Bridge Research Institute, Warsaw, Poland
Lublin University of Technology, Lublin, Poland

J. Ciesla
Road & Bridge Research Institute, Warsaw, Poland

ABSTRACT

Strengthening of existing multi-storey reinforced concrete parking superstructure with the use of composite FRP material is a main objective of the paper.

Slabs of floor of parking are typical reinforced concrete structure, supported by columns, like a reinforced concrete bridge That's the reason, that the parking at the International Frederic Chopin Airport in Warsaw has been designed and strengthened like a bridge structure, especially because it is working under traffic loads. In strengthening the traffic loads on bridges involved with cars, including heavy transport for different purposes, has been taken into account. In that way, procedure of designing according with current bridge standards has been done.

Support points and way and value of loads were changed. It caused, that existing reinforcement would be insufficient in many places in case of construction of new hotel building on the top of parking. In such situation it was a need for essential structural strengthening of parking. For defining possibility and conditions needed for construction of new hotel building and other consequences, especially on top of parking, the structure of parking had to be recalculated. For calculations finite element method was applied with the use of ROBOT MILLENIUM program in version 17.0.

The new factors need to be discussed, taking into account, that condition of exploitation also have to be different, e.g. the fire-proof condition.

That factor has an essential influence for the choice of structural solution. The analysis of ultimate limit states and serviceability limit states have been taken into account, especially in durability and fire resistance aspects.

The idea of the strengthening has been accepted in the form of CFRP composite strips glued to the existing slabs in place of the lacking reinforcement.

All reinforced concrete floor slabs were strengthened toward "x" by means of near surface mounted CFRP strips of the type L (Fig. 1). However the strengthening of slabs toward "y" have been designed by means of CFRP strips the type K, glued to the bottom-surface of the construction.

Such way of strengthening is very cheap and can be executed in short time. Passed analysis and suggested design of the strengthening makes possible leaving of the existing structure without demolition, and significantly reduce of building expenses.

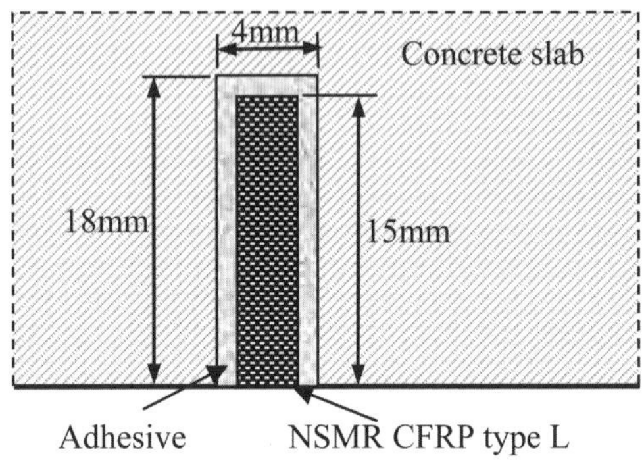

Figure 1. The strengthening of floor slabs toward "x" with inserted tapes CFRP.

REFERENCES

CEB-FIP 1993. *Model Code 1990, Design Code.* Lausanne Switzerland, Comité Euro-International du Béton. Thomas Telford.

EN 1990: 2002. *Basis of structural design.* Brussels, European Committee for Standardization.

EN 1991–2: 2003. *Traffic loads on bridges.* Brussels, European Committee for Standardization.

Lagoda G., Lagoda M. 2011. Wzmacnianie kompozytami ustroju nosnego wielopoziomowego parkingu w Porcie lotniczym im. F. Chopina w Warszawie. *Archives of Insitute of Civil Engineering:* 169–174 Poznan: University of Technology Press.

Bridge Maintenance, Safety, Management, Resilience and Sustainability – Biondini & Frangopol (Eds)
© 2012 Taylor & Francis Group, London, ISBN 978-0-415-62124-3

Comparative behaviour of FRP confined square concrete columns under eccentric loading

X. Lei, T.M. Pham & M.N.S. Hadi
University of Wollongong, Wollongong, NSW, Australia

ABSTRACT

Retrofitting existing columns in bridges and buildings have become an indispensable requirement in recent decades. Strengthening existing reinforced concrete columns using Fibre Reinforced Polymer (FRP) have been demonstrated as an excellent method. Its benefits range from high strength and stiffness to enhanced ductility. This study focuses on introducing a new technique which is aimed to maximise the confinement efficiency of FRP for square columns confined with FRP: circularisation of square columns by bonding four pieces of segmental circular concrete covers. Meanwhile, steel straps confinement is evaluated as an alternative confining material and compared to Carbon Fibre Reinforced Polymer (CFRP). Finally, eccentric loading is incorporated to evaluate the axial load and bending interaction capacity of the columns.

To achieve the aforementioned objectives, an experimental program was carried out. Sixteen reinforced concrete square columns were cast. The reinforcement and strength of the concrete was kept at minimum simulating columns needing retrofitting. All columns were internally reinforced with 4N12 bars (12 mm deformed bars with nominal tensile strength of 500 MPa) as longitudinal reinforcement, and R6 bars (6 mm plain bars with nominal tensile strength of 250 MPa) at 120 mm spacing as ties. Normal strength concrete with nominal compressive strength of 32 MPa was used. The columns were divided into four groups, with each one simulating a specific strengthening technique. Group N was used as a reference group and no external reinforcement was provided. Columns in Group RF were wrapped with three layers of CFRP after the corners were rounded by 20 mm. Groups CF and CS incorporate the process of circularisation, i.e. adding four circular segments of concrete to change the cross section from a square to a circle. After that, columns in Group CF were wrapped with three layers of CFRP and columns in Group CS were confined with steel straps with 30 mm spacing. From each group, one column was concentrically loaded, while the second and the third columns were subjected to axial loading with 15 mm and 25 mm eccentricity, respectively. The fourth specimen was tested under four-point loading as a beam to observe the flexural behaviour.

Results from the experiment showed that (1) all confined columns achieved higher ultimate load and ductility than unconfined columns, (2) Circularisation effectively reduced corner sharpness and thus increase the confinement effectiveness of FRP, (3) For columns under concentric loading, those wrapped with CFRP showed ascending second branch after the columns were yielded while columns confined with steel straps did not, which indicates that the confinement efficiency of steel strapping is relatively lower than CFRP, (4) For eccentric loading tests, all columns showed lower ultimate load and no ascending branch in the second branch was witnessed, which suggests confinement was less effective for eccentrically loaded columns than concentrically loaded columns.

Conclusion from the experimental programme includes: steel straps are an excellent confining material with lower price and are easier to apply in general construction than CFRP. Steel strapping can be used as an effective alternative approach in retrofitting existing columns. Meanwhile, circularisation was proven to be effective and practical. The bonding between the circular segments and columns was reliable and the modified columns can be treated as complete circular columns. It is therefore clear that this method can be practically used to significantly increase the load-carrying capacity of the columns as well as the efficiency of FRP confinement.

Bridge Maintenance, Safety, Management, Resilience and Sustainability – Biondini & Frangopol (Eds)
 ISBN 978-0-415-62124-3

Influence of the axial stiffness of the reinforcement on the FRP-concrete interface's fracture energy

C. Mazzotti
DICAM – Structural Engineering, University of Bologna, Italy

ABSTRACT

The present paper shows the results of an experimental campaign devoted to the investigation of the sensitivity of the bond strength of a FRP-concrete system when changing the number of FRP layers composing the strengthening intervention and the combined effect of the variation of the elastic modulus of the reinforcement. In particular, several prismatic specimens have been strengthened with a number of layers of carbon fiber reinforced polymers (CFRP) sheets ranging from one to three. All the specimens have been subjected to bond test according to a particular version of the single-lap push-pull set-up proposed by the University of Bologna (Mazzotti et al., 2009); the results show the decreasing effectiveness of the FRP reinforcement when increasing the number of layers and, more generally, the overall FRP axial stiffness while keeping constant all the others influencing parameters (the concrete properties, in particular).

The specimens have been instrumented by means of a series of strain gauges placed along the longitudinal axis of the reinforcement. Starting from the strain distribution along the reinforcement, measured at various force levels, shear stress-slip interface behavior has been observed for the different cases considered. A reduction of the peak shear stress and of the fracture energy have been observed when increasing the axial stiffness ($E_f \cdot t_f$) of the reinforcement, thus leading to a more brittle failure mechanism. Though the number of tests is limited, the trend is quite apparent.

Finally, comparison of the experimental results with the design provisions by CNR guidelines (CNR, 2004) showed that actual code formulations can be adopted also when multiple layers of reinforcements are bonded together. Nevertheless, the experimental evaluation of the k_G coefficient showed that, although the mechanical properties of the concrete were constant (since all the specimens are characterized by the same properties of the substrate), a remarkable reduction occurs when increasing the axial stiffness of the reinforcement (Figure 1). This aspect requires further attention since, to date, the fracture energy is thought

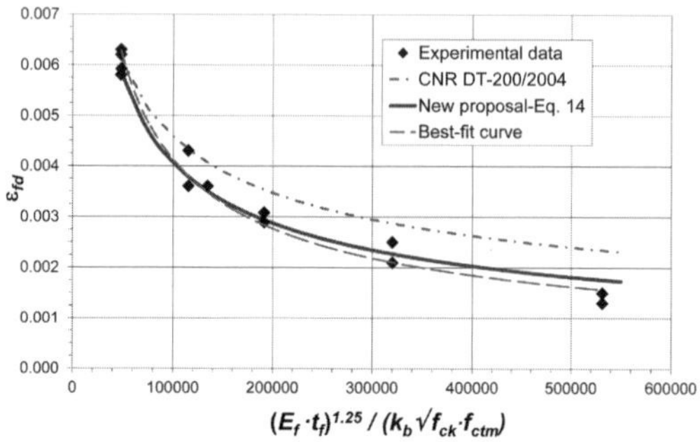

Figure 1. Experimental values of maximum strain at debonding compared with the proposed prediction formula.

to be mainly governed by the mechanical properties of the concrete substrate. In this perspective, a new formulation for the prediction of the maximum strain at debonding has been proposed:

$$\varepsilon_{fd} = \sqrt{\frac{2 \cdot 0.825 \cdot k_b \cdot \sqrt{f_{ck} \cdot f_{ctm}}}{(E_f \cdot t_f)^{1.25}}}, \qquad (1)$$

where the classical k_g parameter is no longer a constant value.

Given the complexity of the problem, nevertheless, to draw final conclusions more tests are necessary considering also various concrete classes and different types of FRP reinforcement.

REFERENCES

Mazzotti C., Savoia M., Ferracuti B., 2009. "A new single-shear set-up for stable debonding of FRP–concrete joints", Construction and Building Materials, 23(4): 1529–1537.

CNR National Research Council, "DT-200: Istruzioni per la progettazione, l'esecuzione, controllo di interventi di consolidamento statico mediante l'utilizzo di compositi fibrorinforzati", 2004.

Bridge Maintenance, Safety, Management, Resilience and Sustainability – Biondini & Frangopol (Eds)
© 2012 Taylor & Francis Group, London, ISBN 978-0-415-62124-3

Finite element modelling of beams strengthened with FRP sheets during short and long-term loads

G. Mazzucco, V.A. Salomoni, C. Majorana & C. Pellegrino
Department of Structural and Transportation Engineering, University of Padova, Italy

ABSTRACT

FRP (Fiber Reinforced Polymer) technique was often diffused to repair damaged reinforced concrete bridges. In this work it is proposed to investigate the complex mechanism of stress-strain evolution at the FRP interface, during short or long-time loadings, until complete debonding. This study has been performed by means of a fully three-dimensional approach within the context of damage mechanics, to appropriately catch transversal effects as well as normal stresses, developing a realistic and comprehensive study of the delamination process. The adhesion properties have been studied through a contact model incorporating an elastic-damage constitutive law, relating inter-laminar stresses acting in the sliding direction. A F.E. research code (FRPCON) has been developed (Mazzucco 2011, Salomoni et al. 2011), incorporating a numerical procedure accounting for Mazars damage law inside the contact algorithm. The code is able to describe the delamination process considering the various surface preparations of the concrete part (Figure 1). Long-time behaviour of this composite structures has been studied by means of two visco-elastic formulations: i) B3 law (Bazant et al. 2000) has been considered in the concrete component, where creep effect is composed by three different terms, i.e. the elastic part, basic creep and drying creep; ii) different long-time behaviour for fibres and matrix have been implemented for FRP component, using a micromechanical approach (Ascione et al. 2003).

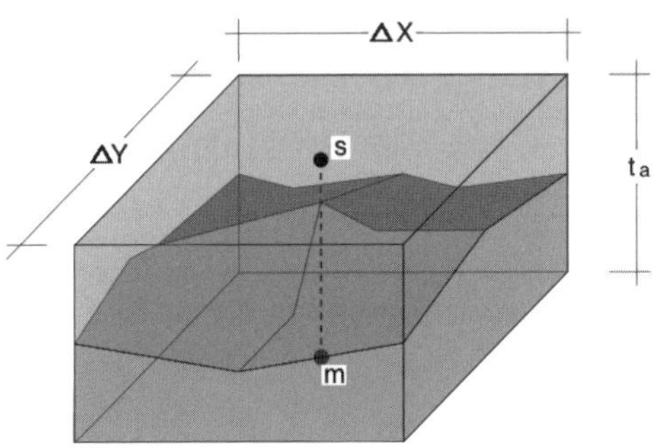

Figure 1. Typical interface volume for every contact pair.

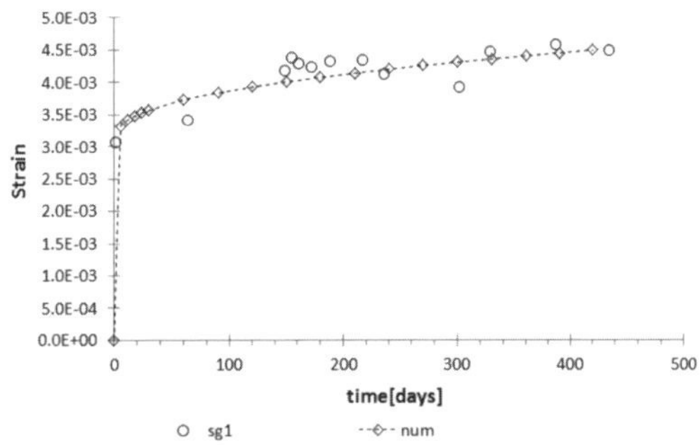

Figure 2. Comparison between experimental and numerical interfacial strains.

The experimental results of long-time bending tests have been used to calibrate and validate the numerical models.

The research F.E. code has demonstrated to be able to simulate delamination processes and long time stress-strain evolutions. By comparing the numerical results with those of the experimental investigations, in terms both of ultimate load and strain vs. time (Figure 2), it has been shown that such an approach is able to catch delamination from a three-dimensional point of view and its evolution during the entire loading process.

REFERENCES

Ascione L. Berardi, V.P. & D'Aponte, A. (2003). Il Comportamento a lungo termine dei materiali compositi fibrorinforzati: un modello micromeccanico, *XXXII Congress AIAS*, Salerno (in Italian).

Bazant, Z. & Baweja, S. (2000). Creep and Shrinkage prediction model for analysis and design of concrete structures: Model B3, Adam Neville Symposium: Creep and Shrinkage – *Structural Design Effects, ACI SP-194, A. Al-Manaseer Ed., Am. Concrete Institute, Farmington Hills, Michigan*, 1–83.

Mazzucco, G. (2011). *Experimental and Numerical analysis of bond behaviour between concrete and FRP*, PhD Thesis, University of Trento, Italy.

Salomoni, V.A., Mazzucco, G. Pellegrino, C. & Majorana, C.E. (2011). Three-dimensional modelling of bond behaviour between concrete and FRP reinforcement, *Eng. Comp.*, 28(1), 5–29.

Bridge Maintenance, Safety, Management, Resilience and Sustainability – Biondini & Frangopol (Eds)
© 2012 Taylor & Francis Group, London, ISBN 978-0-415-62124-3

Effect of FRP retrofit interventions on seismic vulnerability of existing bridges

R. Morbin, E. Casadei, C. Pellegrino & C. Modena
Dipartimento di Costruzioni e Trasporti, Università di Padova, Padova, Italy

ABSTRACT

Italy is one of the Mediterranean Countries with a high seismic risk and previous earthquakes had relevant social and economic effects.

Bridges are the one of the most vulnerable elements in existing road and railway networks during earthquakes, hence the assessment of their seismic vulnerability and the study of the techniques to reduce it are common issues in the general context of the management of the transportation network. One of the most diffused method to evaluate seismic vulnerability of existing bridges is based on the calculation of the fragility curves, which give the probability of a bridge reaching a certain damage state for a given ground motion parameter. By means of fragility curves, damage of critical elements of the bridge can be predicted and consequent economic losses can be estimated. Moreover retrofit interventions can be planned.

In this framework fragility curves for a bridge having a common structural scheme in Italy (multi-span simply supported reinforced concrete bridge) have been calculated by means of non-linear dynamic analyses (Shinozuka et al. 2000, Choi et al. 2003, Morbin et al. 2010). Then the influence of methods to reduce seismic vulnerability of the bridge by pier confinement with fiber-reinforced polymers (FRP) was studied. Two theoretical models (CNR-DT 200 2004, Pellegrino & Modena 2010) for the evaluation of the constitutive law of confined piers were considered to generate the fragility curves of the bridge taking retrofit effects into account.

The following general conclusions can be drawn regarding the modelling strategy for obtaining seismic fragility curves:

- the choice of accelerograms affects fragility curves trend, hence the more accelerograms are considered, the more seismic vulnerability estimation is precise;
- strength values and elastic properties of considered materials can significantly affect results, hence a proper survey, e.g. in-field tests and laboratory analyses, may be recommended;
- non-linear dynamic analysis seems to be suitable for the determination of seismic fragility of the structure, but with a consistent computational effort with respect to other simplified analyses.

The following conclusions can be drawn in relation to FRP retrofitting by means of pier jacketing:

- as expected, seismic vulnerability decreases for retrofitted bridges with respect to the bridge without FRP confinement;

the model proposed by Pellegrino & Modena (2010) shows a reduction of the vulnerability more evident than that obtained with the model proposed adopted in the Italian Code since the former takes into account the contribution of the steel stirrups confinement and the interaction mechanisms between internal reinforcing steel and external FRP strengthening.

REFERENCES

Choi, E.S., DesRoches, R. & Nielson, B. 2003. Seismic fragility of typical bridges in moderate seismic zones. Engineering Structures, 26 (2): 187–199.

Morbin, R., Pellegrino, C., Grendene, M. & Modena, C. 2010. Strategies for seismic vulnerability evaluation of common RC bridges typologies. In *14th European Conference on Earthquake Engineering,* Ohrid, Republic of Macedonia: 30th August – 3rd September 2010.

National Research Council 2004. Guide for the Design and Construction of Externally Bonded FRP Systems for Strengthening Existing Structures, CNR-DT 200.

Pellegrino, C. & Modena, C. 2010. Analytical Model for FRP Confinement of Concrete Columns with and without Internal Steel Reinforcement. Journal of Composites for Construction, 14 (6): 693–705.

Shinozuka, M., Feng, M.Q., Lee, J. & Naganuma, T. 2000. Statistical analysis of fragility curves. Journal of Engineering Mechanics-Asce, 126 (12): 1224–1231.

Bridge Maintenance, Safety, Management, Resilience and Sustainability – Biondini & Frangopol (Eds)
© 2012 Taylor & Francis Group, London, ISBN 978-0-415-62124-3

Ultimate limit state of MF-FRP beams

F. Nardone, G.P. Lignola, A. Prota & G. Manfredi
Department of Structural Engineering (DIST), University of Naples (Federico II), Italy

A. Nanni
Department of Civil, Architetural and Environmental Engineering (CAE), University of Miami, Florida

ABSTRACT

Usually, FRP materials are adhesively bonded to the structural substrate in strengthening applications. An innovative technique is based on the use of mechanical fastening (MF-FRP) by means of steel anchors to attach FRP laminates with enhanced bearing strength to concrete substrate. The benefit of MF-FRP, compared to adhesive bonding for FRP flexural strengthening is due to speed of installation using unskilled labor, minimal or absent surface preparation under any meteorological condition and immediate use of the strengthened structures. Some of the potential shortcomings are: brittle failure modes for members strengthened with the MF-FRP without a proper design; possible concrete damage during the drilling and dense internal reinforcement of the members that could limit the installation. Laboratory testing and a number of field applications have shown the effectiveness of such method. In this paper an analytical model is discussed for reinforced concrete (RC) beams strengthened by using MF-FRP strips. The model accounts for equilibrium, compatibility and constitutive relationships of materials, in particular, it accounts explicitly for the slip between the surface of the substrate and the FRP strip due to the behavior of fasteners. This model represents a refined analytical model, starting from that proposed by Nardone et al. 2011.

The model by Nardone et al. 2011 limits the sustained bearing failure to the first outermost fastener (maximum slip, s_{max}, is equal to slip at the beginning of bearing, s_{b*}) because it focused on the design point of view.

Furthermore, in terms of strengthening assessment, extending the response to the perfectly plastic bearing-slip curve, an increasing number of fasteners (n'_h) can achieve sustained bearing force ($N_{h,max} = N_b$ and $s_{max} > s_{b*}$) and higher loads can be attained (Fig. 1).

A comparison between the analytical predictions and the experimental results has been performed to validate the proposed model on the base of 28 tested specimens (available in scientific literature). The force acting on the fastener-slip relationship proposed by Elsayed et al. (2009) is adopted for the comparison.

In particular, the maximum scatter between experimental results and theoretical predictions in terms of flexural capacity is equal to 42% assuming that only the most solicited fastener reaches the sustained baring stress, and reduces up to 16% assuming that many fasteners could reach the sustained bearing stress. Proposed model provides also strain levels for FRP at beam failure.

The knowledge of the relationship between the force acting on the fastener (N_h) and the slip (s) is fundamental in order to apply the proposed model.

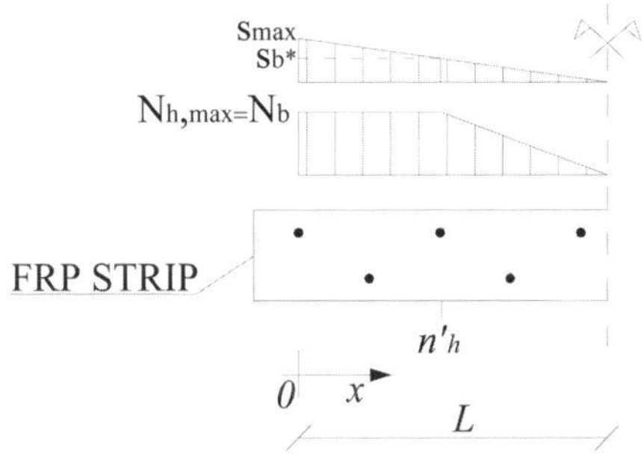

Figure 1. Assumed strain and stress distributions.

REFERENCES

Elsayed, W.E., Ebead, U.A., Neale, K.W. 2009. Studies on mechanically fastened fiber-reinforced polymer strengthening systems. ACI Struct J, 106(1): 49–59.

Nardone, F. Lignola, GP Prota, A Manfredi, G. Nanni, A. 2011. Modeling of flexural behavior of RC beams strengthened with mechanically fastened FRP strips. In Elsevier Composite Structures 93(8): 1973–85.

Bridge Maintenance, Safety, Management, Resilience and Sustainability – Biondini & Frangopol (Eds)
 ISBN 978-0-415-62124-3

Strength and behavior of anchoring devices of CFRP rods for steel girder strengthening

K. Nozaka & A. Tsukiyama
Ritsumeikan University, Shiga, Japan

M. Matsumura
Osaka City University, Osaka, Japan

N. Ochi
Akashi National College of Technology, Hyogo, Japan

T. Ishikawa
Kyoto University, Kyoto, Japan

N. Hisabe
Mitsubishi Plastics Inc., Tokyo, Japan

ABSTRACT

Recently, many of the steel bridges around us are deteriorating due to long term usage, the rapid increase of motor traffic volume and the increased size and speed of vehicles. In order to prolong the life of such structures, repair and strengthening is necessary. The reconstruction of public works structures causes a large economic burden and furthermore has a great influence on the people using the structure. For this reason, making these structures safe to use for the long term is essential from the perspective of developing social capital. Furthermore, it is also necessary to establish construction methods that curb costs, even if only a little.

As identified in research considering the repair and strengthening of steel I-girders, application to the repair and strengthening of bolted connection that have a splice plates is difficult as there are difficulties in bonding the CFRP plate to anything other than flat sections. In the case of steel girder strengthening, the repair and strengthening of the vicinity of the bolted connection is considered necessary if detachment prevention for the CFRP plate and a uniform stress reduction is to be expected, as the CFRP plate is discontinuous at points projecting from the underside of the lower flange, such as the splice plates.

A method that establishes a CFRP rod such that it straddles the splice plates has been proposed as a strengthening method for the vicinity of the bolted connection. In this case, it is necessary to attach the CFRP rod to the steel girders (lower flange) on both sides of the splice plates and transfer the load that was on the CFRP plate to the CFRP rod. A waveform CFRP plate (fixture) is used to fix the rod, the inside of which is filled with adhesive to fix the CFRP rod. Furthermore, the fixture itself is also fixed to the girder by adhesive. Accordingly issues such as the pulling-out or fracturing of the rod, or the detachment of the fixture are conceivable.

In this study, pull-out tests are performed for both high-strength and high-elasticity types of CFRP rod using the fixtures employed in strengthening construction methods that use CFRP rods. The objective is to experimentally verified the pull-out/fracture strength of the rods, the detachment strength and detachment behavior of the fixture, and assemble source material for the future establishment of design and construction guidelines.

In this research, fixtures attached to a steel plate of thickness 9 mm were made, and a load test was performed by installing these specimens in the experimental apparatus in order to verify the effect of differences in shape of fixture and thickness of adhesive layer on strength etc. When manufacturing the specimen, the height from the joining surface that fixes the rod and the thickness of adhesive layer around the rod were varied. A total of nine types of specimen were manufactured, including specimens in which two rods were fixed in parallel in one fixture.

The primary conclusions acquired from this research are as follows.

1) The failure conditions such as pulling out or fracture etc differ according to the elastic modulus of the high-strength type and high-elasticity type rods.
2) The high-strength type was prone to failure by detachment of the adhesive surrounding the rod.
3) For the high-elasticity type, local shear stress of the adhesive acted on the rod is large and is probably the cause of failure for the rod.
4) A thicker adhesive layer can mitigate shear stress and can suppress the detachment of the fixture.
5) As placement location (height) of the CFRP rod becomes high, pulling-out and fracture strength lowers.

Bridge Maintenance, Safety, Management, Resilience and Sustainability – Biondini & Frangopol (Eds)
© 2012 Taylor & Francis Group, London, ISBN 978-0-415-62124-3

Strengthening of bridges with pretensioned FRP laminates: Experimental investigation and a case study

C. Pellegrino
Dipartimento di Costruzioni e Trasporti, Università di Padova, Padova, Italy

G. Giacomin
G&P Intech s.r.l., Altavilla Vicentina, Italy

ABSTRACT

Externally bonded Fiber Reinforced Polymer (FRP) sheets are currently used to repair and strengthen existing Reinforced Concrete (RC) and Prestressed Reinforced Concrete (PRC) structures. A number of experimental programs and analytical studies have been developed in the last few years at the University of Padova on flexural, shear, confinement and bond behaviour of FRP strengthened elements.

The method of strengthening and/or reinforcing RC structures with FRP sheets/laminates as externally bonded reinforcement can be extended when FRP strengthening is prestressed before they are applied to the concrete surface. Following this technique, a unidirectional FRP sheet/laminate is first pre-tensioned and applied to the tension face of the beam, the two far ends of the composite are anchored once the adhesive has fully hardened, and the FRP laminate is then transformed into prestressing element (Triantafillou et al. 1992). Few experimental tests have been developed on specimens strengthened with pre-tensioned FRP laminates and almost all these experiments have been on reduced-scale elements. Furthermore, very few papers are available on the efficiency of various kinds of end-anchorage devices, to develop the strength of the composite laminates and preclude failure by debonding of the composite systems from their concrete support.

In this paper, firstly the results of experimental testing of real-scale RC and PRC beams (Fig. 1) strengthened in flexure with ordinary and pre-tensioned FRP laminates are given (Pellegrino and Modena 2009) and then a practical application of the same FRP pretensioning technique for strengthening of the girders of the Battiferro – Navile viaduct (A14 Highway Bologna-Taranto, Italy) is shown (Fig. 2).

Figure 1. Typical beam before execution of test.

Figure 2. General view of the bridge after the strengthening intervention.

REFERENCES

Pellegrino C., Modena C. 2009. Flexural strengthening of real-scale RC and PRC beams with end-anchored pre-tensioned FRP laminates. *ACI Structural Journal*, 106(3): 319–328.

Triantafillou T. C., Deskovic N., Deuring M. 1992. "Strengthening of concrete structures with prestressed fiber reinforced plastic sheets". *ACI Structural Journal*, 89(3), pp. 235–244.

Bridge Maintenance, Safety, Management, Resilience and Sustainability – Biondini & Frangopol (Eds)
© 2012 Taylor & Francis Group, London, ISBN 978-0-415-62124-3

Open issues in design procedures for FRP strengthening of reinforced concrete bridges

C. Pellegrino
Dipartimento di Costruzioni e Trasporti, University of Padova, Padova, Italy

ABSTRACT

Reinforced concrete bridges are the most common typology in the road transportation network but a number of these structures need rehabilitation or strengthening because of improper design or construction, change of the design loads, damage caused by environmental factors or seismic events.

In this context, externally bonded FRP sheets/plates have found increasingly wide applications in civil engineering due to their high strength-to-weight ratio and high corrosion resistance.

The structural behaviour of FRP-strengthened RC elements and structures has been widely studied over the last decades and some studies have resulted in the first design guidelines for strengthened concrete. American ACI 440.2R-08 (2008), European fib T.G. 9.3 (2001), Italian Recommendations CNR DT 200 (2004), German, British, Japanese and Canadian documents, are examples of such guidelines (currently under revision due to the continuously increasing knowledge on the topic) but their predictions are sometimes contrasting and disagreeing with experimental results related to particular applications. In this context, there are a number of critical issues that are not sufficiently clear. Despite the number of experimental, analytical and numerical works available in literature aimed at investigating the mechanical performances of RC strengthened elements and structures, design rules for the intervention and procedures aimed at checking the efficiency of the strengthening technique are not well defined yet. In particular the efficiency of the intervention in relation to the existing mechanical and geometric characteristics of the RC structural element, particularly the existing steel reinforcement, does not seem deeply studied.

Particular problems/critical issues related to bond between concrete and FRP, shear strengthening of flexural elements, confinement of columns are topics currently under investigation by the scientific community and code predictions related to these problems can lead to unconservative results and have to be improved. Furthermore there are other related problems, for which, at the moment, the knowledge is not sufficiently advanced to propose reliable recommendations e.g. FRP prestressing (particularly interesting for strengthening bridge girders), durability/environmental effects, long term behaviour, cyclic/seismic behaviour, fatigue.

In relation to the above topics, the comparison of the predictions of the current available codes, recommendations and guidelines with selected experimental results will contribute to show or confirm possible critical issues (discrepancies, lacunae, relevant parameters, test procedures, etc.) related to current code predictions or to evaluate their reliability, in order to develop more uniform methods and basic rules for design and control of FRP strengthened RC structures.

In this framework the RILEM Technical Committee "234-DUC – Design procedures for the Use of Composites in strengthening of reinforced concrete structures" has the aim to clarify these general problems/critical issues on the basis of the actual experiences, detect discrepancies in existing codes, lacunae in knowledge and, concerning these identified subjects, provide proposals for improvements.

REFERENCES

ACI Committee 440 2002. Guide for the design and construction of externally bonded FRP systems for strengthening concrete structures (ACI 440.2R-02), American Concrete Institute, Farmington Hills, Michigan, USA.

CNR – Italian Research Council, Italian Advisory Committee on Technical Recommendations for Construction 2004. Guide for the Design and Construction of Externally Bonded FRP Systems for Strengthening Existing Structures. Materials, RC and PC structures, masonry structures (CNR-DT 200/2004), Rome, Italy.

fib Task Group 9.3 2001. Externally bonded FRP reinforcement for RC structures, fib bulletin 14, Lausanne, Switzerland.

Bridge Maintenance, Safety, Management, Resilience and Sustainability – Biondini & Frangopol (Eds)
 ISBN 978-0-415-62124-3

A new composite section for strengthening orthotropic steel decks

R. Sarkhosh & A. Romeijn
Delft University of Technology, Delft, The Netherlands

C.Q. Klap
Ballast Nedam Engineering B.V., The Netherlands

R. Sterkman
Pakor B.V., The Netherlands

ABSTRACT

This study is done to develop a new composite deck system with the aim of strengthening of orthotropic steel bridge decks. The main goal of this research was to obtain a significant improvement of the fatigue performance of the welded components, which means a large reduction of the stresses in the steel deck plate. As an indication, using the so-called S-N curves as given by Eurocode 3 (2005) in case of a number of load cycles up to 5 million, a stress reduction of 50% results into a 8 times increase of fatigue life (Romeijn, 2010).

In order to increase the fatigue strength of welded components and static strength of the bridge deck, the deck plate is strengthened with a number of Glass Fibre Reinforced Polymer (GFRP) laminates and a mixture of epoxy resin with fine aggregates. The number of laminates and thickness of each layer is optimized after performing several experimental tests.

The stress reduction factor (SRF) is expressed as the ratio of the stress in steel plate before and after strengthening. This factor is calculated based on 50% of average elastic stress. As shown in Table 1, a stress reduction factor of 7.0 is expected after strengthening of the beam with 25 mm epoxy filler and 11 mm GRFP sheets. In other words, This research has indicated that the proposed deck strengthening solution reduces bending stresses of steel deck plate under current loads to 15% of the stress before strengthening. The static bending strength capacity and stiffness behaviour is also investigated to evaluate the performance of the composite section in negative and positive bending moments.

Comparing the results of the fatigue tests with fatigue strength of steel plate (detail category 160) shows a significant increase in fatigue strength of deck after applying composite deck. The results of the test of composite sections show a low rate of decrease of strength after a certain number of cycles comparing to a single steel plate.

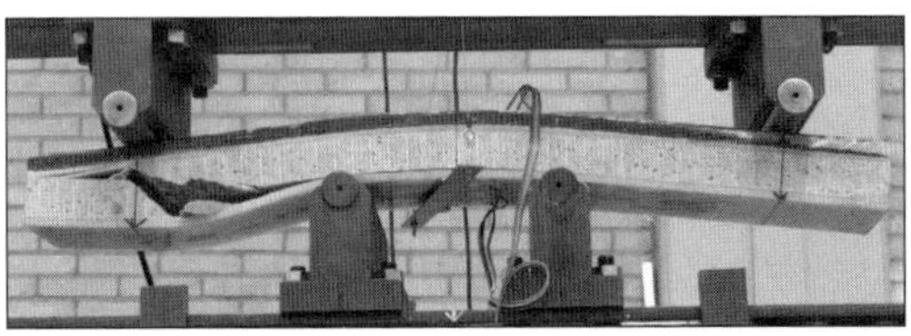

Figure 1. Failure of specimen loaded in positive moment.

Table 1. Stress reduction factor.

	Stress in steel plate		
Specimen	After strengthening	Before strengthening	SRF
EP2	176.3	1258.8	7.1
EP3	180.7	1327.5	7.3
EP4	−58.6	−407.4	6.9
EP5	−59.2	−425.7	7.2
EP6	176.9	549.3	3.1
EP7	−152.6	−540.1	3.5
EP8	184.1	1373.2	7.4
EP9	−63.3	−444.0	7.0
EP13	177.3	1528.9	8.6
EP14	−60.1	−567.6	9.4

REFERENCES

Antonopoulos, C.P. & Triantafillou, T.C. 2003, Experimental Investigation of FRP-Strengthened RC Beam-Column Joints, *J. Compos. Constr.*, 7:1(39): 39–49.

Eurocode 3, Design of Steel Structures, part 1-9 fatigue. EN 1993-1-9; 2005.

Jong, F.B.P. 2006. Renovation techniques for fatigue cracked orthotropic steel bridge decks, *PhD dissertation, ISBN-10: 90-9021412-7*, Delft University of Technology, The Netherlands.

Otte, A. 2009, Proposal for modified Fatigue Load Model based on EN1991-2, MSc-thesis, Delft University of Technology, The Netherlands.

Romeijn, A. 2010a. *Bridge deck strengthening; laboratory research, Technical report 25.5-10-03*, Delft University of Technology, The Netherlands.

Romeijn, A. 2010b. *Bridge deck strengthening; prototype testing, orthotropic steel deck section, Technical report 25.5-10-01*, Delft University of Technology, The Netherlands.

Bridge Maintenance, Safety, Management, Resilience and Sustainability – Biondini & Frangopol (Eds)
© 2012 Taylor & Francis Group, London, ISBN 978-0-415-62124-3

Prediction of the interfacial shear stress with critical stress state criterion for externally bonded FRP-to-concrete substrate

H. Toutanji, S. Ueno & R. Vuddandam
Department of Civil and Environmental Engineering, University of Alabama in Huntsville, Huntsville AL, USA

ABSTRACT

For the externally bonded FRP to concrete structures, the bond strength is evaluated as the maximum transferable load, P_{max}, from single pull-off tests. P_{max} is theoretically predicted using the expression in terms of the FRP properties and the interfacial fracture energy, G_f, for the bond joint. The critical part of the bond is the interface, and the interfacial shear stress along the bonded area is expressed as a function of $\tau(x)$. For the bond behavior, the relationship between the interfacial shear stress and the local slip is important. Non-linear models for the slip-shear relationship capture the non-linear stress descending behavior after reaching the maximum interfacial shear stress, τ_{max}, and the corresponding local slip, s_0. Using the Popovics' expression as the slip-shear model, G_f can be determined as the function of τ_{max} and s_0. For determining τ_{max} and s_0 values, the empirical expressions are previously developed from the regression analysis.

In this work, for predicting τ_{max}, a failure criterion is proposed where the stress state on elements in concrete substrate is limited to the tensile concrete strength. With the assumption that the debonding failure occurs in concrete substrate at t mm below the concrete-to-adhesive interface, the stress state on the concrete substrate element is derived as a function of $\tau(x)$ by applying the surface load transfer method from the elasticity theory. By developing an analytical solution based on the mechanics of materials, the adhesive properties are theoretically adopted to derive the expressions for $\tau(x)$ and s_0. Using the failure criteria and $\tau(x)$ for the state of τ_{max}, the value for τ_{max} is determined, so as s_0. Knowing τ_{max} and s_0, G_f is obtained for predicting P_{max}. The comparison of the predicted P_{max} shows the good results verifying the proposed criterion (Table 1). Due to the effect of the adhesive thickness, the predicted values of G_f have some variations compared with the values of the existing model (Table 2).

Table 1. Predicted values with $t = 1.0$ mm and experimental results for different specimens, and the comparison.

Specimen ID	Predicted load P_{pred} kN	Experiment load P_{exp} (Dai et al. 2005) kN	Ratio P_{exp}/P_{pred}
CR1L1-a	23.16	23.4	1.010
CR1L1-b	22.86	23.1	1.010
CR1L1-b	22.86	24.9	1.089
CR1L2-a	38.65	33.5	0.867
CR1L2-b	38.37	39.3	1.024
CR1L3-a	53.71	42.9	0.799
CR1L3-b	53.48	38.4	0.718
CR1L3-b	53.48	36.9	0.690

*C = carbon fiber; R1 = epoxy type; L1–L3 = number of FRP plies; –a & –b = different adhesive thickness.

Table 2. Comparison of predicted G_f using $t = 1.0$ mm with the existing model for different specimens.

Specimen ID	Predicted G_f Proposed N/mm	Karbhari and Niu (2006) N/mm
CR1L1-a	1.060	1.265
CR1L1-b	1.033	1.265
CR1L2-a	1.476	1.265
CR1L2-b	1.455	1.265
CR1L3-a	1.900	1.265
CR1L3-b	1.884	1.265

REFERENCES

Dai, J., Ueda, T. & Sato, Y. 2005. Development of the nonlinear bond stress-slip model of fiber reinforced plastics sheet-concrete interfaces with a simple method. *Journal of Composites for Construction* 9 (1): 52–56.

Karbhari, M. V. & Niu, H. 2006. Review and comparison of fracture mechanics-based bonded strength models for FRP-strengthened structures. *Journal of Reinforced Plastics and Composites* 25 (17): 1757–1794.

Steel bridge rehabilitation

Organizer: M. Sakano

Bridge Maintenance, Safety, Management, Resilience and Sustainability – Biondini & Frangopol (Eds)
 ISBN 978-0-415-62124-3

Bearing replacement and strengthening of Forth Road Bridge approach viaducts, UK

B. Colford
Forth Road Bridge, Edinburgh, UK

M. Chiarello, C. Hendy & J. Sandberg
Atkins, Epsom, UK

ABSTRACT

The Forth Road Bridge (Figure 1) spans the Firth of Forth and was completed in 1964. The main structure is a three span suspension bridge. At each end of the bridge, there are two multi-span approach viaducts comprising a pair of longitudinal steel box girders with cross girders supporting a concrete deck slab. The approach viaducts carry two carriageways, each with two lanes, and extend from the abutments to the side towers, which are shared with the main suspension bridge. The total width of the structure is 36 metres.

The box girders rest on steel roller or rocker bearings on reinforced concrete portal piers, varying between 11 m and 40 m tall, founded on rock. Locations with roller bearings allow for horizontal movement through movement of the roller while the locations with pinned bearings allow for movement through flexing of the piers.

An initial study of the bearings identified that the rollers had locked up due to corrosion and distortion, and the concrete beneath the bearings and elsewhere on the pier tops had deteriorated due to chloride contamination. Assessment showed that structural deficiencies in the pier were exacerbated by both the concrete deterioration and change in articulation. These factors lead to the decision to replace all the bearings on the viaducts. The original structure was not designed to facilitate replacement of the bearings so the structure had to be strengthened and modified by the addition of jacking stiffeners and corbels to the pier tops.

This paper outlines the design of the strengthening and modifications to the bridge to facilitate bearing replacement, together with a detailed description of the design of the temporary works needed to maintain the bridge's articulation during jacking. Lessons learnt during construction are discussed and the structural benefits of design to Eurocodes are emphasised.

Figure 1. Forth Road bridge, viewed from south.

Figure 2. Typical roller bearing.

REFERENCES

Hendy CR, Sandberg J, Shetty NK, *Recommendations for Assessment Eurocodes for Bridges*, ICE Bridge Engineering March 2011, Thomas Telford, UK

Hendy CR, Smith DA, Strengthening of Irwell Valley Bridge, UK, ICE Bridge Engineering March 2008, Thomas Telford, UK

Bridge Maintenance, Safety, Management, Resilience and Sustainability – Biondini & Frangopol (Eds)
© 2012 Taylor & Francis Group, London, ISBN 978-0-415-62124-3

Study on performance evaluation and maintenance management system of weathering steel bridge

J. He, Y.Q. Liu & A. Chen
Department of Bridge Engineering, Tongji University, Shanghai, China

T. Yoda
Department of Civil and Environmental Engineering, Waseda University, Tokyo, Japan

ABSTRACT

It is well known that the most common types of damage that occur to structural steels during their lifetime are fatigue and rusting. Corrosion of steel not only degrades the performance of steel bridges, but also increases the work of management and maintenance; at the meantime periodic re-painting consume a lot of cost and labor (Liu 2005). Thus, weathering steels are proposed and widely employed in several aspects of construction such as bridges because of their perfect anti-corrosion performance, i.e., a protective layer can develop on the surface of weathering steel exposed to a suitable environment. For example, weathering steel developed and initiated application in bridge engineering in USA at the beginning of last century; large span deck arch bridge- New River Gorge Bridge and cable-stayed bridge- Mississippi River Bridge using weathering steel were constructed in 1977 and 1983 respectively. FHWA published design guideline of Uncoated Weathering Steel in Structures at 1989 (AISC), then weathering steel increasingly used in steel bridges. Weathering steel bridges firstly built in German and England in 1969 and 1970 respectively. And there is about 90% of the new steel bridge constructed using weathering steel in Canada. In 1991, Korea started producing and selling weathering steel, and applying in bridges at 1992.

At 1955, Japan initiated developing and researching weathering steel, and firstly applied in bridge engineering at 1967. The guideline of weathering steel application was initially established in 1993 based on the research done by the Ministry of Construction, Kozai club and Japan steel bridge construction association (JRA 2005). It describes the applicable limit of weathering steel based on the exposure test all around Japan, which is maximum amount of airborne salt is no more than 0.05 mdd (mg/dm^2/day). However, most of the transportation infrastructures are close enough to the ocean, and accordingly, the use of weathering steel is limited. To overcome this difficulty, new weathering steel which contains 3% of Nickel, has higher performance in corrosion protection, especially in the severe environment near the coastal area, was developed. About 70% of steel and concrete composite girder bridge constructed use weathering steel in Japan (Miki 2004). At 1989, Baoji bridge factory fabricated 3 spans weathering steel box girder in China, and put into use at 1991.

Weathering steel is widely applied in bridge structures in Europe, America, Japan and other developed countries due to its cost-effectiveness during whole life cycle and reduction of environmental pollution and energy consumption during construction and maintenance process (He 2006).

This paper describes the corrosion mechanism of weathering steel through previously atmospheric corrosion tests. The time-dependent performance is evaluated by appearance assessment, image processing analysis and thickness measurement of rust layer. According to the evaluation grade, the checking frequency and contents, maintenance proceeding were determined. And maintenance management system was built to provide the reference to the operation and development of weathering steel bridges.

REFERENCES

AISC (American Institute of Steel Construction) 1989. *Cost of Painted A 572 Grade 50 Steel Bridge Versus Uncoated A588 Bridge*. Chicago.

He J., Liu Y.Q., and Chen A.R. 2006. Application of weathering steel in bridge engineering, *Journal of Beijing Jiaotong University*, Vol. 30: 310–315 [In Chinese].

JRA (Japan Road Association). 2005. *Specification for Highway Bridges: III Weathering steels* [In Japanese].

Liu Y.Q. 2005. *Steel-Concrete Hybrid Bridge*. Beijing: China Communication Press [In Chinese].

Miki C., and Ichikawa A. 2004. *Modern bridge engineering – Front technology of un-coating steel bridge*. Saiensu-sha Co., Ltd. Publishers [In Japanese].

Bridge Maintenance, Safety, Management, Resilience and Sustainability – Biondini & Frangopol (Eds)
© 2012 Taylor & Francis Group, London, ISBN 978-0-415-62124-3

Development of fatigue test method and improvement of fatigue life by new functional steel plates for welding of trough rib and deck plate of orthotropic decks

N. Konda, K. Arimochi & M. Nishio
Sumitomo Metal Industries, Ltd. Tokyo, Japan

M. Ichimiya, T. Kasugai & S. Kiyokawai
Yokogawa Bridge Holdings Corp, Chiba, Japan

ABSTRACT

Under heavy traffic conditions, some fatigue cracks have been found at welding joints between U-shaped ribs "trough ribs" and deck plates. Fatigue cracks along thickness direction of deck plate from root tip are important because visual crack detection from outside is limited. In this paper, new functional steel plates with initiation resistance of fatigue crack at weldment and propagation resistance at base material are applied to deck plate to evaluate improvement of fatigue life. We conducted fixed-point fatigue tests with real scale partial models consisted with one trough rib and a transverse girder, and the elastic FEM analysis corresponding to the fatigue test specimen. Based on these results, we studied the fatigue properties which focused on the difference between conventional steel plates and newly developed steel plates, FCA. Fatigue cracks were initiated from root tip of welded joints between a trough rib and a deck plate. It is clarified that FCA steel plates instead of conventional steel plates can improve the fatigue life under the same welding conditions. This improvement effect becomes much clearer in the high cycle region, which is noticeable at the operation and management stage of bridges.

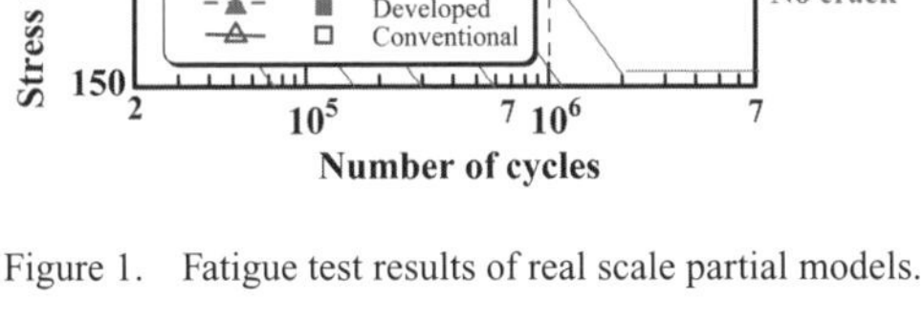

Figure 1. Fatigue test results of real scale partial models.

REFERENCES

Arimochi K. et al. 2006. Study for Improvement of Fatigue Strength at Welded Joints between Deck Plate and U-rib with Fatigue Crack Arrester Steel, Proc. of Symp. on Welding Structures, pp. 233–236. (in Japanese)

Hara J. et al. 2010. Practical Application of High Fatigue-Strength Steel to Weldments (FCA-W) in an LNG Carrier, The 24th Asian-Pacific Technical Exchange and Advisory Meeting on Marine Structures, pp. 263–266.

Ichimiya M. et al. 2008. Improvement of Fatigue Strength by Application of Fatigue Crack Arrester FCA, 63rd Annual Conference of Japan Society of Civil Engineering, pp. 415–416. (in Japanese)

Kurihara & Kawabata 2006. 61st Annual Conference of Japan Society of Civil Engineering, pp. 1087–890. (in Japanese)

Miki C. et al. 2007. New Type of Steel Orthotropic System, Proceedings of the 10th Symposium on Steel Structure and Bridges, pp. 71–91. (in Japanese)

Terao K. & Suzuki K. 1996. A Parametric Study of Details at Weld between Transverse and Longitudinal Ribs in Orthotropic Steel Deck Bridges, Technical Report of Yokagawa Bridge Corp, Vol. 25, pp. 87–95. (in Japanese)

Une T. et al. 2009. Application of Fatigue Crack Arrester FCA to Steel Concrete Combined Bridge, 64th Annual Conference of Japan Society of Civil Engineering, pp. 889–890. (in Japanese)

Yamada K. 2007 Fatigue Damage in Orthotropic Steel Deck under Heavy Wheel Loads around Nagoya, Proceedings of the 10th Symposium on Steel Structure and Bridges, pp. 11–18. (in Japanese)

Bridge Maintenance, Safety, Management, Resilience and Sustainability – Biondini & Frangopol (Eds)
© 2012 Taylor & Francis Group, London, ISBN 978-0-415-62124-3

The analysis on the characteristic of fatigue crack in railway plate girder bridge and its retrofit method

S.J. Lee & K.S. Kyung
Korea Maritime University, Busan, Korea

H.H. Lee & J.C. Jeon
Construction Technology Consultant Co., Ltd., Gunpo, Kyeonggi-do, Korea

1 INTRODUCTION

In case of steel railway plate girder bridges, eccentricity acts in girders because the width between girders is larger than the one of rails. Bending deformation occurs in the upper flange of the girder under the influence of this eccentricity and it is reported that fatigue crack occurs in the welded joints of the upper flange and vertical stiffeners by repeated load. This study carried out structural analysis about service load for steel railway plate girder bridges which are serviced. The structural modeling was verified based on visual inspection and field measurement about the subject bridges and the structural analysis on repair and reinforcement plan of fatigue crack in the welded joints of the upper flange and vertical stiffeners under train load was carried out using the verified structural modeling.

2 THE DIMENSIONS OF THE TARGET BRIDGE

The target bridge of this study is steel railway plate girder bridges of the total span of 32 (the total length: 592.6 m) which was completed in 1945. The span length is divided into the welding section (13.45 m (S1 ~ S7)) and the rivet section (19.74 m (S8 ~ S32)). As for design live load, LS-22 which is the Korean railway design standard was applied and SWS41 is used for steel grade.

3 FIELD TEST AND ITS ANALYSIS

The loading test was carried out for the purpose of obtaining the data to evaluate safety and loading capacity of the structure by grasping the movement of the practical structure about the live load of the target bridge, investigating and revealing the response characteristics and comparing it with the theoretical response according to the structural analysis.

4 STRUCTURAL ANALYSIS

The structural analytical model which integrated the one using 4-node plate element and 8-node solid elements was applied for the target bridge. As for the analytical program, MIDAS Civil which is the general structural analysis program was used. The structural analysis of element size around the flange, vertical stiffener and the welded joints was done by dividing it into 3 mm considering the report with the reliability of the analytical result even though welding throat thickness is not considered if the plate model will be divided into a half element size of each welding zone.

Table 1. Results of fatigue stress of welded joint (Impact factor: design impact factor 50% applies).

Item	Stress range (MPa)			Reduction (%)	allowable stress range
	Before	After	Including impact factor		
Case_1	99.3	13.2	16.3	83.6	64.0
Case_2		15.8	19.5	80.4	
Case_3	192.2	24.8	30.6	84.1	

5 THE RETROFITTING METHOD OF THE WELDED JOINT AND ITS EVALUATION

See Table 1.

REFERENCES

Korea Railroad Corporation (2009), Miryang River Bridge Research, in Korean

Korea Road/Transportation Association (2004), Korean Bridge Design Standard Specification, pp. 7, pp. 160–171, in Korea

S.J. Lee, K.S. Kyung, J.E. Park, J.H. Lee, "Evaluation and Retrofit of Fatigue Crack of Welded Structural Detail for Steel Plate Girder Railway Bridge", The Fall Conference for Korean Society for Railway, 2009, in Korean

Bridge Maintenance, Safety, Management, Resilience and Sustainability – Biondini & Frangopol (Eds)
© 2012 Taylor & Francis Group, London, ISBN 978-0-415-62124-3

Fatigue crack repair using drilled holes and externally bonded CFRP strips

F. Lin, J.G. Sun, H. Nakamura & K. Maeda
Tokyo Metropolitan University, Tokyo, Japan

ABSTRACT

Fatigue cracks in steel bridges are often repaired by constructing drilled holes at the crack tips, because these drilled holes can stop or delay crack propagation. Repair methods using drilled holes have been actively investigated, with techniques such as the tightening of high-strength bolts in the drilled holes used to enhance the repair effect. However, it may be difficult to tighten high-strength bolts in narrow locations.

On the other hand, the results of some studies have demonstrated that Carbon-Fiber-Reinforced Plastic (CFRP) strips, which are lightweight, have high strength, display excellent corrosion resistance, and are easy to use, are useful for the repair of fatigue cracks. Therefore, an improvement in the effect of fatigue-crack repair by combining drilled holes with externally bonded CFRP strips is expected.

In this study, a repair method combining drilled holes with externally bonded CFRP strips is proposed as shown in Figure 1, and tensile tests and fatigue tests are carried out to verify the effects of the fatigue-crack repair using specimens with welded web gusset plates.

As an example of the tensile test results, Figure 2 shows the relationships between the tensile nominal stress and the strain in the drilled holes after several tensile tests. Compared to the repair method using only drilled holes (DH series), the strain concentration factor was greatly reduced by externally bonded CFRP strips (DH5C, DH7C series).

Figure 3 shows *S-N* curves by fatigue test. The fatigue limits were reached with stress ranges of 100 MPa for the DH5C series. Although the durability of the adhesive joint became dominant in higher stress ranges, a significant effect on the fatigue-crack repair was confirmed when combining drilled holes with the application of externally bonded CFRP strips.

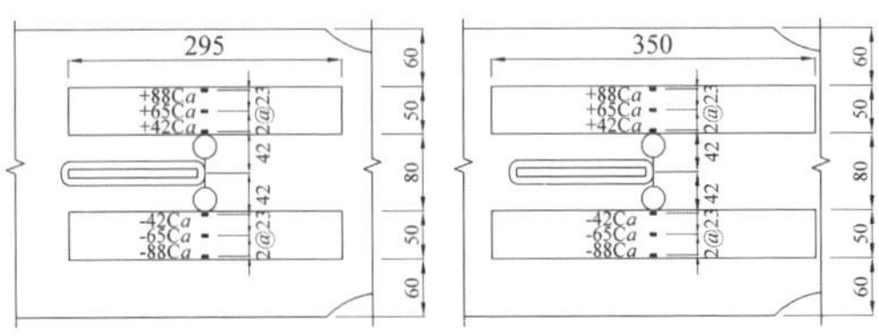

(a) DH5C series (b) DH7C series

Figure 1. Proposed repair method.

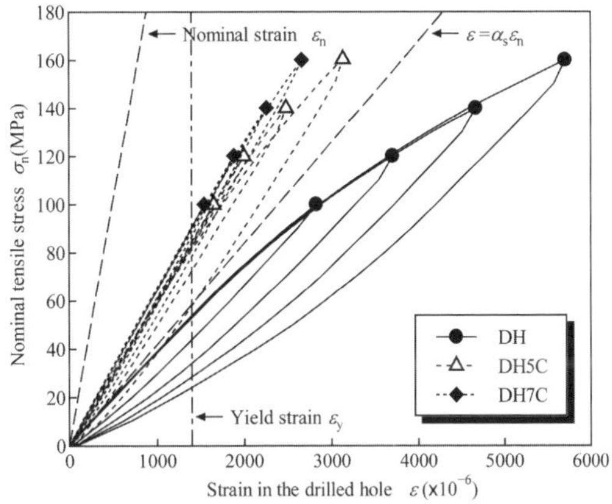

Figure 2. Comparsion of strain in the chamfer of the hole edge.

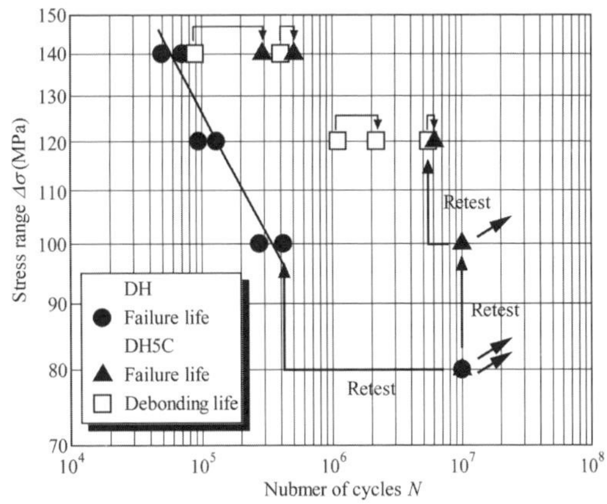

Figure 3. *S-N* curves.

Bridge Maintenance, Safety, Management, Resilience and Sustainability – Biondini & Frangopol (Eds)
© 2012 Taylor & Francis Group, London, ISBN 978-0-415-62124-3

Performance and durability verification tests on rationalized joint of precast steel-concrete composite deck for replacement of deteriorated highway bridge slab

H. Mizuno & H. Kaido
Kawada Industry Inc, Osaka, Japan

S. Matsui
Osaka Institute of Technology, Osaka, Japan

T. Sugiyama
Yamanashi University, Yamanashi, Japan

ABSTRACT

In Japan, as so many highway bridges constructed in the last four decades became superannuated, maintenance of those bridges has become the most serious problem to extend the service lives with appropriate rehabilitation works. The bridges that are over 40 years old and need to be restored are accounted for around 50% of the total and it has been reported that 50–70% of damages on steel bridges are caused by reinforced concrete floor slab deterioration. On the basis of the evaluation above, damaged reinforced concrete floor slabs are replaced with high durability deck slabs so as to prolong the life of the bridges). When deck slabs are replaced on a crowded road such as main national roads or express highways, traffic jam in accordance with traffic suspension leads to social loss and therefore it is desired to decrease the traffic troubles as the minimum.

Considering a general field requirement that (1) site operation period is desirable to be shorter than in the case of cast-in-place concrete deck slab and traffic control can be shortened, and (2) high durable decks are profitable from the viewpoint of maintenance, the authors aimed on the precast steel-concrete composite deck slab (hereinafter called precast composite deck slab) shown in Figure 1 for deck slab replacement. Therefore, rationalized joint structure which shortens the operation period and saves labor for the joint structure of the jointing part is devised and developed.

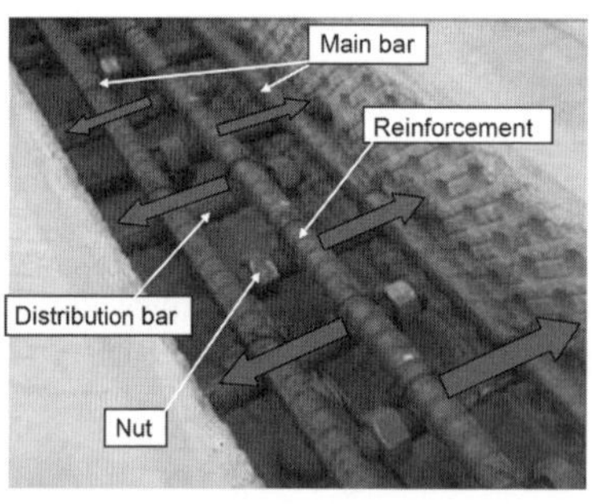

Figure 1. Rationalized joint.

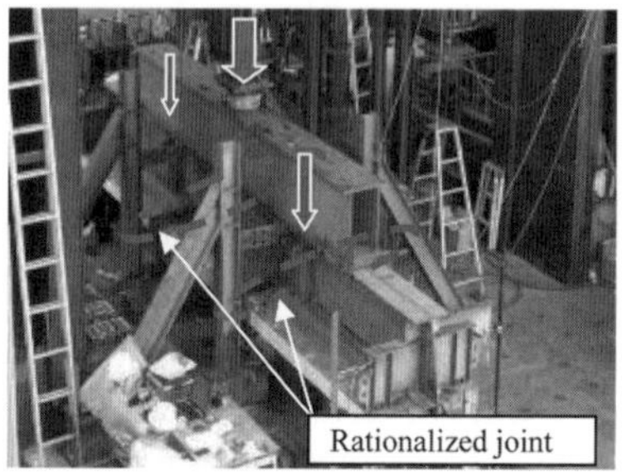

Figure 2. Negative bending test.

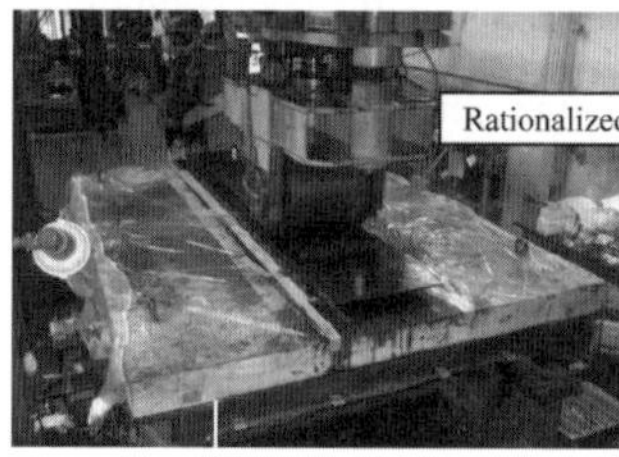

Figure 3. Wheel load running test.

Lap joint has been often applied at jointing of reinforcements between the precast composite deck slabs and it widens the jointing width for concrete to around 650 mm, which increases the volume of concrete used on the site. For this rationalized joint, as shown in Figure 1, distribution bars (hereinafter called distribar) are arranged alternately and their tips are formed as screw top and a nut is arranged for each tip. Stress transfer is facilitated by axial force resistance obtained with nuts, concrete of the jointing part, reinforcement and therefore it largely shortened the jointing length to around 200 mm. Here, the authors investigated the stress transfer properties by a composite girder specimen under negative bending moment as under the intermediate support and fatigue durability for traffic load by wheel load running test using a real-size specimen. In this paper, the results of those loading tests are reported.

Bridge Maintenance, Safety, Management, Resilience and Sustainability – Biondini & Frangopol (Eds)
© 2012 Taylor & Francis Group, London, ISBN 978-0-415-62124-3

Steel plate pre-stressing reinforcement for coped steel girder ends

Y. Nagao
Kansai University, Osaka, Japan

K. Matsumoto
Railtec Co., Ltd., Osaka, Japan

H. Namiki
Kyobashi Mentec Ltd., Osaka, Japan

M. Sakano
Kansai University, Osaka, Japan

ABSTRACT

Recently, it has been extensively reported that fatigue cracks have been detected at coped steel girder ends; such cracks are reinforced through the application of a ribbed steel plate.

However, the conventional method of reinforcement is not always effective because the size of the reinforcing plate is not always sufficient, due to the limited space at the coped steel girder end. And it has also case that sufficient friction grip connection cannot be produced by the reinforcing plate rib when the reinforcing plate is fixed on to the web. So, it has been reported that fatigue cracks were again propagated after reinforcement.

In the previous study, we proposed the steel plate pre-stress reinforcement method for coped girder end as an alternative for conventional method. Steel plate pre-stress reinforcement method fixes angle iron on the web of coped girder end and bottom flange by high-strength bolts, and introduce compressive pre-stress into the web. This reinforcement effect can be expected to the prevention of fatigue crack. And, it is shown to be more effective than conventional reinforcement. However, this reinforcement method cannot show enough reinforcement effects depending on the development direction of the fatigue crack.

In this study, the reinforcing effect of a coped steel girder end reinforced by the steel plate pre-stressing method is investigated through static loading tests.

The principal results obtained through this study are as follow:

1. By steel plate reinforcement of the 45 degrees direction, the maximum principal stress near the coped corner is reduced about 25%. This reinforce effect is same level at previous study. In addition, Principal stress directions are same as no reinforcement.
2. By steel plate pre-stressing reinforcement, the maximum principal stress has change to compressive stress by the introduced pre-stress. It is expected this reinforcing effect that fatigue crack is prevented from propagating to the web.

REFERENCES

M. Sakano, H. Namiki, K. Okabe, T. Sakata, and Y. Osakada (2004): Span Extension of H-Beam Bridges by Steel Plate Pre-stressing, Journal of Structural Engineering. (in Japan), Vol. 50A, pp. 855–863.

K. Matsumoto, H. Takahashi, M. Sakano, and H. Namiki (2006): The Proposal of more effective reinforcement method for coped steel girder ends, JSCE (in Japan).

K. Matsumoto, M. Sakano, and H. Namiki (2007): Steel plate Pre-stressing Reinforcement For Notched Steel Girder ends, Structural Failures, Miedzyzdroje, Poland 1.

M. Sakano, H. Takahashi and H. Namiki (2007): High performance H-beam bridges reinforced by steel plate pre-stressing method , Journal of Structural Engineering. A (in Japan), Vol. 53A, pp. 706–717.

Bridge Maintenance, Safety, Management, Resilience and Sustainability – Biondini & Frangopol (Eds)
© 2012 Taylor & Francis Group, London, ISBN 978-0-415-62124-3

Experimental study on high strength one-side bolted joints

K. Nakajima
Lobtex Fastening System. co., Ltd., Chuo Ward, Tokyo, Japan

H. Suzuki
Science and Engineering, Meisei University, Hodokubo, Hino City, Tokyo, Japan

Y. Kawabe & K. Fujii
Lobtex Fastening System. co., Ltd., Chuo Ward, Tokyo, Japan

ABSTRACT

High strength one-side bolts, which enable construction only from one side, are effective for repairing or strengthening closed-section members or for connecting steel pipes as joints. In this study, experiments were conducted to verify the fatigue strength and other basic characteristics of high strength one-side bolted joints. As a result, it was revealed that high strength one-side bolted joints have fatigue design curve to those of joints using conventional high strength bolts. Also presented in this paper are case studies of application of high strength one-side bolted joints to strengthening of steel truss bridge and arch bridges.

Experiments were conducted to verify the slip factor, relaxation of axial force and fatigue strength, which represent the basic characteristics of OS-Bolted joints. Tensile strength after fatigue experimenting was also verified. The findings are described below.

1. Slip experiments were conducted for friction surface that was subjected either to shot blasting or grit blasting. As a result, a slip factor of 0.40 or higher was obtained. Performance the same level to that of standard high-strength bolted joints can be obtained by properly cleaning the friction surface.
2. Relaxation of axial force was approximately 5%. The relaxation of axial force of OS-Bolt is similar to that for standard high strength bolts. No special consideration is required for the design axial force or the axial force to be introduced.
3. The number of loading cycles for a stress range 1.8 times the stress range for high strength bolted joints was five million, more than 2.5 times two million, the fatigue limit. OS-Bolted joints have fatigue design curve to conventional high strength bolted joints.
4. It was verified in tensile tests before and after fatigue experiments that no strength reduction occurred in OS-Bolt after cyclic loading was applied approximately five million times.

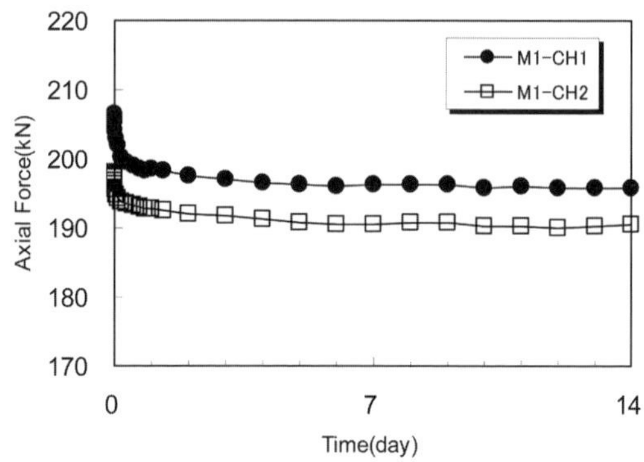

Figure 1. Axial force relaxation.

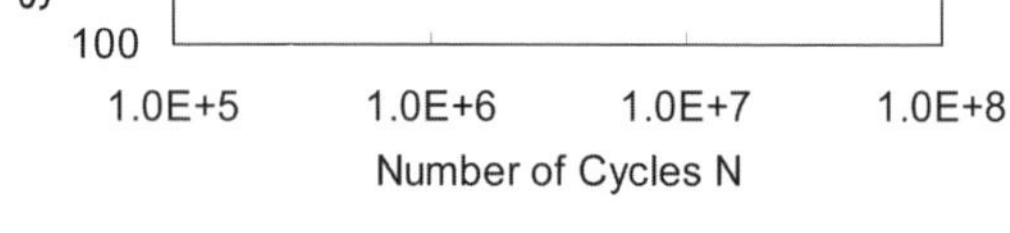

Figure 2. Fatigue experiment results.

REFERENCES

Ohno, T., Natori, T. and Matsumoto, Y.: Friction bonding method using torque control type one-side construction high-strength bolts, Yokogawa Bridge Group technical report, No. 26, pp. 236–240, January 1997 (in Japanese).

Japan Road Association: Fatigue Design Guidance for Steel Highway Bridges, March 2002 (in Japanese).

Japan Society of Civil Engineers: Steel structure series 15 Recommendations on Design, Construction and Maintenance for Friction Type of High strength bolted Connections, December 2006 (in Japanese).

Bridge Maintenance, Safety, Management, Resilience and Sustainability – Biondini & Frangopol (Eds)
© 2012 Taylor & Francis Group, London, ISBN 978-0-415-62124-3

Rehabilitation of steel exprcssway bridge with repeatedly developed fatigue cracks

H. Nakata & Y. Takamura
Hanshin Expressway Engineering, Osaka, Japan

K. Tokumasu, M. Kawamura & Y. Adachi
Hanshin Expressway, Osaka, Japan

M. Sakano
Kansai University, Osaka, Japan

ABSTRACT

Three-span continuous steel plate bridge with steel orthotropic deck of Hanshin expressway was found to have fatigue cracks near the support of the main girder at 1991 after 13 years of service for traffic. The cracks were repaired by patching. However, new fatigue cracks were detected again at the same portion of the girder after 19 years after the repair. The new cracks were also initiated either at the weld bead of the sole plate or at around the bearing set bolt. But some of those cracks penetrated through the bottom flange of the main girder and the end cross beam, and none of them were detected at the resent periodic inspection conducted at four years before. This means the detected cracks quickly have developed within four years, therefore, urgent actions for repair were required to prevent rapidly developments of cracks.

The detailed survey, static load analysis, and stress measurements were carried out for investigation of the cause of those cracks. The detailed survey revealed that the small gap was observed between the bearing and the bottom flange of the girder. The girders were found to move in the vertical direction with every passage of vehicle, generating metallic sound. And also, shoe base mortar was found damaged. It suggested that occurrence of negative reaction. But the static load analysis and stress measurement revealed no occurrence of negative reaction in the support area. On the other hand, the static load analysis revealed that the live load effect is much more than that of the dead load and live load ratio is 1:5 so that the bridge is sensitive to the live load.

Based on these considerations, the damage was not an uplift of the girders due to negative reaction. The assumption was that deterioration of thc shoe base mortar with time caused subsidence of the bearings, which resulted in vertical movements of the main girders, giving impacts to the damaged parts.

There were two plans for emergency repair: (1) patching; and (2) partial replacement. Replacement of bearings was also taken into account because of the damage to the shoe base mortar.

Through a comparative examination, the partial replacement measure was adopted by removing the damaged part of the girder and replacing new one connected by high tension bolts. Figure 1 shows the schematic diagrams of the partial replacement measure.

This measure could eliminate future concerns about developments of cracks by the complete removal of cracks. In addition, this measure provides improved fatigue durability to the structure. And bearing were replaced at the same time with the emergency repair work. This allowed to obtain a sufficient works space in the limited under girder, improving ease of operation.

The emergency repair has been completed and stress measurement was performed in the vicinity of the emergency repair site to evaluate the repair effect.

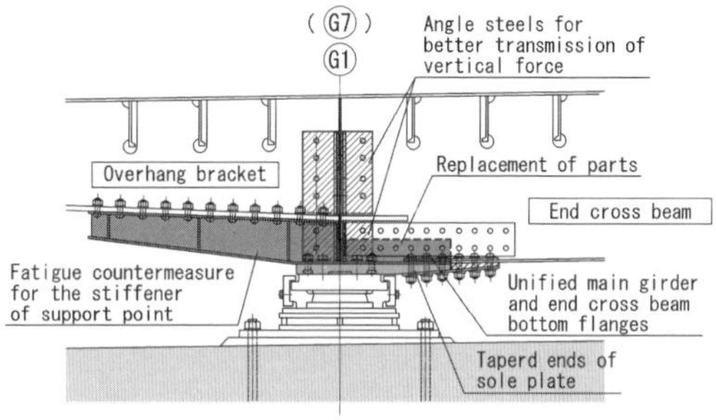

Figure 1. Schematic diagram of the repair (front view).

REFERENCES

Japan Road Association (JRA), March 2002. Specifications for Highway Bridges and Commentary, Part II Steel Bridges.

Bridge Maintenance, Safety, Management, Resilience and Sustainability – Biondini & Frangopol (Eds)
 ISBN 978-0-415-62124-3

Health monitoring via horizontal displacement at the end of steel bridge girders

H. Namiki & T. Kamizono
Kyobashi Mentec Co., Ltd., Osaka, Japan

Y. Otsuka & S. Takahashi
Keihan Electric Railway Co., Ltd., Osaka, Japan

ABSTRACT

Vertical deflection in response to a live load is often used as an integrated indicator when diagnosing the soundness of girder bridge structures. Herein, we show how vertical deflection of a girder can be calculated by measuring the horizontal displacement generated at the upper and lower ends of the girder. The horizontal displacement generated at a girder end can then be used as a new indicator in place of measuring vertical deflection at the span center.

Additionally, we show that the horizontal force acting on aged bearings can be calculated from measured horizontal displacement at the upper and lower ends of a girder, allowing the sliding function of such bearings to be monitored and evaluated. Onsite measurements were conducted using eddy current displacement sensors. The horizontal displacement observed between the girder ends was in good agreement with calculated values.

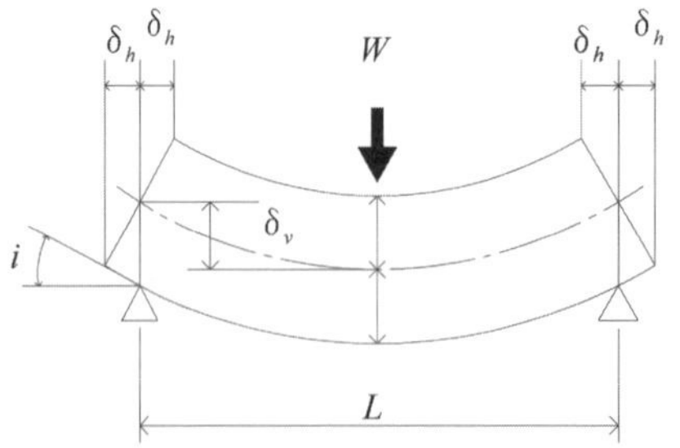

Figure 1. Deflection at girder center and girder end.

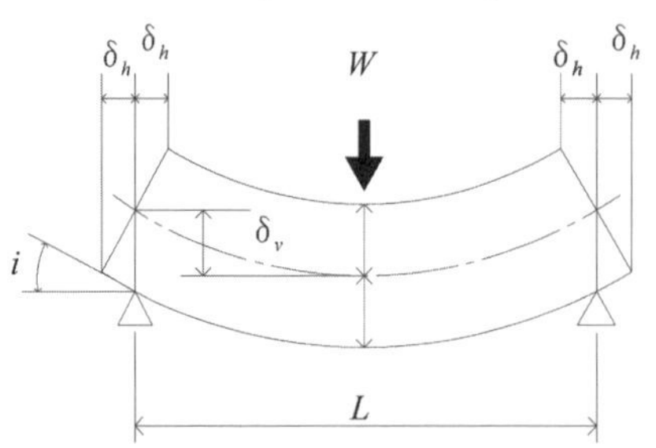

Figure 2. Girder with bearings both sides fixed.

Figure 3. Sensors installed at girder ends.

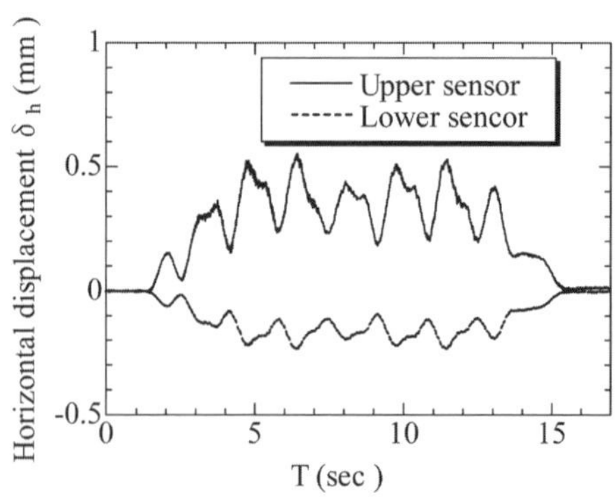

Figure 4. Observed horizontal displacement (after bearing replacement).

Bridge Maintenance, Safety, Management, Resilience and Sustainability – Biondini & Frangopol (Eds)
© 2012 Taylor & Francis Group, London, ISBN 978-0-415-62124-3

Analysis on deck replacement plans of tied arch bridge with composite girder

S. Qing-Tian, Z. Ming-gen, W. Chong & D. Bing
Tongji University, Shanghai, China

ABSTRACT

Cracks in concrete deck are inevitable because the deck carries the vehicle loads directly in steel and concrete composite bridge Local deck damage may occur during operation state by the accidental loads. If the design service life of concrete deck is same as the steel structure in composite girder, the concrete deck must be thick enough, which cause the girder weight, dimension of superstructure and foundation increased, at the same time to enhance the total cost of the project. Therefore, in the view of the economics of the bridge whole life, a plan for replaceable concrete deck is reasonable.

Concrete deck as a component of composite structure takes part in the carrying loads of whole structure. During the deck maintenance and removal, part or all concrete in the section will lose the role of burdening loads. The effects of local deck removal on the mechanical behavior of steel structure need carefully be investigated. Moreover, there are many sequences for concrete deck removal, and the different deck removal sequence may affect the mechanical behavior of steel girder in different results. Through relatively researches, selecting a deck removal plan which is easy operation and safety enough from these removal sequences is very important to ensure the bridge safe during decks removal in the whole service life.

In this paper, combined a project of main bridge of Jiubao bridge in Hangzhou, Zhejiang province of China, several deck removal plans are proposed and the structural response with different removal plans are analyzed.

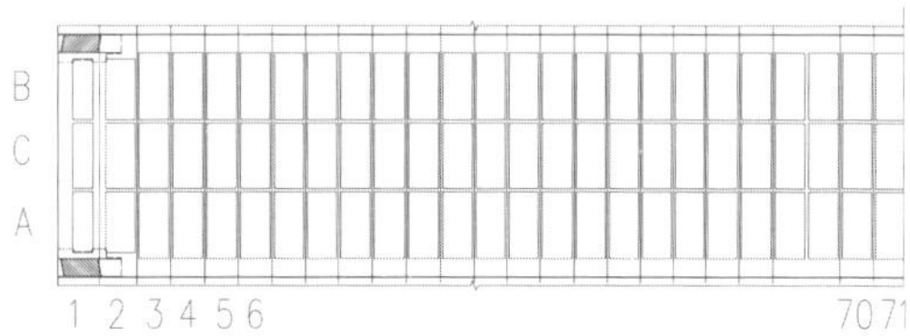

Figure 1. The arrangement of the concrerte bridge decks.

The arrangement of the bridge decks of the half of the bridge is shown in Fig. 1. The numbering of the bridge decks in transverse direction is A, B and C, and the numbering of the bridge decks in longitudinal direction is $1, 2, \ldots, 21$ and 22.

The state of the completion of the bridge, replacements of local decks of side A, replacements of local decks of side C and replacements of local decks of side A, side B and side C are calculated and analyzed. The results show that the replacements local decks have slightly influence on the stresses of the girder at the top and bottom flange and the stress of other concrete decks.

The results of the replacement plans of the whole span show that the schemes of the removal from the mid-span to the side of the span behave better compared to the schemes of the removal from the side of the span to the mid-span. The scheme that the bridge decks are removed from the mid-span to the side of the span symmetrically, firstly the decks of side A and side B and then the decks of side C is the best one.

Bridge Maintenance, Safety, Management, Resilience and Sustainability – Biondini & Frangopol (Eds)
© 2012 Taylor & Francis Group, London, ISBN 978-0-415-62124-3

Investigation of the fracture surface of a cast iron finger joint

K. Sakiya, Y. Takamura & T. Yamagami
Hanshin Expressway Engineering Company Limited, Osaka, Japan

M. Yamanishi
Kobe Material Testing Laboratory, Kako-gun Hyogo, Japan

ABSTRACT

In May 2008 a cast iron expansion joint located in one end of a long span bridge was found to have a broken finger. Visual inspection detected no major problems in the appearance, suggesting a need for detailed causal analysis. The authors carried out fracture surface analysis and material composition analysis on the damaged member to investigate possibility and mechanism of such fracture from the viewpoint of fracture mechanics.

As a result, the fracture found in the expansion joint was likely to have occurred in the following three steps.

(1) Fatigue cracks initially occurred in the flake graphite cast iron portions which were present in the underside of the finger (crack depth: about 5 mm to 8 mm).
(2) The fatigue cracks propagated in the spheroidal graphite portions.
(3) Transition to brittle fracture occurred as the fatigue cracks continued to propagate and reached a point about 30 mm from the underside of the finger, causing the breakage of the finger.

Fracture of an expansion joint was investigated from the viewpoint of fracture mechanics including fatigue cracks and brittle fracture, checking against the actual condition of the fracture surface by precise observation, so that the mechanism of the fracture would be understood. Flake graphite with low fatigue strength was found to be present in the underside of the broken finger, and fatigue cracks were likely to have occurred due to the live load stress exceeding the reduced capacity.

Fracture-mechanical investigation for the cause of the fracture revealed that the cracks which occurred in the flake graphite portions propagated in the spheroidal graphite portions and caused transition to brittle fracture when crack propagation length exceeded its limit. These phenomena were in a close agreement with the actual condition of the fracture surface.

The findings obtained from the present study are expected to be useful for causal investigation in case of fracture in other structures of similar types.

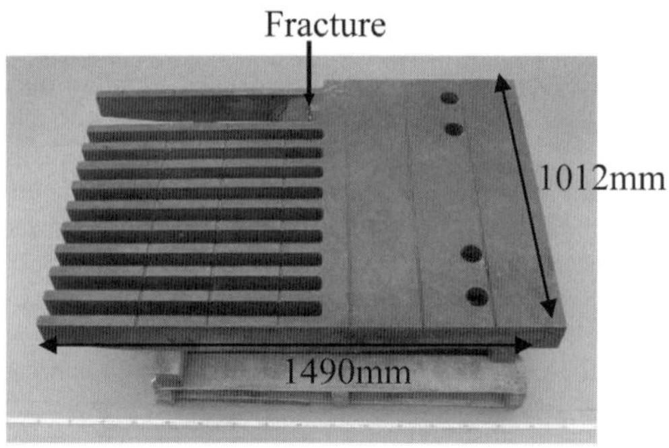

Photo 1. Broken finger plate.

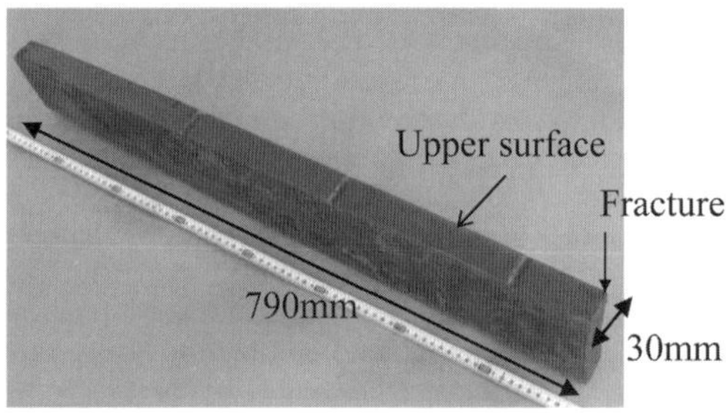

Photo 2. Broken finger (from the top).

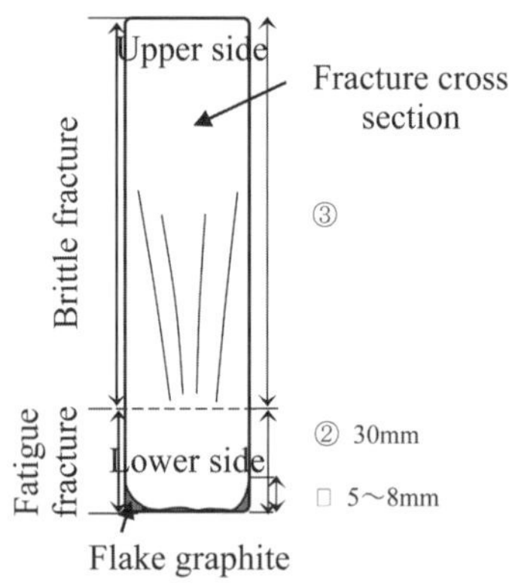

Figure 1. Investigation results and findings.

Bridge Maintenance, Safety, Management, Resilience and Sustainability – Biondini & Frangopol (Eds)
© 2012 Taylor & Francis Group, London, ISBN 978-0-415-62124-3

Rapid emergency replacement of fire-damaged composite bridges using precast decks

C.S. Shim
Chung-Ang University, Seoul, Korea

C.H. Chung
Dankook University, Yongin, Korea

I.K. Kim & Y.J. Kim
Daewoo E&C, Suwon, Korea

ABSTRACT

An expressway line with average daily traffic of 230,000 was blocked by a fire under a five span continuous steel-box girder bridge of 60 m length. Estimated loss was 2.28 billion dollars including indirect social loss and toll fee during removal and replacement of the fire-damaged bridge. Fast construction techniques for the composite bridge were crucial to reduce the loss. Full-depth precast deck system was selected to finish the recovery within 4 months even during winter season. It was impossible to reduce time for the removal of the damaged bridge of 73 m length, 38.6 m width. Producing and fabricating steel plates of the same grade as the existing steel boxes needed absolute time schedule. This paper deals with significant design and construction issues of precast decks without knowing accurate alignment and elevation of steel girders for fast-track process. The existing continuous bridge was cut after supporting the remained girders with temporary supports. Even though the steel box sections were the same as the previous girders, precast decks should accommodate elevation difference and connection details by bedding layer of less than 40 mm. Concrete decks were connected by loop joints with cast-in-place concrete between existing slab and precast decks. Details of stud shear connection were changed to accommodate shear pockets in precast decks. Through the emergency replacement, a fast construction method and connection details were developed.

The most important technical challenge is to replace the concrete deck because it is difficult to cast concrete during cold season. Full-depth precast deck is the only solution, but joints and bedding layers needed to be filled by non-shrink mortar. The steel girder had low temperature and it might cause quality problems of mortar curing.

Precast decks have bedding layers on steel girders and the bedding height should be carefully decided by considering camber of the girders and connection details. Leveling of the decks is normally done after camber adjustment of steel girders. However, the bedding height and spacers needed to be determined

Figure 1. Replacement of fire-damaged bridge.

before the camber adjustment because of time constraint.

In terms of erection plan for the full-depth precast decks, all the erection should be executed above the bridge girders because of detour lines. At both ends of the bridge, small cranes could be utilized, but it was impossible to erect the decks in the middle of the bridge because deadweight of the cranes could not be allowed in the design. New girders should have the same section as the existing parts.

ACKNOWLEDGEMENT

This research was supported by a grant from the Construction Technology Innovation Program (10CTIPB01-Modular Bridge Research & Business Development Consortium) funded by Ministry of Land, Transportation and Maritime Affairs(MLTM) of Korean Government.

REFERENCE

Shim CS, Chang SP, Lee PG. 2001. Design of Shear Connection in Composite Steel and Concrete Bridges with Precast Decks, Journal of Constructional Steel Research, 57(3): 203–219.

Bridge Maintenance, Safety, Management, Resilience and Sustainability – Biondini & Frangopol (Eds)
 ISBN 978-0-415-62124-3

Executive extremely urgent project for the rehabilitation of vehicular and pedestrian traffic of the bridge over Corace river in Gimigliano municipality

E. Siviero
IUAV University of Venice, Italy

A. Stocco
ETCAEH University of Nova Gorica, Slovenia

ABSTRACT

This work is one of the boldest kind of Maillart ever made by Adriano Galli. This is first example of arched road bridge made of reinforced concrete and with stiffening deck in the South of Italy, over 170 meters with two access viaducts consisting of a three-deck continuous beams on four sturdy support and a very solid, as slender central arch of 80 feet.

The disruption that caused the unavailability of the bridge forcefully poses two essential questions: the first one purely functional and "utmost urgency", the second one of cultural and philological nature for the need of seismic rehabilitation increasing its functions.

The urgency needs to be resolved at an early stage are: quickly making the bridge fit for use again, at least for traffic of light vehicles.

The re-functionalization, carried out in a second phase, aimed at adjusting the crossing to the new tasks that the infrastructure has to perform, eliminating the mixing of pedestrians and vehicles and at the same time performing seismic adjustment of the work itself.

Bearing in mind that seismic adjustment of the existing work relies on the need for its temporary closure to traffic for a short period, it is necessary a preventive action, in order to ensure transit at least limited to light vehicles.

The bridge built by Adriano Galli, though damaged, still appears today sleek and majestic as connection of the two top sides of Corace river. The work is characterized by light shape, clear and simple.

The static approach on the structural aspects of architectural design is, in this case, characterized by the binomial form and function, maintaining, also after rehabilitation, those features of clarity and formal lightness that already characterized the work of Galli.

The two phases of general project provide at first step the measures needed to reopen the bridge to traffic, and later a functional enlargement tiling of two walkways to the central lane. The overall intervention is then planned in two coordinated phases of execution, subsequent and linked by a dichotomous relationship within very compelling logic.

Figure 1. Bridge over Corace at the end of works.

Bridge Maintenance, Safety, Management, Resilience and Sustainability – Biondini & Frangopol (Eds)
 ISBN 978-0-415-62124-3

Study of the hybrid structures changed from the steel bridges for railroad which considered construction

N. Taniguchi & M. Hansaka
Railway Technical Research Institute, Tokyo, Japan

N. Koide & K. Ohgaki
Kawasaki Heavy Industries, Tokyo, Japan

F. Ohkubo, S. Satake & Y. Sugino
Taiheiyo Material, Tokyo, Japan

ABSTRACT

There has recently been a requirement for constructive structures to be efficiently maintained and managed. Specifically, steel bridges for railways, many of which are past their design service life, often continue to be used as matters now stand. While steel bridges in main routes already in backboned services that elapsed the nominal aging-limits may sometimes be rene wed through replacement, those existing structures in local lines with less profitability are required just to prolong their lives also at as low cost as possible. As another big issue for steel bridges in railways, on the other hand, there is a subject on how to suppress acoustic noises as well as the life prolongation. Steel bridges for railways, mostly having open-grating structures without floor panels, may develop larger vehicle-noises on train-passage than other road-bridges or concrete bridges because acoustic noises emitted from the rails (rolling noises) and the vehicle itself are directly transmitted to the outside further in addition to noises caused by the steel-members.

As a countermeasure for life-extension and noise-reduction of such structures, revamping of existing steel bridges into composite structures is now under investigation. A shift to a composite structure is to proceed with an action to place a complementary member such as a concrete panel on to the existing steel girder for the purpose of preventing it from having corrosion, improving the beam stiffness, reducing the stress-amplitudes at the time of a response to active forces, and consequently prolonging the fatigue life. Furthermore, consolidating the steel and concrete, on the other hand, acoustic noises caused by the steel member can also be reduced at the same time.

Therefore, this research offers a new composite-revamping method for railway bridges by assuming composite remodeling process of existing steel bridges having no specific fissure damages or serious corrosions and effectively making use of relatively new materials to it. Furthermore, it was confirmed by the construction tests using actual steel bridges that the switching work can be performed well even under strict limitation. Moreover, effects of noise reduction and improved stiffness were further confirmed through hammer-impact tests (vibration-measurement tests, Figure 1) and loading tests (Figure 2) inorder to prove the efficacy of the composite remodeling.

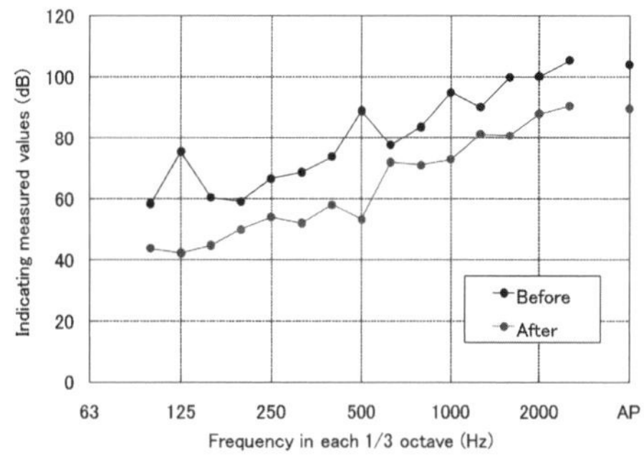

Figure 1. Results of the average of the measurements.

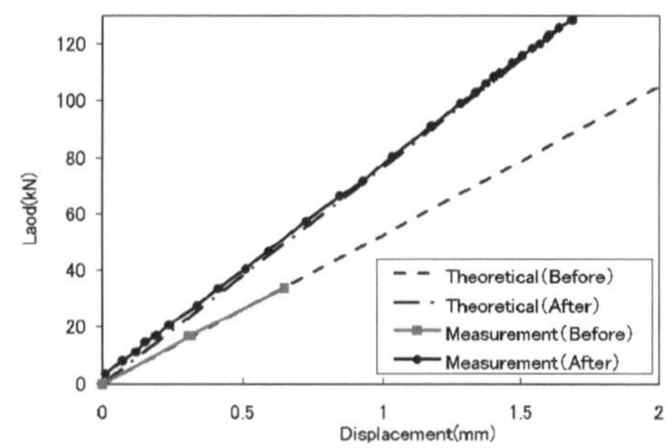

Figure 2. The load versus displacement relationship.

REFERENCE

Nozomu Taniguchi, Masanori Hanzaka, Haruhisa Ikariyama, Takashi Kouzuki, Akiro Tanahashi, Teruhiko Yoda: A study on the development of a low-noise steel bridge with rubber latex mortar coating, Journal of Structural Engineering, Vol. 55A, pp. 1164–1171, 2009.

Bridge Maintenance, Safety, Management, Resilience and Sustainability – Biondini & Frangopol (Eds)
© 2012 Taylor & Francis Group, London, ISBN 978-0-415-62124-3

Repair of fatigue cracks on steel plate deck in highway bridges with heavy traffic

K. Tokumasu & M. Kawamura
Hanshin Expressway Company Limited, Osaka, Japan

T. Nishioka
Hanshin Expressway Engineering Company Limited, Osaka, Japan

ABSTRACT

There are about 1,400 spans, that is 88.2 km, of orthotropic steel deck bridges in Hanshin Expressway. Hanshin Expressway Company Limited is one of the road authorities that manage huge volumes of orthotropic steel deck bridges. They have found numerous fatigue cracks of orthotropic steel deck in recent years since they identified the first one in 1990 (Photo 1). Most of fatigue cracks are detected in the routes with high heavy vehicle ratio.

Inspect of the orthotropic steel deck is the basis of research aimed at determining about "the need for repair or reinforcement" and "the cause of the damage occurred". We understand the strengths and weaknesses about each inspection, and have conducted inspections effectively example combine several test methods depending on the number of targeted inspections and inspection points (Photo 2).

Hanshin Expressway Co. Ltd. grouped by the occurrence of fatigue cracks to orthotropic steel deck.

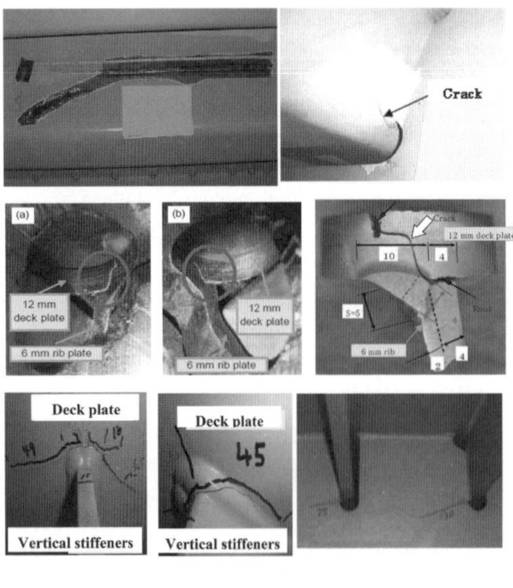

Photo 1. Fatigue cracks on orthotropic steel deck.

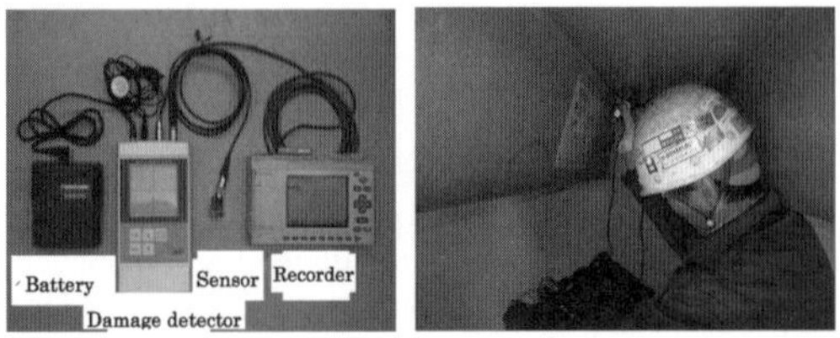

Photo 2. The non-destructive inspection of orthotropic steel deck.

Photo 3. Repair of fatigue cracks.

On that basis, we have determined a method for each type of crack repair (Photo 3). Moreover, we conducted stop-hole construction, grinding crack tip. By performing so, we expect that it will reduce the stress of the crack tip.

We are expected to proceed to consider how inspection and repair an efficient and effective in the future.

Bridge Maintenance, Safety, Management, Resilience and Sustainability – Biondini & Frangopol (Eds)
 ISBN 978-0-415-62124-3

Development of the hot-spot stress sensor and application to orthotropic steel deck

T. Uesugi
Kyowa electronic instruments Co., Ltd., Tokyo, Japan

S. Fujita
NewTec Co., Ltd., Hyogo, Japan

S. Tsukamoto & T.Yamagami
Hanshin Expressway Engineering Company Limited, Osaka, Japan

M. Sakano
Kansai University, Osaka, Japan

ABSTRACT

Fatigue life of welded steel structure, in general, is evaluated by the nominal stress and the fatigue life diagram for the appropriate class. But when we cannot define the nominal stress, we can use the hot-spot stress. In that case, it is necessary to extrapolate by the stress values of two points. But it is hard to paste accurately two strain gages on the weldment neighborhood. In this study we developed the Hot-spot Sensor (HsS) that can measure the hot-spot stress with one gage. It was verified to the orthotropic steel deck for the trial test.

First, for the purpose of validation of HsS, we carried out static load test and fatigue test. Through the static loading tests of welded joint specimens, it was confirmed that HsS can measure the hot-spot stress accurately regardless of the thickness. And through a fatigue test of welded joint specimens, it was confirmed that HsS could predict correctly location of crack initiation and estimate the fatigue life with the adequate accuracy on the conservative side.

Secondly, we applied HsS to trough rib of orthotropic steel deck of highway bridge. We pasted HsS to the trough rib and deck plate of in-service highway bridge steel plate deck, and performed an analysis to evaluate cumulative damage. We adopted two measuring points, the central point of the trough rib and near the end point of the trough rib. Through the measurement of the orthotropic steel deck with the trough ribs, the fatigue life of weld toe was shown to be different between the left side and the right side of a trough rib and between the end and the center of a trough rib.

As a result, it was able to be confirmed to measure the hot-spot stress easily by applying HsS.

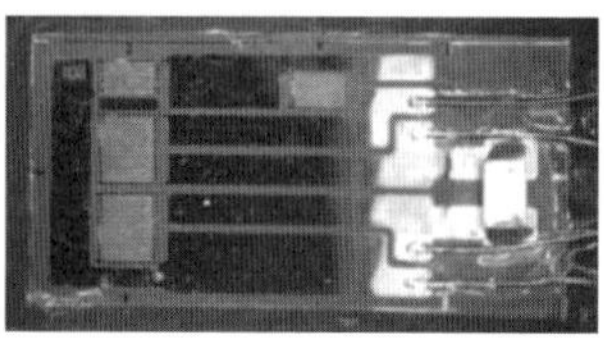

Figure 1. Hot-spot Sensor.

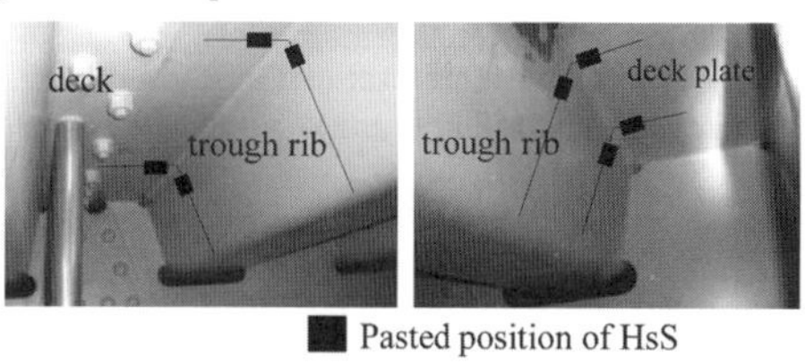

Figure 2. Pasted position of HsS on orthotropic steel deck with trough ribs.

Table 1. Evaluate cumulative damage.

Class of joint Measurement position	E	F
near the end of the trough rib, left side of the trough rib, on the trough lib	∞	∞
near the end of the trough rib, right side of the trough rib, on the trough rib	∞	34
near the end of the trough rib, left side of the trough rib, on the deck plate	∞	∞
near the end of the trough rib, right side of the trough rib, on the deck plate	∞	∞
the center of the trough rib, left side of the trough rib, on the trough lib	460	99
the center of the trough rib, right side of the trough rib, on the trough lib	74	19
the center of the trough rib, left side of the trough rib, on the deck plate	∞	∞
the center of the trough rib, right side of the trough rib, on the deck plate	∞	∞

REFERENCES

Japan Society of Steel Construction. 1993. Fatigue Design Recommendation for Steel Structure. Tokyo: Gihodo shup-pan.

Watanabe, O. 1977. Strain Gage and its Applications. Tokyo: Nikkan Kogyo Shimbun

Suganuma, H. & Miki, C. 2007. Fatigue strength evaluation with effective notch stress of the weld between trough rib and deck plate on orthotropic steel deck, Doboku Gakkai Ronbunshuu A 63(1): 35–42

Takada, Y. Tabata, A. Sakano, M. 2009. Macro analysis of fatigue cracks in existing orthotropic steel decks. Proceedings of Constructional Steel 17: 329–336.

Bridge Maintenance, Safety, Management, Resilience and Sustainability – Biondini & Frangopol (Eds)
© 2012 Taylor & Francis Group, London, ISBN 978-0-415-62124-3

Princess Margaret Bridge rehabilitation

A.R. Zaki
SNC-Lavalin, Montreal, Canada

A.FM Girgis
Ecomstruct, Nebraska, USA

ABSTRACT

In 2010 and 2011, the Princess Margaret Bridge, one of Canada's important bridge underwent a major rehabilitation of its 50-year old structure. Canada's major existing infrastructures are in need for rehabilitation. Dealing with an old steel truss bridge such as the Princess Margaret Bridge required a great deal of creativity and innovation. This is an 80 million dollar rehabilitation project in Fredericton, New Brunswick, Canada. The Princess Margaret Bridge was officially inaugurated in 1959 as one of the largest and most modern bridge in New Brunswick, Canada, to serve as a connecting line between the north and south side of the Saint John River in Fredericton. The bridge measures 1097 meters in length, composed of several structural systems, comprising 12 land piers and 10 water piers. After more than 50 years of operation, the bridge had taken its toll due to the combined effects of age, increase in weight and numbers of trucks, and the extensive use of deicing salts to keep the bridge operational during Canada's harsh winter conditions. The deck was deteriorated beyond repairs, the steel members of the trusses were deficient, and the concrete piers had suffered severe damage. The bridge had to be repaired with the least amount of closure time. A creative solution had been proposed: precast deck panels that are made composite with the steel trusses. To our knowledge, this is the first time trusses were made composite with a precast concrete deck. Not only did this solution speed up construction, but also saved a significant amount on the strengthening of the structural steel in the trusses.

The deck design of most of the bridge length was double tee panels in which the panel ribs oriented transverse to traffic. The double tee ribs were pretensioned in the traffic transverse direction of the bridge. The 180 mm double tee slab was post-tensioned in the longitudinal direction, parallel to traffic. A very unique innovative system of post-tensioning of which eliminated all the duct coupling and possible misalignment, was implemented for the first time (Figure 1).

Figure 1. Deck panel installation.

During the construction of this project, we faced an interesting challenge. The bridge trusses could not support the crane that was to be used to remove the old deck and install the new one. Therefore, a creative electronically controlled machine that performs this removal and replacement work much faster and more effectively than a manually operated crane was developed and used.

This 2 years design build project has been carried out in 2010–2011, and all construction works were completed to ensure that the Bridge achieves the required live load carrying capacity and service life objectives.

REFERENCES

Canadian Highway Bridge Design Code.
NBDOT, Princess Margret Project specifications.
Canadian National Resources Website.
Seismic Retrofit Manual for Highway structures (Part 1).

SPECIAL SESSIONS

Gusset plates in steel truss bridges: Testing, analysis and monitoring

Organizer: D. Duthinh

Bridge Maintenance, Safety, Management, Resilience and Sustainability – Biondini & Frangopol (Eds)
© 2012 Taylor & Francis Group, London, ISBN 978-0-415-62124-3

Structural health monitoring of a steel railway bridge on the river Suaçuí

W.S. Assis
Federal University of Alagoas, Maceió, Brazil

L.M. Trautwein
Federal University of ABC, Santo André, Brazil

T.N. Bittencourt & A.P.C. Neto
University of São Paulo, São Paulo, Brazil

ABSTRACT

This paper describes the monitoring scheme developed to observe the structural behavior of a steel bridge at river Suaçuí. The goal of monitoring performed on the Suaçuí River Bridge is to gather the information needed to quantify and qualify the response of the structure about their safety as well as provide background information for planning and carrying out actions such as: detailed inspections of the structure; theoretical studies on the behavior of the structure and its elements through numerical simulation; decision-making regarding the maintenance and define strategies to strengthen or replace structural components (Bittencourt, 2009).

Located in the Vitória-Minas railway, in Baguari, Brazil, the Suaçuí River Bridge, shown in Figure 1, is a metallic bridge constructed in the 40's and has a span of 41 meters and height of 7.80 meters. Its superstructure is made of two Warren truss supported at the ends by metallic structural bearings. The trusses are interlinked in its lower region by transverse floor beams, with approximately 4.5 meters long, and stringers that connects to transverse floor beams. Over the stringers are supported the wooden sleepers, and over these, the rails. Nodal connections are made of riveted steel plates.

During the monitoring, measurements have been taken for different trains entering the bridge from both edges and crossing at different speeds. This structural monitoring system was comprised of strain gages, optical sensors, displacement transducers and accelerometers. The bridge was also simulated as a 3D finite element model (Figure 2), considering linear elastic material properties. The numerical results obtained from the numerical analysis were then compared with the data obtained in the real monitoring system. Several load cases were analyzed and results in all of them were very consistent with each other.

Figure 1. General view of the Suaçuí River Bridge.

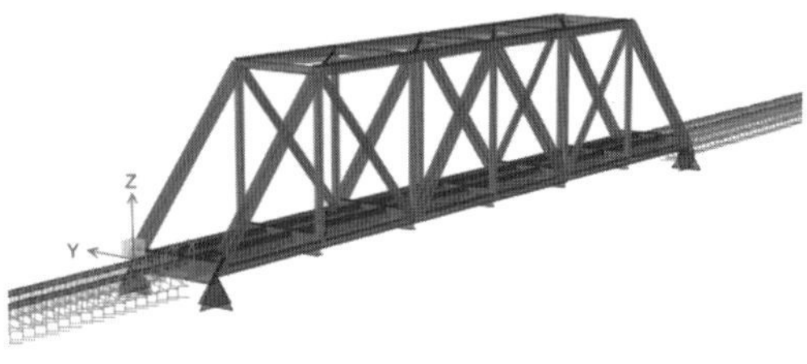

Figure 2. Perspective viewing of the bridge over the river Suaçuí.

REFERENCES

Marecos, VMS. 2008. Bridge safety control in real time. *M.Sc. Dissertation*. National Laboratory for Civil Engineering (LNEC). Lisbon.

Bittencourt TN et al. 2009. Monitoring of the Suaçuí River Bridge. Technical report. São Paulo. 111 p.

Thakkar, SK; Ghosh, G; Singh, Y. 2006. Structural damage identification and health monitoring and damage identification of Bridges. *Advances in Bridge Engineering*, March 24–25: 11–30.

Bridge Maintenance, Safety, Management, Resilience and Sustainability – Biondini & Frangopol (Eds)
© 2012 Taylor & Francis Group, London, ISBN 978-0-415-62124-3

Simplified gusset plate model for failure prediction of truss bridges

C. Crosti
Sapienza University of Rome, Rome, Italy

D. Duthinh
National Institute of Standards and Technology, Gaithersburg, MD, USA

ABSTRACT

The 2007 catastrophic collapse of the I-35 W Bridge in Minneapolis, Minnesota, under ordinary traffic and construction loads, was triggered by the buckling of an undersized gusset plate [National Transportation Safety Board NTSB 2008]. Current procedures [Federal Highway Administration FHWA 2009] for the design and load rating of multi-member gusset plates consist in checking axial, bending and shear stresses along various sections deemed critical, using elastic beam theory. These procedures are intended to ensure a safe and conservative design, but produce results that can be quite different from more realistic finite-element (FE) results, and cannot predict either stiffness or actual behavior. To do so would require highly sophisticated and detailed FE models (FEM), such as the ones used in the investigation of the I-35 W collapse. An intermediate approach is presented here, whereby a simplified connection modeled by nonlinear semi-rigid springs is proposed to improve on current design analysis, which typically is linear and assumes rigid connections

In the present work, we take advantage of the NTSB detailed FEM of gusset plate U10 to establish the equivalent stiffness of springs that model the nonlinear behavior of the connection. The FEM has 5 stub elements attached to a pair of gusset plates and that model is connected to the appropriate members in the global model. For the simplified connection model, the stub elements and gusset plates are replaced by 5 user-defined structural elements called springs for short, that have 6 degrees of freedom. To capture the nonlinear behaviour of the gusset connection up to failure, the detailed FEM of the five-element connection is loaded into the nonlinear range (beyond yielding and small displacements), resulting in force-displacement and moment-rotation curves.

As the global nonlinear analysis may involve buckling and out-of-plane deformation, it is necessary to use a full 3D model of I-35 W (Fig. 1) with a sufficiently accurate representation of lateral bracing

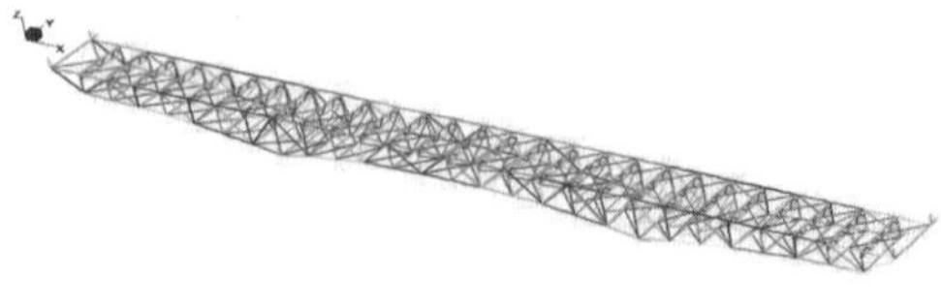

Figure 1. 3D model of I-35 W.

In the model, U10-W joint is semi-rigid, represented by nonlinear connection elements with diagonal stiffness matrix and the five members around the U10-W joint are elasto-plastic. All the other joints are rigid and all the other members are elastic. The model was loaded with the actual uniform loads, but the construction loads were applied gradually, with a load factor of 1.0 corresponding to the actual loads. The model predicts that U10-W begins to fail at a load factor of 0.92, and completely fails at load factor 1.7, leading to the collapse of the bridge.

The simplified connection model is most useful for multiple load cases, or structures that use the same connection repeatedly, or only with changes in plate thickness or material properties that can be accommodated easily in the same detailed FEM. The general procedure is applied to a Howe truss bridge as an example.

The nonlinear connection model provides a simple and affordable way to account for connection performance in global analysis to achieve accurate prediction of bridge failure.

REFERENCES

Federal Highway Administration (2009) "Load rating guidance and examples for bolted and riveted gusset plates in truss bridges" FHWA-IF-09-014, Washington, D.C. 20590

National Transportation Safety Board (2008) "Collapse of I-35 W Highway Bridge, Minneapolis, Minnesota, August 1, 2007" Accident Report, NTSB/HAR 08/03 PB 2008-916213, Washington D.C. 20594

Bridge Maintenance, Safety, Management, Resilience and Sustainability – Biondini & Frangopol (Eds)
© 2012 Taylor & Francis Group, London, ISBN 978-0-415-62124-3

Full scale fatigue testing of original truss members and connections

R. Helmerich
Federal Institute for Materials Research and Testing (BAM), Berlin, Germany

ABSTRACT

The analysis of the I-35 W collapse in Minneapolis, USA, initiated a discussion about gusset plates in ageing truss girders. The cause analysis indicated a range of possible causes contributing to the sudden collapse of the riveted steel truss girder bridge (NTSB 2008). The question to analyze and assess the condition of heritage bridge structures is often raised, since reassessment standards do not exist in most countries. Last year, Switzerland has issued a set of reassessment standards. (Brühwiler et al 2010).

The Federal Institute for Materials Research and Testing (BAM) has performed laboratory fatigue tests of original riveted steel girders in their laboratories dismantled from the Berlin underground line U1 (Helmerich 2005). The Berlin underground was designed by Siemens in the late 19th century as one of the first electrified train lines. The Berlin underground line U1 was partly built over ground as continuous Gerber girders. Although truss elements were designed according to the expected traffic load, the thickness of the connecting gusset plates was estimated according to the diameter of the rivet holes.

10 m long girders were replaced and four original members were tested in the BAM laboratories. In all performed fatigues tests, the gusset plates of the connected first diagonal failed, although load set-up and the load cycles were changed in the different tests.

During the test we have monitored the strains in the truss elements and in an accessible location of the gusset plate (Figure 1). An increase of the load by ~10% was a defined to limit the end of the test. Displacement of the lower chord did not display the initiated fatigue crack propagation sufficient early enough. Before, during and after the fatigue test, nondestructive testing accompanied the visual inspection of the details periodically. Radiography confirmed that a change in the stress condition of the gusset plate was a clear indication for crack initiation long before the changes got visible during simple visual inspection. The failure occurred in all tests in the gusset plates connected to the highest stressed truss members. After the gusset plates failed, the materials characteristics of the gusset plate were analyzed in tension tests, fracture mechanic tests, by means of sulfurous print and eddy current test.

The knowledge from this and other worldwide performed tests was transferred into *Recommendations for the Assessment of Existing Steel bridges* issued by the European Convention for Constructional Steelwork (ECCS) and the European Joint Research Center (JRC) as background document EUR23252EN for the further developement of the Eurocode 3 (Kühn et al 2008).

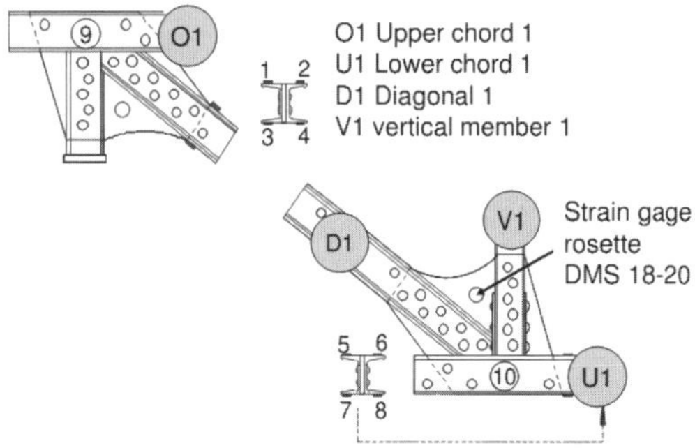

Figure 1. Monitored strain cycles in gusset plates connected to the first diagonal member D1 and truss chords U1 and O1.

REFERENCES

Helmerich, R. 2005. Old steel and steel structures (in German: Alte Stähle und Stahlkonstruktionen) *BAM Report 271*, ISBN3-86509-362-0, http://www.bam.de/de/service/publikationen/publikationen_medien/fb271_vt.pdf

Kühn, B., Nussbaumer, A., Lukic, M., Helmerich, R. et al., 2008. Recommendation for the Assessment of Existing Steel brid-ges, EUR23252EN, ISSN 10185593. JRC and ECCS: http://www.stalforbund.com/Fagboker/EUR23252EN.pdf

National Transportation Safety Board, NTSB (2008). Collapse of I-35 W Highway Bridge, Minneapolis, Minnesota, August 1, 2007, Accident report, NTSB/HAR 08/03 PB 2008-916213, Washington D.C. 20594.

Bridge Maintenance, Safety, Management, Resilience and Sustainability – Biondini & Frangopol (Eds)
© 2012 Taylor & Francis Group, London, ISBN 978-0-415-62124-3

Inspection strategies to prevent fatigue failure of gusset plates in steel truss bridges

J. Herter
BAM Federal Institute for Materials Research and Testing, Berlin, Germany

ABSTRACT

Many old steel bridges have to carry heavy traffic loads during many years. They are suffering amongst others from fatigue related problems. Their resistance against fatigue failure has to be taken into account. Advance methods of assessment are based in addition to calculations on experimental investigations and adjusted inspection strategies.

Two different concepts are used for the assessment of the remaining fatigue life of existing steel bridges. One is based on fracture mechanics, the other on S-N-curves. Both approaches desire the identification of the stress range caused by traffic loads at the "hot spots". The stress range can be calculated based on a assumed static model but much more relevant are results from an adopted model based on load tests on site.

Several studies have been performed to approve the fatigue related safety and the remaining fatigue life of an over 100 year's old riveted steel viaduct of a Berlin subway line (Herter et al. 2002).

In general the following studies are possibly necessary:

- Building survey
- Load tests on site
- Laboratory fatigue tests
- Radiographic inspection on site
- Measurement based analysis
- Material investigation
- Special local investigation
- Assessment based on S-N curves
- Assessment based on fracture mechanic
- Concept for future inspections

The riveted steel viaduct consists of several bridges on steel columns arranged in a row. The double tracked rail track is carried by riveted mean truss girders.

Fatigue tests on original riveted truss girders, performed in the BAM laboratory, showed that the first failure mode was the failure of a gusset plate and not failure of a truss element.

Failure in the gusset plate raises the problem that the crack starts at one of the rived wholes which is covered by the connected truss elements. Radiographic inspection is necessary to see the beginning of the crack or to determine the actual crack length.

An assessment of the gusset plates concerning fatigue failure seems to be only possible on the base of the fracture mechanic approach. This calculation requires the determination of an initial crack length. The result of the calculation, based on fracture mechanic, is a value for an assumed safe further operating time interval. After the end of the first interval successive inspections are needed. The question is how this inspection strategy should be configured.

Due to the importance of the accuracy of the value of the initial crack length we perform presently more investigations to validate the execution process of x-ray inspection. Some ongoing investigation in BAM referring to this will be discussed. More examples of investigations, performed recently by BAM division 7.2 "Safety of Structures", which show the need of both, laboratory tests and experimental investigation on site, are described in Herter et al. 2010.

REFERENCES

Herter, J., Brandes, K. & Helmerich, R. 2002. Assessment of the safety and remaining fatigue life of old steel bridges by nondestructive field tests. *Proc. of the first international conference on Bridge maintenance, safety and management, IABMAS 02, Barcelona 14–17 July 2002.* (CIMNE).

Herter, J., Fischer, M. & Brandes, K. 2002. 100 Jahre U-Bahn in Deutschland. Zur Ermüdungssicherheit eines Stahlviaduktes der Linie 1 in Berlin. In *Stahlbau* 71(2):97–107.

Herter, J., Mehdianpour, M. & Hille, F. 2010. Repair of a railway viaduct in Berlin and inspection concepts to extend the service life. *Structural faults & repair 2010, 13th international conference and exhibition, Edinburgh, UK,* 15th–17th June 2010. On *CD-ROM.*

Bridge Maintenance, Safety, Management, Resilience and Sustainability – Biondini & Frangopol (Eds)
© 2012 Taylor & Francis Group, London, ISBN 978-0-415-62124-3

Quantitative evaluation of digital image correlation as applied to large-scale gusset plate experiments

M.A. Iadicola
National Institute of Standards and Technology, Gaithersburg, Maryland, USA

R.S. Zobel & J.M. Ocel
Federal Highway Administration, McLean, Virginia, USA

ABSTRACT

In response to the 2007 collapse of the I-35W bridge in Minnesota and the findings of the subsequent investigation (NTSB 2008), steel truss bridges designed with gusset plates have been under added scrutiny. The Federal Highway Administration (FHWA) published guidance (Ibrahim 2009) for bridge owners to use in their load ratings. This document was based on known limit-state models that had been developed over time, since scant test data was available on bridge gusset plate connections. However, the guidance document did yield knowingly conservative load ratings that some owners felt were too conservative. This prompted the FHWA to partner with the American Association of State Highway and Transportation Officials (AASHTO) to conduct full-scale testing of steel truss gusset-plated connections to validate and/or expand upon the published FHWA guidance. The large-scale gusset-plate experiments for this study are being performed at FHWA Turner-Fairbank Highway Research Center Structures Laboratory (Ocel et al. 2012). Researchers from the National Institute of Standards and Technology are working in conjunction with FHWA to perform digital image correlation (DIC) measurements during these experiments. The unique combination of measurement methods used in these experiments presents an opportunity to compare multiple techniques each with their own advantages and disadvantages for certain measurements of interest.

In this paper, a comparison is presented of four measurement methods used simultaneously during the mechanical testing of a structural connection (gusset plate). These methods include: traditional foil strain gauges, photoelastic strain measurement, laser displacement tracking, and DIC used to measure the initial plate shape and elastic strains under load. The DIC method is assumed to be the least sensitive of the methods with regard to both the shape and elastic strain measurements. Qualitatively, the results for both shape and elastic strains agree well based on full-field plots of the results. More quantitative analyses show that the agreement for shape is actually not very good, where as the agreement for elastic strains is generally within the uncertainties of each method (Fig. 1). The sensitivity of DIC in this range of shape and elastic strain measurements might be limited, but DIC is able to capture the major behaviors of interest. This in spite of the system being setup primarily to measure plastic strains beyond the limits of the photoelastic method as applied here.

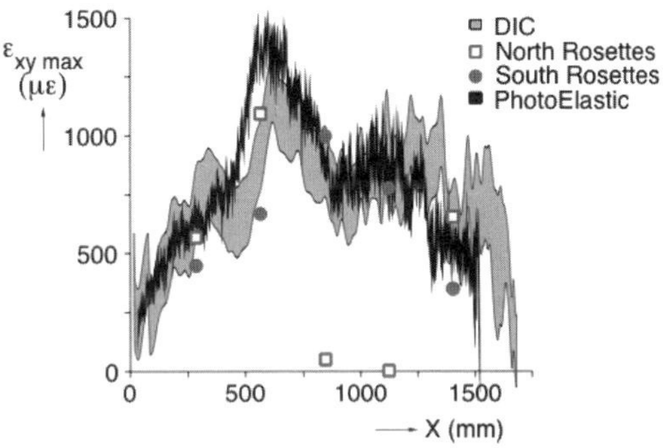

Figure 1. Plot of the measured variation of maximum in-plane shear strain along the width of the plates. Upper and lower bounds of the DIC (North plate) and photoelastic (South plate) results are based on one standard deviation of the full-field data over a 25.4 mm height centered on the strain gauge positions.

REFERENCES

Ibrahim, F.I.S. 2009. *Load Rating Guidance and Examples For Bolted and Riveted Gusset Plates In Truss Bridges*, Federal Highway Administration Publication No. FHWA-IF-09-014, Washington, DC., USA

National Transportation Safety Board 2008. Collapse of I-35W Highway Bridge, Minneapolis, Minnesota, August 1, 2007, *Highway Accident Report NTSB/HAR-08/03*, Washington, DC., USA

Ocel, J. et al. 2012. FHWA Full-Scale testing of Steel Truss Bridge Gusset Plate Connection, to appear in the *Proceedings of IABMAS 2012 Lake Como, Italy*

TEAM: A Marie Curie training network on bridge management

Organizer: C. McNally

Bridge Maintenance, Safety, Management, Resilience and Sustainability – Biondini & Frangopol (Eds)
 ISBN 978-0-415-62124-3

Evaluation of time-dependent chloride parameters for assessing reinforced concrete bridges

A. Attari, C. McNally & M.G. Richardson
School of Civil, Structural and Environmental Engineering, University College Dublin, Ireland

ABSTRACT

One of the main sources of deterioration of reinforced concrete bridges, especially in marine environment is chloride-induced corrosion due to exposure to salt-rich winds and sea-spray. Chlorides penetrate the concrete cover, destroy the passive layer of reinforcement and initiate corrosion, which leads to a decrease in the cross-sectional area of rebar, and reduces the load carrying capacity.

Diffusion coefficient (D_{eff}) is an index representing the permeability of concrete and rate of chloride ingress and is an important input parameter in any model concerning transportation of chloride in concrete. In cementitious systems the diffusion process slows as the diffusion coefficient decreases over time. This is largely due to chloride binding and refinement of the pore structure associated with continued hydration. However many of the currently employed models for chloride transport do not consider this effect, and instead employ a constant diffusion coefficient over the whole life-time of the structure. This leads to overestimated chloride penetration, and consequent conservative durability design. This is particularly the case when ground granulated blast-furnace slag (GGBS) or pulverized fuel ash (PFA) are present in the mix.

The purpose of the current study is to investigate how the diffusion coefficient, and consequently the resistance and durability of concrete changes over time. A novel numerical model is used to determine the degree of hydration, which in turn is used to predict the time-dependent chloride diffusion coefficient of the concrete. This is investigated using concrete mixes with various GGBS replacement levels, with the outputs used in service-life prediction models employed for reinforced concrete structures.

An experiment has been designed involving four different mixes of concrete; these employ limestone aggregates, blended cement (CEM II) and varied GGBS replacement levels. Mix proportions are presented in table 1. Curing of concrete is conducted at a temperature of 21°C. Samples were tested at regular intervals to allow characterisation of the influence of these key mix design properties with respect to time. Diffusion coefficients were determined from the Nordtest rapid migration test while degree of hydration was determined from the heat of hydration produced by the concrete. The correlation between the two methods is discussed, as are the implications for specifying chloride diffusion coefficients for reinforced concrete bridges.

Diffusion coefficients are measured for each mix, using the standard Nordtest method of Rapid Chloride Migration at several points within the first 6 weeks after casting. Curve fitting to the diagrams obtained this way, gives the equations that are describing the changes of Diffusion coefficient with time.

The cumulative amount of heat evolved at time t of the hydration process, divided by the total amount of heat available at 100% hydration, has been used as the measure of degree of hydration. To calculate the cumulative heat of hydration, thermocouples were embedded inside the specimens for monitoring the internal temperature development during the first week after casting. Hydration characterization curves are build based on the results of current experiment and parameters provided in the literature.

Combining the equations obtained for development of the hydration degree of specimens, with those obtained for diffusion coefficient change over time, and eliminating t from both equations, equations are developed for evaluating diffusion coefficient based on hydration degree of the mix.

Table 1. Material proportions for the four different mixes used in the experiment.

Mix	GGBS level (%)	Cement Content (Kg/m^3)	GGBS Content (Kg/m^3)	Fine aggregate (Kg/m^3)	Coarse (5–10 mm) (Kg/m^3)	Coarse (10–20 mm) (Kg/m^3)	water content (Kg/m^3)
1	0	323	0	1003	330	660	163
2	30	222	95	987	325	649	177
3	50	160	160	995	328	655	169
4	70	96	224	994	327	654	170

Bridge Maintenance, Safety, Management, Resilience and Sustainability – Biondini & Frangopol (Eds)
© 2012 Taylor & Francis Group, London, ISBN 978-0-415-62124-3

Prediction of moment redistribution and influence of rotation capacity in reinforced concrete beams

N. Bagge & C. Pedersen
Rambøll Danmark A/S, Copenhagen, Denmark

A.J. O'Connor
Trinity College Dublin, Dublin, Ireland

ABSTRACT

According to standards for concrete structures, such as Eurocode 2, it is allowable to make use of plastic redistribution of sectional forces in continuous beams at the ultimate limit state for the design and assessment of existing structures. Plastic analysis, where applicable, can result in an increased load carrying capacity. Plastic analysis is therefore beneficial for owners/managers of structures, as a consequence of higher load rating can be to reduce maintenance costs through avoidance of unnecessary repair and/or optimize the repairs that are shown to be necessary. An essential factor to determine the degree of plastic redistribution of sectional forces is the rotation capacity of critical regions, expressing the structural ductility, which depends on material, structural and loading factors, CEB (1998). This paper presents a theoretical methodology for plastic analysis of statically indeterminate reinforced concrete structures. The approach is validated against experimental results for 2-span continuous beams, illustrated in Figure 1, with specific emphasis placed upon the ductility of the structure.

The proposed model to analyze the moment distribution, and thereby the allowable moment redistribution, is based on a lower bound (static) method, where cross-section analyses are combined with structural analysis. In addition to the structural analysis the model consist of an explicit check the rotation capacity of critical regions according to the deformable strut-and-and-tie model developed by Michalka (1986). The fundamental condition in the check of the structural ductility is higher or equal rotation capacity in the critical region in comparison to required rotation corresponding to a given redistribution.

Comparison of results from analysis according to the proposed approach and results from experiments performed by Mathiasen & Nielsen (2008) and Hartvig & Trúgvason (2009), it seem to be in good agreement. The analysis of 17 continuous beams indicates an ability to redistribute up to approximately 60% of the bending moment at the intermediate support. However, CEN (2004) allows moment redistribution up to 20% or 30%, depending on the reinforcing steel ductility, regarding the methodology of linear elasticity with limited redistribution.

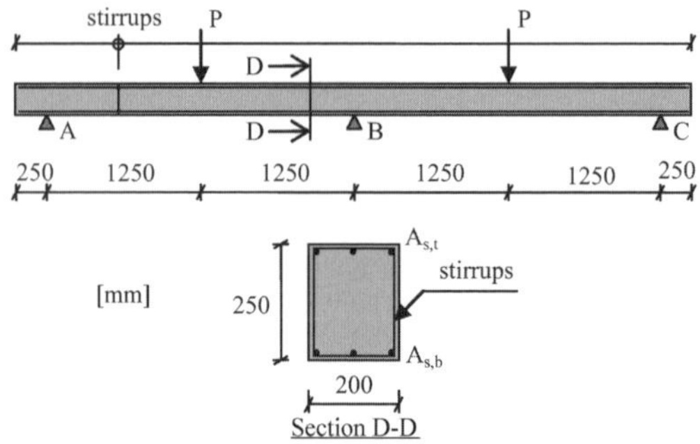

Figure 1. Beam setup.

REFERENCES

Comité Euro-International du Béton (CEB) 1998. *Ductility of Reinforced Concrete Structures*. Bulletin d'Information 252. Lausanne: Comité Euro-International du Béton.

European Committee for Standardization (CEN) 2004. *Eurocode 2: Design of concrete structures – Part 1-1: General rules and rules for buildings*. Brussels: European Committee for Standardization.

Hartvig, E. & Trúgvason, J. 2009. *Experimental Investigation of Moment Redistribution in Continuous Reinforced Concrete Beams* (in Danish). Bachelor thesis. Århus: Engineering College of Århus.

Mathiasen, S. & Nielsen, K.S. 2008. *Experimental Investigation of Moment Redistribution in Continuous Reinforced Concrete Beams* (in Danish). Bachelor thesis. Århus: Engineering College of Århus.

Michalka, C. 1986. *Zur Rotationsfähigkeit von Plastischen Gelenken in Stahlbetonträgern*. PhD thesis. Stuttgart: University of Stuttgart.

Bridge Maintenance, Safety, Management, Resilience and Sustainability – Biondini & Frangopol (Eds)
© 2012 Taylor & Francis Group, London, ISBN 978-0-415-62124-3

Load effect of multi-lane traffic simulations on long-span bridges

C.C. Caprani
Dublin Institute of Technology, Dublin, Ireland

A. Lipari & E.J. OBrien
University College Dublin, Dublin, Ireland

ABSTRACT

It is well acknowledged that long-span road bridges (about 50 m long and more) are governed by congested traffic rather than free-flow conditions. In free flowing traffic, vehicles have large gaps between them, while congestion implies long queues of closely spaced vehicles.

Although long-span bridges are strategic points of the road network, they are not well represented in current design codes. For instance, Eurocode 1 applies only to spans up to 200 m. Common design practice usually relies on conservative assumptions about the traffic and does not consider variability of congestion patterns and driver behaviour. Since bridge maintenance operations are rather expensive, such assumptions may play a decisive role in the assessment of existing bridges. Therefore, a more accurate and site-specific traffic loading may result in significant savings in maintenance operations.

The available models for traffic loading take into account the variability of truck weights, but they often assume full stop traffic with a mix of cars and heavy vehicles at a minimum bumper-to-bumper distance (for instance Ditlevsen & Madsen 1994). However, real-world observations have shown that congestion can take up different forms. Moreover, data is very often collected during free-flowing traffic, due to the fact that it occurs more frequently than congested traffic and the sensor accuracy is generally higher.

An important feature of traffic to bridge loading is that drivers do not usually like staying between larger vehicles and therefore cars typically move out from between trucks, as traffic becomes congested. This results in the formation of truck platoons in the slow lane, thereby changing the car-truck mix during congestion events. This makes the direct use of the widely available (and used) free traffic measurements problematic.

In this paper, an in-house micro-simulation tool is used for generating congested traffic on a two-lane same-way roadway. It is based on an acknowledged car-following model, which is able to replicate most observed congestion patterns (Treiber et al. 2000), as well as on a lane-changing model (Kesting et al. 2007). Since this is the first application of both models for the study of congestion patterns, the results may be of interest for the traffic engineering community as well.

The different congestion patterns are then studied with regards to their effects on the load on a sample 100 m long bridge. It is found that very slow-moving traffic returns the highest loading events, rather than full stop conditions (Figure 1). It is also found that a moderate inflow returns load events of the same order of the ones occurring at the road capacity, thus suggesting that critical load events may also occur outside of rush hours.

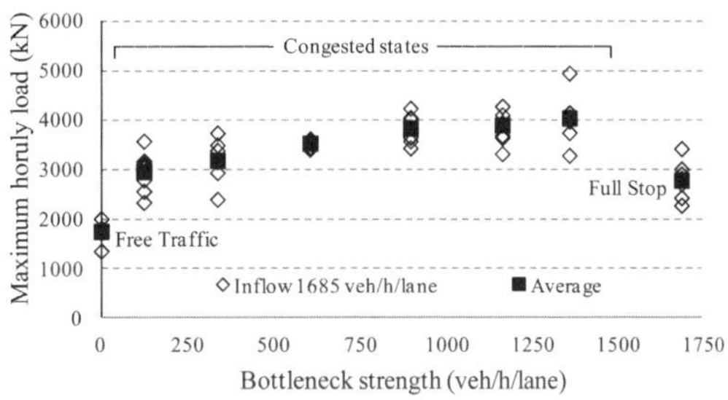

Figure 1. Maximum hourly total load on the sample bridge.

REFERENCES

Ditlevsen, O. & Madsen, H.O. 1994. Stochastic vehicle-queue-load model for large bridges. *Journal of Engineering Mechanics* 120(9): 1829–1847.

Kesting, A., Treiber, M. & Helbing, D. 2007. General lane-changing model MOBIL for car-following models, *Transportation Research Record* 1999: 86–94.

Treiber, M., Hennecke, A. & Helbing, D. 2000. Congested traffic states in empirical observations and microscopic simulation. *Physical Review E* 62: 1805–1824.

Bridge Maintenance, Safety, Management, Resilience and Sustainability – Biondini & Frangopol (Eds)
© 2012 Taylor & Francis Group, London, ISBN 978-0-415-62124-3

Comparison of electromagnetic non-destructive evaluation techniques for the monitoring of chloride ingress in cover concrete

R. du Plooy, S. Palma Lopes, G. Villain & X. Dérobert
LUNAM Université, IFSTTAR, MACS, Bouguenais, France

B. Thauvin & C. Lestréhan
LRPC (Laboratoire Régional des Ponts et Chaussées), Saint-Brieuc, France

ABSTRACT

A major cause of damage in reinforced concrete bridges is steel corrosion resulting from the ingress of aggressive agents, particularly chlorides.

The focus of this paper is to evaluate cover concrete in terms of water and chloride content during the corrosion initiation stage using electromagnetic non-destructive evaluation (NDE) techniques. The electromagnetic (EM) observables obtained from these techniques are direct current resistivity and relative permittivity, material properties known to be sensitive to parameters related to the aggressiveness of the environment such as chloride and volumetric water content (Monfore 1968, Adous 2006).

An experimental study was conducted to compare the capacity of three in-situ electromagnetic techniques to monitor concrete condition: Electrical Resistivity Tomography (ERT), capacimetry and Ground Penetrating Radar (GPR). These techniques were used to indirectly obtain DC-resistivity (from ERT), low frequency permittivity (from capacimetry) and high frequency permittivity (from GPR).

Three slabs were placed in solutions of 0 g/L, 15 g/L and 30 g/L NaCl respectively. The ingress and concentration of the electrolytes were then monitored over a period of three days and again measured at 23 days after a saturated state had been reached.

ERT was found to be a useful tool to track the ingress of a saline electrolyte in concrete. Results correlate well with semi-destructive gammadensimetry tests conducted on cores of identical conditioning. The technique also clearly indicates a decrease in resistivity as the salt content in the slabs increases.

Capacimetry shows promise in the monitoring of property gradients over depth, though the interpretation of results remains difficult. Capacimetry clearly discerns between different chloride contents due to the sensitivity of permittivity to chlorides at low frequencies.

Results from monostatic radar (reflection mode) show good capability of tracking the saturation front. However, the technique requires a clear boundary of contrasting permittivity to be successful. The sensitivity of radar to chloride content is much less than for the other techniques due to the higher frequencies at which it operates.

The sensitivity of the techniques to chloride content is presented in Figure 1, indicating the EM observables for contaminated slabs relative to Slab 1, containing no chlorides (saturated condition).

The three techniques show varying capacity to provide quantitative and qualitative information on the condition of concrete and to identify the presence of property gradients. Though ERT (resistivity) seems to yield the best results for this particular study, capacimetry and radar techniques are quickly developing and have great promise for the future non-destructive characterisation of concrete.

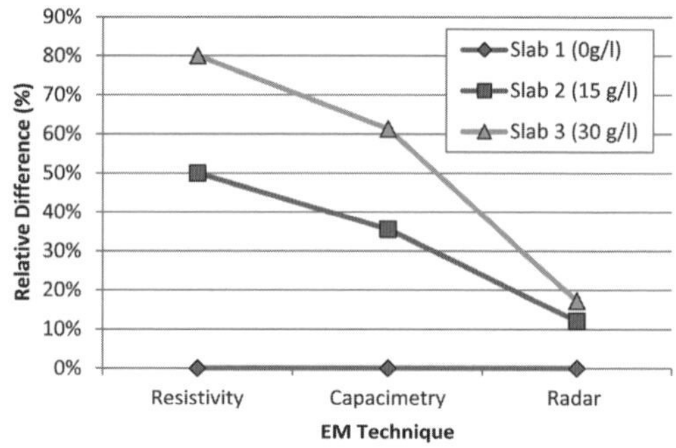

Figure 1. Sensitivity of EM methods to chloride content.

REFERENCES

Adous, M., P. Quéffélec & Laguerre L. 2006. "Coaxial cylindrical transition line for broadband permittivity measurement of civil engineering materials." *Measurement Science and Technology* 17(8):2241–2246.

Monfore, G. E. 1968. "The Electrical Resisitivity of Concrete." *Journal of the PCA Research and Development Laboratories* 10:35–48.

Bridge Maintenance, Safety, Management, Resilience and Sustainability – Biondini & Frangopol (Eds)
© 2012 Taylor & Francis Group, London, ISBN 978-0-415-62124-3

TEAM – a Marie Curie approach to bridge management

C. McNally
School of Civil, Structural & Environmental Engineering, University College Dublin, Ireland

ABSTRACT

Transport demand both for passengers and freight continues to grow strongly across Europe. To meet economic growth targets and facilitate further economic integration of the Member States, European transportation needs to cater for a continuing medium term growth in demand. However transport growth using the technologies of today is unsustainable as the tools do not yet exist to allow optimal management of bridge portfolios. Recent technical advances however offer the potential to significantly change the way that bridges are managed. New technologies and processes are being developed that will deliver more efficient infrastructure management systems. The TEAM project (Training in European Asset Management) is currently exploiting the benefits of new sensor technologies, methodologies, models and algorithms to monitor the condition and safety of bridges. This is providing new sources of asset health data, which is combined with new computer models and algorithms to produce a step change improvement in the accuracy of condition and safety assessments. Knowing exactly the processes, parameters, implications and state of health of a portfolio of bridges, will extend their safe working lives and reduce costs. It will prevent premature and sub-optimal repair, rehabilitation and replacement of assets without compromising safety. The TEAM project is working to achieve this through the implementation of a coordinated research training network targeting these key issues.

Knowledge and understanding is a fundamental requirement for determining the optimal solution to any problem. For transport structures safety continues to remain a critical issue as all bridges, culverts, retaining walls, etc., must have an acceptably low probability of failure and knowledge of the safety is fundamental to an optimal strategy. The TEAM project proposes 14 interlinked PhD projects, all with the goal of extending the lives of transport assets through improved methods of assessment and monitoring. The research is arranged in three Research Clusters as follows:

- Pavement service life optimization
- Enhancing structural capacity
- Traffic loading impacts

A key component of the TEAM project is the implementation of a structured training approach. The coordinated research topics address key issues facing current infrastructure management practice. However for these to be adequately researched, a range of supports, training activities and dissemination events are needed and these are currently being provided by the project partners. These include the provision of:

- Advanced technical skills so as to correctly address the technical challenge
- Research secondments to produce a more integrated project team with a wider appreciation of the technical and cultural issues
- Communication skills to allow the researchers disseminate their results to other engineers and the general public
- Business skills to allow the results be fully exploited by researchers who understand the commercial realities of bridge management
- Transferrable skills to allow the researchers overcome cultural and language barriers

The ambition of the TEAM project is that this approach will lead to a new generation of researchers with the wide range of skills needed to solve the challenges facing bridge engineers today. This will lead to improved bridge management practices internationally, as well as offering a new model for the training of future bridge managers.

Bridge Maintenance, Safety, Management, Resilience and Sustainability – Biondini & Frangopol (Eds)
 ISBN 978-0-415-62124-3

Estimation of lifetime maximum distributions of bridge traffic load effects

E.J. OBrien & D. Hajializadeh
University College Dublin, Dublin, Ireland

E. Sheils
Roughan and O'Donovan, Sandyford, Dublin, Ireland

B. Enright
Dublin Institute of Technology, Dublin, Ireland

ABSTRACT

Repairing or replacing deteriorated bridges is expensive due to the cost of the repair itself but also due to the disruption to traffic and the resulting delays. Large savings are therefore possible by proving that bridges are safe without intervention. To assess the safety of an existing bridge, the traffic loads to which it may be subject in its lifetime need to be accurately quantified. Statistical or probabilistic approaches are commonly adopted to find an appropriate level of load for such an assessment.

In this paper, an extensive Weigh-in-Motion (WIM) database, from Slovakia, is used to calibrate a sophisticated simulation model of two-directional traffic. It is used as a baseline to find the distribution of extreme values for a number of load effects. These distributions are used in turn to find the probabilities of failure. Using the probability of failure as the benchmark, traditional measures of safety – factor of safety and reliability index – are reviewed. Both are found to give inconsistent results.

It is found in the literature (Ayyub et al. 2002, Ayyub & McCuen 2003, Kenshal 2009, Lee & Hwang 2008, Melchers 1999) that the relationship between the reliability index β and the probability of failure, P_f, is given by,

$$P_f = 1 - \Phi(\beta)$$

where $\Phi(\cdot)$ is the cumulative distribution function for the standard Normal distribution. This relationship assumes not only Normal probability distributions for all the random variables in the limit state equation but also linear performance functions. However, in practice, it is common to deal with nonlinear performance functions with relatively low levels of linearity. For a given load effect, the reliability index using the inverse Normal distribution function of the probability of failure is 4.3 for a probability of failure of 10^{-6}. However, as can be seen in Figure 1, the reliability index at this probability varies from 2.0 to 3.3, depending on load effect. It is not surprising that the relationship between reliability index and probability of failure is different from that commonly assumed though the extent of the difference is great. What is perhaps more of concern is the great variability in the reliability index values for a given probability of failure.

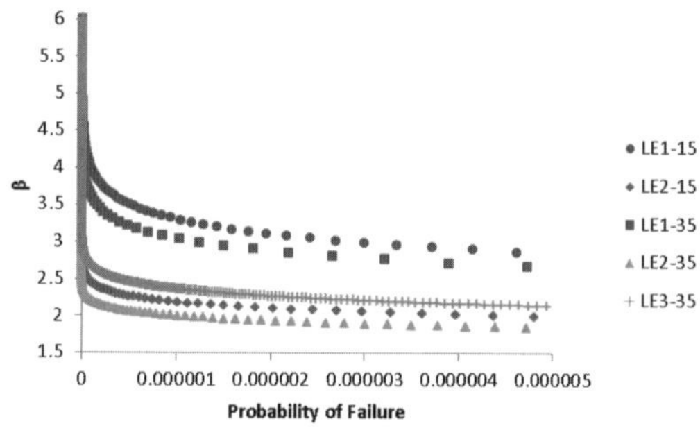

Figure 1. Relationship between reliability index, β, and probability of failure for a range of load effects.

REFERENCES

Ayyub, B. M., Assakkaf, I. A., Beach, J. E., Melton, W. M., Nappi Jr, N. & Conley, J. A. 2002. Methodology for developing reliability-based load and resistance factor design (lrfd) guidelines for ship structures. *Naval engineers journal* 114(2): 23–42.

Ayyub, B. M. & McCuen, R. H. 2003. *Probability, statistics, and reliability for engineers and scientists* Taylor and Francis.

Kenshal, O. 2009. Influence of spatial variability on whole life management of reinforced concrete bridges. Ph.D Thesis, Trinity College Dublin.

Lee, Y. K. & Hwang, D. S. 2008. A study on the techniques of estimating the probability of failure. *Journal of the Chungcheong Mathematical Society* 21(4): 573–583.

Melchers, R. E. 1999. *Structural reliability: Analysis and prediction* John Wiley & Sons New York.

Bridge Maintenance, Safety, Management, Resilience and Sustainability – Biondini & Frangopol (Eds)
© 2012 Taylor & Francis Group, London, ISBN 978-0-415-62124-3

Fatigue assessment of bridges using realistic train models

A. Ottosson & C. Pedersen
Ramboll Group A/S, Copenhagen, Denmark

A.J. O'Connor
Trinity College Dublin, Dublin, Ireland

ABSTRACT

This paper demonstrates the variation in calculated fatigue damage caused by railway traffic mixes defined by the CEN (2003) in comparison to those derived from weigh in motion (WIM) data. The fatigue assessment is carried out according to the cumulative damage method defined in CEN (2005).

The weigh in motion (WIM) data was collected during 2010 by Danish railway operators and contains information on ~6500 freight trains. Eurocode defines twelve train configurations consisting of passenger trains, freight trains and underground trains. The train configurations are combined in three traffic mixes that can be chosen depending on whether the structure carries mixed traffic, predominantly heavy freight traffic (25 t axles) or lightweight passenger traffic (CEN 2003).

The annual traffic volumes differ with almost a factor of three between the Eurocode traffic mixes and the WIM data. In order to satisfactory compare them and their results from the calculations, it was decided to evaluate effects per million tonnes of traffic volume.

The bridges used to investigate the fatigue damage are continuous bridges over 2 spans of equal length, with overall lengths ranging from 20 m to 120 m. Section properties have been optimized and derived for 6 sections by using Eurocode load model 71 (CEN 2003) with dynamic enhancement factors taken at the ultimate limit state.

The use of train models based on WIM data illustrates a large spread in axle loads, see Figure 1, compared to the train models supplied by the Eurocode, making various mixes of heavy and light wagons possible. There are spans of critical length for the Eurocode traffic mixes, which have a strong relation to the lack of mixed wagon weights within the train models, and will result in non conservative results compared to WIM data.

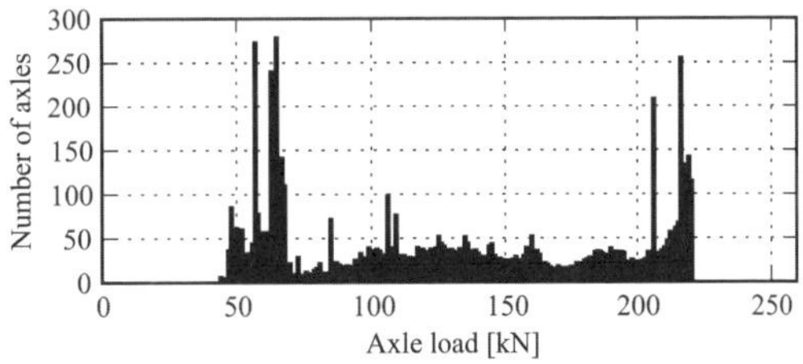

Figure 1. Axle load distribution for WIM data per million tonne.

The critical span length is shown to correspond to the average wagon length for the Eurocode traffic mixes and is most prominent for sections closest to midspan.

Despite the large differences in wagon weight distribution between Eurocode traffic mixes and the WIM data, the resulting fatigue damage show clear similarities for longer spans. Shorter spans show less resemblance due to susceptibility to local effects, caused by the combinations of heavy and light wagons and the individual axle loads.

REFERENCES

European Committee for Standardization 2003. *Eurocode 1: Actions on structures – Part 2: Traffic loads on bridges*. Brussels: European Committee for Standardization.

European Committee for Standardization 2005. *Eurocode 3: Design of steel structures – Part 1–9: Fatigue*. Brussels: European Committee for Standardization.

European Committee for Standardization 2006. *Eurocode 3 – Design of steel structures – Part 2: Steel Bridges*. Brussels: European Committee for Standardization.

Bridge Maintenance, Safety, Management, Resilience and Sustainability – Biondini & Frangopol (Eds)
© 2012 Taylor & Francis Group, London, ISBN 978-0-415-62124-3

A monitoring system for determination of real deck slab behaviour in prestressed box girder bridges

M.A. Treacy & E. Brühwiler
Ecole Polytechnique Fédérale de Lausanne (EPFL), Switzerland

ABSTRACT

The demand for more sustainable infrastructure requires road authorities to maximise the lifetimes of existing bridges whilst limiting resource usage during improvement interventions. This is complicated by substantial increases in the volume and weight of heavy traffic in recent years. For example, estimates show that goods transport by road in tonne-kilometres has almost doubled in the period from 1985 to 2010 in the Western Europe EU-15 countries (Belluci et al. 2011) and this trend is also present in many other countries.

Deck slabs of road bridges are particularly exposed to high numbers of fatigue stresses. Studies such as Schläfli and Brühwiler (1998) found that the reinforcement is the determinant fatigue element in the deck slabs of reinforced concrete bridges. It can be difficult to accurately model their service behaviour as secondary elements such as parapets, kerbs and surfacing layers reduce the stress levels in the steel reinforcement bars. For these reasons a monitoring regime is favoured for the accurate determination of real 'action effects' within the bridge elements.

This paper presents a 1960's prestressed concrete box girder motorway bridge in Switzerland which is instrumented with a remotely accessible structural health monitoring system. The key focus of the study is on the direct measurement of strains in reinforcement bars. The approach involves measuring multiple phenomena at a strategic location in the structure to provide a 'Smart Section'. Latest sensing and data retrieval technologies have been implemented to demonstrate this monitoring approach is a suitable assessment method to reflect the real behaviour within the deck slabs of such bridges.

The installed instrumentation allows for accurate and reliable readings of relevant bridge deck data, in particular the capture of very small strain values. The system has the ability to capture the signature of extreme vehicles crossing event as strain and vibrational response in the structural elements. This recorded data allow for a first evaluation of the fatigue and structural safety of a given element and is a direct approach to answering the question: "is there enough safety?"

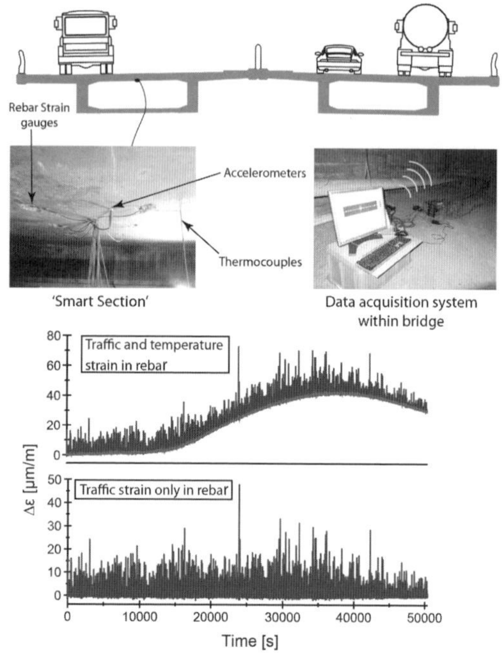

Figure 1. Schematic of monitoring system (Top) and; Strain changes in a deck slab reinforcing bar throughout day (Bottom).

REFERENCES

Bellucci, P., di Roma, C., Fusco, G. & La Monica, S. 2010. FEHRL Vision 2025 for Road Transport in Europe. Available at: www.fehrl.org/?m=32&mode=download&id_file=184

Schläfli, M. & Brühwiler, E. 1998. Fatigue of existing reinforced concrete bridge deck slabs. Engineering structures 20: 991–998.

Bridge Maintenance, Safety, Management, Resilience and Sustainability – Biondini & Frangopol (Eds)
© 2012 Taylor & Francis Group, London, ISBN 978-0-415-62124-3

Extrapolation of traffic data for development of traffic load models: Assessment of methods used during background works of the Eurocode

X.Y. Zhou, F. Schmidt & B. Jacob
IFSTTAR, France

ABSTRACT

Transport demand is growing worldwide, especially the road freight transport has increased in Europe by 45.7% between 1995 and 2008 and this strong growth trend seems likely to keep continuing at the same rate of about 2.7% per year (European Commission 2010). It is important to ensure that the European transport network can sustain this continuing growth in demand by assessing the health of existing bridges exposed to the current and future traffic loads.

The currently used normal load model, LM1, in Eurocode 1, Part 2 was first calibrated with a number of two weeks of heavy traffic dataset from Auxerre (A6 motorway, Paris to Lyon, France) in the late 1980s; it was then re-calibrated by O'Connor et al (2001) with several representative European traffic datasets recorded at France weigh-in-motion (WIM) sites. This model needs to be periodically re-assessed by current traffic, because of the wide changes in traffic volume, composition of traffic, vehicle weights and sizes, to ensure a satisfactory safety level for the design of new bridges; and also the quality of WIM data has increased greatly in the last decade due to improved technologies and the development of specifications regulating accuracy levels. Accurate prediction of the extreme load effects expected during the proposed or remaining lifetime of a structure is a key issue for the design or assessment of highway bridges.

This paper reviews several methods to calculate these extreme effects: the extrapolation methods (fitting a distribution to the upper tail data and Rice's formula) implemented in the background study of the Eurocode EN1991-2, and recently extensively used methods (block maxima method and peaks over threshold method). The methods are then applied on weigh-in-motion (WIM) recorded traffic to extrapolate characteristic values of long return period to reveal the evolution of transportation. Comparisons are made on the extrapolated extreme gross vehicle weight (GVW), total load on lane (TLL) and induced traffic load effect (TLE).

The approach of fitting a Gaussian distribution to the upper tail, block maxima method and peaks-over-threshold method, except for Rice formula, have been used to extrapolate characteristic values of GVW for long return period. The results show that normal distribution based estimates are larger than those based on extreme value theory (EVT), it is surely due to the normal distribution converge to a infinite bound extreme value distribution, while the negative shape parameters of the two extreme value distribution means they have a bounded tail. Because the rare cranes or special trucks were not found in these two set of data, it can be conclude an infinite bound normal distribution is not suitable to be used to extrapolate the characteristic value of GVW for long return period, while the EVT based methods are recommended.

Only block maxima method adopted to extrapolate the characteristic value of TLL, because the TLL is a square-waved process, it avoids the Gaussian process assumption of Rice formula. The extrapolated values are lower than the LM1 design values, which are said that the design load is still enough to ensure structural safety under modern traffic. All methods except the first one are used to extrapolate the characteristic values of TLE.

In this study, various prediction methods are applied to extrapolate the characteristic GVWs, TTLs and TLEs for long return periods. The results show that it is very important to choose the adapted prediction method for the considered effect.

REFERENCES

Eurocode 1: Actions on structures Part 2: Traffic loads on bridges, 2003.

O'Connor, A., O'Brien, E.J., & Prat, M. 1998. Report of studies performed on the normal load model, LM1, ENV 1991-3 Traffic loads on bridges. Paris: Laboratoire Central des Ponts et Chaussees.

Energy harvesting in bridges and transportation infrastructure networks

Organizer: K. Gkoumas

Bridge Maintenance, Safety, Management, Resilience and Sustainability – Biondini & Frangopol (Eds)
© 2012 Taylor & Francis Group, London, ISBN 978-0-415-62124-3

Vibration energy harvesting devices based on magnetostrictive materials

D. Davino & C. Visone
Dipartimento di Ingegneria, Università del Sannio, Benevento, Italy

A. Giustiniani
DIEII, Università di Salerno, Fisciano (SA), Italy

ABSTRACT

Energy harvesting from mechanical vibrations can be a suitable method of powering wireless sensors for bridge health monitoring. This paper tackles the main properties and modeling techniques of magnetostrictive materials and devices for harvesting purposes. A novel possible magnetostrictive harvester is presented too.

Indeed, magnetostrictive materials (Engdahl 2000) have been proposed for vibration energy harvesting applications by exploiting the Villari effect (Zhao & Lord 2006). Vibrations are a source of energy with almost a 24h per day availability in several infrastructures. On the other hand, in civil engineering there is a strong need of structural monitoring the health of bridges (Del Grosso et al. 2002) and other civil infrastructures, by measuring accelerations and resonant frequencies (Lynch 2007). Bridges vibrates because of the wind action and for traffic loadings. The possibility to convert this ambient mechanical energy into electrical energy to feed the sensors is very attractive in those applications. This type of conversion can be performed by means of linear electromagnetic generators or by using magnetostrictive materials. Up to now, three magnetostrictive materials have been studied and employed for energy harvesting applications: Terfenol-D and, more recently, Galfenol and Metglas.

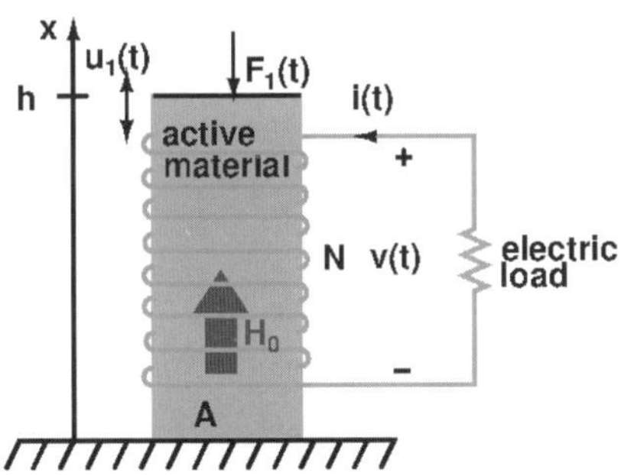

Figure 1. Basic structure of the magnetostrictive harvester. The active material is a magnetostrictive rod with section "A", length "h" and with a N-turns coil closed over an electric load. The rod tip is under a force $F_1(t)$ and moves at a speed $u_1(t)$.

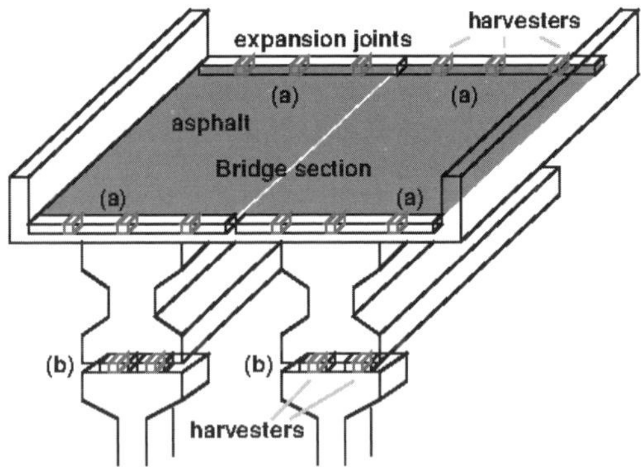

Figure 2. Sketch of a beam-column connection in a bridge with possible harvesters location sites.

The linear modeling of the material characteristics and of a direct force device, as the one represented in Figure 1, lead to a transfer function with a high-pass frequency behavior. This model can be implemented in a numerical simulator to test realistic forces signals. As sketched in Figure 2, such harvesters can be located under suitable asphalt expansion joints (a) or in suitable bridge bearings between beams and columns (b). The average powers of both cases, with suitable dimensions of the various elements, are fully compatible with literature results and are, in some cases, higher with respect to other harvesting techniques, reaching hundreds of mW with voltages far above the volt range.

REFERENCES

Del Grosso, A., Inaudi, D., Pardi, L. 2002, Overview of european activities in the health monitoring of bridges. *Proc. 1st Int. Conf. Bridge Maintenance, Safety Management (IABMAS)*, Barcelona, Spain, 14–17.

Engdahl, G. (ed.) 2000, Handbook of Giant Magnetostrictive Materials, NewYork: Academic.

Lynch, J.P. 2007. An overview of wireless structural health monitoring for civil structures. *Phil. Trans. R. Soc. A* 365: 345–372.

Zhao, X. & Lord, D.G. 2006, Application of the Villari effect to electric power harvesting. *J. Appl. Phys.* 99 (8): 08M 703.

Bridge Maintenance, Safety, Management, Resilience and Sustainability – Biondini & Frangopol (Eds)
© 2012 Taylor & Francis Group, London, ISBN 978-0-415-62124-3

Nonlinear vibration harvesting for extended structures monitoring

L. Gammaitoni, H. Vocca, I. Neri, F. Orfei & F. Travasso
NiPS Laboratory, University of Perugia and INFN, Perugia, Italy

ABSTRACT

The long time continuous monitoring of extended structures via small-scale electronic mobile devices is considered of primary importance for many applications in civil engineering.

These devices all need distributed powering systems. Presently this means wired power-grids, batteries or RF-sources, however all these solutions present some back-draws. Wiring is expensive, add weight and is subjected to high failure rate in devices subjected to repeated motion. Traditional batteries are not a viable solution to the powering of such devices mainly because they have to be replaced once exhausted. Alternative solutions based on micro fuel cells and micro turbine generators are also not suitable: both require refuelling when their supplies are exhausted. Thus the goal is powering such devices with with energy harvested from the ambient has been in recent years the subject of a great research efforts. It has been observed that the harvesting of kinetic energy present in the form of random vibrations is an interesting option due to the almost universal presence of this kind of motion.

Present working solutions for vibration energy harvesting are based on oscillating mechanical elements that convert kinetic energy via capacitive, inductive or piezoelectric methods. These oscillators are usually designed to be resonantly tuned to the ambient dominant frequency. However, in most cases the ambient random vibrations have their energy distributed over a wide spectrum of frequencies, being rich especially at low frequency (like in the vibration from a transit bridge, Figure 1), and frequency tuning is not always possible due to geometrical/dynamical constraints.

In this work we discuss a different method based on the exploitation of the dynamical features of stochastic bistable oscillators (Cottone et al. 2009) employed to model nonlinear piezoelectric harvesters (Gammaitoni et al. 2010).

Such a method is shown to outperform standard linear oscillators and to overcome some of the most severe limitations of present approaches. Digital simulations are shown together with experimental measurements realized on extended structures.

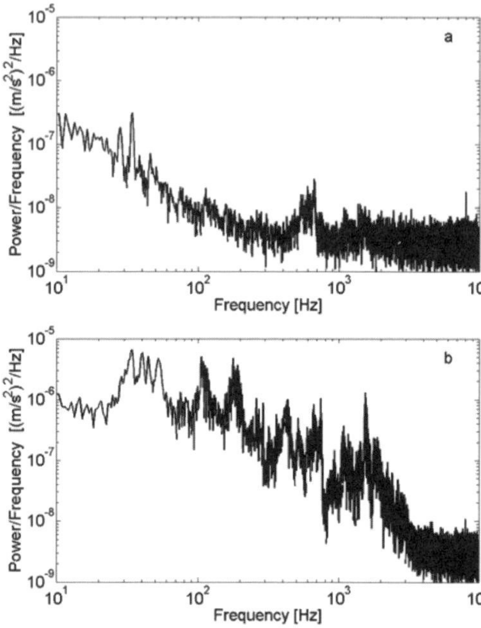

Figure 1. Bridge excitation on the guardrail: a) Power spectrum of the bridge without any running car; b) Power spectrum of the bridge excited by a single car passing through.

REFERENCES

Cottone F., Vocca H., Gammaitoni L., 2009, "Nonlinear Energy Harvesting", Physical Review Letters, Vol 102, 080601.

Gammaitoni L., Neri I., Vocca H., 2010, "The benefits of noise and nonlinearity: Extracting energy from random vibrations", Chemical Physics 375 (2–3), 435–438.

Bridge Maintenance, Safety, Management, Resilience and Sustainability – Biondini & Frangopol (Eds)

Energy harvesting in bridges and transportation infrastructure networks: State of art, recent trends and future developments

K. Gkoumas
Sapienza University of Rome, Italy

ABSTRACT

In the upward trend of renewable energy growth, several proposals have been made concerning energy harvesting devices in transportation infrastructure networks. The objective, concerning higher power extraction, is to supply power to auxiliary systems (e.g. road lights or information panels), thus, satisfying the requirement for sustainable transportation infrastructures. This paper deals with the state of art, the recent trends and the future developments in energy harvesting for bridges and transportation infrastructure networks. A literature review is provided (both for low- and high-energy applications), and after a taxonomy, focus is given to the definition of a broader framework of energy extraction for such systems. In this sense, it is possible to evaluate the system efficiency for different meso-scale energy harvesting applications in the civil engineering field (Table 1). Finally, a survey takes place for the possible research issues and synergies among different research sectors.

Table 1. Classification of the energy production schemes.

RESOURCE Defined in terms of E_H, E_C^{SS} +	From natural resources $E_H = \infty; E_C^{SS} = 0$	From artificial resources $E_H < \infty; E_C^{SS} > 0$	From natural resources, modified by artificial systems $E_H = ?; {}^*E_C^{SS} > 0$ * Greater than if a system of extraction was implemented
EXTRACTION SYSTEM Defined in terms of E_C^{ES} =	Magnetic induction converter; Electrostatic converter; Piezoelectric converter	Photovoltaic converter; Radian energy converter	Thermal energy converter; RF converter; …
COUPLED SYSTEM Defined in terms of E_C, $\hat{E}_H$, ΔE	Example: wind turbines for a wind farm.	Example: piezoelectric converters placed on deformable joints of a flexible structure.	Example: wind turbine on a skyscraper (the source is natural but it is placed in high attitude due to the skyscraper).
ΔE: ACTIVE (ΔE>0)	SUCCESSFUL[1]	SUCCESSFUL[3]	SUCCESSFUL[3]
ΔE: PASSIVE (ΔE<0)	UNSUCCESSFUL	SUCCESSFUL[1] $\hat{E}_H - E_C^{ES} > 0$ UNSUCCESSFUL $\hat{E}_H - E_C^{ES} < 0$	SUCCESSFUL[1] $\hat{E}_H - E_C^{ES} > 0$ UNSUCCESSFUL $\hat{E}_H - E_C^{ES} < 0$
ΔE: BALANCED (ΔE=0)	UNSUCCESSFUL	SUCCESSFUL[2]	SUCCESSFUL[2]

SUCCESSFUL[1]: Production of energy; SUCCESSFUL[2]: Production of an amount of energy equal to the one consumed; SUCCESSFUL[3]: Production of an amount of energy higher than the one consumed
Definitions: E_H ~ Maximum extractable energy; E_C^{SS} ~ Energy cost of the structural system; E_C^{ES} ~ energy cost of the extraction system; E_C ~ Total energy cost of the coupled system; $\hat{E}_H$ ~ Effective extracted energy of the coupled system; ΔE ~ Energy balance of the coupled system; ΔE' ~ Energy balance of the extraction system

Bridge Maintenance, Safety, Management, Resilience and Sustainability – Biondini & Frangopol (Eds)

Adaptive MR damper based control system

X.C. Guan, Y.H. Huang & H. Li
School of Civil Engineering, Harbin Institute of Technology, Harbin, China

J.P. Ou
School of Civil Engineering, Dalian University of Technology, Dalian, China

ABSTRACT

To simplify the magnetorheological (MR) damper based system and enhance its reliability, an adaptive MR damper based control system, combining piezoelectric energy harvesting technology and MR damper based control technology, is proposed. The proposed adaptive MR damper based cable control system and adaptive MR damper based isolation system are respectively shown in Figure 1 and Figure 2.

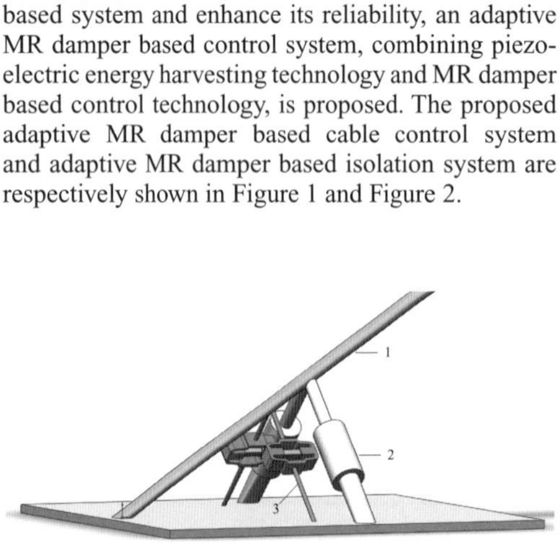

Figure 1. Schematic diagram of adaptive MR damper based cable control system (1. Cable, 2. MR damper, 3. Piezoelectric energy harvester).

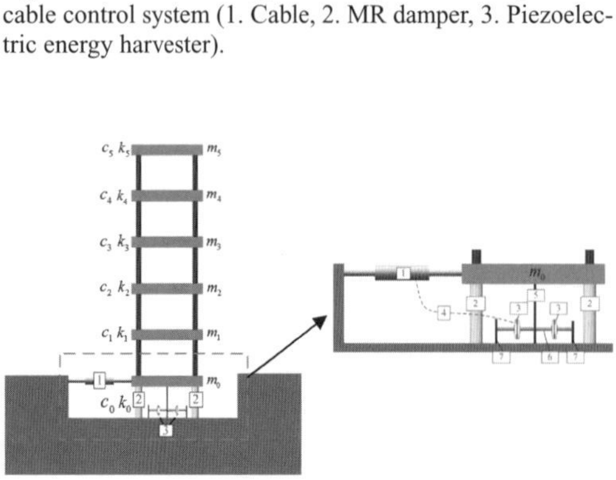

Figure 2. Schematic diagram of adaptive MR damper based isolation system (1. MR damper; 2. Isolator; 3. Piezoelectric energy harvester; 4. Wire; 5. Flexible fixture; 6. Elastic rope; 7. Rigid fixture).

The piezoelectric energy harvester can be designated to contain an external amplification device and a PZT stack. To set an external amplification device is to conveniently connect the PZT stack and isolated building or the cable and to solve the stack deformation-limited problem. In one vibration circle, two sets of piezoelectric energy harvesters can alternately power the damper.

This paper focuses on the determination of PZT stack parameters. To ensure the stack safe, the maximum strain should less than or equal to the limit strain, and then the range of the stack length can be obtained.

The active control force is applied when the controlled structure is away from the balance; otherwise, the minimum force of MR damper is applied. In the whole control process, the demanding energy W_{xq} should be less than or equal to the energy supplied by the harvester W_{sj}, and then the lower limit of the stack area can be got.

To ensure the controllability of the system and the harvester not broken, the ratio of the damping force supplied by harvester to the whole system control force should be small. Assuming the ratio of the force by harvester to maximum MR damping force shouldn't exceed $1/k$, the upper limit of the stack area can be got.

Adaptive MR damper based cable control system and isolation system are respectively devised and simulated based on the N26 cable model from Shandong Binzhou Yellow River Bridge and one five-storey building model. According to the simulation results, it is concluded:

1) For the efficacy of cable control system, adaptive control system is obviously better than the conventional control system with Simple Bang-Bang control strategy and Passive-off strategy; adaptive system is slightly worse than active system, but can be further improved by optimizing the parameters of PZT stack.
2) For the efficacy of isolation control system, adaptive control can more effectively to restrict the base drift and roof relative displacement nearly without increasing the peak isolator and roof accelerations.

Bridge Maintenance, Safety, Management, Resilience and Sustainability – Biondini & Frangopol (Eds)
 ISBN 978-0-415-62124-3

A self-powered vibration monitoring and control system for stay cables: Numerical study

S. Zhu & W.A. Shen
Department of Civil and Structural Engineering, The Hong Kong Polytechnic University, Hong Kong

J. Li
Guangdong Provincial Academy of Building Research, Guangdong, China

ABSTRACT

This paper presents an innovative strategy to integrate damper system and sensor system together so as to drive sensors using the power output from dampers. This strategy not only solves the common power supply problem associated with emerging wireless sensors (WS), but also avoids the overheating problem observed in some conventional fluid dampers. As a key element in the integrated system, electromagnetic dampers (EMD) can convert vibration energy into electricity, and provide both vibration mitigation function and energy harvesting function when connected with appropriate energy harvesting circuits (EHC). A PID compensated buck converter operating in a continuous conduction mode is employed in the EHC to charge a rechargeable NiMH battery with the power output from the EMD. Bridge stay cables, which are vulnerable to excessive vibration due to their low damping and great flexibility, are one of potential applications of the proposed EMD-WS system. A coupled numerical model consisting of stay cable, EMD and EHC is established using Matlab Simulink toolbox. The simulation results reveal that the EMD connected to a PID compensated buck converter can effectively reduce the cable's vibration, and it achieves a slightly better control effect than conventional viscous damper because of the existence of the feedback control in the EHC. The average power harvested in this study is 30.49 W, which is considerable compared to conventional micro energy harvesting devices and is able to power quite a few WS when stay cables are subject to excessive vibrations. The peak energy conversion efficiency of 36.7% was achieved. The performance of the EMD predicted in this study clearly demonstrates the feasibility of developing an integrated vibration monitoring and control system based on EMD and WS, and establishing a smart stay cable with superior self-powered and self-diagnostic features.

REFERENCES

Erickson, R.W. & Maksimovic, D. 2001. Fundamentals of Power Electronics. Norwell, MA: Kluwer.
Mehrabi, A.B. & Tabatabai, H. 1998. Unified finite difference formulation for free vibration of cables. *Journal of Structural Engineering-ASCE* 124 (11): 1313–1322.
Zhu, S., Shen, W.A. & Xu, Y.L. 2011. Linear electromagnetic devices for vibration damping and energy harvesting: Modeling and testing. *Engineering Structures* 34 (1): 198–212.

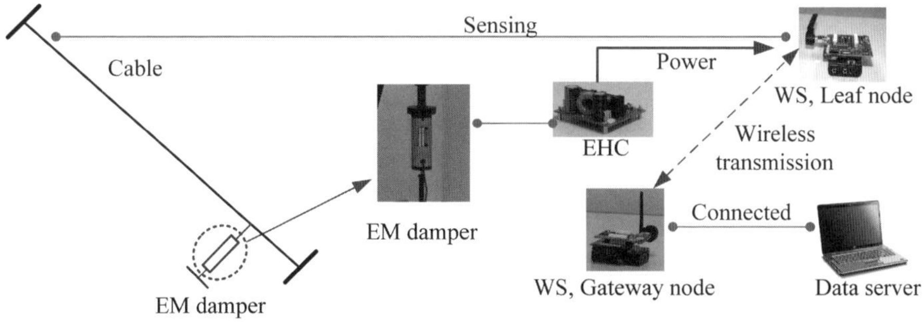

Figure 1. Configuration of SVMC system on a stay cable.

Vulnerability of bridges to fire and explosion

Organizer: L. Giuliani

Bridge Maintenance, Safety, Management, Resilience and Sustainability – Biondini & Frangopol (Eds)
© 2012 Taylor & Francis Group, London, ISBN 978-0-415-62124-3

Consequence-based robustness assessment of bridge structures

F. Brando & L. Cao
Thornton Tomasetti, New York, NY, United States

P. Olmati & K. Gkoumas
Sapienza University of Rome, Italy

ABSTRACT

The aim of this paper is to apply to a steel truss bridge, a methodology that takes into account the consequences of extreme loads on structures, focusing on the influence that the loss of primary elements has on the structural load bearing capacity. In this context, the topic of structural robustness, intended as the capacity of a structure to withstand damages without suffering disproportionate response to the triggering causes while maintaining an assigned level of performance, becomes relevant. In the first part of this study, a procedure for the evaluation of the structural response and robustness of skeletal structures under impact loads is presented and tested in simple structures. In the second part, an application focuses on a case study bridge, the extensively studied I-35W Minneapolis Steel Truss Bridge, a continuous truss bridge over four piers (Figure 1).

The bridge, which has a structural design particularly sensitive to extreme loads, collapsed for a series of other reasons, in part still under investigation. For the analysis, a single lateral span of the bridge is considered (Figure 2, top).

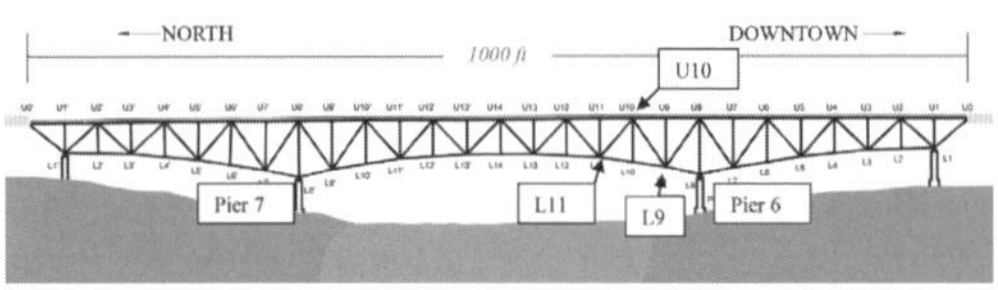

Figure 1. Bridge overview (from Malsch et al. 2011).

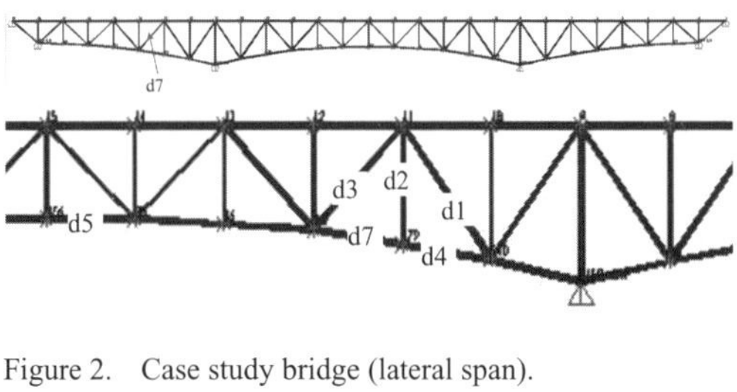

Figure 2. Case study bridge (lateral span).

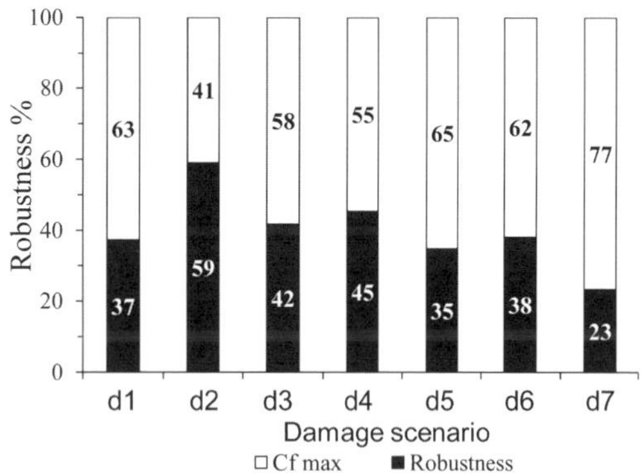

Figure 3. Damage scenario evaluation in terms of C_f for the original configuration of the bridge.

The structure is subjected to a set of damage scenarios (Figure 2, bottom) and the consequence of the damages is evaluated using as metric the so-called "member consequence factor" (C_f) that for convenience can be easily expressed in percentage (Nafday, 2011). The results are shown in Figures 3.

With the aim of increasing the structural robustness of the bridge, and in order to test the sensitivity of the method proposed, an improved variation of the structural system is considered (in the form of a hyper-static steel truss structure). The results of both the original and the enhanced structural schemes, under the same damage scenarios, are compared and commented.

REFERENCES

Malsch, E., Brando, F., Iannitelli, A., Abruzzo, J. & Panariello, G., 2011. The Causes of the I-35 West Bridge Collapse. Proc. 35th Annual Symposium of IABSE / 52nd Annual Symposium of IASS/6th International Conference on Space Structures, London, September 2011.

Nafday, A.M., 2011. Consequence-based structural design approach for black swan events. Structural Safety, 33(1): 108–114.

Bridge Maintenance, Safety, Management, Resilience and Sustainability – Biondini & Frangopol (Eds)

Evaluation of structural risk for bridges under fire

F. Gentili & F. Petrini
Sapienza University of Rome, Rome, Italy

ABSTRACT

One of the most challenging problems of the modern Structural Engineering is connected to the conception and the subsequent analysis and design of constructions able to face Low Probability – High Consequences (LPHC) scenarios. These situations arise for a lot of different and multifaceted reasons, being possibly followed by catastrophic consequences and it's almost impossible to frame them inside any well-recognized probabilistic format. From the design point of view, it has become clear that while a system might have very good LCHP events performances, it can still be very vulnerable to HCLP event induced failures. Then, HCLP and LCHP accidents have generally very different characteristics both from the design and the analysis point of view (Handling Exceptions, 2008).

A specific situation is represented by fire scenarios. In this case, one must follow a) the development of the fire (from the beginning to the spread inside the construction) b) the thermal diffusion inside the construction; c) the structural response linked to the alterations in the material properties due to the change in time of temperature and to the large displacements and deformations that are usually developed. In these situations, it is particularly interesting to follow the progression of failures inside the structural system.

The progression of a fire accident in a construction appears as a really complex phenomenon with possibly unexpected developments. The construction can be considered as composed by a series of firewalls (in the Computer Science meaning) that block the progression of the hazard into a collapse (Reason, 1990). Also if the single shortage is not critical, when an alignment of these weaknesses arises, this can lead to the development of a crisis.

One of the most rational methods to handle structural risk is the Performance-Based Design (PBD). PBD is a modern approach that allows designers to consistently take into account all the aspects related to the serviceability and safety of both existing and new structures without enforcing any limitation to the available design solutions. In case of PBFD of complex structures, such arguments as i) use of the heuristic approach, ii) difficulties encountered in conventional collapse and performance threshold definition, iii) use of advanced models, assume a crucial importance.

The case study presented is the Tsing Ma Bridge (Xu and Ko,1997), located in Hong Kong, China. In this kind of structures a fire in a train may represent a realistic threat. The fire causes a very fast thermal expansion of the upper part of the deck (Fig. 1 top) which is followed by a progressive return to the initial position (Fig. 1 bottom).

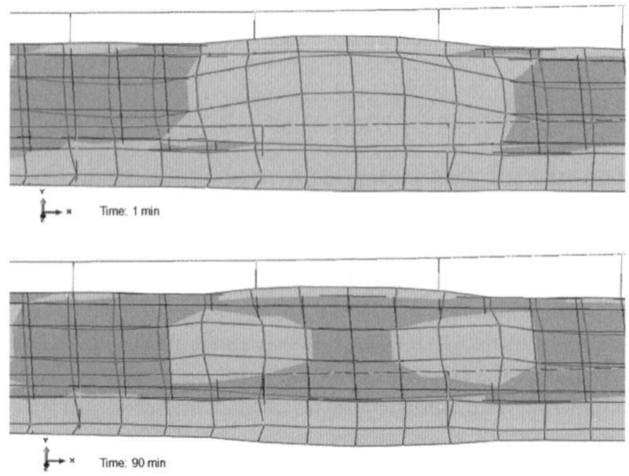

Figure 1. Deformed after 1 min (top) and after 90 min (bottom).

REFERENCES

Handling Exceptions in Structural Engineering 2008, Rome, DOI:10.3267/HE2008.
Reason J. (1990), Human Error, Cambridge University Press.
Xu Y.L., Ko J.M. (1997). "Vibration studies of Tsing Ma suspension bridge", ASCE Journal of Bridge Engineering, vol. 2, no. 4, 149–156.

Bridge Maintenance, Safety, Management, Resilience and Sustainability – Biondini & Frangopol (Eds)
© 2012 Taylor & Francis Group, London, ISBN 978-0-415-62124-3

Vulnerability of bridges to fire

L. Giuliani
Technical University of Denmark, Civil Engineering Department, Lyngby, Denmark

C. Crosti & F. Gentili
Sapienza University of Rome, School of Engineering, Rome, Italy

ABSTRACT

Nowadays a general consensus seems to be achieved among the civil engineering community on the need of taking into account in the design the behavior of structures to exceptional events, such as impacts, fire and explosions (Ellingwood, 2006; Giuliani, in press). However, the design against fire and explosion in particular is mostly limited to the case of buildings and hardly considered in the design of bridges. Nevertheless, not only the vulnerability of bridges to fire may be high, as a consequence of possible high slenderness of the elements and exposed surface to volume ratio of the sections, but also fire-induced failures of bridge, even if not numbered among the first cause of collapses, represent a relatively significant percentage of bridge failures, comparable if not higher to the failure caused by earthquakes (Garlock et al., 2012), which are instead a common concern of bridge designs.

The possible high vulnerability of bridges to fire is witnessed by some recent examples of major collapse of bridges induced by fire, such as the case of the very recent case of the 60 Freeway fire in Montebello, CA, on 14 December 2011, which led to the demolition of the concrete overpass structure, heavily damaged by the intense heat of the fire.

From the review of cases presented in the paper and reported in literature (Astaneh-Asl et al., 2009; Garlok et al., 2012) it is possible to observe that:

- more than 20 bridge fires have been reported since 1995, corresponding to a frequency of more than one occurrence per year in the past 17 years.
- the majority of the fires were caused by a fuel tanker truck accident under a viaduct or falling from it;
- in most case bridges did not collapse but were significantly damaged and high repair costs had to be sustained;
- in some cases where limited structural damages had occurred, high costs deriving from temporary closure and disservice to the traffic had to be sustained.

Figure 1. Truck fire under the Paramount Boulevard bridge (credit: Brian van der Brug/Los Angeles Times).

In view of the considerations above, it seems reasonable to contemplate fire safety design aspects for bridges structures. In particular, the consideration of a fire load case in the form a localized hydrocarbon fire appears sensible for viaducts and overpass, which are mostly subjected to tanker truck fires.

REFERENCES

Astaneh-Asl A., Noble C.R., Son J., Wemhoff A.P., Thomas M.P., McMichael L.D., 2009. Fire Protection of Steel Bridges and the Case of the MacArthur Maze Fire Collapse. *Proc. of the 2009 ASCE Technical Council on Lifeline Earthquake Engineering Conference (TCLEE09)*, Oakland, June 2009

Ellingwood B.R., 2006. Mitigating Risk from Abnormal Loads and Progressive Collapse, *Journal of Performance of Constructed Facilities*, vol. 20, no. 4, pp. 315–324, 2006

Garlock M., Paya-Zaforteza I., Kodur V., Gu L., 2012. Fire hazard in bridges: Review, assessment and repair strategies, *Engineering Structures*, vol.35, pp. 89–98, February 2012

Giuliani L., in press. Structural safety in case of extreme actions, *International Journal of Life-Cycle and Performance Engineering*, accepted for publication in the special issue on Performance and Robustness of Complex Structural Systems on December 2011

Bridge Maintenance, Safety, Management, Resilience and Sustainability – Biondini & Frangopol (Eds)
© 2012 Taylor & Francis Group, London, ISBN 978-0-415-62124-3

Evacuation of mixed populations from trains on bridges

C. Kindler, J.G. Sørensen & A.S. Dederichs
Department of Civil Engineering, Technical University of Denmark, Kgs. Lyngby, Denmark

ABSTRACT

An understanding of human evacuation dynamics and performance are important when designing complex buildings such as bridges and applying performance-based codes in order to reduce the risk of exposing occupants to critical conditions in case of fire. Literature provides a number of case studies of real fire incidents as well as experiments concerning fire and evacuations []. The majority of previous studies deal with the evacuation behavior of homogeneous groups and applies normative standard. However, a significant part of the population is poorly described such as are people with impairments which are about 10%–21% of the world's population (Bendel, 2006), furthermore a mixed population comprehends elderly people, giving an additional 10% (Bendel, 2006). In Denmark 20% of the population are aged below 15 years (Danmarks statistik, 2011). In recent years a series of studies have focused on a broader population for experiments and models (Larusdottir, 2010). The discussion of "equal access" (Steinfeld, 1979) is slowly followed by the demand on "equal ergress" (Proulx, 1996). However, the passengers on trains on bridges are rarely homogeneous mixture. At the same time equal egress is far from assured today (Diament, 2009).

This paper is on the evacuation of mixed populations from trains on bridges. The populations applied in the experiment are mixed corresponding to a composition corresponding to the population of Denmark.

REFERENCES

Bendel, J., Decision Support System for Evaluating Accessibility of Facilities,The Israeli Journal of Occupational Therapy. Vol.15:3, 2006

Diament, M., Disaster plans do not address needs of those with disabilities, FIRE ENGINEERING 2009.

Larusdottir, A.R. and Dederichs, A.S.: *Evacuation of Children: Movement on Stairs and on Horizontal Plane*, Fire Technology: special issue 2010, Springer New York LCC, DOI: 10.1007/s10694-010-0177-6 (2010)

Proulx, G., Pineau, J., Review of Evacuation Strategies for Occupants with Disabilities; National Research Council Canada 1996. http://www.nrc-cnrc.gc.ca/obj/irc/

Steinfeld, E. Access to the Built Environment: A Review of Literature. Washington, D.C., US Dept. of Housing and Urban Development, Office of Policy Development and Research, April,1979

Bridge Maintenance, Safety, Management, Resilience and Sustainability – Biondini & Frangopol (Eds)
© 2012 Taylor & Francis Group, London, ISBN 978-0-415-62124-3

Adapting OpenSees to simulate bridge structures in fire

A. Usmani, Y. Jiang, J. Jiang, L. Jiang & S. Welch
School of Engineering, The University of Edinburgh, Edinburgh, UK

ABSTRACT

OpenSees is an object oriented software framework originally developed at UC Berekeley, McKenna (1997), and currently supported by PEER and Nees. OpenSees has so far been focussed on providing an advanced computational tool for analysing the non-linear response of structural frames subjected to seismic excitations. Given that OpenSees is open source (available for free download at opensees.berkeley.edu) and has been available for best part of this decade, it has spawned a rapidly growing community of users as well as developers who have added to it's capabilities over this period. We have been working to develop this framework to enable simulation of structures subjected to fire, e.g. Usmani (in press), and under compound loadings such a fire following an earthquake or an explosion. This paper will provide a summary of the work carried out so far to apply the software to model the effect of vehicle fires on a highway bridge structure.

Bridge fires are typically rare however a US survey, Battelle (2001), revealed that fires were responsible for three times the number of bridge collapses than earthquakes. The risk of bridge fires arises primarily from vehicular impacts, specifically when they involve trucks carrying flammable materials, however fires in adjacent properties are also possible (e.g. M1 fire near London, 15 April 2011). In addition to fire events under normal conditions, fire may also occur following and earthquake (FFE). Such events have a high probability of occurrence in urban areas primarily caused by electrical short circuiting of gas leaks. Authors have not found any clear data on the risk of post earthquake bridge fires, however given the absolute necessity of ensuring the resilience of transport infrastructure in the aftermath of a disaster it would be prudent to consider this kind of risk for bridges along key routes. Comprehensive safety assessments can be carried out readily based on estimating the risk of fire or FFE events and their severity. A recent paper by Kodur et al. (2010) provides an interesting overview of bridge fire hazard in the United States including a useful case study where they modelled a typical composite steel concrete highway girder bridge. This work extends the work presented in Kodur et al.'s papers to study the performance of the same bridge using a more realistic fire model, based on a localised fire as a result of a moderate size burning vehicle. All the modelling work has been carried out using the author's modified version of OpenSees.

Hasemi's localised fire model is use for the fire (based on Eurocode EN1991-1-2), which is used to determine boundary heat fluxes for the girder and the slab. A pseudo-3D heat transfer analysis is then carried out to determine the evolution of temperatures in the girder and the slab. Finally the structural response is modelled using nonlinear beam-column finite elements and treating the concrete slab and the girder as a single section. The heat transfer analysis showed that the web temperature was considerably higher than the bottom flange temperature throughout the analysis. The maximum deflection after a two hour exposure was less than 300 mm (for a span of 49 m). The vehicle fire chosen was not severe enough to cause significant damage primarily because of the massivity of the steel girder.

REFERENCES

Battelle 2001. Comparative risks of hazardous materials and non-hazardous materials truck shipment accidents/incidents, Technical report for FMCSA, US Department of Transportation.

Kodur et al. 2010. Review and assessment of fire hazard in bridges, *Transportation Research Record*, Journal of the Transportation Research Board, No 2172: 23–19.

McKenna, F. T., 1997. Object-Oriented Finite Element Programming: Frameworks for Analysis, Algorithms and Parallel Computing, PhD thesis, University of California, Berkeley.

SFPE 2002. Society of Fire Protection Engineers Handbook, 3rd Edition, 2002.

Usmani et al. 2012. Using OpenSees for structures in fire, *Journal of Structural Fire Engineering*, 3: 57–70.

Advances on structural robustness and redundancy of bridges

Organizers: F. Biondini & D.M. Frangopol

Bridge Maintenance, Safety, Management, Resilience and Sustainability – Biondini & Frangopol (Eds)
© 2012 Taylor & Francis Group, London, ISBN 978-0-415-62124-3

Evaluation of bridge redundancy under lateral loads

G. Anitori & J.R. Casas
Technical university of Catalonia, Barcelona, Spain

M. Ghosn
The City College of New York, USA

ABSTRACT

The design of bridges has traditionally been performed on a member by member basis and little consideration is usually provided to the remaining capacity of the system after the failure of one structural element. As a consequence of several tragic collapses that followed the failure of single elements, the evaluation of structural redundancy and robustness of bridges has become of primary importance. Since redundancy is related to the overall system behavior, this study attempts to bridge the gap between a component by component design and the system effect. The aim of this work is to numerical evaluate the capability of the structural system to resist loads after the design load.

A systematic analysis method and a set of carefully calibrated criteria are therefore necessary to evaluate the capability of the structural system to resist collapse following extreme events. In this work, a methodology is proposed for the evaluation of the redundancy of bridge systems under lateral loads. A deterministic method is first presented which uses the criteria previously proposed in the NCHRP 458 (Liu et al. 2000) research project. The method is illustrated using as an example a static nonlinear analysis of the 3D space frame model of a typical bridge configuration. Different structural scenarios are considered in the method, in particular, as shown in figure 1, an intact one and a damaged one. The damage scenario is represented by removing the load carrying capacity to one main member from the original configuration. Material nonlinearity is considered to assess the safety of the structure at the ultimate stages.

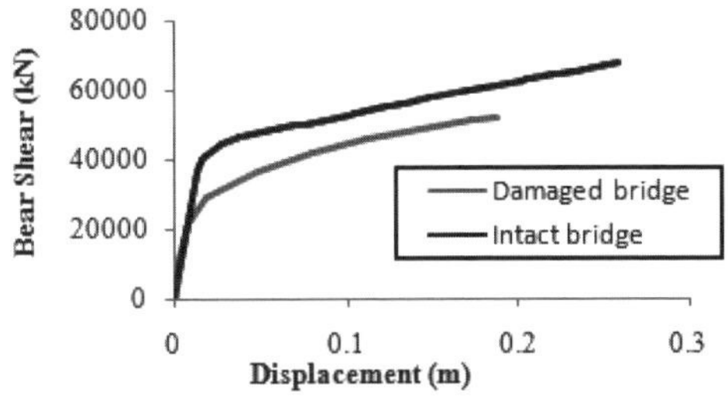

Figure 1. Capacity curves of the intact and damaged bridge.

A probabilistic method is then implemented to verify the accuracy of the deterministic method and to evaluate the redundancy of bridges by means of reliability based measures. The uncertainties in the materials parameters are considered from test results (Casas 1995) and literature review (Val et al. 1997).

REFERENCES

Casas J.R. & Sobrino J.A. 1995. *Geometrical and material uncertainties in reinforced and prestressed concrete bridges.* Proceedings of Structures congress XIII, ASCE, Boston, USA.

Liu, W.D., Ghosn, M., Moses, F. & Neuenhoffer A. 2000. *Redundancy in highway bridge substructures*, NCHRP Report 458, Transportation Research Board, Washington DC.

Val, D., Bljunger F. & Yankelevsky D. 1997. Reliability evaluation in non linear analysis of reinforced concrete structures. In Elsevier (Ed) *Structural Safety:* Vol. 19, N. 2 203–217, 1997, The Netherlands.

Bridge Maintenance, Safety, Management, Resilience and Sustainability – Biondini & Frangopol (Eds)
© 2012 Taylor & Francis Group, London, ISBN 978-0-415-62124-3

Enhancement of bridge redundancy to lateral loads by FRP strengthening

G. Anitori & J.R. Casas
Technical university of Catalonia, Barcelona, Spain

M. Ghosn
The City College of New York, USA

S. Jurado
Technical university of Catalonia, Barcelona, Spain

ABSTRACT

Assessment of bridge redundancy and robustness is receiving increasing attention due to the fact that existing structural assessment methods, which are based on individual member capacity while ignoring the behavior of the bridge as a system, may lead to inefficient decisions regarding the management of bridge networks or the serviceability of particular bridges.

Bridge redundancy has been defined by Ghosn (1998) and Liu et al. (2000) as the ability of a bridge to continue to carry a substantial portion of the design load after the failure of one of its main members, without initiating a progressive collapse mechanism that leads to total failure. In this paper, a general methodology is first presented to evaluate the structural redundancy of bridge systems. Subsequently, the method is applied to examples of bridge systems under lateral loads such as wind or earthquake loads.

The method is based on a non-linear static (pushover) analysis using finite elements. The overall system response of the structure is found by a frame model in which non-linear material properties are taken into account. Realistic stress-strain relationships have been studied to consider the confinement effect of the FRP confinement on the stress strain curve of the material itself as suggested by Eid et al. (2008). These are implemented to carry on sectional analysis first and therefore to define a non-linear structural model by means of concentrated plasticity (plastic hinges).

The method is applied to the case of an integral bridge with rigid connections between the piers and the box girder deck. The first step calculates the redundancy of the original bridge system under lateral load. The second step illustrates how bridge redundancy is influenced by strengthening the bridge piers by the external wrapping of Carbon Fiber Reinforced Plastics (CFRP) sheets which increase the confinement of the concrete and improve its ductility and strength.

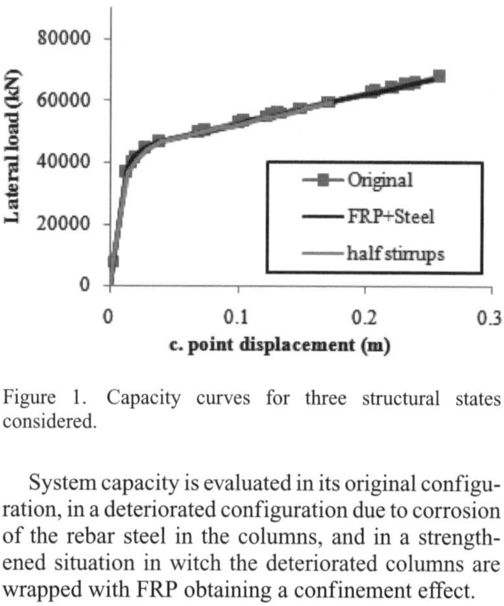

Figure 1. Capacity curves for three structural states considered.

System capacity is evaluated in its original configuration, in a deteriorated configuration due to corrosion of the rebar steel in the columns, and in a strengthened situation in witch the deteriorated columns are wrapped with FRP obtaining a confinement effect.

Based on the results, relevant conclusions concerning the behavior and the development of design criteria for CFRP-strengthened bridge piers are derived.

REFERENCES

Eid, R. & Paultre, P. 2008. *Analytical model for FRP-confined circular reinforced concrete columns.* Journal of composite for construction, ASCE. Vol. 12, N. 5, 541–552.

Ghosn, M. & Moses, F. 1998. *Redundancy in highway bridge superstructures NCHRP report 406*, Transportation research board, Washington DC.

Liu, W. D., Ghosn, M., Moses, F. & Neuenhoffer, A. 2000. *Redundancy in highway bridge substructures*, NCHRP Report 458, Transportation Research Board, Washington DC.

Bridge Maintenance, Safety, Management, Resilience and Sustainability – Biondini & Frangopol (Eds)
© 2012 Taylor & Francis Group, London, ISBN 978-0-415-62124-3

Lifetime structural robustness of concrete bridge piers under corrosion

F. Biondini
Department of Structural Engineering, Politecnico di Milano, Milan, Italy

D.M. Frangopol
Department of Civil and Environmental Engineering, ATLSS Center, Lehigh University, Bethlehem, PA, USA

ABSTRACT

During the last few decades, increasing attention has been focused on the concepts of structural robustness, disproportioned failure and progressive collapse (Ellingwood 2006). In general, the concept of structural robustness is associated to damage suddenly provoked by accidental actions and abnormal loads. However, damage could also arise gradually in time from aging of structures. Depending on the damage propagation mechanism, such kind of damage may also involve disproportionate effects (Biondini and Restelli 2008). These effects are particularly relevant for bridge structures due to their environmental exposure. Notable events of bridge collapses due to the environmental aggressiveness and related phenomena, such as corrosion and fatigue, include for example the Silver Bridge in 1967, and the Mianus River Bridge in 1983. Structural robustness should therefore be considered as key factor for a rational approach to life-cycle design of deteriorating structure and infrastructure systems (Frangopol *et al.* 1997). In this context, it is of great interest to investigate the evolution in time of structural robustness under a progressive deterioration of structural performance.

Recently, the time factor has been explicitly included in a lifetime scale for a time-variant measure of structural redundancy and robustness both in deterministic and probabilistic terms (Biondini 2009, Okasha and Frangopol 2009, Biondini and Frangopol 2010). Based on this approach, the lifetime structural robustness of concrete bridge piers exposed to corrosion is investigated in this study by considering uncertainties. Criteria and methods for the definition of suitable lifetime performance indicators and the quantitative evaluation of structural robustness are presented. The effects of the damage process on the structural performance are evaluated at cross-sectional level by using a proper methodology for life-cycle assessment of concrete structures exposed to diffusive attacks from environmental aggressive agents (Biondini *et al.* 2004, 2006).

The proposed approach is illustrated through the application to a reinforced concrete pier with box cross-section of an existing bridge. The results highlight the important role of the environmental exposure and demonstrate the effectiveness of the proposed time-variant measure in comparing the robustness associated to different damage scenarios. These results may be used to plan repair interventions and maintenance actions to protect, improve and/or restore the lifetime performance of concrete bridges (Frangopol et al. 1997, Okasha and Frangopol 2009).

REFERENCES

Biondini, F., 2009. A Measure of Lifetime Structural Robustness, *Structures Congress 2009*, ASCE/SEI, Austin, TX, USA, April 29-May 3.

Biondini, F., Bontempi, F., Frangopol, D.M., Malerba, P.G., 2004. Cellular Automata Approach to Durability Analysis of Concrete Structures in Aggressive Environments, *Journal of Structural Engineering*, ASCE, **130**(11), 1724–1737.

Biondini, F., Bontempi, F., Frangopol, D.M., Malerba, P.G., 2006. Probabilistic Service Life Assessment and Maintenance Planning of Concrete Structures, *Journal of Structural Engineering*, ASCE, **132**(5), 810–825.

Biondini, F., Frangopol, D.M., 2010. Structural Robustness and Redundancy of Deteriorating Concrete Bridges, *5th Int. Conference on Bridge Maintenance, Safety and Management* (IABMAS 2010), Philadelphia, PA, USA, July 11–15. In *Bridge Maintenance, Safety, Management and Life-Cycle Optimization*, D.M. Frangopol, R. Sause, and C.S. Kusko (Eds.), CRC Press, Taylor and Francis Group.

Biondini, F., Restelli, S., 2008. Damage Propagation and Structural Robustness, *First Int. Symposium on Life-Cycle Civil Engineering* (IALCCE'08), Varenna, Italy, June 10–14. In *Life-Cycle Civil Engineering*, F. Biondini and D.M. Frangopol (Eds.), CRC Press, Taylor and Francis Group.

Ellingwood, B.R., 2006. Mitigating Risk from Abnormal Loads and Progressive Collapse. *Journal of Performance of Constructed Facilities*, ASCE, **20**(4), 315–323.

Frangopol, D.M., Lin, K-Y., and Estes, A.C., 1997. Life-Cycle Cost Design of Deteriorating Structures, Journal of Structural Engineering, ASCE, **123**(10), 1390–1401.

Okasha, N.M., Frangopol, D.M., 2009, Lifetime-oriented multi-objective optimization of structural maintenance considering system reliability, redundancy and life-cycle cost using GA, *Structural Safety*, **31**(6), 460–474.

Bridge Maintenance, Safety, Management, Resilience and Sustainability – Biondini & Frangopol (Eds)
© 2012 Taylor & Francis Group, London, ISBN 978-0-415-62124-3

Robustness assessment of a corroded RC bridge deck

E.S. Cavaco & L.A.C. Neves
Universidade Nova de Lisboa, Civil Engineering Department, Monte de Caparica, Portugal

J.R. Casas
Technical University of Catalunya, Civil Engineering Department, Barcelona, Spain

ABSTRACT

This work is a contribution to the definition and assessment of robustness of deteriorating reinforced concrete structures. Cavaco et al. (2010) proposal is used to assess robustness of a multi-girder bridge deck under corrosion located in the Center of Portugal near the seaside. Robustness is defined as a property of the structure that measures the degree of structural performance loss after damage occurrence.

During a visual inspection carried out to the bridge in 2010, significant corrosion levels where detected in the deck girders increasing the concerns about the bridge safety. Higher corrosion levels were detected on girder 1 more exposed to winds from seaside (see Figure 1).

Robustness assessment accordingly to the proposed methodology allows having a perception on how the bridge safety evolves with the increasing corrosion. Also it shows the importance, for the bridge safety, of considering different corrosion levels for the four girders. The proposed methodology is capable of evaluating important corrosion characteristic phenomena such as reinforcement area reduction and bond (steel-concrete) strength deterioration. This last phenomenon proved to be the most important factor responsible for structural load carrying capacity degradation. The corrosion analysis methodology is then used, coupled with non-linear force based frame finite element models, to predict the load carrying capacity of the corroded bridge. The safety factor for the traffic loads, γ, is considered as the structural performance indicator. Reinforcement corrosion on girder 1, X_{p1}, is adopted as the damage indicator. Several levels of corrosion are defined for the three less corroded girders, considering that the reinforcement loss in these girders is equal to 100%, 75%, 50%, 25% and 0% of that corresponding to girder 1. Results presented in Figure 2 show the competence of the robustness index to predict the bridge safety with corrosion level increase. Results also show the importance of bond strength deterioration and load redistribution on bridge robustness and safety.

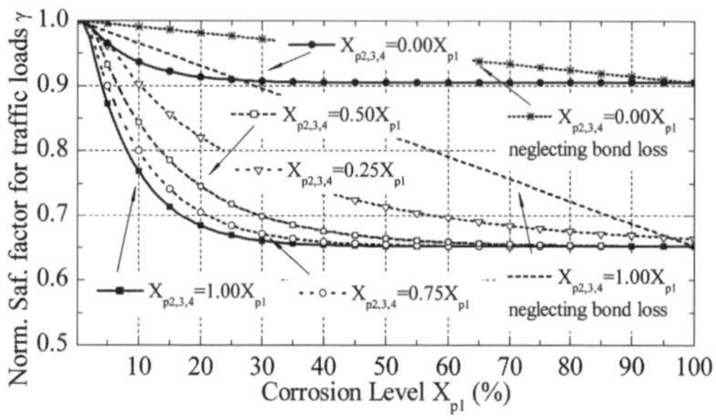

Figure 2. Normalized safety factor for the traffic loads as a function of corrosion level on girder 1.

Figure 1. Bridge deck under view (Jacinto, 2011).

REFERENCES

Cavaco, E. S., et al. (2010). Robustness of corroded reinforced concrete structures – a structural performance approach. *Structure and Infrastructure Engineering.* DOI: 10.1080/15732479.2010.515597.

Jacinto L. (2011). Avaliação da segurança de pontes existentes. Abordagem probabilística bayesiana. Ph.D. Thesis, Universidade Nova de Lisboa, Faculdade de Ciências e Tecnologia.

Bridge Maintenance, Safety, Management, Resilience and Sustainability – Biondini & Frangopol (Eds)
© 2012 Taylor & Francis Group, London, ISBN 978-0-415-62124-3

Robustness assessment of suspension bridges

M. Haberland, S. Haß & U. Starossek
Hamburg University of Technology, Hamburg, Germany

ABSTRACT

Structural robustness – defined as the insensitivity of a structure to initial damage – has been recognized as a desirable characteristic of structural systems because it mitigates their susceptibility to progressive and disproportionate collapse. The robustness of suspension bridges is investigated using a generic suspension bridge design. Based hereupon, recommendations for a robust design of suspension bridges are developed.

The critical components of a suspension bridge are the hangers because they are vulnerable and exposed. The failure of one (or a few) hangers can lead to a progressive collapse of the whole bridge. On the one hand, explicit design measures could be taken to prevent a hanger failure. On the other hand, the collapse progression can be prevented by ensuring robustness.

The robustness of a suspension bridge can be enhanced by providing alternative load paths within the suspension system or by segmenting the bridge girder or the suspension system. These options are discussed in this paper.

The investigation concerning alternative load paths showed that the robustness of a suspension bridge increases when the flexural stiffness of the girder is increased or the hanger spacing is decreased (assuming unchanged girder stiffness). However, girder stiffness and hanger spacing are interdependent parameters. It has further to be investigated which of these parameters is more important. Additionally, it turned out that the failure of long hangers is less critical as the failure of short hangers.

The investigation of the segmentation method showed that introducing hinges in the girder reduces the robustness for small initial damage but increases the robustness for large initial damage. It turned out necessary that the hinges should only develop in the case of imminent collapse progression. Until that point, the segment borders should behave like a monolithic connection.

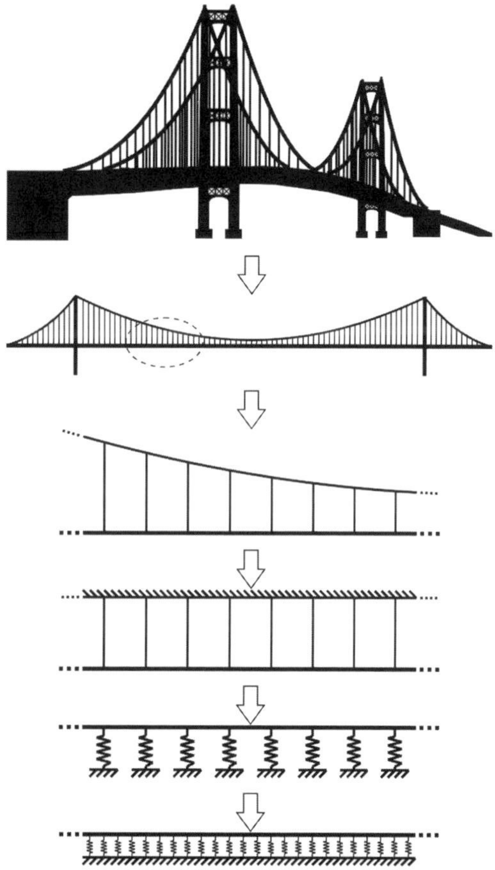

Figure 1. From bridge to model.

Research on the robustness of suspension bridges is at its very beginning. Some issues have been addressed and clarified here, further questions are open for further research.

Light rail bridges in Chongqing, China

Organizer: M.-C. Tang

Bridge Maintenance, Safety, Management, Resilience and Sustainability – Biondini & Frangopol (Eds)
© 2012 Taylor & Francis Group, London, ISBN 978-0-415-62124-3

Design of CaiJia rail bridge over JiaLing river in CongQing

B. Liu & Y. Qi
T.Y. Lin International China, Chongqing, China

L. Lin
Chongqing Rail Transit Design Institute, Chongqing, China

ABSTRACT

This paper describes the design process of the rail bridge crossing the Jialing River in Caijia area.

The paper first introduces bridge scheme choice. From economy, landscape, construction comprehensive, structure dynamic performance, several bridge types have been compared. The cable-stayed bridge with double cable planes is a more suitable bridge type.

Form the vortex vibration characteristics of the main bridge girder section, box girder section is better than triangle section.

Pylon style also has been chosen, The pylon style can use A-shaped or diamond, and diamond style is cheaper than A-shaped style, so diamond style is decided here.

Based on the principle of not affecting the normal traffic and reducing social influence while replacing cables, this cable-stayed bridge chose steel strand stay-cables and adopts the Single cable replacement technology during cable replacement process, which will not interrupt the traffic.

The paper focuses on light rail's functional requirements, structural design of the cable-stayed bridge, construction monitoring, structural health monitoring system, as well as maintenance requirements. In the approach span section, it introduces several features such as tall pier, large span, width change, and horizontal curve. Movable scaffolding system has been used for construction of the approach spans. This bridge provides valuable experience in design and construction for similar projects in the future.

REFERENCES

Yongle Li, Shizhong Qiang, Haili Liao and Y.L. Xu, Dynamics of wind-rail vehicle-bridge systems, Journal of Wind Engineering and Industrial Aerodynamics, 2006, 93: 483–507.

Yongle Li, H.L. Liao and S.Q. Qiang, Simplifying the simulation of stochastic wind velocity fields for long cable-stayed bridges, Computers and Structures, 82 (2004): 1591–1598.

Bridge Maintenance, Safety, Management, Resilience and Sustainability – Biondini & Frangopol (Eds)
© 2012 Taylor & Francis Group, London, ISBN 978-0-415-62124-3

Design of the Chongqing Caiyuanba Yangtze River Bridge for dual highway and rail traffic

A.-S. Liu & Z.-G. Jing
T.Y. Lin International Engineering Consulting China, Chongqing, China

ABSTRACT

The Chongqing Caiyuanba Yangtze River Bridge, located in the core area of the city, is a long-span bridge built for dual highway and rail traffic, the main span of which is a combined rigid frame-tied arch bridge of 420 m long, which is the world's longest span among bridges of the same kind. The bridge carries six lanes of highway and two pedestrian walkways on its upper deck as well as double-line urban light rails on its lower deck. The design of the bridge adheres to the principle of "safe, innovative, cost effective" and features innovative design concepts and technologies, which helps to merge the bridge structure's safety, functionality, economy and the bridge aesthetics into one natural manner. The hybrid structure of combined prestressed concrete rigid frame and steel box tied arch is applied in the bridge scheme to improve both span capacity and efficiency of material use. The combination of orthotropic steel plate and steel truss girder is used to carry highway and rail transit loads. Key innovative technologies in the bridges design and construction include:

- Hybrid bridge structure system combining "Y"-type prestressed concrete rigid frame and steel box tied arch;
- Combined orthotropic steel plate – steel truss girder system for dual highway and rail traffic;
- Industrialized technologies of segmental design, transport and construction;
- Active control technology for main structure and for tie cable system in both main span and side spans.

This paper describes overview of the project, bridge schematic design as well as detailed structural design.

REFERENCES

"Two-stage Construction Drawings of the Caiyuanba Yangtze River Bridge of Chongqing". Design Joint Venture of Chongqing Communications Research & Design Institute Co, Ltd. And TY.LIN International, Nov 2003.

"Construction Organization Design Drawings of the Chongqing Caiyuanba Yangtze River Bridge". China Zhongtie Major Bridge Engineering Group Co, Ltd, Nov 2004.

"Key Construction Technique Research of highway-rail long-span bridge-Innovative Technology Research of Composite Structure and Active Control for long Span Bridge". Chongqing City Construction Investment Corporation/ Chongqing Communications Research & Design Institute Co, Ltd./TY.LIN International) Feb 2008.

Bridge Maintenance, Safety, Management, Resilience and Sustainability – Biondini & Frangopol (Eds)
© 2012 Taylor & Francis Group, London, ISBN 978-0-415-62124-3

Wind-resistant study on Chongqing Chaotianmen Yangtze Bridge – The longest arch bridge in the world

C. Wang, Z. Li & P. Deng
College of Civil Engineering, Chongqing University, Chongqing, China

ABSTRACT

With the rapid development of economy in China, a great number of modern long-span bridges have been built or under construction across large rivers, lakes, sea bays and channels in recent years. Among these bridges, Chongqing Chaotianmen Yangtze River Bridge (Figure 1 and Figure 2) is a semi-supporting, steel-truss, and tied arch bridge with three spans of 190 m + 552 m + 190 m. It is also an extraordinary one not only for it being used for highway traffic and light-rail traffic but also for the world's longest center span of 552 m among the same types.

To ensure the safety of the bridge under winds, wind-resistant behaviors of Chongqing Chaotianmen Bridge over Yangtze River were investigated via wind tunnel tests. The wind-tunnel tests of 1:100 scale-downed full bridge aeroelastic models were carried out in the 8×6 wind tunnel at China Aerodynamics Research & Development Center. The Service Stage and four key construction stages were considered in the tests. The tests were carried out in both smooth and simulated boundary layer wind fields with various wind yaw angles. The areoelastic models were elaborately designed and manufactured (Figure 3), and the dynamic properties in every stage were investigated before the tests.

The results (Figure 4) show that Chongqing Chaotianmen Bridge over Yangtze River has enough aerodynamic stability for all the service and construction stages in both the smooth wind field and the turbulent air flow of boundary layer. During the service stage, the data calculated for the Dieckmann index K, a comfort index, are within the acceptable range. The flutter test results obtained in the full bridge aeroelastic model test fit in perfectly with those attained via the sectional model test. Buffeting response of the bridge to turbulent wind may reach its maximal value in yaw wind case with the most unfavorable yaw angle at 0°. Furthermore, buffeting responses increase with wind speed at approximate conic curves.

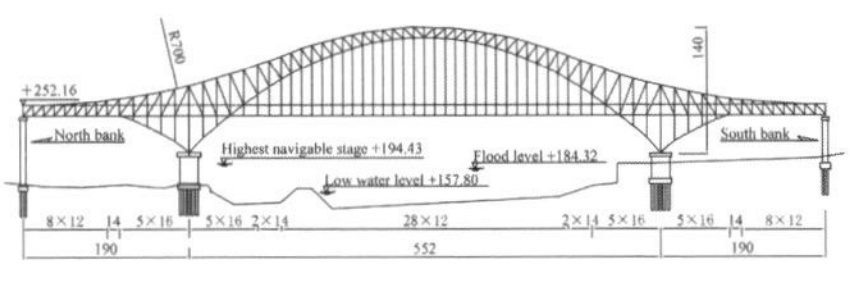

Figure 1. Elevation of bridge (unit: m).

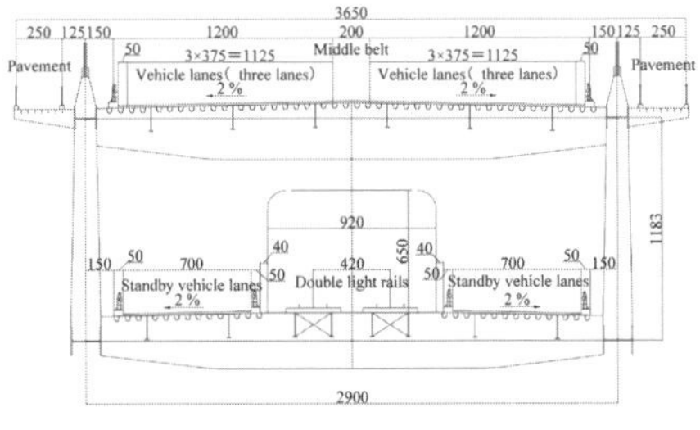

Figure 2. Deck cross-section (unit: mm).

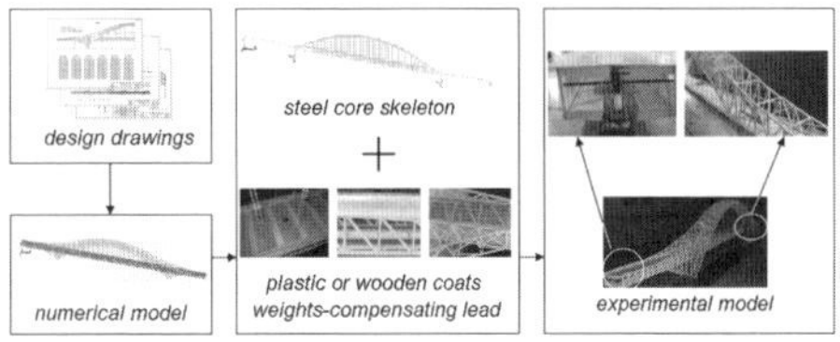

Figure 3. Construction of the experimental model.

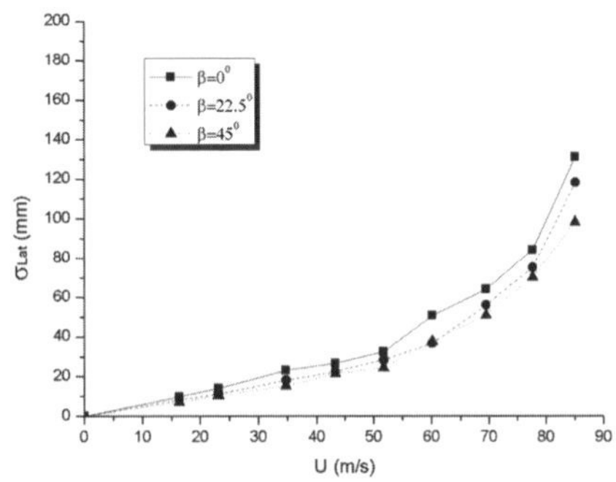

Figure 4. Standard deviations of lateral responses to turbulent wind at deck mid-span in Service Stage.

REFERENCES

Cai, C.S., Albrecht, P., Bosch, H.R., 1999. Flutter and buffeting analysis. Finite-element and RPE solution, J. Bridge Eng. ASCE 4(3),174–180.

Chen, Z.Q., Xiang, H.F., 2005. Wind Engineering for Bridges, China Communication Press, Beijing, pp. 129

D. Dieckmann, (1958). A study of the influence of vibration on man, Ergonomics. 1(4), 347–355

Davenport, Allan Garnett, 1962. Buffeting of A Suspension Bridge by Storm Wind, J. Struct.Div.ASCE 88(3), 233–268 (1962).

Bridge Maintenance, Safety, Management, Resilience and Sustainability – Biondini & Frangopol (Eds)
 ISBN 978-0-415-62124-3

Development of rail transit in mountainous city of Chonqing, China

J.-H. Zhong
Chognqing Railway Transit Group Company

ABSTRACT

As the major metropolitan city of West China, Chongqing has seen rapid economic development in the last two decades. Together with the expansion of the city, rail transit infrastructure has been the focus of overall urban planning and construction from the very beginning. As the backbone of the public transportation system, rail transit has many benefits including energy efficiency, effective use of space, high volume capacity, low environmental impact, and superior level of safety. It is a green and sustainable system and is highly effective in metropolitan regions. For the unique terrain of Chongqing with both mountainous areas and major water ways, the development of rail transit system has overcome many challenges to achieve a rather rapid construction pace.

By 2011, the rail transit system of Chongqing already has three lines actively in service. The No. 2 line, which was put into service on June 18, 2005, is not only the first straddle-type monorail in China, but also the first urban rail transit line in the western region. The No. 1 line is the main east-west bound line while the No. 2 and No. 3 lines are the two north-south lines. Together they form the backbone of the city's rail transit network.

Situated at inland of Southwest China and upper reaches of the Yangtze River, Chongqing is a typical cluster-type city. The entire city is divided into three parts by the Yangtze River and the Jialing River: Yuzhong, Nan'an and Jiangbei. The north, east and south part of Chongqing is surrounded by mountains because it is also located in transition zone between the Qinghai-Tibet Plateau and the plain of middle and lower reaches of the Yangtze River. Chongqing's topography is dominated by hilly terrains. It has been known as the "Mountain City" due to the large amount of steep hills in the urban area. This distinguished feature brings more challenges in construction of rail transportation infrastructure. However, on the other side, it also makes Chongqing's transportation system more unique.

The dual-configuration of the rail transit infrastructure system in Chongqing consists of regular rail track and straddle-type monorail.

There are many points in Chongqing Rail Transit planning where major river crossing bridges are necessary due to the city's unique geographic characters. With the accelerated construction of rail transit infrastructure of Chongqing, these bridges will be built in near future.

Development of rail transit infrastructure in Chongqing is not only a project construction but also a cultural development. The "one theme for each line" principle has been followed as early as in the planning stage. Each line will have its own unique cultural theme, together to form a completed Chongqing impression.

A series of discussion on culture themes and representative elements has been steered by a group of culture specialists from the Rail Transit Ltd. The finalized themes contain culture elements from ordinary life styles, and have gained recognition from most local residents.

According to the development plan of Chongqing, the total rail transit system of Chongqing adds up to 700 km with over 350 stations, which greatly exceeds the originally envisioned budget of 150 billion RMB.

Advances in nondestructive evaluation and monitoring of concrete bridge decks

Organizer: N. Gucunski

Bridge Maintenance, Safety, Management, Resilience and Sustainability – Biondini & Frangopol (Eds)

Verification of advanced electromagnetic measurement techniques for corrosion and fracture detection of bridge tendons

A. Holst, H. Budelmann & H.-J. Wichmann
Institute for Building Materials, Concrete Construction and Fire Protection (iBMB), Department of Civil Engineering, University of Braunschweig, Germany

ABSTRACT

Corrosion induced defects strongly affect the integrity and durability of internal and external tendons of prestressed concrete structures. On this account, the condition assessment of prestressed members should be addressed to the early and non-destructive detection of existing flaws and damages at the steel elements, such as injection faults, corrosion defects and fractures.

This paper reports on the development, testing and application of the following innovative electromagnetic measurement techniques for non-destructive corrosion detection and fracture localization at bridge tendons, cp. also Budelmann et al. (2010):

- Magnetoelastic coil-sensors (cp. figure 1) for force measurements and material defect detection of prestressed tensile elements, where the modification of measured induction voltage is a function of the tensile stress and the cross section of the tendon. By moving the sensor along the cable local signal anomalies indicate a local steel defect.
- The Electromagnetic Resonance Measurement Method (Frequency Domain Reflectometry, FDR) in combination with electromagnetic field strength investigations. When a microwave coupled into the steel is reaching a cable corrosion spot, a fracture or grouting flaws, variations of the characteristic impedance and subsequently a signal modification occur. The identification of surface corrosion is attributed to the skin effect, i.e. the ousting of primary currents with increasing frequency towards the surface due to eddy currents. The amplitude, phase and the shift of the reflected signal in the frequency domain are a function of existing steel defect.

Moreover by moving the fieldmeter-probe externally along the current-carrying tendon as depicted in figure 2, a local drop of the electric or magnetic field strength indicates the position of a steel fracture or of a void.

The thematic focus lays here on discussing the current state of research, essential findings and acquired experiences as well as improvements and further developments of the different sensing methods. It will be also demonstrated that the experimental data and trends could be confirmed by several simulations. Here the microwave reflectometry and the skin-effect as basis for the detection of corrosion spots are discussed, Budelmann et al. (2004).

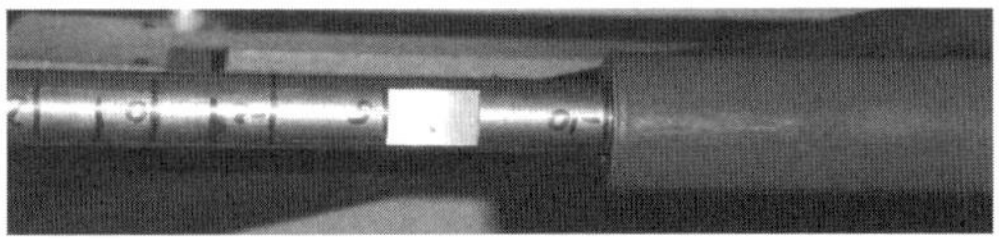

Figure 1. In-line measurement with a magnetoelastic sensor at a steel bar damaged with a local groove.

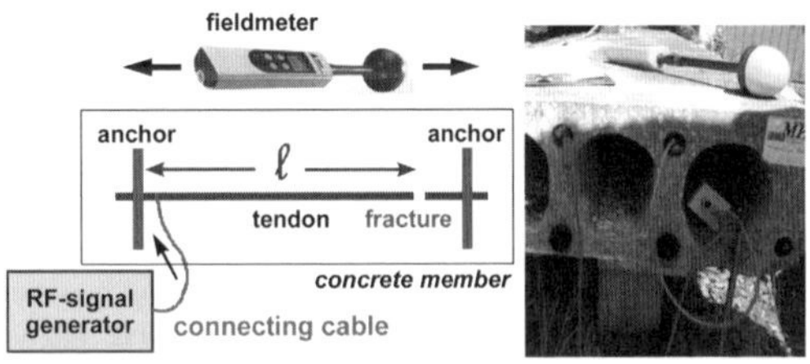

Figure 2. Defect detection by means of electromagnetic field strength measurements by use of a fieldmeter: Schematic diagram and experimental setup on site at a hollow slab.

REFERENCES

Budelmann, H., Hariri, K., Holst, A. & Wichmann, H.-J. 2004. New Monitoring Techniques for Bridge Tendons. In *2nd In-tern. Conf. on Bridge Maintenance, Safety and Management IABMAS 04, Kyoto, 18–22 Oct. 2004*, Leiden: Balkema: 653–654.

Budelmann, H., Holst, A. & Wichmann, H.-J. 2010. Measurement Tool Kit for Corrosion and Defect Monitoring of Bridge Tendons. In *5th Int. Conf. on Bridge Maintenance, Safety and Management – IABMAS 2010, Philadelphia*, 11–15 July 2010, Leiden: Balkema: 364–371.

Bridge Maintenance, Safety, Management, Resilience and Sustainability – Biondini & Frangopol (Eds)
© 2012 Taylor & Francis Group, London, ISBN 978-0-415-62124-3

Damage assessment of reinforced concrete decks due to chloride-induced corrosion of reinforcing bars and fatigue

Y. Ishikawa & M. Aoyama
Central Nippon Highway Engineering Nagoya Co., Ltd., Kanazawa, Japan

Y. Adachi
Central Nippon Expressway Co., Ltd., Kanazawa, Japan

M. Nagai
Nagaoka University of Technology, Nagaoka, Japan

ABSTRACT

Durability of reinforced concrete (here in after RC) decks in highway bridges strongly depends on the corrosion of reinforcing steel bars and fatigue damage due to the traffic. Since the 1990s, the use of deicing salts has dramatically increased in areas that have a high volume of snowfall in Japan.

The remnants of dissolved deicing salts, rainwater and melted snow eventually saturate the bridge expansion joints, as well as the bridge decks. When the deicing salts (i.e. chloride ion) and water reach down the rebars in a concrete deck, an electrolytic action begins, causing the steel to corrode. As the corrosive particles build up, they expand against the concrete cover, eventually exceeding the tensile strength of the concrete. This causes horizontal cracking along the upper area of rebars forming a delaminated area of concrete. Figure 1 shows the delamination of a bridge deck.

The bridge decks are also subjected to high live-load stresses, including impacting vehicle force, and a combination of both damage factors. The weight from the traffic creates potholes in the pavement. Figure 2 shows a pothole in the pavement. These damages lead to reduction of serviceability, safety, and service life of the bridge decks.

This paper presents the investigation of damaged bridge decks caused by deicing salts, through various laboratory tests carried out with some samples from a removed bridge deck. The authors propose a new structural assessment method based on the deflection from the actual bridge decks loading test.

From the field investigation, several findings regarding the corrosion of the rebars due to chloride ion penetration and fatigue mechanism of bridge decks are presented. Chloride ion penetration and corrosion of rebars were concentrated in the local areas, such as the joints of the asphalt pavement and on the line of wheel tracks (Figure 3). Local chloride ion penetration resulted in the reduction of the cross-sectional area of rebars with various reduction patterns. This reduction of the cross-sectional area affects the flexural rigidity of the bridge decks.

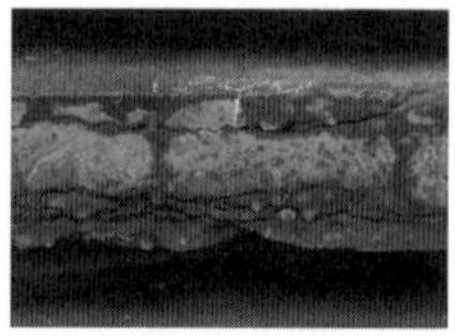

Figure 1. Delamination of a deck (Cross section area).

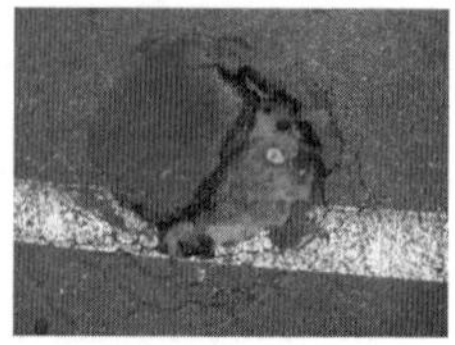

Figure 2. Pothole in a asphalt pavement.

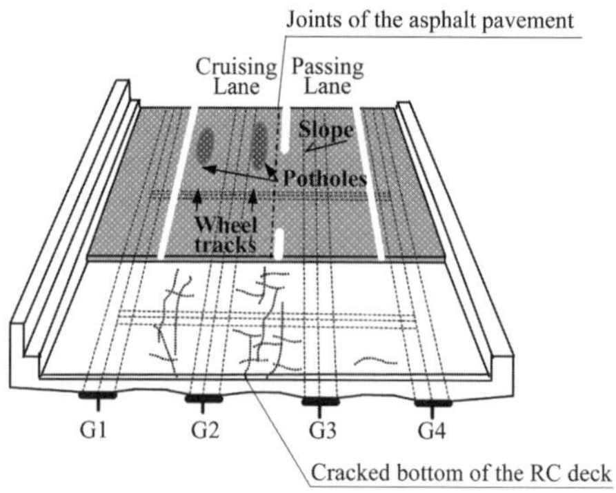

Figure 3. Location of deterioration on the RC decks.

The authors propose a structural assessment method of the degree of bridge decks deterioration, based on the deflection of the bridge decks by test car. By measuring the deflection of the bridge decks of an actual structure and comparing the deflection with the numerical analysis, we can predict the flexural rigidity of the bridge deck and conduct the structural assessment.

REFERENCE

Matsui, S. & Muto, K. 1993. Rating for evaluation of deteriorated reinforced concrete slabs of highway bridges, Proceedings on Bridge Management 2 – Inspection, Maintenance, Assessment and Repair

Bridge Maintenance, Safety, Management, Resilience and Sustainability – Biondini & Frangopol (Eds)
 ISBN 978-0-415-62124-3

Characterization and detection of bridge deck deterioration

K.R. Maser & A.J. Carmichael
Infrasense, Inc., Arlington, Massachusetts, USA

N.M. Martino & R. Birken
Northeastern University, Boston, Massachusetts, USA

ABSTRACT

Bridge decks represent the most expensive bridge component over the bridge life cycle. During the life of a bridge, decks are replaced at least once and rehabilitated multiple times. Properly timed preventive maintenance and rehabilitation can extend the deck's life and reduce life cycle costs. To do this, owners need to decide which decks are in good condition, which need work (and what type of work), and which are not economical to repair and need to be replaced. These decisions require accurate assessment of the deck condition. Traditional methods for deck condition evaluation are not economically feasible to implement on a large number of structures. More recently, ground penetrating radar (GPR) has been implemented to rapidly scan bridge decks and provide the needed information However, the relationship between the electromagnetic response of a reinforced concrete deck and the state of deck deterioration is not well understood, and thus the full acceptance and adaptation of GPR has had limitations. Most bridge decks deteriorate due to corrosion induced delamination, a process that is initiated by the infiltration of de-icing salts. The goal of the work reported in this paper has been to establish a clearer connection between the electromagnetic response detected with GPR equipment and the elements and stages of the deterioration process. This goal was achieved by conducting comprehensive experimental evaluation of specific bridge deck slabs. These include slabs extracted from in-service structures, controlled slabs specially cast and exposed to corrosion, and actual decks in service. Four decks slabs, saw-cut out of a heavily deteriorated deck, were brought into the lab for evaluation. The slabs were subjected to a full battery of tests, including measurements using hammer sounding, ultrasonic impact-echo (IE), half-cell corrosion potentials (HCP), and ground and air-coupled GPR systems ranging from 1.0 to 2.6 GHz. In parallel with this work, 17 specially constructed laboratory slabs exposed to a controlled corrosion environment over an extended period of time were similarly evaluated to examine the correlation between GPR signals and the level of corrosion. Finally, GPR data collected on a new in-service deck was examined and compared to that from the extracted deck to investigate the distribution rebar reflection amplitudes.

The multiple methods applied to the extracted deck slabs show a strong spatial correlation, as depicted in the Table 1 above. The test results on the controlled slabs showed a correlation between average half cell potentials and GPR rebar reflection amplitudes with an R^2 of 0.85. Finally, the rebar reflection amplitude distributions for the new vs. deteriorated decks showed distinct statistical differences, suggesting that these distribution statistics can serve to identify overall deterioration quantities.

Table 1. Correlation between test methods.

Methods Compared	% Match
GPR vs. HCP	90.2%
GPR vs. IE	79.3%
HCP vs. IE	76.4%

REFERENCES

Barnes, Christopher L. & Trottier, Jean-Francois 2004. Effectiveness of Ground Penetrating Radar in Predicting Deck Repair Quantities, *Journal of Infrastructure Systems 10.2: 69–76*, ASCE.

Maser, K.R. 1991. Bridge Deck Condition Surveys Using Radar: Case Studies of 28 New England Decks. *Transportation Research Record No. 1304, TRB National Research Council, 13–17 January, 1991, 94–102*, Washington, DC.

Rhazi, Jamal & Dous, Omar & Laurens, Stephane 2007. A New Application of the GPR Technique to Reinforced Concrete Bridge Decks. *Proc. of 4th Middle East NDT Conference and Exhibition*, Kingdom of Bahrain.

Bridge Maintenance, Safety, Management, Resilience and Sustainability – Biondini & Frangopol (Eds)
© 2012 Taylor & Francis Group, London, ISBN 978-0-415-62124-3

Non-destructive highway inspection methods using high definition video and infrared technology

M. Matsumoto
NEXCO-West USA, Inc., Washington DC, USA

M. Sugimoto
Fuji Engineering, Osaka, Japan

K. Hashimoto
NEXCO-West Engineering Shikoku, Takamatsu, Japan

K. Mitani
NEXCO-West USA, Inc., Washington DC, USA

ABSTRACT

Identifying appropriate applications for technology to assess the health and safety of bridges is an important issue for bridge owners around the world. Traditionally, highway bridge conditions have been monitored by visual inspection with structural deficiencies being manually identified and classified by qualified engineers and inspectors. With traditional on-site inspections, qualified inspectors are performing close-up visual inspections and sounding tests, often from crane suspended lifting cages or built-in inspection staging; arguably putting inspectors at some safety risk. The need for safer inspection methods calls for new innovations in bridge inspection technologies. One of the solutions for this issue is leveraging non-destructive technologies or mechanical sensor technologies as well as experimental approaches for a more advanced and efficient inspection process. Due to advancements in technology and computer science, new technologies are becoming more and more cost effective, and some of the technologies are ready to be applied for on-site bridge assessment practices. Some technologies can contribute to reducing inspection costs by replacing part of the visual inspection or sounding tests, while other technologies can provide additional information for bridge engineers to make a decision on optimum inspection intervals or more cost effective maintenance strategies. If we can improve data collection efficiencies and reduce the time required by inspectors in the field to make general structure condition assessments, more time will be available for these same inspectors to perform detailed hands-on inspections and/or to apply non-destructive testing technologies for pre-screened bridge elements in areas requiring close attention. NEXCO-West, one of major toll road operators in Japan has been working to develop efficient non-destructive highway bridge inspection methods using high resolution image and Infrared (IR) Imagery technologies. High definition video records the surface condition of concrete structures. Recorded data is analyzed by image processing to determine crack widths and length on concrete surfaces. Infrared imaging supports a thermographic assessment of structural integrity. In general, thermographic assessments involve analyzing structural integrity through an analysis of variations in a structural element's temperature at different times of the day; recognizing that damaged or deteriorating structural elements demonstrate different temperature variation gradients than do sound structural elements. Combining high resolution image processing result and IR data enables an accurate assessment of a bridge's structural integrity. The technology can provide a powerful tool for engineers responsible for developing highway maintenance strategies and contribute to efficient and smart infrastructure management.

This paper describes these inspection methods and introduces some examples of practical on-site application to highway bridge superstructures. It was verified from the pilot application results that the accuracy of detection and measurement surface cracks and potential subsurface deterioration using these new technologies provided satisfactory and acceptable results for practical routine and special condition bridge inspections in compliance with recognized inspection practices. It was also demonstrated that new HDV and IR technologies could significantly reduce site inspection times and on-site inspection resource requirements.

REFERENCES

Akashi K, Hashimoto K, Koide H, & Togawa M. 2006.9. Development of Inspection Method using Infrared Thermography Technology and its Technical Consideration, *Proceeding of JSCE 61st Annual Meeting*, 1113–1114.

Florida Department of Transportation. 2008.2. Bridge Inspectors Field Guide Structural Elements.

Bridge Maintenance, Safety, Management, Resilience and Sustainability – Biondini & Frangopol (Eds)
© 2012 Taylor & Francis Group, London, ISBN 978-0-415-62124-3

Rapid seismic scanning for bridge deck NDE

J.S. Popovics & T. Oh
The University of Illinois, Urbana, Illinois, USA

R.W. Arndt
Rutgers University, New Brunswick, NJ, USA

ABSTRACT

Concrete in civil infrastructure systems is susceptible to deterioration caused by various mechanisms. The most serious problem affecting the service life of the reinforced concrete bridge decks is the formation of delaminations, which are thin cracked areas that lie directly above the steel reinforcement mat within the deck. Visual inspections provide only superficial information and, as well as for sounding, the results depend strongly on the experience of the inspectors. Thus nondestructive test (NDT) methods, which can be applied rapidly and ubiquitously across the bridge deck structure, find much utility (ACI committee 228, 1998).

The air-coupled impact-echo test equipment and scheme were developed at the University of Illinois (Zhu and Popovics 2007). The equipment consists of five unshielded dynamic (unpowered) microphones that are mounted onto a small rolling cart with a horizontal mounting frame. The frame positions the microphones at a regular six-inch transverse spacing, and holds them about 12 mm above the surface of the deck. It should be noted that the tests were carried out in a noisy ambient environment without any sort of noise shielding or suppression technology.

The air-coupled impact-echo data are presented in a "cloud plot" image, which was conceived by the research team from the University of Illinois. A cloud plot is a 3-D presentation of the spectral data (individual amplitude spectra) across the two (x-y) spatial dimensions of the bridge deck surface. Frequency is plotted along the vertical z axis, and the spatial dimensions along the x and y axes. In order to clearly interpret the data in this form, the amplitude of the signal is indicated by color and also opacity, where high amplitude signals are represented by opaque "warm" colors, medium amplitude signals are represented by semi-transparent "cold" colors, and low amplitude signals are transparent. A 3-D cloud plot presentation of all of the data is shown in Figure 1. The cloud plot indicates specific regions of interest, where the "cloud" density is high. The density of these data clouds, and the frequency at which they occur, provide clues about the location and nature of defects within the sample.

The interpretation of the x-y cloud plot data is relatively straight-forward: regions underneath opaque "clouds" likely contain some sort of defect, especially if the frequency content of those opaque formations is below 5 kHz.

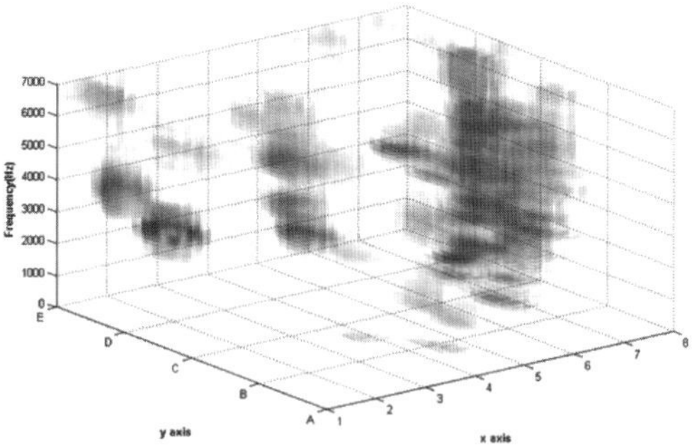

Figure 1. 3-D cloud plot presentation of spectral data collected from the actual bridge deck test section. Spectral data are shown up to 7 kHz.

REFERENCES

ACI committee 228 (1998). "Nondestructive test methods for evaluation of concrete in structures." *Report ACI 228.2R-98*, American Concrete Institute, Farmington Hills, MI.

Zhu, J., and Popovics, J. S. (2007). "Imaging concrete structures using air-coupled impact-echo." *J Eng Mech-ASCE*, 133(6), 628–640.

Bridge Maintenance, Safety, Management, Resilience and Sustainability – Biondini & Frangopol (Eds)

Acoustic testing of concrete bridge decks for detection of delamination

P. Shokouhi
BAM, Berlin, Germany

ABSTRACT

Delamination in a concrete slab is defined as "... a horizontal splitting, cracking, or separation of a slab in a plane roughly parallel to, and generally, near the upper surface; found most frequently in bridge decks and caused by the corrosion of reinforcing steel or freezing and thawing." [*1*]. In concrete bridge decks, delamination is commonly an indirect indication of severe steel rebar corrosion. Therefore, the extent of delamination is often regarded as a measure of the severity of deck deterioration. Delaminations grow in size over time and may eventually develop into large planes of separation resulting in surface peeling and spalling. Early detection of delaminations is a vital task to enable taking timely preventive measures and avoiding the high cost of bridge deck replacement. At present, bridge owners and highway agencies still rely heavily on visual surveys, sometimes augmented by the so-called traditional deck condition surveys (DCS) including Electro-Mechanical Sounding, Hammer Sounding (a.k.a. Metal Tapping) and Chain Drag (ASTM D 4580).

Impact Echo (IE), Ultrasonic (US) Echo (two measurement systems: a single probe and a linear array system) were used to detect (and possibly characterize) delaminations in a concrete specimen. The test specimen is a part of a deteriorated bridge deck preserved from the demolition of a 45-year old prestressed box girder bridge in Berlin, Germany. The IE and US (single probe) data collection was done automatically on a fine grid utilizing a vacuum mounted scanner system developed at BAM, while the testing with US linear array system was conducted manually on a coarser grid. A number of cores were extracted to provide ground-truth information and the basis to assess the effectiveness of each test method in detecting and characterizing delaminations.

In detection of delaminated zones in the test specimen, IE provided a satisfactory overall assessment, but the individual results at single measurement points

Figure 1. Automatic US-testing of the test specimen.

were often difficult to interpret (due to the presence of multiple frequency peaks in the spectra); US could detect and locate deep delaminations, but not the shallow ones; US Linear Array (3D datasets) could locate the extent of deep delaminations and provided indications of shallow ones (through the weakened or vanished reflections from the bottom of the slab as well as distortions in surface waves).

It should be noted that the conclusions made here are bound to the particular characteristics of the test slabs and testing equipment. However, since the test specimens were former parts of a bridge deck, it is reasonable to assume that the attained results were more realistic than those obtained in laboratory experiments on specimens with built-in defects.

REFERENCES

ACI Committee 116R-00. 2003, Cement and concrete terminology (2003), Manual of Concrete Practice, Part I, American Concrete Institute, Farmington Hills, MI.

Bridges for high-speed railways

Organizer: R. Calçada

Bridge Maintenance, Safety, Management, Resilience and Sustainability – Biondini & Frangopol (Eds)
© 2012 Taylor & Francis Group, London, ISBN 978-0-415-62124-3

Analysis of lateral dynamics of railway vehicles on viaducts with coupled models

J.M. Goicolea, P. Antolín & J. Oliva
School of Civil Engineering, Technical University of Madrid, Spain

ABSTRACT

The study of lateral dynamics of running trains on bridges is of importance mainly for the safety of the traffic, and may be relevant for laterally compliant bridges. These studies require 3D coupled vehicle-bridge models, and consideration of wheel to rail contact, a phenomenon which is complex and costly to model in detail.

A nonlinear coupled model is proposed for vehicle-structure vertical and lateral dynamics. Vehicles are considered as fully three-dimensional multibody systems, and the bridge structure is modeled by means of finite elements. It has been implemented within an existing finite element analysis software with multibody capabilities, Abaqus (Simulia Ltd., 2010). Both subsystems (bridge and vehicles) are described with coordinates in absolute reference frames, as opposed to alternative approaches which describe the multibody system with coordinates relative to the base bridge motion. The models are capable of full consideration of geometrical and material nonlinearities. The approach for wheel-rail geometrical interaction and mechanical contact model is fully nonlinear as well, not being limited neither to constant conicity assumptions nor to linearized elastic contact forces. Further details of the model are presented in Antolín et al. (2012).

Two applications are presented, firstly to a vehicle subject to a strong wind gust traversing a continuous beam type bridge, showing the relevance of the nonlinear wheel-rail contact model as well as the interaction between bridge and vehicle (figure 1). The second application is to a real viaduct in a high-speed line, with a long continuous deck and tall piers with high lateral compliance (Millanes et al. (2007), see figure 2). The results show the safety of the traffic as well as the importance of considering nonlinear coupled models, as well as track alignment irregularities.

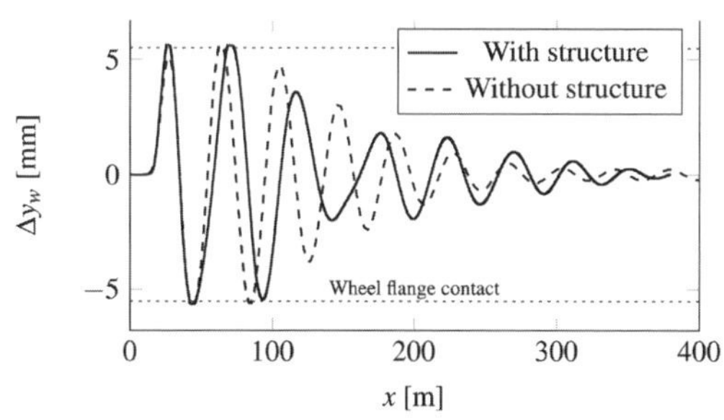

Figure 1. Lateral response of wheelset for vehicle on single track bridge subject to wind gust.

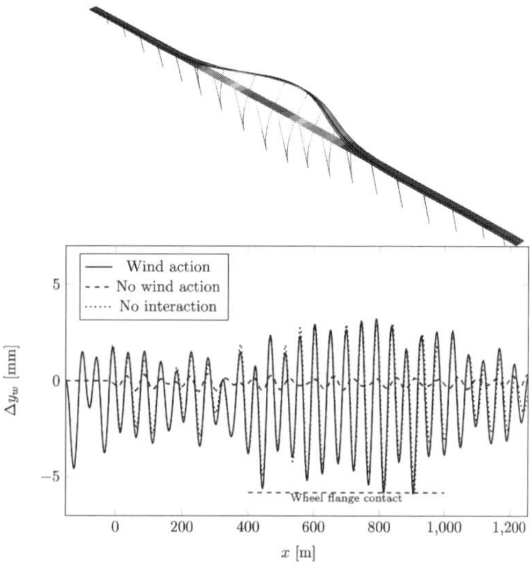

Figure 2. Viaduct "Arroyo las Piedras" in Córdoba–Málaga HS line. Top: first mode of vibration 0.313 Hz. Bottom: relative lateral displacement Δy_w of wheelset for $v = 200$ km/h.

REFERENCES

Antolín, P., J. M. Goicolea, J. Oliva, and M. A. Astiz (2012). *Nonlinear train-bridge lateral interaction using a simpli?ed wheel-rail contact method within a finite element framework*. J. Comp & Nonl Dyn In press.

Millanes, F., J. Pascual, and M. Ortega (2007). *"Arroyo de las Piedras" viaduct: The first composite steel-concrete high speed railway bridge in Spain.* Structural Engineering International 17 (4), 292–297.

Simulia Ltd. (2010). *Abaqus 6.10 User's manual*. Providence, RI: Simulia Ltd.

Bridge Maintenance, Safety, Management, Resilience and Sustainability – Biondini & Frangopol (Eds)
 ISBN 978-0-415-62124-3

Dynamic interaction between rails and structure in a composite bridge of 120 m length

C. Jurado
Polytechnic University of Madrid, Madrid, Spain; Ingecal Ingenieros, S.L.

ABSTRACT

The new steel composite bridge over Verdugo river for the line of high-speed trains in Galicia (Spain) is placed between the localities of Arcade and Pontesampaio.

The total length of the viaduct is 120.00 m with a central pillar arranged in the middle of the river, dividing the total length of the bridge in two spans with 60.00 m each one.

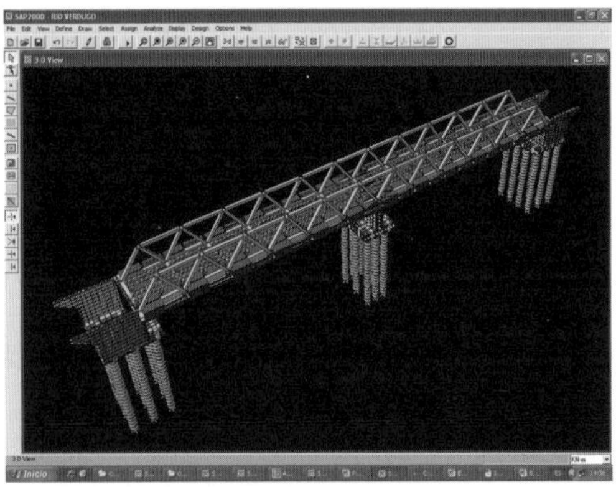

Figure 1. Model of finite elements of the Bridge over Verdugo river.

The structure consists of a composite deck with two metallic main trusses of 1.80 m of height, on both sides of the platform. The composite slab deck has metallic transverse girders of 0.90 m of height arranged every 5.0 m, where are disposed the pre-slabs, that support the pouring concrete of the deck.

The foundation has 6 reinforced concrete piles of 1.75 m of diameter and 21 m of length under the central pillar and 9 of 22 m of length under each abutment.

The project velocity of the line is 200 km/h. Under the rails and over the deck there is ballast and according with the owner, it was needed to study, if it was necessary to dispose expansion devices for the rails.

A 3D finite element model with 7,265 nodes, 6,569 elements type FRAME and 2,003 elements type SHELL has been made, which comprises all the elements of the structure: deck, trusses, girders, central pillar, abutments, mat foundations and concrete piles.

According with the European and Spanish Regulations it is needed to make an extensive model in order to verify three aspects:

- Maximum tensions in rails.
- Maximum displacements.
- Maximum rotations at the ends of spans.

The paper presents all the aspects of the study, compare the different Rules and finally exposes the conclusions.

Bridge Maintenance, Safety, Management, Resilience and Sustainability – Biondini & Frangopol (Eds)
 ISBN 978-0-415-62124-3

Investigation of major dynamic responses in the high-speed railway bridges for KTX

B.S. Kim
SOC Research Institute, Korea Institute of Construction Technology, Goyang, Korea

W.J. Chin, E.S. Choi & J.W. Kwark
Structural Engineering Research Division, Korea Institute of Construction Technology, Goyang, Korea

ABSTRACT

The Gyeongbu High-Speed Railway, the first high-speed railway line of Korea, started its commercial operation in 2004 to link Seoul and Busan. This line completed in two stages connected first Seoul and Daegu, followed later by the completion of the section extending from Daegu to Busan. The first section supports the high-speed train by gravel ballast while the second section adopts concrete ballast. The bridges erected along the Gyeongbu High-Speed Railway applied various bridge types according to the field conditions with 2@40 m PSC box girder bridge as the representative bridge type. This study collects and analyzes major dynamic responses obtained through field measurement performed on 8 bridges of different types crossed by trains in the Gyeongbu High-Speed Railway, and investigates the appropriateness of these responses with reference to the relevant specifications. In addition, the serviceability of the actual bridges vibrating due to the crossing of the high-speed train as well as the riding safety of the train are also investigated. The response margin of the bridges in service was examined. The analysis of the test results revealed that the vibration acceleration and vertical displacement of all the bridges satisfy the limit-values. The comparison of the distribution of the acceleration at the sleepers, ballast and, upper and lower decks verified that the acceleration measured in bridge with concrete ballast satisfied more conservatively the limits than the acceleration measured in bridge with gravel ballast. The vibration serviceability criteria of the bridge considering the vibration duration time is used as the vibration serviceability criteria of the acceleration component related to the vertical acceleration of the train in order to evaluate the vibration serviceability of the high-speed railway bridge considering the comfort. The analysis results confirmed that all the bridges satisfied the vibration serviceability. Accordingly, it was recommended that, in the future, the evaluation should be performed using the vibration serviceability criteria of the bridge structure considering the vibration duration time. Moreover, comparison was done for the measured deflection at mid-span and the deflection converted from the vibration acceleration inside the coach using the transfer function with respect to the allowable deflection limit for the vibration serviceability of a simple bridge. Since all the measured deflection and converted deflection of the selected bridges remained below the allowable deflection limit, the satisfaction of the vibration serviceability by the deflection could be verified.

Table 1. Maximum dynamic responses of selected bridges.

Bridge name	Max. acc.(1) (g)	Max. acc.(2) (g)	Vert. disp.(1) (mm)	Vert. disp.(2) (mm)
Gunghyeon1	0.32	0.04	0.23	0.13
Biryong	0.15	0.02	0.25	0.17
Gomo	0.24	0.01	0.91	0.87
Jitan	0.35	0.05	2.87	2.85
SongSun	0.26	0.02	1.04	1.01
Seongdong	0.26	0.02	1.68	1.49
MeYaGi	0.45	0.03	0.51	0.31
Taegi	0.11	0.02	0.87	0.45

(1): measured; (2) 30 Hz filter.

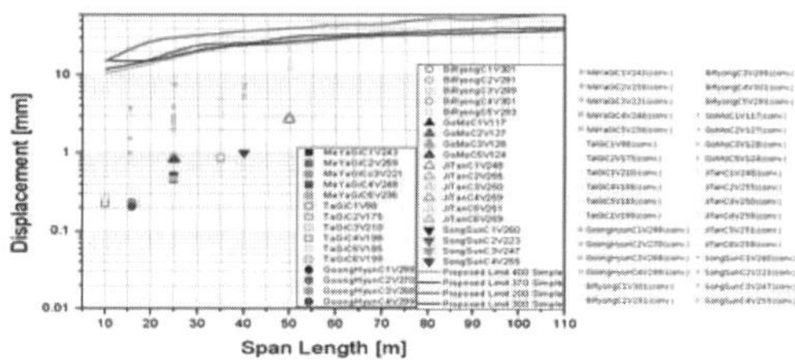

Figure 1. Comparison of deflection limit, measured deflection and converted deflection of simple bridge according to span length and running speed.

REFERENCE

Kim, N.S. et al. 2010. Deflection limit on vibration serviceability of high-speed railway bridges considering the exposed time duration. *Proc. 2010 spring annual confer. of the Korean Society for Railway.*

Bridge Maintenance, Safety, Management, Resilience and Sustainability – Biondini & Frangopol (Eds)
© 2012 Taylor & Francis Group, London, ISBN 978-0-415-62124-3

The vertical acceleration on a bridge deck for riding stability of high-speed train

J.W. Kwark, H.J. Yoon & W.J. Chin
Structural Engineering Research Division, Korea Institute of Construction Technology, Goyang, Korea

B.S. Kim
SOC Research Institute, Korea Institute of Construction Technology, Goyang, Korea

ABSTRACT

Trains crossing a bridge at high speed generate vibrations due to the interaction occurring among the structure, track and train. In the case of resonance between the structure and high-speed train, such vibrations can reach high amplitudes, which may increase the risk of derailment through the weakening of the fastening strength of the rails or loss of wheel load and affect the riding stability of the train. These excessive vibrations should be maintained below the design criteria to secure the riding safety of the high-speed train. In the Honam High-Speed Railway Design Guidelines, currently accepted as the high-speed railway design specifications in Korea, the maximum acceleration at the deck is prescribed unilaterally as 0.35 g in the case of deck with ballast track and 0.5 g in the case of deck with the concrete track. However, this is the acceleration at the ballast which may really affect the riding safety of the train. Moreover, the prescribed acceleration appears in the form of a single peak value. Accordingly, this study intends to propose criteria complementing the limiting values of the vertical acceleration.

In this study, the vertical acceleration signals obtained through field measurement tests performed on high-speed railway bridges currently in service are collected and analyzed. The bridges for field measurement were selected considering the type of track structure since the limit-values of the acceleration vary according to ballast and concrete track.

The vibration acceleration response depends sensitively on the measurement location inside the bridge deck, the consideration or not of the frequency brought by the track system, and the adopted evaluation method.

For a bridge with ballast track, the loosening of the fastening strength of the ballast provoked by excessive vibrations appeared to have direct influence on the riding stability of the high-speed train. Therefore, the acceleration should be imperatively examined at the ballast. Besides, in the case of Taegi Viaduct applying a concrete track, even if the acceleration measured at the bottom of the deck appeared to be larger than that measured at the top of the deck, this acceleration was seen to have no direct effect on track system. Accordingly, the acceleration should not be investigated at the bottom of the deck but should be examined at a location having direct relationship with the track system.

For the evaluation method, the sizes of the acceleration obtained by the peak value and RMS evaluation methods were compared considering the inclusion or not of the frequency brought by the track system. The largest acceleration was obtained by applying the peak value evaluation method on data measured at sampling rate of 250 Hz. The smallest acceleration resulted from the RMS evaluation method applied on signals from which the effect of the track was removed by low pass filtering.

The limiting values of the acceleration suggested by the current Honam High-Speed Railway Design Guidelines apply a single maximum peak value including the frequency generated by the track system. These values were verified to be excessively large for securing the riding stability of the high-speed train. Since the acceleration depends sensitively on the measurement location, the eventual consideration of the frequency of the track system and the evaluation method, need is to limit its value considering these factors.

REFERENCES

European Committee for Standardization. 2003. *EUROCODE 1 Part 2: Actions on Structures – Traffic loads on bridges*, European Committee for Standardization.

European Rail Research Institute. 1999. *ERRI D 214 RP9: Rail Bridges For Speeds v > 200 km/h.*

Korea Rail Network Authority. 2007. *Honam High-Speed Railway Design Guidelines*. Korea Rail Network Authority.

Bridge Maintenance, Safety, Management, Resilience and Sustainability – Biondini & Frangopol (Eds)
© 2012 Taylor & Francis Group, London, ISBN 978-0-415-62124-3

Train-bridge interaction effects on the dynamic response of a small span high-speed railway bridge

J. Rocha, A.A. Henriques & R. Calçada
University of Porto – Faculty of Engineering, Porto, Portugal

ABSTRACT

The main scope of this paper is the analysis of train-bridge interaction effects on the dynamic response of a small span high-speed railway bridge. This type of structures is particularly affected by these effects.

As a case study the Canelas railway bridge was selected. The bridge has six simply supported spans of 12 m each leading to a total length of 72 m. The bridge deck is a composite structure and has two half concrete slab decks with nine embedded rolled steel profiles HEB 500, each supporting one rail track. This is a very common structural solution for small span bridges in the European high speed railway lines, especially in France and Germany.

Small span bridges are also particularly sensitive to the variability of both structural-related parameters and train-related parameters. Since the effects of the structural-related variables had already been studied in previous papers, this paper focused on the train-related variables, providing a full view of this problem.

The sensitivity analysis of the train-related variables on the dynamic response of the bridge showed that the bridge response is practically unaffected by the changes on the train properties, except when the car body mass is changed. However, since the maximum response will always occur for the maximum mass, in subsequent analysis only this scenario was analysed.

Furthermore, a comparison was made between the results obtained using two different methods: the moving loads method and the train-bridge interaction method. This allows assessing the influence of the train-bridge interaction effects on the dynamic response. An analysis was made assigning the mean values to all structural variables and using the two different methods. The obtained results are presented in Figure 1. It can be seen that for non-resonant speeds the results are almost identical. However, in the presence of resonant effects, a reduction of about 10% in the dynamic response is observed when taking into account interaction effects.

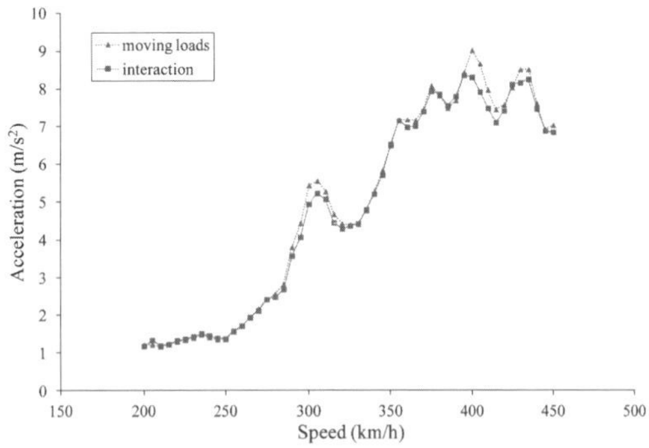

Figure 1. Maximum acceleration comparison for different methods.

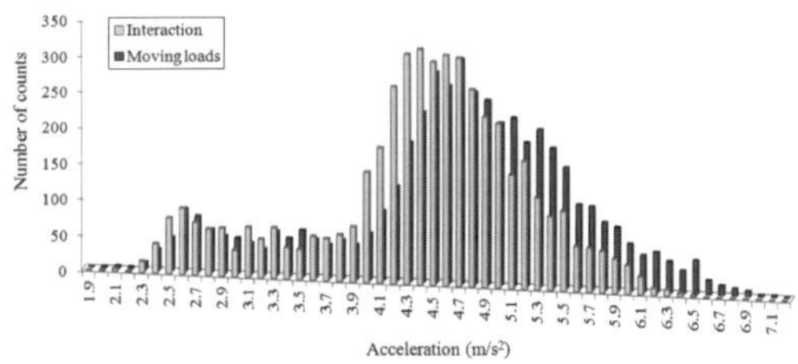

Figure 2. Histogram comparison for different methods (v = 300 km/h).

In order to take into consideration the parameters variability, simulation techniques were used. This allowed assessing the influence that the train-bridge interaction has in the dynamics response.

The obtained results can be seen in Figure 2 and confirmed the results obtained before. These results indicate that train-bridge interaction might play an important role for safety assessment problems limited by resonant effects.

Bridge Maintenance, Safety, Management, Resilience and Sustainability – Biondini & Frangopol (Eds)
© 2012 Taylor & Francis Group, London, ISBN 978-0-415-62124-3

Fatigue analysis of precast girder webs in railway bridge decks

C. Sousa, J. Rocha, R. Calçada & A.S. Neves
Faculty of Engineering, University of Porto, Porto, Portugal

ABSTRACT

Railway bridge decks are usually subjected to important cyclic loads, which makes them sensitive to the fatigue phenomenon. In high speed lines, resonance effects give rise to important dynamic enhancements of the cyclic loads and, consequently, relevant fatigue damage. Therefore, fatigue must be properly taken into account during the analysis and design of these elements. This paper focuses on the specificities of fatigue analyses of box girder webs. This is a complex problem. On the one hand, the calculation of the fatigue damage due to shear forces is more difficult than the analysis of elements subjected to cyclic bending and normal forces. On the other hand, box girder webs are also subjected to important transverse bending moments, which play an important role, namely in the case of slender webs (e.g. box girders made with precast "U" shaped girders, which is a solution employed in the construction of bridges in high speed lines). Moreover, current design codes do not explicitly address this issue.

In this work, a recently developed numerical methodology for fatigue analysis of box-girder webs in railway bridge decks is applied to the analysis of a real structure. This methodology involves: load models that include the influence of the lateral distribution of the train axle loads; dynamic analyses for calculation of the dynamic enhancement of internal forces; a specific model for calculation of stresses caused by the combined effect of shear forces and transverse bending moments; the adoption of proper S-N curves to describe the fatigue behavior of both the steel and the concrete; a linear damage accumulation model for calculation of the fatigue life.

The main specificity of this methodology consists of the algorithm for stress calculations considering the combined effect of in-plane shear and transverse bending. The simplicity of the algorithm for stress calculations makes it feasible for the calculation of stress histories due to the passage of trains, if the histories of internal forces (in-plane shear and transverse bending) are calculated by means of dynamic FEM analyses. Given the complexity of the cyclic behaviour of cracked girder webs subjected to in-plane shear and transverse bending, the calculated stresses on both the steel stirrups and the web concrete depends on a parameter k_m which defines the relative contribution of two resisting mechanisms. Conservative fatigue analyses are obtained if the fatigue strength is verified, not for a single k_m value, but for the plausible range of k_m. A procedure for determination of the plausible range of k_m values was also established.

This methodology was validated by means of the comparison with experimental results reported in the bibliography. This validation revealed that this analysis methodology provides conservative estimates for the fatigue life of the beams tested by various authors.

The practical consequences of employing this methodology in the fatigue analysis of new structures are shown through the application to a real case study, which consists of a continuous railway bridge deck, made with precast "U" shaped girders.

The analysis of this real structure showed that the proposed criteria for fatigue verification of girder webs, considering the combined effect of in-plane shear and transverse bending, is suitable to be applied at the design stage of new structures.

Despite the fact that the numerical methodology was primarily devised for the analysis of girder webs, it can also be applied to the analyses of deck slabs, considering the interaction between in-plane shear forces (resulting from the transference of forces between webs and flanges) and transverse bending.

Bridge Maintenance, Safety, Management, Resilience and Sustainability – Biondini & Frangopol (Eds)
 ISBN 978-0-415-62124-3

Bridge/train interaction analysis of a suspension bridge subjected to seismic loads

K. Sungil
Korea Railroad Research Institute, Uiwang-City, South Korea

K. Jongwon
Korea Institute of Construction Technology, Goyang-City, South Korea

ABSTRACT

The estimation of traffic safety and passenger comfort when the train is running on the bridge should be defined for the railway bridge. The standards for such estimation are included in the design criteria. The items are bridge responses including vertical displacement of bridge, vertical acceleration, and slab twist. However, in principle, a direct estimation based on the train responses (wheel load decrement and acceleration of car-body) has to take place for precise estimation. Especially, the characteristic flexibility of the long-span bridge is essentially unfavorable for dynamic behaviors. Therefore, examination of traffic safety and passenger comfort for such bridges is very important.

Therefore, in the present study we used a 3-dimensional train and bridge model and conducted a bridge/train interaction analysis. Through responses at the bridge and also responses at the train, we were able to conduct examinations on both traffic safety and passenger comfort.

The bridge in the present study is a suspension bridge whose main span is 300 meters long. We conducted analyses for speeds up to 350 km/h at 10 km/h intervals and we also considered road irregularities as a measure for future operation. In addition, train operation on the bridge during earthquakes is also considered. From the dynamic responses of the train, the traffic safety and passenger comforts of the suspension bridge are directly estimated and the speed limit for the bridge can also be verified.

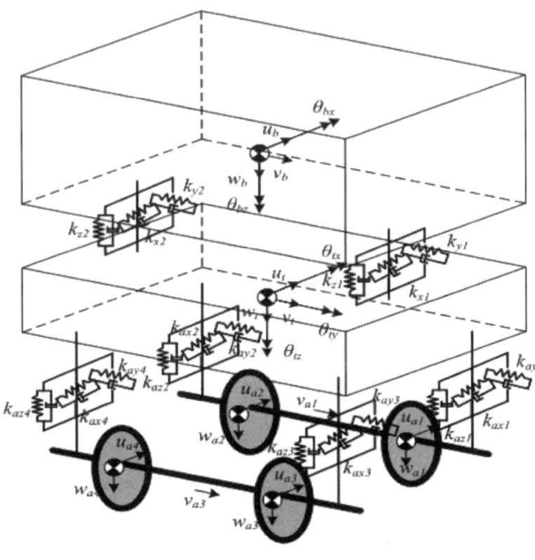

Figure 1. Passenger Car Model for Articulated Bogie.

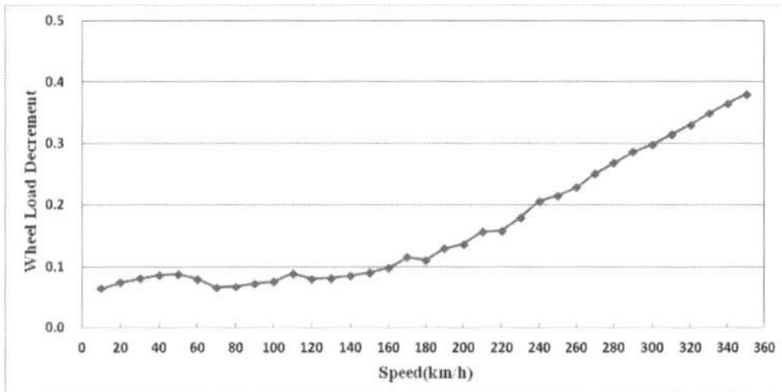

(a) Wheel Load Decrement (with Seismic Load)

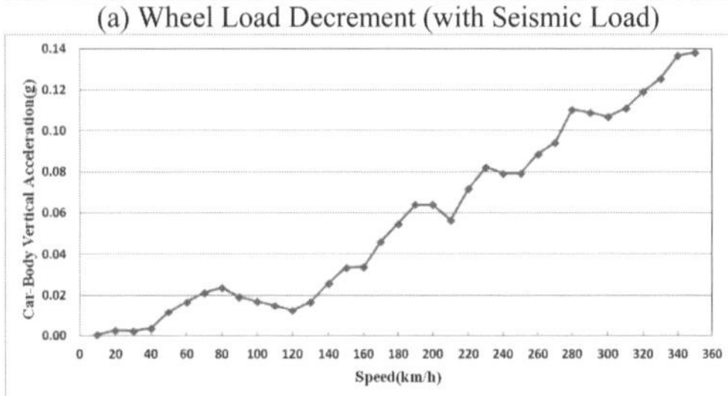

(b) Vertical Accelerations of Train (with Poor Irregularities)

Figure 2. Responses of the Train.

REFERENCES

S.I. Kim. 2000, *Train-Bridge Interaction Analysis of High Speed Railway Bridges*. Seoul. Ph.D. Thesis: Seoul National University

J.D. Yau. 2009. Dynamic Response Analysis of Suspended Beams subjected to Moving Vehicles and Multiple Support Excitations. *Journal of Sound & Vibration* 325: 907–922.

M. Sogabe, M. Ikeda & Y. Yanagisawa. 2006. A Study on Train-Running Quality during Earthquake and its Improvement for Railway Long Span Bridge. *RTRI Report* 20(11): 47–51.

Bridge Maintenance, Safety, Management, Resilience and Sustainability – Biondini & Frangopol (Eds)
© 2012 Taylor & Francis Group, London, ISBN 978-0-415-62124-3

Inspection and evaluation of steel bridges from a high-speed railway network

Ph. Van Bogaert
Ghent University, Ghent, Belgium

ABSTRACT

As some high-speed railway lines have been in service now for several decades, inspection of the bridges and viaducts has become an important issue. Various types of inspection and non-destructive testing can be considered. From these, visual inspection can still be considered as the most reliable, since other types necessarily concern isolated locations, without obtaining data about the general condition of the structure. To be successful, inspections should be based on inspection plans and consistent checklists. The latter must be based on data from the design and from the construction to focus on the critical locations in the structure.

The inspection plan for the high-speed bridge at Halle is being presented. This is a steel tied arch bridge with triangular arrangement of hangers, connection the arches and the lower chord, as well as steel orthotropic plated bridge deck. The inspection plan mainly focuses on the most delicate locations and areas of the bridge. Similar inspection plans may be worked out for other types of structures.

At least once every 2 months a test train is running on various sections of the high-speed network. This train is equipped with measuring devices, which mainly register track quality parameters. Track gauge, cant and levelling are the main quantities being monitored. Another parameter is the vertical train car acceleration. The results from continuous measurements of vertical accelerations of test trains may also constitute a valuable alternative for detecting possible defects in structures. From analyzing these results local acceleration peaks can be explained. Hence, possible abnormal results should be due to defects.

Recent (2011) data from test trains have been compared to older results (1997) from the time of commissioning a section of the Belgian high-speed network. This comparison shows that the average value of maximum accelerations has not changed fundamentally, although the dispersion of the results is large.

However, a more fundamental track-related phenomenon can influence the measured values. The test series of 1997 were carried out in early spring (mostly March), whereas the recent values were collected during summer. This means that rail stress has a larger compression part than at lower temperature. Apparently, rail alignments are less consistent and singular points, such as viaduct piers and consequently also transitions from elastic earth platform to concrete abutments are detected as having larger influence, compared to the condition of having general tension in the rails.

In addition, the test train results allow assessment of the galloping parameter for longer viaducts. This parameter allows determining whether train cars can accumulate kinetic energy during the crossing of longer structures. The results have shown that the galloping effect has decreased with time, which is probably due to spreading of local track discontinuities.

Data from test trains concerning more recent structures, such as the Prester bridge, the Battice viaduct and a bridge at Leuven, confirm these findings, although these results require careful interpretation and local effects from track devices can easily trigger the use of acceleration diagrams as a tool for assessment of structural health.

Artificial intelligence methods in bridge analysis and design

Organizers: E. Garavaglia & L. Sgambi

Bridge Maintenance, Safety, Management, Resilience and Sustainability – Biondini & Frangopol (Eds)
© 2012 Taylor & Francis Group, London, ISBN 978-0-415-62124-3

Prestress optimization of hybrid tensile structures

A. Albertin, P.G. Malerba, N. Pollini & M. Quagliaroli
Department of Structural Engineering, Politecnico di Milano, Milan, Italy

ABSTRACT

The coming of higher tall buildings, of longer span bridges and of wider large structures has been made possible through increasingly lighter structures. The lightest in absolute are the structures made of different arrangements of cables and/or of cables and struts. In this field we distinguish among:

a) pure tensile structures, made of two set of cables, one set carrying the supporting function and the other exerting on the first a stabilization function;
b) tensegrities, made of cables and struts internally self equilibrated through suitable prestress distributions and no requiring boundary containing stiffeners;
c) hybrid tensile structures, made of cable and struts, designed according to engineering intuition and based on many of the concepts of the first two typologies. Widely used hybrid tensile structures are the cable domes.

The cable dome typology was firstly proposed by Geiger and employed in roofing in the Olympic Gymnastics Hall in Seoul. At present, the largest

Figure 1. The Georgia Dome in Atlanta, U.S.A. Source: http://www.columbia.edu/cu/gsapp/BT/DOMES/GEORGIA/geo-21.jpg.

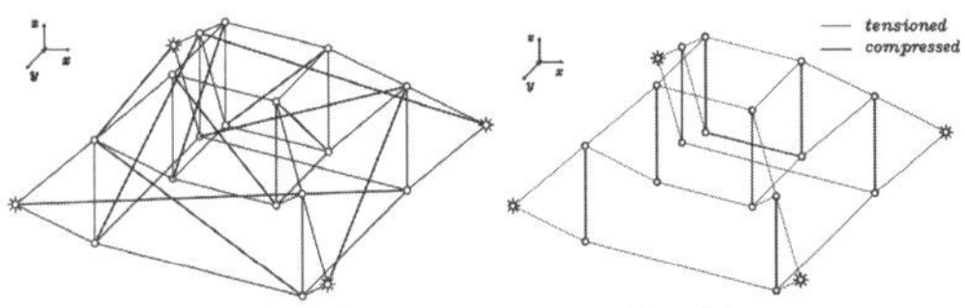

(a) with bracing elements (b) with prestress

Figure 2. The two domes presented in the paper.

existing structure of this type is the Georgia Dome designed for the Atlanta Olympics in 1996, Figure 1.

This paper deals with cable domes design. In particular, it concerns the role of the cable prestress and the optimization of the prestress intensity.

An introductory example is used to show how a simple spatial truss structure, originally born with internal mechanisms, may be made stable not only through additional bracing elements, but, in alternative, by pretensioning certain elements to an adequate pretensioning level and attributing in this way a geometrical stiffness to the whole structure (Figure 2).

In a second part of the paper an optimization procedure, based on a genetic algorithm, is presented. Such a procedure allows searching a solution at the same time of minimum weight and respecting given technological constraints, as shown trough a final example.

REFERENCES

Malerba, P.G., Patelli, M. & Quagliaroli, M. 2012. An extended Force Density Method for the form finding of cable systems with new forms. Structural Engineering and Mechanics, An International Journal 42(2).

Pellegrino, S. & Calladine, C.R. 1986. Matrix analysis of statically and kinematically indeterminate frameworks. International Journal of Solids and Structures 22(4): 409–428.

Schek, H.J. 1974. The force density method for form finding and computation of general networks. Computational Methods in Applied Mechanics and Engineering 3: 115–134.

Bridge Maintenance, Safety, Management, Resilience and Sustainability – Biondini & Frangopol (Eds)
© 2012 Taylor & Francis Group, London, ISBN 978-0-415-62124-3

An expert system for bridge inspection

S. Becker & N. Gebbeken
University of the Bundeswehr Munich, Neubiberg, Germany

ABSTRACT

Due to the age structure of the present bridge stock, condition assessment and the planning of maintenance actions will become more important over the next decades. The national codes in Germany define some boundary conditions but many aspects have to be judged by the engineer. In order to assist the engineers, an Expert System that stores and uses expert knowledge is being developed. The Expert System approach separates the knowledge base which stores the known information from the inference engine which contains the algorithms to infer new knowledge from the knowledge base. In order to adequately describe the relations for such a task an ontology based knowledge representation was implemented instead of a hierarchical one. As a practical example an assessment of a reinforced concrete bridge is presented.

Figure 1. Example bridge with spalling of the concrete cover.

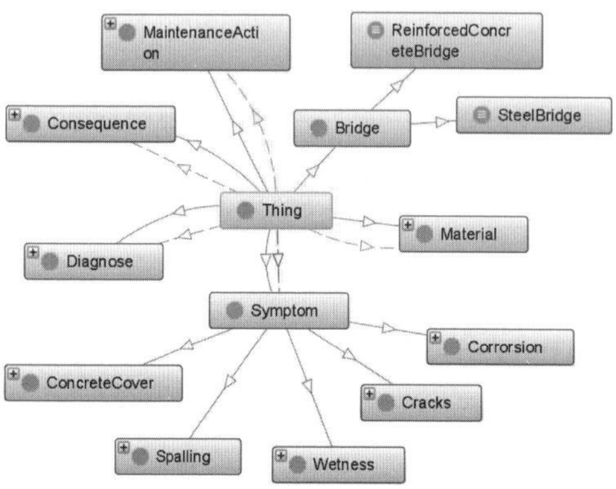

Figure 2. Overview about the developed knowledge base of the Expert System.

Figure 3. Asserted and inferred relations to classes and object properties for the assessed bridge.

Bridge Maintenance, Safety, Management, Resilience and Sustainability – Biondini & Frangopol (Eds)
 ISBN 978-0-415-62124-3

Selective maintenance strategies applied to a bridge deteriorating steel truss

E. Garavaglia, L. Sgambi & N. Basso
Department of Structural Engineering, Politecnico di Milano, Milan, Italy

ABSTRACT

This paper presents a probabilistic model for the service life assessment of bridge deteriorating truss.

Damage evolution of each structural member is dealt as a transition process through different states of performance. The transition process is modeled as a Markovian renewal process which formulation allows considering eventual improvements of the structural performance. In this way, the model is used to set up selective maintenance and/or rehabilitation interventions. Then the preventive maintenance scenarios selected are economically compared.

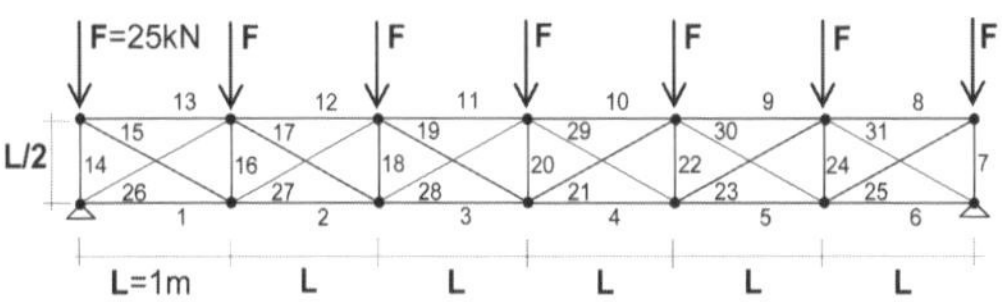

Figure 1. Statically indeterminate truss system with members area $A = 2500\,\text{mm}^2$ and total volume $V = 0.0723\,\text{m}^3$.

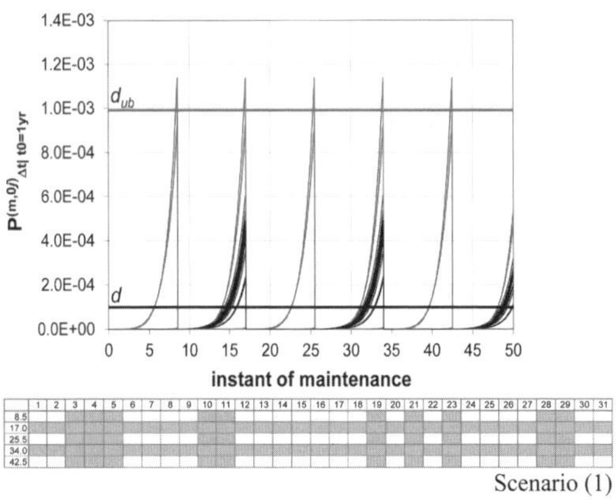

Figure 2. First possible selective maintenance scenarios.

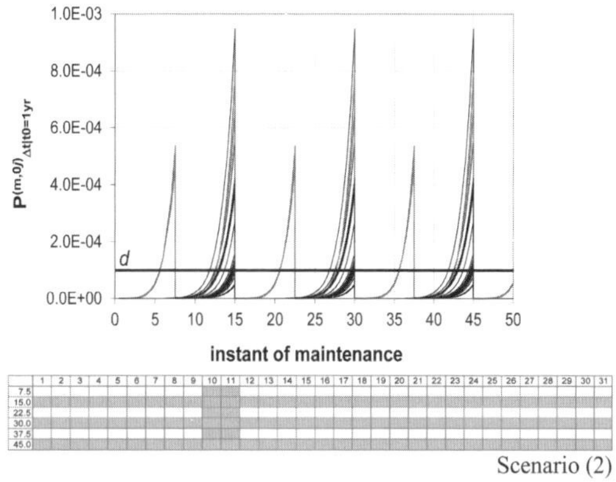

Figure 3. Second possible selective maintenance scenarios.

REFERENCES

Biondini, F., Frangopol, D.M. & Garavaglia, E. 2008 Life-Cycle Reliability Analysis and Selective Maintenance of Deteriorating Structures. *First International Symposium on Life-Cycle Civil Engineering* (IALCCE'08), Varenna, Lake Como, Italy, June 10–14, 2008, F. Biondini & D.M. Frangopol (Eds.), CRC Press, Taylor & Francis Group, London, UK, 483–488.

Garavaglia, E., Gianni, A. & Molina, C. 2004 Reliability of Porous Materials: Two Stochastic Approaches. *J. of Materials*, ASCE, USA., **16**(5), 419–426.

Gupta, A. & Lawsirirat, C. 2005 Strategically optimum maintenance of monitoring enabled multi-component systems using continuous-time jump deterioration models. *International Conference on Simulation and Modeling*, SimMod2005, V. Kachitvichyanukul, U. Purintrapiban, P. Utayopas, eds. Nakornpathom, Asian Institute of Technology, 17–19 January 2005, Bankgok, Thailand.

Kong, J.S., and Frangopol, D.M., 2003 Life-Cycle Reliability-Based Maintenance Cost Optimization of Deteriorating Structures with Emphasis on Bridges. *J. of Structural Engineering*, ASCE **129**(6), 818–824.

Bridge Maintenance, Safety, Management, Resilience and Sustainability – Biondini & Frangopol (Eds)
© 2012 Taylor & Francis Group, London, ISBN 978-0-415-62124-3

Time dependent behaviour of an elementary bridge model in presence of uncertainties

P.G. Malerba, M. Quagliaroli & L. Sgambi
Department of Structural Engineering, Politecnico di Milano, Milan, Italy

P. Baraldi
Energy Department, Politecnico di Milano, Milan, Italy

ABSTRACT

During the last decades the excessive deflections of many long span bridges received wide attention both in the research and maintenance fields. The faults of the original previsions can be attributed to (a) lacks in the models adopted for the structural assessment, (b) weak reliability of some shrinkage and creep formulations used for the analyses and (c) differences between the phases of the actual erection techniques and those planned during the design tasks. In this paper the role of the uncertainty affecting quantities and parameters governing the whole attitude of these structures is outlined.

In the first part, the time dependent behaviour of a simple viscoelastic cantilever beam, suspended at the tip by a pretensioned and variously inclined stay, has been studied (Figure 1). The comparisons among the results obtained for the different stay attitudes through a traditional deterministic approach, outline how creep and shrinkage may modify the purely elastic results.

In the second part of the paper, through a non deterministic approach, analogous comparisons have been carried out by taking into account the uncertainties associated to the relative humidity percentage and to the intensity of the pretensioning force. Two approaches have been considered: a Pure Monte Carlo Approach (Figure 2) and a Hybrid Monte Carlo/Possibilistic Approach. For a given set of data, the creep effects have been handled by means of a step by step computation of the stress history, based on a particular matrix partitioning technique.

From the structural point of view, we can distinguish between two different kinds of structures: those which have a low sensitivity with respect to uncertainties and remain stable with time, and those which are greatly affected both by creep effects and by uncertainties and exhibit time diverging behaviour.

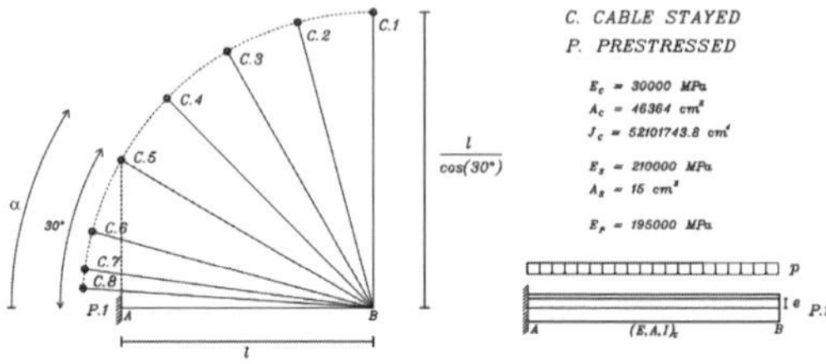

Figure 1. Cable stayed (C.) and prestressed structures (P.1).

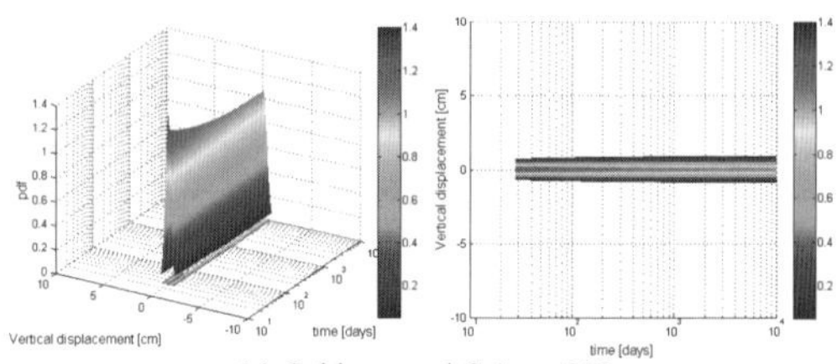

(a) Cable stayed C.1. α=90°.

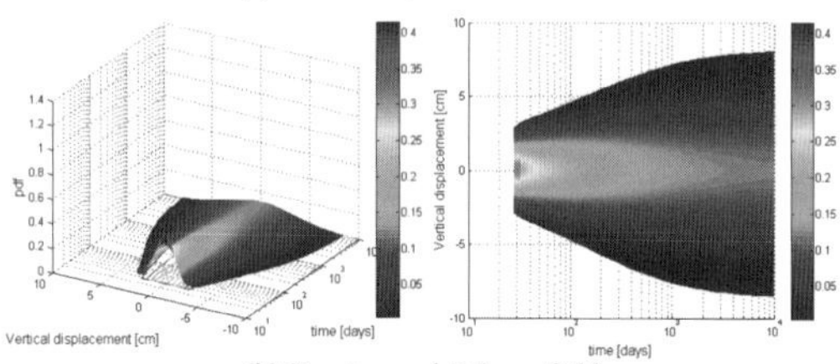

(b) Prestressed P.1. α=0.0°.

Figure 2. Results of the Pure Monte Carlo simulation. Pdf of the vertical displacement in time.

REFERENCES

Ghali, A., Favre, R. & Eldbadry, M. 2002. Concrete Structures, Stresses and Deformation, London and New York, E & FN Spon.

Sustainability assessment of bridges

Organizer: U. Kuhlmann

Bridge Maintenance, Safety, Management, Resilience and Sustainability – Biondini & Frangopol (Eds)
 ISBN 978-0-415-62124-3

Life Cycle Assessment for representative steel and composite bridges

T. Beck & M. Fischer
Department of Life Cycle Engineering, University of Stuttgart, Stuttgart, Germany

M. Pfaffinger
Chair for Metal Construction, Technical University Munich, Munich, Germany

ABSTRACT

Within the construction sector, the topic of sustainability assessment rapidly gains importance. Whereas the origin of the movement lies within the building sector, for the time being, appropriate solutions are searched for also for infrastructure.

Following commonly used sustainability definitions, it becomes evident that to address ecological aspects of an object, a long-term perspective taking into account as many possible current and future impacts of the object as possible is necessary. This objective directly leads to the concept of Life Cycle Thinking and Life Cycle Assessment (LCA), as described by the standards DIN EN ISO 14040 and DIN EN ISO 14044.

Within the approach to a holistic assessment of steel and composite bridges as described in the abstract No. 1399 (NaBrue project), this method is used to assess and to develop reference values for a set of representative bridges, which reflect market relevant options for current steel construction.

Within the assessment system, the LCA indicators addressing the ecological quality are Global Warming Potential, Ozone Depletion Potential, Acidification Potential, Eutrophication Potential and Photochemical Ozone Creation Potential.

The system boundaries are set in line with E DIN EN15978, a European standard currently developed, dealing with the sustainability assessment of buildings and defining a building life cycle to include the product stage, construction stage, the use stage and the end-of-life stage, whereas benefits and loads beyond the system boundary have to be accounted for separately. Sub-stages given by E DIN EN15978 have been adapted with respect to bridge characteristics.

Whereas in most other building LCAs, currently transport to construction site, construction and installation processes as well as deconstruction itself are not regarded, it was decided to include them into the NaBrue analyses. Respective LCA data is compiled in cooperation between LCA experts and bridge engineers.

Modell evaluations will on the one hand show the importance of the different life cycle stages: How important are construction or deconstruction processes or transports in comparison to production and end-of-life? For future assessments, which aspects have to be analyzed more in detail, and which ones can be excluded from system boundaries due to their low influence on results? Where are the important and "low-hanging" environmental improvement potentials? Which maintenance strategies perform best in terms of environmental impacts? A further new aspect is the consideration of environmental impacts caused by external effects, or respectively their comparison to the other environmental impacts of the bridges' life cycles.

Sample data provided in form of the LCA results of the reference bridges will serve as a starting point for the development of benchmark and rating values, and will therefore play an important part in the innovative holistic sustainability assessment system for bridges as presented by abstract No. 1399.

REFERENCES

DIN EN ISO 14040 Environmental Management – Life Cycle Assessment – Principles and Framework, 2006.

DIN EN ISO 14044 Environmental Management – Life Cycle Assessment – Requirements and Guide-lines, 2006.

E DIN EN 15978: Nachhaltigkeit von Bauwerken – Bewertung der Umweltleistungsfähigkeit von Gebäuden – Berechnungsmethode; Deutsche Fassung prEN 15978:2010 (Entwurf). DIN – Deutsches Institut für Normung e.V. Berlin: Beuth 2010.

Bridge Maintenance, Safety, Management, Resilience and Sustainability – Biondini & Frangopol (Eds)
 ISBN 978-0-415-62124-3

Life cycle analysis of highway composite bridges

H. Gervásio & L. Simões da Silva
ISISE, Department of Civil Engineering, University of Coimbra, Portugal

Vera Perdigão & Paulo Barros
BRISA – Auto-estradas de Portugal S.A., Portugal

A. Orcesi
French Institute of Science and Technology for Transport, Development and Networks (IFSTTAR), France

Kim Nielsen
Rambøll Denmark A/S, Denmark

ABSTRACT

Highway bridges are important assets in every infrastructure network. The cost of construction and operation of highway bridges are key issues for bridge engineers. However, the design of bridges requires more than compiling with safety requirements according to safety standards and economic constraints. The construction, operation and ultimately the way the bridge is demolished have a significant impact on the environment, in the surrounding population and in the users of the bridge. The multidimensional perspective of sustainability requires the combination of current design criteria with other important aspects such as society and environment, usually considering a life cycle approach. This paper presents a framework for integrated life cycle analysis of bridges, illustrated in Figure 1, which is being developed within the scope of the European Research Project SBRI "Sustainable Steel-Composite Bridges in Built Environment".

The proposed approach is applied to a case study: a motorway composite bridge. The results of the environmental, economical and social analysis are presented in Figures 2, 3 and 4, respectively. The life cycle analysis covers all stages from raw material acquisition to deposition of demolition waste, through construction and operation of the bridge, and the transportation of materials and equipment between stages.

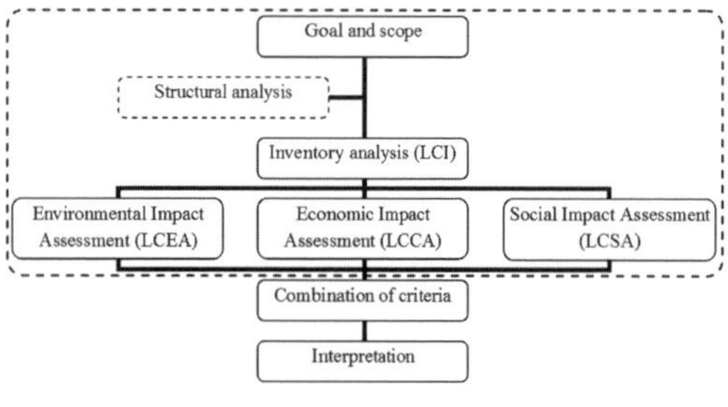

Figure 1. Flowchart of the Life Cycle Integral Analysis.

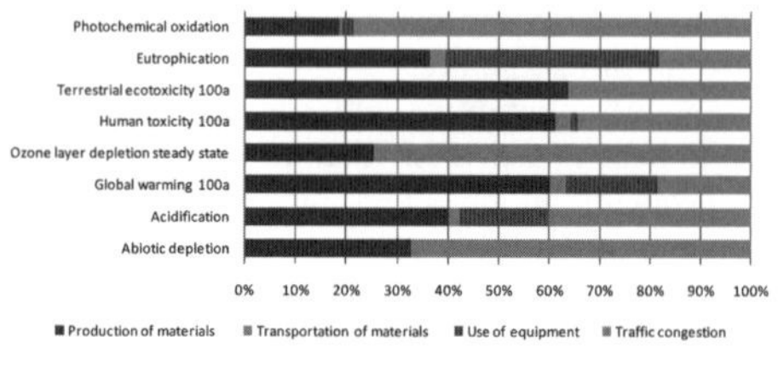

Figure 2. Contribution of each main process to impact category.

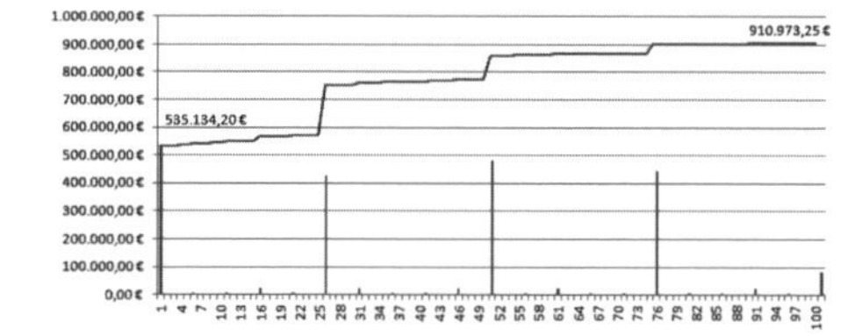

Figure 3. Life cycle costs of the composite bridge and accumulated present value (line in red).

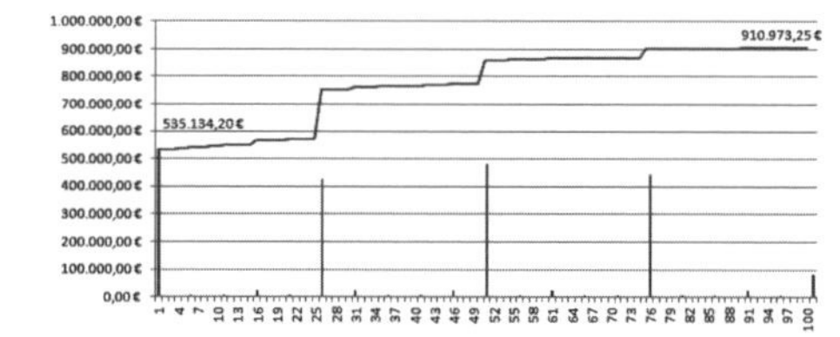

Figure 4. Life cycle social costs of the composite bridge and accumulated present value (line in red).

The life cycle analysis enabled to identify the process(es) with major impacts in the life cycle of the bridge and the causes of those impacts. The stages of material production and operation are the ones with major environmental impacts. The results of the life cycle cost analysis and life cycle social analysis allow to conclude about the importance of minimizing maintenance and rehabilitation activities over the service life of the structure.

Bridge Maintenance, Safety, Management, Resilience and Sustainability – Biondini & Frangopol (Eds)
© 2012 Taylor & Francis Group, London, ISBN 978-0-415-62124-3

Quantification of sustainability principles in bridge projects

C. Hendy & R. Petty
Atkins, Epsom, UK

ABSTRACT

Quantifying civil engineering projects in terms of sustainability and meeting carbon reduction targets is a new challenge for the civil engineering industry. Whilst the development of carbon accounting tools identify areas of bridge design and construction that have the greatest carbon emissions, quantifying sustainability overall has been less well studied. The sustainability index for bridges described in this paper is a significant step towards facilitating the systematic quantification of the sustainability of schemes through a simple and graphical tool.

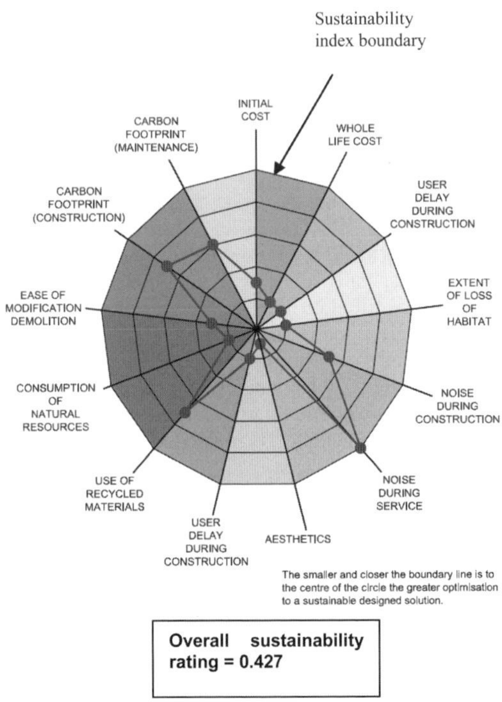

The output identifies where improvements can be made on a design and allows comparison of alternatives. It can be used throughout the design process to monitor decisions, the success of design changes and to inform decisions on future projects thereby improving the sustainability of designs.

The overall sustainability index rating provides a means of enabling targets to be set for the desired sustainability performance of bridges produced by an organisation. The methodology can be adopted by Clients so that, once the key attributes are set and weighted accordingly, designs can be benchmarked across their whole bridge stock. This paper presents the background to the development of the sustainability index, its key features and examples of its use and benefits.

REFERENCES

Report of the World Commission on Environment and Development, General Assembly Resolution 42/187, United Nations, 11 December 1987.

Environment Agency Carbon Calculator for Construction Activities Available at http://www.environment-agency.gov.uk/static/documents/Business/Carbon_calculator_v3_1_1.xls Accessed 8 August 201-

New Model Code 2010, Bulletin 55, International Federation for Structural Concrete (*fib*), March 2010, ISBN 978-2-88394-095-6

New Model Code 2010, Bulletin 56, International Federation for Structural Concrete (*fib*), April 2010, ISBN 978-2-88394-096-3

Bridge Maintenance, Safety, Management, Resilience and Sustainability – Biondini & Frangopol (Eds)
© 2012 Taylor & Francis Group, London, ISBN 978-0-415-62124-3

The proposed damage model and mechanical behaviors of damaged short suspenders in arch bridges

Y.B. Li
Department of Bridge Engineering, Tongji University, Shanghai, China

Q.W. Zhang
State Key Laboratory of Disaster Reduction in Civil Engineering, Tongji University, Shanghai, China

ABSTRACT

Recent bridge condition surveys show that the short suspenders are one of the most vulnerable components for arch bridges, and the in-service disease of arch bridges mainly result from damage deterioration of suspenders. For example, the Caihong Bridge in Chongqing and the Nanmen Bridge in Sichuan, China collapsed in 1999 and 2001 (See Fig. 1), respectively, and both direct causes were the brokenness of short suspenders. Much investigation indicates that damage deterioration of suspenders result from the chemical processes and their interaction with mechanical stress.

It is believed that the corrosion of high-strength steel wires is a quiet complex phenomenon that consists of several different (but interrelated) mechanisms, such as general corrosion, pitting corrosion, stress corrosion, corrosion fatigue, and corrosion-induced cracking. The complex damage deterioration mechanisms of suspenders result from uniform corrosion or pitting and cracking of wires (Riyadh et al. 2001). With the corrosion and actual load history, the mechanical behaviors of wires in suspender will degrade and the bearing capacity of suspender decrease. It becomes clear that the geometry of the wire, as a result of the corrosion process, seems to be the dominating mechanism in the reduction in mechanical behaviors of wires. The changes in geometry of the wire's cross-section are determined by the amount of corrosion and pitting that the wire has undergone. Therefore, in order to investigate the mechanical behaviors of damaged suspenders, the degradation of mechanical properties of wires must be discussed and a novel damage model be proposed by using a monotonic tensile test and finite element model simulation.

A new damage model and bilinear approximation of deteriorated wires in suspenders are suggested based on the mechanical properties of deteriorated wires which examined by experimental tests and FEM simulation in this paper. According to force-displacement envelope of deteriorated wires, a novel damage model and the bilinear approximation are proposed. The model is cumulative and capable of combining the real constitutive relations of wires, strength, stiffness, ductility, corrosion, and loading history. The new bilinear approximation of deteriorated wires is advanced based on the exact solutions of integral method or S. M. Elachachi constitutive method (Elachachi et al. 2006). Furthermore, the damage indicator of suspender is established combined with the corrosion distribution and damage indicators of wires. Finally, the rationalization of the proposed model is validated and the sensibility of parameters is illustrated (Park et al. 1975). It is found that the mechanical properties of wires degrade significantly with the corrosion and actual load history. The damage indicator of suspender in this paper is reasonable which taking account of degradation in strength, stiffness, and limit strain.

Figure 1. Short suspenders collapsed of Nanmen Bridge in Sichuan, China.

REFERENCES

Elachachi S. M., Breysse D., Yotte S., and Cremona C. 2006. A Probabilistic Multi-scale Time Dependent Model for Corroded Structural Suspension Cables. *Probabilistic Engineering Mechanics* 21:235–245.

Park R. and Paulay T. 1975. *Reinforced Concrete Structures*. New York: John Wiley and Sons.

Riyadh A. H. 2001. *A Proposed Damage Model for R/C Bridge Elements under Cyclic Loading*. Canada: The University of British Columbia.

Bridge Maintenance, Safety, Management, Resilience and Sustainability – Biondini & Frangopol (Eds)
© 2012 Taylor & Francis Group, London, ISBN 978-0-415-62124-3

Steel-composite bridges – Holistic approach applied to European case studies

P. Maier & U. Kuhlmann
Institute of Structural Design, University of Stuttgart, Germany

Y. Tardivel, N. Robert & J. Raoul
Service d'études sur les transports, les routes et leurs aménagements, Paris, France

V. Perdigão, N. Martins & P. Barros
BRISA Engenharia e Gestão SA –BEG, São Domingos de Rana, Portugal

H. Friedrich & J. Krieger
Federal Highway Research Institute (BASt), Bergisch Gladbach, Germany

ABSTRACT

In the European funded research project SBRI a holistic approach is applied to steel-composite bridges by combining analysis of Lifecycle Assessment (LCA), Lifecycle Cost (LCC) and Lifecycle Performance (LCP). Under the perspective of sustainability an entire lifespan, from the construction to the demolition of a bridge, is regarded. As German and Portuguese bridge authorities are involved a valuable collection of data in view of existing bridges has been obtained. With the bridge management system BMS and SIB-Bauwerke a lifecycle scenario was described including maintenance strategies. By postponing and pre-scheduling of measurements optimization towards cost-effectiveness and low environmental impact can be achieved.

Assumptions were made for the lifecycle scenario and maintenance actions. These influence not only the lifecycle assessment but also lifecycle and user costs. Optimized crack detection during inspections results in minimization of maintenance needed. The application of post-weld treatment may increase the fatigue resistance of critical details remarkably. The way of construction and materials used are an essential element to act on performances during the different phases of the lifecycle of bridges. Durability can be increased, the frequency of maintenance and the quantities of construction materials reduced. High performance materials, such as high strength steels, can reduce the weight of the structure and optimize the quantities used. In terms of maintenance, the use of specific materials like weathering steels can reduce the vulnerability of the structure to external environmental aggressions. Therewith lower lifecycle costs as well as less environmental impacts are achieved as no renewals of the painting are needed.

The new sustainable design process is evaluated on steel-composite bridges with a complete design of realistic case studies. Following the experience in the different European countries of the project partners, typical road bridges and traffic situations are selected. The functionality and therewith the span length of the bridges is the criteria by which the bridges are divided into three main types. A differentiation between small motorway bridges, crossings of motorways and big motorway bridges is made.

To the case studies the holistic approach of combining LCA, LCC and LCP is applied and optimized. Over the entire lifespan of the bridges none of the factors can be neglected but an optimization of the overall analyses must be found. A most specific description of the lifecycle is therefore a basis for the analyses. In the case studies realistic bridge situations are analysed.

Bridges are "living" structures, however changes during service are difficult and costly in terms of economic and environmental impacts. In a changing environment, these bridge structures must therefore have capacity or reserve of evolutions otherwise they will not satisfy the contemporary needs of users and society over their whole lifespan.

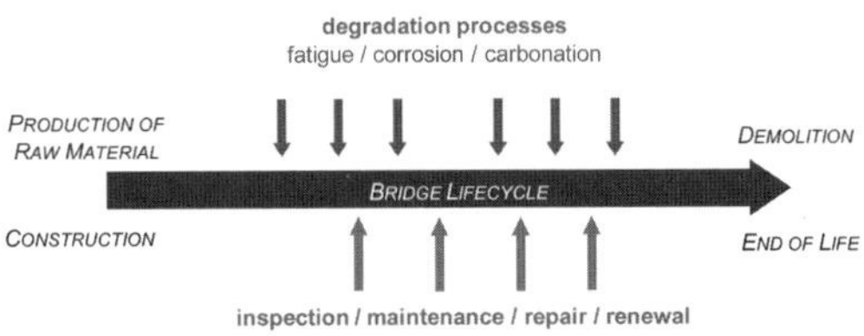

Figure 1. Schematic representation of the lifecycle of a bridge.

REFERENCE

Kuhlmann, U. et al. 2009, *SBRI Sustainable steel-composite bridges in built environment*, research project, RFSR-CT-2009-00020.

Bridge Maintenance, Safety, Management, Resilience and Sustainability – Biondini & Frangopol (Eds)
 ISBN 978-0-415-62124-3

Optimizing bridge design by improved deterioration models through fatigue tests

P. Maier & U. Kuhlmann
Institute of Structural Design, University of Stuttgart, Germany

N. Popa
ArcelorMittal Long Products Research & Development, Esch/Alzette, Luxembourg

R. Willms
AG der Dillinger Hüttenwerke, Dillingen/Saar, Germany

ABSTRACT

Bridges are essential structures for public, bounded to a significant investment volume over several decades. Managing these structures is an important task for engineers. When designing bridges the issue of fatigue cannot be neglected. Steel and steel-composite bridge structures experience deterioration especially due to fatigue. Fatigue can lead to major defects implying loss of structural capacity and safety. The basis of the description of the lifecycle of a specific detail is to capture the process of fatigue in a precise way. Describing the lifecycle of a detail is therefore the precondition for an optimized design of the entire bridge structure.

In the frame of the European funded research project SBRI two typical bridge details are investigated under fatigue and described here.

In the first part of this paper own fatigue tests on steel bridge girders are described. The transverse stiffener is an often used detail in steel and steel-composite bridges under fatigue load. The resistance against fatigue can be improved by optimizing the detail itself in an efficient way. Not changing the geometry but looking at the fabrication of the welded detail promises a sustainable extension of the fatigue resistance. Therefore the influence of post-weld treatment is compared to a standard fabrication in fatigue tests.

The fact of crack detection is a crucial issue when bridges are inspected. Several methods are applied during the own fatigue tests. A comparison regarding application on site, existing anti-corrosion coating and certainty of detection is made for the non-destructive inspection methods of magnetic particle and penetration testing. As innovative method the ultrasonic technique of phased array is used to detect cracks which are not close to the surface and therefore allow optimized detection of fatigue deterioration in the bridge structure.

In the second part of this paper own fatigue tests on composite girders under transversal fatigue loading are described. Horizontally lying shear studs are used for the shear connection. In a complex test set-up in addition to the transversal cyclic loading an axial force is applied leading to a cracking of the concrete. The behaviour of horizontally lying shear studs is described and based on own fatigue tests a S-N-curve is developed and compared to tests without any axial force.

An improved knowledge of the fatigue behavior of typical bridge details is reached and should be taken into account for a sustainable lifecycle design.

Figure 1. Set-up of fatigue tests on transverse stiffeners and horizontally lying shear studs.

REFERENCES

EN 1993-1-9- Eurocode 3. Design of steel structures, part 1–9: Fatigue strength of steel structures, European Committee for Standalization (CEN), Bruxelles.

Kuhlmann, U., Dürr, A. & Günther, H.-P. 2006. *Application of post welded treatment methods to improve the fatigue strength of high strength steels in bridges*. In: Proceedings of the Third International Conference on Bridge Maintenance, Safety and Management, IABMAS '06, Porto, Portugal.

Kuhlmann, U. & Raichle, J.; 2008b. *Ermüdungsverhalten liegender Kopfbolzendübel infolge Querschub*; final report FE 15.407/2004/CRB. Stuttgart, Germany.

Kuhlmann, U. et al. 2009, *SBRI Sustainable steel-composite bridges in built environment*, research project, RFSR-CT-2009-00020.

Bridge Maintenance, Safety, Management, Resilience and Sustainability – Biondini & Frangopol (Eds)
© 2012 Taylor & Francis Group, London, ISBN 978-0-415-62124-3

German approach to a holistic assessment of steel and composite bridges

P. Maier & U. Kuhlmann
Institute of Structural Design, University of Stuttgart, Germany

H. Friedrich & J. Krieger
Federal Highway Research Institute (BASt), Bergisch Gladbach, Germany

M. Pfaffinger & M. Mensinger
Chair for Metal Structures, Technische Universität München, Munich, Germany

ABSTRACT

Bridges are fascinating structures, not only for engineers, and possess a central role as an indispensable component of the infrastructure network as they do not only cover a large investment volume, but also do guarantee the smooth transport of goods and passengers. Compared to buildings bridges are especially long-living structures with a scheduled service life of 100 years. Since an increase of freight and passenger traffic is to be anticipated, adaptations to these rising requirements are necessary. However orders for a new bridge construction are usually given to the tenderer demanding the lowest price. Such a "cheap" bridge policy does usually not lead to especially long living structures characterized by high durability and flexibility in use. With the new concept of sustainable design developed in the German research project *NaBrue* a holistic assessment over the entire lifecycle should allow to introduce other criteria than only the initial costs. Particularly for infrastructure, such as bridges, which are publicly financed and maintained, the concept of sustainability is of special importance, for in this way, in addition to ecological qualities also socio-cultural and functional aspects can be integrated into decision making. The overarching goal is a holistic assessment for bridge structures, especially for steel and composite bridges. For this purpose, the three aspects of ecological quality, economical quality and socio-cultural/functional quality are substantiated with precise criteria specific for bridges.

In order to guarantee a real holistic view, analyses on different levels are necessary from detail investigations on critical points of the construction via investigations on the general bridge design to the effects on the directly affected traffic routes as e.g. the determination of the economic benefit in the transregional relationship or the assessment of the ecological impact. The development of methods for a holistic assessment of the sustainable design of steel and composite bridges is realized for three specific prototypes of bridges. For these bridges the entire lifecycle is regarded from the material production stage and the construction over the operation of the bridge (including maintenance etc.) till the demolition at the end of life. Considering the whole lifecycle of these bridges, identifying the decisive details and describing their degradation allows an optimization in view of sustainability already at the design stage. Comparing variants of bridge design for realistic conditions enables to identify the main influences on the various aspects.

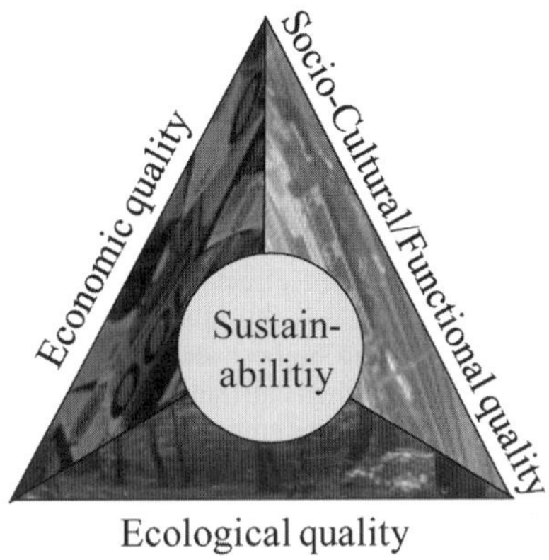

Figure 1. Holistic assessment of three qualities of sustainability.

REFERENCES

BASt: Bundesanstalt für Straßenwesen. *Federal Highway Research Institute of Germany*. website: http://www.bast.de

Beck, T. et al. 2011. *Life Cycle Assessment for representative steel an composite bridges*. In: proceedings of the 6th International Conference on Bridge Maintenance, Safety and Management, Lake Como, Italy, July 8–12, 2012.

Kuhlmann, U. et al. 2011. *Ganzheitliche Bewertung von Stahl- und Verbundbrücken nach Kriterien der Nachhaltigkeit*. In Stahlbau, 80(10): 703–710.

NaBrü, 2010. *Holistic Evaluation of Steel and Composite Bridges according to Criteria of Sustainability*, research project, FOSTA P843.

Zinke, T. et al. 2011. *The social dimension of bridge sustainability assessment – Impacts on users and the public*. In: proceedings of the 6th International Conference on Bridge Maintenance, Safety and Management, Lake Como, Italy, July 8–12, 2012.

Bridge Maintenance, Safety, Management, Resilience and Sustainability – Biondini & Frangopol (Eds)
© 2012 Taylor & Francis Group, London, ISBN 978-0-415-62124-3

Experimental and analytical studies on fatigue strength of corroded bridge wires

S. Nakamura
Tokai University, Tokyo, Japan

K. Suzumura
Nippon Steel Engineering, Futtu, Japan

ABSTRACT

Fatigue strength of corroded galvanized steel wires was studied in four steps in this paper.

In the first step, corroded wires on three corrosion levels (Levels-1, 2, and 3) were produced at laboratory and fatigue tests were conducted under dry and wet environments to find the effects of corrosion fatigue of corroded wires. It was found that fatigue strength of new wires was the highest and severer corrosion decreased fatigue strength further. There is also a tendency that the fatigue strength under wet environments is lower than that under dry environments.

In the second step, surface roughness or pits were measured on the corroded galvanized wires to find the relation between corrosion levels and pit sizes. An average maximum depth of Level-2 wires was 0.18 mm and that of Level 3 was 0.34 mm, which means severer corrosion produces deeper pits. The maximum depth was 0.56 mm among all of the data. All of the pits with the depth of 0.4 mm to 0.6 mm were within the area with a width of 10 mm and a length of 10 mm. On the other hand, some of the pits with the depth less than 0.4 mm were outside of this area. Relatively shallow corrosion pits spread widely and deeper corrosion pits concentrated on narrower area.

In the third step, cyclic tests were conducted for the wire specimens with artificial pits whose sizes were decided based on the measured data. To clarify the effect of pits, the specimens with three different shapes, round (group-P) and triangle (group-Q) and triangle with a notch (group-R), and with three different lengths of 3.5, 6.0 and 10 mm were used. Then, the fatigue strength of the wires with artificial pits was compared with that of actually corroded wires. However, the fatigue tests were conducted with a stress range of 400 MPa and further data with stress ranges such as 200 MPa and 300 MPa should be required to confirm this tendency.

The wire specimens were cyclically loaded with a stress range from 500 to 900 MPa. All of the group-P specimens with round shape pits did not break until one million cycles. The group-Q specimens with triangle shapes broke at fewer cycles for shorter pit length. The critical cycles did not depend on pit length for the group-R specimens with triangle shapes with a notch, and the pit with a notched triangle shape had the lowest fatigue strength. This is because stresses concentrate on a notch at the sharp edge of a triangle and initiates a fatigue crack from there. The triangle shape pit also decreased fatigue strength but it depended on the ratio of pit length to width. This ratio also related with stress concentration factor. The round shape pit is safer because stress concentration is smaller than those of two others. The S-N data of the triangle pit specimens (group-Q) and the notched triangle pit specimens (group-R) were on the extended S-N line of the Level-3 corrosion wires.

In the fourth step, the stress concentration factor was evaluated by two methods: measurement by strain gauges and FEM analysis. Four wire models with different artificial pit shapes were tensioned and strains were measured at the pit bottom and at the parallel parts, from which stress concentration factors were obtained. The same models were analyzed by FEM and the maximum stresses are calculated, by which the stress concentration factor was obtained. Both methods showed that stress concentration factor was higher for sharper triangle shape pits with shorter pit length, and it was smaller for round shape pits.

REFERENCES

Nakamura S. and Suzumura K. 2009. Hydrogen embrittlement and corrosion fatigue of corroded bridge wires, *Journal of Constructional Steel Research*, Elsevier 65: 269–277.

Suzumura K. and Nakamura S. 2004. Environmental Factors Affecting Corrosion of Galvanized Steel Wires, *Journal of Materials in Civil Engineering*, ASCE 16(1): 1–7.

Bridge Maintenance, Safety, Management, Resilience and Sustainability – Biondini & Frangopol (Eds)
 ISBN 978-0-415-62124-3

The social dimension of bridge sustainability assessment – Impacts on users and the public

T. Zinke & T. Ummenhofer
Karlsruhe Institute of Technology, Karlsruhe, Germany

M. Pfaffinger & M. Mensinger
Technische Universität München, München, Germany

ABSTRACT

In the past, bridge design usually concentrated on structural aspects. The cultural significance of a bridge as well as the interaction with the surrounding environment were frequently neglected. Within the scope of a growing demand for a holistic assessment of building activities, many evaluation criteria going beyond the pure manufacturing cost come to the fore. They are, however, not completely developed for bridges. While methods like life-cycle costing and life-cycle assessment are already applied for infrastructures, especially in the field of social assessment a lot of inaccuracies, assessment problems and methodological difficulties exist.

Up to now, just a few approaches for sustainability assessment systems are available that unite all relevant aspects within the life cycle of a bridge. Especially in the field of the social assessment a lot of different criteria have been developed. Dependent on the origin of the study, an emphasis on different topics have been devoted. For example, studies dealing with macroeconomic aspects of traffic normally use monetization methods to quantify different impacts. On the other hand, assessment systems created for a practical application consist of a big number of qualitative indicators which are united by using weighting factors. Two problems when dealing with social aspects have to be highlighted:

- Dependent on the country different classification schemes are used and different sets of indicators are often not directly comparable (front side of the cube in figure 1). Therefore, a close look at the classification structure is necessary.
- Secondly, research in the field of social aspects (especially for infrastructures and bridges) is still in its infancy. About 20 commonly used social aspects are part of different approaches. These aspects are often split up into a lot of single criteria which are described qualitatively.

To achieve a future development of social aspects for bridges, an application of external cost calculation should take place. This methodology allows to monetize the preferences of users and the public so that external costs and direct costs can be compared directly. As a consequence, the problem of finding weighting factors in assessment schemes can be avoided. Simultaneously, the acceptance of social assessment could be strengthened if valid methods for assessing social aspects exist. Depending on the boundary condition (traffic intensity and type of traffic route), external costs can exceed direct costs by far. In the example presented in the paper, delay costs, fuel costs and environmental costs (because of additional air pollution) are about 9 times higher than the direct costs in the construction phase. These results highlight the enormous importance of social aspects and external costs, respectively.

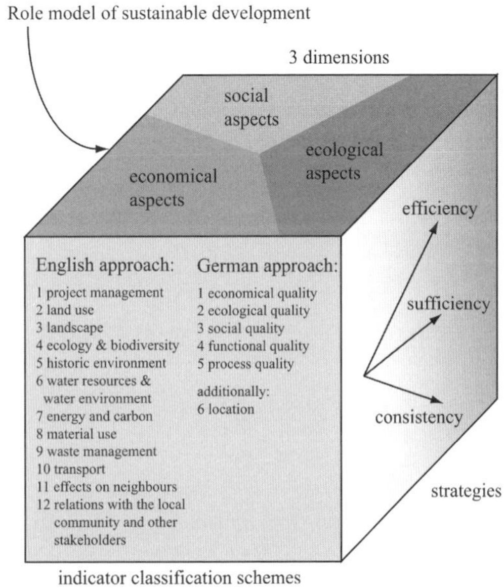

Figure 1. Classification possibilities for sustainable development (role model of sustainable development defined by the Brundtlandt commission and three different classification schemes at each side of the cube).

Numerical simulation of durability of concrete bridges

Organizer: A. Chen

Bridge Maintenance, Safety, Management, Resilience and Sustainability – Biondini & Frangopol (Eds)
© 2012 Taylor & Francis Group, London, ISBN 978-0-415-62124-3

Modeling corrosion-induced longitudinal crack width and its effect on corrosion rate

C. Cao & B.Y.B. Chan
The Hong Kong University of Science and Technology, Hong Kong, China

M.M.S. Cheung
The Hong Kong University of Science and Technology, Hong Kong, China
Tongji University, Shanghai, China

ABSTRACT

Chloride-induced corrosion of steel reinforcement in concrete may cause severe damage to RC structures. Longitudinal cracks may form during the corrosion rust expansion process and these cracks may enhance the transport processes of chloride, moisture, and oxygen in concrete, which may result in a significant increase the corrosion rate.

This paper examines the typical localized corrosion process by employing coupled micro- and macro-cell corrosion modeling technique. The reinforcing steel bar is divided into two parts: active zone and passive zone. In each zone, the corrosion processes are modeled by numerical polarization equations along with experimentally obtained parameters. The modeling of coupled oxygen distribution and potential distribution is accomplished by utilizing finite element method. The modeling scheme for corrosion propagation is shown in Figure 1.

Also, a simple uniform cracking model has been utilized to analyze the crack width propagation process. Smeared crack assumption is made in order to study the problem in the regime of continuum mechanics.

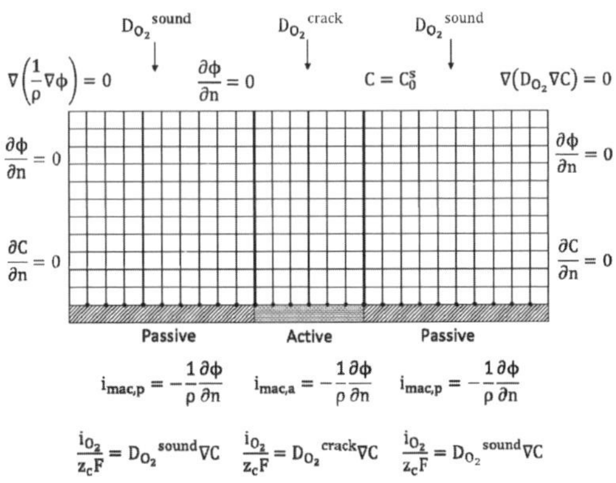

Figure 1. 2-D numerical model for corrosion propagation.

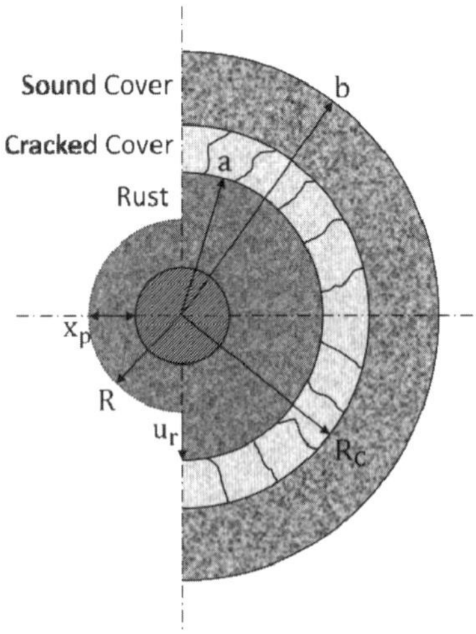

Figure 2. Uniform cracking model.

The modeling scheme for crack propagation is shown in Figure 2.

The coupled process between corrosion and crack propagation has been accomplished. It has been found that after crack appears, oxygen permeability will increase, which may cause an increase in microcell corrosion rate in the active zone. The microcell corrosion rate may increase a lot as the crack width increases while the macrocell corrosion rate may not change so much. As a result, the total corrosion rate at the active zone may increase substantially and this may lead to accelerated damage.

REFERENCES

Yuan, Y., Jiang, J. & Peng, T. 2010. Corrosion process of steel bar in concrete in full lifetime. *ACI Material Journal* 107(6): 562–567.

Wang, X. & Liu X. 2005. Bond strength modeling for corroded reinforcements. *Construction and Building Materials* 20(3): 177–186.

Bridge Maintenance, Safety, Management, Resilience and Sustainability – Biondini & Frangopol (Eds)
© 2012 Taylor & Francis Group, London, ISBN 978-0-415-62124-3

Bond slip model for generalized excitations

E. Mazzarolo
Università IUAV di Venezia, Venice, Italy

T. Zordan
Tongji University, Shanghai, China

B. Briseghella
Fuzhou University, Fuzhou, China

ABSTRACT

Correct evaluation for bond–slip effects is a crucial point when investigating progressive damage of reinforced concrete structures under random or earthquake excitations. For bridges, this aspect affects in particular the seismic response of piles' base and pier–deck joint in Integral Abutment Bridges as well. The need for a bond model, more accurate than those currently available in literature, without renouncing to ease of implementation, suggested to develop a new one. This model is defined by summing the effects of different bond resistance contributes (namely mechanical bond, friction bond and virgin bond) defined by means of continuous functions (Yankelevsky et al. (1992). This allows to fit, with reasonable precision, experimental monotonic and cyclic bond–slip paths, even along reloading branches. New relationships have been provided for updating the main law parameters at each load reversal. Moreover a specific progressive damage rule is introduced, able to account for generalized excitation. This model should be adopted in both numerical or analytical studies, when a realistic bond stress–slip relationship is required. Actually the model deals with short anchorage length and good confinement hypothesis. This means that splitting cracks should be neglectable and ultimate bond failure should be caused by rebar pull–out

Proposed bond-model analytical response is compared against several cyclic loading tests performed by Eligehausen et al. (1983) (test series 2). It appears that predicted curves fit reasonably well with experimental ones. Also reloading response is correctly simulated, independently from cycle number or maximum and minimum excursion slip. The damage parameter seem to capture well the progressive degradation of rebar-to-concrete interface even if in some cases a 10% overestimation of strength is revisable during first load cycle (Figure 2). Further work is however required to verify the goodness of the proposed approach when dealing with more general load excitations.

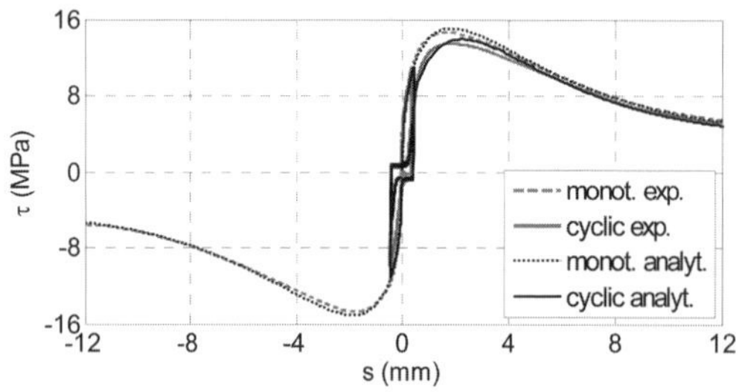

Figure 1. 10 cycles between ± 0.44 mm.

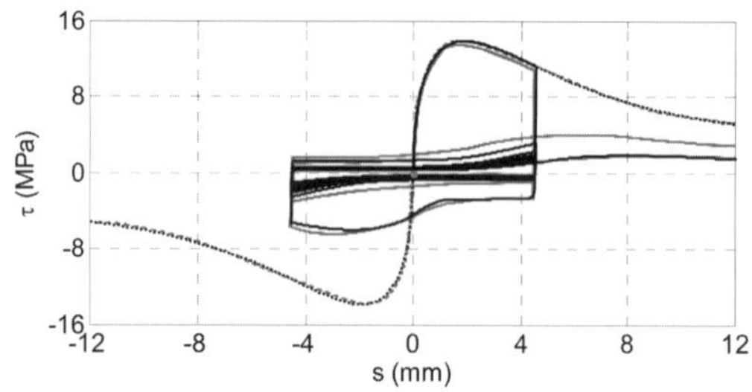

Figure 2. 10 full cycles between ±4.56 mm.

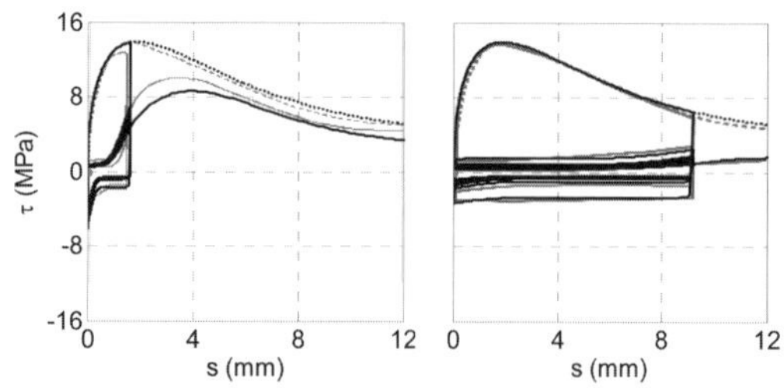

Figure 3. Half cycles with excursion 1.65 mm and 9.2 mm.

REFERENCES

Eligehausen R, Popov EP, Bertero VV, 1983. Local bond stress-slip relationships of deformed bars under generalized excitations: Experimental results and analytical model. Report No UCB/EERC-83/23, University of California.

Yankelevsky DZ, Adin MA, Farhey DN, 1992. Mathematical model for bond-slip behavior under cyclic loading. ACI Structural Journal 89(6).

Bridge Maintenance, Safety, Management, Resilience and Sustainability – Biondini & Frangopol (Eds)
© 2012 Taylor & Francis Group, London, ISBN 978-0-415-62124-3

Optimization of maintenance planning for deteriorating RC bridges. I: Theory

H. Tian
Zhejiang Institute of Communications, Hangzhou, Zhejiang, China
Zhejiang Scientific Research Institute of Communication, Hangzhou, Zhejiang, China

D.M. Frangopol
Department of Civil and Environmental Engineering, ATLSS Engineering Research Center, Lehigh University, Bethlehem, PA, USA

A. Chen
Department of Bridge Engineering, Tongji University, Shanghai, China

ABSTRACT

In this first part of a two-part paper, a novel approach for maintenance optimization of deteriorating reinforced concrete (RC) bridges is developed. The performance of these bridges is modeled using time-dependent reliability. Specifically, the component reliabilities for bending and shear are used as performance indicators. Firstly, component time-dependent reliabilities are evaluated using Latin Hypercube Sampling, Monte Carlo simulations and the computer program CBDAS. Secondly, maintenance models for RC bridges are presented. Finally, a novel approach for optimization of maintenance of RC bridges is proposed. In this approach, the original optimization problem is divided into a series of sub-optimization problems based on the possible sequences of maintenance types. Then the optimal solution of each sub-optimization problem is sought. The main steps involved are as follows:

(1) Time-dependent performance
An approach for evaluating the time-dependent reliability of deteriorating RC bridges is developed. On one hand, the computer program CBDAS was written to assess the variation of structural performance during service life; on the other hand, the Latin Hypercube Sampling and Monte Carlo Simulations are used to compute the time-dependent reliability.

(2) Effects of maintenance actions
According to the structural state during service life, the maintenance actions for deteriorating RC bridges can be classified in four types: Preventive type I (i.e. delay of corrosion initiation), Preventive type II (i.e. postponement of corrosion after its initiation), Strengthening, and Replacement. The quantitative effects of these maintenance types on structural performance are investigated.

(3) Optimization of maintenance
The original optimization problem is divided into a series of sub-optimization problems. The possible sequences of optimization types are assumed. Then the optimal solutions related to each sub-optimization problem are sought.

REFERENCES

FORTRAN 95. 1995. The Fortran Company, 6025 N. Wilmot Rd, Tucson, AZ 85750, USA.

Frangopol DM and Liu M. Maintenance and management of civil infrastructure based on condition, safety, optimization and life-cycle cost. Structure and Infrastructure Engineering, 2007;3(1):29–41.

Frangopol DM. Life-cycle performance, management, and optimization of structural systems under uncertainty: Accomplishments and challenges. *Structure and Infrastructure Engineering*, 2011;7(6):389–413.

Neves LC, Frangopol DM and Petcherdchoo A. Probabilistic lifetime-oriented multi-objective optimization of bridge maintenance: Combination of maintenance types. *Journal of Structural Engineering*, 2006;132(11):1821–34.

Neves LC, Frangopol DM and Cruz PJS. Probabilistic lifetime-oriented multi-objective optimization of bridge maintenance: Single maintenance type. *Journal of Structural Engineering*, 2006;132(6):991–1005.

Okasha NM and Frangopol DM. Novel approach for multicriteria optimization of life-cycle preventive and essential maintenance of deteriorating structures. *Journal of Structural Engineering*, 2010;136(8):1009–22.

Tian H., Frangopol DM and Chen A. Optimization of maintenance planning for deteriorating RC bridges. II: Application. Proceedings IABMAS 2012, Biondini F. and Frangopol DM, eds., CRC Press, 2012.

Tian H. Research on Performance Evolution of Concrete Bridges in Given Service Life[doctoral dissertation]. Shanghai: Tongji University, 2009.

Bridge Maintenance, Safety, Management, Resilience and Sustainability – Biondini & Frangopol (Eds)
© 2012 Taylor & Francis Group, London, ISBN 978-0-415-62124-3

Optimization of maintenance planning for deteriorating RC bridges. II: Application

H. Tian
Zhejiang Institute of Communications, Hangzhou, Zhejiang, China
Zhejiang Scientific Research Institute of Communication, Hangzhou, Zhejiang, China

D.M. Frangopol
Department of Civil and Environmental Engineering, ATLSS Engineering Research Center, Lehigh University, Bethlehem, PA, USA

A. Chen
Department of Bridge Engineering, Tongji University, Shanghai, China

ABSTRACT

This paper, as the second part of a two-part paper, presents the application of the proposed maintenance optimization approach to a reinforced concrete bridge. Chloride penetration-induced reinforcing-steel corrosion is used as the aggressive environment. Silane treatment, cathodic protection, bonded steel plate attaching, and replacement are selected as specific maintenance actions. The following simulation results are demonstrated and discussed: (a) the point-in-time component reliabilities throughout the service life and the associated time-dependent reliabilities, (b) the quantitative effect of bonding steel plate when Strengthening is used, and (c) the optimal solutions of the maintenance with respect to various specified conditions.

(1) Time-dependent reliability
The time-dependent reliability is nearly invariant at the beginning of service life, since corrosion didn't initiate in the reinforcing steels wrapped in the concrete section. After about 20 years, however, the time-dependent reliabilities begin to decrease due to the corrosion of reinforcing steels and reduction of concrete section. Thus, appropriate appropriate maintenance interventions are necessary to preserve the time-dependent reliability at acceptable level over the life-cycle of a structure.
(2) Quantitative effect of bonding steel plate
In the case of Strengthening, bending and shear reliabilities are improved
(3) Optimal solution of maintenance strategy
The optimal solutions of maintenance strategies with respect to various specified conditions are investigated. Four cases are discussed: (a) the threshold value of reliability index is 3.5, (b) the available maintenance budget is ¥1,800,000 (i.e. 1.8 Million Chinese RMB), (c) the threshold value of reliability index is 4.0, and (d) the bridge is assumed to have operated for 25 years and the threshold value of reliability index is defined as 3.5.

REFERENCES

FORTRAN 95. 1995. The Fortran Company, 6025 N. Wilmot Rd, Tucson, AZ 85750, USA.

Frangopol DM and Liu M. Maintenance and management of civil infrastructure based on condition, safety, optimization and life-cycle cost. Structure and Infrastructure Engineering, 2007;3(1):29–41.

Frangopol DM. Life-cycle performance, management, and optimization of structural systems under uncertainty: Accomplishments and challenges. *Structure and Infrastructure Engineering*, 2011;7(6):389–413.

Neves LC, Frangopol DM and Petcherdchoo A. Probabilistic lifetime-oriented multi-objective optimization of bridge maintenance: Combination of maintenance types. *Journal of Structural Engineering*, 2006;132(11):1821–34.

Neves LC, Frangopol DM and Cruz PJS. Probabilistic lifetime-oriented multi-objective optimization of bridge maintenance: Single maintenance type. *Journal of Structural Engineering*, 2006;132(6):991–1005.

Okasha NM and Frangopol DM. Novel approach for multicriteria optimization of life-cycle preventive and essential maintenance of deteriorating structures. *Journal of Structural Engineering*, 2010;136(8):1009–22.

Tian H., Frangopol DM and Chen A. Optimization of maintenance planning for deteriorating RC bridges. I: Theory. Proceedings IABMAS 2012, Biondini F. and Frangopol DM, eds., CRC Press, 2012.

Tian H. Research on Performance Evolution of Concrete Bridges in Given Service Life[doctoral dissertation]. Shanghai: Tongji University, 2009.

Bridge Maintenance, Safety, Management, Resilience and Sustainability – Biondini & Frangopol (Eds)
© 2012 Taylor & Francis Group, London, ISBN 978-0-415-62124-3

Diffusion process and life-cycle analysis of concrete structures

X. Tu & A. Chen
Department of Bridge Engineering, Tongji University, Shanghai, China

ABSTRACT

Durability is an important issue for long term operation of concrete bridge structures, such as continuous bridges, arch bridges, the pylon of cable stayed bridges and various concrete components. The capacity degradation of bridges relies on the corroded rebar and related damages within concrete attacked by aggressive agent from environment, being proved by numerous convictive experiments for natural corrosion and accelerated corrosion. Numerical simulation approach was adopted due to its currently practicality and applicability to evaluate the durability of concrete structures.

The relationship between pitting and average penetration was simplified into deterministic law by mathematical fitting and research experience, suggested as below:

$$R = \left(0.0263\delta_{As} + 0.000528\right)^{-0.308} - 2.048 \left(0 \le \delta_{As} \le 1\right)$$

where, δ_{As} is average section area reduction of rebar and R is pitting factor for area.

Based on Finite Element Method (FEM), the general methods for analyzing structural performance and diffusion of aggressive agent were developed and verified.

As was recognized as a unique feature of diffusion, inconsistent diffusion, indicating the diffusion process in two types of medium, was carefully studied (Figure 1).

As the classification of diffusion cracks, Major Crack and Micro Crack, were defined, Interactive Durability Analysis (IDA) was carried out based on the combination of structural analysis program and 3D diffusion program. A simply supported beam (Figure 2) was studied for researching on the time evolution of diffusion and effective section of rebar. Result of diffusion was obtained as shown in Figure 3.

IDA was proved to be a practical way to obtain the diffusion result within concrete structure under external loading. The simplified classification of crack introduced was a preliminary way to define the influence of loading on structures and provided be basis of considering the interaction between diffusion and structural performance.

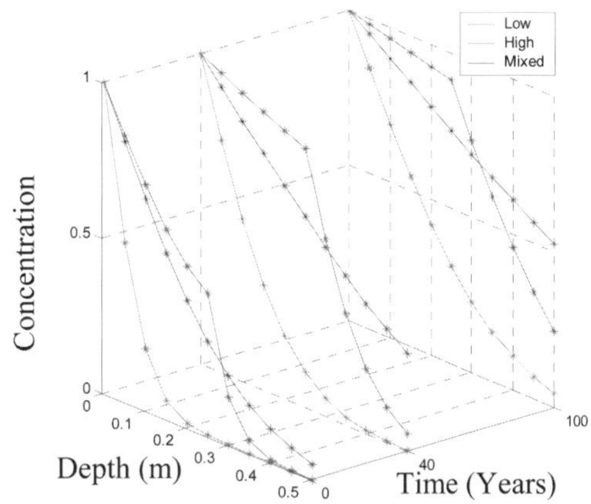

Figure 1. Time Evolution of Concentration Profile.

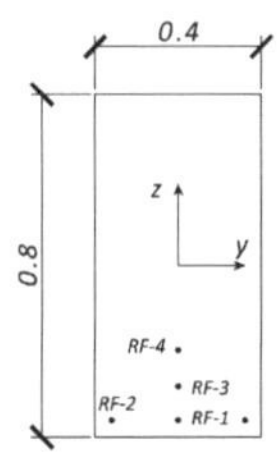

Figure 2. Section and Layout of Reinforcement (Unit: m).

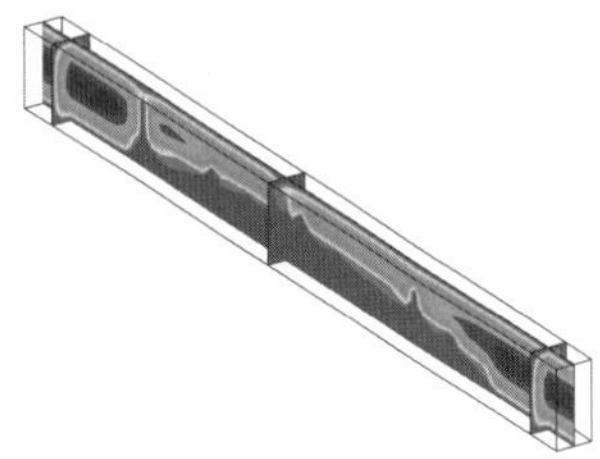

Figure 3. 3D Distribution of Concentration.

REFERENCES

Biondini, F. Bontempi, F., Malerba P.G. Fuzzy reliability analysis of concrete structures, *Computers & Structures*, Volume 82, Issues 13–14, Advances in Probabilistic Mechanics and Structural Reliability, May 2004, Pages 1033–1052.

Biondini, F. Bontempi, F., Frangopol, D.M., Malerba, P.G., Probabilistic service life assessment and maintenance planning of concrete structures, *Journal of Structural Engineering*, ASCE 132 (5) (2006), pp. 810–825.

Coronelli, D. & Gambarova, P. Structural Assessment of Corroded Reinforced Concrete Beams: Modeling Guidelines, *Journal of Structural Engineering*, Vol. 130, No. 8, August 2004, pp. 1214–1224.

Analysis, design and testing of road timber bridges

Organizers: A. Palermo, K. Crews & R. Kliger

Bridge Maintenance, Safety, Management, Resilience and Sustainability – Biondini & Frangopol (Eds)
 ISBN 978-0-415-62124-3

Non-linear analysis of a stress-laminated-timber bridge loaded to failure

K. Ekholm
Chalmers University of Technology, Steel and Timber Structures, Gothenburg, Sweden
Vectura Consulting, Gothenburg, Sweden

I.R. Kliger
Chalmers University of Technology, Steel and Timber Structures, Gothenburg, Sweden

R. Crocetti
Lund University, Structural Engineering, Lund, Sweden

ABSTRACT

The general assumption in stress-laminated-timber (SLT) bridge design is that the structural response is linear both in the serviceability limit state (SLS) and in the ultimate limit state (ULS). However, this has been shown not to be the case according to a full-scale test performed in Sweden where the SLT deck was subjected to a failure load (Ekholm et al. 2011).

When an SLT deck is loaded to failure, non-linear behaviour must be considered when a structure of this kind is analysed. Both horizontal and vertical slip occur in the interlaminar interface between the stressed glulam laminations, as shown in Figure 1. This behaviour does not occur in a solid timber plate. Once slip has occurred, the stresses are redistributed between the laminations.

The interlaminar slip occurs when the interlaminar horizontal rolling shear stress τ_{ZX} or the interlaminar vertical rolling shear stress τ_{ZY} becomes larger than the coefficient of friction (COF) multiplied by the resisting compressive pre-stress (σ_Y). The three distortions shown in Figure 1 are generated by:

1. Twisting moment M_{XY}
2. Transversal shear force V_{YZ}
3. In-plane twisting moment M_{XZ}

A rectangular SLT deck, $5.40 \times 7.98 \times 0.27\,m^3$ (length, width and thickness), with two patch loads

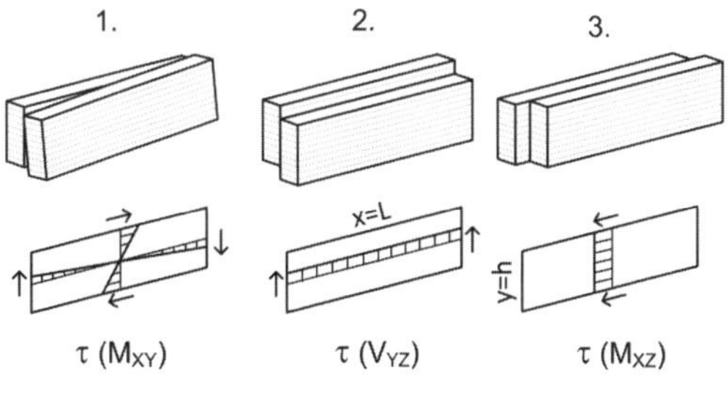

Figure 1. Distortions of laminates in a SLT deck due to interlaminar slip.

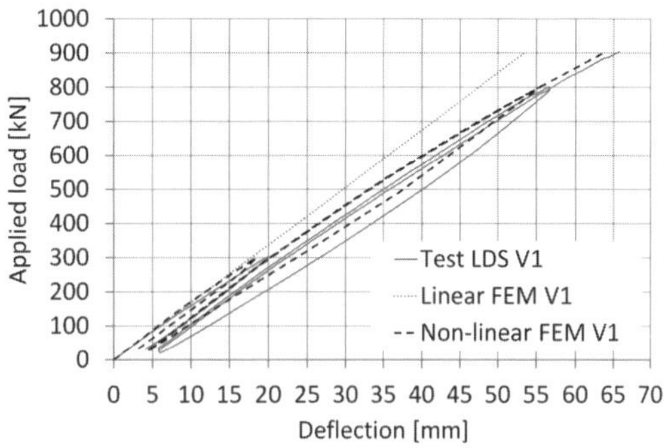

Figure 2. Comparison between the experimental test, linear and non-linear FE simulation. Deflection values taken at the edge of the deck, 0.4 m from mid-span. The load sequence $0 \rightarrow 300 \rightarrow 30 \rightarrow 800 \rightarrow 30 \rightarrow 900$ kN is shown.

positioned close to the edge was tested. Deflection values for both ultimate- as well as non-destructive loads were compared against finite element (FE) models of the SLT deck in order to evaluate the behaviour of the deck.

A non-linear model was created using the FE software Abaqus. Nonlinearity was obtained by introducing contact based friction properties in between the laminations.

Figure 2 shows a comparison between the deflection values from the test, a linear FE model and a non-linear FE model. The maximum deflection from the test was 65.1 mm. The non-linear model predicted a deflection of 64.0 mm (1.7% lower than the test), while the linear model predicted 53.7 mm (17.5% lower than the test).

REFERENCE

Ekholm, K., Crocetti, R. & Kliger, I. R. 2011. Full-scale ultimate-load test of a stress-laminated-timber bridge deck. Accepted for publication in ASCE Journal of Bridge Engineering posted online ahead of print 2 September 2011.

Bridge Maintenance, Safety, Management, Resilience and Sustainability – Biondini & Frangopol (Eds)
© 2012 Taylor & Francis Group, London, ISBN 978-0-415-62124-3

Timber bridges in Sweden – On-going research and steadily expanding market

I.R. Kliger
Chalmers University of Technology, Steel and timber Structures, Gothenburg, Sweden

K. Ekholm
Chalmers University of Technology, Steel and timber Structures, Gothenburg, Sweden
Vectura Consulting, Gothenburg, Sweden

R. Crocetti
Lund University, Structural Engineering, Lund, Sweden

ABSTRACT

The research on timber bridges both in the past in Scandinavia and on-going in Sweden is summarised. The on-going, more recent research at Chalmers University of Technology focuses on improving our understanding of the mechanical behaviour of stress-laminated-timber (SLT) decks loaded until failure. Three full-scale tests have been performed during the last couple of years within the framework of this project. All the SLT decks were made of glulam laminations.

The following parameters were studied: simply-supported decks versus continuous deck, type of load (two-patch or single load), load position (centre and close to the edge), level of pre-stress (three levels 300 kPa, 600 kPa and 900 kPa), and non-destructive tests and tests to failure. The non-destructive tests were all loaded up to five cycles. All the full-scale tests were comprehensively described by Ekholm (2011).

Some most important results from these tests presented in this paper are as follow:

- For the continuous deck over two spans, the deck maintained its ability to withstand a load corresponding to 70% of the ultimate failure load for almost half an hour. However, the stiffness reduction was significant. No stiffness reduction was observed for loads of up to approximately 30% of the ultimate load.
- For the simply-supported deck, there was a significant increase in the deflection of the deck when the pre-stress was reduced from 600 kPa to 300 kPa. A reduction of this magnitude was not observed when the pre-stress was reduced from 900 kPa to 600 kPa.
- The irreversible horizontal interlaminar slip began to appear at a load level corresponding to 17%–25% of the ultimate load.
- When the load was positioned at the very edge of the deck, the deflection values were as much as 110% higher than they were for a load positioned centrically.
- The performance of an SLT deck is highly dependent on its ability to withstand pre-stressing force in SLT decks, which needs to be relatively large and these forces creates compression perpendicular to the grain. Normally, it is common practice in Scandinavia to use high-strength steel rods with a diameter of d = 20 mm in order to pre-stress SLT bridges. The reinforcement of compression perpendicular to the grain by means of fully-threaded self-tapping screws increases the capacity of the anchorage system by 50 to 85% for a screw diameter of 10 mm and 12 mm respectively, (Crocetti & Kliger 2010).

Two case-study bridges completed in 2011 are also reported as examples of innovative solutions when it comes to the industrialised production and assembly of very large sections as examples of more spectacular projects performed in Sweden. Timber producers in Sweden are very positive when it comes to the future of timber bridges. However, many people at the authorities are still very sceptical, especially when it comes to questions such as service length, durability and maintenance costs.

REFERENCES

Crocetti, R. & Kliger, I. R. 2010. Anchorage systems to reduce the loss of pre-stress in stress-laminated timber bridges. International Conference Timber Bridges, ICTB2010. K. A. Malo, O. Kleppe and T. Dyken. Lillehammer, Norway, ICTB 2010 & Tapir Academic Press, Trondheim 2010. Procedings of the ICTB 2010: 359–370.

Ekholm, K. 2011. Stress-laminated-timber bridge decks subjected to ultimate loads – Experimental tests, analytical and numerical modelling. Department of Civil and Environmental Engineering, Division of Structural Engeineering. Goteborg, Sweden, Chalmers University of Technology. Licentiate of Engineering: 134.

Bridge Maintenance, Safety, Management, Resilience and Sustainability – Biondini & Frangopol (Eds)
© 2012 Taylor & Francis Group, London, ISBN 978-0-415-62124-3

Capacity of compression members in heritage timber truss bridges

A. Nicholas & W. Ariyaratne
NSW Roads and Maritime Services, Australia

K. Crews
University of Technology, Sydney, Australia

ABSTRACT

New South Wales Roads and Maritime Services are responsible for maintaining a number of timber bridges on the road network. Many of these are truss bridges which were designed over 100 years ago, and are listed on the State Heritage Register. The timber truss bridges were built when vehicles were much lighter, slower and less numerous. Since 1900, the design loadings for bridges in Australia have increased by almost ten percent every ten years.

Since the introduction of the limit state design method for Australian bridge design in the Austroads Bridge Design Code in 1992, there has not been an associated limit states timber bridge design code in Australia. Compression members in many of New South Wales' timber truss bridges take the form of spaced columns, consisting of two timber flitches separated by timber spacers. Although there are provisions for spaced columns in the Australian Standard for Timber Structures, they do not cover the kinds of columns used in timber truss bridges.

A research and testing program was conducted at the University of Technology, Sydney in 2010 and 2011 in order to develop guidelines to allow reasonable prediction of the capacity of members in these heritage timber truss bridges. The four primary areas that were studied were buckling modes of column assemblies, properties of Australian hardwoods used in timber bridges, shear capacity of timber spacers with bolts along the grain loaded perpendicular to grain, and stress relaxation of timber members loaded in bending with permanent deflections.

Spaced columns in many of New South Wales' timber truss bridges have larger spacers towards the centre so that a 25 mm bow is created in the timber flitches. This means that bending stresses must be considered in combination with compressive stresses. From the creep relaxation test, it was clear that a reduction factor can be applied to the bending stresses due to fabrication in order to take into account the fact that this stress will decrease with time.

From the experiments on the properties of bridge timbers, it was concluded that there is a wide variability in the modulus of elasticity of timber, which will affect both the design bending moment and the design compressive capacity. It is therefore important that a number of values of modulus of elasticity are checked to determine which is critical.

From the buckling mode experiment and the experiment on the shear capacity of timber spacers, valuable information was obtained to use in computer modelling of spaced column assemblies.

From the theoretical analysis and literature review, it was reasoned that the duration of load modification factor should not be applied to the buckling capacity of slender columns. This is because for slender columns, it is not the material strength that governs, but the stiffness, and there is no significant deflection until elastic buckling occurs.

From the testing and analysis described in this paper, new guidelines for the design and assessment of heritage timber truss bridges in New South Wales have been developed. The guidelines make use of a critical elastic buckling analysis, as well as a simple interaction formula for bending and compression. In order to make the guidelines useful and practical to those who may use them, they have been developed to be consistent with the Australian Standard for Timber Structures, AS 1720.1-2010.

The method developed and described in this paper permits a logical and transparent design process leading to improved reliability in the design and assessment of timber truss bridges in New South Wales. This work substantiates the existing sizes of truss members in many cases for the increased loads they now have to carry. It also allows a more accurate understanding of the actual behaviour of compression members in timber truss bridges, so that future modification or strengthening work can be undertaken with a sound engineering basis.

Bridge Maintenance, Safety, Management, Resilience and Sustainability – Biondini & Frangopol (Eds)
© 2012 Taylor & Francis Group, London, ISBN 978-0-415-62124-3

Simplified fatigue verification for timber-concrete composite bridges considering notched connections

K. Stephan & U. Kuhlmann
University of Stuttgart, Institute of Structural Design, Stuttgart, Germany

P. Aldi
Hess Timber GmbH & Co. KG, Kleinheubach, Germany

ABSTRACT

In order to be used as road bridge the fatigue behavior of the interface connection of a timber-concrete composite bridge has to be considered. Some knowledge has been gathered on the ultimate load capacity and stiffness of different shear connector types, however only little is known about the fatigue strength. As due to high stiffness and strength notch connectors are especially appropriate for bridges, fatigue tests have been realized for them. To investigate the connection characteristics and the failure mechanism at the interface, symmetrical push-out tests and tests on composite beams have been carried out (Kuhlmann & Aldi 2010). The S-N-line derived from push-out tests is compared with indications given in EN 1995-2 (2010) for shear timber fatigue strength.

Besides the fatigue strength of the connection, the considered fatigue load model is of importance for the service life prediction of a TCC bridge. The fatigue verification and utilization of a TCC bridge with notches according to the simplified fatigue load model FLM 3 of EN 1991-2 (2004) and the linear damage accumulation hypothesis with FLM 4 are introduced. The application of the simplified FLM 3 shows a conservative estimation of the damage rate due to traffic loads. In this case, the introduction of damage equivalent factors seems to be appropriate to adjust the applied load spectrum according to the real traffic situation. These factors take into account the length of the critical influence line, traffic volume, design life of the bridge and traffic lanes. But λ-factors given in EN 1993-2 (2010) are calibrated for steel S-N-lines and therefore different for timber and TCC structures.

Within this paper the developed S-N-line related to test results and investigations on the fatigue verification of a bridge with notched connections are presented. Both, S-N-line and a simplified fatigue concept, are required for the economical application of TCC road bridges.

Figure 1. Push-out specimen under load.

REFERENCES

EN 1993-2 2010. *Eurocode 3: Design of steel structures – Part 2: Steel Bridges*. European Communities and the European Committee for Standardisation (CEN).

EN 1995-2 2010. *Eurocode 5: Design of timber structures – Part 2: Bridges*. European Communities and the European Committee for Standardisation (CEN).

EN 1991-2 2004. *Eurocode 1: Actions on structures – Part 2: Traffic loads on bridges*. European Communities and the European Committee for Standardisation (CEN).

Kuhlmann, U. & Aldi, P. 2010. *Ermüdungsfestigkeit von Holz-Beton-Verbundträgern im Straßenbrückenbau*. Universität Stuttgart, Institut für Konstruktion und Entwurf: Forschungsbericht DGfH/iVTH 15052N.

Risk-based and disaster resilience analysis of bridge systems and networked infrastructures under multiple hazards

Organizers: G.P. Cimellaro & L. Dueñas-Osorio

Bridge Maintenance, Safety, Management, Resilience and Sustainability – Biondini & Frangopol (Eds)
© 2012 Taylor & Francis Group, London, ISBN 978-0-415-62124-3

Road network's disaster resilience assessment methodology

V. Arcidiacono, G.P. Cimellaro & A. Infuso
Politecnico di Torino, Turin, Italy

A.M. Reinhorn
The State University of New York, Buffalo, New York, USA

ABSTRACT

In the recent years numerous catastrophic events have occurred that show how the communities are vulnerable and have not sufficient knowledge about how to manage critical events. Our intent is to create a new methodology that can evaluate damages, functionality, and resilience of a road network during an extreme event.

A critical review of existing indices, such as ones proposed by Chang & Nojima (2001) and Bocchini & Frangopol (2011), is provided by a comparison between the indices. The strengths and the weaknesses are evaluated with respect to the new proposed methodology.

In order to model the road network that is composed by bridges, tunnels, major roads, districts (secondary roads), and traffic sources it has been used the graph theory. The methodology was inspired by the human circulatory system, where the traffic sources are comparable to the heart, while the network is equivalent to the blood vessels. The capacity of the road network has been considered as the parameter that defines the functionality and the behavior of the system. Therefore, the main parameters of entire methodology are: (i) the interdependences between road network and building units; (ii) the rearrangement of the network after the disaster to achieve the maximum functionality; (iii) the accessibility of the road network from the traffic sources; and (iv) the development of a recovery plan that is able to maximize the resilience index. The results are expressed in term of resilience index GIR and of time to complete the works TEW.

The methodology has a light computational weight, making the methodology a useful tool to be implemented in software. In fact, it was developed into a self-standing software tool (ArciResilience 2.0) based on PEOPLES framework. The software provides a recovery plan that maximizes the resilience index with respect to physical, social, and economic limitations. It is proposed with a user-friendly graphical interface working into the Google Earth, which it is used to provide and display the input and output of the data for/of the analysis, allowing the user to see the geospatial distribution of the given hazard scenario.

Finally, the proposed model has been tested using two practical cases: (i) the transportation network of L'Aquila town (including highways and primary roads) subjected to the 2009 Italian earthquake; and (ii) the road network of Treasure Island in San Francisco Bay affected by an earthquake with a return period of 2450 yrs.

REFERENCES

Arcidiacono, V., Cimellaro, G.P., and Reinhorn, A.M. (2011). "A software for measuring disaster community resilience according to the PEOPLES methodology". Proceedings of COMPDYN 2011 – 3rd International Conference in Computational Methods in Structural Dynamics and Earthquake Engineering, Corfu', Greece, May 26–28, 2011.

Cimellaro G.P., and Arcidiacono V. (2011). "Resilience-Based Design for urban cities". Proceedings of "How the concept of resilience is able to improve urban risk management? A temporal and spatial analysis", November 3rd–4th, 2011 Paris.

Cimellaro, G.P., Reinhorn, A.M., and Bruneau, M. (2006). "Quantification of seismic resilience, in 'Proceedings of the 8th U.S. National Conference of Earthquake Engineering." San Francisco, California, USA.

Cimellaro, G.P., Reinhorn. A.M., and Bruneau, M. (2010). "Seismic resilience of a hospital system." Structure and Infrastructure Engineering 6(1), pp. 127–144.

Bridge Maintenance, Safety, Management, Resilience and Sustainability – Biondini & Frangopol (Eds)
 ISBN 978-0-415-62124-3

Probabilistic functionality recovery model for resilience analysis

P. Bocchini, A. Decò & D.M. Frangopol
Department of Civil and Environmental Engineering, Lehigh University, Bethlehem, PA, USA

ABSTRACT

The topic of structural resilience is certainly one of the most rapidly emerging research trends in Civil Engineering. The aim of resilience analysis is to emphasize the importance for structures, infrastructure systems, and entire communities, to withstand and recover efficiently from extreme events (Bruneau et al. 2003). For this reason, the concept of resilience can be applied to preventive assessment as well as to disaster management (Bocchini & Frangopol 2012). In this latter case, the small amount of information usually available after the occurrence of an extreme event and the need for quick effective tools that assist the decision making process have led to the development of resilience analyses characterized by a deterministic approach. On the other hand, when the focus is on the resilience assessment of existing systems (e.g. structure, infrastructures, communities) for design and maintenance management purposes, the information available is usually more complete and the time constraints are more flexible. Therefore, in these cases resilience is treated more properly in a probabilistic context (Chang & Shinozuka 2004).

The most popular analytical definition of resilience is rooted in the concept of functionality. A general definition of functionality is necessarily elusive, because different structures and systems should have their functionality defined in very different ways. Indeed, even for the same system, usually there are multiple metrics that can be used to represent the functionality. For instance, the functionality of a hospital system can be measured as the normalized waiting time for a patient; the traffic flow capacity can be used as a metric for the functionality of a bridge. By definition, resilience depends entirely on the residual functionality of a system after the occurrence of an extreme event and on the functionality recovery over time. Therefore, a proper model for the functionality recovery is paramount for any resilience assessment.

In this paper, a novel model for the functionality recovery of a generic system is proposed. The aim is to develop a model versatile enough to be adapted to the typical recovery profiles of many components of the civil infrastructure. In fact, in most studies that deal with resilience, complex systems are involved and the recovery of very different entities (e.g. an individual bridge, an individual tunnel, and an entire transportation network) must be combined. Moreover, even for the same component, the functionality recovery profile can be very different depending on the magnitude, type, and location of the damage caused by the extreme event. A general model that can adapt to fit all these needs facilitates the adoption of a unified and coherent framework for resilience analyses. Moreover, the proposed model introduces six clearly defined parameters that can be considered random variables to account for the uncertainties involved in the problem. Depending on the specific application and the initial damage conditions, the probability distributions of the random parameters define the probabilistic characteristics of the time-dependent functionality recovery, which is a stochastic process.

In this paper, a review of the most popular recovery functions is provided. Then, the proposed model is presented in detail. A numerical example is used to illustrate the application of the proposed model using a highway bridge.

REFERENCES

Bocchini, P. & Frangopol, D.M. 2012. Restoration of bridge networks after an earthquake: multi-criteria intervention optimization. *Earthquake Spectra* 28(2).

Bruneau, M., Chang, S.E.,. Eguchi, R.T., Lee, G.C., O'Rourke, T.D., Reinhorn, A.M., Shinozuka, M., Tierney, K., Wallace, W.A. & Winterfeldt D.V. 2003. A framework to quantitatively assess and enhance the seismic resilience of communities. *Earthquake Spectra* 19(4): 733–752.

Chang, S. & Shinozuka, M. 2004. Measuring improvements in the disaster resilience of communities. *Earthquake Spectra*, 20(3): 739–755.

Bridge Maintenance, Safety, Management, Resilience and Sustainability – Biondini & Frangopol (Eds)
© 2012 Taylor & Francis Group, London, ISBN 978-0-415-62124-3

Seismic vulnerability of shallow buried rectangular structures

E. Debiasi, A. Gajo & D. Zonta
DIMS – University of Trento, Trento, Italy

ABSTRACT

Compared to bridges, underground structures are unjustly regarded as non-critical components of the road infrastructure, due to their supposed low seismic vulnerability. Conversely, experimental observations indicate that shallow buried rectangular structures can be affected by shaking failure. In literature there are no reliable models to calculate fragility curves for this kind of buried structure. In this paper we suggest vulnerability models for shallow buried rectangular structures. Cut and cover structures belong to this category and are made with an open excavation and later placing the backfill over the finished structure. This structure is typical of shallow tunnels in highway and railway transportation systems, of small bridges (box culvert) crossing streams or roads, as well as of railway subways. In seismic zones the support of underground facilities must be designed for static overburden loads and for the additional forces due to soil-structure interaction under seismic conditions. In the paper we first introduce a seismic response model for shallow buried 3 and 4-sided rectangular structures. In the formulation proposed here we rely on simplified methods described in the literature, and in particular on the method proposed by Wang (1993). Next, we propose a simplified formulation to estimate their capacity A_i based on knowledge of their geometry, where with capacity we mean the median peak ground acceleration that produce a determinate damage state on the structure. This quantity is the basis for estimating the structural seismic risk using the fragility curve approach as proposed in HAZUS (Basöz & Mander, 1999). Particularly, capacity is defined with respect to an Operational Limit State (OLS) and a Collapse Limit State (CLS), for the scope of this work assumed equivalent to the slight damage and the complete damage of HAZUS. Next, we show with the aid of two examples how the fragility curves are derived and the seismic risk is calculated for individual case studies, representative of 3 and 4-sided boxes located in the Italian Province of Trento. In Table 1 the results of the two cases are summarized, where P[D > OLS|PGA] and P[D > CLS|PGA] are the probability of exceeding the OLS and CLS respectively, with an earthquake return period of $T_R = 475$ years.

Last, in our conclusion, we take the same approach to estimate the vulnerability of a complete infrastructural system, the bridge stock of the Italian Autonomous Province of Trento (APT). In Fig. 1 we plot the probability of exceeding the CLS with an earthquake return period of $T_R = 475$ years for this bridge stock. Every dot corresponds to one structure and shows the position of the structure. The colors of the dots depend on the value of the probability P of collapse; in particular if this probability is lower than 10^{-5} the color is green; yellow if $10^{-5} < P < 10^{-4}$; orange if $10^{-4} < P < 10^{-3}$; and red if greater than 1/1000.

Table 1. Results of the two case studies.

Buried structure	A_{OLS}	P[D > OLS| PGA]	A_{CLS}	P[D > CLS| PGA]
4-sided	0.92 g	$1.3 \cdot 10^{-5}$	1.39 g	$5.0 \cdot 10^{-7}$
3-sided	0.29 g	$6.7 \cdot 10^{-2}$	1.28 g	$3.0 \cdot 10^{-5}$

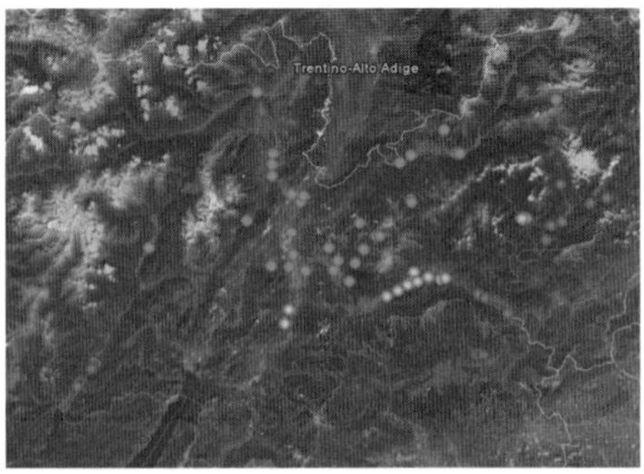

Figure 1. Probability to exceeding the CLS with $T_R =$ 475 years

REFERENCES

Basöz, N., Mander, JB., 1999. Enhancement of the Highway Transportation Lifeline Module in HAZUS. *Final Pre-Publication Draft (#7) prepared for the National Institute of Building Sciences,* March 31, 1999.

Wang, J., 1993. Seismic Design of Tunnels: A Simple State-of-the-art Design Approach, *Monograph 7, Parsons, Brinckerhoff, Quade and Douglas Inc.,* New York.

Bridge Maintenance, Safety, Management, Resilience and Sustainability – Biondini & Frangopol (Eds)
© 2012 Taylor & Francis Group, London, ISBN 978-0-415-62124-3

Accounting for bridge condition and correlation estimates in the seismic reliability analysis of aging transportation networks

J. Ghosh, K. Rokneddin, J.E. Padgett & L. Dueñas-Osorio
Rice University, Houston, TX

ABSTRACT

The system level reliability of spatially distributed systems is affected by correlated response of their components. In highway bridge networks, correlations among the seismic failure probabilities of bridges may stem from geographical proximity and spatial orientation of bridges; similarities in design, material, and construction methods; environmental agents; network, topology, traffic loading, etc. Moreover, recent advances in bridge sensing and health monitoring techniques highlight the opportunity to account for in-field measurements when estimating the seismic reliability of aging bridges and transportation networks. Such in-field measurement data and correlation sources at network level are often absent in the bridge network reliability literature. A framework is presented in this paper to account for the combined effect of known and unknown correlation sources while improving bridge fragility estimates via field condition data and Bayesian updating for an improved network reliability assessment.

Kriging, a spatial interpolation technique, is applied to estimate deterioration parameters for all bridges in the network from measurements of a sample set of bridges. At the bridge reliability level, deterioration parameters either measured in the field or inferred from Kriging are used to update prior estimates of the seismic fragility of bridges conditioned upon such degradation parameters. The correlation structure among seismic failure probabilities of bridges is established using data sources for bridge condition ratings, types of roads in the network, and the network layout. The Dichotomized Gaussian Method is presented to generate correlated binary random variables that are compatible with the correlation structure. These samples are used to evaluate the network connectivity reliability by Monte Carlo simulations. The proposed methodology is applied to an 18-bridge

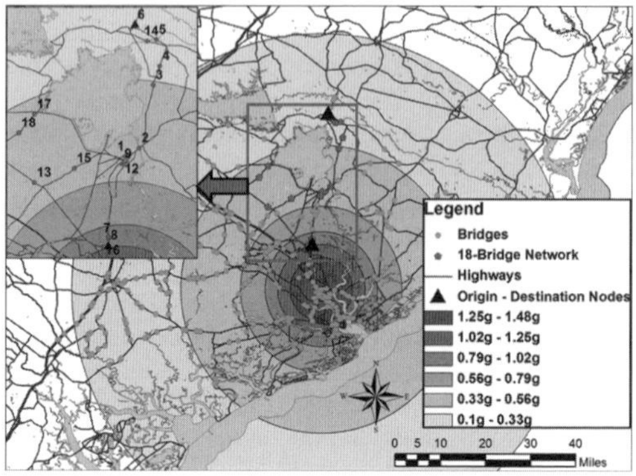

Figure 1. Case-study region of the SC transportation network.

aging transportation network in South Carolina, USA, revealing a significant change in the transportation network reliability after accounting for the bridge damage correlation structure.

REFERENCES

Bocchini, P., and Frangopol, Dan M. (2011). "Generalized bridge network performance analysis with correlation and time-variant reliability." *Structural Safety*, 33(2), 155–164.

Ghosh, J., and Padgett, J. (2010). "Aging Considerations in the Development of Time-Dependent Seismic Fragility Curves." *Journal of Structural Engineering*, 136(12), 1497–1511.

Kiremidjian, A., Stergiou, E., and Lee, R. (2007). "Issues in seismic risk assessment of transportation networks." *Earthquake Geotechnical Engineering*, 461–480.

Rokneddin, K., Ghosh, J., Dueñas-Osorio, L., and Padgett, J. (2011). "Bridge retrofit prioritization for aging transportation networks subject to seismic hazards." *Structure and Infrastructure Engineering*, In Review.

Advanced technologies in standard bridge structures – from research to implementation

Organizer: M. Saiid Saiidi

Bridge Maintenance, Safety, Management, Resilience and Sustainability – Biondini & Frangopol (Eds)
 ISBN 978-0-415-62124-3

Application of Shape Memory Alloys (SMAs) for prevention of bridge deck unseating during hurricane wave and surge loading

N. Ataei, E. McCarthy & J.E. Padgett
Department of Civil and Environmental Engineering, William Marsh Rice University, Houston, TX, USA

ABSTRACT

Often, high profile bridges uniquely employ innovative and complex mechanisms for management of anticipated dynamic loads. Natural hazards however are unmindful of design and bring common bridges prematurely to failure through a-typical conditions. Tidal surge for example routinely unseats coastal bridge decks during hurricanes resulting in a break down in a transportation network, excessive repair and economic costs, and potential loss of life. Through the distinctive recentering and energy dissipating characteristics inherent to shape memory alloys (SMAs), some deficiencies seen in bridges vulnerable to deck unseating can be addressed. Idyllic shape memory alloy restraining and damping abilities are mechanically defined by a long stress plateau which allows for the accommodation of large vertical displacements without the transference of detrimental loads to the adjoining structure, a strain hardening behavior which limits undesirable structural displacements, and a series of transformation states which provide high strength and energy dissipation. The superelasticity effect is descriptive of these SMA characteristics which are stress induced and strain recoverable up to approximately 6–8%. By analyzing a standard coastal bridge which implements SMA devices as a means of deck restraint during a surge wave event, the viability of application in-situ is determined. The OpenSEES framework is used for monitoring such a bridge's response both with and without SMA restrainers and thus facilitates monitoring of potential bridge response modification. Figure 1 shows the case study bridge model with restrainer SMA cables.

Fragility surfaces, which show the vulnerability of the structure under hazard intensity, were developed as part of this research to provide a better comparison of the retrofitted bridge with the reference model. The result of this study reveals the potential of SMA cables for application as a retrofit measure for coastal bridges to prevent the deck unseating mode of failure.

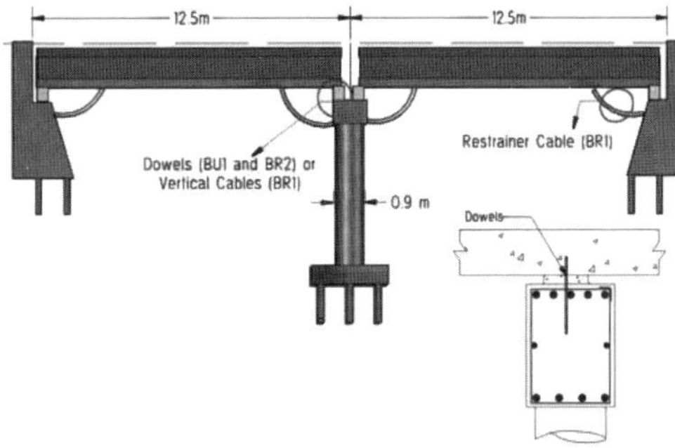

Figure 1. Case study bridge configuration.

Bridge Maintenance, Safety, Management, Resilience and Sustainability – Biondini & Frangopol (Eds)
© 2012 Taylor & Francis Group, London, ISBN 978-0-415-62124-3

Assessment of a historical railway bridge toward traffic regulation requirements

N. Boumechra
University of Tlemcen, Tlemcen, Algeria

F. Casciati
University of Pavia, Pavia, Italy

S. Casciati
University of Catania, Siracusa, Italy

ABSTRACT

Not far away from the Algerian town of Tlemcen, an arch bridge was built, in 1889, with the purpose of enabling the railway link to cross a rather sharp mountain area. The structure was named the El Ourit bridge, and its design was conceived by the world-famous French engineer Gustave Eiffel. It consist of a metallic truss structure forming two parallel arches, which are stiffened together by the deck and by the horizontal and vertical braces. The span is 68 m, but the entire deck length is 70 m. Each arch is actually made of two parts joined by a hinge at mid-span and two prestressed diagonal elements are placed across the link to improve the stiffness. At the edges, the main support consists of a bolted I-section beam of variable height, which is hinged in the masonry abutement. The deck shows a small horizontal curvature of radius 300 m, which causes a train speed limit of 60 km/h.

The bridge is now 123 years old and it is demanded to satisfy the present safety requirements, resulting from the new train traffic features and from the updated seismic code. A performance evaluation of the bridge is therefore pursued in this paper. The results of preliminary studies are reported in (Boumechra & Hamdaoui, 2008; Boumechra & Hamdaoui, 2010).

First, the different components of the bridge were deeply inspected by a geometric, topographic and pathological survey. No corrosion at all was detected thanks to the dry environment and the maintenance plan of the network manager, SNCF

Specimens of wrought iron were tested in the laboratory to characterize the mechanical behavior of the bridge constitutive material. such a material is characterized by a peculiar metallographic structure, with inclusions of variable density oriented along the lamination direction. These inclusions are made of iron oxides, silica and phosphor. In this particular case, the iron mineral comes from the French region of Lorain with more than 1% of phosphor; this component improves the anti-corrosion characteristics of the material.

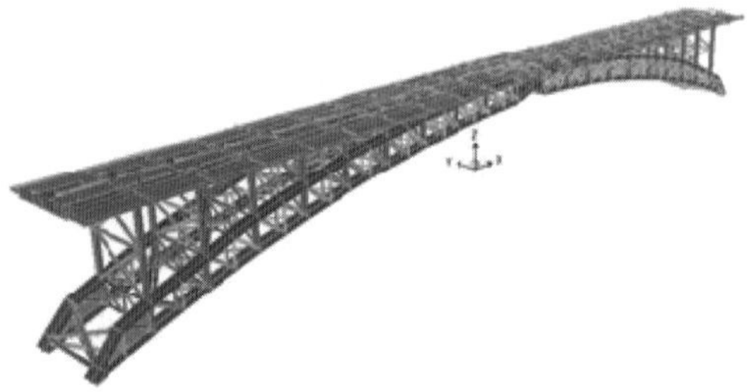

Figure 1. Finite element model of the El Ourit Bridge.

A finite element model of the bridge (Figure 1) was then created on the basis of its actual geometry. The modal analysis of the bridge was a conducted, and a historical ground motion record was used to carry out the dynamic analysis. The structural safety was checked with reference to both the last version of the Algerian seismic code (MTP, 2008) and the Eurocode 8 (Afnor, 2000). It was possible to conclude that the original design is consistent even with the seismic hazard update.

REFERENCES

Afnor, 2000, Eurocode 8: Design of structures for earthquake resistance – Part 2: Bridges.

Boumechra N. and Hamdaoui K., 2008, Dynamic and Fatigue Analyses of an 18th Century Steel Arch Bridge, *Proceedings of the 2008 Seismic Engineering International Conference (MERCEA'08)*, July 08–11, Reggio Calabria, Italy, pp. 69–75.

Boumechra N. and Hamdaoui K., 2010, Dynamic Analysis of Steel Truss Bridges with Uncertain Parameters – A SABRA's Iron Bridge Case, in *Proceedings of the 5th EWSHM*, Casciati F. and Giordano M. (eds.), DEStech Publications, Lancaster, PA, 238–245.

M.T.P., 2008, *Document technique reglementaire D.T.R. – Règles parasismiques applicables au domaine des ouvrages d'art* (in French) – R.P.O.A. 2008, Ministère des Travaux Publics.

Bridge Maintenance, Safety, Management, Resilience and Sustainability – Biondini & Frangopol (Eds)
© 2012 Taylor & Francis Group, London, ISBN 978-0-415-62124-3

Research and application of precast segmental concrete bridge columns in regions of high seismicity

K.-C. Chang & M.-S. Tsai
Department of Civil Engineering, National Taiwan University, Taipei, Taiwan

Y.-C. Ou
Department of Construction Engineering, National Taiwan University of Science and Technology, Taipei, Taiwan

P.-H. Wang
National Center for Research on Earthquake Engineering, Taipei, Taiwan

G.C. Lee
Department of Civil, Structural and Environmental Engineering, University at Buffalo, Buffalo, USA

ABSTRACT

Precast segmental concrete bridge column construction by reducing on-site construction activities and time has been proven to be an effective method to reduce traffic disruption and environmental impact. However, conventional segmental columns typically have much lower seismic energy dissipation capacity than monolithic columns due to the discontinuity of mild steel reinforcing bars at precast joints. This paper presents a series of research activities designed to investigate methods to improve the seismic behavior of segmental columns for use in regions of high seismicity. The methods include the use of mild steel reinforcing bars continuous across precast joints, the use of cast-in-place plastic hinge regions, and the use of seismic isolations. Results of cyclic tests of large-scale columns demonstrated that the proposed methods can greatly enhance the seismic behavior of segmental columns. The applications of the proposed segmental columns with cast-in-place plastic hinge method and seismic isolation are also presented.

REFERENCES

ASTM A955-04a. 2004. Standard specification for deformed and plain stainless steel bars for concrete reinforcement. ASTM International, West Conshohocken, PA, U.S.A.,

FHWA. 2010. Prefabricated bridge elements and systems – get in, get out, stay out, Federal Highway Administration, Washington, DC, U.S.A. Available from: http://www.fhwa.dot.gov/bridge/prefab/index.cfm

Ou, Y.-C., Tsai, M.-S., Chang, K.-C., and Lee, G. C. 2010. Cyclic behavior of precast segmental concrete bridge columns with high performance or conventional steel reinforcing bars as energy dissipation bars. *Earthquake Engineering and Structural Dynamics*, 39(11):1181–1198.

Ou, Y.-C., Wang, P.-H., Tsai, M.-S., Chang, K.-C., and Lee, G. C. 2010. Large-scale experimental study of precast segmental unbonded post-tensioned concrete bridge columns for seismic regions. *Journal of Structural Engineering, ASCE*, 136(3): 255–264.

Bridge Maintenance, Safety, Management, Resilience and Sustainability – Biondini & Frangopol (Eds)
© 2012 Taylor & Francis Group, London, ISBN 978-0-415-62124-3

Damping system for stay cables

P. Egger & J. Kollegger
Vienna University of Technology, Vienna, Austria

ABSTRACT

Vienna University of Technology has developed a method to increase the structural damping of stay cables. Advantages in regard to conventional dampers are the simple and robust technology and the fact that all components required for the new method are located inside the structural components (cables). The new method works like a vibration absorber attached along the whole stay cable in order to decrease the dynamic hazards in various eigenfrequencies. It is a kind of a tuned mass damper which was studied by Den Hartog and, in its simplest form, it can be modeled as a secondary mass-spring dashpot attached to the cable (Den Hartog 1956). The stay cable can be modeled as single degree of freedom. Kovacs mentions in his paper that a punctual tuned mass damper attached to a stay cable could increase the damping ratio user-defined, but the technical realization seems tricky (Kovacs 1982). A more precise approach could be to have a taut cable with a multipendulum attached along the cable. The impulses of the impact damper lead to a momentum transfer on one side and due to friction, energy dissipation on the other side. The damping system consists of a tube attached to the stay cable where a strand is located inside the tube. Under dynamic excitation, the energy dissipation leads to an increased structural damping. The strand is connected to the structure at one point only. In our method, the required components (strand, tube) are placed inside the structural system and energy is dissipated along the total length of the structure, whereas in conventional damping systems the damping devices are only attached to certain points of the structure. To place the damping system inside the structural system is advantageous for stay cables because it facilitates the inspection along the cable length and no additional corrosion protection, as in the case of external viscous dampers or orthogonal external cable, has to be provided, and thus the appearance of a cable stayed structure is not impaired by external damping measures.

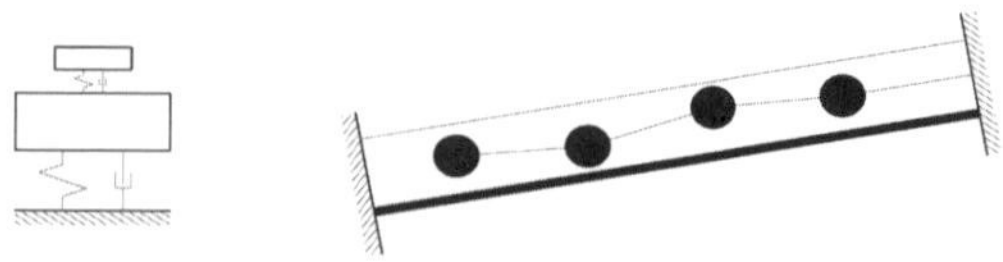

Figure 1. Model of the damping system.

Figure 2. Setup of the field experiment.

In full scale field tests on stay cables with a length of 31.2 m the damping ratio of a conventional stay cable and a stay cable equipped with the damping system could be compared. The damping ratio of the simple cable has a damping ratio of approximately 0.1%. By mounting the damper we achieved damping ratios up to 0.8%.

REFERENCES

Den Hartog, J.P., (1956), "Mechanical Vibrations" 4th Edition, McGraw-Hill, NY.

Kovacs, I. (1982). "Zur Frage der Seilschwingungen und der Seildämpfung." Bautechnik, vol. 10, 325–332.

Kollegger, J. & Egger, P. "Tragkonstruktion". Patent application T12456 Austria, March 18, 2009.

Bridge Maintenance, Safety, Management, Resilience and Sustainability – Biondini & Frangopol (Eds)
© 2012 Taylor & Francis Group, London, ISBN 978-0-415-62124-3

Shake table testing of a quarter-scale 4-span bridge with composite piers

F. Kavianipour & M. Saiidi
Department of Civil and Environmental Engineering, University of Nevada, Reno, NV

ABSTRACT

The target performance in the current bridge seismic design practice for standard bridges is collapse prevention under severe earthquakes, thus post-earthquake serviceability is not generally considered. Because of the high cost of repairing bridges and also the need for their availability for emergency response transportation after the earthquake, a new approach to seismic design and construction of bridges is needed. Funded by the National Science Foundation through the Network for Earthquake Engineering Simulation (NEES) research program, a major multi-university research project has been in progress at the University of Nevada, Reno. This abstract describes one of the three large-scale bridge models that were tested to failure on three shake tables system. This model was supported on fiber-reinforced polymer (FRP) composite piers implementing accelerated bridge construction (ABC) techniques.

The bridge was a quarter scale model of a 4-span bridge with continuous reinforced concrete superstructure and a drop cap two-column pier design. Figure 1 shows the assembled bridge model on the shake tables. Each pier utilized different unconventional FRP details. The purpose of using these innovative details was to improve the seismic performance of the bridge. The first pier consisted of cast-in-place concrete-filled glass FRP tubes with ±55 degree fibers. The third pier had the same configuration but the columns and footing were precast. The top connections in both piers consisted of pipe-pin joints to facilitate ABC and provide hinge behavior. The middle bent was a segmental reinforced concrete column wrapped with layers of unidirectional carbon FRP sheets to provide confinement and shear reinforcement. Only nominal hoops were used to hold the longitudinal reinforcement, as FRP jacket and tube were sufficient in providing confinement and shear required reinforcement.

Figure 1. Overview of the bridge set in UNR lab.

Shake table testing was conducted using modified 1994 Northridge, CA ground motion recorded in Century City and it was applied to the bridge model in ten runs with increasing amplitudes. Compared to conventional reinforced concrete bridges, experimental results showed superior performance under extreme seismic loading even under lateral drift ratios exceeding 9%. The modeling of different piers was done using program Open System for Earthquake Engineering Simulation (OpenSees). This paper summarizes design, construction and setup of the testing, the results of the experiments, and performance of different details.

Bridge Maintenance, Safety, Management, Resilience and Sustainability – Biondini & Frangopol (Eds)
© 2012 Taylor & Francis Group, London, ISBN 978-0-415-62124-3

Fatigue behaviour of bridge deck slab elements strengthened with reinforced UHPFRC

T. Makita & E. Brühwiler
Laboratory of Maintenance and Safety of Structures (MCS), École Polytechnique Fédérale de Lausanne (EPFL), Lausanne, Switzerland

ABSTRACT

With the occurrence of higher and more frequent axle loads on roads, in particular bridge deck slabs are more severely solicitated by fatigue loading. To avoid heavy interventions for strengthening of bridge deck slabs, an improved building material is used, namely Ultra-High Performance Fibre Reinforced Concrete containing steel rebars (=R-UHPFRC). UHPFRC has eminent properties such as relatively high compressive and tensile strength, low permeability, strain-hardening behaviour in tension.

The objectives of the present paper are to describe the bending fatigue behaviour of R-UHPFRC reinforced concrete composite member (RU-RC member) and demonstrate the effectiveness of fatigue strengthening of existing RC bridge deck slabs using R-UHPFRC. The experimental set-up represents a strip of RC bridge deck cantilever strengthened with R-UHPFRC.

Bending fatigue tests were conducted on RU-RC slab-like beams of 1900 mm length with a cross section of 400 mm × 220 mm. Thickness of R-UHPFRC layer is 50 mm using an in-house developed UHPFRC mix called HIFCOM 13. The average elastic limit and ultimate strength of UHPFRC were determined from quasi-static uniaxial tensile tests to be around 10 MPa and 12 MPa each. Four steel rebars of 8 mm diameter were arranged in the UHPFRC layer with a spacing of 100 mm. Concrete in the RC part was C30/37 grade with a maximum aggregate size of 16 mm. Four steel rebars of 16 mm and 10 mm diameter were arranged in the top and bottom of the RC part with a spacing of 100 mm. All of steel rebars used in the experimental tests were of B500B grade and had a nominal yielding strength of 500 MPa. The RC part was first cast, and 28 days later the UHPFRC layer was cast on the top surface of the RC part which was roughened with hydro-jetting to obtain monolithic bond between UHPFRC and concrete. (Neither any adhesion products nor any shear connector was used for the bonding.).

S-N diagram was obtained from bending fatigue tests on RU-RC beams. Although some scatter was observed in the test results, fatigue limit at 10 million cycles was estimated to be at a solicitation level of about 50% of the ultimate static strength of the RU-RC beam.

All the RU-RC beams showed similar behaviour until fatigue failure. Differences in deformation of R-UHPFRC layer between calculation and measurement is explained to be due the variation of UHPFRC material properties. Growth of the deformation of R-UHPFRC was attributed to stiffness degradation of UHPFRC caused by micro-cracking in the hardening domain.

The four rebars in R-UHPFRC layer fractured one after the other and the RU-RC beam failed and lost its load bearing capacity soon after the fracture of R-UHPFRC layer. The R-UHPFRC layer, in particular the steel rebars in the layer determine the behaviour of the RU-RC beam under bending fatigue. Stress amplitude in the steel rebars in R-UHPFRC layer is thus the most relevant factor for the fatigue behaviour of the RU-RC beam. UHPFRC carries stress and contributes to reduction of the stress in the rebars; however, stress in UHPFRC transfers to the rebars gradually as deformation (crack opening) of UHPFRC increases due to fatigue.

On the basis of the present experimental test results, design rules for RU-RC member under bending fatigue are proposed for fatigue safety check with respect to the fatigue limit. Fatigue safety needs to be checked for RU-RC member as well as for steel rebars and UHPFRC fatigue resistances.

Finally, an application of this novel technology is briefly described demonstrating that improvement of bridge deck slabs using UHPFRC is a relatively gentle intervention with limited intervention costs. There is a potential inherent with this novel construction method to limit the duration of the working site and thus to reduce the user costs as well as life cycle costs.

Bridge Maintenance, Safety, Management, Resilience and Sustainability – Biondini & Frangopol (Eds)
© 2012 Taylor & Francis Group, London, ISBN 978-0-415-62124-3

Gradient anchorage method for prestressed CFRP strips – Principle and application

J. Michels & C. Czaderski
Structural Engineering Research Laboratory, Empa, Switzerland

R. Brönnimann
Electronic, Metrology and Reliability Laboratory, Empa, Switzerland

M. Motavalli
Structural Engineering Research Laboratory, Empa, Switzerland
Department of Civil Engineering, University of Tehran, Iran

ABSTRACT

This paper presents the development and application of an innovative anchorage method for prestressed carbon fiber reinforced polymer (CFRP) strips.

At Empa, Professor Urs Meier and Ywan Stöcklin developed a purely epoxy-based anchorage system for prestressed CFRP strips without any remaining mechanical anchor plates or bolts. The system is based on the epoxy's ability to cure faster under elevated temperature and subsequently being able to carry significant shear stress after a short heating duration. The concept foresees a segment-wise heating of the epoxy at the strip end followed by a partial prestress force release in the hydraulic jack. This procedure is repeated until the prestress force is decrease to zero at the very end, whereas it is kept constant at the initial value over the free strip length outside the anchorage gradient. With this procedure, the total prestress force to be anchored is divided into several individual bond lengths, each one responsible of carrying a part of the total load.

Preliminary analysis showed that the short-term heating of the epoxy under elevated temperature has an accelerating effect on its cure and furthermore offers a sufficiently low stiffness in order to evenly distribute the shear stresses over the bond length.

In the present paper, the development of the necessary equipment (see Figure 1) for practical application is presented. Whereas the clamping and prestressing devices remained almost identical to an existing system with mechanical anchorage, the core innovation lies in the production of an electronic device connected to a special type of heating elements. These elements (eight in total, individually controllable by means of a developed software) are used for the accelerated curing of the underlying adhesive layer.

In the manuscript, the complete installation as well as the prestress and heating procedures are presented and commented in detail. It is shown that the installation of such a gradient anchorage is considerably faster than conventional solutions with mechanical systems.

Figure 1. Prestressing and heating device for the gradient anchorage installation.

REFERENCES

Czaderski, C., Motavalli, M. 2007. 40-Year-old full-scale concrete bridge girder strengthened with prestressed CFRP plates anchored using gradient method, *Composites Part B: Engineering* 38: 878–886.

Czaderski, C., Martinelli E., Michels, J., Motavalli, M. 2011. Effect of curing conditions on strength development in an epoxy resin for structural strengthening, *Composites Part B: Engineering*, in press. DOI: 10.1016/j.compositesb.2011.07.006

Meier U., Stöcklin, Y. 2005. A Novel Carbon Fiber Reinforced Polymer (CFRP) System for Post-Strengthening, *International Conference on Concrete Repair, Rehabilitation and Retrofitting (ICCRRR)*, Cape Town, South Africa.

Michels, J., Czaderski, C., El-Hacha, R., Brönnimann, R., Motavalli, M. 2011b. Temporary bond strength of partly cured epoxy adhesive for anchoring prestressed CFRP strips on concrete, *Composite Structures*, submitted.

Stöcklin, Y., Meier, U. 2003. Strengthening of concrete structures with prestressed and gradually anchored CFRP strips, FRPRCS-6, Singapore.

Design and seismic analysis of long span bridges – Case studies

Organizer: A.H. Malik

Bridge Maintenance, Safety, Management, Resilience and Sustainability – Biondini & Frangopol (Eds)
 ISBN 978-0-415-62124-3

Seismic assessment of long curved bridges using modal pushover analysis: A case study

M.S. Ahmed & C.C. Fu

Bridge Engineering Software & Technology (BEST) Center, Department of Civil and Environmental Engineering, University of Maryland, College Park, MD, USA

ABSTRACT

The assessment of existing bridge structures against earthquake threat has become a major issue lately, motivated by the maturity of seismic design of new structures, on one side, and by the recognition of the inadequate level of seismic protection, the aging and the constant need of maintenance of the existing ones, on the other.

While nonlinear time history analysis (NL-THA) is the most rigorous procedure to compute seismic demands, many seismic-prone countries, such as United States, New Zealand, Japan and Italy, have recently released standards for the assessment of buildings, all of which include the use of the non-linear static analysis procedure, the so-called pushover. Non-linear static pushover analysis (SPA) is a widely used analytical tool for the evaluation of the structural behavior in the inelastic range and the identification of the locations of structural weaknesses as well as of failure mechanisms.

Nevertheless, the method is limited by the assumption that the response of the structure is controlled by its fundamental mode. Extension of the SPA to consider higher modes effects has attracted attention. Chopra & Goel (2002) proposed the modal pushover analysis (MPA) where pushover analyses are carried out separately for each significant mode and the contributions from individual modes are combined using an appropriate combination rule. While many studies are available dealing with the application of pushover analysis to building structures, the situation is quite different when bridges are considered. The number of studies is very limited, among those are Kappos et al. (2004) and Pinho et al. (2005).

This study is intended to evaluate the accuracy of the modal pushover analysis (MPA) procedure in estimating seismic demands for the case of an actual long span curved bridge after proposing some modifications that would render the MPA procedure applicable for bridges. Key elements of applying the MPA procedure for the case of bridges are: Definition of the control node, developing of the pushover curve and transforming it to capacity curve and number of modes that should be considered in the analysis. Definition of the monitoring point was presented and different appropriate locations; deck mass center, equivalent SDOF location, or most critical pier location, were investigated. It was concluded that:

- Most critical pier location was found to be the most appropriate location to be considered as the control point.
- The MPA procedure introduced was found to yield better results when the level of earthquake excitation was increased and more inelasticity developed in the structure.
- Maximum Demand displacements, total base shear and plastic rotations obtained from SPA and MPA were compared with the corresponding values resulting from the NL-THA. Comparison shows a good agreement between MPA and NL-THA results and MPA is deemed to be accurate enough for practical use.

REFERENCES

Chopra, A. K., & Goel, R. K. 2002. A Modal pushover analysis procedure for estimating seismic demands for buildings. *Earthquake Engineering and Structural Dynamics, 31*(3), 561–582.

Kappos, A. J., Paraskeva, T. S., & Sextos, A. G. 2004. Seismic assessment of a major bridge using modal pushover analysis and dynamic time-history analysis. *Advances in Computational and Experimental Engineering and Science*, 673–680.

Pinho, R., Antoniou, S., Casarotti, C., & Lopez, M. 2005. A displacement-based adaptive pushover for assessment of buildings and bridges. *NATO International Workshop on Advances in Earthquake Engineering for Urban Risk Reduction*. Istanbul, Turkey.

Bridge Maintenance, Safety, Management, Resilience and Sustainability – Biondini & Frangopol (Eds)
© 2012 Taylor & Francis Group, London, ISBN 978-0-415-62124-3

Innovative methodology towards the design of long span bridges

A.H. Malik
Private Consultant, Pinestump Road, Albany, New York

ABSTRACT

Long span bridges present architectural design and are used as a signature artifact for the local area. The advancements in bridge technology and the associated software have encouraged the designers to meet record-breaking challenges throughout the world. Cable supported bridge technology has provided the designers to bridge across large waterways, deep valleys and unsuitable subsurface conditions. At the same time these exotic looking structures have become the landmark symbol for many regions around the world.

During the past two decades, the enhanced engineering challenges were very well tackled by the remarkable developments in the bridge technology. The advance methodologies have been implemented on many infrastructures throughout the world. The increased demand of transport loads and the traffic volume requires innovative approach towards design, construction, and maintenance to meet this challenge. The designers, researchers, contractors and software engineers responded by demonstrating their technical skills by adopting creative approach towards the record breaking structures. The bridge design and analysis software has a significant role in the emergence of various conceptual architectural ideas into realities.

The conventional method of trial and error to arrive at the optimum design works very well, based on experience, for various non-cable supported structures. Whereas, for cable supported structures, that tedious process has been replaced with the current software's analytical and modeling techniques to provide the engineers the optimal tensioning strategy while applying the prestressing forces of the cables of a cable-stayed bridge.

This paper presents two case studies involving comprehensive software for the seismic and non-seismic analysis reflecting the modeling of various phases of design and construction utilized for the long span cable stayed bridges. The software should investigate the various phases of construction in the design which includes cable sagging, p-delta effects, and large displacements as well as wind-induced vibrations and optimization of cable stressing sequence.

Bridge Maintenance, Safety, Management, Resilience and Sustainability – Biondini & Frangopol (Eds)

Seismic design of the San Francisco – Oakland Bay Bridge self anchored suspension bridge

M. Nader & M.-C. Tang
T.Y. Lin International, San Francisco, California

B. Maroney
CALTRANS, Sacramento, California

ABSTRACT

The seismically vulnerable East Span of the San Francisco-Oakland Bay Bridge will be replaced with a dual east bound and west bound 3.6 km long parallel structure. The cost of the Replacement Bridge is estimated at six billion dollars and the bridge will be constructed by the year 2013. The Bay Bridge lies between the Hayward and the San Andreas faults which can generate magnitude 7.5 M and 8.1 M earthquakes, respectively. Performance criteria require that the bridge must be operational immediately following a 1500-year return period earthquake from either of these two faults. Four distinct structures will make up the bridge crossing: a low rise post-tensioned concrete box girder near the Oakland shore; a 2.4 km long segmental concrete box girder; a self-anchored suspension signature span; and a post-tensioned concrete box girder that connects to the east portal of the Yerba Buena Island tunnel.

The single tower asymmetric self-anchored suspension bridge was selected from a total of four design alternatives that were developed for the signature main span; these included two cable stayed bridges and two self-anchored suspension bridges (each bridge type included single tower and dual portal tower alternatives). Each design alternative was evaluated based on its seismic response, construction cost and aesthetic properties.

Analytically, single tower asymmetric self-anchored suspension bridge was selected from a total of four design alternatives that were developed for the signature main span; these included two cable stayed bridges and two self-anchored suspension bridges (each bridge type included single tower and dual portal tower alternatives). Each design alternative was evaluated based on its seismic response, construction cost and aesthetic properties.

Bridge Maintenance, Safety, Management, Resilience and Sustainability – Biondini & Frangopol (Eds)
 ISBN 978-0-415-62124-3

Structural design and analysis of long span bridges

S. Wang & C.C. Fu
Bridge Engineering Software and Technology (BEST) Center, Department of Civil and Environmental Engineering, University of Maryland, College Park, MD

ABSTRACT

The structural design and analysis of long span bridges involve detailed and sophisticated methods for various stages of design and construction. The advancements in modeling and simulation tools provided by various computer software made record breaking long-span bridges possible. This paper presents structural analysis examples of a recent-built long-span cable-stayed bridge in China and an existing suspension bridge in the US, using current computer tools.

The geometric nonlinear effects have to be considered in the structural analyses of long-span cable-stayed and suspension bridges. These effects include 1) initial stress, 2) large displacements and 3) sag of long cables. The latter is derived from large displacements and needs to be considered only when sub-meshing of cables is not affordable. In that case, Ernst formula (Ernst 1965) can then be used to deduct the axial stiffness of a cable that is meshed as one element between its two anchor points. Thus, all these nonlinear analyses can be performed by using Newton-Raphson method. Due to these nonlinearities, theoretical extreme live loads are not able to be obtained. It is more practical taking the same extreme loads and positions obtained by linear analysis to perform one round of nonlinear analysis to adjust the results.

The "ideal state", which is defined as minimizing the total bending energy under dead loads in girder is minim, is unique to cable-stayed bridges. Cable stresses in "ideal state" can be obtained by using influence matrix method. These cable stresses are their ideal state when superimposed dead loads are applied. Backward analyses, in which each girder segment and cables are removed step by step, are conducted to predict the control stresses of cables in each erection stage.

For a better understanding of how to apply these nonlinearities in structural analyses and to define the ideal state of a long-span bridge, two examples are introduced. The first one is a feasibility study of a steel cable-stayed bridge in China with a main span of 1088 m. Methodologies for defining the ideal state, backward and forward analyses for construction simulation and control, and the geometric nonlinearities together with static stability analysis are discussed. This example shows 1) initial stresses accumulated in flat arch-like girder will enhance girder stiffness if large displacement is considered in live load analyses and 2) both geometric and material nonlinearities have to be included in stability analysis.

The second case is to analyze LRFD live loads under existing condition for a cost allocation study. The modeled bridge is a typical suspension bridge with a center span of 487.68 m and two 205.74 m side spans. It is modeled in 3D. The method to simulate cable saddle, results of dead loads, live loads and temperature loads analyses are presented. Static blast load cases analyses are also briefly introduced.

REFERENCES

ASCE 1992. *Guidelines for the Design of Cable-stayed Bridges*. American Society of Civil Engineers

CSI 2005. SAP2000 Version 10. Berkeley, CA: Computers and Structures, Inc.

Ernst, J. H. 1965. Der E-Modul von Seilen unter berucksichtigung des Durchhanges. *Der Bauingenieur* 40(2):52–55

Fleming, J. F. 1978. Nonlinear static analysis of cable-stayed bridge structures. *Computer and Structures* 10(4): 621–635

Tang, M.C. 1971. Analysis of cable-stayed girder bridges *Journal of the Structural Division, ASCE* 97(ST5): 1482–1495

Tang, M.C. 1976. Bulking of cable-stayed bridges. *Journal of the Structural Division, ASCE* 102(ST7):1675–1684

Suthar, Kunal 2007. The Effect of Dead, Live and Blast Loads on a Suspension Bridge. Thesis submitted to the Faculty of the Graduate School of the University of Maryland, College Park

O.C. Zienkiewicz, Olgierd Cecil 1977. *The Finite Element Method (Third Edition)*. McGraw-Hill Book Company (UK) Limited

Advances in modeling and analysis for the performance-based design of bridge structures subjected to multiple hazards

Organizers: F. Petrini & A. Palmeri

Bridge Maintenance, Safety, Management, Resilience and Sustainability – Biondini & Frangopol (Eds)
© 2012 Taylor & Francis Group, London, ISBN 978-0-415-62124-3

Structural response of bridges to fire after explosion

C. Crosti, P. Olmati & F. Gentili
Sapienza University of Rome, Rome, Italy

ABSTRACT

The aim of this paper is to focus the attention on the structural response that steel truss bridges have when involved by two particular hazards: fire and explosion. The choice of those two hazards was made according to the fact that many bridges are potential terrorist targets, so hazards like fire and explosion can unfortunately be occurred in their lifetime.

The bridge under study is the I35-W Bridge that collapsed on August 1st, 2007, due to the buckling of an under-designed gusset plate. After the collapse of this bridge, the FHWA focused its attention on all the 465 steel deck truss bridges present in the National Bridge Inventory (NTSB, 2008). They classified this bridge to be a fracture critical system, which means that if one main component of a bridge fails, the entire structure could collapse.

Focusing the attention on a particular zone of this bridge, four hypothetical scenarios of damage due to an explosion were considered. The procedure presented is based on the assumption of a certain damage level caused by a generic explosion load which is able to instantaneously cut off the contribution of a structural element to the load bearing capacity of the system. To point out the difference between the other damage scenarios, in Figure 1 the deformed shape of the upper chord is drawn by connecting the vertical displacements of each node. Those displacements are measured at the time corresponding to the reaching of the maximum displacement for each different scenario. Scenario 2, 3 and 4 show almost the same deformed shape, for scenario 1 it is instead different. In fact looking at Figure 1 it is possible to assess that a collapse happened in that zone when the element is removed. This consideration points out that scenario 1 involves the damage in one of the key elements of the structure.

Running nonlinear static analyses on the damaged bridge following the scenarios already described, the critical temperatures are evaluated. Scenario 0 has $T_{crit} = 200°C$, for scenario 2 T_{crit} is 310°C, scenario 3 and 4 have T_{crit} of 190 and 210°C. It is not accurate to think that over those critical temperatures the structure collapses, but it is reasonable to think that over this temperature limit, the bridge suffers a modification of stiffness and resistance that, in a performance approach, highlights the possibility that the safety of the structure is not longer guaranteed (Crosti, 2009). All of that matches with the considerations made in the beginning about the importance of some elements.

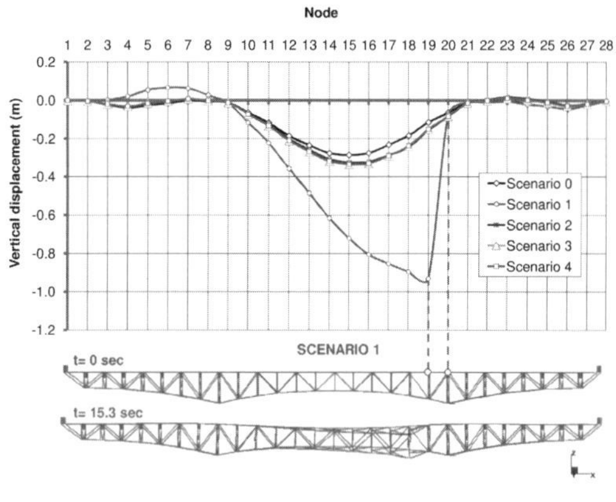

Figure 1. Deformed shape of the nodes of the upper chord at the time of the maximum vertical displacement. Scenario 0 is the one with no level of damage.

REFERENCES

Crosti C., 2009. Structural Analysis of Steel Structures Under Fire Loading. Acta Polytechnica. Vol. 49, No. 1/2009.

National Transportation Safety Board, 2008. Collapse of I-35 W Highway Bridge, Minneapolis, Minnesota, August 1, 2007. Accident Report, NTSB/HAR 08/03 PB 2008-916213, Washington D.C. 20594.

Bridge Maintenance, Safety, Management, Resilience and Sustainability – Biondini & Frangopol (Eds)
 ISBN 978-0-415-62124-3

The performance-based evaluation of kinematic pile response due to lateral spread at an historic bridge in Costa Rica

K.W. Franke & K.M. Rollins
Brigham Young University, Provo, Utah, USA

ABSTRACT

Lateral spreading is a common phenomenon associated with soil liquefaction during earthquakes. Commonly associated with cyclic mobility, lateral spreading has repeatedly been observed to cause significant damage to bridge structures, and many case histories of such damage have been documented during the past 50 years. One such case history was documented following the M7.6 earthquake that struck the Limon Province in Costa Rica on April 22, 1991. The Rio Bananito Railway Bridge (Figure 1) was constructed prior to 1890 and served as both a critical rail and highway bridge at the time of the earthquake. The single-truss bridge is supported on elliptically-shaped caissons that are approximately 1.46 meters by 2.16 meters across the major axes. Liquefaction during the 1991 earthquake caused lateral spreading displacements estimated to range from 2.0 meters to 2.5 meters beneath the north abutment of the bridge. These soil displacements caused significant deformations of the bridge caissons (4.3 meters to 5.7 meters), pushing them out from beneath the seating plates of the bridge. As a results, the bridge fell off its foundation and tipped eastward at an approximate angle of 15 degrees. The damage sustained at the bridge rendered it unsafe for use until repair efforts returned the bridge to an operational status in the months following the earthquake.

A recent study (Rollins & Franke 2011) was performed to evaluate the performance of the Rio Bananito Railway Bridge following the 1991 Limon earthquake. A new performance-based procedure for the evaluation of deep foundations under kinematic lateral spreading loads is presented. This procedure utilizes the performance-based probabilistic framework developed by the Pacific Earthquake Engineering Research Center (PEER) (Cornell & Krawinkler 2000; Krawinkler 2002; Deirlein et al. 2003) and incorporates analysis techniques familiar to most practicing geotechnical design professionals. The procedure is applied to the Rio Bananito Railway Bridge and produces probabilistic estimates of kinematic response of a bridge caisson to lateral spreading displacements. Such estimates could be useful in the site-specific lateral spreading performance evaluation of deep foundations from both new and existing bridges, and could eventually aid owners and engineers in making informed and economical decisions regarding seismic design and/or remediation.

Figure 1. Rio Bananito Railway Bridge in April, 2010.

REFERENCES

Cornell, C.A. & Krawinkler, H. 2000. Progress and challenges in seismic performance assessment. *PEER News:* 1–3. PEER.

Deirlein, G.G., Krawinkler, H. & Cornell, C.A. 2003. A framework for performance-based earthquake engineering. *2003 Pacific Conference on Earthquake Engineering, Proc.*

Krawinkliner, H. 2002. A general approach to seismic performance assessment. *International Conference on Advances and New Challenges in Earthquake Engineering Research, Proc., Hong Kong.* ICANCEER.

Rollins, K.M. & Franke, K.W. 2011. Subsurface investigation and analysis of bridges in Costa Rica damaged by lateral spread to evaluate and improve design procedures for kinematic loading of pile. Final Report for Award No. G10AP00047, 188p. USGS.

Bridge Maintenance, Safety, Management, Resilience and Sustainability – Biondini & Frangopol (Eds)
 ISBN 978-0-415-62124-3

Numerical simulation of bridges remodeling

A. Marí, J. Bairán, R. Moreno & E. Oller
Universitat Politècnica de Catalunya, Barcelona, Spain

J.J. Álvarez
Dragados, Madrid, Spain

ABSTRACT

Many infrastructures of the current roadway system need to be upgraded to accommodate them to new traffic loads or to increase the capacity of roads. Examples of modifications needed are widening of decks or movement of piers, among others.

In order to ensure the required levels of serviceability and safety of the retrofitted bridge during the execution and along its service life, it is necessary to realistically evaluate the structural response before and after the intervention, taking into account the previous state of the structure. Therefore, sequential non-linear analysis models become useful tools for design of efficient strengthening solutions.

In this paper, a numerical model, capable to reproduce the effects of the non-linear and time-dependent materials behaviour and those due to the construction sequence and to those modifications needed to adapt the actual bridges to future needs, is presented. The model allows capturing the service and ultimate capacities of the bridges providing valuable information when comparing the efficiency of different alternative solutions. In order to show the capabilities of the model a freeway overcrossing subjected to piers movement is analysed.

Due to the need of increasing the number of lanes of AP-7 freeway near Barcelona (Spain), the intermediate piers of several overcrossing bridges had to be moved 2 m. towards the abutments, without interrupting the traffic. In this work, the developed analytical model has been used to simulate the strengthening operations carried out over one of the bridges affected (OF.29-3), in order to assess the serviceability and safety of the bridge after the intervention. The overcrossing consists on a 5 span continuous prestressed concrete bridge with a slab deck. The solution adopted consisted in placing two longitudinal steel beams under the side cantilevers, connected to the concrete deck over the piers position before they were moved, and supported over the new piers constructed. The reactions from the old piers to the previously deformed steel beam were transferred by means of hydraulic jacks, so the deck did not suffer any change in stresses under permanent loads.

An alternative solution studied consisted on placing an exterior prestressing system under the side cantilevers with an adequate layout to introduce upward forces at the intermediate old piers and downward forces at the new piers, transferring to them the reactions. Comparisons between the structural response of the bridge with the two strengthening systems showed that the prestressed solution was more efficient, although more difficult to built, so it was not the one chosen at the design stage.

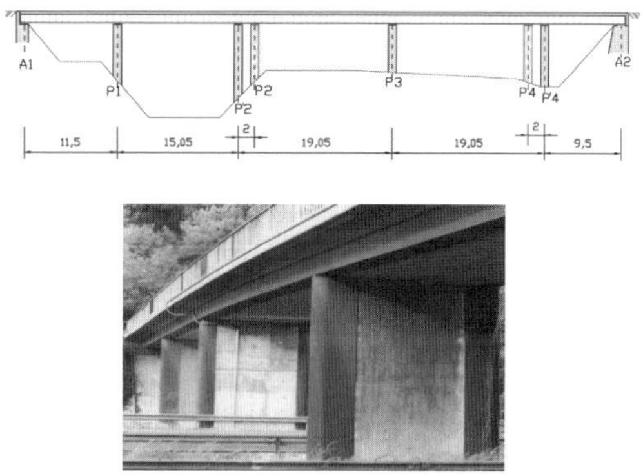

Figure 1. Frontal view and picture of the remodeled bridge.

REFERENCES

Marí, A. 2000. Numerical Simulation of the Segmental Construction of Three Dimensional Concrete Frames. *Engineering Structures*, 22: 585–596.

Martínez Calzón, J. 1994. Ampliación de las luces de 14 pasos superiores continuos postensados, para conseguir el ensanchamiento a seis carriles de la autopista AP-7 Barcelona-La Junquera, Tramo: Hostalric-Maçanet. *Informes de la Construcción*, 46(431): 7–38.

Bridge Maintenance, Safety, Management, Resilience and Sustainability – Biondini & Frangopol (Eds)
© 2012 Taylor & Francis Group, London, ISBN 978-0-415-62124-3

Performance-based design of bridge structures subjected to multiple hazards: A review

F. Petrini
Sapienza University of Rome, Rome, Italy

A. Palmeri
Loughborough University, Loughborough, UK

ABSTRACT

Performance-Based Design (PBD) is a modern and efficient framework to conceive and assess complex structural systems, which allows designers to consistently take into account both natural and man-made hazards. Initially formalised and applied for earthquake engineering applications (FEMA 1997), PBD has been recently extended to cope with other design situations, like blast, fire, tsunami and wind scenarios (e.g. Hamburger and Whittaker 2003). Given its versatility, PBD appears to be a viable strategy for a more reliable design of bridges. As a matter of fact, these structures may play a critical role in the aftermath of large-scale natural disasters, and therefore their level of structural safety must be rigorously evaluated. PBD is usually carried out by taking into account each single threat individually, therefore neglecting complex scenarios with simultaneous and/or cascaded hazards. In recent years, however, researchers and practitioners have started to pay an increasing attention to more advanced approaches enabling multi-hazard exposures for the design of structures. Nevertheless, five key issues arise when the structural problem is consistently addressed in a multi-hazard context:

- first, the level of knowledge reached in different fields has to be joined into a unified framework of risk assessment (joining knowledge problem);
- second, interactions between different hazards are intrinsically difficult to model, both for lack of raw data and unavailability of concurrent hazards models (hazards interaction problem);
- a third issue is represented by the necessity of considering uniform hazard levels for the different threats (uniform hazard problem);
- the fourth main task consists in balancing the design in order to furnish similar safety levels with regard to different multi-hazard scenarios (uniform risk problem);
- finally, design philosophies for different hazards lead very often to opposites strategies for bridge structures, e.g. either reducing or increasing the flexibility and/or the redundancy (opposite design strategy problem).

These five main issues have been sparsely addressed in the technical literature, mainly in the context of risk assessment and referring to pairs of concurrent hazards (e.g. flooding and wind, wind and earthquakes, etc.). Moreover, even if PBD can be nowadays assumed as a complete risk assessment procedure (Augusti and Ciampoli 2008, Ciampoli et al. 2011), a rigorous PBD framework for structural multi-hazard scenarios is not yet available. To the best of the authors' knowledge, this paper represents a first step in this direction, as a multi-hazard extension is proposed for the theoretical framework originally developed by the Pacific Earthquake Engineering Research (PEER) centre (e.g. Kunnath 2007) for the case of seismic input. Since the present work has to be intended as a preliminary investigation in this research topic, most of it is focused on reviewing the existing literature covering the previously introduced issues for bridge structures, starting with some historic cases of bridges that have been stroked by multiple hazards.

REFERENCES

FEMA (Federal Emergency Management Administration) 1997. NEHRP Guidelines for the Seismic Rehabilitation of Buildings, Report No. FEMA 273, prepared by the Applied Technology Council for FEMA, Washington, D.C

Hamburger, R.O. & Whittaker, A.S. 2003. Considerations in performance-based blast resistant design of steel structures. *Proc. AISC-SINY Symposium on Resisting Blast and Progressive Collapse, American Institute of Steel Construction, New York, USA.*

Kunnath, S.K. 2007. Application of the PEER PBEE Methodology to the I-880 Viaduct, PEER Report 2006/10 (I-880 testbed committee). Available on line at http://peer.berkeley.edu/publications/peer_reports.html [Accessed 2 Febraury 2010].

Bridge Maintenance, Safety, Management, Resilience and Sustainability – Biondini & Frangopol (Eds)

 ISBN 978-0-415-62124-3

Finite element analysis of innovative solutions of precast concrete beam-column ductile connections

A.Saviotti, P. Olmati & F. Bontempi
Sapienza Università di Roma, Rome, Italy

ABSTRACT

The experience of hazards, like earthquakes and even accidental situations, has shown that reinforced concrete bridges are subject to beam-column joint failure. If connection design is inadequate, for example for the shear load that develops under earthquake excitation, the connection may exhibit deteriorating stiffness and strength. This behavior may result in inadequate structural performance.

Especially to precast concrete structure connections are one of the most essential parts. Connections transfer forces between precast members, so the interaction between precast units is obtained. They are generally the weakest link in the structure. An acceptable performance of precast concrete structure depends especially on the appropriate kind of connections choice, adequate detailing of components and design of the connections is fundamental. It is interesting to study the behavior of connecting elements and to compare different solutions of ductile connections for precast concrete structures in case of horizontal applied force and vertical imposed displacement, as well as those produced by hazards situation, like that earthquake and explosion, whereby topics of structure robustness are carried out.

Structural continuity (Fig.1) is an important problem with regard to the strength of connections between precast elements. In this paper the mechanical behavior of the beam-column connections developed by B-S Italia Styl – Comp Group (www.bsitaliagroup.com) (Fig. 2) is examined by a finite element analysis. To develop the numerical analysis is mainly used DIANA software, modeling the nonlinear behavior of concrete and mortar using total strain crack model. The reinforcing steel is modeled by a bilinear plasticity model. A detailed geometry of the system is been meshed and a non linear constitutive law of the material is been adopted. The full load capacity of the bars is developed without the failure of the concrete and the mortar, therefore the connection system is well performing because a brittle failure do not occurs. The progress of the cracking of the concrete is well reproduced.

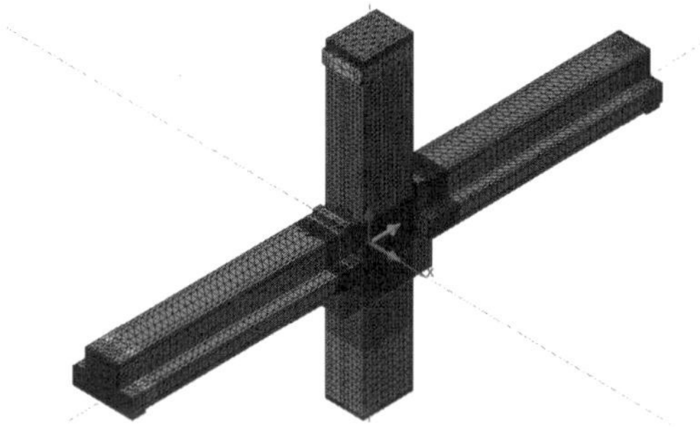

Figure 1. 3D Model "A": Mesh.

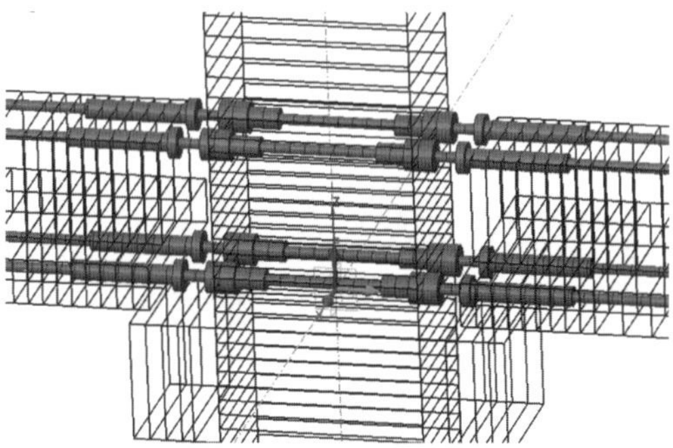

Figure 2. Reinforcing Steel of Model 3D: Zoom at the section of innovative solutions for ductile connections.

An important aspect of the present research regards the soundness of the results of the numerical simulations. To this aim, and independent model developed by ASTER is compared with the one developed with DIANA. The similarity obtained between the two modeling and the sensitivity analysis developed to test the difference in behavior between different connection configurations make the authors confident in the research outcome.

Advances in engineering structure management in Finland

Organizer: M.-K. Söderqvist

Bridge Maintenance, Safety, Management, Resilience and Sustainability – Biondini & Frangopol (Eds)
© 2012 Taylor & Francis Group, London, ISBN 978-0-415-62124-3

Guidelines for calculating the life cycle costs

R. Kiviluoma
WSP Finland Ltd, Helsinki, Finland

P. Korhonen
Finnish Transport Agency, Helsinki, Finland

ABSTRACT

This paper describes the life-cycle-cost (LCC) guideline taken in use by the Finnish Transport Agency 2011 (Tirkkonen et al. 2001). The guideline aims to harmonise preparation of the LCC-estimates at the design stage of new bridges or at the renovation stage of the old ones. With harmonisation, and publishing unit data to be used for the estimates, one seeks possibility to compare design alternatives and to promote general development and awareness of LCC-efficiency.

LCC-estimate is targeted to be prepared by the bridge engineer, who as part of the task, has to plan and schedule major maintenance operations according to the service-life unit data given in the guideline.

The guideline combines conventional LCC of the owning agency with the indirect costs including risks, users' costs and society's costs. To this end, monetary values are assigned for non-monetary quantities like traffic delays, environmental burden and risks. Environmental costs include global stressors, whose magnitude is assessed based on quantity take-off for construction and maintenance operations. Thus, a simplified Life Cycle Assessment (LCA) is included. Unit data for the LCA is tabulated based on worked case studies from more rigorous LCA methods, especially those developed in the Nordic ETSI project (ETSI 2007, ETSI 2009).

The same breakdown of structures is used than in the stranded quantity take-off and cost estimation of new bridges; the LCC-estimate includes as bases the conventional cost estimate of the bridge.

In order to support different scenarios, present value calculation with multiple discount rates is applied, including 0%, 1%, 2% and 5%. Depending on scope of the analysis, recommendations are given for the review period (typically 100 years), discount rate and weighting of the cost types. Here, the present value for cost C_j; with subscript j denoting the scope of analysis; is given by

$$C_{\mathrm{j}} = \sum_{i=1}^{N_{\mathrm{A}}} w_{\mathrm{A,i}} C_{\mathrm{A,i}} + \sum_{i=1}^{N_{\mathrm{U}}} w_{\mathrm{U,i}} C_{\mathrm{U,i}} + \sum_{i=1}^{N_{\mathrm{S}}} w_{\mathrm{S,i}} C_{\mathrm{S,i}} \quad (1)$$

	Direct costs	Indirect costs
Agency (N_A = 3)	• Construction ($C_{A,1}$) • Maintenance ($C_{A,2}$) – routine maintenance – operating – repairing – dismantling	• Risks ($C_{A,3}$)
Users (N_U = 2)		• Traffic delays ($C_{U,1}$) • Risks ($C_{U,2}$)
Society (N_S = 2)		• Environmental ($C_{S,1}$) – noise & vibration – waste & contamination – global stressors • Risks ($C_{S,2}$)

Figure 1. Cost types included in the guideline.

where $w_{A,i}$ = weighting factor for the present value of agency costs; $w_{U,i}$ = weighting factor for the present value of users' costs; $w_{S,i}$ = weighting factor for the present value of society's costs; and other symbols are as given in Figure 1.

Guideline has been subjected to comment circulations for bridge engineers and material industry; and LCC-estimates have been prepared for roughly ten pilot bridges. The general methodology has been found promising. Unit data in the guideline is voluminous, and this part of the guideline should evidently be updated regularly.

REFERENCES

ETSI Project (Stage 1), 2007, Bridge Life Cycle Optimisation. Eds. Jutila, A. & Sundquist, H., TKK-SRT-37. 157 pp.

ETSI Project (Stage 2), 2009, Bridge Life Cycle Optimisation. Ed. Salokangas, L., TKK-R-BE3. 140 pp.

Tirkkonen, T. & Korhonen, P. & Piispanen, M. and Kiviluoma, R., 2011. Finnish bridge life-cycle-cost guideline. PIARC XXIVth World Road Congress; Mexico City, Mexico, 10 pp.

Bridge Maintenance, Safety, Management, Resilience and Sustainability – Biondini & Frangopol (Eds)
© 2012 Taylor & Francis Group, London, ISBN 978-0-415-62124-3

Management of inspection data quality of the Transport Agency's structures

M.-K. Söderqvist
The Finnish Transport Agency, Helsinki, Finland

ABSTRACT

The Transport Agency's bridge unit is the mentor of all the country's bridge activities. In the 1980's work started to develop a bridge management system including an inspection system, an inspector examination system, inspection data quality assessment, a data inventory, a group of reference bridges, repair programming and versatile reporting. Today the Finnish Transport Agency manages together 14,700 road and 2300 railway bridges and numerous piers, quays and tunnels with the help of the management system.

Because of a well organised inspection system is the key element of a successful engineering structure management it forms the basis and an integral part of the whole management system. The system results are as reliable as the data in the database. Hence, the experience of the engineers and the use of the management system have shown that the available condition and damage data must continuously be improved and completed. The improvement of data quality is emphasised in the revised inspection training and in the Bridge Inspector Qualifications, which were taken in use in spring 2000. The inspection system, which runs since 1970s, the quality improving methods, their effects on and meaning for bridge maintenance, repair and rehabilitation (MR&R) and bridge age behaviour modelling are discussed in the paper.

Bridge inspector qualifications

The bridge inspector certificate includes the Bridge inspector examination and the basic course in Bridge Register use. It is also required that the inspector – to preserve his/her inspection competence and the certificate – participates the Advanced yearly Training Day. The day works as "a calibration" of inspectors, to get the inspection data more comparative and standardised and even in quality.

A quality report has been published yearly in the Transport Agency's internal report series since 2002. The development of data quality during the last years has been good. The results show that the inspection data quality has improved clearly after the year 2002.

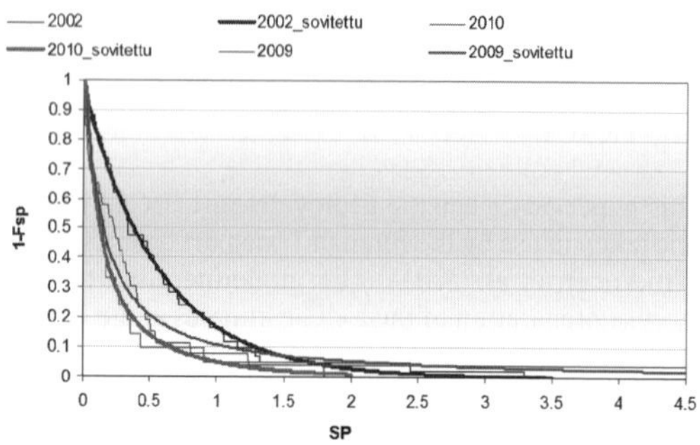

Figure 1. Quality improvement of inspection data, distribution function of damage points relative deviation SP.

REFERENCES

The references can be found on the Transport Agency's web pages: www.liikennevirasto.fi/sillat

Finnish Road Administration 2004. Sillantarkastusohje (Bridge Inspection Guidelines) 4th renewed edition in Finnish, TIEH 2000008-04 Helsinki. Edita Prima Ltd.

Finnish Road Administration 2006. Sillantarkastuskäsikirja (Bridge Inspection Manual, The Directives for Bridge Inspection Procedures. English 1st edition 1989) 6th renewed edition in Finnish, TIEH 2000020-06 Helsinki. Edita Prima Ltd.

Finnish Road Administration 2009. Siltojen yleistarkastusten laatuvaatimukset (Quality requirements of bridge general inspection). TIEH 2000010-v-09 Helsinki. Web publication.

Finnish Road Administration 2009. Siltojen ylläpito, toimintalinjat (Guideline and Policy for Bridge MR&R Operation), TIEH 1000217-09 Helsinki. Edita Prima Ltd.

Finnish Road Administration. 2003. Bridge Repair Manual (SILKO), General Guidelines 1, Repair Instructions 2, Product File 3, Work Equipment File 4. English editions TIEH 2230095 to 2230098. Tampere 2003. Ecuprint Ltd.

Transport Agency 2010. Laiturien tarkastusohje (Pier and Quay Inspection Manual). LO 2/2010. Helsinki, Edita Prima Ltd.

Bridge Maintenance, Safety, Management, Resilience and Sustainability – Biondini & Frangopol (Eds)
© 2012 Taylor & Francis Group, London, ISBN 978-0-415-62124-3

The new management system of engineering structures in Finland

M.-K. Söderqvist
Finnish Transport Agency, Helsinki, Finland

M. Veijola
Morro Kb, Esbo, Finland

ABSTRACT

The Finnish Transport Agency manages 15.000 road bridges. In the 1980's work started to develop a bridge management system including an inspection system, an inspector examination system, inspection data quality assessment, a data inventory, a group of reference bridges, repair programming and versatile reporting. Over the years most bridges have been inspected several times, several modelling methods have been used to model bridge behaviour and repair action effects, several software versions have been implemented and handbooks published. Lately road tunnels and road related piers and quays have been added to the system. A total overhaul and redesign of the whole bridge management system became necessary to manage many additional types of engineering structures: rail bridges, rail tunnels, pile decks and retaining walls. An overhaul is also called for by the need for new software features, such as agency-wide integrated reporting and flexible use of maps. Rail bridges and tunnels are the most important new structures to be included and this is related to organisational changes within the state administration. As a part of this effort a project was started to design a general framework for a management system for engineering structures using the best available methods and practices making the most of experiences gained over the years. The paper describes this project, which includes the following tasks: 1) comparison of the suitability of different multi-objective optimisation methods for the management of bridges and other engineering structures, 2) choosing the most suitable multi-objective optimisation method, 3) finding out which optimisation criteria should be used for each category of engineering structures, 4) finding out how to integrate the multi-objective optimisation into the repair programming process of the management system, 5) finding out how to integrate the condition prediction models, based on deterioration and repair action models of the present management system, into the management system, the data storage and the reporting system, 6) finding out how to integrate the bridge-specific life-cycle management (analysis and costs, LCA & LCC) into the management system, 7) finding out how to integrate the goals (management by objectives) of the Transport Agency into the management system, 8) finding a method to establish network level performance targets for the management system.

CONCLUSIONS

Challenges of building a good management system:

- understanding multi-objective optimisation
- producing deterioration and repair action models
- specifying what data to collect
- testing models and criteria
- testing the overall system
- adjusting to organisational processes and changes
- finding and extending software system support
- integration with the agency's system architecture
- implementing and deploying with step-by-step projects
- financing the projects

REFERENCES

Atostek Oy. 2010. Vaatimusmäärittely (System requirements specification), unpublished.

Atostek Oy. 2012. Taitorakennerekisterin ja taitorakenteiden hallintajärjestelmän toiminnallinen määrittely (Functional Specification of the Engineering Structure Register and Management System), January 2012, unpublished.

Tiehallinto. 2008. Siltarekisterin ja Hanke-Sihan tekninen tarveselvitys (Requirements analysis), internal report, unpublished.

Virtala, Pertti, Thompson, Paul D, Ellis, Reed. 2011. Taitorakenteiden hallintajärjestelmän periaatteellisen toiminnan määrittely. Liikenneviraston tutkimuksia ja selvityksiä 28/2011. Helsinki 2011, published on Transport Agency web site.

Virtala, Pertti, Thompson, Paul D, Ellis, Reed. 2011. Multi-objective optimization of engineering structures, paper to be presented at IABMAS 2012, Como, Italy.

Bridge Maintenance, Safety, Management, Resilience and Sustainability – Biondini & Frangopol (Eds)
© 2012 Taylor & Francis Group, London, ISBN 978-0-415-62124-3

Bridge life cycle optimisation, the Nordic ETSI project

M. Torkkeli & M. Piispanen
Finnish Transport Agency, Helsinki, Finland

ABSTRACT

ETSI project is a Nordic co-operation project, which aim was to create appropriate tools of information related to a bridge and its life cycle for designers, constructors and bridge owners. In project were aspired to be optimize a bridge considering all aspects: functionality, technique, economics and aesthetics in the lifetime of the bridge, produce measuring methods for comparing and evaluation of different items for judging the life cycle of a bridge in a meaningful way, bring the principle of sustainability into practice in bridge engineering and emphasize the importance of cultural values in bridge design, construction and maintenance. The project was a good communication channel between the Nordic countries in bridge life cycle engineering.

The main objective of ETSI was to develop programs to estimate bridge life cycle costs (LCC), life cycle assessment (LCA) as well as bridge aesthetics and cultural values. Tools which be developed will be in such a form that they are relevant and directly applicable to practical problems or decision-making.

The participants of ETSI project were Finnish, Swedish, Norwegian and Danish Road Administrations, Universities of Technology from Sweden (KTH Royal Institute of Technology), Norway (NTNU Norwegian University of Science and tehcnology) and Finland (Aalto University, School of Engineering) as well as COWI Consultants from Denmark, WSP Finland and VR Track from Finland.

ETSI project divided into three stages. In the stage I (2006–2007) was made a study about different aspects of life cycle analysis. In the stage II (2007–2009) was developed the tools for life cycle analysis (WebLCC and BridgeLCA) as well as simple method for evaluating aesthetic and cultural values. The reports of earlier ETSI stages could be found from webpage: http://etsi.aalto.fi/

The main focus of the ETSI stage III (2010–2012) was to continue developing the programs further, updated the used emission values in LCA with current Nordic values and gather a Nordic database of service life assessment of structural parts of bridges. The database was included information for different structural properties and protections, such as exposure classes, service life, interval of reparations and renovation. In the studies the used way to acquire repair interval estimates was combined the result of Delphi-interview (an expert interview) and the data of bridge management systems from all Nordic countries. The new Excel-based programs for LCA and LCC calculations was developed well as program to estimate aesthetics and cultural values of bridge and bridge environments. The developed programs and reports could be found from webpage: http://etsi.aalto.fi/

REFERENCES

Jutila, A. and Sundquist, H. (Editors) 2007. ETSI PROJECT (Stage 1) Bridge Life Cycle Optimisation. Helsinki University of Technology, Publications in Bridge Engineering TKK-SRT-37, 165 p.

Salokangas, L.. (Editor) 2009. ETSI PROJECT (Stage 2) Bridge Life Cycle Optimisation. Helsinki University of Technology, TKK Structural Engineering and Building Technology Publications B, TKK-R-BE3. 136 p.

Bridge Maintenance, Safety, Management, Resilience and Sustainability – Biondini & Frangopol (Eds)
© 2012 Taylor & Francis Group, London, ISBN 978-0-415-62124-3

Guidelines and policy for maintaining and managing all engineering structures of the Traffic Agency

M. Torkkeli & J. Lämsä
Finnish Transport Agency, Helsinki, Finland

ABSTRACT

The Finnish Transport Agency (FTA) is a government agency operating under the Ministry of Transport and Communications. The FTA, which started operating from 1 January 2010, combines the Finnish Rail Administration, the Finnish Maritime Administration and the Finnish Road Administration. The Finnish Transport Agency is an expert organization tasked to develop the transport system to meet the needs of citizens and businesses. In addition to common transport route maintenance skills, road and rail administration and shipping each also require a specific type of in-depth professional competence, which the new Agency makes possible to maintain and developed further.

The Finnish Transport Agency manages great number of different engineering structures like road and railway bridges, wharves, canals with locks, aids to navigation, tunnels, noise barriers, pile slabs, canopies, rock cutting etc. altogether more than twenty thousand of different engineering structures.

New guidelines of maintaining Engineering Structures will be completed and it will be published at the end of 2012. The new guidelines of maintaining will cover primarily bridges (road and railway), wharfs and railway tunnels because inspection data of the other engineering structures is insufficient. Anyway all new guidelines for designing (e.g. Eurocodes), construction or maintaining will always concern all structures managed by FTA. Also all valid guidelines have to reform as well as overlaps and contradicts of guidelines have to remove. The best practice of maintaining and managing will copy and spread out for other engineering structures.

The managing of engineering structures of FTA will base on new Management System of Engineering Structures. The condition of different structures vary from critical to good condition however the structures of the previous Road Administration are in lightly better condition than the structures of Rail Administration because of the different age distribution of structures. At present the all structures of the Maritime Administrations are out of any managing systems. The main subject will be to find method and management system to compare conditions, safety and money-value of repairing of different structures because of budget scenario and uncertain economic situation in Finland. The Finnish Transport Agency's basic principle for everything is "Less money, more to do".

REFERENCES

Economic and strategy plan of Finnish Transport Agency, years 2013-2016. Helsinki 2012, 64 p. Not published.

Tiesillat 1.1.2010. Liikenneviraston tiesillaston rakenne, palvelutaso ja kunto (Road Bridges of Finnish Transport Agency 1.1.2010. Structure, service level and condition of the bridge stock) Helsinki 2010. Finnish Transport Agency, Statistics of the Finnish Transport Agency 3/2010.90 p.

Transport conditions in 2035, Finnish Transport Agency, Helsinki 2011. 32 p. http://portal.liikennevirasto.fi/sivu/www/e/fta/operating_philosophy/transport_conditions 2035

Virtala, P., Thompson, P. D., Ellis, R.: Conceptual Definition of the Management System for Engineering Structures in Finland (in Finnish). Finnish Transport Agency, Helsinki 2011. Research reports of the Finnish Transport Agency 28/2011. 229 p.

Bridge Maintenance, Safety, Management, Resilience and Sustainability – Biondini & Frangopol (Eds)
© 2012 Taylor & Francis Group, London, ISBN 978-0-415-62124-3

Multi-objective optimization of engineering structures

P. Virtala
Destia Ltd, Vantaa, Finland

P.D. Thompson
Consultant, Bellevue, WA, USA

R.M. Ellis
Stantec Inc, Edmonton, AB, Canada

ABSTRACT

The management of transportation infrastructure in Finland has recently been restructured with a multimodal vision and a new allocation of responsibilities. Highways, railways, and seaways, previously separate agencies, have been combined. However, planning and implementation of infrastructure projects are in agencies under separate ministries. Finland is developing a new multi-objective optimization system for structure management, to enable its Transport Agency to incorporate condition, safety, functionality, cost, and environmental concerns in a more flexible system for accommodating and reconciling the goals and objectives of two separate Ministries. The system uses up-to-date concepts of utility theory and capital program optimization, designed to be scalable to handle responsive interactive modeling for an inventory of 20,000 structures. The system will employ a combination of proven algorithms and innovative new forecasting models, designed to make optimal use of Finland's unique damage-oriented visual inspection process.

The three layer decision support system has a database as the ground level of the system where general data, functional data and inspection data are stored and updated annually. A very important part of the system are models, rules, and decision trees so that the system uses the data effectively and makes predictions and comparisons using mathematical tools.

Utility theory is a technique that is commonly used to construct the preference order of alternatives, either by directly eliciting evidence of the decision maker's preferences, or by deducing the preference structure by means of the decision maker's choices of cost and performance outcomes.

Evaluation of performance can be done at the levels of assets, projects, or network-wide. Utility computations are needed for comparing and prioritizing projects, and for describing, tracking, and setting targets for network-wide performance.

Recent research conducted in the USA (Patidar et al 2007) investigated several exact algorithms and approximate heuristics to solve the knapsack problem for problems similar to the Finnish application. It found that a procedure known as the incremental utility/cost (IUC) heuristic could produce near-optimal solutions in reasonable time for a 20,000-structure inventory (McFarland et al 1983). Therefore the new Finnish system will use this method.

REFERENCES

Hawk, Hugh. 2003. *Bridge Life Cycle Cost Analysis*. National Cooperative Highway Research Program Report 483. Washington: Transportation Research Board of the National Academies.

McFarland, W., J. Rollins and R. Dheri. 1983. *Documentation for Incremental Benefit-Cost Technique*, Technical Report prepared for the Federal Highway Administration by the Texas Transportation Institute, Austin: Texas A&M University System, College Station Texas.

Mild, Pekka. *Applications of Multicriteria Portfolio Models in Project Evaluation and Resource Allocation*. 2007. Licentiate Thesis, Helsinki University of Technology.

Patidar, Vandana, Samuel Labi, Kumares Sinha, and Paul D. Thompson. 2007. *Multi-Objective Optimization for Bridge Management Systems*. National Cooperative Highway Research Program Report 590. Washington: Transportation Research Board of the National Academies.

Thompson, Paul D., and John O Sobanjo. 2010. "Estimation of enhanced Pontis deterioration models in Florida." *Proceedings of the Fifth International Conference on Bridge Maintenance, Safety, and Management*, Philadelphia.

Virtala, Pertti, Thompson, Paul D, Ellis, Reed 2011: *Taitorakenteiden hallintajärjestelmän periaatteellisen toiminnan määrittely*. Liikenneviraston tutkimuksia ja selvityksiä 28/2011. Helsinki.

Operation and maintenance of major landmark bridges

Organizer: J.S. Jensen

Bridge Maintenance, Safety, Management, Resilience and Sustainability – Biondini & Frangopol (Eds)
© 2012 Taylor & Francis Group, London, ISBN 978-0-415-62124-3

Optimal maintenance of major bridges

M.L. Bloomstine
COWI AS, Denmark

ABSTRACT

Major bridges are often iconic structures and of great interest to the general public. They are also complicated and unique structures. They most often carry busy roadways and sometimes railways as well and are vital transportation links with great value for society. Traffic disruptions should generally be avoided. Due to the enormous size, height and complexity of the bridges, as well as traffic considerations, access is complicated and very expensive. Special access equipment is generally required and access periods are generally restricted. For all these reasons and in order to maintain structural integrity it is essential that maintenance is carried out in an optimal manner.

The main goal of maintenance is to keep the bridge open at full traffic capacity at all times with the proper safety level. Secondary goals that help achieve the main goal are:

- Ensure that the structural integrity of the bridge is satisfactory at all times
- Ensure that there are no disruptions due to maintenance work
- Ensure that other safety features are fully functional (e.g. signs, warnings, roadway surfacing, etc.)

Another important goal for bridge owners/operators is that the maintenance required to fulfill the above mentioned goals is carried out in the most economical manner possible.

The paper covers recommendations for the following topics concerning optimal maintenance of major bridges:

- The maintenance parties; owner/operator, consultants and contractors. A description of recommended qualifications.
- Maintenance activities including planning. regular inspections, special inspections, investigations, reliability based inspection, regular maintenance, reliability centered maintenance, repair design and repairs works.
- Maintenance design of new bridges.
- Upgrading existing bridges

And finally there are four case stories to illustrate a number of aspects of optimal maintenance.

Fulfilling maintenance for major bridges is a complicated multidisciplinary task. If the guidelines in this paper are followed it is however possible to perform maintenance in an optimal manner to the benefit of the bridge users and the owner/operator and at a much lower life cycle cost.

REFERENCES

Bloomstine, M. et al 2001, Operation Management and Systems for Cable-Supported Bridges, IABSE Conference, Cable-Supported Bridges – Challenging Technical Limits, Seoul, Korea June 2001

Bloomstine, M. 2011, Main Cable Corrosion Protection by Dehumidification – Experience, Optimization and New Development, 6th New York City Bridge Conference 2011, Modern Techniques in Bridge Engineering

Linneberg, P. et al 2011, Messina Bridge – Reliability Based Inspection and Reliability Centered Maintenance – The Challenge, IABSE-IASS Symposium 2011, London, England

Linneberg, P. et al 2012, Reliability Based Inspection and Reliability Centered Maintenance, The 6th International Conference on Bridge Maintenance, Safety and Management (IABMAS), Lake Como, Italy, July 2012

Bridge Maintenance, Safety, Management, Resilience and Sustainability – Biondini & Frangopol (Eds)
© 2012 Taylor & Francis Group, London, ISBN 978-0-415-62124-3

Maintenance of bridge cable systems

J. Laigaard, O. Sørensen & N. Bitsch
COWI, Kgs. Lyngby, Denmark

ABSTRACT

This paper describes the challenges of maintenance of cable systems. This is systems, which are difficult to access and inspect.

Cable systems are a fundamental element in major span bridges. A cable system provides operation and maintenance challenges with respect to:

- Access – as the cable system typical involves works at height with interfaces to bridge traffic.
- Inspection – as the maintenance condition of the cable often is very difficult to inspect, if not impossible. Often only indirect indications may be used to assess the maintenance conditions of a cable system.
- Maintenance – as the access conditions generally is difficult together with the fact that the access into every corner of the cable system for the maintenance of each individual cable element often is more than difficult.
- Replacement – as the structural redundancy can be small, and because the cable in itself provides the access to the element of the cable.

The access, inspection, maintenance and replacement challenges of cable systems stress the need for applying effective approaches to operation and maintenance of cable systems. This need is born in the design phase, long before operation but lacks very frequently the right level of attention. This common deficiency makes the challenges of inspection of maintenance of cable systems even larger (Sørensen et al. 2006; Sørensen et al. 2006).

Due to these challenges, the effectiveness in inspection and maintenance very much depend on the Owner taking a risk based approach in inspection and maintaining his cable systems seeking to optimize his efforts (Lindeberg et al. 2010; Laigaard et al. 2009).

The paper illustrates these problems by the experiences and the careful follow-up by the owners of major Danish and Swedish bridges such as Great Belt Bridge, New Little Belt Bridge, Farö Bridge, Uddevalla Bridge and Tjörn Bridge. Problems with wind induced cable vibrations impose discomfort for traffic and risk of fatigue to the cable systems. The most severe fatigue problems typically are related to the anchorages/fixation points with bending and water ingress.

Less spectacular but equally important, water ingress into cables provides a major maintenance challenge in terms of the risk of corrosion of strands, e.g. into lower anchorages of cables. This aspect requires pro-active maintenance of drainage and water tightness arrangements. In this connection special consideration are required for bridges in regions, where water freezes into ice and block drainages.

REFERENCES

Bitsch, N., Laigaard Jensen, J., Ernst Laursen & Andersen, J.E. 2010. Fatigue Monitoring Systems Great Belt Bridge, IABMAS, 5th International Conference on Bridge Maintenance, Safety and Management.

Laigaard Jensen, J., Bitsch, N. & Laursen, E., 2007. Fatigue Risk Assessment of Suspenders on Great Belt Bridge. 7th International Cable Dynamic Conference, Wienna.

Laigaard Jensen, J., Bitsch, N. & Zinck, M., 2009. Design and Tendering Strait Crossing for Life. Strait Crossing Conference, Trondheim.

Linneberg, P., 2011. Messina Bridge – Reliability Based Inspection and Reliability Centred Maintenance – The Challenge, IABSE-IASS Symposium, London.

Raof, M., 1995. Characteristics of locked coil strands under free bending, Int. J. Offshore and Polar Engineering, Vol. 5m No. 2, pp. 153–160.

Sørensen, O. & Bloomstine, M.L., 2006. Rehabilitation of Main Cables. IABSE Conference, 2006 Copenhagen.

Sørensen, O., Bloomstine, M.L. & Hommel, D.L., 2006. Investigation of Corrosion of Cables, IABSE Conference, Copehagen.

Stoltzner, E., Jensen, B. B. & Zinck, M., 2008. Maintenance of the Faro Bridges, ISCBOC, Japan.

Bridge Maintenance, Safety, Management, Resilience and Sustainability – Biondini & Frangopol (Eds)
© 2012 Taylor & Francis Group, London, ISBN 978-0-415-62124-3

Reliability based inspection and reliability centered maintenance

P. Linneberg & M. Zinck
COWI A/S, Kongens Lyngby, Denmark

K.V. Christensen
Rail Net Denmark, Fredericia, Denmark

ABSTRACT

The common purpose of Reliability Based Inspection (RBI) and Reliability Centred Maintenance (RCM) is to ensure that bridge maintenance and inspection are carried out in a systematic and consistent manner that will ensure bridge availability and meet levels of reliability that are defined by the client or other serviceability limits.

RBI addresses elements and systems of the bridge with predictable failure pattern, where degradation rate can effectively be measured before actual failure. Inspection based updating of the degradation model for the element or system in question allows for prediction of safe maintenance interval. For example, bridge girders exposed to corrosion as a failure, and the protection system in place to mitigate corrosion such as paint. The RBI programme is a detailed description of type of inspection, the location and frequency of inspection. The RBI programme is optimised based on criticality and vulnerability criteria for each element or system.

RCM addresses elements and systems with a predictable failure pattern, where degradation cannot effectively be measured before failure. Such systems are often mechanical, hydraulic and/or electrical in nature. The RCM programme is optimised based on criticality and vulnerability criteria for each element or system and proposes detection method, mitigation measures, maintenance frequency, and the number of spare parts needed to ensure that a required level of reliability is maintained for the element or system under consideration.

During the RBI and RCM analyses, each system or element is analysed and the optimal individual strategy is determined. The individual strategies are then combined and adjusted to give an overall optimal inspection and maintenance strategy for the entire bridge. The results of these analyses are bench marked against Life Cycle Cost analysis (LCC).

During operation RBI and RCM plans should be evaluated and optimised in order to prolong service life, optimise inspection and maintenance frequency and prevent escalation of damages, i.e. reduce maintenance costs.

Inspired by the above methodology Rail Net Denmark has performed the first phase of a systematic and consistent optimization of the inspection and maintenance regime on the superstructure of the Little Belt Bridge from 1935. The Little Belt Bridge of 1935 is a 825 m long, five span, riveted steel lattice bridge carrying two rail tracks and a two road lanes, i.e. fatigue of riveted connections and degradation of paint where some of the key issues analyzed. Concrete approach spans were not part of this study.

The coming phase is detailing of I&M planning for selected elements and systems based on actual degradation and combining all information in an overall I&M plan for the steel superstructure.

REFERENCES

Linneberg, P. et al 2011. Messina Bridge – Reliability Based Inspection and Reliability Centred Maintenance – The Challenge, IABSE-IASS Symposium 2011 (CD-ROM)

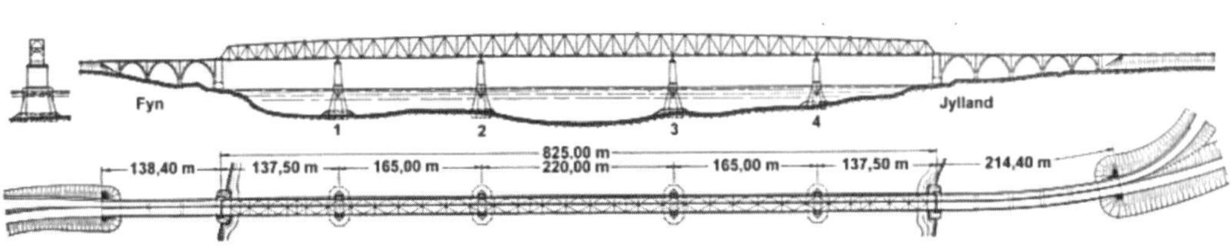

Figure 1. Little Belt Bridge of 1935. Plan and elevation.

Bridge Maintenance, Safety, Management, Resilience and Sustainability – Biondini & Frangopol (Eds)
 ISBN 978-0-415-62124-3

Maintenance of long span bridges

D.K. MacKenzie
Flint & Neill Limited, London, England

B. Colford
Forth Estuary Transport Authority, Edinburgh, Scotland

ABSTRACT

Modern bridges are expected to last over 100 years. But this expectation cannot be met without a detailed and long lasting commitment to maintenance. It also cannot be met without a recognition that the design of a new bridge needs to consider maintenance as a critical design case, alongside strength and serviceability criteria.

In the UK, it has been recognized that the design philosophy of minimizing initial capital cost was having a significant effect on durability. Design standards were changed around 2001 to promote measures to improve durability. These changes have focused on the more genetic issues of improving cover and concrete strength to the specific problems of poor detailing at piers supports by the use of continuous decks over intermediate supports. In conjunction with this, the use of integral bridges with abutments connected directly to the deck has been promoted alleviating some of the shared problems that arise when water gets trapped in expansion joints and under bearings. Other advances such as the use of non-corrodible reinforcement in concrete has some considerable benefit along with the even simpler solution of the use of mass concrete in abutments and wing walls to avoid the problem of corrosion all together. Allied to the technical improvements is a greater recognition of the need to adequate provision is now made for access for maintenance; cleaning and painting; bearing replacement and inspection of closed cell and box members.

The operation of major bridges provides a clear insight into some of the limitations within the design process, where the designers' emphasis on strength and stiffness often neglect important considerations of maintenance. Difficulties occur primarily with access and the replacement of components that do not have the same design life as the bridge, resulting in significant increases to operation costs and even the inability to carry out maintenance. Although modern design of cable supported bridges embraces planned cable replacement as a standard design case, issues still abound with items such as the replacement of bearings, joints and holding down bolts and safety fences. In the case of cable replacement, the load case is considered but not the impact on traffic, the use of temporary works and the safety of operators carrying out the works. In the design and build framework where performance specifications work well for strength and stiffness constraint, less attention is paid to the equally important maintenance design case. This paper looks at the difficulties in maintaining major bridges and the examines how designers can consider access and maintenance issues as design cases alongside conventional strength and displacement criteria, resulting in greater reliability of the structure.

Case studies are presented that show how the cost of maintenance of major bridges can be reduced by setting down a maintenance strategy for the bridge at an early stage. The strategy should consider the major repair/replacement work expected in the lifetime of the bridge and assesses the cost benefit of the impact of these operations. With suitably low discount rates on valuing future work and the use of DBFO forms of contract rather than just design and build contracts owners can avoid the short term approach towards sacrificing maintainability at the altar of lowest initial cost.

Bridge Maintenance, Safety, Management, Resilience and Sustainability – Biondini & Frangopol (Eds)
ISBN 978-0-415-62124-3

A structural health monitoring systems for long span bridges

D.K. MacKenzie
Flint & Neill Limited, London, England

N.M. Apaydin
KGM 1st Division, Istanbul, Turkey

O. Akkol
Orient Research, Istanbul, Turkey

ABSTRACT

The Bosphorus Bridge in Istanbul, Turkey, was one of the earliest examples of a major box girder suspension bridge. The structure, designed by Freeman Fox and Partners, was completed in 1973[1]. With a main span on 1074 m the crossing demonstrates a significant stage in the development of long span bridges. A Structural Health Monitoring (SHM) system was installed on the Bosphorus Bridge in Istanbul in 2009. This Paper discusses the practical applications of strain gauge data from the orthotropic steel deck and how this data can be used to assess the residual fatigue life on the deck.

The complex geometry of many fatigue sensitive details and uncertainties regarding the actual application of loading (a problem that is compounded by the cube rule relationship of fatigue stresses) means that there is still considerable difficultly in analytically predicting the fatigue life of new and existing structures.

It is likely that the in-service performance of major bridge crossings will continue to be significantly hampered by the onset of fatigue damage; therefore it is critically important to understand the reduction in fatigue life in a structure and whether the predicted effects match the actual effects. Furthermore, knowledge of the likely occurrence of damage will influence inspection requirements – favorably if the predictions are higher than the actual stresses. For existing structures, where the fatigue design may be based on wholly inappropriate standards, knowledge of the actual performance is extremely useful in planning the future maintenance requirements of the structure.

The structure remains a critical piece of infrastructure within Turkey, connecting the Asia and European shores. Over 100,000 vehicles cross the bridge in either direction each day.

Due to evidence of fatigue damage cracking of the orthotropic deck shortly after construction, the bridge has been closed to heavy goods vehicles for many years; however, the fatigue performance of the bridge remains a concern to the Turkish General Directorate of Highways (KGM).

In 2007 KGM employed Flint & Neill to undertake a detailed inspection and assessment of the overall bridge. As part of this work a Structural Health Monitoring (SHM) system was installed.

This paper provides an overview of the SHM system and then discusses how the strain gauge data is being used to inform the fatigue assessment of the deck. The paper then goes on to compare the measured data against results obtained from a FE analysis and considers how the development of cracks in the troughs affect the local load distribution in the deck.

The observed data has been shown to agree with local finite element modeling.

It is found that the fatigue capacity of the deck plate is low. Despite the fact that heavy goods are prevented from using the crossing there is still evidence of new fatigue cracking.

The bridges operators, KGM, have adopted a responsible and proportionate maintenance program in which the deck is regularly inspected and any new cracks swiftly repaired to ensure the continuous service of this landmark bridge.

Recent advances in bridge health monitoring

Organizers: C.F. Cremona & A. Orcesi

Bridge Maintenance, Safety, Management, Resilience and Sustainability – Biondini & Frangopol (Eds)
 ISBN 978-0-415-62124-3

Perturbation based stochastic model updating methods for the evaluation of structural modifications

R. Biyoghe, D. Clair & M. Fogli
Clermont Université, Université Blaise Pascal, LaMI-Campus des Cézeaux, Aubière Cedex, France

ABSTRACT

In Structural Health Monitoring of civil engineering structures, the construction of mathematical models that are expected to produce accurate predictions of the behavior of the system of interest is an important issue. During the construction of such predictive models, errors due to imperfect modeling and uncertainties due to incomplete information about the system and its environment always exist. Test data are inherently exposed to uncertainties due, for example, to measurement errors or modal identification procedures. These uncertainties can be appropriately accounted for by using a stochastic approach in which the model parameters are considered as random variables. In the recent years the problem of "Stochastic Model Updating" has been addressed by numerous authors (Govers, Y & Link, M. 2010, Hua, X & al. 2008, Khodarapast, H.H & al. 2008, Mottershead, J.E & al. 2006). This article presents three approaches of this problem using perturbation techniques based on different linearization strategies. These approaches are tested and compared on an academic application concerning a beam supported by random elastic foundations. Procedure A relies on a deterministic updating procedure coupled with a posteriori statistical estimations. It is the most accurate, very simple to implement but suffers from the major drawback to be very time consuming and cannot be reasonably used to update complex finite element models. Procedures B and C are built to directly estimate the mean and covariance matrix of the parameters. Their performances on the proposed applications are very close. This is mainly due to the fact that they use the same formula for the estimation of the mean. This may also be associated with the simplicity of the model parameter covariance used in the applications. Further investigations are needed to clarify this point. The application of Procedure C to the case of a railway bridge demonstrates its efficiency in updating a finite element model using modal frequencies identified from experimental measurements (Table 1). The calculation of the natural frequencies using the updated values of the parameters shows a clear improvement of the model prediction (Table 2).

Table 1. Initial and updated model parameter values.

	$S^{(0)} = (I_{xx} M_l)^{(0)T}$		$S_{up} = (I_{xx} M_l)^T_{up}$	
Parameter	$I_{xx}(\mathrm{m}^4)$	$M_l(\mathrm{kgm}^{-1})$	$I_{xx}(\mathrm{m}^4)$	$M_l(\mathrm{kgm}^{-1})$
Mean	5.50	12700	5.005	9500
Std Deviation	0.55	1270	0.18	1400

Table 2. Identified and computed natural frequencies (Hz).

Mean Frequency	F_1	F_2	F_3	F_5
Identification	8.5	12.78	14.64	18.55
Initial Model	8.27	12.97	13.86	18.80
Updated Model	8.73	12.80	14.65	18.56

REFERENCES

Govers, Y & Link, M. 2010. Stochastic model updating-Covariance matrix adjustment from uncertain experimental modal data. *Mechanical Systems and Signal Processing* 24(3): 696–708.

Hua, X, Ni, Y, Chen, Z & Ko, J. 2008. An improved perturbation method for stochastic finite element model updating. *International Journal for Numerical Methods in Engineering* 73(13): 1845–1864.

Mottershead, J.E., Mares, C., James, S. & Friswell, M.I. 2006. Stochastic model updating: part 2 Application to a set of physical structures. *Mechanical Systems and Signal Processing* 20(8): 2171–2185.

Bridge Maintenance, Safety, Management, Resilience and Sustainability – Biondini & Frangopol (Eds)
© 2012 Taylor & Francis Group, London, ISBN 978-0-415-62124-3

Supervised learning algorithms for damage detection and long term bridge monitoring

C. Cremona
Technical Centre for Bridge Engineering, SETRA, Sourdun, France

A. Cury
Federal University of Ouro Preto (UFOP), Ouro Preto, Brazil

A. Orcesi
University Paris-Est, IFSTTAR, Paris, France

ABSTRACT

In the past few years, numerous methods for damage assessment for structural health monitoring were proposed in the literature. Several problems are raised for making these approaches practical for the engineer. The first concern is to determine whether a structure presents an abnormal behavior or not. Statistical inference is concerned with the implementation of algorithms that analyze the distribution of extracted features in an effort to make decisions on damage diagnosis. Learning algorithms have extensively been applied to classification and pattern recognition problems in the past years and deserve to be used for structural health monitoring. In addition, data acquisition campaigns of civil engineering structures can last from several minutes to years.

Dealing with large amounts of data is not an easy task and suitable tools are required to correctly extract important features from them: symbolic data analysis (SDA) is such an approach. In this paper, some supervised learning methods (Bayesian decision trees, neural networks and support vector machines) are introduced to discriminate structural features and are developed within the concept of symbolic data analysis in order to compress data without losing its inherent variability. To highlight the different features of these techniques for structural health monitoring, this paper focuses attention on the monitoring of a railway bridge belonging to the high speed track between Paris and Lyon (Fig. 1). During the month of June 2003, a strengthening procedure was carried out in this bridge. In so doing, vibration measurements were recorded

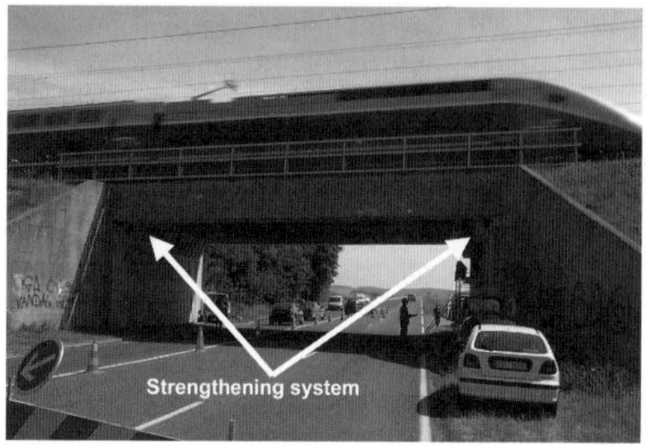

Figure 1. PK 075 + 317 bridge with its strengthening system.

under three different structural conditions: before, during strengthening. In the following years (2004, 2005 and 2006), new tests were performed to observe how the dynamic behavior of the bridge evolved, especially for the case of frequency changes. The objective was to verify whether the strengthening procedure was still effective or not, in other terms if the new data could be still assigned to the condition "after strengthening". This paper reports the major results obtained and shows how the supervised learning techniques can be applied to cluster structural behaviors and classify new data. An original assignment approach is also presented: based on dissimilarity measures this approach shows that in fact the structural behavior of the bridge seems totally different of the initial behaviors used for training the supervised learning methods.

Bridge Maintenance, Safety, Management, Resilience and Sustainability – Biondini & Frangopol (Eds)
© 2012 Taylor & Francis Group, London, ISBN 978-0-415-62124-3

Bridge characterization and structural health monitoring: A suspension bridge case study

A. Houel
CETE of Lyon, Laboratory Department of Lyon, Bron, France

A. Orcési
University Paris-Est, IFSTTAR, Paris, France

R. Leconte
DIADES, Lyon, France

ABSTRACT

Detecting and assessing the severity of structural damage is one major objective when monitoring a bridge. In particular, when some degradation phenomena threaten the bridge safety, the goal can be to assess monitoring critical components, which allows checking a safety threshold that is not reached until some repair/rehabilitation actions are performed (F08b 2009, Sohn et al. 2004, Worden & Farrar 2007).

This paper describes an overall monitoring-based decision process on structures with or without pathologies (Houel & Leconte 2011).

In particular, when pathologies are susceptible to question the bridge safety, the bridge owner can decide to put in place a continuous structural health monitoring program so that immediate countermeasures (defined in advance) can be decided in case of imminent danger (LCPC 2009).

The objective of continuous monitoring can be to maintain in service the damaged structure (restricted or not), to postpone a repair or a reinforcement action, or to detect an irreversible damage. It can be implemented only after having led a complete structural analysis with in-depth, damage and special inspections/investigations, and confirmation of inputs to calculation (Ministère de l'Equipement, des Transports et du Logement – Direction des Routes 1998, LCPC 2005).

This decision process is illustrated with the case of a suspension bridge in France, the Teil Bridge near Montelimar. Following the sudden fracture of a cable near the anchorage, the bridge was totally closed to traffic. Several investigations were carried out to better understand the reason of the fracture occurrence and assess the performance of other anchorages (Houel & Gonod 2008). The bridge safety was ensured by strengthening each anchorage, starting an acoustic monitoring program (during one year), and defining warning thresholds (Tessier 2007). Meanwhile, several other experimental investigations (in situ load tests, chemical analysis of concrete samples) were conducted to estimate the residual structural service life and define the bridge repair actions (Dierkens 2008, Germain 2009). Finally, a major rehabilitation program with the change of all cables and anchorages was decided. The example of the Teil Bridge illustrates the progress of monitoring information integration in a bridge management system.

REFERENCES

Dierkens, M. 2009. *Teil Bridge – Concrete diagnosis*. CETE of Lyon report [in french].

F08b 2009. *Guideline for structural health monitoring, SAMCO final report 2006.*

Germain, D. 2009. *Teil Bridge – Fatigue study of the deck, bridge robustess and fatigue crack propagation.* CETE of Lyon report [in french].

Houel, A. & Gonod, B. 2008. *Teil Bridge – partial detailed inspection*. CETE of Lyon report [in french].

Houel, A. & Leconte, R. 2011. *Caracterization of structural health monitoring*. report GT5-3. Research project Structural Health Monitoring [in french].

LCPC, 2005. Technical Guide: *Telemonitoring system for bridges under high surveillance*. [in french]

LCPC, 2009. Technical Guide: *Investigations and dynamic evaluations on bridges*. [in french]

Ministère de l'Equipement, des Transports et du Logement – Direction des Routes, 1998. *Technical instruction for monitoring and maintenance of bridges – part 2, fascicule 03: Diagnosis, reinforced monitoring, high monitoring, decision-making.* [in french]

Sohn, H., Farrar, C. R., Hemez, F. M., Shunk, D. D., Stinemate, D. W., Nadler, B. R. & Czarnecki J. J., 2004. *A review of Structural Health Monitoring Literature: 1996–2001.*

Tessier, C. 2007. *Teil Bridge – Investigation and surveillance after cable break*. LCPC report [in french].

Worden, K. & C. R. Farrar, 2007. *The fundamental axioms of structural health monitoring.* Proceedings of the Royal Society of London, Series A (Mathematical, Physical and Engineering Sciences) 463 (2082): 1639–64.

Bridge Maintenance, Safety, Management, Resilience and Sustainability – Biondini & Frangopol (Eds)
© 2012 Taylor & Francis Group, London, ISBN 978-0-415-62124-3

DynaMo – Software for vibration based structural health monitoring

F. Magalhães, S. Amador, Á. Cunha & E. Caetano
University of Porto, Faculty of Engineering (FEUP), Porto, Portugal

ABSTRACT

This paper presents an innovative software package, called DynaMo, for continuous dynamic monitoring of relevant civil infrastructures, which was developed with the goal of extracting useful information from continuously collected acceleration time series. It contains 3 components: the main software that includes routines for data and results management and data processing, a graphical user interface to provide access to the results database and a watchdog software to insure the DynaMo is always 'alive'.

Having in mind the identification of stiffness changes due to damage, it was followed an approach that is based in the continuous on-line automatic identification of the structure modal parameters, using its response under operation and adopting state-of-the-art identification algorithms.

Therefore, the data processing includes 3 main steps: (i) automatic identification of modal parameters, (ii) elimination of environmental and operational effects on modal parameters, (iii) calculation of a suitable index to flag relevant frequency shifts.

These steps are illustrated in Figure 1 adopting results from a monitoring application on concrete arch bridge that is presented in the paper to illustrate the use of the software. It shows that from the collected acceleration time series it was possible to track the time evolution of 12 natural frequencies. Then, with adequate statistical tools, the effects of the traffic over the bridge and of the ambient temperature on each natural frequency were minimized. After this, it was possible to obtain a control chat that is able to identify abnormal frequency shifts that might be associated with the occurrence of damage (Magalhães et al. 2011).

Figure 2 shows a screenshot of the graphical user interface (DynaMo Viewer) customized for the same bridge application. The paper includes several plots produced by this interface that demonstrate the quality of the information that can obtain with the processing routines included in the DynaMo software.

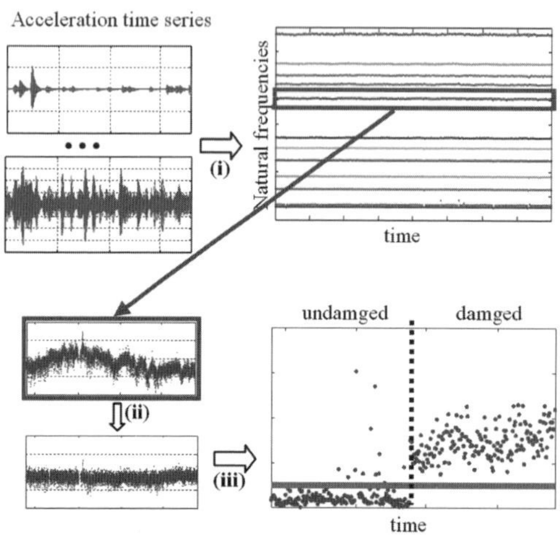

Figure 1. Main processing steps of a vibration-based health monitoring system.

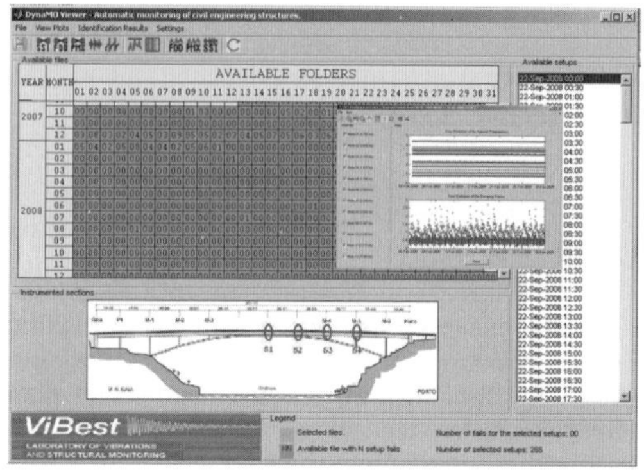

Figure 2. Screenshot of the DynaMo Viewer software.

REFERENCE

Magalhães, F., Cunha, Á. & Caetano, E. 2011. Vibration based Structural Health Monitoring of an arch bridge: from automated OMA to damage detection. *Mechanical Systems and Signal Processing*, 28, 212–228.

Bridge Maintenance, Safety, Management, Resilience and Sustainability – Biondini & Frangopol (Eds)
 ISBN 978-0-415-62124-3

Structural monitoring of the Tacony-Palmyra Bridge using video and sensor integration for enhanced data interpretation

M.T. Yarnold
Intelligent Infrastructure Systems (IIS) a division of Pennoni Associates, Philadelphia, PA, USA

F.L. Moon & A.E. Aktan
Drexel University, Philadelphia, PA, USA

B. Glisic
Princeton University, Princeton, NJ, USA

ABSTRACT

A comprehensive structural health monitoring (SHM) system has been implemented on the Tacony-Palmyra Bridge in Philadelphia, Pennsylvania (USA). The Tacony-Palmyra Bridge is an over 80 year old structure that includes a two-leaf, rolling lift, bascule system (79 m span). The health monitoring system was developed using the structural identification process along with a historic review and vulnerability assessment. The underlying objective of the project is to indefinitely preserve the structure by leveraging modern technology. As a result, the primary focus has been placed on addressing engineering, operation, and maintenance needs of the bridge. This was executed with the use of distributed data acquisition systems. The instrumentation includes electrical resistance and vibrating wire strain gages, tilt sensors, and a weather station. Cameras were also placed at selected locations on and around the structure. The integration of data and video was accomplished through the development of a live web portal and a customized playback program. The live web portal allows for real-time remote viewing of the data and video over the internet. The structural monitoring software includes the ability to record events such as bascule openings and overloaded vehicle passage. These events can be viewed in the playback program (Figure 1). This program allows the data to be viewed both spatially and temporally so as to maximize data interpretation and benefit for the end user of the SHM. The system is also equipped with trigger and alert functionality for specified events. The goal is to provide the bridge owner with an effective and reliable decision making tool to better manage their structure.

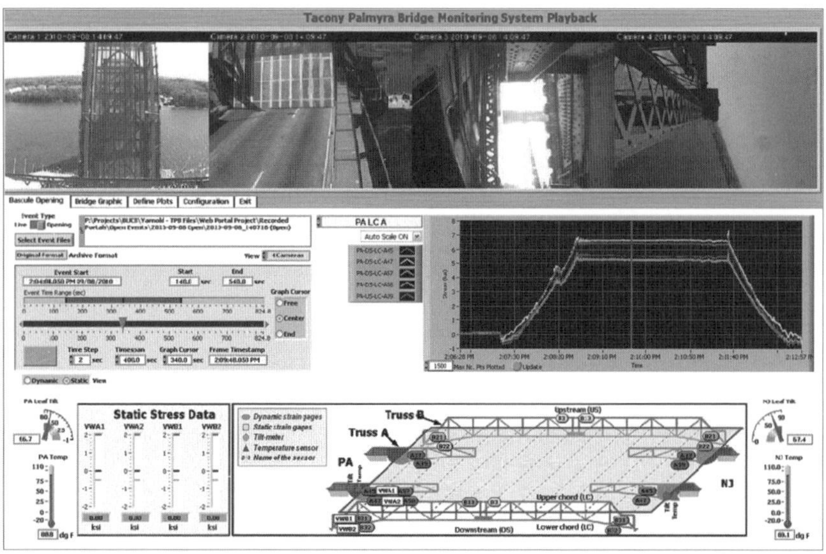

Figure 1. Playback program for detailed review of recorded events.

Optical monitoring techniques for bridge maintenance and safety

Organizers: P. Sumitro & H. Matsuda

Bridge Maintenance, Safety, Management, Resilience and Sustainability – Biondini & Frangopol (Eds)

Field loading measurement of post-tension PC girder bridge with line sensor scanner

A. Demizu & H. Matsuda
Nagasaki University, Nagasaki, Japan

Y. Ito
Saga University, Saga, Japan

K. Hida
K&T Consultant Co., Tokyo, Japan

T. Okamoto
Keisoku Research Consultant Co., Hiroshima, Japan

M. Uchino
Fukuoka Industry Science & Technology Foundation, Fukuoka, Japan

P.S. Sumitro
Smart Structures, Illinois, USA

ABSTRACT

Recently, the research on the strain measurement that uses the optical full field measurement is done. There is digital image correlation method (DICM) in one of the optical full field measurements. DICM can easily calculate displacement, the strain distribution within the range of the measurement, and the direction by analyzing the digital image of which it takes a picture with CCD camera and CMOS camera, etc.

However, a highly accurate strain measurement cannot be achieved in outdoor by ①change in the illuminance in the measurement, ②fluctuation of the air between objects, ③blur of the image by the vibration, ④various aberrations of the lens, and ⑤lack of the number of pixels for CCD(CMOS), etc.

Then, to solve the problem of the above-mentioned, authors developed an applicable line sensor type full field strain measurement device (scanner) in outdoor.

In this thesis, the outline of the scanner and the strain accuracy experiment that uses the steel material are described. And, the strain measurement that uses the scanner is described in the field loading test on the PCT beam bridge.

In the strain measurement accuracy experiment that used the steel material, when the strain gage was compared with the measurement of the scanner, the error margin was the greatest – 1.3%. The average of the absolute value of the error margin in each strain level was 0.59%. The standard deviation of the measurement of five times in the same strain level with the scanner was 25 μ or less. In a word, high reproducibility was confirmed.

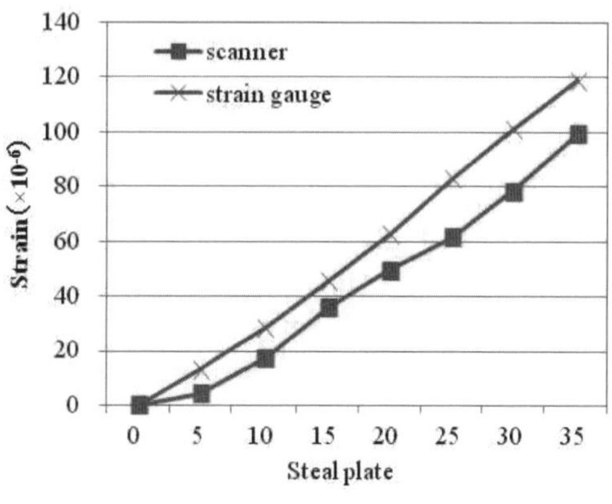

Figure 1. Measurement result.

As for the scanner, the result of the same accuracy as the strain gage was confirmed.

In the field loading test on the PCT beam bridge, the strain of the main girder lower flange was measured. The scanner indicates the small value from the strain gage, and the difference of the strain value has grown with a load increase.

The average error of the scanner and the strain gage was 15 μ. As for the scanner, the result of the same accuracy as the strain gage was confirmed in outdoor.

REFERENCE

Jsndi, 2010, Feature: Stress and Strain Analysis, Journal of the Japanese Society for Non-destructive inspection, Vol. 59 No. 7.

Bridge Maintenance, Safety, Management, Resilience and Sustainability – Biondini & Frangopol (Eds)
 ISBN 978-0-415-62124-3

Application of full-field non-contact measurement technology to clarification of deterioration mechanism on constructional material

H. Goda & M. Hibino
Kyushu Institute of Technology, Kitakyushu, Japan

Y. Kawabata
Port and Airport Research Institute (PARI), Yokosuka, Japan

M. Uchino
Fukuoka Industrial Technology Center, Kitakyushu, Japan

H. Matsuda
Nagasaki University, Nagasaki, Japan

ABSTRACT

This study is the basic research for proposing a maintenance method for the structures using full visual field non-contact measurement technique. In this paper, examples of deterioration of concrete structures due to alkali silica reaction (ASR) have been reported, and methods used to diagnose such structures have become controversial issues. Regarding the diagnosis of ASR affected the structures, it is difficult to evaluate induced strain/stress of concrete due to ASR expansion. Therefore it is necessary to develop a method to evaluate the induced strain in reinforced concrete (RC) member. One of the conventional methods is to measure the strain with strain gauges before and after releasing induced stress by cutting and/or coring the surface of concrete. This requires, however, the attachment of many strain gauges in order to evaluate multidirectional strains since ASR expansion shows anisotropic behavior.

Meanwhile, digital image correlation Method (DICM), a full-field optical measurement, gives strain distribution of concrete with high resolution. By using this technique, the authors developed a method to evaluate induced strain in RC member caused by ASR expansion.

Figure 1. Specimen and camera for DICM.

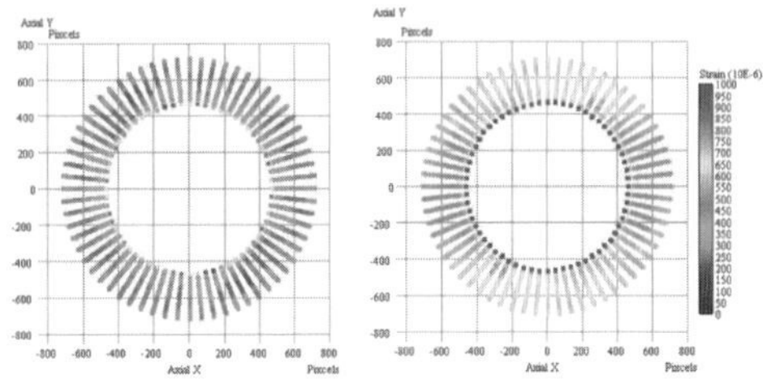

Figure 2. Strain contributions on stress release method using by multi rossete analysis.

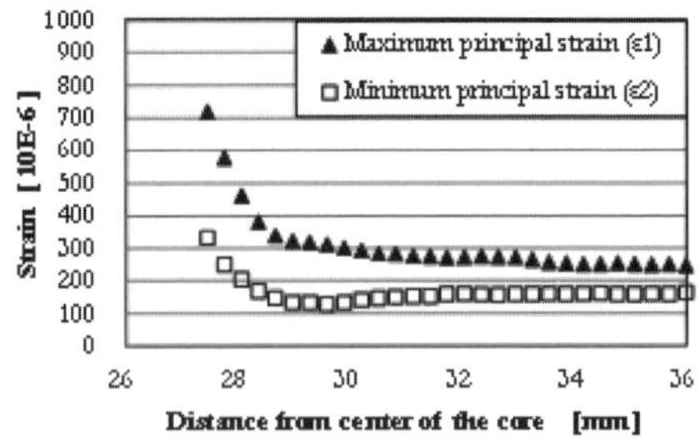

Figure 3. Relationship between the distance from center of the core and strain on the surface of the specimen.

This paper describes the new method to evaluate induced strain by using DICM. After expansion, deformation of concrete before and after coring the concrete was measured. The results showed one possibility that the new method can evaluate the induced strain in RC member caused by ASR expansion.

REFERENCE

Uchino, M., et al, Strain analysis method using multi-rosette analysis by digital image correlation method, Proceeding of the International Conference on Bridge Maintenance, Safety and Management, (2010), vol. 5, pp. 2564–2569.

Bridge Maintenance, Safety, Management, Resilience and Sustainability – Biondini & Frangopol (Eds)
© 2012 Taylor & Francis Group, London, ISBN 978-0-415-62124-3

Existing bridge structural identification by vibration measurements using laser doppler velocimeter

K. Makino, C. Morita & H. Matsuda
Structural Engineering, Nagasaki University, Japan

P.S. Sumitro
Smart Structures, United States

S. Yanai
K&T Consultant Co. LTd, Japan

ABSTRACT

Many civil engineering structures have been constructed in the 1970s at a period of high economic growth in Japan. Most of the bridges, built at that time, have aging problems. The importance of maintenance technology is more and more important.

As inspection method of structures, visual inspection and hammering testing are widely used. How-ever, these inspections need a lot of experience and the results are depend on inspectors. In recent years, vibration measurements are utilized to examine structures. For evaluate the structural characteristics of the structures, various vibration measurements have been conducted. Most of methods have been used natural frequencies and vibration mode shapes as parameters. In such vibration measurement, the sensor must be placed in structures. But such setting is not easy.

In our study, Laser Doppler Velocimeter is conducted for vibration measurement of existing structures. Laser Doppler Velocimeter is an instrument that can measure long distance and the non-contact. First experiment using concrete specimens, the effects of the material damage to vibration characteristics is evaluated. Then, the vibration measurement of existing bridge to estimate the natural frequencies and vibration mode shapes of superstructure is con-ducted using two Laser Doppler Velocimeters. Vibration characteristics of the bridge are evaluated from the natural frequencies and vibration mode shapes. In addition, the three-dimensional FEM analysis is performed. The result of the analysis was compared with measurement result, and evaluated range of the analysis model. We have measured an existing bridge column, built in 1976, before and after strengthening with additional reinforced concrete layer. From the results of natural frequencies, we can estimate the stiffness change good enough.

Table 1 and Figure 1 show a comparison of measurements and analysis of the superstructure. The measurement and analysis are mostly consistent. From these results it is considered that the vibration characteristics of existing bridges are able to evaluate using Laser Doppler Velocimeter.

Table 1. Comparison of natural frequency.

Mode	Measurements (Hz)	Analysis (Hz)	Difference ratio (%)
1	3.25	3.31	1.85
2	6.23	6.13	1.61
3	8.2	7.92	3.41

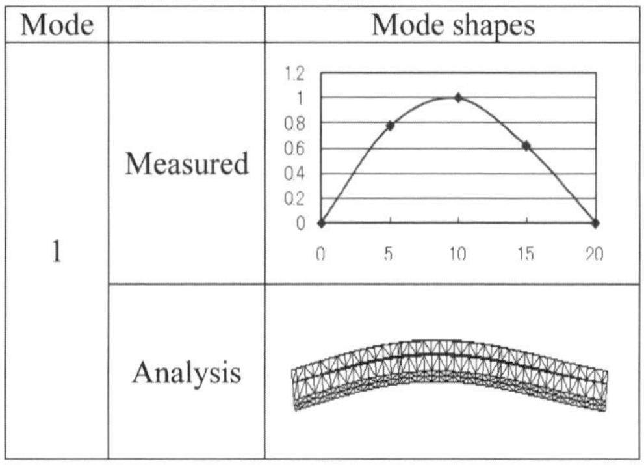

Figure 1. Comparison of vibration mode shape.

REFERENCES

JECE. 2010. Guidelines for Monitoring of Bridge Vibration. Maruzen: 5–15.

JSCE. 2005. Standard Specifications for Concrete Structures-2007 Test Methods and Specifications. JSCE: 394–398.

Kotsubo, S. 1996. Introduction to Construction Vibration. Morikita Syuppan: 150–159.

Miyashita, T. 2006. Identification of Vibration Mode of Bridge Using Two Accelerometers Wireless LAN. Proc. 61th Annual Conference of the Japan Society of Civil Engineers, 1–496: 989–990.

Uehan, H. 2007. Development of Noncontact Vibration Measurement System "U-Doppler" for Structure Diagnosis. RTRI Report, 21–12: 17–22.

Uehan, H. 2010. Noncontact Measuring of Structural Vibration. Railway Research Review, 64-4(2007): 22–25.

Bridge Maintenance, Safety, Management, Resilience and Sustainability – Biondini & Frangopol (Eds)
 ISBN 978-0-415-62124-3

Development of a hybrid camera system for bridge inspection

S. Nishimura & K. Kimoto
Keisoku Research Consultant Co., Ltd, Hiroshima, Japan

S. Kusuhara & S. Kawabata
Honshu-Shikoku Bridge Expressway Co., Ltd, Kobe, Hyogo, Japan

A. Abe & T. Okazawa
Bridge Engineering Co., Ltd, Kobe, Hyogo, Japan

In these days, acquisition of high resolution digital images and processing techniques have been developed, which are applied to such as the tunnel wall inspection. In this study, we applied these techniques to huge bridge maintenances and management. However, we predicted that it is difficult to specify the taking a picture parts only with the individual detailed images, because huge bridges, especially truss bridge, is complicated with various components, compared with tunnel walls.

By using sphere images and detailed images taken at the same time, we built Images Research System which use sphere images as an index, and examine application possibility.

We applied experimentally this system to the Akashi Kaikyo Bridge with a higher efficiency and laborsaving in image arrangement work. This paper describes the effectiveness of this System and remained problems.

We took pictures to investigate various structural members of stiffing girder of the Akashi Kaikyo Bridge by four methods shown in Fig 1. From wide inspection way, we took pictures, running vehicle which has a camera for taking sphere image (Ladybug3) and 16 industrial measurement cameras for taking detailed image (Grasshopper) at 15 km/h. Moreover, we experimentally set another one Grasshopper which has 5 mega pixels and synchronized it with Ladybug3. Thus, we identified shooting position of detailed image which use sphere images as an index.

Table 1 shows images which the part located in 20 m from camera position with 4 kinds of cameras.

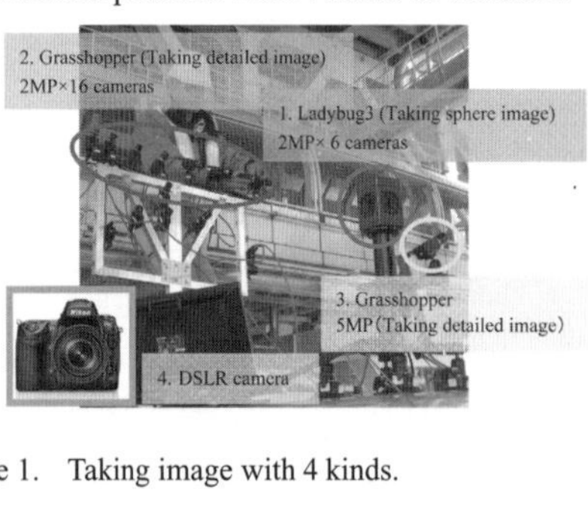

Figure 1. Taking image with 4 kinds.

Table 1. The comparison table of the each image.

Camera	Ladybug3	FULLCAP
Pixels	2MP	2MP
Taking image	Sphere Image	Detailed Image
Full Image		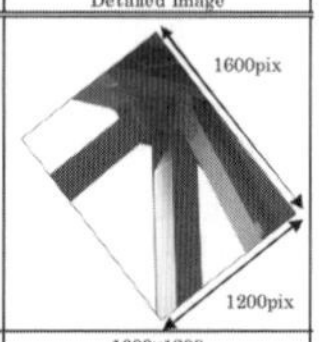
Size	1616×1232	1600×1200
Extended Image	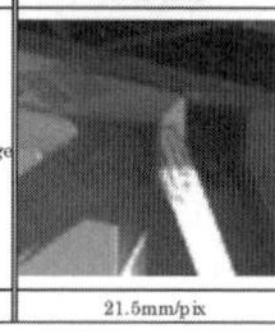	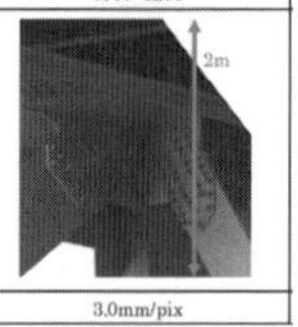
Resolution	21.5mm/pix	3.0mm/pix

As a result, we verified that this system which takes pictures by sphere image camera-Ladybugs3, specify the inspection parts, and then takes detailed image (FULLCAP) is able to apply for inspecting such as long regular structural forms bridges. Additionally, in such bridge, it is necessary to adjust camera angle and lens focus were adjusted sufficiently because it takes pictures while running. However, accurate adjustment makes it possible to take images effectively.

REFERENCES

Hirata. K et al. 2010: Investigation use of the Photogrammetry which utilized the digital technology JAST vol. 21, (2010) (CD-ROM) (in Japanese).

Kurashige. H et al. 2010: Development of On-Board Image Measurement System for Actual Running and Application to Wall Surface Surveys of Structures, IABMAS2010.

Kusuhara. S et al. 2011: Development of a Hybrid Camera System for Bridge Inspection, Proceedings of the 66th Annual Conference of the Japan Society of Civil Engineers, (CD-ROM) I-535 (in Japanese).

Nishimura. S et al. 2010: Development of On-Board Image Measurement System for Actual Running and Application to Wall Surface Surveys of Structures, JAST vol. 21, (2010) (CD-ROM) (in Japanese).

Nishimura. S et al. 2011: The Measurement and Monitoring at Gunkanjima-Island by using 3D laser scanner and Sphere Camera, J. JSEM (in Japanese).

Bridge Maintenance, Safety, Management, Resilience and Sustainability – Biondini & Frangopol (Eds)
© 2012 Taylor & Francis Group, London, ISBN 978-0-415-62124-3

Monitoring of short & long term cable force on a cable stayed bridge using package type FBG sensors

D.Y. Park, D.G. Kim, S.H. Kim & S.W. Lee
Daelim Industrial Co., Ltd, Seoul, Korea

ABSTRACT

Stay cables are the main load-bearing components of a stayed-cable bridge. The positive and negative moments of the decks are also generally controlled by adjusting the cable force under construction and the safety of the bridge is determined by measuring the cable force in service. Especially, in case of a cable-stayed bridge with a concrete edge girder constructed using the free cantilever method, cracks in the girder easily arise depending on the cable force when bridge is the under construction. A cable is also easily damaged due to factors such as environmental corrosion, fatigue, materials aging and stress redistribution, with the result being that the cable can no longer guarantee the designed cable force. Therefore, monitoring of the cable force is very important during the construction and maintenance of the bridge.

Currently, strain gauges, accelerometers and electro-magnetic sensors are widely used to measure the cable force in many applications. However, existing sensors such as a strain gauge cannot easily evaluate stress or damage state of a cable because they are difficult to install and they contain electrical noise. In addition, the transmission capacity actually declines as the length of the cable increases. For accelerometers that measure MS-type cables, these accelerometers are installed on HDPE pipe which serves to protect the cable, they contain a gap between the HDPE pipe and cable. Therefore, an accuracy of the measurement is also reduced. Moreover, the degree of cable force adjustment must be determined by reflecting the force distribution of the cables adjacent to the cable which is to be adjusted. In a cable stayed bridge with MS-type cables, the cable force is commonly determined by a lift-off test as shown Figure 1. However, this test requires a great deal of time to measure the cable force. To solve the problem as above, the sensor using an optical fiber was recently developed. However, most FGB sensors were developed for an application to the cable of parallel wire strand (PWS) cables. In contrast, FGB sensors for MS type cable are rare. Thus, a real-time monitoring system using a FBG sensor required to measure the exact cable force of a MS-type cable.

In this paper, a monitoring system which consists of a package type FBG sensor, two types of jigs with and

Figure 1. Lift-off test.

without internal ribs and a data logger is developed and a field test is carried out on Second Dolsan cable-stayed bridge, MS-type cable, in Korea. The variation in both the short and long-term cable forces are stored by the data logger and the field test serves to verify that the developed sensor system with the FBG sensor is suitable for real-time cable force measurements during the construction and service phases.

In conclusion, the field test in the Second Dolsan cable-stayed bridge shows that there are no significant differences between the FBG sensor and load-cell in term of the cable forces which result in that the developed FBG sensor system is suitable for measuring the cable force in real time.

REFERENCES

Kim, J.M. 2008. A Methodology for Monitoring Prestressed Force of Bridges Using OFS-embedded Strand. Computational Structural Engineering Institute 21(3): 287–294.

Park, D.Y. et al. 2010. A development of smart monitoring system using a sensor based on IT. Seoul: Daelim Industrial Co., Ltd.

Zhou, Z. et al. 2004. A new kind of smart bridge cable based on FBG sensors. SPIE Proceddings 5502: 196–199.

Zhou, Z. et al. 2005. Applications of FRP-OFBG sensors on bridge cables. SPIE Proceddings 5765: 668–677.

Bridge Maintenance, Safety, Management, Resilience and Sustainability – Biondini & Frangopol (Eds)
 ISBN 978-0-415-62124-3

Strain visualization sticker using Moiré fringe for remote sensing

T. Takaki, K. Fujii & I. Ishii
Hiroshima University, Hiroshima, Japan

S. Umemoto, H. Ohata, N. Miyamoto & T. Okamto
Keisoku Research Consultant Co. Ltd., Hiroshima, Japan

ABSTRACT

Evaluation of the strain on a bridge's members is required to maintain its safety (Chang, S.J. et al. 2010) (Ni, Y. Q. et al. 2010) (Uchino, M. et al. 2010), and a large amount of research has been carried out by measuring such strain using electrical sensors, such as strain gauges. When such sensors are attached to measurement points, it is necessary to supply electrical power to these points. This implies that external power must be supplied via wiring or that an internal electrical power source must be fitted into the sensor. Furthermore, wired or wireless methods are needed to transmit the measurement data from the sensor. This tends to make the overall systems larger and more expensive.

To solve this problem, measurement techniques that do not require electrical power to be supplied to measurement points have been developed, utilizing optical elements or ultrasound, for example. In an earlier paper, we proposed a mechanism that utilizes moiré fringe patterns to visualize a physical force without the need for an electrical power supply at the measurement point (Takaki, T. et al. 2008, 2010).

In this study, we propose a strain visualization sticker that can display characters and fringe patterns that correspond to the magnitude of strain without the use of electronic elements, such as amplifiers, strain gauges and wires. We focus on the slight displacement produced by the strain and visually magnify and display its displacement with a moiré fringe. Figure 1 shows the concept of how to use the sticker. The sticker is simply attached to the bridge's members, and it provides strain information by displaying characters which can be seen with the naked eye. An accurate numerical value of the strain can be obtained from the image of the sticker through image processing. When using a telescopic lens, this value can be remotely measured. The structure of the sticker is simple, and its fabrication is inexpensive; therefore, our proposed measurement method using the sticker can be employed for multipoint measurement.

We assume that a slight displacement is generated by the strain, and results show that the sticker provides displacement information in the form of readable characters, giving accurate numerical value of displacement of less than 1 μm using image processing.

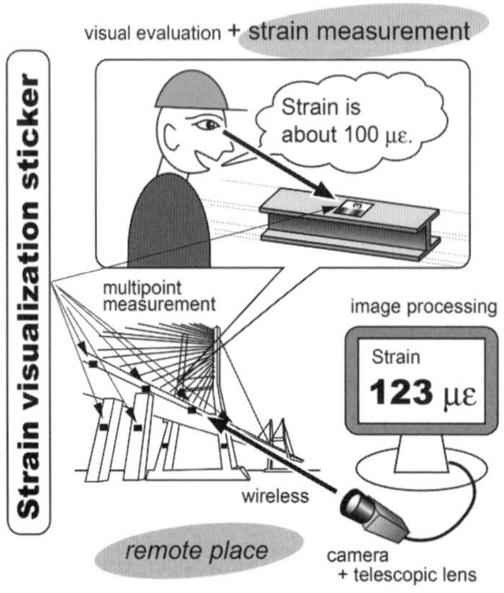

Figure 1. Concept of a proposed strain visualization sticker.

REFERENCES

Chang, S.J., Kim, N.S. & Kim, H.K. 2010. Prediction of displacement response of a suspension bridge using FBG strain sensors, Proceedings of the Bridge Maintenance, Safety, Management and Life-cycle Optimization, pp. 139–144, 2010.

Ni, Y. Q., Xia, H. W., Ko, J.M. & Wong, K.Y. 2010. Condition assessment of bridge deck truss using in-service monitoring data of strain, Proceedings of the Bridge Maintenance, Safety, Management and Life-cycle Optimization, pp. 799–805, 2010.

Uchino, M., Okamoto, T., Hida, K., Ito, Y., Sumitoro, P. & Matsuda, H. 2010. Strain analysis method using multi-rosette analysis by digital image correlation method, Proceedings of the Bridge Maintenance, Safety, Management and Life-cycle Optimization, pp. 2557–2562, 2010.

Takaki, T. Omasa, Y. & Ishii, I. 2008. Force Visualization Mechanism using moiré Fringe for Robot Grippers, Proceedings of the 28th Annual Conference of the Robotics Society of Japan, 1K3-07, 2008. (in Japanese)

Takaki, T. Omasa, Y. Ishii, I. Kawahara, T. & Okajima, M. 2010. Force Visualization Mechanism Using a moiré Fringe Applied to Endoscopic Surgical Instruments, Proceedings of the 2010 IEEE International Conference on Robotics and Automation, pp. 3648–3653, 2010.

Bridge Maintenance, Safety, Management, Resilience and Sustainability – Biondini & Frangopol (Eds)
© 2012 Taylor & Francis Group, London, ISBN 978-0-415-62124-3

Strain measurement of bridge members using strain visualization sticker

S. Umemoto, H. Ohata, N. Miyamoto & T. Okamoto
Keisoku Research Consultant Co Ltd., Hiroshima, Japan

T. Takaki, K. Fujii & I. Ishii
Hiroshima University, Hiroshima, Japan

ABSTRACT

Rational, cost-effective maintenance and control of aging social infrastructure such as bridges and other structures has now become an important issue, heightening the need for efficient, effective techniques for investigating the soundness of structures. The main type of periodic inspection used in investigations of structural soundness is visual inspection of external appearance. If the stresses acting on structural members due to external forces could be determined at the same time during these routine inspections, this would provide useful information for assessing structural soundness.

Stress measurements of structural members are generally made using electrical sensors such as strain gauges. When electrical sensors are installed at a number of points, it is necessary to supply external power to each measurement point via the signal cable, or alternatively, to provide an internal power supply in the sensors. The measured data from the sensors must also be transmitted by a cable or wireless method. These requirements tend to increase the scale and cost of measurement systems. In order to solve this problem, the authors propose a "Strain Visualization Sticker" which is capable of displaying characters corresponding to the amount of strain without using conventional electrical elements such as amplifiers, signal cables, and strain gauges. Based on the principle of the Moiré fringe, the device displays a visible enlarged pattern showing the micro-displacement caused by strain, thereby providing strain information in characters which are visible to the naked eye. Precise values of strain can also be obtained from sticker images by image processing, and strain values can be measured remotely using a camera with a telescopic lens.

A tensile test of a test specimen and bending test of a bridge member were carried out using the Strain Visualization Sticker. Accurate strain information was obtained non-contact in both tests. In the tensile test, error was less than 10 $\mu\varepsilon$, showing accuracy equal to that of the conventional strain gauge method, while error in the bending test was as small as 30 $\mu\varepsilon$. The visibility of strain information provided by the Strain Visualization Sticker was also investigated in the bending test, demonstrating that approximate strain information can be read with the unassisted eye. These results demonstrated that the Strain Visualization Sticker can provide an effective tool for evaluations of structural soundness.

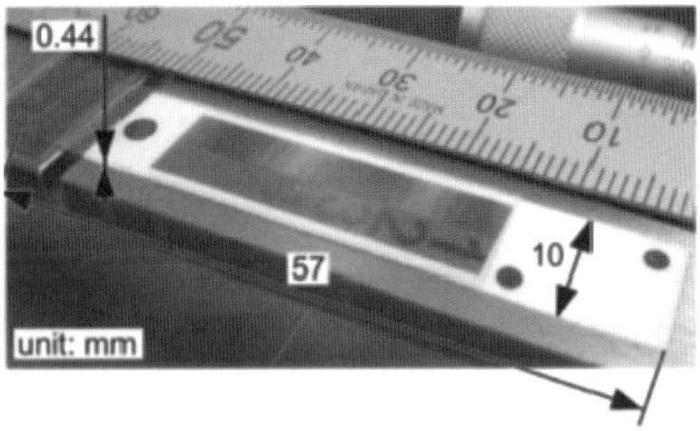

Figure 1. The developed strain visualization sticker.

REFERENCES

Takaki, T. Omasa, Y. & Ishii, I. 2008. *Force Visualization Mechanism using moiré Fringe for Robot Grippers*, Proceedings of the 28th Annual Conference of the Robotics Society of Japan, 1K3-07, 2008. (in Japanese)

Takaki, T. Omasa, Y. Ishii, I. Kawahara, T. & Okajima, M. 2010. *Force Visualization Mechanism Using a moiré Fringe Applied to Endoscopic Surgical Instruments*, Proceedings of the 2010 IEEE International Conference on Robotics and Automation, pp. 3648–3653, 2010.

New developments on the bridge safety, maintenance and management in Mexico

Organizer: D. De León

Bridge Maintenance, Safety, Management, Resilience and Sustainability – Biondini & Frangopol (Eds)
ISBN 978-0-415-62124-3

Time variation of bridges structural reliability due to corrosion in Mexico

D. De León, C.A. González-Pérez, S. Díaz, D. Delgado & J.C. Arteaga
Autonomous University of Mexico State—Faculty of Engineering, Toluca, Mexico State, Mexico

ABSTRACT

Risk is defined in terms of the bridge's probability to attain or exceed a specified limit state that characterizes moderate damage (life safety or damage control), severe damage (collapse prevention), or collapse, to identify bridge vulnerability, and the cost of failure consequences. In Mexico, there are bridges where the corrosion attack plays an important role on the bridge capacity and, therefore, its structural reliability. In this paper the time variation of the Cornell's reliability index for a simply supported beam of a vehicles bridge is presented. This is made through the behavior analysis on the changes in the cracking moment capacity of the prestressed concrete box cross section, obtained from its bending moment-curvature curve. The only damaging factor considered here is the corrosion on the beams prestressed steel. In the statistical analysis, Monte Carlo simulation techniques were used considering the concrete strength, the prestressed steel area and the live load as random variables whereas the members geometry and dead load was deterministic. At the corrosion initiation of prestressed steel, it was observed that the variation on the annual reliability index was not significant within the first four years. However, in the subsequent years, the reliability index decreased to values under 1.75 at the six years after corrosion initiation, which is not considered as acceptable for the structure operation and it is necessary to recommend restoration works to increase the reliability index to the level of at least 3.

Further research may lead to the development of optimal inspection and maintenance schedules for bridges under corrosion attack in Mexico.

REFERENCES

Frangopol, D., Furuta, H., Kameda, T., Ankara, K. and Takahashi, Y., 2006. Optimal Bridge Maintenance Planning using Improved Multi-Objective Genetic Algorithm. *Structure and Infrastructure Engineering*, Ed. Taylor & Francis, 2(1): 33–41.

Ang, Alfredo H-S. and De Leon, D., 2005. Modeling and analysis of uncertainties for risk-informed decisions in infrastructures engineering, *Journal of Structure and Infrastructure Engineering*, 1(1): 19–31.

Torres Acosta, A. A. and Martinez Madrid, M., 2001. *Design of Concrete Structures with Durability Criterias.* Secretary of Communications and Transportation. Mexican Institute Transportation. *Technical Publication 181*, Sanfandila, Qro., 2001, 87 (in Spanish).

Bridge Maintenance, Safety, Management, Resilience and Sustainability – Biondini & Frangopol (Eds)
© 2012 Taylor & Francis Group, London, ISBN 978-0-415-62124-3

Parametric study of bridges with substructure irregular conditions

M.C. Gómez-Soberón & D.J. Salas-Mengchún
Materials Department, Universidad Autónoma Metropolitana

ABSTRACT

Some of the proposed preliminary evaluation methods for bridges consider irregularity of the substructure as a relevant parameter (Gómez *et al.*, 2002). To determine the different irregularity conditions influence in bridge substructures, parametric analyses of monolithic, continuous, and simple-supported linear elastic models and simple-supported no-linear models were conducted. In these models, irregularity was considered by varying center and extreme pier height in a simple bridge (Priestley *et al.*, 1996). The percentages of variation of pier height were of +25%, +50%, +75%, −25%, −50%, and −75%. Bridges were subjected to a database of 53 earthquakes recorded in one of most seismic activity zones in Mexico. These analyses were used to record maximum displacements and mechanical elements (elastic) and a damage index (no-linear), with them the normalized difference in percentage, between regular and irregular structures, was obtained.

Based on the results, graphs were plotted showing the normalized differences between regular and irregular models by quartiles, which represent 25, 50, 75, and 100% of the data, respectively. The obtained results show that increasing central pier height increases the variation between displacements for the irregular models, compared with the regular bridge. That is, bridges are more vulnerable when the difference between their piers height is greater, although increasing the length produces greater dispersion than reducing it, as it can be observed in Figure 1. It is also shown that the displacements variation is not linear with the variation in the center or extreme pier height, but approaches a quadratic polynomial function. It is possible to say that a bridge with variations in the central pier height is more vulnerable, than other with variations of near-the-abutment pears, as can be deduce of the comparison of Figures 1 and 2. Trends are similar for monolithic, continuous and simple-supported bridges. With no-linear analysis similar trends for normalized damage index were defined, so linear analysis represents an adequate procedure to define the influence of substructure irregularity.

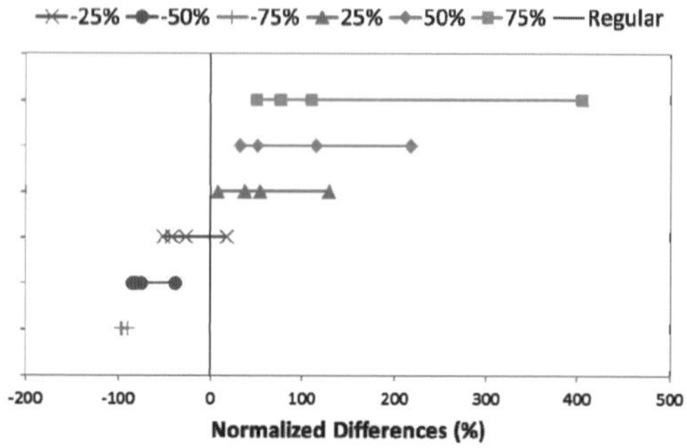

Figure 1. Displacement normalized differences by quartiles. Monolithic model, variation of the central pier.

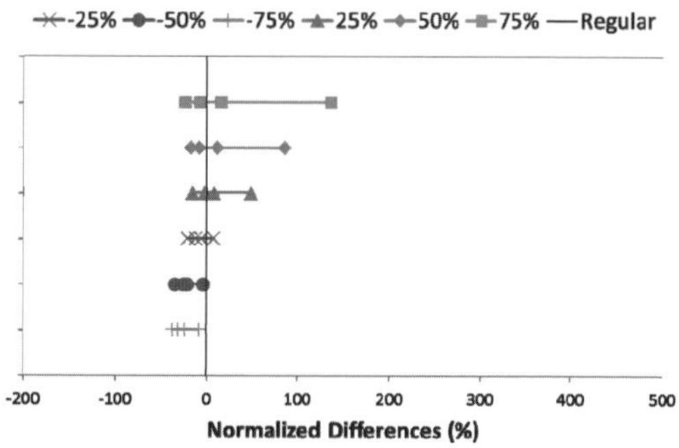

Figure 2. Displacement normalized differences by quartiles. Monolithic model, variation of the extreme pier.

REFERENCES

Gómez, C, Barbat A H & Oller, S, 2002. Seismic vulnerability of bridges using simplified models. *Monograph Series IS-47 International Center of Numeric Methods in Engineering, CIMNE*, Barcelona, Spain.

Priestley, M J, Seible, N & G M Calvi, 1996, *Seismic design and retrofit of bridges*. John Wiley & Sons, New York.

Bridge Maintenance, Safety, Management, Resilience and Sustainability – Biondini & Frangopol (Eds)
© 2012 Taylor & Francis Group, London, ISBN 978-0-415-62124-3

Simplified revision of bridge structural types on seismic zones. Specific cases on Oaxaca, Guerrero, Michoacan, Colima and Mexico State

H. Hernández-Barrios
Engineering School, Universidad Michoacana de San Nicolás de Hidalgo, Michoacan State, Mexico

D. de León-Escobedo
Engineering School, Universidad Autonoma del Estado de Mexico, Mexico State, Mexico

ABSTRACT

In Mexico, the free toll federal highway network has a length of more than 45,000 km and it is being operated by the Communications and Transportation Secretary (SCT). Oaxaca, Guerrero, Michoacan, Mexico State and Colima are located on a potentially seismic zone and some bridges there have been designed without any seismic consideration.

Oaxaca State count with a free toll highways system distributed in 8 routs and with 475 total bridges and it is the Mexican State with most bridges in the country; Guerrero State has a 6 routs with 386 total bridges; Michoacan State has a 12 routs with 473 total bridges; Mexico State has a 286 total bridges, and Colima State has a 81 total bridges (Table 1). All of them are important for the social and economical development of the region and almost of them are on a seismically active zone.

Table 2 shows the period of construction and number of bridges by State. It can see that 1125 bridges (66%) were built before the 70's and were designed according to load and seismic provision from AASHO. The AASHO for many years included in their loading the HS20 truck, and semi-trailer combination.

In this paper, a bridge classification based on construction year, substructure type, number of spans and substructure material is presented. Visually inspected damages are used for a preliminary assessment of safety conditions and repair/maintenance actions are proposed for each class of bridge importance and damage level.

Table 1. Number of bridges by State.

State	Number of bridges
Oaxaca	475
Guerreo	386
Michoacan	473
Mexico State	286
Colima	81
Total	1701

Table 2. Period of construction and number of bridges by State.

State	Period of construction (year)				
	1910	1920	1930	1940	1950
Oaxaca	1	0	12	66	31
Guerrero	0	1	1	3	4
Michoacan	0	0	2	63	62
Mexico State	0	0	2	11	16
Colima	0	0	2	0	7
Total number	1	1	17	143	120
	1960	1970	1980	1990	2000
Oaxaca	54	189	16	41	65
Guerrero	229	73	11	38	26
Michoacan	63	105	52	100	26
Mexico State	15	81	16	72	73
Colima	8	26	7	8	25
Total number	369	474	102	259	215

In the future, a probabilistic damage assessment and life-cycle economic consequences appraisal will serve to improve the evaluation procedures and to set optimal maintenance strategies.

REFERENCES

Alfawakhiri, F. & Bruneau, M. 2000. Flexibility of superstructures and supports in the seismic analysis of simple bridges. *Earthquake Engineering and Structural Dynamics*, 29: 711–729.

Hernandez, H. & De León N. 2010. Seismic vulnerability of free toll bridges in Michoacan state, Mexico. *Bridge Maintenance, Safety, Management and Life-Cycle Optimization*, Frangopol, Sause & Kusko (eds), Taylor & Francis.

Hernández, H. & Hurtado F. 2010. Caracterización de los puentes vehiculares del estado de Guerrero con fines de evaluación sísmica. *XVII Congreso Nacional de Ingeniería Estructural*, León, Guanajuato, México (in Spanish).

Bridge Maintenance, Safety, Management, Resilience and Sustainability – Biondini & Frangopol (Eds)
© 2012 Taylor & Francis Group, London, ISBN 978-0-415-62124-3

Dynamic characterization of highway bridges

M.E. Ruiz Sandoval
Universidad Autónoma Metropolitana, Mexico

G. Martínez, R. Rojas, A.R. Sánchez, E.O. Navarro, J.M. Jara & B.A. Olmos
Universidad Michoacana de San Nicolás de Hidalgo, Morelia, Mexico

J. de la C. Tejeda
Universidad de Colima, Mexico

ABSTRACT

A nation's infrastructure is one of the pillars for its economic development. An adequate supervision is required to anticipate problems and to implement schedule maintenance. The monitoring processes are not very common in our country. In addition to its supervision, it is suggested to carry out seismic vulnerability studies, that in an seismic event, would allow to determine which structures need to be inspected in first instance.

As part of Mexican federal effort, a group of three public institutions were given the task to carry out the construction of fragility curves for 3 highway bridges. This will take place in different stages which are: the instrumentation and dynamic characterization, risk assessment and earthquake vulnerability and finally fragility curves of selected structures.

The aim of this paper is to present only the first stage of the work, that consist in the instrumentation and dynamic characterization of 3 road bridges, along with their mathematical models.

The bridges selected were on the Morelia – Guadalajara highway, one of the most important of the country. In the first part, the bridges are described both in its geometry and structural configuration, as well as the characteristics of their construction materials. It can be said that, in general, the structures are supported by intermediate columns, and have more than 2 spans. The vast majority are skew. All of the bridges studied are composed of AASHTO girders, and allocate two traffic lines. With this basic information, mathematical models of the structures were constructed in commercial software.

The following part of the paper describes the different data acquisition systems and sensors that were used over the structures, as well as their location. A complete description of the type of amount and type of data is presented. Different types of data signal processing were used to determine the dynamic characteristics (periods and modal parameters) of the bridges. Results are presented and compared.

Finally, it is presented the calibration of the mathematical models. Detailed descriptions of the parameters adjusted, and under what assumptions were changed.

REFERENCES

Computers and Structures, Inc. (2010) http://www.csiberkeley.com

Gómez R, Muria-Vila D, Sanchez-Ramirez R, Escobar JA (1997). "Non destructive structural evaluation of two bridge superstructures" Conference Information: Mouchel Centenary Conference on Innovation in Civil and Structural Engineering. pp. 121–132

Kinemetrics Inc. (2010) http://www.kinemetrics.com

PCB Piezotronics, Inc (2010) http://www.pcb.com

Manual de Obras Civiles de la Comisión Federal de Electricidad "Diseño por sismo" (2008)

Morassi A, Tonon S (2008) "Dynamic Testing for Structural Identification of a Bridge" Journal of bridge engineering 13(6) 573–585.

Muriá, D. y R. González. (1995). "Propiedades dinámicas de edificios de la ciudad de México". Revista de Ingeniería Sísmica. (51) 25–45.

O. Ortiz, J. Téllez, F.J. Burgos, A. Patrón, E. Reyes, V. Robles, C. Cremona y M.E. Ruiz-Sandoval (2008) "Load capacity assessment of "Antonio Dovali Jaime" bridge using static and dynamic tests", Proceedings of International fib Symposium 2008 (FIB), Amsterdam, Holland.

Non deterministic schemes for structural safety and reliability of bridges

Organizer: S. Arangio

Bridge Maintenance, Safety, Management, Resilience and Sustainability – Biondini & Frangopol (Eds)
© 2012 Taylor & Francis Group, London, ISBN 978-0-415-62124-3

Redundancy of highway bridge decks

G. Anitori & J.R. Casas
Technical University of Catalonia, Barcelona, Spain

M. Ghosn
The City College of New York, USA

ABSTRACT

Structural redundancy is defined, according to Ghosn (1998), as the capability of a structure to continue to carry load after the failure of one main member.

Typical design is based on a member by member basis and little attention is provided to the behavior of the structure after the failure of one member.

A methodology, based on Ghosn (1998) and Liu (2000), is applied to evaluate the redundancy of bridge system, representative of a typical design solution and calibrate system factors. System factors can be used in design equations to take into account the system effect and take advantage of the capability of the structure to redistribute loads beyond the design capacity.

The structural typology considered herein is the simply-supported span with precast prestressed box-beams (figure 1).

Figure 1 also shows the numerical approach to the problem that consists in a plane grillage based on Hambly (1991). Material non-linearities are taken into account by using realistic relationships for the material constituents and therefore obtain realistic moment curvature curves for principal and secondary members of the deck.

Damage scenarios are also investigated, to consider the behavior of the system after the failure of one main member due to exceptional reasons. The damage scenarios are represented by the partial failure of the most loaded external beam.

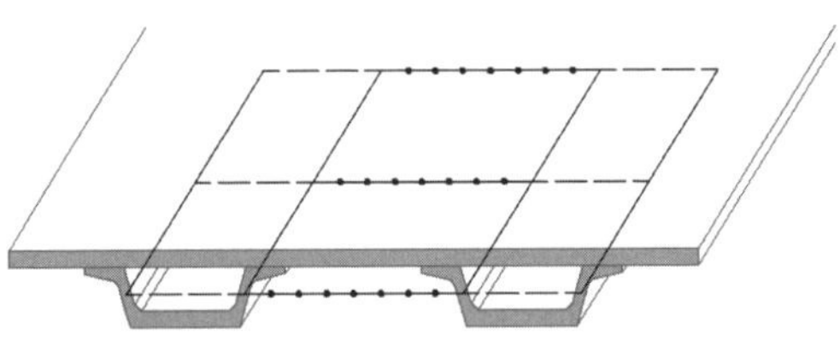

Figure 1. Structural typology and frame elements configuration.

Table 1. System factors.

		# of beams	
Span (m)	Beam spacing (m)	3	2
36	4.50	1.29	1.00
	5.25	1.19	0.93
	6.00	1.13	0.87
30	4.50	1.37	1.18
	5.25	1.28	1.15
	6.00	1.21	1.21
21	4.50	1.69	1.27
	5.25	1.28	1.25
	6.00	1.25	1.09

The probabilistic analysis used to characterize the resistance of the structural systems are based on statistical data of the materials. In order to calculate the reliability indices a FORM model has been implemented, and then the results have been used in an iterative algorithm needed to calculate the system factors.

Table 1 shows the system factors calculated for the different parameters defining the bridge. Linear interpolation can be used to fit a specific bridge configuration.

REFERENCES

Ghosn, M. & Moses, F. 1998. *Redundancy in highway bridge superstructures NCHRP report 406*, Transportation research board, Washington DC.

Hambly, E.C. 1991. *Bridge deck behavior*, Taylor & Francis, New York.

Liu, W. D., Ghosn, M., Moses, F., & Neuenhoffer A. 2000. *Redundancy in highway bridge substructures*, NCHRP Report 458, Transportation Research Board, Washington DC.

Bridge Maintenance, Safety, Management, Resilience and Sustainability – Biondini & Frangopol (Eds)

Bayesian neural networks for damage identification of a cable-stayed bridge

S. Arangio & F. Bontempi
Sapienza University of Rome, Rome, Italy

ABSTRACT

In recent years there has been a growing interest on the application of soft computing methods for processing the large quantity of data coming from long term structural health monitoring. In particular, this work deals with the applicability of Bayesian neural networks for damage identification of a cable-stayed bridge. Bayesian neural networks come from the optimization of the neural networks model by using Bayesian inference at four hierarchical levels (Arangio and Beck, 2010).

The selected structure is a real bridge proposed as benchmark problem by the Asian-Pacific Network of Centers for Research in Smart Structure Technology (ANCRiSST). They shared data coming from the long term monitoring of the bridge with the Structural Health Monitoring community in order to assess the current progress on damage detection and identification methods with a full scale example. The dataset includes vibration data before and after the bridge was damaged; this is a rare case of an instrumented bridge that has been damaged.

The monitoring system include 14 unixial accelerometers on the bridge deck. The available data include two dataset: the first one (health structure) consist of 1 hour registrations that have been repeated for the 24 hours by the 14 sensors on January 17, 2008. The sampling frequency is 100 Hz. The second one includes measurements in the same points with the same instruments carried out some months later.

The available data have been used to test a Bayesian neural networks-based damage detection strategy and this work summarizes the preliminary results of the analyses. The proposed strategy was originally proposed in (Arangio & Beck, 2010, Arangio & Bontempi, 2010). It consists in building different neural networks, one for each measurement point and for each hour of measurements. The neural network models are built and trained using the time-histories of the accelerations of the selected points in the undamaged situation. The purpose of these models is to approximate the behavior of the health bridge. The trained models are then tested with new time histories of the response, both in health and damaged conditions; if the error in one or more points is large, the presence of an anomaly (that can represent or cannot represent damage) is detected.

The proposed method is able to detect anomalies on the behavior of the structure, which could be related to the presence of damage but at the moment it is not able to identify clearly the damaged zone. Current studies are finalized at the optimization of the strategy in this sense.

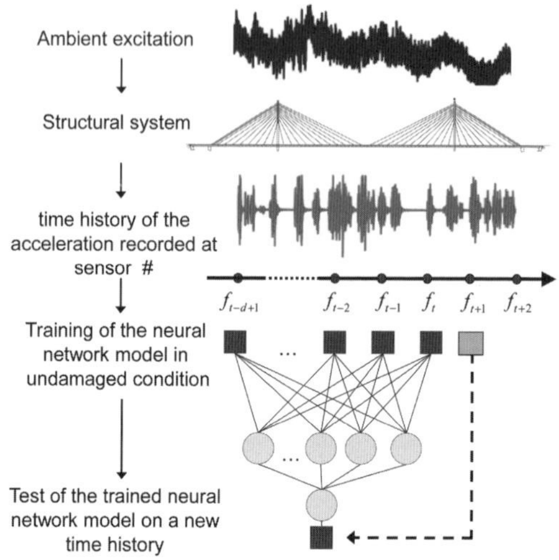

Figure 1. Scheme of the proposed damage detection strategy.

REFERENCES

Arangio S., Beck J.L. 2010. Bayesian neural networks for bridges integrity assessment, *Structural Control & Health Monitoring*, published online from November 2010, DOI: 10.1002/stc.420.

Arangio S., Bontempi F. 2010. Soft Computing based Multilevel Strategy for Bridge Integrity Monitoring, *Computer-Aided Civil and Infrastructure Engineering*, 25, 348–362.

Bridge Maintenance, Safety, Management, Resilience and Sustainability – Biondini & Frangopol (Eds)
© 2012 Taylor & Francis Group, London, ISBN 978-0-415-62124-3

Dynamic load allowance for capacity rating of prestressed concrete girder bridges based on reliability studies

L. Deng
College of Civil Engineering, Hunan University, Changsha, China

C.S. Cai & M. Barbato
Department of Civil and Environmental Engineering, Louisiana State University, Baton Rouge, Louisiana, USA

ABSTRACT

The current highway bridge design in the United States follows the AASHTO Load and Resistance Factor Design (LRFD) specifications (AASHTO 2004), which employs a dynamic load allowance, *IM*, equal to 0.33 for the dynamic effect of vehicular loading. Previous studies have shown, through both field tests and numerical simulations, that the *IM* value of 0.33 may underestimate the vehicle dynamic effect under poor road surface conditions (RSCs).

In this paper, the *IM* is explicitly modeled as a random variable with different statistical properties (taken from Deng and Cai 2010) for different RSCs. The reliability indices of a selected group of prestressed concrete girder bridges, designed using the AASHTO LRFD code, are calculated for both moment and shear strength limit states. The results relative to the moment strength limit state are shown in Fig. 1. It is found that while the calculated bridge reliability indices are usually above the target reliability index value of 3.5 for above average RSCs, they can be significantly below the target value of 3.5 when the RSCs are below average.

By selecting a group of *IM* values for each RSC and calculating the reliability indices of the selected bridges with respect to the selected *IM* values, an appropriate *IM* value is suggested for each RSC to achieve a consistent target reliability index. The suggested *IM* values are shown in Table 1, together with the minimum values needed to ensure a reliability index larger or equal to 3.5 for all cases considered.

The results presented in this paper are particularly valuable for rating of existing prestressed concrete girder bridges, for which the actual RSCs can be directly evaluated. The RSCs need to be taken into account properly in order to accurately estimate the actual safety of the considered bridge.

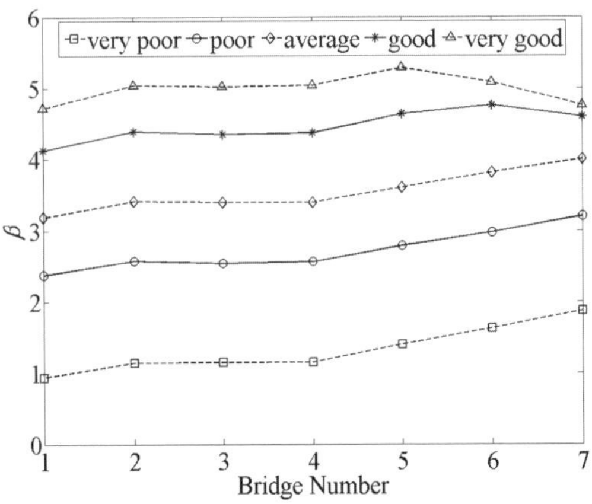

Figure 1. Reliability indices relative to the moment strength limit state the for the seven benchmark bridges designed following the LRFD AASHTO specifications and according to the five road surface conditions defined in ISO 1995.

Table 1. Dynamic load allowance *IM*s obtained in this study.

Road surface condition	Theoretical minimum *IM*s	Proposed *IM*s
Very poor	2.50	2.40
Poor	1.00	0.90
Average	0.50	0.33
Good	0.15	0.20
Very good	0.05	0.10

REFERENCES

American Association of State Highway and Transportation Officials (AASHTO) 2003. *Manual for Condition Evaluation and Load and Resistance Factor Rating (LRFR) of highway bridges.* AASHTO, Washington, DC.

American Association of State Highway and Transportation Officials (AASHTO) (2004). *LRFD Bridge Design Specifications.* AASHTO, Washington, DC.

Deng, L. & Cai, C.S. 2010. Development of dynamic impact factor for performance evaluation of existing multi-girder concrete bridges. *Engineering Structures* 32:21–31.

International Organization for Standardization (ISO) 1995. Mechanical vibration-Road surface profiles-Reporting of measured data. *ISO 8068: 1995 (E).* ISO, Geneva.

Bridge Maintenance, Safety, Management, Resilience and Sustainability – Biondini & Frangopol (Eds)
© 2012 Taylor & Francis Group, London, ISBN 978-0-415-62124-3

Application of orthogonal decomposition approaches to long-term monitoring of infrastructure systems

E. Kallinikidou & S.F. Masri
University of Southern California, Los Angeles, USA

A.K. Jain & H.-B. Yun
University of Central Florida, Orlando, USA

L.-H. Sheng
California Department of Transportation, CA, USA

ABSTRACT

The ability to monitor the condition of bridges and other civil infrastructure systems is of primary interest to various agencies and the engineers responsible for the maintenance of such civil infrastructure. The development of advanced structural health monitoring (SHM) technologies is an active, ever growing research field, which is being pursued by numerous researchers world-wide. The intent of the SHM implementation is to facilitate the early detection of a possible structural deterioration, with the goal to extend the life-span of structures. Timely information gathered from such SHM technologies help in the corrective and urgent decision making process to ensure a safe operational condition for the monitored structures. Currently, state of the art technologies applied to SHM applications, enable the collection and dissemination of large amounts of data; however, the development of analytical tools and data management methodologies, with the capacity to analyze and extract useful information, for the condition assessment of monitored structures, poses an emerging and challenging research area. There is a need for new analytical tools designed to enable data archiving and measurement "stacking" for data compaction.

The long-range monitoring of civil infrastructure systems monitored with dense sensor arrays that are capable of generating voluminous amounts of data from continuous online monitoring requires the implementation of a proper data processing and archiving scheme to maximize the benefits of structural health monitoring operations. This paper focuses on the areas of data-management, data quality control, and feature-extraction of meaningful parameters to describe large-scale infrastructure systems' response to ambient excitation in the context of SHM. Recordings from the monitoring system installed on the Vincent Thomas Bridge (VTB) in San Pedro, California form the database of the proposed data-management and archiving methodology.

The data processing methodology for the VTB is based on the calculation of the sensor array acceleration covariance matrices for every hour of available data, and the subsequent orthogonal decomposition of the covariance matrices. The dominant proper orthogonal modes of the bridge are determined and their statistical variations over an extended observation period covering several months of continuous data are quantified and analyzed. The empirical probability density functions for the mean daily bridge accelerations are computed and used to compare the statistical variations in different periods of operation of the bridge (working days, weekends, and holidays). It is shown that the computed statistical distributions of the bridge response can provide a quantitative baseline through which to facilitate the early detection of any anomalies indicative of a possible structural deterioration due to either fatigue (service loads), or extreme loading events, i.e., earthquake, man-made, or any other natural hazard.

REFERENCES

Abdel-Ghafar, A. M. et al., 1992. Seismic performance evaluation of suspension bridges. *Proceedings of the Tenth World Conference on Earthquake Engineering:* 4845–4850.

Smyth, A.W., Pei, J.-S., and Masri, S. F. 2003. System identification of the Vincent Thomas Suspension Bridge using earthquake records. *Earthquake Engineering and Structural Dynamics*, 32: 339–367.

Yun, H., Nayari, R., Tasbihgoo, F., Wahbeh, M., Caffrey, J., Wolfe, R., Nigbor, R., Masri, S., Abdel-Ghaffar, A., and Sheng, L., 2008. Monitoring the collision of a cargo ship with the vincent thomas bridge. *Structural Control and Health Monitoring*, (15): 183–206.

Bridge Maintenance, Safety, Management, Resilience and Sustainability – Biondini & Frangopol (Eds)
 ISBN 978-0-415-62124-3

Strength and reliability of FRP-reinforced concrete beams

S.E.C. Ribeiro
Department of Materials and Construction, Federal University of Minas Gerais, Belo Horizonte, Brazil

S.M.C. Diniz
Department of Structural Engineering, Federal University of Minas Gerais, Belo Horizonte, Brazil

ABSTRACT

Reinforced concrete (RC) bridges are often subjected to deicing salts or in a marine environment. As such, a major problem to the durability of these structures is the corrosion of reinforcing steel. In this light, fiber reinforced polymers (FRP), as noncorrosive materials, provide a promising prospect for use as reinforcement in concrete bridges construction. Although the use of FRP as structural reinforcement shows great promise in terms of durability, the characteristics of these materials have led to new challenges in the design of FRP-reinforced concrete (FRP-RC) components.

In recent years, there has been a growing interest on high performance materials such as high-strength concrete and FRP, among others. These materials may offer not only greater durability but also higher resistance and, as a consequence, potential gains throughout the life-cycle of the structure. Moreover, the specific advantages of each material may be combined in innovative systems as in the case of FRP-RC beams and slabs. For instance, it has been reported that the high tensile strength of FRP is most efficiently used when paired with high-strength concrete (Nanni, 1993).

In conventional RC beam design, failure is dictated by yielding of steel, thus resulting in a ductile failure. In the case of FRP-RC beams where two brittle materials are involved, a fragile failure is unavoidable. As a result, a change in the RC beam design paradigm (under-reinforced beams) is necessary. Therefore, due to differences between the mechanical properties of steel and FRP, the reliability of FRP-RC beams shall be evaluated.

Since most of the variables involved in the project (mechanical properties of concrete and FRP, geometric characteristics, loads, etc.) are random, probabilistic methods are required to assessing the reliability of these structural elements. Most of the suggestions proposed for the design of FRP-RC beams are based on a deterministic point of view; only few works have used probabilistic methods in the reliability checking of design recommendations for FRP-RC structures, e.g. Diniz (2007) and Shield et al. (2011). However, it should be observed that the work presented in Shield et al. is limited to 28 MPa specified concrete compressive strength, and obviously not addressing the consequences of using high-strength concrete.

In this study, a contribution to the development of semi-probabilistic design recommendations for FRP-RC beams is reported. The strength and reliability of 81 FRP-RC beams designed according to ACI-440 (2006) are assessed. Monte Carlo simulation is used in the probabilistic description of the beam strength and failure mode, and in the computation of the probability of failure of the designed beams with respect to the ultimate flexural strength. Special attention is given to the deterministic procedure for the computation of FRP-RC beam resistance, thus extending the investigation presented in Diniz (2007).

The results of this study point to the possibility that reduced levels of reliability can be achieved, in particular for the combination of higher concrete compressive strength, lower FRP tensile strength and smaller dead to live load ratios.

REFERENCES

ACI Committee 440 (2006). *Guide for the Design and Construction of Structural Concrete Reinforced with FRP Bars*. ACI 440.1R-06.

Diniz, S. M. C. (2007). Reliability Evaluation of FRP Reinforced Concrete Beams. *Computational Stochastic Mechanics – CSM5*, Rhodes, Greece, June 21–23, 2006, Mill Press, pp. 215–225.

Nanni, A. (1993). Flexural Behavior and Design of Reinforced Concrete Using FRP Rods, *Journal of Structural Engineering*, 119(11), pp. 3344–3359.

Shield, C.K., Galambos, T.V., and Gulbrandsen, P. (2011). On the History and Reliability of the Flexural Strength of FRP Reinforced Concrete Members in ACI 440.1R, *SP-275: Fiber-Reinforced Polymer Reinforcement for Concrete Structures*.

Bridge Maintenance, Safety, Management, Resilience and Sustainability – Biondini & Frangopol (Eds)
© 2012 Taylor & Francis Group, London, ISBN 978-0-415-62124-3

Back analysis for earthquake damaged bridges. Part I: General procedure

P.E. Sebastiani, P. Franchin, F. Petrini & F. Bontempi
Department of Structural and Geotechnical Engineering, Sapienza University of Rome, Rome, Italy

ABSTRACT

This paper and its companion (*Sebastiani et al.*, 2012) present a back analysis method for the interpretation of earthquake damage pattern of bridges. The a posteriori damage evaluation is based on a number of finite element analyses carried out with the aim of bounding the structural and geotechnical configurations. This first part presents the methodological aspects of the procedure (Fig. 1), comprised of a series of investigations (Fig. 2) and modeling (Fig. 3) steps. Regarding the investigations, they consist in: i) geological and geotechnical surveys to evaluate soil properties, ii) studies to characterize the occurred ground motion, iii) structural investigations to establish the structural conditions of the bridge before and after the occurred earthquake. Afterwards, a set of models to evaluate the local site effects, the soil-structure interaction and the structural behavior by means of nonlinear time history analysis is employed (Fig. 4).

All necessary steps to reproduce the damage scenario are discussed, with flowcharts representing the various analysis phases.

The methodology does not only aim at identifying the causes of damage, but also at bounding the occurred damage within the range of scenarios corresponding to the problem uncertainties.

To this end, different structural, seismic and geotechnical models are compared in order to evaluate the influence of uncertainties on the results.

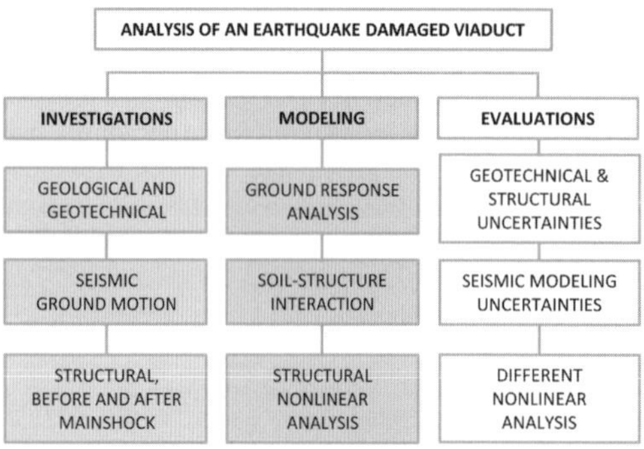

Figure 1. Flowchart of general procedure.

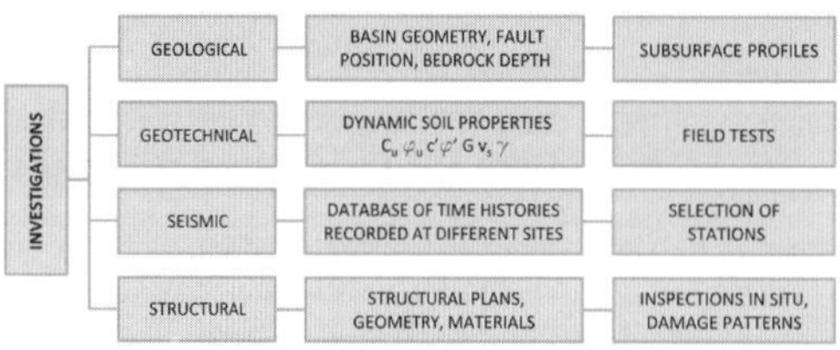

Figure 2. Flowchart of surveys.

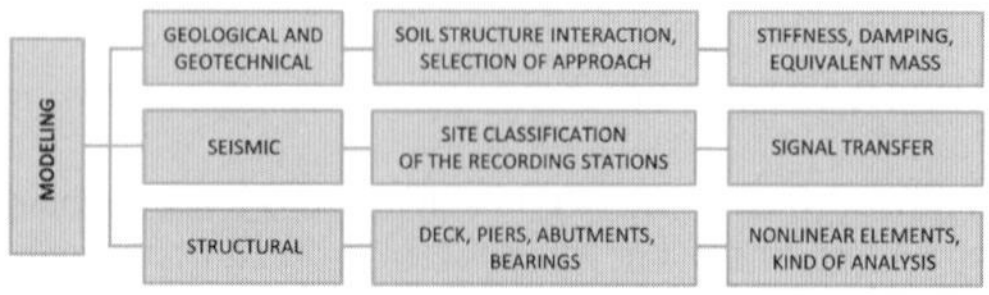

Figure 3. Flowchart of modeling.

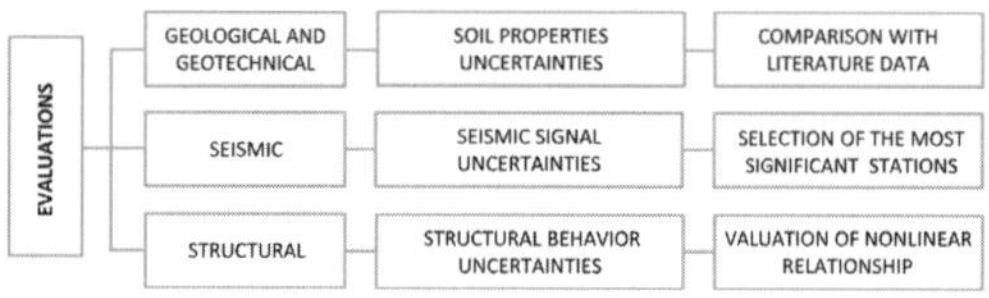

Figure 4. Flowchart of evaluations.

REFERENCES

Pinto P.E., Franchin P. (2010) Issues in the Upgrade of Italian Highway Structures.

Sebastiani P.E., Franchin P., Petrini F., Bontempi F. (2012) Back Analysis for Earthquake Damaged Bridges. Part II: application to a viaduct damaged in the April 6th, 2009 L'Aquila earthquake.

Taherzadeh R., Clouteau D., Cottereau R. (2009) Simple formulas for the dynamic stiffness of pile groups.

Bridge Maintenance, Safety, Management, Resilience and Sustainability – Biondini & Frangopol (Eds)
© 2012 Taylor & Francis Group, London, ISBN 978-0-415-62124-3

Back analysis for earthquake damaged bridges. Part II: Application to a viaduct damaged in the April 6th, 2009 L'Aquila earthquake

P.E. Sebastiani, F. Petrini, P. Franchin & F. Bontempi
Department of Structural and Geotechnical Engineering, Sapienza University of Rome, Rome, Italy

ABSTRACT

In order to verify the applicability of the methodology illustrated in the companion paper (*Sebastiani et al.*, 2012) to practical engineering problems, a case study is presented.

The examined structure is an existing curved highway viaduct, named "Popoli viaduct", located in the Abruzzo region, central Italy (Fig. 2). The bridge was built during the decade from 1970 to 1980 and it wasn't equipped with seismic protection devices, so the decks are simply supported by steel cylindrical bearings (Fig. 3). The viaduct is situated in an alluvial basin and it was damaged by the L'Aquila earthquake on April 6th, 2009.

Starting from the observed damage suffered by the bridge, this paper presents an example of the Back Analysis method presented in the companion paper, with a comparison between damage patterns and model results (Fig. 1–4)

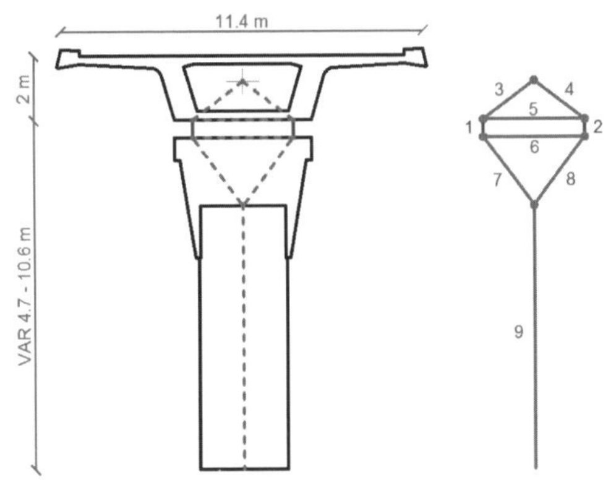

Figure 1. Pier-deck connection modeling.

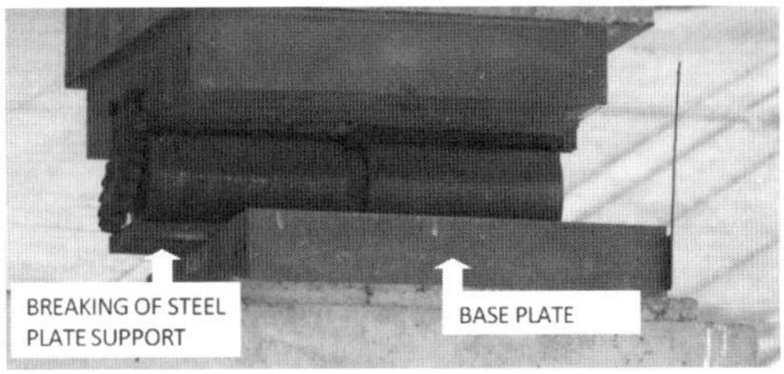

Figure 3. Breaking of bearings.

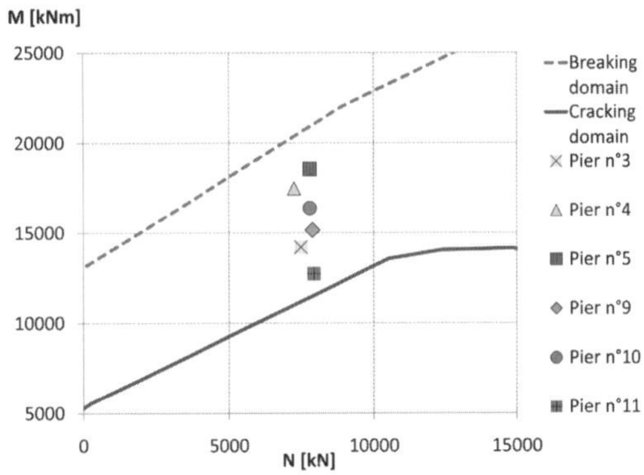

Figure 4. Domains of resistance for the piers.

REFERENCES

Pinto P.E., Franchin P. (2010) Issues in the Upgrade of Italian Highway Structures.

Sebastiani P.E., Franchin P., Petrini F., Bontempi F. (2012) Back Analysis for Earthquake Damaged Bridges. Part I: general procedure.

Taherzadeh R., Clouteau D., Cottereau R. (2009) Simple formulas for the dynamic stiffness of pile groups.

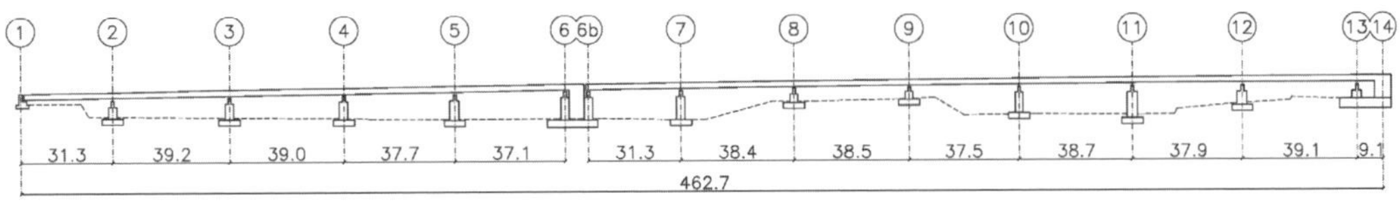

Figure 2. Popoli viaduct – side view.

Structural control of bridges and footbridges: Extreme and every-day events

Organizers: L. Martinelli & M. Domaneschi

Bridge Maintenance, Safety, Management, Resilience and Sustainability – Biondini & Frangopol (Eds)
 ISBN 978-0-415-62124-3

Seismic protection of the ASCE updated cable-stayed bridge benchmark with RNC passive devices

G. Carusone, M. Domaneschi & L. Martinelli
Politecnico di Milano, Milan, Italy

M. Ismail
Structural Engineering Department, Zagazig University; Egypt Postdoctoral researcher, Universitat Politècnica de Catalunya-BarcelonaTech, Barcelona, Spain.

J. Rodellar
Department of Applied Mathematics III, Control, Dynamics and Application Laboratory, School of Civil Engineering, Universitat Politècnica de Catalunya-BarcelonaTech, Barcelona, Spain

ABSTRACT

Seismic protection solutions for large structures are supposed to meet high standards of performance, feasibility and safety. When addressing civil structures, such as long-span bridges, there is a considerable uncertainty, including nonlinearity, associated with both physical properties and disturbances such as earthquakes and winds.

In this work a refined version of the international ASCE benchmark for a controlled cable-stayed bridge is considered as a case study in order to evaluate the performance of the RNC (roll-n-cage) protection system, which is a recently developed passive control device, which synergistically takes the advantage of two simple and high effective approaches that are: base isolation and passive energy dissipation (Ismail et al. 2008).

The benchmark structure is developed in a commercial finite element code (ANSYS) including enhanced aspects in the simulation of the stay cables dynamics, in the implementation of the seismic excitation and in the soil-structure interaction (Domaneschi & Martinelli 2011).

The RNC isolator incorporates isolation, energy dissipation, buffer in a single unit. It can also provide adequate wind resistance, a great range of horizontal flexibility, always exhibits no uplift during its lateral motion while being intrinsically versed to carry also tensile forces up to some extent.

The main goal of this work is to perform numerical mechanical characterization and efficiency assessment of such type of passive control devices considering the aforementioned benchmark problem of cable stayed bridges.

The positive contribution of the new dissipative devices in the mitigation of the seismic effects is assessed in terms of internal actions and displacements.

REFERENCES

Ismail M, Rodellar J, Ikhouane F. A Seismic Isolation System for Supported Objects. Spanish Patent No. P200802043, Spanish Office of Patents and Marks; 2008.

Domaneschi M. & Martinelli L. 2011. Performance Comparison of Passive Control Schemes for the Numerically Improved ASCE Cable-Stayed Bridge Model. Accepted in *Earthquakes and Structures*. Techno-Press.

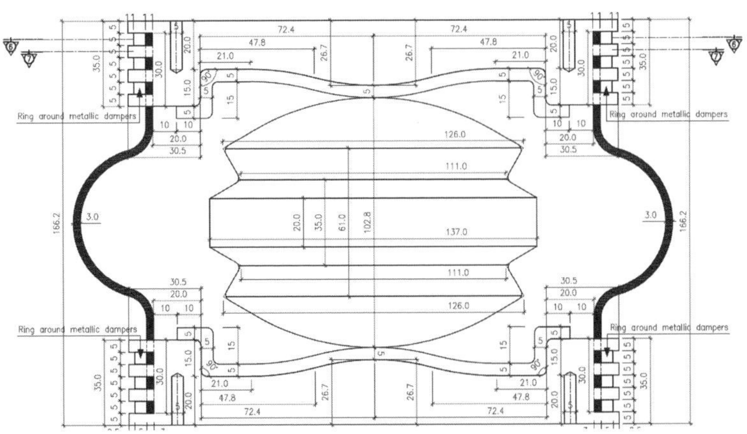

Figure 1. Design table of the RNC-c scaled prototype.

Bridge Maintenance, Safety, Management, Resilience and Sustainability – Biondini & Frangopol (Eds)
 ISBN 978-0-415-62124-3

Seismic performance of a wind designed control strategy on a suspension bridge

M. Domaneschi & L. Martinelli
Politecnico di Milano, Milan, Italy

ABSTRACT

There is today a growing interest in the control of large structures, as suspension bridges are. The modern design of this category of complex structures pays a greater attention to performance, while safety, reflecting the advances in knowledge accumulated in time, must be explicitly assessed under an increasing number of conditions. An important contribution can be sourced from structural control which can provide solutions to satisfy the required high standard of performance, feasibility and safety of today codes.

Herein, a model version of an existing suspension bridge is developed at the numerical level in the ANSYS Finite Element (FE) code, starting from original data, and used to simulate the structural response under wind and seismic excitation.

The main goal of this study is the evaluation of passive and semi-active control strategies, designed and proven effective for mitigating the effects of the wind action (Domaneschi & Martinelli 2010), when the bridge structure is subjected to seismic excitation.

The suspension bridge model is inspired by the Shimotsui-Seto Bridge, in Japan, spanning from the side of Mt. Washu to the Hitsuishijima Island. Figure 1a shows the main geometry of the bridge with a total span of about 1000 m and towers height of 150 m. The transversal section of the bridge deck is depicted in Figure 1b (30 m width, 13 m thickness).

The seismic input is represented by the Kobe earthquake, applied herein as a synchronous signal at all the bridge-soil contact points. Scaled inputs have been also evaluated by increasing and reducing the recorded one.

The implemented schemes are able to mitigate the seismic effects in the bridge deck reducing both the lateral displacements and the bending moments for the recorded intensity of the adopted signal. The towers internal forces appear equivalent for all the bridge configurations.

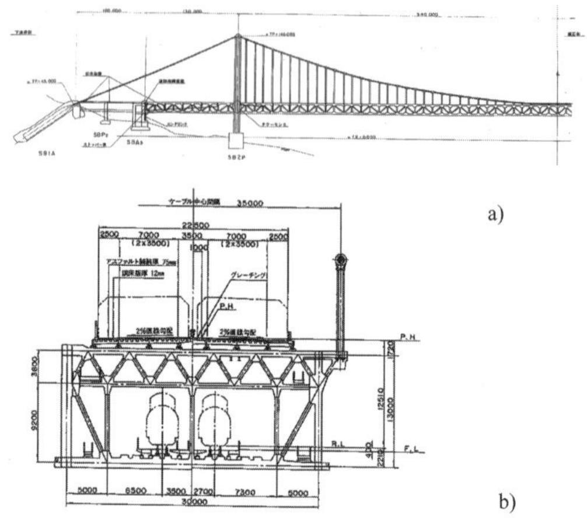

Figure 1. Main geometry of the bridge (a). Transversal section of the bridge deck (b).

REFERENCES

ANSYS, v. 12.0 2010. Ansys Inc., Canonsburg, PA, United States.

Domaneschi, M. & Martinelli, L. 2010. Optimal Passive and Semi-active Control of a Wind Excited Suspension Bridge. Accepted (November 19, 2010) in *Structure and Infrastructure Engineering*. Taylor and Francis Group Ltd.

Bridge Maintenance, Safety, Management, Resilience and Sustainability – Biondini & Frangopol (Eds)
© 2012 Taylor & Francis Group, London, ISBN 978-0-415-62124-3

Structural control of a wind excited suspension bridge model accounting for motion induced wind forces

M. Domaneschi & L. Martinelli
Politecnico di Milano, Milano, Italy

ABSTRACT

Modern design of long suspension bridges must satisfy at the same time performance and safety under several different conditions. An important contribution to the solution of this difficult design task can be given by the Structural Control discipline.

Pursuing this goal, the authors have in the past studied the feasibility of controlling wind induced vibration in long span suspension bridges through passive and semi-active control systems. These studies were based on a model of an existing suspension bridge, developed at the numerical level inside the ANSYS FE framework. A representation of the wind loading, which is considered the main dynamic excitation, through drag forces only applied on the towers, the cables and the deck of the suspended bridge. A simulation of wind velocity as a spatially correlated process acting in the horizontal direction, transversal to the deck. The performance of the control system, which was designed for a specific wind intensity was studied at different levels of wind intensity to make the results general.

The attention was initially focused on passive control systems for the Shimotsui-Seto Bridge under wind excitation; semi-active control strategies were later also studied in a decentralized configuration on the suspended bridge model. These developments adopted as a starting point the operating parameters for the devices associated to the passive control configuration previously identified as optimal.

In this paper, the optimization process for passive control systems, previously applied within a simplified model of the buffeting excitation which considers only the drag forces, is reevaluated in association with a more refined version of description the wind-structure interaction forces. This refined description in time domain is based on modeling the drag force as completely non linear, within the quasi-steady theory, while adopting for the lift force and the aerodynamic moment a linearized form with corrections for frequency dependent loading using indicial functions. The field of turbulent wind velocities has been simulated as a spatially correlated process.

The results obtained do not change the results already obtained with the simplified wind forces model. The optimal parameters for the passive control arrangement, previously identified, retain their validity. Further studies are underway to confirm that this result can be due the aerodynamic stability characteristics of the bridge cross-section.

Additional validation work is anyway needed, since the results are based on a single realization of wind turbulent velocity of predefined average speed.

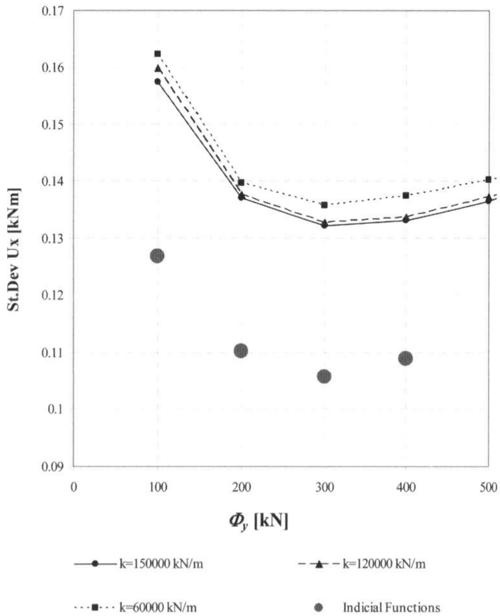

Figure 1. Standard deviations of the displacement Ux, in direction x, at the deck mid-span as function of Φ_y and k (control device yield force and stiffness, respectively).

REFERENCE

Domaneschi, M. and Martinelli, L., 2011. Optimal Passive and Semi-active Control of a Wind Excited Suspension Bridge. *Structure and Infrastructure Engineering*, Taylor and Francis Group Ltd. (DOI:10.1080/15732479.2010.542467).

Bridge Maintenance, Safety, Management, Resilience and Sustainability – Biondini & Frangopol (Eds)
 ISBN 978-0-415-62124-3

Detailed numerical and experimental dynamic analysis of long-span footbridges to optimize structural control measures

C. Meinhardt
GERB Vibration Control Systems, Berlin, Germany

ABSTRACT

Footbridges in urban areas are more and more meant to set architectural highlights and not just to connect two sides. According to that, the design becomes more and more spectacular by realizing slender long-span structures. Despite the extreme lightness, bridge structures span more than 30 meters, cable stayed structures more than 100 meters while maintaining a filigree support structure. These complex structures display several natural frequencies distinctively below 5 Hz so they can be easily excited to perturbing vibrations by pedestrians or joggers due to low structural damping. Since the modal parameters of the bridge can't get arbitrarily adapted, the application of structural control devices such as Tuned Mass Dampers (TMDs) is a common measure to ensure comfort and safety of footbridges. The effectiveness of the applied systems hereby strongly depends on the specification and tuning of the TMDs. The application of TMDs is already state of the art and has been realized at many projects, for example the Millennium Bridge in London. So this contribution will focus on the challenge to apply TMD systems to be effective for several modes that are identified to be susceptible by broadening the frequency range in which the system is effective without necessarily increasing the effective mass of the TMDs to avoid too much additional loading. This approach requires a detailed dynamic analysis to assess the dynamic response of footbridges under realistic loading according to defined comfort criteria. It also requires to modify the commonly known optimization criteria by analyzing the TMD effect for a more realistic assumption of human induced vibrations and for interaction effects. The following paper describes calculation approaches for detailed dynamic analyses of footbridges to assess relevant susceptible vibration modes considering realistic loading group loading and synchronization effects. It also shows examples for adapted Tuned Mass Damper System specifications which are optimized to provide sufficient damping for several susceptible modes by abandoning the optimum specification for just one single susceptible mode. In addition project examples will be introduced for which the preliminary theoretical analysis and methods for an experimental dynamic analysis will be presented.

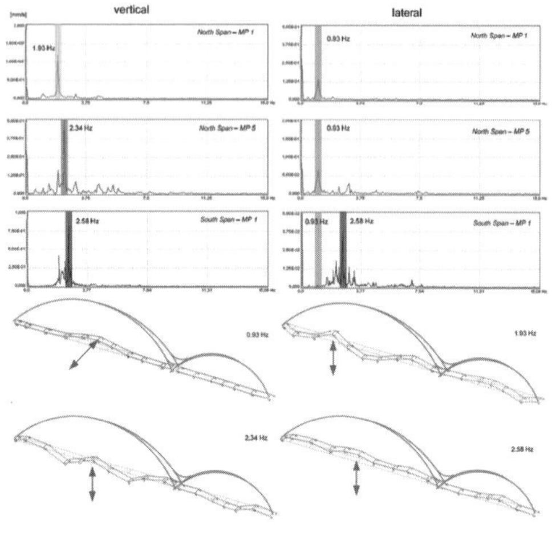

Figure 1. Experimentally determined natural frequencies and corresponding mode shapes – layout of measurement points.

REFERENCES

Bachmann, H., Lively Footbridges – a real challenge, *Footbridge 2002 – Design and dynamic behavior of Footbridges*, 2002

Butz, C., Sedlacek, G., Bemessungskonzept für fußgängerinduzierte Brückenschwingungen, *Stahlbau 76, Ernst und Sohn Verlag*, 2007

Den Hartog, J.P., Mechanische Schwingungen, *2. Edition, Springer Verlag*, 1952

Petersen, C., Dynamik der Baukonstruktionen', *Vieweg Verlag*, 1996

Matsumoto, Y., Nishio, T., Shiojiri, H, Matsuzaki, K., Dynamic Design of Footbridges, *Proceedings P17/18, IABSE*, 1978

Meinhardt, C., Eusani, R., Detailed Analysis of large-span footbridges, *Proceedings IABSE-IASS 2011 London Symposium Report*, 2011

Wise, C., Weir, A., Harris, T., Dynamic Analysis of the Infinity Bridge, *Internal Report Expedition Engineering Ltd.*, 2007

Bridge Maintenance, Safety, Management, Resilience and Sustainability – Biondini & Frangopol (Eds)
© 2012 Taylor & Francis Group, London, ISBN 978-0-415-62124-3

Elaboration of the vibration comfort criteria for footbridges during vibrations induced by pedestrians

M. Pańtak
Cracow University of Technology, Cracow, Poland

ABSTRACT

New structural materials characterize by better and better resistance parameters what in contemporary designed structures leads to: greater spans, smaller dimensions of cross sections of structural elements, smaller vertical and horizontal stiffnesses of the structure, smaller masses and lower values of damping parameters. For this reasons contemporary footbridges become susceptible to different dynamic actions. Dynamic response of these footbridges, correctly designed for static loads, can exceed established comfort limits. Users can feel unpleasant vibrations of the footbridge deck.

In the paper, specificity of elaboration of vibrations comfort criteria are presented. The proposals of vibration comfort criteria taking into account frequency of vibrations occurrence (frequent, rare and intentional events) are characterized and compared with recommendations of different standards, guidelines and propositions of other authors.

Presented criteria were elaborated on the basis of series of in situ experimental investigations performed on over 30 footbridges of different structural schemes (15 truss footbridges, 7 cable-stayed footbridges, 8 suspension footbridges and 1 ribbon footbridge, with spans ranging from 20 m to 110 m). The propositions can be characterized as follows:

- criteria are related to men-induced vibrations i.e. more or less vibrations of harmonic character,
- criteria define comfort levels in case of vibrations sensed by walking users,
- criteria are related separately to vibrations in vertical and horizontal direction,
- criteria are related to peak acceleration (a_{max}) as a function of vibrations frequency,
- criteria taking into account frequency of vibrations occurrence: frequent event (base curve M1), rare events (curve M1.7) and intentional (vandal) actions (curve M10).

In case of vertical vibrations proposed criteria are shown and characterized in relations to obtained research results (Fig. 1) and are compared with requirements of ISO 10137.

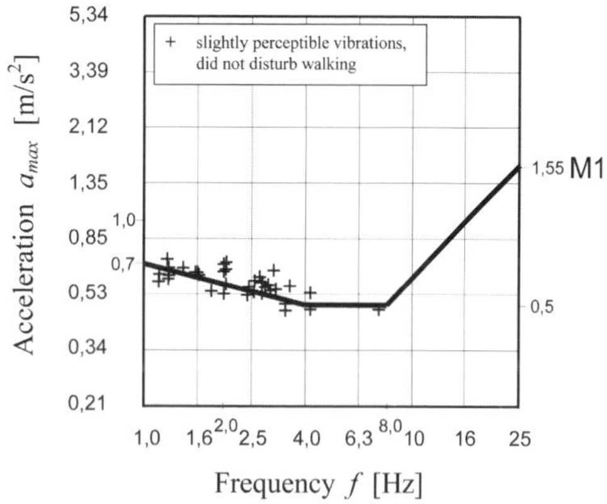

Figure 1. Proposed vibration comfort criteria for men-induced vertical vibrations in comparison with experimental results – vibrations slightly perceptible and did not disturb walking.

The criteria can be used for design of passive, active, semi-active and hybrid structural control systems.

REFERENCES

Bachmann, H. & Ammann, W. 1987. *Vibration in structures induced by man and machine.* Zürich: International Association of Bridge and Structural Engineering (IABSE).

BS 1978. Steel, Concrete and Composite Bridges: Specification for Loads. Vibration Serviceability Requirements for Foot and Cycle Track Bridges. *BS 5400 – Part 2, Appendix C.* London: British Standards Association.

Flaga, A. & Pantak, M. 2008. Vibration comfort criteria for pedestrians on footbridges. *Footbridges 2008; Proc. Third Inter. Conf.* Portugal, 2–4 July 2008, Porto: Faculdade de Engenharia Universidade do Porto.

ISO 2007. Bases for Design of Structures. Serviceability of buildings and walkways against vibrations. *ISO 10137.* Geneva: International Organization for Standarization.

SÉTRA 2006. Assessment of vibrational behaviour of footbridges under pedestrian loading. *Technical guide.* Paris: Technical Department for Transport, Roads and Bridges Engineering and Road Safety (Service d'études techniques des routes et autoroutes – Sétra).

Integral bridges: Design and technological issues

Organizers: P.G. Malerba & V. Kristek

Bridge Maintenance, Safety, Management, Resilience and Sustainability – Biondini & Frangopol (Eds)
 ISBN 978-0-415-62124-3

Integral bridge design solutions for Italian highway overpasses

L. Ferretti Torricelli, A. Marchiondelli, R. Pefano & R. Stucchi
Bridge Design Department, SPEA Ingegneria Europea S.p.a., Milan, Italy

ABSTRACT

The present paper highlights some of the peculiar aspects characterizing integral bridges, which emerged from an extensive study carried out in view of their implementation in Italy as a standard solution for highway overpasses. The advantages offered by integral bridges in terms of durability by the elimination of joint and bearing devices are well known, since the presence of deck-abutment joint has always represented a "weak" point, prone to water and aggressive agents attack, that is one of the primary causes of premature bearing replacement and frequent maintenance interventions. In addition, the presence of the connection between deck and abutments, which transforms the system in a rigid or semi-rigid portal frame, clearly improves the global structural performance. By considering these aspects, two main span configurations have been considered, characterized by one and three spans arrangement respectively, with maximum lengths of the main span equal to 38 and 45.50 m depending on the width of the overpassed highway carriageway. In dependence of the type of interesting roads, four different overpass width has been designed, calling for a total slab width equal to 5.40, 9.0, 12.0 and 13.50 m respectively. Finally, depending on the morphology of the surrounding territory, two different types of abutment have been provided.

The preliminary part of the study has been dedicated to the choice and evaluation of a proper simplified analysis, in order to overcome the difficulties involved in non linear soil-structure interaction problem. Discarded methods based on limit equilibrium theory, which usually bring to a too large overestimation of stresses in abutments, the attention has been focused on methods based on the splitting of the global model into two sub-models formed by deck only, and soil-abutment system only, respectively. In each model, the presence of the "mutual" part has been considered by applying appropriate restraints or loading conditions located at the abutment/deck connection. By adopting this approach, the structural analysis has been carried out by recurring to dedicated softwares in each model.

Finally, the attention has been then focused on the potential advantages, in terms of structural performance, involved in the frame behavior which characterizes the jointless construction. To this aim, a direct comparison between integral and simply supported schemes has been attempted by highlighting the main differences in terms of internal actions and material exploitation. By means of case studies, the paper has shown the benefits derived from the adoption of the integral configuration for highway overpasses, in spite of a more complex design process. In particular, in case of one span configuration, a great economic convenience in adopting integral scheme has been made evident, since it allowed to lighten both deck and abutments. In addition, in case of integral bridge, joint and bearing devices can be avoided, and this aspect assures a great saving in terms of both initial and maintenance costs. It is also worth remembering that it influences durability aspects, by adding value to bridge quality.

REFERENCES

Biddle, A.R., Iles, D.C. & Yandzio, E., 1997, Integral steel bridges: Design guidance. (P163), The Steel Construction Institute.

Burk, Jr. & Martin P., 1996, An Introduction to the Design and Construction of Integral Bridges, FHWA, West Virginia DOT and West Virginia University, Workshop on Integral Bridges, November 13–15, 1996.

England, G.L., Tsang, N.C.M. & Bush, D.I., 2000, Integral Bridges, Thomas Telford Publishing, London.

Furlanetto, G., Ferretti Torricelli, L. & Marchiondelli, A., 2010, Towards Eurocodes: an Italian experience. A comparative analysis between the former and the new Eurocode-based codes, 2010 joint IABSE-fib Conference, May 3–5, 2010, Dubrovnik, Croatia.

Wasserman, E.P. & Walker, J.H., 1996, Integral Abutments for Steel Bridges, Highway Structures Design Handbook, Tennessee Department of Transportation for the American Iron and Steel Institute, October 1996.

Bridge Maintenance, Safety, Management, Resilience and Sustainability – Biondini & Frangopol (Eds)
© 2012 Taylor & Francis Group, London, ISBN 978-0-415-62124-3

Structure-soil interaction of buried arch bridges

M. Foglar & V. Kristek
Czech Technical University in Prague, Czech Republic

ABSTRACT

Arch is the oldest structural shape that mankind has invented. Already the Romans constructed a large number of stone arch bridges on roads or water-supply routes (aqueducts), many of which survived to present.

Arch bridges represent a major part of the bridge population in Europe, including more than 200,000 masonry arch bridges. They comprise railway bridges built in the 19th century, medieval and Renaissance bridges which have survived along the centuries and nowadays represent relevant landmarks around Europe. Most of these bridges are considered as historical and belong to the architectural cultural heritage (Cultural Heritage Bridges). In this respect arch bridges represent a fundamental value as they symbolize important structural achievements, providing a tangible experience on past construction technologies.

Buried arch bridges are built on both roads and railways, both as underbridges or overbridges. Due to low construction costs, great durability and endurance, they are favoured by contractors and owners. Buried arch bridges are built with spans from 2 m up to 40 m.

This paper discusses the structure-soil interaction of buried concrete arch bridges. On the basis of an analytical derivation which is developed first, a method of centre-line optimization of buried arch bridges is proposed and boundaries of this approach discussed.

The process of centre-line optimization of buried arch bridges leads to a centre-line which is a resultant of forces affecting the structure, thus only minor bending moments are present; see equations for vertical and horizontal direction:

$$H y'' + H' y' - \gamma y = h \gamma \tag{1}$$

$$H' + k \gamma y y' + k h \gamma y' = 0 \tag{2}$$

Then a specific example is studied, where the beneficiary effect of the proposed method is showed.

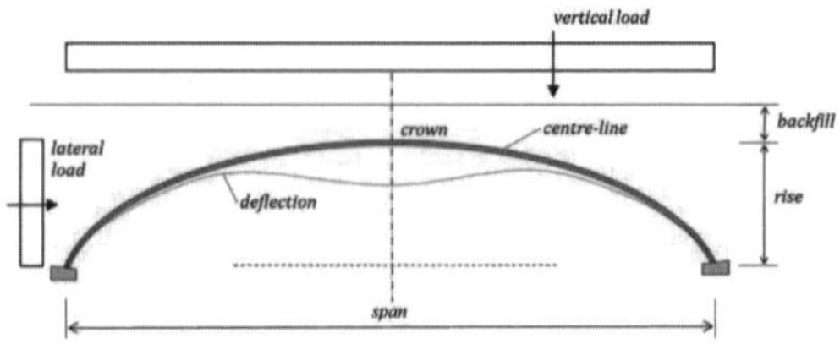

Figure 1. Loading and deflection of a buried arch bridge.

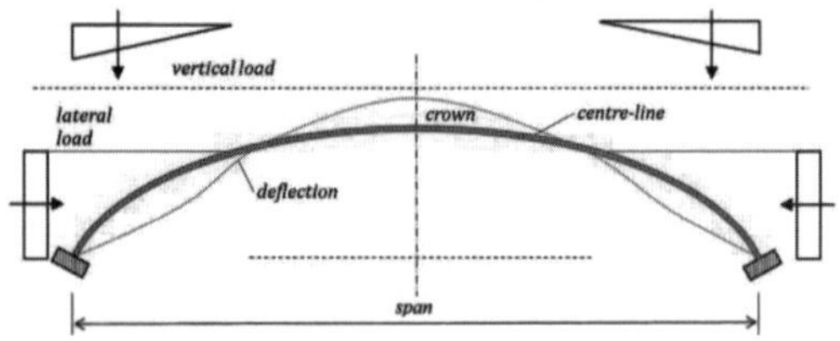

Figure 2. Loading and deflection of a buried arch bridge during construction.

Finally, the effect of construction phases is studied, compare Fig. 1 and 2.

REFERENCES

Foglar, M. & al. (2010). The use of fiber reinforced concrete in buried arch bridges (in Czech). *Beton TKS* **10**, No. 2, 92–95.

Kristek, V. & al. (2004). D11 Praha – Hradec Kralove – Design of arches on rural roads over D1105-1 (in Czech). *Proc. of Int'l Conf. Betonarske dny*, CBS Servis, Praha, 477–482.

Malina D. *Design and analysis of a buried structure in unfavourable site conditions* (in Czech), MSc Thesis, supervisor Foglar M. Czech Technical University, Prague, 2011.

Bridge Maintenance, Safety, Management, Resilience and Sustainability – Biondini & Frangopol (Eds)
 ISBN 978-0-415-62124-3

Two integral bridges connecting the runways of the Milano Malpensa Airport

P.G. Malerba
Department of Structural Engineering, Politecnico di Milano, Milan, Italy

G. Comaita
Studio Malerba, Milan, Italy

ABSTRACT

The construction of the H ring is part of a general project to promote Malpensa Airport as an international hub. The project was started in the 90s and includes the construction of a new terminal, a freight area, called Cargo City, and a new apron capable of hosting more than 100 aircraft at one time.

The two bridges which complete the ring are located to the west (toward Malpensa) and to the east (toward Milan) of the longitudinal axis of the main runway (Fig. 1). The bridges decks have straight axes intersecting the curved railway track at a skew angle of about 45°, and are made of pre-tensioned, precast beams, made continuous by cast-in-place transverse beams and upper slab. The geometry of these structures differs from that of a conventional bridge having length greater than its width. In this case, the West Bridge has a 146 m wide deck and a 21.50 m long effective span, while the East Bridge has a 134 m wide deck and a 19.50 m long effective span.

The vertical section of each bridge is a portal frame (Fig. 2). The area enclosed by the bridges was determined by the underlying rail platform, which consists of three tracks (two operating, one upcoming) and a service road used for inspection and maintenance operations, for a total width of 21.5 m for the West Bridge and 19.5 m for the East Bridge. The vertical walls supporting the decks have variable heights of about 6 m and also serve as retaining walls for the embankments on the two sides of the trench. The two bridges are supported on spread foundations.

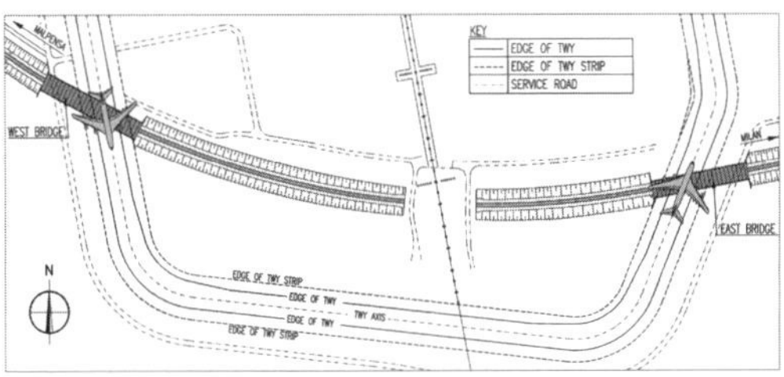

Figure 1. A plan of the H ring.

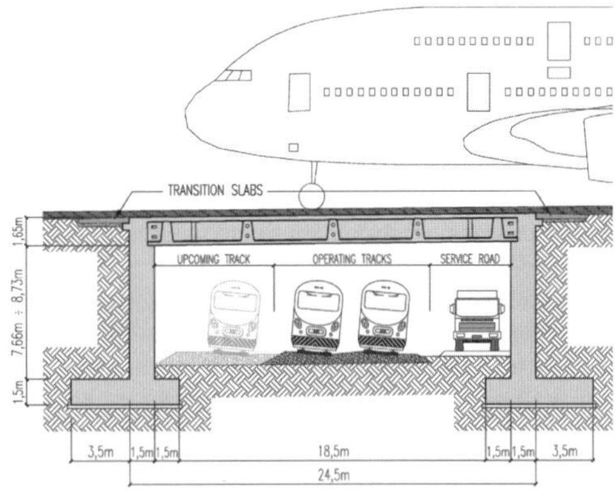

Figure 2. Cross section of West Bridge (the depicted aircraft gives an idea of the bridge proportions).

Table 1. Maximum loads for AIRBUS A380/800 Series aircraft.

AIRBUS 380/841-861		[kN]
Maximum Ramp Weight	MRW	5620
Maximum Takeoff Weight	MTOW	5600
Maximum Landing Weight	MLW	3860
AIRBUS 380/843F-863F	(F = freight)	[kN]
Maximum Ramp Weight	MRW	5620
Maximum Take Off Weight	MTOW	5600
Maximum Landing Weight	MLW	4270

One of the features of these bridges is the magnitude of the service loads, on account of the transit of AIRBUS 380 Class aircrafts. The load intensities are summarised in Table 1.

This paper briefly describes how the above requirements were dealt with in designing the two integral bridges.

Bridge Maintenance, Safety, Management, Resilience and Sustainability – Biondini & Frangopol (Eds)
© 2012 Taylor & Francis Group, London, ISBN 978-0-415-62124-3

Integral bridge: A review on its behaviour under earthquake loads

M. Masrilayanti & L. Weekes
University of Salford, UK

ABSTRACT

Integral bridges, also called jointless bridges are bridges which do not use expansion joints and bearings. Their decks are continues and connected monolithically to the abutment with a moment resisting connection. This cause the structure will be acting as one unit.

Integral bridges or jointless bridges have shown a good record of initial cost savings in economical use of material and maintenance. Because of do not have any expansion joints at abutment and bridge deck. This condition leads on reducing both of construction and continual maintenance costs. Because of this, year after year many engineers are interested on using integral bridges and it was stated that this kind of structure will be more often constructed throughout the world. However, there are still many problems should be overcome regarding this monolithic structure.

During the last several decades, integral bridges have been constructed and used for a considerable time, especially in Europe and America. In the United States of America, there are recently more than a thousand integral bridges built in several states, such as Tennessee and California. These kinds of bridges have shown good structural performance due to their redundancy and are durable. Meanwhile in United Kingdom, this bridge has become more popular. In fact, there are regulations from the Highway Agency of United Kingdom which now state that for a new bridge with a length of less than 60 meters, where possible, must be in the form of an integral bridge.

Soil structure interaction is acknowledged as the most significant problem of integral bridge. The additional loads due to pressure of soil behind the wall or abutment are considered gives an important impact to the structure as a whole.

This paper will describe the characteristic of integral bridge in term of design, construction and its effectiveness in overcoming critical load such as earthquake. However, its complicated issue in terms of soil structure interaction is going to be described as well to make a balance consideration.

REFERENCES

Burke, M. P. 2009. *Integral and Semi Integral Bridge*. Newyork: Willey-Blackwell Publishing.

Connal, J. 2004. Integral Abutment Practice: Australian and US Practice, Proc. 5th bridge conference, 19–21 May

Faharani, V. R. 2010. Seismic Analysis of Integral Abutment Bridge in Tennessee, Including Soil–Structure Interaction. *Bridge Engineering* (2).

Frosh, R. J. et al. 2009. *Earthquake Resistance Of Integral Abutment Bridges*. Purdue Libraries.

Goel, R. K. 1997. Earthquake Characteristics of Bridges with Integral Abutments. *Journal of Structural Engineering* 11(123): 1435–1443

Iles, D. C. 2006. Integral Bridges in the UK, *Technical Report of International Workshop on the Bridges with Integral Abutments*, Sweden

Karantzikis & Spyrakos. 2000. Seismic Analysis of Bridge including Soil Abutment Interaction, *Proceeding in 12 World Conference on Earthquake Engineering, no 2471, New Zealand Earthquake*, Auckland

Pekcan, G. et.al. Seismic Design Recommendations for Steel Girder Bridges with Integral Abutments TRB 2010 Annual Meeting CD-ROM, available online at: http://docs.trb.org/prp/10-0747.pdf accessed in 21st January 2011

Shamsabadi & Kapuskar. 2006. Nonlinear Seismic Soil Abutment Structure Interaction Analysis of Skewed Bridge. *Fifth National Seismic Conference on Bridges & Highways: Innovation in Earthquake Engineering for Highway Structures*, San Fransisco, September 18–20

Tandon, M. 2005. Economical Design of Earthquake Resistant Bridge. *Journal of Earthquake Technology*. 1(42). pp. 13–20

White, H. 2007. Integral Abutment Bridges: Comparison of Current Practice Between European Countries and The United States of America. *Special Report 152 Transportation Research And Development Bureau*, Report Fhwa/Ny/Sr-07/152

Management and preservation of long span historic bridges

Organizers: A.E. Aktan, F. Moon & D.S. Lowdermilk

Bridge Maintenance, Safety, Management, Resilience and Sustainability – Biondini & Frangopol (Eds)
© 2012 Taylor & Francis Group, London, ISBN 978-0-415-62124-3

Preservation and management of historic landmark bridges that remain essential as critical infrastructure elements

I. Bartoli & A.E. Aktan
Department of Civil, Architectural and Environmental Engineering, Drexel University, Philadelphia, PA, USA

ABSTRACT

A survey of the recent US and international workshops and conferences on maintenance and preservation of bridges confirmed continuous struggles of many agencies, organizations and societies when dealing with the growing challenges of managing aging infrastructures. Advances in research have introduced promising technologies to assess and monitor the condition of infrastructures (Terry et al., 2010; Frangopol, 2007). NSF, FHWA, NIST, DOTs and several other US and foreign agencies have vigorously supported these progresses recognizing the potential of Structural Health Monitoring paradigm to provide long term bridge performance monitoring (Aktan, 1995; Frangopol, D.M., 2007), simultaneously serving as an effective tool to perform objectively informed preservation management.

An obvious goal of the agencies, owners and the engineering community is to guarantee and possibly enhance the safe performance of bridges (Aktan, 2008) as well as the entire transportation infrastructure. Bridges serve as critical links within the highway and railway networks, and in times of hazard they become essential for evacuation, response and recovery. Consequently, aging bridge populations represent a major concern for safety and security (Doyle and Betti, 2009). Several international conferences over the years have helped increase the sensitivity of the worldwide engineering communities to bridge safety and security while exploring the potential of innovative technologies to help their maintenance (Chang and Guemes, 2007).

While recognizing the importance and accomplishments made by all previous efforts, it is noticed that rarely the proactive maintenance and management of aging historic infrastructures is seen as an integrated paradigm inclusive of the political, social, economic, organizational and historic constraints. Asset management of major long-span highway bridges in a holistic manner by identifying not only engineered but also human, organizational and natural elements as assets, and by defining and tracking collective system-level and individual asset level performance measures and metrics is a promising approach that deserves a closer study for customized applications to major historic bridges that cannot be removed from service even for short durations. We have to note that the economic impact of removing one of the East River Bridges in New York City from service may be comparable to the economic impacts of a major earthquake in California.

The paper will first focus on the present critical situation of transportation infrastructure investment in United States and how policy, planning, and financing are reacting and are influenced by the current US economic condition. Secondly, a discussion on Performance and Asset Management and Preservation of historic landmark bridges will be provided. Finally, the issue of leveraging the technology as a mean for condition evaluation of infrastructures to assist objective bridge maintenance and repair will be explored.

REFERENCES

Aktan, A.E., 1995. North American Workshop for Bridge Instrumentation and Vibration Analysis, Cincinnati, Ohio.
Chang, F.K., Guemes J.G. 2007. IWSHM, 7th International Workshop on Structural Health Monitoring: Session on Health Monitoring of Bridges for Asset Management, Stanford University.
DeLony E. and Klein T.H. 2003. Historic Bridges: a Heritage at Risk. A Workshop on the Preservation and Management of Historic Bridges, Washington, D.C.
Doyle, C. and Betti, R. 2009. Aging Infrastructures Workshop, New York City, NY.
Frangopol, D.M., 2007. Bridge Long Term Performance Program FHWA/NSF Workshop on Future Directions for Long-Term Bridge Performance Monitoring, Assessment, and Management, Las Vegas, Nevada.
Terry, J., Siewert, T., McColskey, D., Luecke, W., Iadicola, M. and Duthinh,, D. 2010. Workshop on Quantitative Tools for Condition Assessment of Aging Infrastructure, Boulder, Colorado.

Bridge Maintenance, Safety, Management, Resilience and Sustainability – Biondini & Frangopol (Eds)

Knowledge management for aging infrastructure

E. Jackson, E. Richter, P.L. Gurian, A. Pradhan, A.E. Aktan & F.L. Moon
Drexel University, Philadelphia, PA, USA

ABSTRACT

This paper describes the basic principles and techniques of knowledge management, documents preliminary progress in applying these approaches to a Bridge Owner, and outlines the remaining steps that would be required to fully implement knowledge management practices. Organizational knowledge can be divided into categories of explicit (theoretical understanding) and tacit (unconscious competence). There are four core processes that make up the framework for integrating tacit and explicit knowledge into a knowledge management system – socialization, externalization, combination, and internalization. Socialization is the sharing of tacit knowledge from person to person and is essential for the retention and transfer of organizational knowledge. Externalization is the process of formalizing tacit knowledge into an explicit form, typically through structured interviews. Combination is the sorting and organizing of explicit knowledge into a more easily accessible form, and could include information mapping or development of a metadata system. Internalization is the process that develops tacit knowledge out of compiled explicit knowledge, typically by training.

Socialization efforts conducted as part of this study consisted of the development of a stakeholder diagram to identify and map the social structure around the Bridge Owner along with each stakeholder's role in the bridge management process. Future socialization work should include the development and sharing of case studies, along with employee shadowing.

Externalization proceeded through project scoping interviews that were conducted as a starting point for the capture of tacit knowledge that is critical for job performance. An additional protocol for a structured interview was developed to determine key sources of organizational knowledge, identify existing strengths along with areas where management systems could improve the organization, and capitalize on unexploited ideas and innovations.

The combination process began with the development of information flow diagrams and use case analysis. Information flow diagrams identified existing information flows along with areas where information flows could be facilitated through the development of a knowledge management system. Use-case analysis was then used to illustrate how a knowledge management system would address these gaps. Work on these efforts should continue as more information is gathered from additional structured interviews.

Efforts have not yet begun on the internalization component, but could include the development of training modules based on the case studies elicited during the socialization process.

Several commercially available software systems that can be customized to meet the use-case needs have been identified for the Bridge Owner. However, the effort should not be confined to the development of new software tools but should include the development of procedures to identify and transfer implicit knowledge through monitoring, shadowing, and reflection on case studies. The Bridge Owner now has the opportunity to undertake a knowledge management program to improve the efficiency of current operations and aid in the storage and transfer of knowledge for the future. This program would extend the initial interviews that have been undertaken to identify the existing organizational structure and management processes. The effort would propose an architecture for the knowledge management system that mirrors existing management structures and can accommodate the use cases identified here.

Bridge Maintenance, Safety, Management, Resilience and Sustainability – Biondini & Frangopol (Eds)
© 2012 Taylor & Francis Group, London, ISBN 978-0-415-62124-3

Maintaining and preserving long span signature structures

D.S. Lowdermilk & G. Nyikita
Pennoni Associates Inc., Philadelphia, Pennsylvania, USA

J. Jeffers
Burlington County Bridge Commission, Palmyra, New Jersey, USA

ABSTRACT

It is no secret that the infrastructure of the United States is rapidly deteriorating as it continues to age. The American Society of Civil Engineers (ASCE) has been tracking the continuing aging of roads, bridges and other important infrastructure since first developing and releasing their "Infrastructure Report Card" in 1990's. Table 1 shows the difference in bridge infrastructure from 2001 and 2005 as published in the 2005 ASCE Report Card.

Note that the number of structurally deficient or functionally obsolete bridges again dropped about 1% during the four year span from 2005 and 2009, but it is important to note that the anticipated cost for improving that same bridge infrastructure increased from \$9.4 billion in 2005 to \$17.5 billion in 2009... nearly double the annual investment. Likewise the total financial need for the infrastructure is now \$2.2 trillion in 2009, an increase of nearly one trillion dollars. The United States is clearly losing the battle of successfully funding the maintenance and preservation of the nation's infrastructure, most notably its aging bridge inventory.

This paper mainly focuses on outlining challenges that bridge owners face in managing and preserving their assets, especially the aging infrastructure in the United States. In specific, this paper describes the Burlington County Bridge Commission (BCBC) in Palmyra New Jersey, USA as an example and the challenges they are facing in managing their assets, especially their inventory of signature movable bridges, and how BCBC has hired technology to create safer bridges for their customers. In fact, the technology has helped the BCBC to prioritize their budget and infinitely increase the life of their bridges without increasing the tolls for over a decade. In addition, the paper briefly describes the monitoring systems installed on two of the BCBC's movable bridges (Tacony-Palmyra Bridge and Burlington-Bristol Bridge), and how they have been helping the BCBC to monitor their bridges during

Table 1. ASCE Report Card for Bridges.

Subject	2001 Grade	2005 Grade	Comments
Bridges	*C*	*C*	Between 2000 and 2003, the percentage of the nation's 590,750 bridges rated structurally deficient or functionally obsolete decreased slightly from 28.5% to 27.1%. However, it will cost \$9.4 billion a year for 20 years to eliminate all bridge deficiencies. Long-term underinvestment is compounded by the lack of a Federal transportation program.

construction and other extreme events such as earthquake and hurricane, specifically Hurricane Irene that hit the East coast USA in summer of 2011. Based on the findings of the monitoring projects on the aging bridges, it is concluded at the end that condition assessment of such bridges should involve more than just the visual inspection of the elements, and should take into account the redundancy of the system, different load carrying paths, integrity of the system, remaining life, and other factors that require implementation of technology.

REFERENCES

ASCE 2009 Report Card for America's Infrastructure, January 2009, https://apps.asce.org/reportcard.

History of Tacony Palmyra Bridge (TPB), 75th Anniversary of the Opening of TPB, August 2004, Pennoni Associates and the Burlington County Bridge Commission.

"Tacony-Palmyra Bridge over the Delaware River between Philadelphia PA and Palmyra NJ," Final Engineering Report Submitted by Majeski, Masters and Chase Engineers to the President and Directors of Tacony-Palmyra Bridge Co., 1931 (Courtesy of the Burlington County Bridge Commission).

Bridge Maintenance, Safety, Management, Resilience and Sustainability – Biondini & Frangopol (Eds)
 ISBN 978-0-415-62124-3

iCOMPASS: An integrated approach in performance-based management of infrastructures

E. Minaie, M. DePriest, N.C. Dubbs, F.L. Moon & A.E. Aktan
Intelligent Infrastructure Systems Division of Pennoni Associates Inc., Philadelphia, PA, USA

P. Adams & S. Ozalis
Burlington County Bridge Commission (BCBC), NJ, USA

ABSTRACT

The Burlington County Bridge Commission (BCBC) has been investing in implementing technology into their operation and management by adding technology as a value added to tradition and culture. Pennoni associates, the engineering program manager of BCBC, has been helping BCBC with the design of an Integrated Centralized Operation, Management, Performance, and Security System (iCOMPASS). This system encompasses different modules such as maintenance, capital planning, inspection, performance monitoring, security, and digital archive. Figure 1 illustrates how these modules interact with each other and with the digital archive. Each of these components has a role in adding value and improving the safety of the BCBC assets and productivity of the agency in general. The stakeholders (bridge owners, bridge managers, maintenance managers and staff, financial officers, security managers and officers, resident engineers, program and project managers, etc.) will have different levels of access to the modules of iCOMPASS, depending on their role defined in the Concept of Operations (ConOps) which is prepared earlier in the project in the concept studies phase. This paper briefly describes the design of the system architecture, role of each module, interconnectivity of different modules, and the integration process of iCOMPASS.

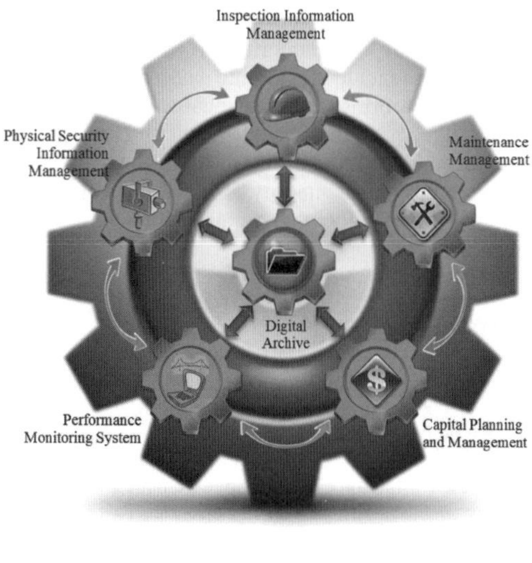

Figure 1. Modules of iCOMPASS.

REFERENCES

Deshan S. and L. Qiao (2009), "Development of a Smart-Client Based Bridge Management and Maintenance System fir Existing Highway Bridges," Proceedings of the International Conference on Transportation Engineering, 2009, China.

Liu C. and Y. Itoh, (2001) "Information technology applications for bridge maintenance management", Logistics Information Management, Vol. 14 Iss: 5/6, pp. 393–400.

Lounis Z. (2005), "Network-Level Bridge Management Using a Multi-Objective Optimization Decision Model", Proceedings of 1st CSCE Specialty Conference on Infrastructure Technologies, Management and Policy, Toronto, Ont., June 2–4, 2005.

Patidar V. (2007), "Multi-Objective Optimization for Bridge Management Systems", NCHRP Report 590, Transportation Research Board of the National Academies, Washington DC, USA.

Valenzuela S., Soliminihad H., Echaveguren T. (2010), "Proposal of an Integrated Index for Prioritization of Bridge Maintenance," ASCE J. Bridge Eng. 15, 337 (2010)

Bridge Maintenance, Safety, Management, Resilience and Sustainability – Biondini & Frangopol (Eds)
 ISBN 978-0-415-62124-3

Evaluation of a long-span steel tied arch bridge using temperature-based structural identification

M.T. Yarnold, F.L. Moon, N.C. Dubbs & A.E. Aktan
Intelligent Infrastructure Systems (IIS) a Division of Pennoni Associates, Philadelphia, PA, United States
Drexel University, Philadelphia, PA, USA

ABSTRACT

The Tacony-Palmyra Bridge steel tied arch span is currently being evaluated through a new approach termed Temperature-Based Structural Identification (TBSI). TBSI uses temperature as the forcing function for long-span bridge evaluation. Most bridge structures experience significant daily and seasonal temperature variations thus causing relatively large strains and displacements. As a result, TBSI takes advantage of this for long-span bridge assessment. The Tacony-Palmyra Bridge arch span has been instrumented for measurement of local temperatures (input) along with local strains and global displacements (output response). The objectives of the evaluation include (1) finite element model calibration for refined structural ratings, (2) evaluation of long-term performance criteria, and (3) development of automated alert criteria for a real-time structural health monitoring (SHM) system. The results of the project provide the bridge owner with reliable information to better manage the structure.

Figure 1. TPB Steel Tied Arch Span.

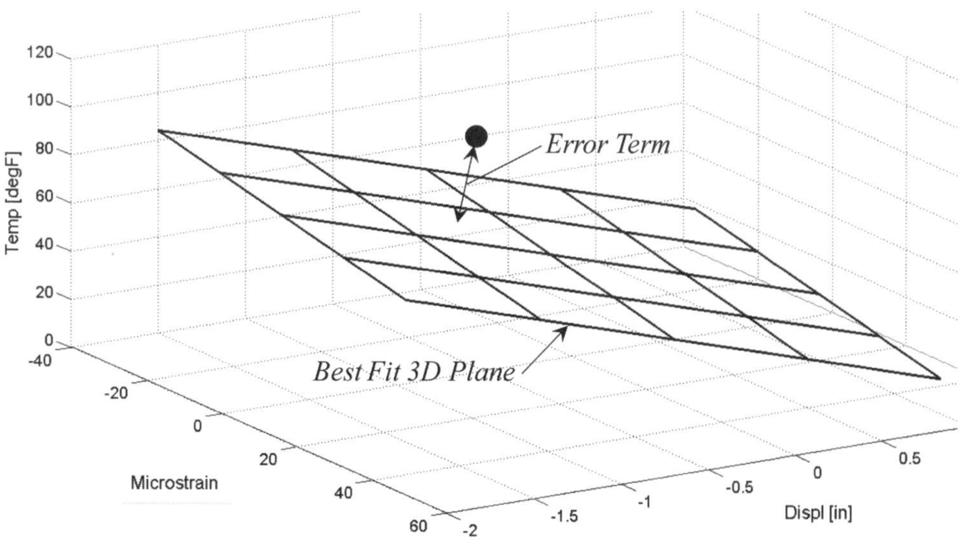

Figure 2. SHM System Alert Criteria using TSBI.

Many bridges aren't straight – Investigations of curved and skewed structures

Organizer: D. Linzell

Bridge Maintenance, Safety, Management, Resilience and Sustainability – Biondini & Frangopol (Eds)
© 2012 Taylor & Francis Group, London, ISBN 978-0-415-62124-3

An evaluation of lateral flange bending in straight and skewed short-span steel bridges

K.E. Barth & G.K. Michaelson
West Virginia University, Morgantown, West Virginia, USA

N.Y. Galindez
Modjeski and Masters, Inc., Philadelphia, Pennsylvania, USA

ABSTRACT

During construction, exterior steel I-girders are required to withstand torsional effects from deck overhang loads, which results in the development of lateral flange bending (LFB). AASHTO proposes simplified models to estimate LFB in exterior girders, which were principally adopted from LFB estimates in straight bridges.

Current AASHTO limits are intended to ensure that these maximum LFB stresses do not exceed the girder's flexural capacity. However, these models have not been calibrated to predict the effects offered by structural elements such as cross frames, interior girders and deck forms. Additionally, these models do not directly account for the skew angle effects which can produce differential displacements and/or out-of-plane rotations.

This paper presents the results of a comprehensive parametric study of the levels of LFB during deck placement on short-span steel bridges. 48 bridges were analyzed using the commercial software package Abaqus/CAE (Dassault Systèmes, 2010). S4R elements were used to simulate the girder webs and flanges; B33 elements were used to simulate the cross-frame members and the stiffeners. Figure 1 shows a screen capture of a finite element model analyzed in this study.

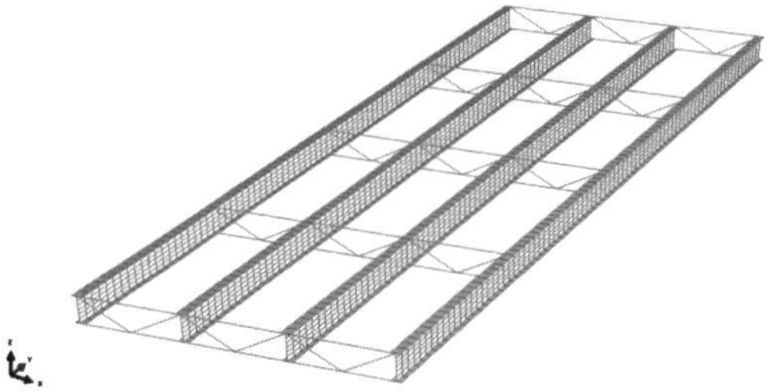

Figure 1. Abaqus screen capture of parametric model.

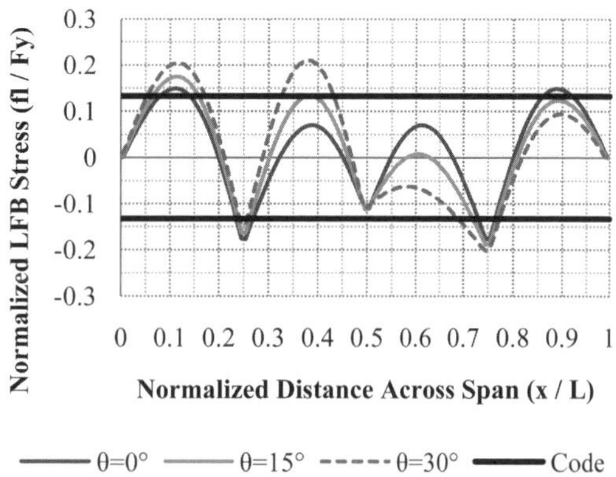

Figure 2. Method comparison for $L = 36.58$ m, $N_b = 5$, $S = 2286$ mm, Case 1 loading condition, plate girder design.

Throughout the parametric matrix of bridges analyzed, AASHTO procedures proved quite accurate in estimating LFB in flanges at the center of the unbraced lengths. However, for all bridges, these approximations underestimated the LFB at the cross-frame locations, which is the location where the maximum LFB is present. In addition, current AASHTO Specifications do not account for the effect of skew. This is apparent when observing Figure 2; as the skew angle increases, AASHTO estimations of LFB prove to be less accurate.

REFERENCES

American Association of State Highway and Transportation Officials. 2010. *AASHTO LRFD Bridge Design Specifications, 5th Edition*. Washington, DC: AASHTO.

Dassault Systèmes. 2009. *Abaqus/CAE Users Manual (Version 6.9)*. Providence, RI: Dassault Systèmes Simulia Corp.

Galindez, N. 2009. *Levels of Lateral Flange Bending in Straight, Skewed and Curved Steel I-girder Bridges during Deck Placement (Ph.D. Dissertation)*. Morgantown, WV: West Virginia University.

Bridge Maintenance, Safety, Management, Resilience and Sustainability – Biondini & Frangopol (Eds)
© 2012 Taylor & Francis Group, London, ISBN 978-0-415-62124-3

Behavior of skewed concrete box girder bridge under static and dynamic loading

X.-H. He & X.-W. Sheng
Central South University, Changsha, China

A. Scanlon & D.G. Linzell
The Pennsylvania State University, Pennsylvania, USA

ABSTRACT

Skewed bridges are widely used in China, with one prominent application being for high speed railways. This paper summarizes some aspects of a research project that encompassed 1:8 scale model testing and analyses of a three-span continuous, prestressed concrete (PC), box girder bridge having a 45° skew. Along with summarizing the design and construction details and the experimental procedure, experimentally obtained displacements and stresses, natural frequencies, mode shapes and damping ratios are presented and compared against those obtained from FE analyses of the tested structures. The influence of skew on the skewed bridge's static and dynamic behavior is investigated.

The overall length of the model bridge was 19 m, consisting of an 8.75 m main span and two side spans of 5 m each. The width of the model girder flanges were 1.65 m at the top and 0.85 m at the bottom. The bridge depth at mid-span of the center span was 0.375 m and at the piers the depth was 0.625 m. The top flange thickness was 70 mm, the bottom flange thickness varied between 70 mm and 90 mm and the web thickness was 110 mm. The model girder bridge was fully prestressed in the longitudinal direction to ensure that the structure was always in compression under the combination of the prestressing, dead and live load. Ten unbonded, continuous prestressed tendons symmetrically placed across the bridge section were used. Concrete used for the model bridge had an elastic modulus of 36 GPa and a compressive strength of 50 MPa.

Static tests were performed on the model bridge under two arrangements of dead load and live load designated as Loading 1 and Loading 2. A third loading combination, Loading 3, was used for computational analysis. Loading 1 represents a symmetrical loading case while Loading 2 and Loading 3 are unsymmetrical, intended to induce torsional effects in the box girder bridge depending on possible locations of trains on the bridge. This paper focuses on elastic static and dynamic performance. Information on inelastic tests and analyses can be found in Sheng and Xin (2005).

Static finite element analyses were performed using a 3-dimensional (3D) finite element (FE) model created in SAP that contained both 3D solid and 1D linear elements.

Ambient vibration testing of the model bridge was completed to measure its dynamic characteristics. The random decrement technique (RDT) method (He et al. 2011) was used for modal extraction and identification. The dynamic tests included two phases, the first being completed with the bridge under an initial, non-skewed support system and the second carried out after the structure was supported by the final, skewed, bearing system. An additional computational examination of a third, non-skewed support system was also completed.

Conclusions are drawn from the results of the study. For example, as box girder bridge skew angles increase, vertical deformations decrease. However, torsional stresses and deformations increase as well as differential reaction levels. Consequently, large skew angles (above 45°) were not recommended for skewed bridges on the high speed railway.

REFERENCES

Sheng, X. W., and Xin, X. Z., (2005). "Nonlinear analysis of the prestressed concrete skew box continuous girder." *Zhongguo Tiedao Kexue*, 26(2), 64–68.

He, X. H., Hua, X.G., Chen, Z.Q., Huang, F.L., (2011). "EMD-based random decrement technique for modal parameter identification of an existing railway bridge". *Engineering Structures*, 2011, 33(4): 1348–1356.

Bridge Maintenance, Safety, Management, Resilience and Sustainability – Biondini & Frangopol (Eds)
 ISBN 978-0-415-62124-3

Skewed steel bridge cross-frame live load performance

S.D. Murphy & D.G. Linzell

Department of Civil and Environmental Engineering, The Pennsylvania State University, University Park, Pennsylvania, USA

ABSTRACT

Skewed bridges are a common and necessary part of American infrastructure. They are often required due to difficult roadway alignments and site constraints. Although there is already a vast inventory of skewed bridges in service, uncertainty still exists with respect to their behavior. Specifically, the interaction between girders and bracing members, such as cross-frames, cause load distributions which are not easily predicted. Cross-frames, which are historically required to provide stability to the girders during construction prior to the hardening of the concrete deck and in negative moment areas where the bottom flange is in compression, are also relied upon to distribute both lateral loads, such as wind, and vertical wheel loads across the bridge. The distribution of load through cross-frames, coupled with the occurrence of differential girder deflections and rotations, common in skewed bridges, could warrant an increased emphasis on cross-frames members as they may attract higher loads.

Research involving skewed steel bridges has considered a variety of components in their behavior using both experimental and analytical methods. However, no studies using field testing have been located which examine typical cross-frame forces in skewed steel bridges under live load. It is the goal of this paper to present cross-frame field testing results of an in-service, highly skewed, steel, bridge and investigate the live load force levels in the cross-frames.

The tested bridge is a 60° skew, slab on steel girder bridge located in central Pennsylvania. The structure has a single span of 55 feet (16.7 m), with nine rolled American Institute of Steel Construction (AISC) W27×91 beams, spaced at 34″ (863.6 mm) center-to-center, compositely acting with a $7^1/_2$″ (190.5 mm) concrete slab connected to the girders via shear studs. Cross-frames consisted of two back-to-back L3″×3″×1/4″ angle members in "X" type frames. The cross-frames are in a staggered orientation running parallel to the abutments at a typical spacing of 17 feet (5.18 m).

Strains on individual angle members in selected cross-frames were recorded during controlled field tests. Field testing was completed using two loaded, three axle trucks.

It was determined from the field tests that cross-frames members at the obtuse corners of the tested bridge were the most critical, and experienced the highest compressive live load forces. Cross-frames near supports generally experienced greater compressive forces than intermediate cross-frames. Intermediate cross-frame members typically had higher tensile forces but those forces were considerably lower than the compressive axial force magnitudes recorded in members near the supports. Cross-frame member bending had considerably less disparity with respect to magnitudes between cross-frames that were monitored, though the highest load was again calculated for a member at the obtuse corner. No axial forces or bending moments were significant compared to member capacities considering only live load.

An analytical parametric study was performed to determine the effect of skew angle and parapet combinations on cross-frame forces. A total of 16 models were created for four skews of 0°, 20°, 40°, and 60°. Of the four models for each skew, one had no parapets and the other three had parapets with construction joints at the half and third points of the span or with no construction joints.

The parametric study indicated that parapet stiffness reduced member forces in cross-frames near the supports. As the skew was reduced, tensile forces became prevalent and compressive forces were reduced. Parapets were shown to increase the tensile forces and reduce compressive forces in intermediate cross-frame members. It was generally indicated that increased skew angle reduced the impact of parapets on cross-frame axial forces.

Bridge Maintenance, Safety, Management, Resilience and Sustainability – Biondini & Frangopol (Eds)
 ISBN 978-0-415-62124-3

Special considerations in curved segmental post-tensioned bridges

A.J. Schokker
University of Minnesota Duluth, USA

ABSTRACT

Segmental concrete bridges have many design and construction considerations that are specific to their specific geometry and assembly. Time dependent staging, complex geometry, and multiple distinct types of construction for curved segmental bridges make them a unique challenge for designers and contractors. This paper covers the specifics unique to curved concrete segmental bridges, including long-term durability. Cast-in-place and precast bridges are discussed for each of the major types of segmental construction methods, including balanced cantilever and span-by-span.

Figure 1. Balanced cantilever construction (I-35 inter-change, Austin, Texas).

Segmental post-tensioned concrete bridges can be extraordinarily durable and aesthetic bridges when properly constructed. Their unique construction considerations (segmental, geometry control, specialized equipment) make their construction more of a challenge for a designer and/or contractor not experienced in segmental concrete construction. This paper provides an overview of the most common construction types (based on U.S. experience, but also typical across the world). The discussion will also include best practices for achieving long-term durability in these structures. A section is devoted to grouting of post-tensioned bridges since this area has received the majority of the attention over the last decade of segmental concrete construction.

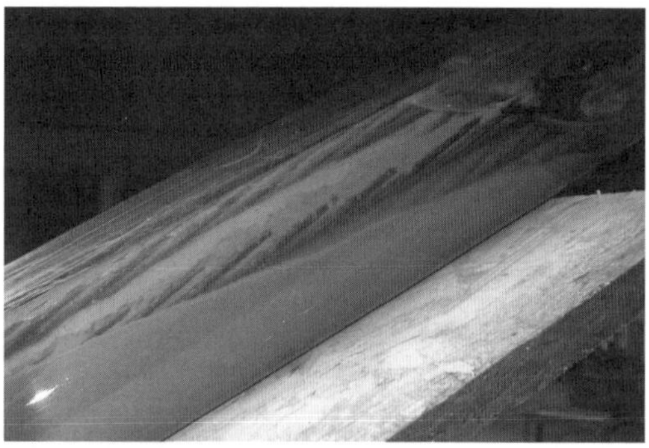

Figure 2. Bleed water in a clear test dust.

Lessons learnt from the Canterbury earthquakes: Assessment, testing and analysis of New Zealand bridges

Organizers: A. Palermo & L.M. Wotherspoon

Bridge Maintenance, Safety, Management, Resilience and Sustainability – Biondini & Frangopol (Eds)
© 2012 Taylor & Francis Group, London, ISBN 978-0-415-62124-3

Observed and predicted bridge damage following the recent Canterbury earthquakes: Toward the calibration and refinement of damage and loss estimation tools

M. Brando
Politecnico di Milano, Milan, Italy

S.L. Lin, S. Giovinazzi & A. Palermo
University of Canterbury, Christchurch, New Zealand

ABSTRACT

Earthquake loss estimation procedures, which attempt to predict damage to structures and infrastructures, casualties and losses for deterministic or probabilistic hazard scenario, can be very powerful tools for supporting the post-event emergency management and pre-event mitigation planning. The availability of a comprehensive database of the damage sustained by the bridge stock in the most affected areas of the Canterbury Region, represent a unique occasion to validate and identify potentialities and criticalities of existing damage and loss estimation tools and procedures toward their calibration and refinement.

In order to offer an unbiased method of assessing the performance and cause of damage to bridges in the Canterbury Region in New Zealand following the 4th September 2010 Darfield earthquake and 22nd February Christchurch 2011 earthquakes (and followings aftershocks), the New Zealand Natural Hazard bridge group, under the leadership of Dr Alessandro Palermo, prepared a Bridge Damage Database [Palermo et al. 2011]. The database covers the area were strongest shaking and land damages were experienced and include 223 bridges. It gathers together both typological caracheristics and constructive details of the bridges together with the damage observed to different bridge components, including deck, bearing and pavement. An overall damage level has been attributed to each bridge combining damages observed to the bridge's components (Figure 1).

The observed damage to the bridges resulting from the processing of the Bridge Damage Database was compared with the damage predicted using of MAEviz platform (MAE, 2007) for a hazard scenario interpolating the peak ground accelerations recorded by the instrumentation available in the Canterbury Region and Christchurch City (GeoNet, 2012). The predicted damage distribution differed from the observed ones (Figure 1), as follow: the percentage of bridges slightly damaged increased resulting in 86% (being 61% the observed one) and, as a consequence, the amount of bridges belonging to the other categories, decrease.

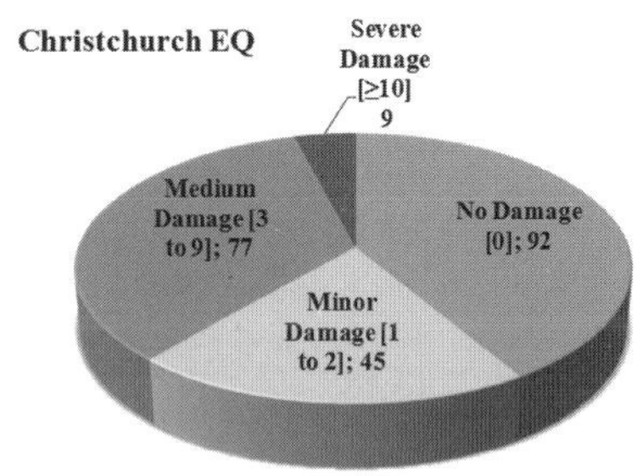

Figure 1. Damage severity to the Canterbury Bridge stock following 22nd February 2011 Christchurch earthquake.

However the percentage of severely damaged bridges resulted almost the same, being down by only 1% on respect to observed damage.

It is worth highlighting that many necessary assumptions were made when conducting loss estimation via MAEviz on Christchurch bridge data, including: 1) fragility curves specific for US; 2) impossibility to include a represent liquefaction induced land-damage; 3) differences in the typological and damage classification taxonomies. Different possible options for improving tools and procedures for predicting damage and induced-impact on the bridge stock have been therefore discussed and suggested as part of the paper.

REFERENCES

GeoNet. 2012. GeoNet Strong Motion Database. *www.geonet.org.nz/resources/basic-data/strong-motion-data/*

MAE, 2007, Mid-America Earthquake Centre. Mid-America Earthquake Centre Seismic Loss Assessment System, MAEviz v3.1.1. University of Illinois at Urbana-Champaign, Urbana, Illinois.

Palermo A., Wotherspoon, L., Wood, J., Chapman, H., Scott, A., Hogan, L., Kivell, A., Camnasio, E., Yashinsky, M., Bruneau, M., and Chouw, N., 'Lessons Learnt from 2011 Chrischurch Earthquakes: Analysis and Assessment of Bridges.', *Bulletin of the New Zealand Society for Earthquake Engineering*. Vol. 44 (2011), 319–33.

Bridge Maintenance, Safety, Management, Resilience and Sustainability – Biondini & Frangopol (Eds)
© 2012 Taylor & Francis Group, London, ISBN 978-0-415-62124-3

Lateral spreading interaction with bridges during the Canterbury earthquakes

E. Camnasio
Politecnico di Milano, Milan, Italy

M. Le Heux
Rose School of Pavia

A. Palermo
University of Canterbury, Christchurch, New Zealand

L.M. Wotherspoon
University of Auckland, Auckland, New Zealand

ABSTRACT

In less than six months the city of Christchurch, New Zealand, experienced two major earthquakes, occurring on September 4, 2010 and February 22, 2011. Since Canterbury soils are susceptible to liquefaction [Bartlett & Youd 1995], the damage that bridges suffered was mainly due to the lateral spreading of the river banks throughout the city of Christchurch and in the rural areas in the surroundings. The soil spreading and the consequent settlement of the surroundings affected the foundation system of the bridges and damaged the approaches in general. As a consequence, bridge serviceability was often disrupted, especially considering the cumulative effect that both earthquakes had in some cases.

Lateral spreading led to pile damage in the form of plastic hinging below grade due to the imposed displacement. Significant liquefaction settlement led to the exposure of the piles. Apart from ground shaking, bridge piers were sometimes damaged by lateral spreading imposing a large displacement at the bottom of piers and inducing a large moment at the pier-deck interface. Flexural cracks developed and rotations at the bottom of the pier tilted the columns [UoC 2011].

In the present work, the three-layer lateral spreading model [Cubrinovski & Ishihara 2004] is adopted in order to determine the effect of lateral spreading forces on bridge structures. This methodology models the soil as equivalent springs, which have stiffness and strength accordingly to the non liquefied/liquefied soil properties. The numerical analysis is then carried out in the framework of a p-y analysis, where displacements are imposed to the foundation system reproducing the lateral spreading of the river banks.

Two partial pile-abutment models, i.e. Pages Road Bridge and ANZAC Drive Bridge, and a global bridge structure, i.e. Dallington Pedestrian Bridge, were herein investigated. These integral, precast and arch concrete bridges are chosen among the most damaged bridges to show how the lateral spreading affects different structural typologies.

As far as the pile-abutment subassemblies, particular attention was put in the modeling of the deck-abutment connection to simulate the interaction between the foundation system and the superstructure. Rotational springs with properly calibrated stiffness were introduced. The results provided a better understanding of the damage that likely occurred to the piles below the ground level. The analyses showed that plastic hinging usually develops at the interface between liquefied and non-liquefied layers or at the pile-abutment connection.

The case study of the Dallington Pedestrian Bridge [Le Heux 2012] gave an example of the global performance of a bridge subjected to lateral spreading. The compression forces passing through the bridge were evaluated, as the bridge acted as a strut spanning across the river and bearing against the river banks. The analysis results compared well with the observations made at the site and led to the provision of recommendations for the selection of certain parameters required when determining the forces that acts on a bridge due to lateral spreading.

REFERENCES

Bartlett, S.F., & Youd, T.L. 1995. Empirical prediction of liquefaction-induced lateral spread. *Journal of Geotechnical Engineering*, 121(4), 316–329.

Cubrinovski, M., & Ishihara, K. 2004. Simplified method for analysis of piles undergoing lateral spreading in liquefied soils. *Soils and Foundations*, 44(25), 119–133.

Le Heux, M. 2012. Darfield Earthquake 2010 Bridge Performance & Lateral Spreading Design Lessons. *Masters Dissertation*. Rose School, Italy.

UoC, University of Canterbury 2011. Database of Bridge Performance to Significant Events. Department of Civil and Natural Resource Engineering. Contact: alessandro.palermo@canterbury.ac.nz

Bridge Maintenance, Safety, Management, Resilience and Sustainability – Biondini & Frangopol (Eds)
 ISBN 978-0-415-62124-3

Forced vibration testing of bridge damaged in the 2010 Darfield earthquake

L.S. Hogan, L.M. Wotherspoon, S. Beskhyroun & J.M. Ingham
Civil and Environmental Engineering, University of Auckland, Auckland, New Zealand

ABSTRACT

Abutments provide the large interfaces between the bridge superstructure and the surrounding soil. This interface contributes significantly to the overall stiffness and damping of a bridge system when loaded seismically (El-Gamal and Siddharthan 1998; Kotsoglou and Pantazopoulou 2009). While abutments contribute greatly to the dynamic response of a bridge, there has been little testing investigating the interaction between the abutment and the remaining structures of in situ bridges.

The opportunity to test the interaction between the abutments and the rest of the bridge structure became available in South Canterbury, New Zealand. Following the 2010 M7.1 Darfield Earthquake, the Davis Road Bridge, located 5 km southeast of Lincoln, New Zealand, sustained significant damage to the western approach. The single span bridge consisted of six double hollow core precast concrete beams with six square precast concrete piles and a 2.0 m long friction slab at each abutment. During the earthquake, lateral spreading caused the western approach to subside approximately 0.5 m. The subsidence of the approach effectively removed the stiffness provided by the abutment and provided a unique opportunity to directly measure the influence of this resistance on the system.

The span was subjected to shaking along both axes from a large eccentric mass shaker (Figure 1) and a benchmark system identification was made of the bridge in the damaged state. Soil was then recompacted, and the road repaved at the approach. Once the bridge was reinstated, another round of shaking was performed, and differences in mode shapes and natural periods were compared between the damaged and reinstated state. Testing was able to detect one mode in each direction for the bridge both with and without both abutments intact (Figure 2).

Figure 1. Setup of shaker and accelerometers for forced vibration tests.

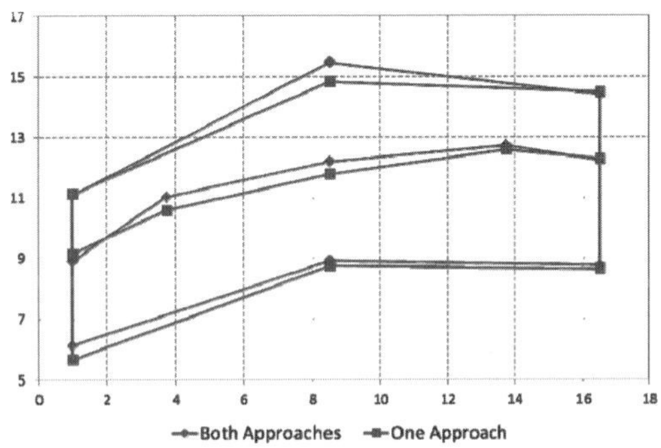

Figure 2. Comparison of first mode shapes in transverse direction of the bridge with and without both approaches intact.

REFERENCES

El-Gamal, M. and R. V. Siddharthan (1998). "Stiffnesses of Abutments on Piles in Seismic Bridge Analyses." *Soils and Foundations* **38**(1): 77–87.

Kotsoglou, A. N. and S. J. Pantazopoulou (2009). "Assessment and Modeling of Embankment Participation in the Seismic Response of Integral Abutment Bridges." *Bulletin of Earthquake Engineering* **7**(2): 343–361.

Bridge Maintenance, Safety, Management, Resilience and Sustainability – Biondini & Frangopol (Eds)
© 2012 Taylor & Francis Group, London, ISBN 978-0-415-62124-3

Overview of bridge performance during the 2011 Christchurch earthquake

A. Palermo & A. Kivell
University of Canterbury, Christchurch, New Zealand

L.M. Wotherspoon & L.S. Hogan
University of Auckland, Auckland, New Zealand

M. Yashinsky
CALTRANS, Earthquake Engineering Office, Sacramento, USA

M. Bruneau
University of Buffalo, Buffalo, United States

E. Camnasio
Politecnico di Milano, Milan, Italy

ABSTRACT

The M_W 6.3 February 22, 2011 Christchurch earthquake was centred 10 km south east of central Christchurch on the edge of the city at a depth of 5 km. Peak Ground Accelerations (PGA) in the area of South-East Christchurch were much higher than the design level in the period range of New Zealand road and highway bridges, with exceptional values of vertical acceleration being registered.

Most of the damage was a result of liquefaction and lateral spreading of the river banks (Figure 1), with very few examples of significant bridge damage on non-liquefiable sites. A number of bridges suffered non-structural damage such as slumping of abutment aprons and fracture of deck drainage pipes. Overall, bridges suffered little structural damage compared to other structures such as residential houses and commercial buildings. Damage was mainly confined to the bridges spanning the Avon River in the Central/Eastern part of the city, nearest to the earthquake epicentre. However, despite the moderate structural damage, extensive disruption to traffic occurred due to the closure of bridges critical to the network.

Following the earthquake, the bridge stock was inspected by the network consultants and researchers to establish safety conditions, repairs that were required to enable traffic to flow and document damage. A site walk-over was carried out at each of the inspected bridges with particular attention focused on checking for evidence of movements at the piers and abutments.

This paper presents a summary of observations from the field on a selection of the most severely damaged bridges in the Central Business District (CBD) and East and Southern suburbs, subdivided by typology, *i.e.* road bridges and highway bridges.

(a) (b)

Figure 1. (a) Backwards rotation of the abutments of Anzac Drive Bridge [Photo by A. Palermo]; (b) Plastic hinging at the pier caps of Anzac Drive Bridge [Photo by E. Camnasio].

REFERENCES

Chapman H.E., Lauder M.K., Wood J.H. (2005). Seismic assessment and retrofitting of New Zealand state highway Bridges. *Proc. of the New Zealand Society Earthquake Engineering Conference (NZEES),* March, 11–13, 2005, Wairakei (New Zealand), CD-ROM.

GeoNet (2011) "Strong motion FTP site, available from ftp://ftp.geonet.org.nz/strong/processed/Proc/2011/02_Christchurch_mainshock_extended_pass_band/ (Accessed August 12, 2011)

GNS Science (2011). http://www.geonet.org.nz/earthquake/historic-earthquakes/ GeoNet – Historic Earthquakes. (Accessed September 1, 2011).

NZTA (2003). Bridge Manual, Second Edition, Including Amendments to December 2004. New Zealand Transport Agency, Wellington.

Palermo, A., Wotherspoon, L., Hogan, L., Le Heux, M., Camnasio, E. (2011a). Seismic Performance of Concrete Bridges during Canterbury Earthquakes. Structural Concrete (accepted).

Palermo, A., Wotherspoon, L., Wood, J., Chapman, H., Scott, A., Hogan, L., Kivell, A., Camnasio, E., Yashinsky, M., Bruneau, M., Chouw, N. (2011c). Lesson learnt from 2011 Christchurch earthquakes: analysis and assessment of bridges. *NZSEE Bulletin* (in review).

Standards New Zealand, NZS 1170.5:2004. Structural Design Actions. Part 5: Earthquake actions – New Zealand.

Transit New Zealand (1998). Manual for seismic screening of bridges, SM110, Wellington, New Zealand.

Transit New Zealand (2004). New Zealand Bridge Design Manual, second edition.

Wood, J.H., Chapman, H.E., Brabhaharan, P. (2011). Performance of Highway Structures During the Darfield and Christchurch Earthquakes. Report to New Zealand Transport Agency. (In preparation).

Bridge Maintenance, Safety, Management, Resilience and Sustainability – Biondini & Frangopol (Eds)
© 2012 Taylor & Francis Group, London, ISBN 978-0-415-62124-3

Performance of bridges during the 2010 Darfield earthquake

L.M. Wotherspoon & L.S. Hogan
Civil and Environmental Engineering, University of Auckland, Auckland, New Zealand

A. Palermo
Civil and Natural Resources Engineering, University of Canterbury, Christchurch, New Zealand

M. Le Heux
Rose School, Pavia, Italy

M. Bruneau & M. Anagnostopoulou
Civil, Structural and Environmental Engineering, University of Buffalo, Buffalo, USA

ABSTRACT

The M_W7.1 Darfield earthquake occurred on September 3 2010 (UTC) with an epicentre approximately 40 km west of central Christchurch. It was felt throughout the South Island of New Zealand, and the maximum felt intensity was approximately MM 9. The duration of strong shaking was between 8 and 15 seconds in the Canterbury region (Gledhill et al. 2011). In the region near the epicentre, horizontal PGAs up to 0.76 g were recorded. Within Christchurch, the typical horizontal PGAs ranged from 0.15–0.25 g, while to the north of Christchurch in the town of Kaiapoi the PGA was 0.34 g. In general spectral accelerations were below the 1000 year return period design levels in this region, the typical return period used for bridge design.

The immediate districts surrounding the epicentre and associated fault rupture contain more than 800 road, rail and pedestrian bridges. Five authorities own and operate the majority of the bridges within the area where the felt intensity was above Modified Mercalli (MM) 6. The initial response of the different authorities was to carry out preliminary visual inspections on all bridges in the region to determine the level of damage to the road and rail networks. Visual inspections of fifty five bridges were carried out by the authors, which was combined with information provided by other institutions involved in the post-earthquake inspections. Most bridges sustained little to no damage and were operational immediately after the event. Those bridges which were moderately to severely damaged were limited to areas of extensive liquefaction and lateral spreading.

Out of the total bridge stock, eight road bridges required closure following the event, mainly due to approach damage caused by liquefaction induced lateral spreading. Five of these road bridges remaining closed for at least five days after the earthquake while temporary repairs were completed. With the exception of one of the eight closed bridges, none suffered major damage to the superstructure, with damage to the approach the main reason for closure. Remediation efforts were therefore focused on the repair and building up of the approaches to allow traffic to pass. More permanent repairs were made in the months following the earthquake. Highways bridges generally performed well, with only one case of closure for more than a day where the approach subsided and cracked.

Six pedestrian bridges suffered severe structural damage as a result of lateral spreading. The superstructure stiffness of these bridges was much less than the road bridges and they were therefore unable to resist compressive loads induced by lateral spreading. Because of this, they were damaged to the extent that replacement will be necessary.

Bridges near the epicentre suffered no damage at locations were liquefaction was not observed. For example, the Selwyn River Bridge on State Highway 1, situated less than 5 km from the epicentre, was operational immediately after the earthquake.

The two major factors that contributed to the good performance of the bridge stock were first, most bridges in the Canterbury area had small to moderate spans that generally exhibit a sturdy seismic response, due largely to their symmetry and limited reactive mass. Secondly, newer bridges were generally designed to resist forces larger than the demands imparted by this earthquake.

REFERENCE

Gledhill, K. Ristau, J. Reyners, M. Fry, B. and Holden, C. 2011. The Darfield (Canterbury, New Zealand) Mw 7.1 earthquake of September 2010: A preliminary seismological report. *Seismological Research Letters* 82(3): 378–386.

Corrosion detection in cables and concrete bridges by magnetic methods

Organizers: A. Ghorbanpoor & B. Hillemeier

Bridge Maintenance, Safety, Management, Resilience and Sustainability – Biondini & Frangopol (Eds)
 ISBN 978-0-415-62124-3

Magnetic inspection of adjacent box-beam girders

B. Fernandes, D.K. Nims & V. Devabhaktuni
The University of Toledo, Toledo, Ohio, USA

ABSTRACT

Load rating of aging prestressed concrete adjacent box-beam bridges (PCBB) requires an accurate assessment of the condition of the prestressing strands. Compared to other nondestructive evaluation techniques, magnetic methods are the most effective for detecting strand corrosion (Jones et al. 2010). This signal obtained by magnetizing a steel body, such as a steel rope, can be of two types – induced magnetic field (IMF) due to the main flux in the magnetic circuit and leakage flux generated due to flaws in the steel body under inspection (see Fig. 1). The amount of material lost due to corrosion, strand breaks or dislocations can thus be detected by magnetizing the strand and analyzing its magnetic signals. Presently, magnetic flux leakage (MFL) detection system has been used to inspect prestressed strands with good success (Jones et al. 2010). The perturbation in the MFL signal gives information about flaws in the strand, such as cracks, fractures and position of section loss. Apart from that the magnitude of the signal can be used to estimate the cross-sectional area of the strand using a correlation method. The authors have been working on development of an electromagnet-sensor system based on measuring the IMF due to main flux. This system has been tested in the laboratory as proof-of-concept. It was also tested in the field and was able to predict loss of cross-sectional area to a good level of accuracy.

A comprehensive practical magnetic inspection system (MIS) based on detection of main flux and leakage flux has been proposed. Work is in progress to resolve the challenges in applying the magnetic flux leakage and induced magnetic field methods. Both methods produce a signal which is visually interpreted. By consideration of the MIS as a whole and addressing the weak links in the devices and processes, significant strides in developing a practical inspection system are being made. To scan the smooth soffit of box-beam bridges, a semi-automated mechanical system that can maneuver the sensor unit under the bridge is being designed. A computer based simulation model to generate magnetic field signals from noncorroded strands is discussed. Such a model can be useful tool to make estimations of corrosion by comparison with data obtained in the field.

Too much detail in inspection is costly and does not result in a more accurate load rating; too little information leads to inaccurate load rating. The research has shown a magnetic system can extract the necessary data. However, several steps need to take place before an MIS is practical for PCBB. The paper discusses the background of magnetic inspection of strand and the steps to make it practical.

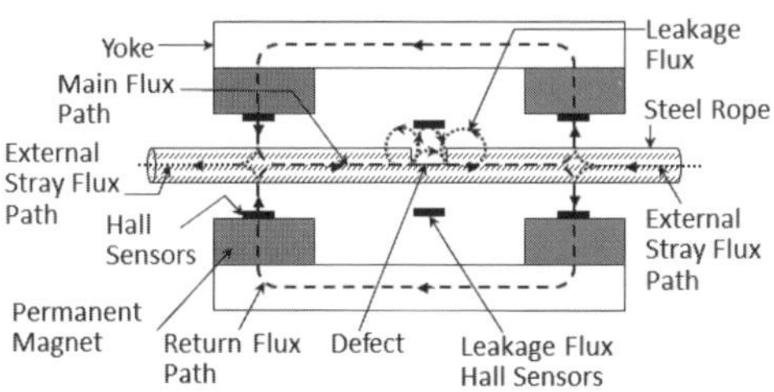

Figure 1. Magnetic field inspection of a steel rope (Weischedel 2011).

REFERENCES

Jones, L. et al. 2010. Inspection methods & techniques to determine non visible corrosion of prestressing strands in concrete bridge components, Task 2 – Assessment of candidate NDT methods. ATLSS Report No. 09-09, Lehigh University, Bethlehem, PA and Pennsylvania Department of Transportation, Harrisburg, PA.

Weischedel, H.R. The magnetic flux leakage inspection of wire ropes. NDT Technologies, Inc. Available at: http://www.ndttech.com/Papers/FluxLeakageWireRopes.pdf. Last accessed November 14, 2011.

Bridge Maintenance, Safety, Management, Resilience and Sustainability – Biondini & Frangopol (Eds)
 ISBN 978-0-415-62124-3

Fast and innovative detection of fractures in prestressing tendons on German highway-bridges

C. Flohrer
HOCHTIEF Solutions AG, Mörfelden-Walldorf, Germany

ABSTRACT

A historical development of the test method and the application of various investigations and test equipment in case studies will be presented. Case studies of pre-tensioned and post-tensioned concrete structures and bridges in Germany and Hungary, and the application of the scanning remanent magnetism method to detect wire fractures in lateral tendons of bridge decks, show the possibilities of non-destructive testing methods. Logistics, traffic management, economics and methods for rehabilitation also will be discussed.

The method of magnetic detection of reinforcement fractures in pre-stressed concrete elements is based on the effect, that a dipole is formed at the fracture point of the magnetized wire. The stray magnetic field at the fracture point of a magnetized wire is physically comparable to a broken bar magnet. The basics are presented in (Hillemeier et al., 2012).

At the end of the nineteen eighties the stray magnetic field method was developed in the construction industry by HOCHTIEF, specifically for the assessment of the structural stability of Bavarian cowsheds. The aim was to locate non-destructive steel fractures in pre-stressed concrete beams, Flohrer (1990). The method was applied successfully over 20 years.

The measurement of single transverse tendons is complex. First the tendons have to be located with radar and then the remanent magnetic measurement, described above, must be performed. This does not necessarily require a rail system; however, to measure the road pavement, it has to be removed.

The measuring process can be accelerated by using a larger magnet and the use of magnetic sensors which detect larger areas. At the TU Berlin, Department of Building materials and construction materials testing, a bridge-inspection system for checking the transverse tendons was developed (Hillemeier et al., 2012).

The application of magnetic field measurements with the large bridge scanner requires preliminary conditions and preparations.

Preliminary conditions are:

- accessibility to the building,
- for bridge investigations under traffic, minimum two lanes must be blocked for traffic for each measurement line (because of the dimensions of the magnetic scanner and a necessary working area two lanes are required),
- the area under the coils cannot be investigated because the magnetic field is too strong to identify fracture signals; results can be obtained at about 40 cm beside the border of the measuring unit,
- the number of lanes to be tested; the client has to determine how many lanes are to be investigated,
- Type and thickness of the pavement structure on the concrete surface; preferably one should measure directly on the surface of the concrete slab,
- Depth of the transverse tendons; a maximum depth of the cables of about 25 cm is acceptable.

For economical bridge investigations with the magnetic field scanning unit, the works on site have to be prepared meticulously. The magnetic yoke with a mass of approximately 2.0 t is operated with a controllable DC power supply with a maximum power output of approximately 16 kW. Because of the high starting current, a highly efficient power unit is required.

REFERENCES

Flohrer, C. (1990), Zerstörungsfreie Ortung von Spanndrahtbrüchen in Spannbeton-Decken-trägern, VDI-Berichte 1990, S. 425–433

Hillemeier, Bernd, Pak, Chol-I.: Magnetic Localization of Fractures of Broken Wires in Pre-stressing Cables of Bridges and Parking Decks; IABMAS 2012

Bridge Maintenance, Safety, Management, Resilience and Sustainability – Biondini & Frangopol (Eds)
 ISBN 978-0-415-62124-3

Corrosion detection in tendons of segmental concrete bridges

A. Ghorbanpoor & E. Abdel-Salam
Department of Civil Engineering, University of Wisconsin-Milwaukee, USA

ABSTRACT

Segmental concrete bridge structures primarily rely on prestressing steel that are post-tensioned to carry all relevant bridge loads. During recent years, many bridge owners throughout the world have experienced deterioration, and sometimes complete fracture, of prestressing steel within post-tensioned (P-T) tendons due to corrosion. Since prestressing steel within P-T tendons is normally hidden from view, routine inspection practices do not allow for effective evaluations and detection of corrosion in such tendons.

Recent developments in the field of magnetic-based NDE have shown significant improvements in corrosion detection capabilities when used for NDE of external tendons in post-tensioned segmental concrete bridges. One successful magnetic-based system that is developed by the authors of this paper is based on the concept of Magnetic Flux Leakage (MFL). The system consists of a set of strong permanent magnets, a series of Hall-Effect sensors, and a number of electronic and mechanical devices that are required for its operation (Ghorbanpoor et al. (2000)). This MFL system has been evaluated and optimized for field applications through conducting extensive laboratory and field tests. It has been demonstrated that the current MFL system is capable of detecting corrosion, or section loss, as small as less than 1 percent of the cross sectional area of external P-T tendons in segmental bridges. The MFL system has been used by the authors on a number of major cable supported bridges in the USA and other countries. Figure 1 shows an example of detected surface corrosion in a P-T tendon of a segmental concrete box bridge.

One major shortcoming of the current MFL system has been its limitation to inspect the P-T cables at the ends inside the anchorage region. Normally, the P-T cable in a segmental concrete box is extended at each end from the face of the concrete a few feet into the anchorage region so it can be stressed and anchored in place at the opposite face of the concrete. Accordingly, it has not been possible to use the MFL system to inspect the cable within the anchorage region. To overcome this shortcoming, the authors have developed a new system that integrates the capabilities of the current MFL system with those of a magnetostrictive (MS) transceiver. The magnetostrictive transceiver includes a transmitting coil that introduces a guided wave in the cable and a receiver that captures cable corrosion or fracture signals from reflections of the transmitted guided wave from such anamolies inside the anchorage region. Figure 2 shows a recorded MS signal that includes a reflection of the guided wave from the fracture end of a single strand that is a part of a P-T cable with 19 strands.

This paper presents the results of NDE investigations to evaluate the condition of external tendons in major post-tensioned segmental concrete bridges in the United States using the MFL system as well as laboratory results for the use of the MS transceiver in conjunction with the MFL to detect corrosion or fracture of the cable.

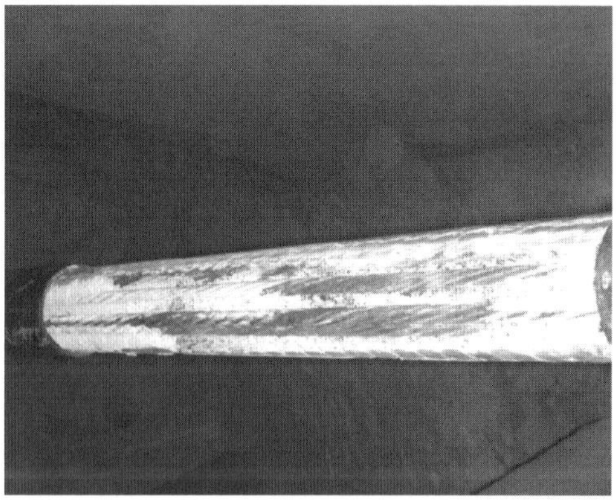

Figure 1. An example of detected corrosion in a P-T tendon of a segmental concrete box bridge.

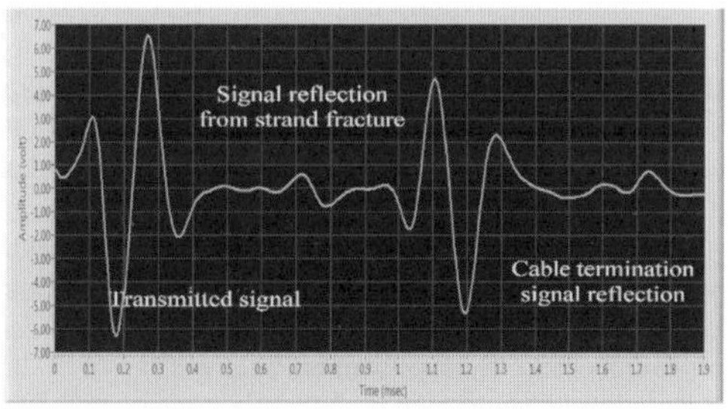

Figure 2. MS signal showing a reflection of the guided wave from the fracture end of a single strand in a P-T tendon.

REFERENCE

Ghorbanpoor, A., Borchelt, R., Edwards, M. & Abdel-Salam, E. 2000. Magnetic-based NDE of prestressed and post-tentioned concrete members – The MFL system. Final *Report No. FHWA-RD-00-026; Federal Highway Administration, US Department of Transportation.*

Bridge Maintenance, Safety, Management, Resilience and Sustainability – Biondini & Frangopol (Eds)
© 2012 Taylor & Francis Group, London, ISBN 978-0-415-62124-3

Magnetic localization of fractures of broken wires in pre-stressing cables of bridges and parking decks

B. Hillemeier
Technische Universität Berlin, Germany

C.-I Pak
Paektusan Architectural Academy Pyongyang, Korea, DVR

ABSTRACT

The Remanent Magnetism Method (RMmethod) allows the identification of potentially unsafe conditions in pre-tensioned and posttensioned concrete structures by locating fractures in the pre-stressing steel. This nondestructive method identifies fractures in single wires, even when they are bundled with intact wires. The magnetic field of the tendons is measured at the concrete surface, once they have been premagnetized with an electromagnet. Fractures produce characteristic magnetic leakage fields, which can be measured with appropriate sensors at the concrete surface.

The parameters associated with fractured wires have been quantitatively identified in the laboratory and have been confirmed in the field. The knowledge of these parameters allows to draw conclusions about the reduction of the crosssectional area or the number of fractured wires in a tendon.

The method has been independently evaluated by the German Federal Highway Research Institute (Bundesanstalt für Straßenwesen, BASt) and won the Innovation Prize Berlin-Brandenburg 2006.

Together with the Fraunhofer Institute for Non-destructive Testing, Saarbrücken, a large rotating magnetic field scanner has been built.

The latest development is an array of 120 Hall effect sensors in a straight line of 3.00 m measuring as well in a remanent as in an active field. The test results are indicated just in time in a gray scaled magnetic laptop picture.

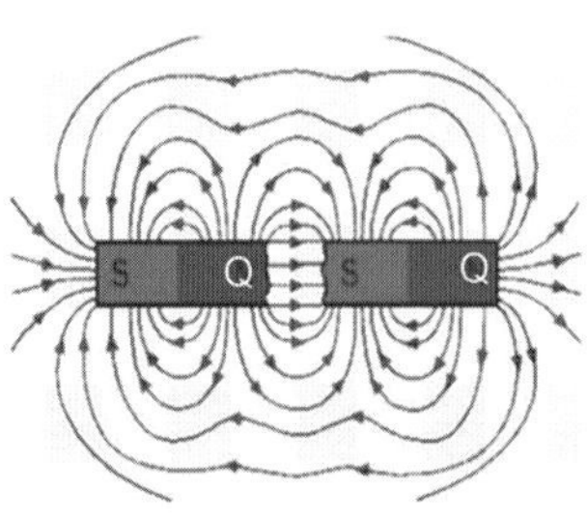

Figure 1. At the fracture point of a bar magnet (pre-stressing wire), a magnetic north (N) pole and south (S) pole are located directly next to each other.

Figure 2. Detecting fractures in pre-stressing tendons on a parking deck with the big magnetic testing device.

$$\Delta B_{Bruch}(x,y,z) = -\frac{B_{R,S}\cdot z\cdot d^2}{16}\frac{3\cdot y\cdot l_B}{\left(x^2+y^2+z^2\right)^{5/2}} + O(l_B^3)$$

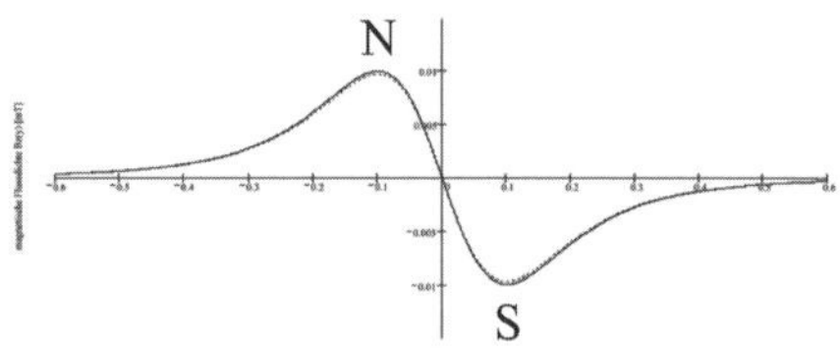

Figure 3. Graph shows that the fracture signal can be approximated very well with the formula mentioned above, derived by Maxwell theory and simplified by Taylor series development.

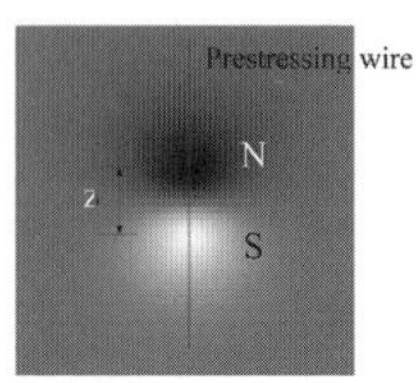

Figure 4. Typical fracture signal measured two-dimensionally and visualized in a gray scaled magnet picture with south pole (white) and north pole (black).

REFERENCE

Hillemeier, B. & Walther, A. & Pak, Chol-I. 2008. Fast Non Destructive Localization of Pre-stressing Steel Fractures in Post-Tensioned Concrete Bridges. *Accelerated Bridge Construction-Highway for Life, Conference, Baltimore, Maryland* 2008: 409–410

Bridge Maintenance, Safety, Management, Resilience and Sustainability – Biondini & Frangopol (Eds)
 ISBN 978-0-415-62124-3

Application of line scanner in remanent and active field compared with the big magnet impulse magnetization

S. Knapp & B. Hillemeier
IFDB, Hillemeier & Knapp GmbH, Berlin, Germany

ABSTRACT

The contribution describes case studies concerning the application of the remanent magnetism method (RM-method). The development is shown with the examination of roofs, halls, bridge buildings and tank constructions. Generally one can say that every non-destructive test procedure passes through different stages of development. At first it is necessary to recognize a problem situation. Fractures in prestressed tendons must be detected non-destructively in prestressed concrete structures. The physical effect used in this case is based on the magnetic flux leakage (MFL-method). Both the remanent magnetic field and the active magnetic field are used. Both magnet procedures have their specific advantages (Hillemeier et al. 2008, Taffe et al. 2010).

The success is always pleasant when a new measurement procedure works in the laboratory. With the success, however, the problems also grow. The data acquisition and the data evaluation permanently must be improved and optimized. The equipment has to be developed further for practical applications. Organizationally care must be taken to access to the structure surface safe and secure. The operation of the equipment requires the expert. This should be a civil engineer with profound knowledge in physics, mathematics, electrical engineering and computer software

Figure 1. Magnet-Sensor-Unit, which was adapted by a substructure at the curvature of the double-curved HP-shells.

Figure 2. In-situ measurements with the large magnet and the newly developed line scanner on a highway bridge in Germany.

and last but not least he must work like a construction worker. The greatest difficulty is finally finding the suitable staff. The better a non-destructive test procedure works, the less demanding is the work for the staff. If the procedure can be used successfully, the wish grows to increase the speed and to reduce the costs. In the beginning the quality counts the speed at the end. Every non-destructive test procedure can be developed further in all stages of development. The computer software, the equipment engineering, the increase of the ruggedness, the personnel training and the making of manipulators to avoid the erection of scaffolding they all have specific development potential. The following two examples show the application of the procedure in its stages of development.

REFERENCES

Hillemeier, B., Walther, A. & Pak, C. 2008. Fast Non-Destructive Localization of Pre-stressing Steel Fractures in Post-Tensioned Concrete Bridges, In *2008 Accelerated Bridge Construction – Highway for Life, Conference, Baltimore – Maryland*, 409–410.

Taffe, A., Hillemeier, B. & Walther, A. 2010. Condition Assessment of a 45-year old pre-stressed concrete bridge using NDT and verification of the results, In *NDE/NDT for Highways and Bridges: Structural Materials Technology (SMT), New York, 16–20 August 2010.*

Hybrid composite bridge system

Organizers: H. Furuta & S.-H. Kim

Bridge Maintenance, Safety, Management, Resilience and Sustainability – Biondini & Frangopol (Eds)

Partial interaction analyses of composite steel-concrete girders subjected to combined bending and shear

P. Ansourian & G. Ranzi
The University of Sydney, Sydney, Australia

A. Zona
University of Camerino, Ascoli Piceno, Italy

ABSTRACT

Earlier studies on the behaviour of steel-concrete composite beams highlighted that the relative displacement between the steel beam and the reinforced concrete slab (partial interaction) requires to be included in a composite beam model to properly estimate the structural stiffness as well as the structural capacity, as shear connection can also be responsible for the structural collapse. Such considerations are now widely accepted (Spacone and El-Tawil 2004) and even included in some modern structural codes.

One of the first papers dealing with the analysis of composite beams with partial interaction is the one by Newmark et al. (1951) where a beam model that couples two Euler-Bernoulli beams, i.e. one for the reinforced concrete slab and one for the steel beam, by means of a deformable shear connection distributed along their interface, is presented. This shear connection enables longitudinal relative movement to occur between the two components while preventing their vertical separation. The Newmark kinematic model has been widely applied for static linear elastic analyses, and various formulations were presented for nonlinear static analysis under monotonic and cyclic loadings, and more recently for nonlinear dynamic analysis under earthquake ground motions. Some specific applications to steel-concrete composite girders involved probabilistic nonlinear analysis in order to assess different design approaches in continuous girders (Zona et al. 2010).

Modifications to the Newmark kinematic model were recently proposed in order to include the combined effect of bending and shear in the composite beam components. A beam model (EB-T model) including the shear deformability of the steel component only was introduced by Ranzi and Zona (2007). This model was obtained by coupling an Euler-Bernoulli beam (only flexural deformability and flexural failure mode of each beam component) for the reinforced concrete slab with a Timoshenko beam (flexural and shear deformability as well as flexural and shear failure mode) for the steel member. The composite action was provided by a continuous shear connection which, as in the Newmark model, enabled longitudinal relative displacements while preventing vertical separation. Successively, Zona and Ranzi (2011) introduced a beam model (T-T model) including the shear deformability of both component of the composite beams, i.e., model derived by coupling two Timoshenko beams with composite action was provided by a continuous shear connection enabling longitudinal relative displacements while preventing vertical separation.

In this paper the results obtained at service and ultimate conditions based on the Newmark model, the EB-T model and the T-T model are compared, summarizing the results presented by Ranzi and Zona (2007) and Zona and Ranzi (2011), in order to give a comprehensive evaluation of the effects of the shear deformability of the steel and slab components at various load levels. Results show that differences between the three models are significant in many cases, both at service state and at collapse.

REFERENCES

Newmark, N.M., Siess, C.P. & Viest, I.M. 1951. Tests and analysis of composite beams with incomplete interaction. *Proceedings of the Society of Experimental Stress Analysis* 9(1): 75–92.

Ranzi, G. & Zona, A. 2007. A steel-concrete composite beam model with partial interaction including the shear deformability of the steel component. *Engineering Structures* 29(11): 3026–3041.

Spacone, E. & El-Tawil, S. 2004. Nonlinear analysis of steel-concrete composite structures: state-of-the-art. *Journal of Structural Engineering* 130(2): 159–168.

Zona, A. & Ranzi, G. 2011. Finite element models for nonlinear analysis of steel-concrete composite beams with partial interaction in combined bending and shear. *Finite Elements in Analysis and Design* 47(2): 98–118.

Zona, A., Barbato, M., Dall'Asta, A. & Dezi, L. 2010. Probabilistic analysis for design assessment of continuous steel-concrete composite girders. *Journal of Constructional Steel Research* 66(7): 897–905.

Bridge Maintenance, Safety, Management, Resilience and Sustainability – Biondini & Frangopol (Eds)
© 2012 Taylor & Francis Group, London, ISBN 978-0-415-62124-3

Study on crack inspection of in-service steel structure by EPDM

K.H. Chang, D.N.V. Vuong & S.H. Hyun
Department of Civil and Environmental Engineering, Chung-Ang University, Seoul, Korea

C.H. Lee
Structures and Bridges Research Division, Korea Institute of Construction Technology, Seoul, Korea

M. Hirohata
Graduate school of Engineering, Nagoya University, Nagoya, Japan

Y.C. Kim
Joining and Welding Research Institute., Osaka University, Osaka, Japan

ABSTRACT

Recently, steel construction such as desalination plant, steel bridge and vessels is currently increasing on account of several merits of steel structures. Due to numerous environmental factors and fatigue phenomena by cyclic loading, many cracks and extensive corrosion occur in steel structures, which cause the deterioration of the performance and life cycle of such systems. Thus, maintenance of steel structures is strongly demanded for safety control. However, the inspection methods that are currently being used are very limited and can detect only local defects in steel structures. It also takes large amount of time to detect and incur high maintenance costs. Moreover, such method cannot be applied to huge steel structures, which people can't be unapproachable. They also require much time because of local point measurement. Such method is needed the periodic inspection and may lead to cost loss due to stopping of service. Therefore, the development of an inspecting technology that can detect defects of whole structures and can reduce the repair and strengthening costs at early stage is very much needed. In this study, inspection method of cracks or corrosion is newly developed by using the electric-potential-drop method (EPDM) which is the non-destructive inspection method to detect corrosion and cracks of steel structure system. EPDM technique is also the way to evaluate the fatigue life, the direction generation and propagation of cracks could be specified. Based on the obtained results between the analysis and experiment, it was verified that this technique can be detected with high accuracy and extensively.

REFERENCES

Lee, JaeSun, Chang. KyongHo, Hwang. JeeHoon, Park. Hyun Chan, Jeon. Jun Tai, Kim. YouChul, Study on Crack Monitoring System in Steel Structure, KSSC Journal, 23(2) (2011) , pp.159–167.

Oku, K., Arita, K., and Kim, Y.C. (2006) Monitoring of Initiation and Propagation of Fatigue Crack on Steel Plate Deck by FSM, Steel Structure ISSS'06, Vol. 1, pp.173–179

Tae. Hwan. Um, Kyong Ho Chang, You Chul Kim, Sung Ho Hyun, The Study on Damage Diagnosis Method for Storage Tank, KWJS 2011-Spring, (2011), 133

Bridge Maintenance, Safety, Management, Resilience and Sustainability – Biondini & Frangopol (Eds)
© 2012 Taylor & Francis Group, London, ISBN 978-0-415-62124-3

Seismic performance evaluations of bridge-pier system with uncertainty

T. Iida
Osaka Sangyo University, Osaka, Japan

D. Lim
Kawada Construction Co., Ltd., Osaka, Japan

K. Kawano
Kagoshima University, kagoshima, Japan

ABSTRACT

Deterioration of Japan's social infrastructure due to aging has progressed with time and is now a challenging issue. Establishment of a method for evaluating seismic performance incorporating damage and aging during the service life of structures is extremely important for proper maintenance. In particular, in performance design-based structural design, it is necessary to clarify nonlinear response when a structure is subjected to severe seismic motions exceeding its yield level.

Therefore, in the present research, a nonlinear response analysis considering aging situation in the properties of structural materials was performed using a vibration model of a nonlinear system with 3 degrees of freedom (Fig. 1), and a damage evaluation of the structure was carried out using Park and Ang's damage index D. In addition, the seismic safety of the structure was examined under aging situation in structural properties using the cumulative probability distribution of damage index D and reliability index β for each year obtained by Monte Carlo Simulation (MCS), treating the maximum acceleration of the input seismic motion as an uncertain parameter.

The following results are obtained through the present study:

(1) For aging of the structure, the damage index increases and reliability index decreases (Fig. 2); however, the evaluations is closely affected by the natural period of the structure and the dynamic characteristics of the seismic motion and similar factors.
(2) The allowable degree of damage in structures where nonlinear response is permissible is an important issue. Taking into accounts the uncertainty of the maximum acceleration of the seismic motion and the aging of structural properties in the performance evaluation of the structure, it is available for the reliability index to evaluate performance of the structure with respect to the design values of the structure.

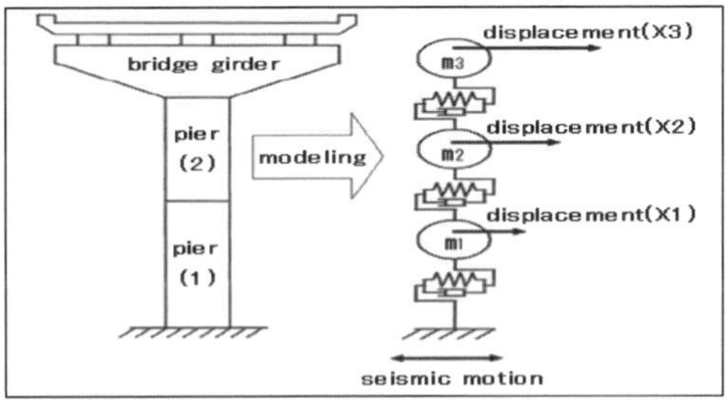

Figure 1. Idealized vibration model of bridge pier system.

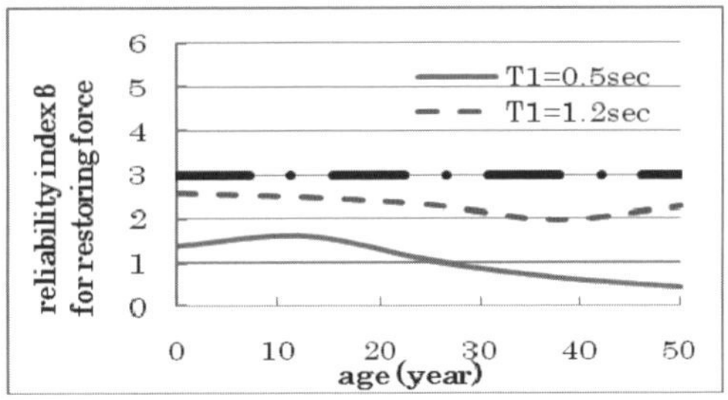

Figure 2. Reliability evaluation with respect to restoring force.

(3) Damage index and reliability index can provide useful indicators for seismic performance evaluations under aging effects in structural properties.

REFERENCES

Guan, X.L. and Melchers, R.E. 2000. A comparison of some FOSM and Monte Carlo results, Applications of Statistics and probability, *ICASP 8*, Vol. 1, pp. 65–71.

Kawano, K., et al. 2007. Performance based evaluation of offshore structure to wave and seismic forces with uncertainties, *Proc. of Seventeenth International Offshore and polar engineering conference*, pp. 3554–3561.

Park, Y.J. and Ang, A.H.S 1985. Mechanistic seismic damage model for reinforced concrete, *Journal of Structural Engineering*, Vol. 111, No. 4, pp. 722–739.

Bridge Maintenance, Safety, Management, Resilience and Sustainability – Biondini & Frangopol (Eds)
© 2012 Taylor & Francis Group, London, ISBN 978-0-415-62124-3

Applications of hybrid system with Perfobond rib shear connectors

S.-H. Kim, C.-Y. Jung & J.-Y. Joung
School of Civil and Environmental Engineering, Yonsei University, Seoul, Republic of Korea

J.-H. Ahn
Department of Civil Engineering, Kyushu University, Fukuoka, Japan

ABSTRACT

New hybrid systems have been recently spotlighted in bridge construction. These systems are especially attractive to bridge engineers since they maximize the structural and economical advantages of the steel and concrete components. Figure 1 shows the concept of hybrid bridge system. The most important factor in designing the steel-concrete hybrid system is the joint which should effectively transfer member forces between steel and concrete members. Thus, these systems need a great deal of shear connections in order to improve the composite action between the steel and concrete. Several types of joints have been studied to develop an appropriate feature for steel-concrete connections. The most classical feature of steel-concrete connection is the head stud; however, it needs complicated rebar arrangement and these connectors should be welded to the cramped area. To overcome such problems, studies on large stud shear connectors and grouped stud connectors are widely investigated and those connectors are discovered to have better performance than the general forms of the head stud. However, the head stud may cause cracks in the concrete slab and suffer from fatigue under service loading conditions. Therefore, various types of shear connection systems with sufficient shear capacity, ductility, and high fatigue resistance have been presented to provide better alternative for steel-concrete connection which may efficiently replace the head stud connection.

Recently, a perfobond rib shear connection was proposed for the connection of the hybrid steel-concrete structural system. The perfobond rib shear connector is originated in Germany, and its classical feature consists of a flat steel plate containing a number of holes. This connector resists horizontal shear and vertical uplift forces by using concrete end-bearing zone, concrete dowel, and transverse rebars in the rib holes. The perfobond rib shear connector has several advantages; it has high shear resistance capacity and high fatigue resistance and its shape allows easy installation. In this study, the new hybrid bridge systems using

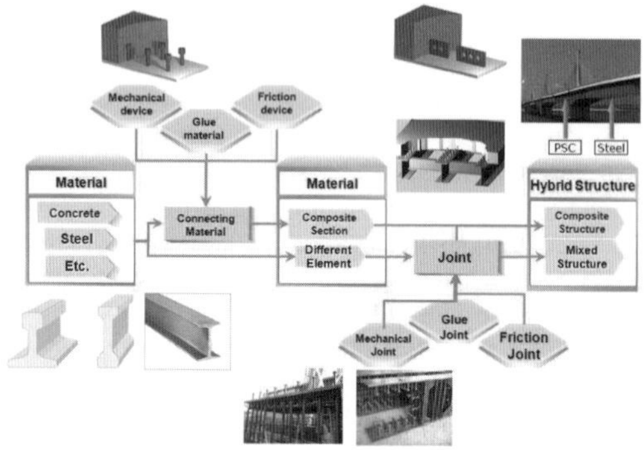

Figure 1. Sample of a figure caption.

perfobond rib shear connector, a flat plate containing a number of holes through which the reinforcements are passed, imparting high shear and fatigue capacities and allowing simple installation. The perfobond rib connectors can be applied to various hybrid to improve the behavior of composite structures or constructability. In this paper, the experimental evaluation of some new hybrid bridge systems using various connector are presented: 1) a spliced PSC-Steel-PSC hybrid I-girder, 2) the partially corrugated web PSC girder with Y-type perfobond rib, and 3) the abutment-pile connection in integral abutment bridge, etc.

REFERENCES

Ahn, J.H., Kim, S.H., and Jeong, Y.J. 2008. Shear behaviour of perfobond rib shear connector under static and cyclic loadings. Magazine of Concrete Research, 60(50): 347–357.

Dunai, L. 2004. Experimental and numerical studies on the cyclic behavior of steel and composite joints. International Journal of Steel Structures, 4(4): 197–208.

Dunai, L., Fukumoto, Y. and Ohtani, Y. 1996. Behavior of steel-to-concrete connections under combined axial force and cyclic loading. Journal of Constructional Steel Research, 36(2): 121–147.

Bridge Maintenance, Safety, Management, Resilience and Sustainability – Biondini & Frangopol (Eds)
© 2012 Taylor & Francis Group, London, ISBN 978-0-415-62124-3

Fundamental study on rigid connection detail of steel-concrete composite rigid frame bridge using bearing plate

K. Tani , T. Yamada & D. Sagou
Takadakiko Co.,Ltd, Osaka, Japan

T. Yamaguchi & Y. Kawamoto
Osaka City University, Osaka, Japan

ABSTRACT

Recently, the number of steel-concrete composite rigid frame bridges are increasing for short and medium span length in Japan. The features of this type of the bridge are the rigid connection between the steel girder and the RC abutment, and lack of expansion joints. Application of this type of the bridge can expect enhancement of the service life and reduction of the life cycle cost due to no bearings and expansions, and improvement of earthquake resistance. To make this type of the bridge more rational, it is important to ensure a good performance of the rigid connection securely, which can transfer the load from the steel girder to the abutment. However, it is reported that such a detail has some problems as follows: 1) High bearing and tensile stress occurs at the contact surface between the steel girder and the concrete section. 2) Maintenance of the inside portion of the connection where the steel girder installed is difficult. 3) Arranging reinforcement bars, filling concrete at the connection might be difficult.

Therefore, a new type of the connection detail with a bearing plate has been proposed by the authors as shown in Figure 1. In this study, effectiveness and applicability of the proposed rigid connection detail with a bearing plate was examined by elasto-plastic finite displacement analysis varying the thickness of the bearing plate.

It is found that the load can be securely transferred from the steel girder to the abutment through the bearing plate as shown in Figure 2. It is concluded that the proposed connection detail can be one of the practical and effective rigid connection detail considering easiness of construction.

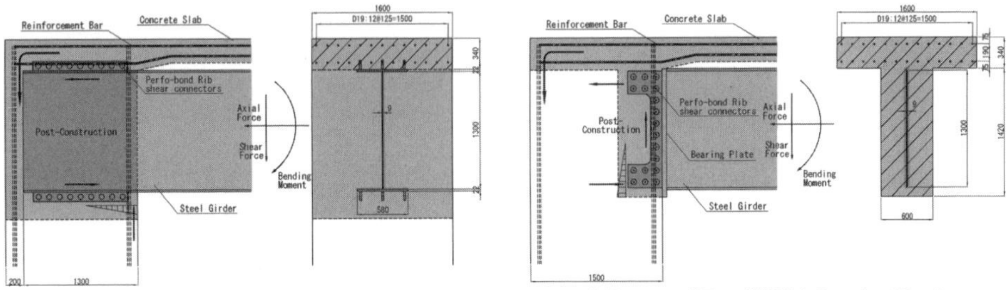

(a) Typical Current Detail (b)Proposed Detail(With Bearing Plate)

Figure 1. Proposed New Rigid Connection Detail (unit: mm).

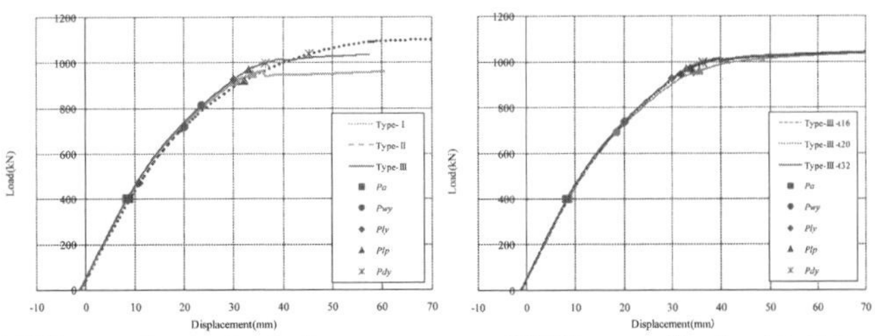

(a) Differences of connection detail (b) Differences of bearing plate thickness

Figure 2. Load-Displacement Relationship.

Bridge Maintenance, Safety, Management, Resilience and Sustainability – Biondini & Frangopol (Eds)
© 2012 Taylor & Francis Group, London, ISBN 978-0-415-62124-3

Fatigue crack detection of steel truss bridge by using mechanoluminescent sensor

N. Terasaki, C.-N. Xu, C. Li, L. Zhang, Y. Sakata & N. Ueno
National Institute of Advanced Industrial Science and Technology (AIST)

C.-N. Xu
CREST, Japan Science and Technology Agency (JST)

K. Yasuda
Registered consulting Engineer Road group, NEW JEC Inc.

L.H. Ichinose
Japan Industrial Testing Corp

ABSTRACT

Recently there have been innovative mechanoluminescent (ML) particle made available, each of which emits light repeatedly in response to small stresses applied, such as deformation, friction, or impact. When dispersedly coated on a structure, each particle acts as a sensitive mechanical sensor, while the 2-dimentional emission pattern of the whole assembly reflects well the dynamical stress distribution. This stress visualization technique provides a novel way of diagnosing the structural health, far more advantageous over the conventional point-by-point measurement method. It is possible to detect invisible defects and microcracks in small parts of machinery as well as to monitor the safety of huge constructions. In this study, we have investigated the detection of cracks and defects on an old bridge by using the ML sensor. The target bridge for ML monitoring was Torikai great bridge (Osaka, Japan, Fig. 1), 9 spans continuous Gerber truss type steel bridge (through bridge, Length 550 m), and have been provided for use by the public since 1954. After 56 years of the providing, fatigue crack, degradation and corrosion had become prominent in the course of increasing traffic, then the alternative new bridge was constructed beside the old one and finally the providing was finished in 2010.

Furthermore, one diagonal member on the steel truss bridge had already been cut in advance for the other analytical experiment to clarify the influence of the diagonal failure. In addition, we can pass across a pseudo-overloaded vehicle even in the failure condition for research purpose. Namely, the bridge was convenience for our applying test of our ML sensor against such damaged failure bridge.

As the results on application of the ML sensing technique to the aged bridge with fatigue crack and diagonal failure, we successfully measured stress concentration at the tip of the crack and crack mouth opening displacement, at the point from a to c (Fig. 2), accompanied by dynamical loading test with the heavy vehicle, in particular undetected and unexpected cracks at point d in spite of safety measures just like a stop-hole. We also succeeded to distinguish the influence between before and after the diagonal failure through the ML intensity and pattern.

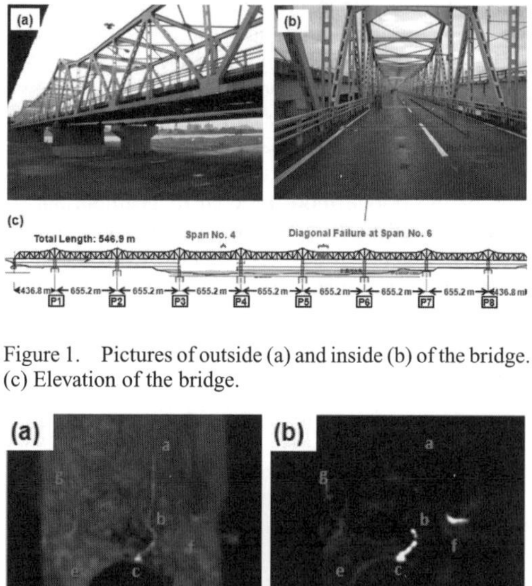

Figure 1. Pictures of outside (a) and inside (b) of the bridge. (c) Elevation of the bridge.

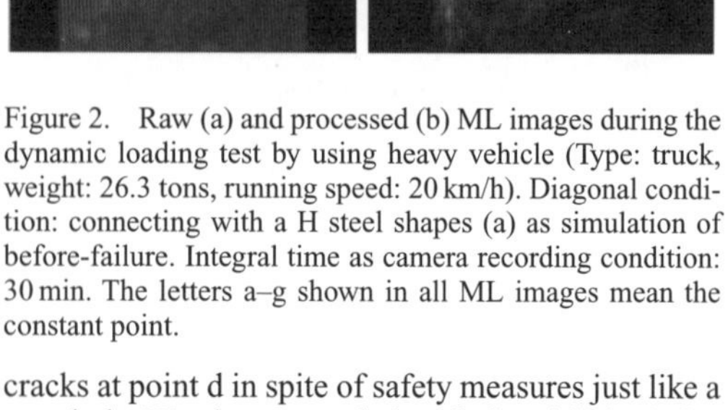

Figure 2. Raw (a) and processed (b) ML images during the dynamic loading test by using heavy vehicle (Type: truck, weight: 26.3 tons, running speed: 20 km/h). Diagonal condition: connecting with a H steel shapes (a) as simulation of before-failure. Integral time as camera recording condition: 30 min. The letters a–g shown in all ML images mean the constant point.

REFERENCE

Xu, C.N. 2002. Coatings. *Encyclopedia of Smart Mterials*. 1-2: 190–201.

Understanding and enhancing bridge performance

Organizer: J.M. Hooks

Bridge Maintenance, Safety, Management, Resilience and Sustainability – Biondini & Frangopol (Eds)
© 2012 Taylor & Francis Group, London, ISBN 978-0-415-62124-3

Monitoring bridges with wireless sensor networks: A critical assessment

G. Feltrin
Empa, Swiss Federal Laboratories for Materials Science and Technology

ABSTRACT

In the past decade, wireless sensor networks have emerged as an innovative structural monitoring technology. Their advantages such as rapid deployment, ease of scalability, and self-configuration allow to reduce the costs, to increase the flexibility, and to simplify the operability of monitoring. However, since wireless sensor networks are battery powered, the advantages are weakened by the energy limitations that impose low power hardware with limited resources and carefully chosen data acquisition, data processing and data communication policies.

Currently wireless sensor networks provide competitive solutions for monitoring slowly changing physical quantities like temperature, humidity, etc. Most deployments have proven to run with reasonable stability and reliability for several months to a year.

Structural monitoring scenarios, however, frequently deal with capturing dynamic quantities that involve measuring time series of data with high sampling frequencies. These tasks consume significantly more energy and reduce dramatically the battery lifetime of monitoring system based on wireless sensor networks. Therefore, data intensive long-term monitoring still remains a very challenging task.

The experiences of long-term installations show that a replacement period of batteries of several months cannot be achieved exclusively with low power hardware and optimized sensing and data acquisition policies. Since data transmission is very expensive in terms of energy, an important improvement is achieved by significantly reducing the raw data by means of in-node data processing.

Monitoring tasks with relatively complex in-node data processing (tracking of natural frequencies, monitoring of strain cycles, monitoring of vibration amplitudes) could be operated for up to several months with high stability and reliability in in- and outdoor environments.

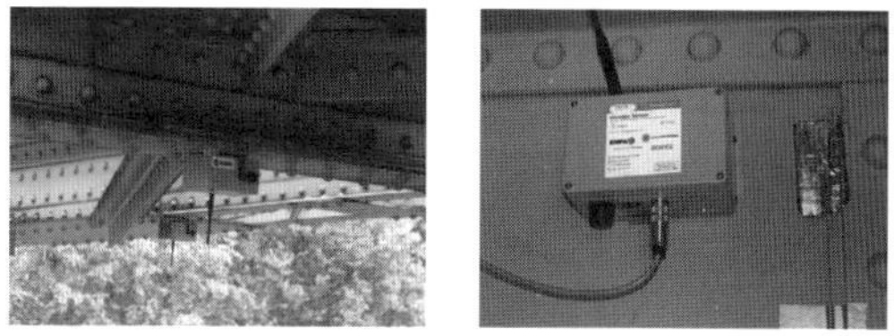

Figure 1. Sensor nodes mounted on a bridge.

Although the hardware limitations of wireless monitoring systems do not allow to achieve the accuracy of wired systems, the data quality generally matches the requirements of structural assessment.

Triggered data acquisition, which is often applied in practice, is still on a primitive level. The wired systems approach based on measuring permanently and checking for the trigger condition is expensive in terms of energy since the sensors as well as the processor are permanently switched-on. Innovative solutions are required to make wireless sensor networks competitive.

Furthermore, many common sensors, which were developed for wired monitoring systems assuming no power restrictions, are not suitable for wireless sensor networks in terms of fast deployment and power consumption and need to be optimized.

Despite the still existing deficiencies, wireless sensor networks have achieved a grade of maturity, which allows end-user to reliably perform many complex monitoring tasks.

REFERENCE

Feltrin, G., Saukh, O., Meyer, J., Bischoff, R. and Motavalli, M. 2011 Structural monitoring with wireless sensor networks: Experiences from field deployments. Proceedings of the First Middle East Conference of Smart Monitoring, Assessment and Rehabilitation of Civil Structures, Dubai, UAE, 8–10 February 2011.

Bridge Maintenance, Safety, Management, Resilience and Sustainability – Biondini & Frangopol (Eds)
© 2012 Taylor & Francis Group, London, ISBN 978-0-415-62124-3

Evaluating and forecasting bridge performance under uncertainty

D.M. Frangopol & D. Saydam
Lehigh University, Bethlehem, PA, USA

ABSTRACT

The methodologies and tools for evaluating and forecasting bridge performance under uncertainty are presented. The uncertainties associated with the lifetime performance are summarized. Several structural performance indicators that are used in bridge performance evaluation are presented. A review of the methodologies for system analysis of bridges is provided. Furthermore, the role of structural health monitoring in evaluating and forecasting bridge performance is described.

The major sources of uncertainty in engineering problems are divided in two groups: aleatory uncertainties and epistemic uncertainties. The performance indicators reviewed include probability of failure, time to failure, survivor function, reliability index, redundancy index, vulnerability, robustness, resilience, life-cycle cost and risk. The interaction of the vulnerability index, redundancy index and reliability index is illustrated qualitatively in Figure 1. It is possible to evaluate and predict the entire bridge structural system reliability by making appropriate assumptions (e.g., series, parallel and combined system modeling assumptions) regarding the interaction of individual components.

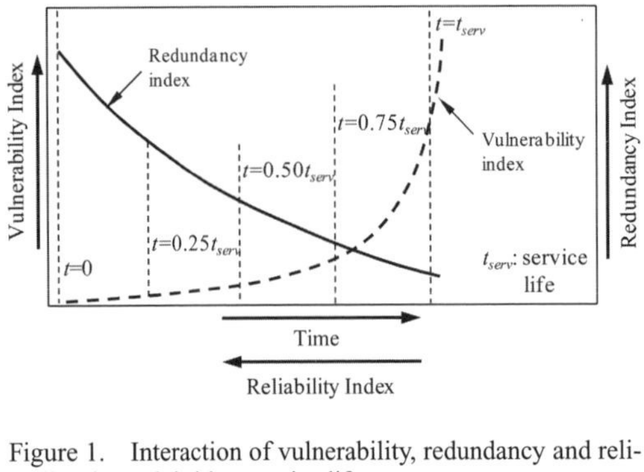

Figure 1. Interaction of vulnerability, redundancy and reliability through bridge service life.

REFERENCES

Akgül, F. & Frangopol, D.M. 2004. Computational platform for predicting lifetime system reliability profiles for different structure types in a network. *Journal of Computing in Civil Engineering* 18 (2): 92–104.

Cimellaro, G.P., Reinhorn, A.M., Bruneau, M. 2010. Seismic resilience of a hospital system. *Structure and Infrastructure Engineering* 6(1): 127–144.

Estes, A.C. & Frangopol, D.M. 1999. Repair optimisation of highway bridges using system reliability approach. *Journal of Structural Engineering* 125(7): 766–775.

Frangopol, D.M. & Bocchini, P. 2011. Resilience as optimization criterion for the rehabilitation of bridges belonging to a transportation network subject to earthquake. *Proceedings of the 2011 Structures Congress*, ASCE, Las Vegas, NV, USA, April 14–16, 2011.

Ghosn, M., Moses, F., Frangopol, D.M. 2010. Redundancy and robustness of highway bridge superstructures and substructures. *Structure and Infrastructure Engineering* 6(1–2): 257–278.

Okasha, N.M. & Frangopol, D.M. 2009. Lifetime-oriented multi-objective optimization of structural maintenance considering system reliability, redundancy and life-cycle cost using GA. *Structural Safety*, 31(6), 460–474.

Saydam, D. & Frangopol, D.M., 2011. Time-dependent performance indicators of damaged bridge superstructures. *Engineering Structures* 33(9): 2458–2471.

Bridge Maintenance, Safety, Management, Resilience and Sustainability – Biondini & Frangopol (Eds)
© 2012 Taylor & Francis Group, London, ISBN 978-0-415-62124-3

Studying, understanding & enhancing the performance of bridges in the United States

J.M. Hooks
John M. Hooks & Associates, Cape Coral, FL, USA

A. Foden
Parsons Brinckerhoff, Princeton, NJ, USA

M.C. Brown
Virginia Center for Transportation Innovation & Research, Charlottesville, VA, USA

ABSTRACT

There are approximately 600,000 bridges in the United States and the characteristics of these bridges very widely in terms of age, type, construction materials, span length, design load, local climate and environment, traffic loadings, etc. Many aspects of the performance of these bridges at the level of the structural system, the component level, and at the material level are not well understood. Consequently the search for cost-effective solutions to performance issues typically involves significant uncertainties about the best method and materials and best time for application of the solution. This problem of uncertainty is particularly widespread and consequential when considering solutions for problems with concrete bridge decks.

The US Federal Highway Administration has launched a 20-year research program to collect high quality data on bridges, to use this data to better understand critical aspects of bridge performance and to create the knowledge and tools necessary to significantly enhance the long-term performance of bridges. During this program the research team will deploy numerous investigative and testing methodologies and technologies to collect data on the condition and behavior of bridge materials, components and systems. The team will collect important information on service environment, traffic loadings and maintenance histories in order to create a complete performance picture for different types of bridges.

Experimental studies will be conducted using non-destructive testing, material testing, live loads and dynamic testing, periodic and long-term monitoring and finite element modeling. State-of-the-art techniques for data fusion and data visualization will be used to maximize the value of the data collected. The results of the program will include improved deterioration and lifecycle cost models and guidance on improvements to bridge design, construction, inspection, maintenance and rehabilitation. The findings from the program are expected to create critically needed knowledge on the effectiveness of various approaches to preservation and maintenance of bridges. A second benefit will be knowledge about more effective ways to periodically assess the condition of bridges – a practice that is now typically done on a fixed 2-year cycle using visual inspection procedures as the primary tool of assessment.

Most Departments of Transportation in the United States conduct ongoing programs to manage the assets in their highway systems including bridges. Within those programs, the preservation, maintenance and rehabilitation of bridge decks consumes a large percentage of the total time and financial resources available for bridge programs. Most of the input to decision-making on actions to be taken on bridge decks is the result of visual inspections, occasionally enhanced by more sophisticated inspection technologies. This paper discusses the strategies and tactics envisioned for the Long Term Bridge Performance program with the emphasis on how deploying advanced methods of testing can greatly improve the effectiveness of programs and actions aimed at preservation, maintenance and rehabilitation of bridge decks.

Bridge Maintenance, Safety, Management, Resilience and Sustainability – Biondini & Frangopol (Eds)
© 2012 Taylor & Francis Group, London, ISBN 978-0-415-62124-3

A model-free data-interpretation approach for long-term monitoring of bridges

I. Laory, T.N. Trinh & I.F.C. Smith
Applied Computing and Mechanics Laboratory (IMAC), Swiss Federal Institute of Technology Lausanne (EPFL), Switzerland

ABSTRACT

Long-term monitoring of bridges has the potential to improve structural sustainability through early detection, thereby avoiding unnecessary replacement and leading to more efficient maintenance. However, for long-term bridge monitoring, interpreting continuous measurement data to assess bridge conditions remains a challenge.

In general, based on the presence of physics-based behavioral model, there are two classes of data interpretation methods: model-based and model-free methods (ASCE 2011). Model-free data-interpretation is the analysis of measurements without using behavioural models of the structures. These methods are purely statistical and data driven; thus, they are well suited for simultaneous long-term monitoring of structures. Posenato et al. (2010) performed a comparative study of model-free data interpretation methods and demonstrated that the performances of Moving Principal Component Analysis (MPCA) and Robust Regression Analysis (RRA) for anomaly detection were superior to other methods (Wavelet packet transform, Discrete wavelet transform, ARIMA, Box-Jenkins, Instance based method, Short Term Fourier Transform and correlation anomaly scores analysis) when dealing with civil-engineering challenges such as significant noise, missing data and outliers. Laory et al. (2011) valuated the performance of MPCA and RRA under traffic and temperature variations. The study showed that while MPCA is better than RRA in terms of damage detectability, RRA is better than MPCA in terms of time to detection. Hence, both methods were considered to be complementary and it was noted that synergies between both methods may result in a better overall methodology for damage detection.

This paper presents a new model-free data-interpretation approach for damage detection using long-term monitoring data. The approach combines two model-free methods: MPCA and RRA. The objective is to integrate complementary advantages of these methods in order to improve damage detectability and time to detection. To illustrate the applicability of the approach, a railway truss bridge in Zangenberg (Germany) is selected as a case study. A numerical model inspired from the bridge is used. The performance of the proposed approach is compared with that of using MPCA and RRA alone. Table 1 shows the result of the comparative study. The combined approach is able to detect lower damage level than the minimum detectable damage-level when using MPCA and RRA alone. A scenario with 50% damage level is selected to study time to detection. As shown in Table 1, while MPCA and RRA require more than ten days to detect damage, the new methodology detects damage instantly.

Table 1. Performance of three data-interpretation approaches.

Methodology	Minimum detectable damage-level (%)	Time to detection (days)
MPCA	10	29
RRA	35	17
MPCA-RRA	2	<1

Results demonstrate that combining Moving Principal Component Analysis (MPCA) and Robust Regression Analysis (RRA) exploits the complementary advantages of these methods for damage detection during long-term monitoring of bridges.

REFERENCES

ASCE 2011. Structural Identification of Constructed Facilities. Structural Identification Comittee, American Society of Civil Engineers).

Laory, I., Trinh, T. N. and Smith, I. F. C. 2011. Evaluating two model-free data interpretation methods for measurements that are influenced by temperature. *Advanced Engineering Informatics,* 25(3), 495–506.

Posenato, D., Kripakaran, P., Inaudi, D. and Smith, I. F. C. 2010. Methodologies for model-free data interpretation of civil engineering structures. *Computers & Structures,* 88(7–8), 467–482.

Bridge Maintenance, Safety, Management, Resilience and Sustainability – Biondini & Frangopol (Eds)
 ISBN 978-0-415-62124-3

Extracting knowledge from structural response data

F.L. Moon & J.S. Weidner
Drexel University, Philadelphia, PA, USA

N.C. Dubbs
Intelligent Infrastructure Systems, Philadelphia, PA, USA

ABSTRACT

Over the past two decades there has been significant attention paid to the use of long-term health monitoring systems to aid in the management, maintenance and preservation of individual bridges. Although such systems are quite promising, the economic reality is that for the foreseeable future they will continue to be cost-prohibitive to deploy in a comprehensive manner. As a result, there is a need for methods capable of extracting generic knowledge about bridge performance from the long-term monitoring of a relatively small number of bridges.

The aim of this paper is to present a framework that achieves this through the integration and leveraging of competing data interpretation approaches, specifically, direct and comparative methods.

The direct methods of data interpretation include the use of statistical and pattern recognition approaches. Once properly identified, these models are employed to forecast future performance and identify when the performance of the structure has deviated from the expectation based on past performance (e.g. damage detection). The primary advantages of direct data interpretation approaches is that they are highly efficient, easily automated (so they may be employed on-line), and capable of handling structural complexities such as nonlinearity. The primary drawback of these approaches is that they cannot provide insight into the cause or the importance of any identified change.

In contrast to direct methods, comparative methods of data interpretation employ independent information and use the comparison with such information to provide insight into the measure data. The more robust and versatile comparative interpretation approach is termed structural identification (St-Id) and involves the correlation of a simulation model with experimental observations. The simulation model (commonly an FE model) essentially represents the expected nominal behavior of the structure. By correlating this model with the observed behavior, the analyst can gain insight into the validity of the underlying assumptions and idealizations employed, and in turn learn a great deal about the structure of interest. Given their explicit inclusion of kinetic and kinematic mechanisms, such comparative methods hold potential for identifying physically meaningful explanations for the observed responses. The primary drawback of these approaches is that they require significant user interaction and rely heavily on the intuition and heuristics of the user. As a result such methods do not lend themselves to automation and can be cumbersome and inefficient.

Considering the complementary nature of these two perspectives, the following data interpretation framework is proposed:

(1) Employ short-term testing and comparative methods to identify the key mechanisms and vulnerabilities of the structure
(2) Employ comparative methods together with parametric studies to select a robust set of response quantities (including spatial/temporal resolutions, etc.) that are sensitive to the vulnerabilities or performances of interest
(3) Train and validate a set of direct data interpretation methods to ensure that the inherent complexities of operational responses are properly embedded
(4) Implement the direct data interpretation methods to track the response quantities of the structure and identify when the behavior has changed in a statistically significant manner
(5) Employ comparative methods to diagnose identified changes and perform prognosis to establish the likely impact that the changes may have on future performance

Extending bridge life through industry academic partnerships

Organizer: E.J. OBrien

Bridge Maintenance, Safety, Management, Resilience and Sustainability – Biondini & Frangopol (Eds)
 ISBN 978-0-415-62124-3

Attenuating resonant behavior of a tied arch railway bridge using increased hanger damping

A. Andersson & R. Karoumi
Division of Structural Engineering and Bridges, KTH Royal Institute of Technology

ABSTRACT

In this paper, dynamic analyses and field measurements of a tied arch railway bridge is presented. The bridge is simply supported with a span of 45 m, carrying a single track supported directly by stringer beams. A view of the bridge is illustrated in Figure 1.

Excessive vibrations of the hangers were obtained, caused by resonance during train passage. Due to very low damping of the hangers, increased stress levels and number of stress cycles were shown to decrease the fatigue service life significantly. The most critical section is a threaded turnbuckle connection of the hangers, illustrated in Figure 2. At the most critical point measured, more than 50% of the fatigue service life was due to stresses induced during free vibrations.

Passive dampers were installed in 2005 to attenuate the vibrations by means of increased hanger damping. Based on free vibration tests, the critical damping ratio of the hangers was increased from an average of 0.2% to 3.5%.

A 2D FE-model of the bridge was used to compare measured frequencies, displacements and stresses of the hangers. Assuming the hangers to be fully clamped at the connection with the main beam and hinged at the connection with the arch generally produced the best agreement with measured data. The pre-stress due to permanent load was included.

The stress components in one hanger are presented in Figure 3, based on results from the 2D FE-model. The resonant behavior inducing high bending stresses is also shown from the measurements without dampers. With increased hanger damping, the FE-model predicts a significant decrease of the bending stresses.

Figure 1. View of the Ljungan railway bridge.

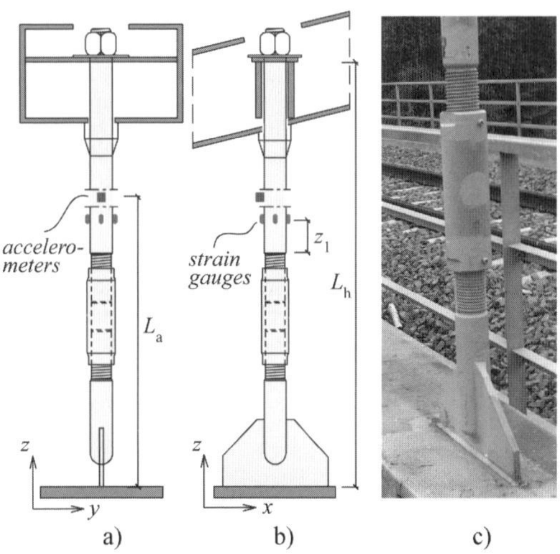

Figure 2. Detail of the hangers, a) section across the bridge, b) section along the bridge, c) detail of the turn buckle connection.

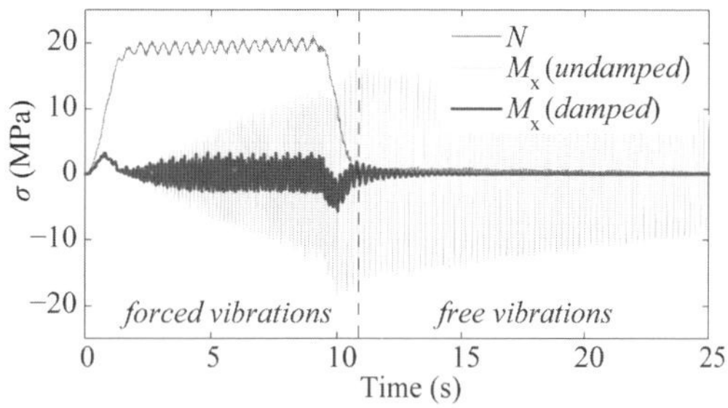

Figure 3. Stress components in hanger 6 during passage of a freight train, influence of increased hanger damping (based on a 2D FE-model).

Bridge Maintenance, Safety, Management, Resilience and Sustainability – Biondini & Frangopol (Eds)
 ISBN 978-0-415-62124-3

Probabilistic approach to fatigue

J. Baussaron & T. Yalamas
PHIMECA Engineering, Cournon d'Auvergne, France

ABSTRACT

In structures submitted to repeated variable loads, initiation and growth of cracks in the material are the most common causes of the deterioration of structures. When it happens after a long time, this failure mode is called fatigue. Under variable amplitude and repeated loads, some local cracks initiate on critical points of the structures. The fatigue phenomenon is the cause of different well-known accidents over the last hundred years. The complicated phenomenon has been discussed in many articles and books, in order to observe, explain and model it. All fields of engineering have been confronted to failures of structures incurring fatigue. Many parameters used for predict times to failure of structure are variable and their variations have a big influence on the real lifetime. This paper focus on a global methodology to take into account main sources of variability in fatigue life prediction. This approach is summarized on Figure 1. The first step of this methodology is to determine the variability of each parameter. Loading is one of the most important sources of variability. Rainflow counting are used to transform a time variant signal into constant amplitude cycles. Then, a fatigue equivalent load is defined using cumulative damage rules; it produces the same damage as the initial signal. Another important source of variability is the strength of the structure. Probabilistic S-N curves are fitted to represent the failure probability of the structure. An accurate reliability assessment of the structure can be performed when the main variability sources are considered. Reliability methods also allow to rank the influence of each variable and optimize parameters in order to reach product reliability target.

An European project named "Long Life Bridges" has started in September 2011. One aim of this project is to apply and extend this methodology on bridge to assess risk in fatigue damage. A probabilistic approach will be developed to the analysis and design of fatigue critical details in cable stayed bridges. This will include the identification of fatigue critical details in cable stayed bridges (typically in welded details), stochastic modeling of the fatigue loads (long term description of stress ranges and uncertainty related assessment of the stress ranges) and stochastic modeling of the fatigue strength.

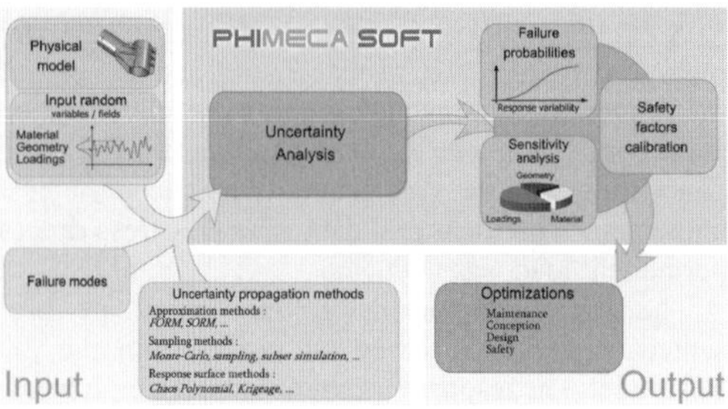

Figure 1. Reliability analysis methodology.

REFERENCES

Dowling, N. E., 1972. Fatigue failure predictions for complicated stress-strain histories. Journal of Materials, JMLSA, p. 71–87.

Endo T., Mitsunaga, K. and Nakagawa H., 1967. Fatigue of metals subjected to varying stress – predic-tion of fatigue lives. Preliminary Proceedings of the Chu-kogu-Shkoku District Meeting, The Japan Society of Mechanical Engineering, p. 45–46.

Lemaire M. and Pendola, M., 2006. Phimeca-soft. Structural Safety, 28, 130–149.

Miner, M. A., 1945. Cumulative damage in fatigue. Journal of Applied Mechanics, 12:A159–A164

Stephens, R. I., Fatemi, A. ,Stephens, R. R. and Fuchs, R. O., 2001. Metal Fatigue in Engineering. Wiley Interscience

Bridge Maintenance, Safety, Management, Resilience and Sustainability – Biondini & Frangopol (Eds)
 ISBN 978-0-415-62124-3

Improved bridge response evaluation based on dynamic testing

I. González, R. Karoumi & A. LLorens
KTH Royal Institute of Technology, Stockholm, Sweden

ABSTRACT

In this study the dynamic behavior of the portal frame Rössjö railway bridge is studied by means of a 2D finite element model. Furthermore parameters in the proposed model are updated to make the predicted response match the acceleration response measured in site.

The Rössjö Bridge is a reinforced concrete bridge that carries a single ballasted track over a span of 15.25 m with a free height of 4.7 m. The bridge is founded in concrete footings.

The model used was a 2D Timoshenko beam model that used a constant cross-section for the slab and a varying cross-section in the walls to account for the mass and stiffness contribution of the wing walls.

The measurements were carried on while the bridge was crossed by the Swedish Green Train a prototype train develop to study the feasibility of high speed tracks. Eight accelerometers of MEMS type were used, all placed around the mid-point. In this study two accelerometers placed 1.5 m away from the center line, on the edge beams, were used to obtain the acceleration response used for updating and validation.

The parameters updated were mainly the support stiffness. The density of the slab was also modified to account for the weight of the ballast and track, but the stiffness contribution of the ballast and track was not considered.

The main parameters used to judge the proximity between the modeled response and the predicted response, were the eigen-frequencies.

The support stiffness was set to $1 \cdot 10^8$ N/m since this value gave the best agreements between simulated and measured response. The degree of agreement was obtained using rather subjective criteria. How much the deterministic simulations fit the measured data that will always include important scattering is an engineer's judgment. The model as to be safe and thus "underperform" the real bridge, therefore a minimum square fitting, or other similar techniques, is not applicable.

A more mathematical rigorous approached is suggested in the form of Bayesian updating that might take into account even previous information about the variables to be updated. Different strategies existing in the literature are offered as possible updating techniques.

Bridge Maintenance, Safety, Management, Resilience and Sustainability – Biondini & Frangopol (Eds)
© 2012 Taylor & Francis Group, London, ISBN 978-0-415-62124-3

Traffic load models for long span bridges

A.A. Harapetova, A.J. O'Connor & E.J. OBrien
Roughan and O'Donovan Innovative Solutions, Sandyford, Dublin, Ireland

ABSTRACT

Long-span traffic loads, particularly for bridges in excess of 200 m, are not well accounted for by the codes of practice. Most of the earlier load models represent real traffic in a way that assumes that all vehicles behave in a consistent manner. In reality, vehicle drivers behave in a wide range of ways. The human factor is traditionally ignored, which weakens the ability of these models to replicate congested scenarios.

This paper contributes to the development of traffic load models for long-span bridges through the application of micro-simulation modelling. The latter is used to simulate site specific traffic flow conditions and the effects of vehicle interactions, taking account of driver behaviour. Analysis based on micro-simulation can result in a significantly more accurate assessment of traffic loading on long-span bridges. More accurate assessments in turn can result in the retention of existing bridges for longer lives by proving that they are safe for the load carried.

In the micro-simulation approach, traffic load on a bridge is considered as a multi-vehicle system where each vehicle is an agent with its own set of driver behaviour characteristics, such as acceleration/deceleration and lane changing. Heterogeneity across drivers is captured through the driver's target headway and reaction time distributions. The lane changing decision process considers a sequence of three steps: decision to consider a lane change, choice of the target lane, and gap acceptance.

Real traffic observations and weigh-in-motion data were used in this study to develop an accurate traffic micro-simulation models, taking account of driver behaviour as traffic changes from free-flowing to congested. The latter governs in the assessment of long span bridges. Driver behaviour under congested conditions results in changes to the lane distribution of the car-truck mixes. This is critical as it causes 'platooning' whereby long platoons of heavy trucks can form in critical positions on a long span bridge.

The final result of the micro-simulation models is snapshot records of traffic on a bridge. The movement of the recorded vehicles over the bridge model generates a history of traffic load effects, which is used to perform the probabilistic assessment of a structure. The extrapolation of the extremes of these effects is incorporated through the Normal distribution theory for the definition of the characteristic values of interest.

REFERENCES

Bank, J.H. 2003. Average time gaps in congested freeway flow. *Transportation Research Part* A37: 539–554.

Daganzo, C.F. 2002. A behavioural theory of multi-lane traffic flow. Part I: Long Homogeneous freeway sections. *Transportation Research Part* B 36: 131–158.

Hayrapetova, A.A. 2006. Micro-simulation of traffic load models on long span bridges. *PhD thesis, University College Dublin*. Ireland.

Kerner, B.S. 2004. Three-phase traffic theory and highway capacity. *Physica* A 333: 379–440.

Lee, D-H., Yang, X. 2000. Parameter Optimisation for Paramics Using Generic Algorithm. *Proceedings of the 6th ASCE AATT International Conference*. Singapore.

OBrien, J.E., Hayrapetova, A., Walsh C. 2010. The use of micro-simulation for congested traffic load modelling of medium- and long span bridges. *Structure and Infrastructure Engineering*. Taylor & Francis.

Paramics 2002. Reference Manual. SIAS.

Bridge Maintenance, Safety, Management, Resilience and Sustainability – Biondini & Frangopol (Eds)
© 2012 Taylor & Francis Group, London, ISBN 978-0-415-62124-3

Procedures for calibrating Eurocode traffic Load Model 1 for national conditions

E.J. OBrien, A.J. O'Connor & J.E. Arrigan
Roughan and O'Donovan Innovative Solutions, Sandyford, Dublin, Ireland

ABSTRACT

Since April 2010 Eurocode Load Model 1 (LM1) is the prescribed traffic load model to be employed in the design of highway bridges in the European Union (EU). Uniquely, the code permits member states to calibrate the load model, through the application of 'α-factors' to allow for national or regional conditions. Some countries with high volumes of very heavy traffic may find that they require α-factors in excess of unity whilst other less heavily trafficked road networks may require much lesser values. The importance of accurate calibration of the α-factors is clear from a safety and economic point of view. This paper describes procedures for calibration of α-factors using Weigh in Motion (WIM) data, illustrated in Figure 1. WIM data allows classification of the traffic loads in individual countries, enabling the specific Gross Vehicle Weights (GVWs), axle loads and frequencies of heavy trucks to be taken into account. Simulations calibrated using this data, for a wide range of structural forms (i.e., influence lines, spans and numbers of lanes) and scenario types (i.e., free flowing, congested and mixed traffic conditions), allow comparison of the load effects generated by the site-specific traffic

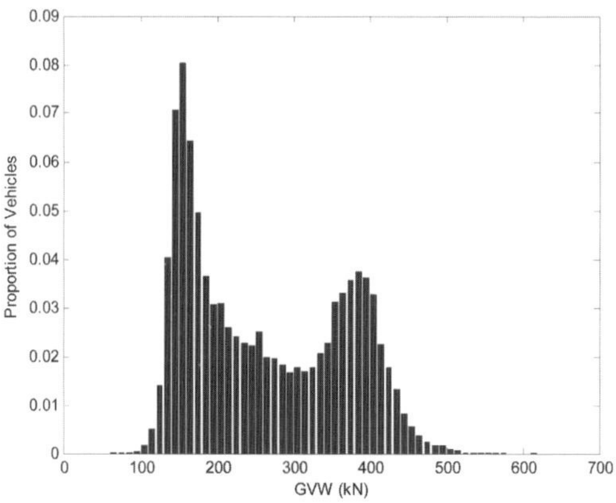

Figure 1. GVW histogram, 5 axle trucks (note the two peaks, corresponding to empty and full trucks).

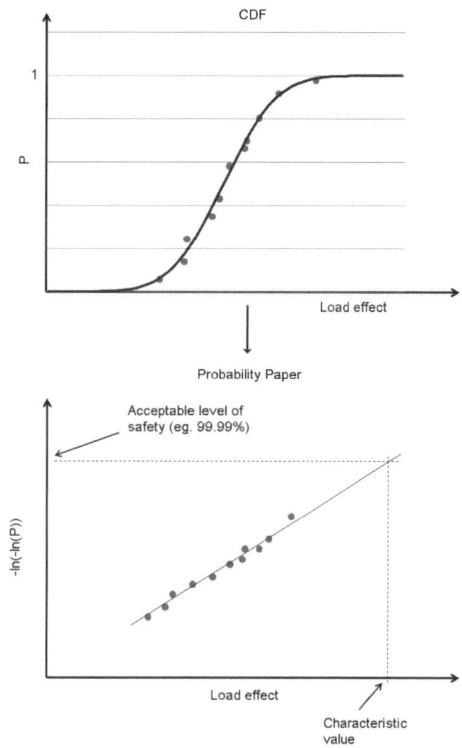

Figure 2. Gumbel probability paper plot.

to those obtained when employing LM1. Statistical Extreme Value Distributions (EVDs) are fitted to simulated results, Figure 2, to determine characteristic load effect values using the same methodology as was employed in the calibration of LM1 itself. Appropriate α adjustment factors are then determined to cater for variation in predicted characteristic extreme load effects on a network by network basis. Where $\alpha < 1.0$, the prescribed approach delivers significant savings by preventing unnecessary overdesign of bridges. On the other hand, for cases where $\alpha > 1.0$ it allows bridge designers to design bridges with adequate levels of safety.

Bridge Maintenance, Safety, Management, Resilience and Sustainability – Biondini & Frangopol (Eds)
© 2012 Taylor & Francis Group, London, ISBN 978-0-415-62124-3

Introduction to the Long Life Bridges project

E.J. OBrien, A.J. O'Connor & J.E. Arrigan
Roughan and O'Donovan Innovative Solutions, Sandyford, Dublin, Ireland

ABSTRACT

Long Life Bridges is a European 7th Framework research project funded under the Marie Curie Industry Academia Partnerships and Pathways programme (IAPP). The project has an estimated budget of €890,000, shared between four partners consisting of two universities and two Small/Medium Enterprises (SMEs). The goal of the IAPP programme is to transfer technology between universities and industry. This is achieved through a number of secondments between the two sectors.

The Long Life Bridges project has three tasks in its research work package, as illustrated in Figure 1: Railway Bridge Dynamics, Life Cycle Evaluation and Fatigue Evaluation. Each arrow indicates a staff secondment. These vary in length from three months to one year. Research staff in the SMEs will spend up to a year in the universities to learn from leading experts in the field and to bring that knowledge and understanding back to the SME. Post-doctoral staff and undergraduate students from the universities will travel in the other direction, to gain industrial experience in the SMEs and to develop valuable international contacts.

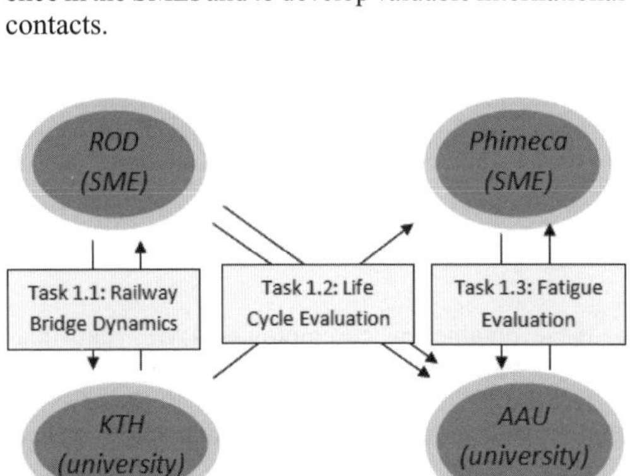

Figure 1. Staff transfers of *Long Life Bridges* Project.

The *Long Life Bridges* consortium consists of two SMEs and two universities. It is led by Roughan and O'Donovan (ROD), an SME firm of consulting engineers based in Dublin. A subsidiary company, Roughan and O'Donovan Innovative Solutions, focuses on the application of recent research developments in advanced bridge assessment and loading. Staff have developed software and published extensively on probabilistic approaches to bridge safety assessment, bridge dynamics, bridge traffic loading and Weigh-In-Motion.

Phimeca is an SME in France with expertise in uncertainty, particularly in the nuclear industry in the quantification of extremely remote risks. Their expertise will allow the development of a probabilistic approach to the analysis and design of fatigue critical details in cable stay bridges.

The two university partners, Aalborg University (AAU) in Denmark and The Royal Institute of Technology (KTH) in Stockholm both conduct world class research in areas key to the success of the project. AAU performs research in structural reliability and risk analysis with experience in application to industrial sectors such as offshore structures, bridges and wind turbines. This also includes the development of reliability and risk based methods for life-cycle assessment and optimal planning of inspection, operation and maintenance of structures. KTH conduct research into the analysis and design of bridges, with particular expertise in dynamic behaviour including the instrumentation and monitoring of several bridges across Sweden.

The research goal of the project is to extend the lives of bridges. This will be achieved through the development of more accurate models of railway bridge loading and dynamics. Total life cycle costs will be minimized by developing optimal maintenance strategies. The life cycle approach will be applied to fatigue damage and deterioration due to environmental effects.

Bridge Maintenance, Safety, Management, Resilience and Sustainability – Biondini & Frangopol (Eds)
© 2012 Taylor & Francis Group, London, ISBN 978-0-415-62124-3

Reliability-based assessment of fatigue life for bridges

H.S. Toft & J.D. Sørensen
Aalborg University, Denmark

ABSTRACT

Reliability assessment of both new and existing bridges is becoming increasingly more important in many countries due to degradation of existing bridges along with increase in the traffic flow.

Existing bridges are normally designed for a limited service life which in many cases will expire in the near future. Additionally, the weight of the individual vehicles is increasing and few but very heavy special transportation vehicles are observed. Both existing and new bridges should be capable of resisting the present amount of traffic and weight of the individual vehicles with a satisfactory reliability level, see e.g. (Righiniotis & Chryssanthopoulos 2004) and (Fisher & Roy 2011).

In the present paper the reliability level in the structural standards and recommendations from JCSS, ISO, NKB and Eurocodes are compared. In general a consistent reliability level is observed between the different references. For fatigue critical details local failure does often not lead to structural collapse because there is still reserve capacity in the structure. The reliability level for fatigue design of these details can therefore be reduced especially if the probability of collapse given local failure is small.

One way of increasing the reliability level for bridges is to apply inspections or monitoring in order to detect local damage from especially fatigue at an early stage, see e.g. (Chung et al. 2003). Detection and repair of cracks which can lead to local failure and later collapse of the structure can also lead to higher allowed traffic loads without increasing the probability of failure.

A method for reliability analysis of welded steel details is described using both an SN-approach and a Fracture Mechanics approach by which the information from inspections can be taken into account. The uncertainty related to the inspection method is modelled by Probability Of Detection (POD) curves.

In an illustrative numerical example a typical welded detail relevant for steel bridges is considered.

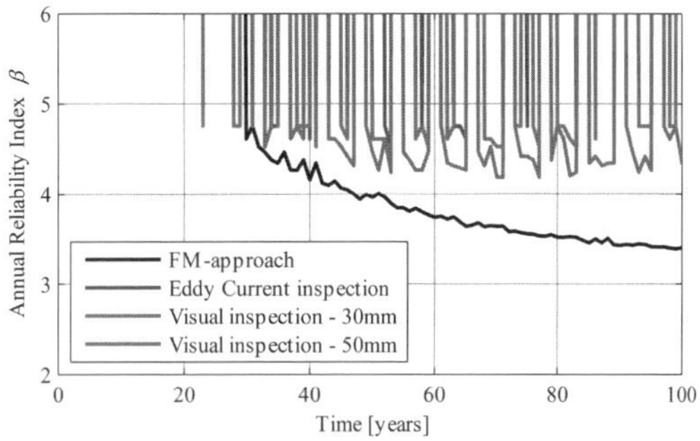

Figure 1. Annual reliability index with inspections at 6 years intervals.

The reliability is estimated both with and without regular inspections, see figure 1. The results show that the annual reliability level can be significantly increased by applying inspection, especially if the inspection method has a high quality.

The methodology described in the present paper can be further developed for other types of fatigue critical details and for other types of inspections and monitoring techniques providing direct or indirect information about the condition of the bridges.

REFERENCES

Chung, H.-Y., L. Manuel & K.H. Frank: Reliability-based optimal inspection for fracture-critical steel bridge members. Proc. TRB 2003 Annual Meeting, Washington, D.C., 2003.

Fisher, J.W. & S. Roy: Fatigue of steel bridge infrastructure. Structure and Infrastructure Engineering, Vol. 7, 2011, pp. 457–475.

Righiniotis, T.D. & M.K. Chryssanthopoulos: Fatigue and fracture simulation of welded bridge details through a bi-linear crack growth law. Structural Safety, Vol. 26, 2004, pp. 141–158.

Extreme events of long span bridges: Design, assessment and management

Organizer: A. Chen

Bridge Maintenance, Safety, Management, Resilience and Sustainability – Biondini & Frangopol (Eds)
© 2012 Taylor & Francis Group, London, ISBN 978-0-415-62124-3

Analysis on applicability of health monitoring techniques on a curved cable stayed bridge

B. Briseghella
Fuzhou University, Fuzhou, China

A. Chen, X. Li & T. Zordan
Tongji University, Shanghai, China

C. Lan & E. Mazzarolo
Università IUAV di Venezia, Venice, Italy

ABSTRACT

The investigated bridge – erected in the Commercial Harbor of Marghera (Venice, Italy) – was opened to traffic on January 2007 and includes six generally curved spans, where the two main (105 m + 126 m) spans are suspended by cable-stays, connected to the centerline of the deck. As a consequence of the spatial structural arrangement, the dynamic behavior of the deck appears very complex in both bending and torsion.

In this paper, based on the experimental dynamic tests on a curved cable-stayed bridge carried out during construction phase in May 2006 and during public service in July 2010, finite element model was built and tuned accurately according to comparisons between experimental and theoretical analyses. According to the analysis of tuned finite element model, the characteristics differences between two dynamic tests were caused mainly by the asphalt pavement completion, while the temperature differ-ence of two tests has little effect on both the static and dynamic behaviors of the bridge.

From the finite element simulation on the damage on the cable, being compared to monitoring measurement precisions, the limitations of monitoring techniques were found, that is, the existing measurement system have difficulty in identify the early damage of the bridge. Based on that, the recommendations on future monitoring were presented.

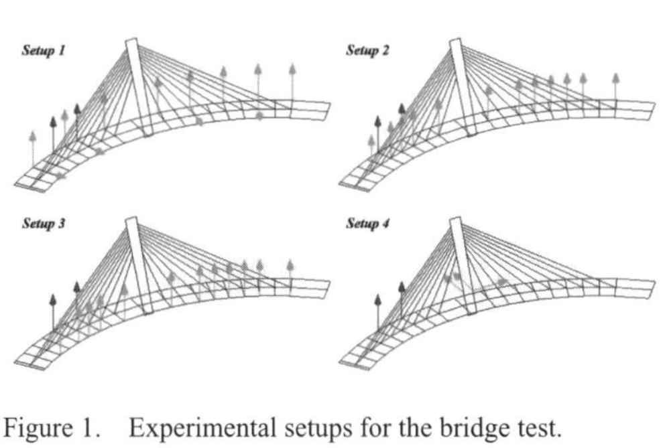

Figure 1. Experimental setups for the bridge test.

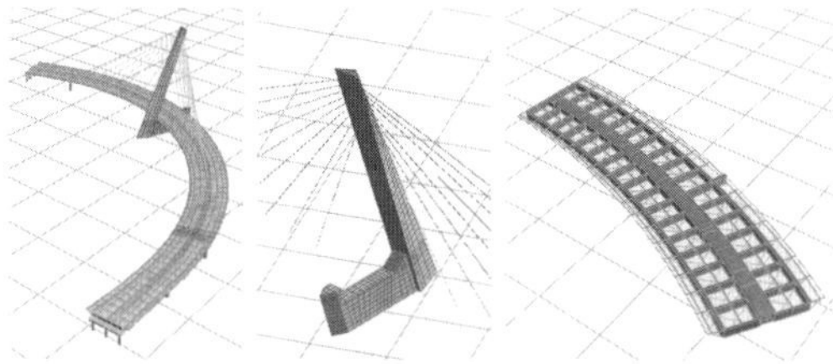

Figure 2. General arrangement and details of the FE model.

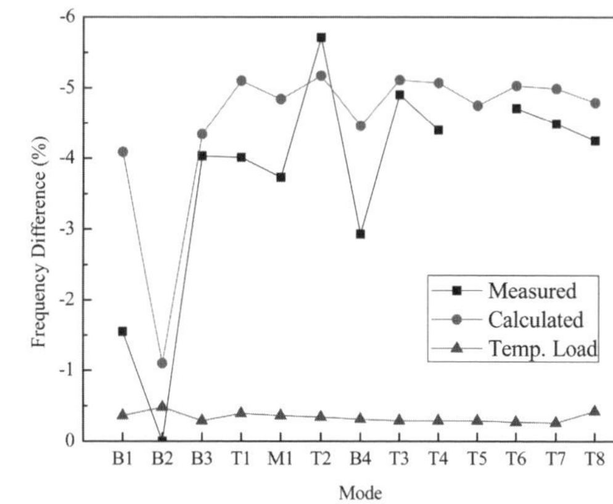

Figure 3. Frequency differences comparison.

REFERENCES

Briseghella, B., Zordan, T., Lan, C., Mazzarolo, E., Busatta, F. & Gentile, C., 2010. Safety Monitoring of the Cable Stayed Bridge in the Commercial Harbor of Venice. *IABMAS 2010 Conference, The Fifth International Conference on Bridge Maintenance*, Philadelphia, USA.

Gentile, C. 2006. Modal and structural identification of a R.C. arch bridge, *Structural Engineering and Mechanics*, Vol. 22, No. 1, pp. 53–70.

Maia, N.M.M. & Silva, J.M.M. 1997. *Theoretical and Experimental Modal Analysis*, Research Studies Press Ltd., England.

Bridge Maintenance, Safety, Management, Resilience and Sustainability – Biondini & Frangopol (Eds)
© 2012 Taylor & Francis Group, London, ISBN 978-0-415-62124-3

Computational approach to predict transporting possibility of concrete in long-distance pumping

S.D. Jo, S.G. Lee, K.P. Jang & S.H. Kwon
Myongji University, Yongin, South Korea

C.K. Park, J.H. Jeong & S.H. Lee
Samsung C&T Corporation, Seoul, South Korea

ABSTRACT

Concrete pumping is a major issue in the construction of high-rise buildings, long-span bridges, and other structures that require long-distance transport of concrete. For example, knowledge of the pumping flow rate (that is, the amount of concrete transported per hour) is required for the construction of a high-rise building of several hundred meters. This variable is useful for controlling the casting speed and for determining the total duration of the construction period, which is directly related to the construction cost.

The existing studies on predicting the pumping speed showed that the slip layer or lubricating layer that forms near the interface of the concrete and the inner wall of the pipe is a dominant factor in facilitating pumping. However, there is no way to directly measure the thickness of the lubricating layer or to experimentally determine how the rheological properties of the layer vary near the wall.

The objective of this study is to use numerical analysis to estimate the thickness and rheological properties of a lubricating layer. Numerical analysis was performed to estimate the lubricating layer considering shear-induced particle migration (SIPM) which was considered as a formative mechanism for the formation of the layer. A parametric study was undertaken to understand the effects of changes to the pumping pressure, and the rheological properties of concrete on the formation of the lubricating layer. From the analysis, the particle distribution and the viscosity varying over the cross section were obtained. Based on the results, the thickness and the rheological properties of the layer were estimated.

Bridge Maintenance, Safety, Management, Resilience and Sustainability – Biondini & Frangopol (Eds)
© 2012 Taylor & Francis Group, London, ISBN 978-0-415-62124-3

Time-dependent reliability of carbonation process for concrete component with surface coating protection

X. Liu, X. Ruan & A. Chen
Department of Bridge Engineering, Tongji University, Shanghai, China

ABSTRACT

Durability deficiency reduces the serviceability and safety performance of a reinforced concrete structure, and increases the social and economic costs in the future maintenance. The corrosion of reinforcement in atmospheric environment is mainly caused by concrete carbonation. Applying surface coating to concrete component is an effective measure to prevent concrete carbonation.

The surface coatings prevent concrete carbonation by reducing the diffusion of carbon dioxide into the concrete matrix. According to their materials, the coatings can be mainly divided into two kinds: 1) the materials can react with carbon dioxide, such as plaster, cement paste, etc. The alkaline substances in the coating will react with carbon dioxide first to reduce the carbon dioxide concentration on the concrete surface and delay the time carbon dioxide reaches concrete surface (Liu et al. 1997). 2) the materials can not react with carbon dioxide, such as asphalt, organics, etc. The coatings prevent the carbonation process by reducing permeation and diffusion speed of carbon dioxide in the coating and the water content on the concrete surface (Park 2008).

Based on existing carbonation models, a diffusion-reaction numerical carbonation model taking account of surface coating protection is presented in this work. And this model has been validated by comparing results obtained from the model with accelerated carbonation experiments using coated concrete specimens (Liu et al. 1997, Park 2008).

The carbonation front reaching reinforcement is taken as limit state of the durability of concrete cover. Sophisticated prediction model for computing the carbonation depth leads to the non-linear implicit performance function. An improved response surface method, the first order reliability method (FORM) together with response surface method (RSM), is adopted to estimate reliability index of concrete component.

Finally, an example is given to illustrate the probabilistic approach and the effects of degradation of organic coatings. The results show that organic coating is better in preventing concrete carbonation than cement paste coating.

The degradation of organic coating will obviously decline its prevention to carbonation. The recoating at fixed period is required to maintain high durability of concrete component (Figure 1).

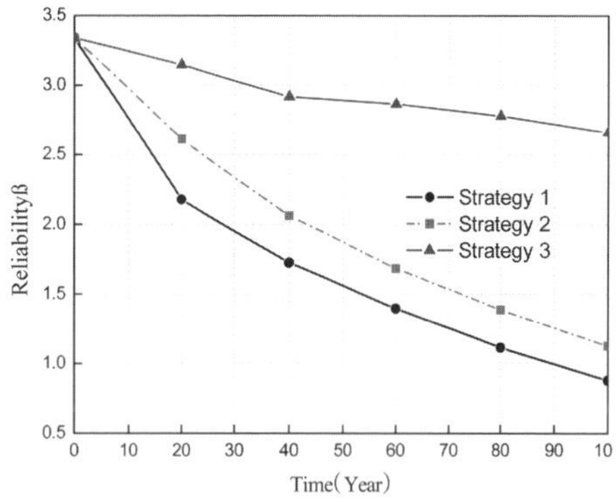

Figure 1. Comparison of three maintenance strategies.

ACKNOWLEDGEMENT

This work is supported by National Natural Science Foundation of China (Grant NO.50878145), the Fundamental Research Funds for the Central Universities and Traffic technology Foundation of Guizhou Province, China (Grant NO.2010-122-018)

REFERENCES

Liu Y.Q., Zhang Y., Zhang W.P., Wang J. 1997. Influence and calculation of surface coatings on carbonation of concrete. Industrial Construction 27:41–45.

Park D.C. 2008. Carbonation. of concrete in relation to CO2 permeability and degradation of coatings. Construction and Building Materials 22: 2260–2268.

Bridge Maintenance, Safety, Management, Resilience and Sustainability – Biondini & Frangopol (Eds)
 ISBN 978-0-415-62124-3

Risk based management in Minpu Bridge

X. Ruan, Z.Y. Yin & Z.G. Yan
Department of Bridge Engineering, Tongji University, Shanghai, China

ABSTRACT

Minpu Bridge is located in Shanghai, connecting both sides of the city. Minpu Bridge is the first long-span double-layer steel cable-stayed bridge in China, with a 708-metre main span (Figure 1). Double cable plane is designed with 88 stayed cables on each side in total symmetric arrangement. Main girder in main span and side span of Minpu Bridge adopts steel truss and truss typed diagonal composite form separately. Both layer of the girder are served for traffic purpose. The upper layer is designed as expressway with 8 lanes and a width of 40.5 m, while the lower layer is designed for local traffic with 6 lanes and a width of 26.0 m.

A risk based management research on Minpu Bridge, which consists of risk identification, risk analysis, risk assessment and risk management, is delivered in order to overcome the management caused by various risk scenarios during operation period.

A comprehensive analysis on the potential risk factors during operation is implemented, and four risk types, named as natural disaster, traffic accident, terrorism attack and other, are identified.

Detailed risk analysis including component vulnerability and structure robustness analysis is introduced. Risk analysis of Minpu Bridge is implemented according to the types of components, individually the pylon, side piers, main girder and stayed cables.

Risk analysis results indicates that 60 risk scenarios are in need of further risk assessment and risk matrix method is chosen. Assessment results demonstrate that no risk scenario in the operation period of Minpu Bridge is categorized in unacceptable level, 11 risk scenarios in strict control, 25 in reasonable control, 9 in acceptable and 15 in neglectable.

A risk management guidebook is presented in a risk scenario based form, in which both specific description and management strategies are provided for each risk scenario.

Risk based management achievements proposed in this paper have been applied in the operation period management of Minpu Bridge and received a satisfying feedback. For further application in other engineering objects, similar process can be adopted.

REFERENCES

Park, Y.J. & Ang, A.H. 1985. Mechanistic Seismic Damage Model for Reinforced Concrete. *Journal of Structural Engineering*. 111(4): 722–739.

Ruan, X., Chen, A.R. & Shi, X.F. 2008. *Risk Assessment of Bridge Engineering*. Beijing: China Communications Press

Ruan, X., Xu, D.Z., Li, L.J. & Zhang, Z.Y. 2008. Risk Assessment and Management System for Bridge Construction. *Information and Communication Technology(ICT) for Bridges, Building and Construction Practice. IABSE Symposium Report, Helsinki, June, 2008.*

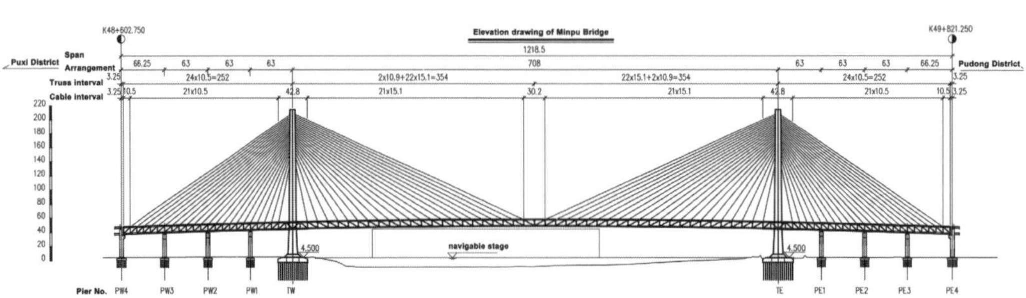

Figure 1. Elevation drawing of Minpu Bridge.

GENERAL SESSIONS

Organizers: F. Biondini & D.M. Frangopol

Bridge Maintenance, Safety, Management, Resilience and Sustainability – Biondini & Frangopol (Eds)
 ISBN 978-0-415-62124-3

Negative moment region composite action of steel-concrete girders with grouped studs

I.-S. Ahn, S.S. Chen & A.J. Aref
Department of Civil, Structural and Environmental Engineering, University at Buffalo, Buffalo, New York, USA

J.A. Carpenter
Birdair, Inc., Amherst, New York, USA

M. Chiewanichakorn
Degenkolb Engineers, Los Angeles, USA

ABSTRACT

Many bridges utilize concrete deck slabs composite with steel girders. A steel-concrete composite girder is a structural member where a steel beam is connected to a concrete slab by a shear transfer mechanism. Shear studs are typically used to provide a sound shear transfer mechanism between these two components. Thus, a composite girder behaves as one member resisting applied loads. During historical development of composite girder design, the composite action of simple-supported girders became well understood. That understanding was extrapolated to use of composite girders in negative as well as positive moment regions of continuous girders. The current AASHTO LRFD Bridge Design Specifications (AASHTO 2010) incorporated the results of research conducted by the authors (Chen et al. 2007) in the definition of the effective width: girder spacing can be used as the effective flange width in most cases in both positive and negative moment regions. In particular, Aref et al. (2007) investigated composite actions and corresponding effective flange width in negative moment regions of a continuous girder.

There is some ambiguity and confusion in the determination of whether girders are composite or noncomposite if only grouped shear studs are provided at contra-flexure points instead of distributing them longitudinally throughout the negative moment region. In this paper, ultimate behavior of two types of composite construction are compared based on analytical and experimental investigations: girders with distributed shear studs and girders with studs grouped at the contra-flexure points. Figure 1 shows the force-displacement relationship from the specimen having distributed shear studs with finite element method (FEM) and line-girder analysis results.

From the comparison of test results and accompanying analysis, it is observed that the global behavior of the specimen with grouped studs is similar to that of the specimen with distributed studs, including the ultimate limit state. However, the specimen with grouped studs

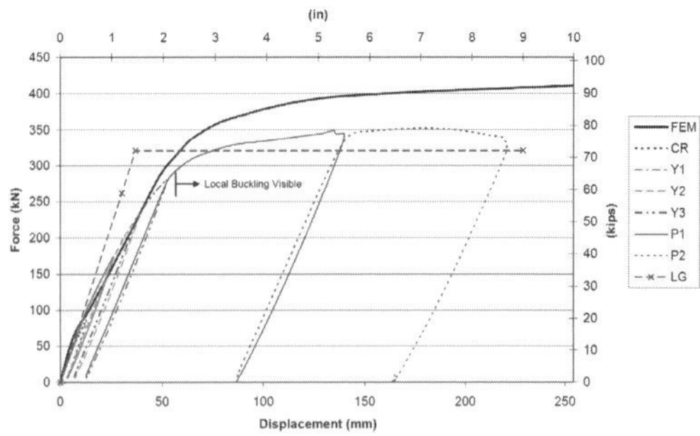

Figure 1. Load vs. Displacement in Negative Moment Subassemblage.

indicates some slippage and separation of the concrete deck from the steel girder as it reaches the ultimate capacity.

It can be concluded that girders with grouped studs develop significant composite action that makes them behave like composite girders having distributed studs. If girders with grouped studs are designed assuming non-composite behavior, then actual composite action would increase the compression at the bottom flange and move the neutral axis up in the web. Consequently, the section can change from compact to non-compact, the sectional capacity can be substantially reduced, and local buckling can happen in the web or the bottom flange.

REFERENCES

AASHTO. 2010. *AASHTO LRFD Bridge Design Specifications (5th ed.)*, American Association of State Highway and Transportation Officials.

Aref, A. J., M. Chiewanichakorn, S. S. Chen and I.-S. Ahn. 2007. Effective Slab Width Definition for Negative Moment Regions of Composite Bridges. *Journal of Bridge Engineering-ASCE* 12(3): 339–349.

Chen, S. S., A. J. Aref, M. Chiewanichakorn and I.-S. Ahn. 2007. Proposed Effective Width Criteria for Composite Bridge Girders. *Journal of Bridge Engineering-ASCE* 12(3): 325–338.

Bridge Maintenance, Safety, Management, Resilience and Sustainability – Biondini & Frangopol (Eds)
© 2012 Taylor & Francis Group, London, ISBN 978-0-415-62124-3

Structural analysis of bridges with time-variant modulus of elasticity under moving loads

H. Aied & A. González
School of Civil, Structural and Environmental Engineering, University College Dublin, Dublin, Ireland

ABSTRACT

This paper investigates the response of a one-dimensional simply supported bridge model to a moving loading when allowing for a varying modulus of elasticity according to the strain rate experienced by the structure. This dynamic (time-variant) modulus is used throughout the simulation instead of the traditional constant modulus used in vehicle-bridge interaction problems. The change in dynamic modulus with time is estimated based on the equation proposed by CEB-FIB (Rowe & Rene, 1991). Each point on the cross-section of the beam has an associated strain rate and a value of dynamic modulus that varies with longitudinal section under investigation and the location of the moving load on the beam. The moment of inertia of the cross-section is adjusted at each point in time to ensure that the bridge model has the same distribution of stiffness as the one using a dynamic modulus for each elementary beam. Aied & González (2011) have found that strain rates in structures subject to moving loads can be sufficiently high to justify the use of a dynamic modulus. For example, Figure 1 shows how the dynamic modulus of the bottom fiber at the mid-span section changes with strain rate as a 100 kN load crosses a 10 m beam at 25 m/s.

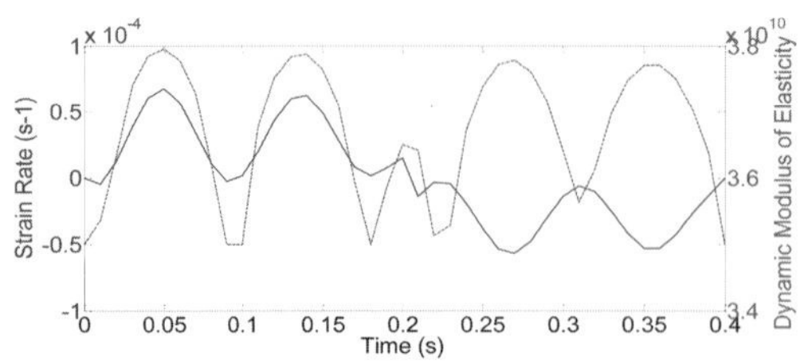

– – – Time-variant modulus of elasticity (N/m^2)

—— Strain rate (s^{-1})

Figure 1. Time-variant modulus with strain rate versus time at mid-span at lowest section of 10 m long bridge model subject to moving load (velocity = 25 m/s and load = 100 kN).

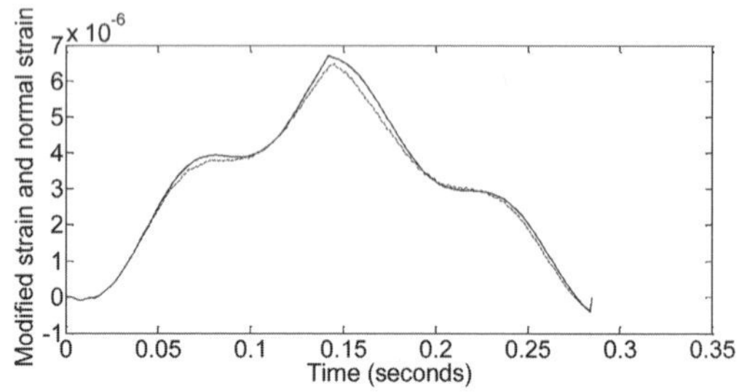

—— Strain due to constant modulus

– – – Strain due to time-variant modulus

Figure 2. Time-variant modulus with strain rate versus time at mid-span.

This paper analyses how these changes in stiffness affect the overall structural response when subject to a relatively moderate moving load. Structural responses are obtained for a range of bridge spans, typical vehicle highway speeds and bridge locations. It is found that strain and displacement can be about 5% smaller when considering a time-variant modulus compared to a constant modulus, although this percentage will vary depending on the properties of the moving load, the bridge and the section under investigation. Figure 2 illustrates this phenomenon for the case of a 100 kN load travelling at 35 m/s over a 10 m bridge.

REFERENCES

Aied, H. & González, A. 2011. The Effect of Strain Rate on the Total Response of a Structure subject to a Moving Load. *In:* B.H.V. Topping, Y. T. (ed.) *Civi.* Stirlingshire, UK: Civil-Comp Press.

Rowe, R. E. & Rene, W. 1991. Bulletin D'Information No. 203 CEB-FIP Model Code 1990- Final Draft *In:* (CEB), C. E.-I. D. B. (ed.) *Comite Euro-International du Beton (CEB).*

Bridge Maintenance, Safety, Management, Resilience and Sustainability – Biondini & Frangopol (Eds)
© 2012 Taylor & Francis Group, London, ISBN 978-0-415-62124-3

A BMS development project with an integrated inspection program

F. Akgül
Department of Engineering Sciences, Middle East Technical University, Ankara, Turkey

ABSTRACT

BMSs are developed in various countries and the existing systems are being continuously improved. Some examples of existing BMSs include PONTIS and BRIDGIT in US, DANBRO in Denmark (Bjerrum and Jensen 2006), and BMSs for prefectures in Japan (Furuta & Watanabe 2010), T-BMS in Taiwan (Liao *et al.* 2008), SBMS for steel bridges in Korea (Kong *et al.*) and others. The BRIME study proposed a framework for a European BMS and considerable progress has been made for BMSs in US since the 90s.

This paper presents the details of an inspection and data collection program devised as part of the development of a BMS in Turkey. Figure 1 shows the number of bridges built at different times in Turkey. A pilot bridge group consisting of 200 bridges is selected for inspections which are located along the state and provincial roads. In addition to visual inspections of the pilot bridges, detailed inspections for 10 deteriorated bridges will also be performed using NDT equipments. Testing equipments to be used include probes for resistance penetration (for strength evaluation), ultrasonic testing equipment, concrete reinforcement locator and a corrosion measurement device for concrete reinforcement. The web-based BMS is partially developed and the bridge inventory is constructed for the pilot bridge group. Matrices containing definitions of Bridge types vs. Element types and Element types vs. Damage types have been constructed and implemented in the BMS. The remaining tasks include the determination of the condition states and transfer of inspection results to the system. The system is expected to have a prioritization module and an optimization module for bridge maintenance-repair-replacement decisions. The project duration is 2.5 years with a budget of $1,120,000. Project details are presented including budget and time schedules. Project consists of personnel employment, technical and safety training of the personnel, procurement, software and database development, development and integration of an optimization algorithm for maintenance repair and replacement activities, development of the inspection program, field-to-database data transfer, and final testing and implementation. The developed Management System will be used by the General Directorate of Highways to plan and monitor the maintenance and repair of the bridges.

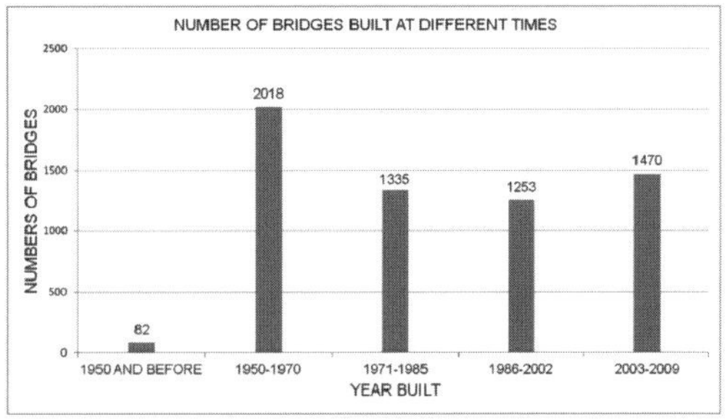

Figure 1. Number of bridges built at different times.

REFERENCES

Bjerrum, J & Jensen, F.M. 2006. Internet-based management of major bridges and tunnels using the Danbro+ system. *Proc. of IABMAS2006 the 3rd int. conf. on bridge maint., safety, manag., life-cycle perf. and cost, Porto, Portugal.* CRC-Press-Balkema-Taylor & Francis Group.

Furuta, H. & Watanabe, E. 2010. Bridge maintenance and practical bridge management systems in Japan. *Proc. of IABMAS2010 the 5th int. conf. on bridge maint., safety, manag. and life-cycle optimization. Philadelphia, PA, USA*

Kong, J.S., Park, S.H., Kim, S.H., Park, K.H. & Frangopol, D.M. 2006. *Proc. of IABMAS2006 the 3rd int. conf. on bridge maint., safety, manag., life-cycle perf. and cost, Porto, Portugal.*

Liao, H.-K, Yen, C.-I. & Yau, N.-J. 2008. Development of a bridge maintenance decision support module for Taiwan Bridge Management System. *Proc. of IABMAS2008 the 4th int. conf. on bridge maint., safety, manag., health monit. and informatics.*

Bridge Maintenance, Safety, Management, Resilience and Sustainability – Biondini & Frangopol (Eds)
© 2012 Taylor & Francis Group, London, ISBN 978-0-415-62124-3

Design of externally plated RC beams in bridging applications

S. Aliamiri & M. Raoof
School of Civil and Building Engineering, Loughborough University, Loughborough, UK

ABSTRACT

The need for structural strengthening (in preference to the environmentally undesirable and economically often much more expensive alternative of demolition and reconstruction) may arise for several possible reasons, including, for example, poor or outdated initial design; deterioration of structures due to both age and environmental factors and changing loading conditions (especially relevant to road bridges with increasing traffic loads over time).

Steel and/or FRP plate bonding to the tension face of RC beams is a relatively simple and cost-effective method of upgrading such structural members. Moreover, in addition to being less labor intensive, the plate bonding method causes (cf. certain other strengthening methods) minimal influence on the strengthened member's dimensions, and can be carried out with minimal disruptions to the users of the structure: obviously, this is particularly of importance in the case of bridging applications. Over the last three decades, this method has been used extensively for upgrading both buildings and bridges in a large number of countries.

Bearing the above in mind, it is however noteworthy that, despite extensive worldwide research on the various characteristics of externally plated RC beams over the last two to three decades, a generally accepted method to predict the failure load of plated beams is not yet available. This is particularly true in the case of widely observed undesirable premature (plate peeling and interface debonding) failures of such elements which are often of a largely brittle nature. It is noteworthy that interface debonding failure is associated with those instances when there is a bond failure occurring at the plate/glue/concrete interface whereas the peeling failure involves the plate and concrete cover separating as a unit from the underside of the main reinforcing bars.

In a series of publications, Raoof & his associates have already reported details (as well as extensive experimental verification) of two distinctly different semi-empirical models for predicting both the peeling and the debonding failure loads of steel as well as FRP plated RC beams; hereafter referred to as the Tooth and the Interface debonding models, respectively. Most importantly, in a recent publication, Raoof and one of his associates argued that, in contrast to the previous widely held view in the literature, in their extensive tests on steel plated RC beams, even when the occurrence of premature plate peeling failure was successfully prevented by using effective plate end anchorages (in the form of sufficiently long prestressed bolts), the full flexural capacity of the steel plated beam was still not achieved and, instead, an interface debonding mode of premature failure was found to occur at the plate/glue/concrete interface with its consequent associated drastic reductions in the ultimate strength (cf. the full flexural capacity).

In the present paper, using an extensive set of test data relating to 484 FRP and 203 steel plated RC beams (i.e. total of 687 plated specimens), covering a very wide range of first order beam design parameters, ample support is provided for the general reliability of a simple design method against occurrence of premature flexural failures in externally plated simply supported RC beams with the steel and/or FRP plates glued to their soffits. In particular, the proposed design method is amenable to simple hand calculations, using a pocket calculator, aimed at practicing engineers for their use in everyday designs. In addition, by using the same extensive set of test data, it has been demonstrated that the plate premature failure moment can, in some cases, be significantly lower than the ultimate failure moment of the original RC beam- hence, the warning that it is absolutely necessary to guard against occurrence of such premature failures by using a reliable model (such as the one proposed in the present paper) for designing against such potentially very dangerous premature failures. Finally, the present paper also provides a complete numerical worked example for a typical steel plated RC beam, based on the presently proposed design approach.

Bridge Maintenance, Safety, Management, Resilience and Sustainability – Biondini & Frangopol (Eds)
 ISBN 978-0-415-62124-3

Network bridge management with life-cycle cost optimization

J.O. Almeida
Instituto Politécnico de Viana do Castelo, Viana do Castelo, Portugal

R.M. Delgado
Faculdade de Engenharia da Universidade do Porto, Porto, Portugal

P.F. Teixeira
Instituto Superior Técnico, Lisboa, Portugal

ABSTRACT

This paper presents a strategic medium/long-term intervention planning methodology for managing bridges and other similar structures networks, during their working periods of life, assuring the required safety and serviceability levels and taking the available budget into consideration, useful to minimize the associated high life-cycle costs and also to pre-pare an eventual extra-budget request argumentation. The methodology processes a set of a different kind of information to achieve some support decision indexes like the minimum life-cycle cost intervention plan on a set of bridges during a medium/long period of time, for different performance levels and respecting all the user's local and global restrictions. The methodology's main models are those presented on Figure 1 where the bridge performance deterioration is predicted with a probabilistic process and the life-cycle cost estimation is based on some expected economical rates and some other considerations grounded on a Portuguese expert judgment base.

From the presented work it's possible to conclude that the selected degradation model could have a huge importance on the final results. So, is very important to choose the most suitable degradation model for each bridge, taking in consideration its specific kind of structure.

As usually referred Ekberg et al. (2010), the actualization rate variation is a factor with a great influence on the final results, and it's especially relevant for longer analysis periods of time.

However, it's important to note that the main purpose of this kind of analysis is to predict the necessary medium/long time future needs, in terms of intervention and budget. However, a specific verification before the real implementation of the improvement works is always needed. Moreover, the analysis should be redone periodically in order to actualize bridges' condition states with last inspections results and to redefine other kind of parameters that were better known in the meantime.

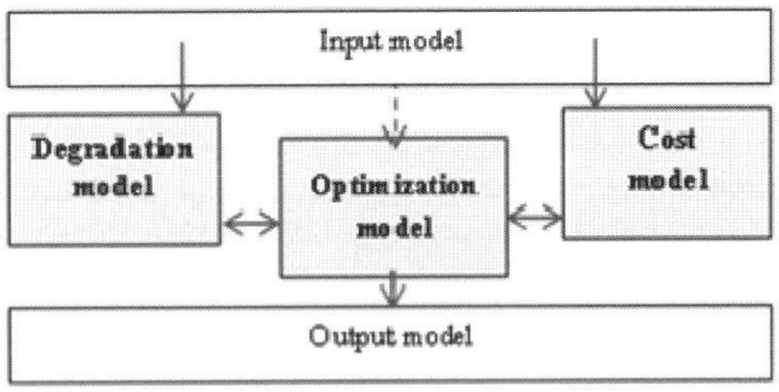

Figure 1. Methodology's main modules.

The results obtained on the methodology's factors sensitive analysis are also presented in order to highlight the most relevant parameters so that a more precise characterization and calibration should be done in future developments and applications.

The genetic algorithm showed to be suitable for this kind of process optimization and the main attention on that model should be the definition of the number of individuals by population according to the number of structures in analysis.

So, suitable results could be obtained for estimate network bridges medium/long term future needs, if the referred especial care in the model degradation choice is taken, the correct genetic algorithm parameterization is done and a good determination of the previously identified methodology's main parameters is made. The presented methodology could be easily used by bridges administration to minimize their high total life-cycle cost and to justify future investment needs.

REFERENCE

Ekberg, A. and B. Paulsson (2010). InnoTrack – Concluding Technical Report, UIC.

Bridge Maintenance, Safety, Management, Resilience and Sustainability – Biondini & Frangopol (Eds)
 ISBN 978-0-415-62124-3

Application of the operational modal analysis and modal updating methods for the characterization of the longitudinal modulus of an ancient reinforced concrete truss bridge in Almeria (Spain)

J.F.J. Alonso & A.S. Pérez
Department of Continuum Mechanics and Structural Analysis, University of Seville, Seville, Spain

ABSTRACT

As part of the project for the repair, reinforcement and extension of an old reinforced concrete bridge, designed by isostatic double-truss girders and built on the outskirts of the city of Almeria (Spain) in 1927, it was necessary to characterize the actual behavior of the structure, provided its apparent advanced state of deterioration and the proximity of the structure to a marine environment.

Since the Property did not allow completing any type of destructive testing, two types of nondestructive tests were considered for this purpose.

On the one hand, a powerful rebound hammers test campaign in order to get a first estimate of the modulus of elasticity of the structure.

On the other hand, a test was conducted to characterize the dynamic behavior of the structure by performing an environmental test bt means of modal operational technique under a normal traffic flow on the structure.

This test was preceded by the development of a detailed structural finite element (FE) model. In order to correlate the experimental dynamic results and the analytical FE model, such model was adjusted using the modal updating techniques to actualize the calculation models from dynamic measurements.

The parameters considered in this setting were the modulus of elasticity of the structure and the coefficient of horizontal stiffness of the substructure.

Given the high existing levels of cracking in the lower wing of the concrete truss, a sensitivity study was performed by considering reductions in the inertia of these elements to values of about 30%. Finally, results obtained by both methods were compared, exhibiting a very good correlation.

The longitudinal modulus estimated in this way was used as the base for the development of the subsequent repair and reinforcement structural project.

Figure 1. Image of Los Molinos Bridge (Almería, Spain).

REFERENCES

Caetano E., Cunha A., Hoorpah W. and Raoul J., "Footbridge Vibration Design", CRC Press, Taylor & Francis Group, 2009, ISBN 978-0-415-49866-I.

Ewins, D.J. (2000) Modal Testing: theory and practice. Research Studies Press, U.K.

Friswell, M.I., Mottershead, J.E., "Finite Element Model Updating in Structural Dynamics, Kluwer Academic Publishers. ISBN 0-7923-3441-0, 1995.

Koh Ghee C., Perry M.C., "Structural Identification and Damage Detection using Genetic Algorithms" CRC Press, Taylor & Francis Group, 2010, ISBN 978-0-415-46102-3.

Magalhaes, F. (2010) Operational Modal Analysis for Testing and Monitoring of Bridges and Special Structures. PhD Thesis, Faculdade de Engenharia Universidade do Porto, Portugal.

Peeters, B. (2000) System Identification and Damage Detection in Civil Engineering. PhD Thesis, Katholieke Universiteit Leuven, Belgium.

Teughels, A., "Inverse Modelling of Civil Engineering Structures Based on Operational Modal Data", Ph. D. Thesis, Katholieke Universiteit Leuven, 2003.

Zivanovic, S., Pavic, A. Reynold, P., "Finite element modelling and updating of a lively footbridge: The complete process", Journal of Sound and Vibration, 2006.

Bridge Maintenance, Safety, Management, Resilience and Sustainability – Biondini & Frangopol (Eds)
© 2012 Taylor & Francis Group, London, ISBN 978-0-415-62124-3

Seismic vulnerability and retrofitting of "Gioieni bridge" in Catania using innovative materials

L. Anania, A. Badalà, S. Costa & G. D'Agata
University of Catania, Italy

ABSTRACT

This paper deals with a large series of in situ investigations as well as some modern techniques of retrofitting for maintenance and restoration of road bridges in urban areas. The paper reports the study of the evaluation of the materials degradation and of the seismic vulnerability of the bridge known as "Tondo Gioeni bridge" built in the early '60s. The structural typology of this bridge represents the typical structural architecture adopted in the city of Catania at that time, besides it also plays a key role in the urban road network, so for its historical, architectural and functional importance, now in its 50th anniversary, we have considered it necessary to undertake the process of analysis of the status quo in order to evaluate the possibility of the restoration and retrofitting to modern standards required by structural rules.

After a first phase that covered the control of the deterioration of reinforced concrete (obtained by measuring the concrete degree of carbonation, the chromatography of the concrete and the measure of the degree of oxidation of the steel reinforcement by the electrochemical potential, etc.), we proceeded to the determination of the real materials resistance also on the basis of the data obtained from previous investigations in compliance with the new rules. The analysis of seismic vulnerability of the structure was carried out according to new structural Italian codes by means of a 3-D modeling, in order to identify the most deteriorated areas. It is necessary to strengthen the bridge, both to increase the margin of safety (actually very low), and to remedy the absence of concrete cover for reinforcement mainly due to two reasons: an installation error and the degradation of the concrete coating. The PBO-CFRP reinforcement of the slab deck is thus provided for the only purpose of increasing the safety margin of the structure, giving flexibility and a residual strength capacity typical of PBO-CFRP composites.The effectiveness of the proposed technique is confirmed by experimental tests conducted on six models of reinforced concrete plate. The plates are the same as the dimensional characteristics of the deck slabs under consideration.

The comparison between the two tests in terms of "load – vertical displacement curves shows that the ultimate load of reinforced plate system is about 2.5 times greater than that of the unreinforced plate, since the tensile strength is, respectively, of 160 kN and 64 kN for the plate reinforced with PBO FRCM and for unreinforced plate.

Figure 1. "Aereal view of "Tondo Gioieni" bridge in Catania.

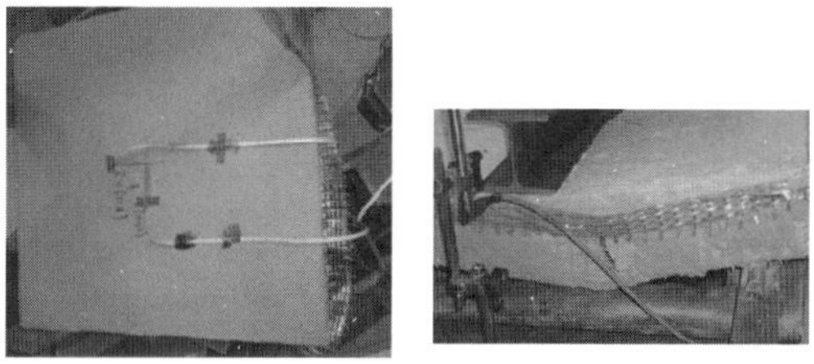

Figure 2. Tested plate, configuration at the collapse.

REFERENCES

In situ test report: "Indagini preliminari sulle strutture in cemento armato del Ponte di Piazza Gioeni – Elaborati 1–4".

Italian Code D.M. 14/01/2008: "Norme tecniche per le costruzioni".

Ruredil S.p.A., 2009a. Istruzioni per la Progettazione di Interventi di Consolidamento Statico mediante l'utilizzo di Compositi Fibrorinforzati a matrice cementizia-FRCM.

Bridge Maintenance, Safety, Management, Resilience and Sustainability – Biondini & Frangopol (Eds)
© 2012 Taylor & Francis Group, London, ISBN 978-0-415-62124-3

Exploring system interdependencies via a multi-disciplinary modeling approach: Application to bridge management

E. Andrijcic, S. Chase, Z. Guo & S. Hwang
University of Virginia, Charlottesville, VA, USA

ABSTRACT

Highway bridges are essential, and in some cases critical, elements of regional and national transportation networks. Bridge infrastructure failures (defined as the inability of a bridge to provide planned-for levels of service) propagate through interdependent socio-economic systems across multiple regions. Our inability to understand the critical connection between the bridge system as such and the regional socio-economic system within which the bridge system is embedded presents a great challenge and is explicitly addressed in this paper. To address this challenge, this paper presents a conceptual multi-disciplinary modeling approach that is grounded on the tools and principles of systems engineering, and is used to model and assess the broader implications of bridge reliability and management.

We describe the US bridge infrastructure as a complex system of systems, encompassing all physical and engineered components of a bridge, commercial and non-commercial users, and local-, state-, and federal-level decision makers. Such regional bridge systems cannot be modeled by a single model, as no single model can capture all the dimensions necessary to adequately represent and manage the structural viability of a bridge system and evaluate the efficacy of risk assessment and management activities associated with it. This paper considers the concurrent use of a multi-disciplinary set of tools and models to explore the interdependencies between the various elements of the bridge system of systems, with the goal of discovering how current bridge maintenance practices affect, and are affected by, social and economic dimensions. The conceptual modeling framework includes finite element analysis and girder line approaches, statistical deterioration models, input-output economic models, and a conceptual agent-based simulation model. Results of the traditional bridge models can be used as inputs into the agent-based model, which can be built from credible and publicly available databases, and which can model travel behavior changes caused by reduced bridge capacity. The proposed agent-based simulation model outputs (i) the changes in non-commercial bridge users activity participation patterns, and (ii) changes in commodity deliveries by freight carriers. By relating the changes in the travel behavior of bridge users to changes in supply and demand of regional economic sectors, the results of the agent-based simulation model can augment existing results provided by the economic input-output models.

The proposed use of multi-disciplinary models allows us to examine the operation of the bridge infrastructure from various perspectives, and might help us gain insight into robust bridge infrastructure management and emergent risks better than any single-model approach can. Through the integrated use of models that are traditionally used in bridge management, and new models that enable us to explore the socio-economic effects of bridge maintenance decisions, we propose a holistic, albeit currently conceptual, approach to bridge management, which supports the formulation of bridge management policies by improving the transparency and explicitness of the system interdependencies.

REFERENCES

Aktan, A.E., Faust, D. 2003. A holistic integrated systems approach to assure the mobility, efficiency, safety and integrity of highway transportation. Published in *Structural Health Monitoring and Intelligent Infrastructure – Volume 1*. Eds. Wu, Z. & Abe, M. Sweets and Zeitlinger.

DeLaurentis, D., Callaway, R.K. 2004. A System-of-Systems Perspective for Public Policy Decisions. *Review of Policy Research* Vol. 21, No. 6: 829–837.

DeLaurentis, D. 2008. Appropriate modeling and analysis for systems of systems: Case study synopses using a taxonomy. *IEEE International Conference on System of Systems Engineering, June 2008.*

Thissen, W.A.H., Herder, P.M. 2009. System of Systems Perspective on Infrastructures. Published in *System of Systems Engineering*, Ed. Jamshidi, M. New York: John Wiley & Sons.

Bridge Maintenance, Safety, Management, Resilience and Sustainability – Biondini & Frangopol (Eds)
© 2012 Taylor & Francis Group, London, ISBN 978-0-415-62124-3

Earthquake retrofit campaign for large scale bridges in Istanbul

N.M. Apaydın
Division of General Directorate of State Highways, Istanbul, Turkey

ABSTRACT

This study describes the seismic retrofit and structural strengthening campaign for large scale bridges in Istanbul. Two main motorways at east-west axis connect two continents and two parts of the city. These motorways serve domestic trade as well as international trade between Europe and Middle East countries and they serve to the city traffic at the same time. In the event of a possible destructive earthquake, the earthquake losses may increase due to failure of the transportation network. Any broken link due to the bridge failures may totally paralyze the whole transportation system in the city. For this reason, earthquake protection of the bridges has great importance for the transportation network of Istanbul.

There are totally 730 bridges located within the same transportation network. 120 of them are considered as large scale bridges. The motorways O-1 and O-2 in the metropolitan area of Istanbul contain 165 bridges; two of them are long-span suspension bridges and 16 of them are large scale steel and concrete bridges.

The aim is to keep the bridges open to traffic at strategic locations without causing any interruption for emergency services (fire, health, search-rescue, communications, etc.). The most critical issues in assessing the retrofitting needs of existing bridges were the identification of the strategic importance of the bridge (prioritization) and the assessment of the structural deficiency. The first issue was treated by means of determining real needs of the people and the city. When we look at the city plan, we can see that the settlements areas are separated from each other with waterways. Considering the unusual geographical location of Istanbul, a lot of bridges have very important strategic position. For prioritization the whole Marmara Region was taken into consideration and a wide range study was performed according to importance-based optimization of the resources. For evaluation of bridges initial structural capacity assessments were made by using only earthquake resistant design parameters. For this purpose, the bridges were grouped with respect to their highways and on the basis of the outcome of the preliminary evaluation. For the importance-based prioritization and optimization of the resources to be allocated the criteria were taken to account. As a result of the selected criteria and the related analysis, three main bridge network groups were determined. After this prioritization phase, the seismic retrofitting campaign was performed following steps: 1-Preliminary Evaluation and Ranking at in each group, 2-Determination of Seismic Performance Criteria, 3-Determination of Retrofit Design procedure and Details for Retrofitting.

The seismic performance target levels were determined according to needs of transportation. Target performance objectives for each group of bridges were assessed according to followings; Safety against earthquake, Functionality after earthquake and Repair to be needed after earthquake.

To reduce earthquake losses, performing seismic retrofitting on the right time is very important. Seismic retrofitting of bridges in such a huge city like Istanbul is a challenging study. Success is depends on sufficient funds, using time efficiently, applying effective retrofitting methodologies and well organization in heavy traffic. All critical bridges in the city center of Istanbul have been strengthened against earthquakes and this challenging work has been accomplished.

REFERENCES

Apaydın Memisoglu N. 2010. Earthquake performance assessment and retrofit investigations of two suspension bridges in Istanbul. Soil Dynamic Earthquake Engineering, doi:10.1016/j.soildyn.2010.02.011

Erdik, M., Aydinoglu, N., Fahjan, Y., Sesetyan, K., Demircioglu, M., Siyahi, B., Durukal, E., Ozbey, C., Biro, Y., Akman, H., &Yuzugullu, O. 2003. "Earthquake Risk Assessment for Istanbul Metropolitan Area", Earthquake Engineering and Engineering Vibration, Vol. 2, No. 1, 1–27.

Bridge Maintenance, Safety, Management, Resilience and Sustainability – Biondini & Frangopol (Eds)
© 2012 Taylor & Francis Group, London, ISBN 978-0-415-62124-3

Enhanced analytical method of predicting residual strength capacities of corroded steel bridge plates

J.M.R.S. Appuhamy, M. Ohga & P. Chun
Ehime University, Matsuyama, Japan

S. Furukawa
West Nippon Expressway Engineering Shikoku, Japan

P.B.R. Dissanayake
University of Peradeniya, Peradeniya, Sri Lanka

ABSTRACT

Efficient maintenance, repair and rehabilitation of existing bridges require the development of a methodology that allows for an accurate evaluation of the load carrying capacity and prediction of the remaining life. Steel girder bridges, like other structures, deteriorate over time due to environmental effects, material fatigue, and overloading. Corrosion becomes one of the major causes of deterioration of steel bridges and there have been many damage examples of older steel bridge structures due to corrosion around the world during past few decades. Therefore, designing of bridge infrastructure systems for a particular service life and maintaining them in a safe condition during their entire service life have been recognized as very critical issues worldwide.

Usually, regular and detailed inspections are necessary in order to be acquainted with the present condition of the infrastructure, assure the adequate safety and determine maintenance requirements, in bridge infrastructure management. Several experimental studies and detailed investigations were done by some researchers during the past few decades, in order to introduce methods of estimating the remaining strength capacities of corroded steel plates. But, to develop a more reliable strength estimation technique, only experimental approach is not enough as actual corroded surfaces are different from each other. Further, due to economic constraints, it is not possible to conduct tests for each and every aged bridge structure within their bridge budgets. And hence, numerical simulation is being used to replace the time-consuming and expensive experimental work and to comprehend on the lack of knowledge of mechanical behavior, stress distribution, ultimate behavior and so on.

This paper proposes two equations for the corrosion condition modeling (CCM) parameters, based on the results of many tensile coupon tests conducted on corroded specimens cut from a steel girder used for about hundred years with severe corrosion.

Then a simple and brisk analytical methodology is developed with CCM parameters and Figure 1 shows

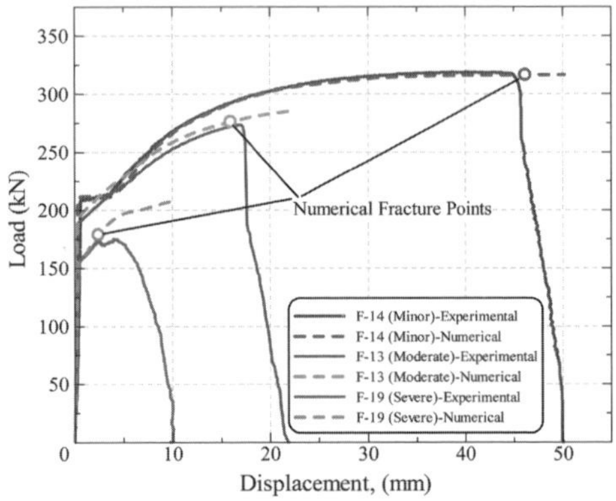

Figure 1. Comparison of load-displacement curves of experimental and proposed analytical model.

that a very good comparison of load-displacement behavior can be seen for all three classified corrosion types with the proposed analytical model. Therefore this analytical model, developed by using only the measurement of maximum corroded depth, can be used as a reliable and brisk method for the maintenance management of corroded steel infrastructures.

REFERENCES

Appuhamy, J.M.R.S., Kaita, T., Ohga, M. and Fujii, K. 2011. Prediction of residual strength of corroded tensile steel plates, *International Journal of Steel Structures*, 11(1): pp. 65–79.

Matsumoto, M., Shirai, Y., Nakamura, I. and Shiraishi, N. 1989. A proposal of effective thickness estimation method of corroded steel member, *Bridge Foundation Engineering*, 23(12): pp. 19–25.

Muranaka, A., Minata, O. and Fujii, K. 1998. Estimation of residual strength and surface irregularity of the corroded steel plates, *Journal of Structural Engineering*, 44A: pp. 1063–1071

Bridge Maintenance, Safety, Management, Resilience and Sustainability – Biondini & Frangopol (Eds)
© 2012 Taylor & Francis Group, London, ISBN 978-0-415-62124-3

Thermography for the inspection of infrastructures

R.W. Arndt & H. Parvardeh
Rutgers University, Piscataway, NJ, USA

ABSTRACT

The bridge inspector guidelines and – techniques in the United States were first introduced in 1971 with the National Bridge Inspection Standards (NBIS). They were triggered by the collapse of the Silver Bridge over the Ohio River at rush hour in 1967, where 46 people died. According to the NBIS requirements, the vast majority of bridges in the US that are 20 feet (6 meters) or longer are inspected at least once every 24 months.

Today, these inspections are no longer limited to visual inspection and simple tools such as hammers and chains as in the beginnings of the standardized bridge inspection. They are increasingly complemented by the use of non-destructive testing methods and other advanced technologies like structural health monitoring to improve the reliability of the inspection. One of these methods complementing visual inspection especially of concrete bridge decks is the so called Infrared Thermography (IRT) where temperature differences are monitored to detect e.g. debonding and delamination in reinforced concrete structures. Already since 1988 this type of bridge inspection with passive IRT is performed standardized according to ASTM standard D4788 by using the heating affect of the sun (ASTM 2005). Other applications for bridge inspection using IRT include the quality control of the bonding in composite materials (CFRP laminates) or paintings.

For highways IRT is used for some years now effectively for the quality assurance during manufacturing of hot mix asphalt pavements (Muench & Willoughby 2006). Here IRT offers new perspectives. It allows a very accurate and complete coverage of the surface temperature conditions at a very high point density with the added benefit of rapid statistical analysis over any parts of the overall temperature of any thermogram (temperature contrast image). Thus new insights into important conditions such as uniformity of temperature distribution and cooling speeds can be gained.

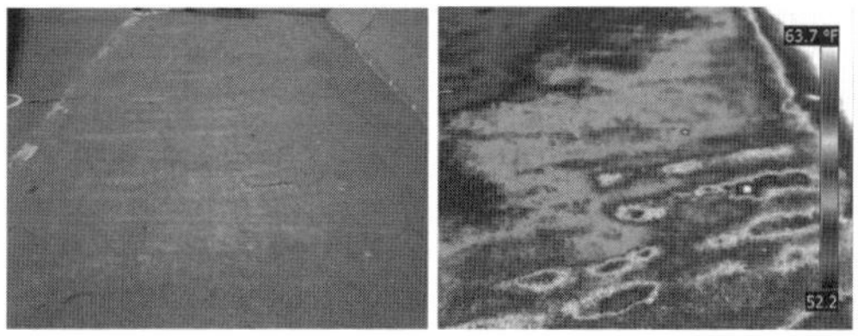

Figure 1. Digital photo with markers and visible spalling (left) and IR image of $3\frac{3}{4}$ hrs after sunrise with clearly visible debonded transversal reinforcement (right).

So, e.g. the State of Washington developed a uniform quality assurance concept for the fabrication of asphalt by implementing among other things the systematic use of thermography monitoring thus extending the life of highway overlays by an estimated 20–80% (20 to 80 years) (Muench & Willoughby 2006).

The paper gives a general introduction into the application of IRT in civil engineering followed by practical IRT applications for the inspection of infrastructures. Results presented and discussed include the inspection of a concrete bridge deck using the heating affect of the sun (Fig. 1) and winter IRT applications on a concrete bridge substructure and an artificial reinforced concrete specimen using nightly cooling effects. The paper is completed by results of an IRT based quality control of the production process of an asphalt highway test bed.

REFERENCES

ASTM 2005. Standard Test Method for Detecting Delaminations in Bridge Decks Using Infrared Thermography. West Conshohocken: ASTM.

Muench, S. & Willoughby, K. 2006. Preventing Pavement Failure Caused by Hot-Mix Asphalt Temperature Differentials – Washington State's Systematic Approach. TR News 246: 26–28.

Bridge Maintenance, Safety, Management, Resilience and Sustainability – Biondini & Frangopol (Eds)
© 2012 Taylor & Francis Group, London, ISBN 978-0-415-62124-3

The broad impact of disaster risk mitigation based on IT solutions

G.M. Atanasiu
The Multidisciplinary Center of Structural Engineering & Risk Management, Department of Structural Mechanics, Technical University "Gheorghe Asachi" of Iaşi, Romania

F. Leon
Department of Computer Science & Engineering, Technical University "Gheorghe Asachi" Iaşi, Romania

ABSTRACT

The challenge of disaster management is to reduce the harm that disasters cause to society, economy, and the lives of individuals and communities. Therefore, disaster managers must reduce uncertainty, calculate and compare costs and benefits, and manage resources. This must often be accomplished more extensively and much faster than it is possible by traditional problem solving methods. Information Technology provides capabilities that can help people understand the dynamic realities of a disaster more clearly by simulations of risk scenarios and can allow them to make better decisions more quickly by providing intelligent support for decision making. Also, IT can help record the multitude of details involved in all phases of disaster management. Beside extensive experimental research, one of the important roles of laboratories consists in opening the opportunities for using IT solutions in enhancing knowledge in earthquake engineering, for the development of human resources, especially young researchers, and for virtually interconnecting European laboratories with the global world and the already existing operating networks in the USA and Japan. The research studies carried out within the EFAST Project, which is part of the European FP7 Program, have shown direct and indirect benefits of IT-based seismic risk mitigation. This involves offering IT solutions with telepresence capabilities for European and world-wide researchers, organization of virtual seminars, videoconferences, training courses for researchers, developing real-time experiments in connection with other networks such as NEES and E-Defence, services such as consultancy for disaster mitigation in Europe and other active seismic regions. Some of the possible applications of IT solutions, suitable to be used in Earthquake Engineering research laboratories are also presented in a SWOT analysis, including their economic impact before, during, and after a seismic event.

REFERENCES

Atanasiu, G. M., Leon, F., Gâlea, D. (2006). GIS-Based Monitoring of Seismic Vulnerability for a Class of Concrete Structures during Lifetime Serviceability, Proceedings of the European Symposium on Service Life and Serviceability of Concrete Structures, ESCS-2006, pp. 183–188, Concrete Association of Finland.

EFAST (2010). Economic Impact Evaluation, Technical Report.

EFAST (2011). "Design Study of a European Facility for Advanced Seismic Testing", [online] http://efast.eknowrisk.eu.

Rao, R. R., Eisenberg, J., Schmitt, T., eds. [2007], National Research Council, Committee on Using Information Technology to Enhance Disaster Management, Improving Disaster Management: The Role of IT in Mitigation, Preparedness, Response, and Recovery, The National Academies Press, Washington, D.C.

Bridge Maintenance, Safety, Management, Resilience and Sustainability – Biondini & Frangopol (Eds)
© 2012 Taylor & Francis Group, London, ISBN 978-0-415-62124-3

Noncontact bridge deformation monitoring using laser tracking technology

U.B. Attanayake, A. Servi & H.M. Aktan
Western Michigan University, Kalamazoo, MI, USA

ABSTRACT

Monitoring of bridge deformations under static and temperature loads is vital to understanding structural behavior and performance. Use of traditional technology such as LVDTs, DCDTs, and Potentiometers requires having a fixed reference. This requirement is satisfied by connecting a string to the ground or to a temporary structure. This entire operation is tedious and requires time to mount the sensors and provide power supply and data acquisition. This operation becomes complicated when longer spans or wider bridges are to be monitored. Also, when the bridge is over a river or a highly congested road use of traditional sensors becomes challenging. Today's focus is to identify technologies that can be used in the field for bridge monitoring with confidence and minimum or no disturbance to the public while developing no hazardous conditions to the crews. Further, it should be a mobile technology that can be easily operated under field conditions, during day or night. With this understanding, a laser based metrology equipment with micrometer accuracy in distance measurement, *Laser Tracker*, was identified. Capabilities and limitations of the technology were evaluated under various field conditions. Field implementation of this technology includes monitoring of a 42 degree skew steel girder bridge behavior under thermal loads.

The objective of this study is to evaluate the capabilities and limitations of the technology and demonstrate field implementation.

This study yields the following conclusions,

1) The *Laser Tracker* can be used in the field for monitoring structural deflection/deformation at multiple locations under static loads. There is no limitation to number of measurement locations on a structure. Generally, the decision is governed by the objective, scope, and the project budget.
2) The superiority of the equipment with compared to other technologies is that the entire bridge superstructure profile can be measured from the side of the structure, as opposed to beneath the structure.
3) Laser is very sensitive to exposure conditions and the quality of air it travels through. System is capable of performing few basic adjustments based on measured in-situ weather conditions (temperature, humidity, and pressure). However, adequate measures are needed to avoid interference of heat waves. Performing measurements at night helps avoiding heat waves and minimizing impact to the public
4) Measurements performed within a radius of 250 ft. yield highly accurate results (within 5% with respect to a potentiometer).
5) Laser beam divergence limits measurement range to about 150 ft when 0.5 in. reflectors are used. When RRR are used measurements can be made well beyond the 300 ft.
6) Additional research is needed to overcome challenges due to equipment/accessory limitations and in-situ exposure conditions and to develop procedures and guidelines for field implementation of the technology.

REFERENCES

Aktan, A. E., Catbas, F. N., Grimmelsman, K. A., and Pervizpour, M. 2003. *Development of a Model Health Monitoring Guide for Major Bridges*, Drexel Intelligent Infrastructure and Transportation Institute, Drexel University, Philadelphia, PA.

Chiang, F.P., and Yu, J. 2010. Full field mapping of bridge deformation using digital speckle Photography, Bridge Maintenance, Safety, Management and Life-Cycle Optimization, 591–595.

Ja'uregui, D. V., White, K. R. P.E., Woodward, C. B., and Leitch, K. R. 2003. Noncontact Photogrammetric Measurement of Vertical Bridge Deflection, Journal of Bridge Engineering, Vol. 8, No. 4, 212–222.

Yoneyama, S., Kitagawa, A., Iwata, S., Tani, K., and Kikuta, H. 2010. Bridge Deflection Measurement Using Digital Image Correlation, Experimental Techniques, 34–40.

Bridge Maintenance, Safety, Management, Resilience and Sustainability – Biondini & Frangopol (Eds)
© 2012 Taylor & Francis Group, London, ISBN 978-0-415-62124-3

Assessing full-depth deck joint durability using embedded sensors and FE simulations

U.B. Attanayake, O. Abudayyeh, H.M. Aktan, I. Abdel-Qader, C. Mansiz & E. Almaita
Western Michigan University, Kalamazoo, MI, USA

ABSTRACT

Precast full-depth deck panels are used in new bridge construction and during deck repair activities. Effectiveness of precast component joints, in terms of durability, is questioned by looking at the past performance. Currently, a number of research projects are being conducted evaluating effectiveness of various connection details through laboratory testing. The paper will describe a bridge with full-depth deck panels that was constructed by implementing ABC techniques. The deck panels of the bridge are instrumented with more than 180 vibrating wire strain gages (VWSG) to monitor panel stresses at the middle and along the joints. Currently, temperature and strain data are collected every 10 minutes since the bridge was opened to traffic in 2008. The data is evaluated to identify the dominant loads on the bridge and to develop stress envelops representing maximum and minimum values observed during a period of more than two years. In parallel to data analysis, a detailed finite element (FE) model is developed representing the instrumented bridge superstructure. The model is simulated with as-is conditions and with embedded deteriorations under traffic and thermal loads to develop stress signatures specific to the simulated deterioration. The paper will detail the bridge configuration, sensor network, FE model and simulated distresses, stress envelops, and the use of stress signatures from FE analysis in conjunction with sensor data to identify joint durability issues as they stem from the existing structure to develop effective and efficient bridge maintenance activities.

The Parkview Bridge was designed with four spans and three traffic lanes, with all its major bridge elements including substructure prefabricated off site. The structural health monitoring system was implemented following the completion of the construction. This enabled the remote collection of continuous strain and temperature data at ten-minute intervals. Strain and temperature measurements were chosen in this project for efficiency and cost effectiveness. To effectively monitor the bridge performance under varying load conditions, sensors were grouped, depending on their locations, to address the structural monitoring.

The objectives of finite element modeling to present design details of the Parkview Bridge superstructure, to display and discuss the finite element (FE) modeling of components and their interactions, to show model calibration using sensor data, and to elaborate upon the simulation of identified distress types to develop stress/strain contours. The analysis results, in conjunction with sensor data, are used to identify signatures of potential performance issues of the full-depth deck panel system.

A detailed finite element model was developed and the model was first calibrated using load test data. However, due to the dominance of thermal loads, it was required to calibrate the model using stresses developed in the structural system due to thermal loads. This was a great challenge due to a lack of thermocouples along the depth of bridge superstructure cross-section. A model was identified from literature that is capable of representing the gradient profile from noon to 6 p.m. in a summer day. The FE model was calibrated using sensor data and the temperature gradient profile during this specific duration. Stress signatures were developed by simulating the debonding of a joint between two deck panels. The signatures can be used to identify the on-set of deterioration of the joints to make necessary maintenance decisions.

The process presented in this paper can be further developed to use for real-time structural monitoring to capture onset of deterioration to plan and execute an effective and efficient bridge management and maintenance program.

REFERENCES

Abudayyeh, O. Y., Barbera, J., Abdel-Qader, I., Cai, H., and Almaita, E. (2010). *Towards Sensor-Based Health Monitoring Systems for Bridge Decks: A Full-Depth Precast Deck Panels Case Study*, Advances in Civil Engineering, Hindawi Publishing Corporation.

Abudayyeh, O., Aktan, H., Abdel-Qader, I., and Attanayake, U. (2012). *Development and Validation of a Sensor-Based Health Monitoring Model for the Parkview Bridge Deck*, the Michigan Department of Transportation, Lansing, MI.

Bridge Maintenance, Safety, Management, Resilience and Sustainability – Biondini & Frangopol (Eds)
© 2012 Taylor & Francis Group, London, ISBN 978-0-415-62124-3

Seismic assessment of monolithic vs. pin column top connections in R/C skewed bridges

N. Attarchian, A. Kalantari & A.S. Moghadam
International Institute of Earthquake Engineering and Seismology, The Iran

ABSTRACT

Bridge failures, especially unseating or displacement of superstructure in skewed bridges, have been observed in recent earthquakes. Even when the centers of mass and stiffness coincide; skewed bridges tend to develop rotation. Based on increased stiffness of piers to deck connections, with monolithic relative to pin connections, reduction of displacement demand is predictable. However it is anticipated that, there would be a combined force demand, in monolithic connections of skewed bridges under multi-directional loading of earthquake which can affect the strength and deformation capacity of piers.

Therefore in this paper, the effect of monolithic versus pin connections at the top of the pier, has been investigated on the dynamic displacement and force fields that develop in a seismic response of a class of reinforced concrete (R/C) skewed bridges. The deck displacement demand, force fields in both ends of the piers and shear demand on shear keys of the abutments are compared in two options, under bidirectional earthquake excitations.

Following a preliminary design, three dimensional spine models of a three-span continuous concrete box girder, single column bridge with skew angles from 0 to 60 degrees are analyzed under bidirectional excitations in OpenSees. Ten different variations of the bridges are investigated to evaluate the effects of (1) skew angle; (2) continuous versus pin connection on displacement and force demands. The abutments are seat-type with elastomeric bearings under the web of each girder. In the longitudinal direction, movement of the superstructure is free without deck-abutment interaction and, in transverse direction, shear keys prevent the movement. The strong ground motion suite is selected among, standardized set of strong ground motions, recently published in Pacific Earthquake Engineering Research center in order to facilitate comparative evaluations in transportation research programs. The studies offer an insight in the complex seismic behavior of skewed bridges and help to establish the required design seat width, the strength of the shear keys and capacity of columns under combined loading.

REFERENCES

Akbari, R. & Maalek, Sh., 2010. Adequacy of the Seismic Analysis Method for Single-column-bent Viaducts Considering Regularity and higher modes effects. *Journal of vibration and control* 16: 827–852.

Baker, J.W., Lin, T., Shahi, S.K. & Jayaram, N., 2011. *New Ground Motion Selection Procedures and Selected Motions for the PEER Transportation Research Program* PEER Report 2011/03.

Caltrans Seismic Design Criteria, June 2006. Version 1.4.

Calvi, G.M. & Pinto, P.E., 1996. *Experimental and numerical investigations on the seismic response of bridges* ECOEST-PrEC8, Report No. 4, Brussels.

Casarotti, Ch., Pinho, R., 2006. Seismic response of continuous span bridges through fiber-based finite element analysis *Earthquake Engineering and Engineering Vibration* 5: 119–131.

Elnashai, A., Gencturk, B., Kwon, K., Al-Qadi, I., Hashash, Y., Rosesler, J.R., Kin, S.J., Jeong, S.H., Dukes, J. & Valdivia, A., 2010. *The Maule (Chile) Earthquake of February 27, 2010, Consequence Assessment and Case Studies*, Report No. 10-04, MAE Center, USA.

Irainian code of practice for seismic design of buildings Standard No. 2800-05 3rd Edition.

Isakovi'C, T., Lazaro, M., & Fischinger, M., 2008. Applicability of pushover methods for the seismic analysis of single column bent viaducts. *Earthquake Engineering & Structural Dynamics* 37: 1185–1202.

Kappos, A.J., Manolis, G.D. & Moschonas, I.F., 2002. Seismic Assessment and design of R/C bridges with irregular configuration, including SSI effects *Engineering Structures* 24: 1337–1348.

Mander, J.B., Priestley, M.J.N. & Park, R., 1988. Theoretical Stress-strain Model for confined Concrete. *Journal of Structural Engineering, ASCE* 114:1804–1826.

Meng, J.Y. & Lui EM., 2000. Seismic Analysis and Assessment of a Skew Highway Bridge. *Engineering Structures* 22: 14333–1452.

Mohti, A.A. & Pekcan, G., 2008. Seismic Response of Skewed RC box-girder Bridges. *Earthquake Engineering and Engineering Vibration* 7(4): 415–426.

Bridge Maintenance, Safety, Management, Resilience and Sustainability – Biondini & Frangopol (Eds)
 ISBN 978-0-415-62124-3

A comprehensive guide for designing bridges for service life

A. Azizinamini
Civil and Environmental Engineering Department, Florida International University, Miami, Florida, USA

E. Power
HDR Engineering, Inc., Virginia, USA

G. Myers
Atkins, Miami, Florida, USA

ABSTRACT

The second Strategic Highway Research Program (SHRP 2) was authorized by U.S. Congress to address some of the most pressing needs related to the U.S. highway system. Four areas were identified for research: Safety, Renewal, Reliability and Capacity. The Renewal area was tasked to develop technologies and institutional solutions to support systematic rehabilitation of highway infrastructure in a way that is rapid, presents minimal disruption to users, and results in long-lasting facilities. Four projects were funded in the bridge area. As part of SHRP2 R19A project, within Renewal are and entitled "Bridges for Service Life beyond 100 Years: Innovative Systems, Subsystems, and Components", the research team led by the author is developing a comprehensive document, referred to as "Guide for Bridges for Service Life", hereinafter referred to as the Guide. The objective of this paper is to present the outline of the Guide and its content.

The Guide for Bridges for Service Life is developed to serve as a comprehensive and stand-alone reference document, devoted to service life and durability of bridge systems, subsystems, components and elements. The Guide complements existing AASHTO specifications and includes subjects in an integrated manner crucial to the development of bridge systems capable of providing enhanced service life. The Guide addresses design, fabrication, construction, operation, maintenance, repair and replacement related issues and considerations at both individual and Department of Transportation (DOT) policy levels. Materials presented in the Guide are applicable to both existing and new bridges. The Guide also introduces several new and innovative systems and approaches. New bridge systems introduced in the Guide includes, a) self-stressing method for producing compression in bridge decks, b) new systems, where expansion joints are completely eliminated, even at the end of the approach slab, applicable to both flexible and rigid pavements, c) new pile head details to expand application of jointless bridges to longer bridges, and d) extending the application of jointless bridges to curved girder bridges. Further, the Guide provides complete design, construction and maintenance provisions for jointless bridges. This paper concentrates only on the development of Guide.

The Guide will be the main product of Project SHRP2 R19A of the Strategic Highway Research Program 2 (SHRP2), entitled "Bridges for Service Life beyond 100 Years: Innovative Systems, Subsystems, and Components".

SHRP 2 is administered by the Transportation Research Board of the National Academies under a Memorandum of Understanding with the Federal Highway Administration and the America Association of State Highway and Transportation Officials.

Bridge Maintenance, Safety, Management, Resilience and Sustainability – Biondini & Frangopol (Eds)
© 2012 Taylor & Francis Group, London, ISBN 978-0-415-62124-3

Seismic risk assessment and retrofit design of existing concrete bridges for the Italian highway Savona-Ventimiglia

C. Bafaro, G. Massone & G. Pasqualato
SINECO, Milano, Italy

G. Massa & F. Lenti
Autostrada dei Fiori, Imperia, Italy

ABSTRACT

The seismic behaviour of existing bridges can be assessed with different tools which range from simplified linear elastic calculations to more refined 3D linear or non-linear finite element analyses. The paper describes the seismic analysis and structures vulnerability of a important number of bridges and viaducts on the highway from Savona to Ventimiglia in Italy, placed over a high seismic hazard area. The activity performed over the last ten years, was the understanding the bridges seismic behaviour and the road safety in an earthquake scenario. The method adopted for the seismic risk assessment is the classic Response Spectrum Analysis and in some particular cases the 3D non linear; regarding the Eurocode 8 (2005) and the new Italian seismic design guidelines (NT 2008). All the structural elements were analysed: from the bearing to the pier, abutment and the foundation system, in order to apply a real risk analysis (the probability of failure) and the corresponding vulnerability of the highway bridges.

The retrofit design were also described for the more relevant and typological structures. Full details of the retrofit given in the paper, basically consisting in the repair of local structural damages and the bearing substitution.

The seismic isolation was also consider as a solution that could bring the viaducts to acceptable levels of seismic safety. Finally in the paper, some case studies are presented during the construction phases.

Figure 1. Lifting system and new bearing under construction.

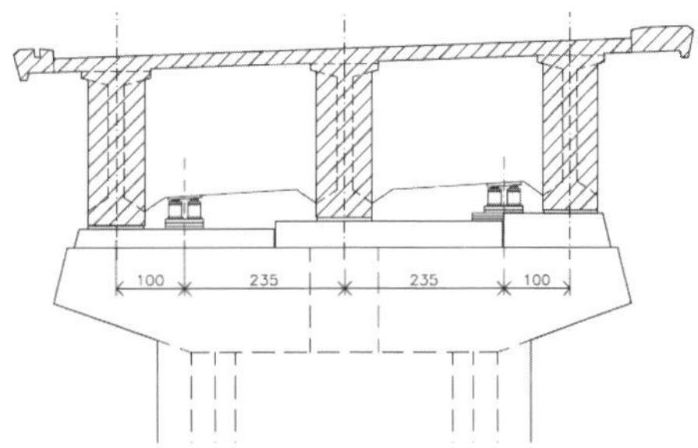

Figure 2. Design of the lifting system.

REFERENCES

Ministero delle Infrastrutture e dei Trasporti 2008. *Nuove Norme Tecniche per le Costruzioni, DM 14 Gennaio 2008.* Italy.

Circolare del Dipartimento della Protezione Civile del 21 ottobre 2003.

Circolare del Dipartimento della Protezione Civile del 21 aprile 2010.

UNI EN 1998-1 2005. *Eurocode 8: Design of structures for earthquake resistance – Part 1: General rules, seismic actions and rules for buildings.*

prEN 1998-2, 2005. *Eurocode 8 – Design of structures for earthquake resistance – Part 2: Bridges.*

Wai-Fan Chen, Lian Duan 2003. *Bridge Engineering Seismic Design.* CRC Press.

APPLIED TECHNOLOGY COUNCIL, ATC-6-2 1983. *Seismic retrofitting guidelines for highway bridges.* Palo Alto, CA: ATC.

PAUL E. MONDORF 2006. *Concrete bridges.* USA: Taylor & Francis.

R. L. MAYERS 1993. *Dynamic isolation systems.* Seminar notes on the seismic isolation of bridges. Berkeley, California.

A. CASTELLANI 1983. *Costruzioni in zona sismica* Masson Editore.

A. DUSI, A. MARCELLINI 2006. *Valutazione dei segnali sismici per i viadotti Prino e Impero. Sineco Internal Report.*

AAVV *"SIOS Sistema ispezioni opere Sineco".*

Bridge Maintenance, Safety, Management, Resilience and Sustainability – Biondini & Frangopol (Eds)
© 2012 Taylor & Francis Group, London, ISBN 978-0-415-62124-3

Towards a load rating methodology for concrete-encased pre-stressed steel girder bridges based on US standards

U. Barajas-Valdes
Department of Civil Engineering and Surveying, University of Puerto Rico, Mayagüez, Puerto Rico

G. Portela-Gauthier
Department of Engineering Science and Materials, University of Puerto Rico, Mayagüez, Puerto Rico

R.N. González, G. Velázquez & W. Varela-Ortiz
Engineer Research and Development Center, US Army Corps of Engineers, Vicksburg, MS, USA

ABSTRACT

A very unique type of bridge is in service on a few installations within the US Army garrisons (Barajas et al. 2010). These bridges utilize steel main girders that are pre-stressed and encased in concrete. Beyond these few, no others are known to exist in North America, and likewise no US load rating standards exist for analytical evaluation of their safety. This study is aimed toward the development of an analytical methodology for this purpose by combining AASHTO rating factors recommended for pre-stressed concrete bridges and steel-girder bridges. In this study, load ratings of a 125-ft long pre-stressed composite (steel beam/concrete encased) bridge are obtained. The variable parameters were: bridge condition, steel section loss, and concrete compressive strength. Details of load distribution and construction stage analysis are discussed in Portela et al. 2010. The pre-stressed composite system designed with modifications to current AASHTO provisions exhibited favorable rating results, even when affected by steel section loss and concrete strength reductions. The main contribution to the flexural capacity of the pre-stressed composite bridge appears to be attributable to the steel section embedded in the concrete slab.

The LRFR method seems to adequately represent the current conditions of the bridge selected for this work. However, specifically for this bridge, condition factors were estimated, which represent a reduction in the order of 3.5% for poor conditions and an increment in the order of 4.2% for fair conditions established by AASHTO LRFR. These condition factors seem to account for reductions in the steel sectional areas. The results also suggest the use of four condition factors instead of three based directly in four sectional area percentages, i.e. 85%, 90%, 95%, and 100%.

The results show that although the capacity of the concrete in compression is essential for the stability of the system during the construction stages, once the bridge is in service, this does not contribute substantially to its capacity. In addition, after the final stage of construction, the tension capacity of concrete in the bottom flange is neglected, assuming that the steel resists all the tension, similar to the analysis of conventional reinforced concrete girders. Because of this, losses in concrete strength do not represent a significant reduction in the capacity of the CEPS girder. Even the detection of cracks in the concrete bottom does not necessarily represent a decrease in capacity.

These effects should be considered in the rating method to be developed in further stages of this study. It is recommended to the inspector to consult the structural engineer for the corresponding analysis. It is recommended to develop load and condition factors based on section loss and load increments (live and dead), respectively. Since in this preliminary study we noticed that the two factors that contribute substantially to the reduction in safe load carrying capacity are the steel section loss and load increments.

This study represents an initial stage in the development of a methodology for load rating of CEPS bridges. It is recommended an experimental study of these systems and the use of a more representative sample with variations in span length, vehicle type and load distribution.

REFERENCES

Barajas U., Portela G., Stanton T., Varela W. & Velazquez G. 2010. "Método de Análisis para Vigas Pre-flexionadas basado en Estándares de Norteamérica". In Spanish, *Revista Internacional de Desastres Naturales e Infraestructura Civil,* Vol. 10(2): 95–105.

Portela G., Barajas U. & Varela W. 2010. Parametric Study of Long-Span Pre-Cambered Bridge Girders. *3rd fib International Congress*, Washington, D.C., USA.

Bridge Maintenance, Safety, Management, Resilience and Sustainability – Biondini & Frangopol (Eds)

Newest development of concrete safety barriers for bridges and the need to harmonize national collision force regulations

A. Bares, T. Edl & G. Ferner
DELTA BLOC International GmbH, Sollenau, Austria

A. Barnaš
KIRCHDORFER Fertigteilholding GmbH, Wöllersdorf, Austria

ABSTRACT

In the past decade vehicle restraint systems became increasingly important as the awareness of consequences of fatalities drew attention to more passive road safety. This is especially true for bridges as accidents on special structures tend to have more serious consequences.

Subsequently, we notice a strong trend towards passive road safety as the economic impact and cost of fatalities were analyzed in a broader context. In several studies the huge social costs of accidents are mentioned which amount to more than 1% of the GDP (130 billion EUR) in Europe 2009.

The creation of the harmonized EN 1317 and the redeployment of this harmonized standard via the Construction Products Directive (CPD) lead to safer restraint systems and added economic value by introduction of space limitations via working width.

Bridges always played an important role in the further development of new technologies as boundary conditions are stricter and more important. Therefore applications for bridges indicate the direction of future development.

As the significance of passenger safety became more immanent the ASI value became the most important factor describing the safety of a restraint system. The existence of a correlation between the safety level of a system and the deformation has been repeatedly demonstrated. Until recently a low ASI value was considered to be in conflict with small deformation of safety barriers.

Aiming to find a system with a containment level of H4b and working width W1 with ASI A seemed the ultimate goal. Several research was done to permit minimal deformation and reduce the impact energy in a smooth way to obtain safety barriers with low deformation and low ASI values.

Anchoring the system seemed a solution that was tested in different ways to minimize the deformation. Several tests performed were not satisfactory as either the anchoring was too expensive or the anchorage did not bring the effect as wished.

It should take years finding a system with containment level H2, working width W1 and still receive an ASI value of B. Due to a new anchoring system a restraint system was developed and tested with impressive result in fall 2010. The objective to find a system withstanding heavy impacts that fulfills both aspects, passenger safety and minimal deflection, was finally met.

The key of the system is the so called "energy absorption link" that connects a parapet to the bridge in a very efficient way.

The research process for this new anchoring system (energy absorption link) made it obvious that harmonization of static loading cases is absolutely necessary for the bridge design. It should be aimed to consider specific testing methods and the system for more accurate design loads.

REFERENCES

EN 1317-1, Road restraint systems – Part 1 & 2: Terminology and general criteria for test methods, October 2010

EN 1317-2, Road restraint systems – Part 2: Performance classes, impact test acceptance criteria and test methods for safety barriers, October 2010.

Hellmich K., Stella J., Stengl E., Piringer S., Heimel H., Plomer J. 2002, Restraint systems on bridges – acceptance tests with force measurements, Final report of the research contract of the Austrian Ministry of Traffic, Innovation and Technology, Road research contract Nr. 3554.

RVS 15.04.71, 2005: Brückenausrüstung, Rückhaltesysteme auf Brücken

ENV 1991-2, 2004, Eurocode 1– Actions on structures, Part 2: General actions – Traffic loads on bridges.

DIN-Fachbericht 101, 2009: Einwirkungen auf Brücken

COMMUNICATION FROM THE COMMISSION TO THE EUROPEAN PARLIAMENT, THE COUNCIL, THE EUROPEAN ECONOMIC AND SOCIAL COMMITTEE AND THE COMMITTEE OF THE REGIONS – Towards a European road safety area: policy orientations on road safety 2011–2020

Council Directive 89/106/EEC of 21 December 1988

REGULATION (EU) No 305/2011 OF THE EUROPEAN PARLIAMENT AND OF THE COUNCIL of 9 March 2011

Bridge Maintenance, Safety, Management, Resilience and Sustainability – Biondini & Frangopol (Eds)
© 2012 Taylor & Francis Group, London, ISBN 978-0-415-62124-3

Assessment of the levels of load-path redundancy in short span steel truss bridges

K.E. Barth, G.K. Michaelson, J.M. Stains & K.P. Mertens
West Virginia University, Civil Engineering, Morgantown, West Virginia, USA

ABSTRACT

Owner agencies and state transportation departments have expressed concern over costly inspection requirements for short-span steel truss bridges due to the potential designation of members as fracture-critical. This designation has a direct impact on future implementation of these structures as well as perceived inherent safety in existing bridges of this type. Many engineers and researchers have classified bridges with fracture-critical members as nonredundant, as failure of a single member/element caused failure, or in these cases, collapse, of the entire bridge.

This paper presents the results of a highly accurate, nonlinear finite element investigation of the load-path redundancy, or the means of redistributing loads that does not exceed acceptable levels of stresses, deflections, etc. in the presence of failure/damage of a member/component, of a representative short-span steel truss bridge. This finite element investigation is benchmarked against physical load testing of a representative bridge. The bridge chosen was a steel through truss over Little Mill Creek on County Route 5/18 in Jackson County, West Virginia. Figure 1 shows an isometric view of the bridge.

Modeling was done using the commercial finite element software package, Abaqus/CAE Version 6.9 (Dassault Systèmes, 2010). Element selection for these finite element models included a 4-node, doubly-curved, finite-membrane-strain, general-purpose shell with reduced integration, which were used to simulate all structural elements with the exception of lateral rods, which were modeled with 2-node linear beam elements. Figure 2 shows a screen capture of the finite element model.

After benchmarking the model, an example redundancy analysis was performed. It was shown that the bridge possesses sufficient levels of load-path redundancy in a damaged state when evaluating a typical built-up fracture critical member; in this case, a bottom chord double channel section.

Figure 1. Jackson County Little Mill Creek bridge.

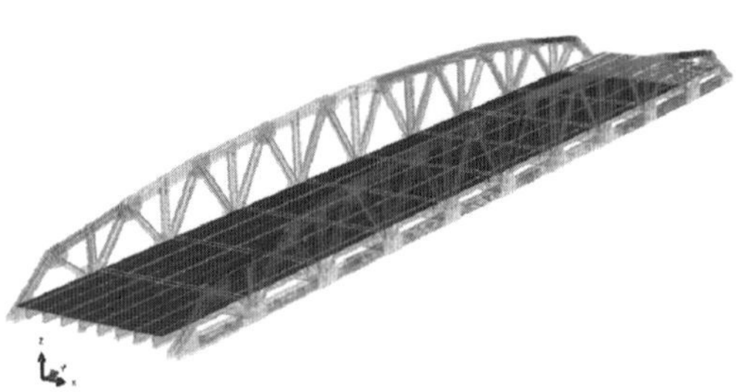

Figure 2. Abaqus screen capture.

REFERENCES

American Association of State Highway and Transportation Officials. 2010. *AASHTO LRFD Bridge Design Specifications, 5th Edition*. Washington, DC: AASHTO.

Dassault Systèmes. 2009. *Abaqus/CAE Users Manual (Version 6.9)*. Providence, RI: Dassault Systèmes Simulia Corp.

Ghosn, M. & Moses, F. 1997. Redundancy in Highway Bridge Superstructures. *National Cooperative Highway Research Program, Report 406*. Washington, DC: Transportation Research Board.

Bridge Maintenance, Safety, Management, Resilience and Sustainability – Biondini & Frangopol (Eds)
© 2012 Taylor & Francis Group, London, ISBN 978-0-415-62124-3

Structural assessment of Bullona 1929 railway bridge station to double span by external post-tensioning

C. Beltrami, C. Silvestri & G. Pedrazzi
Lombardi-Reico Engineering s.r.l., Milan, Italy

ABSTRACT

Bullona F.N. is an old liberty style railway bridge station built in 1929 for the Northern Line Railway Company in the city of Milan. This building have the peculiarity to cross a large trench with two railways lanes to Cadorna F.N. main station. The structure of Bullona F.N. building is composed by reinforced concrete beams and columns with masonry infill and main walls. The building is built on a I-girder deck composed by a symmetric two span frame structure with two external retaining walls and a central support given by four columns.

Recent works for quadrupling railways lanes induced the enlargement of the railways trench up to 18 m internal span and the roofing of the entire railway lane in town. At the same time, in 2003 after 75 years of service, the Bullona F.N. railway station have been closed and declared historical heritage by Italian Ministry of Culture.

Large structural engineering task come out: the critical evolution of Bullona F.N. building station because of the demolition of two of the three existing supports, the widening of the building span and the joining of a new segment to support the traffic load of the near street.

The concept design developed is founded on the idea that the structure should tolerate as less as deformations as possible during all the construction phases and obviously to the final configuration.

This design object have been successfully reached by the application of a post-tensioning tendons/bars layout able to form an orthotropic pre-stressed deck and to give the necessary contribution in terms of strength and stiffness to the existing reinforce-concrete deck and stand up to new spans and move/reduction of supports. External prestressing was an unavoidable choice, not only for compatibility with geometry of the original reinforced concrete structure, but also to introduce through cables deviations new vertical forces instead of missing boundary conditions.

More than 30 construction phases have been developed while the railway lane has been kept fully operating. One of the most important and preliminary phase consist in a deep traditional concrete and reinforcement strengthening of the deck structure, mainly in shear capacity, carried out with the original static and boundary configuration of the building.

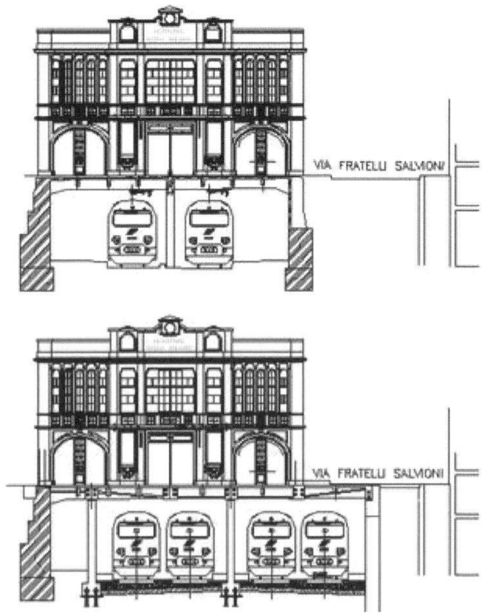

Figure 1. Structural scheme evolution of building bridge railway station by external pre-stressing and more than 30 construction stages.

REFERENCES

C.E.B. Model Code 1990, Bulletin de information n.203, July 1991

D.M. 04.05.1990: "Aggiornamento delle norme tecniche per la progettazione, la esecuzione e il collaudo dei ponti stradali" and instructions (Circ. Min. LL. PP. 25.02.1991, n. 34233).

Bridge Maintenance, Safety, Management, Resilience and Sustainability – Biondini & Frangopol (Eds)
© 2012 Taylor & Francis Group, London, ISBN 978-0-415-62124-3

Structural assessment of bridges and health monitoring programs based on dynamical tests

F. Benedettini
University of L'Aquila, L'Aquila, Italy

A. Morassi
University of Udine, Udine, Italy

F. Vestroni
University of Roma "La Sapienza", Roma, Italy

ABSTRACT

The assessment of the structural conditions of bridges is a very important problem and a great amount of papers are devoted to this item. Usually the bridge conditions are analyzed by means of visual inspections and standard controlling procedures. However, this approach is heavy, time consuming and too much dependent from the inspectors. Another approach, potentially able to give more refined information on the healthy of bridges, is based on dynamic tests of the structure; in this work the use of systematic dynamical tests is discussed as indispensable mean to correctly quantify the level of maintenance and to collect data to update a consistent *FE* model. Moreover, the proposed path is the only one acceptable when structural modifications are needed to improve the performance of the structure under certain loading conditions (moving loads, earthquake, and so on).

This dynamical characterization is of particular importance in Italy because of the high seismicity of several regions. A certain number of bridges belonging to the most representative and common typologies have been selected for dynamic monitoring, seismic evaluation, model updating and damage assessment purposes.

The possibility to reach good results depend on many factors, such as reliable experimental results, a suitable quantities of sensors and well positioned with respect to unknown damage location, not influent ambient conditions.

Notwithstanding experimental campaigns on the dynamic behavior of bridges are very diffused, so far, very effective results are not yet reached, a lot of aspects concerning excitations, sensors layouts, observed quantities, damage indexes, must be optimized, but the confidence in their potentiality and utility in view of structural conditions monitoring has increased.

In the paper two case studies are presented, where dynamic tests have been used with different goals.

The first case is concerned with a damage analysis on an artificially damaged bridge during the demolition phase and the steps of *damage detection* and *damage localization* are accurately described. Modifications of modal shapes, and in particular of their curvatures, have been shown to be effective in damage detection.

The second case describes the calibration of an accurate *FE* model of a bridge to be used to assess its actual seismic performance and to forecast the level of improvement connected with specific restoration works. The dynamic properties experimentally determined evidences an unexpected behavior of the bridge in the out-of-plane direction which makes critical its seismic resistance.

The two case studies are well representative in view of enlightening the wide range of possible applicability of dynamical tests in SHM programs.

REFERENCES

Benedettini, F., Zulli, D. & Alaggio, R. 2009. Frequency-veering and mode hybridization in arch bridges. Proceedings of IMAC XXVII, Orlando, USA.

Brownjohn, J.M.W., Moyo, P., Omenzetter, P. & Lu, Y. 2003. Assessment of highway bridge upgrading by dynamic testing and finite-element model updating. *Journal of Bridge Engineering* 8(3): 162–172.

Cruz, P. & Salgado, R. 2008. Performance of vibration-based damage detection methods in bridges. *Computer-Aided Civil and Infrastructure Engineering* 24: 62–79.

Dilena, M. & Morassi, A. 2011. Dynamic testing of a damaged bridge. *Mechanical Systems and Signal Processing* 25(5): 1485–1507.

Salawu, O.S. & Williams, C. 1995. Review of full-scale dynamic testing of bridge structures. *Engineering Structures* 17(2): 113–121.

Whelan, M.J. & Janoyan, K.D. 2010. In-service diagnostics of a highway bridge from a progressive damage case study. *Journal of Bridge Engineering* 15(5): 597–607.

Vestroni, F. & Benedettini, F. 2010. Dynamical tests and analysis for the assessment of structural conditions of bridges. Proceedings of 34th IABSE Int. Symposium on Bridge and Structural Engineering, Venice, Italy

Bridge Maintenance, Safety, Management, Resilience and Sustainability – Biondini & Frangopol (Eds)
© 2012 Taylor & Francis Group, London, ISBN 978-0-415-62124-3

Anchorage capacity of naturally corroded reinforcement in an existing bridge

F. Berg, D. Johansson, K. Lundgren & M. Plos
Chalmers University of Technology, Gothenburg, Sweden

K. Zandi Hanjari
Swedish Cement and Concrete Research Institute (CBI), Borås, Sweden

ABSTRACT

Corrosion of reinforcement is one of the most common causes of deterioration in reinforced concrete bridges. Anchorage, prior to shear and bending moment resistance, is the main uncertainties in the evaluation of the structural behavior of corroded reinforced concrete bridges. Thus, to assess the remaining load-bearing capacity of deteriorated existing bridges, models to estimate the remaining bond and anchorage capacity are needed. Most of our knowledge on the structural behavior of corroded reinforced concrete structures is based on experimental investigations of artificially corroded concrete specimens.

In this study, the anchorage capacity of naturally corroded steel reinforcement was investigated experimentally. The test specimens were taken from edge beams of a bridge, Stallbackabron, in Sweden. Since the dimensions and the amount of reinforcement were given on beforehand, it was only the test set-up which could be chosen freely. A test set-up consisting of a four point bending test indirectly supported with suspension hangers was considered to be the best alternative with the least disturbance and influence of the natural damages. Detailed design was done by using a non-linear finite element method, for more information see Berg & Johansson (2011). It was seen that the edge beams needed to be strengthened with transverse reinforcement, else they would have failed in a local failure at the suspension hole or in shear. The technique adopted for the strengthening was an internal mounting of steel reinforcement using epoxy as adhesive.

The bond and anchorage behavior was examined in tests through measurements of applied load, free-end slip and mid-span deflection. A first test showed that additional measures were needed to ensure anchorage of the strengthening bars. In subsequent tests, they were therefore anchored at the top of the beam with nuts and flat steel plates. In two following tests, the beams failed in a splitting induced pull-out failure, i.e. anchorage failure was achieved as wanted. The crack pattern in a test with a high corrosion level causing cover spalling is shown in Fig. 1.

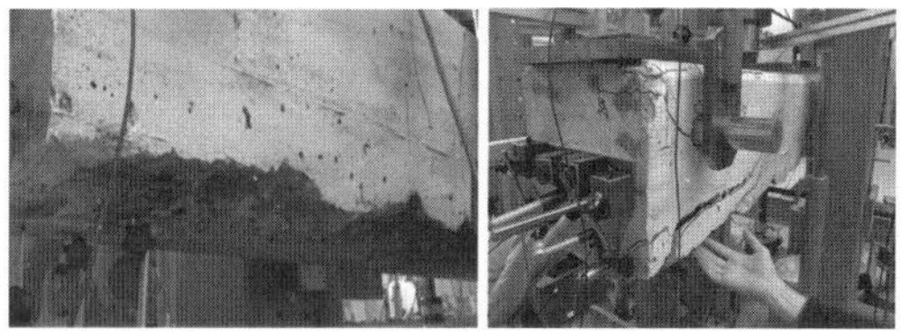

Figure 1. Severly corroded test specimen with cover spalling; before testing and at failure.

This work has opened for further experiments on edge beams from Stallbackabron. In a first series, test specimens are taken from the south part of the bridge; eight specimens in addition to the three described in this paper will be tested. A second series of tests are planned for with specimens from the north part of the bridge; these are planned to be tested in 2012. These tests will produce benchmark data of anchorage of naturally corroded reinforcement in bridges. They will be evaluated with detailed nonlinear finite element modeling, using the bond and corrosion model developed in Lundgren (2005) and further developed in Zandi Hanjari *et al.* (2011).

REFERENCES

Berg F. and Johansson D. 2011. Design of test set-up using FEM: Pilot tests on anchorage of naturally corroded reinforcement. Master of Science Thesis, Civil and Environmental Engineering, Chalmers University of Technology, Gothenburg, 2011.

Lundgren K. 2005. Bond between ribbed bars and concrete. Part 2: The effect of corrosion. *Magazine of Concrete Research*, Vol. 57, No. 7, pp. 383–395.

Zandi Hanjari K., Lundgren K., Plos M. and Coronelli D. 2011. Three-dimensional modelling structural effects of corroding steel reinforcement in concrete. Accepted for publication in *Structure and Infrastructure Engineering*. Published online 2011, http://dx.doi.org/10.1080/15732479.2011.607830.

Bridge Maintenance, Safety, Management, Resilience and Sustainability – Biondini & Frangopol (Eds)
© 2012 Taylor & Francis Group, London, ISBN 978-0-415-62124-3

Static test for three span highway bridge in Romania

G. Boaca & C.C. Comisu
"Gheorghe Asachi" Technical University of Iaşi, Faculty of Civil Engineering and Building Services, Department of Bridges

ABSTRACT

The tests are an essential requirement provided into Romanian standards and regulations and also important for the quality assessment process.

Tests perform on bridges is required when a road bridge is important, when the composition is less common, or when are used new technologies.

The execution of connections that will reduce transport time has become a very topical issue. Therefore the design and implementation of highway infrastructure is an important point at this moment and is a direction which allocates considerable funds.

The purpose of the static test is to obtain technical information regarding the behaviour of the superstructure, to establish the state of viability, according to STAS 12504-86 "Railway, road bridges and foot-bridges. Testing of superstructures with action test." Assessing the state of viability, functionality and behaviour of the passage under the action of the test loads will be considered under the following aspects:

- Resistance and structural stability of the superstructure and infrastructure.
- Elastic structural behaviour.
- State of deformations and displacements in characteristic sections.
- Compliance with the design calculation.

The study took into account refers to a bridge on the first traffic path from one of the four major bridges on the bypass of Sibiu, bypass made to the rank of highway.

The test load consists of two rows of trucks with a total mass of 43.7 tons, distributed as follows: 12.6 + 7.1 + 8.0 + 8.0 + 8.0 distance from 1.40-3, 60-4, 00-1, 30-1, 30-1, 20 meters.

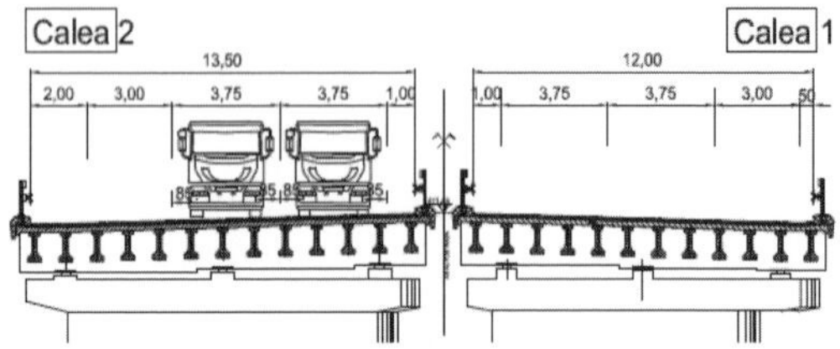

Figure 1. Transversal displacement.

Load test program consist, first of all, was realised a depth study of documents and design. After that the bridge was inspected and measured

The bridge was loaded in two steps (marginal and central spawn). All four trucks loaded the bridge spawn one after another.

Analytical results by the computerized model were realized in SAP2000. The trucks convoy position was established with influence line.

Experimental consideration consist by deformations were measured before starting the test, every loading and after unloading the structure, deformation values are presented in tables for stage I and stage II, measurements were made on leveling points place outside of structures, and sections established.

REFERENCE

Boaca, G. 2011, *Structural assessment of bridges through static and dynamic tests*, PhD Thesis – Iasi, October.

Bridge Maintenance, Safety, Management, Resilience and Sustainability – Biondini & Frangopol (Eds)
© 2012 Taylor & Francis Group, London, ISBN 978-0-415-62124-3

Life time extension of prestressed beams using cathodic protection

R. Brueckner, C.P. Atkins & P. Lambert
Mott MacDonald, Altrincham, UK

ABSTRACT

Cathodic protection (CP) is an electro-chemical repair technique which can be used to significantly extend the life time of reinforced concrete structures. This technique is able to reduce the corrosion activity at the reinforcement and its affects to the structure to a minimum without the necessity to remove all chloride-contaminated concrete. CP is established as a viable alternative to standard repair techniques, being easier and faster to install. It reduces health and safety issues to the contractors, is less disruptive and also reduces the carbon footprint as contaminated but undamaged concrete can remain.

Chlorides commonly derive from de-icing salts and/or were formerly used as concrete admixture. Chloride contaminated reinforced concrete structures mainly suffer from cover concrete delamination and loss of reinforcement cross-sections. By using CP there is no need to remove all of the chloride-contaminated concrete reducing the time for access, repair and traffic management.

Many structures, in particular bridges, are constructed using reinforced and prestressed concrete elements. The combination of the two structural forms can result from the original design or from later construction works involving widening, extension or strengthening works. The application of CP to prestressed concrete elements is not routinely considered for two reasons. The first is that prestressed construction is generally not as old as in-situ reinforced concrete and therefore is consequently in a better condition. However, even a small reduction in cross-sectional area can result in a brittle and catastrophic failure because the steel is typically loaded at 80% of its capacity. The second concern is the risk of hydrogen embrittlement which may occur naturally or be a result of inappropriate applied cathodic protection.

The earliest example of cathodic protection of prestressed concrete is referred to in a paper produced in 1964 by Franquin (1964), and makes reference to the first application by the authors being in 1946. This may well make it older than CP of conventional concrete, and it certainly predates the widespread adoption of the technique. Several other case studies from a range of different environments are discussed in the NACE State-of-the-Art report (2002) on CP of prestressed structures and the COST 534 (2009).

This paper discusses the design and application of a CP system to the retro-fitted prestressed beams on a bridge in the North West of England. It also reviews the advantages and disadvantages to be considered as novel technique to prevent catastrophic failure. The beam soffits were protected using a galvanic system as it was found to be relatively straightforward to install and safe with respect to hydrogen embrittlement of the prestressing wires. However, installation practicalities also generated significant concerns.

Prestressed elements are very sensitive to changes of concrete cross-sectional area, bond of tendons and welding procedures during installation. All remaining issues are similar to CP of in-situ reinforced concrete. Provided the system has been designed and installed competently and is operated with the normal degree of care required then cathodic protection provides an acceptable method of preventing corrosion of prestressing steel. The alternative approach is to remove the chloride contaminated concrete, which in many cases would involve beam replacement.

REFERENCES

European Cooperation in Science and Technology – COST 534, 2009. New Materials, Systems, Methods and Concepts for Prestressed Concrete Structures – Final Report.

Franquin, J. 1964. The Corrosion Protection of Prestressed Concrete Pipes, presented at the Corrosion Conference, Olympia, Dec 2–4, 1964.

NACE International Publication 01102, 2002. State-of-the-Art Report: Criteria for Cathodic Protection of Prestressed Concrete Structures, NACE International.

Bridge Maintenance, Safety, Management, Resilience and Sustainability – Biondini & Frangopol (Eds)
© 2012 Taylor & Francis Group, London, ISBN 978-0-415-62124-3

Updating existing railway bridges based on monitoring data

E. Brühwiler, M. Rocha & M.A. Treacy
École Polytechnique Fédérale de Lausanne (EPFL), Switzerland

T.N. Bittencourt & A.P.C. Neto
University of São Paulo, Brazil

ABSTRACT

Due to ever increasing traffic demands the fatigue safety and service life of the bridges of a railway line in Brazil need to be examined. Conventional assessment methods using load models and approaches as suggested in codes lead to conservative results resulting in significant strengthening interventions. Due to the important direct and indirect costs of the intervention, more detailed examination methods based on data as obtained from monitoring is suggested. This paper reports on an ongoing study to examine the fatigue safety of one standard bridge type, i.e. riveted steel truss structure. First results show that all bridge members are safe under the various limit states. The level of stress ranges found in the truss members due to fatigue loading are low such that only two members experience fatigue damage. The bridge structure has thus significant reserves in capacity which makes a future increase in axle loads feasible.

REFERENCES

Brühwiler, E., Lüchinger, P., Lang, T. & Vogel, T. 2011. *Swiss standards for existing structures*, Structural Engineering International, IABSE, Zurich.

EN 1991-2:2003. Eurocode 1: *Actions on structures – Part 2: Traffic loads on bridges*. European Committee for Standardisation.

Standard SIA 269, 2011. *Existing structures – Basis for examination and interventions*. Swiss Standard SN 505 269, Swiss Society of Engineers and Architects (SIA), Zurich.

Standard SIA 269/3, 2011. *Existing structures – Steel structures*. Swiss Standard SN 505 269/3, Swiss Society of Engineers and Architects (SIA), Zurich.

Stephens, R.I., Fatemi, A. Stephens, R.R. & Fuchs H.O. 2001. *Metal Fatigue in engineering*. Wiley, New York.

Bridge Maintenance, Safety, Management, Resilience and Sustainability – Biondini & Frangopol (Eds)
 ISBN 978-0-415-62124-3

Dynamic interaction of cable supported bridges with traffic loads including the effect of an accidental failure in the cable system

D. Bruno, F. Greco, P. Lonetti & P. Nevone Blasi
Department of Structural Engineering, University of Calabria, Rende, Cosenza, Italy

ABSTRACT

During the last decades, with new developments on the field of high performance materials and construction techniques, cable supported bridges have received much attention and many applications have been proposed to overcome increasing spans. However, new structural problems related to bridge deformability arise due to extreme loading conditions. As a matter of fact, long span bridges are widely subjected to highway or railway loads, which are, frequently, comparable with those involved by the self-weight ones (Bruno et al., 2008; Giamsing, 1997). Additional complexities arise when the bridge structure is affected by an accidental failure in the bridge components. The existing codes on this topic, suggest to take into account this failure condition by means of a quasistatic analyses taking into account the dynamic effects of the sudden failure of the cable by means of a fictitious amplification factor chosen in the range between 1.5 and 2.0.

In the proposed work, the dynamic behavior of long span bridges is analyzed by using a generalized formulation based on the finite element method, in which both in plane and out-of plane deformation modes have been accounted for undamaged and accidentally damaged bridge structures. Cable-stayed bridges based on both "*H*" and "*A*" shaped typologies with a double layer of stays have been considered. The cable system is analyzed by using an improved formulation of the cable system in which local vibration of the stays cables are taken into account. Moreover, the stay failure mechanism is defined, consistently with a Continuum Damage Mechanics approach (Lemaitre and Chaboche, 1990), by means of time dependent degradation functions, which control the constitutive behavior of the failure mechanisms and thus the inertial characteristics of the failure in the cable system.

The analysis focuses attention on the influence of the inertial characteristics of the moving loads, by accounting coupling effects arising from the interaction behavior between bridge deformations and moving system parameters. Comparisons in terms of dynamic amplification factors are proposed for both damage and undamaged bridge structures (Fig. 1).

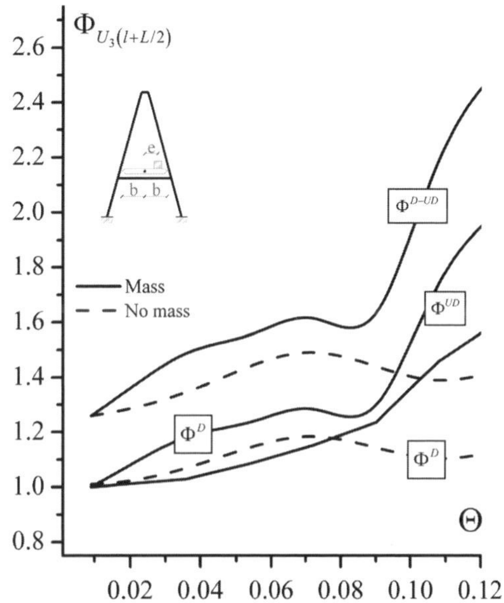

Figure 1. Dynamic amplification factors of the vertical midspan displacement for A-shaped tower topology for undamaged and damaged bridge structures due to the failure of the anchor stay.

REFERENCES

Bruno D, Greco, F, Lonetti, P, 2008. Impact analysis of cable stayed bridges, Enginnering Structures, Vol. 30, pp. 1160–1177.

Gimsing NJ, 1997, Cable supported bridges: concepts and design. John Wiley & Sons Ltd.

Lemaitre J and Chaboche JL, 1990, Mechanics of Solid Materials, Cambridge University Press, Cambridge, UK.

Bridge Maintenance, Safety, Management, Resilience and Sustainability – Biondini & Frangopol (Eds)
© 2012 Taylor & Francis Group, London, ISBN 978-0-415-62124-3

Cameras as displacement sensors to get the dynamic motion of a bridge: Performance evaluation against traditional approaches

G. Busca, A. Cigada, P. Mazzoleni & E. Zappa
Politecnico di Milano, Dipartimento di Meccanica, Milano, Italy

M. Franzi
Ferrovie Nord, Milano, Italy

ABSTRACT

Nowadays there is a strong interest towards new measurement techniques for structural health monitoring, especially related to bridges. The cost and complexity of any new measurement system are the main restraints or might constitute the reason for success. On the other side the performances of any new approach must be known and granted, accounting for facts like resolution, sensitivity, uncertainty. That is why recently a testing activity has been carried out on a bridge by comparing several state of the art approaches against traditional ones. Each of them had advantages and drawbacks if compared against the usual approaches; in this paper the use of cameras to monitor the bridge response during trains transit is being considered and some results will be presented. A camera is a non contact measurement device, having the great advantage to be simple, as an off-the-shelf camera can be used coupled to targets: therefore no cables are to be laid on the bridge, no power supply must be provided to sensors, reliability is higher with respect to that of common instrumentation. Cameras measure displacements (so limited in bandwidth), but offering a high level of redundancy, as a single image can provide many information. The leading idea of the paper is to use cameras moving from the actual measurements of small region rigid displacements towards wide area deformable motions, fully exploiting the huge information content provided by a single camera. The extreme final solution would be having wide area measurements with no targets, to make measurements very easy. Several solutions have been tried and compared along this path, in both terms of measurement strategy and vision analysis algorithms, always having some reference measurements to validate the camera results. The main tested approaches are: circular targets with pattern matching approach, camera normal to the bridge, with different mm/pixel ratios (Figure 1); random pattern for digital image correlation analysis, viewing the bridge from one side to compress the view of the bridge span (Figure 2); Markerless approach, just working on a pattern matching performed on the upper beam, having a good contrast against the sky (Figure 3).

Tests, though being preliminary ones, seem to be promising and offer good improvement margins as well as a radical new design of bridge monitoring strategies.

Figure 1. The pattern for the analysis and two targets on the bridge.

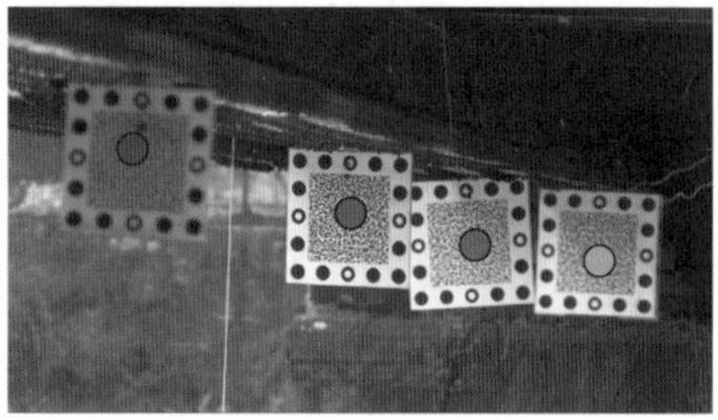

Figure 2. Targets with random texture mounted below the bridge.

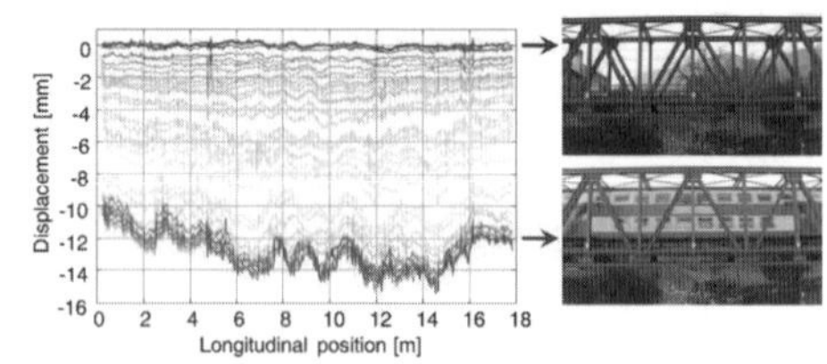

Figure 3. Vertical displacement of a bridge section obtained using a "dense" and targetless approach.

REFERENCES

Giorgio Busca, Alfredo Cigada, Marcello Vanali, Emanuele Zappa, Vision-based vibration monitoring of a large steel structure, Experimental Vibration Analysis for Civil Engineering Structures (EVACES), Varenna, Italy 3–5 October 2011.

Jong Jae Lee, Masanobu Shinozuka, 2006. A vision-based system for remote sensing of bridge displacement, NTD&E International 39 (2066) 425–431.

Bridge Maintenance, Safety, Management, Resilience and Sustainability – Biondini & Frangopol (Eds)
 ISBN 978-0-415-62124-3

Lane changing control to reduce traffic load effect on long-span bridges

C.C. Caprani, B. Enright & C. Carey
Department of Civil & Structural Engineering, Dublin Institute of Technology, Ireland

ABSTRACT

Long-span bridges represent a major infrastructural investment and many are reaching the end of their design lives. To keep these bridges in service whilst ensuring public safety, methods to curtail the exposure of these bridges to increasing traffic loads must be investigated.

The governing form of traffic loading on long span bridges is congested traffic (Buckland 1981). Cars changing lane increases the size and frequency of truck platoons, and these truck platoons dominate the characteristic loading events for long span bridges (Enright et al. 2012). This work investigates how the control of lane-changing events can help reduce the traffic load effects on long span bridges.

Traffic was recorded using Weigh-In-Motion equipment over a six month period in 2008 on the A4 (E40) motorway at Wroclaw, Poland. This traffic was used as an input to a traffic micro-simulation model, based on the Intelligent Driver Model (IDM), a microscopic car-following model developed by Treiber et al. (2000) and MOBIL, a lane-changing model developed by Kesting et al. (2007)

On a 10 km virtual road in the traffic micro-simulation a bottleneck is introduced to induce congestion. Various lane-changing restrictions, see Table 1, are implemented on the virtual road, and the associated traffic loading on a 200 m, 500 m, and 1000 m bridge, located at 8.25 km, is compared to the typical case of unrestricted lane changing.

Table 1. Scenarios of lane change control considered.

Scenario	Lane Change Restriction
1	None
2	Complete
3	1.0 km*
4	2.5 km*
5	5.0 km*

*Ends at the start of the bridge at 8.25 km

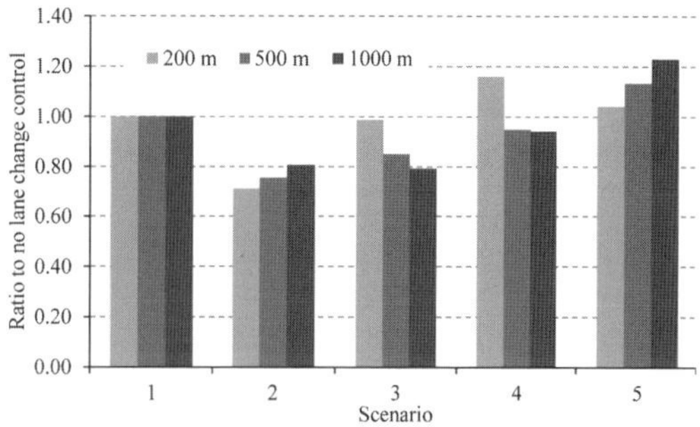

Figure 1. Ratios of 1000-year load effect resulting from lane change control scenarios (2–5), compared to no lane change control (1).

Figure 1 presents the results. It is found that restriction of lane changing is an effective means of reducing long-span bridge traffic load effect (Scenario 2). However, it is also found that the results are sensitive to the length of the lane changing control zone (Scenarios 3–5). This works explores the reasons for this. The results may assist bridge owners in adopting easily-implemented measures to prolong the life of existing infrastructure.

REFERENCES

Buckland, P.G., 1981. Recommended design loads for bridges, *Journal of the Structural Division, ASCE*, 1161–213.

Enright, B., Carey, C. & Caprani, C.C. 2012. The Effect of Lane Changing on Long-Span Highway Bridge Traffic Loading. *Proc. 6th Intl. Conf. on Bridge Maintenance, Safety and Management*, IABMAS, Lake Como, Italy, July.

Kesting, A., Treiber, M. & Helbing D. 2007. General lane-changing model MOBIL for car-following models. *Transportation Research Record: Journal of the Transportation Research Board*, 1999: 86–94.

Treiber, M., Hennecke, A., & Helbing, D. 2000. Microscopic Simulation of Congested Traffic. in: *Traffic and Granular Flow '99*, Eds. D. Helbing, H.J. Herrmann, M. Schreckenberg, & D.E. Wolf, Springer, Berlin: 365–376.

Bridge Maintenance, Safety, Management, Resilience and Sustainability – Biondini & Frangopol (Eds)
© 2012 Taylor & Francis Group, London, ISBN 978-0-415-62124-3

Reliability analysis and in-field investigation of a r.c. bridge over river Adige in Verona, Italy

F. Carturan, K. Islami, C. Pellegrino & C. Modena
Dipartimento di Costruzioni e Trasporti, Università degli Studi di Padova, Padova, Italy

ABSTRACT

In transportation networks the reliability of existing infrastructures is a key issue since the operability of entire network is related to each single bridge.

To assess reliability of bridges three approaches can be found in literature: structural reliability assessment, field-testing and technical judgment.

In this context, a reinforced concrete bridge over the river Adige in Verona (Italy), Ponte Nuovo del Popolo (Fig. 1), has been used as case study; a parametric reliability study has been performed to evaluate its traffic operability and then a field test with dynamic identification has been applied. The bridge is located in the city center of Verona and is characterized by three spans with a total length of over 90 m. It was built in 1946, and it was not subjected to significant interventions. The bridge recently revealed severe damage on lateral girders at the middle of the spans. The concrete from main and cross beams is highly deteriorated and carbonated (Fig. 2).

A reliability analysis has been performed by means of OpenSees Reliability package. The reliability index in critical cross sections has been calculated to assess safety of entire bridge. In a second stage, data about materials and geometry have been obtained from in-situ testing and the analysis has been updated.

The bridge was subjected to a system of investigation for modal testing and the principal natural frequencies have been identified. The response of the bridge was recorded using high-sensitivity piezoelectric acceleration transducers (Fig. 3). A number of destructive and non-destructive tests were performed to use best-fitted material characteristics in every element of the structure. A finite element model was updated and adjusted according to the results of investigation; this model was then used for static and seismic parametric studies for the assessment of the structural elements.

Figure 1. A photo of Ponte Nuovo del Popolo.

Figure 2. Damage due to steel corrosion on the longitudinal edge girders.

Figure 3. Setup scheme, a typical 3-axes accelerometer.

Bridge Maintenance, Safety, Management, Resilience and Sustainability – Biondini & Frangopol (Eds)
 ISBN 978-0-415-62124-3

Service life assessment of steel riveted railway bridges: A case study

L. Cascini, M. D'Aniello, F. Portioli & R. Landolfo
Department of Constructions and Mathematical Methods in Architecture, Naples, Italy

ABSTRACT

In this paper a service life assessment of an aged steel riveted railway bridge is presented. Nowadays, such bridges show structural deficiencies in the riveted connections due to the increased traffic demand, the current code requirements, the deterioration due to fatigue and corrosion, etc. To this end, a life-cycle evaluation of structural performance over time is envisaged. The assessment of fatigue capacity is carried out considering the influence of variation of stress intensity related to the thickness loss by corrosion over time. The residual service life is assessed for several design scenarios, each defined as a combination of the maintenance performed during the service life, the train loads acting on the structure and the corrosion of bridge components, taking into account the fatigue and corrosion effects. The time-dependent assessment of residual life of steel riveted bridge developed in this work is a step-by-step procedure, which is based on combination of codified approach for fatigue verification and for the evaluation of corrosion decay for a set of operating scenarios. Each operating scenario is defined on the basis of a fixed loading condition and a fixed maintenance programme, which are generally stipulated by the infrastructure owner.

The procedure is framed into the following steps:

1. *Structural analysis.*
2. *Evaluation of Stress spectra.*
3. *Evaluation of environmental class and relevant degradation law.*
4. *Definition of maintenance programme.*
5. *Life-time discretization.*
6. *Calculation of thickness loss due to corrosion.*
7. *Evaluation of damage due to the combined presence of fatigue and corrosion at each time-range.*
8. *Evaluation of cumulative damage.*
9. *Evaluation of residual life-time.*

The presented methodology has been applied to a metal riveted railway bridge located on the Italian Foggia-Cervaro-Napoli railway. It consists of a symmetric structure about the middle length of the intermediate span. Per span, the main girders are two riveted trusses 3.50 m high and 3.30 m apart, consisting of combined plates and L and C sections. In addition, the main trusses are connected with riveted built-up transverse secondary beams, a horizontal bracing system, and longitudinal secondary trussed beams, where the railway superstructure transmits the train loads. The study highlighted the important influence of corrosive phenomenon on the residual life. Indeed, the study revealed that the presence of corrosion could significantly reduce the fatigue life, because of the increase of stress intensity due to the thickness loss.

REFERENCES

D'Aniello A., Mannara G. (2006). Monitoraggio dinamico dei ponti ferroviari: sperimentazione di un sistema per l'analisi in continuo dei parametri strutturali e della loro evoluzione nel tempo. Ingegneria Ferroviaria N. 5, 23–25 May (in Italian).

D'Aniello M., Portioli F., Fiorino L., Landolfo R. (2010). Experimental investigation on shear capacity of riveted connections in steel structures. Engineering Structures (doi:10.1016/j.engstruct.2010.11.010).

Landolfo R., Cascini L., D'Aniello M., Portioli F. (2011). Gli effetti del degrado da fatica e corrosione sui ponti ferroviari in carpenteria metallica: un approccio integrato per la valutazione della vita residua. Rivista Italiana della Saldatura, N. 3, p. 367–377.

Rete Ferroviaria Italiana (RFI). Istruzioni 44F-Verifiche a fatica dei ponti ferroviari me-tallici. Roma 30.1.1992.

Bridge Maintenance, Safety, Management, Resilience and Sustainability – Biondini & Frangopol (Eds)
© 2012 Taylor & Francis Group, London, ISBN 978-0-415-62124-3

Design of energy harvesting bridge considering practical traffic conditions

M.S. Choi & S.-H. Kim
Yonsei University, Seoul, Korea

H.E. Youn
SK Construction and Engineering, Seoul, Korea

ABSTRACT

The piezoelectric system is an innovative energy system that uses mechanical energy as a source of generating electricity. Because of its cleanliness, it is spotlighted as one of the prominent energy harvesting systems which faithfully satisfies the aim of low-carbon development. Infrastructures are designed to sustain various external loads and some of these loads have potential to be used as a source of piezoelectric energy generation. Because piezoelectric modules are small and does not need complicated energy generating device, it can be easily applied to the civil structures where the repetitive mechanical stress is being loaded. In case of bridge, traffic load can be a good source of energy generation, as it continuously applied to bridge throughout the service life.

This study mainly focused on the quantitative evaluation of energy generation from the highway bridge which installed the piezoelectric modulus. In order to estimate the energy generation characteristic from traffic load, traffic headway distance and heavy traffic ratio are considered as the analysis parameters.

Application of piezoelectric energy generating module into a bridge is the first trial and there is no piezoelectric module that is qualified for bridge environment. This study focused on the energy harvesting characteristics from the bridge, by using a typical piezoelectric module.

In order to maximize the generation of electricity the bridge, it is necessary to place the piezoelectric module into the bridge member where the change of strain is most severe. This study assumed that piezoelectric modules are placed in the secondary members of steel plate bridge.

This study performed the analysis using a steel plate girder bridge. The bridge has a steel deck which is reinforced by U-ribs and crossbeams. The bridge is modelled by the commercial finite element method software, MIDAS Civil 2006. 3-dimensional frame elements are used for modelling the bridge. The piezoelectric modules are assumed to be placed in three different places in the bridge; the bottom of deck plates, the bottom of u-ribs, and the lower flange of crossbeams.

Two parameters are considered for the analysis. One is the traffic headway distance and the other is heavy traffic ratio.

In order to estimate the energy harvesting characteristics regarding to various traffic headway distance, nine analysis cases are considered, changing the vehicle types and headway distances. The results indicated that narrower vehicle headway distance will produce more electricity than that from wide headway distance. Such result indicates that constantly and slowly entering traffics which keep narrow headway distance amplify the electricity generation. This study also randomly generated 10,000 vehicles having different heavy traffic ratio, in order to estimate the change of energy harvesting characteristics regarding to various heavy traffic ratios. The results showed that increase of heavy traffic ratio significantly raise the electricity generation. Similar tendencies are drawn from all three headway distances, although the narrower headway distance still have advantage in energy generation. From the results, it can be noticed that the effect of consecutively running vehicle should be carefully considered when estimating the energy generation characteristic.

Bridge Maintenance, Safety, Management, Resilience and Sustainability – Biondini & Frangopol (Eds)
© 2012 Taylor & Francis Group, London, ISBN 978-0-415-62124-3

Health monitoring system of bridges network in Romania

C.C. Comisu & G. Boaca
"Gheorghe Asachi" Technical University of Iaşi, Faculty of Civil Engineering and Building Services, Department of Bridges, Iaşi, Romania

A. Ianos
S.C. POD-PROIECT S.R.L, Iaşi, Romania

ABSTRACT

The team's research Laboratory of Bridges of the Faculty of Civil Engineering and Building Services from Iasi, Romania, develop a modern program of health monitoring system of bridges in Romania. The system must interface and integrate the actual practice mainly based on visual inspections and combines the response of a number of different reliable sensors, installed on the structure to monitor the progress and identify the process of damage, with enhanced realistic deterioration models. The program of health monitoring system is developed to cover the parameters for the most important deterioration mechanisms: corrosion of reinforcement in bridges, carbonation of concrete, freeze-thaw cycles, alkali-silica reaction and mechanical damage, as well as the changes in the structures behavior and safety: static deformation, strains; crack widths and vibrations (frequencies, amplitudes, accelerations and vibration modes).

The bridges network in Romania will be equipped with a system of sensors which will permanently determine their technical condition, indicating the temperature, the humidity and implicitly the moment of the apparition of the white frost, the PH and also the corrosion state of the metal fitting and of the concrete, the condition of crack, of deformation, displacement and vibrations (frequent and amplitudes).

This information will be transmitted through a wireless system to a zone center of monitoring, where they will be centralized, registered, processed, and analyzed with the help of software dedicated to this scope.

The information regarding the temperature, the humidity, the apparition of the white frost, the PH, the degree of corrosion, the degree of cracking, the degree of deformation, the displacements and vibrations, will be visualized in a direct and continuous manner, in a real time and on monitors placed in the zone monitoring center, which will be placed at the beginning in a research lab from the Faculty of Constructions from Iasi, the technical assistance will be ensured on the entire duration of the exploitation by the research collectives consisting of teachers, the students, those from the master and the doctoral candidate will be directly involved in this research program.

The research project foresees the elaboration of an *"integrated system of active monitoring and structural identification of an intelligent transportation substructure"*. The scope of the monitoring system is the establishment of the moment in which the degradation process starts to manifest on the transportation structure (road, railway, bridge, bed, supporting walls, troughs).

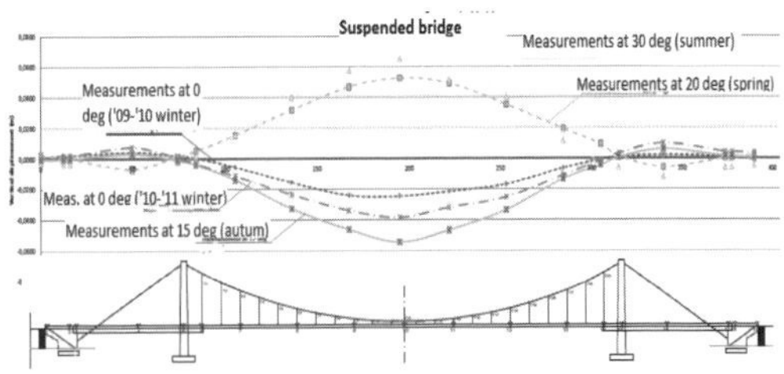

Figure 1. Bridge deformation over four measurements in one year.

REFERENCE

Comisu C.C. 2006. Mobile lab for dynamic testing and diagnosis of bridges for highways. *Research program CEEX 309/2006*, Faculty of Civil Engineering of Iasi – CNCSIS, Romania.

Bridge Maintenance, Safety, Management, Resilience and Sustainability – Biondini & Frangopol (Eds)
© 2012 Taylor & Francis Group, London, ISBN 978-0-415-62124-3

Numerical durability analysis of reinforced concrete bridges with focus on hygro-thermal behavior

F. Cramer, U. Kowalsky & D. Dinkler
Institute for Structural Analysis, TU Braunschweig, Germany

ABSTRACT

At reinforced concrete bridges operated nowadays damage affects primarily the durability. The detection of the durability during the design process is very difficult in comparison with the load-carrying capacity. There are a lot of guidelines and recommendations for the design engineer, but avoidable damages occur repeatedly. Many of these defects result from corrosion of steel reinforcement due to weather and chemical attacks.

In order to include this, a numerical model is proposed, which allows the coupled description of the most important processes deformation, transport and reaction during the life cycle of the reinforced concrete bridges.

The numerical model consists of three parts, the principal variables, the principal processes and the material models. The principal variables are variables, which characterize the general state of construction during the whole life cycle. Examples of these are the deformation, relative humidity, temperature or concentration of toxic substances. The change of these variables may be computed with help of the proposed model at every time step.

The principal processes describe the changing of the principal variables at every time step. Transport, reaction and deformation behavior could be identified as the most important principal processes. Different principal processes are used according to the external hazards during the life cycle.

Additionally to the formal treatment, constitutive models dependent on the principal variables and the material have to be considered. These models describe the development of internal variables as well.

The description of the degradation of the structure due to extreme environmental influence requires the description of moisture and heat transport as well as transport of toxic substances by the principal variables relative humidity, temperature and the concentration of the toxic substances related to the content of cement.

The deformation is the determining process for the assessment of the structural response. Therefore different material models, for example viscous or multi surface plasticity models are incorporated.

Beside the transport and the deformation, the chemical reactions are the third principal process. For each process the reaction kinetics has to be formulated. So far, the hydration process, dehydration, chloride binding and carbonation are considered, see Tacke 2002, Ostermann 2007 & Steffens et al. 2002.

The principal processes, have to be described by mathematical and physical models. The theoretical-numerical model is based on the theory of porous media using the conservation laws of mass, energy and linear momentum. The description of the material behavior takes place in the context of non-local damage theory and considers the mechanical and the chemical concrete damage.

Finally, the results of a numerical analysis of the durability of a reinforced concrete bridge are discussed. The presented model allows a detailed analysis of several degradation mechanism as well of the durability during the whole life cycle of a reinforced concrete bridge in a monolithic closed algorithm. Measures for the maintenance can be deduced from the results.

REFERENCES

Ostermann, L. & Dinkler, D. 2007. Numerical Simulation of the transport and the mechanical behavior of concrete at high temperatures. In Proceedings, *EURO:TUN 2007*. Vienna, Austria.

Steffens, A., Dinkler, D. & Ahrens, H. 2002. Modeling carbonation for corrosion risk prediction of concrete structures. In Elsevier, *Cement and Concrete Research* 32: 935–941.

Tacke, R. 2002. Feuchte- und Festigkeitsentwicklung hydratisierenden Betons – Modellierung und numerische Analyse. PhD Thesis, TU-Braunschweig.

Bridge Maintenance, Safety, Management, Resilience and Sustainability – Biondini & Frangopol (Eds)
© 2012 Taylor & Francis Group, London, ISBN 978-0-415-62124-3

Application of risk analysis for the preservation of post-tensioned girder bridge decks

C. Cremona & G. Lacoste
Technical Centre for Bridge Engineering, SETRA, Sourdun, France

P. Corfdir
Bridge Management Group, DIR East, Metz, France

S. Neiers
Bridge Division, CETE East, Metz, France

C. Aubagnac
LRPC Autun, CETE Lyon, Autun, France

E. Delahaye & C. Marcotte
Bridge Division, CETE North-Picardy, Lille, France

P. Paillusseau
Bridge Division, CETE South-West, Bordeaux, France

ABSTRACT

The VIPP bridges (Viaducs à travées Indépendantes à Poutres préfabriquées en béton Précontraint par post-tension) are simple span viaducts made of precast concrete girders prestressed by post-tension. This construction technique was very popular in the 60s in France (Fig. 1) but is no longer used today, competed by cheaper and safer ones. The enthusiasm that prevailed at the time of the construction of the first VIPPs' generation led to a blind faith in the absence of concrete cracking (total prestressing) and in the perfect tightness of compressed concrete. In the oldest VIPP bridges, it is very common to see longitudinal cracks following the cables' paths in the webs and heels of the girders. These cracks start from the anchors, grow when the cables are tensioned and stops afterwards. From a structural point of view, these cracks do not present any particular risk, but from a durability aspect, they play preferential water gateways inside the structure, especially for edge beams when drainage systems are also deficient; it is common to see efflorescence along some of these cracks.

Figure 1. Example of a VIPP bridge.

Deficiencies in waterproofing, bad injections of prestressing ducts and defective sealing of the anchors are the main defects leading to the corrosion of prestressing tendons. At last, the lack of inspections and corrective maintenance actions constituted the two main maintenance shortcomings versus these corrosion problems. In addition, the operation conditions led to an increasing use of de-icing salts that have been able to increase the aggressiveness of the water seeping on the structures.

All these problems do that the VIPP bridges form a very sensitive family of bridges which has received a lot of attention during the past twenty years. For this reason, the Technical Department for Transport, Roads and Bridges (SETRA) and the Technical Centres for Public Works (CETE) of the Ministry of Ecology, Sustainable Development, Transport and Housing were appointed by the Directorate for the Transportation Infrastructures (DIT) of the Ministry to develop a specific methodology for analyzing the criticality of the VIPPs' asset. The proposed approach relies on a risk-based analysis for identifying and qualifying hazards and quantifying vulnerability and consequences in order to state proposals for further investigations, detailed structural assessment and corrective actions. This paper presents the major results obtained with this risk-based methodology and the related consequences in terms of management for this family of bridges.

REFERENCES

SETRA 2010. *Simplified risk analysis for simply-supported post-tensioned girder bridges*, (in French), SETRA press, Bagneux, France.

SETRA 2011d. *LAGORA: Bridge management system*, (in French), SETRA press, Bagneux, France.

Bridge Maintenance, Safety, Management, Resilience and Sustainability – Biondini & Frangopol (Eds)
© 2012 Taylor & Francis Group, London, ISBN 978-0-415-62124-3

The reconstruction of the Williamsburg Bridge in New York City

R.D. Csogi
Greenman-Pedersen, Inc., Babylon, New York

ABSTRACT

The Williamsburg Bridge was opened in 1903, making it the second bridge to cross the East River, connecting the Lower East Side of Manhattan with Williamsburg in Brooklyn. Its main suspended span of 1,600 feet was four feet and six inches longer than the Brooklyn Bridge, making it the longest bridge in the world.

The Williamsburg Bridge was designed to carry four streetcar tracks, two elevated railway tracks (all of which were later converted to carry subway trains), two-lane roadways on each side (cantilevered outside the trusses), and footpaths and bicycle paths (above the trolley tracks). The stiffening trusses are 40 feet deep, designed to support the rail traffic on the Bridge, and the overall width of the bridge is 118 feet (See Figure 1).

On April 12, 1988, the bridge was closed to motor vehicles and subways for safety concerns after an inspection revealed that some of the steel beams in the roadway structure were severely corroded, the result years of deferred maintenance during the city's fiscal crisis. Emergency repairs were conducted and the Bridge was eventually re-opened to traffic. An Open Design Competition was held to foster ideas as to how to replace the crossing, but ultimately other studies were conducted to determine if the bridge could be rehabilitated. It was ultimately decided to keep the bridge and a 20-year reconstruction program began in 1991 with a cost of $1.4 billion.

The first major reconstruction Contract targeted the rehabilitation of the four Main Cables and the replacement of the Suspenders to a redundant suspension system. This Contract was known as Contract 4, with the three prior Contracts being conducted in the 1980's and were technically not considered Reconstruction Contracts since they addressed emergency and maintenance related work.

The next three reconstruction Contracts (Contracts 5, 6 and 7) were directed at the roadway, subway, and superstructure areas for both the Approaches and the Suspended Spans. In its current configuration, the Williamsburg Bridge has four lanes carrying vehicular traffic into Manhattan via two lanes that are cantilevered off of the Main Outer Truss, and two lanes that are framed between the Inner and Outer Trusses. These are designated as the South Outer and South Inner Roadways, respectively. These Roadway areas were completely replaced, from end to end under a major reconstruction Contract known as Contract 5.

Contract 6 addressed the complete replacement of the BMT subway areas on the Bridge. This included the entire sub and superstructure on the Approach Spans, framing replacement on the Suspended Spans, and all of the track, signals and communication systems from end to end of the Bridge.

Contract 7 was a mirror image of Contract 5 and addressed the replacement of entire North Outer and North Inner Roadways Roadway areas on the Approaches, and reconstruction of these Roadways on the Suspended Spans.

Contract 8 was the final reconstruction contract on the Williamsburg Bridge. This contract addressed the Main Tower Stiffening and Strengthening, the final adjustment of the suspender lengths for the final bridge geometry, and all other miscellaneous repairs at various areas on the Bridge.

REFERENCES

Ahmed H., 2002. Innovations & Accomplishments. *2002 Bridges and Tunnels Annual Condition Report*: 39–40.

Ahmed H., 2002. Innovations & Accomplishments. *2006 Bridges and Tunnels Annual Condition Report*: 58–62.

Koch, R.W., 2007. Williamsburg Bridge Rehabilitation – Con-tract 8. *Steel Institute of New York*: 37–41.

Bridge Maintenance, Safety, Management, Resilience and Sustainability – Biondini & Frangopol (Eds)
© 2012 Taylor & Francis Group, London, ISBN 978-0-415-62124-3

Bridge dynamic response due to truck load sequence

S. Darjani & M.A. Saadeghvaziri
Department of Civil and Environmental Engineering, New Jersey Institute of Technology, University Heights, Newark, NJ, USA

N. Aboobaker
New Jersey Department of Transportation, Trenton, New Jersey, USA

A. Khan
Ali Khan & Associates – Moorestown, NJ, USA

ABSTRACT

Through the development and usage of high-strength materials, the design of more flexible bridges is unavoidable. More flexibility causes more static deflection which may exceeds optional AASHTO deflection criteria.

AASHTO Standard Specifications (1996) limit live load deflection to L/800 for general bridges and L/1000 for bridges that are used by pedestrians. These limits were originally employed to avoid undesirable structural and psychological effects of vibration due to moving trucks. However, results of prior studies indicate that deflection limits do not necessarily control vibration. Existing limits do not prevent damages in structures because they check global deflection, while the damages are more attributed to local deformations such as connection rotations and twisting of cross beams relative to support stringers (Roeder 2001). Furthermore, human susceptibility is more influenced by derivatives of deflection (e.g., acceleration or velocity) rather than the deflection itself (Wright and Walker 1971, Griffin 2004). Thus, there is a need for development of a more rational serviceability criterion that reliably controls bridge vibration while enhancing the use of HPS.

As evidenced by the reasons stated above, deflection limits have been made optional in AASHTO LRFD since 1998. However, they are still being used by transportation agencies and designers mainly due to the lack of an appropriate and rational guideline that can limit bridge vibration. Thus, there is a need for a more rational bridge vibration control guidelines that enhances structural performance and human comfort while allowing for application of high strength materials. That is the purpose of this ongoing study to develop the knowledge base with the aim of attaining such a guideline.

Important to this is availability of a versatile analytical model that can reliably and accurately determine vibrational response of bridges subjected to moving loads. This paper presents an analytical study that employs 2-D and 3-D Finite Element (FE) models to evaluate dynamic response of bridges under moving truck load. The effect of vehicle length, vehicle speed, bridge frequency, and span length are investigated.

REFERENCES

AASHTO (1996). Load Factor Design: Bridge Design Specifications, (16th edition). American Association of State Highway and Transportation Officials, Washington, D.C.

AASHTO (1998), Load Resistance and Factor Design: Bridge Design Specifications, (first edition), American Association of State Highway and Transportation Officials, Washington D.C.

Griffin, M. J. (2004). "Handbook of Human Vibration." Elsevier Academic Press, ISBN 0-12-303041-2.

Roeder, C. W., Barth, K. and Bergman A. (2002). "Improved Live Load Deflection Criteria for Steel Bridges." National Cooperative Highway Research Program Transportation Research Board of the National Academies, November.

Wright, R. N. and Walker, W. H. (1971). "Criteria for the deflection of steel bridges", American Iron and Steel Institute, Bulletin No. 19, November.

Bridge Maintenance, Safety, Management, Resilience and Sustainability – Biondini & Frangopol (Eds)
© 2012 Taylor & Francis Group, London, ISBN 978-0-415-62124-3

Reinforcement of structural elements by the use of composite materials and external prestressing

E. De Angelis, F. Incelli, B. Rinaldi & S. Mancini
Spea Ingegneria Europea, Rome, Italy

ABSTRACT

In the last years the use of composite materials for structural strengthening has arisen considerable interest.

One of the most usual application is the improvement of existing bridges resistance for live loads in order to fulfill the requirements of new construction standards.

An example of this kind of technology is the enlargement of the deck of the bridge over Adda River.

The intervention was mainly concerned with FRP flexural strengthening of the slabs and both flexural-shear improvement of the beams. In order to allow the reinforcement reach an effective stress configuration with very low strain, the best design decision was the one to use High-Module FRP (CFRP_HM) with ultimate cracking stress $f_{tk} \geq 3000$ MPa and ultimate cracking strain $\varepsilon_{fu} = 0.8\%$. This solution therefore provides the fiber to contribute immediately with existing materials

However the reduced strain range of CFRP_HM reinforcement due to delamination between resine matrix and fibers, is the main responsible of the growth in stiffness of the reinforced elements. As a consequence further analysis have been done to verify that concrete failure or delamination failure are avoided and in particular that the strength parameters used during design are respected in construction on site.

Table 1.

Mechanical parameters of CFRP_HM	
Failure stress	$f_{tk} \geq 3000$ MPa
Failure strain	$\varepsilon_{fu} = 0.8\%$
Fabric thickness	$t_f = 0.165$ mm
Elastic Module	$E_f = 390$ GPa
Relative density	$\rho = 1.9$ g/cm^3
Cracking energy	$\Gamma_{Fk} = 82$ kJ/m^3

It is important to mention that the use of carbon fiber is particularly suitable for reinforcing elements made with ordinary concrete, in spite of prestressed concrete element in which the strengthening given by the fiber is not able to reach the required performance levels. In this case it is recommended to use external post-stressing with high resistance steel rods or strands. As a replacement of the existing strands is often unsuitable and extremely dangerous for the integrity of the beam itself in this case the best solution is the one to anchor two clamping steel heads on the beam ends and a steel deflector at midspan connected by post-stressed rods. The anchorages transmit the axial load of the rods to the beam increasing its axial compression while the deflector develops a vertical upward load that fades the effects of live load bending moment.

An example is the strengthening of lateral girders of the viaduct Vallonalto I of the Highway A16. It was noticed a loss of pre-stressing (30%) non-recoverable with FRP, therefore it was necessary to integrate the loss through external high-strength rods with appropriate pre-stressed load that restores the original serviceability and ultimate resistance of the beam; in that case the average axial tension of the rods was 86000 kgf at initial steps, 71800 kgf after tipical loss of tension in post-stressing technology.

REFERENCES

Arduini, M., Nanni, A., 1997. Behavior of Precracked RC beams strengthened with carbon FRP Sheets. ASCE Journal of Composites for Construction 1 (2), 63–70.

Chajes, M.J., Thomson, T.A., Januszka, T.F., Fin, W., 1994. Flexural strengthening of concrete beams using externally bonded composite materials. Construction and Building Materials 8 (3), 191–201.

Mahmoud, T.E.M., Joseph, W.T., 2000. Analysis of reinforced concrete beams strengthened with FRP laminates. ASCE Journal of Structural Engineering 126 (6), 684–691.

Bridge Maintenance, Safety, Management, Resilience and Sustainability – Biondini & Frangopol (Eds)
© 2012 Taylor & Francis Group, London, ISBN 978-0-415-62124-3

Monitoring during large construction projects

H. De Backer, A. Outtier, K. Schotte, D. Stael, W. Nagy & Ph. Van Bogaert
Ghent University, Ghent, Belgium

ABSTRACT

When studying the long-term behaviour of structures, it often becomes necessary to resort to monitoring of certain structural parameters, even during construction. This research paper discusses several such monitoring projects where monitoring data is used to verify design assumptions.

The first example deals with an integral bridge that is being constructed in several separate building stages, continuously allowing traffic on parts of the bridge. In order to verify the actual behavior of the bridge, strain gauges have been applied on the beams of the superstructure, while ground pressure sensors have been attached to the back of the abutment walls. Strain and ground pressure are being monitored continuously during a one year period wherein the bridge is constructed, post-tensioned and fully opened for traffic. This project necessitated the use of a fully autonomous acquisition and data storage system since no reliable power supply was available.

As a second example, an important tunneling project is described, consisting of 2 single-track tunnels beneath the airport's runway and taxi-lanes. During construction of both tunnel shafts, measurements have been carried out regarding the tunnel lining at several cross-sections. The monitoring was carried out from at the moment of each segment's installation until the completion of the tunnel works.

A smaller example deals with the monitoring of the forces in struts during repair works on a lock door. It was important to monitor the buildup of pressure on the walls surrounding the lock doors because of the buildup of groundwater pressure.

This article gives an overview of these experiences and on the lessons learned concerning power supply and possible electromagnetic interference in harsh construction site condition. Furthermore, the most important results of both test cases are discussed in short, including the effectiveness off the integral bridge and the actual ovalisation of the tunnel.

Figure 1. Downloading of strain measurement data in the Liefkenshoek Tunnel.

REFERENCES

De Backer H., De Corte W. and Van Bogaert Ph. 2003. "A case study on strain gauge measurements on large post-tensioned concrete beams of a railway support structure" *Insight*. Dec 2003 Vol 45 Nr 12. Ed. British Inst of Non-Destructive Testing – Northampton UK pp 822–825.

De Backer H., De Pauw B., De Corte W. and Van Bogaert Ph. 2004. "*Precast HPC post-tensioned alternatives for a railway crossing viaduct*", in Proc. of the 3rd Int. Symposium on HPC, 2004, CD-ROM.

De Backer H., Outtier A., Van Bogaert Ph. 2007. "High precision strain gauge measurements in areas of high stress concentrations of orthotropic plated bridge decks" *Insight*, Vol 49, No. 7, July 2007, Ed. British Inst of Non-Destructive Testing – Northampton UK, pp 384–389.

De Pauw B., De Backer H. De Corte W. and Van Bogaert Ph. 2004. "*Two Alternatives for a Railway Viaduct in Densily Populated Area*", in Proc. Conference Int. Ass. For Bridge and Struct. Eng. IABSE. Role of Structural Engineers towards Reduction of Poverty. Ed IABSE Zürich. pp 455–462.

Bridge Maintenance, Safety, Management, Resilience and Sustainability – Biondini & Frangopol (Eds)
© 2012 Taylor & Francis Group, London, ISBN 978-0-415-62124-3

Lifetime risk assessment of bridges affected by multiple hazards

A. Decò & D.M. Frangopol
Lehigh University, Bethlehem, PA, USA

ABSTRACT

In this paper, life-cycle risk associated with traffic and seismic hazards is investigated for a group of bridges within a transportation network. In a life-cycle perspective, optimal bridge intervention schedule can be based on risk ranking and risk importance factor. Bridge vulnerability with respect to earthquakes and traffic is investigated. Consequence analysis is performed according to different levels of bridge functionality including direct and indirect consequences. The proposed approach accounts for different types of cost-related consequences, including rebuilding costs, rehabilitation costs, material damage costs, running costs, time loss costs, loss of human life costs, and injury costs. Fragility analysis is performed for the evaluation of the effects induced by earthquake events. Probabilities of exceeding specific damage states (slight, moderate, extensive, and complete) are computed.

Random earthquakes are generated by using Latin Hypercube sampling according to a truncated exponential magnitude-frequency relationship. Traffic hazard is assessed considering Weibull time-to-failure probability distribution. Both vulnerabilities and consequences vary over time.

Finally, time-dependent risk profiles are obtained in the case of traffic and seismic hazards. Direct and indirect consequences are evaluated for a group of existing bridges.

REFERENCES

Akgül, F. & Frangopol, D.M. 2004. Computational platform for predicting lifetime system reliability profiles for different structure types in a network. *Journal of Computing in Civil Engineering*, 18(2): 92–104.

Akiyama, M., Frangopol, D.M. & Matsuzaki, H. 2011. Life-cycle reliability of RC bridge piers under seismic and airborne chloride hazards. *Earthquake Engineering & Structural Dynamics*, 40(15): 1671–1687.

Basöz, N. & Mander, J. 1999. *Enhancement of the highway transportation lifeline module in HAZUS*. NIBS, Washington, DC.

Bocchini, P. & Frangopol, D.M. 2011a. Optimal resilience- and cost-based post-disaster intervention prioritization for bridges along a highway segment. *Journal of Bridge Engineering*, doi:10.1061/(ASCE)BE.1943-5592.0000201.

Bocchini, P. & Frangopol, D.M. 2011b. Generalized bridge network performance analysis with correlation and time-variant reliability. *Structural Safety*, 33(2): 155–164.

Campbell, K.W. & Bozorgnia, Y. 2007. *Campbell-Bozorgnia NGA ground motion relations for the geometric mean horizontal component of peak and spectral ground motion parameters*. PEER Report 2007/02, Berkeley, CA.

Decò, A. & Frangopol, D.M. 2011a. Risk assessment of highway bridges under multiple hazards. *Journal of Risk Research*, 14(9): 1057–1089.

Decò, A. & Frangopol, D.M. 2011b. Life-cycle risk assessment of spatially distributed aging bridges under seismic and traffic hazards. (submitted).

FHWA 2010. *NBI database*. Federal Highway Administration (FHWA), Washington, DC.

Frangopol, D.M. & Bocchini, P. 2011. Bridge network performance, maintenance and optimization under uncertainty: accomplishments and challenges. *Structure and Infrastructure Engineering*, doi:10.1080/15732479.2011.563089.

Frangopol, D.M. 2011. Life-cycle performance, management, and optimization of structural systems under uncertainty: accomplishments and challenges. *Structure and Infrastructure Engineering*, 7(6): 389–413.

Ghosh, J. & Padgett, J.E. 2010. Aging considerations in the development of time-dependent seismic fragility curves. *Journal of Structural Engineering*, 136(12): 1497–1511.

Liu, M. & Frangopol, D.M. 2006. Probability-based bridge network performance evaluation. *Journal of Bridge Engineering*, 11(5): 633–641.

Okasha, N.M. & Frangopol, D.M. 2010. Advanced modeling for efficient computation of life-cycle performance prediction and service-life estimation of bridges. *Journal of Computing in Civil Engineering*, 24(6): 548–556.

Orcesi, A.D. & Frangopol, D.M. 2011. Use of lifetime functions in the optimization of nondestructive inspection strategies for bridges. *Journal of Structural Engineering*, 137(4): 531–539.

Padgett, J.E., DesRoches, R. & Nilsson, E. 2010. Regional seismic risk assessment of bridge network in Charleston, South Carolina. *Journal of Earthquake Engineering*, 14(6): 918–933.

van Noortwijk, J.M. & Klatter, H.E. 2004. The use of lifetime distributions in bridge maintenance and replacement modelling. *Computers and Structures*, 82(13–14): 1091–1099.

Yang, S.-I., Frangopol, D.M. & Neves, L.C. 2004. Service life prediction of structural systems using lifetime functions with emphasis on bridges. *Reliability Engineering & System Safety*, 86(1): 39–51.

Bridge Maintenance, Safety, Management, Resilience and Sustainability – Biondini & Frangopol (Eds)

Further study of chloride penetration in a RC slab sustaining in-service loads

A. Deif
Jubail University College, Jubail, Kingdom of Saudi Arabia

B. Martín-Pérez
University of Ottawa, Ottawa, Canada

B. Cousin
University of Montpellier, Montpellier, France

ABSTRACT

Reinforced concrete structures exposed to de-icing salts and marine environments are susceptible to chloride-induced reinforcement corrosion, which is considered to be one of the major durability problems that affect reinforced concrete infrastructure. This paper presents additional results of an experimental program wherein a reinforced concrete slab was subjected to wet/dry cycles (90 days each) with a saturated chloride solution while sustaining mechanical static loads (Fig. 1). Concrete cores were taken at the end of the tenth wet/dry cycle after total time of 30 months. The chloride penetration front at different locations where the strains vary was determined by spraying freshly concrete cores with silver nitrate.

The purpose of this work is to compare these results with that obtained after first three months in order to establish relationships between the rate of chloride penetration in concrete and the strain induced in the concrete deck sustained under mechanical loads.

Figure 1. Completed RC slab.

From the results, it is observed that the measured chloride penetration depths are affected by the w/c of the concrete as well as the level and type of sustained load. A correlation between structural parameters, as given by the magnitude of the applied bending moment, and chloride transfer properties was first attempted. Future measurements of chloride concentrations by standard titration procedures will give more information on the nature of this relationship.

REFERENCES

Deif, A., Martín-Peréz, B. and Cousin, B. 2008. Experimental study of chloride penetration in a RC slab sustaining in-service loads. *8th International Conference on Creep, Shrinkage and Durability of Concrete and Concrete Structures,* 30th September – 2nd October 2008. Ise-Shima, Japan, 7 pp.

De Schutter, G. 1999. Quantification of the influence of cracks in concrete structures on carbonation and chloride penetration. *Magazine of Concrete Research* 51(6): 427–435.

Gowripalan, N., Sirivivatnanon, V. and Lim, C.C. 2000. Chloride diffusivity of concrete cracked in flexure. *Cement and Concrete Research* 30(5): 725–730.

Konin, A., Franfois,R. and Arliguie, G. 1998. Penetration of chlorides in relation to the microcracking state into reinforced ordinary and high strength concrete. *Materials and Structures*, **31**, 310–316.

Lim, C.C., Gowripalan, N. and Sirivivatnanon, V. 2000. Microcracking and chloride permeability of concrete under uniaxial compression. *Cement and Concrete Composites* 22(5): 353–360.

Lindquist, W.D., Darwin, D., Browning, J. and Miller, G.G. 2006. Effect of cracking on chloride content in concrete bridge decks. *ACI Materials Journal* 102(6): 467–473.

Otsuki, N., Nagataki, S. and Nakashita, K. 1992. Evaluation of $AgNO_3$ solution spray method for measurement of chloride penetration into hardened cementitious matrix materials. *ACI Materials Journal* 89(6): 587–592.

Bridge Maintenance, Safety, Management, Resilience and Sustainability – Biondini & Frangopol (Eds)
© 2012 Taylor & Francis Group, London, ISBN 978-0-415-62124-3

Controlled demolition of damaged bridge decks

L. Della Sala
Università degli Studi di Basilicata, Potenza, Italy

ABSTRACT

The present paper describes an innovative technique used for the demolition of two strongly damaged decks of "Fiumara of Tito" viaduct on the highway RA005 Sicignano-Potenza in Basilicata – Italy.

The structures resulted affected by high corrosion of the steel tendons of the prestressing cables. In a first phase in situ tests and numerical analysis with the aim of evaluating the residual bearing capacity of the decks and their global safety factor were carried out. The results obtained indicated that the repairing of the structures was not convenient in economic terms and so a specific procedure for the demolition of the decks was studied.

The method consisted in the lifting of the entire damaged deck, in the lateral translation on the parallel twin deck and in its successive demolition in safe conditions.

With the aim of reducing the entity of the forces in the lifting phase, before translation the decks to be demolished were previously lightened by removing the slabs using WIDIA cutting wheels.

After the lightening operations, each deck was reduced to a grid composed of the two principal beams and the four cross beams.

Figure 1. Shifted grid and first beam ready for demolition.

Then, each deck was lifted and then translated for its complete demolition by cutting.

All the operative phases were conducted in such a way as to secure full control of the works and to preserve the external parts of the viaduct to be maintained. The structure has been monitored using electronic displacement transducers to control the two directional translations of the beams of the decks. The works were partially performed under traffic.

REFERENCES

Cerasi, E., Banfi, L. 2009. Relazione n. ROP 099/09 – Intervento di sollevamento e traslazione delle campate 6 e 8 per successiva demolizione. Sistral s.r.l. (MI)

Della Sala, L., Cerone, R., Chioini, N., Gennari Santori, A. 2002. Caratterizzazione dinamica del viadotto strallato Carpineto I. *Giornate AICAP 2002: Le grandi opere di ingegneria civile: la progettazione globale, le tecniche esecutive e le procedure di accettazione – Il comportamento delle strutture in calcestruzzo in fase di servizio*. Bologna (IT).

Della Sala, L., Cerone, R., Gennari Santori, A. 2002. Effetti dei fenomeni viscosi sulle strutture a sbalzo in c.a.p. – Un caso di studio. *Giornate AICAP 2002: Le grandi opere di ingegneria civile: la progettazione globale, le tecniche esecutive e le procedure di accettazione – Il comportamento delle strutture in calcestruzzo in fase di servizio*. Bologna (IT).

Della Sala, L., Cerone, R., Gennari Santori, A., Tommasini, M. 2009. Assessment, Strengthening and Validation of Prestressed Damaged Beams (Pietrastretta Viaduct). Damage Assessment of Structures VIII – *Key Engineering Materials* Volumes 413–414: 359–366. Switzerland: Trans Tech Publications.

Giangreco, E. 1992. Teoria e tecnica delle costruzioni. Napoli: Liguori Editore.

Suntharavadivel, T.G. & Aravithan, T. 2005. Overview of external post-tensioning in bridges. Southern Engineering Conference

Virlogeux, M. 1993. External prestressing – Historical and modern applications. *Proceedings of External Prestressing in Structures*. Saint Rémy lès Chevreuse.

Bridge Maintenance, Safety, Management, Resilience and Sustainability – Biondini & Frangopol (Eds)
 ISBN 978-0-415-62124-3

Assessment of bridge expansion joints using long-term displacement measurement under changing environment

Y. Deng
School of Civil Engineering and Architecture, Changsha University of Science and Technology, Changsha, Hunan Province, China

Y.L. Ding, A.Q. Li & G.D. Zhou
School of Civil Engineering, Southeast University, Nanjing, Jiangsu Province, China

ABSTRACT

In recent years, damage alarming and safety evaluation using long-term monitoring data is an area of significant research activity for long-span bridges. This paper attempts to extend the research in this field through the application of long-term monitoring data to the damage alarming problem for bridge expansion joints.

First, the effects of temperature, traffic condition and wind on the expansion joint displacement are analyzed and interpreted. The results reveal that temperature and traffic condition are the dominant environmental factors to influence displacement and the correlation between displacement and wind speed is very week.

Next, multiple linear regression models are obtained to describe the correlation between displacements and the dominant environmental factors. Based on the multiple regression models the damage alarming index is defined.

Then, the X-bar control chart is utilized to detect the abnormal change of the displacements. The false-positive tests of the control chart reveal that X-bar control chart is robust against false-positive indication of damage and can effectively reduce the probability of misjudgment by the multi-sample hypothesis test.

At last, the damage sensitivity of the proposed method is discussed according to a simulated example. The results reveal that the proposed method can provide a good capability for detecting the damage-induced slight change of the expansion joint displacement.

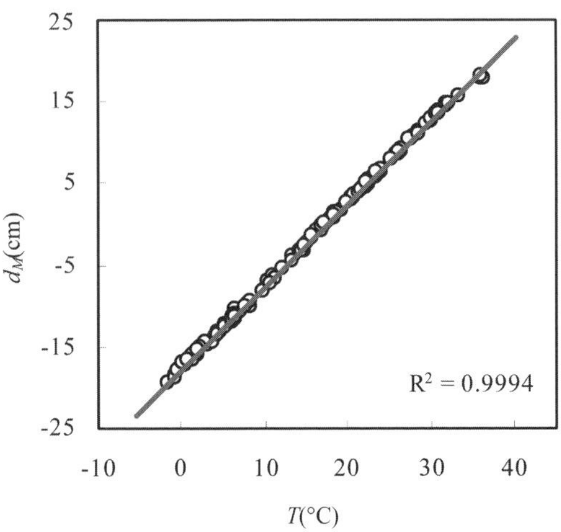

Figure 1. Correlation between the north displacement and temperature (daily averaged).

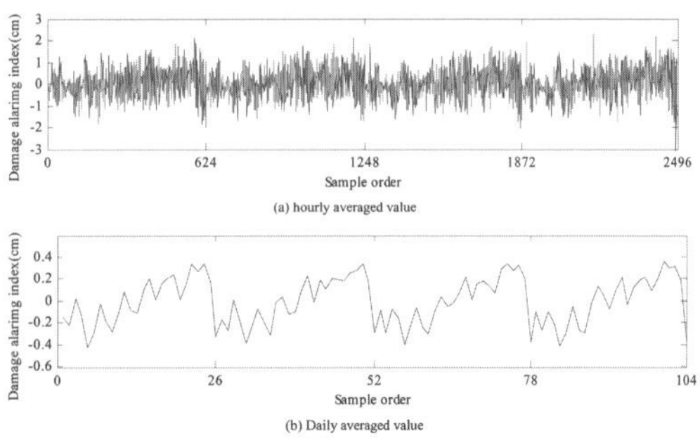

Figure 2. Damage alarming indices of the training data from the north abutment.

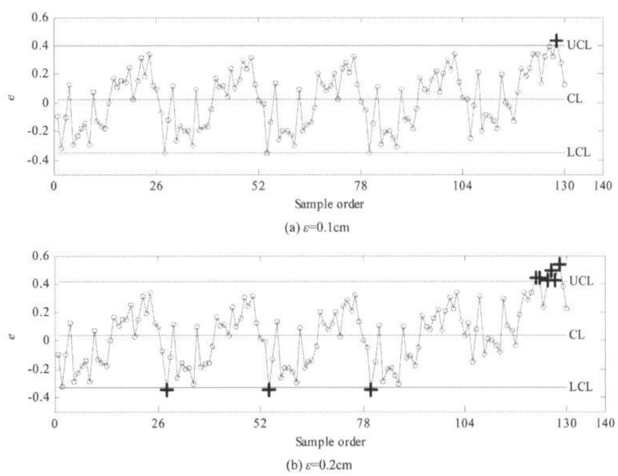

Figure 3. Control charts of damage expansion joints.

Bridge Maintenance, Safety, Management, Resilience and Sustainability – Biondini & Frangopol (Eds)

MMA polymer concrete materials for aging bridge rehabilitation and sustainability

A.M. Dinitz & M.S. Stenko
Transpo Industries Inc., New York

ABSTRACT

Rehabilitation and Sustainability for Aging Bridges require materials that heal, seal, patch, set rapidly, protect the structure from deterioration and can also be applied under a wide variety of conditions, temperatures and application/construction requirements.

Methyl Methacrylate (MMA) materials for joint headers, precast slab pockets, pier cap replacement, setting of structural beams, proactively sealing and healing concrete structures reducing chloride permeability and protecting them from further deterioration is a cost effective way to extend the serviceable life of both new and existing bridges.

Precast Polymer Concrete safety shape panels are manufactured with materials that result in bridge rails and tunnel linings with high strength and resistance to corrosion. In addition to their superior physical properties, the unique retro reflective Visi-stripe and the bright white surface offer added safety features for motorists. The ability to be manufactured in virtually any shape make them ideal for fast track projects and retro fitting existing structures without expensive demolition.

The paper will present four successful installations of sustainable rehabilitation using Polymer Concrete materials:

1. An emergency full depth repair of an expressway bridge in $2^1/_2$ days in cold weather.

Table 1. Physical Properties of Polymer Concrete

Viscosity	10–12 mPa · s (cP)	Brookfield
Pot Life	@70°F 24 min.	AASHTO T237
Compressive Strength	>34 MPa (5000 psi) – 1 hour	ASTM C109
	>51 MPa (7500 psi) – 3 hours	ASTM C109
	>58 MPa (8500 psi) – 24 hours	ASTM C109
Flexural Strength	13 MPa (2000 psi)	ASTM D790
Bond Strength	>1.7 mPa (250 psi) – 2 hours	ACI 503R

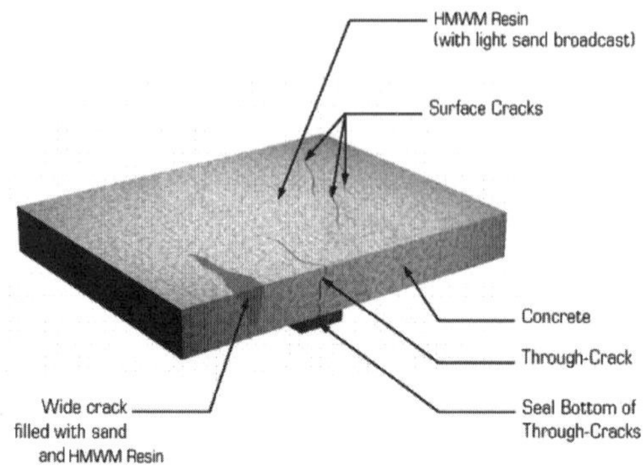

Figure 1. Concrete sealed with HMWM Resin.

2. The use of a Methacrylate based resin to seal cracks in a new thin concrete bridge overlay (Fig. 1).
3. The rehabilitation of a bridge deck that included extensive spall repair, thin wearing surface overlay and concrete crack sealing all done with Methyl Methacrylate based materials.
4. A retrofit of a deteriorated bridge barrier with precast polymer concrete safety shape panels eliminating the need to remove the existing cast in place concrete barrier.

REFERENCES

New York City DOT, Belt Parkway Bridges. *Failing pier caps and bearing seats*: Brooklyn and Queens, NY USA.

Pennsylvania DOT. Fort Dusquesne Bridge. *Bridge Deck 320,000 sq ft. (29,700 m²)*: Pittsburgh, PA USA.

New Jersey DOT, Dorland J.Henderson Memorial Bridge/ Manahawkin Bay Bridge. *Steel girder bridge rehabilitation 130,000 sq ft (12, 077 m²)*, Manahawkin, NJ USA.

Louisiana DOTD, Interstate 20 Bridges. *Barrier Retrofit, over 500 linear feet (152 m) of barrier rehabilitated*. Shreveport and Monroe, Louisiana, USA.

Bridge Maintenance, Safety, Management, Resilience and Sustainability – Biondini & Frangopol (Eds)

Push-out tests of straight shear connectors based on steel-concrete adherence

H.J.F. Diógenes, A.L.H.C. El Debs & M.K. El Debs
University of Sao Paulo, Sao Carlos, Brazil

ABSTRACT

This paper presents the results of an experimental program investigating the behavior of an innovative technology for connections by adherence in composite structures. Connections have a strategic importance for precast concrete and steel-concrete composite structures, since they affect the whole production process, like execution, assemblage and other services on site, besides determining the obtained structural behavior. Currently, however, steel-concrete composite connections are not completely adapted for the use of prefabricated slabs. In this way, the development of new types of connections is clearly necessary, especially to be used in composite bridges, where connections by adherence seem quite promising. To improve the performance of the steel-concrete interface, this research proposes a mechanical treatment of surface associated to the use of a high performance grout. Preliminary push-out tests were performed, and their results are presented and discussed. A satisfactory behavior of the proposed connection was observed, justifying further studies on the subject.

REFERENCES

Associação Brasileira de Normas Técnicas. *NBR 8800: 2008 – Projeto de estruturas de aço e de estruturas mistas de aço e concreto de edifícios*. Rio de Janeiro: ABNT.

CEN – EN 1994-1-1: 2004: Design of composite steel and concrete structures. *European Committee for Standardization* (CEN).

Dauner, H.-G., Antille, S. 2002. Ponts mixtes à construction rapide, *rapport OFROU 567.* Berne: Office fédéral des routes.

Hino, S. et al. 2000. Composite girder bridge with channel-shaped precast PC deck slabs, Composite and Hybrid Structures, In: *6th Asscs International Conference On Steel-Concrete Composite Structures*, pp. 181–188, Los Angeles.

Kitoh, H., Sonoda, K. 1996. Bond characteristics of embossed steel elements, In: *Composite Construction in Steel and Concrete III, Proceedings of an Engineering Foundation Conference, pp. 909–918.* New York: American Society of Civil Engineers.

Leonhardt, F. et al. 1987. Neues, vorteilhaftes Verbundmittel für Stahlverbund-Tragwerke mit hoher Dauerfestigkeit, *Beton und Stahlbetonbau, v. 12, pp. 325–331.* Berlim: Ernst & Sohn.

Shim, C.-S., Lee, P.-G., Chang, S.-P. 2001. Design of shear connection in composite steel and concrete bridges with precast decks. *Journal of Constructional Steel Research, Vol. 57, No 3, pp. 203–219.* Oxford: Elsevier Science Ltd.

Sonoda, K., Kitoh, H., Uenaka, K. 2000. Embossed steel plate-concrete composite member subjected to bending and shear. In: *Composite Construction in Steel and Concrete IV, Proceedings of an Engineering Foundation Conference*, pp. 355–367. New York: American Society of Civil Engineers

Thomann, M. 2005. *Connexions par adherence pour les ponts mixtes acier-béton*. PhD. Thèse, Ècole Polytechnique Fédérale de Lausanne, Faculté Environnnement Naturel, Architectural et Construit. Lausanne.

Vanderlei, R. D. 2004. *Análise Experimental do Concreto de Pós Reativos: Dosagem e Propriedades Mecânicas.* Tese (Doutorado), Escola de Engenharia de São Carlos – EESC/USP, São Carlos.

Veríssimo, G. S. 2006. *Desenvolvimento de um conector de cisalhamento em chapa dentada para estruturas mistas de aço e concreto e estudo do seu comportamento.*Tese (Doutorado) Universidade Federal de Minas Gerais. Escola de Engenharia. Belo Horizonte.

Yamane, T. et al. 1998. Full depth precast/prestressed concrete bridge deck system, *PCI Journal, Vol. 43, No 3, pp. 50–58.* Chicago: Precast/Prestressed Concrete Institute.

Bridge Maintenance, Safety, Management, Resilience and Sustainability – Biondini & Frangopol (Eds)
 ISBN 978-0-415-62124-3

The study on stability of bridge on which heavy military vehicle passes

A. Do Kyoun Kim, B. Jae Uk Lee & C. Young Gu Kang
Korea Infrastructure Safety & Technology Co, Goyang, South Korea

ABSTRACT

The aim of this paper center around identifying the stability of the Bridges on which Heavy Military Vehicle passes on general national road of Korea, with an emphasis on putting restrictions on its passage. Because it has not been controlled on Heavy Military Vehicle such as HET(Heavy Equipment Transporter), Caterpillar vehicle and Wheel vehicle due to the deficiency of regulation on restriction of vehicle operation in Korea Road Act While it has been cracked down on truck and construction machinery exceeding the weight limit. It is not only to protect a bridge from structural hazard during its passing, but also to prevent accident by lack of bridge stability. Regulation on restriction of vehicle operation in Korea Road Act, which range is limited to only heavy vehicle under axial load 10 ton or total load 40 ton.

But the range of vehicle by Korea Road Act covers vehicles by Automobile Management Act and Construction Machinery Management Act besides vehicle by Act on the Management of Military Supplies. As a result, Tank or HET(Heavy Equipment Transporter) has been excluded from object on restriction of vehicle operation so that result in being excluded from object on operation permission by present act, which is logical inconsistency.

According to the safety evaluation by numerical analysis and loading test, Stress occurred by two Heavy Military Vehicle paralleling driving on bridge(two-lane road) at the same time, which is 60 Vehicle Classification Number($60\,\text{Ton} \times 0.907 = 54.42\,\text{tonf}$) is less than allowable stress as well as the bridge bears enough the stress by bending and shearing, with the result that the bridge has a capacity for 60 Vehicle Classification Number during two vehicles paralleling driving on bridge.

Bridges in this study turn out to be safe when heavy military vehicle passes but also show the difference of the safety depending on the position of members

As a consequence, Load-carrying capacity assessment in quality is required to be thoroughly carried out on Road bridge in use that has various of form, the difference of design load class and specially structure deterioration as time passes.

Appropriate measures is required to be thoroughly taken before passing of the heavy military vehicle including HET(Heavy Equipment Transporter) for the stability of bridge and the safe passage during ordinary day, not during war.

For these reasons, Adjustment of current Korea Road Act is needed for rational operation of military heavy vehicle. And close cooperation between public official in charge of maintenance for bridge and transport army is necessary to minimize the damage of bridge.

REFERENCES

Kang Young Chul & Lee Pil Jae(2009), "Feasibility Study on the Road Bridge Passed by Heavy Equipment Transporter"

Korea Infrastructure Safety and Technology Corporation(2004), "The study on the safety assessment and guidelines for the passage when the special equipment vehicle passes on facilities in Gyung-gi province."

Park Byung Hee & Song Jae Ho & Jang Il Young(2006), "Feasibility Study on the Road Bridge Passed by Military Heavy Vehicle"

Bridge Maintenance, Safety, Management, Resilience and Sustainability – Biondini & Frangopol (Eds)
© 2012 Taylor & Francis Group, London, ISBN 978-0-415-62124-3

Damage detection in a suspension bridge model using the interpolation damage detection method

M. Domaneschi, M.P. Limongelli & L. Martinelli
Politecnico di Milano, Milano, Italy

ABSTRACT

Bridges and roads are strategic structures for emergency activities for example in the aftermath of an earthquake or a tornado. Traditional methods based on visual inspection carried out by experts take a long time to be performed and sometimes are not able to detect low levels of damage Furthermore the lack of uniform standards makes the evaluation of damage somehow dependent on the human judgment of the expert. For this reason the rapid and accurate detection of severe damage affecting the structure is essential in order to evaluate the operability and to schedule maintenance operations.

In this paper the use of the IDDM (Interpolation Damage Detection Method) is proposed for a speedy damage detection and localization, able to provide a reliable real-time evaluation of the structural condition without the need of any user interaction after an initial phase of fine-tuning.

The method is based on the definition of damage defined in terms of the accuracy of a spline function in interpolating the operational mode shapes (ODS) of the structure recovered form responses to wind excitation. At a certain location, a decrease (statistically meaningful) of accuracy, with respect to a reference configuration, points out a localized variation of the operational shapes thus revealing the existence of damage.

The method has been applied to the numerical model of a cable suspended bridge derived from the ANSYS model of the Shimotsui- Seto Bridge in Japan (940 m length of the main span). The wind excitation is simulated as a spatially correlated process acting in the horizontal direction, transversal to the deck.

Results show that the IDDM is able to provide with a good level of accuracy the correct location of damage for the several damage scenario herein considered.

Table 1. Damage scenarios for noise level 2%.

Setting	Location Node	Duration s	Damage %
N43_n20_35%	43	19800	35
N43_n20_50%	43	19800	50
N43_n20_60%	43	19800	60

Table 2. Damage scenarios for noise level 5%.

Setting	Location Node	Duration s	Damage %
N43_n40_15%	43	39600	15

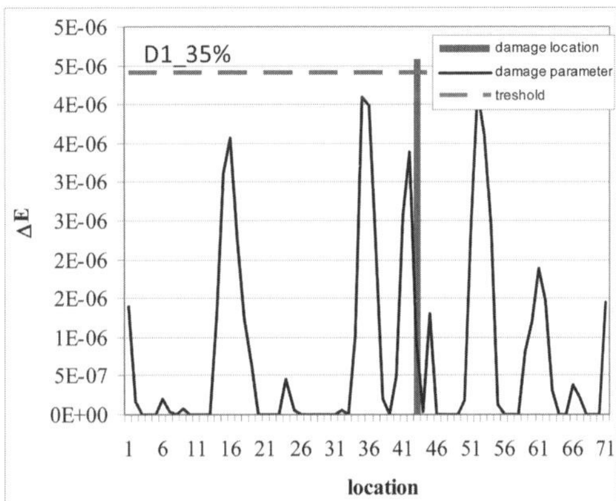

Figure 1. Damage parameter for a damage scenario at noise level 2% and threshold corresponding to 1% probability of false alarm.

The accuracy decreases with the level of noise hence an initial fine-tuning of parameters related to range of frequency corresponding to a reasonable signal to noise ratio and to the false and missing alarms rate is required.

REFERENCES

Domaneschi, M. and Martinelli, L., 2010. Optimal Passive and Semi-active Control of a Wind Excited Suspension Bridge. *Structure and Infrastructure Engineering*, Taylor and Francis Group Ltd. (in press)

Limongelli, M.P., 2010. Frequency Response Function Interpolation for Damage Detection under Changing Environment. *Mechanical Systems and Signal Processing*, 24(8), 2898-2913.

Bridge Maintenance, Safety, Management, Resilience and Sustainability – Biondini & Frangopol (Eds)
 ISBN 978-0-415-62124-3

Using internal electrical resistivity measurements as a tool for structural health monitoring

N.H. El-Ashkar, M.I.S. Elmasry & M.F.A. Alasadi
Arab Academy for Science, Technology and Maritime Transport, Alexandria, Egypt

ABSTRACT

Worldwide, vital and strategic structures have special importance. These structures, such as bridges sky scarpers, power utilities, *etc*, age and deteriorate with time. Consequently in recent years, Structural Health Monitoring has been studied and used by the industry as a possible method to obtain continuous real-time data about the health, and subsequently the safety and serviceability of the vital structures, and thereby reducing their operational cost. When a vital structure abruptly collapses, *e.g.* what happened in the I-35 Mississippi Bridge in Minnesota, the collapse could lead to huge risks on human lives as well as undesirable economic losses. Accordingly, it is necessary to detect hidden damage and take remedial repair actions. Electrical resistivity measurements method is a low cost, simple and efficient which has received relatively little attention in terms of structural health monitoring (SHM). The objective of this research is to study a suitable technique using Inner Electrical Resistivity Measurement (IERM) that can be used in damage detection for structural health monitoring (SHM) of RC structures. In this study, an experimental program is conducted to investigate the effect of crack initiation, propagation, location and orientation on IERM. Two specimens are fabricated, one is made of mortar and the other is made of concrete. Directly after casting the specimens, a plastic piece, of 1.25 mm thickness and 15x10 cm height and width, is placed in the central part of the blocks. This plastic piece resembles crack effect in the specimen. The electrical resistivity is estimated from the measured resistance by earth resistance tester using probes embedded inside the specimens. The results of this study demonstrate the sensitivity of the IERM technique to crack initiation and propagation. Also, the crack location and orientation effect on IERM is investigated and found to be a vital parameter that affects the IERM. It was shown that when the crack is initiated and propagated between the voltage drop probes, the IERM values are significantly increased as shown in Figure 1. While, it was also observed that IERM values are decreased when the crack is initiated and propagated between the current and voltage drop probes or parallel to the line between the voltage drop probes. Thus, the proposed technique can be successful in detecting damages in concrete structures. This technique when compared to other SHM techniques in concrete structures such as ultrasonic, acoustic emission and other methods can be cheaper, simpler, and more efficient.

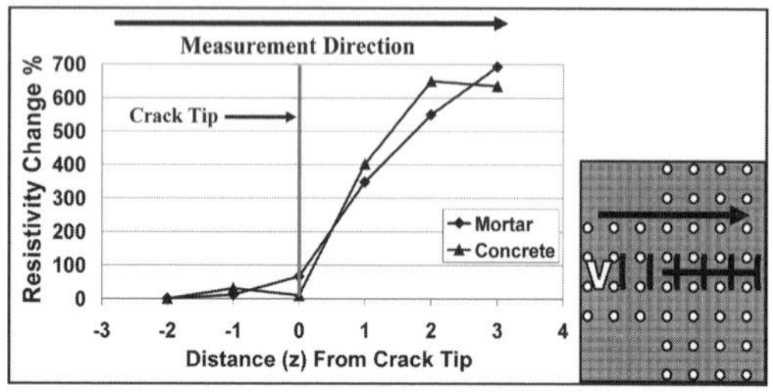

Figure 1. Effect of crack presence on Square Inner Electrical Resistivity Measurement (SIERM) resistivity change.

REFERENCES

Ahlborn, et al., 2010. *The State-of-the-Practice of Modern Structural Health Monitoring for Bridges: A Comprehensive Review*. Michigan Tech Transportation Institute, Michigan Tech Research Institute, Michigan Technological University.

Beard, et al., 2005. *Practical issues in real world implementation of structural health monitoring systems*. Industrial and commercial applications of smart structure technologies. Conference, San Diego CA, vol. 5762, pp. 196-203.

Sun, M., Staszewski, W. J., & Swamy, R. N., 2010. *Smart Sensing Technologies for Structural Health Monitoring of Civil Engineering Structures*. Advances in Civil Engineering journal, Vol. 2010, Article ID 724962, 13 pages.

Bridge Maintenance, Safety, Management, Resilience and Sustainability – Biondini & Frangopol (Eds)
© 2012 Taylor & Francis Group, London, ISBN 978-0-415-62124-3

Strengthening of box girders using adaptive "tube-in-tube" concepts

M. Empelmann & D. Busse
Institut für Baustoffe, Massivbau und Brandschutz, Fachgebiet Massivbau, TU Braunschweig, Germany

S. Hamm, M. Girmscheid & T. Zedler
HOCHTIEF Solutions AG, Essen, Germany

ABSTRACT

Many existing prestressed concrete bridges show deficiencies with regard to material, construction, design and life load rating. In addition, the traffic volumes on German motorways have increased rapidly and a further increase is forecasted (Naumann, BMVBS). Strengthening of existing bridges and adaptation methods of new bridges for future modified boundary conditions is a key aspect to secure mobility.

In this context the Federal Ministry of Transport, Building and Urban Transportation in Germany (BMVBS) has introduced an innovation program promoting the development of new products, technologies and methods for maintenance and strengthening of infrastructure. The innovation project introduced in this paper aims to develop adaptive bridge structures that can react flexible on varying boundary conditions in the future such as increased loads or modified geometric requirements.

State of the art strengthening of bridges comprises inter alia external prestressing, the addition of concrete and epoxy-glued reinforcement. While strengthening for bending using external prestressing has become rather common, strengthening for shear is rather complex and usually requires intensive workmanship and bridge closures. Hence, a strengthening method combining the strengthening of bending as well as shear and torsion capacities that can be applied without bridge closures is envisaged.

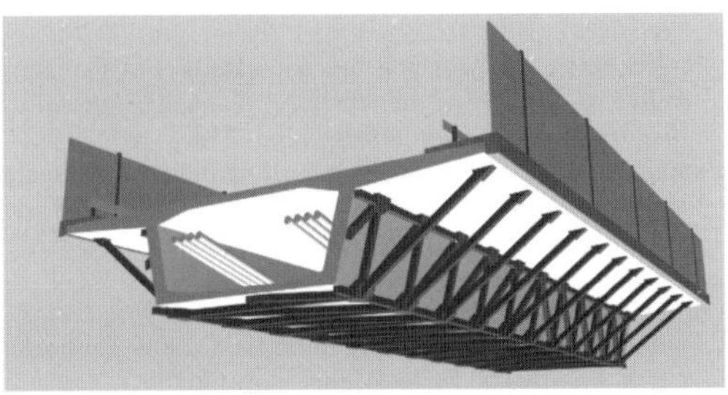

Figure 1. Tube-in-tube strengthening concept.

The innovative idea is to create new "tube-in-tube" structural concepts for box girder bridges combining a primary tube (box girder) with a secondary tube (e.g. truss structure). Adaptation methods using tube-in-tube concepts are developed for increased effects in transverse direction as well as for increased effects in longitudinal direction of the structure. The strengthening concepts are developed under the premise of minimum disruption of traffic.

The adaptation concepts in the transverse direction comprise diagonal compression struts, additional downstand beams, struts inside the box and additional beams between the girder webs stressed against the existing top slab.

In longitudinal direction of the box girder bridge post-tensioned truss structures have been investigated (Figure 1). The post-tensioning is applied to the bottom chord of the truss before the truss joints are connected to the box girder. This results in reduced shear and bending demands in the original structure. For load transfer different alternative joint options are investigated in detail in this study and are discussed in this paper.

The effectiveness of the strengthening method for the transverse and longitudinal direction is shown using detailed design calculations for a representative bridge structure.

It is highlighted that the developed strengthening method has positive effects on the bending, shear and torsion capacity of the bridge. The strengthening concept is in principal applicable to existing structures as well as to new structures.

REFERENCES

Naumann, J. 2010. Bridges and heavy goods vehicle traffic – a strategy for strengthening existing bridges on federal trunk roads. In Springer Verlag (ed.), *Bauingenieur* 85: 210–216

BMVBS 2006. Bericht über die Qualität, Dauerhaftigkeit und Sicherheit von Spannbetonbrücken. www.bmvbs.de

Bridge Maintenance, Safety, Management, Resilience and Sustainability – Biondini & Frangopol (Eds)

The effect of lane changing on long-span highway bridge traffic loading

B. Enright, C. Carey & C.C. Caprani
Department of Civil & Structural Engineering, Dublin Institute of Technology, Dublin, Ireland

E.J. OBrien
School of Civil, Structural & Environmental Engineering, University College Dublin, Dublin, Ireland

ABSTRACT

The governing form of traffic loading on long-span bridges is induced by congested traffic (Buckland 1981). As the number of freight trucks travelling on European highways rises (Eurostat 2009), the frequency of occurrence of truck platoons also rises, consequently increasing the traffic loading on bridges. This may cause an increase in the regularity and severity of bridge repair programs. The cost associated with bridge repair is twofold – the actual repair cost and the economic cost due to traffic delays while repairs are being carried out. Reduced rehabilitation needs could therefore have significant associated costs savings. This paper investigates how the lane-changing behavior of cars affects the formation of truck platoons, and the effect these platoons have on the traffic loading on long-span bridges. Traffic on the A4 (E40) motorway at Wroclaw, Poland, was recorded using Weigh-In-Motion equipment over a six month period in 2008. The traffic at the measured site is free-flowing, and in this study it is passed through a traffic micro-simulation model to induce congestion (Caprani 2010), and the platooning behavior in this congested traffic is studied, see Figure 1.

From Figure 1 it can be seen that the probability of the next vehicle being a truck increases as the number of trucks traveling in front increases. This suggests that, for this traffic, the occurrence of a truck or car is not independent of the number of similar type vehicles travelling in front, in contradiction to conventional vehicle arrival processes for long-span bridges.

A Monte Carlo simulation model in which individual car drivers probabilistically decide, based on a lane-changing bias probability, whether or not to change lane is developed. In the model, various lane-changing bias probabilities for cars travelling behind trucks and other cars are investigated in order to replicate as closely as possible the platooning behavior found in the congested traffic.

The sensitivity of bridge loading to this system is investigated for different bridge lengths and traffic compositions. This research concludes that the lane-changing behavior of car drivers has an effect on bridge loading for long-span bridges, and the magnitude of this effect is quite sensitive to the percentage of trucks in the traffic.

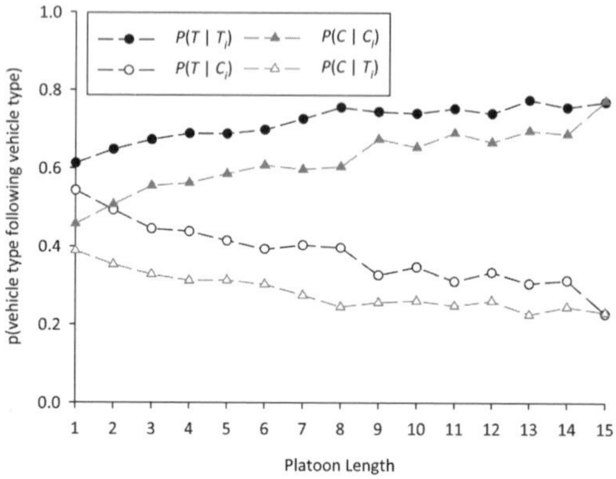

Figure 1. Platoon dependency plot for congested traffic from micro-simulation, where $P(T)$ is the probability that the next vehicle is a truck and $P(C)$ is the probability of the next vehicle being a car, T_i is a platoon of i trucks in length and C_i is a platoon of i cars in length.

REFERENCES

Buckland, P.G., 1981. Recommended design loads for bridges (Committee on loads and forces on bridges of the committee on bridges of the structural division), *Journal of the Structural Division, ASCE*, 1161–213.

Caprani, C.C. 2010. Using Microsimulation to Estimate Highway Bridge Traffic Load. *Proc. 5th Intl. Conf. on Bridge Maintenance, Safety and Management*, IABMAS, Philadelphia, July.

European Commission Directorate-General for Energy, Transport and Environment Indicators, 2009. European Union Energy and Transport in Figures 2009.

Bridge Maintenance, Safety, Management, Resilience and Sustainability – Biondini & Frangopol (Eds)
© 2012 Taylor & Francis Group, London, ISBN 978-0-415-62124-3

Exploiting linear system behaviour to determine structure level costs based on element condition states

D. Fernando & B.T. Adey
Institute of Construction and Infrastructre Management (IBI), Department of Structral, Environmental and Geomatic Engineering (D-BAUG), ETH Zürich, Switzerland

S. Walbridge
Department of Civil and Environmental Engineering, University of Waterloo, Ontario, Canada

ABSTRACT

Bridges play a key role in the functionality of road infrastructure. Due to the deterioration of the bridges, interventions are required to ensure that they will continue to provide adequate level of service over a specified time period. The determination of preservation intervention strategies that maximize benefit to all stakeholders requires consideration of not only the condition of the elements of which a bridge is comprised, but also the performance of the structure as a whole. This is required because although some costs can be related to the condition of the elements, others, such as the costs of accidents for those travelling on the bridge, cannot. The costs related to the performance of a bridge can in many cases be related to specific performance indicators. For example, travel time costs which are related to the travelling speed over the bridge, can be related to the longitudinal unevenness of the bridge.

One methodology to predict these structure level costs is to determine structure performance states from the element condition states, to estimate the values of the relevant performance indicators for each of these performance states, and to associate costs to each value of the performance indicator. Although this methodology will allow cost predictions that facilitate the determination of optimal preservation intervention strategies, when structure level costs are to be considered, even a bridge with a moderate number of elements and element condition states results in an unwieldy number of possible bridge performance states, when the values of the performance indicators are to be estimated directly for each performance state. The amount of work required to estimate these values, however, can be drastically reduced by exploiting the almost linear system behavior of the bridge that occurs between many performance states.

In this paper, a methodology to be used to relate the structure performance to element condition states for the purpose of determining structure level costs is presented. An exact method, which yields highly accurate results but is computationally demanding, as well as an approximate method, which yields slightly less accurate results but it computationally much less demanding, are presented and demonstrated using examples. It is seen that the use of the approximate method for estimation of the values of the performance indicators results in only a small discrepancy in the estimate of total costs that would be incurred over the expected life time of the example bridge when compared with the use of the exact method.

Bridge Maintenance, Safety, Management, Resilience and Sustainability – Biondini & Frangopol (Eds)
© 2012 Taylor & Francis Group, London, ISBN 978-0-415-62124-3

A Timoshenko–based structural model for the analysis of bridges

M. Ferraz, R. Faria & J. Figueiras
LABEST, Faculty of Engineering of the University of Porto, Portugal

ABSTRACT

The objective of this work is to present all the phases of a process that lead to the development of object oriented software for structural analysis of bridges (Ferraz 2010). The objectives and motivation for the process are presented, as well as the Object Oriented Programming paradigm and the platform concept. The structural analysis model was developed to perform 3D analyses of evolutive structures, from the building phase up to the final application, using a discretization with FE based on the Timoshenko beam formulation. The model allows for changes of the static system to be made over time, incorporating specific FE for the discretization of rebars, prestress (both embedded and exterior) and stay cables, different types of external links (including sliding and unidirectional supports), time dependent material behavior laws (such as concrete ageing, shrink-age and creep, as well as prestress relaxation) and other nonlinear constitutive laws (such as the elastoplastic behavior of steel and the concrete cracking). The model includes incremental and iterative tools to solve the nonlinear problem involved.

The developed model to numerically predict the behavior of bridges and viaducts proved to be efficient and accurate, as demonstrated by the analysis of the structural behavior of the Corujeira Viaduct in Oporto (Fig. 1) during the construction phase, the load tests and for maintenance operations.

The Timoshenko FE formulation (3D), combined with a fiber discretization of the cross-section, allowed

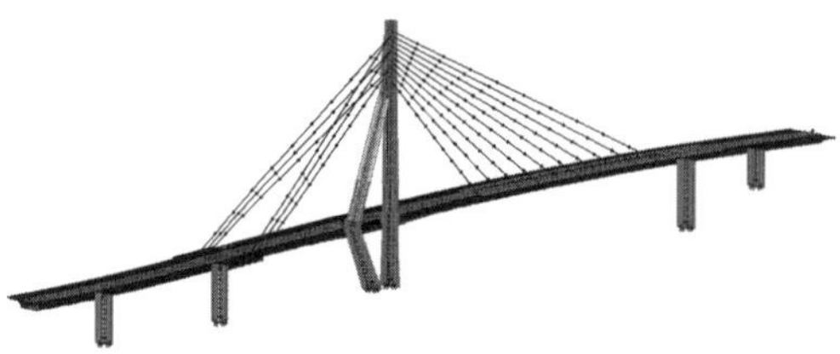

Figure 1. Corujeira Viaduct model view.

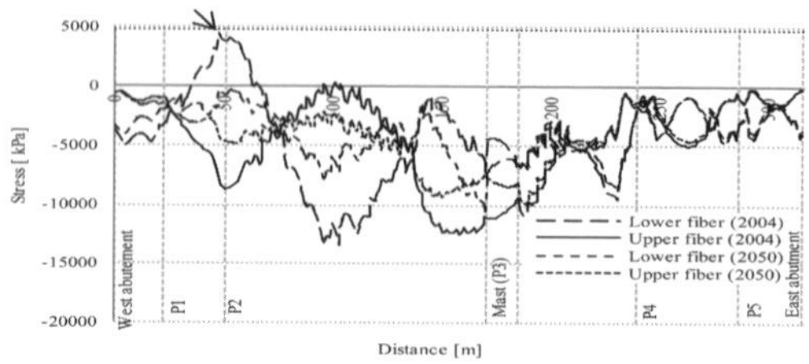

Figure 2. Normal stresses on concrete deck after construction and at 2050.

the use of independent material laws for each fiber, and by this way a great simplification in the simulation of complex phenomena like creep, shrinkage and aging of concrete, and relaxation of prestress. Also the specific formulations used for the modeling of embedded prestress and stay cables showed to be very useful and accurate.

The existence of a monitoring system on the Corujeira Viaduct allowed an extensive and rigorous comparative analysis of the numerical results with the in situ measurements. This permitted also to make realistic predictions of the future structural behavior of the viaduct (Fig. 2), including the modeling of maintenance operations like the replacement of the bearings.

The extensive list of implemented and validated features specifically targeted for structural bridge analysis allows claiming the developed model as a useful tool for project management and structural analysis of bridges built using different construction methods.

REFERENCE

M. Ferraz, 2010, Modelo para Avaliação do Comportamento Estrutural de Obras de Arte, PhD theses, Faculdade de Engenharia da Universidade do Porto

Bridge Maintenance, Safety, Management, Resilience and Sustainability – Biondini & Frangopol (Eds)
© 2012 Taylor & Francis Group, London, ISBN 978-0-415-62124-3

Truck weight limits on concrete bridges regarding ultimate limit bending moment using reliability theory

L.M. Ferreira, M.K. El Debs & A.L.H.C. El Debs
University of Sao Paulo, Sao Carlos, Brazil

ABSTRACT

The increase in gross weight limits allowed by Brazilian legislation and the soaring number of new truck con-figurations on national highways has called for greater attention regarding the structural safety of bridges when submitted to real traffic.

The objective of the present paper is to propose truck weight limits for reinforced and prestressed concrete bridges, designed according to the latest version of Brazilian code regarding bridge loading, using the reliability theory and equation known as bridge formula.

The employed procedure involves initially the selection of typical bridges that reliably represents the existing structures constructed over the country. Looking for attending the most common structural systems and obtaining envelopes of actual situations, even through fictitious cases, 60 typical bridges were selected for the calculation of the bridge formula, as shown in Table 1.

Once determined the typical bridges, each critical cross section is designed and its statistical resistance parameters are obtained through Monte Carlo simulations. The procedure includes the analysis of the compressive strength of concretes with several design specifications.

The next step was to develop a live load model which was based on weightings data. A live load model is developed based on trucks weightings carried out by a private administrator of some public highways segments at the state of São Paulo. The data analyzed was that ranging from January 2001 to October 2002, totaling 184,603 weightings, 126,389 of which were during the first year and 58,214 in 2002.

Making use of the calculated resistance and loading statistical parameters, the performance of the selected cross sections is measured in terms of the reliability index β. Based on the analysis of the results a target reliability index is fixed, understood as the minimum value that the effects of considered safe trucks must respect in all structures.

The gross weight limits are presented by an equation, known as bridge formula, which are applicable to any group of two or more consecutive axles. Based on the developed study, the following equation can be obtained:

$$W_{ULS} = 26.2 \cdot B + 233.9 \quad (1)$$

where W = maximum gross weight in kN on any group of two or more consecutive axles; B = length of the axle group in meters.

The use of this weight limits indicates that there is restriction to the circulation for LCV (Long Combination Vehicles) with 19.8 meters length and 740 kN.

Table 1. Typical brides analyzed.

Structural system	Number of typical bridges
Slab	8
2 girders	23
5 girders with crossbeams:	11
5 girders without crossbeams	8
box section (1, 2, 3 or 4 cells):	10

REFERENCES

Brazilian Association of Technical Standards (ABNT). 1984. *NBR 7188 – Live load in bridges and footbridges*. (in Portuguese). Rio de Janeiro.

Ferreira, L. M. 2006. *Development of truck weight limits for concrete bridges using reliability theory* (in Portuguese). PhD thesis, São Carlos School of Engineering, São Carlos. University of São Paulo.

Ferreira, L. M., Nowak, A. S, El Debs, M. K. 2008. Development of truck weight limits for concrete bridges using reliability theory. *Ibracon Structures and Materials Journal*, 1(4): 421–450.

Ghosn, M. 2000. Development of truck weight regulations using bridge reliability model *Journal of Bridge Engineering*, 5(4): 293–303.

Nowak, A.S. & Collins, K. R. 2000. *Reliability of structures*. McGraw-Hill, Boston.

Bridge Maintenance, Safety, Management, Resilience and Sustainability – Biondini & Frangopol (Eds)
© 2012 Taylor & Francis Group, London, ISBN 978-0-415-62124-3

Analysis of normative approaches to service life design for carbonation induced reinforcement depassivation: Fib MC-SLD, by50 and LNEC E465

R.M. Ferreira
C-TAC, University of Minho, Guimarães, Portugal

E. Vesikari
VTT Technical Research Centre of Finland, Espoo, Finland

ABSTRACT

Normative approaches to service life design in concrete codes are relatively recent. No design approach has been generally accepted, however, the awareness provided by EU projects like LifeCon and DuraCrete has influenced the implementation of design approaches in concrete codes.

In this paper three approaches to service life design are analysed. They are distinguishable, not only by the theoretical background, but also the mathematical manner in which they are implemented. The design approaches analysed are the fib Model Code for Service Life Design (full probabilistic), the Finnish Concrete Code by50 (factor approach), and the Portuguese specification LNEC E-465 (partial safety factor approach).

Design scenarios for the carbonation exposure classes are used to illustrate the implementation of the design approaches for the serviceability limit state of reinforcement depassivation.

The requirements of the design approaches as well as the ease of use and design outcome are critically analysed, individually and comparatively. The analysis reveals the fundamental differences of the different design approaches, their limitations and the implication on the outcome of service life design is obtained. The models outcome are analysed directly and through the service life distributions.

One of the main difficulties was to define a sufficiently detailed case study with realistic information on concrete performance and environmental loading that would satisfy the individual model's needs. For this study, the following considerations where made:

- service life = 50 years;
- two exposure classes XC3 (moderate humidity) and XC4 (Cyclic wet and dry) – the Finnish interpretation of class XC3 is that the concrete is sheltered from rain was adopted in this study because it gives the maximum for carbonation and minimum for initiation time of corrosion;
- concrete with two distinct cement types according to carbonation performance: beneficial effect (CEM I or CEM II/A) and non-beneficial effect (CEM II/B to V), and two minimum strength classes (C20/25 and C35/45);
- concrete cover of 25 mm used for all models;
- durability performance of the concrete measured as the resistance to carbonation is based on published values and identical for each model.

In this paper, the results of the calculation using all three models are analysed in two formats: i) direct model outcome; and, ii) Service life cumulative and probability distribution functions (See Fig. 1).

The FIB model is very flexible (prone to manipulation [9]) because there is freedom to choose the type of distribution and the parameter values. On the other hand, the E465 and the BY50 are less flexible – by restricting design choices to what is considered to be within acceptable practice.

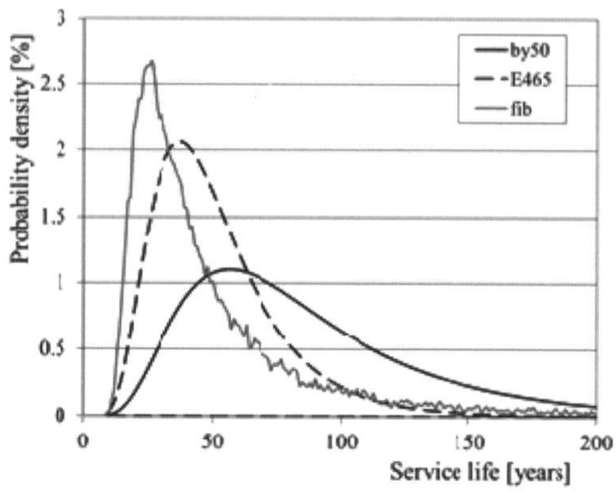

Figure 1. Service life PDF curves for the XC3 exposure class for: C20/25 concrete with CEM I or II/A cement.

Bridge Maintenance, Safety, Management, Resilience and Sustainability – Biondini & Frangopol (Eds)
© 2012 Taylor & Francis Group, London, ISBN 978-0-415-62124-3

Bridges and viaducts of "Variante di Valico" project

L. Ferretti Torricelli & A. Marchiondelli
Bridge Design Department, SPEA Ingegneria Europea S.p.a., Milan, Italy

ABSTRACT

The A1 Milan-Naples Highway, the strategic link which connects the North and South parts of Italy, was built in the '50 and is nowadays inadequate to sustain the continuously increasing traffic flow. For this reason an intervention aimed to improve clients' safety and comfort is currently under realization.

After a long design process, made difficult due to the great morphological complexity of territory, the project called "Variante di Valico" was set up by SPEA Ingegneria Europea, the design company of Autostrade per l'Italia, the main National transportation agency. The alignment develops for 59 km between Sasso Marconi and Barberino di Mugello, by crossing Appennines along Emilia Romagna and Tuscany.

First of all the project consists of the modernization of the existing alignment between Bologna and Incisa Valdarno by means of the widening of the present highway and the realization of local alternative routes. Secondarily, a new highway is realized in order to cross the Apennines by strongly reducing the actual altimetric rise. A great effort has been spent in order to minimize environmental and antropic impact, and to reduce interferences with local road network. As a consequence, the alignment of the "Variante di Valico" develops mainly underground in the respect of the high morphological complexity of the territory. From the structural point of view, the project provides 22 tunnels, developing for 29,3 km in total, and 22 bridges and viaducts, for a total length of 10,5 km.

After a description of the adopted design criteria, the paper presents the most used bridge design solutions, highlighting all the aspects and constrains related to the design choices. A final chapter is dedicated to the most remarkable works, designed to represent the symbol of the whole project, that are the Aglio Viaduct, with its elegant slender profile and 90 m high piers (Figure 1), and the Reno Bridge, characterized by two couples of elegant inclined piers aimed to sustain the central span of 135 m (Figure 2).

Figure 1. View of the Aglio viaduct.

Figure 2. View of the Reno bridge.

REFERENCES

Ferretti Torricelli, L., Furlanetto, G. & Marchiondelli, A., 2006, *Durability design criteria for the Reno bridge*, IABMAS'06 Conference, July 16–19, 2006, Porto, Portugal.

Ferretti Torricelli, L., Furlanetto, G. & Marchiondelli, A., 2007, *A1 Milan-Naples Motorway. The Variante di Valico project*, IABSE Symposium, September 19–21, 2007, Weimar, Germany.

Ferretti Torricelli, L., Furlanetto, G. & Marchiondelli, A., 2009, *The new Aglio bridge on the A1 Milan-Naples Motorway*, fib Symposium, June 22–24, 2009, London, England.

Bridge Maintenance, Safety, Management, Resilience and Sustainability – Biondini & Frangopol (Eds)
© 2012 Taylor & Francis Group, London, ISBN 978-0-415-62124-3

Experimental testing of blast resistance FRC and RC bridges

M. Foglar, M. Kovar & A. Kohoutkova
Czech Technical University in Prague, Czech Republic

ABSTRACT

According to recent publications, from 2005 to 2008 there were more than 13000 terrorist attacks around the world, which took more than 73000 human lives. The attacks were targeted mainly on the technical and civic infrastructure, like governmental buildings, bridges, etc.

Due to improved ductility, fibre-reinforced concrete (FRC) shows better performance under blast and impact loading compared to conventionally reinforced concrete.

Field tests of FRC and reinforced concrete specimens were performed in cooperation with the Czech Army corps and Police of the Czech Republic in the military training area Boletice.

The paper presents conclusions from two sets of tests and results of their numerical evaluation. The tests were performed in cooperation with the Czech Army corps and Police of the Czech Republic at the military training area Boletice using real scale precast slabs and 25 kg of TNT charges placed in distance from the slab for better simulation of real in-situ conditions.

Dimensions of the specimens were designed in real scale of a small span bridge as concrete slabs, 6 m long, 1.5 m wide and 0.3 m thick. The specimens were designed to fulfill the design bridge loading according to EN 1991-2 design code.

Five specimens were made in total, where three of them were made of C30/37 grade concrete (specimen No. 1, 2 and 5), two of C55/67 grade concrete (No. 3 and 4). Polypropylene 54 mm long synthetic fibers were used in three of the specimens. The fiber dosage was following: specimen No. 1 0, No. 2 4.5 kg/m^3, No. 3 0, No. 4 4.5 kg/m^3 and No. 5 9 kg/m^3. The dosage of the fibers was kept low as it can be achieved on-site.

Figure 1. Layout of the experiments.

Figure 2. Damaged specimen No. 3 after blast – bottom view.

REFERENCES

Coughlin, A.M. & al. (2010) Behavior of portable fiber reinforced concrete vehicle barriers subject to blasts from contact charges. *International Journal of Impact Engineering*, V. 37, No. 5, pp. 521–529.

Schenker, A. & al. (2008) Full-scale field tests of concrete slabs subjected to blast loads. *International Journal of Impact Engineering*, V. 35, No. 3, pp. 184–198.

Seible, F. & al. (2008) Protection of our bridge infrastructure against man-made and natural hazards, *Structure and Infrastructure Engineering*, Vol. 4, No. 6, pp. 415–429.

Bridge Maintenance, Safety, Management, Resilience and Sustainability – Biondini & Frangopol (Eds)
© 2012 Taylor & Francis Group, London, ISBN 978-0-415-62124-3

Bridge management system implementation in Italy: Pontis® and other BMS application in Italy

E. Frugulietti & G. Pasqualato
Sineco S.p.A., Italy

E. Spallarossa
Archimede srl, Italy

ABSTRACT

The activities described in this paper have been performed within the collaboration between Archimede s.r.l. and Sineco s.p.a. Since 2002 the two companies have been cooperating in Pontis implementation, intended to choose the optimal bridge network maintenance or improvement policies, consistent with italian agency's usual policies, long term targets and budget constraints on the basis of inspection data recorded by Sineco. A study was performed to set the bridge management systems (BMS) that best fitted Sineco needs. Sineco chose Pontis because it assessed its own needs (different agencies with different needs might opt for a different system), being developed by AASHTO, that is a non-profit association, and being widespread among nearly all the United States, grants a systematic implementation. Furtemore using Pontis provides a systematic approach for a bridge management system application. Once the system was identified, data collected by Siceco during inspections and used within its maintenance system called SIOS were converted into CoRe elements condition state language, used by Pontis system. The two different approaches in considering structures are explained and the conversion method is described, also providing a numerical example. The paper also describes and makes a comparison between SIOS and Pontis maintenance needs evaluationor, in other words, preservation policy. Based on inspection data collected through past years, SIOS method is able to forecast deterioration evolution on the basis of statistical criteria. Needs definition, intended as work priorities, is obtained, both for the entire structure, both for each element composing the bridge, from the statistical analysis of the changes in the condition states during time. This predictive method in efficient for short-medium term analysis (up to ten years planning horizons) thank to the great amount of data collected performing annual inspections since 1994. Preservation Policy, for Pontis, is a set of feasible actions associated to each structural element that, maintaining existing bridges in operation at their current level of service, assumes that operations must continue and deterioration must be detected and remedied before operations are affected, at minimum cost. Concluding remarks underline that integration of the two systems allows different simulation scenarios for different planning horizon: short, medium and long period, from five up to 30 years. Moreover, as one of Pontis output is the prioritized work list, on the basis of the benefit cost ratio, the works identified by Pontis are used as input data for the Asset management tool in use within Sineco company.

REFERENCES

Aldayuz José L. & Marshall Allen R. 2004. Issues in deploying a bridge management system internationally.

Astudillo Pastor, R., Arrieta Torrealba, J.M., C. Velando Cabañas, C. & Lozano Bruna, C. 1999. Decision on repair/replacement. BRIME (Bridge Management in Europe) PL97-2220 Deliverable D7.

Cambridge Systematics, Inc., 2004. Pontis Release 4.4 Technical Manual.

Marshall, A., Robert, W., Anderson, K., Floyd, R. & Corso, F. 1999. Comparison of Pontis Bridge Project Recommendations to Programmed Work for Three U.S. Transportation Agencies. 8th International Bridge Management Conference, Transportation Research Circular 498.

Robert, W., Marshall, A., Shepard, R. & Aldayuz, J. 2003. Pontis Bridge Management System: State of the Practice in Implementation and Development. 9th International Bridge Management Conference, Transportation Research Circular E-C049.

Sineco, Inc., 2004. "Manuale di riconoscimento e valutazione anomalie"

Spallarossa, E., Robert, W., Aldayuz, J., 2005. National Application of Pontis BMS models. ICOSSAR'05.

Spallarossa, E., 2004. Evaluation of candidate bridge management systems for use with italian highway bridges. IABMAS'04

Bridge Maintenance, Safety, Management, Resilience and Sustainability – Biondini & Frangopol (Eds)
© 2012 Taylor & Francis Group, London, ISBN 978-0-415-62124-3

"The Maintenance Manual" in important infrastructural project, from the design up to the implementation after construction

E. Fruguglietti & G. Pasqualato
Sineco S.p.A., Italy

R. Sagula
Point S.r.l., Italy

ABSTRACT

The management of infrastructural asset is a very complex activity, because is necessary to be able to maintain structures, the bridges, the galleries and the systems installed in an acceptable condition and in a cost effective way. This involves numerous multidisciplinary tasks such as: planning, inspection, maintenance, optimization of resources, preparation of procedures and instructions, coordination and cooperation with external bodies and provision of reliable information to the public about the on serviceability of the infrastructure. Since the design phase, it has been clear that a particular strategy for optimizing management, operation and maintenance of an asset, a bridge, for example, shall be a fundamental requirement both to the dimension and technical complexity of the structure, and to its economic importance for the national transport system.

This paper describes the methods applied to produce the "Maintenance manual" in particular is presented the technical approach based on FMEA/FMECA analysis, a common method in the industrial production world but relatively new in the civil construction world, that usually is based on approaches less sophisticated. The manual is oriented to: breakdown of the structure in order to analyze the function of each component, in relation on its reliability and the effects of its unreliability; Identify the critical failure in order to define the correct maintenance policy; Definition of the maintenance program, as sum of the single maintenance tasks and the correlated technical documentation.

The approach proposed is also to reach the valorization of the RPN (Risk priority number) that together with the experience and the know how help to define the quality of the service of maintenance. The Maintenance Manual together with a number of additional documents with supplementary information combined will be a tailored Asset Management or more Bridge Management (BMS), capable of storing inventory & inspection data, documents, correspondence and budgets in a transparent way. The Inspection and Maintenance Manual comprises a large number of Technical Procedures and Instructions which are based on the information from design and execution phases, from suppliers, etc. Finally the paper describes the main steps followed by Sineco to apply this technique to an important infrastructural project in Italy. These Technical Procedures and Instructions must be complied with by those involved in inspection and maintenance after the construction, and are a sort of practical guidelines to the software implementation.

REFERENCES

Ponte sullo stretto di Messina "Specifiche tecniche per l'ingegneria di manutenzione e dei sistemi di controllo e gestione – Linee guida", 2004 – Stretto di Messina S.p.A.

Ponte sullo stretto di Messina "Opere ferroviarie connesse – Prescrizioni di manutenzione", 2004 – Stretto di Messina S.p.A.

Ponte sullo stretto di Messina – COWI "Preliminary Inspection and Maintenance Manual", 2010 – Stretto di Messina S.p.A.

Ponte sullo stretto di Messina – COWI "Reliability Based Inspection (RBI) and Reliability Centred Maintenance (RCM)", 2010 – Stretto di Messina S.p.A.

Furlanetto L., Garetti M., Macchi M., 2006. "Principi generali di gestione della manutenzione" – Franco Angeli

Circolare 19/07/1967 n.6736/61A1 del M.LL.PP. "Controllo delle condizioni di stabilità delle opere d'arte stradali"

Istruzione 44c "Visite di controllo ai ponti, alle gallerie, ed alle altre opere d'arte del corpo stradale" ed. 1994 – Ferrovie dello Stato

UNI EN 13306:2003, "Manutenzione – Terminologia", 2003

UNI 10147:2003, "Manutenzione – Termini aggiuntivi alla UNI EN 13306 e definizioni", 2003

MIL-STD 1629A, "Procedures for performing a Failure Mode, Effects and Criticality Analysis (FMECA)", 1984

SAE J1739, "Potential Failure Mode and Effects Analysis in Design and Potential Failure Mode and Effects Analysis in Manufacturing and Assembly Processes", 2002

Sineco, Inc., 2007. " Manuale di riconoscimento e valutazione anomalie", "Manuale per la gestione dei dati ispettivi", "Manuale per le ispezioni alle opere d'arte". Sineco S.p.A.

Bridge Maintenance, Safety, Management, Resilience and Sustainability – Biondini & Frangopol (Eds)
© 2012 Taylor & Francis Group, London, ISBN 978-0-415-62124-3

Strength of corroded steel structure bonded with steel cover plate

M. Fukuda & K. Fujii
Graduate School of Engineering of Hiroshima University, Hiroshima, Japan

T. Nakayama
West Japan Railway Company, Osaka, Japan

H. Horii
Konishi Co., Ltd, Osaka, Japan

ABSTRACT

This paper presents a repair method to bond the steel plates to the corroded structure with adhesive. The effects of this method are investigated experimentally through the loading tests; 1) tensile test of corroded plates and 2) bending tests of plate girders with corroded flange.

Firstly, we made six specimens whose both surfaces are dug by ball-end-mill. Axial tensile test specimens had a length of 1000 mm and a width of 100 mm, and the corroded surface was dug on the plate with 400 mm × 100 mm area near the center. And, Table 1 shows the size of the cover plate and the maximum load.

The load-elongation curves obtained from tests are shown in Figure 1. After exfoliation of cover plates, the axial strength of Type B-1~B-4 decreased. Thereafter, it can be noticed that the shape of their curves after exfoliation is almost the same as that of Type B-0. Also as shown in Table 1 and Figure 1, the thicker cover plate does not necessarily improve the deteriorated strength by the reason of larger stress concentration at the edge of it. If engineer expects to use thicker cover plate effectively, remarkable the bond length shall be necessary.

Table 1. The size of the cover plate and results.

	Thickness	Length	Max. Load
A	–	–	485.8
B-0	–	–	276.7
B-1	3.1	500	352.2
B-2	3.1	200	276.1
B-3	2.3	500	379.2
B-4	2.3	200	285.5

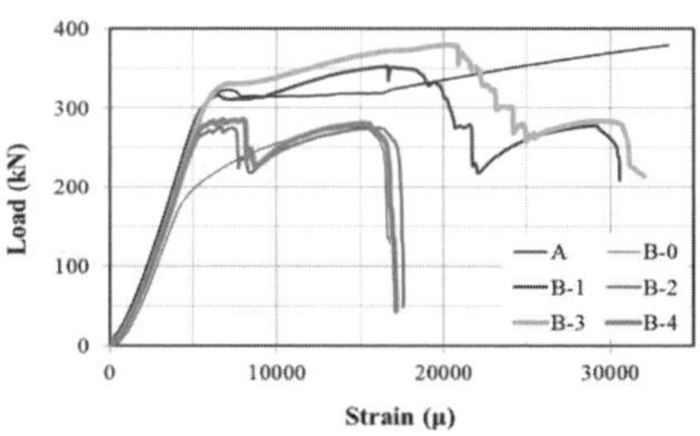

Figure 1. Load-elongation curve.

Table 2. The size of the cover plate and results.

	Corrosion length	C.P. length	Max. Load M_u (kN · m)
G-0	–	–	213.8
G-1	200	–	167.7
G-1c	200	500	209.6
G-2	600	–	158.4
G-2c	600	900	180.2

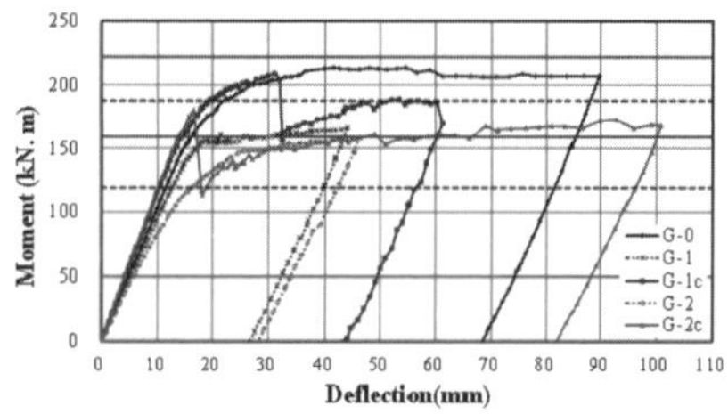

Figure 2. Load-deflection curve.

Secondly, we made five specimens by using the I-girders and the steel cover plates. For the steel cover plate, its width and thickness are 150 mm and 18.6 mm, respectively. All girders have the same dimensions, but the corroded and reinforced conditions are different for each girder (Table 2). The corroded flange is located in pure bending region.

The specimen of G-0 which has not corrosion didn't appear buckling, but the load of the specimens: G-1c and G-2c declined due to debonding adhesive together with the big sound. The ultimate strength of these girders was decided by debonding adhesive. In the tests of plate girder subjected to bending moment, the cover plate can improve its flexural rigidity and diminished strength due to corrosion, if the cover plate is bonded firmly (Figure 2).

REFERENCE

Morishita, T. and Fujii, K. 2009. On Rehabilitation of Corroded Steel Plates by Using Adhesive, *Proc. of 10th Japan-Korea Joint Symposium on Steel Bridge*, Vol.10, CD-ROM.

Bridge Maintenance, Safety, Management, Resilience and Sustainability – Biondini & Frangopol (Eds)
 ISBN 978-0-415-62124-3

Approach for the life-cycle management of structures including durability analysis and maintenance planning

P. Furtner & R. Veit-Egerer
Vienna Consulting Engineers, Vienna, Austria

ABSTRACT

The paper describes a case study elaborated for the Austrian Federal Highway Company ASFINAG in 2011. A 25 km long highway section (see Figure 1) comprising 102 structures on the Austrian S6 highway was analysed in order to derive a maintenance concept for the upcoming 30 years.

This maintenance concept was intended to give a long-term outlook for maintenance measures (heavy and routine maintenance) during the upcoming service life of the analysed structures.

To derive tailored maintenance plans a life cycle model was developed which utilizes state-of-the-art information from literature as well as VCE's experience from performing monitoring and bridge inspections worldwide. The multi-level procedure is based on the entire lifecycle of a structure and considers all available information gained from visual inspections, the applied design concept, field measurements and the relevant loading exposure.

Probabilistic methods are used for the service life calculations of the whole structure as well as for individual items delivering lower and upper bounds of life expectancy.

The calculated lifetime prognosis represents estimations at the time of investigation. This means it is necessary to update the incorporated ageing curves periodically, using the latest knowledge from on-site inspections and assessment in further succession.

Based on the results of the life cycle analysis maintenance instructions were elaborated for every structure and structural member in order to ensure the demanded structural service life and operability.

In times of reduced maintenance budgets cost optimization is an important issue in the field of infrastructure management. In the present case study the cost and availability optimization considering the existing pavement management concept was one of the key demands. Furthermore the case study includes proper maintenance concepts tailored to different budget scenarios given by the client. The result can be used as a basis for decision making in the long run. As the analysed structures consist of bridges, culverts, flyovers, access ramps, gantries and tunnels (see Table 1) cross asset harmonization was another task to be considered.

Figure 1. The S6 highway and the given analysis section.

Table 1. Analysed structures.

Structural Type	Amount
Bridges	76
Tunnels	8
Gantries	18
Total	102

In a final step a parameter study was conducted comparing different asset specifications in order to detect possible correlations between certain structural characteristics and life expectancy on the one hand and life cycle costs on the other hand.

REFERENCES

Forschungsgesellschaft für das Verkehrs- und Straßenwesen 1995. *RVS 13.03.11 Überwachung, Kontrolle und Überprüfung von Kunstbauten – Straßenbrücken*. Wien.

Veit-Egerer R. & Widmann M. 2010. *Life Cycle Methodology & Durability Analysis with regard to relevant heavy-maintenance instructions*. VCE-WD 220; IRIS (Integrated European Industrial Risk Reduction System – IP: SP3 & 7: Workers Safety & Online Monitoring), European Commission FP7, Project No.: FP7-NMP-2007-LARGE-1.

Wenzel H., Veit-Egerer R., Widmann M. 2011. Life Cycle Analysis addressing Maintenance and Repair Options and its affections on Remaining Lifetime. In "*Proceedings of the ACI Convention Fall 2011 – Bridging Theory and Practice*", Cincinnati: ACI.

Bridge Maintenance, Safety, Management, Resilience and Sustainability – Biondini & Frangopol (Eds)

Low cost dynamic structural identification system for extensive bridge monitoring

J. García-Palacios, A. Araujo, O. Nieto-Taladriz & A. Samartín
Technical University of Madrid, Madrid, Spain

E. Reynders & G. de Roeck
Department of civil Engineering K.U. Leuven Belgium

1 INTRODUCTION

A complete system for dynamic structural identification test under ambient loads is presented. The main objective of these improvements is to achieve significant cost reductions. In this way the system can be suitable for application to a large number of structures within the framework of a comprehensive optimal maintenance program of structures. To achieve this cost reduction of the tests, the following improvements have been introduced: (a) a complete wireless system to acquire sensor data, (b) a wireless localization system that permits referring sensors to a local coordinate system. This localization system has been the object of recent patent, and (c) finally, further developments of software aimed at diminishing human intervention in the structural identification process.

In case of standard bridge monitoring the application of the above improvements can achieve the generation of an automatic identification report. This goal is to be attained eventually, and to which the described developments contribute. Therefore these improvements can reduce engineer's intervention mainly to propose general guidelines for a suitable sensors placement.

ACKNOWLEDGEMENTS

The authors acknowledge financial support from the Spanish Ministry of Economy and Competitiveness (project reference P8/08) within the National Plan for Scientific Research, Development and Technological Innovation 2008–2011.

Bridge Maintenance, Safety, Management, Resilience and Sustainability – Biondini & Frangopol (Eds)
© 2012 Taylor & Francis Group, London, ISBN 978-0-415-62124-3

Blow-up oscillating solutions to some nonlinear fourth order differential equations describing oscillations of suspension bridges

F. Gazzola & R. Pavani
Dipartimento di Matematica, Politecnico di Milano, Milano, Italy

ABSTRACT

Under suitable boundary and initial conditions, the following nonlinear beam equation was proposed by Lazer & McKenna (1990) as a model for a suspension bridge, when $t > 0$

$$u_{tt} + u_{xxxx} + \gamma u^+ = W(x,t) \quad x\epsilon(0,L), \tag{1}$$

where $L > 0$ denotes the length of the bridge, $u^+ = max\{u,0\}$, γu^+ represents the force due to the cables which are considered as a spring with a one-sided restoring force (equal to γu if u is downward positive and to 0 if u is upward negative), and W represents the forcing term acting on the bridge (including its own weight per unit length and the wind or other external sources). The solution u represents the vertical displacement when the beam is bending. Normalizing (1) by putting $\gamma = 1$ and $W \equiv 1$, and seeking traveling waves $u(x,t) = 1 + w(x - ct)$ to (1) gives the equation

$$u''''(s) + ku''(s) + [w(s) + 1]^+ - 1 = 0 \quad (s \in \mathbb{R}, k = c^2) \tag{2}$$

We wish to suggest here a variant of this model and to consider the more general equation

$$w''''(s) + kw''(s) + f\big(w(s)\big) = 0, \tag{3}$$

where $s \in \mathbb{R}$, $k \in \mathbb{R}$, and f is a locally Lipschitz function satisfying the sign condition

$$f(t)t > 0 \quad \text{for all } t \in \mathbb{R}\backslash\{0\}.$$

This assumption reflects the fact that the nonlinearity has the same sign as the vertical displacement w of the beam. Since the parameter k equals the squared velocity of the traveling wave, one usually assumes that $k > 0$. Similar equations with a nonlocal term and with $k < 0$ are considered by Como-Del Ferraro-Grimaldi (2005). Our main theoretical result considers a particular situation.

Fix an integer $n \geq 5$, let $k = n^2 - 4n + 8/2 < 0$ and let

$$f(t) = \left(\frac{n(n-4)}{4}\right)^2 t + |t|^{8/(n-4)}t \tag{4}$$

so that f is superlinear. In this case, we prove there exists a solution $w = w(s)$ to the equation

$$w''''(s) - \frac{n^2-4n+8}{2} w''(s) + \left(\frac{n(n-4)}{4}\right)^2 w(s) + \\ +|w(s)|^{\frac{8}{n-4}} w(s) = 0 \tag{5}$$

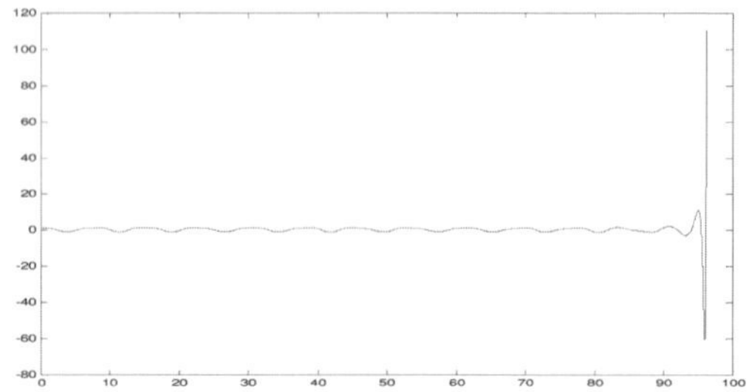

Figure 1. The computed solution to (3), with $f(t) = t + t^3$, $k = 3.6$, $[w(0), w'(0), w''(0), w'''(0)] = [0.9, 0, 0, 0]$

which is defined in a neighborhood of $s = -\infty$ and such that $\lim_{s\to R} \inf w(s) = -\infty$, $\lim_{s\to R} \sup w(s) = +\infty$, for some finite $R \in \mathbb{R}$. Note that if $n = 8$, equation (5) simply becomes

$$w''''(s) - 20w''(s) + 64w(s) + w(s)^3 = 0. \tag{6}$$

This result proves that fourth order equations such as (3) may exhibit finite time blow up with wide oscillations. The proof of this result is based on three main ingredients: a Liouville-type nonexistence result for critical growth biharmonic partial differential equations, the radial version of this pde, and a suitable change of variables in the corresponding ode. For more general nonlinearities, our numerical results display solutions behaving as in Figure 1. As our theoretical and numerical results seem to suggest, traveling waves with superlinear nonlinearities f blow up in finite time after wide oscillations.

Is this the explanation of the Tacoma collapse?

REFERENCES

M. Como, S. Del Ferraro, A. Grimaldi, 2005. A parametric analysis of the flutter instability for long span suspension bridges, *Wind and Structures*, 8, 1–12

P. Drábek, G. Holubová, A. Matas, P. Necesal, 2003. Nonlinear models of suspension bridges: discussion of the results, *Appl. Math.*, 48, 497–514

A.C. Lazer, P.J. McKenna, 1990. Large-amplitude periodic oscillations in suspension bridges: some new connections with nonlinear analysis, *SIAM Rev.*, 32, 537–578

P.J. McKenna, 2006. Large-amplitude periodic oscillations in simple and complex mechanical systems: outgrowths from nonlinear analysis, *Milan J. Math.*, 74, 79–115

Bridge Maintenance, Safety, Management, Resilience and Sustainability – Biondini & Frangopol (Eds)
 ISBN 978-0-415-62124-3

Spectral analysis of dynamic response of footbridges to random pedestrian loads

M. Gladysz & W. Zielichowski-Haber
Wroclaw University of Technology, Wroclaw, Poland

ABSTRACT

Contemporary footbridges are usually very sensitive to the dynamic loads due to limited stiffness of their structure and low mass. In such case the dynamic moving loads induced by pedestrians are crucial for safety design of the structures and for structural vibration parameters important for the comfort of the users. In this paper the random train of moving forces has been assumed to be a Poisson process of moving load and consequently vibrations of a structure subjected to this kind of load became stochastic process, Lin (1995), Sniady (1989). An analytical technique is developed to determine spectral density function of structure response. Rectangular (Figure 1) and sinusoidal (Figure 2) shape functions are used for modeling of the loads during pedestrian walk and run, respectively.

According to the Poisson process, the appearance of impulses in time is characterized by its intensity – parameter of Poisson process λ [1/s] describes the number of footstep impulses in time unit. The second important parameter is T_o which means the time duration of every impulse at the considered point. The third parameter of every impulse is its magnitude A_k.

Due to the fact that the impulses are random, they have to be characterized by the probabilistic characteristics i.e. the expected value $E[A_k]$ and the covariance $Cov[A_kA_k]$.

Proposed procedures confirm usefulness of the spectral analysis method for the analysis of stationary

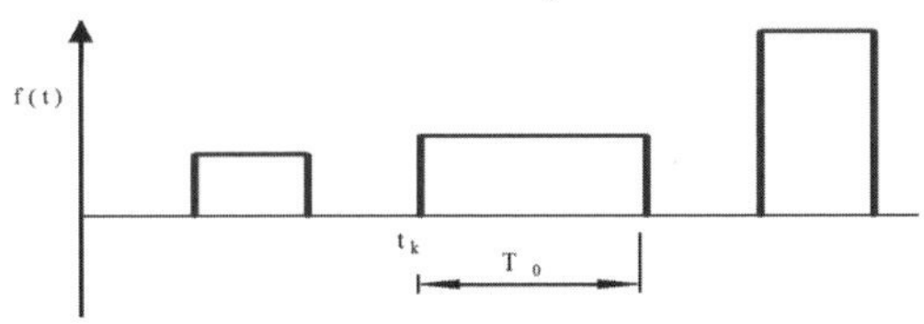

Figure 1. Shape function of the rectangular pulses.

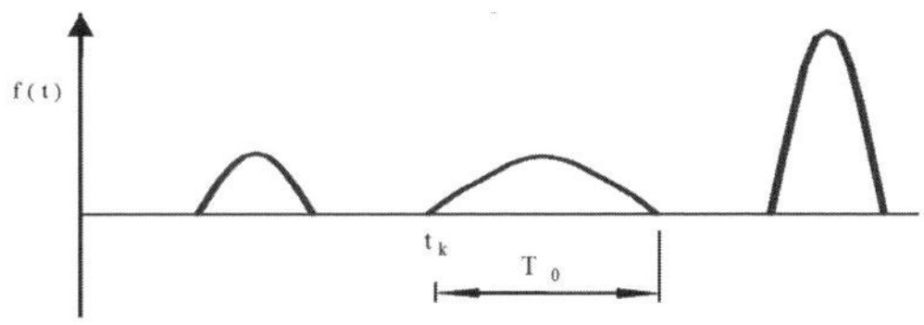

Figure 2. Shape function of the sinusoidal pulses.

response of a linear dynamic footbridge system. Dynamic response of linear system under random train of pulses driven by a Poisson process constitutes a filtered Poisson process. In this case the spectral analysis of the response of footbridge system cannot be obtained using the complex frequency response function. To overcome this difficulty the dynamic influence function and the frequency influence function have been introduced. Taking advantage of these both functions we can obtain a simple algebraic relation between the input and the output spectrum. The relationship describing the spectral density function has the simple form easy to use in the numerical analysis.

In this paper the influence of shape function of human-induced forces with specific parameters for walk and run on the spectral density response function is considered. One can see that in the case of presented load model the footbridge response is a narrow-band process with clear peaks. This type of stochastic response is one of the factors which has significant influence on fatigue in footbridge structures.

REFERENCES

Lin, Y.K. & Cai, G.Q. 1995. *Probabilistic Structural Dynamics. Advanced Theory and Applications*. New York: McGrow-Hill.

Sniady, P. 1989. Dynamic response of linear structures to a random stream of pulses. *Journal of Sound and Vibration* 131: 91–102.

Bridge Maintenance, Safety, Management, Resilience and Sustainability – Biondini & Frangopol (Eds)
 ISBN 978-0-415-62124-3

Bridge condition assessment based on long-term monitoring data and finite element model updating

H.C. Gomez & Maria Q. Feng
Department of Civil & Environmental Engineering, University of California, Irvine, USA

ABSTRACT

This paper presents a simple approach for the identification of structural parameters and condition assessment of highway bridges based on long-term vibration monitoring data and Finite Element (FE) model updating. In the proposed approach the purpose is to identify changes in the structural physical parameters i.e. Young's modulus and boundary stiffness to assess the condition of bridge structures. The physical parameters of the FE model are identified by means of FE model updating. The updating is based on the optimization of the difference between measured and calculated bridge modal parameters and, if available, seismic response. The proposed approach was applied to three existing prestressed bridge structures, Figure 1.

During a period of eight years, approximately 3,500 ambient vibration and seismic records were collected and processed using a variety of signal processing techniques. Disregarding seismic events, the analyses of the acceleration data reveals a continuous reduction, of approximately 5–8%, in the first three natural frequencies over the period of study (see Table 1).

Linear finite element models were generated using OpenSees® software and used for model updating. After FE model updating it was found the value assigned to the Young's modulus of deck (E_d) and columns (E_c) decreased approximately 9% during a period of five years (see Table 1). The reduction in frequencies and stiffness is attributed to aging factors such as creep, shrinkage and relaxation of tendons. In the case of seismic records it was found the reduction in stiffness is a function of the peak ground acceleration (PGA) in transverse direction. This reduction is in accordance to the formation of a cracked section and it justifies the use of an effective moment of inertia for the columns in design practice (Caltrans 2006; ACI 2008).

The updated FE models were enhanced by providing nonlinear attributes to the column elements and they were used to generate fragility curves in order to assess bridge damageability as a function of ground motion intensity, e.g. PGA. In general, as a bridge gets older, its vulnerability increases due to the aforementioned aging factors.

Figure 1. Three Instrumented Bridges.

Table 1. Approximate percentage change of identified bridge parameters for a 5 year monitoring period.

Parameter	3-Span straight	3-span curved	4-span
*f*1	−2.25	−4.91	−2.85
*f*2	0.00	−3.43	−2.82
*f*3	0.00	−3.58	−2.10
Ed	−9.40	−9.00	0.00
Ec	9.40	−9.00	−9.40

REFERENCES

ACI 318-08 (2008). Building Code Requirements for Structural Concrete and Commentary. *American Concrete Institute*, Farmington Hills, MI.

Caltrans amendments to AASHTO LRFD Bridge Design Specifications (2008). 4th edition. *California Department of Transportation (Caltrans)*, Sacramento, CA.

Bridge Maintenance, Safety, Management, Resilience and Sustainability – Biondini & Frangopol (Eds)
 ISBN 978-0-415-62124-3

Innovative numerical modeling to investigate local scouring problems induced by fluvial structures

R. Guandalini & G. Agate
Environment and Sustainable Development Dept. – RSE SpA, Milan, Italy

S. Manenti, S. Sibilla & M. Gallati
Hydraulics and Environmental Eng. Dept., University of Pavia, Pavia, Italy

ABSTRACT

The local scouring problems induced by fluvial structures have been traditionally investigated by means of scaled physical models that allow to compare on a quality level different design solutions and to search for the better configuration from the scouring point of view. In addition to this approach, also investigations and forecasting based on numerical tools have been successfully recently considered for this type of problems. The numerical models allow to analyze scenarios without domain scaling and to perform a number of sensitivity analyses of different layouts at very low costs with respect to physical ones, also supporting the design of a corresponding experimental facility for validation purposes. In the frame of the Italian Electric System Research Projects a work devoted to develop innovative numerical models, based both on a classic CFD and on a new SPH approaches, has been carried out with the goal of improving the accuracy of the results and to support very detailed analyses.

The CFD model is based on fully 3D finite volume hydrodynamic module, accounting for the turbulence effects, coupled with a solid transport module using a moving bed technique, while the innovative SPH model is based on a meshless smoothed particle hydrodynamics for describing the coupled dynamics of two different materials (the water and the sediment) and their interactions with the solid structures.

The paper shows the application of the two models to simulate the local scouring effects induced by a barrage, comparing the accuracy of the results with experimental measurements and discussing advantages and limits of each model. The Figure 1 shows a qualitative comparison from upstream of the depressed areas resulting from CFD model and physical model. In Figure 2, the results of the 2D SPH simulation is shown. Preliminary 3D results are also discussed in the paper.

The results show that both the CFD model and SPH innovative model can be applied successfully to investigate local scouring problems when fixed structures are involved and they can be used in conjunction with

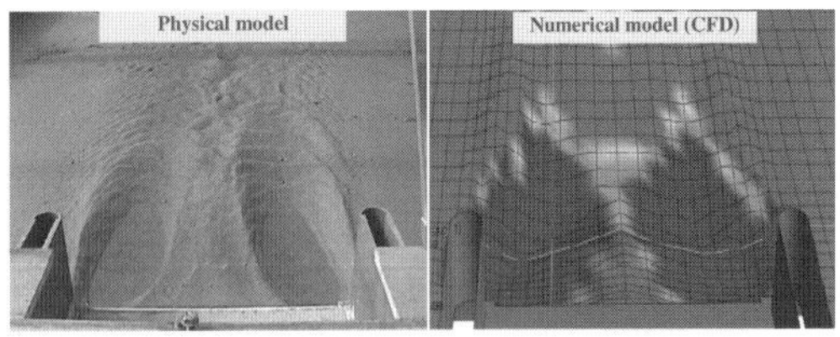

Figure 1. Comparison between CFD model and physical model.

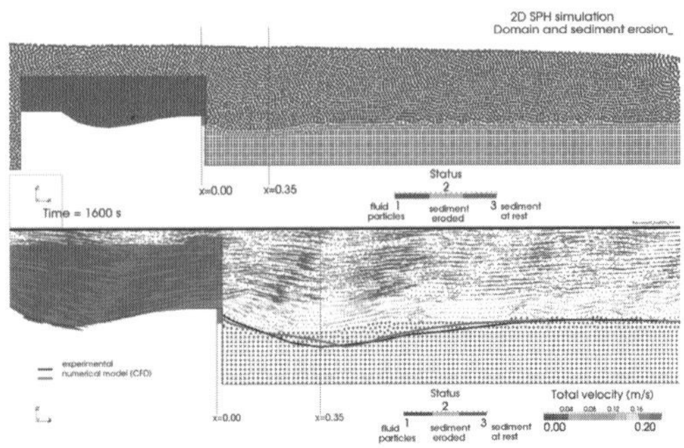

Figure 2. 2D results from SPH numerical model.

the physical models to investigate safety aspect and to support optimization design of fluvial structures.

REFERENCES

Maffio A., Agate G. 2005. Modelli matematici per la valutazione del fenomeno dell'erosione localizzata in alvei fluviali, *Rapporto di Ricerca di Sistema* (secondo periodo) prot. A5055694, (progetto SISET-SITER).

Manenti, S., Sibilla, S., Gallati, M., Agate, G., Guandalini, R., 2010. Prediction of sediment scouring through SPH. *Proc. 5th Int. Spheric Workshop*, Manchester (Uk), pp. 56–60

Manenti, S., Sibilla, S., Gallati, M., Agate, G., Guandalini, R., 2011. SPH Simulation of Sediment Flushing Induced by a Rapid Water Flow. *Journal of Hydraulic Engineering ASCE* (in press).

Monaghan, J.J. (1992) Smoothed particle hydrodynamics, *Ann. Rev. Astron. Astrophys.* Vol. 30, 543–574

Bridge Maintenance, Safety, Management, Resilience and Sustainability – Biondini & Frangopol (Eds)
© 2012 Taylor & Francis Group, London, ISBN 978-0-415-62124-3

3D Numerical simulation of soil-structure interaction effect: The Acquasanta, Genoa, railway bridge

R. Guidotti & I. Mazzieri
Politecnico di Milano, Milan, Italy

M. Stupazzini
Munich RE, Munich, Germany

ABSTRACT

On the Genoa-Ovada railway, North-West Italy, is located the Acquasanta bridge (Fig. 1), in the Genoa district. The bridge represents a typical structure the ancestor of which can be dated back to the Roman Age and, thanks to the excellent design achieved no less than two thousand years ago, this structural type did not change significantly along the centuries.

Referring to this challenging case study, remarkable both for the site features and the local geological and geo-morphological conditions, the soil-structure interaction problem is addressed by three-dimensional numerical simulation, based on the Spectral Element formulation (Faccioli et al. 1997, Stupazzini 2004, Stupazzini et al. 2009). The wave propagation problem is handled in its 3D complexity and in this way it is possible to examine several features that play a major role in seismic hazard assessment studies, usually analysed under restrictive and simplified assumptions: (i) soil-structure interaction; (ii) topographic amplification; (iii) soft soil amplification caused by the superficial alluvium deposit; (iv) complex tectonic and geological structures; (v) deterministic evaluation of the spatial variability of ground motion. The comparison between displacement time-histories of the structure considering or disregarding the soil-structure interaction effect is provided, pointing out that the three-dimensional site effects contribution to the dynamic response of the structure is significant and cannot be neglected.

Relying on non-conforming techniques (Antonietti et al. 2012), new numerical advancements in meshing strategy are introduced. Through independently generated meshes and different spectral degrees in different sub-domains, it is possible to refine the mesh where it is necessary, i.e. in the structural model of the bridge, without the generation of a huge number of transition elements. This allows us to tackle efficiently the typically multi-scale wave propagation problem, from the far-field to the near-field, and from the near-field to the soil-structure interaction effect, reducing the overall computational burden of the analysis.

Figure 1. The Acquasanta (Genoa) railway bridge and its three-dimensional numerical modellization.

REFERENCES

Antonietti, P., Mazzieri, I., Quarteroni, A. & Rapetti, F. 2012. Non-conforming high order approximations of the elastodynamics equation. *Computer Methods in Applied Mechanics and Engineering*, 209–212: 212–238.

Faccioli, E., Maggio, F., Paolucci, R. & Quarteroni, A. 1997. 2D and 3D elastic wave propagation by a pseudospectral domain decomposition method. *Journal of Seismology* 1(3): 237–251.

Stupazzini, M. 2004. A spectral element approach for 3D dynamic soil-structure interaction problems, *Ph.D. Thesis*, Politecnico di Milano, Department of Structural Engineering, Milan, Italy.

Stupazzini, M., Paolucci, R. & Igel, H. 2009. Near-fault earthquake ground-motion simulation in Grenoble Valley by high-performance spectral element code. *Bulletin of the Seismological Society of America*, 99(1): 286–301.

Bridge Maintenance, Safety, Management, Resilience and Sustainability – Biondini & Frangopol (Eds)
© 2012 Taylor & Francis Group, London, ISBN 978-0-415-62124-3

Fatigue reliability analysis of steel bridge details based on field-monitored data and linear elastic fracture mechanics

T. Guo & Y.W. Chen
Key Laboratory of Concrete and Prestressed Concrete Structure, Ministry of Education, Southeast University, Nanjing, P.R. China

ABSTRACT

Fatigue and fracture are among the most critical forms of damage that occur to metal structures subjected to cyclic loads. Such damage often accumulates in an invisible way and may eventually trigger a sudden fracture in critical components of the structure. Therefore, fatigue behaviours of steel structures have received extensive attention for decades. In fatigue analyses, one crucial step is to determine the fatigue stress spectra. The traditional way of obtaining the fatigue stress spectra on bridges is based upon the stress analysis with a traffic load model and a structural model. The recent development of structural health monitoring (SHM) systems enables authentic long-term stress spectra can be obtained, which provides a reliable and efficient tool for the fatigue evaluation on steel bridges.

To consider the uncertainties that abound in fatigue assessments and the SHM, a fatigue reliability approach is proposed in this paper, based on the linear elastic fracture mechanics (LEFM) and the long-term stress monitoring. Application is made in the fatigue reliability analysis of welded details of a forty-year old steel box-girder bridge. To be more specific, fatigue crack growth is described by using the LEFM, based on which the ultimate limit state is established as a function of the crack size. Using the daily stress range histograms, the effective stress range S_{re} is calculated, and it is observed that the derived S_{re} during the measurement period can be represented by the lognormal PDFs. The mean value and the standard deviation of $\ln S_{re}$ are then used in reliability analyses to consider the uncertainty of stress ranges.

Details at discontinuous backing bar splices, which have not been explicitly listed in the AASHTO specification are evaluated on their fatigue reliabilities. It is found that the backing bar splice detail may have high fatigue reliability,as shown in Figure 1, considering that low stress ranges and small number of cycles occurred in the monitoring period.

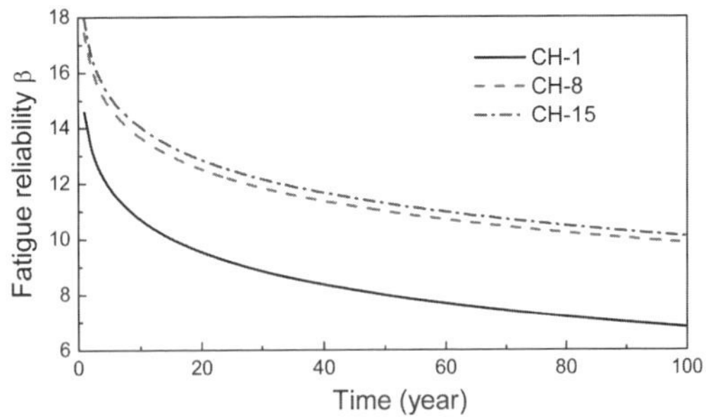

Figure 1. Fatigue reliability evaluation of backing bar splice detail.

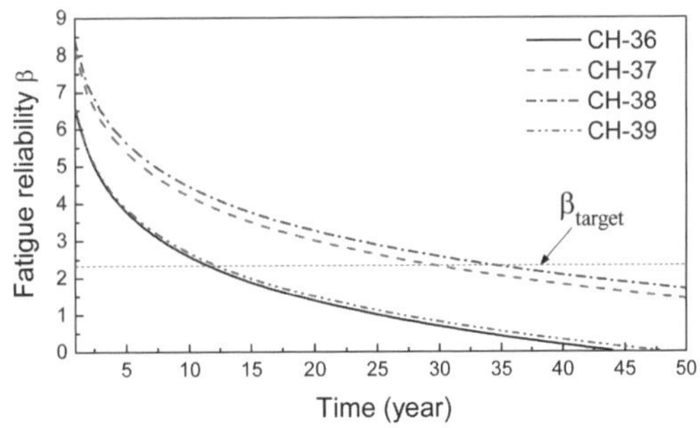

Figure 2. Time-dependent fatigue reliabilities of the details at the base of Bent NB13S.

For the weld at the base of Bent NB13S, the details have high fatigue reliabilities ranging from 6.5 to 8.5 at the beginning of service, as shown in Figure 2, while a rapid decrease in the reliability indexes occurs as time goes by. Currently, the reliability indexes of these details are all below the target value, so that maintenance or retrofitting actions may be necessary. For the details at the upper base of Bent SB13S, high reliability indexes are expected during 100 years of service.

Bridge Maintenance, Safety, Management, Resilience and Sustainability – Biondini & Frangopol (Eds)
© 2012 Taylor & Francis Group, London, ISBN 978-0-415-62124-3

Prediction of service performance for RC bridge by considering the coupling effect of load-environment in service cycle

T. Guobin
Henan Transportation Research Institute, Zhengzhou, Henan Province, China

X. Yiqiang & W. Qiangqiang
Department of Civil Engineering, Zhejiang University, Hangzhou, Zhejiang Province, China

ABSTRACT

Since the concrete bridges are subjected to the effect of various load acting on the bridges and environment factors surrounding them in service cycle of the structures, the defects such as reinforcement bar corrosion, concrete carbonation, splitting and cracking appear in these structures, which will result in deterioration of service performance in the service bridges. In general, these defects or deterioration phenomena always and simultaneously occur in the service bridges. However, for most concrete bridge, it's difficult to predict long term performance under load and environment effect by experimental tests, because of complex structure and boundary conditions. A more useful assessment tool is numerical method. With a reasonable physical model for the corroded reinforcing steel in concrete, coupled with concrete nonlinear theory, the long term performance of bridge is able to work out based on a numerical model.

In this study, idealized model is used for studying corrosion of reinforcing steel under a sustaining load. Then the corrosion of reinforcing steel under a sustaining load can be reflected by the changing of elastic module. The nominal elastic module is adopted to indicate the corrosion propagation of reinforcing steel as the Eq. (1).

$$E(t) = E_0\left[1-\eta(t)\right] \tag{1}$$

where the η(t) is the rate of area loss of the steel bar η(t) at time t. Then the strain of steel bar can be subdivided into two parts: one due to instant strain ε_{in}, and one due to corrosion of steel ε_{co}. Then the corrosion factor $\varphi(t)$ is defined which indicates the ratio between corrosion strain and instant strain. Using corrosion factor the time-depended constitutive relationship of corroded steel bar can be constructed.

Based on the time-depended constitutive relationship, the finite element equilibrium of reinforced concrete can be induced at any time during the service life. When the embedded model of reinforcement in concrete are used in 3D numerical simulation, at any instant, the required equilibrium equation to be satisfied is given by

$$\boldsymbol{K}^{\tan}\Delta\boldsymbol{U} = \Delta\boldsymbol{R} + \Delta\boldsymbol{R}^{\text{co}} \tag{2}$$

where $K^{\tan}$ is the updated tangent stiffness in the load step t_{n-1}, which may be obtained by the displacement in the last step and the corrosion status of reinforcement. ΔR is the load increment of step due to applied load, ΔR^{co} is the load increment of step due to corrosion of reinforcement. In the nonlinear algorithm they are defined by

$$\boldsymbol{K}^{\tan} = \iiint_\Omega \boldsymbol{B}^{\mathrm{T}} D_{\mathrm{c}}^{\tan} \boldsymbol{B}\mathrm{d}\Omega + \int_l \boldsymbol{B}^{\mathrm{T}}\boldsymbol{T}^{\mathrm{T}} E(\tau_0)\boldsymbol{T}\boldsymbol{B}\mathrm{d}l \tag{3}$$

$$\Delta\boldsymbol{R}^{\text{co}} = \int_l \boldsymbol{B}^{\mathrm{T}}\boldsymbol{T}^{\mathrm{T}} E(\tau_0)\Delta\varepsilon_{\mathrm{s}}^{\text{co}}\mathrm{d}l \tag{4}$$

where B is the strain-displacement matrix of 3D element, T is the transformation matrix. Based on the above equilibrium, the long-term performance of reinforced concrete beam can be predicted incorporated with the elastic-plastic theory for concrete.

Finally, a numerical example is given and analyzed. The results approve the applicability and validity of the presented theory and procedure, which can be used in the prediction of service performance for concrete bridge.

REFERENCE

Mehta, P.K., Schiessl, P. and Raupach, M.(1992). "Performance and Durability of Concrete Systems." Proceedings of the 9th International Congress on the Chemistry of Cement, New Delhi.

Bridge Maintenance, Safety, Management, Resilience and Sustainability – Biondini & Frangopol (Eds)
© 2012 Taylor & Francis Group, London, ISBN 978-0-415-62124-3

An in-depth analysis of I35W Bridge collapse

S. Hao
ACII, INC. Wilmette, IL, USA

ABSTRACT

This article briefs the analysis and major conclusions obtained in an independent investigation of the highway I35W Bridge's collapse, discussing the lessens learned in the following three respects: (i) structural design, (ii) issues for maintances and load-rating, (iii) health monitoring.

August 1st of 2007, 6:05 PM, the 1000-foot long, 8-lane wide, three-span steel-truss deck of the interstate highway I35W Bridge in the City Minneapolis, Minnesota, collapsed suddenly, resulted in 13 fatalities and 145 injuries; see the figure below.

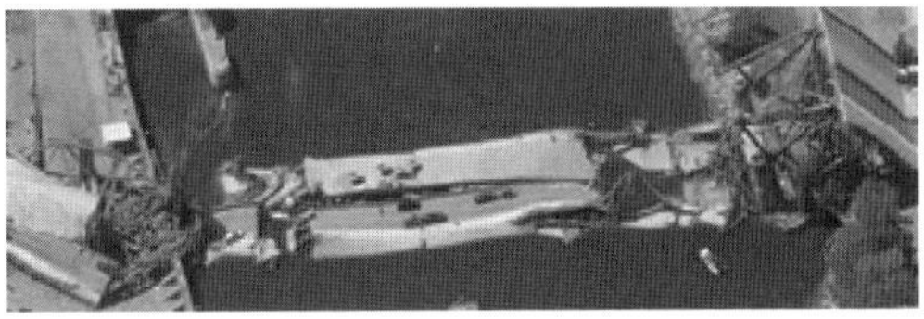

Figure 1. Minneapolis I35W Bridge, August 1st, 2007.

Designed in 1964 and opened to traffic at 1967, the I35W Bridge stretches from north to south crossing the Mississippi River, with a main unit of three-span steel deck-truss directly over the river and other eleven steel-girder concrete-slab approach-spans. In 2004 the measured daily-average traffic volume over the bridge was 141,000/day, made it among the list of the busiest bridges in United States.

After the collapse, National Transportation Safety Board (NTSB) investigation team collected the bridge's wreckages from the river's bed, lined up on an opening-ground for examination. The investigation has identified the undersized gusset plates U10 and L11, concluded that the failure of U10 triggered the collapse of the entire bridge (NTSB 2008).

Based on the information released to public and the materials' evidences disclosed by NTSB, this independent investigation has evaluated all original design-drawings and conducted a completed load-ratings of the bridge under different conditions and limit states through a series of fully-scaled computations and detailed analysis of key structural components. The results were organized into four reports submitted to the related government agencies (Hao, 2007,2008). The major conclusions are summaried as follows:

(1) The load path of the shallow arch deck-truss structured bridge determines the gusset plates with the distance to a nearby pier from 1/6 to 1/3 of a span length, for example, the U10 gusset in I35W, are the "pivots" that transfer live load and superstructure's weight to the pier.
(2) This load path results in high amplitude force flows through diagonal members and gusset plates connected. By contrast, in conventional one-dimensional influence line model, the solved bending moment around this location is zero or very small. This coincidence may mislead to an underestimate of the stress level for the structural components within this area in past.
(3) A consistency has been found between the I35W bridge's gusset plates' thicknesses, its upper and lower chords' thicknesses, and the bending moment distribution of a classical influence line solution for such a three-span bridge, which reveals that the NTSB disclosed "undersized" gusset plates are the consequences of the one-dimensional model-biased original design that did not give sufficient consideration of the forces from diagonal trusses.
(4) Stress concentration may significantly accelerate localized corrosion-assisted fatigue and damage evolution in both tension and compression structural members; whereas structural instability is a growing risk for those compression-dominated members when additional lateral force presents after deterioration. Bending moment, which causes both additional tension and compression on a gusset plate, should be considered in design and rating.
(5) Bearing condition is crucial for maintaining an aged bridge's safety operation. Corrosion or other deterioration induced roller bearing lock may result in significant thermal stresses.
(6) To avoid a progressive collapse that occurred in I35W Bridge, an additional safety factor in ASD-Allowable Stress Design (or η_R in LRFD) may be necessary for those key single load-path structural members stress concentration, especially, for those aged multi-span bridges.
(7) Effective health-monitoring relies on the understanding of a bridge's load-path and structural details.

REFERENCES

NTSB, 2008, Report NTSB/HAR-08/03, Washington, D. C.
Gee, King W., 2008, FHWA Technical Advisory T 5140.29
Hao, S., 2010, ASCE, J. Bridge Engineering, Sept/Oct, pp. 608–618
Hao, S., 2007–2009, http://www.suhao-acii.com/I35Wa.html

Bridge Maintenance, Safety, Management, Resilience and Sustainability – Biondini & Frangopol (Eds)

Fatigue design of plated structures using structural hot spot stress approach

M. Heshmati & M. Al-Emrani
Chalmers university of technology, Gothenburg, Sweden

ABSTRACT

The nominal stress method has been the most widely used fatigue assessment method in the field of structural engineering for decades. This method is generally referred to as the classic fatigue assessment method. The ease of use for versatile structural details and the acceptable accuracy, compared to the work effort, are some advantages of this approach. However, the implementation of more complicated details in steel structures on the one hand, and the increasing demand for more efficient and accurate design methods on the other hand, has caused new limitations for designers to use the nominal stress method.

The hot spot stress concept was first introduced for the fatigue design of tubular structures decades ago. Over the years, the advantages of this approach compared to the traditional nominal stress method provoked design associations to introduce guidelines and instructions regarding fatigue design of plated structures using hot spot stress as well. By definition, the structural hot spot stress approach (SHSS) designates the basic stress, including stress concentration effects caused by geometrical variations of the detail at the expected fatigue crack initiation area (hot spot). SHSS disregards the notch effect caused by the weld profile and comprises all other geometric parameters. Hence, one hot spot stress S-N curve can be associated to several details.

The results of several three-dimensional finite element analysis have shown that certain instructions regarding the element types and meshing techniques should be followed in order to obtain comparable results.

In this paper, a large number of fatigue test data of frequently used details in plated steel structures are collected and used to produce nominal stress S-N curves. Figure 1 illustrates an example of studied details and Table 1 presents a summary of the test data. The results are discussed and compared with the IIW and Eurocode 1993-1-9 recommendations.

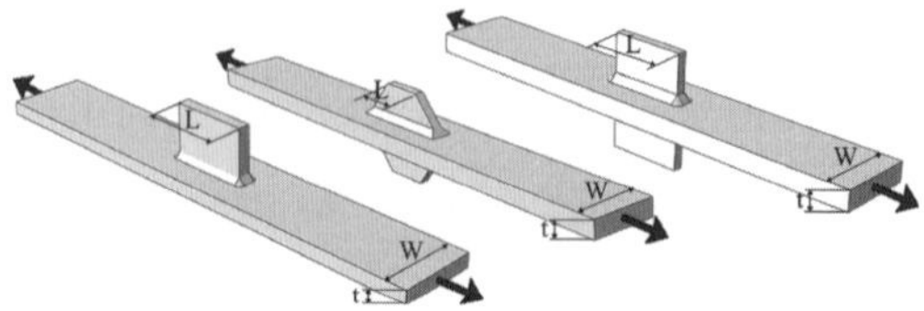

Figure 1. Schematics of investigated longitudinal non-load-carrying attachments.

Table 1. Evaluated fatigue details.

Structural detail	Number of specimens
Longitudinal none load carrying attachments	286
Over-lapped joints with crack at main plate	10
Over-lapped joints with crack at cover plate	9
Beams with cope-holes	29
Beams with cover-plates	183

The FEA instructions given by international welding institute (IIW) are adapted in order to create three-dimensional solid models with fine meshes. Both linear and quadratic extrapolation methods are exploited in order to obtain hot spot stress S-N curves. Eventually, an equivalency between these two approaches with reference to the fatigue strengths of the studied details is established.

REFERENCES

Hobbacher, A. 2008. IIW document IIW-1823-07 ex XIII-2151r4-07/XV-1254r4-07.

European standard, 2005. Eurocode 3: Design of steel structures – Part 1–9: Fatigue, Brussels: European Committee for Standardization.

Bridge Maintenance, Safety, Management, Resilience and Sustainability – Biondini & Frangopol (Eds)
 ISBN 978-0-415-62124-3

Durability of corrosion protecting materials under sleepers of railway steel bridges

H. Higashiyama
Kinki University, Higashiosaka, Osaka, Japan

T. Nakayama & M. Kimura
West Japan Railway Company, Osaka, Japan

T. Sakamoto
Railway Technical Research Institute, Kokubunji, Tokyo, Japan

S. Matsui
Osaka Institute of Technology, Yawata, Kyoto, Japan

ABSTRACT

Corrosion of the upper flange of railway steel girders directly in contact with railway sleepers is a unique deterioration phenomenon. Many open-floor-type railway steel bridges have been constructed in the service area of West Japan Railway Company. Railway sleepers are directly mounted on the upper flange of main steel girders and/or steel stringers. The upper flange under the railway sleepers is easily exposed to wet condition with rain and is thus under corrosive environment. Furthermore, blowing dust containing small particles of sand affects the corrosion because the small particles function as a polisher on railway bridges across a river. Under this condition, the deterioration of a paint coating as well as the corrosion of the upper flange under railway sleepers is accelerated. Visual inspection of a paint coating on an actual railway bridge conducted after one and two years reveals that the paint coating on the upper flange under railway sleepers was partially rubbed and cracked and was locally worn away (Nakayama, et al. 2010). To overcome this problem, durable corrosion protecting materials applied on the upper flange under railway sleepers are needed to insure a longer life of railway steel bridges.

In this study, two test methods have been used to simulate the deterioration of corrosion protecting materials in the laboratory. With the first test method, the repetitive load was applied to railway sleepers by a hydraulic servo jack as the actual railway sleeper system was simulated. However, it was complicated to measure the coating thickness of materials with this test method. Then, the authors developed the new test method, which allows easy handling and an accelerated degrading test, to quickly evaluate the durability of corrosion protecting materials. The testing machine shown in Figure 1 is composed of a hydraulic motor, an eccentric rotary cam and springs. To simulate the severe condition of actual railway bridges, water and silica sand were supplied from both sides of the railway sleeper into the interface between the specimen and the railway sleeper.

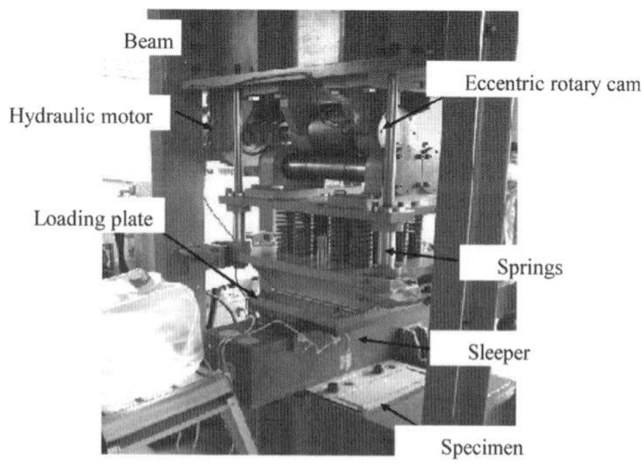

Figure 1. A specially developed loading machine.

In this study, three types of corrosion protecting material were investigated: a paint coating called "B-7" generally employed on railway steel bridges; an epoxy coated glass fiber sheet; and an epoxy coated carbon fiber sheet.

The results of the loading tests conducted by this machine indicate that the characteristic deterioration of B-7 matches the deterioration pattern observed in an actual railway bridge. The epoxy coated glass fiber sheet and the epoxy coated carbon fiber sheet have significantly higher durability than the B-7. When the reduction ratio of the coating thickness of the epoxy coated glass fiber sheet and the epoxy coated carbon fiber sheet increased up to around 33% and 39%, respectively, an exposed part of the steel plate surface was detected in this study. This ratio is determined as the serviceability limit state of the epoxy coated fiber sheets.

REFERENCE

Nakayama, T., et al. 2010. Deterioration Characteristics of Upper Flange Paint Contact with Railway Sleepers. *JSSC Proc. Construction Steel* 18: 633–638 (in Japanese).

Bridge Maintenance, Safety, Management, Resilience and Sustainability – Biondini & Frangopol (Eds)
 ISBN 978-0-415-62124-3

Pre-assessment of existing road bridges – New procedure for a rough but quick estimation of the capacity of existing road bridges

M. Hofmann & B. Kühn
Verheyen-Ingenieure GmbH & Co. KG, Bad Kreuznach, Germany

H. Frießem, B. Winkler & M. Hoffmann
LBM, Landesbetrieb Mobilität Rhineland-Palatinate, Koblenz, Germany

ABSTRACT

In many countries an increasing age of the existing bridge stock can be noticed, on the other hand the traffic impact on these bridges increased, too. Studies predict that road traffic and transport of goods in particular will increase considerably in near future. During the recent 50 years design loads for road bridges had to be raised five times in Germany. Yet another rise is currently discussed.

Is the aging bridge stock able to carry increasing traffic loads?

The German Federal Ministry of Transport, Building and Urban Development (BMVBS) wants this question to be answered by the highway- and bridges authorities of the federal states.

From the engineering point of view this is an interesting question. Several new recommendations and guidelines issued recently will provide detailed information on how to assess existing bridges. So assessment itself should not pose a problem. The big problem is the huge number of bridges.

There are more than 38,423 bridges in federal highways in Germany with an asset value of about 50 billion €. The average age of those bridges is between 30 and 50 years, as shown in figure 1.

Of course, detailed calculation could answer the question of BMVBS, but this would take years with estimated costs of up to 300 million €.

So the idea was to develop a system of simple criteria and short calculations which would enable a quick pre-assessment. In 2010 the LBM, state highway authority of Rhineland-Palatinate (a state of Germany), assisted by an engineering office (Verheyen Engineers), started to develop a new method for a rough but quick pre-assessment:

1. To verify, which of the existing bridges probably will be able to carry the new Eurocode design traffic loads.
2. To give hints on bridges which failed the test, whether a complete re-building or a strengthening of the existing structure should be preferred, from the technical as well as from the economical point of view.

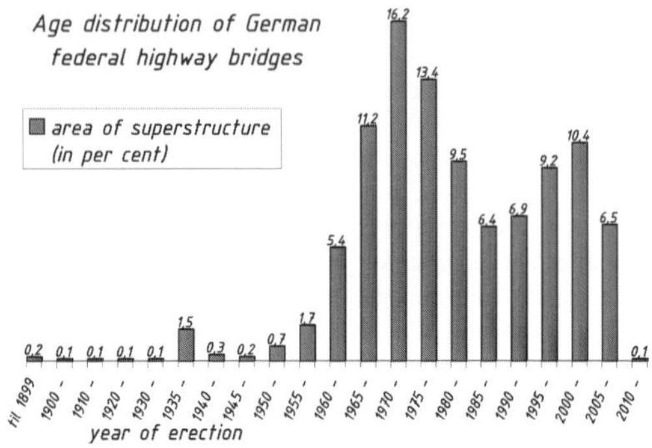

Figure 1. Age distribution of bridges, BMVBS 2011.

As there was no time for inspection of drawings, only the following data were available: Year of erection, type of bridge, spans, width of the road surface, total width, total area of the deck, design load classification.

A system of ranking was developed, starting by considering the sensitivity of the bridges to vehicle loads by use of especially calculated load indices. Afterwards ranking was done by regarding the condition state and the history of German design codes and the effect on special points of weakness.

Information was given on which option (maintenance, strengthening, rebuilding superstructure (and) substructure) seemed to be most cost effective.

In a first step 134 highway bridges were analyzed. The whole process finally led to a table, showing all the 134 bridges sorted after pre-assessment. This procedure was carried out by a single engineering office within only 3 weeks.

The system was developed to suit German conditions and design codes, but can be applied – after modification – in other countries as well.

REFERENCE

Bundesministerium für Verkehr, Bau und Stadtentwicklung, (BMVBS) 2011. *Deutscher Bundestag Drucksache 17/5325*

Bridge Maintenance, Safety, Management, Resilience and Sustainability – Biondini & Frangopol (Eds)
 ISBN 978-0-415-62124-3

Humber bridge A-frame refurbishment/replacement

S.R. Hornby & J.H. Collins
Ove Arup and Partners Limited, Leeds, UK

P.G. Hill & J.R. Cooper
Humber Bridge Board, Hessle, East Riding of Yorkshire, UK

ABSTRACT

Opened in 1981, the Humber Bridge carries the A15 dual carriageway over the Humber estuary between the East Riding of Yorkshire and North Lincolnshire. The main span is 1410 m, the side spans 280 m and 530 m.

The deck box is discontinuous at the towers. At each end of the side span and main span decks are pairs of A-frame bearing rockers. The A-frames are made up of welded steel cruciform sections arranged in an "A" shape. At each apex of the A-frames is a pinned bearing. The A-frames provide vertical and lateral restraint to the deck box. Pairs of A-frames provide deck box torsional restraint. The cruciform cross sections of the A-frames result in low plan torsion resistance, hence providing little restraint to deck end plan rotations. The pins allow free longitudinal movement to the deck box. The original design accommodates approximately ±1 m of longitudinal deck movement sourced from:

- Traffic actions, ±0.7 m. Vehicular passage changes the shape of the main cable, pulling the deck longitudinally underneath it.
- Dynamic (buffeting) wind actions, ±0.2 m.
- Thermal effects (±0.3 m).

In recent years the adequacy of capacity and performance of the main span A-frames has been questioned. The paper discusses the development of refurbishment/replacement options for the A-frames as far as scheme design stage.

To understand the constraints on the refurbishment/replacement exercise, a study of the existing A-frames behaviour and condition was undertaken. This included a review of historic assessments and monitoring, site inspections and setting up a monitoring system using Moiré tell-tales. It was found that:

- The main span A-frame bearing arrangements appear worn with a resulting drop in level of the A-frames.
- The main span longitudinal accumulated sliding distance is 4.5 km/year.
- A typical 400 kN vehicle crossing the bridge results in a deck movement of 20 to 25 mm at a peak velocity of 2 to 3 mm/s.
- The existing A-frames holding down arrangement is heavily loaded.
- Two of the main span A-frames were seized against live load movements.
- A-frame reactions are dominated by the provision of a torsional reaction to the deck box.

Under the original design loads, it was calculated that reversible axial loading on the A-frames is in the region of 8 MN (compression) and 5 MN (tension).

Six feasible refurbishment/replacement options were considered:

1. Replace pins only and reset A-frames.
2. Replace with new like-for-like A-frames.
3. Replace with new A-frames with spherical bearing housed pins.
4. Replace with vertical pendels and wind shoe.
5. Replace with deck ballast, sliding bearings and wind shoe.
6. Replace with "C" framed bearing arrangement.

Three dimensional models using SolidWorks® software were generated for all options to help identify key areas of interest.

The options were assessed against weighted criteria. Option 4 was identified as the most preferable option. Amongst other favourable distinguishing features, awkward and impractical construction methods would be required for the other options.

The inclusion of a longitudinal buffer to limit displacements was considered. However, the reaction generated by restraint of such imposed deformations were calculated as very high. This was attributed to the truss action of the inclined hangers present at the Humber Bridge.

Scheme design has progressed with the pendel and wind shoe option. Construction is forecast to commence in 2012.

Bridge Maintenance, Safety, Management, Resilience and Sustainability – Biondini & Frangopol (Eds)
© 2012 Taylor & Francis Group, London, ISBN 978-0-415-62124-3

The implementation of a bridge management system in Portugal

C.S. Horta & E.C. Lopes
Estradas de Portugal, S.A., Almada, Portugal

ABSTRACT

The Bridge Maintenance Management System (BMMS) of "Estradas de Portugal, SA" (EP) – *the general concessionaire of the Portuguese Road Network* – is a decision support tool whose main purpose is the definition of priorities, mainly based on a cost-benefit analysis of interventions on its bridges. It provides the technical staff involved in the managing of bridges under EP jurisdiction the possibility of taking consistent and structured decisions concerning their preservation and maintenance, aiming at optimizing the available resources. It is in great measure based upon the assessments made by the technical staff involved in the process.

There is currently a combination of a set of activities which are at different development stages: Inventory/History; Principal/ Underwater Inspections; Prediction Model/Decision Support.

The bridge Inventory is composed by registration data, technical data and constitution data. History compiles important data regarding inspections, projects and interventions.

The Principal Inspection aims to evaluate the structural safety of bridges under EP jurisdiction. It is conducted by a civil engineer with the support a helper and it is mainly a visual inspection to all accessible parts of the structure allowing to evaluate it with an indicator designated by "Maintenance Status" that goes from 0 (excellent) to 5 (very poor).

The structural complexity of bridges makes it difficult to the general use of prediction models linked to the deterioration of structures. In order to ensure the necessary following of situations that deserve greater attention, structural monitoring procedures are implemented with regular reporting.

In December 31st 2010, in the system-supplied data, 5724 bridges under EP jurisdiction (or direct interference in their roads) were inventoried. In this total there is a predominance of Hydraulic Crossings (40%).

In 2010, the routine inspection programme to bridges under EP jurisdiction (and bridges with direct

Table 1. Maintenance Status (MS) Distribution.

Maintenance Status	% From Total
MS0	1
MS1	23
MS2	50
MS3	18
MS4	7
MS5	1

interference on EP road network) scheduled for the biennium 09/10 was concluded. It was found that 33% of its components were in poor condition, particularly by lack of cleaning, as well as anomalies regarding the road surface and guard rails. Continuing the strategic option taken in 2008, in mid-2010 EP began to perform Principal Inspections using its internal means. In 2009 was taken the decision to organize the plan for Underwater Inspections in cycles of four years, divided in two separate periods, each one corresponding to 50% of bridges that present immersed foundations.

In 2012 the financial investment is estimated to increase both financially and by number of bridges repaired, trying to avoid greater costs in the future.

Concluding, in the last 10 years of activity, EP bridges MS has come to be acknowledged, which has allowed taking strategic actions consistent with the needs in a prioritized manner and with a gradual rehabilitation of the assets under EP management.

From the experience gained it is now important to plan the future strategic actions proved required, in particular the updating of the computer application that serves as base to the management system, the deepening of the bases for diagnosis by performing tests and to continue with the training of inspectors. The strategic planning also comprises the programming of preventive interventions (3–5 years), gradually reducing the need for reactive interventions (1–2 years), as long as the technical and financial means necessary to achieve these objectives are provided.

Bridge Maintenance, Safety, Management, Resilience and Sustainability – Biondini & Frangopol (Eds)
© 2012 Taylor & Francis Group, London, ISBN 978-0-415-62124-3

Live load factors for serviceability limit state of prestressed concrete girder stresses

E.-S. Hwang, S.-M. Kim & S.H. Nguyen
Kyung Hee University, Yongin, Korea

ABSTRACT

This paper focuses on reliability analysis of stresses in PSC girder at initial stage and under service live load. Reliability indices for ULS of prestressed concrete girders are also calculated to compare with those for SLS. Since the first reliability-based design code is under development now in Korea, reliability indices for current live load model and proposed live load model are calculated and compared. To calculate the stresses in prestressed concrete girder, actual design should be carried out and section dimensions must be determined. In this study, typical sections already designed according to design code (MLTM, 2005) are used in the reliability analysis. However, to propose the relevant live load factors for SLS load combinations, the reliability indices embedded in the code should be calculated and new design is performed to meet the minimum required design criteria (Figure 1 and 2). Reliability indices for minimum sections are compared with target reliability indices of international standards, as shown in Figure 3 and 4. Live load factors related to tension in prestressed concrete girders with objective of crack control are proposed corresponding to target reliability indices of international standard. The live load factor 0.7 to 0.8 for both I-shaped and box type prestressed concrete girder may be appropriate to meet the international standards.

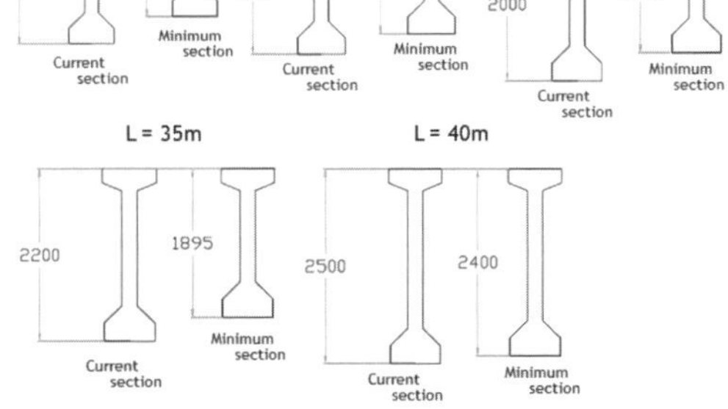

Figure 1. Dimension of minimum section of I-shape girder.

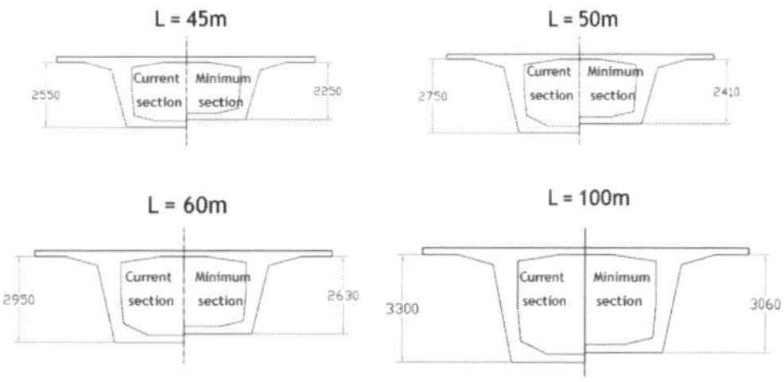

Figure 2. Dimension of minimum section of box type girder.

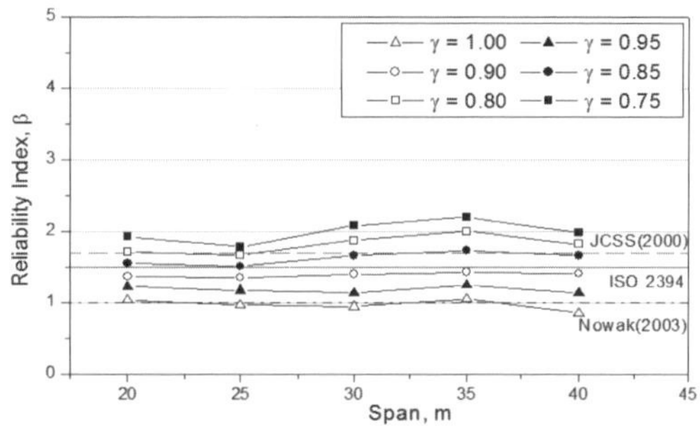

Figure 3. Reliability Indices for minimum section PSC I-shape girder by proposed live load model.

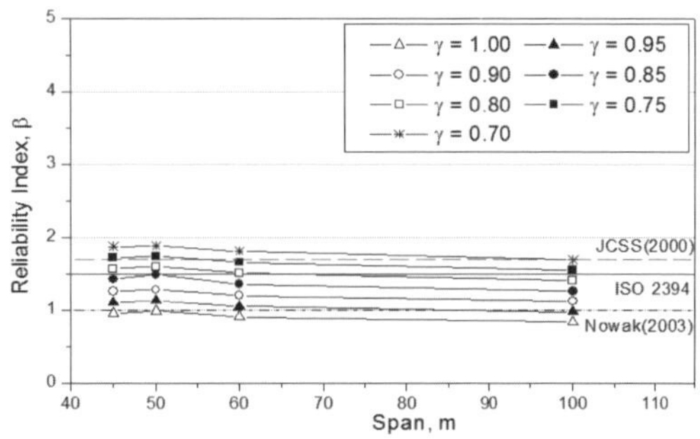

Figure 4. Reliability Indices for minimum PSC box type girder by proposed live load model.

REFERENCE

MLTM. 2005. *Korean Bridge Design Code*, Ministry of Land, Transportation, and Marine Affairs, Seoul, Korea.

Bridge Maintenance, Safety, Management, Resilience and Sustainability – Biondini & Frangopol (Eds)
© 2012 Taylor & Francis Group, London, ISBN 978-0-415-62124-3

The benefits and use of FE modelling in bridge assessment and design

P. Icke
LUSAS, Kingston upon Thames, UK

C. Margheriti
LUSAS (Italia), Terni, Italy

ABSTRACT

For the evaluation and assessment of existing bridges, and also for new design, FE modelling allows for a more rigorous analysis approach to be adopted that can often lead to more accurate and economical results with respect to codified methods.

A load capacity study by Atkins for a pair of braced steel beams shows that a nonlinear FE analysis can provide a significantly larger collapse load factor than that obtained from a design code.

When the diaphragms of a steel box bridge do not comply with assessment code criteria, FE analysis will allow a detailed analysis to be performed in order to help prove the integrity of the design. An assessment of the diaphragms on the Midlands Link Viaducts, in the UK is given as an example.

Two, US steel truss bridges – the US-2 over Cut River, and the M-55 over Pine River, and a typical UK steel 'through' bridge at Hackbridge, are given as examples of how load capacity can be proved on structures even if the latter was 'seen' to fail according to a design code.

The use of measured structural data to verify, calibrate and sometimes even fine-tune an FE model can lead to greater accuracy in the results obtained. Mention is made of a study by Catbas & Gokee into the use of AASHTO distribution factors, and of a study by Das of a variety of in-service bridges in North Carolina which saw controlling load rating factors for particular rating vehicles increase by up to 75%. The West Gate Bridge Strengthening project also saw calibrated models used.

Bridge assessments requiring retrofit solutions, are another area where FE analysis can assist greatly with what-if scenarios. A two span deck truss bridge with a lower lateral bracing system that vibrated excessively under truck crossings and required remedial action is mentioned.

The role that FE plays in creating innovative new designs is illustrated by the Gateshead Millennium Bridge and the initial proposed New Mississippi River crossing by Modjeski & Masters.

When a re-design is undertaken using FE analysis there can be occasions when great savings can be made and the Estero Parkway Flyover is cited as an example of this.

Methods of customising FE software and carrying out automated modelling and loading is described for rail track-structure interaction analysis to UIV774-3 and for Eurocode traffic loading. Benefits of using specialist design checking software in conjunction with FE analysis are stated.

For erection and demolition engineering, the Bagley Street cable stayed footbridge is used to show how critical pylon geometry could be checked using FE to ensure key design criteria were not exceeded during construction and the use of FE modelling to assist with the disassembly of the self-anchored Paseo Bridge in Kansas City, Missouri is cited.

REFERENCES

Catbas, N & Gokee B. 2009. A Novel Approach to Analyse Existing Bridges Efficiently. ASCE Structures Congress.

Das, S. 2010. Refined load rating analysis of in0-service bridges in North Carolina. Bridge Maintenance, Safety, Management and Life-Cycle Optimization.

Ferretti Torricelli, L. et al. 2010. Towards Eurocodes: an Italian experience. A comparative analysis between the former and the new Eurocode-based codes. Proc. of Joint IABSE-fib Conference 2010, Cavtat, Croatia

Hendy, C.R. 2008. The implications of the change to Eurocodes for bridge design. Proceedings of the ICE – Bridge Engineering Vol 1. Issue 1.

LUSAS. 1997–2011. Finite element analysis case studies. See http://www.lusas.com/case/bridge/index.html

Taylor, S. et al. 2009. West Gate Bridge Strengthening. Austroads 2009.

Zhou, E. & Biegalski, A.M. 2008. Problem Diagnosis and Retrofit of Lateral Bracing System of a Truss Bridge. Structures Congress, Canada.

Bridge Maintenance, Safety, Management, Resilience and Sustainability – Biondini & Frangopol (Eds)
© 2012 Taylor & Francis Group, London, ISBN 978-0-415-62124-3

Monitoring of a prestressed high performance concrete bridge from construction through service using an embedded optical fiber sensor system

R.L. Idriss
New Mexico State University, Las Cruces, New Mexico, USA

ABSTRACT

An optical fiber monitoring system was designed and built into the I-25 bridge at the Dona Ana exit in Las Cruces NM. The bridge is a simple span, high performance prestressed concrete girder bridge. The girders are six BT-63 high performance concrete (HPC) girders with a span length of 112.5 ft (34.2 m). Fiber Bragg Grating (FBG) optical fiber deformation sensors along with thermocouples were embedded in the girders during fabrication. Sensors were installed along the top and bottom flanges, at mid-span and quarter spans. Pairs of crossed sensors in a rosette configuration were embedded in the webs at the supports. The bridge was monitored from transfer of the prestressing force through service. The sensor data was analyzed to evaluate shear and moment girder distribution factors, in-situ material properties, prestress losses, camber, dynamic load allowance, and the bridge performance under traffic loads. This paper focuses on the lateral load distribution in the bridge. Shear and moment girder distribution factors are obtained from a finite element model, sensor measurements under a live load test as well as regular traffic loading and compared to the values specified by the AASTHO Standard Specifications (2002) and the AASHTO LRFD Specifications (2007).

Moment and shear GDFs from both codes were conservative when compared to the FE model.

Under regular traffic, the maximum moment GDF was found to be well below the moment GDF given by the AASHTO Standard Specifications and the AASHTO LRFD Specifications as shown in Figure 1. As shown in figure 2, the maximum shear GDF measured under regular traffic was very close, within 4% to the shear GDF specified by the AASHTO Standard Specifications, but still well below the AASHTO LRFD Specifications value.

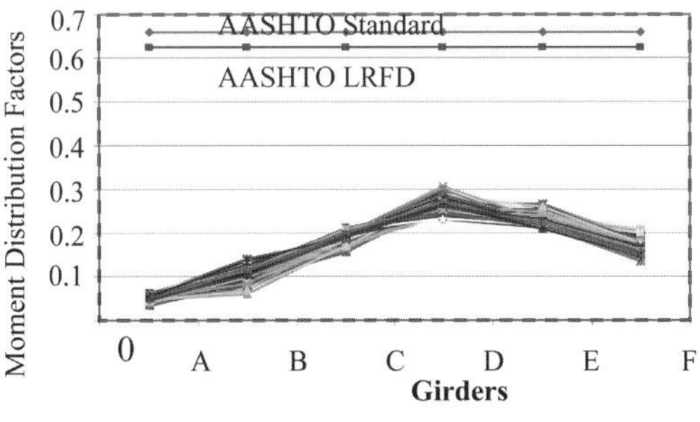

Figure 1. Moment distribution factors under regular traffic.

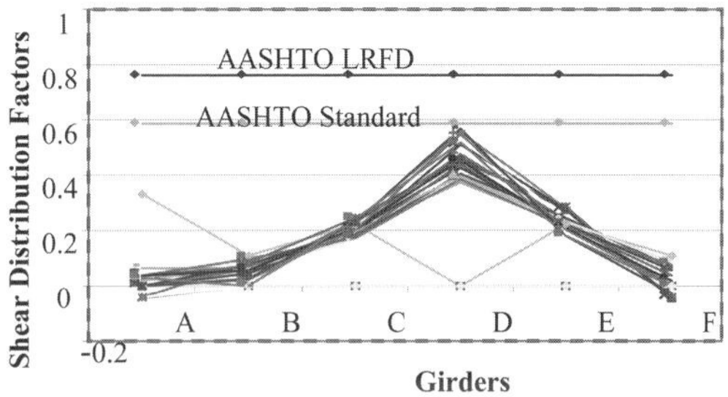

Figure 2. Shear distribution factors under regular traffic.

This project was funded under the FHWA Innovative Bridge Research and Construction (IBRC) Program.

REFERENCES

AASHTO 2002. Standard Specifications for Highway Bridges, Bridge Design Specifications., Washington, D.C., 2002.

AASHTO 2007. LRFD Bridge Design Specifications., Washington, D.C.

Hughs E. & Idriss R.L. 2006. Live-Load Distribution Factors for Prestressed Concrete, spread box-girder bridge. *Journal of Bridge Engineering*, 11(5), 573–581.

Liang Z. 2008. Monitoring Of A Prestressed High Performance Concrete Bridge Using An Embedded Optical Fiber Sensor System In Dona Ana County, New Mexico. PhD Dissertation. New Mexico State University, Las Cruces, NM.

Nassif H. & Nowak A. 1995. Dynamic Load Spectra for Girder Bridges. *Transportation Research Record* 1476, TRB, National Research Council, Washington, D.C., pp. 69–83.

Nowak A. & Hong Y.K. 1991. Bridge Live Load Models. *ASCE Journal of Structural Engineering*, Vol. 117, No. 9, 1991, pp. 2757–2767.

Bridge Maintenance, Safety, Management, Resilience and Sustainability – Biondini & Frangopol (Eds)
 ISBN 978-0-415-62124-3

Fatigue assessment of a railway bridge detail using dynamic analysis and probabilistic fracture mechanics

B.M. Imam & G. Kaliyaperumal
Faculty of Engineering and Physical Sciences, University of Surrey, Guildford, UK

ABSTRACT

The objective of this paper is to present a generic methodology for the use of PFM within the context of bridge loading for the fatigue design and assessment of steel railway bridges and to provide detailed guidance on how to use the proposed methodology in order to carry out a PFM-based fatigue assessment. The problem is set in a probabilistic context to take into account material, loading as well as modelling uncertainties. Guidance is given on how to calibrate a constant amplitude PFM analysis against an S-N curve. Finally, as a case study, a cracked welded bridge detail is considered. Dynamic finite element (FE) analysis of the bridge is carried out and the reliability of the resulting stresses is established through comparison with field measurements to account for modeling uncertainties. By using the fatigue load spectra developed from typical railway traffic, fatigue life estimates are obtained via the PFM methodology.

The first part of the paper deals with the S-N based FM model calibration. This is required in order in order to verify the model assumptions of the FM method employed. Figure 1 shows the calibration results. The comparison of the S-N and FM curves demonstrated that, unless unrealistically small initial crack sizes are used for the calibration, a crack initiation time N_{in} would need to be introduced in the analysis as a basis of calibration. However, the use of very small values of a_{in} is not recommended for two reasons. Firstly, because very short crack growth cannot be described by the Paris-Erdogan crack growth rule and secondly because calibration on a_{in} will, in general, affect inspection outcomes under bridge loading. Therefore, the use of N_{in} as a calibrating parameter appears to be particularly attractive and is suggested here for steel bridge details.

Following the calibration procedure, the fatigue reliability of the case study stiffener detail is estimated. The bridge is analysed under the passage of typical train traffic and the annual stress range histogram is developed. By using the histogram and through Monte Carlo analysis, the time-dependent probability of fatigue failure of the connection is determined and shown in Figure 2. The effect of modeling uncertainties has also been investigated. A factor taking into account the differences between actual stresses, obtained from field measurements, and stresses obtained from FE analysis has been developed. Furthermore, the effect of neglecting the cut-off stresses in fatigue damage calculations, suggested by EN1993-1-9 (2005), has been shown to be negligible on fatigue reliability.

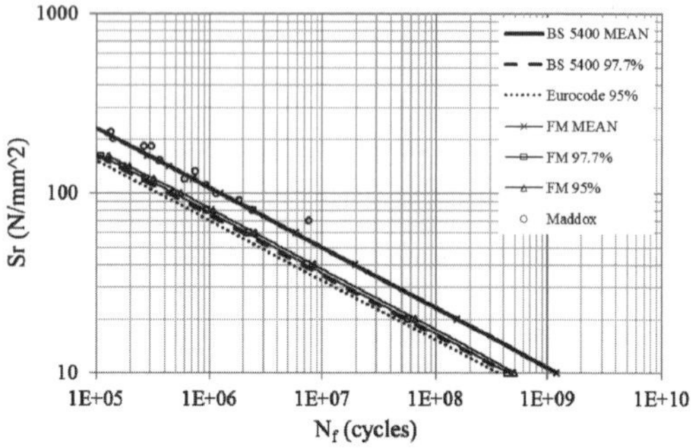

Figure 1. Calibration of the FM lines against the BS 5400 lines.

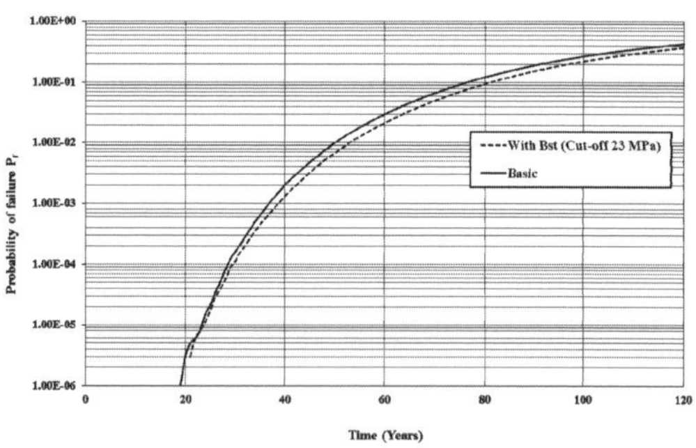

Figure 2. Probability of fatigue failure over time for case study bridge detail.

Bridge Maintenance, Safety, Management, Resilience and Sustainability – Biondini & Frangopol (Eds)
© 2012 Taylor & Francis Group, London, ISBN 978-0-415-62124-3

Highway A24 in Italy: Improvement of seismic performance

F. Incelli, B. Rinaldi, E. De Angelis & S.Mancini
Spea Ingegneria Europea, Rome, Italy

ABSTRACT

After the earthquake that struck the city of L'Aquila in April 2009, SPEA has developed projects for seismic retrofitting of the main viaducts of the Highway A24. The seismic study has begun with a preliminary SDOF analysis that has shown the need to improve performance against the seismic loads of the whole project works. Then a non-linear dynamic MDOF analysis has been performed. The next step was the management of a series of targeted intervention in order to reduce the loads transmitted to the substructures by the excited masses and to consolidate and strengthen the structural elements. As the decks are all simply supported, the main aim is to transfer the entire longitudinal seismic force to contrast artifacts behind the abutment of the bridges, and transverse seismic action both to piers and abutments.

In the following a detailed list of the required interventions is remarked:

- structural union of the slab of bridge deck
- replacement of the bearings
- installation of seismic devices
- construction of contrast element on the back of the abutments
- transversal union of the bridge decks
- reinforcement of the piers

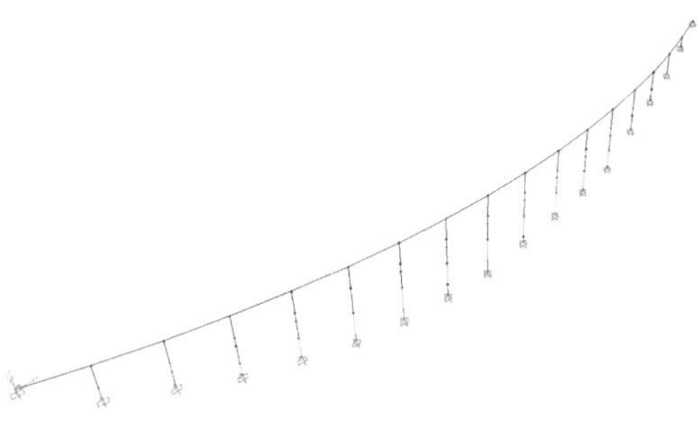

Figure 1. MDOF model for dynamic analysis.

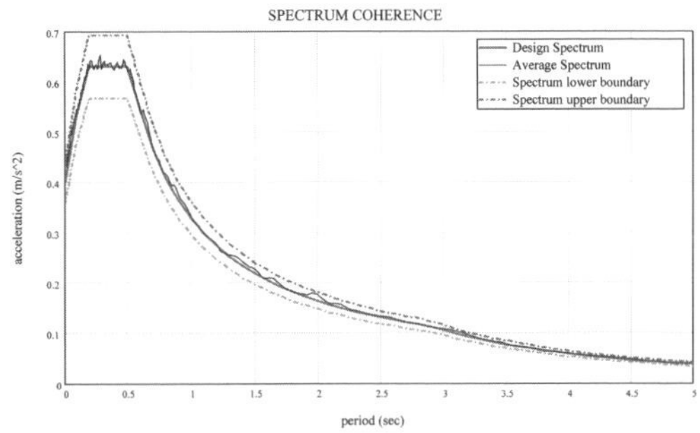

Figure 2. Example of Spectrum Coherence.

In longitudinal direction the entire deck is bound to the abutment by means of horizontal dampers between the bottom of the girders and the contrast element behind the abutments. In transverse direction the transmission of the seismic force is provided by dampers on the piers and concrete blocks on the abutment to fix transverse displacements. All the bearings are multidirectional to allow those displacements needed by the damping systems. A union of the decks has been done to avoid hammering of themselves in case of transversal seismic loads.

Due to the lack of natural accelerograms for the dynamic analysis, 50 simulated accelerograms have been developed. Seven accelerograms have been selected in complete coherence with design spectra referred to the site. The lower variance from design spectrum has been fixed in a range of 10%. Thus from each accelerogram the respective spectrum has been generated and combined with the others to obtain the average spectrum used to lead the coherence analysis.

REFERENCES

Pinto, P.E., Giannini, R., Franchin, P., 2004. *Seismic Reliability Analysis of Structures*. IUSS Press, Pavia.

Priestly, M.J., Seible, F., Calvi, G.M., 1996. *Seismic Design and Retrofit of Bridges*, John Wiley & Sons, N.Y.

Bridge Maintenance, Safety, Management, Resilience and Sustainability – Biondini & Frangopol (Eds)
© 2012 Taylor & Francis Group, London, ISBN 978-0-415-62124-3

Imperfection sensitivity of hanger of a suspension bridge for different hanger arrangements

M. Inoue
IHI Infrastructure Systems, Co., Ltd., Tokyo, Japan

ABSTRACT

The imperfection sensitivity of hangers of a suspension bridge for three hanger arrangements has been studied in this study. The difference among 1-hanger, 2-hangers and inclined hanger models was discussed in detail through the analysis results by using a bridge model with a main span of 3,000 m.

When the fabrication length error of 5 mm is considered to one of hangers, the increase ratio of hanger tension compared with that in the reference condition (dead weight condition) is obtained as shown in Table 1 for the shortest hanger in the main span.

As conclusion of this study, the notable findings are shown below;

< Hanger arrangement type >

2-hangers model, where a bridge has two or more hanger ropes in the longitudinal direction at each hanger location per each cable plane, shows more imperfection sensitivity than others. In addition, the unbalanced forces and static/dynamic wind effects shall also be taken into account. Except for the situation that 1-hanger model is not feasible for example because of too large diameter of hanger, it seems to be better not to adopt 2-hangers model as a hanger arrangement type. Or the use of this type of arrangement shall be limited for longer length hangers, e.g. at the tower location where larger design hanger tension per one hanger location is mostly required.

Table 1. Additional hanger tension

	1-hanger	2-hangers	Inclined
Direct effect*	12%	96%	35%
Adjacent effect**	28%	28%	42%
Total	40%	124%	77%

*additional due to a length error of designated hanger
**additional due to a length error of adjacent hangers

< The requirement for fabrication >

The allowable fabrication error for unstressed hanger length shall be distinguished between the hanger arrangement types. For 1-hanger model, a similar concept to past experiences can be suitable but the upper limit for long hanger rope needs to be defined for a suspension bridge with a main span longer than 2,000 m. The upper limit of error of 15 mm seems to be practical since the error of long hangers located near tower does not much affect on the additional hanger tension and the deformation (imperfection to a bridge geometry). For 2-hangers model, more severe requirement and/or concept than past experiences which were likely based on 1-hanger model need to be defined.

It is of importance to discuss an appropriate length in detail at the early stage of design with consideration of a suitable balance with other factors such as cost and workability to find the best solution for each project.

REFERENCES

HONSYU-SHIKOKU Bridge Authority, 1994. Structure of cable band and suspender on the Akashi Kaikyo Bridge (part. 1), *Honshi Technical report* Vol. 18 No. 70 pp. 22–34.

HONSYU-SHIKOKU Bridge Authority, 1998. Design of suspender rope and cable band of Kurushima Bridge, *Honshi Technical report* Vol. 22 No. 85 pp. 56–64.

HONSYU-SHIKOKU Bridge Authority, 2000. Aerodynamic stabilization for hanger ropes of the Akashi Kaikyo Bridge, *Honshi Technical report* Vol. 24 No. 93 pp. 18–25.

IHI Corporation, 1989. Construction report of Second Bosporus Bridge.

Bridge Maintenance, Safety, Management, Resilience and Sustainability – Biondini & Frangopol (Eds)
 ISBN 978-0-415-62124-3

Shear strengthening of columns in existing bridges

T. Isaković, M. Fischinger & Z. Vidrih
University of Ljubljana, Faculty of Civil and Geodetic Engineering, Ljubljana, Slovenia

A. Bevc
Slovenian National Building and Civil Engineering Institute

ABSTRACT

Bridges, which were built before the modern principles of the earthquake engineering were established, often comprise construction details that are nowadays considered inappropriate for the seismic regions. These deficiencies are mostly related to the transverse reinforcement in columns, which typically cannot ensure adequate shear strength and the confinement of concrete core and cannot prevent the buckling of the longitudinal flexural bars.

Many older bridges in Central Europe (Slovenia) are supported by hollow box columns, which comprise specific substandard construction details that are not considered in any literature. Therefore, the experimental and analytical investigation of such columns has been performed at University of Ljubljana and Slovenian National Building and Civil Engineering Institute.

The cyclic experimental investigation demonstrated that, in spite of the poor construction details, these columns had relatively good displacement ductility capacity, which was acceptable for moderate seismic demand regions (e.g. Central Europe). This ductility was provided by the favorable hollow box cross-section with its large compression zone, by the low axial forces, and by the relatively high strength of the concrete. The mixed shear-flexural type of failure was observed during the cyclic test of the as-built column.

Therefore the shear strengthening was performed using two techniques: concrete jacketing and CFRP wrapping. Since the shear demand in the regions of moderate seismicity was not considerably larger than the shear capacity of the columns, the minimum amount of strengthening was provided.

Both strengthening techniques successfully increased the shear strength of the investigated column. As far as the sufficient shear strength was provided, the other unfavorable types of failure induced by other column's construction deficiencies were activated. Particularly critical was buckling and rupture of the longitudinal bars as well as the bar slip at the column base.

The concrete strengthening was more efficient considering enhancement of column ductility and energy dissipation capacity. To improve the ductility capacity of the investigated type of column using CFRP wrapping, larger amount of CFRP strips was needed (note that the absolute minimum amount of CFRP strips was used).

The analytical estimation of the shear strength of as-built and strengthened columns was also performed. Different methods, reported in the literature (UCSD method, – Priestley et al., 1996) and standards (EC2 – CEN 2004 and EC8/3 – CEN 2005) were employed.

The shear strength estimated according to the standard EC2 was too low, because the contribution of the concrete to shear strength was neglected when the demand exceeds its value. However, as far as this contribution was taken into account, the shear strength of the column was comparable with the results of other two methods, but only to the values that corresponded to larger displacement.

A more suitable procedure for the estimation of the shear strength of investigated column is that proposed in the standard EC8/3. Using this procedure the shear strength of the bottom part of the column was estimated quite accurately. However, the standard underestimated the shear strength of the top of the column, where the ductility demand was low.

Good estimation of the type and the location of failure were obtained by using the UCSD method.

Different types of macro models were successfully used to model the response of the investigated as-built and columns strengthened by CFRP strips.

REFERENCES

CEN 2004, Eurocode 2: Design of concrete structures – Part 1-1: General rules and rules for buildings. EN 1992-1-1. Comité Européen de Normalisation, Brussels, 2004.

CEN 2005, Eurocode 8: Design of structures for earthquake resistance. Part 3: Strengthening and repair of buildings. EN 1998-3, Euro Commit for Stand, Brussels.

Priestley, M.J.N., Seible, F. & Calvi, G.M. 1997. *Seismic Design and Retrofit of Bridges*, John Wiley & Sons.

Bridge Maintenance, Safety, Management, Resilience and Sustainability – Biondini & Frangopol (Eds)
© 2012 Taylor & Francis Group, London, ISBN 978-0-415-62124-3

Expected seismic performance of irregular isolated bridges

J.M. Jara, M. Jara & B.A. Olmos
Civil Engineering School, Universidad Michoacana de San Nicolás de Hidalgo, Morelia, Michoacán, México

D. Villanueva
Civil Engineering School, Universidad de Sonora, Hermosillo, Sonora, México

H. Varum
Department of Civil Engineering, Universidade de Aveiro, Aveiro, Portugal

ABSTRACT

Bridge structures are usually built on irregular topographical surfaces which create substructures with pier heights of different lengths. Three height irregularity types of typical RC medium length bridges are analyzed aimed at determining the best strength and stiffness parameters of an isolation system. The models were located in a high seismicity zone of Mexico. The isolation system is composed by lead rubber bearings (LRB) located on each pile and abutment. The bridge and isolation parameters conducted to the nonlinear time history analysis (NLTHA) of 169 models. Ten seismic records representative of the subduction zone in the Pacific Coast in Mexico were chosen to carry out the study. The maximum pier demands of drifts, bending moments and shear forces were analyzed to identify the best isolation properties for improving the bridges' structural behavior. Additionally, the seismic response of the bridges supported on traditional neoprene bearings was carried out.

Many seismic bridge damages in different countries have been identified with force demand concentrations on certain piers of substructures composed by piers with different lengths which conduct to strong lateral stiffness irregularities. This study was motivated by the increasing use of isolation systems in bridges with these characteristics and the lack of studies quantifying the influence of the isolation parameters on the seismic response of bridges with height irregularities. Bridges closely located to active seismic sources are very attractive to incorporate seismic isolation systems, mainly due to the high seismic activity and the typical frequency content of the accelerograms recorded in these areas. The selection of the bridge parameters intends to characterize typical schemes of variations in pier heights representing a large number of medium length isolated bridges. The substructure irregularity was considered by using a five span simple supported bridge with three pier height configurations.

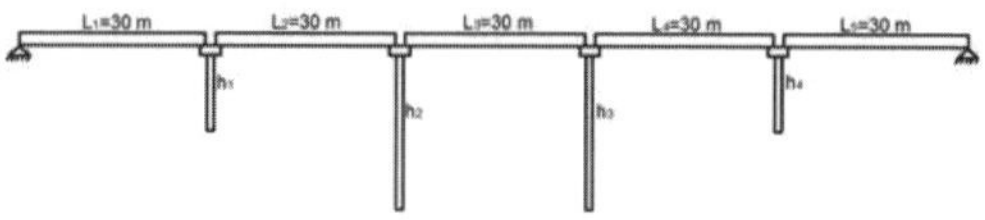

Figure 1. Bridge irregularity type I.

Table 1. Pier heights of the type I bridge models.

Model	h_1 (m)	h_2 (m)	h_3 (m)	h_4 (m)	h_2/h_1
MODI_1	5	5	5	5	1.0
MODI_2	5	7.5	7.5	5	1.5
MODI_3	5	10	10	5	2.0
MODI_4	5	15	15	5	3.0

The bridge models are supported on four piers and two fixed abutments (figure 1) and all of them have the same superstructure type.

For each irregularity type, it was created several models varying the pier heights (table 1). Model I_1 is the regular bridge model whereas model I_4 is the model with more pronounced difference between the central and lateral pier heights. The last column of the table displays the ratio between the pier heights.

The analyses of two additional irregularity types of bridge substructures, allowed identifying general trends of behavior and a specific range of the isolator strength and stiffness parameters, to have a uniform substructure seismic response.

Bridge Maintenance, Safety, Management, Resilience and Sustainability – Biondini & Frangopol (Eds)
© 2012 Taylor & Francis Group, London, ISBN 978-0-415-62124-3

Great marquee for high-speed trains in the new railway station of Málaga

C. Jurado
Polytechnic University of Madrid, Madrid, Spain; Ingecal Ingenieros, S.L.

ABSTRACT

The new railway station of María Zambrano for AVE (Spanish high-speed trains) located in Malaga, has been inaugurated in November 2006, just on the site of the former railway station. The new railway station with an investment of 134.7 million Euros occupies a surface of 51,377 m^2, five times the surface of the former station. The enclosure is the biggest intermodal transport and commercial center of Spain which comprises a parking of 21,000 m^2 for 1,300 parking places, one commercial area and a hotel of 35 m, with a total extension constructed of approximately 100,000 m^2. The spaces of leisure contain cinemas, shops, restaurants, bowling, gymnasium, swimming pool and zones for passenger's traffic.

Before going the station in the sense of arrival trains, there is a great marquee over the train platforms with a huge span of 78 meters, without any intermediate pillar and a length in the direction of the railway rails of 72 meters.

Under the marquee there have been constructed six platforms that give service to 6 trains, four with an international width and two with an Iberian width.

Figure 1. Aerial sight of the new commercial centre and railway station of the AVE, in Málaga.

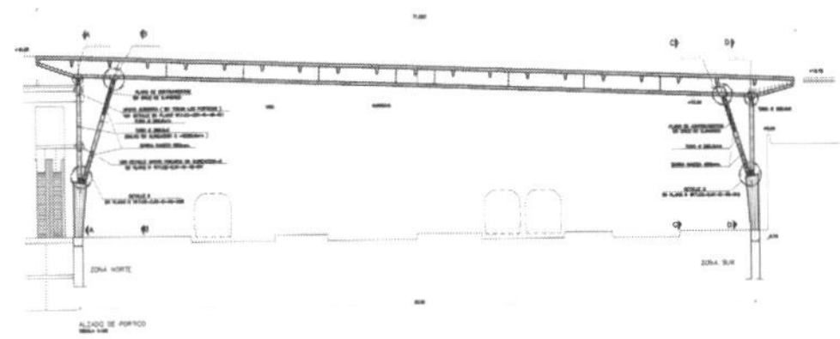

Figure 2. Transverse section of the marquee of platforms.

Figure 3. View of the marquee from the zone of parking in construction.

The portals of the marquee with different height in his ends of 16.50 m (left hand) and 13.72 m (right hand) have on the top a metallic girder beam of 1.60 m of height.

The structure is constituted by 10 metallic portals separated 8.00 m each from the following one, covering a zone of 72.00 × 71.475 m^2. The roof is rigidity by means of San Andres's crossings on the two initial and end portals of the marquee. The Project of the marquee has been realized by means of a three dimensional model of finite elements with the program SAP2000N, considering static and dynamic loads.

The marquee is singularity placed on another underground station of trains.

Bridge Maintenance, Safety, Management, Resilience and Sustainability – Biondini & Frangopol (Eds)
© 2012 Taylor & Francis Group, London, ISBN 978-0-415-62124-3

Effect of skewness on shear force applied to shear keys in skewed highway bridges

A. Kalantari & S.M.J. Foroughi
International Institute of Earthquake Engineering and Seismology (IIEES), Tehran, Iran

ABSTRACT

In this study effect of skewness angle on amount of shear force applied to external shear keys in skewed highway bridges is investigated. Skewed bridge is a bridge that unlike a straight bridge, direction of the piers is not perpendicular to the longitudinal direction of the bridge deck. Sacrificial shear keys are used at abutments to provide transverse support for bridge superstructures under seismic loads. In addition, sacrificial shear keys serve as structural fuses to control damage in abutments and the supporting piles under transverse seismic loads. Sacrificial shear keys may be interior or exterior. Exterior shear keys are usually recommended for new construction because they are easier to inspect and repair (Megally et al. 2001).

A set of bridge models with different skewness angle are subjected to three ground motion in three direction in OpenSEES software. According to free vibration analysis, natural periods of second and third mode of the bridge that are in lateral and longitudinal direction in straight bridge, respectively, increase by increasing skewness angle. But natural period of first mode, that is deck rotation, in straight bridge, does not change clearly. Shear force applied to shear keys, increase with increasing skewness angle in shear keys that are in acute corners of the deck. But this force reduces in shear keys that are in obtuse corners (Fig. 1).

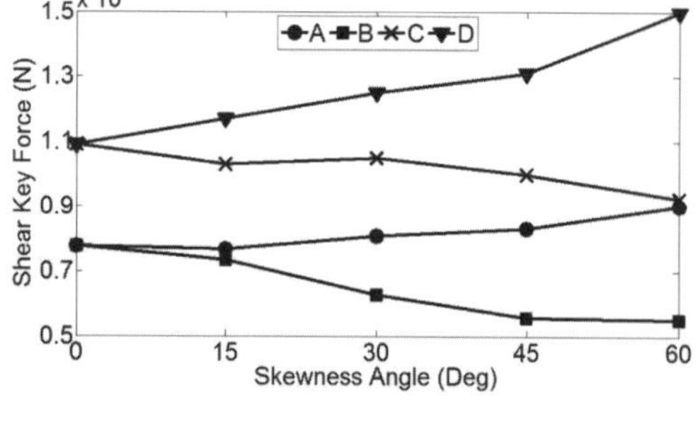

Figure 1. Shear Key Force in Unsymmetrical models.

Symmetrical bridges are more stiff than unsymmetrical bridges, as in symmetrical models, center of mass and stiffness, coincide. Therefore, shear keys suffer smaller forces in symmetrical bridges. Shear force of the shear keys in symmetrical straight bridge are nearly equal in four corners of the deck with small differences. But as skewness angle increases, this differences between forces become more. As shear force applied to shear keys in skewed bridges, is different in shear keys that are in acute and obtuse corners, this issue should be considered in design procedure of external shear keys, in skewed highway bridges.

REFERENCES

Bozorgzadeh, A., Megally, S.H., Ashford, S. & Restrepo, J.I. 2004. Seismic response of sacrificial exterior shear keys in bridge abutments. *Structural systems research report, Department of structural engineering, university of California San Dieg, La Jolla, CA, Oct 2007* (SSRP-04/14).

California Department of Transportation (CALTRANS) 2006. *Seismic design criteria.* Sacramento, California. http://www.dot.ca.gov

Ghobarah, A.A. & Tso, W.K. 1974. Seismic analysis of skewed highway bridges with intermediate supports. *Earthquake Eng. Struct, Dyn*, 2: 235–248.

Jennings, P.D. (ed) 1971. Engineering features of the san Fernando earthquake. *Rep. No EERL 71-02*, California Institute of Technology, Pasadena, Calif.

McKenna, F. & Fenves, G. 2001. *The OpenSEES command language manual.* Version 1.2, Pacific Earthquake Center, Univ of California, Berkeley, California.

Megally, S.H., Silva, P.F. & Seible, F. 2001. Seismic response of sacrificial shear keys in bridge abutments. *Structural systems research report, Department of structural engineering, university of California San Dieg, La Jolla, CA, May 2002* (SSRP-2001/23)

Meng, J.Y. & Lui, E.M. 2002. Refined stick model for dynamic analysis of skewed highway bridges. *Journal of bridge engineering* ASCE, 7(3): 1–11.

Ngoc L.A., Tirasit P. & Kawashima K. 2008. Seismic performance of a skewed bridge considering flexure and torsion interaction. *14th world conference on earthquake engineering (WCEE), Beijing, China, October 2008.*

Bridge Maintenance, Safety, Management, Resilience and Sustainability – Biondini & Frangopol (Eds)
 ISBN 978-0-415-62124-3

Ductility behavior of deteriorating reinforced concrete members

A.N. Kallias & M.I. Rafiq
University of Surrey, Guildford, Surrey, UK

ABSTRACT

Rebar corrosion is the most commonly observed deterioration cause in reinforced concrete (RC) bridges, due to the extensive salt applications during winter. Rebar corrosion is associated with the loss of steel area and the formation of expansive rust at the steel-concrete interface, which eventually leads to concrete cracking/spalling, loss of bond and degradation of rebar mechanical properties due to pitting formation. Corrosion damage may adversely affect the behavior of the RC element; this, however, depends on the location and amount of corrosion loss.

Numerous studies in literature, designed to investigate the influence of the different corrosion effects, on the performance of RC beams. Most studies examined the residual load-capacity and serviceability performance of corroded beams. Only a few studies examined the influence of corrosion damage on the ductile behavior of under-reinforced beams. These investigations do not allow for sound conclusions to be made on ductility behavior due to the significant scatter in the observations. For example, Maaddawy et al. (2005) observed an increase of beam displacement ductility factor, μ_δ, for steel area losses (Q_{cor}) up to approximately 30% in their tension rebars. In this study, μ_δ started to decrease for $Q_{cor} > 22\%$ but it all cases it remained above the value of μ_δ of the control beam. The gradual reduction of ductility was also accompanied by a change in beam behavior at ultimate, causing a shift in the failure mode from ductile towards brittle modes. In contrary, Du et al. (2007) reported loss of beam's ductility for $Q_{cor} = 11\%$ in the tension rebars (T282 beam). The loss of beam's ductility was due to the premature rupture of the tension rebars which occurred prior to concrete crushing. Rodriguez et al. (1996) reported ductility losses in under-reinforced beams for increasing corrosion damage. These beams, however, have all their rebars corroded; hence it is difficult to quantify the influence of each of the corrosion effects on the evolution of beam's ductility. Hence, it is important to understand the factors influencing the observed structural behavior.

Simplified sectional analysis is, in many cases, inadequate in capturing the evolution of complex behavior of corroding RC beams. Non-linear finite element analysis (NLFEA) can be considered as a suitable assessment tool for the analysis of corroding RC beams, since the different effects of corrosion can be explicitly incorporated in the analysis.

This paper examines the effect of corrosion on the ductile behavior of under-reinforced beams. Numerical models of uncorroded/corroded beams are initially developed and validated. The numerical models are then used in parametric studies to examine the influence of different variables and modeling assumptions on the evolution of ductility in the corroding beams. The results are discussed in relation to key experimental and numerical studies. This in turn assists the interpretation of the existing experimental results. Parameters which affect the ductile behavior of corroding beams can be identified and set the basis for further laboratory investigations.

REFERENCES

Du, Y.G., Clark, L.A., Chan, A.H.C. 2007. Impact of reinforcement corrosion on ductile behaviour of reinforced concrete beams. ACI Structural Journal 104(3):285–93.

Maaddawy, T.E., Soudki, K., Topper, T. 2005. Long-term performance of corrosion-damaged reinforced concrete beams. ACI Structural Journal 102(5):649–56.

Rodriguez, J., Ortega, L.M., Casal, J., Diez, J.M. 1996. Assessing structural conditions of concrete structures with corroded reinforcement. Concrete Repair, Rehabilitation and Protection. Dhir, R.K. & Jones, M.R. (Eds). E&FN Spon. ISBN 0419214909. pp. 65–78.

Bridge Maintenance, Safety, Management, Resilience and Sustainability – Biondini & Frangopol (Eds)
 ISBN 978-0-415-62124-3

Improvement in tensile performance of steel fiber reinforced high strength concrete: Influence of fiber shape and sand to aggregate ratio

S.T. Kang
Department of Civil Engineering, Daegu University, Gyeongsan, Korea

H.W. Cho & J.H. Lee
Korea Institute of Construction Technology, Goyang, Korea

D.J. Kim
Department of Civil and Environmental Engineering, Sejong University, Seoul, Korea

ABSTRACT

As well known, brittleness of high strength concrete limits its applications. Short-fiber reinforcement, in general, is incorporated into concrete in order to improve its mechanical properties. The enhanced properties include tensile strength, post-cracking tensile behavior, toughness, ductility, energy absorbing capacity, etc. Properties of fiber reinforced concrete are influenced by various factors such as the type, aspect ratio, volume fraction of fiber, and mix proportion, aggregate size of concrete and so on. Recently, several brand-new shapes of fibers have been suggested in order to enhance pullout resistance governing the tensile performance especially of fiber reinforced strain hardening cementitious composites.

As a partial study to develop a high performance steel fiber reinforced concrete (SFRC) with better mechanical properties, the efficiency of adopting the twisted fibers was investigated. In addition, in order to improve the mechanical performance more by adding higher fiber contents as much as possible without detrimental reduction in workability and fiber balling phenomenon, the effect of sand to aggregate ratio on the performance of fresh and hardened state. Workability of fresh concrete was estimated by slump test and the mechanical performance was tested by means of compressive test, direct tensile test, and four points bending test.

The test results indicate that twisted fiber resulted in better mechanical performance than smooth and hooked fiber. The sand to aggregate ratio was beneficial to improve workability of concrete and fiber dispersion but had a tendency to decrease the strength and the tensile behavioral performance.

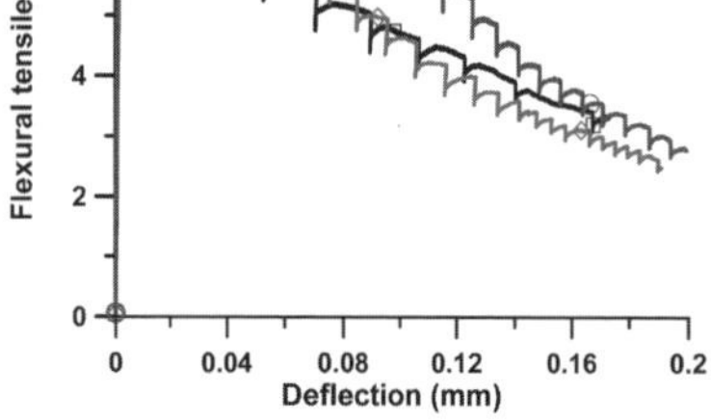

Figure 1. Flexural tensile behavior with different fiber types.

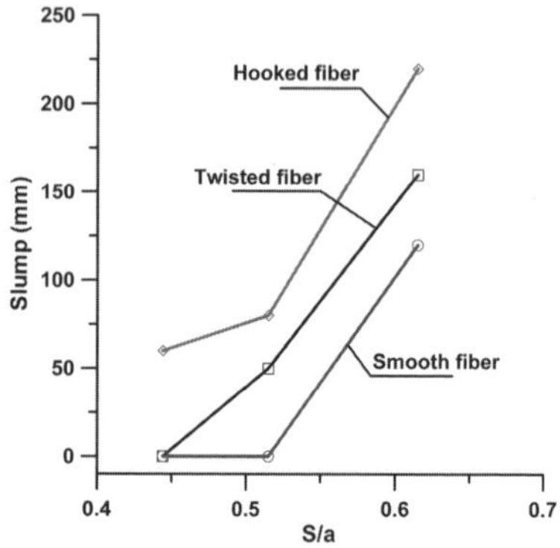

Figure 2. Influence of S/a on concrete slump.

REFERENCES

American Concrete Institute 544 Committee 1996. Report on Fiber Reinforced Concrete (ACI 5441R-96). American Concrete Institute

Song, P. S. & Hwang, S. 2004. Mechanical properties of high-strength steel fiber-reinforced concrete. *Construction and Building Materials* 18(9): 669–673.

Barros, J. A. O. & Sena Cruz, J. 2001. Fracture Energy of Steel Fiber-Reinforced Concrete. *Mechanics of Composite Materials and Structures* 8(1): 29–45.

Naaman, A. E. 2003. Engineered Steel Fibers with Optimal Properties for Reinforcement of Cement Composites. *Journal of Advanced Concrete Technology* 1(3): 241–252.

Bridge Maintenance, Safety, Management, Resilience and Sustainability – Biondini & Frangopol (Eds)
© 2012 Taylor & Francis Group, London, ISBN 978-0-415-62124-3

Reliability analysis of footbridge serviceability considering crowd loading

J. Keogh & C.C. Caprani
Department of Civil & Structural Engineering, Dublin Institute of Technology, Ireland

P. Archbold
Structures & Materials Research Group (STRAIT), Athlone Institute of Technology, Ireland

P. Fanning
School of Civil, Structural & Environmental Engineering, University College Dublin, Ireland

ABSTRACT

Recent developments in the design of structures and structural materials along with pressure on designers to deliver more aesthetically pleasing structures have led to longer and lighter footbridges. Increasingly, these (typically) low-frequency structures are experiencing serviceability problems.

The characteristic vertical response of flexible footbridges subjected to single pedestrian and crowd loading is examined in this paper. Vibrations of a bridge's walking surface can be expected if its natural frequency is within the pedestrian pacing frequency range, due to the dynamic nature of pedestrian load application. If these vibrations are large enough they will lead to discomfort for pedestrians crossing the bridge, thus exceeding the serviceability limit state. The vibration response due to crowd loading is very difficult to predict and is typically estimated using enhancement factors applied to the effect caused by a single pedestrian.

Eurocode 5 uses an equation to predict the single pedestrian response in which the pedestrian is assumed to be walking at the same frequency as the bridge. Use of this equation, means that regardless of the natural frequency of the bridge, once it is below 2.5 Hz, the single pedestrian response is the same. This equation neglects the sensitivity of vibrations of the deck to the pacing frequency found by Pedersen and Frier (2010).

In this paper a moving force model is used in Monte Carlo simulations of non-homogeneous samples of single pedestrians and crowds to estimate characteristic vertical vibration levels. Pedestrian mass, stride length and pacing frequency were all assigned statistical distributions obtained following a literature survey.

Also in this work, statistical distributions of the bridge parameters obtained from literature are considered. These include flexural rigidity, mass and damping ratio. In addition rotational stiffness at the supports is accounted for, as shown in Figure 1. This

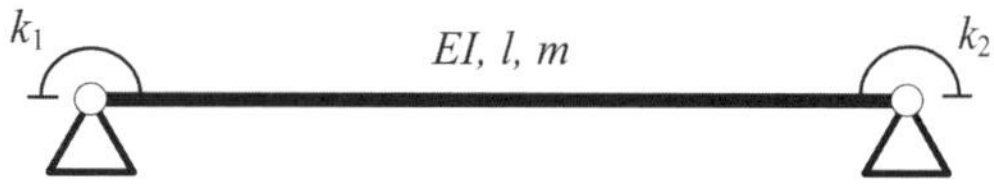

Figure 1. Considered beam with rotational spring supports.

is done to reflect the reality of construction forms, in which the assumption of perfectly free rotation is not realized in practice. The rotational spring stiffness allowance makes the bridge slightly stiffer, increasing its natural frequency by a small amount.

As incorporating statistical ranges of bridge parameters leads to variations in the natural frequency of the bridge, the bridge deck acceleration limit is determined using two limit state functions (LSF). LSF1 uses the design mean value of the bridge parameters and so respects the likelihood that the design bridge will fail. LSF2 uses the natural frequency of the bridge realized in each simulation. From this it was found that the addition of statistical ranges for bridge parameters has only a small effect on the acceleration response of the bridge deck.

Also in this paper the model is compare to currently available design codes and guidelines and compares well with the findings of other authors, including Pavic (2011). The work reported herein also highlights the differences in approach to this issue adopted in some common design codes.

REFERENCES

EN 2004. Eurocode 5: Design of timber structures – Part 2: Bridges, EN 1995-2:2004, *European Committee of Standardization*, Brussels.

Pavic, A. 2011. Vertical Crowd Dynamic Action on Footbridges: Review of Design Guidelines and Their Application, *Proceedings of the Fourth International Conference Footbridge 2011*, Wroclaw, 6–8 July.

Pendersen, L. & Frier, C. 2010. Sensitivity of footbridge vibrations to stochastic walking parameters, *Journal of Sound and Vibration*, 329(13), 2683–2701.

Bridge Maintenance, Safety, Management, Resilience and Sustainability – Biondini & Frangopol (Eds)
© 2012 Taylor & Francis Group, London, ISBN 978-0-415-62124-3

Shear performance of long-term corroded reinforced concrete beams

I. Khan & R. François
Université de Toulouse; UPS, INSA; LMDC (Laboratoire Matériaux et Durabilité des Constructions), Toulouse, France

A. Castel
Université de Nice-Sophia Antipolis; Laboratoire J.A. Dieudonné, Nice, France

ABSTRACT

Bridges are affected by reinforcement corrosion and corrosion causes deterioration of slabs, piers and main beams of sub-standard bridges. The common modes of failure in these elements are flexure and general shear failure. This paper presents the experimental results of two highly corroded shear critical beams subjected to chloride environment to assess the shear behavior of long-term corrosion damaged beams. The beams were kept in aggressive chloride environment. (Salt fog 35 g/l of NaCl corresponding to the salt concentration of sea water) under sustained loading for 16 years then unloaded. The control beams have the same concrete composition and lay-out of the reinforcement, but were stored in a 50% of R.H. and 20°C laboratory room.

A 26 year old corroded reinforced concrete beam exposed to chloride environment was cut into two small shear critical beams which were tested along with a control beam of same age and same length and cross-section (115 × 28 × 15 cm) under three point bending until failure. Concrete cover of A beams is 40 mm for stirrups which correspond to 48 mm for longitudinal bars.

Cracking maps were drawn with the locations of flexural transverse cracks and longitudinal corrosion cracking at four stages, at the age of 28 days, 6, 14, 17 and 26 years. Corrosion maps of corroded beams were drawn after the steel was recuperated from concrete. Corrosion maps of both longitudinal and transverse reinforcement were drawn. It was noted that the longitudinal bars as well as the stirrups were highly corroded and the mass loss was more important in the areas where the corrosion cracks were of larger size.

The force displacement curves for the corroded beams as well as control beam were drawn. From flexure load responses of all three beams, three stages was observed from the load deflection curve, the first stage represent the behavior of beam before flexure cracking and the second stage represents the behavior of beam after flexure cracking and in third stage flexure cracks are accompanied by a large diagonal shear crack up to failure which was basically the behavior change from beam action to arch action. Third stage seems to lead to a plateau which could be the indication of the

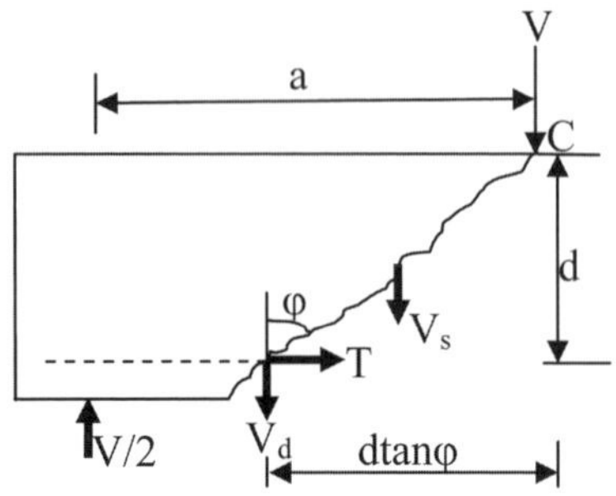

Figure 1. Forces on free body diagram of deep beam at failure.

beginning of yielding of tensile bar or a continuous slip of re-bars.

After testing the beams until failure the main steel bars along with stirrups were extracted from the beams and loss of mass for both longitudinal and transverse reinforcement was measured and plotted. It was noted that the loss of mass of transverse reinforcements plays an important role on the failure mode of beam and it reduces the ductility of beam. From the results it appeared that corrosion of stirrups has a significant impact on shear capacity of beams.

The experimental shear capacity of beams was compared by theoretical shear capacity by using a theoretical model presented by Prodromos (2003). Cracked concrete members were analysed and all the internal forces at diagonal shear cracks were taken into account to determine the ultimate shear capacity of deep beam as shown in figure 1.

The model was used for both control and corroded beams. In case of corroded beams, the reduced steel areas of longitudinal and traverse reinforcement were used to calculate the ultimate shear capacity. The shear capacity of beams calculated by the model had a close agreement with the experimental results.

REFERENCE

Prodromos D. Zararis (2003), Shear Compression Failure in Reinforced Concrete Deep Beams, *Journal of Structural Engineering*, Vol. 129, No. 4, 544–553.

Bridge Maintenance, Safety, Management, Resilience and Sustainability – Biondini & Frangopol (Eds)

Effect of soil-structure-interaction on the reliable seismic retrofit design of an existing highway bridge

S.R. Khavari
Sama Technical and Vocational Training College, Islamic Azad University (IAU) – Roudehen Branch, Roudehen, Iran

M. Hosseini
Structural Engineering Research Center, International Institute of Earthquake Engineering and Seismology (IIEES), Tehran, Iran

M. Taghipour, E. Davoodi, M. Zohrehvandi, A. Esmaeilifar & S. Tahmasebifard
Sama College, Islamic Azad University, Roudehen, Iran

ABSTRACT

In this study the effect of inclusion of SSI in seismic retrofit design of a 2-span Reinforced Concrete (RC) slab bridge with round-column piers, by three various techniques, has been investigated by Nonlinear Time History Analysis (NLTHA) by using the three-component accelerograms of some earthquakes compatible to the site soil of the bridge. The used retrofit techniques include: 1) increasing the bridge's strength and stiffness by adding the fixity of deck connections to abutments in lateral direction and using diagonal bracing elements beside the bridge abutments in longitudinal direction, 2) adding some diagonal braces between columns in transverse direction and beside abutments in longitudinal direction, and 3) using dampers in diagonal members between deck and piers and abutments. The high seismic vulnerability of the bridge has been illustrated in a previous work of the authors (Hosseini et al 2008), and later, some more detailed studies have been performed on the seismic retrofit of this bridge (Khavari & Hosseini 2010).

To find out how the inclusion of SSI affects the seismic retrofit designs of the bridge, the seismic responses of the bridge, retrofitted by each one of the three mentioned techniques, subjected to three-component accelerograms of some selected earthquakes have been investigated, by NLTHA, once without, and once with considering the SSI, and the results have been compared. The chosen set of accelerograms relate to Chi-Chi, Taiwan earthquake of 1999, Northridge earthquake of 1994, and Palm Spring earthquake of 1986, and all are compatible with the site soil of the bridge and have the near fault features. Furthermore, the records have been chosen in such as way that their Peak Ground Acceleration (PGA) values have been very close to the predicted PGA value of the bridge site, and therefore, there has been no need to scale the records. For modeling the soil surrounding the bridge foundations and locating behind the abutments the nonlinear Winkler springs have been used in a three dimensional setting. The response values, considered for investigation, include the total base reaction of bridge in both longitudinal and lateral directions, as well as the displacements of the middle point of the bridge deck in both main directions. Figure 1 shows a sample of the results.

Results show that consideration of soil-structure-interaction can change the retrofit design specifications to some extent. The change level is not the same for different retrofit techniques, and can be significant in some cases. Therefore, it is recommended that the soil-structure-interaction is taken into account in seismic retrofit design of existing bridges to make it both reliable and economical.

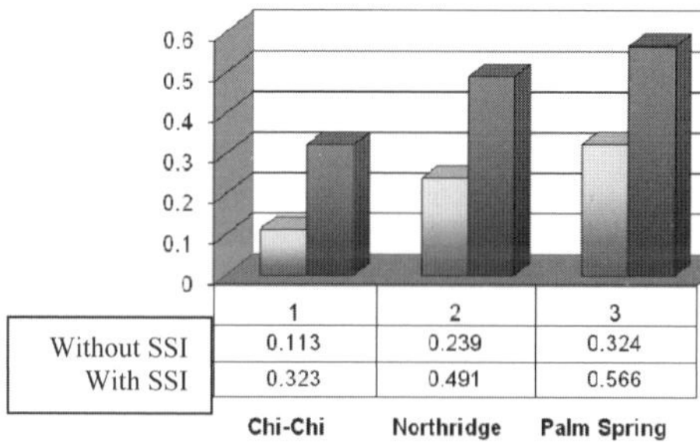

Figure 1. Lateral maximum displacements (cm) of the bridge deck, in case of second retrofit design, subjected to the three selected earthquakes with exclusion and inclusion of SSI effect.

REFERENCES

Hosseini, M., Banaeizadeh, Y., and Khavari, S. R., 2008. Vulnerability Assessment of an Existing Highway Bridge by Push-Over and 3-D Nonlinear Time History Analyses for its Seismic Retrofit Design, The 6th US National Seismic Conf. on Bridges and Highways, Charleston, SC, USA, 27–30 July.

Khavari, S. R. and Hosseini, M., 2010. Seismic Vulnerability Assessment and Comparing Various Retrofit Designs for an Existing Highway Bridge, Proceedings of IABMAS 2010, Philadelphia, Pennsylvania, USA, 11–15 July.

Bridge Maintenance, Safety, Management, Resilience and Sustainability – Biondini & Frangopol (Eds)
© 2012 Taylor & Francis Group, London, ISBN 978-0-415-62124-3

Remote monitoring of suspension bridge cables as calibrated in the laboratory and tested in the filed

D. Khazem & K. Serzan
Parsons Transportation Group, New York, NY, USA

R. Betti
Department of Civil Engineering and Engineering Mechanics, Columbia University, NY, USA

ABSTRACT

Suspension bridges are essential links in the transportation networks especially of large metropolitan areas, such as New York City, and their serviceability is extremely important for the economic and societal growth. The safety of such structures is closely linked to the safety of the cable supported system and, in particular, of their main cables.

Current state of the art approach for cable evaluation, with visual inspection at a few locations was found to be deficient in uncovering the worst condition along the main cable as discovered during cable rehabilitation projects where the entire cable length was unwrapped and in depth inspected.

In this 5 year study, sponsored by the Federal Highway Administration (FHWA) and led by a research team consisting of Columbia University, Parsons Transportation Group and Mistras Group developed a corrosion detection and remote monitoring system for suspension bridge cables that can be considered as an effective tool in major bridge management system.

An integrated methodology that uses state-of-the-art sensing capabilities and NDE technologies to assess the cable conditions has been developed. A smart sensor system that would map the entire length of the cable was developed which provides an accurate method for reliably assessing the condition of suspension bridge cables. In addition to direct sensing technologies, an indirect sensing system consisting of a network of 75 miniature sensors embedded deep within the 7 m long, 0.5 m in diameter cable mock up to monitor the internal environment of the cable and provide information that can be used to assess the potential for deterioration and progression was assembled tested in both the laboratory and in the field.

Sensor readings showed excellent correlation between the environmental conditions and the corrosion rate. This system was also successfully installed and tested on the main cable of one of the major suspension bridges in New York City and data about the internal environment of the cable was remotely collected for one year. This paper discusses the findings of this research and the technology used.

Figure 1. Full scale cable mock-up, 6 m long and 0.5 m in diameter under tension encased within a custom made corrosion chamber.

REFERENCES

American Society for Testing and Materials (ASTM 1994a). "Standard practice for modified salt spray (fog) testing." *G85-02*, West Conshohocken, Pa.

Betti, R., West, A. C., Vermaas, G., and Cao, Y. (2005). "Corrosion and embrittlement in high-strength wires of suspension bridge cables." *J. Bridge Eng.*, 10(2), 151–162.

Betti, R., and Yanev, B. (1999). "Conditions of Suspension Bridge Cables: The New York City Case Study." *TRB*, 1654, 105–112.

Kwun, C. J. Dynes, S. Kim, K. A. Bartels, and D. A. Khazem, "Application of the Magnetostictive Sensor (MsS) for NDE of Cables in Suspension Bridges," *Final Report for Federal Highway Administration, SwRI Project No. 17-7602, Grant No. DTFH61-95-X-00044, Mod. 4 (August, 2000).*

Bridge Maintenance, Safety, Management, Resilience and Sustainability – Biondini & Frangopol (Eds)
© 2012 Taylor & Francis Group, London, ISBN 978-0-415-62124-3

Ultimate strength interaction of unstiffened steel box members subjected to bending and torsion

K. Kim
Department of Civil Engineering, Cheongju University, Cheongju, Korea (ROK)

ABSTRACT

Enclosed boxes are widely adopted for both curved and straight structural members due to their superb torsional properties. When horizontally curved members subjected to vertical loads as shown in Fig. 1, the primary actions are bending about its major axis and torsion. Due to the interaction, the ultimate bending strength is affected by the torsional action, and vice versa. It is inevitable to conduct nonlinear analyses as the combination of linear elastic theories is not adequate for predicting the failure of members subjected to the combined vertical and torsional loading.

The circular form interaction equation incorporating full plastic moment capacity was also suggested for straight members under combined bending and torsion by several researchers. However, it may be inadequate to use such interaction equation for strength design of box members against combined bending and torsion actions. The reason is attributable to the fact that the full plastic bending and torsional moments of the cross section could not be used as the reference value of bending and torsional moments as the ultimate strengths are definitely much lower than those from full plastic bending and torsional moments of the cross-section. The strength computed form full plastic section represent the upper limit value of strength, thus, the interaction equation with full plastic capacity must overestimate the interaction strength. Similarly, in practice, ultimate strength of steel box girder cannot reach the upper limit value because elastic or inelastic buckling occurs in the plates subjected to compressive or shear stresses before full plastic cross-section is formed. It is reasonable that ultimate interaction equation between bending and torsion has the following form:

$$\left(\frac{M}{M_u}\right)^{\alpha}+\left(\frac{T}{T_u}\right)^{\beta}\leq 1$$

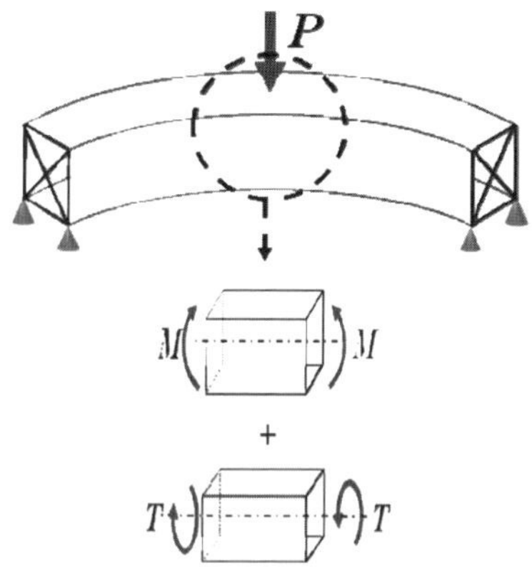

Figure 1. Combined action of bending and torsion in horizontally curved members subjected to vertical loads.

where M_u and T_u = ultimate bending and torsional strengths, respectively. α and β = exponents. Ultimate strength interactions for rectangular steel box members are examined by nonlinear incremental analysis using a commercial FEA program, ABAQUS. The results from numerical investigation are compared to those from other researchers' experimental investigation for verification. Ultimate strength analyses for straight steel box beams subjected to various combinations of vertical and torsional loads were carried out incorporating the effect of both residual stress and initial imperfection. Based on the data collected from the numerical analyses, an ultimate strength interaction equation between bending and torsion was constructed. The suggested equation that is simple and on very good agreements with the experimental test results well represent the lower limits of expected ultimate strengths.

Bridge Maintenance, Safety, Management, Resilience and Sustainability – Biondini & Frangopol (Eds)
© 2012 Taylor & Francis Group, London, ISBN 978-0-415-62124-3

Effects of superstructures on seismic behavior of steel bridge frame piers with circular columns

K. Kinoshita
Gifu University, Gifu, Japan

ABSTRACT

In the author' previous research, seismic performance of existing steel bridge frame piers with circular columns against large earthquakes in the transverse direction was accurately evaluated by using an elasto-plastic FEM pushover analysis and earthquake response analysis (Kinoshita et al. 2008). This research found that existing frame piers could satisfy seismic performance for large earthquakes (Kinoshita et al. 2008). On the other hand, there are influential parameters, such as foundation and superstructures. It is important to grasp effects of their parameters on seismic behavior of existing frame piers. If these effects are expected to be not significant or have good advantage for evaluating seismic performance of existing frame piers, it is thought that seismic performance was simply evaluated by not considering these parameters.

In order to grasp effects of superstructures on seismic behavior of existing steel bridge frame piers, frequency analysis and elasto-plastic FEM earthquake response analysis by considering superstructures in the FEM models were carried out. Based on the author' previous research (Kinoshita et al. 2008), a 2-story type frame pier was used as object frame piers.

Figure 1 shows FEM model of the object steel bridge frame pier considering superstructures. A span length between the object frame pier and neighbor frame piers is 26.5 m. The object frame pier was modeled by shell elements. The neighbor frame piers were modeled by beam elements.

As a result, effects of superstructures on seismic behavior in the longitudinal direction of the 2-story frame pier were beneficial as shown in Figure 2(a). However, their effects on the seismic behavior in the transverse direction are not so significant as shown in Figure 2(b). Thus, seismic performance of existing frame piers can be applicably evaluated by seismic performance evaluation in the transverse direction.

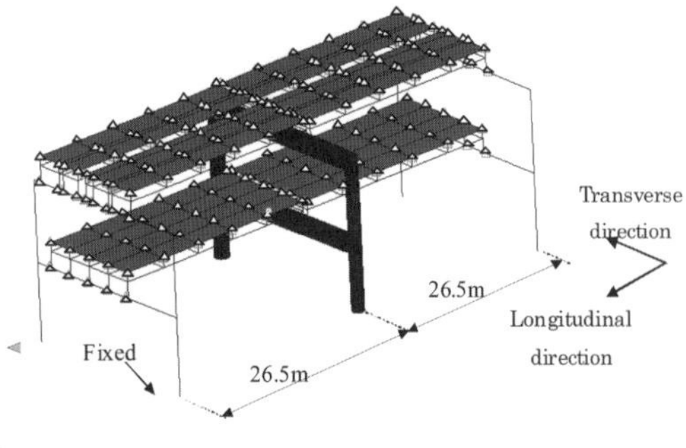

Figure 1. FEM model.

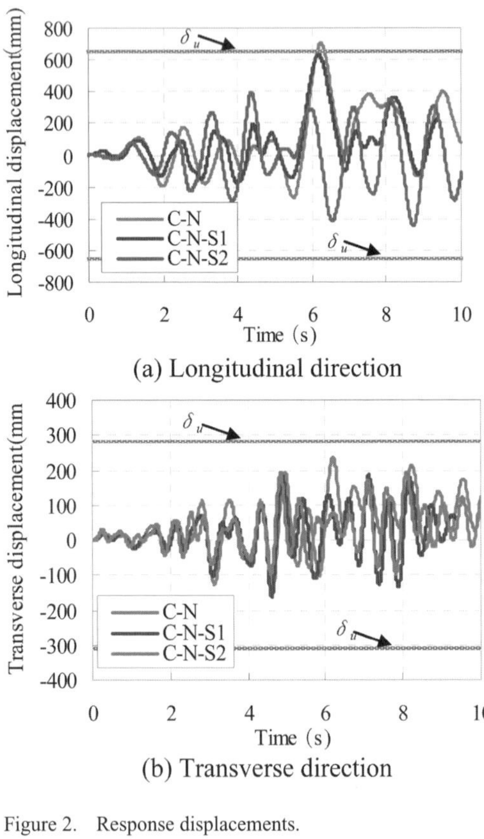

Figure 2. Response displacements.

REFERENCE

Kinoshita K., Chitoshi, M. & Ichikawa, A. 2008. Study on seismic performance of existing steel bridge frame piers with circular column. *Doboku Gakkai Ronbunshuu A* 64(3): 571–587. (in Japanese)

Bridge Maintenance, Safety, Management, Resilience and Sustainability – Biondini & Frangopol (Eds)
© 2012 Taylor & Francis Group, London, ISBN 978-0-415-62124-3

Wind shielding on long span bridges

S. Kite
Arup, UK

M. Carter
Hong Kong, UK

E. Ozkan
Arup, UK

ABSTRACT

Long span bridges are often located in exposed areas subject to strong winds. In recent years the advantages of installing wind shields to protect traffic have been recognised, and there are many examples.

Generally there are two types of wind shields – continuous ones along the deck edges and those which smooth out changes in air flow locally around towers. The incorporation of wind shields results in higher design loads on the bridge and a cost premium - not just the components themselves, but the effect they have on the overall structure. For new bridges, these effects can be designed for and a suitable balance found between performance and cost. Retrofit of existing bridges with the addition of wind shields to improve traffic safety is often compromised by the changes in structural and aerodynamic behaviour that they would induce, and usually only limited measures are possible.

Light-weight high-sided vehicles, particularly empty curtain-walled trucks and double-decker buses, are the most prone to cross winds. Determining the design criteria for wind shields is not straightforward, as accidents are often the result of a combination of effects, including driver reaction.

To eliminate as many unknowns as possible, criteria can be established that the conditions on the bridge should be no worse than on the surrounding road network. Hence, if a vehicle can get to the bridge it should be able to cross. Typically this may lead to a requirement to reduce the overturning moments on vehicles on the bridge deck by 50%.

Numerical simulation can provide a useful indication of the changes in air flow caused by wind shields, but wind tunnel testing is necessary to fully evaluate their effects. Shielding is often formed of continuous horizontal slats, or louvres, with gaps in between. A greater concentration of louvres near the top can have an enhanced reduction in the overturning effects. The use of transparent louvres is favoured to limit the detrimental effect on views from the bridge deck.

Local effects around towers should also be considered. Vertical wind shields rising in height towards the tower can be effective in reducing the sudden changes in wind environment. But the 3-dimensional environment should also be considered, as vertical wind flows resulting from the down-draft effect at towers can have an impact on the likelihood of a vehicle overturning, and may need further mitigation.

Figure 1. Smoke visualisation at edge wind shield (tested at Politecnico di Milano, Dipartimento di Meccanica).

Figure 2. Possible configuration for tower canopy detail, integrated with sign gantry, to limit downdraft effect.

Bridge Maintenance, Safety, Management, Resilience and Sustainability – Biondini & Frangopol (Eds)
© 2012 Taylor & Francis Group, London, ISBN 978-0-415-62124-3

Use of structural monitoring in simulation of train-bridge interaction

R. Kiviluoma
WSP Finland Ltd, Helsinki, Finland

H. Yli-Villamo
Finnish Transport Agency, Helsinki, Finland

ABSTRACT

This paper deals with possibilities and a case study to complement dynamic Finite-Element (FE) simulations of railway bridges with structural monitoring. In simulation, it is straightforward to model vertical axle loads as moving static loads to pick up bridge's resonances depending on span-lengths and train speed, as in EN1991-2 (2003). Developing and assuming a dynamic load model for lateral forces is more demanding, and might benefit on calibration with structural monitoring. From measurements it is known that notable lateral vibration may occur also if the track is perfectly straight on the bridge.

A case study using this approach is conducted for the railway bridge shown in Figure 1. This single track steel bridge has continuous girder of 186 m in length; the main span length is 42 m. Bridge has no ballast and has a lightweight superstructure.

Lateral vibrations have enforced the owner to set speed limits on the bridge (30 km/h and 80 km/h for freight and passenger trains, respectively) and doubted the structural integrity. Vibration has been harmful to overhead contact lines by causing fatigue in droppers. To overcome the speed restrictions and fatigue problems, a detailed FE-simulation has been made. Dynamic load model for the lateral excitation has been deduced from theoretical bases, originating from railway vehicle dynamics on the strait track. The wavelength of the kinetic oscillation (Hunting motion) of the wheel set, whose mathematical formulation has been credited 1883 for Klingel (Wickens 2006), is

$$L = 2\pi\sqrt{\frac{re}{2\lambda}} \tag{1}$$

where L = wavelength of kinematic oscillation, r = nominal radius of the wheel, e = lateral distance between the wheel contact points and λ = conicity of the wheel set. To consider modern railway vehicles and inclined rails, λ is to be replaced with effective conicity λ_e. Due to the inertia of cars and bogies, Equation (1) implies lateral excitation on frequency f_H proportional to the speed V of train ($f_H = V/L$). Parameters r and λ_e varies slightly from car and wheel set to another, and are treated as random variables. To shortcut the complexity that might exist in actual wheel-rail interaction, the resulting lateral load model includes one open parameter, which has been calibrated by structural monitoring on the bridge.

Simulation and monitoring revealed that, due to the tall piers, the fundamental lateral natural frequency bridge is low; 1.08 Hz. The relatively small forces due to the Hunting motion are enough to excite the bridge, and especially the overhear contract lines. It was concluded that the bridge is structurally safe; and the best alternative to mitigate the harmful vibrations is to install dampers to electrification portals. The study strengthens the authors view formed in SB (2007) that for clearly defined target structural monitoring could be useful, and could be recommended for actual problem solving.

Figure 1. Saimaa Canal Railway Bridge, Finland.

REFERENCES

EN1991-2, 2003. Eurocode 1: Actions on structures, Part 2: Traffic loads on bridges, 164 pp.

SB8.2, 2007. Demonstration of bridge monitoring. Sustainable Bridges, EU FP6. Deliverable 8.2, 129 pp.

Wickens, A.H., 2006. A history of railway vehicle dynamics. In Handbook of Railway Vehicle Dynamics, Ed. Iwnicki, S. Taylor & Francis. pp. 5–38.

Bridge Maintenance, Safety, Management, Resilience and Sustainability – Biondini & Frangopol (Eds)
© 2012 Taylor & Francis Group, London, ISBN 978-0-415-62124-3

Probability-based design of spun concrete bridge piers

R. Kliukas & R. Vadlūga
Vilnius Gediminas Technical University, Vilnius, Lithuania

A. Kudzys
Kaunas University of Technology, Kaunas, Lithuania

ABSTRACT

The object of this paper is giving highway and structural engineers a possibility to assess the bearing capacity and reliability index of rational spun concrete bridge piers by unsophisticated but fairly exact probability-based method and to encourage them to use reliability index concept in their design, erection and maintenance practice.

Precast spun (centrifugally cast) concrete members of annular cross-sections reinforced by hot rolled or cold worked steel bars uniformly distributed throughout their parameters satisfy economical and constructive requirements for pier shafts and aesthetical features for short-span road bridges and footbridges.

In spite of efforts to improve and modify semi-probabilistic approaches, it is inconceivable to fix correct values of reliability indices of members and their systems. Therefore, the reliability level of spun concrete piers of analogous bridges designed by these concepts may differ considerably. It is clear that the actual reliability level of piers or their reliability index may be defined only by probability based concepts and approaches.

The means and standard deviations of shaft resistances, representing the ratio between their actual and predicted values, may be defined as: $\theta_{RMm} = 1.02$, $\sigma\theta_{RM} = 0.08$ and $\theta_{RNm} = 0.99$, $\sigma\theta_{RN} = 0.08$ for the shafts of bracing and braced piers, respectively (Vadlūga 1985, Kudzys and Kliukas 2008).

The probability distributions of spun concrete shaft resisting action effects M_R or N_R and permanent load effects M_G or N_G of bridges are very close to a normal distribution. An application of a lognormal distribution is conventional for live load effects M_Q and N_Q due to the road traffic consisting of the sum of identically distributed independent lorries, cars and special vehicles.

Therefore, the survival probabilities of pier shafts within 50-year reference time may be calculated by fairly simple equations written in the single integral forms:

$$\mathbf{P}_{SM} = \mathbf{P}(\beta_M) = \int_0^\infty f_{MRc}(x) F_{Mc}(x) dx \tag{1}$$

$$\mathbf{P}_{SN} = \mathbf{P}(\beta_N) = \int_0^\infty f_{NRc}(x) F_{Nc}(x) dx \tag{2}$$

where f_{MRc} and f_{NRc} are the density functions of normally distributed conventional variables $M_{Rc} = \theta_{RM} M_R - \theta_M M_G$ and $N_{Rc} = \theta_{RN} N_R - \theta_N N_G$;

$$\beta_M = \mathbf{\Phi}^{-1}\mathbf{P}_{SM} \text{ and } \beta_N = \mathbf{\Phi}^{-1}\mathbf{P}_{SN} \tag{3}$$

are the reliability indices of pier shafts. For eccentrically loaded spun concrete shafts of bridge piers, the target reliability index $\beta_T = 3.8$ for 50 years reference period.

The analysis results corroborated that the resistance concept of spun concrete shafts may be based on their ultimate resisting bending moments (for bracing piers) or compressive forces (for braced piers). Thus, the structural safety of bracing and braced shafts may be assessed and predicted by the generalized reliability index β_M or β_N from Equation (3), respectively.

It is shown that the statistical parameters of resisting action effects of shafts of annular cross-sections and their second order destroying forces may be expressed in a fairly simple and easily perceptible manner. They may stimulate engineers having minimum appropriate skills to use full probabilistic approaches in bridge design and maintenance practice more courageously.

REFERENCES

Kudzys, A. and Kliukas, R. 2008. The resistance of compressed spun concrete members reinforced by high-strength steel bars. *Journal of Materials and Structures* 41: 419–430.

Vadlūga, R. 1985. The evaluation of strength of reinforced concrete members of ring cross-section. *Concrete Structures*. Proceedings 14: 97–102.

Bridge Maintenance, Safety, Management, Resilience and Sustainability – Biondini & Frangopol (Eds)
© 2012 Taylor & Francis Group, London, ISBN 978-0-415-62124-3

Analytical and experimental study for flexure of composite bridges with CFT girder

H.J. Ko & H.-E. Lee
School of Civil, Environmental and Architectural Engineering, Korea University, Seoul, Korea

J. Moon
Civil & Environmental Engineering, University of Washington, Seattle, State of Washington, USA

ABSTRACT

Concrete-filled tubes (CFTs) have been widely used across the world as a building column and bridge pier which mainly resist against axial load. Key benefit of CFTs is that the concrete infill is confined by the steel tube, resulting in a tri-axial state of compression that increases the strength and strain capacity of the concrete. In addition, the concrete infill restrains buckling deformation of the steel tube.

Recently, several researchers have attempted to use CFT in the girder bridge system. By applying CFT to the girder bridge system, noise and vibration induced by cars or trains, which is the major disadvantage of the steel I-girder bridge, can be reduced and significant increase in flexural strength is expected. Despite of these benefits, CFTs have limited to use in girder bridge system. This is mainly because deign method has not been well established. Thus, it is crucial to thoroughly understand the flexural strength of the CFT girder bridge system.

In this study, flexural strength of the CFT girder bridge system was investigated. Based on plastic-stress distribution method, simple equation to evaluate the flexural strength of the CFT girder bridge system was derived for various location of the neutral axis.

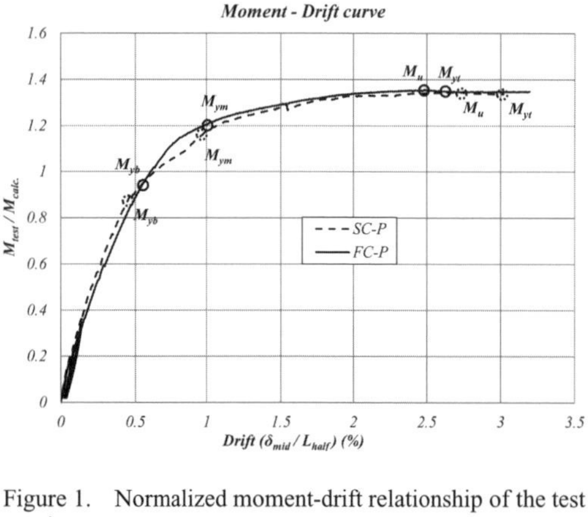

Figure 1. Normalized moment-drift relationship of the test specimens.

Experimental study was then conducted to evaluate the proposed equations and to investigate the effect of internal shear connector inside the steel tube. From the results, it is found that proposed equations underestimated flexural strength of the CFT girder reasonably as shown in Figure 1. Also, it can be seen that internal shear connector has insignificant effect on the global response of the CFT girder.

Finally, FE model for composite girder with CFT was developed to enhance understanding of the test results and further parametric study. From the comparative results, developed FE model provided good prediction of global response of the CFT girder as well as local behavior such as crack pattern and slip.

REFERENCES

ABAQUS. 2010. Abaqus Analysis User's Manual version 6.10. Dassault Systèmes Simulia Corp.

American Concrete Institute (ACI). 2008 Building code requirements for structural concrete and commentary. ACI 318-08 American Concrete Institute. Farmington Hills, MI.

American Institute of Steel Construction (AISC). 2005. Specification for structural steel buildings. AISC 2005. Chicago, IL.

Baltay, P. & Gjelsvik, A. 1990. Coefficient of friction for steel on concrete at high normal stress. *J. Mat. Civ. Eng.* 2(1): 46–49.

Hsu, T.T.C. Mo, Y.L. 2010. Unified theory of concrete structures. John Wiley & Sons, Ltd.

Hu, H.T. & Schnobrich, W.C. 1989. Constitutive modeling of concrete by using nonassociated plasticity. *J. Mat. Civ. Eng.* 1(4): 199–216.

Roeder, C.W. et al. 2010. Strength and Stiffness of Circular Concrete-Filled Tubes. *J. Struct. Eng.* 136(12): 1545–1553.

Saenz, L.P. 1964. Discussion of 'Equation for the stress-strain curve of concrete' by P. Desayi, and S. Krishnan. *ACI J.* 61; 1229–1235.

Shima, H. & Watanabe, S. 2010. Formulation for load-slip relationships of headed stud connector. *34th Int Sym on Bridge and Struct Eng.* Venice.

Bridge Maintenance, Safety, Management, Resilience and Sustainability – Biondini & Frangopol (Eds)
© 2012 Taylor & Francis Group, London, ISBN 978-0-415-62124-3

Life-cycle cost estimation of a new metal spraying system for steel bridges

T. Kondo
Institute of Technologists, Saitama, Japan

S. Okuno
Nitto Engineering Co., Ltd., Osaka, Japan

ABSTRACT

Japan often experiences high humidity and hot temperatures, especially during the summer. Therefore, the use of techniques to prevent metallic corrosion is important in steel construction projects. With this in mind, a new metal spraying system has been developed. An outline of this recently developed metal spraying system and its performance for zinc-aluminum complex sprayed steel bridges was previously reported in IABMAS '04. In that paper, the efficacy of the system in preventing the corrosion of steel bridges was discussed. In IABAMAS '06 and IABMAS 2010, the corrosion prevention mechanism was discussed on the basis of an analysis of results from actual meta-sprayed steel structures. The durability of the sprayed film was also discussed.

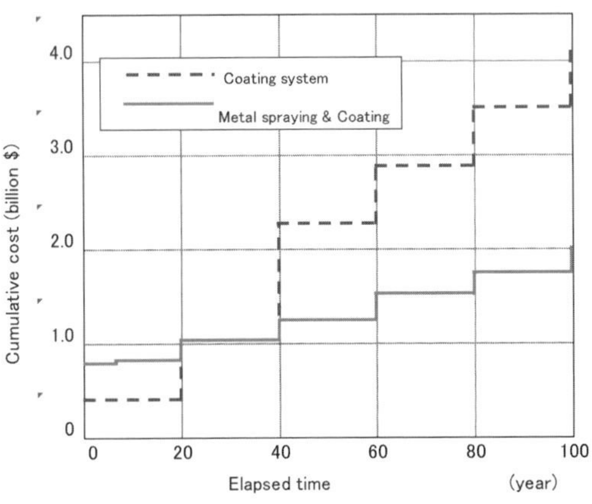

Figure 1. Estimated life-cycle cost in severe conditions.

In addition, results of an examination of several steel structures that had been sprayed by using the system about 10 years before IABMAS'08 were presented at IABMAS'08, and the reference service life of the system was discussed on the basis of these results.

This paper is continuation of the IABMAS'08 report. Results are presented from an examination of the above-mentioned steel structures and the reference service life and life-cycle cost estimation of the system are discussed on the basis of these results.

The zinc-aluminum sprayed film retained corrosion inhibition effects for a long time.

In this paper, the reference service life of the metal spraying system is discussed on the basis of investigations of the actual condition of steel structures with a metal sprayed film. Consequently, the life of this product is related to the location and part of the structure. If investigations are repeated after the completion of construction, it will be possible to estimate more accurate product service life numbers.

Furthermore, the LCC is estimated on the basis of above results. It is found that using the metal spraying system in severe conditions is economically advantageous compared to the coating system. While, using the metal spraying system in mild conditions is not so economical compared to coating system.

REFERENCES

T. Kondo et al. 2004: *Application of a new metal spraying system for steel bridge Part 1 & 2*, Proceedings of IABMAS '04, Kyoto Japan.

T. Kondo et al. (2006): *Application of a new metal spraying system for steel bridge Part 3*, Proceedings of IABMAS '06, Porto Portugal.

T. Kondo et al. (2008): *Application of a new metal spraying system for steel bridge Part 4*, Proceedings of IABMAS '08, Seoul Korea.

Bridge Maintenance, Safety, Management, Resilience and Sustainability – Biondini & Frangopol (Eds)
© 2012 Taylor & Francis Group, London, ISBN 978-0-415-62124-3

Static and dynamic load tests of a long-span cable-stayed bridge over Odra River in Wrocław

M. Kużawa, J. Bień, P. Rawa, T. Kamiński & J. Zwolski
Wroclaw University of Technology, Poland

ABSTRACT

Procedures and results of the preliminary load static and dynamic tests of the long-span cable-stayed bridge along the Wroclaw Motorway Bypass are presented and discussed in the paper. The tested bridge consists of two parallel decks connected with the central pylon by means of cable stays. Each construction is a four-span continuous box beam with the following spans: 49 m + 256 m + 256 m + 49 m. The pylon (122 m high) is constructed as a RC structure and the decks are made of prestressed concrete.

Four schemes of static load tests were performed by means of 40-ton trucks. Maximum load consisted of 40 trucks of 1600 tons total weight (Figure 2).

Dynamic load consisted of three series of truck passages with various configurations of the passage routes and at the speeds from 5 km/h to 80 km/h. Special tests with the artificial bumps modelling defects of the road surface were also performed (Figure 3).

Experimental results were compared with effects of the numerical analysis achieved by means of advanced finite element model (Figure 4).

Figure 1. General view of Rędziński Bridge.

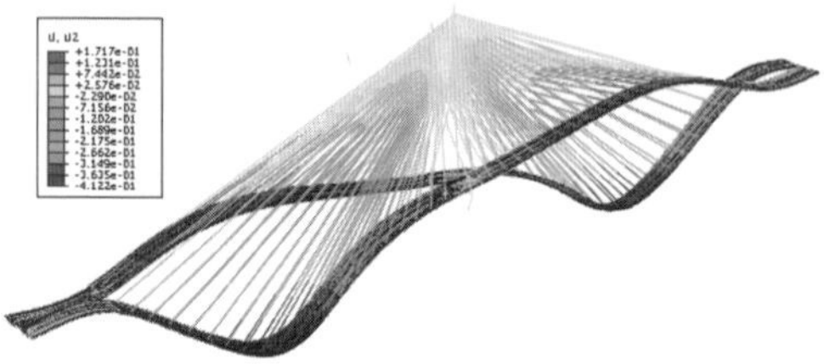

Figure 2. Deformations of the bridge under 40 trucks (scheme S3).

Figure 3. Dynamic tests with artificial bumps.

Figure 4. Axonometric top view of details of FEM model used in numerical analysis.

Bridge Maintenance, Safety, Management, Resilience and Sustainability – Biondini & Frangopol (Eds)
© 2012 Taylor & Francis Group, London, ISBN 978-0-415-62124-3

Composite concrete encased steel beam-column design in AASHTO specifications

L.L.-Y. Lai
Specialty Engineering, Inc., Bristol, PA, USA

ABSTRACT

A comparison of the ACI (2011), AISC (2011), and AASHTO (2010) design and detailing requirements for composite concrete encased steel beam-columns reveals that there are discrepancies among these design specifications.

The most significant discrepancy is the design methods used in the three codes. The ACI Codes have long adopted the strain compatibility method for the design of reinforced concrete and composite beam-columns. The AISC Manual (2011) has also revised the design criteria recommending either the strain compatibility method or plastic stress distribution method for composite beam-columns. A rigid plastic simplified method is also proposed in the commentary. However, the AASHTO LRFD Specifications (2010) are still using the tri-linear equation.

A MATHCAD procedure was developed to compute the P-M resistance curves using the strain compatibility, plastic stress distribution, and rigid plastic methods (Leon and Hajjar 2008, Lai and Chang 2010). The results are compared to the resistance curves obtained using the AASHTO simplified criteria. The design criteria used in the AASHTO Specifications result in over-conservative estimate of the P-M resistance curves, particularly the flexure resistance. This conservativeness is due to the use of a simplified nominal flexure resistance equation, which is not used by any other design codes. One example shows that the flexure resistance computed using the strain compatibility method could be 3.1 times of that obtained using the criteria in the AASHTO Specifications.

The simplified nominal flexure resistance equation in the AASHTO Specifications is believed to be derived for bending about the major axis only and it results in less and inconsistent conservativeness if it is applied to bending about the minor axis. A different equation for bending about the minor axis is needed to provide consistent conservativeness and safety factor.

REFERENCES

AASHTO 2010. *AASHTO LRFD Bridge Design Specifications. 5th Edition*, Washington, D.C.: American Association of State Highway and Transportation Officials.

ACI 2011. *Building Code Requirements for Structural Concrete (ACI 318-11),* Farmington Hills, MI: American Concrete Institute.

AISC 2001. *Manual of Steel construction, Load and Resistance Factor Design. 3rd Ed.* Chicago: American Institute of Steel Construction.

AISC 2011. *Manual of Steel construction. 14th Ed.* Chicago: American Institute of Steel Construction.

Galambos, T. V. (ed.) 1998. *Guide to Stability Design Criteria for Metal Structures, 5th Ed.* New York, NY: John Wiley & Sons.

Lai, L. L-Y & Chang, W-S, 2010. Rehabilitation of Market and Chestnut Street. *International Bridge Conference Proceedings*, Pittsburgh, PA.

Leon, R. T. & Hajjar, J. F. 2008. Limit State Response of Composite Columns and Beam-Columns, Part II: Application of Design Provisions for the 2005 AISC Specification. *Engineering Journal, First Quarter, Vol. 45, No. 1, 21–46*, American Institute of Steel Construction, Inc.

Ziemian, R. D. (ed.) 2010. *Guide to Stability Design Criteria for Metal Structures, 6th Ed.* New York, NY: John Wiley & Sons.

Bridge Maintenance, Safety, Management, Resilience and Sustainability – Biondini & Frangopol (Eds)
© 2012 Taylor & Francis Group, London, ISBN 978-0-415-62124-3

Control of traffic loads on Great Belt Bridge

J. Laigaard, N. Bitsch & H. Gjelstrup
COWI, Kgs. Lyngby, Denmark

K.A. Nielsen
A/S Great Belt, Korsør, Denmark

ABSTRACT

Since inauguration of the Great Belt Link in June 1998 the traffic growth has been significantly above the forecasts. In order to determine the risk of fatigue damages in the orthotropic bridge deck, the influence of the increased traffic on the fatigue capacity of the orthotropic steel deck has been investigated by structural monitoring. An automatic monitoring system has been installed to continuously sample stresses and calculate the accumulated fatigue impact. This is carried out for two bridge deck sections in the suspension span, a section approximate 1590 meters from the anchor block and a section close to an expansion joint.

The system provides the bridge owner, A/S Great Belt with a tool for controlling the wear of the pavement and the orthotropic steel deck, and thus the maintenance costs. From the system, the owner can conclude on the influence of traffic patterns, heavy vehicles and temperature. This allows A/S Great Belt to control wear from traffic. For instance, this may provide A/S Great Belt with argument for increasing or reducing toll fares for heavy vehicles, possible with a dependency on the hour of the day for driving across the bridge. The monitoring system has proven that the applied orthotropic deck design with pavement steel composite effect is robust against fatigue.

The influence of high temperatures has attracted some attention as the orthotropic steel deck has been designed by taking the composite effect of the asphalt pavement into account. This means that traffic loads transferred to the deck depends on the pavement temperature, and that wear and fatigue contribution will reach its maximum during hot sunny days where the asphalt reaches the highest temperatures. This has been assessed by the monitoring system from which, it is possible to estimate the remaining fatigue life by taking the temperature influence into account.

For the Great Belt Bridge, the results show that there is no indication of fatigue problems within the service life of the bridge with the current expectations to traffic. However, although there is no apparent fatigue risk, the fatigue monitoring system has demonstrated that it provides a tool for documenting and controlling the impact of heavy vehicles.

Generally, the bridge deck measurements on the Great Belt Bridge implicate that traffic loads are likely to influence the operational expenditure of the bridge deck and its service life in terms of fatigue and wear.

Although there is no fatigue risk, an optimized and cost effective design as for the Great Belt orthotropic steel deck means that it will be more sensitive to traffic loads. The sensitivity of the deck structure means that the owner has a large interest in monitoring and controlling the traffic loads. This becomes further important given the impact of temperature, which is outside the control of owner.

The investigations and experiences from the Great Belt shows that temperature as well as heavy vehicles play a key role to the fatigue stresses and wear of the steel bridge deck. Bearing in mind that this also will be true for the pavement, owners should be encouraged to focus not only on controlling service life of the steel deck but also on the pavement.

In the future, if the combined impact of high temperature and heavy vehicles becomes too large, the owner may have the opportunity to control and optimize the traffic pattern and thereby extending the service lives and keeping maintenance costs down by using higher toll fares during high day temperature and low tolls fares at low night temperatures.

Bridge Maintenance, Safety, Management, Resilience and Sustainability – Biondini & Frangopol (Eds)
 ISBN 978-0-415-62124-3

Residual capacity from aggregate interlock of cracked concrete slab bridge

E.O.L. Lantsoght, C. van der Veen & J.C. Walraven
Delft University of Technology, Delft, The Netherlands

ABSTRACT

In the Netherlands, 600 reinforced concrete slab bridges, most of which were built before 1976, are under discussion as a result of the increased traffic loads and volumes. A large research program to determine the actual shear bearing capacity was started. (Walraven 2010).

This paper shows a case study which was performed on a 50-year-old reinforced concrete slab bridge with large cracking in the southern concrete approach bridge. After repair, it was uncertain at which crack width the traffic loads on the bridge should be further restricted. The shear capacity was calculated by counting on the aggregate interlock capacity of a supposedly fully cracked cross-section (Walraven 1981). The tension acting on the cross-section as a result of the restraint of deformation was subtracted from the clamping force of the reinforcement to determine a reduced effective reinforcement ratio ρ_{clamp}. Comparing the resulting shear capacity shows that the capacity of the section with a through crack is higher than the capacity of a section without a through crack calculated according to the Dutch Code NEN 6720.

Furthermore, an aggregate interlock relation between shear capacity and crack width based on an unreinforced section (based on Walraven 1981) was used to find the crack width at which the shear capacity V_{u_unr} of section with a through crack becomes smaller than the shear capacity V_{VBC} of a member without a through crack, Figure 1. Limits for crack widths at which load restrictions should be imposed were derived.

Since the aggregate interlock model regards both axial and shear forces, a sensitivity analysis of the influence of the restraint of deformation was carried out. The resulting axial force from the tension in the concrete cross-section was compared to the remaining axial force delivered by the top reinforcement after using the required clamping capacity necessary for equilibrium resulting from the aggregate interlock expressions. This comparison showed the importance of the equilibrium in the axial direction. Equilibrium was obtained for 71% restraint or less. For higher levels of restraint, no equilibrium was found. Therefore, measurement of the actual restraint level were proposed.

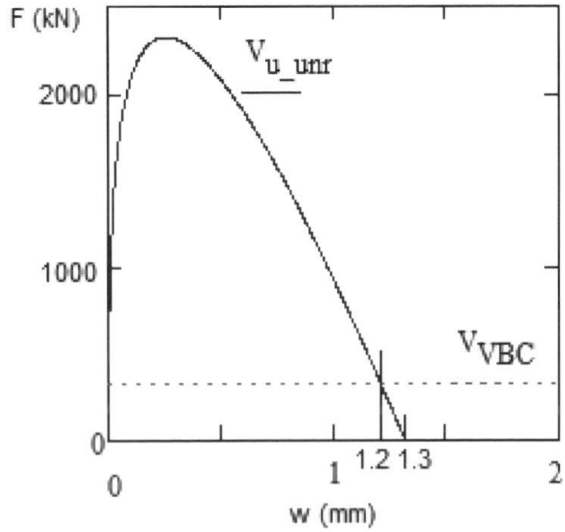

Figure 1. Plot of shear capacity from NEN 6720 (V_{VBC}, dashed line) and from aggregate interlock based on an unreinforced section (V_{u_unr}, solid line) as a function of the crack width w.

The large structural capacity of the through-cracked concrete section shows that the residual capacity based on the aggregate interlock capacity of reinforced concrete slab bridges with existing cracks offers a practical and easy-to-implement method to determine the residual bearing capacity.

REFERENCES

Normcommissie 351001. 1995. *Voorschriften Beton TGB 1990 – Constructieve Eisen en Rekenmethoden (VBC 1995, NEN 6720)*. Delft: Nederlands Normalisatie-instituut.

Walraven, J.C. 1981. Fundamental analysis of aggregate interlock. *Journal of the Structural Division – ASCE* 107(11): 2245–2270.

Walraven, J.C. 2010. Residual shear bearing capacity of existing bridges. In *fib* Bulletin 57, *Shear and punching shear in RC and FRC elements; Proceedings of a workshop held on 15–16 October 2010 in Salò, Lake Garda, Italy*. Lausanne: International Federation for Structural Concrete (*fib*).

Bridge Maintenance, Safety, Management, Resilience and Sustainability – Biondini & Frangopol (Eds)
© 2012 Taylor & Francis Group, London, ISBN 978-0-415-62124-3

The renewal of the Burtscheidt Viaduct in Aachen Germany

A. Laubach
Schüßler-Plan Consulting Engineers, Düsseldorf, Germany

ABSTRACT

The Burtscheidt Viaduct is located at the eastern entrance to the Aachen main station and was built in the years 1838 to 1840. It is one of the oldest railway bridges still in service in Germany and is therefore a listed monument. The viaduct is 251 m long and up to 20 m high, it has 8 large arches of 12, 2 m width, a middle section with 4 arches in 2 vertical arrays and 11 small arches of 6 m width. The structure was completely made of brickwork.

The viaduct carries a two-track railway line which is part of the European link from Paris over Brussels, Liege and Aachen to Cologne. It was therefore absolutely necessary to maintain both tracks in service during the complete construction phase of the renewal works. The load bearing capacity of the viaduct was examined by a combination of displacement and strain measurements and finite element calculations with diverse models. In result it was found that the calculated load bearing capacity was not any longer sufficient due to the deterioration overage and condition of the more than 170 year old structure. A renewal of the railway viaduct was necessary.

Due to the boundary conditions in the centre of the 2000 year old city of Aachen it was impossible to build a new railway bridge beside the existing viaduct. Therefore it was necessary to strengthen the existing structure while the railway viaduct was in service. The solution of Schüßler-Plan to this design challenge was a separate concrete structure below the existing arches which is designed to carry all dead and live loads of the existing viaduct and itself.

Concrete arches with a thickness of 50 cm are placed under the brickwork viaduct and existing hollow spaces in the viaduct are filled with concrete. The concrete arches are connected with each other by anchors drilled trough the existing brickwork pillars. The supporting concrete structure is situated on a drilled pile foundation in case of the large arches 4 to 11 respectively a raft foundation in case of the small arches 16 to 26. The arch 5 is situated over thermal water resources; therefore it was not allowed to influence the thermal water by drilled piles. A 3 m high concrete girder takes the loads of the arch 5 to the piles of the arches 4 and 6.

The face side of the brickwork viaduct is still visible and contains therefore the elevation of the existing viaduct. The face side is secured by stainless steel anchors and the edges of the concrete arches.

At the moment, the existing viaduct will still carry all loads by itself. In the future the supporting structure will take more and more of the loading from the existing viaduct with increasing deterioration of the old structure. The renewed structure complies with all serviceability and ultimate limit state requirements like a new railway bridge and can therefore be kept in service for future times.

REFERENCES

D.B. Netz, A.G. Magdeburg, 2000. Messung und Nachrechnung des Burtscheider Viadukts in Aachen zur Ermittlung seiner Tragfähigkeit, Bericht 99233, NPZ 25 Brückenmessung und – bewertung.

Heitfeld, K.-H., Heitfeld, M., Rosner, P., Schetelig, K. & Steinmetz, E., Mitt. Ing.- u. Hydrogeol., Heft 91, S. 119–128, Aachen Erneuerung Burtscheider Viadukt – Gründungsmaßnahmen im Thermalwasserzug.

Bridge Maintenance, Safety, Management, Resilience and Sustainability – Biondini & Frangopol (Eds)
 ISBN 978-0-415-62124-3

Rate of convergence of measured stress range spectra

J. Leander & R. Karoumi
Division of Structural Engineering and Bridges, KTH Royal Institute of Technology, Stockholm, Sweden

ABSTRACT

The subject treated is the determination of stress rage spectra for fatigue assessment using monitoring. In several guidelines for fatigue assessment, monitoring is mentioned as a mean to reduce the uncertainties in the response of a structure. The duration of the monitoring is however rarely treated.

Due to the exponential relation between fatigue life and the stress range, small changes in the stress range can have a substantial influence in the predicted service life. For existing bridges that are desirable to keep in service, a monitoring campaign can be motivated by an expected decrease of the stress ranges. Monitoring does, however, give rise to additional questions such as the quality of the response and the required duration of measurements. Apart from saving time, lower cost and fewer data to analyze are incentives to keep the duration of the measurements short. The monitoring period should, however, have a sufficient length to capture a representative stress range spectra for the structure. Variation in traffic between days, weeks and perhaps even months should be taken into consideration in the decision concerning the monitoring duration.

A case study of the Söderström Bridge, one of the most important bridges in Sweden, is incorporated in the study. It is a part of the main railway line in Stockholm City and joins the north and south main railways for passenger trains through Sweden. Theoretical assessments have indicated that the fatigue life of the bridge is exhausted.

The contribution of the paper is the closed form model for convergence progress. As an alternative to a predetermined monitoring duration, the end time could be a variable dependent on the convergence of one or several stress range spectra.

The model for convergence is based on the accumulated damage according to the Palmgren–Miner rule, and the assumption of an exponential shape of the convergence progress as

$$f_D(n) = a\exp(-bn) \tag{1}$$

where a and b are coefficients determined by a least squares fit and n is the measurement period in a suitable unit, e.g. a day or a week. The function value f_D is fitted against the normalized value of the accumulated damage as

$$d_D = \left|1 - \frac{D(n)}{D(N)}\right| \tag{2}$$

where $D(n)$ is the accumulated damage after period n and $D(N)$ is the accumulated damage for all available periods N. Figure 1 shows the normalized accumulated damage for two gauges from the case study.

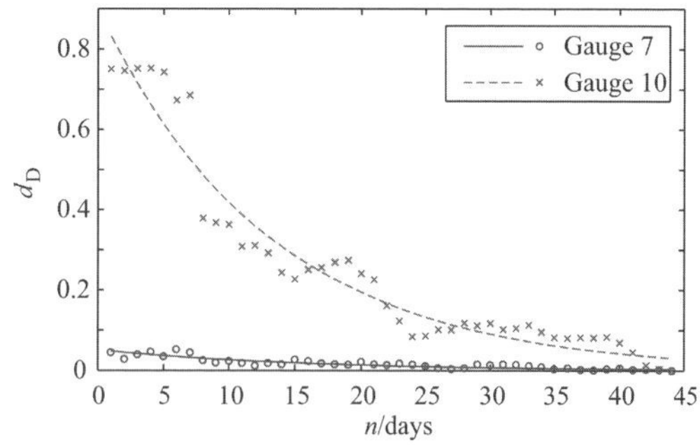

Figure 1. Normalized accumulated damage for two gauges from the case study. The values are normalized according to (2) with $N = 44$ days.

The required duration of monitoring differs between different structural parts in the bridge. Stress range spectra for members with a short influence line seem to converge faster than for members with a longer influence line. For some gauge locations, a diverging shape of the spectrum has been found due to non-linear behavior of the structure.

Bridge Maintenance, Safety, Management, Resilience and Sustainability – Biondini & Frangopol (Eds)
© 2012 Taylor & Francis Group, London, ISBN 978-0-415-62124-3

Effectiveness of multiple unseating prevention devices for bridges under extreme earthquakes

T.Y. Lee
Department of Civil Engineering, National Central University, Taiwan

D.W. Chang & D.C. Dzeng
CECI Engineering Consultants, Inc., Taiwan

ABSTRACT

In the past major earthquakes, such as the 1995 Japan Kobe earthquake and the 1999 Taiwan Chi-Chi earthquake, numbers of bridges suffered damage as well as unseating of superstructures. Whenever unseating failure occurs again, the importance of unseating prevention strategy is emphasized repeatedly. Multiple unseating prevention devices then have been used to increase the safety in modern bridge seismic design. Lately, the seismic performance design on entire bridges as well as components thereof has been highlighted. Unseating prevention is the major requirement to achieve the safety goal. Since unseating prevention devices generally work in extreme situations, it is questionable whether the unseating prevention devices are effective during major earthquakes and whether multiple unseating prevention devices provide safer protection. Actually, it is quite difficult to conduct a shaking table test by using a proto-model, especially for multi-span or long-span bridges to recognize the function of unseating prevention devices under major earthquakes. Therefore, this study is aimed to clarify the effectiveness of multiple unseating prevention devices for bridges under strong near-field ground motions by numerically predicting the ultimate situation of bridges.

The Vector Form Intrinsic Finite Element (VFIFE), a new computational method developed by Ting et al. (2004), is superior in managing the engineering problems with material nonlinearity, discontinuity, large deformation, large displacement and arbitrary rigid body motions of deformable bodies. Once reaching the ultimate state, a bridge underwent nonlinear and hysteresis behavior, progressive failure, fragmentation and collapse. Thus, the VFIFE is selected to be the analysis method in this study. Many elements and analysis methods, such as hysteresis elements, pounding elements, gap and hook elements, friction elements, in the VFIFE have been developed for ultimate analysis. A six-span simply-supported bridge and a continuous-span bridge are analyzed and compared. Two types of unseating prevention devices, tension type and compression type, are separately or simultaneously installed on the bridges. A parametric study on the gap length is performed. In simulation, the bridges are subjected to near-field ground motions recorded at JMA Kobe in the 1995 Japan Kobe earthquake. The ground acceleration is amplified from 100% to 300% at an increment of 10%.

Through numerical simulation of two bridges for three cases with different installation of unseating prevention devices, the ultimate states are demonstrated and compared. It is observed that the simply-supported bridge suffers unseating of the superstructure as the ground motion is amplified equal to and larger than 280% while the continuous-span bridge do not unseat even under 260% ground motion when only steel tendons are installed. However, when both steel tendons and stoppers are installed, the simply-supported and continuous bridges suffer unseating 60%–70% earlier than the case only with steel tendons and the case only with stoppers.

The deck unseating is attributed to column failure because the stoppers transmit larger force to the columns. Actually, the collapse of the simply-supported bridge with only steel tendons or stoppers is due to insufficient unseating prevention length. The simulation results of the continuous bridges show that the failure can be also attributed to column collapse except the end span unseat due to insufficient seating length. The results demonstrate that multiple unseating prevention devices do not increase the safety of the studied bridges as expected, even deteriorate the safety of bridges.

REFERENCES

Ting, E. C., Shih, C. and Wang Y. K. 2004. Fundamentals of a vector form intrinsic finite element: Part I. basic procedure and a plane frame element. *Journal of Mechanics* 20: 113–122.

Bridge Maintenance, Safety, Management, Resilience and Sustainability – Biondini & Frangopol (Eds)
© 2012 Taylor & Francis Group, London, ISBN 978-0-415-62124-3

Static and dynamic windproof efficiency evaluation for bridge cross section considered transmission of fairing

H. Lee, S. Oh, N. Chun & H.-E. Lee
School of Civil, Environmental and Architedtural Engineering, Korea University, Seoul, Korea, South

ABSTRACT

In recent years, length of bridge is longer and longer according to develop new material and construction technology. To Evaluate effect of behavior under wind induced has became main issue as the length of bridge progress. Evaluating effect of behavior under wind induced is divided into static and dynamic efficiency. The static efficiency is a factor which determine wind load and evaluated by drag coefficient. The dynamic efficiency evaluates a harmful vibration like vortex shedding, flutter and buffeting. Through the elevation of the bridge cross section can improve the static and dynamic windproof performance. Recently there is a case that a twin-box section improved by many researches is constructed, and a section of triple-box shape is constructing. Also, researches about fairing and railing have been progressing constantly. In this study, under the change of transmission of fairing, effect of behavior under wind induced is evaluated. Through Performing wind tunnel test, check the drag coefficient, vortex shedding, flutter, and performing CFD(Computational Fluid Dynamics), check the drag coefficient, flutter derivatives. At this time, the fairing maintains constant angle, length. And the transmission is realized from idealizing fairing part. The wind tunnel test carried out 3 cases about transmission 0%, 25%, 50% in fairing. In this test, we got drag coefficient according to attack angle and dynamic performance according to wind velocity each cases. CFD analysis carried out 3 cases about transmission 0%, 25%, 50% in fairing. In this analysis, we also got we got drag coefficient according to attack angle and flutter derivatives using moving mesh code according to wind velocity each cases. From extracted flutter derivatives, we found out flutter stability. Finally it is confirmed that the windproof efficiency of the bridge cross section is concerned about the transmission of fairing through compare test with CFD result.

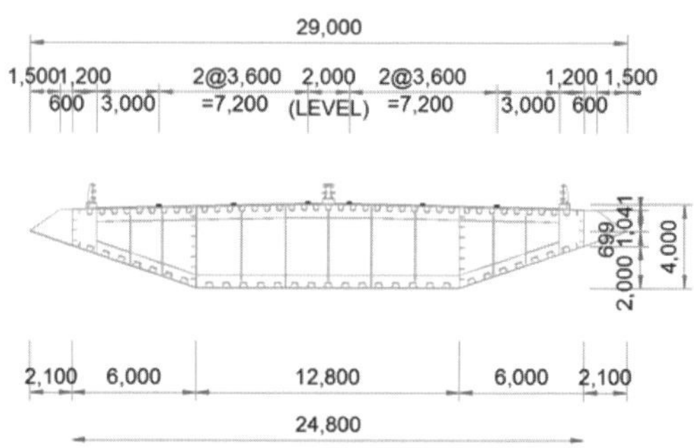

Figure 1. Bridge cross section.

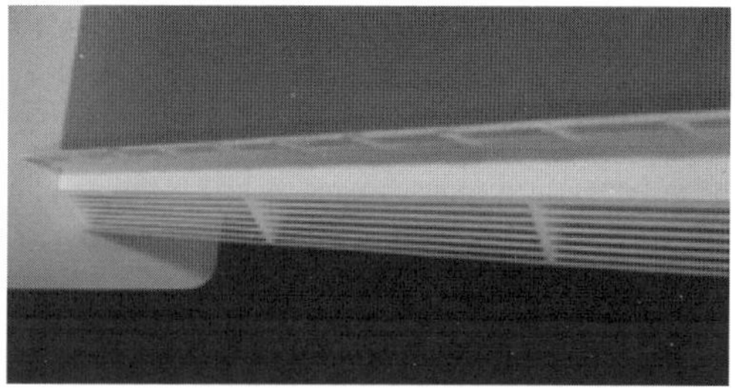

Figure 2. Idealizeing fairing part (Transmission).

REFERENCES

Cermak, J. E., 1975, "Applications of Fluid Mechanics to Wind Engineering", A Freeman Scholar Lecture, ASME
Fluent Inc., 1995, "Fluent Theory Manual", Fundamentals of Fluid Mechanics 4th, Wiley, pp. 533–606
Niemann, H., 1998, "The boundary layer wind tunnel: and experimental tool in building aerodynamics and environmental engineering," J. of Wind Engineering and Industrial Aerodynamics, Vol. 74–76, pp. 826–838
Simiu, E. and Scanlan, R. H., 1996, Wind Effects on Structures, 3rd edition, John Wiley and Sons, Inc.
Tanaka, H. 1992, Similitude and modelling in bridge aerodynamics, Proceedings of the First International Symposium on Aerodynamics of Large Bridges, Copenhagen, Denmark, pp. 83–94

Bridge Maintenance, Safety, Management, Resilience and Sustainability – Biondini & Frangopol (Eds)
 ISBN 978-0-415-62124-3

Assisting routine inspection of highway bridges with IFC-based 3D models

S.-H. Lee, M.G. Huang & B.-G. Kim
School of Civil and Environmental Engineering, Yonsei University, Seoul, South Korea

ABSTRACT

In this paper, an approach for integrating bridge routine inspection information gathered by means of handwritten notes/sketches, photographs, and field measurements with a 3D bridge model is proposed for the integrated management of bridge inspection information and design/construction information.

The 3D bridge model, which is component-wise and open standard based 3D models built through bridge information modeling technique, plays the central role of the information management during routine inspections in this approach. Industry Foundation Classes (IFCs), which is an open standard data model developed by buildingSMART International to describe building and construction industry data, is adopted for constructing the 3D bridge model. In order to integrate gathered inspection information with IFC-based 3D bridge model, the authors developed a faceted classification method where inspection information is classified into three facets according to the material, element role, and construction type of a bridge element, respectively.

The techniques for positioning and visualizing of specific damages on a bridge element are also identified. Positioning of bridge damages is facilitated through generating Cartesian grids on surfaces of 3D objects in the bridge model. Damages can be located three dimensionally and their shape can be precisely modeled using the proposed method, while the traditional routine inspection could only describe the damage linguistically and assign it to a bridge element through predefined element number system. Visualization of bridge damages are realized by texture mapping, which is a method for adding detail, surface texture, or color to a computer-generated graphic or 3D model; the damage types (e.g. corrosion, concrete spalls, rebar exposure) could be identified intuitively. Then, IFC model requirements are identified based on the proposed faceted classification and additional IFC property sets are proposed.

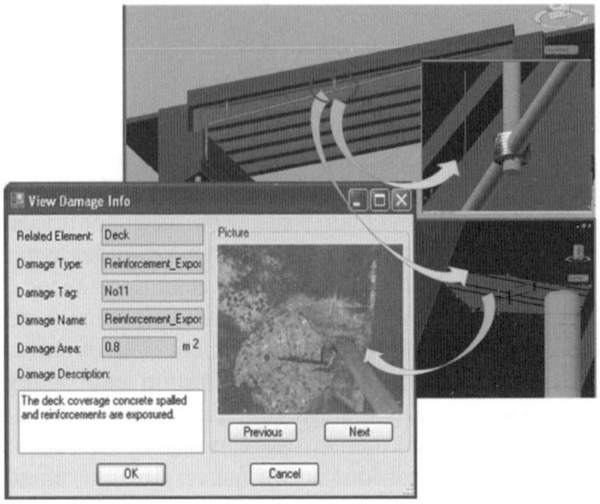

Figure 1. 3D visualization of bridge routine inspection results.

Finally, a prototype tool for assisting 3D bridge model based bridge routine inspection is developed based on the method provided by Lee and Jeong (2006). A case study is conducted for Kae-Peong Bridge (Fig. 1), which is a one span prestressed concrete bridge in Korea. The case study indicates the applicability and efficiency of the approach in supporting intuitionally and quick collection, recording and retrieving of bridge inspection data. The further research will focus on applying the method in mobile devices such as iPad for assisting field inspector to collect high quality bridge condition data. More tools (e.g. matrix barcode and RFID) will be integrated with the system.

REFERENCES

Lee, S.-H. & Jeong, Y.-S. 2006. A system integration framework through development of ISO 10303-based product model for steel bridges. *Automation in Construction* 15: 212–228.

Bridge Maintenance, Safety, Management, Resilience and Sustainability – Biondini & Frangopol (Eds)
ISBN 978-0-415-62124-3

Impact effect statistic investigation of concrete filled steel tube arch bridge under moving vehicles based on the field test and simulation analysis

Y. Li, Y. Liu & H. Sun
School of Traffic Science and Engineering, Harbin Institute of Technology, Harbin, P.R. China

ABSTRACT

The dynamic effect of moving vehicles on bridges was generally treated as a dynamic impact factor in many design codes. Due to the road surface deterioration of existing bridges, studies have shown that the calculated impact factors from field measurements could be higher than the values specified in design codes that mainly target at new bridge designs. This paper developed a 3D vehicle-bridge coupled model to simulate vehicle-bridge coupled vibration and investigated the impact factor for a typical half through concrete filled steel tubular arch bridge. The dynamic field test was carried out on the bridge under moving vehicle and ambient vibration. The dynamic response of the selected points was recorded for the investigation of the bridge's dynamic behavior. It was found that the numerical analysis matched well with the experimental data. The stochastic effect of surface roughness on the impact factors for the bridge was thoroughly investigated by using these models. Then vehicle-bridge dynamic response analysis was carried out based on multi-group random roughness samples. Impact effect of the bridge was studied based on statistics analysis method under different road surface conditions and vehicle speeds. Chi-square tests were then performed on the impact factors and it was found that the impact factors obtained under the same road surface condition followed the Log-Normal distribution. The analysis results showed that the statistic characteristic parameters of the maximal dynamic response tended to be stable when the roughness samples number exceeded 40, the impact factors changed greatly under different confidence levels.

The comparison of impact factors under different RSC and assurance rate were shown in Figure 1. From the Figure 1 it was clear the impact factor changed greatly with assurance rate increasing. The difference between various RSC tended to be greater with assurance rate increasing. Obviously the China codes underestimate the impact factor seriously, as a result that may bring some potential risks on the bridge.

For the bridge example, when RSC was "good", the authors suggested that the reasonable assurance rate range should be 90%–99%, the corresponding impact factor was 0.225–0.294 based on the present study. On the other hand, the analysis result also provided useful information for bridge engineers and administers when it came to the performance evaluation and maintenance with bridge getting old. The revision coefficient of impact factor (defined as the impact factor divide baseline impact factor under same assurance rate) under different assurance rates were showed in the paper. The revision coefficients range were 1.25–1.55, 1.95–2.50, 3.60–5.40, corresponding the average, poor and very poor RSC respectively. The impact factor value can be determined considering actual RSC and required assurance rate. The research results were used to evaluate and maintain the Chaobai River Bridge.

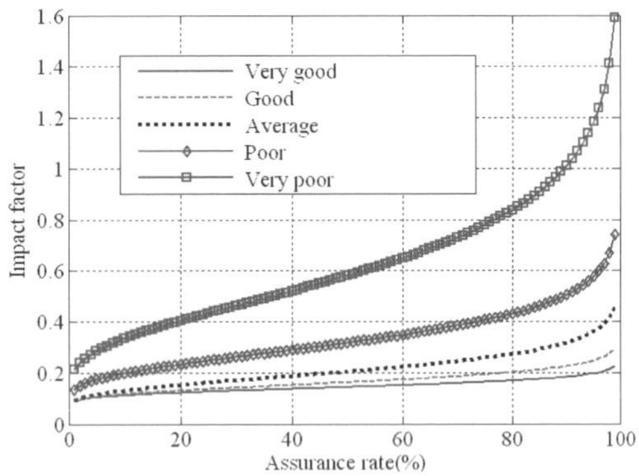

Figure 1. Impact factor comparison under different RSC.

REFERENCES

Huang, D.Z., 2005. Dynamic and impact behavior of half-through arch bridges. Journal of Bridge engineering, 10(2): 133–141.

Li, Y., Chen, S.F. & Sun, H., 2009. Assessment of safety and service status of Chaobai River Bridge based on the static and dynamic field test, ISEV2009, Beijing.

Bridge Maintenance, Safety, Management, Resilience and Sustainability – Biondini & Frangopol (Eds)
© 2012 Taylor & Francis Group, London, ISBN 978-0-415-62124-3

Temperature effect on cable frequencies and evaluation of cable corrosion

X.R. Li, X.C. Chen, Z.Z. Bai & D.W. Chen
Department of Bridge Engineering, Tongji University, Shanghai, China

X.C. Chen & F.T.K. Au
Department of Civil Engineering, The University of Hong Kong, Hong Kong, China

ABSTRACT

Since cables are important to the structural safety of cable-stayed bridges, it is essential to identify and monitor cable forces throughout their design lives. Changes in temperature may also cause variations in cable forces. In structural health monitoring systems, the cable forces are often identified by analyzing the cable frequencies according to the cable vibration theory (Irvine & Caughey 1974, Zui *et al.* 1996). Based on the structural health monitoring system of Hengfeng Road Bridge in Shanghai, China, the temperature-induced variations in cable frequencies are studied.

A linear regression model has been developed to analyze the relationship between the cable frequencies and temperature measurements. Data analysis demonstrates that the measured cable frequencies and temperatures are highly correlated linearly as demonstrated by the coefficients of determination R^2 that are above 0.8; the first five cable frequencies decrease by about 0.16% when the temperature increases by 1°C; and the linear change in cable frequency of mode n with respect to the temperature, namely γ_n, satisfies approximately the proportion $\gamma_1 : \gamma_2 : \gamma_3 : \gamma_4 : \gamma_5 \approx 1:2:3:4:5$ as n varies.

By taking account of the factors, including temperature-induced changes in cable forces and cable geometry, a formula to quantify the temperature-induced variation in cable frequencies is derived as

$$\gamma_n = \frac{n}{2} \cdot \sqrt{\frac{F_0}{4Ml_0}} \cdot \left[\left(\frac{1}{F_0} - \frac{1}{E_0 A_0} \right) \theta - \alpha \right] \tag{1}$$

where F_0 = cable force at reference temperature (in N); M = total mass of the cable (in kg); l_0 = effective length of cable at reference temperature (in m); E_0 = Young's modulus of cable at reference temperature (in Pa); A_0 = effective cross sectional area of cable at the reference temperature (in m^2); θ = linear change in cable force with respect to the change in the temperature (in N/°C); and α = coefficient of thermal expansion (in /°C). The formula shows good agreement with results from data analysis.

Rearranging Equation 1 gives the expression of linear change in cable force, θ, as

$$\theta = \left(\frac{2}{n} \cdot \gamma_n \cdot \sqrt{\frac{4Ml_0}{F_0}} + \alpha \right) \Bigg/ \left(\frac{1}{F_0} - \frac{1}{E_0 A_0} \right) \tag{2}$$

In Equation 2, the value of γ_n can be obtained from data analysis and the values of other parameters are given. Substituting these values into Equation 2 can give the value of θ.

In this paper, a method based on the mass variation of the cable is proposed to evaluate the corrosion of cable. The value of γ_n decreases as the mass of cable increases as shown in Equation 1. Hence, according to the value of γ_n calculated from measurements at different stages, the degree of corrosion can be evaluated. The ratio ν of the corroded cable mass M_{cor} (in kg) to the initial cable mass at the reference time M_0 (in kg) is used as an index to represent the degree of corrosion. The theoretical relationship between ν and γ_n has also been developed as

$$\nu = \frac{M_{cor}}{M_0} = \frac{1}{\chi} \left(\frac{\gamma_{n,0}^{\,2}}{\gamma_{n,t}^{\,2}} - 1 \right) \tag{3}$$

where $\gamma_{n,0}$ and $\gamma_{n,t}$ = linear changes in cable frequency of mode n with respect to temperature at reference time and time t respectively (in Hz/°C); and χ = ratio of increased mass ΔM (in kg) to the mass of corroded steel. Because the main ingredient of rust is $Fe_2O_3 \cdot H_2O$ (Broomfield 1997), χ is approximately taken as

$$\chi = \frac{\Delta M}{M_{cor}} \approx \frac{4 \times \mathrm{O} + 2 \times \mathrm{H}}{2 \times \mathrm{Fe}} = \frac{4 \times 16 + 2 \times 1}{2 \times 56} = 0.589 \tag{4}$$

REFERENCES

Broomfield, J.P. 1997. *Corrosion of steel in concrete: understanding, investigation and repair*. London: London E & FN Spon.

Irvine, H.M. & Caughey, T.K. 1974. The linear theory of free vibration of a suspended cable. *Series A, Mathematical and Physical Sciences: Proceedings of the Royal Society of London, England, 1974*. 341(1626): 299–315.

Zui, H., Shinke, T. & Namita, Y. 1996. Practical Formulas for Estimation of Cable Tension by Vibration Method. *Journal of structure Engineering* 122(6): 651–656.

Bridge Maintenance, Safety, Management, Resilience and Sustainability – Biondini & Frangopol (Eds)
ISBN 978-0-415-62124-3

A comparative study of bridge traffic load effect using micro-simulation and Eurocode load models

A. Lipari & E.J. OBrien
University College Dublin, Dublin, Ireland

C.C. Caprani
Dublin Institute of Technology, Dublin, Ireland

ABSTRACT

It is generally accepted that short span bridges are governed by small numbers of vehicles in free flowing traffic, with an allowance for dynamic amplification. Long span bridges on the other hand are governed by congested conditions, when a greater number of vehicles are present at much closer spacing, but no allowance for dynamics is appropriate.

Since April 2010, the Eurocodes are the legal standards for structural design throughout the European Union. Eurocode 1 – Part 2 (2003) deals with bridge loading, but its provisions apply only to the design of new bridges with spans up to 200 m. No provisions are made regarding the traffic loading for bridge safety assessment.

In the absence of code provisions, common design practice about traffic loading on long-span bridges usually relies on conservative assumptions about the traffic, typically assuming a queue of vehicles at a minimum bumper-to-bumper distance. In fact, maintenance operations for long-span bridges are rather expensive and therefore such conservative assumptions may play a decisive role, possibly resulting in unnecessary expenditure or disruption.

Also Eurocode traffic load models have been calibrated considering a standstill queue for longer spans (Prat 2001).

In reality, observations have shown that congestion can take up different forms. Micro-simulation (i.e. simulating the motion of individual vehicles) is a powerful tool for effectively generating different observed congestion patterns (Treiber et al. 2000) and for applications to traffic loading on long-span bridges (OBrien et al 2010).

In this paper, we generate heavy congestion on a single-lane road by means of an in-house micro-simulation tool. Traffic data comes from weigh-in-motion measurements collected at a Polish site.

The load effects on three sample long-span bridges are calculated for 1000 hours of congestion, deemed to represent 1 year of traffic. The results are then extrapolated to 75-year (for bridge assessment) and 1000-year return period (for bridge design, in order to draw comparison with the Eurocode load model).

It is found that, for the site under consideration, the Eurocode model may be reduced by 19 to 29%. An additional 5% of reduction can be allowed for the bridge assessment. The ratio of the extrapolated values through micro-simulation and the Eurocode load model are shown in Figure 1.

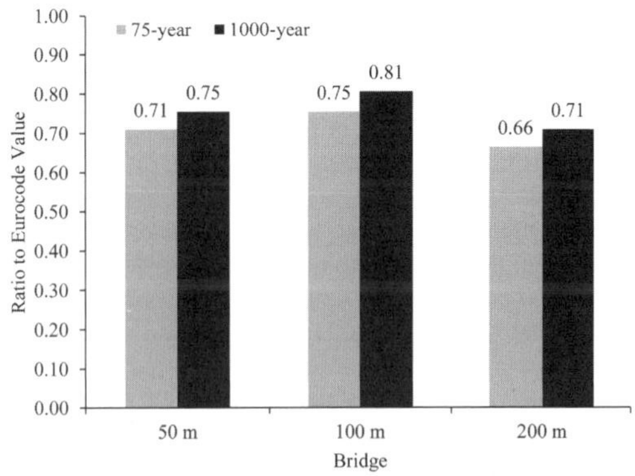

Figure 1. Ratio of microsimulation to Eurocode values for 75- and 1000-year return periods.

REFERENCES

EN 1991-2:2003, Eurocode 1: Actions on structures, Part 2: Traffic loads on bridges.

OBrien, E.J., Hayrapetova, A., Walsh, C. 2010. The use of micro-simulation for congested traffic load modelling of medium- and long-span bridges. *Structure and Infrastructure Engineering* 7(3): 1–8.

Prat, M. 2001. Traffic load models for bridge design: recent development and research. *Prog Struct Engng Mater* 3: 326–334.

Treiber, M., Hennecke, A., Helbing, D. 2000. Congested traffic states in empirical observations and microscopic simulation. *Physical Review E* 62: 1805–1824.

Bridge Maintenance, Safety, Management, Resilience and Sustainability – Biondini & Frangopol (Eds)
 ISBN 978-0-415-62124-3

Variability in dynamic characteristics of the Sutong cable-stayed bridge under routine traffic conditions

J. Liu
Department of Bridge Engineering, Tongji University, Shanghai, China

Q. Zhang
State Key Laboratory for Disaster Reduction in Civil Engineering, Tongji University, Shanghai, China

ABSTRACT

In the field of structural health monitoring, techniques based on vibration monitoring often evaluate the possible degradation or destruction of the structure by measuring changes of modal parameters. But an important problem the techniques have to face is that modal parameters may vary under different environmental conditions or operation conditions, such as traffic loads, wind loads, temperature and so on. The normal variation of structure under routine operation is prone to mask or confuse the variation caused by structural changes in physical properties. Therefore, it is essential for bridge health monitoring to quantize the variability in dynamic characteristics of routine operation and distinguish it from the changes caused by structural damage. Recently many studies have been carried out for the variation of dynamic parameters under different environmental conditions. But investigation of ultra long span cable-supported bridges is still lacking.

The Sutong Bridge is the first-built cable-stayed bridge with a thousand-meter scale main span. Its variability in dynamic characteristics under routine traffic conditions is focused and investigated. Before the bridge was open to the traffic, field vibration test was carried out. Acceleration responses of the main girder and the two towers were measured and precise dynamic characteristics of the bridge were identified as a reference. Then in the subsequent different seasons of the operational phase, vibration monitoring of the bridge was performed and dynamic characteristics of the bridge under different environmental conditions were estimated by using the enhanced frequency domain decomposition method and the stochastic subspace identification technique. The variability in dynamic characteristics was discussed and some valuable information was provided for the application of the vibration-based health monitoring techniques.

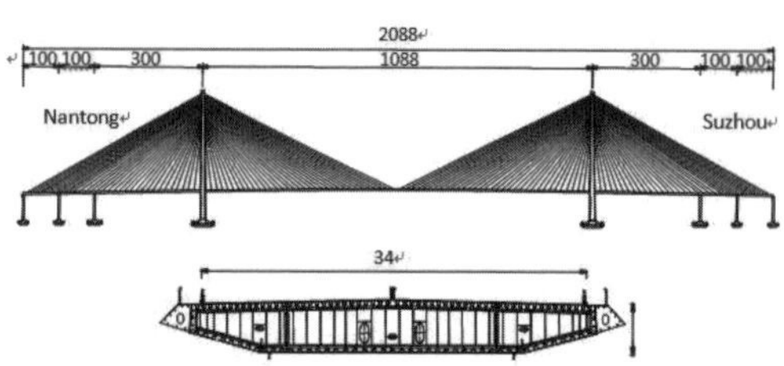

Figure 1. General Layout of the Sutong Bridge and typical section (Dimensions in meter).

REFERENCES

Doebling S W, Farrar C R, Prime M B. A summary review of vibration-based damage identification methods[J]. The shock and Vibration Digest, 1998, 30: 91–105.

Sohn H., Dzwonczyk M., Straser E.G., Kiremidjian A.S., Law K.H., Meng T., "An experimental study of temperature effect on modal parameters of the Alamosa Canyon Bridge," Earthquake Engineering and Structural Dynamics, 28, pp. 879–897, 1999.

Abdel Wahab M., De Roeck G., "Effect of temperature on dynamic system parameters of a highway bridge," Structural Engineering International, 7, pp. 266–270, 1997.

Ni Y Q, Fan K Q, Zheng G, et al. Automatic modal identification and variability in measured modal vectors of a cable-stayed bridge[J]. Structural Engineering and Mechanics, 2005, 19(2): 123–139.

Peeters B, Roeck G D. One-year monitoring of the Z24-Bridge: Environmental effects versus damage events[J]. Earthquake Engineering an Structural Dynamics, 2001, 30: 149–171.

Fujino Y, Abe M, Shibuya H, et al. Monitoring of Hakucho Suspension Bridge using ambient vibration[A]. Proceedings of the Workshop on Research and Monitoring of Long Span Bridges[C]. Hong kong: [s. n.], 2000. 142–149.

Zhang Qiwei. Variability in dynamic properties of cable-stayed bridge under routine traffic conditions [J]. Jounal of Tongji University, 2002, 30(6): 661–666.

Bendat, JS, Piersol, AG. "Engineering applications of correlation and spectral analysis," 2nd Ed., *Wiley Interscience*, New York, 1993.

Peeters B, De Roeck G. "Reference-based stochastic subspace identification for output-only modal analysis," Mechanical Systems and Signal processing, 13(6): pp. 855–878, 1999.

Bridge Maintenance, Safety, Management, Resilience and Sustainability – Biondini & Frangopol (Eds)
© 2012 Taylor & Francis Group, London, ISBN 978-0-415-62124-3

Estimation of Markovian deterioration models for bridge management

T. Liu & J. Weissmann
University of Texas at San Antonio

ABSTRACT

A key element for the use of Elemental Bridge Inspections in Bridge Management is the development of forecasts of bridge deterioration rates by element. Previous studies focused on the prediction of bridge performance using overall inspection values as specified by the US Federal Highway Administration (FHWA) National Bridge Inventory (NBI) coding guide, which provides global ratings for items as deck, superstructure and substructure. In this paper, we propose methodologies for modeling the deterioration of individual elements of a bridge. These individual element deterioration curves will be key elements in the development of future BMS modules that forecast bridge condition and optimize limited funding. Individual element deterioration curves will also be a key element in the prediction of overall bridge network condition as summarized by indices such as the Bridge Health Index (BHI).

REFERENCES

AASHTO Guide Manual for Bridge Element Inspection, American Association of State Highway and Transportation Officials, 2011.

Basheer P. A. M., S. E. Chidiact and A. E. Long, "Predictive models for deterioration of concrete structures", Department of Civil Engineering, The Queen's University of Belfast, UK, 1995.

Delwiche Lora D., Susan J. Slaughter, "The Little SAS Book 3rd Edition" SAS Institute Inc., Cary, NC, 2003.

Federal Highway Administration (FHWA) "Recording and Coding Guile for the Structure Inventory and Appraisal of the Nation's Bridges", FHWA-PF-96-001, Washington D.C, 1995.

Liu Min, Dan M. Frangopol," Optimal bridge maintenance planning based on probabilistic performance prediction", Engineering Structures, Volume 26, Issue 7, 991–1002, Jun 2004.

Liu Thaoshin, "Markov-Based Bridge Deterioration Model in Texas", Masters Thesis, Department of Civil and Environmental Engineering, University of Texas at San Antonio, 2011.

Morcous G., "Performance Prediction of Bridge Deck Systems Using Markov Chains", ASCE Journal of Performance of Constructed Facilities, Vol. 20, No. 2, pp. 146–155, May 2006.

Pontis Bridge Management User's Manual, Release 4.4. American Association of State Highway and Transportation Officials, Washington, D.C., January 2005.

Son Jung Baeg, Michael Blumenstein, Yew-Chaye Loo, Hong Guan, Kriengsak Panuwatwanich, "Generating Historical Condition Ratings for the Reliable Prediction of Bridge Deteriorations", Transportation Research Record, Volume 96, Transportation Research Board, 2009.

Texas Department of Transportation, "'Elements' Field Inspection and Coding Manual", Texas, 2001.

Texas Department of Transportation, "'Elements' Field Inspection and Coding Manual", Texas, 2010.

Bridge Maintenance, Safety, Management, Resilience and Sustainability – Biondini & Frangopol (Eds)
 ISBN 978-0-415-62124-3

Comparisons of four time-dependent reliability approaches for safety assessment of deteriorated concrete bridges

D.G. Lu, X.P. Fan & W. Jiang
School of Civil Engineering, Harbin Institute of Technology, Harbin, China

ABSTRACT

The reliability of existing reinforced concrete bridges is one of the world-wide concerned problems in the field of structural engineering. Reliability assessment for existing reinforced concrete structures can help make rational decisions to repair or demolition; moreover, it can also be directly applied to structural design. Based on the importance of reliability assessment, in this paper, firstly four computation methods of time-dependent reliability are studied in detail, which include time-integrated approach, time-discretized approach, time-discretized and integrated approach and first-passage probability approach; and then by analyzing the practical examples, the computational accuracy and shortcomings of the four methods are compared carefully; finally load probability models of the bridges are analyzed in detail, and based on the load probability models, the time-dependent reliability indices of the safety is solved with the four time-dependent methods about the flexural members of concrete continuous beam bridge, and the safety-based time-dependent reliability of the components has been evaluated.

REFERENCES

Melchers R.E. 1999. Structural reliability: analysis and prediction [M]. 2nd Edition, John Wiley, New York, NY.

Niu D.T. 2003. Durability and service life prediction of concrete structures [M]. Beijing: Science Press. (in Chinese)

Li T., Liu X.L. 1999. Durability analysis and design of concrete structures [M]. Beijing: Science Press. (in Chinese)

Li Z.N. 1996. Structural reliability analysis on the condition of truncated distribution [J]. Wuhan: Journal of Wuhan University of Technology, 18(4): 43–45. (in Chinese)

Zhang A.L., Zhao G.F., Wang G.Y. 1996. Engineering structural reliability based on many cases [J]. Journal of Dalian University of Technology, 36(6):771–775. (in Chinese)

Bridge Maintenance, Safety, Management, Resilience and Sustainability – Biondini & Frangopol (Eds)
© 2012 Taylor & Francis Group, London, ISBN 978-0-415-62124-3

Integrated bridge management from 3D-model to network level

K. Lukas & A. Borrmann
Computational Modelling and Simulation Group, Technische Universität München, Germany

ABSTRACT

In bridge management, namely in complex networks like e.g. in urban areas, it is of equal importance to consider the condition of the individual building as well as the effects of maintenance measures on network level. In this paper we introduce a concept to integrate a 3D-model based management of single bridges with a maintenance schedule optimization on network level.

The use of 3D-models in building management has some advantages to the mere textual based approach used in most common management systems: The orientation of building inspectors and maintenance planning engineers at navigation through the data is facilitated; the possibility of misplacements and misinterpretation of data is reduced, as all data is attached directly to a 3D representation of the building (Figure 1). Further the current condition (gained from inspection data), as well as prognoses for the future condition trends can be visualized on this 3D-model thus helping to identify weak points in the construction to be subject of special attention in later inspections.

The condition prognosis obtained on building level is input data for finding an ideal schedule of maintenance on network level. This schedule is not only subject to the wish to keep all bridges in the network under good condition and therefore safe, but also to some additional considerations by the building manger: For example, the manager may have a limited budget for maintenance measures each year or may want to steady the amount of money spent on maintenance over the years. It may also be of interest to plan maintenance and thus involved (partial) road closures in such a way, that the impact on traffic flow is as low as possible. Additionally, there may be synergies with maintenance measures by other parties, e.g. streetcar operators, which should also be considered. All these considerations make the task of finding a good schedule a constrained multi-objective optimization problem.

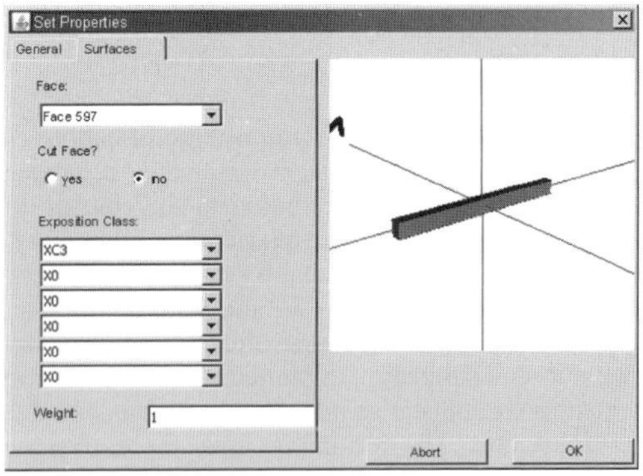

Figure 1. Adding data to the 3D-model.

For solving this problem we are using ant colony optimization (Dorigo, 1992). The basic Ant Colony system (Dorigo & Gambardella, 1997) for the single-objective problem and Pareto Ant Colony Optimization (Doerner et al. 2004) for the multi-objective problem are adapted to fulfil the special requirements of this optimization problem. These adapted algorithms are presented in this paper.

REFERENCES

Dorigo, M., 1992. *Optimization, learning and natural algorithms*. Milan: Politecnico di Milano.

Dorigo, M. & Gambardella, L. M., 1997. Ant colony system: a cooperative learning approach to the traveling salesman problem, *IEEE Transactions on Evolutionary Computation*, 1(1): 53–66.

Doerner, K., Gutjahr, W. J., Hartl, R. F., Strauss, C., Stummer, C., 2004. Pareto Ant Colony Optimization: A Metaheuristic Approach to Multiobjective Portfolio Selection, *Annals of Operations Research* 131: 79–99.

Bridge Maintenance, Safety, Management, Resilience and Sustainability – Biondini & Frangopol (Eds)
© 2012 Taylor & Francis Group, London, ISBN 978-0-415-62124-3

Evaluation of Eurocode damage equivalent factor based on traffic simulation

N. Maddah & A. Nussbaumer
EPFL, Lausanne, Switzerland

ABSTRACT

The goal of this paper is to propose modifications to improve the damage equivalent factors (λ) for fatigue design of road bridges according to the concept of Eurocodes. The factor λ can be decomposed into several partial factors, i.e. $\lambda = \lambda_1 \cdot \lambda_2 \cdot \lambda_3 \cdot \lambda_4$. This study especially focuses on $\lambda_1, \lambda_2, \lambda_4$ and λ_{max}.

Our study on damage equivalent factors was carried out with simulation of real continuous traffic, in a manner as realistic as possible, on different bridge static systems. The results are thus more accurate than the former simulations which have been done within writing of the Eurocode. Based on these simulations, a fatigue load model with a single axle is proposed, because it leads to a significant decrease in the dispersion of the values of λ_1 as well as λ_{max} obtained for different static systems. In addition, a new method for determining the fatigue equivalent length is proposed as:

$$L_\lambda = \frac{A_{\text{inf}}}{\Delta_{\text{inf}}} \tag{1}$$

where A_{inf} is absolute sum of area under influence line, as shown in Figure 1, and Δ_{inf} is difference between maximum and minimum values of influence line. This expression provides a simpler method for uniquely determining the fatigue length for all influence lines.

In addition, λ_2 is evaluated by performing simulations for different single lane traffic conditions. The results show that λ_2 adapts the average gross weight of heavy vehicles in different traffic conditions; however, the annual number of heavy vehicle traffic cannot properly be modified by λ_2. It can be explained because λ_2 does not take into account the probability of having several heavy vehicles on a bridge which depends on the annual number of heavy vehicle traffic.

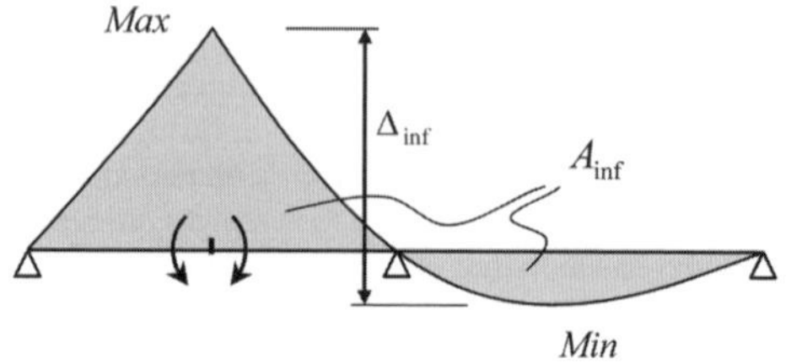

Figure 1. Schematic influence line to determine parameters of fatigue equivalent length.

For λ_4, similar simulations have been done to study the effect of double lane traffic. Two cases of bidirectional traffic and three cases of unidirectional traffic are studied. The obtained results for box cross section, which has uniformed transverse distribution, show that the effect of overtaking or crossing is important and its effect cannot be neglected in λ_4; however, λ_4 in the Eurocode only considers the damage accumulation due to volume of traffic in the additional lanes. Consequently, a new λ_4 is proposed as:

$$\lambda_4 = \left[(1-c) + \left(\frac{N_2}{N_1} - c \right) \cdot \left(\frac{\eta_2 Q_{m2}}{\eta_1 Q_{m1}} \right)^5 + c \cdot \left(1 + \frac{\eta_2 Q_{m2}}{\eta_1 Q_{m1}} \right)^5 \right]^{1/5} \tag{2}$$

where c is crossing (or overtaking) ratio. This Equation adds up damage due to crossing (or overtaking) as well as damage due to traffic volume on fast lane. In addition, based on the traffic simulations for highways, the maximum crossing ratio in bidirectional highway traffic condition is proposed 20 percent, and the maximum overtaking ratio in unidirectional traffic, assuming the annual heavy vehicle traffic of the fast lane is 20 percent of the slow lane, is proposed 2.5 percent.

Bridge Maintenance, Safety, Management, Resilience and Sustainability – Biondini & Frangopol (Eds)
© 2012 Taylor & Francis Group, London, ISBN 978-0-415-62124-3

Dynamic characterization of multiple identical spans of a steel girder bridge

R. Maestri, E.V. Fernstrom & K.A. Grimmelsman
Department of Civil Engineering, University of Arkansas, Fayetteville, Arkansas, USA

ABSTRACT

Full-scale dynamic testing of in-service bridges is an effective approach for quantitatively characterizing their condition and performance from a global perspective.

Ambient vibration testing is one of the most commonly used testing approaches for identifying the dynamic properties of an in-situ bridge structure. In ambient vibration testing, unmeasured and uncontrolled dynamic excitation provides the input to the structure, and the measured vibration responses of the structure are used to identify its dynamic characteristics. The natural sources of dynamic excitation include both the environment and the service live loads operating on the bridge. The unmeasured ambient dynamic excitation is assumed to have broadband white noise characteristics.

The results of any full-scale characterization program for bridge structures may be subject to uncertainty arising from several possible sources. These sources of uncertainty can be broadly classified as those related to experimental considerations, environmental effects, and structural characteristics.

The uncertainty associated with the structure can be an especially important consideration for guiding the design and execution of experimental characterization programs for multi-span bridge structures. The obvious question that arises is that when a multi-span bridge structure is to be characterized, how many of the individual spans should be evaluated. It may be reasonable to assume that the characterization of an individual span from a multiple-span bridge crossing would be representative of the remaining spans that share the same design characteristics.

A field investigation on a multiple span concrete deck on steel girder bridge was performed to evaluate the consistency of the experimental characterization results for the individual spans of the bridge crossing. The guiding objective for the research was to test and evaluate whether the dynamic characteristics identified for a single span of the structure can be reliably assumed to be representative of the other spans sharing the same superstructure design details. In this case, the bridge evaluated consisted of ten identical simply-supported bridge spans each sharing the same superstructure designs and detailing, and each having no observable damage or deterioration associated with any of the individual spans. A secondary objective of the research was to evaluate if the uncertainty associated with the field-constructed nature of bridges would be significant for such a structure that was built using modern construction quality control methods.

The individual spans of the structure where characterized using two different variations of ambient vibration testing: actual ambient dynamic excitation and pseudo ambient dynamic excitation from a small shaker. The characterization results obtained from the study generally indicated that the identified dynamic properties were consistent for the individual spans of this multiple span bridge crossing. The only significant difference identified was associated with an end span that had different substructure from the internal spans. The uncertainties associated with the field constructed nature of bridges were not significant for this structure as evidenced by the consistency of the identified dynamic properties for the interior spans evaluated simultaneously. The substructures were observed to affect the dynamic properties of the spans, but differences associated with the foundation conditions for the tested spans were not observed.

REFERENCES

Allemang, R.J. 2003. The Modal Assurance Criterion (MAC) – Twenty Years of Use and Abuse. *Sound and Vibration*, 37(8):14–21.

Brincker, R. Zhang, L. Andersen, P. 2001. Modal identification of output-only systems using frequency domain decomposition. *Smart Materials and Structures*, (10): 441–445.

Bridge Maintenance, Safety, Management, Resilience and Sustainability – Biondini & Frangopol (Eds)
© 2012 Taylor & Francis Group, London, ISBN 978-0-415-62124-3

Uncertainty evaluation of the behavior of a composite beam

J.C. Matos, I.B. Valente & P.J.S. Cruz
ISISE – Institute for Sustainability and Innovation in Structural Engineering, Universidade do Minho, Guimarães, Portugal

L.A.C. Neves
Civil Engineering Department, Universidade Nova de Lisboa, Lisboa, Portugal

ABSTRACT

Concrete and steel are the most used materials in civil engineer structures. Their combination takes advantage on each material's behavior. By making use of each property it is then possible to obtain slender structural elements with a higher stiffness, guarantying a better performance and economy. The use of lightweight concrete on composite structures presents additional advantages, such as the weight reduction. Despite the intensive experimental and numerical research focused on performance of steel and lightweight concrete composites, there are still significant uncertainties related to its behavior.

One of studied problems lies in the connection between these two materials. Valente (2007) developed several experimental tests to study different connection types. From all tested types it is important to point out the good results obtained with headed studs in lightweight concrete slabs. The study presented in this paper intends to complement those tests with a probabilistic numerical analysis in which the main uncertainty sources are considered. It starts with a numerical model calibration that is based on a composite beam previously tested up to failure (Figure 1). A sensitivity analysis is then developed to identify the parameters that present higher influence on the overall structure behavior. Probabilistic distribution laws are respectively attributed to such parameters based on a literature review, being the numerical simulation and statistical processing of results further developed.

Figure 1. Failure mode, Valente (2007).

Table 1. Resistance distribution curves.

Numerical model	μ [kN]	σ [kN]	Index-p [%]
Without Bayesian inference	18.997	2.500	89.598
With Bayesian inference	22.763	2.214	96.627

One advantage of using this type of analysis lies in the possibility of considering uncertainty on numerical model parameters. The evaluated structure presents uncertainties in material (e.g. lightweight concrete), geometric and physical (e.g. connection) parameters. Additionally, characterization tests were developed by Valente (2007) to study material and connection behavior. The results obtained in these tests are now used for numerical model adjustment. In addition, a Bayesian inference procedure is considered to perform it (Bernardo and Smith, 2004). A resistance curve is obtained before and after this procedure. An index-p is then presented to evaluate the reliability of such curves. This evaluation is done by comparing numerical and experimental results. Table 1 indicates the mean (μ) and standard deviation (σ) values obtained for these two models.

By analyzing obtained results it becomes clear that such procedure results are an improvement on the reliability of probabilistic model. Also, it allows the identification of a resistance capacity reserve that was not firstly considered. It is so recommended that engineers, when evaluating composite structures behavior, decide by a full probabilistic analysis and, if additional tests are executed, consider them into model adjustment.

REFERENCES

Bernardo, J., Smith, A. 2004. Bayesian theory. New York: Wiley.

Valente, M. I. B. 2007. Experimental studies on shear connection systems in steel and lightweight concrete composite bridges, PhD Thesis, Minho University.

Bridge Maintenance, Safety, Management, Resilience and Sustainability – Biondini & Frangopol (Eds)
 ISBN 978-0-415-62124-3

Dynamic response of isolated viaduct considering knocking-off effects of displacement restrainers

M. Matsumura
Department of Civil Engineering, Osaka City University, Osaka, Japan

M. Yoshida
Kawakin Core-tech, Tochigi, Saitama, Japan

ABSTRACT

Displacement restrainers are usually set asides isolation rubber bearings to restraint transverse displacement of superstructure for damage protection of expansion joints during earthquake (JRA 2002). Here, these can be utilized as a trigger device to differ vibration modes before/after their breaks. The authors have been developing displacement restrainers with knocking-off function, called as CSB, to mitigate damage of bridge pier and/or its foundation against a strong earthquake. 3 types of the displacement restrainers with the knocking-off function can be treated as shown in Fig. 1; bar/bolt type, cantilever column type and beam type. After the break, the restrainer itself is required to be replaced to new one.

Targeted in this paper among the 3 types of the displacement restrainers with the knocking-off function is the cantilever column type where the slit is inserted to the column root, as illustrated in Fig. 1(b). In this type of the displacement restrainer with knocking-off function, called CSB, both the vertical stress and the shear stress are subjected to the breaking part, then the breaking part is designed by focusing on predominant shear stress to the part.

According to the tested results already obtained through static and dynamic breaking tests of the CSB, the depth of the slit and the loading speed greatly influence on the breaking load of the CSBs (Asada et al. 2008). Also fluctuations of the breaking load experimentally obtained, are small and take about 1.1 times the design breaking load, calculated by using tensile strength of steel material.

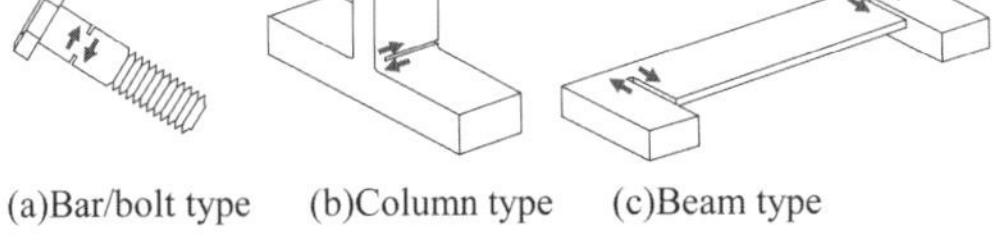

(a)Bar/bolt type (b)Column type (c)Beam type

Figure 1. Types of displacement restrainers with slit.

In this paper 5 specimens are additionally prepared to examine the effects of the loading position, h_l and the height of the slit, h_s on the breaking load of the CSBs. Also installation effects of the CSBs asides the rubber bearings are verified through shaking table tests. Moreover the installation effects of the CSBs in a viaduct are revealed through dynamic response analysis. The followings are conclusions;

1. The breaking load can be controlled by the change of the cross sectional area. The slit height influences on the breaking displacement a bit but not on the breaking load. The loading height of horizontal load, that is, the distance be-tween the loading position and the breaking part, greatly influences on the breaking load and correction factor is presented based on tested results.
2. Based on the shaking table test results, after the CSBs breaks, the vibration mode of the system shifts into the isolated one promptly and smoothly. However, setting gap between the mass and the CSBs will fluctuate the breaking timing, that is, a larger spacing arises a faster loading speed at the contact and the spacing should be checked in the design carefully.
3. Dynamic responses of a viaduct with rubber bearings and with the CSBs can be simulated by the seismic response analysis. Simulated results indicate that the breaking load of the CSB should be approximated to a minimum requirement for damage mitigation of the substructures.

REFERENCES

Asada, N., Matsumura, M., Kitada, T., Sakaida, M. and Yoshida, M., 2008. Effectiveness of rupture controllable steel side blocks for elevated girder bridges with isolation bearings, *Proceedings of the Second International Conference on Bridge Maintenance, Safety and Management, IAB-MAS08, Seoul, Korea*, 8 pages.

Japan Road Association, 2002. Japanese Specifications for Highway Bridges (JSHB), Part 5, Seismic design. (in Japanese).

Bridge Maintenance, Safety, Management, Resilience and Sustainability – Biondini & Frangopol (Eds)
© 2012 Taylor & Francis Group, London, ISBN 978-0-415-62124-3

An asset management approach to bridge barrier retrofits

P.S. McCarten
Opus International Consultants Limited, Napier, New Zealand

N. Lloyd
New Zealand Transport Agency, Wellington, New Zealand

ABSTRACT

Road bridges are an important and critical component within transportation network infrastructure. They are required to convey road users over objects or obstructions and hence raise the risk of falling for all road users and significantly raise the damage severity for all errant vehicles. Bridge designers control the risks to road users by providing a side protection system commensurate to the site specific conditions.

The New Zealand Transport Agency (NZTA) has recognised that semi-rigid bridge barrier retrofit design needs to be from a first principle approach due to the lack of tested systems and the range of different site constraints and bridge configurations and has developed a guide. The aim of the guide is to overcome variability in design standard, to achieve consistency in the evaluation of the existing structures and ensure consistency in the development of the bridge barrier retrofit design. This paper introduces the guide, outlines the guide development and provides details of the guide content.

In New Zealand the historic bridge side protection consisted of a Kerb, used as the dominant vehicle re-direction device, with a handrail behind. A range of kerb types were available consisting of different materials and of two main forms, an ordinary kerb and a safety kerb, and these could be combined with a range of handrail types. It is interesting to note the historic philosophy was to protect vehicles from colliding with the handrail, with the safety kerb having a greater kerb width than the ordinary kerb. This historic philosophy is the converse of the current vehicle collision design philosophy where the barrier must be hit and have sufficient strength and stiffness to provide the vehicle re-direction required.

In 1998 the NZTA adopted North American practice for safety performance evaluation of New Zealand highway features and the principles also apply to bridge barriers. The guideline report NCHRP 350 (1993) sets out the details for evaluating many highway features including longitudinal barrier performance.

Interrogation of the NZTA Bridge Data System undertaken in 2007 showed 52% of bridges had at least one kerb. The 'rigid' concrete handrail and kerb system for typical bridge deck designs has been found to only provide TL2 collision performance. Recent risk studies for barrier selection have shown that TL2 barrier collision performance is inadequate for current traffic volumes, vehicle weights and vehicle travel speeds on state highway bridges and this provides further justification for barrier upgrading. Typical bridge barrier retrofits using semi-rigid rail systems must consider combinations with an existing concrete kerb.

As the project evolved there was a change from presenting a range of 'best practice' examples to developing more robust processes for bridge barrier selection, existing bridge evaluation for collision loads and for bridge barrier retrofit design, as well as developing a range of acceptable solutions. The paper outlines the verification and calibration work that was undertaken through the project as well as setting out the justification for the development of new, TL3 High and TL4 High, intermediate bridge barrier systems. The interim guide, recently released for industry use and feedback, has received good reviews to date. It is concluded the bridge barrier retrofit project undertaken has been worthwhile. The Government Policy Statement (2008) requirement to make best use of the existing networks will be met and the aims to achieve industry wide standardisation and value-for-money are expected to be realised.

REFERENCES

New Zealand Transport Agency, 2004. "Bridge Manual".
Standards Australia and Standards New Zealand, 1999. "AS/NZS 3845: Road Safety Barrier Systems".
Transportation Research Board, 1993. "NCHRP Report 350". National Research Council, TRB, Washington, USA.

Bridge Maintenance, Safety, Management, Resilience and Sustainability – Biondini & Frangopol (Eds)
© 2012 Taylor & Francis Group, London, ISBN 978-0-415-62124-3

An approach to evaluating the influences of aging on the system capacity of steel I-girder bridges

J. McConnell, G. McCarthy & D. Wurst
University of Delaware, Newark, DE, USA

ABSTRACT

Limited financial resources for infrastructure require ever more thoughtful approaches to bridge management. One significant opportunity for better prioritizing the limited funding available for this purpose is through better quantification of the existing strength of aging structures. It has been shown in the first author's and others' previous work that bridges have significant reserve capacity above what is predicted from modern analysis and design methods predicated on the design of individual members. In contrast, by considering the system level interactions of all structural components in three dimensional analyses, the true strength of a bridge structure can be more accurately determined. This strength may be much higher than what is calculated based on current design and rating procedures.

However, in such an analysis approach, the transverse load distribution mechanisms become more influential than in current methods. Furthermore, the reliability of these transverse load distribution mechanisms in aging structures with various states of deterioration is currently unknown, significantly limiting the practicality of performing system level analyses. Thus, the goals of the research that is the subject of this paper are to quantify: (1) the transverse load distribution provided by the concrete deck of highway bridges as a function of the deterioration of the deck and (2) the resulting system capacity of steel-concrete composite bridges resulting from these changes in load distribution characteristics.

This is accomplished by first synthesizing the existing literature on deterioration of concrete, particularly due to the influences of deicing salts and freeze-thaw processes, to quantify changes in concrete member performance as a function of deterioration and aging. This review showed a linear degradation in member strength as a function of time exposed to chloride. Extrapolating the available data for a 25 year service life of a concrete bridge deck led to selecting 50% strength reduction as a characteristic of poor condition bridge decks. The existing literature also reports an increase in ultimate deflection, although no change in elastic stiffness, as reinforced concrete specimens are exposed to more chlorides. Again extrapolating the available data, an 82% increase in ultimate deflection was the second criteria targeted to represent poor condition bridge decks.

A finite element analysis (FEA) calibration study is then used to determine changes in FEA input that produce the targeted changes in member performance. Changes in concrete tensile strength, concrete failure strain, concrete shear failure strain and rebar area have been explored, but are not sufficient to produce the desired strength loss and deflection increase. Thus, current work is focusing on changing the method in which the rebar and concrete interact in the model to decrease the constraint between the rebar and concrete elements and thus more realistically represent the decreased bond between corroded rebar and the surrounding concrete. Additionally, the existing literature will be evaluated to determine a precise a value to input for the rebar section loss.

The results of this calibration study will be applied in combination with the analysis methodology validated in previous work to compare the system strength of bridges with good and poor condition bridge decks to the strength predicted by current bridge specifications. It is anticipated that, even with consideration for potentially deteriorated conditions of concrete bridge decks, system capacities much greater than the code-predicted girder capacities can be reliably achieved. Confidence in this hypothesis is based on the large, 250%, increase in system capacity compared to code-predicted girder capacity for the preliminary example evaluated. Thus, even though the system capacity may be dramatically reduced due to deterioration of the concrete bridge deck, system analysis is still likely to provide significantly more capacity than the current approach of considering the capacities of only individual girders.

Bridge Maintenance, Safety, Management, Resilience and Sustainability – Biondini & Frangopol (Eds)
© 2012 Taylor & Francis Group, London, ISBN 978-0-415-62124-3

Seismic upgrade of steel curved highway viaducts with isolation bearings and cable restrainers

C. Mendez Galindo, G. Moor & R. Berger
Mageba SA, Bulach, Switzerland

T. Hayashikawa
Hokkaido University, Sapporo, Japan

ABSTRACT

This paper investigates the effectiveness of the use of seismic isolation devices on the overall 3D seismic response of curved highway viaducts with an emphasis on expansion joints. Furthermore, an evaluation of the effectiveness of the use of cable restrainers is presented. For this purpose, the bridge seismic performance has been evaluated on four different radii of curvature, considering two cases: restrained and unrestrained curved viaducts. Depending on the radius of curvature, three-dimensional non-linear dynamic analysis shows the vulnerability of curved viaducts to deck unseating and joint residual damage.

The highway viaduct considered in this study is composed by a three-span continuous seismically isolated bridge section connected to a single simply supported non-isolated span. The bridge alignment is horizontally curved in a circular arc. Four different radii of curvature are taken into consideration measured from the origin of the circular arc to the centerline of the bridge deck: 100 m, 200 m, 400 m and 800 m. The total viaduct length of 160 m is divided in equal spans of 40 m. The bridge superstructure consists of a concrete deck slab that rests on three I-shape steel girders (Fig. 1). The deck weight is supported on five hollow box section steel piers of 20 m of height. Two cases have been considered, when the superstructure is supported on steel bearings (SB), and the second which has been seismically isolated by lead rubber bearings (LRB).

The use of LRB's provides a significant reduction of seismic damage. Furthermore, even though the differences on the radii of curvature among the viaducts, the application of cable restrainers reduces the possibility damage. In this analysis, the effectiveness of seismic isolation combined with the use of restrainers on curved highway viaducts is demonstrated not only by reducing in all cases the possible damage but also by providing a similar behavior in the viaducts despite of curvature radius.

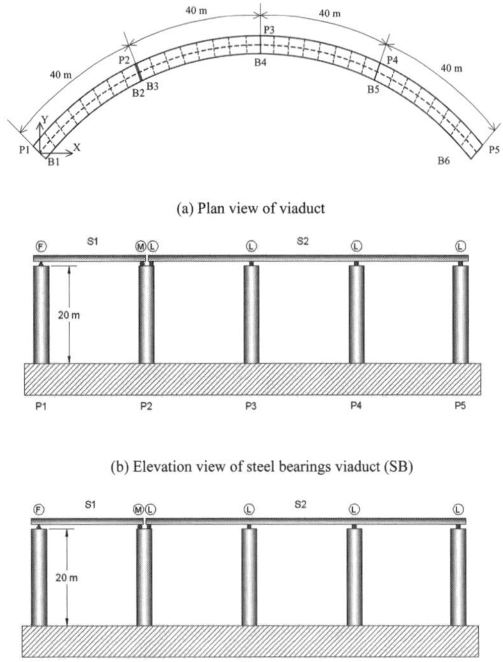

(a) Plan view of viaduct

(b) Elevation view of steel bearings viaduct (SB)

(c) Elevation view of LRB supports viaduct (LRB)

Figure 1. Model of curved highway viaduct.

REFERENCES

Japan Road Association (JRA), *Specifications for Highway Bridges – Part V Seismic Design*, Maruzen, Tokyo, 2002.

Kawashima, K., and Shoji, G. Effect of restrainers to mitigate pounding between adjacent decks subjected to a strong ground motion, *12th World Conference on Earthquake Engineering*, Paper No. 1435, 2000.

Mendez Galindo C., Hayashikawa T., Ruiz Julian F.D., Pounding and deck unseating damage of curved highway viaducts with piers of unequal heights, *Journal of Constructional Steel*, JSSC, Vol. 15, pp. 285–292, 2007.

Bridge Maintenance, Safety, Management, Resilience and Sustainability – Biondini & Frangopol (Eds)
 ISBN 978-0-415-62124-3

Design and testing of seismic protection devices for bridges according to EN 15129

C. Mendez Galindo, T. Spuler, G. Moor & R. Berger
Mageba SA, Bulach, Switzerland

ABSTRACT

Although Europe is not as seismically active as other parts of the world, the design of critical structures to withstand the effects of earthquakes continues to gain importance on the continent. This was underlined by the publication of the new European norm for anti-seismic devices, EN 15129 on August 2010. This norm regulates the design, production and testing of most existing types of anti-seismic devices, and crucially, also allows the development of new devices, as long as they fulfill the established performance criteria. From August 2011, only manufacturers certified to supply seismic devices with the CE label will be able to provide these devices in Europe (CEN 2009).

This is a significant development for the bridge industry in Europe, due to the critical role bridges play as lifelines in the aftermath of an earthquake – enabling access for emergency services and the evacuation of the affected population. The cost associated with repair or replacement of damaged bridges is likely to be small compared with the economic impact caused by disruption to traffic after an earthquake and during the long reconstruction phase.

In order to assure functionality of bridges, they must be designed to safely withstand the devastating forces of seismic ground movements. Past earthquakes have served as full-scale tests and the often tragic results have forced engineers to reconsider design principles and philosophies. Recent earthquakes have repeatedly demonstrated, for example, that during an earthquake, adjacent spans of multi-span bridges often vibrate out-of-phase, causing significant damage to the structures (Moor et al. 2011).

Among the great variety of anti-seismic devices, seismic isolators such as lead rubber bearings (LRB) and curved surface sliders (CSS) have found wide application in bridge and building structures. In order to reach the CE Certification according to EN 15129, a complete testing campaign has been carried out as specified by the European norm.

Despite the demanding testing requirements in the new European norm, the results proved that the proposed design of the isolators successfully fulfilled the performance required for the CE Certification.

Figure 1. Lead rubber bearing.

Figure 2. Curved surface slider type *Pendulum Mono*.

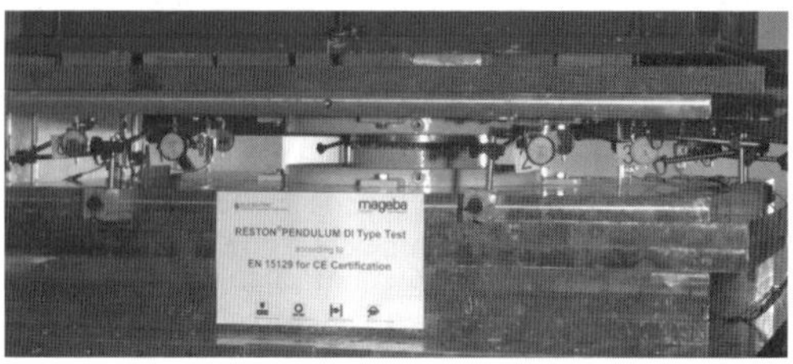

Figure 3. Curved surface slider type *Pendulum Duplo*.

REFERENCE

European Standard EN 15129:2009 for Anti-seismic Devices, European Committee for Standardization, November 2010.

Bridge Maintenance, Safety, Management, Resilience and Sustainability – Biondini & Frangopol (Eds)
© 2012 Taylor & Francis Group, London, ISBN 978-0-415-62124-3

Fatigue damage assessment of railway steel bridges based on short-term monitoring data

L.C. Meneghetti, R.M. Teixeira, R. Oliveira, T.N. Bittencourt & A.P. Conceição Neto
Structural Concrete Modeling Group, Polytechnic School, São Paulo University, São Paulo, Brazil

ABSTRACT

Long-term monitoring for assessment of a structural system generally requires the adoption of solutions with high costs. Alternatively, short-term monitoring when applied to structures of railway bridges may be used as tools to characterize the load spectra, since they are samples of the main loads and the confidence levels of models for extrapolation to longer periods are verified.

In this paper, the fatigue damage of a steel railway bridge was determined using the measured load spectrum obtained from the one week short-term monitoring, considering stress-life approach according AREMA (2009). The fatigue life of riveted member is also evaluated based linear elastic fracture mechanic (He & Wu 2011, Kühn et al. 2008).

The steel bridge considered in this study consists of two steel units of Warren truss with a span length of 30 m and 41 m supported by a concrete column in the middle of the bridge and by abutments at the ends. Three representative structural members located in steel truss with span length 41 m and height 7.49 m was chosen to estimate the fatigue life due to variable train loading. The members analyzed are pointed in Figure 1 as M3 (central post), D2 (intermediate diagonal) and Bi3 (central bottom cord).

In all cases, the average daily traffic (ADT) considered was equal 24 trains/day. Ore trains loaded and empty were used to calculate the damage and to estimate the fatigue life. The fatigue life forecast of vertical post is 82 years considering the set of the loaded ore trains with ADT 24 trains/day. When the analysis is made to unloaded set of ore trains the fatigue life increase to 156 years.

In the members that fatigue life is infinite (intermediate diagonal D2) was calculated the safety index γ to guarantee the security in terms of corrosion or others types of damage. The safety factor to infinite fatigue of these members can be considered satisfactory if good condition is maintained.

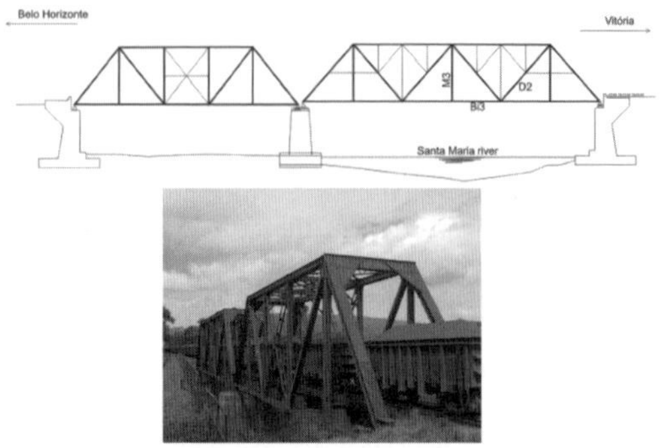

Figure 1. Overall profile of the Santa Maria bridge.

Based on linear elastic fracture mechanics, the fatigue life of vertical post is also evaluated, which is slightly smaller than that obtained by S-N curves and Miner rule. Considering the ADT equal 24, the remaining fatigue life can be estimated in 53 years. There is a difference between the results obtained by S-N curve and linear elastic fracture mechanics. This difference is related to life consumed during the crack initiation stage since the stress-life approach does not distinguish between the initiation and propagation stages.

REFERENCES

American Railway Engineering and Maintenance-of-way Association, AREMA. 2009. Manual for railway engineering: Washington.

He, X. & Wu, H. 2011. Fatigue life evaluation of existing railway riveted steel bridges. *Advanced Materials Research* 243–249:387–395.

Kühn et al. 2008. Assessment of existing steel structures: recommendations for estimation of remaining fatigue life. In Sedlacek et al. (eds.), *JRC Scientific and Technical Reports EUR 23252 EN:* Luxembourg.

Bridge Maintenance, Safety, Management, Resilience and Sustainability – Biondini & Frangopol (Eds)
© 2012 Taylor & Francis Group, London, ISBN 978-0-415-62124-3

A simplified procedure to evaluate seismic vulnerability of R.C. circular bridge piers

M. Mezzina, F. Palmisano & D. Raffaele
Technical University of Bari, Bari, Italy

ABSTRACT

In this paper a simplified procedure to evaluate the seismic vulnerability of R.C. circular piers of bridges with simply supported deck is presented.

The seismic response of this kind of structures essentially depends on the behaviour of its piers. Therefore the vulnerability analysis of the most exposed pier, often coincides with that of the whole structure.

According to the Italian Technical Standards (MIT 2008) the seismic intensity for a construction in a given site is defined by the reference return period T_R of the seismic action. This is correlated to the probability of exceedance P_{VR} in a reference period V_R by the relation

$$T_R = \frac{V_R}{\ln\left(1 - P_{VR}\right)} \tag{1}$$

where V_R depends on the importance class of the construction.

Thus structural seismic vulnerability can be evaluated determining the reference period consistent with a given performance (i.e. a given limit state *SL*).

In particular, the proposed methodology, starting from the assumption of the characteristics of the site (i.e. location, soil stratigraphy, topography) and of the pier (i.e. material, geometry, reinforcement, seismic masses) aims at evaluating the reference return periods $T^C_{R,SLY}$, $T^C_{R,SLV}$, $T^C_{R,SLC}$ consistent with the elastic limit state (SLY), the limit state of life safe (SLV) and that of near collapse (SLC) respectively.

The proposed procedure is based on the following steps:

- rapid evaluation of the pier capacity curve by using normalised curves;
- definition of a procedure, based on non-linear static analysis, that starting from the capacity curve is able to evaluate the reference return period T^C_R;
- definition of normalised curves that are able to quantify the contribution of each single parameter to T^C_R.

In the definition of the capacity curve of the pier the following collapse modes are taken into account: ultimate curvature, lap-splice failure of longitudinal bars, buckling of longitudinal bars, shear failure, unseating of the deck, second order effects.

Once that $T^C_{R,SL}$ has been evaluated for a specific limit state it has to be compared with the return period $T^D_{R,SL}$ prescribed by technical standard for the case under study. Hence it is possible to define the following seismic vulnerability index

$$I_{V,SL} = \frac{T^C_{R,SL}}{T^D_{R,SL}} \tag{2}$$

Taking into account that the probability of exceedance P_{VR} of the capacity spectrum has to be the same of the demand spectrum, from relations (1) and (2) it is possible to obtain

$$I_{V,SL} = \frac{T^C_{R,SL}}{T^D_{R,SL}} = \frac{\dfrac{V^C_{R,SL}}{\ln\left(1 - P^C_{VR,SL}\right)}}{\dfrac{V^D_R}{\ln\left(1 - P^D_{VR,SL}\right)}} = \frac{V^C_{R,SL}}{V^D_R} \tag{3}$$

In the work the effectiveness of the proposed procedure is shown by an application to a case study.

Moreover it is highlighted that this procedure is also able to analyse the influence of a single parameter (e.g. transversal confinement, longitudinal bars, seismic mass, slenderness) on the vulnerability index in order to define the best structural intervention necessary to achieve the design seismic performance

REFERENCE

MIT (Ministero delle Infrastrutture e dei Trasporti) 2008. *Decreto del Ministero delle Infrastrutture 14.01.2008 "Approvazione delle nuove norme tecniche per le costruzioni".*

Bridge Maintenance, Safety, Management, Resilience and Sustainability – Biondini & Frangopol (Eds)
© 2012 Taylor & Francis Group, London, ISBN 978-0-415-62124-3

Stress analysis method for steel plate multilayered CFRP under uniaxial loading

T. Miyashita & M. Nagai
Nagaoka University of Technology, Nagaoka, Japan

ABSTRACT

The objective of this study is to develop a new stress analysis method for a steel plate with multi-layered CFRPs under uniaxial loading to repair or reinforcement of bridge. In the method, strains in each layer are considered to be state variables in the first order simultaneous differential equations deriving from force equilibriums in infinitesimal regions. A general solution for strain is numerically calculated from eigenvalue analysis for a system matrix making from the differential equations. For validation of the proposed method, comparisons with analytical solutions, laboratory experiment and FEA were carried out. As a result, it was confirmed that the proposed method included conventional analytical solutions, and made possible to analyze stress distributions of a steel plate with multi-layered CFRPs.

Application of CFRP (Carbon Fiber Reinforced Polymer) to steel structures has been studied actively for rational repair and reinforcement. Herein, debonding of CFRP is a problem despite a steel member in elastic. For a countermeasure of the problem, the method setting steps at the edges of CFRPs has been proposed in order to increase the maximum load at debonding.

For clarifying the condition maximizing debonding load, a steel plate layered by two CFRPs with a step under uniaxial loading was theoretically analyzed. In the study, the forth order differential equation on stress of the plate was derived. However, a steel plate multi-layered by more than two CFRPs has never been theoretically analyzed. Since the order of differential equation on stress of the plate multi-layered by n CFRPs becomes 2n, the equation becomes complicated as the number of CFRPs increases.

As mentioned above, in a stress analysis for a steel plate multi-layered by more than two CFRPs, there is no analytical solution and the cost of numerical calculation such as FEM is large although structure itself is very simple. The objective of this study is to develop a new stress analysis method for a steel plate with multi-layered CFRPs under uniaxial loading.

Application scope of the proposed method is limited in simple structures. However, comparing with FEA, the proposed method significantly reduces the modeling effort and affords analytical solutions by numerical calculation. In order to verify the validity of the proposed method, it is compared with analytical solutions, laboratory experiment and finite element analysis.

It was confirmed that the proposed method included existing analytical solutions and gave the same accurate solution as FEA. Therefore, the proposed method makes possible to analyze a steel plate layered by more than two CFRPs without analytical solutions in addition to the reduction of modeling effort in conventional numerical method such as FEA.

REFERENCES

Al-amery R. and Al-mahaidi R. (2006), Numerical analysis of multilayered CFRP retrofitted RC beams with partial interaction, *Composite Structures*, Vol. 75, pp. 479–488.

Ishikawa T., Okura I. and Komuro K. (2009), Theoretical analysis on increase in debonding load by stepping ends of CFRP strips, *JSCE Journal of Structural and Earthquake Engineering*, Vol. 65, No. 2, pp. 362–367 (in Japanese).

Sugiura H., Ohgaki K., Inaba N., Tomita Y., Nagai M. and Kobayashi A. (2008b), Experimental study on repair effect using carbon fiber sheets for damaged steel members due to corrosion, *JSCE Journal of Structural Engineering A*, Vol. 54, pp. 548–554 (in Japanese).

Sugiura H., Kobayashi A., Inaba N., Homma A., Ohgaki K. and Nagai M. (2009), Experimental study on seismic retrofitting of circular steel bridge piers stiffened circumference direction by CFRP sheets, *JSCE Journal of Construction Engineering and Management*, Vol. 65, No. 1, pp. 106–118. (in Japanese).

Bridge Maintenance, Safety, Management, Resilience and Sustainability – Biondini & Frangopol (Eds)
 ISBN 978-0-415-62124-3

Partial safety factors for existing reinforced concrete structures

T. Moser
Schimetta Consult, Vienna, Austria

A. Strauss & K. Bergmeister
Department of Civil Engineering an Natural Hazards, University of Natural Resources and Life Sience, Vienna, Austria

ABSTRACT

There is a continuously increase in the traffic during the last decades and thus in the requirements of new and existing structures. In consequence the codes and regulations are in permanent development. The current standards allow the calculation of structures using probabilistic methods. These probabilistic methods for reliability evaluation allow the consideration of uncertainties in planning, design, execution and maintenance of new and historic reinforced concrete structures, JCSS (2000) & Braml (2010). The implementation of the probabilistic method used for the reliability evaluation which is allowed according to the standard is in very limited use and requires some modifications.

A higher acceptance of probabilistic verification procedures can be achieved – at the same time as making simplified procedures available for determining reliability – through the systematic use of stochastic models for key properties (e.g. material properties, resistances, loading models, etc.) of new and existing structures.

A method of adapting the partial safety coefficients of semi-probabilistic verification procedures based on measured or specified stochastic models would be of great interest – especially for engineers – while maintaining the established standardized proofing routines.

The purpose of this paper is to present a method to adapt the partial safety factors according to the current standards in case of the reduction of the reliability index β. The reduction of the reliability index is based on the acquired stochastic models of characteristics of the structure and on the expected remaining lifetime of the building. By a mathematical combination of the probabilistic limit state function and the mechanical calculation models according to the codes the partial safety factors are interwoven in the probabilistic analysis, Moser et al. (2011). Using a probabilistic inverse calculation with the help of special statistics – software packages the partial safety factors for a reduced reliability level can be calculated. In this contribution especially the shear failure of the shear reinforcement is carried out.

By determining the stochastic models of the structural properties and the effects the calculation system can be standardized. This normalization allows a practical determination of the reduced partial safety factors in case of a declared minimum reliability index.

The method will be demonstrated on a typical three-span slab bridge used by the Austrian Federal Railways, see Fig. 1.

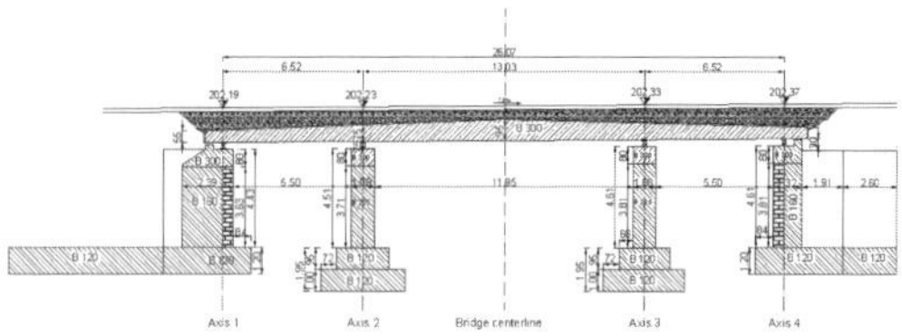

Figure 1. Ringstraßenbridge longitudinal section of bridge axis.

REFERENCES

Braml T. 2010. *Zur Beurteilung der Zuverlässigkeit von Massivbrücken auf Grundlage der Ergebnisse von Überprüfungen am Bauwerk.* München: Dissertation an der Universität der Bundesweh

Joint Committee on Structural Safety (JCSS). 2000. *Probabilistic Model Code 12th draft.* http://www.jcss.ethz.ch

Moser T., Strauss A., Bergmeister K. 2011. *Teilsicherheitsbeiwerte für bestehende Stahlbetontragwerke – Nachweis der Schubtragfähigkeit*; Paper in Review

Bridge Maintenance, Safety, Management, Resilience and Sustainability – Biondini & Frangopol (Eds)
 ISBN 978-0-415-62124-3

Research on bond between non-metallic reinforcement bars and concrete for bridge applications

P. Mossakowski & W. Radomski
Warsaw University of Technology, Warsaw, Poland

ABSTRACT

Research presented below concerns two imprtant problems observed currently in the building industry, in bridge engineering in particular. First of them is of more specific nature and concerns bridge durability, while the second one is connected with the general concept of sustainability.

One of the newest methods to improve the bridge durability is application of non-metallic reinforcement for structural concrete instead of the traditional steel one. The main idea of the use of non-metallic reinforcing bars results from the fact that they are not sensitive to corrosion. They are mostly made of polymer materials reinforced with glass or carbon fibers, and named glass fiber reinforced polymer (GFRP) bars or carbon fiber reinforced polymer (CFRP) bars, respectively. Although GFRP and CFRP bars are produced relatively long since (more or less since two decades), their structural applications are very limited as yet. Moreover, some problems of their engineering use are not sufficiently recognized and, therefore, require to be investigated. One of such problems is bond between non-metallic reinforcing bars and structural concrete, especially concrete of new type. Above problem is a main subject of this paper.

Concept of sustainability in its construction aspect is connected among others with the rational utilization of waste materials in the building industry. One of them is fly ash obtained from circulating fluidized bed combustion (CFBC) of coal. CFBC is a new technology developed in production of energy and has several advantages compared with the traditional one, especially concerning lower emissions of SO_2, CO_2, NO_x, and C_xH_y. However, fly ash from CFBC boilers differs considerably from the fly ash obtained from traditional installations. Moreover, CFBC fly ash as a waste material is characterized by high variation of its properties and cannot meet the current standard requirements for components of structural concrete. Therefore, an extensive research project has been undertaken and realized in the years 2007–2010 in Poland by six research teams and coordinated at the Institute of Fundamental Technological Research, Polish Academy of Sciences, to determine of conditions and limits for application of that kind of non-standard fly ash in structural concretes. This paper is based on the part of the above project, in which the authors have been personally involved.

The use of CFBC fly ash, which is a waste material, for partial replacement of Portland cement is of prime importance taking into account environmental requirements in particular and the strategy of sustainability in general. Similarly, the use of non-metallic reinforcement to improve bridge durability is in accordance with the above strategy.

The paper presents the author's experimental investigations on bond between smooth and ribbed Glass Fibre Reinforced Polymer (GFRP) reinforcing bars with diameters 8, 12, and 16 mm and concrete with the addition of CFBC fly ash with amount of 15% and 30% of the Portland cement mass. The bond between traditional steel bars and the above concretes has been also tested for comparison. The reference bond between steel and GFRP bars ad concrete without addition of CFBCV fly ash has been investigated too. The experiments have been carried out using somewhat modified RILEM procedure. It has been found that the bond in general is very good in all the tested cases and the bond strengths are above those specified in the bridge design codes. However, the best bond has been observed in case of concrete with the addition of fly ash with amount of 15% of the cement mass. It has been also found that there are some differences in the form of bond loss in case of steel and GFRP bars.

Several examples of structural uses of non-metallic reinforcement are given, including the first its applications in Poland. Economical aspects of the applications of non-metallic reinforcement in concrete structures are discussed.

Bridge Maintenance, Safety, Management, Resilience and Sustainability – Biondini & Frangopol (Eds)

Intelligent bridges – Adaptive systems for information and holistic evaluation in real time

T. Neumann & P. Haardt
Bundesanstalt fuer Straßenwesen, Bergisch Gladbach, Germany

ABSTRACT

The German highway network has to face new challenges in the near future, e.g. increasing traffic density and loads, climate change effects and new quality requirements regarding sustainability. The current maintenance management is damage-based and reactive (DIN 1076, 1999; RI-EBW-PRÜF, 2007). This approach is only useful when the structure indicates that it is about to fail, when the load situation does not change over time and when sufficient maintenance funds are available. Since one cannot rely on the adherence of these reconditions, it is necessary to come up with foresighted concepts in the present to be prepared for these challenges. Therefore it is important to adapt and enhance innovative attempts which take changing impacts, underdimensioning and insufficient and defective materials into account.

The objective is to go further than the monitoring does so far and set the starting point before a potential danger is manifested. Therefore it is planned to implement an adaptive system that provides not only information but also a holistic evaluation in real time. The corresponding software system – the expert system – derives intelligently warning messages for the user or owner, which might even lead us to self-organized reaction in the future (Krüger & Große, 2009; Zilch & Straub, 2010).

Within the complex structure of the adaptive system of an intelligent bridge the variety of components and aspects affect each other decisively. Most of these components can be classified into one of the three component groups seen in Fig. 1.

All these decisive elements have to be considered and developed precisely, since they are mutually depended. The three outlined component groups undertake different tasks and functions.

The hard- and software retrieves the required information: innovative, wireless sensors and sensor networks with an intelligent data management and a sufficient energy supply are existential. The measured data feeds the evaluation models, which on the other hand indicate the catalogue of requirements regarding e.g. measured parameters or sampling rates.

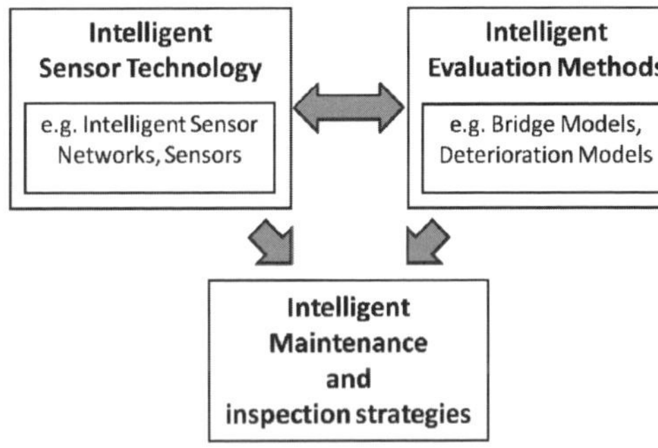

Figure 1. Component groups within the adaptive system of an intelligent bridge.

Both component groups have to be implemented into one adaptive system, which is to be integrated in the progressive inspection strategies. The objective is to support the existing (mostly deterministic) strategies by the input of the intelligent bridge especially by probabilistic and statistic means. The modular system advances the evaluation in respect of the development of condition and damage.

The paper describes the recent research and developments on a system for information and holistic evaluation in real time.

REFERENCES

DIN 1076 1999. Ingenieurbauwerke im Zuge von Straßen und Wegen; Überwachung und Prüfung, Berlin: Beuth Verlag.

Krüger, M. & Große, C. 2009. Einsatz von Sensorik an Brückenbauwerken. Final report FE 88.001/2009. *Funded by the Innovation Research Programme of the Bundesanstalt für Straßenwesen.*

RI-EBW-PRUEF 2007. Richtlinie zur einheitlichen Erfassung, Bewertung, Aufzeichnung und Auswertung von Ergebnissen der Bauwerksprüfung nach DIN 1076. BMVBS.

Zilch, K. & Straub, D. 2010. Entwicklung von Verfahren einer risikobasierten Bauwerksprüfung. *Final report FE 88.0002/2009. Funded by the Innovation Research Programme of the Bundesanstalt für Straßenwesen.*

Bridge Maintenance, Safety, Management, Resilience and Sustainability – Biondini & Frangopol (Eds)
 ISBN 978-0-415-62124-3

Repair and dynamic-based condition assessment of impact damage to a freeway overpass bridge near Mossel Bay, South Africa

A.A. Newmark
BKS (Pty) Ltd, Cape Town, South Africa

P. Moyo
Civil Engineering Department, University of Cape Town, Cape Town, South Africa

E.J. Kruger
The South African National Roads Agency SOC Limited

ABSTRACT

This freeway overpass bridge is located near Mossel Bay, South Arica. It consists of a four span beam and slab type structure. Vehicle impact damage to the bridge occurred as a result of an illegally loaded vehicle passing underneath. The impact caused significant damage to an edge beam during which the functionality of this beam was largely destroyed (Figure 1). As a result the beam was mainly supported by the transverse diaphragm beams connected to adjacent beams.

It was concluded from a detailed assessment and further design evaluations that repairs would require reconstruction of a section of the beam and the reinstatement of longitudinal tension capacity over the full extent of the beam. The final repair method included positive load application to ensure that the longitudinal and transverse load characteristics of the structure were reinstated. This was done by utilizing post tensioned tendons which were cast into a new in-situ beam web section attached to the inside face of the damaged beam. A new steel plate which would also act as edge armouring provided the optimum solution to reinstate part of the original capacity.

On instruction of the owner, the South African National Roads Agency SOC Limited (SANRAL), the effectiveness of the installation was assessed by means of dynamic based condition assessment before and after completion. These tests were performed by the University of Cape Town, Department of Civil Engineering. Following the dynamic testing and strain measurements during the retro-fitment of this bridge it was concluded that the damage to the tendons of the outer beam did have a significant impact on global stiffness in the longitudinal direction and in the transverse direction as reflected in change of natural frequencies and redistribution factors. This is in agreement with the analysis carried out by the consulting engineers and the testing confirmed that the rehabilitation intervention was successful in restoring the interaction between the damaged beam and the rest of the bridge.

Figure 1. Extent of damage to web and tendons.

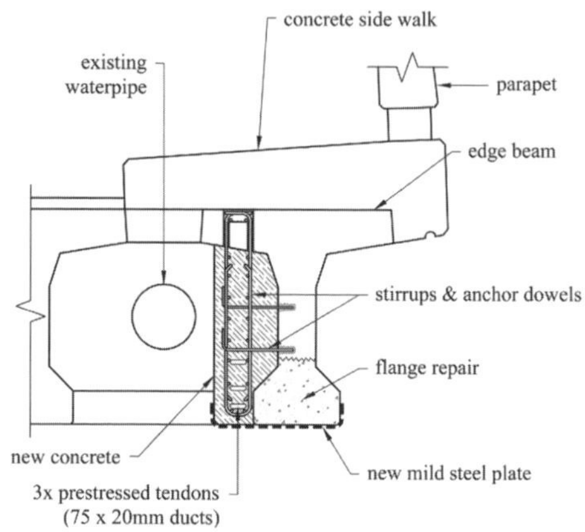

Figure 2. Final design proposal.

REFERENCES

Maia, N.M.M., et al, 1997. Theoretical and experimental modal analysis.

Nuno, M.M, Maia, J.M. and Silva, M. (eds) *1st Ed.* Research Studies Press Ltd, Baldock, England.

Bridge Maintenance, Safety, Management, Resilience and Sustainability – Biondini & Frangopol (Eds)
 ISBN 978-0-415-62124-3

Optimal seismic retrofit strategy selection of deteriorating concrete structures

N.A. Ni Nuallain
Division of Structural Engineering and Bridges, KTH Royal Institute of Technology, Stockholm, Sweden

A.J. O'Connor
Department of Civil, Structural & Environmental Engineering, Trinity College Dublin, Ireland

ABSTRACT

Seismic retrofitting techniques have undergone significant advances in recent years, considering that the first effort to seismically retrofit bridges was in the 1970's. Today, there are many possible retrofit strategies, making use of traditional and/or innovative materials. This presents designers with many options (Caterino et al. 2008). The problem, however, is that the selection of the most suitable strategy for a particular structure is not an easy task. Nonlinear methods of assessment, though accurate, are computationally very expensive. Thus, to compare a number of strategies for one bridge, using nonlinear methods of analysis, can be extremely time consuming.

Furthermore, there are many inherent uncertainties in the seismic vulnerability assessment process. Ground motion definition, seismic demand calculation, capacity calculations and material and modeling assumptions are all sources of uncertainty. To overcome these uncertainties, many researchers have combined seismic assessment with probabilistic methods of analysis in the form of fragility curves. A limitation at the moment, however, is that fragility curves combined with nonlinear methods of analysis can be time consuming.

One of the objectives, therefore, of this paper is to present an approach for selecting an optimal retrofit strategy for a particular bridge structure that overcomes the above limitations.

In addition, the second aim of this paper is to investigate the influence that deterioration due to corrosion will have on the retrofit strategy selection. The objective is to investigate if failing to include deterioration, when necessary, could result in a retrofit selection that is not optimal and not adequate.

To achieve the aims of the paper, a bridge in California has been selected. The selected bridge, Bixby Creek Bridge, is a concrete arch bridge. The results of a deterministic analysis revealed the vulnerabilities in the bridge and this information is used to design a number of options.

The paper presents an approach for comparing these options that is objective and efficient. Fragility curves are employed to account for the inherent uncertainties and the fragility curves are constructed using the elastic spectral response method so that a number of options can be compared without excessive time and computational expense.

After constructing the fragility curves assuming an un-deteriorated condition, the fragility curves are regenerated assuming deterioration is present. This is achieved by modifying the capacity models to account for deterioration.

The approach is summarized as follows:

1. Identify the vulnerable members from a deterministic analysis
2. Define appropriate limit states
3. Identify a number of retrofit options
4. Choose an appropriate suite of ground motions for the bridge site and create the demand models using response spectrum analysis for each option
5. Create capacity models for each member considering an un-deteriorated and deteriorated state
6. Generate fragility curves for each of the critical members and hence using joint probability theory; generate the fragility curve for the overall bridge sys-tem for each retrofit option
7. Compare the retrofit options and identify the optimal strategy for the bridge

The results show that the approach provides a good graphical method for comparing a number of options.

REFERENCE

Caterino, N., Iervolino, I., Manfredi, G. & Cosenza, E. 2008. Multi-Criteria Decision Making for Seismic Retrofitting of RC Structures. Journal of Earthquake Engineering 12(4): 555–583

Bridge Maintenance, Safety, Management, Resilience and Sustainability – Biondini & Frangopol (Eds)
 ISBN 978-0-415-62124-3

Flexural tests on GFRP RC slabs: Experimental results and numerical simulations

E. Nigro, A. Bilotta, G. Cefarelli, G. Manfredi & E. Cosenza
Department of Structural Engineering, University of Naples Federico II, Italy

ABSTRACT

The excellent resistance to corrosion of Fiber Reinforced Polymer (FRP) Composite Materials undoubtedly is important when they are used to reinforce concrete structures, due to a great reduction of maintenance costs that are particularly relevant when referring to structures constantly exposed to severe weather conditions such as bridges (Bakis et al., 2000). For these reasons many research activities are aimed to validate theoretical models for the design of concrete members reinforced with FRP bars (Pecce et al., 2000). In this framework, experimental tests were recently performed on concrete slabs reinforced with Glass FRP.

The tests herein presented are a part of a more wide experimental program concerning 11 full-scale tests in a four-point bending scheme (two at room temperature and nine in fire situations) on concrete slabs reinforced with GFRP bars. Nine fire tests were shown elsewhere (Nigro et al., 2011a,b) with the aim of providing informations on the FRP-RC slabs behavior in fire condition; the results of two tests carried out at room temperature, instead, are shown and discussed in this paper.

Deflection prediction provided by Italian, American and Canadian codes for design of FRP reinforced concrete structures are in a good agreement with the experimental results at the serviceability limit state, and some times more conservative than prediction provided by more sophisticated procedure, such as integration of curvature along the member. However, a more depth investigation on the prediction provided by relationships suggested in Italian code indicated that deflection predicting relationships should be better calibrated, being design of FRP-RC members more governed by serviceability limit state (SLS) than ultimate one (ULS).

Experimental results were also simulated by a FEM model analysis which properly takes into account the effect of tension stiffening in FRP reinforced concrete members. The load vs. displacement curve was in good agreement with experimental ones.

A more refined validation will be performed by comparing stresses and strains simulated with those recorded in FRP and concrete during the tests.

The model should be used to refine simplified formula suggested in design code for FRP reinforced

Figure 1. Test set-up from top.

concrete members as well as to simulate the behaviour in fire condition.

REFERENCES

Bakis, C.E., Bank, L.C., Brown, V.L., Cosenza, E., Davalos, J.F., Lesko, J.J., Machida, A., Rizkalla, S.H. and Triantafillou, T.C. (2002) Fiber-Reinforced Polymer Composites for Construction – State-of-the-Art Review. *Journal of Composites for Construction*, Vol. 6, No. 2, May 1, 2002, ASCE. ISSN 1090-0268/2002/2-73-87

Nigro, E., Cefarelli, G., Bilotta, A., Manfredi, G. and Cosenza, E. (2011a). "Fire resistance of concrete slabs reinforced with FRP bars. Part I: experimental investigations on the mechanical behavior". *Composites: Part B*, 42 (2011), 1739–1750.

Nigro, E., Cefarelli, G., Bilotta, A., Manfredi, G. and Cosenza, E. (2011b). "Fire resistance of concrete slabs reinforced with FRP bars. Part II: experimental results and numerical simulations on the thermal field". *Composites: Part B*, 42 (2011), 1751–1763.

Pecce, M., Manfredi, G., and Cosenza, E. (2000). "Experimental response and code models of GFRP RC beams in bending." *Journal of Composites for Construction*, 4(4), 182–190.

Bridge Maintenance, Safety, Management, Resilience and Sustainability – Biondini & Frangopol (Eds)
© 2012 Taylor & Francis Group, London, ISBN 978-0-415-62124-3

Development of an advanced orthotropic steel deck system for long span bridge

C.K. Oh, K.J. Hong & D. Bae
Kookmin University, Seoul, Republic of Korea

ABSTRACT

Orthotropic steel deck systems have been widely used especially for long span bridge construction due to great advantages such as (1) light self-weight compared with that of concrete deck systems (2) reliable quality control at manufacturing factories (3) rapid construction and so on. As the welding and construction techniques become advanced, design service life of the long span bridges is increased, e.g., 200 years nowadays, and fatigue is one of the major factors to govern the design.

The complicated welding connection of the orthotropic steel deck system, however, may cause fatigue cracks at welding intersection regions such as longitudinal rib and cross beam connections and so on. Therefore, it requires to develop an advanced orthotropic steel deck system that has smaller amount of maximum principal stresses at the fatigue-vulnerable region of interest, because the concentrated stresses can initiate fatigue cracks.

In this study, three major factors are taken into consideration. First of all, bulkhead plates, also known as internal diaphragm, are suggested to be installed inside longitudinal ribs. Then, the optimal scallop shapes are proposed through parametric studies. Finally, stiff surfacing effects are investigated based on computational analysis which is verified by experimentation.

Several different shapes of bulkhead plates are installed inside longitudinal ribs and their performance is examined in terms of concentrated stress reduction. The optimal shape turns out to be a curved one having 300 mm and 200 mm radii of curvature for upper and lower parts of bulkhead plates, respectively. When the optimal ones are installed, maximum principal stresses are reduced by about 30% (Oh et al. 2011).

As for scallop, two kinds of traditional scallop shapes specified on KSCE (2011)/MOTC (2006) and Eurocode are compared with another newly suggested one. When the new one is utilized along with the optimal bulkhead plates, the concentrated stresses can be decreased by about 35%.

Then, a stiff surfacing effect is examined through ultra high performance concrete overlay (Harris 2004) onto the steel deck. Maximum stresses decrease up to 30% only with a 40 mm UHPC layer compared those obtained from a deck overlaid by 80 mm traditional asphalt surfacing (Han 2011).

Finally, a static load test is carried out to a 3-dimensional full-scale experimental structure and the analytical findings demonstrated so far are verified through the experimentation.

Based on the achieved results, an advanced orthotropic steel deck system that combines the aforementioned three factors such as installation of the optimal bulkhead plates inside longitudinal ribs, using the optimal scallop configuration and the stiff surfacing, is undoubtedly very effective as well as efficient to decrease the magnitude of concentrated stresses which may cause fatigue cracks.

REFERENCES

AASHTO. 2007. *AASHTO LRFD Bridge Design Specifications*, American Association of State Highway and Transportation Officials, Washington D.C., U.S.A.

Han, T. 2011. Analytical study on optimal connection detaila and behavior for 3-dimensional full-scale orthotropic steel deck. M.S. thesis in Kookmin University, Seoul, Korea. [in Korean]

Harris, D.K. 2004. Characterization of punching shear capacity of thin UHPC plates, M.S. thesis in Virginia Polytechnic Institute and State University, V.A., U.S.A.

KSCE. 2010. Design specifications for highway bridges, Korean Society of Civil Engineers, Seoul, Korea. [in Korean]

MOCT. 2006. Guide lines for the design of details of steel highway bridges, Ministry of Construction and Transportation, Kyonggi-do, Korea. [in Korean]

Oh, C.K., Hong, K., & Bae, D. 2011. Analytical and experimental studies on optimal details of orthotropic steel decks for long span bridges. International Journal of Steel Structures 11(2): 227–234.

Bridge Maintenance, Safety, Management, Resilience and Sustainability – Biondini & Frangopol (Eds)
© 2012 Taylor & Francis Group, London, ISBN 978-0-415-62124-3

Displacement fragility curves for bridges with medium length

B.A. Olmos, J.M. Jara & M. Jara
Civil Engineering School, Universidad Michoacana de San Nicolás de Hidalgo, Morelia, Michoacán, México

ABSTRACT

One fertile and important area of research is the study of the vulnerability of both new and existent bridges located on the Pacific Coast of Mexico. In this work we evaluated the vulnerability of ten reinforced concrete bridges modeled with typical geometry, and two existing bridges, as well. The bridges' capacity was evaluated based on the FEMA recommendations, and their demand was evaluated using a group of more than one hundred accelerograms recorded close to the subduction zone of Mexico. The vulnerability of the bridges was assessed determining displacements fragility curves for several limit states.

Most of the strong earthquakes that occur in Mexico have epicenter situated along the Pacific Coast on the borders of the Pacific Coast and the North American plates. The frequent occurrence of earthquakes and the high number of bridges in this zone of the country are issues that must be considered in the estimation of the bridges' seismic vulnerability. The topic has especial relevance because of the age of the bridges and the non-seismic considerations during the design of some of them. This makes particularly vulnerable a group of bridges located on highly seismic zones of Mexico. In addition, the construction of longer bridges with higher substructures located on highly seismic zones continues, becoming more important the estimation of the vulnerability of these type of structures.

The bridges considered in this work are representative of the ones build along the Mexican highway system: Two spans reinforced concrete (RC) bridges with span lengths of 20 and 40 m, piers with regular elevation and highs of 6 and 10 m. Two substructures types are analyzed: the first one is a frame type structure with three piers and the other is a one column substructure. The analytical model was developed using the SAP2000 program as a tridimensional (3D) structure. The model for the superstructure consists of a RC deck resting on six simple supported prestressed girders AASHTO type, RC diaphragms located at the spans' ends and at 10 m apart each one along the span. The substructure is constituted by RC piers located at each of the support axes with columns of circular cross section. The bent caps were modeled as RC elements with a rectangular cross section. The elements used in the SAP2000 to define the analytical model were beam elements with six degrees of freedom at each end for the girders, diaphragms, piers and bent caps whereas the deck was represented with finite thin shell elements. Most of the girders' bearings used in Mexico correspond to a combination of steel and rubber plates that were modeled with link type elements. The abutments were modeled as simple supported bearings neglecting the flexibility of the surrounding soil based on the important contribution in the lateral stiffness given by the bridge. The piers' support was assumed fixed for the six degrees of freedom at each node since most of the bridges' foundations are built on a rigid stratum.

Figure 1. Two existing bridges studied.

Additionally, two real bridges located on the Pacific Coast of Mexico are studied (figure 1).

The results of fragility curves allow concluding that it is possible to estimate the seismic vulnerability of existing bridges by interpolating the curves obtained for the representative group of bridges, provided that the geometry of the cases of interest is similar to one of the cases studied.

Bridge Maintenance, Safety, Management, Resilience and Sustainability – Biondini & Frangopol (Eds)
© 2012 Taylor & Francis Group, London, ISBN 978-0-415-62124-3

Mode shape estimation of a bridge using the responses of passing vehicles

Y. Oshima, K. Yamamoto & K. Sugiura
Kyoto University, Kyoto, Japan

ABSTRACT

Traditionally the inspection of bridges has been performed visually, especially in daily inspection, and then if something wrong happens in the inspection, detail investigations will be conducted at the suspected area. As an alternative way to the traditional inspection, screening technologies by using vehicle responses to assess bridges roughly but rapidly for early identification of damage has been needed recently, but no study has been reported on damage detection on the basis of mode shapes. Thus the purpose of this study is to propose a method to estimate bridge mode shapes based on the responses of passing vehicles.

The estimation theory consists of the sequence of the following four steps: (1) estimation of the incoming displacement to monitoring vehicles, (2) extraction of bridge vibration component from the incoming displacement, (3) conversion of the data obtained by moving vehicles to those by fixed observations, and (4) estimation of bridge mode shapes based on the data obtained in (3).

For verification of the proposed method, this study adopts simple numerical simulation in which the monitoring vehicles are modeled by sprung-mass system and a bridge is modeled by finite beam elements. The number of monitoring vehicles is four and they runs between two heavy vehicles as shown in Figure 1.

As a result, it was found that mode shapes can be identified as shown in Figure 2, but higher order than 3rd cannot be identified since the accuracy of estimation decreases as the number of vehicles increases due to shorter intervals. Thus, it is difficult to estimate higher mode shapes even though many monitoring vehicles are used. There are also other difficulties in measuring position, exciting higher modes, and deduction of road unevenness.

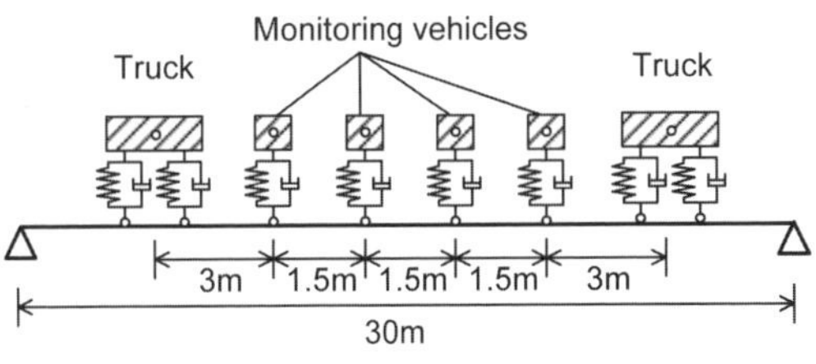

Figure 1. Assumed bridge and vehicles in simulation.

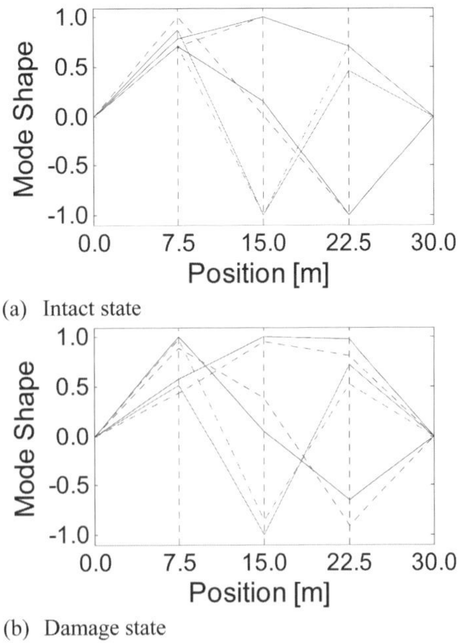

Figure 2. Estimated and true mode shapes.

REFERENCES

Y.B. Yang, C.W. Lin. 'Vehicle-bridge interaction dynamics and potential applications,' *Journal of Sound and Vibration*, 284, 205–226. 2005

Y.B. Yang, K.C. Chang,'Extracting the bridge frequencies indirectly from a passing vehicle: Parametric study,' *Engineering Structures*, 31, 2448–2459, 2009.

Bridge Maintenance, Safety, Management, Resilience and Sustainability – Biondini & Frangopol (Eds)
© 2012 Taylor & Francis Group, London, ISBN 978-0-415-62124-3

Assessing impact-echo test variables for detecting loss of bond in RC bridge columns

A. Pagnotta & P. Gardoni
Texas A&M University, College Station, TX, USA

D. Trejo
Oregon State University, Corvallis, OR, USA

Q. Huang
The University of Akron, Akron, OH, USA

ABSTRACT

The premature cracking of some reinforced concrete (RC) bridge columns has been investigated by the Texas Department of Transportation (TxDOT). This cracking has been attributed to Alkali-Silica Reaction (ASR) and to a lesser extent, Delayed Ettringite Formation (DEF). These two reactions create expansive byproducts that lead to distributed cracking of concrete. Petrographic analyses of cores removed from affected specimens show a build-up of these products at the steel-concrete interface of longitudinal reinforcement. Under certain loading conditions, areas where steel-concrete bond is critical (lap-splice regions, anchorage zones, etc.) may experience bond failure, resulting in structural failure. A practical tool is needed to determine the state of steel-concrete bond in these critical areas such that accurate assessments of structural reliability and rehabilitation plan can be made.

Previous studies have identified the impact-echo method as a viable technique for detecting voids around steel reinforcing bars in reinforced concrete (RC) members. This paper investigates the limitations of impact-echo scanning with respect to multiple experimental variables. A small-scale experimental setup for making this determination is designed. Variables in the experiment include void thickness, void length, and the interval at which impact-echo tests are performed. Impact-echo results from a small-scale experimental design are presented and interpreted to determine detectable void sizes given the interval at which tests are performed.

Several conclusions are made based on the results of this study. With regards to observing voids of varying thickness, reducing the interval at which impact-echo tests are performed from 25 mm (1 in) to 13 mm (0.5 in) does not necessarily improve the ability to detect voids when observing amplitudes at the frequency corresponding to reflections from the full depth of the member. However, when examining the mean transfer amplitude at the steel frequencies, reducing the length interval at which tests are performed from 25 mm (1 in) to 13 mm (0.5 in) improves the ability to qualitatively detect voids at the steel-concrete interface. This may be attributed both to a reduction in the granularity of the image and a reduction in statistical uncertainty that arises from an increase in the amount of data available. A minimum void size that can be qualitatively identified with impact echo testing is voids that are 0.19 mm (0.075 in) thick and 203 mm (8 in) in length.

With regards to observing voids of varying length, reducing the interval at which impact-echo tests are performed from 25 mm (1 in) to 13 mm (0.5 in) does not improve the ability to detect voids shorter than 102 mm (4 in). When examining displacement amplitudes at the frequency of P-wave reflections through the full-depth, the minimum void length that can be identified is between 51 (2 in) and 102 mm (4 in) when the void is 0.25 mm (0.01 in) thick. When making similar observations using the frequency of P-wave reflections from the concrete-steel interface with 0.25 mm (0.01 in) thick voids, the minimum detectable length is between 102 mm (4 in) and 203 mm (8 in).

REFERENCES

Cheng, C.-C. & Yu, C.-P. 2000. An investigation of the transfer function of the impact-echo response and its application. *Proceedings of the 15th World Conference on Nondestructive Testing (WCNDT), Roma, Italy, 2000.*

Cheng, C.-C., Yu, C.-P., & Chang, H.-C. 2004. On the feasibility of deriving transfer function from Rayleigh wave in the impact-echo displacement waveform. *Key Engineering Materials* 270–273: 1484–1492.

Hsu, K.-T., Cheng, C.-C., & Lin, Y. 2008. Use impact-echo method to evaluate bond of reinforced concrete subjected to early-age vibration. *Journal of Solid Mechanics and Materials Engineering* 2(12): 1528–1538.

Sansalone, M. & Streett, W. B. 1997. *Impact-Echo: Nondestructive Evaluation of Concrete and Masonry.* Ithaca: Bullbrier.

Bridge Maintenance, Safety, Management, Resilience and Sustainability – Biondini & Frangopol (Eds)
© 2012 Taylor & Francis Group, London, ISBN 978-0-415-62124-3

Direct and probabilistic interrelationships between half-cell potential and resistivity test results for durability ranking

V. Pakrashi
Department of Civil and Environmental Engineering, School of Engineering, University College Cork, Ireland

J. Kelly
Roughan and O'Donovan Consulting Engineers, Sandyford, Dublin, Ireland

A.J. O'Connor
Roughan & O'Donovan Innovative Solutions, Sandyford, Dublin, Ireland

ABSTRACT

Tests related to durability studies on structures often feature half-cell potential and resistivity data. An approximately linear relationship between Half-Cell Potential (HCP) testing and resistivity data has been discussed and well-researched. Despite criticisms related to environmental sensitivity of resistivity tests it remains as a popular choice for investigations into durability of structures. HCP and resistivity tests were carried out on different components of a number of bridge structures. The direct and the probabilistic empirical correlations of the two tests were investigated based on the collected data. The exposure condition, age and the distance of the bridges from each other were very significantly varied.

Generally, a linear correlation was observed to exist between the HCP and the resistivity tests from the absolute values and the percentile values. The scatter of the test results indicated that the variability of the levels of correlation among bridges and among different components should be investigated using absolute and percentile values. Correlation levels of the tests among the bridges varied. Computation of percentile values rescales the HCP and resistivity data differently and leads towards a marginal increase in scatter. However, the correlations observed from absolute values are changed little through the use of percentiles.

The calibration values for relating the two tests varied significantly from a rule of thumb indicating an approximate relationship of $10\,\mathrm{mV/k\Omega \cdot cm}$. For tests with good correlations, the deviation of this calibration value from the general rule of thumb is less. A calibration value obtained from the correlations using percentile values of tests results was observed to be within 3–5 percentile of HCP result variation for a unit percentile variation of resistivity. Investigation on different components of the bridges corroborated the findings presented in the previous sections.

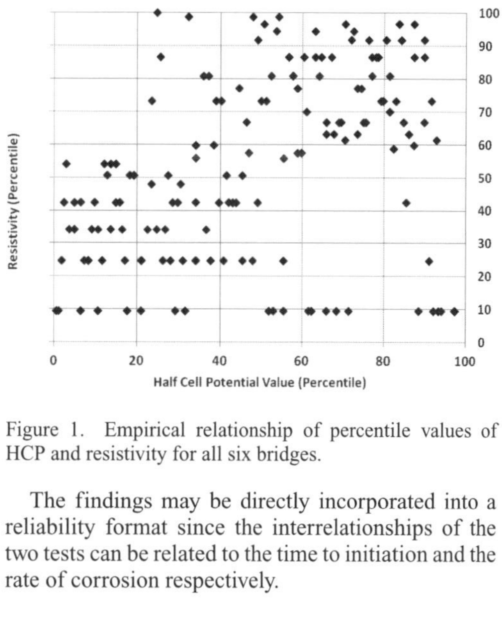

Figure 1. Empirical relationship of percentile values of HCP and resistivity for all six bridges.

The findings may be directly incorporated into a reliability format since the interrelationships of the two tests can be related to the time to initiation and the rate of corrosion respectively.

REFERENCES

ASTM Standard, C876-09. 2009. Standard Test Methods for *Half—Cell Potentials of Uncoated Reinforcing Steel in Concrete.*

Estes A.C. & Frangopol D.M. (2003). Updating Bridge Relia bility Basaed on Bridge Management Systems Visual Inspection Results. *ASCE Journal of Bridge Engineering*, 8(6): 374–382.

Gulikers J. 2005. Theoretical Considerations on the Supposed Linear Relationship between Concrete Resistivity and Corrosion Rate of Steel Reinforcement. *Materials and Corrosion*, 56(6): 393–403.

Gulikers J & Elsener B. 2009. Development of a Calcula tion Procedure for the Statistical Interpretation of the Results of Potential Mapping Performed on Reinforced Concrete Structures. *Materials and Corrosion*, 60(2): 87–92.

Bridge Maintenance, Safety, Management, Resilience and Sustainability – Biondini & Frangopol (Eds)
© 2012 Taylor & Francis Group, London, ISBN 978-0-415-62124-3

A displacement-based procedure for seismic assessment of reinforced concrete bridges

A.M. Paksoy
European School for Advanced Studies in Reduction of Seismic Risk (ROSE School), Pavia, Italy

L. Petrini
Politecnico di Milano, Department of Structural Engineering, Milano, Italia

ABSTRACT

In Italy, as well as in other seismic countries, many viaducts along the highways were built in the Sixties without antiseismic criteria. Moreover, some of these bridges are fundamental after a seismic event for allowing the civil protection interventions and first aid organizations. Therefore, they are needed to be assessed against seismic risk.

Among the displacement-based design approaches developed in the last 20 years, the Direct Displacement-Based Design (DDBD) procedure, proposed by Priestley and his co-workers has been fully developed and successfully applied on structures with different typologies. Based on the principles of DDBD a new assessment methodology for multi-span reinforced concrete (RC) bridges, named Direct Displacement-Based Assessment (DDBA), has been recently proposed (Sadan 2009). The procedure basically consists of 4 main steps: i) acquisition of the structural information; ii) derivation of the displacement shape; iii) definition of the equivalent SDOF properties, iv) evaluation of the structural performance. As shown by the published results (Petrini et al 2009) the methodology is reliable and easily implementable. However, at each step it requires the development of different models and the use of different numerical codes: this aspect limits the diffusion of the methodology out of the research world. To promote the use of DDBA also between the practition engineers and Department of Transportation Agencies of different countries with the aim of giving them a useful tool for taking rational and quick decisions about bridge assessment, an important work of revision of the procedure and implementation of it in a software (named *DDBA toolbox*) able to automatically run the DDBA has been performed.

The *DDBA toolbox* was used to study the transverse behavior of 4 different RC bridge configurations. All the configurations refer to the A16 Italian highway (Fiumarella viaduct), described in the work of Cardone et al (2011). They have 4-span simply-supported isostatic deck of 135 m total length, featuring two seat-type abutments and three frame-type piers characterized by five RC columns with 1.2 m diameter circular cross section. Configurations C1SH and C1N (Fig. 1 top) have pier heights H = 3.2 m and steel bearings or neoprene pads, respectively. Configurations C2SH and C2N (Fig. 1 bottom) have pier heights equal to H-2H-3H and steel bearings or neoprene pads, respectively. The agreement between DDBA toolbox and NLTHA results, in terms of displacement shapes and Capacity/Demand ratio shows the good accuracy of the numerical tool and implemented procedure.

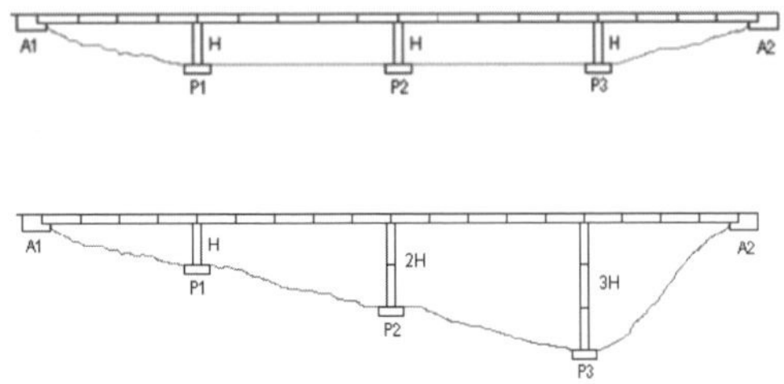

Figure 1. Geometrical configurations of the studied cases: top configuration C1, bottom configuration C2.

REFERENCES

Cardone, D., Capone, V., Perrone, G. and Petrini, L., 2011. Direct displacement-based seismic assessment of multi-span simply supported deck bridges. Proc. of COMPDYN 2011 – III ECCOMAS Thematic Conference on Computational Methods in Structural Dynamics and Earthquake Engineering, Corfu, Greece, 26–28 May 2011, on CD-rom: 1–20.

Petrini, L., Sadan, O.B. and Calvi, G.M., 2009. Direct Displacement-Based Seismic Assessment Procedure for Multi-Span Reinforced Concrete Bridges, Proc. of 13th Italian conference of earthquake engineering, 28 June–2 July 2009, on CD-rom: 1–8.

Sadan, O.B., 2009. Direct Displacement Based Seismic Assessment. Procedure for Multi-Span Reinforced Concrete Bridges, Phd Thesis, Politecnico di Milano.

Bridge Maintenance, Safety, Management, Resilience and Sustainability – Biondini & Frangopol (Eds)
© 2012 Taylor & Francis Group, London, ISBN 978-0-415-62124-3

A practical overweight permit analysis system in Seoul

A.H.M. Park
IAM Corporation & Korea University, Seoul, South Korea

B.J.K. Lim, C.Y.M. Chon & D.S.H. Kim
IAM Corporation, Seoul, South Korea

ABSTRACT

The Korean government limits exceeding total weight upto 40 ton and axle weight upto10ton for ensuring safety in facilities. But illegally pass of vehicles exceeding 40 tons on bridges has occurred frequently, because overweight vehicle users should prepare evaluation report which is structural analysis of facilities on the permit route for permission of overweight vehicles and overweight vehicles like self-propelled mobile crane have difficulties in dissembling also.

Therefore, Seoul Metropolitan Government (SMG) has been faced with the impending need to develop solutions associated with the permit of overweight vehicles.

Whenever overweight vehicle users apply for permit, implementing structural analysis is impossible. Because length, height, total weight, axle weight and the requested permit route are varied to the type of heavy vehicles.

So we have developed the system that calculating the permit ratings automatically by inputting the length, height, total weight, axle weight and axle distances based on influence line analysis. This system (Overweight PERmit Analysis System, OPERAs) is web-based system that users are not needed to visit for permission and implement structural analysis.

We have performed influence line analysis for all facilities on permit routes to calculate live load effect by specified overweight live load.

Influence line analysis could make to calculate permit ratings without additional structural analysis.

In this study, we have considered 146 facilities to develop 830km permit routes in Seoul.

Almost all road and facilities were evaluated for permit routes in Seoul.

And we have studied reliability-based permit analysis of bridges that is mainly considered AASHTO LRFR Guide Manual, also calibrated permit load factors based on reliability concept and Korean live load characteristics.

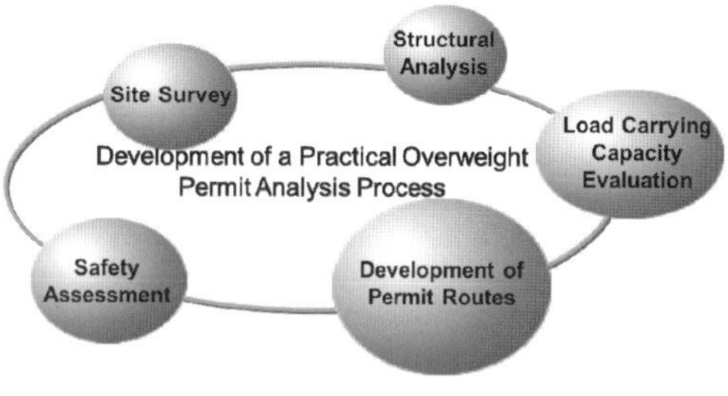

Figure 1. Research Scope.

It may be concluded that a practical overweight permit analysis system is developed in order to simplify the overweight vehicle permit process and provide accurate evaluation results. This system should be further expanded throughout the nation, thus providing an advanced heavy vehicle permit services.

It will provide rational and practical solution to the permit overweight vehicles in Korea

REFERENCES

2003(a). AASHTO LRFD Bridge Design Specifications.: American Association of State Highway and Transportation Officials.

2003(b). Guide Manual for Condition Evaluation and Load and Resistance Factor Rating (LRFR) of Highway Bridges: American Association of State Highway and Transportation Officials.

1996. Highway Bridge Design Standard Specifications (Appendix, Load-Resistance Factor Design): Korea Road and Transportation Association.

Minervino, C. & Sivakumar, B. 1999. Load Rating and Permit Review Using Load and Resistance Factor Philosophy. Transportation Research Circular 498, k6/1-8, Transportation Research Board:National Research Council.

Bridge Maintenance, Safety, Management, Resilience and Sustainability – Biondini & Frangopol (Eds)
 ISBN 978-0-415-62124-3

Estimation of diffusion coefficient of chloride ions for concrete durability design

B.S. Park, J.W. Seok, J.M. Park, J.-Y. Cho & J.Y. Kim
Seoul National University, Seoul, Korea

ABSTRACT

In general, concrete is known as an excellent material in terms of durability. But, degradation of durability by physical and chemical factors as time passes is inevitable. Particularly, steel corrosion caused by chloride ion penetration into concrete is one of the most important factors leading to durability degradation. Steel corrosion caused by chloride ion penetration into concrete decreases steel area and causes cracking due to expansion by corrosion by-products. This deterioration may threaten the functionality and safety of concrete structures during their service lives. So, consideration of chloride ion penetration through concrete should be performed for durability design of concrete structures.

In this study, a new method is developed to estimate chloride diffusion coefficient of concrete specimen by using the decrease of chloride ion concentration in the solution containing chloride ions.

First, the analytical solution is developed from Fick's second law using initial and boundary conditions. And a new experimental method is introduced to get the parameters to calculate analytical solution. In this experiment, any equipment is not required to accelerate chloride ions penetration. Therefore, the diffusion coefficient estimated by this method is more accurate than the other methods that use special equipments to accelerate chloride ion penetration. And this approach can make it possible to get the diffusion coefficient of chloride in concrete specimen within 3∼4 days.

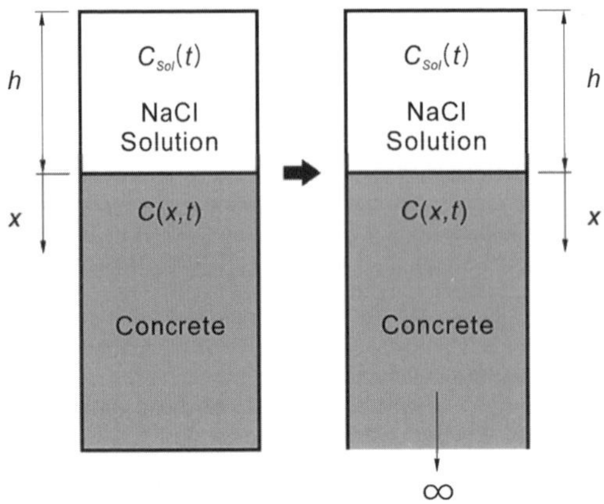

Figure 1. Conceptual diagram of mathematical model.

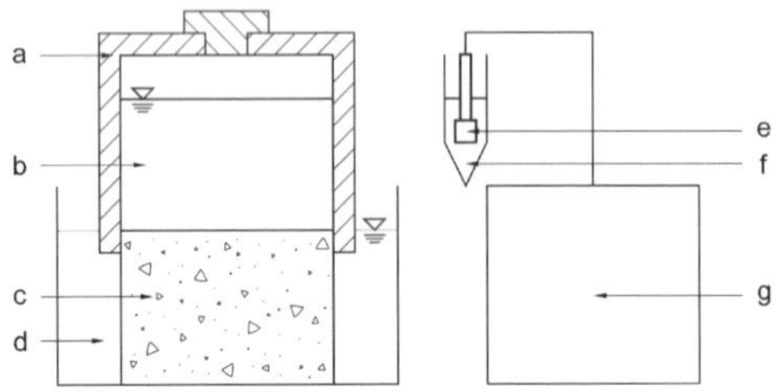

Figure 2. Experimental arrangement of the proposed method.

REFERENCES

Seo, B. R. 2003. A study on mass transfer of volatile organic compounds through geomembranes, Master thesis, Publication: 53552, *Dept. of Civil, Urban & Geosystem Engineering, Seoul National University*, Seoul, Korea.

Tang, L. 1996. Chloride Transport in Concrete – Measurement and Prediction, Division of Building Materials, Doctoral thesis, Publication P-96:6, *Dept. of Building Materials, Chalmers University of Technology*, Goteborg, Sweden.

Bridge Maintenance, Safety, Management, Resilience and Sustainability – Biondini & Frangopol (Eds)
© 2012 Taylor & Francis Group, London, ISBN 978-0-415-62124-3

A study on temperature variation of steel box girder for construction of key-segment closure of partially Earth-anchored cable-stayed bridges

S.-J. Park, Y.-G. Kim & S.-H. Kim
School of Civil & Environmental Engineering, Yonsei University, Seoul, Korea

ABSTRACT

Partially earth-anchored cable stayed bridge can be constructed using the free cantilever method (FCM) which is the most widely used technique for bridge construction. However, it is difficult to apply the set back & reset back work, because of the existence of the earth-anchored cables. Due to such difficulty, an innovative key-segment closure technique was proposed using the thermal prestressing method (TPSM). This study aims to propose a prediction method for temperature variation and examined the construction details of key-segment closure.

The proposed method considers solar exposure and wind velocity during the prediction of temperature variation, which are the most momentous parameters for the temperature distribution in the steel-box girder. The proposed method improved the accuracy of the temperature distribution after the solar exposure, and suggested the model that injecting hot air into the steel box. To verify proposed method, the experimental tests are conducted under various temperature conditions. As a result of the tests, difference between the measured temperature value and the expected value is evaluated to be 1.3%(in average). From such result, it can be concluded that the proposed method gives reliable result. To examine the heat transfer effect caused by thickness of the steel box girder, the reduced-scale model experimental tests were performed. There were no clinolimnion problems occurred during the test. Such result indicates that heat in the inside of the flange is sufficiently transferred to the outside of the flange. It is also verified that the ribs in the inside of the flange does not obstruct heat transfer.

In order to reach target temperature of the steel box girder, the analyses are performed considering seasonal effects with different target temperatures. The results showed that the proposed method released accurate results and maintain the target temperature consistently. It can be concluded that the proposed method can be effectively used in predicting heating temperature, time and the time required of heating during a closure work in cable-stayed bridges. However, it is necessary to have solar exposure and wind velocity data in order to get accurate result.

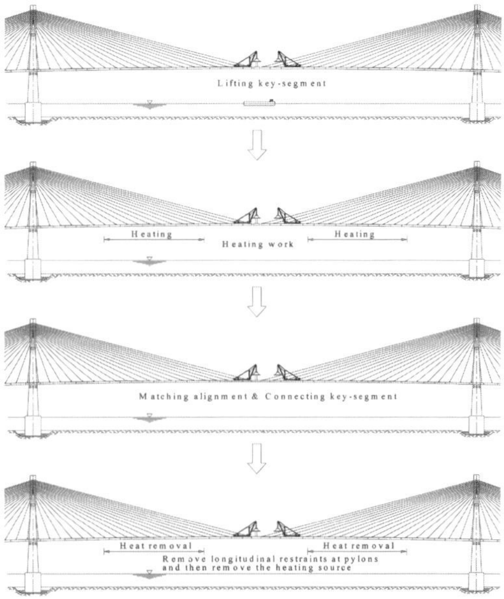

Figure 1. Process of the key-segment closure (Won, 2008).

REFERENCES

Orgill and Hollands, 1977, "Correlation equation for hourly diffuse radiation on a horizontal surface", Solar energy, Volume 19, Issue 4, pp. 357–358.

Stewart, R.W., 1979, "The atmospheric boundary layer", World Meteorological Organization, pp. 64–65.

Won, J.-H., Cho, K.-I., Yoon, J.-H., and Kim, S.-H. (2008), "Innovative key-segment closing methoad using thermal prestressing technique for partially earth-anchored cable-stayed bridges", Advances in Structural Engineering, Vol. 11, No. 5, pp. 549–569.

Bridge Maintenance, Safety, Management, Resilience and Sustainability – Biondini & Frangopol (Eds)
© 2012 Taylor & Francis Group, London, ISBN 978-0-415-62124-3

Estimation of elastic modulus of reinforcement corrosion products using inverse analysis of digital image correlation measurements for input in corrosion-induced cracking model

B.J. Pease, A. Michel, A.E.A. Thybo & H. Stang
Technical University of Denmark, Department of Civil Engineering, Kgs. Lyngby, Denmark

ABSTRACT

A combined experimental and numerical approach for estimating the elastic modulus of reinforcement corrosion products is presented. Deformations between a steel reinforcing bar and the surrounding mortar were measured using digital image correlation (DIC) during accelerated corrosion testing at 100 μA/cm^2 (~1.16 mm/year). Measured deformations were compared to a numerical corrosion-induced damage model that considers electrochemical, transport, and mechanical processes. Previous investigations utilizing x-ray imaging of the corrosion process indicate corrosion products penetrate into the surrounding cementitious matrix (Michel et al., 2011); therefore, the model allows a portion of the corrosion products to penetrate into a 'corrosion-accommodating region,' or *CAR*, provided by the mortar's capillary porosity. As illustrated in Figure 1 the CAR expands during filling of the capillary pores, likely due to microcracks forming and providing connection between previously isolated pores. Figure 2 provides a comparison of modeled and experimental deformations between the steel and the cementitious matrix. The order of magnitude of the elastic modulus of corrosion products was varied as a model input. A corrosion product stiffness of 2.0 GPa yielded the best fit to the experimental data.

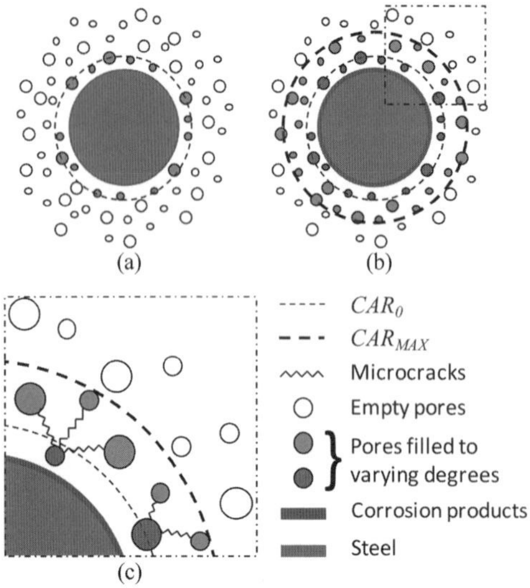

Figure 1. Conceptual schematic of idealized filling process of capillary porosity with corrosion products: (a) shows the initial *CAR*, CAR_0, (b) the subsequent increase in *CAR* size to a maximum, CAR_{MAX} and filling of additional pores due to (c) formation of microcracks between pores allowing movement of corrosion products.

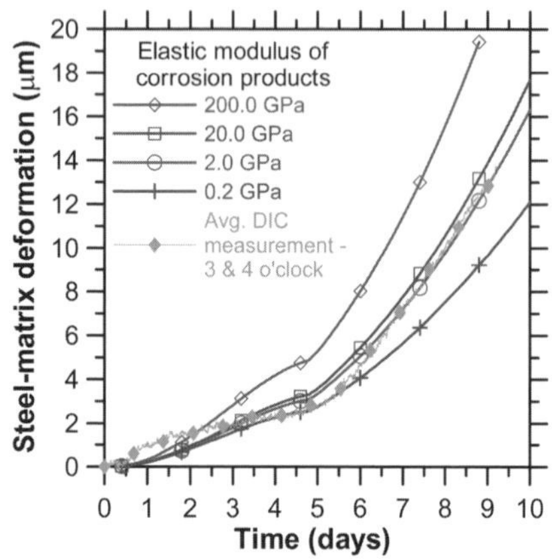

Figure 2. Modeled corrosion-induced deformations between steel and cementitious matrix for varying elastic modulus of corrosion products compared to DIC measured deformations.

REFERENCE

Michel, A., Pease, B., Geiker, M., Stang, H. & Olesen, J. 2011. Monitoring reinforcement corrosion and corrosion-induced cracking using non-destructive x-ray attenuation measurements. *Cement and Concrete Research* 41(11): 1085–1094

Bridge Maintenance, Safety, Management, Resilience and Sustainability – Biondini & Frangopol (Eds)

Assessment procedures and strengthening of an existing metal bridge

A. Pipinato
Department of Construction and Transportation, University of Padova, Padova, Italy

ABSTRACT

Metal historical bridges represent a relevant category of the international cultural heritage, being the evidence of the modern industrial technology, particularly those intended to accommodate activities of an industrial or transport infrastructure. Many of these structures require particular rehabilitation due to design defects, basic elements deterioration, variation of use or change of the intensity of the imposed loads. With regard to Italy, the historical heritage is rich of significant metal structures, which played an essential role in the growth of industrial civilization: the most part of this heritage is represented by bridges, and the 60 per cent of Italian railway steel bridges has about one hundred years, as they were built between 1900th–1920th. A step level approach and practical issues on strengthening of the same structure is presented in order to evaluate the structural integrity of historical and deteriorated metal bridges, incorporating analytical, mechanical and structural topics. Critical regions of hot-spot members were identified by structural finite element analysis, and remaining fatigue life estimation has been performed. In this paper a typical arched railway of the last-eighteenth century is studied according to a step-level assessment procedure proposed in recent studies (Pipinato et al. 2010; Pipinato A. 2011; Pipinato et al. 2009). First the bridge is geometrically described and a literature material investigation is carried out. Then, a linear FEM model is used to find out critical hot spot stress. Finally hot spot stress data are used in order to perform the reliability fatigue assessment. The case study is represented by a single track steel arch, the Garabit Viaduct: it was 565 metres (1853 ft) long and weighed 3587 tons.

Figure 1. The Garabit bridge.

REFERENCES

Pipinato A. (2011), Assessment of existing bridges: safety and security issues – Problemi di sicurezza nelle valutazioni strutturali di ponti esistenti, Ingegneria Ferroviaria, n. 4/2011, pp. 355–371, ISSN 0020-0956, CIFI-Collegio Ingegneri Ferroviari Italiani, Roma.

Pipinato A., Pellegrino C., Modena C. (2010), Fatigue assessment of highway steel bridges in presence of seismic loading, Engineering Structures, ISSN: 0141-0296, Elsevier, London, DOI: 10.1016/j.engstruct.2010.10.008, in print.

Pipinato A., Molinari M., Pellegrino C., Bursi O.S., Modena C. (2009), Fatigue tests on riveted steel elements taken from a railway bridge, Structure and Infrastructure Engineering, Maintenance, Management, Life-Cycle Design & Performance, First published on: 21 August 2009 (iFirst), Doi: 10.1080/15732470903099776, Print ISSN: 1573-2479, Online ISSN: 1744-8980, Taylor & Francis, London.

Pellegrino C., Pipinato A., Modena C. (2009), A Simplified Management Procedure for Bridge Network Maintenance, Structure and Infrastructure Engineering, Maintenance, Management, Life-Cycle Design & Performance, First published on: 22 January 2009 (iFirst), DOI: 10.1080/1573247080265908, Print ISSN: 1573-2479, Online ISSN: 1744-8980, Taylor & Francis, London.

Bridge Maintenance, Safety, Management, Resilience and Sustainability – Biondini & Frangopol (Eds)
 ISBN 978-0-415-62124-3

The manteinance of bridge structures: The case of the Soleri Viaduct in Cuneo

G. Pistone & A.D. Cavallo
Politecnico di Torino, Dipartimento di Ingegneria Strutturale DISTR, Torino, Italy

R. Enrici & S. Nicola
Provincia di Cuneo, Settore Viabilità, Cuneo, Italy

ABSTRACT

In the month of April 2010, the Provincia di Cuneo commissioned the Politecnico di Torino with the job of designing the unscheduled maintenance works of the Soleri viaduct. The viaduct is a renowned public structure, located at the north entrance to the city of Cuneo. It is an infrastructure of elevated historical-cultural value and of great logistical importance for the city: in the western section, it is used both for road and railway traffic, while on the eastern side the railway part is separated from the road section.

The first phase of the work involved the search for documentary information on the structure, an activity that was carried out in the main archives of the cities of Cuneo and Turin. The aim of this part of the work was that of clearly defining the different historical stages that led to the construction of the structure, this being motivated by needs of an infrastructural nature and, no less so, by local political requirements. The historical investigations were followed by a survey campaign which was conducted to establish the main problems and instabilities the structure suffered from: the different types of degrade were reported in specific tables. In depth technical-structural studies were also carried out together with the survey campaign. In particular, reference was made to finite element models of the structure in order to define the behaviour of the structure under the loads established by law in service conditions

The second phase of the work was first concentrated on the definition of the planning of the unscheduled maintenance requirements of the structure. The main guidelines were drawn up on the basis of this planning. The Politecnico di Torino drew up a document of a general nature, the *Design Guidelines,* which acted as a reference during the drafting of the project.

The third phase had the objective of performing an in depth planning of a specific element of the road decking: the flagstones of the pedestrian pavement. As the road decking will have a new bicycle path on both sides of the roadway, the pedestrian pavement requires re-dimensioning. The present pavement is covered in

Figure 1. Image of the Soleri Viaduct.

normal large, heavy, concrete flagstones. These elements will be substituted with other smaller and less dense ones. It was decided to make these elements in reinforced concrete with polymer structural fibres (FRC). The current state of the research is oriented towards the formulation of different types of flagstones (with only lightened concrete, with the addition of polymer structural fibres, with only metallic reinforcement, with the addition of structural fibres and metallic reinforcement) in order to identify the best structural response. After having accurately defined the mix designs, flagstone samples were prepared in order to conduct characterisation tests on the materials. The current state of the research foresees the definition of the behaviour on collapse of the different samples and the definition of an experimental constitutive bond for each type of mix design.

REFERENCES

Bosio Oreste, 1991–92, Tesi di laurea in Architettura, Facoltà di Architettura, Il ponte viadotto sul fiume Stura a Cuneo. Politecnico di Torino, Relatore Ing. Vittorio Nascè, Correlatore Arch. Roberto Roccati.

Bosio Oreste, 2001–02, Tesi di laurea in Ingegneria Civile, I Facoltà di Ingegneria, Verifica del ponte viadotto sul fiume Stura a Cuneo. Politecnico di Torino, Relatore Ing. Giuseppe Mancini.

Bridge Maintenance, Safety, Management, Resilience and Sustainability – Biondini & Frangopol (Eds)
© 2012 Taylor & Francis Group, London, ISBN 978-0-415-62124-3

Practical aspects of imposed autocorrelation and probabilistic nonlinear modeling

J. Podrouzek
Department of Waterresources, Vienna University of Technology, Austria

A. Strauss
Department of Civil Engineering and Natural Hazards, University of Natural Resources and Life Science, Vienna, Austria

ABSTRACT

Stochastic simulations in the civil engineering context represent a challenge in terms of computational cost and statistical data acquisition. Ideally such effort should be recompensed by the ability to formulate the probabilistic characteristics of analyzed system in objective and rational way. In order to do so, one must not only control the conciseness of the correlated random input variables but also relate this information to a specific dimension, i.e. characteristic length, otherwise an important feature of the probabilistic considerations would be lost and the output variance would depend on the system's discretization, often treated as a black-box parameter. Persistence of such aspect is demonstrated on selected case study utilizing data from an existing bridge and stochastic nonlinear finite element analysis.

The need of uncertainty consideration in engineering design has become imperative. In order to quantify the reliability of an engineering system, it is crucial to evaluate the probabilistic characteristics of response variables given those of the input. Generally, environmental loads and structural properties exhibit random spatial and temporal fluctuations. While the temporal fluctuations can be suitably described in terms of random process, the spatial fluctuations are better described by random fields (Bucher 2006). The scope of this paper is to address the effect of adopting various auto-correlated input scenarios on the model output (response) in the form of reliability index β (Basler and Cornell). Since closed form solution exists only for some simple/special performance functions, stochastic simulations are generally used in the civil engineering context (Novak et al. 1997). Despite the important implications, current software tools seem to neglect the option of user-controlled autocorrelation or random field concept, partly legitimately, since real material-based descriptors are rarely available.

This paper provides an insight into implementing such concept within an existing SARA (Pukl et al. 2007) software package for probabilistic nonlinear finite element analysis of concrete and reinforced concrete structures.

Performed numerical simulations of random responses of models utilizing various autocorrelation scenarios demonstrate the necessity of controlling the dimensionality of randomized input variables.

The presented study very well documents the implications of large numbers theory. In order to objectively utilize the statistical properties of input variables, one must tie such information with additional descriptive feature based on characteristic dimension. Otherwise, depending on a number of independent or cross-correlated random variables, the calculated coefficient of variation of structural response and consequent safety factor β would be a consideration based on heuristic trial-error judgment. Problematic are also non-uniformly discretized models due to patchy sensitivity to randomization. More research is required in the field to provide adequate descriptors and test the compatibility of such descriptors with current engineering practice.

REFERENCES

Bucher, C., 2006. Applications of random field models in stochastic structural mechanics. In M. Pandey, W.-C. Xie, and L. Xu (Eds.), Advances in Engineering Structures, Mechanics and Construction, pp. 471–484, Springer.

Novák, D., Teplý, B., Keršner, Z., 1997. The role of Latin Hypercube Sampling method in reliability engineering; In: Structural Safety and Reliability: 7th International Conference on Structural Safety and Reliability, Kyoto.

Pukl, R., Cervenka, V., Cervenka, J., 2007, Virtual reliability assessment of structures with damage, SARA – part II. SHMII-3, Vancouver.

Bridge Maintenance, Safety, Management, Resilience and Sustainability – Biondini & Frangopol (Eds)
© 2012 Taylor & Francis Group, London, ISBN 978-0-415-62124-3

Efficient solution for bridge reconstruction

V. Popa
S.C. Consitrans, Bucharest, Romania
correspondent member of the Romanian Academy of Technical Sciences

ABSTRACT

The bridge rehabilitation or reconstruction is a difficult and expensive work, but sometimes a well-inspired idea could lead to an efficient solution. An efficient solution means not only a low investment cost but also the best way to reach the necessary performances of the construction.

The paper presents two cases of bridge reconstruction, where interesting ideas led to important investment savings.

The first case describes the bridge over Campinitza river at Lunca Cornului.

The old bridge was constructed in 1930. It had three spans of 10.30 m + 10.40 m + 10.30 m and a total length of 31.00 m. The structure consisted in beams with cantilevers and hinges – type Gerber. The carriageway width was only 5.00 m. The foundation of the bridge consisted of wood piles and concrete pile caps. The bridge was rehabilitated in 2001, corresponding to the current life loads.

The superstructure was widened and strengthened with a new reinforced concrete overslab and two edge beams added to the existing edge beams. The infrastructure foundation was consolidated by adding two lateral concrete blocks which widened and deepened the existing foundation.

The floods from 2005 sternly affected the bridge and led to its collapse. The bridge collapse took place not only due to the inadequate foundation, but also because of the pier foundation widening that arose as an increased obstacle in the way of the flowing water.

The solution for the reconstruction was a new bridge with a single span of 40.00 m length.

The new structure consists in simple supported composite girders (six units in the cross section). The infrastructure consists in two abutments founded on piles that reached a good soil for foundation.

The most important advantage of this solution is upgrading the flowing water under the bridge and taking out the causes that affect the foundation stability.

The second case is referring to the bridge over the Ialomitza river at Albesti.

The bridge was completed in 1970. It had four spans of about 20.80 m each and a total length of 92.00 m, including the length of the wingwalls. The structure consisted in simple supported precast prestressed concrete girders. The foundation of the bridge consisted of precast concrete driven piles (25 piles for each pier foundation). The length of the piles is 9.00 m.

The pier no. 2 placed in the middle of the river bed was settled in time, during the frequent floods.

The carriageway was lowering with about 50 cm.

The adopted solution for reconstruction consisted in execution of a new superstructure instead of the two central ones, having a span of 41.50 m and removing the central pier. In order to have the same construction height, a composite structure was adopted for the superstructure.

The new composite structure has 6 steel girders and a reinforced concrete slab in the cross section. The remained piers no. 1 and no. 3 were consolidated by two drilled piles having a diameter of 1.50 m and a length of 22.00 m each and a reinforced concrete pile cap which incorporates the old one.

Sometimes the bridges are damaged due to different causes, especially because of the strong floods.

For rehabilitation or reconstruction of a damaged bridge it is very important to adopt those solution which eliminate entirely the causes which led to bridge problems.

A well-inspired solution could solve all these problems and allow to ensure the durability and safety of the construction.

Bridge Maintenance, Safety, Management, Resilience and Sustainability – Biondini & Frangopol (Eds)
© 2012 Taylor & Francis Group, London, ISBN 978-0-415-62124-3

Low-power wireless monitoring of fracture-critical bridges

D.L. Potter
National Instruments Corporation, Austin, TX, USA

J.D. Fasl, T.A. Helwig, S.L. Wood, V.A. Samaras, A.A. Yousef & K.H. Frank
University of Texas at Austin, Austin, TX, USA

R.E. Lindenberg
Wiss, Janney, Elstner Associates, Inc., Northbrook, IL, USA

ABSTRACT

There are currently over 600,000 bridges in the national bridge inventory, many which are reaching the end of their design life. Labor-intensive visual inspections remain the primary tool for identifying possible safety problems in bridge populations. However, recent developments in wireless technology, low power electronics, and embedded graphical software make real-time monitoring of bridges feasible. A low-power wireless data acquisition node that works with conventional resistive strain gages has been developed to perform real-time strain monitoring and fatigue analysis. The low-power wireless nodes are intelligent devices that can be configured with graphical LabVIEW programs that run embedded on the nodes and can be used to efficiently handle measurement data and minimize power consumption. The research team envisions using the wireless system in conjunction with current methods to detect damage that might escalate between visual inspections. With such a system, transportation officials will have the tools to better allocate inspection resources while providing greater safety to the public.

Figure 1. Steel trapezoidal box girder bridge at US 290 E and IH 35 in Austin, TX.

A complete low-power wireless sensing system utilizing the wireless strain gage wireless data acquisition devices was installed on a bridge in Austin, TX. The bridge, a connector ramp from IH-35 to US 290 in Austin, is a twin trapezoidal box girder bridge and is therefore categorized as a fracture critical bridge.

The wireless strain node continuously collects live load data, performs a local rainflow counting algorithm, programmed in LabVIEW, and periodically transmits the rainflow data to the system network controller, or gateway. The ability of the wireless node to run custom LabVIEW programs, such as the rainflow counting algorithm, enables the node to conserve the majority of energy that would otherwise be required to transmit large amounts of raw strain gage data.

This research was funded by the National Institute of Standards and Technology (NIST) as part of the Technology Innovation Program (TIP).

REFERENCES

Downing, S., & Socie, D. 1982. Simple Rainflow Counting Algorithms. *International Journal of Fatigue*, 4 (1), 31–40.

Fasl, J., Helwig, T., Wood, S., Potter, D., Samaras, V., & Frank, K. 2009. Development of Rapid, Reliable, and Economical Methods for Inspection and Monitoring Fracture Critical Steel Bridges. World Steel Bridge Symposium; San Antonio, TX, 18–20 November 2009. National Steel Bridge Alliance.

Helwig, T., Herman, R., & Li, D. 2004. Behavior of Trapezoidal Box Girders with Skewed Supports. *Research Report 4148-1, Report for Texas Department of Transportation.*

USDOT. 2004 Status of the nation's highways, bridges, and transit: conditions & performance. Retrieved from http://www.fhwa.dot.gov/policy/2004cpr/ (2006).

Bridge Maintenance, Safety, Management, Resilience and Sustainability – Biondini & Frangopol (Eds)
 ISBN 978-0-415-62124-3

Computer simulation of concrete bridges

R. Pukl, V. Červenka & J. Červenka
Cervenka Consulting, Praha, Czech Republic

D. Novák
Brno University of Technology, Brno, Czech Republic

ABSTRACT

Nonlinear computer simulation can support efficient design of concrete bridges. Verification of structural safety assisted by the nonlinear numerical simulation and based on the global safety format represents enhancement of the standard design practice which is based on elastic analysis and partial safety factors. Computer simulation provides additional and refined information about the structure, its behaviour and safety, and is especially suited for complex structures. It has been recently applied during the design of a pre-stressed concrete bridge on the south highway ring around Prague in Czech Republic. The nonlinear analysis was used to verify the bridge stability and ultimate limit state during the construction phases. The segmented bridge during construction is shown in Figure 1.

The global safety was verified for various load cases. The realistic simulation consisted of the modeling of the whole construction process including the gradual attachment of the individual segments and the pre-stressing process. After that the loads for the investigated load case were applied and gradually increased up to failure. This overloading ratio provides the information about the global safety factor. Figure 2 shows selected results from the computer simulation: deformed shape and concrete crushing at the bottom face of the bridge segments close to the pier are presented.

The lowest value of the global safety factor was obtained for the case of the combined loading was equal to 1.7, which lies above the value required by EN 1992-2.

Figure 1. View of the bridge during construction.

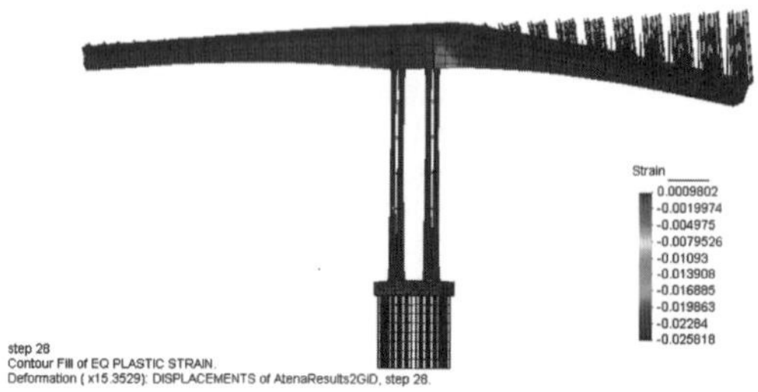

Figure 2. Plastic hinge formation at ultimate limit state.

REFERENCES

Červenka, J., Červenka, V., Janda, Z. & Pukl, R. 2008, Safety assessment of railway bridges by non-linear and probabilistic methods. In: *IABMAS 2008*, Taylor & Francis Group, London, ISBN 978-0-415-46844-2: 3708–3715.

Červenka, V. 2008, Global Safety Format for Nonlinear Calculation of Reinforced Concrete. *Beton- und Stahlbetonbau* 103, Special Edition, Ernst & Sohn: 37–42.

Pukl, R., Červenka, V., Teplý, B., Novák, D., Strauss, A. & Bergmeister, K. 2010, Advanced tools for numerical analysis and life-cycle assessment of reinforced concrete structures. In: *IALCCE 2010*, Taipei, Taiwan,: 211–216.

Novák, D., Pukl, R., Teplý, B. & Strauss, A. 2012, Reliability assessment of concrete bridges, *submitted to: IABMAS 2012*, Stresa, Italy.

Bridge Maintenance, Safety, Management, Resilience and Sustainability – Biondini & Frangopol (Eds)
© 2012 Taylor & Francis Group, London, ISBN 978-0-415-62124-3

High damping curved surface sliding isolators for bridges

V. Quaglini
Politecnico di Milano, Milan, Italy

ABSTRACT

Curved surface sliders or pendulum systems, firstly introduced at the end of the '80s (Zayas et al. 1987) (Mokha et al. 1996) for the protection of buildings from earthquakes, are today reliable seismic isolators with increasing popularity worldwide.

However strict and often conflicting requirements are posed on the properties of the materials of the sliding surfaces when the system is used for isolation of bridge and viaduct structures, subjected to large and frequent motions during their lifetime. The durability of the materials is essential to decide whether the bearing is capable of sustaining high velocities and a large amount of service movements without any deterioration. The coefficient of friction shall allow dissipation of seismic energy during the earthquake and do not produce excessive constraint to the slow service motions induced by the structure. Finally also the wear endurance is a key parameter for the long term serviceability of the bearing.

The paper presents the tests performed to investigate the effectiveness and durability of a curved surface sliding isolator designed for a viaduct structure in a high speed railway line in Turkey. The design requirements are to provide an isolation period of 2.2 seconds and an effective damping of 22% in combination with a seismic movement of ±0.45 m; the characteristics of the isolator shall remain stable over 5 consecutive cycles at the maximum design displacement. The isolator shall also be able to sustain movements corresponding to at least 30 years of expected thermal and live load displacements.

A special PTFE composite had been proposed as the self-lubricating material of the sliding surfaces of the isolator (Quaglini et al. 2011). Characterization tests revealed an exponential dependence of its coefficient of friction on the sliding speed, with an activation velocity of between 25 and 50 mm/s, which is expected to provide the isolator with distinct behaviors under either service or seismic movement. The Wear test confirmed the durability of the special material with no significant deterioration of the coefficient of friction after more than 2400 m of sliding.

In order to account for the increase in temperature at the sliding interfaces occurring in presence of large friction values and high velocities, the isolator was designed with two pairs of curved sliding surfaces with comparable radii, as to share the total movement and limit sliding velocity and heat generation at each surface. The design was validated in prototype tests conducted on full-size devices according to the American code (AASHTO 2010).

The tests showed that the isolator provides high dissipation at seismic velocities (damping >20%), while it is able to accommodate slow movements of the structure with minimum resistance. Despite of the large amount of energy dissipated at the sliding interface, the isolator behaves very stably during cyclic loading, with variations of its characteristics within 10% for the stiffness and 20% for the damping with respect to the theoretical values.

The assessment confirms that the proposed design and manufacture of isolator in combination with the use of the PTFE composite is promising in terms of upgrading the seismic resistibility of bridge structures under near-fault motion.

REFERENCES

AASHTO 2010. Guide Specifications for Seismic Isolation Design, 3rd Edition. Washington DC: American Association of State Highway and Transportation Officials.

Mokha, A.S., Avin, N., Constantinou, M.C., Zayas, V. 1996. Seismic isolation retrofit of a large historical building. ASCE Journal of Structural Engineering, 122: 298–308.

Quaglini, V., Dubini, P., Motta, R., Ferroni, D., Cuminetti, D., Poggi, C. 2011. Sliding isolation bearings with high damping capacity. In ANIDIS 2011; Proc. Italian conf., Bari, 22–24 September 2011. Bari: Digilabs (on CD rom).

Zayas, V.A., Low, S.S., Mahin, S.A. 1987. The FPS earthquake protection system: experimental report. Report No. UCB/EERC-87/01. Berkeley: Earthquake Engineering Research Center, University of California.

Bridge Maintenance, Safety, Management, Resilience and Sustainability – Biondini & Frangopol (Eds)
© 2012 Taylor & Francis Group, London, ISBN 978-0-415-62124-3

The improvement of the seismic response of a concrete bridge by using isolation devices

I.R. Răcănel & D.I. Creţu
Technical University of Civil Engineering of Bucharest, Romania

M. Conţiu
S.C. Consitrans S.R.L., Bucharest, Romania

ABSTRACT

The last seismic events produced in Chile last year 2010 and more recently this year in Japan have led to major damages on many structures including buildings and bridges. These catastrophic events have had a major influence regarding the development of efficient methods and systems for the protection of the structural systems.

Generally, in the early stages of the design, for the estimation of the response of a structure under seismic action, force based methods were used. These methods, based on the use of the design response spectra and modal analysis, led often to the increase of the dimensions of the cross sections of the structural elements in order to lower the obtained stresses below an allowable limit.

In the last decade the principles of the design are changing fast and new innovative methods, based on the displacement demand of the structure for a certain level of seismic action, began to be used. These methods are called "peformance based seismic design methods". Using finite element software together with nonlinear algorithms the inelastic response of a structure subjected to a natural recorded or artificially generated ground motion can be more realistic estimated. Most of the existing software includes for this type of analysis the "pushover" method assuming that the behaviour of the structure is governed by its fundamental mode of vibration and considering a simplified distribution of the seismic forces on the height. Recent studies have shown the fact that higher modes can significantly influence the structural response. Moreover the occurrence of the plastic hinges in the structural elements (plasticization of the cross sections) can change the form of the mode shapes. Thus, the concepts as "multi-modal push over analysis" and "adaptive pushover analysis" are suggested to be more effective in estimate the structural response.

The bridge analyzed in the paper is situated on the national road DN13 on the sector Braşov-Târgu Mureş. It comprises eight spans in the following succession: $26.00 + 6 \times 31.50 + 26.00$ m and has a total length of about 250 m. The bridge superstructure consists in eight prestressed concrete girders covered at the top part by a concrete slab with variable thickness and supports two traffic lanes having a width of 3.90 m each and a 2.50 width footway this being placed asymmetrically, only at one side of the bridge. The bridge substructure includes two abutments and seven piers. The piers elevation was designed with two circular columns of 1.50 m diameter which are connected at the top part through concrete supporting saddles. All substructure elements have deep foundations on piles 1.08 m diameter. The link between super and substructure of the bridge is made using neoprene bearings.

In order to improve the seismic behaviour of the bridge on the seismic action, the standard neoprene bearing devices were replaced by special devices as isolators and viscous dampers. Assuming a target period of the structure, these devices are calibrated in order to obtain the desired stress level on the cross section of the substructure elements, but also a limitation of the displacement at the superstructure level.

The behaviour of the structure is investigated through nonlinear time-history analyses. Also, performing nonlinear pushover analyses, the performance level of the structure could also be estimated.

REFERENCES

Elnashai, A.S., 2001. Advanced inelastic static (pushover) analysis for earthquake applications, Structural Engineering and Mechanics, Vol. 12, No. 1: 51–69.

Kalkan, E. & Kunnath, K., 2004. Method of modal combinations for pushover analysis of buildings, 13-th Conference on Earthquake Engineering, Vancouver, Canada, Paper no. 2713.

Naeim, F. & Kelly J.M., 1999. Design of seismic isolated structures. From theory to practice, Wiley & Sons Inc.

Bridge Maintenance, Safety, Management, Resilience and Sustainability – Biondini & Frangopol (Eds)

Stochastic model of continuously measured vertical pedestrian loads

V. Racic, J.M.W. Brownjohn & A. Pavic
University of Sheffield, Sheffield, UK

ABSTRACT

Stochastic and narrow band nature are the two essential features of vertical walking loading not addressed adequately in the existing design guidelines for pedestrian structures, such as footbridges, long-span floors and staircases (Racic et al. 2011). One of the reasons for this is the lack of a comprehensive database of walking forces in the form of continuously recorded time series that can be used for development of statistically reliable characterisation of these forces for application in the civil engineering context. This paper has addressed the issue by establishing a large database of measured walking time series recorded by a state-of-the-art instrumented treadmill at the University of Sheffield (Figure 1).

Another reason is the lack of an adequate modelling strategy which can simulate reliably the actual forcing signals (Racic & Brownjohn 2011). In this paper, a data-driven mathematical model has been developed to generate synthetic force signals with realistic temporal and spectral features (Figures 2 & 3).

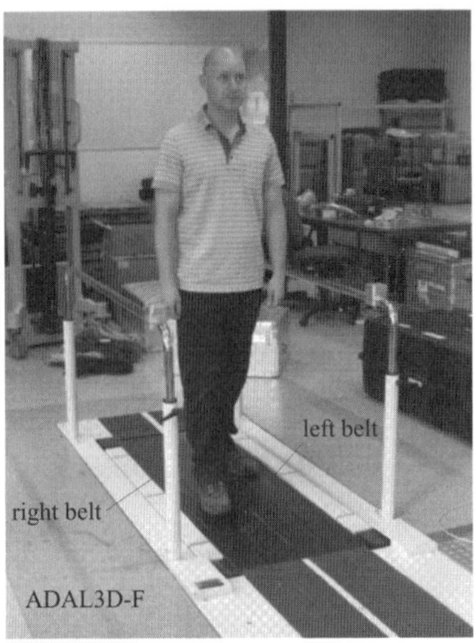

Figure 1. Experimental setup.

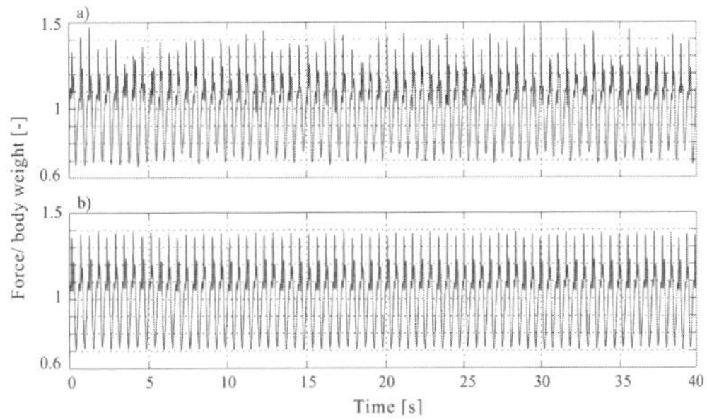

Figure 2. (a) Measured and (b) an example of synthetic walking time series.

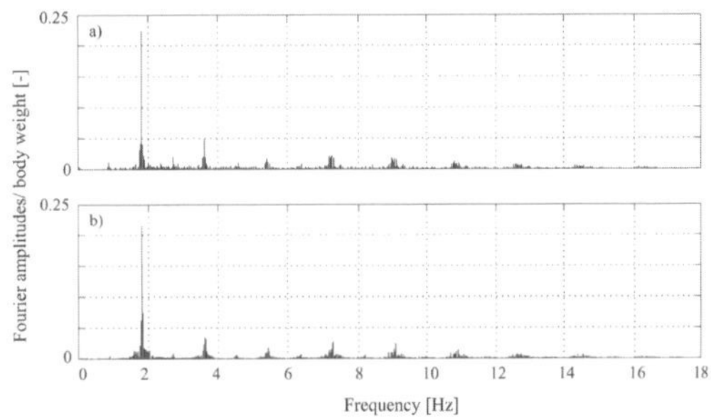

Figure 3. Discrete Fourier amplitudes of (a) measured and (b) synthetic walking time series. Figures 9 and 10 represent the same data.

The modelling strategy takes a complex numerical approach, thus can be coded as a user-friendly software which can be adopted in the vibration serviceability design practice.

REFERENCES

Racic, V., Pavic, A. & Brownjohn, J.M.W. 2009 Experimental identification and analytical modelling of human walking forces: Literature review. *Journal of Sound and Vibration* 326: 1–49.

Racic, V. & Brownjohn, J.M.W. 2011 Stochastic model of near-periodic vertical loads due to humans walking. *Andvanced Engineering Informatics* 25(2): 259–272.

Bridge Maintenance, Safety, Management, Resilience and Sustainability – Biondini & Frangopol (Eds)
© 2012 Taylor & Francis Group, London, ISBN 978-0-415-62124-3

Finite element modelling of Humber Bridge

A.R. Rahbari & J.M.W. Brownjohn
Department of Civil Engineering, University of Sheffield, Sheffield, UK

ABSTRACT

Humber Bridge, built in 1981, is the fifth longest single span bridge in the world, with main span of 1410 m and total length 2220 m. A Structural Health Monitoring (SHM) system is currently in operation to help operators identify significant changes in structural performance and inform maintenance decisions. The Humber Bridge SHM uses experimental data to diagnose performance with the help of a physics-based representation in the form a finite element model. This needs to be validated against prior experimental observations of dynamic performance.

This paper describes the modelling processes of Humber Bridge using ANSYS software. The bridge was divided into five different parts which was modelled separately: main cables, deck structure, towers, suspenders and supports. At the end all parts were attached together and made the model as it shown in figure 1.

The modelling considered two levels of resolution for the deck model. The initial model was made with equivalent plate to decrease the model complexity. However because of differences between the analytical and experimental measurements it was decided to model the bridge with the new equivalent box section and demonstrated that model works properly.

Finally, figure 2 shows the relationship between measured mode frequencies and those from equivalent box. The agreement is remarkable good. For each degree of freedom one colour/symbol is used (Vertical, Torsional and Lateral).

This validated model gives an opportunity to do parametric studies on the bridge and understanding the bridge behaviour in different situation and also can help to use as a part in bridge health monitoring.

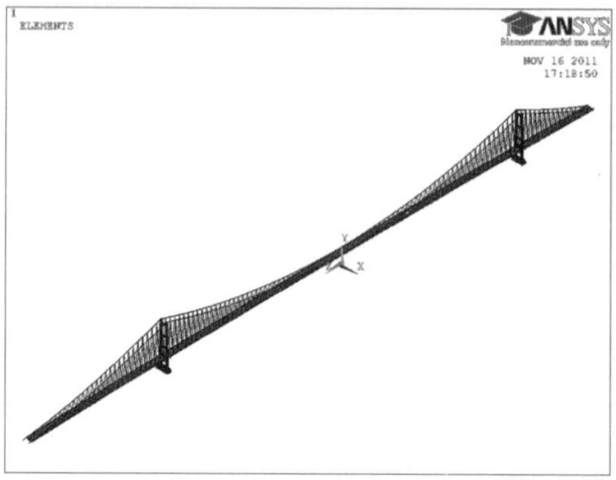

Figure 1. Bridge model with equivalent box section deck.

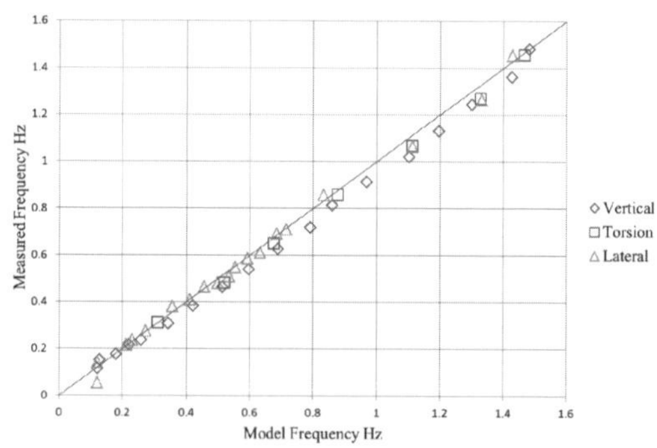

Figure 2. Model frequencies for equivalent box model.

Bridge Maintenance, Safety, Management, Resilience and Sustainability – Biondini & Frangopol (Eds)
 ISBN 978-0-415-62124-3

Jacking of bridge girders for bearing replacement

A.P. Ranasinghe
Dewberry Inc., USA

W.L. Gottshall
Area Engineering Inc., USA

ABSTRACT

As our existing bridges age and deteriorate, it is necessary to replace the existing bearings. Also, the bearings are replaced as a seismic retrofit measure. Old rocker bearings are replaced with other types of bearings such as elastomeric or seismic isolation bearings that improve the seismic performance of the bridge.

In order to replace the bearings, it is necessary to temporarily jack up the bridge girders. Normally, this is done by placing the jacks at one of three locations: under a jacking stiffener placed in front of the bearing, under the girder overhang behind the bearing, or under the end diaphragms.

In many bridges all three locations may not be available due to site conditions or the nature of the existing bridge. The first location may not be suitable or economical if there is not enough bridge seat available, or when the abutments and piers are too high, or if the bridge is above water. The second location is not suitable when there is insufficient girder overhang or for very shallow bearings. The third location may not be suitable when there are utilities passing through the end diaphragms or when there are inspection openings in the end diaphragms. Therefore, it is important to select the most appropriate and economical location for the jacking operation.

This paper discusses how the jacking details can be designed for the above three locations. Details for the design of the innovative method of jacking from girder overhangs are provided. Design considerations, such as geometric constraints and capacity constraints, are included. The paper describes the use of the jacking details in various bridge rehabilitation projects. Furthermore, new findings for jacking from end diaphragms with openings are included. The importance of safety, maintenance and protection of traffic, and economy when preparing a jacking procedure are emphasized. The paper suggests considering future bearing replacements and jacking procedures when designing new bridges.

REFERENCES

AISC, 2003. *Steel Design Guide Series 2: Steel and Composite Beams with Web Openings*, American Institute of Steel Construction. Chicago, USA.

Gottshall W.L. and Ranasinghe A.P., 2003. Jacking Configuration for the Retrofit and Rehabilitation of Bridge Bearings on Steel Girder Bridges. *Proceedings of the 10th International Conference on Structural Faults and Repairs, July 2003.* London, UK.

Bridge Maintenance, Safety, Management, Resilience and Sustainability – Biondini & Frangopol (Eds)
© 2012 Taylor & Francis Group, London, ISBN 978-0-415-62124-3

An innovative experimental approach for nonlinear dynamic physical simulation using a reconfigurable test setup

A. Rank & H.-B. Yun
University of Central Florida, Orlando, Florida, USA

J.P. Caffrey
California State Polytechnic University, Pomona, California, USA

S.F. Masri
University of Southern California, Los Angeles, California, USA

ABSTRACT

Experimental nonlinear dynamics is an important area of study in the field of structural engineering and structural health monitoring. A great challenge is the inability of a physical device to have complete freedom to simulate a wide range of nonlinearities. As a result, many researchers have longed for a versatile, but accurate, test apparatus capable of validating nonlinear modeling, system identification, and stochastic analysis studies.

The objective of this study is to develop a reconfigurable test setup as a tool to be used in a wide range of nonlinear dynamic studies. The main components include a moving mass whose restoring force can accurately be controlled and reprogrammed (with software) based upon directly measured displacement and velocity readings at each time step. The device offers control over nonlinear characteristics (coefficients) and the equation of dynamic motion. The advantage of having such an experimental setup is the ability to simulate various types of nonlinearities with the same test setup. As a result, data collected can be used to help validate dynamic models or system identification methods.

To display a practical application of the device, a case study is presented to physically simulate an orifice viscous damper, commonly used in vibration mitigation in bridges and buildings. For a large-scale viscous damper, physical testing is needed to ensure proper working functionality; however it is a very expensive process and few laboratories have the equipment needed to properly test these large devices. Conversely with the use of the compact reconfigurable test setup, the dynamic signature of the large-scale viscous damper can be simulated with pre-collected data. Because the device is reconfigurable, the controlled deterioration of the dynamic properties of a viscous damper is also possible.

The development of the aforementioned device is a challenging endeavor and despite a number of calibration efforts performed in this study, further analysis is needed to achieve a reliable test-setup. A number of practical difficulties that must be overcome for a reliable real-time simulation device are also discussed in this project.

The apparatus introduced in this study is a useful tool for researchers and designers, allowing for physical data collection for system identification or uncertainty quantification purposes.

This methodology also shows great promise for future studies, including its utilization in a physical deterioration study. Large-scale viscous dampers are becoming increasingly popular to implement into the design of large civil structures due to their ability to mitigate dynamic forces. However should these devices deteriorate, little information is known regarding its effects upon surrounding structural members. For experimental quantification and analysis, the reconfigurable test setup can be installed upon an actual structure. Using the design criteria from the installed viscous dampers, the reconfigurable test setup can physically simulate its attributes and deteriorate in a controlled manner. In this way, the stochastic effects of complex nonlinear dynamic components that propagate through adjacent components can be experimentally quantified. Thus, critical elements can be identified for design consideration and bridge maintenance plans may be able to be modified, allowing for maintenance budgets to be more efficient.

REFERENCES

Arup Corporation. 2011. The Millennium Bridge Project Page.

HITEC. 1996. Guidelines for the Testing of Seismic Isolation and Energy Dissipating Devices. CERF Report No. 96-02. American Society of Civil Engineers, cerf report edition.

Yun, H.-B. and Masri, S. F. (2008). "Stochastic change detection in uncertain nonlinear systems using reduced-order models: system identification." *Smart Materials and Structures*, 17(1), 015040.

Bridge Maintenance, Safety, Management, Resilience and Sustainability – Biondini & Frangopol (Eds)
© 2012 Taylor & Francis Group, London, ISBN 978-0-415-62124-3

Bridge maintenance planning using cross-entropy and non-stationary Markov chains

T.M. Reale & A.J. O'Connor
Trinity College Dublin, Dublin, Ireland

ABSTRACT

As infrastructure, and bridges in particular age, and environmental concerns and sustainability issues come more and more to the fore, bridge maintenance and management has become a significant challenge. To deal with this problem, Bridge Management Systems (BMS) have been implemented effectively in many countries worldwide (Ireland, US, Switzerland, Denmark etc). In 2008, the IABMAS Bridge Management Committee decided to conduct an investigation on the bridge management systems of the world to be issued in conjunction with the 2010 IABMAS conference. This report was based on the completed questionnaires from 18 bridge management systems in 15 countries. The report found that the majority of these systems have only come into place in the last 10 to 20 years (Adey, Klatter et al. 2010 and that of the 18 systems, deterioration is taken into account in 12 of them, although the level of complexity varies widely. The German 'Bauwerk Management System' and the Finnish Bridge Management System appear to be among the most technically advanced, with sophisticated deterioration models which allow for age-behaviour models in the Finnish case and particular deterioration models for specific deterioration mechanisms in the German case. Ireland, as yet does not utilize such deterioration models and so consequently this paper outlines the development of a Markovian-based model suitable for predicting structural degradation due to chloride-ion ingress on RC structures on the Irish road network.

Ideally, given sufficient quantities of inspection data, the deterioration modules of Bridge Management Systems would be calibrated directly from the raw data obtained from inspections. In the meantime however future bridge condition needs to be estimated using predictive models. Naturally, the predictive powers of the system then are only as good as the underlying models describing the physical deterioration processes themselves. RC structures deteriorate due to a number of different mechanisms acting in tandem but for application purposes, the deterioration model developed here is based on that of chloride-induced corrosion. In order to simulate total structural degradation with time, carbonation and other forms of attack could also be modelled and the overall total deterioration aggregated.

The model is Markovian in nature, due to the fact that the predefined states of condition assigned to structures upon inspection lend themselves very well to the discrete states needed for a Markov process.

Since an age-dependent deterioration model is involved, non-homogeneous chains are used to represent different time periods and phases of deterioration. This is expected to bring greater accuracy to the prediction model than homogeneous chains would. This choice of a Markovian basis also ensures that the developed procedure could be incorporated into a discrete-state Bridge Management System.

In particular, Cross-Entropy was investigated as an optimisation algorithm for these Markovian transition probabilities due to the fact that many solvers out there are unsuitable for the purposes of transition probability estimation as they can tend to converge to solutions on the bound constraints of the problem.

Pontis for instance, the US system, frequently obtains negative and larger than unity transition probabilities (Fu and Devaraj, 2011). By smoothing each iteration of the algorithm with the previous generation, Cross-Entropy tends to remain below the bound constraints, thereby making it an appropriate choice in this context.

It is envisaged that that through the reliable prediction of future structural condition, the model could be used for maintenance planning by allowing for the conjecture of future budgets and the development of comprehensive rehabilitation policies.

Bridge Maintenance, Safety, Management, Resilience and Sustainability – Biondini & Frangopol (Eds)
© 2012 Taylor & Francis Group, London, ISBN 978-0-415-62124-3

Rehabilitation of the suspension bridge over Zambezi River in Mozambique

A. Reis
GRID Consulting Engineers, Lisbon, Portugal
Technical University of Lisbon, Portugal

C. Baptista
GRID Consulting Engineers, Lisbon, Portugal

ABSTRACT

The suspension bridge across Zambezi River at Tete in Mozambique originally designed by Professor Edgar Cardoso was built between 1965 and 1970.

It is a multiple span suspension bridge with a total length of 720 m with a triangular hanger system and a flexible deck composed by 72 isostatic modules 10 m long and 11.6 m wide. Each deck module is composed by nine longitudinal beams of variable depth transversally connected by reinforced cross diaphragms and by a reinforced concrete deck slab 15 cm thick. This structural grillage is supported by concrete main transversal beams which are suspended by hangers each 10 m.

This multiple span suspension bridge is unique in its structural scheme. The aim of the designer was to demonstrate the potential of multiple suspension bridges with a flexible concrete deck, suspended from a triangular hanger system.

Inspections during 1999 and 2005 have identified 3 major problems: unbalanced load on hanger system with major misalignments on deck geometry, hanger failure by fatigue and corrosion and concrete damage due to excessive rotations of main transverse beams suspended by hangers.

A strengthening and rehabilitation design was carried out between 2006 and 2008 by the author's, leading to full hanger replacement and concrete deck strengthening carried out between 2009 and 2011 under heavy traffic conditions.

The main part of rehabilitation works did consist in restraining the cross beam rotations, repairing concrete damages and replacing the hanger system (288 hangers).

A preliminary investigation to find out the most adequate structural scheme for the strengthening of the bridge was carried out. The main problems were related to lack of stability of the cross beams that experienced excessive rotations under unbalanced traffic forces at the bearings. To correct concrete damages due to excessive rotations of main transverse beams a stability system was designed, tested during works and finally executed. Hanger force measurements were performed in order to access the force distribution in the hanger system.

A full geometric non-liner analysis was performed and fatigue on hangers was found to be related to the special inclined hanger configuration that go under zero stress conditions under traffic. For fatigue load definition, traffic surveys were performed. Special anchorage systems for hangers were required due to restriction on concrete deck geometry. A fatigue test at $2 \cdot 10^6$ cycles was performed on real scale locked-coil hanger including sockets, confirming detail category $\Delta\sigma_c = 150$ MPa from EN1993 1-11.

The upgraded bridge was opened to traffic without any restrictions, since January 2011.

Figure 1. General view of the bridge.

Bridge Maintenance, Safety, Management, Resilience and Sustainability – Biondini & Frangopol (Eds)
© 2012 Taylor & Francis Group, London, ISBN 978-0-415-62124-3

A live load control procedure for long-span bridges

N.S. Renehan & C.C. Caprani
Department of Civil & Structural Engineering, Dublin Institute of Technology, Dublin, Ireland

ABSTRACT

Many long-span suspension bridges that exist today were built in the early 20th century. Many of these bridges have undergone some form of degradation in strength. The cost of repairing these bridges may be substantial, especially when disruption is considered. As traffic is one of the most variable sources of loading on such bridges, and has typically increased year-on-year, the control of traffic load offers great potential in extending the service life of many of these bridges. Such a system could be combined with existing health monitoring systems that exist on many of these bridges.

Atkinson (2000) introduced lane closures if the load on the two lanes of the bridge exceeded an allowable limit. Kim (2010) examined a system of vehicle selection where oversized vehicles were weighted, extremely heavy vehicles over 40 tonnes were restricted access to the bridge and vehicles under 40 tonnes were allowed into the traffic stream once an acceptable position was found.

Figure 1 shows the proposed system. It consists of a WIM sensor to weigh vehicles, strain monitoring, a bridge safety algorithm that predicts the load on the bridge in the very near future, and a barrier system to restrict access to the bridge, once certain criteria are reached. In this paper, a solely reactive subset of this system was examined for its effectiveness for a range of bridge lengths, load effects, and system geometry. Microsimulation of traffic is used, as well as a novel queue dispersion model to replicate driver behaviour.

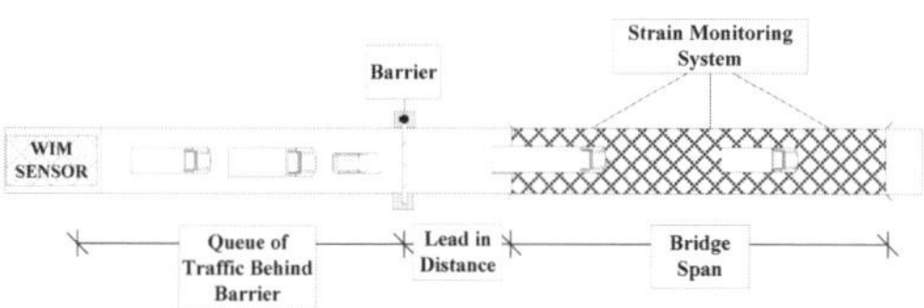

Figure 1. Proposed traffic load control system.

Table 1. Percentage reduction in 10-year return period characteristic load effect for total load on the bridge.

	0	50	100	150	200	250
50	3.55	1.31	5.78	8.69	3.44	1.35
100	5.17	1.16	6.30	7.12	3.92	3.43
200	8.32	1.37	0.85	2.03	4.18	2.35
500	11.69	3.07	2.81	2.39	1.59	1.78
1000	10.94	8.67	6.17	4.70	3.51	2.31
2000	9.01	7.52	6.32	5.93	4.46	3.87

The effect of the barrier system has on reducing the total load on the bridge is given in Table 1, for total load on the bridge length. For bridge lengths over 200 m the reduction is found to be greatest for shorter lead-in distances (See Figure 1 for definition). The closer the barrier is to the bridge, the fewer vehicles enter the bridge once the barrier is closed. Conversely, the further the barrier is placed from the bridge, the lower the reduction, as more vehicles pass the barrier before it is closed, and these vehicles enter the bridge, contributing to load effect.

The reactive system examined does not deliver a large reduction in load effects, but enough that may be important for bridges marginally in need of rehabilitation. The effectiveness of the system, is very sensitive to the shape of the influence line. A pre-emptive control system may prove to be more effective than the solely reactive one considered here.

REFERENCES

Atkinson T., Brown P., Darby J., Ealey, T., Smith J.W. & Zheng, Y. 2000. Active Control of Bridge Loads. In J Ryall, GAR Parke and J E Harding (eds.), *Bridge Management 4*. 475–482. London, April.

Kim S., Cho K., Choi, M. 2010. Reduction of design live load for bridges by controlling the heavy vehicles. In Furuta, Frangopol & Shinozuka (eds.), *Safety, Reliability and Risk of Structures, Infrastructure and Engineering Systems*, ICOSSAR, 13–17 September 2009, Osaka, Japan, Taylor.

Bridge Maintenance, Safety, Management, Resilience and Sustainability – Biondini & Frangopol (Eds)
© 2012 Taylor & Francis Group, London, ISBN 978-0-415-62124-3

Prediction of fatigue life of reinforced concrete bridges using fracture mechanics

M. Rocha & E. Brühwiler
École Polytechnique Fédérale de Lausanne (EPFL), Switzerland

ABSTRACT

With the occurrence of higher and more frequent axle loads, bridges are more solicited by fatigue loading. Bridge elements like deck slabs are subjected to a high number of stress cycles at relatively small stress magnitudes. The application of Fracture Mechanics as a useful tool for the analysis of fatigue crack growth in steel elements was demonstrated by Paris et al. in the early 1960s. With respect to reinforced concrete, the fatigue strength of the steel reinforcement is determinant. The fatigue behaviour of the steel reinforcement is similar to that of structural steel. The fatigue relevant parameters are the stress ranges, the number of fatigue cycles and stress concentrations. This paper presents a study to predict the fatigue life of steel reinforcement based on the Paris law. The method will be validated by a case study.

REFERENCES

ASM Handbook 1996. *Fatigue and Fracture.* Ohio: ASM International.

BS 7910 1999. *Guide on methods for assessing the acceptability of flaws in metallic structures.* London: British Standards Institution.

Dowling, N.E. 1993. *Mechanical Behavior of Materials.* New Jersey: Prentice Hall.

EN 1991-2 2003. Eurocode 1: *Actions on structures – part 2: Traffic loads on bridges.* Brussels: CEN.

EN 1992-1-1 2004. Eurocode 2: *Design of concrete structures – Part 1-1: General rules and rules for buildings.* Brussels: CEN.

Herwig, A. 2008. Reinforced concrete bridges under increased railway loads – fatigue behavior and safety measures. *PhD Thesis no. 1040. ENAC – MCS. École Polytechnique Fédérale de Lausanne.* Lausanne, Switzerland.

Hirt, M.A., Bez, R. & Nussbaumer, A. 2006. *Construction métallique: notions fondamentales et méthodes de dimensionnement, nouvelle édition revue et adaptée aux nouvelles normes de structures.* Traite' de Ge'nie Civil de l'Ecole Polytechnique Fe'de'rale de Lausanne, vol. 10. Lausanne, Switzerland.

Paris, P.C., Gomez, M.P. & Anderson, W.E. 1961. A rational analytic theory of fatigue. *The Trend in Engineering*, 13, 9–14.

Plos, M., Gylltoft, K., Lundgren, K., Elfgren, L., Cervenka, J., Brühwiler, E., Thelandersson, S. & Rosell, E. 2006. Structural assessment of concrete railway bridges: non-linear analysis and remaining fatigue life. *IABMAS'06: 3rd International Conference on Bridge Maintenance, Safety and Management, Porto, Portugal, July 16–19.*

Ritchie, R.O. & Murakami, Y. 2003. *Comprehensive Structural Integrity: Fracture of materials from nano to macro: cyclic loading and fatigue.* Oxford: Elsevier.

Schijve, J. 2001. *Fatigue of Structures and Materials.* Dordrecht: Kluwer.

Schijve, J. 2003. Fatigue of Structures and Materials in the 20th Century and the State of the Art. *International Journal of Fatigue*, 25, 679–702.

Stephens, R.I., Fatemi, A., Stephens, R.R. and Fuchs, H.O. 2001. *Metal Fatigue in engineering.* New York: Wiley Interscience.

Bridge Maintenance, Safety, Management, Resilience and Sustainability – Biondini & Frangopol (Eds)
© 2012 Taylor & Francis Group, London, ISBN 978-0-415-62124-3

Contribution of the FBG based monitoring to the rehabilitation of a centenary steel bridge

C. Rodrigues
LABEST, Faculdade de Engenharia da Universidade do Porto; NewMENSUS, Lda., Portugal

F. Cavadas
LABEST, Faculdade de Engenharia da Universidade do Porto, Portugal

C. Félix
LABEST; Instituto Superior de Engenharia do Porto; NewMENSUS, Lda., Portugal

J. Figueiras
LABEST, Faculdade de Engenharia da Universidade do Porto; NewMENSUS, Lda., Portugal

ABSTRACT

The corrective maintenance of old structures is becoming a cost efficient alternative to the building of new structures. In structurally defective bridges, requiring a more effective approach than a visual inspection or a numerical analysis, the short-term structural monitoring can play an essential role supporting the rehabilitation project. This allows a consistent appraisement of the effective structural behavior and condition.

The Eiffel Bridge (Fig. 1) is a centenary double-deck rail/road steel bridge, located in the north of Portugal, designed by Gustave Eiffel. During the last five years, the structure has been subjected to imperative strengthening and rehabilitation works aiming at the improvement of its transport capacity and the extension of its residual service life.

The most critical stages of the rehabilitation required the evaluation of the structure performance and its changes. In this context, the structural behavior was assessed by means of short-term monitoring campaigns based on FBG strain sensors (Fig. 2).

Structural measurements were performed: i) during the rehabilitation works; ii) at a load test, after the rehabilitation; iii) for the assessment of the premature pavement deterioration (Fig. 3). The most relevant aspects of the monitoring system and their results (Fig. 4) are presented in this paper. The structural behavior is discussed and the effects of the rehabilitation process are clarified.

The utility of the structural monitoring information for the participants in the rehabilitation is highlighted. Concerning the overall results, the presented application was a success in terms of the employment of the structural monitoring serving the rehabilitation process of an old metallic bridge.

Figure 1. General views of the Eiffel Bridge.

Figure 2. FBG strain sensors installation.

Figure 3. Monitoring campaigns carried out during trains and road vehicles crossings.

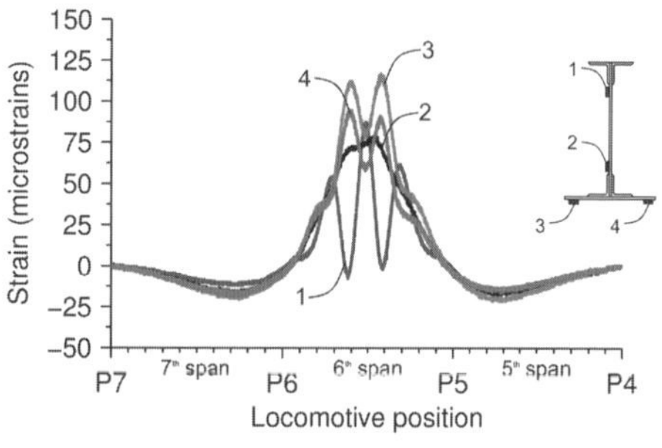

Figure 4. Experimental influence lines obtained during the crossing of a locomotive train.

Bridge Maintenance, Safety, Management, Resilience and Sustainability – Biondini & Frangopol (Eds)
© 2012 Taylor & Francis Group, London, ISBN 978-0-415-62124-3

Laboratory fatigue evaluation of replacement orthotropic deck for a signature bridge

S. Roy, J.W. Fisher, N.K. Manandhar, R.S.D. Alapati & Y.C. Park
ATLSS Engineering Research Center, Lehigh University, Bethlehem, Pennsylvania, USA

ABSTRACT

Steel orthotropic decks are natural choice for replacing decks in long span bridges because of light weight that helps in reducing dead load stresses and increasing the life cycle of the main supporting elements. After a comprehensive study of the site specific loading, it was decided to replace the 45 year old concrete filled steel grid deck at the upper level of a signature suspension bridge with a steel orthotropic deck that was integral with the floor system and the stiffening truss. To accommodate the limitations on vertical clearance, the deck consisted of relatively shallow closed trapezoidal ribs and sub-floor beams that were made integral with the existing stringers, making them load bearing in the transverse direction. Cutouts were provided in the sub-floor beams at the rib intersections for the pass-through ribs and the ribs were provided with full depth internal bulkheads at the sub-floor beam intersections. One of the primary reasons for selecting an orthotropic deck was that if properly designed, the deck system would provide more than 100 years' service life with minimum maintenance (Fisher and Roy 2010). However, the major concern related to this deck was the relatively high initial cost owing to intensive fabrication, and a higher possibility of in-service fatigue cracking from a large number of welded connections. The most severe of these welded details were the bulkhead/sub-floor beam-to-rib connections and the rib-to-deck welds that were subjected to complex in-plane and out-of-plane deformations under localized wheel loads. The cutouts in the load bearing sub-floor beams introduced additional stress concentrations. Due to lack of sufficient validation against experimental data, local stress based fatigue design methods using advanced analyses are not well established in the infinite life regime. As such, fatigue performance of the proposed replacement orthotropic deck was evaluated at the multi-directional testing facility in the ATLSS Engineering Research Centre, Lehigh University. A full-scale prototype consisting about a quarter of the deck between the panel points of the stiffening truss, and including one floor beam and two stringers was tested under simulated passage of AASHTO fatigue truck. Based on the first phase fatigue test results improvements were made in the deck design based on multi-level 3D Finite Element Analyses of the deck. The refurbished deck was further fatigue tested in the second phase to demonstrate infinite life performance. The studies provided critical information on issues related to fabrication and deck installation, and facilitated a life cycle cost-effective design for more than 100 years' service life under site specific loading.

Figure 1. Full scale laboratory testing of orthotropic deck.

REFERENCE

Fisher, J.W., and Roy, S. 2010. Fatigue of Steel Bridge Infrastructure. Structure and Infrastructure Engineering, http://dx.doi.org/10.1080/15732479.2010.493304

Bridge Maintenance, Safety, Management, Resilience and Sustainability – Biondini & Frangopol (Eds)
© 2012 Taylor & Francis Group, London, ISBN 978-0-415-62124-3

Probabilistic modeling of reinforced concrete bridge repair deterioration in marine environments

P.C. Ryan & A.J. O'Connor
Trinity College Dublin, Dublin, Ireland

ABSTRACT

It is widely accepted that chloride induced corrosion is one of the most serious problems encountered in reinforced concrete structures (McPolin, Basheer et al. 2005; Angst, Elsener et al. 2009). This paper focuses on the initiation phase of chloride induced reinforcement corrosion through the presentation of the results of an extensive experimental study performed to identify the characteristics of deterioration of a repaired concrete bridge located in a Marine environment on the south east coast of Ireland. The testing involved exposing concrete samples, constructed using five different repair options, to cyclic salt spray mist and drying in a salt (fog) spray chamber. After eight months of exposure the samples were analyzed to investigate the relative merits of the various repair materials in resisting chloride ion ingress. The repair alternatives investigated were; Ordinary Portland Cement (OPC), OPC with Ground Granulated Blastfurnace Slag (GGBS) as a partial cement replacement, OPC with Pulverised Fuel Ash (PFA) as a partial cement replacement, OPC with a silane treatment applied and OPC with increased cover. The paper presents the details and findings of the laboratory experiments.

The results of the experimental study serve to inform input parameters for a probabilistic deterioration model. The probability based deterioration model is presented and the parameters of the model are discussed in detail. The model is then utilized to study the relative merits of the various repair options in a robust stochastic manner.

The cumulative distribution function plot for time to initiation of corrosion for each repair method obtained from the probabilistic deterioration model is shown in Figure 1. It can clearly be seen from the plot that the standard OPC repair performs poorest of the five repair options studied. There is little to separate the three best performing repair options which can be seen in Figure 1 to be OPC + GGBS, OPC + Silane repair and OPC with increased cover. The OPC + PFA concrete's cumulative probability of initiation of corrosion lies somewhere between the OPC concrete and the three best performing repair options.

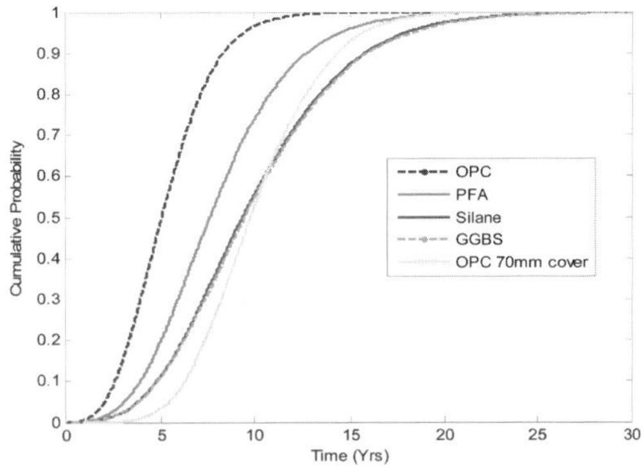

Figure 1. CDF plot of time to initiation of corrosion for the five repair options.

The paper presents the probabilistic model results in detail and discusses the relative merit of each repair option studied. For the experiments conducted and the probabilistic deterioration model utilized it was found that when compared to the standard OPC repair option the time to initiation of corrosion was approximately doubled if either OPC + GGBS, OPC with cover increased by 20 mm or OPC + Silane treatment were utilized. This indicates that the choice of concrete repairs can thus have considerable implications for the whole life cost of a deteriorated marine structure.

REFERENCES

Angst, U., B. Elsener, et al. (2009). "Critical chloride content in reinforced concrete – A review." Cement and Concrete Research 39(12): 1122–1138.

McPolin, D., P. A. M. Basheer, et al. (2005). "Obtaining progressive chloride profiles in cementitious materials." Construction and Building Materials 19(9): 666–673.

Bridge Maintenance, Safety, Management, Resilience and Sustainability – Biondini & Frangopol (Eds)
© 2012 Taylor & Francis Group, London, ISBN 978-0-415-62124-3

Zambia bridge and culvert inspection and management system

M.C. Schellhase, J.K. Shaffer & B.D. Witter
InspectTech Systems, Inc., Pittsburgh, PA, USA

ABSTRACT

The nation of Zambia is located in southern Africa. The Road Development Agency (RDA) manages over 5,000 bridges and culverts on its national road network. Similar to many agencies worldwide, the RDA's inspection processes and procedures were previously based on paper forms and a scattered collection of databases.

The RDA recently embarked on a plan to inventory and inspect all of the bridges and culverts in Zambia. After an extensive investigation process, the RDA selected BridgeInspect™ software from InspectTech for a customized, all-inclusive bridge and culvert inspection and management system. During inspections, multiple teams spent months traveling Zambia's roads and collecting the information utilizing tablet computers with interactive inspection forms. Inspectors were able to quickly enter in condition and inventory information on each structure.

In the field, inspectors use collection software on ruggedized tablet computers. Inspections are performed on-site, and data is then entered into the tablet computers. Inspectors return to the office and automatically synchronize with the server, which is running the second part of the software: a web-based inspection software that enables reports to be continued and finalized from anywhere with a secure internet connection. Reports are reviewed at this stage along with QA/QC processes to ensure a complete final report.

The software allowed for the quick upload and attachment of photographs and other files. Users are able to generate individual inspection reports or summary reports for a roadway network from any tablet or networked computer. The software's management component allows for the meeting performance measures, prioritization of needs, and other asset management based on specific Zambia RDA business processes. The management software also provides full searching across any data field, historical trending, dashboards, summary reporting, and full structure history with photos, sketches, and other attachments.

This project demonstrates how customized software was able to provide the foundation for future inspection and management decisions in Zambia. Both inspection and management processes were streamlined significantly, leading to strategically linking together structure information in a single location as Zambia's bridges and culverts continue to age.

Figure 1. Zambia RDA manages 5,000 bridges and culverts.

REFERENCES

Rankin Engineering. 2011. http://www.RankinEngineering.com/.

Road Development Agency. 2011. http://www.rda.org.zm/.

Schellhase, M.C. 2008. Improving Rating Accuracy in Visual Inspection via Software. In Proceedings of *International Bridge Conference*, Pittsburgh, PA.

Bridge Maintenance, Safety, Management, Resilience and Sustainability – Biondini & Frangopol (Eds)
© 2012 Taylor & Francis Group, London, ISBN 978-0-415-62124-3

Fatigue life time assessment of structural steels by use of ductility parameters

I. Schendel & U. Peil
Technische Universität Carolo-Wilhelmina, Institute of Steel Structures, Braunschweig, Germany

ABSTRACT

The infrastructure is a substantial factor for any national economy. Consequently, an adequate service life prediction is of major interest. The condition of steel structures is commonly determined by visible inspection in combination with an estimation of possible fatigue damage. Modeling load histories and damage accumulation is a widely used approach but also relatively inaccurate because uncertainties are multiplied. The load histories are usually only roughly estimated, since recordings of the former load sequences are seldom available, e.g. train schedules, traffic counts (Schendel and Machledt-Michael, 2009). An approach is presented here to directly determine the inherent fatigue damage of structural steels. Inherent damages are those that accumulate before a visible crack forms (Medgenberg, 2008).

The method is based on the idea of a reduced ductility in the steel matrix microstructure. In order to determine complete S/N-curves, specimens of construction steel S355J2 with multiple symmetric notches are subjected to single-stage cyclic loading. Once a short macro crack of $a = 0.5$ mm is detected in the root of one notch, the corresponding number of load cycles N is defined as damage $D = 1$. Fractions of fatigue damage D (e.g. 0.5 D; 0.9 D) are then applied to further specimens of each series to generate damage $D = n/N$ in up to six steps. All specimens are then individually Charpy-impact-tested at specific temperatures in the upper and lower shelf and in the steep drop.

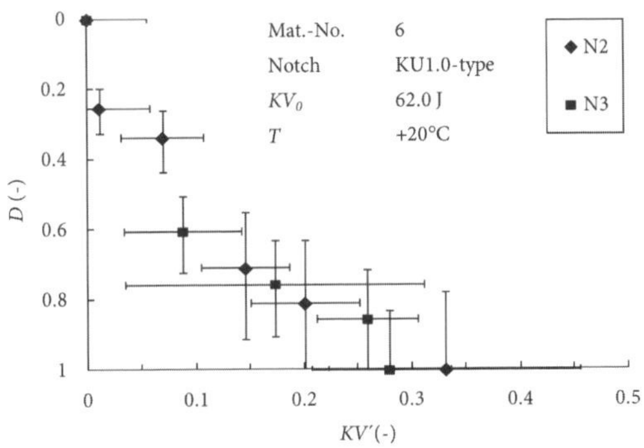

Figure 1. Fatigue damage D, normalized ductility KV' 0.1/0.9-fractiles.

Early results are shown in Figure 1. The fatigue damage D is applied on the ordinate. The abscissa shows the normalized impact energy KV' defined by Equation 1

$$KV' = \frac{KV_D - KV_0}{KV_0} \tag{1}$$

where KV' = normalized ductility reduction, KV_D = notch energy $(n/N > 0)$, KV_0 = initial notch energy $(n/N = 0)$. Fractile values of fatigue tests and of impact tests are displayed by range tracers. Experimental parameters such as the cast of steel (Mat.-No.) and radius of the notch, as well as the testing temperature T were varied in between the testing series.

Aim of this research is a direct assessment of the service life of a structure by testing small material samples (Peil et al., 2004), so grades of ductility reduction from different locations can be compared. Thus, the commonly used error-prone modeling of load histories and damage accumulation could be avoided.

REFERENCES

Medgenberg, J. 2008. Investigation of localized fatigue properties in unalloyed steels by infrared thermography, PhD thesis, TU-Braunschweig.

Peil, U., Frenz, M. & Weilert, K. 2004. Determinination of the inherent damage of older structures. In: Proc. Europ. Workshop on Struct. Health Monitoring, München.

Schendel, I. & Machledt-Michael, S., 2009. Beanspruchungen der Vergangenheit, interner Bericht Historie der Verkehrslasten, Institut für Stahlbau & Institut für Verkehr und Stadtbauwesen, TU Braunschweig.

Bridge Maintenance, Safety, Management, Resilience and Sustainability – Biondini & Frangopol (Eds)
© 2012 Taylor & Francis Group, London, ISBN 978-0-415-62124-3

Rapid non-contact tension force measurements on stay cables

M. Schmieder, A. Taylor-Noonan & R. Heere
Metro Testing Laboratories Ltd, Burnaby, British Columbia, Canada

ABSTRACT

Vibration to Tension Force calculations are known in the engineering community through the Taut String Theory. Precise calculations require accurate identification of the natural frequencies and incorporation of correction factors. Additionally, the duration of individual measurements must be optimized to allow for sufficient data acquisition. The work presented in this paper describes a cost effective and optimized approach to determine the tension force in stay cables. This optimization is necessary to accommodate any funding restrictions.

Cable Stay bridges gain more and more popularity in the structural engineering world. The reasons for the popularity of these structures include the cost efficiency of construction and the pleasing appearance of cable stay bridges. However, the main structural elements holding the deck in place are the stay cables which are in most cases not thoroughly inspected over the lifetime of the bridge. The life span of the structure may be shortened due to corrosion, slippage, settlement of the structure or part of the structure, resulting in load imbalances in the cables.

The costs of repairing significant structural damages may exceed the replacement cost, especially when the repairs involve full or partial closure. Therefore, maintaining the structure and fixing the issues when they arise will be most cost efficient in the long run. Periodic inspection plans for these bridges often include a measurement of the cable tension forces. Traditional methods to measure cable tension forces in stay cables can be classified into 2 main categories:

Lift-off:
direct measurement by the lift-off method, using hydraulic jacks.

Frequency-based:
indirect measurement via the natural frequency of the cable.

The natural frequency-based method centers on the Taut String Theory, where the tension force T in a cable of known length L and unit weight w can be calculated from the n-th natural frequency of vibration f_n by

$$T = \frac{4\omega L^2 f_n^2}{n^2 g}$$

where g is the gravitational constant.

An experimental validation of the determination of tension force through natural frequencies was published in Russell and Lardner.

As the size of bridges presents challenges for accelerometer installation, remote sensing methods have been explored. A Laser Vibrometer is one such device, in which a laser beam is directed at the location of interest and the velocity of the vibration is measured.

Systems of this kind are capable of conducting measurements from distances of up to 200 m.

This paper describes an application of stay cable tension determination through laser-based vibration measurements. First, the main methods of bridge cable tension assessment are compared. Next, the authors present the results of original research on the importance of measurement duration. Three methods for identifying the natural frequency from the frequency plot are compared for varying measurement durations. Then, a literature review on the effects of sag-extensibility and bending stiffness is conducted, and a system of correction factors is presented.

Bridge Maintenance, Safety, Management, Resilience and Sustainability – Biondini & Frangopol (Eds)
© 2012 Taylor & Francis Group, London, ISBN 978-0-415-62124-3

Laser vibrometry for bridge post-repair investigation

M. Schmieder, A. Taylor-Noonan & R. Heere
Metro Testing Laboratories Ltd, Burnaby, British Columbia, Canada

S.E. Chen
University of North Carolina at Charlotte, USA

ABSTRACT

Bridges are important elements of our traffic infrastructure. Structural failure of such elements may have devastating consequences. Therefore, monitoring the condition of such structures is of importance. Traditional monitoring (beyond visual inspection) requires the inspector or a measuring instrument to directly access the structural component to be evaluated. For locations difficult to access, Laser-based vibration monitoring may now be of assistance. The authors have gathered field experience using a Laser interferometer to determine the natural frequency of structural components, using ambient excitation only. Because the natural frequency is related to the stiffness of a component, some structural damages, as well as the effectiveness of structural repairs and strengthening, may be detectable. Further, repeat routine measurements at regular intervals (vibrational fingerprinting) may be able to flag structural changes which will prompt more detailed investigations by traditional means. Further, the authors have demonstrated that localized delamination of fibre composite patches on a concrete bridge can be detected using Laser interferometry in combination with remote acoustic excitation. The authors are currently investigating whether this method can detect concrete de-laminations, failing steel connections and other localized damages to structural elements.

The paper also will address the potential of using Laser vibrometry for post-repair structural system evaluation as a validation of the rehabilitation procedure. Results from several field tests are presented. It is observed that local frequency due to voided contacts seem to be promising for damage detection.

The current process of bridge health investigation (current practice) requires a fair amount of equipment and manpower which leads to a significant cost to the operating agencies. Traffic flow is often impeded by lane closures to accommodate snooper trucks. Environmental conditions, insufficient lighting, and the inherent subjectivity of visual inspection and hammer sounding may lead to incomplete or incorrect results. Vibration and modal analysis is an emerging field, with the aim of providing detailed information on the condition and performance of a structure. Accelerometers are used in most implementations. However, placement of contact sensors poses several logistic and economic limitations. Due to the long and complicated setup and wiring process of an array of accelerometers, most in-service monitoring systems are permanently installed on the structure, while portable vibration measurement systems are used mostly for academic research projects. Permanent systems can be installed during construction when access to the structure is easy, and then left unattended to gather vast amounts of data. These systems are not without maintenance requirements: computer systems may crash and cables and sensors exposed to the weather may corrode or crack.

Scanning Laser Doppler Vibrometer (SLDV) provide a very flexible solution of remotely measuring vibration patterns without the requirement to attach instruments to a structure. This paper explores the practicality of such instruments in the process of Structural Health Investigation.

SLDV could provide a more cost-effective and convenient way of collecting vibrational data on existing structures than accelerometer setups, either permanent or temporary. The system is portable, self-contained, and does not require access to the structure or lane closures. A single portable SLDV can monitor many bridges because the long life span of a bridge means that measurements are not needed every day. Instead, measurements taken with the SLDV at intervals (every year, two years, 5 years) would provide vibrational snapshots or fingerprints of the structure that can be compared. If these snapshots are taken before and after a repair project, the effectiveness of the repair could be quantified.

Bridge Maintenance, Safety, Management, Resilience and Sustainability – Biondini & Frangopol (Eds)
 ISBN 978-0-415-62124-3

Shear strength for corroded plate girder bridge

T. Shimozato, Y. Tamaki, Y. Arizumi & T. Yabuki
Department of Engineering, Ryukyu University, Okinawa, Japan

S. Ono
Japan Construction Mechanization Association, Shizuoka, Japan

ABSTRACT

Structural function abilities of steel plate girder bridges with corrosion damages are not clear yet sufficiently. This paper reports the results examined analytically for girder shear strengths by referring to the data of web plate thicknesses measured for the steel plate girder bridge damaged severely by chloride attack with high humidity and high temperature during 28 years. The models of the web plate with the corrosion aspects measured were analyzed using elastic-plastic finite element approach with experimental results on corroded specimens. As the results, it became clear that it is necessary to evaluate the shear strength not only for corroded de-crease web thickness but also for the distribution of the corroded decrease thickness depending on structural parts in plate girders.

Figure 1. Panorama bridge (Okinawa in Japan).

Figure 2. Specimen of corroded plate girder.

Figure 3. Set-up with specimen.

Figure 4. Shear buckling mode.

REFERENCE

Y. Tamaki, T. Shimozato, Y. Arizumi, T. Yabuki and S. Ono, "Evaluation of corrosion deterioration of weathering steel bridge under the environmental corrosiveness", Proceeding of the Fifth International Conference on Bridge Maintenance,Safety and Management, Philadelphia, Pennsylvania, USA, 1 1–15 July 2010, p. 678.

Bridge Maintenance, Safety, Management, Resilience and Sustainability – Biondini & Frangopol (Eds)
© 2012 Taylor & Francis Group, London, ISBN 978-0-415-62124-3

Structural assessment for high concrete pier with a vertical construction error and suggestion of the improvement measurement

J. Sim, K. Lee, H. Kim & A. Hanif
Department of Civil & Environmental Engineering, Hanyang University, Ansan, South Korea

G. Kim, H. Ju & M. Ju
Bridge Inspection Division, KISTEC, Ilsan, South Korea

ABSTRACT

Slip-form construction is a construction method for tall structures, such as buildings, bridges, towers, offshore platforms and dams, in which concrete is poured into a continuously moving form. These forms serve as supporting walls strong enough to bear the concrete weight poured over top of it. Slip-form has been used for the construction of tall piers of long span bridges not only to reduce the construction period but also to ensure the safety The slip-form method, however, sometimes has showed several problems, such as vertical errors under construction or deflections of forms, from irregular loads by construction materials, workers or its weight, or unsettled friction or bond between concrete and forms. These include, but not limited to, problems in mechanical and hydraulic operations due to greater weight of steel form, problems in ensuring that the slip-form remains plumb i.e. the horizontal deviations do not exceed specified limits and certain other handling problems arising due to the weight of the form itself. This paper presents a case of construction errors that aroused during vertical slip-forming in

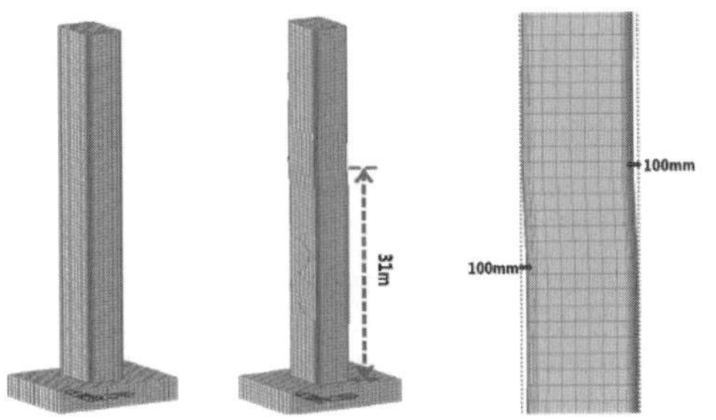

(a) Normal model (b) Model with error (c) Details of error

Figure 1. Structural analysis of P4 using detail modeling.

Table 1. Comparison of allowable stresses (MPa).

		Sections (h = 20 m ~ 40 m)			
		Normal model	Model with error	Allow. bending compressive and bending tensile stresses	
Normal	Max. compressive	1.851	3.872	$0.4f_{ck} = 9.60$	OK
	Max. tensile	0.014	1.148	$0.63\sqrt{f_{ck}} = 3.08$	OK
Seismic	Max. compressive	4.368	4.372	$0.4f_{ck} = 9.60$	OK
	Max. tensile	1.707	1.762	$0.63\sqrt{f_{ck}} = 3.08$	OK

the construction of bridge pier using steel as the slip-form. Moreover, it also includes field inspection, shape survey, structural assessment through the structural analysis and the measures taken against these errors.

The most reasonable causes of these construction errors that may occur were estimated to weight and friction force of steel form. So, using of FRP form was suggested which is efficient in reduction of weight and friction force, therefore proving to be an excellent alternative to the steel slip-form.

REFERENCES

ACI Committee 117, 2008, Specification for tolerances for concrete construction and materials and commentary (ACI-117-06).

Cho, S.K., Kim, H.B., Park, Y.K. & Joo, T.S. 2005.Case study of slip form method to pier construction, Korean society of civil engineers (KSCE) annual convention, 3099–3109.

Lee, S.Y., Park, H.S., Lee, J.H. & Woo, S.M. 2004. An application of slip form system on high pier casting, Magazine of the Korea concrete institute, Vol. 16. No. 2: 52–56.

Korean society of civil engineers (KSCE), 2010, Korean specification for highway bridges.

Bridge Maintenance, Safety, Management, Resilience and Sustainability – Biondini & Frangopol (Eds)
 ISBN 978-0-415-62124-3

Determination of the mean fatigue limit of a French railway bridge puddle iron by self-heating measurements under cyclic loadings

S. Sire, C. Doudard & S. Calloch
LBMS (Brest Laboratory of Mechanics and Systems), Brest, France

ABSTRACT

This study deals with the puddle iron of the viaduct of Lambézellec (Brest, France) built between 1891 and 1893. In order to determine the fatigue properties of this material several samples have been collected during its complete restoration. Indeed this hot riveted structure was damaged during the Second World War and it was much corroded because of the often rainy weather in this region of Brittany.

Puddle iron is a heterogeneous material which contains lots of non-metallic inclusions. This feature makes difficult the mechanical characterization; indeed, the scatter in the results is very high. Indeed, the undertaken tensile tests show that the yield stress ranges between 280 and 240 MPa and that the elongation at failure is between 6 and 12%. Charpy impact tests (V notch) show the same scatter.

The determination of the fatigue properties of the puddle iron is thus very difficult. In order to make a fast characterization and to avoid a long and expensive experimental campaign, self-heating measurements under cyclic loadings have been carried out (Doudard 2009). Indeed, recent research works on the fast characterization of High-Cycle Fatigue (HCF) properties show that this method gives very good results compared to the long classic one.

The self-heating method consists in measuring the change of the specimen mean temperature under cyclic loadings. For the puddle iron, this temperature reaches a steady-state quickly. Plotted as a function of the stress amplitude applied to the specimen, the steady-state temperature increases significantly from some loading levels which can be explained by the assumption of a dissipation induced by the microplasticity. Empirically, the mean fatigue limit can be estimated with the asymptotical behavior of the last plotted points. As shown in Figure 1, the self-heating tests carried out on the puddle iron give an upper and a lower value of its mean fatigue limit σ_D. The mean of σ_D is close to 210 MPa.

A two-scale probabilistic model to analyze the self-heating measurements is also proposed. A set of

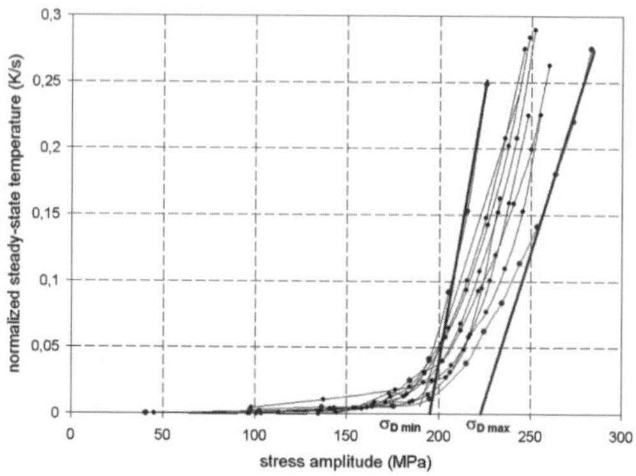

Figure 1. Self-heating experimental set of curves and determination of the mean fatigue limit range from 195 to 225 MPa.

elasto-plastic sites are considered to be randomly distributed within an elastic matrix. The distribution of sites (with a random yield stress) where the microplasticity occurs is assumed to be described by a Weibull law (Weibull 1951). By integrating the heat conduction equation, we show that the scatter observed in the thermal behavior of the puddle iron specimens can be relied to the Weibull shape parameter. Moreover, this developed numerical model is able to predict a scatter of the steady state temperature as a function of the stress amplitude, which corresponds to the experimental observations.

REFERENCES

Doudard C., Calloch S. 2009. Influence of hardening type on self-heating of metallic materials under cyclic loadings at low amplitude, *European Journal of Mechanics – A/Solids*, 28(2), 233–240.

Weibull W. 1951. A statistical distribution function of wide applicability, *ASME Journal of Applied Mechanics*, 18, 293–297.

Bridge Maintenance, Safety, Management, Resilience and Sustainability – Biondini & Frangopol (Eds)
© 2012 Taylor & Francis Group, London, ISBN 978-0-415-62124-3

The design and construction of bridge structure erected by balanced cantilevers method situated on the Prague bypass

M. Šístek, V. Engler & F. Hanus
Novák&Partner, Ltd., Prague

R. Lenner
Valbek, Ltd.

L. Vráblík
Novák&Partner, Ltd., Prague
Czech Technical University in Prague, Faculty of Civil Engineering, Department of Concrete and Masonry Structures

ABSTRACT

The bridge across the Berounka river valley forms a part of the new south-western part of Prague's by-pass and with its total length of more than two kilometers is one of the biggest infrastructure projects in the Czech Republic and has a great importance in calming of the traffic congestion in the capital. Three different construction methods took a part in the building of the bridge – fixed scaffolding, the overhead movable scaffold system and the balanced cantilever method using of up to four pairs of form travelers respectively, with the longest span of 114 m. The main contractor for the No. 204 and No. 205 section was Boegl and Krysl. The bridge was designed by Novak & Partner consulting company as a continuous span with the box girder section of a variable height of 2.6 to 6.5 m.

Over 1500 tons of post-tensioning (PT) were installed during the course of construction in years 2008–2009. The PT contractor, VSL Systems has submitted a modified alternative of the PT in order to achieve the material savings and overall simplification. The construction process was speeded up further by using of prefabricated tendons made in the construction yard. The data about tendon forces and elongations during the stressing works were collected by the Adapt electronic system. The records were then utilized to verify the design; the theoretical friction values could be subsequently compared with the actual measurements.

Prague Ring Road (SOKP) is part of a network of highways and roads of the Czech Republic. SOKP is one of the most significant road constructions of capital city of Prague and Central Bohemia; after its completion it will become one of the busiest traffic roads in the Czech Republic. It is part of the fourth Multimodal Corridor Trans-European Transport Network TEN-T.

The bridge over the Berounka Valley – a part of Prague Ring Road – is a monolithic structure of prestressed concrete with box cross section. Separate structures are built for either direction of traffic. The overall length of the bridge is divided into five separate expansion units – that differ in their arrangements, span sizes and employed technology.

Figure 1. Overall view of the structure.

The main part of the structure of 557 m length passes over the city district of Radotín and the railway line Prague – Pilsen at the height of up to 40 m above ground surface.

The structure has 6 spans of lengths of up to 114 m (72 m + 84 m + 101 m + 2 × 114 m + 72 m) with the variable depth of its box cross sections (3 m at midspans and above terminal abutments; the box depth increases through parabolic variation up to the height of 5.2 m above piers of the shorter spans and to 6.5 m above piers of the long spans).

The process of bridge erection was very complex both in the course of design works and as well as during the construction. The process was changed several times due to unsettled ownership problems it was impossible to get access to a part of the land below the area of cantilever carriage concreting and to carry out the construction according to the planed time schedule.

Bridge Maintenance, Safety, Management, Resilience and Sustainability – Biondini & Frangopol (Eds)
© 2012 Taylor & Francis Group, London, ISBN 978-0-415-62124-3

The whole life costing of bridge deck replacement – A case study

T.W. Siwowski
Rzeszow University of Technology, Rzeszow, Poland

ABSTRACT

The most vulnerable element of a bridge is its deck. Bridge deck deterioration of older bridges is a significant problem in aging of the highway system. Use of an advanced material bridge deck system is viewed as a potential long-term solution for the concrete deck deterioration. But new construction materials for bridge decks must be selected with great care and foresight over the conventional construction materials. The whole life cost analysis (or life-cycle cost analysis, LCCA) allows the engineer to determine which alternative is cost effective over its intended life. The goal of the LCCA performed in this study was to assess the life-cycle cost-effectiveness associated with three bridge deck material options for a steel truss bridge, including new advanced aluminium solution (Siwowski 2009) and two conventional solutions as used typically for bridge redecking in Poland.

The simple and flexible life-cycle cost (LCC) model consistent with the standard method for performing life-cycle costing has been used in this study (Ehlen 1997). The model uses a life-cycle costing methodology based on the ASTM practice for measuring the life cycle costs of buildings and building systems and a cost classification scheme for comparing life cycle costs of alternatives. It is based on the LCCA methodology for new-technology materials in construction sector. The conventional cost categories have been included in the LCC model, i.e.: initial construction costs, operation, maintenance and repair costs and finally disposal costs. The LCC of each alternative is computed as the sum of individual project cost items, each cost discounted to base-year, present-value Euros.

The subject of this study is the bridge over Vistula river, a key component of the National Road No. 9 in Poland. After 50 years of service the bridge needed comprehensive rehabilitation along with enhancing its carrying capacity up to the highest class according to Polish bridge code. Apart from strengthening and anticorrosion protection of steelwork, the rehabilitation also included the total replacement of deteriorated concrete deck slab. Three deck replacement alternatives were considered: new reinforced concrete deck slab (base case), steel orthotropic deck and aluminium deck made of extruded shapes.

Each alternative deck's LCC is the sum of all costs that are incurred over the life of the bridge, i.e. over 60 years. The aluminium deck has the lowest LCC (€ 5,917,030), making it the cost-effective bridge deck. Table 1 shows the unit costs on 1 m^2 of deck area and total cost savings in comparison to base case (RC deck alternative).

Table 1. Unit costs of 1 m^2 of deck and total cost savings (€)

	Deck replacement alternative		
Parameter	RC deck	Steel deck	Aluminium deck
Unit costs of 1 m^2 of deck area	2 269	1 963	1 843
Total net cost saving	0	982 119	1 366 257
Cost saving on 1 m^2 of deck area	0	306	426

The conclusions reached in this study are based on the relative performance of the three systems. Despite of the simplifications used in analysis, the presented LCCA model could serve as a tool for evaluating a bridge deck material from a holistic sustainability perspective – integrating environmental, social and economic indicators. Such a material selection procedure could enhance robust investment decisions and infrastructure sustainability.

REFERENCES

Ehlen M.A. 1997. Life-cycle costs of new construction materials. *Journal of Infrastructure Systems* 3(4):129–133.

Siwowski T. 2009. Structural behaviour of aluminium bridge deck panel. *Engineering Structures* 31(7):1349–1353.

Bridge Maintenance, Safety, Management, Resilience and Sustainability – Biondini & Frangopol (Eds)
© 2012 Taylor & Francis Group, London, ISBN 978-0-415-62124-3

Monitoring integrity and corrosion damage on cable stayed bridge "Jaime Dovali" Mexico

R. Soto-Espitia
Center for Research of Engineering and Applied Sciences (CIICAp), Autonomous University of Morelos State (UAEM), Cuernavaca, Morelos, Mexico

J.R. Vazquez Gonzalez
Federal Toll Roads and Bridges and Related Services (CAPUFE), Cuernavaca, Morelos, Mexico

L. Martinez-Gomez
Institute of Physical Sciences (ICF), National Autonomous University of Mexico (UNAM), Cuernavaca, Morelos, Mexico

ABSTRACT

Cable Stayed Bridges have been built in rapidly increasing numbers since 1950 and have been found to be especially economical for medium to long-span bridges from 100 to 1000 meters, where technical and economic considerations dictate a cable stay solution. The use of cable stayed bridges in Mexico has been expanding in the last 20 years; today already exist more than 10 stay cable in Mexico. Corrosion of steel is complex phenomenon affected by many environmental factor, the location of this bridges are around of the country (sea zones, high humidity zones, etc). Many bridges over 20 years shown deterioration due to corrosion, in 2010 the average age of cable-stayed bridges in the Mexico was 17.5 years. As these bridges age, they need for effective inspection and maintenance methods and tools becomes more acute. The aim of this study is to examine the effects of the corrosion phenomenon on the cable-stayed bridge Jaime Dovali Brigde evaluating by Non Destructive Test using electrochemical techniques and structural parameters like: Detailed visual inspection, Detection of hidden corrosion defects by sounding, Electric continuity verification of steel reinforcement, Electrical resistivity evaluation of the concrete, Corrosion potential mesuarement, Sampling of concrete core for analysis, Evaluation of chloride penetration, Evaluation of carbonation depth and Characterization of concrete samples.

Dovali Bridge has a total length of 1170 m and a main span length is 288 m, this bridge where built in 1984, and is located about 20 kilometers from the Mexican Gulf.

REFERENCES

Frangopol, D. 2011. Life-cycle performance, management, and optimisation of structural systems under uncertainty: accomplishments and challenges, Structure and Infraestructure Engineering. Vol. 7, No. 6, June 2011, 389–413.

Glišić, B. & Inaudi, D. 2007. Fibre Optic Methods for Structural Health Monitoring. John Wiley & Sons, Ltd.

Maya, E., Martínez de la Escalera, L. M., Canto, J., Ascencio, L., De Silva-Munoz, J. A., Morales, M. V. & Martínez Gómez, L. 2011. Inspection, diagnosis, materials and processing methods to repair the commercial dock in Puerto Quetzal Guatemala, NACE Corrosion, Houston. Paper 13497.

Soto-Espitia, R. Vazquez, Enriquez, J. & Martinez-Gómez, L. 2011. Monitoring Integrity and Corrosion Damage on Stay Cable Bridge System in Mexican Highways. Proc. XXIV World Road Congress. Paper 0439.

Bridge Maintenance, Safety, Management, Resilience and Sustainability – Biondini & Frangopol (Eds)
© 2012 Taylor & Francis Group, London, ISBN 978-0-415-62124-3

Simplified and detailed calculations of long-term stress redistributions in continuous precast bridge decks

C. Sousa, M. Fonseca, R. Calçada & A.S. Neves
Faculty of Engineering, University of Porto, Porto, Portugal

ABSTRACT

Precast beams have been widely employed in the construction of continuous bridge decks. These structures undergo significant stress changes over the time, owing to the concrete creep and shrinkage deformations and the influence of phased construction. The service behavior of these structures strongly depends on the adequate calculation of the long-term stress redistributions, so that adequate geometries and prestress forces are adopted at the design stage.

The detailed calculation of long-term effects can be carried out through finite-element phased visco-elastic analyses. If so, the accuracy of the calculation results depends on the reasonability of both the adopted creep and shrinkage models and the assumed construction timings. However, long-term analyses are often performed by using the simplified effective modulus method and, in this case, additional uncertainties are introduced, due to the simplified nature of these calculations and the eventual in-adequate choice of calculation parameters.

This presentation addresses the possible errors associated to long-term predictions through the effective modulus method. The accuracy of the simplified calculations is herein evaluated by taking the results of detailed finite-element analyses (which have been validated through the comparison with experimental results) as reference values. The consequences of in-accurate predictions, in terms of incorrect estimation of prestressing forces and unintended cracking are quantified and discussed by means of a case-study, which consists of a continuous precast bridge deck with 40 m spans.

The structure under analysis consists of a hypothetical railway bridge deck, which carries two rail-way tracks. This structure is built with two "U"-shaped, 2.70 m deep, precast girders. The interior and the extreme span lengths amount to 40 m and 32 m respectively. The total number of spans is equal to 7.

The results presented in this paper show that the quantification of the long-term distribution of concrete stresses strongly depends on the variability associated to the quantification of creep deformations and to the estimate for the girder ages at the time that the continuity connection is established.

The case study shown in this paper provides quantitative information about the parameters to be used in simplified long-term analyses based on the age-adjusted effective modulus method.

The results of deterministic finite-element analyses in terms of long-term concrete stresses installed at the bottom of mid-span cross sections could only be obtained in simplified calculations based on the EMM when the K coefficient took reduced values ($\sim$0.2). On the other hand, the results obtained for the upper fiber of the cross sections located above the continuity supports suggest the adoption of higher values for the K coefficient ($\sim$0.4).

For design purposes, the methodology used in the calculation of long-term stresses caused by permanent actions, should not overestimate the compression stresses in critical locations (bottom of mid-span cross sections and upper fibers of cross sections above continuity supports). By comparing the results of the calculations based on the EMM and the ones obtained in probabilistic analyses, it could be concluded that:

- in the case of mid-span cross sections, the higher the adopted K coefficient, the more conservative the calculated stresses are;
- in the case of cross sections above the continuity supports, the lower the adopted K coefficient, the more conservative the calculated stresses are.

Bridge Maintenance, Safety, Management, Resilience and Sustainability – Biondini & Frangopol (Eds)
© 2012 Taylor & Francis Group, London, ISBN 978-0-415-62124-3

Renovation of a heritage protected suspension bridge with replacement of key components and provision of seismic protection

T. Spuler & G. Moor
Mageba SA, Bulach, Switzerland

ABSTRACT

The restoration of a historic footbridge in Germany is described, with particular focus on the upgrading of key mechanical parts and the retrofitting of seismic protection. The Argen Bridge, with span of 68.4 m, was erected in 1897, making it the second oldest suspension bridge in Germany (Berger et al. 2010).

The renovation work included the replacement or addition of bearings, expansion joints and seismic dampers. It also included extensive concrete reconstruction at each abutment, with access chambers and shafts provided for inspection and maintenance purposes – in particular for the deck's tie-down anchorages. In order to facilitate this construction work, temporary columns were placed under the deck close to each end and ballast was placed on the deck, allowing the structure's existing anchorages to be removed and the deck of the bridge to be lifted using hydraulic jacks on top of the temporary columns.

This also allowed the bridge's bearings to be replaced. The bearings under the edge girders are now free-sliding spherical bearings, chosen for their high rotational capacity and small size, while the bearings under the central axis girder are elastomeric bearings with transverse movement restraints. These also secure the deck longitudinally, allowing some limited movement but recentering it due to the restoring forces which result from their elastic deformation.

Seismic dampers were newly installed at both abutments to resist uncommonly large and sudden longitudinal forces, such as seismic loading or exceptionally high braking forces from traffic. These permit longitudinal movement under normal circumstances, but block and transmit forces to the bridge's abutments under dynamic loading conditions.

Following completion of these works, the Argen Bridge was re-opened in August 2010, now safer, more durable, and requiring less maintenance in the future. This, and in particular the extension of the life of the historic structure, is of great value to the surrounding localities – not only for the bridge's transportation uses, but also for its value in helping make the region an attractive place to live and to visit. The successfully completed project thus demonstrates how modern technology can be used sensibly and sensitively to protect our cultural heritage.

Figure 1. The Argen Bridge in southern Germany.

Figure 2. View of one abutment, showing bearing under central axis girder and tie-down anchorage of one edge girder. A new extension to the central axis girder passes through the abutment wall and connects to a new seismic damper.

REFERENCE

Berger, R., Flietner, M. & Metzger, W. 2010. Old and new combined, Bridge Design & Engineering, Issue 61.

Bridge Maintenance, Safety, Management, Resilience and Sustainability – Biondini & Frangopol (Eds)
© 2012 Taylor & Francis Group, London, ISBN 978-0-415-62124-3

Renewal of small movement expansion joints with minimum break-out and time requirements

T. Spuler & G. Moor
Mageba SA, Bulach, Switzerland

ABSTRACT

Failures of bridge expansion joints, quite often premature, present a danger for traffic and are a heavy burden for bridge owners, maintenance crews and bridge users. This paper presents a robust type of expansion joint which offers substantial benefits when used to replace an existing joint at the end of its life.

A bridge's expansion joints will almost certainly have to be rehabilitated or renewed several times during the course of the bridge's life, and the choice of expansion joint type to be used depends on various factors, which are likely to include:

- disruption to traffic on the structure during the works should be minimised;
- the amount of deck structure which must be broken out should be minimised; and
- ways of improving the performance of the joint should be considered.

The *TensaCrete* expansion joint offers a solution which may satisfy all such needs in many circumstances. With its anchorage in fast-curing, high strength polymer concrete, the depth of the anchorage can be limited to approximately 80 mm. This minimises the effort required to break out the existing expansion joint and structure to create an adequate block-out, and allows the bridge to be quickly re-opened to traffic – for example, after a weekend closure. The joint can also be fitted, if desired,

Figure 1. A low-quality, small movement expansion joint in urgent need of replacement.

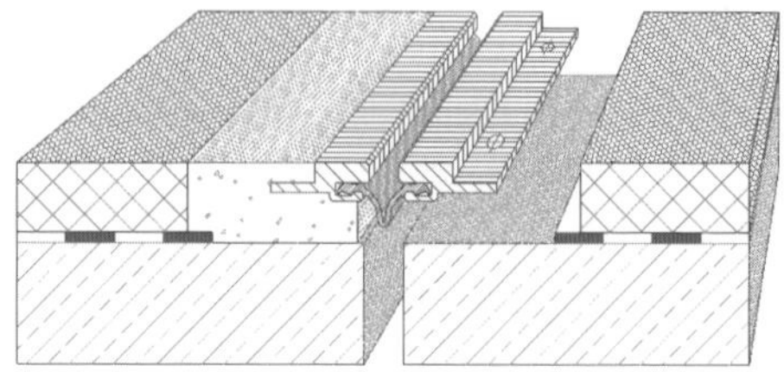

Figure 2. Schematic detail of *TensaCrete* expansion joint.

with noise-reducing surface plates which remove the impacts of wheels on the joint and thus greatly reduce the noise of traffic crossing the joint.

Installation of the joint is relatively simple, and requires little or no construction equipment. The use of this type of expansion joint can also optimise the use of associated machinery on a bridge construction site, especially if an asphalt surface is to be applied to the structure. For instance, following completion of removal of the existing joint, the asphalt can be placed across the entire length of the bridge and the asphalting plant can be redeployed to another site. At a later stage, the asphalt can be cut and removed in the area of the joint, and the joint installed within the depth of the asphalt, with no more asphalting work required. The polymer concrete can then be placed to precisely match the level of the new asphalt, and thanks to its quick-curing properties, the joint can be driven over within a matter of hours of completion.

This paper describes the joint, its advantages, and its installation as a replacement for an existing joint in an important structure – presenting a system which has potential to play an important role in the rehabilitation of many of the countless bridges around the world which require renewal of small movement expansion joints.

REFERENCE

Spuler, T., Moor, G. & O'Suilleabhain, C. 2008. Expansion joints with low noise emission. *Proceedings of the International fib Symposium, Amsterdam, May 2008.*

Bridge Maintenance, Safety, Management, Resilience and Sustainability – Biondini & Frangopol (Eds)

Renewal of bridge expansion joints with minimal disruption to traffic – A solution using modularised sliding finger joints

T. Spuler & G. Moor
Mageba SA, Bulach, Switzerland

ABSTRACT

A solution to the demands to be addressed in the specification, design, installation and replacement of expansion joints on bridges is presented: the *TensaFlex* sliding finger joint and the associated *Mini-Fly-Over* traffic management system. The expansion joint is quiet and durable, with long-term benefits for bridge owners and local residents, while the traffic management system minimises the effort and traffic disruption associated with its installation as a replacement for an existing expansion joint.

The *TensaFlex* sliding finger joint consists of interlocking finger plates, with the plates at one side of the bridge gap providing sliding support to the finger plates which span the gap from the other side. The finger plates that span the gap are pre-tensioned downwards, enabling vertical movements of one side of the joint relative to the other to be facilitated and ensuring that the plates will remain in contact with the sliding surface below at all times. Thanks also to this design, the joint does not require major anchoring to the structure; the "simply supported" design of the fingers which span the bridge gap means that no significant moment loading on the support structures will result from the anchorage of the joint.

The modular design of the system allows the individual elements of the joint to be replaced very quickly (e.g., in one night) on a lane-by-lane basis (see Figure 2). It is also possible to only replace the joint in the lane with heaviest traffic, should this section of joint require replacement earlier than the rest.

Figure 1. The *TensaFlex* sliding finger joint.

Figure 2. Installation of *TensaFlex* sliding finger joint.

Figure 3. Installation of *TensaFlex* sliding finger joint "under traffic" using Mini-Fly-Over traffic management system.

If an existing expansion joint is to be renewed when it reaches the end of its lifetime, the *Mini-Fly-Over* traffic management system can be used to allow traffic to cross the site during the daytime, while the construction works are carried out at night-time on a lane-by-lane basis. In this way, unhindered traffic flow during peak times can always be facilitated. The use of the system to install the joints on a busy bridge carrying 100,000 vehicles per day (Berne, 2007), with almost no impact on traffic, is described.

REFERENCE

Berne, 2007. Information Office of the Swiss canton of Berne. Press release, Berne, 29th March 2007.

Bridge Maintenance, Safety, Management, Resilience and Sustainability – Biondini & Frangopol (Eds)
© 2012 Taylor & Francis Group, London, ISBN 978-0-415-62124-3

Structural reliability of cable stayed bridges based on analysis of deformations

V. Straupe & Ai. Paeglitis
Riga Technical University, Riga, Latvia

ABSTRACT

Behavior of a cable system under uniformly distributed load as well as under variable loads can be analyzed by studying deformations of their components. These correlations can be mathematically derived using the differential equations of deformed shape of girder and the tensioning forces of cables caused by their extensions.

Method presented in this paper shows mathematical correlation between tension forces in cables and fluctuations of stresses in stiffening girder due to variable loads. Study of uncertainties of variable loads allows to find the influence of stiffness and geometry of cables on structural reliability of stiffening girder, cables and whole construction of the bridge.

This approach allows to find influence of the load-bearing element stiffness on any of the optimized parameters – the allowable deflection, permissible stress, consumption of materials, structural reliability index. The proposed analytical method gives reasonable initial assumptions about the optimal cable stayed bridge geometry, as well as predictions about the influence of the individual element stiffness on the overall structural behavior.

The analysis of functional correlations show that by reducing the second moment of area of the stiffening girder, the cross-sectional area of cables can be reduced without changing the bending moment diagram caused by uniformly distributed load. In this case the values of stiffness of components must be chosen depending on the allowable deformations of the stiffening girder.

The bending moment with least possible extreme values caused by variable loads can be achieved by introduction of an "intelligent" cable adjusting system – a system acting as a group of mechanisms monitoring displacements of some nodes and adjusting separate cables depending on the location and acting of the variable loads.

Benefit of such system is more significant in cases with low value of η ratio – percentage of strains caused by permanent loads. The "intelligent" system enables economy of construction materials as well as improves the structural reliability by reducing strain fluctuations in main elements of cable stayed bridge.

It is interesting to investigate how the tensile forces in cables should change when the point load is moving over the bridge in order to secure the desired bending moment diagram. Curves given in Figure 1 represent the tensile force of each cable. Such action of cables is required for an "intelligent system".

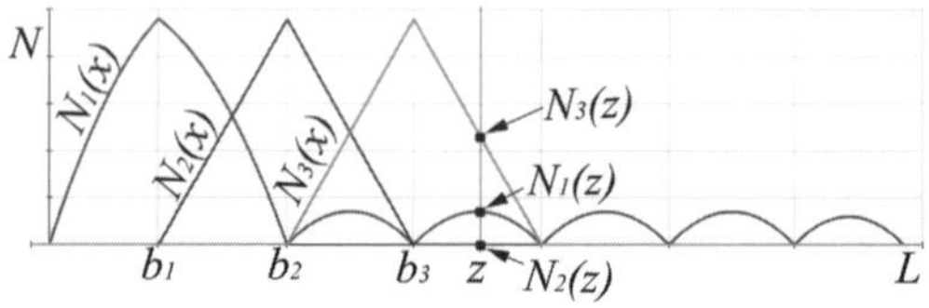

Figure 1. The adjusting forces N_1, N_2 and N_3 of cables depending on location x of point load P_0.

REFERENCES

Achkire, Y. 1997. *Active tendon control of cable-stayed bridges*. Brussels: Universite Libre de Bruxelles

Bruer, A. et al. 1999. Computer based optimizing of the tensioning of cable-stayed bridges. In *IABSE Conference proceedings. Cable-stayed bridges – past, present and future*: 65–74. Malmo: Congrex Sweden AB.

Cruz, J.S. et al. 1999. A new model for cable-stayed bridges control and adjustment. In *IABSE Conference proceedings. Cable-stayed bridges – past, present and future*: 200–209. Malmo: Congrex Sweden AB.

Di Bernardo, S. 1998. *Motion based design of cable-stayed bridges*. Massachusetts: Massachusetts Institute of technology.

Kachurin, V.K. & Bragin, A.V. 1971. *Design of suspension and cable stayed bridges*. Moscow: Press Transport.

Straupe, V. & Paeglitis, A. 2011. Mathematical bases of interaction between elements in cable stayed bridges. In *IABSE-IASS symposium proceeding. Taller, longer, lighter*: 250. London: Hemming Group Ltd.

Walther, R. et al. 1999. *Cable stayed bridges*. London: Thomas Telford Ltd.

Bridge Maintenance, Safety, Management, Resilience and Sustainability – Biondini & Frangopol (Eds)
© 2012 Taylor & Francis Group, London, ISBN 978-0-415-62124-3

Conflicting policies with CWR on open deck bridges

R.A.P. Sweeney
Modjeski & Masters, Montreal, Canada

ABSTRACT

There are 7 Class 1 (major) Railroads, 23 Regional Railroads, and 533 Local Railroads operating over approximately 225 000 km of track. Amtrak operates approximately 37000 km of passenger railroad. The seven Class 1 Railroads in the United States account for 67.5% of the trackage and 93.3% of the freight revenue (Sorgenfrei et al., in press).

Because of conflicting polices with regard to anchoring of Continuous Welded rail on open deck railway bridges, a program of testing was initiated by the writer through the Association of American Railroads' Transportation Technology Center Inc. (TTCI) to try and resolve the conflicts.

Since the requirements of the AREMA Manual are recommended practice and are not mandatory a number of administrations choose not to follow the recommendations. In some cases it seems the policy adopted depended on whether the track engineer or the bridge engineer set the requirements.

Some of the policies do not agree with theoretical studies of the expected behavior but are based on many years of anecdotal observations. For example some railroads anchor the rail very tightly to ensure that if a rail breaks the gap will be restrained. Other railroads allow the rail to be totally unanchored to reduce the stress in the rail to avoid rail breaks.

The test results indicate that full anchorage similar to that on open track is not appropriate on open deck bridges if the tie to structure interface is restrained by rivets or similar restraints. Test results also indicate that minimal restraint reduces considerably the stress in the rail that would cause the rail to break during cold weather. Test results at high temperatures gave clear results but further testing will be required to fully understand cold weather behavior.

This paper explains the recent changes to the AREAM Manual for Railway Engineering concerning Continuous Welded Rail on Open Deck bridges that allow for seemingly conflicting solutions. The changes are based on an analytical model developed based on tests conducted by TTCI at Fast and on several member railroads and data from tests done in China.

Seven tests are described as follows, a long bridge with expansion rails joints and unanchored rail, a long bridge with fully anchored rail, the thermal performance of a short bridge with fully anchored rail, a simulated cold weather break on an open deck bridge, a simulated cold weather rail break in open track, the bridge at eastern mega site, and a cold weather bridge test that is just starting.

Full scale testing is slowly reducing the need to base recommended practice for the anchorage of continuous welded rail to open deck bridges based on opinion and anecdotal evidence that in the past is seemingly contradictory.

REFERENCE

Sorgenfrei, D.F., Marianos, W.N., Sweeney, R.A.P., Handbook of Bridge Engineering, 2nd Edition, Chapter 23 Railroad Bridges, CRC Press, Dec 2012.

Bridge Maintenance, Safety, Management, Resilience and Sustainability – Biondini & Frangopol (Eds)
© 2012 Taylor & Francis Group, London, ISBN 978-0-415-62124-3

Damage detection in suspension bridges from wind response measurements and automatic mode selection: A feasibility study

F. Ubertini & A.L. Materazzi
Università degli Studi di Perugia, Perugia, Italy

ABSTRACT

We explore the effectiveness of dynamic methods for damage detection in the case of wind-excited suspension bridges. The considered damage is represented by a sectional loss in the main cables, with arbitrary intensity, location and extension.

The wind excited vertical response of the structure is analytically modeled using a continuum formulation for the damaged suspension bridge, derived in a separate study (Materazzi & Ubertini 2011), and suitable expressions for the wind forces acting on the deck, modeling the wind velocity field as a multivariate Gaussian stochastic process. Then, a realistic damage scenario is considered and the possibility of observing the resulting frequency variations from response data is examined.

A sketch of the considered structural health monitoring configuration is shown in Figure 1. Particularly, vertical accelerations are collected in 8 equidistant points placed along the deck of the bridge. The damage is assumed to be placed at one support, where fretting fatigue phenomena are usually more significant, and it is assumed to affect 5% of the total length of the cable. Pseudo experimental wind-excited response data are then generated, using the analytic model, for levels of cross-section reduction varying from 0 (undamaged case) to 0.1 (10% reduction of the whole cable section). In the analysis, the model is specialized to the case of the New Carquinez Bridge, that was studied in a previous work by one of the authors (Hong et al. 2011).

Natural frequencies are extracted by pseudo-experimental data using a sophisticated, fully automatic, system identification (SI) technique, recently developed by the authors elsewhere (Ubertini et al. 2011). The observed variations of the natural frequencies caused by damage are compared to the correct values. This comparison is shown, for the most sensitive mode (second in-plane mode) in Figure 2. The results demonstrate that the considered SI technique is able to track the frequency variations due to damage, already at an early stage, thus showing a promise towards its application for detecting main cables damage in suspension bridges.

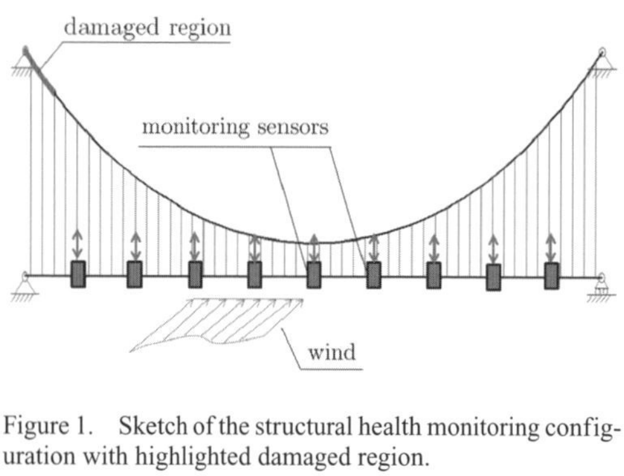

Figure 1. Sketch of the structural health monitoring configuration with highlighted damaged region.

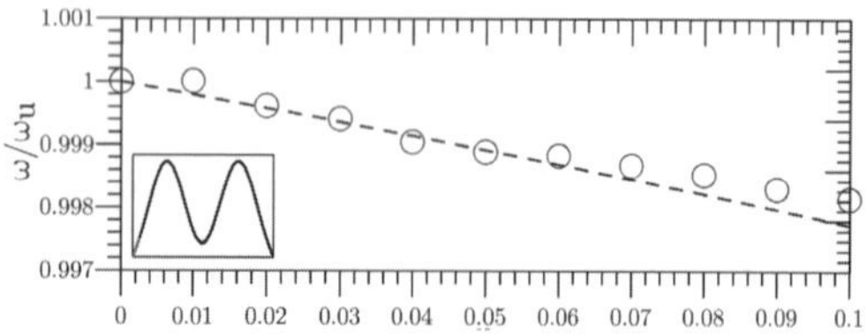

Figure 2. Ratios between damaged, ω, and undamaged, ω_u, circular frequency of the second symmetric mode vs. cross section reduction (correct values: discontinuous line; SI results: circles).

REFERENCES

Materazzi, A.L. & Ubertini, F. 2011. Eigenproperties of suspension bridges with damage. Journal of Sound and Vibration 330: 6240–6434.

Hong, A.L., Ubertini, F. & Betti, R. 2011. Wind analysis of a suspension bridge: identification and finite-element model simulation. *ASCE Journal of Structural Engineering* 137(1): 133–142.

Ubertini, F., Gentile, C. & Materazzi, A.L. 2011a. On the automatic identification of modal parameters by subspace methods. In Carmelo Gentile (ed.), *Proceedings of the EVACES 2011 Conference, Varenna, 3–5 October 2011.*

Bridge Maintenance, Safety, Management, Resilience and Sustainability – Biondini & Frangopol (Eds)
© 2012 Taylor & Francis Group, London, ISBN 978-0-415-62124-3

Analysis and verification of existing bridge structures

C. Unger & M. Empelmann
Institut für Baustoffe, Massivbau und Brandschutz, Fachgebiet Massivbau, TU Braunschweig, Germany

ABSTRACT

To provide uniform standards for the assessment of existing road bridges the German Federal Ministry of Transport, Building and Urban Development (BMVBS) issued the "Guideline for recalculation of existing road bridges (short: Recalculation Guideline)" (BMVBS 2011). In this guideline current and former standards are reconciled with respect to load-bearing capacity, serviceability and durability. This paper gives a brief overview of the potential for the reliability assessment of bridges, which is now regulated more transparent and comprehensive then in the individual guidelines before.

The present article deals with the probabilistic reliability analysis of bridge structures in the framework of the Recalculation Guideline. The application of this procedure is allowed within step 4 of the investigation hierarchy in the Recalculation Guideline. The major advantage of the probabilistic method is that the operational failure of probability can be determined specifically. In comparison to the deterministic calculation the probabilistic method requires more computing time, but the reliability reserves of a bridge can not only be estimated, but also be quantified and used, in case the recalculation with other methods shows deficiencies of the bridge structure and limitations in the use of the bridge have to be specified.

The principle procedure is explained by using the example of an ordinary two-span bridge (post-tensioned T-beam). Firstly, a comparison between the results of the semi-probabilistic safety concept, e.g. according to Eurocode, and the probabilistic assessment is done, indicating a higher reliability index in regard to the bending capacity of the bridge within the probabilistic analysis.

In a further step, a parametric study for the failure of tendons with a constant remaining lifetime of $t = 50$ years is carried out. It shows that, even in case of a failure of 20% of the tendons (2 tendons out of 10), for the chosen bridge example the reliability index does not fail below the required value of $\beta = 3.8$ specified in the current applicable standards.

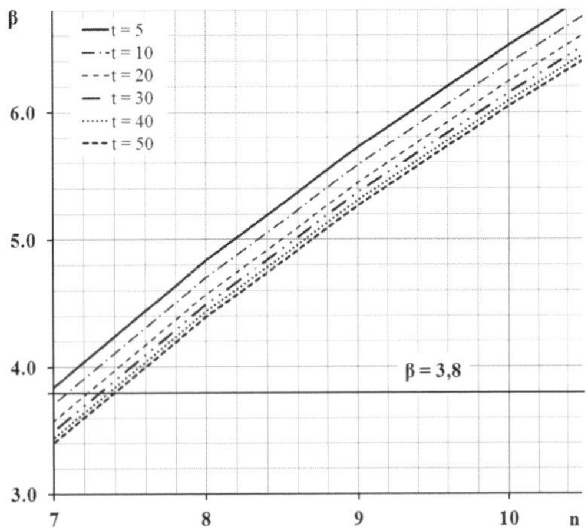

Figure 1. Reliability index β in dependence of the number of tendons n and a reduced remaining service life t.

Finally, the consideration of a reduction of the remaining service life time has been investigated (Figure 1). On the basis of adjusted traffic loads for the varied observation periods an increase of the reliability index can be noted. Furthermore figure 1 illustrates, that a failure of 30% of the tendons (3 out of 10) in combination with a reduced remaining service life of $t = 5$ years still fulfills the required value of $\beta = 3.8$ according to the current applicable standards. In some cases this correlation may help, if the target reliability index cannot be verified for a service life of $t = 50$ years, and a reduced service life has to be assessed.

REFERENCE

BMVBS (Bundesministerium für Verkehr, Bau und Stadtentwicklung, Federal Ministry of Transport, Building and Urban Development): Richtlinie zur Nachrechnung von Straßenbrücken im Bestand (Guideline for recalculation of existing road bridges); May 2011.

Bridge Maintenance, Safety, Management, Resilience and Sustainability – Biondini & Frangopol (Eds)
© 2012 Taylor & Francis Group, London, ISBN 978-0-415-62124-3

Bridge monitoring by fiber optic deformation sensors: A case study

G. Uva, D. Raffaele, F. Porco & A. Fiore
Dipartimento ICAR, Politecnico di Bari, Italy

G. Porco
Dipartimento di Strutture, Università degli Studi della Calabria, Cosenza, Italy

ABSTRACT

In the last few years, the real-time monitoring of civil infrastructures has become an essential tool for the safety inspection, the design and planning of maintenance. In this context, the implementation of fibre optic sensors within the structural elements is particularly useful in order to check strains and displacements and assess the structural safety level.

In this paper, it is presented a methodology aimed at the control of the safety and serviceability level for a Pre-stressed Reinforced Concrete viaduct. The high vulnerability of the site of this structure, related to a severe hydro geological instability occurred in 2003, required in fact the installation of a complex system for the structural safety control, including a optic fibre monitoring system called "SOFO", which belongs to the category of the "long base" sensors.

The system was installed on the mild steel of a pier and on the strands of some deck beams, before the casting of the concrete. Through the processing of the data provided by the sensors at different times of the execution (t_0 – launching of the beams; t_1 – casting of the concrete slab; t_2 – final state of the bridge), the actual response of the structure during the main constructive phases is monitored. The strain variations related to the load increments and to the stress loss have been evaluated at predefined time steps and compared with the expected values (according to the design data), as briefly shown in Fig. 1. Thence, the periodical monitoring of the structure allows to appraise if the actual behaviour reflects the theoretical prediction. Moreover, the proposed methodology is also able to identify possible abnormal behaviour determined by degradation or instability of the structural elements.

The proposed methodology provides many advantages for the planning of maintenance programs and represents an effective tool for checking the safety level of the bridges during the their service life.

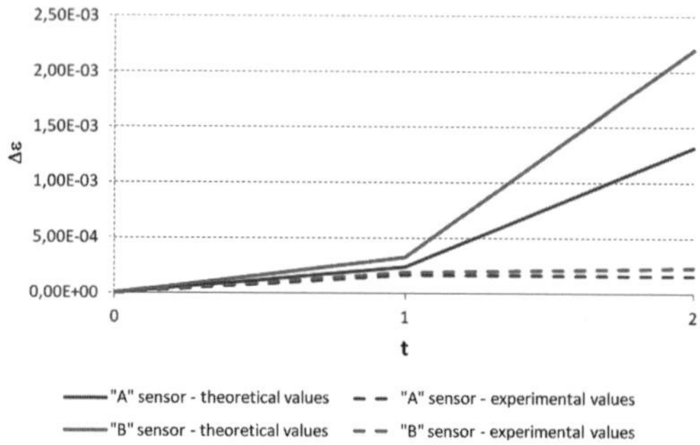

Figure 1. Comparison among theoretical and recorded values of the strain variation.

REFERENCES

Glišić, B & Inaudi, D. 2007. *Fiber optic methods for structural health monitoring*. Chichester, West Sussex: John Wiley & Sons, Ltd.

Idriss, R.L. & Liang, Z. 2006. Monitoring an Interstate Highway Bridge with a Built-In Fiber-optic Sensor System. *IABMAS2006 3rd International Conference on Bridge Maintenance, Safety and Management*, Porto, Portugal

Idriss, R.L. 2001. Monitoring of a high performance prestressed concrete bridge with embedded optical fiber sensors during fabrication, construction and service. *Struct faults repair conference*, London, England.

Inaudi, D. & Del Grosso, A. 2008. Fiber optic sensor for structural control. *The 14th World Conference on Earthquake Engineering* October 12–17 2008. Beijing, China.

Mezzina, M. 2001. *Costruire con il cemento armato*. Torino: UTET Libreria srl.

Bridge Maintenance, Safety, Management, Resilience and Sustainability – Biondini & Frangopol (Eds)
 ISBN 978-0-415-62124-3

Bridge strengthening by network arch: Structural performance and design criteria

M.A. Valenzuela
Public Work Department, Santiago, Chile & Technical University of Catalonia, Barcelona, Spain

J.R. Casas
Technical University of Catalonia, Barcelona, Spain

ABSTRACT

The problems of erosion and scour at bridge piers caused by run-off of rivers have led to an alternative strengthening method, attempting to minimize the processes of monitoring, repair and strengthening. This situation is extremely important in countries with high seismic activity (Figure 1).

The alternative method proposed is a process of modernization of bridges by arches with tight hangers – verticals and network –, capable to lift the deck from its intermediate supports. This allows removing the piers and thus to eliminate future problems of erosion and scouring (see Figure 2).

The present paper describes the behavior of the strengthening method at construction's stage and later on the structural performance of the strengthening bridge in the service state. It also validates the structural technique and establishes recommendations and design criteria for its practical implementation.

The methodology consists in a parametric study of the variables influencing the process of construction method: structural changes and tight hangers' sequence. It uses a structural model based on the bridge San Luis, located in Chile. The generalization of the process and its optimization is performed by interaction between a Finite Element code and genetic algorithms.

Additionally, an analysis with live loads and the factors influencing the dynamics of the bridge is developed, with careful attention to the seismic response, related to the recommendations in countries with high seismic hazard (Chile).

A systematization of this alternative strengthening method, which includes: structural and technological research; feasibility of the construction process; and preliminary design criteria, is provided.

Finally, the study brings to engineers, a set of criteria and recommendations of the basic steps to implement this technique, becoming a competitive and sustainable alternative compared to the conventional method of foundation strengthening.

Figure 1. Scour and erosion in San Luis bridge.

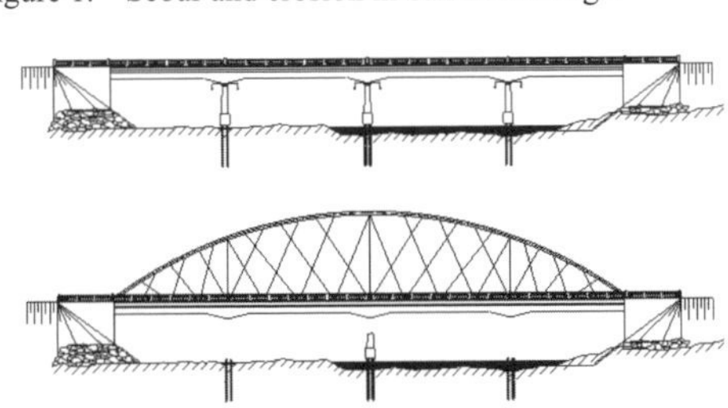

Figure 2. Original and strengthened bridge.

REFERENCES

Schanack, F. 2008, *Network Arch Bridges,* Ph.D Thesis, University of Cantabria, ACHE. Spain.

Tveit, P. 2007, *The Network Arch. Bits of Manuscript after Lectures in 44 Countries*, Internet Edition, http://pchome.grm.hia.no/~pchome/

Valenzuela, M. & Casas, J.R. 2010, *Bridge strengthening by structural change: from continuous beam to network arch.* Proceedings of IABMAS 2010, Philadelphia, United States.

Valenzuela, M. & Casas, J.R. 2011, *Bridge strengthening by structural change: Optimization via genetic algorithm.* Proceedings of IABSE 2011, London, United Kingdom.

Valenzuela. M. 2011, *Refuerzo de puentes de luces medias por conversión en arcos atirantados tipo network.* Ph.D. Thesis. In progress. Technical University of Catalonia. Civil Engineering Department, Barcelona.

Bridge Maintenance, Safety, Management, Resilience and Sustainability – Biondini & Frangopol (Eds)
© 2012 Taylor & Francis Group, London, ISBN 978-0-415-62124-3

Modelling synergistic effects of carbonation/chloride penetration and frost attack for service life design of concrete bridges

E. Vesikari & H. Kuosa
VTT Technical Research Centre of Finland, Espoo, Finland

J. Piironen
Aalto University, Espoo, Finland

R.M. Ferreira
C-TAC, University of Minho, Guimarães, Portugal

ABSTRACT

The traditional approach for the service life design (SLD) of a reinforced concrete structures is typically based on the modelling of a dominating degradation mechanism or several degradation mechanisms separately. However, in real-life exposure, reinforced concrete structures are subject to the simultaneous effect of several degradation mechanisms. The Finnish DuraInt research project was setup to understand the combined effect of degradation mechanisms such as, frost attack, carbonation and chloride penetration, on concrete performance. In this paper the effect of frost attack on carbonation and chloride penetration is analysed. The influence may be substantial and should be taken into account when concrete is exposed to frost attack.

Based on the results of laboratory and in situ exposure tests, the effects of frost attack on the rate of carbonation and chloride penetration as an interacted degradation were developed. A procedure for integrating the synergistic effect of both internal frost attack and frost scaling on carbonation and chloride penetration for the service life design of reinforced concrete structures is presented based on the use of interaction factors.

Results showed that, if frost attack is rapid, it is usually the dominating degradation mechanism, marginalising reinforcement corrosion initiated by carbonation or chloride penetration. If frost attack proceeds slowly, reinforcement corrosion can become dominate and the interaction effects should be considered in the SLD of reinforced concrete.

The results of the calculation of interaction factors are presented here. A detailed description of the calculation process can be found in the full paper. For the calculation of the interaction factors, several serviceability limit states are defined:

- Freeze-thaw scaling: depth scaling = 15 mm;
- Freeze-thaw internal damage: relative dynamic modulus = 66.6%;
- Corrosion initiation due to carbonation: depth of carbonation front = reinforcement cover;
- Corrosion initiation due to chloride penetration: chloride content at reinforcement cover > critical content.

Table 1. Interaction factor for the effect of frost scaling on the initiation time of carbonation induced corrosion for a concrete cover depth of 25 mm, $I_{ca,FS}$.

	$t_{L,FS}$						
t_{ca}	40	50	60	70	80	90	100
40	0.70	0.74	0.78	0.81	0.83	0.84	0.85
50	0.65	0.71	0.76	0.78	0.80	0.82	0.84
60	0.61	0.66	0.71	0.74	0.76	0.79	0.81
70	0.57	0.62	0.67	0.71	0.74	0.78	0.81
80	0.53	0.58	0.63	0.67	0.71	0.75	0.78
90	0.49	0.55	0.61	0.64	0.67	0.72	0.76
100	0.46	0.52	0.58	0.62	0.65	0.68	0.71

Table 2. Interaction factor for the effect of internal cracking on the initiation time of carbonation induced corrosion, $I_{ca,IF}$.

	$t_{L,IF}$						
t_{ca}	40	50	60	70	80	90	100
40	0.83	0.85	0.88	0.89	0.90	0.91	0.93
50	0.82	0.85	0.88	0.89	0.90	0.91	0.92
60	0.80	0.82	0.85	0.86	0.88	0.89	0.90
70	0.77	0.80	0.83	0.84	0.86	0.87	0.88
80	0.75	0.78	0.81	0.83	0.85	0.86	0.87
90	0.73	0.76	0.79	0.81	0.83	0.84	0.85
100	0.71	0.74	0.78	0.80	0.82	0.83	0.85

Carbonation and surface scaling. The interaction factor is influenced by the depth of concrete cover. In Table 1 the interaction factors for the concrete cover of 25 mm, for service life varying between 40 and 100 years, are presented.

Carbonation and internal damage. In Table 2 the interaction factors for service life varying between 40 and 100 years, are presented.

The full paper also presents the interaction factors for chloride penetration and freeze-thaw damage.

Bridge Maintenance, Safety, Management, Resilience and Sustainability – Biondini & Frangopol (Eds)

Experimental modal analysis and fatigue assessment on the Lagoscuro viaduct

L. Vincenzi
DIMEC – University of Modena and Reggio Emilia, Modena, Italy

M. Savoia
DICAM – University of Bologna, Bologna, Italy

W. Salvatore
Consorzio Pisa Ricerche, Pisa, Italy

ABSTRACT

The functionality maintenance of infrastructures like bridges is acquiring more and more importance due to the huge economic losses related to the interruption of their regular service. In particular, fatigue represents one of the more diffused failure modes occurred in existing steel and composite steel-concrete bridges. Several studies were performed in the past in order to assess the fatigue resistance of steel bridges; such studies were the base of modern codes and standards. Despite of these efforts, the fatigue assessment of railway bridges both considering the design of new bridges and the assessment of existing ones is one of the main issues in current practice. In fact, phenomena like "vibration induced" and "distortion induced" fatigue are still partially uncovered by actual design codes and represent critical aspects for the assessment of existing bridge remaining life and for the design of new bridges.

In this context, the FADLESS research project aims to define innovative technical guidelines for the assessment and control of existing and new bridges with regard to fatigue phenomena induced by vibrations and distortions produced by train passage. The project will employ experimental and numerical techniques in order to identify the most typical details frequently subjected to high fatigue effects and to draft technical guidelines suitable to control such effects during bridge design or assessment.

In this paper, the preliminary results concerning the fatigue assessment of Lagoscuro railway viaduct will be described. The Lagoscuro viaduct is composed of 2 steel railway viaducts running parallel one to each other and crossing the Po river. The first viaduct was built in 1948 and is composed by 9 single span truss-girder bridges; members of the main truss girders are composed of 4 L-shaped elements, riveted together by means of steel plates; stringers and additional elements supporting the railway lines are also riveted. The new Lagoscuro viaduct was recently built (2005) in order to potentiate the railway line; the same geometry is adopted but, differently to the old viaduct, the truss girders are composed by H-shaped elements welded together in the joints.

Experimental dynamic tests have been performed in order to identify the modal properties of both viaducts. Results show frequency in the range 1.5–15 Hz. Then, the FE models are corrected with a model updating procedures, where the uncertain model properties are adjusted in order to have the numerical predictions as close as possible to the measured data, in term of modal frequencies and mode shapes. The identification parameters are the equivalent density of steel members, the equivalent moment of inertia of the diagonals, the additional masses on the deck and the bearing stiffness. To obtain the unknown parameters, a modified genetic algorithm has been used [Storn and Price 1997, Vincenzi and Savoia 2009]. After the updating procedure, the numerical frequencies are very close to the experimental values, with errors never greater then 4%.

Making use of the FE model, a preliminary fatigue assessment is carried out following the procedure defined in EN1993 – Eurocode 3, part 1–9. Results show that some elements (stringers and longitudinal beams) are characterized by high value of Damage Index. This is mainly due to the presence of the bending moment induced by the freight train passage. A comparison between stress history obtained from the analysis and experimental ones will carried out. Local model of the lower bracing system will be also studied in order to evaluate local vibrations, "vibration induced" and "distortion induced" phenomena.

REFERENCES

Storn, R. and Price, K., 1997. Differential Evolution – a simple and efficient heuristic for global optimization over continuous spaces. *J. of Global Optimization*, 11(4): 341–359.

Vincenzi, L. and Savoia, M., 2009. Coupling of Differential Evolution algorithm and Quadratic Approximation for dynamic identification, *ICOSSAR – International conference of structural safety and reliability*, Osaka, Japan.

Bridge Maintenance, Safety, Management, Resilience and Sustainability – Biondini & Frangopol (Eds)
© 2012 Taylor & Francis Group, London, ISBN 978-0-415-62124-3

Monitoring of bridges – Detection of traffic loads

C. von der Haar, S. Marx & M. Hansen
Institute of Concrete Construction, Leibniz Universität Hannover, Hanover, Germany

ABSTRACT

The forecast of the remaining service life of existing bridges is an economic as well as safety relevant aspect of an effective life cycle management. Among others the service life addicts to the fatigue stresses influenced by traffic loads in combination with thermal stresses. Particularly the heavy goods vehicles cause remarkable fatigue stresses. It is expected that the frequency as well as the total weights of these heavy goods vehicles will increase further. Many existing, aged bridge structures are not designed for the expected stresses. Therefore, the remaining service life of these bridge structures is unknown. Based on the current traffic loads realistic forecasts of the remaining service life can be performed. The current traffic loads can be extrapolated to determine the past, present and forecasted fatigue stresses. The remaining service life results from these prognosticated fatigue stresses.

In the present research project, the identification of the current traffic loads of a bridge structure is focused. The identification is carried out by a structural monitoring program on a box girder structure (Fig. 1). For this, measurement sensors are installed in three cross-sections as they are MQ-0 at the expansion joint and MQ-2 and MQ-3 at the midspan of the second and third field of the bridge. The measurement sections MQ-2 and MQ-3 are used for the detection of the vehicles and to define further vehicle attributes. The number of axles and the vehicle type is determined at the expansion joint. The measurement system consists of strain gauges, displacement transducers and thermal sensors. The measured data are used to explore the appearance of passing heavy goods vehicles and to determine their total vehicle weights. The total vehicle weights and the recorded data will be combined by a numerical model of the structure and specific algorithm based on influence lines.

Furthermore, a possibility to determine the fatigue stresses influenced by the explored traffic loads will be shown. Time series of the structural stresses influenced by the detected vehicles will be calculated. This generated time series data can be counted by the Rainflow Counting Method to determine the regarding load spectra. The fatigue verification and the remaining service life can be determined by these resulting load spectra.

Figure 1. Instrumented box girder bridge.

REFERENCES

DIN EN 1991-2. 2010. Eurocode 1: Einwirkungen auf Tragwerke – Teil 2: Verkehrslasten auf Brücken.

DIN EN 1992-1-1. 2011. Eurocode 2: Bemessung und Konstruktion von Stahlbeton- und Spannbetontragwerken – Teil 1-1: Allgemeine Bemessungsregeln und Regeln für den Hochbau.

Liebig, J.P. & Grünberg, J. & Hansen, M. 2009. Monitoring an Spannbetonbrücken – Auswirkungen zunehmenden Schwerverkehrs. *Beton- und Stahlbetonbau* 104: 368–371.

Lubasch, P. & Schellenbach-Held, M. & Buschmeyer, W. 2011. Computergestützte Identifikation von Verkehrslasten – Ergänztes Bauwerksmonitoring. *Bauingenieur* 86: 433–442.

WAVE. 2001. Weighing-in-Motion of Axles and Vehicles for Europe (WAVE). RTD project, RO-96-SC, 403.

Bridge Maintenance, Safety, Management, Resilience and Sustainability – Biondini & Frangopol (Eds)
 ISBN 978-0-415-62124-3

Numerical simulations of prestress loss due to creep and shrinkage in singular regions of concrete

L. Vráblík, J. Losko & V. Kristek
Czech Technical University in Prague, Faculty of Civil Engineering, Department of Concrete and Masonry Structures

ABSTRACT

Reliable design of prestressed concrete elements and structures must be based on the accurate determination of prestressing force (including losses and their time development) to satisfy stress as well as deformation limits. Commonly used obsolete computational routines and methods neglect or incorrectly describe real effects and the results obtained by analysis based on these methods are thus necessarily wrong – great differences from results obtained by measurements on real structures have been found. The paper is directed to long term prestressing losses due to creep, shrinkage and cross section warping.

The long-term deflection behavior of long-span prestressed concrete box girder bridges has often deceived engineers monitoring the deflections. A survey of many bridges monitored in various countries showed that all of them have experienced similar deflection histories. It has frequently been experienced that the box girders of many prestressed concrete bridges deflected far more than predicted in design. The deflection evolution has often been counterintuitive, with slowly growing deflections in the early years, followed later by a rapid and excessive deflection growth.

Reliable design of prestressed concrete elements and structures must be based on the accurate determination of prestressing force (including losses and their time development) to satisfy stress as well as deformation limits. Commonly used obsolete computational routines and methods neglect or incorrectly describe real effects and the results obtained by analysis based on these methods are thus necessarily wrong – great differences from results obtained by measurements on real structures have been found.

Results of experimental studies performed on real structures and their analyses confirm that present methods for evaluation of prestressing losses due to concrete creep and shrinkage are not realistic, prestressing losses are extremely underestimated. These methods ignore lots of important factors:

- The true behavior of concrete prestressed structures is three-dimensional – the simplified presumption that cross sections remain plane after deformations is quite far from reality. It is inevitable (when applying advanced computational analyses) to take into account (without any problems for computational process) 3D effects as shear lag and shear deformations of webs resulting in the cross section warping.
- The real development of concrete creep and shrinkage – for its prediction have to be used apposite and calibrated mathematical models; due to this fact, application of the B3 model, which has a strong theoretical basis, is proved as the best choice.

The rheological non-homogeneity of cross sections – creep and shrinkage development is significantly affected by cross section slabs and walls thicknesses. Application of creep and shrinkage models that realistically describe the moisture diffusion process, which causes that the shrinkage and drying creep is significantly affected by cross section slabs and walls thicknesses, is essential.

REFERENCES

Bažant, Z.P. & Baweja, S. 1995 Creep and shrinkage prediction model for analysis and design of concrete structures: ModelB3; *Materials and Structures 28*

Bažant, Z.P. & Li, G. H. & Yu, Q. & Klein, G. & Krístek, V. 2008 Explanation of Excessive Long-Time Deflections of Collapsed Record-Span Box Girder Bridge in Palau; *CONCREEP8, Ise Shima, Japan*

Křístek, V. & Vráblík, L. & Bažant, Z. P. 2008 Misprediction of long-time deflections of prestressed box girders: Causes, remedies and tendon layout; *CONCREEP8, Ise Shima, Japan*

Vráblík, L. & Loško, J. & Křístek, V. 2009 To determination of prestressing in prestressed concrete elements and structures (in Czech), *BETON No. 4, July 2009*

Bridge Maintenance, Safety, Management, Resilience and Sustainability – Biondini & Frangopol (Eds)
© 2012 Taylor & Francis Group, London, ISBN 978-0-415-62124-3

A study on the durability performances for bridge expansion joints

D. Wakabayashi & T. Asai
Nippon Expressway Research Institute Co., Ltd., Machida-shi, Tokyo, Japan

S. Ono
Japan Construction Method and Machinery Research Institute, Fuji-shi, Shizuoka-ken, Japan

ABSTRACT

Performances of bridge expansion joints have been investigated for the introduction of appropriate design methods and maintenance techniques for highway bridges. In this study, the damage cases of expansion joints have been analyzed, and the estimation method for water leak-resistance performance of expansion joints has been discussed. The outline of the current status of expansion joints for highway bridges and the confirmation method of the watertight performance for bridge expansion joints of expressways will be described in this paper.

Fig. 1 shows a water leakage from expansion joint to bridge abutment. According to the result of field survey, a number of water leakage has been found as shown in Fig. 2.

Based on the actual condition of existing expansion joints, confirmation method to assess the watertight performance was proposed as shown in Fig. 3. The outline of this testing method is described below.

(1) Rubber insert type and sealant type are applicable in this test.
(2) Test specimen must be fabricated as similar condition of actual joint.

Figure 1. Water leakage from expansion joints.

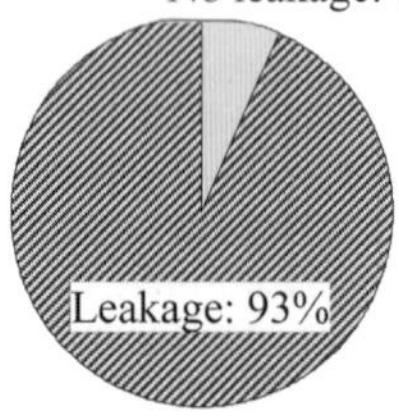

Figure 2. Ration of water leakage in the prefabricated joint (metal type).

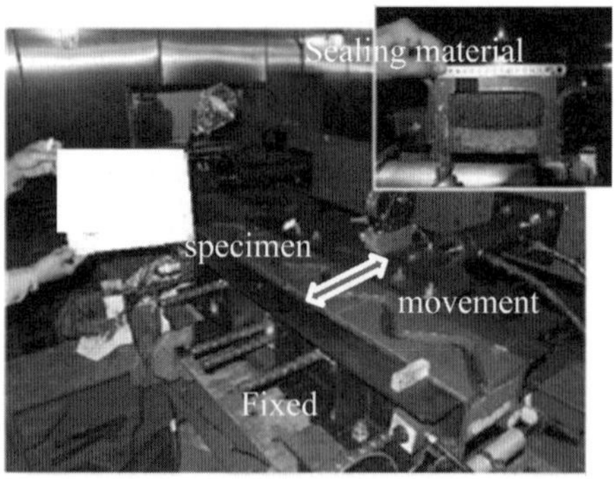

Figure 3. Test method of watertight performance.

(3) Connection of expansion joint including watertight material must be set at the center of the specimen.
(4) Test movement based on the design movement must be expanded and contracted continuously under the prescribed temperature conditions.
(5) After the repetitive movement, water-tightness must be kept for 24 hours under the condition of filled water bath of 10 cm depth

REFERENCES

Nippon Expressway Company, "Design specifications", 2011.7. (in Japanese)

Bridge Maintenance, Safety, Management, Resilience and Sustainability – Biondini & Frangopol (Eds)
© 2012 Taylor & Francis Group, London, ISBN 978-0-415-62124-3

Modelling inspection and fatigue retrofitting by post-weld treatment in bridge management systems

S. Walbridge
University of Waterloo, Waterloo, ON, Canada

D. Fernando & B.T. Adey
Swiss Federal Institute of Technology (ETHZ), Zurich, Switzerland

ABSTRACT

Bridge management systems (BMSs) are routinely used to model the deterioration of bridge components due to manifest processes, such as deck surface wear and corrosion. Fatigue deterioration and retrofitting are generally not modelled in BMSs, however, primarily due to the difficulties associated with detecting fatigue damage at its early stages.

In many existing BMSs the condition of a bridge or bridge element subjected to wear or corrosion is characterized in terms of a distinct set of possible condition states (CSs) and transitions from one CS to another. The time period that the element will stay in each CS is then estimated based on prediction models for the applicable deterioration mechanisms, available inspection data, or expert opinions.

Most existing BMSs identify fatigue damage of critical sections based on inspection results and use criticality identification mechanisms (e.g. smart flags in PONTIS) to highlight the necessity for actions against fatigue damage. Some advanced BMSs make predictions of the deterioration of elements due to fatigue damage by specifying discrete CSs based on fatigue cracking and using mechanistic or historical data-based models to predict the fatigue deterioration. The fatigue damage models used in existing BMSs are normally applied at the element level and are deterministic in nature. Probabilistic models would be an improvement over these deterministic ones, in particular if they can be used within the modelling framework of existing BMSs (e.g. KUBA 2005, Thomson et al. 1998).

For the retrofitting of fatigue damaged welds, bridge managers have a range of options available, including member replacement, reinforcement, load restrictions, and increased monitoring. Post-weld treatments such as grinding, dressing, and peening are increasingly being considered as another possibility for the retrofitting of fatigue damaged bridge welds. To date, however, there have only been limited attempts to model post-weld treatment retrofits in a BMS framework (e.g. Orcesi et al. 2010).

In this paper, a simple Markov chain fatigue model, similar to those commonly used in BMSs to model other deterioration processes, is described and evaluated by comparison with a previously validated probabilistic mechanistic (strain-based fracture mechanics) model. Several alternative fatigue management strategies employing various combinations of weld inspection and retrofitting by post-weld treatment (needle peening) are analyzed.

Based on the work presented in this paper, it is concluded that simple Markov chain models, similar to those used in bridge management systems (BMSs) to model deterioration due to other mechanisms are capable of predicting general trends for fatigue management strategies employing inspection and post-weld treatment by needle peening.

When comparing the investigated fatigue management strategies, it is seen that the benefits of employing post-weld treatments to improve the fatigue performance of bridge welds increase as the ratio between the treatment and the repair/replacement costs increases. This benefit also depends on the applied equivalent stress range and traffic volume.

REFERENCES

KUBA-MS-Ticino User's Manual: Release 3.0. 2005. Federal Department of Highways, Bern, Switzerland.

Orcesi, A.D., Frangopol, D.M. & Kim, S. 2010. Optimization of bridge maintenance strategies based on multiple limit states and monitoring. *Engineering Structures* 32: 627–640.

Thompson, P.D., Small, E.P., Jonhson, M. & Marshall, A.R. 1998. The pontis bridge management system. *Structural Engineering International* 8(4): 303–308.

Bridge Maintenance, Safety, Management, Resilience and Sustainability – Biondini & Frangopol (Eds)
© 2012 Taylor & Francis Group, London, ISBN 978-0-415-62124-3

Study on long-term wind data recorded at Sutong Bridge site

H. Wang, J. Niu, T. Guo & Z. Zong
Education Ministry Key Laboratory of Concrete & Prestressed Concrete Structure, Southeast University, Nanjing, Jiangsu, China

ABSTRACT

The Sutong Bridge is a cable-stayed bridge with the longest main span in the world. The bridge is vulnerable to Pacific typhoons since it is located in the east coastal area of China. The dynamic action on the bridge induced by the turbulence wind therefore needs special consideration.

In order to obtain the turbulent characteristics at bridge site, 3D ultrasonic anemometers in the Structural Health Monitoring System (SHMS) are employed to collect wind data. The wind data used for this study cover a period of 15 months, from 1 May 2008 to 31 July 2009. Seasonal monsoon winds and two typhoons, Kalmaegi and Fung-Wong, occurred during this period.

In this paper, the real-time recorded wind data are analyzed in detail to obtain the wind-rose diagram, mean wind speed and direction, turbulence intensity, turbulence integral scale and power spectral density. Figure 1 shows the relationship between turbulence intensities of the long-term wind and the wind speed, and variations of turbulence intensities scatter.

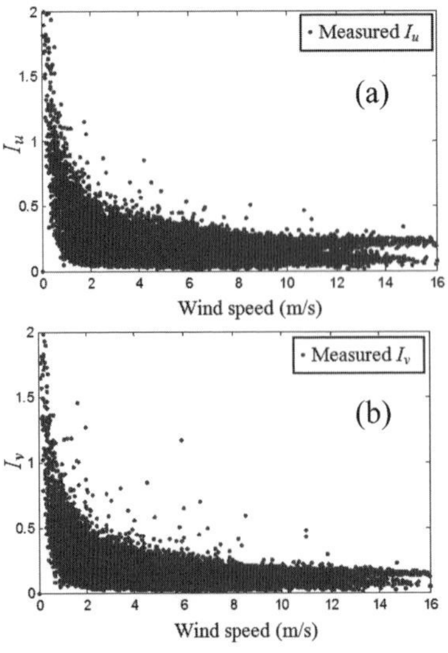

Figure 1. Relationship between turbulence intensity of long-term wind and wind speed: (a) Along-wind I_u; (b) Across-wind I_u.

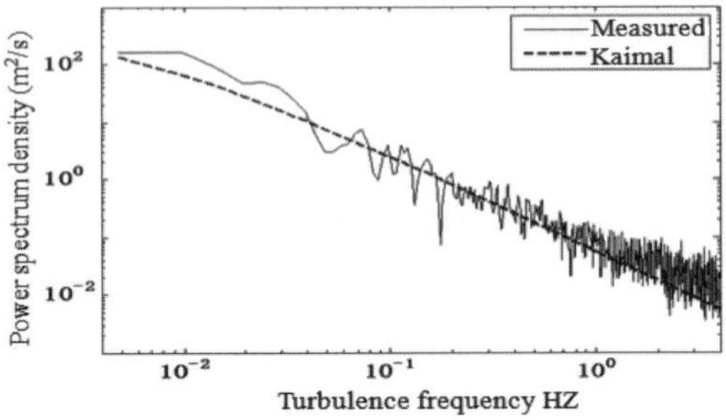

Figure 2. Kaimal and Fung-Wong measured along-wind turbulence power spectra.

Comparison analyses are conducted among inhomogeneous wind characteristics of the northern wind, Typhoon Kalmaegi and Typhoon Fung-Wong, and comparisons concerning turbulence power spectra in along-wind direction between Fung-Wong and Kaimal are plotted in Figure 2 as an example. In addition, calculated values and those recommended in the current specification are compared.

Results show that turbulence intensities decrease as the mean wind speed increases, except the domain of the superposition of general winds and typhoons; meanwhile, turbulence intensities of the typhoons are different from the recommended values. In addition, the measured along-wind power spectra of the northern winds and Typhoon Kalmaegi do not agree well with the recommended Kaimal spectrum, whereas the along-wind power spectrum of Typhoon Fung-Wong is similar to Kaimal spectrum. The conclusions can be used to determine the wind characteristic parameters of the east coastal area of China, and provide references for wind-resistant evaluation of bridges.

REFERENCES

Kaimal, J.C. et al. 1972. Spectral characteristics of surface-layer turbulence. Q. J. R. Meteorol. Soc. 98: 563–589.

Simiu, E. and Scanlan, R.H. 1996. Wind effects on structures. John Wiley and Sons, INC, New York.

Wang, H. et al. 2010, Characteristics of strong winds at the Runyang Suspension Bridge based on field tests from 2005 to 2008, J. Zhejiang Univ-Sci. A. (Appl. Phys. & Eng.), 11(7): 465–476.

Bridge Maintenance, Safety, Management, Resilience and Sustainability – Biondini & Frangopol (Eds)

Targetless precision monitoring of road and rail bridges using video cameras

P.W. Waterfall
Research Engineer at Imetrum, Bristol University Systems Centre, UK

J.H.G. Macdonald
Bristol University, UK

N.J. McCormick
National Physical Laboratory, UK

ABSTRACT

Video cameras have been used within the research community for a number of years for assessing the static and dynamic displacements, including vibration frequencies, of bridges (decks and cables). Researchers have had considerable success at generating measurements on bridges that correlate well with traditional sensors, demonstrating the potential of a technique that could offer substantial cost savings over an installed network of strain gauges, accelerometers and displacement sensors.

This paper undertakes a brief survey of the background and principles of camera based measurement techniques from a historical perspective. It then presents three case studies using a recently developed system based on Digital Image Correlation (DIC) principles, where accuracy has been improved by an order of magnitude over other published algorithms. The system enables multi-point real time dynamic measurement without the need for targets.

The first case study involves cameras at three separate locations on a post-tensioned concrete road bridge. Measurements were taken at 3 Hz & 15 Hz for the purposes of a detailed investigation of vertical displacement at a half-joint. The majority of data was taken from a car park 45 m from the bridge with a resolution of 0.02 mm. Comparison data with a Laser Tracker was also taken, at a distance of approximately 10 m, and is shown below.

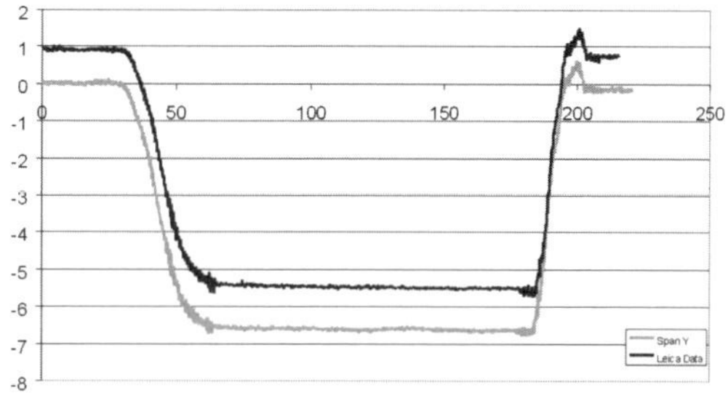

Figure 1. Comparative trace of video gauge system and laser tracker. Laser tracker offset by 1 mm for clarity. Note similar traces, including vibration as lorries stop and start.

The second case study looks at vertical displacement on a riveted railway bridge during the passage of goods trains and high speed passenger trains. Data is compared with pole-mounted potentiometers that were erected for the purposes of the test. The system is placed approximately 10 m from the areas being measured, and data recorded at 117 Hz and fed directly into a data logger on site. Again, a sample trace is shown here:

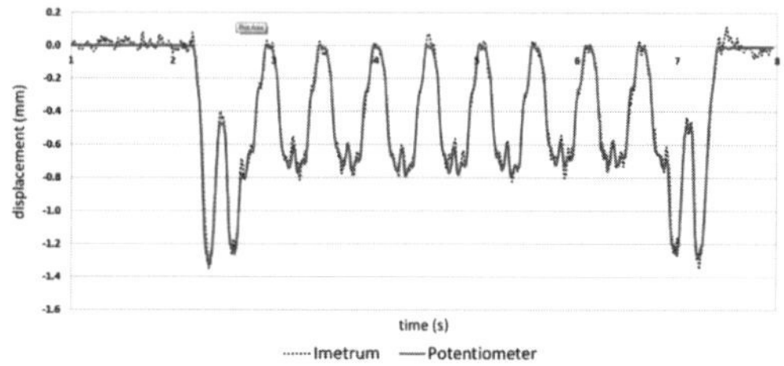

Figure 2. Comparative trace of video gauge system and potentiometer during passage of high speed train.

The third case study looks at the overall movement of a 164 m span motorway structure that has been reconfigured to allow 4 lanes of traffic in each direction instead of 3. Deck torsion is examined, as well as vertical displacement relative to pier movement, using two synchronised cameras. Rotational and displacement data is obtained over the course of a few hours that was not possible to achieve in any other way, enabling the asset owner to study the impact of the lane reconfiguring.

Measurements differ by less than 30 micrometres on 1 mm displacements for relatively close (10 m) range and field of view of around 4 m^2. Rotations are measured to a resolution of 0.001 degrees. These are all without the addition of any targets to the bridges in question, enabling truly remote monitoring. Overall, the paper demonstrates the potential of the video measurement system as a reliable, cost-effective alternative for consultants, maintenance contractors and asset owners looking to understand the dynamic performance of structures.

Bridge Maintenance, Safety, Management, Resilience and Sustainability – Biondini & Frangopol (Eds)

Monitoring of field performance of longitudinally cracked concrete bridge deck

G.W. William, S.N. Shoukry & M.Y. Riad
West Virginia University, Morgantown, West Virginia, USA

ABSTRACT

Empirical bridge deck design has been utilized by many state highways due to the economic benefit of less reinforcement requirements. However uncharacteristic cracking patterns appeared on several bridges that were designed based on the empirical deck design method. The most alarming cracks identified on those newly constructed bridges are longitudinal cracks that are full depth proliferate and extended through the entire length of the bridge decks. These cracks also were considered open of relatively uniform width and confined to the proximity of the interior girders locations. Empirical design method is based on slab resistance to wheel load, however there is lack of evidence that the slab will resist combined loading configurations such as thermal loading, shrinkage stresses, and torsional shear stresses due to slab curvature (Barker and Puckett, 2007). The problem is further complicated by the fact that in many bridges, where longitudinal cracks were observed, the LRFD code produces a steel superstructure that seems to lack sufficient transverse stiffness and is critically designed so that it would not allow for a significant increase in the dead weight of a new elastically designed bridge deck.

The Star City Bridge is a lightweight structure that was designed according to the LRFD specifications with empirical deck. This bridge was instrumented during its construction with over 750 sensors to provide a real life set of data that would demonstrate the behavior of long-span lightweight continuous bridge superstructure since the early age of construction (Shoukry et al., 2009). In 2005, visible longitudinal cracks were developed. Therefore, having this bridge deck instrumented made it unique as it could be used for further investigate the cracking problem and the response of the bridge response post to the development of the longitudinal cracks. Therefore, additional twelve sensors were installed across the cracks in June 2007 to monitor continuously their opening and closure as well as the differential settlement across the cracks. Over two years, the collected data indicate that some cracks have continued to open. In particular the crack along Girder 7 at the edge bay developed a permanent opening of 0.25 mm and a permanent settlement of 0.08 mm as shown in Figure 1.

The opening of the longitudinal cracks could be correlated with the lateral bending of the steel girders. This was visually seen on the steel webs and was found to adversely affect their flexural capacity. Up to 40 percent in capacity could be estimated based on the strains measured on the steel girders. The cracks also caused the axial forces in the diaphragms and cross frames to increase. In addition, the presence of the longitudinal cracks induce shear stress in the transverse steel rebar due to the passage of traffic loading that would affect their fatigue life.

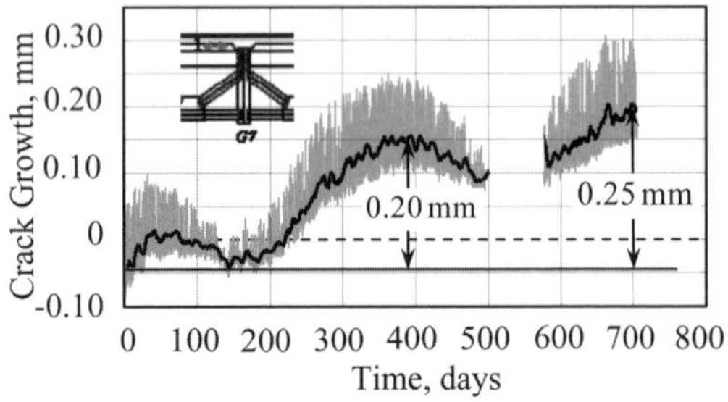

Figure 1. Crack Opening over First Interior Girder (G7).

REFERENCES

Barker, P.M. and Puckett, J.A., 2007. *Design of Highway Bridges An LRFD Approach*, Second Edition, John Wiley and Sons, Hoboken, New Jersey, USA.

Shoukry, S.N., Riad, M.Y. & William, G.W. 2009. Longterm Sensor-Based Monitoring of an LRFD Designed Steel Girder Bridge, *Engineering Structure* 31(12): 2954–2965.

Bridge Maintenance, Safety, Management, Resilience and Sustainability – Biondini & Frangopol (Eds)
© 2012 Taylor & Francis Group, London, ISBN 978-0-415-62124-3

Localized bending fatigue behavior of high-strength steel monostrands

J. Winkler
Department of Civil Engineering, Technical University of Denmark, Kongens Lyngby, Denmark
ATKINS Denmark A/S, Copenhagen, Denmark

G. Fischer & C.T. Georgakis
Department of Civil Engineering, Technical University of Denmark, Kongens Lyngby, Denmark

ABSTRACT

Increasing bridge stock numbers and a push for longer cable-supported span lengths have led to an increased number of reported incidents of damage and replacement of stay cables due to wind and traffic-induced fatigue and corrosion. Limited work has been undertaken to thoroughly assess the fatigue characteristics of bridge cables subjected to cyclic transverse deformations as the cables are in principle not expected to experience bending. Furthermore, the commonly applied qualification tests for the fatigue resistance of stay cables, as outlined in *fib* and PTI, do not specifically address fatigue issues related to transverse cable vibrations and therefore do not require testing for bending. However, recent bending fatigue tests on grouted and PE coated stay cable monostrands have shown that monostrands can experience fatigue failure due to high localized deformations (Winkler et al. 2011) and fretting (Wood & Frank 2010). Although a number of theories describing the failure critera for multilayered strands have been proposed (Hobbs & Raoof 1996), none of these can be applied to monostrands comprised of one layer of wires. As the majority of modern stay cables are comprised of a number of individual high-strength steel monostrands, the understanding of the bending characteristics and failure mechanism of the individual monostrand has become more relevant.

To date, information about local deformations and the resulting strain, often needed to evaluate the failure criteria, has been measured with strain gauges located in the vicinity of the cable anchorage (Miki et al. 1992). However, the critical region of the strand in terms of fatigue is located in the vicinity of the fixation point (wedge) where placement of the gauges is problematic and the strain information obtained with gauges is limited to discrete locations along the strand (i.e. where the strain gauges were placed). In this paper, the localized bending fatigue behavior of pretensioned high strength steel monostrand is investigated. Furthermore, a new methodology employing

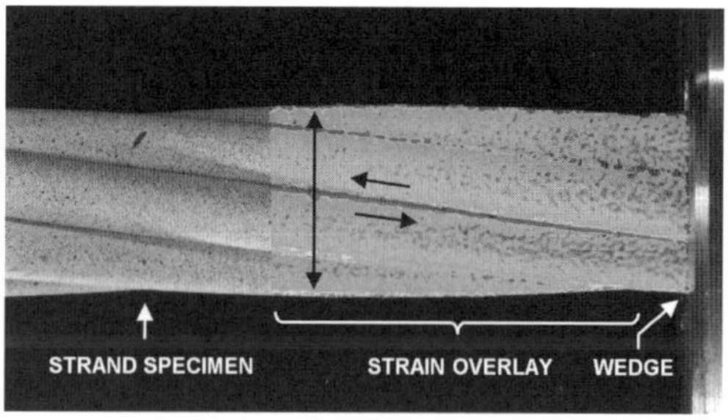

Figure 1. Strand specimen with strain overlay.

an optical photogrammetry system which can track coordinates and calculate deformations and strain is presented. The system enables a measurement of the strain distribution in the strand and helps to identify potential failure mechanisms along the strand and at the wedge location (Fig. 1).

The results reported in this paper show that the bending fatigue life of the monostrand may be controlled either by the local bending strains or by the relative movement of the helically wound wires. The results obtained with the new methodology are a step towards a better understanding of the governing fatigue failure criterion for monostrand cables.

REFERENCES

Hobbs, R.E. & Raoof, M. 1996. Behaviour of cables under dynamic or repeated loading, *Journal of Constructional Steel Research*, 39, 31–50.

Miki C., Endo T. & Okukawa A. 1992. Full-Size Fatigue Test of Bridge Cables, IABSE Reports, Length Effects on Fatigue of Wires and Strands, 66, 167–178.

Winkler, J., Fischer, G., Georgakis, C.T. & Kotas, A. 2011. A Preliminary Bending Fatigue Spectrum for Steel Monostrand Cables, *Journal of IASS*, 52 (2011) No. 4, December n. 170.

Wood, S. & Frank, K.H. 2010. Experimental investigation of bending fatigue response of grouted stay cables, *Journal of Bridge Engineering*, 15, 123–130.

Bridge Maintenance, Safety, Management, Resilience and Sustainability – Biondini & Frangopol (Eds)
© 2012 Taylor & Francis Group, London, ISBN 978-0-415-62124-3

High performance computing for damage detection of civil infrastructural systems

Z.Y. Wu
Bentley Systems, Incorporated, Watertown, CT, USA

T. Mi and J. Zhao
University of Connecticut, Storrs, CT, USA

G. Xu
Microsoft Inc., Seattle, USA

ABSTRACT

Previously developed solution method by authors, based on finite element model updating or calibration, has proved to be effective at locating structure damage elements. The model is formulated to simultaneously search for the given number of the damaged elements and the corresponding damage indicators. It is a mixed integer and continuous optimization problem that is solved by applying a competent genetic algorithm to minimize the discrepancy between the field monitored and model analyzed responses. It is computationally costly to apply it to a large structure system. In this paper, a high performance computing (HPC) framework has been presented and applied to efficiently solve the problem. The HPC framework takes advantage of well-developed FE analysis models and software design patterns. The method is implemented by parallelizing the solution evaluations on a cluster of many-core machines. The parallel optimization essentially speeds up the computation and enables fast convergence of the integrated method. The developed method has been tested on the identification of the damage scenario for a benchmark system. The results obtained show that the proposed method is efficient and effective at detecting damage in a large infrastructure system.

REFERENCES

ASCE (2009). <http://www.infrastructurereportcard.org/>.

Baruch, M., & Bar Itzhack, I. Y. 1978. Optimal weighted orthogonalization of measured modes. *AIAA Journal*, 17(8), 927–928.

Bentley Systems, Inc. 2007. *STAAD.Pro 2007 technical reference manual*, Exton, PA, <http://www.bentley.com/>.

Doebling, S. W., & Farrar, C. R. 1995. Computation of Structural Flexibility For Bridge Health Monitoring Using Ambient Modal Data. *Proceedings of the 11th ASCE Engineering Mechanics Conference.*

Goldberg, D. E., Deb, K., Kargupta H. & Harik, G. 1993. Rapid, accurate optimization of difficult problems using fast messy genetic algorithms. *IlliGAL Report No. 93004*, 1993, Illinois Genetic Algorithms Laboratory, University of Illinois at Urbana-Champaign, Urbana, IL 61801, USA

Goldberg, D. 2002. *The Design of Innovation: Lessons from and for Competent Genetic Algorithms*, Kluwer Academic Publishers: Dordrecht.

Hao, H., & Xia, Y. 2002. Vibration-based Damage Detection of Structures by Genetic Algorithm. *Journal of Computing in Civil Engineering*, 16(3), 222–229.

Pandey, A. K., & Biswas, M. 1994. Damage Detection in Structures Using Changes in Flexibility. *Journal of Sound and Vibration*, 169(1), 3–17.

Sanayei, M. 1997. *PARIS-PARameter Identificafication System*, Tufts University. Medford, MA.

Sanayei, M., Bell, E. S., Javdekar, C. N., Edelmann, J. L., & Slavsky, E. 2006. "Damage Localization and Finite-Element Model Updating Using Multiresponse NDT Data." *Journal of Bridge Engineering*, 11(6), 688–698.

Soh, C. K., & Dong, Y. X. 2001. Evolutionary Programming for Inverse Problems in Civil Engineering. *Journal of Computing in Civil Engineering*, 15(2), 144–150.

Wei, F.-S. 1990. Mass and stiffness interaction effects in analytical model modification. *AIAA JOURNAL*, 28(9), 1686–1688.

Wu, Z. Y. & Xu, G. 2011a. Integrated Evolutionary Optimization Framework for Finite Element Mode Identification. *The First Middle East Conference on Smart Monitoring, Assessment and Rehabilitation of Civil Structures, Feb. 8–10, 2011, Dubai, UAE*

Wu, Z. Y. & Xu, G. 2011b. Effective method for Locating Damage Elements by Parallel Optimization of Model Updating. *The 5th International Conference on Structural Health Monitoring of Intelligent Infrastructure (SHMII-5), Sept. 11–15 2011, Cancun, Mexico*

Wu, Z. Y., Wang, Q., Butala, S. & Mi, T. 2011. Generalized Framework for High Performance Infrastructure System Optimization. *International Conference on Computing and Control for the Water Industry, Sept. 5–7, 2011, Exeter, UK*

Wu, Z. Y. & Simpson, A. R., 2001. Competent genetic algorithm optimization of water distribution systems. *Journal of Computing in Civil Engineering*, ASCE, 15(2), 89–101.

Bridge Maintenance, Safety, Management, Resilience and Sustainability – Biondini & Frangopol (Eds)
 ISBN 978-0-415-62124-3

Monitoring and conservation system design of historic bridge based on the internet of things

Y.Q. Xiang, Q.P. Li, K. Cheng & Q.Q. Wu
Department of Civil Engineering, Zhejiang University, Hangzhou, China

ABSTRACT

The Internet of Things(IOT) refers to uniquely identifiable objects (things) and their virtual representations in an Internet-like structure. In other words, it is a global network infrastructure which links physical and virtual objects through the exploitation of data capture and communication capabilities. This infrastructure is actually a huge network combined with Internet, in which sensors, radio frequency identification technology, global positioning systems, infrared sensors, laser scanners, gas sensor, etc. are used to collect information such as sound, light, heat, electricity, mechanics, chemistry, biology, location and so on. The term Internet of Things was first proposed and used by Kevin Ashton of Massachusetts Institute of Technology in 1999.

Historic bridge, which is considered as an outstanding cultural heritage of ancient buildings, is not only a transport infrastructure, but also the nation's cultural heritage and the representative of its social development. Just like other bridges, the historic bridges have been suffering from action of surrounding environments and various loads (earthquake, river erosion, etc.). However, if maintenance and repairing of the bridges can not be timely obtained, the heritage value will loss, the sudden destruction of the bridge will probably happen. Therefore, the effectively monitoring and assessing historic bridges is a measure to ensure their safety, and preserve its historical value.

Therefore, based on the Internet of Things, the monitoring and conservation system of a typical historic bridge is designed by integrating the modern bridge health monitoring, sensing technology and information technology in this paper.

A typical process of bridge monitoring system is firstly presented, which is divided into 5 modules including online testing, real-time analysis, damage detection, condition assessment and decision-making of maintenance.

Then, the sensor subsystem, data acquisition and transmission subsystem, data processing and analysis subsystem, and control and evaluation subsystem in the monitoring system of historical bridge are given in detail.

The monitoring methods of the relative physical parameters including displacement and strain, and available sensor types for selecting are also described.

What we have done in this paper may be served as the digital platform of the monitoring, assessment and conservation system design of historic bridge. It has a significantly scientific sense and application value for realizing long-distance monitoring and conservation of historic bridge. And also, it is a new management model of culture heritage conservation.

REFERENCES

Ding N. et al. 2010. Ancient Architecture Protection Analysis and Prospects Based on Three-dimensional Laser Scanning Technology. Shandong Architectural University 25(3): 274–276.

Shen S.B. et al. 2010. Conceptual Model and Structure of The Internet of Things. *Nanjing University of Posts and Telecommunications (Natural Science)* (4): 1–8.

Wu H.C. & Xiang Y.Q. 2004. Condition Assessment System of Long-span Prestressed Concrete Cable-stayed Bridge. *Proceedings of the 3rd China-Japan-US Symposium on Structural Health Monitoring and Control*, Oct. 13–16, 2004. Dalian: Dalian University of Technology Press.

Xiang Y.Q. et al. 2003. Health Monitoring and Evaluation Management System of Wenhui Bridge, *First International Conference on Structural Health Monitoring and Intelligent Infrastructure*, Nov. 13–15, 2003 (Tokyo) London: Taylor & Francis.

Yao W.H. 2010. The Concept of the Internet of Things and Basic Content. *China Information* (5): 22–23.

Zhao J. et al. 2010. The Structure and Development of the Internet of Things. *Communications Technology* (9): 106–108.

Zhu Y. & Huang M.K. 2010. The Development and Application of the Internet of Things. *Telecom Network Technology* (11): 34–37.

Bridge Maintenance, Safety, Management, Resilience and Sustainability – Biondini & Frangopol (Eds)
 ISBN 978-0-415-62124-3

Wireless interrogation of passive crack sensor

Z. Xu, Y. Jia & R. Valentín
Department of Mechanical Engineering, University of Puerto Rico, Mayagüez, Puerto Rico

G. Portela
Department of Engineering Science and Materials, University of Puerto Rico, Mayagüez, Puerto Rico

ABSTRACT

This study aims to apply passive wireless (PW) sensors for structural bridge inspections. In this paper, an innovative passive wireless sensor consisting of an interdigital capacitor (IDC) and a loop inductor antenna is proposed for crack monitoring (Figure 1). The crack sensing element is fabricated using photolithography processes consisting of a polyimide composite film on a substrate and rolled-annealed copper foil. The proposed inductive coil consists of a coated copper wire with circular geometry. The abovementioned components work together as an LC resonator whose resonant frequency was designed to change correspondingly with the size, shape, orientation, and propagation of the monitored crack. The capacitance calculation was made using conformal mapping (Igreja & Dias 2004).

Table 1 shows the design parameters of the IDC sensor. The capacitance value of the IDC sensor depends on the crack initial position, orientation, and crack length. The LC frequency reader detects the frequency variation by monitoring the impedance across the terminals of the wide bandwidth reader antenna. Thereafter, the crack growth is retrieved and interpreted based on the received signal from the reader by the resonant frequency shift.

An experimental test was performed to show that the IDC sensing system can measure and record the resonance frequency change due to the propagation of a crack. The sensor was experimentally validated showing an increase in its resonant frequency with an increase in crack size. The comparison with physical damage of a steel specimen showed that the system is capable to wirelessly monitor crack propagation.

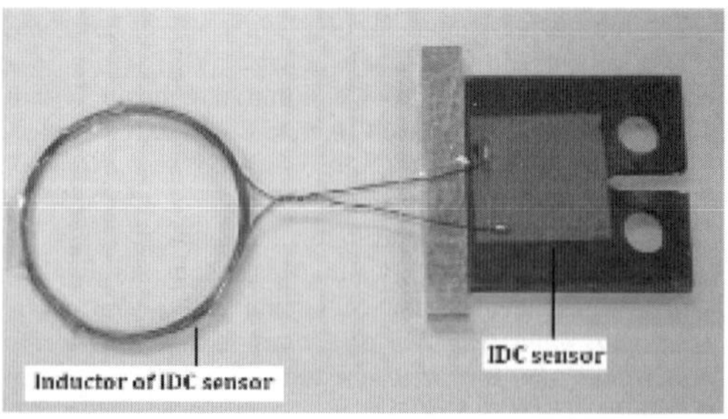

Figure 1. Sensor prototype.

Table 1. IDC design parameters.

Design Parameter	Symbols	Value
Number of total electrodes	N_C	26
Length of fingers	L_C	23 mm
Thickness of fingers	t	35 μm
Thickness of substrate	h	100 μm
Width of fingers	w_C	500 μm
Gaps between electrodes	g_C	800 μm

REFERENCE

Igreja, R. & Dias, C. J. 2004. Analytical evaluation of the interdigital electrodes capacitance for a multi-layered structure. *Sensors and Actuators*: 291–301.

Bridge Maintenance, Safety, Management, Resilience and Sustainability – Biondini & Frangopol (Eds)
© 2012 Taylor & Francis Group, London, ISBN 978-0-415-62124-3

Dynamic response analyses for human-induced lateral vibration on footbridges

M. Yoneda
Kinki University, Osaka, Japan

ABSTRACT

The phenomenon of synchronous lateral excitation caused by pedestrians walking on footbridges such as the London Millennium Bridge has increasingly attracted public attention. This excitation phenomenon for example the triggering of the lock-in phenomenon and its self-limited nature has not been fully understood or modelled. Therefore, in this paper, a neural-oscillator model which is consisted of two simulated neurons arranged in mutual inhibition is investigated to grasp some useful information for human-induced lateral vibration on congested pedestrian bridges.

The new evaluation formula which is proportional to square value of the deck velocity is proposed in order to convert the neural-oscillator output to the pedestrian lateral force (see Figure 1). Dynamic response analyses taking into account the neural oscillator proposed by Matsuoka (Matsuoka, 1985) is also carried out for the pedestrian bridge model with the centre span length of 50 m, placing much emphasis on the lock-in phenomenon (see Figure 2).

The oscillator (pedestrian walking on the bridge) frequency can be obtained by zero-crossing method for the time history oscillator output. It is found that the oscillator frequency varies gradually, and is entrained to be equal to 1.000 Hz which is the lateral natural frequency of the bridge. It can be said that the synchronization (the lock-in phenomenon) is analytically explained.

The results are summarized below:

1. The oscillator behavior is investigated and the oscillator's entrainment property is clarified.
2. A simplified method to evaluate the frequency of the neural-oscillator model is proposed.
3. The new evaluation formula which is proportional to square value of the deck velocity is proposed in order to convert the oscillator output to the pedestrian lateral force.
4. It is confirmed that the neural-oscillator might be one of the effective models to explain the synchronization (the lock-in phenomenon) of a fairly high part of all pedestrians being on the bridge.

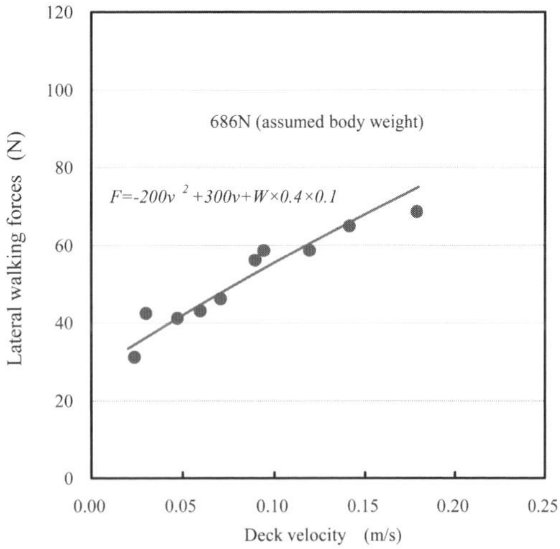

Figure 1. Relationship between newly proposed dynamic load factor and deck velocity as measured at the Imperial College tests.

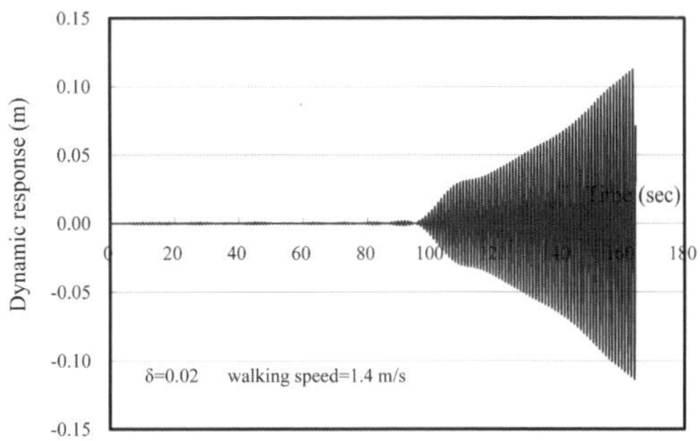

Figure 2. Time history displacement at the centre of the span obtained by the dynamic response analysis.

REFERENCES

Matsuoka, K. 1985. Sustained Oscillations Generated by Mutually Inhibiting Neurons with Adaptations. *Biological Cybernetics* 52:367–376.

Bridge Maintenance, Safety, Management, Resilience and Sustainability – Biondini & Frangopol (Eds)
© 2012 Taylor & Francis Group, London, ISBN 978-0-415-62124-3

Basic creep study and formulation of a new model

M. Zeineddine & W. Raphael
Ecole Supérieure d'Ingénieurs de Beyrouth (ESIB), Saint-Joseph University, Lebanon

A. Chateauneuf
LGC/CUST – Université Blaise Pascal, France

ABSTRACT

Bridge performance undergoes time-varying changes when exposed to aggressive environments. While much work has been done on bridge reliability under corrosion, little is known about the effects of creep and shrinkage on the reliability of concrete bridges. Time-dependent effects of creep and shrinkage of concrete introduce time-dependent variations in the forces. In prestressed members, the prestressing may be lower than expected and cracks may appear in critical areas, leading to concrete, steel and pavement deterioration. In segmental construction bridges, inaccurate creep analysis may cause excessive deflections, difficulties with closure, or un-aesthetic permanent deflections.

We know that concrete is a very important structural material, mainly due to its variety in properties, wide range of applicability and reasonable price. During the last decades, concrete gained large interest and became more and more involved in constructions. But the basic creep of concrete can have prejudicial consequences on structures. These uncontrolled strains appear and can be hardly predicted with precision due to the fact that these strains are highly sensitive to all involved parameters. Obvious factors that increase basic creep at a given time after fabrication are temperature, loading age and mean radius noting that the relative humidity has no effect on it.

In this work, a large database has been constituted for basic creep testing in many European research centers. At first, a comparison is performed between the experimental results and the BPEL design code. This comparison shows that the basic creep deformation is highly overestimated in the BPEL. The error reaches sometimes more than 500%.

In order to provide reliable creep model, half of the experimental database is randomly considered to develop and calibrate a new model, known as FP1, for the prediction of basic creep in RC structures; while the second half is used to validate the results of the new model FP1. Using the collected database, a new model is formulated to reach our goal. The statistical analysis using the M_{CEB} method shows that the new model FP1 is much more accurate than the BPEL design code.

So the new model FP1 is very important and can be used in bridge engineering to avoid the above-mentioned problems.

REFERENCES

Raphael W., Faddoul R., Chateuneuf A., 2009. " Information-based formulation for Bayesian updating of the Eurocode 2 Creep Model " Structural Concrete Journal, FIB Thomas Telford Edition, Volume 10, Issue 2.

Raphael W., Faddoul R., Daher S., Chateuneuf A. – 2007. "Information-based formulation for probabilistic concrete shrinkage models updating" Proceedings of the International Conference on applications of statistics and probability in civil engineering, ICASP 10, Tokyo, Japan

Bridge Maintenance, Safety, Management, Resilience and Sustainability – Biondini & Frangopol (Eds)
© 2012 Taylor & Francis Group, London, ISBN 978-0-415-62124-3

The real-time alarming technique of ship-collision to long-span bridges based on the displacement data of expansion joints

Y.F. Zhang, S.C. Zhu & L.T. Zhang
Jiangsu Transportation Research Institute co., LTD, Nanjing, Jiangsu, China

ABSTRACT

With the social and economic development, all countries in the world have built a large number of long-span bridges with the main form of suspension bridge and cable-stayed bridge over the sea or the river. However, as the buildings acrossing the channel, those bridges will undoubtedly cause some obstacles. The investigation of IABSE shows that ship-collision accidents have been happening in recent years though scientific development brings a significant progress to the equipment level of ships, safety monitoring and management level and the establishment and management level of bridges. What's more, the number of accidents have been confirmed to show a rising trend Since the 1980s.

Currently, the structural health monitoring system is applied in a great deal of bridges to carry out real-time monitoring upon the bridge structure. It plays an important role in how to take the advantage of the system to carry out the real-time alarming when strong wind or ship-collision happens, distinguish the accident styles, track down the accident ship, and prompt the maintaining to the management upon long-span bridges.

One of the important functions of the structural health monitoring system is to monitor environmental loads by recording the structural response in real time when happening ship collision and activating the alarming in time. This helps the bridge management and operation department to control the risk with significant social and economic benefits. In order to deal with the problem of ship-collision alarming, the displacement at expansion joints was choosed as the preferred data source, and furthermore the data was processed by comparing the different sources. According to the vibration features of the beam angle during ship-collision, three features were selected as the characteristic indicators, i.e. the amplitude, the horizontal peak value in the first vibration mode and the amplitude of short-time power spectral density. Based on the three features, relevant thresholds were decided according to the results of ship-collision alarming using measured data of Jiangyin Bridge. The results of calculation showed that using the proposed algorithm, the two ship-collision events were picked out accurately from the measured data of Jiangyin Bridge and the alarming delay was less than one minute.

Though the proposed method achieves ship-collision alarming accurately, the physical meaning of some the non-ship-collision alarming signal is still not entirely clear. Therefore, the interpretation work will continue. At the same time, the measured data of displacement at expansion joints were choosed as the only signal source. In actual system, therefore, the integrated techniques of multi-sensors measured data need to be researched and adopted further to make reasonable judge.

REFERENCES

Chen G.Y. & Wang L.L. 2006. Research on ship-collison and prevention techniques of bridegs. *Railway Press of China*. Beijing.

Liu Y.Z., Chen W.L. & Chen L.Q. 1998. Mechanics of vibrations. *China Higher Education Press*. Beijing.

Manen, S.V. & Frandsen A.G. 1998. Ship collision with bridges, review of accidents. *Ship Collision Analysis*: 3–11. Gluver and Olsen.

Bridge Maintenance, Safety, Management, Resilience and Sustainability – Biondini & Frangopol (Eds)
 ISBN 978-0-415-62124-3

The wind statistical characteristics analyse of long-span bridge based on long-term field measurement data

G.D. Zhou, Y.L. Ding, A.Q. Li & Y. Deng
School of Civil Engineering, Southeast University, Nanjing, China

ABSTRACT

The fluctuating wind induced vibration is one of the most important factors which must be taken into account in the design of long-span bridges, because of the low stiffness and low natural frequency. The sustaining fluctuating wind will result in high fatigue stress amplitude and shorten the structural serves life, so its effects can not be ignored. To calculate the buffer response of long-span bridges, the analysis of sustaining fluctuating wind characteristics is the prior work. Field measurement characteristics of sustaining wind on structure site can provide accurate wind load parameters for structural wind resistance design and wind field simulation.

The continuously wind environment field measurement both in mid-span and on tower top is executed from 2005 up to now by the structural health monitoring system deployed on the Runyang Suspension Bridge (RSB) with 1490 m main span, which has high sensitivity to fluctuating wind, located in the east of China. Based on the recorded data, the wind characteristic parameters, including mean wind speed, wind direction, the turbulence intensity, the gust factors, the turbulence integral length, power spectrum and spatial correlation, are analyzed and the coherence function of those parameters are evaluated using statistical method in this paper. The following conclusions can be deduced:

a) The 10-min mean wind speed is low all the year except some typhoon occasions. The average wind speed is 10.196 m/s. The predominating wind comes from the southeast coast in China.
b) The sustaining wind contains high fluctuating components on the RSB site, which induce stronger vibration response to the bridge. So the fatigue stress induced by long-term wind load needs special concerning. The longitude component and the lateral component of the turbulence intensity have intensive correction, so does the different components of the gust factor. The linearity equation of the turbulence intensity and the gust factor can be expressed as $I_v = 0.921 I_u$ and $G_v = 0.903 G_u - 0.601$ respectively. Furthermore, the turbulence intensity increases proportionally with the gust factor ascending. With the airflow slowing down, the intensity of turbulence wind has the increasing tendency.
c) As a parameter describes the scale of the turbulence eddies, the longitude turbulence integral length is larger than the value suggested by the design specifications of long span bridge and the lateral part is smaller than the recommend vales, which means this item defined in the bridge design specification can not suit to the practical condition of the RSB very well. The Gauss distribution is demonstrated to fit describing the statistical property of the turbulence integral length on the RSB site. The turbulence integral length descends with the turbulence intensity increasing and ascends with the mean wind speed increasing.
d) Both the Simiu spectrum and the von-Karman spectrum can not accurately describe the wind energy distribution over frequency on the bridge site. The special energy distribution of the sustaining wind needs to further study based on large number of field measurement data.

The characteristics of sustaining wind have much difference with the extreme strong wind, the research on the sustaining wind need to be pay enough attention. The SHMS provides an important way to obtain long-term field measurement sustaining wind data. The results obtained in this study can be employed to evaluate the long term reliability of the RSB and provide reference values for wind resistant design of other structures.

Bridge Maintenance, Safety, Management, Resilience and Sustainability – Biondini & Frangopol (Eds)
© 2012 Taylor & Francis Group, London, ISBN 978-0-415-62124-3

Minnesota Department of Transportation new structure information management system

J. Zink
Minnesota Department of Transportation, Oakdale, MN, USA

J.K. Shaffer, M.C. Schellhase & B.D. Witter
InspectTech Systems, Inc., Pittsburgh, PA, USA

ABSTRACT

The Minnesota Department of Transportation is responsible for over 13,000 bridges greater than 20 ft (~6 m) in length. The Department has completely replaced a program of scattered databases with a new system to inspect and manage all of its structures. This system greatly enhances the capabilities of both inspection and management personnel. Multiple personnel on fracture critical inspection teams can also simultaneously collect information on different parts of the bridges and merge the data together into a single report on the server. The system also manages the hundreds of pictures taken on large bridges and directly links to appropriate location (bearing, pier, or other component). The SIMS software now serves as the foundation for the department and links in multiple systems including Pontis, SI&A data, and Electronic Records Management files.

This new approach has allowed the state to improve its analysis and accuracy of data as well as communication between each of the districts' inspection teams, central office, and other personnel. Additional enhancements have included integrated photo references for each field, relevant coding manuals, and drop down options. A number of built-in quality assurance error checks also assist in preventing mistakes.

Figure 1. Mn/DOT maintains over 13,000 bridges, from simple to highly complex structures.

The new system allows the state to engage in much more in-depth management programs. Some of the common features include the ability to automatically spot trends, quantify maintenance needs and priorities, and an advanced scheduling module. The system serves as the complete repository for all bridge data for Mn/DOT. This is allowing for unprecedented access to the data from multiple sites via secure web interfaces.

This presentation shows experiences and lessons learned by a DOT that has sought to integrate the latest technology to improve its inspection and management process.

REFERENCES

Laird, J. H. 2009. Merging and Moving Forward with New Jersey Turnpike Authority's Bridge Inspection Program. In Proceedings of *New York City Bridge Conference*, New York, NY.

Robert, W.E and Marshall, A.R. 2003. The Pontis Bridge Management System: State of the Practice in Implementation and Development. In Proceedings of 9th *International Bridge Management Conference*, Orlando, FL

Schellhase, M.C 2008. Improving Rating Accuracy in Visual Inspection via Software. In Proceedings of *International Bridge Conference*, Pittsburgh, PA

Shaffer, J.K. 2008. Integrated Management and Inspection System for Maryland Counties and Cities. In Proceedings of the *Transportation Research Board 10th International Bridge and Structure Management Conference*, Buffalo, NY.

Thompson, P.D. and R.W. Shepard 1994. Pontis. In Transportation Research Circular 423 *TRB National Research Council.*

United States Federal Highway Administration 2002. *Bridge Inspector's Reference Manual.* U.S. Federal Highway Administration.

United States Federal Highway Administration 2005. *National Bridge Inspection Standards, 23 CFR Part 650.* U.S. Federal Highway Administration.

United States Federal Highway Administration 1995. *Recording and Coding Guide for the Structure Inventory and Appraisal of the Nation's Bridges.* U.S. Federal Highway Administration.

Bridge Maintenance, Safety, Management, Resilience and Sustainability – Biondini & Frangopol (Eds)
© 2012 Taylor & Francis Group, London, ISBN 978-0-415-62124-3

Lessons learned from the Little Lake Harris Bridge settlement restoration project

M. Hassan
Florida Department of Transportation, Deland, FL, USA

A. Sallam
Terracon Consultants, Inc., Winter Park, FL, USA

D. Thompson
Kisinger Campo & Associates Corp., Tampa, FL, USA

ABSTRACT

The Little Lake Harris (LLH) Bridge was constructed in 1950 as a replacement for the east-west County Road 48 Bridge connecting the towns of Howey-in-the-Hills and Astatula in Lake County, Florida, United States (See Figure 1).

The 78-span bridge carries two lanes of traffic, is 3,130 feet long, and 36.1 feet wide. The concrete deck is supported by steel I-girders carried by one row of 18-inch square precast concrete driven piles. In the late 1980's, slight settlement was noted in several bents. In 1990, Florida Department of Transportation (FDOT) initiated annual survey of deck elevations during bridge inspections to monitor settlement. Due to the slow nature and magnitude of settlement, substructure instability was not a major concern. However, public concern over the settlement and rough ride of the bridge continued to rise. FDOT contracted Kisinger Campo & Associates Corp. (KCA) to provide the needed design to correct the dips in the riding surface and minimize future settlement. KCA's initial design was based on the concept of "helper bents" at the settled locations in order to restore the bridge to its original elevation. Estimated construction cost for seven helper bents was about $3.3M. FDOT proposed evaluating other alternatives including jacking the bridge on its existing foundations and using shims to restore the deck grades. After discussions with the FDOT, KCA involved the geotechnical consultant, Nodarse, A Terracon Company (Terracon), to evaluate the proposed alternative. The jacking and shimming option was determined to be both technically viable and economical with a total cost of less than $300K. The design was completed in March 2010 and construction was completed in February 2011. This paper provides an overview of the LLH Bridge history, summarizes the evolution of the design methodology including the findings of the geotechnical study and presents lessons learned during the design and construction of the project.

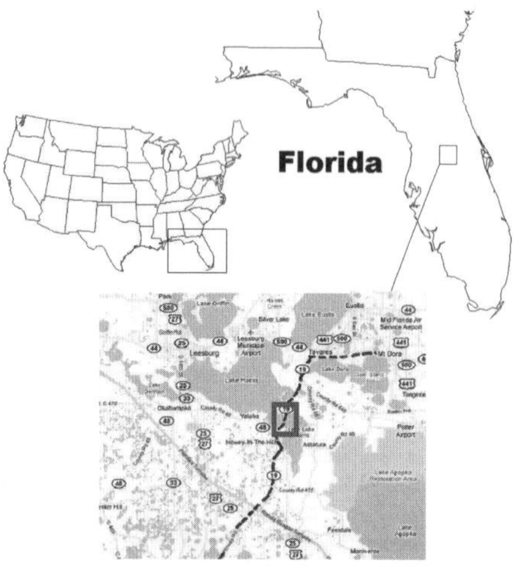

Figure 1. Bridge location map.

REFERENCES

Braja M Das, "Principles of Foundation Engineering", seventh edition, 2010, a book by Cengage Learning.

Schmertmann, J.H., Hartman, J.P., and Brown, P.R. (1978). "Improved Strain Influence factor diagram", J. Geotechnical Engineering Division, ASCE, 104(8), 1131–1135.

Soils and Foundation Handbook, 2000, Florida Department of Transportation.

AASHTO LRFD Bridge Specifications Fifth Edition 2010, American Association of State Highway and Transportation Officials.

Bridge Maintenance, Safety, Management, Resilience and Sustainability – Biondini & Frangopol (Eds)
© 2012 Taylor & Francis Group, London, ISBN 978-0-415-62124-3

Author index